Texts in Computational Science and Engineering

11

Editors

Timothy J. Barth
Michael Griebel
David E. Keyes
Risto M. Nieminen
Dirk Roose
Tamar Schlick

For further volumes:
http://www.springer.com/series/5151

Walter Gander   •   Martin J. Gander   •   Felix Kwok

Scientific Computing

An Introduction using Maple and MATLAB

 Springer

Walter Gander
Departement Informatik
ETH Zürich
Zürich
Switzerland

Martin J. Gander
Felix Kwok
Section de Mathématiques
Université de Genève
Genève
Switzerland

ISSN 1611-0994
ISBN 978-3-319-38158-9   ISBN 978-3-319-04325-8 (eBook)
DOI 10.1007/978-3-319-04325-8
Springer Cham Heidelberg New York Dordrecht London

Mathematics Subject Classification (2010): 65-00, 65-01

This book is dedicated to
Professor Gene H. Golub
1932–2007

(picture by Jill Knuth)

The three authors represent three generations of mathematicians who have
been enormously influenced by Gene Golub.
He shaped our lives and our academic careers through his advice, his
leadership, his friendship and his care for younger scientists.
We are indebted and will always honor his memory.

Preface

We are conducting ever more complex computations built upon the assumption that the underlying numerical methods are mature and reliable.
When we bundle existing algorithms into libraries and wrap them into packages to facilitate easy use, we create de facto standards that make it easy to ignore numerical analysis.

John Guckenheimer, president SIAM, in *SIAM News, June 1998: Numerical Computation in the Information Age*

When redrafting the book I was tempted to present the algorithms in ALGOL, but decided that the difficulties of providing procedures which were correct in every detail were prohibitive at this stage.

James Wilkinson, The Algebraic Eigenvalue Problem, Oxford University Press, 1988.

This book is an introduction to *scientific computing*, the mathematical modeling in science and engineering and the study of how to exploit computers in the solution of technical and scientific problems. It is based on mathematics, numerical and symbolic/algebraic computations, parallel/distributed processing and visualization. It is also a popular and growing area — many new curricula in *computational science and engineering* have been, and continue to be, developed, leading to new academic degrees and even entire new disciplines.

A prerequisite for this development is the ubiquitous presence of computers, which are being used by virtually every student and scientist. While traditional scientific work is based on developing theories and performing experiments, the possibility to use computers at any time has created a third way of increasing our knowledge, which is through modeling and simulation. The use of simulation is further facilitated by the availability of sophisticated, robust and easy-to-use software libraries. This has the obvious advantage of shielding the user from the underlying numerics; however, this also has the danger of leaving the user unaware of the limitations of the algorithms, which can lead to incorrect results when used improperly. Moreover, some algorithms can be fast for certain types of problems but highly inefficient for others. Thus, it is important for the user to be able to make an informed decision on which algorithms to use, based on the properties of the problem to be solved. The goal of this book is to familiarize the reader with the basic concepts of scientific computing and algorithms that form the workhorses of many numerical libraries. In fact, we will also emphasize the effective implementation of the algorithms discussed.

Numerical scientific computing has a long history; in fact, computers were first built for this purpose. *Konrad Zuse* [154] built his first (mechanical) computer in 1938 because he wanted to have a machine that would solve systems of linear equations that arise, e.g., when a civil engineer designs a bridge. At about the same time (and independently), *Howard H. Aiken* wanted to build a machine that would solve systems of ordinary differential equations [17].

The first high quality software libraries contained indeed numerical algorithms. They were produced in an international effort in the programming language ALGOL 60 [111], and are described in the handbook "Numerical Algebra" [148]. These fundamental procedures for solving linear equations and eigenvalue problems were developed further, rewritten in FORTRAN, and became the LINPACK [26] and EISPACK [47] libraries. They are still in use and available at `www.netlib.org` from Netlib. In order to help students to use this software, Cleve Moler created around 1980 a friendly interface to those subroutines, which he called MATLAB (Matrix Laboratory). MATLAB was so successful that a company was founded: MathWorks. Today, MATLAB is "the language of technical computing", a very powerful tool in scientific computing.

Parallel to the development of numerical libraries, there were also efforts to do exact and algebraic computations. The first computer algebra systems were created some 50 years ago: At ETH, Max Engeli created SYMBAL, and at MIT, Joel Moses MACSYMA. MACSYMA is the oldest system that is still available. However, computer algebra computations require much more computer resources than numerical calculations. Therefore, only when computers became more powerful did these systems flourish. Today the market leaders are MATHEMATICA and MAPLE.

Often, a problem may be solved analytically ("exactly") by a computer algebra system. In general, however, analytical solutions do not exist, and numerical approximations or other special techniques must be used instead. Moreover, computer Algebra is a very powerful tool for deriving numerical algorithms; we use MAPLE for this purpose in several chapters of this book. Thus, computer algebra systems and numerical libraries are complementary tools: working with both is essential in scientific computing. We have chosen MATLAB and MAPLE as basic tools for this book. Nonetheless, we are aware that the difference between pure computer algebra systems and numerical MATLAB-like systems is disappearing, and the two may merge and become indistinguishable by the user in the near future.

How to use this book

Prerequisites for understanding this book are courses in calculus and linear algebra. The content of this book is too much for a typical one semester course in scientific computing. However, the instructor can choose those sections that he wishes to teach and that fit his schedule. For example, for an introductory course in scientific computing, one can very well use the least squares chapter and teach only one of the methods for computing the QR decomposition. However, for an advanced course focused solely on least squares methods, one may also wish to consider the singular value decomposition (SVD) as a computational tool for solving least squares problems. In this case, the book also provides a detailed description on how to compute the SVD in the chapter on eigenvalues. The material is presented in such a way that a student can also learn directly from the book. To help the reader navigate the volume, we provide in section 1.2 some sample courses that have been taught by the authors at various institutions.

The focus of the book is algorithms: we would like to explain to the students how some fundamental functions in mathematical software are designed. Many exercises require programming in MATLAB or MAPLE, since we feel it is important for students to gain experience in using such powerful software systems. They should also know about their limitations and be aware of the issue addressed by John Guckenheimer. We tried to include meaningful examples and problems, not just academic exercises.

Acknowledgments

The authors would like to thank Oscar Chinellato, Ellis Whitehead, Oliver Ernst and Laurence Halpern for their careful proofreading and helpful suggestions.

Walter Gander is indebted to Hong Kong Baptist University (HKBU) and especially to its Vice President Academic, Franklin Luk, for giving him the opportunity to continue to teach students after his retirement at ETH. Several chapters of this book have been presented and improved successfully in courses at HKBU. We are also thankful to the University of Geneva, where we met many times to finalize the manuscript.

Geneva and Zürich, August 2013

Walter Gander, Martin J. Gander, Felix Kwok

Contents

Chapter 1. Why Study Scientific Computing?

> *Computational Science and Engineering (CS&E) is now widely accepted, along with theory and experiment, as a crucial third mode of scientific investigation and engineering design. Aerospace, automotive, biological, chemical, semiconductor, and other industrial sectors now rely on simulation for technical decision support.*
>
> Introduction to the First SIAM Conference on Computational Science and Engineering, September 21–24, 2000, Washington DC.

The emergence of scientific computing as a vital part of science and engineering coincides with the explosion in computing power in the past 50 years. Many physical phenomena have been well understood and have accurate models describing them since the late 1800s, but before the widespread use of computers, scientists and engineers were forced to make many simplifying assumptions in the models in order to make them solvable by pencil-and-paper methods, such as series expansion. With the increase of computing power, however, one can afford to use numerical methods that are computationally intensive but that can tackle the full models without the need to simplify them. Nonetheless, every method has its limitations, and one must understand how they work in order to use them correctly.

1.1 Example: Designing a Suspension Bridge

To get an idea of the kinds of numerical methods that are used in engineering problems, let us consider the design of a simple *suspension bridge*. The bridge consists of a pair of ropes fastened on both sides of the gorge, see Figure 1.1. Wooden supports going across the bridge are attached to the ropes at regularly spaced intervals. Wooden boards are then fastened between the supports to form the deck. We would like to calculate the shape of the bridge as well as the tension in the rope supporting it.

1.1.1 Constructing a Model

Let us construct a simple one-dimensional model of the bridge structure by assuming that the bridge does not rock side to side. To calculate the shape of the bridge, we need to know the forces that are exerted on the ropes by the supports. Let L be the length of the bridge and x be the distance from one

W. Gander et al., *Scientific Computing - An Introduction using Maple and MATLAB*, Texts in Computational Science and Engineering 11, DOI 10.1007/978-3-319-04325-8_1,

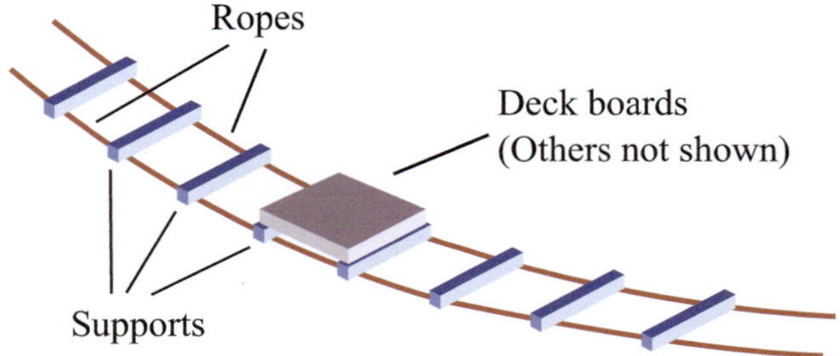

FIGURE 1.1. *A simple suspension bridge.*

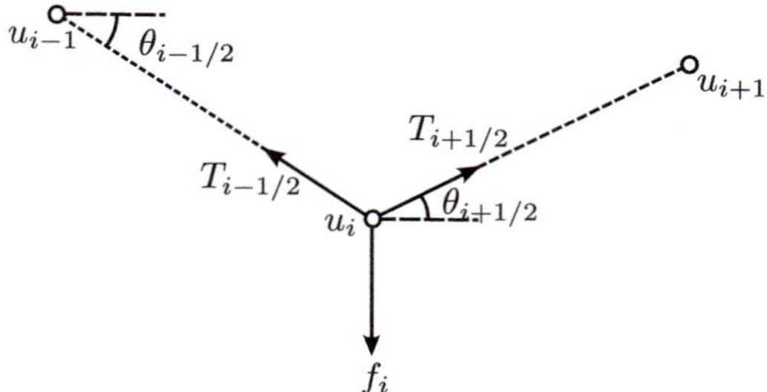

FIGURE 1.2. *Force diagram for the bridge example.*

end of the bridge. Assume that the supports are located at x_i, $i = 1, \ldots, n$, with h being the spacing between supports. Let $w(x)$ be the force per unit distance exerted on the deck at x by gravity, due to the weight of the deck and of the people on it. If we assume that any weight on the segment $[x_{i-1}, x_i]$ are exerted entirely on the supports at x_{i-1} and x_i, then the force f_i exerted on the rope by the support at x_i can be written as

$$f_i = \left(\int_{x_{i-1}}^{x_i} w(x)(x - x_{i-1})\, dx + \int_{x_i}^{x_{i+1}} w(x)(x_{i+1} - x)\, dx \right). \qquad (1.1)$$

We now consider the rope as an elastic string, which is stretched by the force exerted by the wooden supports. Let u_i be the height of the bridge at x_i, $T_{i-1/2}$ be the tension of the segment of the rope between x_{i-1} and x_i, and $\theta_{i-1/2}$ be the angle it makes with the horizontal. Figure 1.2 shows the force diagram on the rope at x_i.

Since there is no horizontal displacement in the bridge, the horizontal forces must balance out, meaning

$$T_{i-1/2} \cos(\theta_{i-1/2}) = T_{i+1/2} \cos(\theta_{i+1/2}) = C,$$

where K is a constant. Vertical force balance then gives

$$T_{i+1/2}\sin(\theta_{i+1/2}) - T_{i-1/2}\sin(\theta_{i-1/2}) = f_i,$$

or

$$C\tan(\theta_{i+1/2}) - C\tan(\theta_{i-1/2}) = f_i.$$

But

$$\tan(\theta_{i+1/2}) = \frac{u_{i+1} - u_i}{h},$$

so we in fact have

$$\frac{K(u_{i+1} - 2u_i + u_{i-1})}{h} = f_i, \qquad i = 1, \ldots, n, \tag{1.2}$$

where u_0 and u_{n+1} are the known heights of the bridge at its ends and $u_1, \ldots, u_n$ are the unknown heights.

1.1.2 Simulating the Bridge

Now, if we want to compute the shape of the bridge and the tensions $T_{i-1/2}$, we must first calculate the forces f_i from (1.1), and then solve the system of linear equations (1.2). To calculate the f_i, one must evaluate integrals, which may not be analytically feasible for certain weight distributions $w(x)$. Instead, one can approximate the integral numerically using a *Riemann sum*, for instance:

$$\int_{x_{i-1}}^{x_i} w(x)(x - x_{i-1})\,dx \approx \frac{1}{N}\sum_{j=1}^{N} w(x_{i-1} + jh/N) \cdot \frac{h}{j}.$$

For large N, this converges to the exact value of the integral, but the error behaves like $1/N$; this means if we want to have three decimal digits of accuracy in the answer, one would need approximately 10^3 points. There are other formulas that give more accurate values with fewer number of points; this is discussed in more detail in Chapter 9.

The next step is to solve (1.2) for the u_i. This can be rewritten as

$$A\boldsymbol{u} = \boldsymbol{f},$$

where $A \in \mathbb{R}^{n \times n}$ is a matrix, $\boldsymbol{u} \in \mathbb{R}^n$ is the vector of unknowns, and $\boldsymbol{f}$ is the vector of forces we just calculated. This system can be solved by *Gaussian elimination*, i.e., by row reducing the matrix, as taught in a basic linear algebra course. So for $n = 4$, a uniform distribution $w(x) = 1$, and

$u_0 = u_{n+1} = 0$, we can calculate

$$
\begin{pmatrix}
-2 & 1 & 0 & 0 & \vline & 1 \\
1 & -2 & 1 & 0 & \vline & 1 \\
0 & 1 & -2 & 1 & \vline & 1 \\
0 & 0 & 1 & -2 & \vline & 1
\end{pmatrix}
\longrightarrow
\begin{pmatrix}
-2 & 1 & 0 & 0 & \vline & 1 \\
0 & -\frac{3}{2} & 1 & 0 & \vline & \frac{3}{2} \\
0 & 1 & -2 & 1 & \vline & 1 \\
0 & 0 & 1 & -2 & \vline & 1
\end{pmatrix}
$$

$$
\longrightarrow
\begin{pmatrix}
-2 & 1 & 0 & 0 & \vline & 1 \\
0 & -\frac{3}{2} & 1 & 0 & \vline & \frac{3}{2} \\
0 & 0 & -\frac{4}{3} & 1 & \vline & 2 \\
0 & 0 & 1 & -2 & \vline & 1
\end{pmatrix}
\longrightarrow
\begin{pmatrix}
-2 & 1 & 0 & 0 & \vline & 1 \\
0 & -\frac{3}{2} & 1 & 0 & \vline & \frac{3}{2} \\
0 & 0 & -\frac{4}{3} & 1 & \vline & 2 \\
0 & 0 & 0 & -\frac{5}{4} & \vline & \frac{5}{2}
\end{pmatrix}.
$$

Back substitution gives $\boldsymbol{u} = \frac{h}{K}(-2, -3, -3, -2)^{\top}$. However, one often wishes to calculate the shape of the bridge under different weight distributions $w(x)$, e.g., when people are standing on different parts of the bridge. So the matrix A stays the same, but the right-hand side $\boldsymbol{f}$ changes to reflect the different weight distributions. It would be a waste to have to redo the row reductions every time, when only $\boldsymbol{f}$ has changed! A much better way is to use the *LU decomposition*, which writes the matrix A in factored form and reuses the factors to solve equations with different right-hand sides. This is shown in Chapter 3.

In the above row reduction, we can see easily that there are many zero entries that need not be calculated, but the computer has no way of knowing that in advance. In fact, the number of additions and multiplications required for solving the generic (i.e., full) linear system is proportional to n^3, whereas in our case, we only need about n additions and multiplications because of the many zero entries. To take advantage of the *sparse* nature of the matrix, one needs to store it differently and use different algorithms on it. One possibility is to use the *banded matrix format*; this is shown in Section 3.6.

Suppose now that the people on the bridge have moved, but only by a few meters. The shape of the bridge would have only changed slightly, since the weight distribution is not very different. Thus, instead of solving a new linear system from scratch, one could imagine using the previous shape as a first guess and make small corrections to the solution until it matches the new distribution. This is the basis of *iterative methods*, which are discussed in Chapter 11.

1.1.3   Calculating Resonance Frequencies

A well-designed bridge should never collapse, but there have been spectacular bridge failures in history. One particularly memorable one was the collapse of the *Tacoma Narrows bridge* on November 7, 1940. On that day, powerful wind gusts have excited a natural resonance mode of the bridge, setting it into a twisting motion that it was not designed to withstand. As the winds continued, the amplitude of the twisting motion grew, until the bridge eventually collapsed[1].

It turns out that one can study the resonance modes of the bridge by considering the *eigenvalue problem*

$$Au = \lambda u,$$

cf. [37]. Clearly, a two-dimensional model is needed to study the twisting motion mentioned above, but let us illustrate the ideas by considering the eigenvalues of the 1D model for $n = 4$. For this simple problem, one can guess the eigenvectors and verify that

$$u^{(k)} = (\sin(k\pi/5), \sin(2k\pi/5), \sin(3k\pi/5), \sin(4\pi/5))^\top, \qquad k = 1, 2, 3, 4$$

are in fact eigenvectors with associated eigenvalues $\lambda^{(k)} = -2 + 2\cos(k\pi/5)$. However, for more complicated problems, such as one with varying mass along the bridge or for 2D problems, it is no longer possible to guess the eigenvectors. Moreover, the *characteristic polynomial*

$$P(\lambda) = \det(\lambda I - A)$$

is a polynomial of degree n, and it is well known that no general formula exists for finding the roots of such polynomials for $n \geq 5$. In Chapter 7, we will present numerical algorithms for finding the eigenvalues of A. In fact, the problem of finding eigenvalues numerically also requires approximately n^3 operations, just like Gaussian elimination. This is in stark contrast with the theoretical point of view that linear systems are "easy" and polynomial root-finding is "impossible". To quote the eminent numerical analyst Nick Trefethen [139],

> Abel and Galois notwithstanding, large-scale matrix eigenvalue problems are about as easy to solve in practice as linear systems of equations.

1.1.4 Matching Simulations with Experiments

When modeling the bridge in the design process, we must use many parameters, such as the weight of the deck (expressed in terms of the mass density ρ per unit length) and the elasticity constant K of the supporting rope. In reality, these quantities depend on the actual material used during construction, and may deviate from the nominal values assumed during the design process. To get an accurate model of the bridge for later simulation, one needs to estimate these parameters from measurements taken during experiments. For example, we can measure the vertical displacements y_i of the constructed bridge at points x_i, and compare it with the displacements u_i predicted by the model, i.e., the displacements satisfying $Au = f$. Since both A and f

---

[1] http://www.youtube.com/watch?v=3mclp9QmCGs

depend on the model parameters ρ and K, the u_i also depend on these parameters; thus, the mismatch between the model and the experimental data can be expressed as a function of ρ and K:

$$F(\rho, K) = \sum_{i=1}^{n} |y_i - u_i(\rho, K)|^2. \tag{1.3}$$

Thus, we can estimate the parameters by finding the optimal parameters ρ^* and K^* that minimize F. There are several ways of calculating the minimum:

1. Using multivariate calculus, we know that

$$\frac{\partial F}{\partial \rho}(\rho^*, K^*) = 0, \qquad \frac{\partial F}{\partial K}(\rho^*, K^*) = 0. \tag{1.4}$$

   Thus, we have a system of two *nonlinear equations* in two unknowns, which must then be solved to obtain ρ^* and K^*. This can be solved by many methods, the best known of which is *Newton's method*. Such methods are discussed in Chapter 5.

2. The above approach has the disadvantage that (1.4) is satisfied by all stationary points of $F(\rho, K)$, i.e., both the maxima and the minima of $F(\rho, K)$. Since we are only interested in the minima of the function, a more direct approach would be to start with an initial guess (ρ^0, K^0) (e.g., the nominal design values) and then find successively better approximations (ρ^k, K^k), $k = 1, 2, 3$, that reduce the mismatch, i.e.,

   $$F(\rho^{k+1}, K^{k+1}) \le F(\rho^k, K^k).$$

   This is the basis of *optimization algorithms*, which can be applied to other minimization problems. Such methods are discussed in detail in Chapter 12.

3. The function $F(\rho, K)$ in (1.3) has a very special structure in that it is a sum of squares of the differences. As a result, the minimization problem is known as a *least-squares problem*. Least-squares problems, in particular linear ones, often arise because they yield the best unbiased estimator in the statistical sense for linear models. Because of the prevalence and special structure of least-squares problems, it is possible to design specialized methods that are more efficient and/or robust for these problems than general optimization algorithms. One example is the *Gauss–Newton method*, which resembles a Newton method, except that second-order derivative terms are dropped to save on computation. This and other methods are presented in Chapter 6.

1.2   Navigating this Book: Sample Courses

This book intentionally contains too many topics to be done from cover to cover, even for an intensive full-year course. In fact, many chapters contain enough material for stand-alone semester courses on their respective topics. To help instructors and students navigate through the volume, we provide some sample courses that can be built from its sections.

1.2.1 A First Course in Numerical Analysis

The following sections have been used to build the first year numerical analysis course at the University of Geneva in 2011–12 (54 hours of lectures).

1. Finite precision arithmetic (2.1–2.6)

2. Linear systems (3.2–3.4)

3. Interpolation and FFT (4.2.1–4.2.4, 4.3.1, 4.4)

4. Nonlinear equations (5.2.1–5.2.3, 5.4)

5. Linear and nonlinear least squares (6.1–6.8, 6.8.2, 6.8.3, 6.5.1, 6.5.2)

6. Iterative methods (11.1–11.2.5, 11.3.2–11.3.4, 11.7.1)

7. Eigenvalue problems (7.1, 7.2, 7.4, 7.5.2, 7.6)

8. Singular value decomposition (6.3)

9. Numerical integration (9.1, 9.2, 9.3, 9.4.1–9.4.2)

10. Ordinary differential equations (10.1, 10.3)

A first term course at Stanford for computer science students in 1996 and 1997 ('Introduction to Scientific Computing using MAPLE and MATLAB, 40 hours of lectures) was built using

1. Finite precision arithmetic (2.2)

2. Nonlinear equations (5.2.1–5.2.3, 5.2.5, 5.2.7, 5.4)

3. Linear systems (3.2.1, 3.2.2, 3.2.3, 11.2–11.2.3, 11.3.2, 11.3.3, 11.4, 11.7.1)

4. Interpolation (4.2.1–4.2.4, 4.3.1)

5. Least Squares (6.2, 6.5.1, 6.8.2)

6. Differentiation (8.2, 8.2.1)

7. Quadrature (9.2, 9.2.4, 9.3.1, 9.3.2, 9.4.1–9.4.2)

8. Eigenvalue problems (7.3, 7.4, 7.6)

9. Ordinary differential equations (10.1, 10.3, 10.4)

1.2.2   Advanced Courses

The following advanced undergraduate/graduate courses (38 hours of lectures each) have been taught at Baptist University in Hong Kong between 2010 and 2013. We include a list of chapters from which these courses were built.

1. Eigenvalues and Iterative Methods for Linear Equations (Chapters 7, 11)

2. Least Squares (Chapter 6)

3. Quadrature and Ordinary Differential Equations (Chapters 9 and 10)

At the University of Geneva, the following graduate courses (28 hours of lectures, and 14 hours of exercises) have been taught between 2004 and 2011:

1. Iterative Methods for Linear Equations (Chapter 11)

2. Optimization (Chapter 12)

1.2.3   Dependencies Between Chapters

Chapter 2 on finite precision arithmetic and Chapter 3 on linear equations are required for most, if not all, of the subsequent chapters. At the beginning of each chapter, we give a list of sections that are prerequisites to understanding the material. Readers who are not familiar with these sections should refer to them first before proceeding.

Chapter 2. Finite Precision Arithmetic

In the past 15 years many numerical analysts have progressed from being queer people in mathematics departments to being queer people in computer science departments!

George Forsythe, What to do till the computer scientist comes. Amer. Math. Monthly 75, 1968.

It is hardly surprising that numerical analysis is widely regarded as an unglamorous subject. In fact, mathematicians, physicists, and computer scientists have all tended to hold numerical analysis in low esteem for many years – a most unusual consensus.

Nick Trefethen, The definition of numerical analysis, SIAM news, November 1992.

The golden age of numerical analysis has not yet started.

Volker Mehrmann, round table discussion "Future Directions in Numerical Analysis," moderated by Gene Golub and Nick Trefethen at ICIAM 2007.

Finite precision arithmetic underlies all the computations performed numerically, e.g. in MATLAB; only symbolic computations, e.g. MAPLE, are largely independent of finite precision arithmetic. Historically, when the invention of computers allowed a large number of operations to be performed in very rapid succession, nobody knew what the influence of finite precision arithmetic would be on this many operations: would small rounding errors sum up rapidly and destroy results? Would they statistically cancel? The early days of numerical analysis were therefore dominated by the study of rounding errors, and made this rapidly developing field not very attractive (see the quote above). Fortunately, this view of numerical analysis has since changed, and nowadays the focus of numerical analysis is the study of algorithms for the problems of continuous mathematics[1]. There are nonetheless a few pitfalls every person involved in scientific computing should know, and this chapter is precisely here for this reason. After an introductory example in Section 2.1, we present the difference between real numbers and machine numbers in Section 2.2 on a generic, abstract level, and give for the more computer science oriented reader the concrete IEEE arithmetic standard in Section 2.3. We then discuss the influence of rounding errors on operations in

---

[1] Nick Trefethen, The definition of numerical analysis, SIAM News, November 1992

W. Gander et al., *Scientific Computing - An Introduction using Maple and MATLAB*,
Texts in Computational Science and Engineering 11,
DOI 10.1007/978-3-319-04325-8_2,
© Springer International Publishing Switzerland 2014

Section 2.4, and explain the predominant pitfall of catastrophic cancellation when computing differences. In Section 2.5, we explain in very general terms what the condition number of a problem is, and then show in Section 2.6 two properties of algorithms for a given problem, namely forward stability and backward stability. It is the understanding of condition numbers and stability that allowed numerical analysts to move away from the study of rounding errors, and to focus on algorithmic development. Sections 2.7 and 2.8 represent a treasure trove with advanced tips and tricks when computing in finite precision arithmetic.

2.1 Introductory Example

A very old problem already studied by ancient Greek mathematicians is the *squaring of a circle*. The problem consists of constructing a square that has the same area as the unit circle. Finding a method for transforming a circle into a square this way (*quadrature of the circle*) became a famous problem that remained unsolved until the 19th century, when it was proved using Galois theory that the problem cannot be solved with the straight edge and compass.

We know today that the area of a circle is given by $A = \pi r^2$, where r denotes the radius of the circle. An approximation is obtained by drawing a regular polygon inside the circle, and by computing the surface of the polygon. The approximation is improved by increasing the number of sides. Archimedes managed to produce a 96-sided polygon, and was able to bracket π in the interval $(3\frac{10}{71}, 3\frac{1}{7})$. The enclosing interval has length $1/497 = 0.00201207243$ — surely good enough for most practical applications in his time.

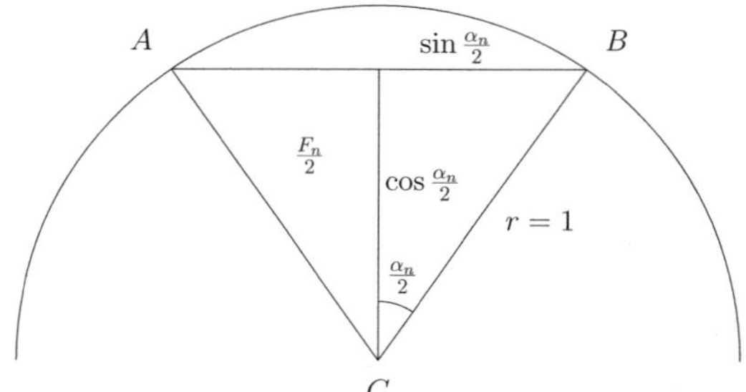

FIGURE 2.1. *Squaring of a circle*

To compute such a polygonal approximation of π, we consider Figure 2.1. Without loss of generality, we may assume that $r = 1$. Then the area F_n of the isosceles triangle ABC with center angle $\alpha_n := \frac{2\pi}{n}$ is

$$F_n = \cos\frac{\alpha_n}{2}\sin\frac{\alpha_n}{2},$$

and the area of the associated n-sided polygon becomes

$$A_n = nF_n = \frac{n}{2}\left(2\cos\frac{\alpha_n}{2}\sin\frac{\alpha_n}{2}\right) = \frac{n}{2}\sin\alpha_n = \frac{n}{2}\sin\left(\frac{2\pi}{n}\right).$$

Clearly, computing the approximation A_n using π would be rather contradictory. Fortunately, A_{2n} can be derived from A_n by simple algebraic transformations, i.e. by expressing $\sin(\alpha_n/2)$ in terms of $\sin\alpha_n$. This can be achieved by using identities for trigonometric functions:

$$\sin\frac{\alpha_n}{2} = \sqrt{\frac{1-\cos\alpha_n}{2}} = \sqrt{\frac{1-\sqrt{1-\sin^2\alpha_n}}{2}}. \tag{2.1}$$

Thus, we have obtained a recurrence for $\sin(\alpha_n/2)$ from $\sin\alpha_n$. To start the recurrence, we compute the area A_6 of the regular hexagon. The length of each side of the six equilateral triangles is 1 and the angle is $\alpha_6 = 60°$, so that $\sin\alpha_6 = \frac{\sqrt{3}}{2}$. Therefore, the area of the triangle is $F_6 = \sqrt{3}/4$ and $A_6 = 3\frac{\sqrt{3}}{2}$. We obtain the following program for computing the sequence of approximations A_n:

ALGORITHM 2.1. *Computation of π, Naive Version*

```
s=sqrt(3)/2; A=3*s; n=6;            % initialization
z=[A-pi n A s];                     % store the results
while s>1e-10                       % termination if s=sin(alpha) small
   s=sqrt((1-sqrt(1-s*s))/2);       % new sin(alpha/2) value
   n=2*n; A=n/2*s;                  % A=new polygon area
   z=[z; A-pi n A s];
end
m=length(z);
for i=1:m
   fprintf('%10d %20.15f %20.15f %20.15f\n',z(i,2),z(i,3),z(i,1),z(i,4))
end
```

The results, displayed in Table 2.1, are not what we would expect: initially, we observe convergence towards π, but for $n > 49152$, the error grows again and finally we obtain $A_n = 0$?! *Although the theory and the program are both correct, we still obtain incorrect answers.* We will explain in this chapter why this is the case.

2.2 Real Numbers and Machine Numbers

Every computer is a finite automaton. This implies that a computer can only store a finite set of numbers and perform only a finite number of operations. In mathematics, we are used to calculating with real numbers $\mathbb{R}$ covering the continuous interval $(-\infty, \infty)$, but on the computer, we must contend with a

n	A_n	$A_n - \pi$	$\sin(\alpha_n)$
6	2.598076211353316	−0.543516442236477	0.866025403784439
12	3.000000000000000	−0.141592653589794	0.500000000000000
24	3.105828541230250	−0.035764112359543	0.258819045102521
48	3.132628613281237	−0.008964040308556	0.130526192220052
96	3.139350203046872	−0.002242450542921	0.065403129230143
192	3.141031950890530	−0.000560702699263	0.032719082821776
384	3.141452472285344	−0.000140181304449	0.016361731626486
768	3.141557607911622	−0.000035045678171	0.008181139603937
1536	3.141583892148936	−0.000008761440857	0.004090604026236
3072	3.141590463236762	−0.000002190353031	0.002045306291170
6144	3.141592106043048	−0.000000547546745	0.001022653680353
12288	3.141592516588155	−0.000000137001638	0.000511326906997
24576	3.141592618640789	−0.000000034949004	0.000255663461803
49152	3.141592645321216	−0.000000008268577	0.000127831731987
98304	3.141592645321216	−0.000000008268577	0.000063915865994
196608	3.141592645321216	−0.000000008268577	0.000031957932997
393216	3.141592645321216	−0.000000008268577	0.000015978966498
786432	3.141592303811738	−0.000000349778055	0.000007989482381
1572864	3.141592303811738	−0.000000349778055	0.000003994741190
3145728	3.141586839655041	−0.000005813934752	0.000001997367121
6291456	3.141586839655041	−0.000005813934752	0.000000998683561
12582912	3.141674265021758	0.000081611431964	0.000000499355676
25165824	3.141674265021758	0.000081611431964	0.000000249677838
50331648	3.143072740170040	0.001480086580246	0.000000124894489
100663296	3.137475099502783	−0.004117554087010	0.000000062336030
201326592	3.181980515339464	0.040387861749671	0.000000031610136
402653184	3.000000000000000	−0.141592653589793	0.000000014901161
805306368	3.000000000000000	−0.141592653589793	0.000000007450581
1610612736	0.000000000000000	−3.141592653589793	0.000000000000000

TABLE 2.1. *Unstable computation of* π

discrete, finite set of *machine numbers* $\mathbb{M} = \{-\tilde{a}_{min}, \ldots, \tilde{a}_{max}\}$. Hence each real number a has to be mapped onto a machine number $\tilde{a}$ to be used on a computer. In fact a whole interval of real numbers is mapped onto one machine number, as shown in Figure 2.2.

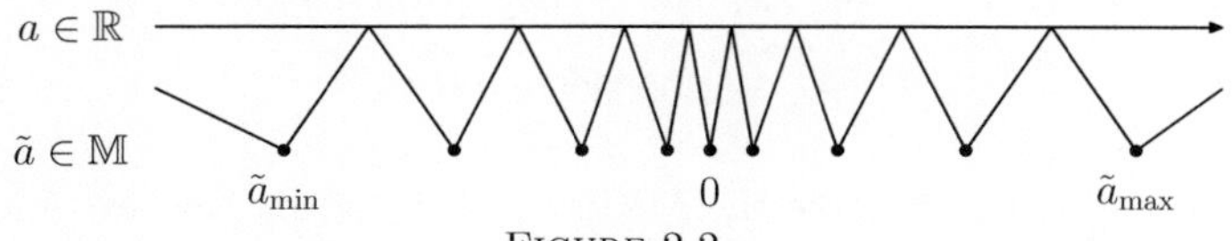

FIGURE 2.2.
Mapping of real numbers $\mathbb{R}$ onto machine numbers $\mathbb{M}$

Nowadays, machine numbers are often represented in the *binary system*. In general, any *base* (or *radix*) B could be used to represent numbers. A real machine number or *floating point number* consists of two parts, a *mantissa* (or *significant*) m and an *exponent* e

$$
\begin{aligned}
\tilde{a} &= \pm m \times B^{e} & \\
m &= D.D \cdots D & \text{mantissa} \\
e &= D \cdots D & \text{exponent}
\end{aligned}
$$

where $D \in \{0, 1, \ldots, B-1\}$ stands for one *digit*. To make the representation of machine numbers unique (note that e.g. $1.2345 \times 10^3 = 0.0012345 \times 10^6$), we require for a machine number $\tilde{a} \neq 0$ that the first digit before the decimal point in the mantissa be nonzero; such numbers are called *normalized*. One defining characteristic for any *finite precision arithmetic* is the number of digits used for the mantissa and the exponent: the number of digits in the exponent defines the *range of the machine numbers*, whereas the numbers of digits in the mantissa defines the *precision*.

More specifically [100], a finite precision arithmetic is defined by four integer parameters: B, the base or radix, p, the number of digits in the mantissa, and l and u defining the exponent range: $l \leq e \leq u$.

The *precision of the machine* is described by the real machine number *eps*. Historically, *eps is defined to be the smallest positive $\tilde{a} \in \mathbb{M}$ such that $\tilde{a} + 1 \neq 1$ when the addition is carried out on the computer.* Because this definition involves details about the behavior of floating point addition, which are not easily accessible, a newer definition of *eps* is simply *the spacing of the floating point numbers between* 1 *and* B (usually $B = 2$). The current definition only relies on how the numbers are represented.

Simple *calculators* often use the familiar decimal system ($B = 10$). Typically there are $p = 10$ digits for the mantissa and 2 for the exponent ($l = -99$ and $u = 99$). In this finite precision arithmetic, we have

- *eps*$= 0.000000001 = 1.000000000 \times 10^{-9}$,

- the largest machine number

$$
\tilde{a}_{max} = 9.999999999 \times 10^{+99},
$$

- the smallest machine number

$$\tilde{a}_{min} = -9.999999999 \times 10^{+99},$$

- the smallest (normalized) positive machine number

$$\tilde{a}_{+} = 1.000000000 \times 10^{-99}.$$

Early computers, for example the MARK 1 designed by Howard Aiken and Grace Hopper at Harvard and built in 1944, or the ERMETH (Elektronische Rechenmaschine der ETH) constructed by Heinz Rutishauser, Ambros Speiser and Eduard Stiefel, were also decimal machines. The ERMETH, built in 1956, was operational at ETH Zurich from 1956–1963. The representation of a real number used 16 decimal digits: The first digit, the q-digit, stored the sum of the digits modulo 3. This was used as a check to see if the machine word had been transmitted correctly from memory to the registers. The next three digits contained the exponent. Then the next 11 digits represented the mantissa, and finally, the last digit held the sign. The range of positive machine numbers was $1.0000000000 \times 10^{-200} \leq \tilde{a} \leq 9.9999999999 \times 10^{199}$. The possibly larger exponent range in this setting from -999 to 999 was not fully used.

In contrast, the very first programmable computer, the Z3, which was built by the German civil engineer Konrad Zuse and presented in 1941 to a group of experts only, was already using the binary system. The Z3 worked with an exponent of 7 bits and a mantissa of 14 bits (actually 15, since the numbers were normalized). The range of positive machine numbers was the interval

$$[2^{-63},\ 1.11111111111111 \times 2^{62}] \approx [1.08 \times 10^{-19},\ 9.22 \times 10^{18}].$$

In MAPLE (a computer algebra system), numerical computations are performed in base 10. The number of digits of the mantissa is defined by the variable `Digits`, which can be freely chosen. The number of digits of the exponent is given by the word length of the computer — for 32-bit machines, we have a huge maximal exponent of $u = 2^{31} = 2147483648$.

2.3 The IEEE Standard

Since 1985 we have for computer hardware the *ANSI/IEEE Standard 754 for Floating Point Numbers*. It has been adopted by almost all computer manufacturers. The base is $B = 2$.

2.3.1 Single Precision

The IEEE *single precision* floating point standard representation uses a 32-bit word with bits numbered from 0 to 31 from left to right. The first bit S is the sign bit, the next eight bits E are the exponent bits, $e = EEEEEEEE$, and the final 23 bits are the bits F of the mantissa m:

$$
\begin{array}{c c c}
 & \overbrace{}^{e} & \overbrace{}^{m} \\
S & EEEEEEEE & FFFFFFFFFFFFFFFFFFFFFFF \\
0 \quad 1 & \quad 8 \quad 9 & 31
\end{array}
$$

The value $\tilde{a}$ represented by the 32 bit word is defined as follows:

normal numbers: If $0 < e < 255$, then $\tilde{a} = (-1)^S \times 2^{e-127} \times 1.m$, where $1.m$ is the binary number created by prefixing m with an implicit leading 1 and a binary point.

subnormal numbers: If $e = 0$ and $m \neq 0$, then $\tilde{a} = (-1)^S \times 2^{-126} \times 0.m$. These are known as *denormalized (or subnormal) numbers.*

If $e = 0$ and $m = 0$ and $S = 1$, then $\tilde{a} = -0$.

If $e = 0$ and $m = 0$ and $S = 0$, then $\tilde{a} = 0$.

exceptions: If $e = 255$ and $m \neq 0$, then $\tilde{a} = $ NaN (*Not a number*)

If $e = 255$ and $m = 0$ and $S = 1$, then $\tilde{a} = -$Inf.

If $e = 255$ and $m = 0$ and $S = 0$, then $\tilde{a} = $ Inf.

Some examples:

```
0 10000000 00000000000000000000000 = +1 x 2^(128-127) x 1.0   =   2
0 10000001 10100000000000000000000 = +1 x 2^(129-127) x 1.101 =   6.5
1 10000001 10100000000000000000000 = -1 x 2^(129-127) x 1.101 = -6.5

0 00000000 00000000000000000000000 =   0
1 00000000 00000000000000000000000 = -0

0 11111111 00000000000000000000000 =   Inf
1 11111111 00000000000000000000000 = -Inf

0 11111111 00000100000000000000000 = NaN
1 11111111 00100010001001010101010 = NaN

0 00000001 00000000000000000000000 = +1 x 2^(1-127) x 1.0 =   2^(-126)
0 00000000 10000000000000000000000 = +1 x 2^(-126) x 0.1  =   2^(-127)

0 00000000 00000000000000000000001
  = +1 x 2^(-126) x 0.00000000000000000000001 = 2^(-149)
  = smallest positive denormalized machine number
```

In MATLAB, real numbers are usually represented in *double precision*. The function `single` can however be used to convert numbers to single precision. MATLAB can also print real numbers using the hexadecimal format, which is convenient for examining their internal representations:

```
>> format hex
```

```
>> x=single(2)
x =
   40000000
>> 2
ans =
   4000000000000000
>> s=realmin('single')*eps('single')
s =
00000001
>> format long
>> s
s =
   1.4012985e-45
>> s/2
ans =
     0
% Exceptions
>> z=sin(0)/sqrt(0)
Warning: Divide by zero.
z =
   NaN
>> y=log(0)
Warning: Log of zero.
y =
  -Inf
>> t=cot(0)
Warning: Divide by zero.
> In cot at 13
t =
   Inf
```

We can see that x represents the number 2 in single precision. The functions
`realmin` and `eps` with parameter `'single'` compute the machine constants
for single precision. This means that s is the smallest denormalized number
in single precision. Dividing s by 2 gives zero because of underflow. The
computation of z yields an undefined expression which results in NaN even
though the limit is defined. The final two computations for y and t show the
exceptions Inf and -Inf.

2.3.2 Double Precision

The IEEE *double precision* floating point standard representation uses a 64-
bit word with bits numbered from 0 to 63 from left to right. The first bit
S is the sign bit, the next eleven bits E are the exponent bits for e and the
final 52 bits F represent the mantissa m:

$$
\begin{array}{c c c}
 & \overbrace{e} & \overbrace{m} \\
S & EEEEEEEEEEE & FFFFF \cdots FFFFF \\
0 \quad 1 & 11 & 12 \qquad\qquad 63
\end{array}
$$

The value $\tilde{a}$ represented by the 64-bit word is defined as follows:

normal numbers: If $0 < e < 2047$, then $\tilde{a} = (-1)^S \times 2^{e-1023} \times 1.m$, where $1.m$ is the binary number created by prefixing m with an implicit leading 1 and a binary point.

subnormal numbers: If $e = 0$ and $m \neq 0$, then $\tilde{a} = (-1)^S \times 2^{-1022} \times 0.m$, which are again *denormalized* numbers.

    If $e = 0$ and $m = 0$ and $S = 1$, then $\tilde{a} = -0$.

    If $e = 0$ and $m = 0$ and $S = 0$, then $\tilde{a} = 0$.

exceptions: If $e = 2047$ and $m \neq 0$, then $\tilde{a} = \texttt{NaN}$ (*Not a number*)

    If $e = 2047$ and $m = 0$ and $S = 1$, then $\tilde{a} = -\texttt{Inf}$.

    If $e = 2047$ and $m = 0$ and $S = 0$, then $\tilde{a} = \texttt{Inf}$.

In MATLAB, real computations are performed in IEEE double precision by default. Using again the hexadecimal format in MATLAB to see the internal representation, we obtain for example

```
>> format hex
>> 2
ans =    4000000000000000
```

If we expand each hexadecimal digit to 4 binary digits we obtain for the number 2:

```
 0100 0000 0000 0000 0000 0000 ....   0000 0000 0000
```

We skipped with .... seven groups of four zero binary digits. The interpretation is: $+1 \times 2^{1024-1023} \times 1.0 = 2$.

```
>> 6.5
ans =    401a000000000000
```

This means

```
 0100 0000 0001 1010 0000 0000 ....   0000 0000 0000
```

Again we skipped with .... seven groups of four zeros. The resulting number is $+1 \times 2^{1025-1023} \times \left(1 + \frac{1}{2} + \frac{1}{8}\right) = 6.5$.

From now on, our discussion will focus on double precision arithmetic, since this is the usual mode of computation for real numbers in the IEEE Standard. Furthermore, we will stick to the IEEE Standard as used in MATLAB. In other, more low-level programming languages, the behavior of the IEEE arithmetic can be adapted, e.g. the exception handling can be explicitly specified.

- The *machine precision* is $eps = 2^{-52}$.

- The largest machine number $\tilde{a}_{max}$ is denoted by `realmax`. Note that

```
>> realmax
   ans = 1.7977e+308

>> log2(ans)
   ans = 1024

>> 2^1024
   ans = Inf
```

This looks like a contradiction at first glance, since the largest exponent should be $2^{2046-1023} = 2^{1023}$ according the IEEE conventions. But `realmax` is the number with the largest possible exponent and with the mantissa F consisting of all ones:

```
>> format hex
>> realmax
ans =   7fefffffffffffff
```

This is

$$
\begin{aligned}
V &= +1 \times 2^{2046-1023} \times 1.\underbrace{11\ldots 1}_{52\,\text{Bits}} \\[2mm]
&= 2^{1023} \times \left(1 + \left(\frac{1}{2}\right)^1 + \left(\frac{1}{2}\right)^2 + \cdots + \left(\frac{1}{2}\right)^{52} \right) \\[2mm]
&= 2^{1023} \times \frac{1 - \left(\frac{1}{2}\right)^{53}}{1 - \left(\frac{1}{2}\right)} = 2^{1023} \times (2 - eps)
\end{aligned}
$$

Even though MATLAB reports `log2(realmax)=1024`, `realmax` does not equal 2^{1024}, but rather $(2 - eps) \times 2^{1023}$; taking the logarithm of `realmax` yields 1024 only because of rounding. Similar rounding effects would also occur for machine numbers that are a bit smaller than `realmax`.

- The *computation range* is the interval $[-\texttt{realmax}, \texttt{realmax}]$. If an operation produces a result outside this interval, then it is said to *overflow*. Before the IEEE Standard, computation would halt with an error message in such a case. Now the result of an overflow operation is assigned the number $\pm\texttt{Inf}$.

- The smallest positive normalized number is $\texttt{realmin} = 2^{-1022}$.

- IEEE allows computations with *denormalized numbers*. The positive denormalized numbers are in the interval $[\texttt{realmin} * \texttt{eps}, \texttt{realmin}]$. If an operation produces a strictly positive number that is smaller than $\texttt{realmin} * \texttt{eps}$, then this result is said to be in the *underflow range*. Since such a result cannot be represented, zero is assigned instead.

- When computing with denormalized numbers, we may suffer a loss of precision. Consider the following MATLAB program:

```
>> format long
>> res=pi*realmin/123456789101112

res =  5.681754927174335e-322

>> res2=res*123456789101112/realmin

   res2 = 3.15248510554597

>> pi   = 3.14159265358979
```

The first result `res` is a denormalized number, and thus can no longer be represented with full accuracy. So when we reverse the operations and compute `res2`, we obtain a result which only contains 2 correct decimal digits. We therefore recommend avoiding the use of denormalized numbers whenever possible.

2.4 Rounding Errors

2.4.1 Standard Model of Arithmetic

Let $\tilde{a}$ and $\tilde{b}$ be two machine numbers. Then $c = \tilde{a} \times \tilde{b}$ will in general not be a machine number anymore, since the product of two numbers contains twice as many digits. The computed result will therefore be rounded to a machine number $\tilde{c}$ which is closest to c.

As an example, consider the 8-digit decimal numbers

$$\tilde{a} = 1.2345678 \quad \text{and} \quad \tilde{b} = 1.1111111,$$

whose product is

$$c = 1.37174198628258 \quad \text{and} \quad \tilde{c} = 1.3717420.$$

The *absolute rounding error* is the difference $r_a = \tilde{c} - c = 1.371742e{-}8$, and

$$r = \frac{r_a}{c} = 1e{-}8$$

is called the *relative rounding error*.

On today's computers, basic arithmetic operations obey the *standard model of arithmetic*: for $a, b \in \mathbb{M}$, we have

$$a \tilde{\oplus} b = (a \oplus b)(1 + r), \tag{2.2}$$

where r is the relative rounding error with $|r| < eps$, the machine precision. We denote with $\oplus \in \{+, -, \times, /\}$ the exact basic operation and with $\tilde{\oplus}$ the equivalent computer operation.

Another interpretation of the standard model of arithmetic is due to Wilkinson. In what follows, we will no longer use the multiplication symbol $\times$ for the exact operation; it is common practice in algebra to denote multiplication without any symbol: $ab \Longleftrightarrow a \times b$. Consider the operations

Addition: $a \tilde{+} b = (a + b)(1 + r) = (a + ar) + (b + br) = \tilde{a} + \tilde{b}$

Subtraction: $a \tilde{-} b = (a - b)(1 + r) = (a + ar) - (b + br) = \tilde{a} - \tilde{b}$

Multiplication: $a \tilde{\times} b = ab(1 + r) = a(b + br) = a\tilde{b}$

Division: $a \tilde{/} b = (a/b)(1 + r) = (a + ar)/b = \tilde{a}/b$

In each of the above, the operation satisfies

Wilkinson's Principle

The result of a numerical computation on the computer is the result of an exact computation with slightly perturbed initial data.

For example, the numerical result of the multiplication $a \tilde{\times} b$ is the exact result $a\tilde{b}$ with a slightly perturbed operand $\tilde{b} = b + br$. As a consequence of Wilkinson's Principle, we need to study the effect that slightly perturbed data have on the result of a computation. This is done in Section 2.6.

2.4.2 Cancellation

A special rounding error is called *cancellation*. If we subtract two almost equal numbers, leading digits will cancel. Consider the following two numbers with 5 decimal digits:

$$
\begin{array}{rcl}
1.2345e0 & & \\
-1.2344e0 & & \\
\hline
0.0001e0 & = & 1.0000e{-}4
\end{array}
$$

If the two numbers were exact, the result delivered by the computer would also be exact. But if the first two numbers had been obtained by previous calculations and were affected by rounding errors, then the result would at best be $1.XXXXe{-}4$, where the digits denoted by X are unknown.

This is exactly what happened in our example at the beginning of this chapter. To compute $\sin(\alpha/2)$ from $\sin \alpha$, we used the recurrence (2.1):

$$
\sin \frac{\alpha_n}{2} = \sqrt{\frac{1 - \sqrt{1 - \sin^2 \alpha_n}}{2}}.
$$

Since $\sin \alpha_n \to 0$, the numerator on the right hand side is

$$
1 - \sqrt{1 - \varepsilon^2}, \quad \text{with small} \quad \varepsilon = \sin \alpha_n,
$$

and suffers from severe cancellation. This is the reason why the algorithm performed so badly, even though the theory and program are both correct.

It is possible in this case to rearrange the computation and avoid cancellation:

$$\sin\frac{\alpha_n}{2} = \sqrt{\frac{1-\sqrt{1-\sin^2\alpha_n}}{2}} = \sqrt{\frac{1-\sqrt{1-\sin^2\alpha_n}}{2}\frac{1+\sqrt{1-\sin^2\alpha_n}}{1+\sqrt{1-\sin^2\alpha_n}}}$$

$$= \sqrt{\frac{1-(1-\sin^2\alpha_n)}{2(1+\sqrt{1-\sin^2\alpha_n})}} = \frac{\sin\alpha_n}{\sqrt{2(1+\sqrt{1-\sin^2\alpha_n})}}.$$

This last expression no longer suffers from cancellation, and we obtain the new program:

ALGORITHM 2.2. *Computation of π, Stable Version*

```
oldA=0;s=sqrt(3)/2; newA=3*s; n=6;      % initialization
z=[newA-pi n newA s];                   % store the results
while newA>oldA                         % quit if area does not increase
    oldA=newA;
    s=s/sqrt(2*(1+sqrt((1+s)*(1-s)))); % new sine value
    n=2*n; newA=n/2*s;
    z=[z; newA-pi n newA s];
end
m=length(z);
for i=1:m
   fprintf('%10d %20.15f %20.15f\n',z(i,2),z(i,3),z(i,1))
end
```

This time we do converge to the correct value of π (see Table 2.2). Notice also the elegant *termination criterion*: since the surface of the next polygon grows, we theoretically have

$$A_6 < \cdots < A_n < A_{2n} < \pi.$$

However, this cannot be true forever in finite precision arithmetic, since there is only a finite set of machine numbers. Thus, the situation $A_n \geq A_{2n}$ must occur at some stage, and this is the condition to stop the iteration. Note that this condition is *independent of the machine*, in the sense that the iteration will always terminate as long as we have finite precision arithmetic, and when it does terminate, it always gives the best possible approximation for the precision of the machine. More examples of *machine-independent algorithms* can be found in Section 2.8.1.

A second example in which cancellation occurs is the problem of *numerical differentiation* (see Chapter 8). Given a twice continuously differentiable

n	A_n	$A_n - \pi$
6	2.598076211353316	-0.543516442236477
12	3.000000000000000	-0.141592653589793
24	3.105828541230249	-0.035764112359544
48	3.132628613281238	-0.008964040308555
96	3.139350203046867	-0.002242450542926
192	3.141031950890509	-0.000560702699284
384	3.141452472285462	-0.000140181304332
768	3.141557607911857	-0.000035045677936
1536	3.141583892148318	-0.000008761441475
3072	3.141590463228050	-0.000002190361744
6144	3.141592105999271	-0.000000547590522
12288	3.141592516692156	-0.000000136897637
24576	3.141592619365383	-0.000000034224410
49152	3.141592645033690	-0.000000008556103
98304	3.141592651450766	-0.000000002139027
196608	3.141592653055036	-0.000000000534757
393216	3.141592653456104	-0.000000000133690
786432	3.141592653556371	-0.000000000033422
1572864	3.141592653581438	-0.000000000008355
3145728	3.141592653587705	-0.000000000002089
6291456	3.141592653589271	-0.000000000000522
12582912	3.141592653589663	-0.000000000000130
25165824	3.141592653589761	-0.000000000000032
50331648	3.141592653589786	-0.000000000000008
100663296	3.141592653589791	-0.000000000000002
201326592	3.141592653589794	0.000000000000000
402653184	3.141592653589794	0.000000000000001
805306368	3.141592653589794	0.000000000000001

TABLE 2.2. Stable Computation of π

function $f : \mathbb{R} \to \mathbb{R}$, suppose we wish to calculate the derivative $f'(x_0)$ at some point x_0 using the approximation

$$f'(x_0) \approx D_{x_0,h}(f) = \frac{f(x_0 + h) - f(x_0)}{h}.$$

This approximation is useful if, for instance, $f(x)$ is the result of a complex simulation, for which an analytic formula is not readily available. If we expand $f(x)$ by a Taylor series around $x = x_0$, we see that

$$f(x_0 + h) = f(x_0) + hf'(x_0) + \frac{h^2}{2}f''(\xi)$$

where $|\xi - x_0| \leq h$, so that

$$\frac{f(x_0 + h) - f(x_0)}{h} = f'(x_0) + \frac{h}{2}f''(\xi). \tag{2.3}$$

Thus, we expect the error to decrease linearly with h as we let h tend to zero. As an example, consider the problem of evaluating $f'(x_0)$ for $f(x) = e^x$ with $x_0 = 1$. We use the following code to generate a plot of the approximation error:

```
>> h=10.^(-15:0);
>> f=@(x) exp(x);
>> x0=1;
>> fp=(f(x0+h)-f(x0))./h;
>> loglog(h,abs(fp-exp(x0)));
```

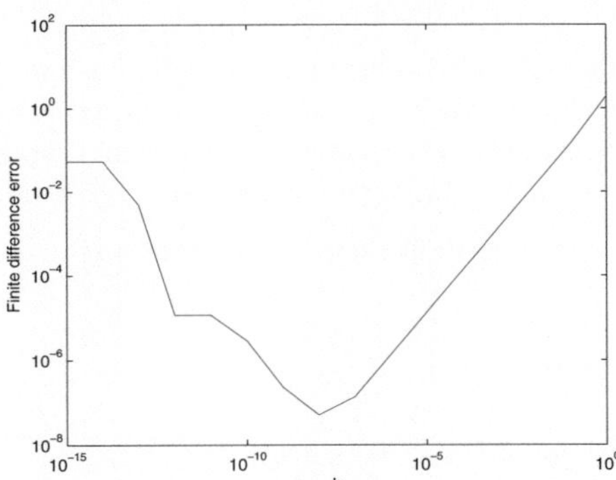

FIGURE 2.3. *Results of numerical differentiation for* $f(x) = e^x$, $x_0 = 1$

Figure 2.3 shows the resulting plot. For relatively large h, i.e., for $h > 1e{-}8$, the error is indeed proportional to h, as suggested by (2.3). However, the plot clearly shows that the error is minimal for $h \approx 1e{-}8$, and then the error increases again as h decreases further. This is again due to *severe cancellation*: when h is small, we have $f(x_0 + h) \approx f(x_0)$. In particular, since $f(x_0) = f'(x_0) = 2.71828...$ is of moderate size, we expect for $h = 10^{-t}$ that $f(x_0 + h)$ differs from $f(x_0)$ by only $|\log_{10}(eps)| - t$ digits, i.e., t digits are lost due to finite precision arithmetic. Thus, when $h < 10^{-8}$, we lose more digits due to roundoff error than the accuracy gained by a better Taylor approximation. In general, the highest relative accuracy that can be expected by this approximation is about $\sqrt{eps}$, which is a far cry from the *eps* precision promised by the machine.

We observe that in the first example, we obtain bad results due to an unstable formula, but a better implementation can be devised to remove the instability and obtain good results. In the second example, however, it is unclear how to rearrange the computation without knowing the exact formula for $f(x)$; one might suspect that the problem is inherently harder. In order to quantify what we mean by easy or hard problems, we need to introduce the notion of *conditioning*.

2.5   Condition of a Problem

Intuitively, the *conditioning* of a problem measures how sensitive it is to small changes in the data; if the problem is very sensitive, it is inherently more difficult to solve it using finite precision arithmetic. In order to properly define "small" changes, however, we need define the notion of distance for $\mathbb{R}^n$.

2.5.1   Norms

A natural way to measure distance in higher dimensions is the *Euclidean norm*, which represents the distance of two points we are used to in daily life. There are however many other ways of measuring distance, also between matrices, and these distance characterizations are called norms.

DEFINITION 2.1. (VECTOR NORM) *A vector norm is a function $\|\boldsymbol{x}\|$: $\mathbb{R}^n \longrightarrow \mathbb{R}$ such that*

1. $\|\boldsymbol{x}\| > 0$ *whenever* $\boldsymbol{x} \neq 0$.

2. $\|\alpha\boldsymbol{x}\| = |\alpha|\,\|\boldsymbol{x}\|$ *for all* $\alpha \in \mathbb{R}$ *and* $\boldsymbol{x} \in \mathbb{R}^n$.

3. $\|\boldsymbol{x} + \boldsymbol{y}\| \leq \|\boldsymbol{x}\| + \|\boldsymbol{y}\|$ *for all* $\boldsymbol{x}, \boldsymbol{y} \in \mathbb{R}^n$ *(triangle inequality)*.

Note that vector norms can also be defined for vectors in $\mathbb{C}^n$, but we will mostly concentrate on real vector spaces, with the exception of Chapter 7 (Eigenvalue problems). Frequently used norms are

The *spectral norm* or *Euclidean norm* or *2-norm*: It measures the Euclidean length of a vector and is defined by

$$\|\boldsymbol{x}\|_2 := \sqrt{\sum_{i=1}^{n} |x_i|^2}.$$

The *infinity norm* or *maximum norm*: It measures the largest element in modulus and is defined by

$$\|\boldsymbol{x}\|_\infty := \max_{1 \le i \le n} |x_i|.$$

The 1-norm: It measures the sum of all the elements in modulus and is defined by

$$\|\boldsymbol{x}\|_1 := \sum_{i=1}^{n} |x_i|.$$

DEFINITION 2.2. (MATRIX NORM) *A matrix norm is a function* $\|A\|$: $\mathbb{R}^{m \times n} \longrightarrow \mathbb{R}$ *such that*

1. $\|A\| > 0$ *whenever* $A \ne 0$.

2. $\|\alpha A\| = |\alpha|\, \|A\|$ *for all* $\alpha \in \mathbb{R}$ *and* $A \in \mathbb{R}^{m \times n}$.

3. $\|A + B\| \le \|A\| + \|B\|$ *for all* $A, B \in \mathbb{R}^{m \times n}$ *(triangle inequality)*.

When the matrix $A \in \mathbb{R}^{m \times n}$ represents a linear map between the *normed linear spaces* $\mathbb{R}^n$ and $\mathbb{R}^m$, it is customary to define the *induced matrix norm* by

$$\|A\| = \sup_{\|\boldsymbol{x}\|=1} \|A\boldsymbol{x}\|, \tag{2.4}$$

where the norms $\|\boldsymbol{x}\|$ and $\|A\boldsymbol{x}\|$ correspond to the norms used for $\mathbb{R}^n$ and $\mathbb{R}^m$ respectively. Induced matrix norms satisfy the *submultiplicative property*,

$$\|AB\| \le \|A\|\, \|B\|. \tag{2.5}$$

However, there are applications in which A does not represent a linear map, for example in data analysis, where the matrix is simply an array of data values. In such cases, we may choose to use a norm that is not an induced matrix norm, such as the norm

$$\|A\|_\Delta = \max_{i,j} |a_{ij}|,$$

or the Frobenius norm (see below). Such norms may or may not be submultiplicative: for instance, the Frobenius norm is submultiplicative, but $\| \cdot \|_\Delta$ is not.

For the vector norms we have introduced before, the corresponding *induced matrix norms* are

The spectral norm or 2-norm: (see Chapter 6)

$$\|A\|_2 := \sup_{\|\boldsymbol{x}\|_2=1} \|A\boldsymbol{x}\|_2.$$

It can be shown (see Problem 2.6) that $\|A\|_2^2$ is equal to the largest eigenvalue of $A^T A$ (or, equivalently, to the square of the largest singular value of A, see Chapter 6). It follows that the 2-norm is invariant under orthogonal transformations, i.e., we have $\|QA\|_2 = \|AQ\|_2 = \|A\|_2$ whenever $Q^\top Q = I$.

The *infinity norm* or *maximum row sum norm*:

$$\|A\|_\infty := \sup_{\|\boldsymbol{x}\|_\infty=1} \|A\boldsymbol{x}\|_\infty \equiv \max_{1 \le i \le n} \sum_{j=1}^{n} |a_{ij}|.$$

The last identity comes from the observation that the vector $\boldsymbol{x}$ which maximizes the supremum is given by $\boldsymbol{x} = (\pm 1, \pm 1, \ldots, \pm 1)^\top$ with the sign of the entries chosen according to the sign of the entries in the row of A with the largest row sum.

The *1-norm* or *maximum column sum norm*:

$$\|A\|_1 := \sup_{\|\boldsymbol{x}\|_1=1} \|A\boldsymbol{x}\|_1 \equiv \max_{1 \le j \le n} \sum_{i=1}^{n} |a_{ij}|.$$

The last identity holds because the supremum is attained for the value $\boldsymbol{x} = (0, 0, \ldots, 0, 1, 0, \ldots, 0)^\top$ with 1 at the column position of A with the largest column sum.

Note that all induced norms, including the ones above, satisfy $\|I\| = 1$, where I is the identity matrix. There is another commonly used matrix norm, the *Frobenius norm*, which does not arise from vector norms; it is defined by

$$\|A\|_F := \sqrt{\sum_{i,j=1}^{n} |a_{ij}|^2}. \tag{2.6}$$

The square of the Frobenius norm is also equal to the sum of squares of the the singular values of A, see Chapter 6.

In the finite dimensional case considered here, all norms are equivalent, which means for any pair of norms $\|\cdot\|_a$ and $\|\cdot\|_b$, there exist constants C_1 and C_2 such that

$$C_1 \|\boldsymbol{x}\|_a \le \|\boldsymbol{x}\|_b \le C_2 \|\boldsymbol{x}\|_a \quad \forall \boldsymbol{x} \in \mathbb{R}^n. \tag{2.7}$$

We are therefore free to choose the norm in which we want to measure distances; a good choice often simplifies the argument when proving a result, even though the result then holds in any norm (except possibly with a different constant). Note, however, that the constants may depend on the dimension n of the vector space, which may be large, see Problem 2.27.

2.5.2 Big- and Little-O Notation

When analyzing roundoff errors, we would often like to keep track of terms that are *very* small, e.g., terms that are proportional to eps^2, without explicitly calculating with them. The following notations, due to Landau, allows us to do just that.

DEFINITION 2.3. (BIG-O, LITTLE-O) *Let $f(x)$ and $g(x)$ be two functions. For a fixed $L \in \mathbb{R} \cup \{\pm\infty\}$, we write*

1. *$f(x) = O_{x \to L}(g(x))$ if there is a constant C such that $|f(x)| \leq C|g(x)|$ for all x in a neighborhood of L. This is equivalent to*

$$\limsup_{x \to L} \left| \frac{f(x)}{g(x)} \right| < \infty.$$

2. *$f(x) = o_{x \to L}(g(x))$ if $\lim_{x \to L} |f(x)|/|g(x)| = 0$.*

When the limit point L is clear from the context, we omit the subscript $x \to L$ and simply write $O(g(x))$ or $o(g(x))$.

The following properties of $O(\cdot)$ and $o(\cdot)$ are immediate consequences of the definition (see Problem 2.7), but are used frequently in calculations.

LEMMA 2.1. *For a given limit point $L \in \mathbb{R} \cup \{\pm\infty\}$, we have*

1. *$O(g_1) \pm O(g_2) = O(|g_1| + |g_2|)$.*

2. *$O(g_1) \cdot O(g_2) = O(g_1 g_2)$.*

3. *For any constant C, $C \cdot O(g) = O(g)$.*

4. *For a fixed function f, $O(g)/f = O(g/f)$.*

The same properties hold when $O(\cdot)$ is replaced by $o(\cdot)$.

Note carefully that we do not have an estimate for $O(g_1)/O(g_2)$: if $f_1 = O(g_1)$ and $f_2 = O(g_2)$, it is possible that f_2 is much smaller than g_2, so it is not possible to bound the quotient f_1/f_2 by g_1 and g_2.

EXAMPLE 2.1. *Let $p(x) = c_d(x-a)^d + c_{d+1}(x-a)^{d+1} + \cdots + c_D(x-a)^D$ be a polynomial with $d < D$. If c_d and c_D are both nonzero, then we have*

$$p(x) = O\left(x^D\right), \qquad as\ x \to \pm\infty,$$
$$p(x) = O\left((x-a)^d\right), \qquad as\ x \to a.$$

Thus, it is essential to know which limit point L is implied by the the big-O notation.

EXAMPLE 2.2. *Let $f : U \subset \mathbb{R} \to \mathbb{R}$ be n times continuously differentiable on an open interval U. Then for any $a \in U$ and $k \leq n$, Taylor's theorem with the remainder term tells us that*

$$f(x) = f(a) + f'(a)(x - a) + \cdots + \frac{f^{(k-1)}(a)}{(k-1)!}(x-a)^{k-1} + \frac{f^{(k)}(\xi)}{k!}(x-a)^k$$

with $|\xi - a| \leq |x - a|$. Since $f^{(k)}$ is continuous (and hence bounded), we can write

$$f(x) = f(a) + \cdots + \frac{f^{(k-1)}(a)}{(k-1)!}(x-a)^{k-1} + O\left((x-a)^k\right)$$

where the implied limit point is $L = a$. For $|h| \ll 1$, this allows us to write

$$f(a+h) = f(a) + f'(a)h + f''(a)\frac{h^2}{2} + O(h^3),$$

$$f(a-h) = f(a) - f'(a)h + f''(a)\frac{h^2}{2} + O(h^3).$$

Then it follows from Lemma 2.1 that

$$\frac{f(a+h) - f(a-h)}{2h} = f'(a) + O(h^2)$$

whenever f is at least three times continuously differentiable.

In MAPLE, *Taylor series expansions can be obtained using the command* `taylor`, *both for generic functions f and for built-in functions such as* `sin`:

```
>p1:=taylor(f(x),x=a,3);
```

$$p1 := f(a) + \mathrm{D}(f)(a)(x - a) + \frac{1}{2}(\mathrm{D}^{(2)})(f)(a)(x - a)^2 + O((x - a)^3)$$

```
> p2:=taylor(sin(x),x=0,8);
```

$$p2 := x - \frac{1}{6}x^3 + \frac{1}{120}x^5 - \frac{1}{5040}x^7 + O(x^8)$$

```
> p3:=taylor((f(x+h)-f(x-h))/2/h, h=0, 4);
```

$$p3 := \mathrm{D}(f)(x) + \frac{1}{6}(\mathrm{D}^{(3)})(f)(x)h^2 + O(h^3)$$

```
> subs(f=sin,x=0,p3);
```

$$\mathrm{D}(\sin)(0) + \frac{1}{6}(D^{(3)})(\sin)(0)h^2 + O(h^3)$$

```
> simplify(%);
```

$$1 - \frac{1}{6}h^2 + O(h^3)$$

Here, the $O(\cdot)$ in MAPLE *should be interpreted the same way as in Definition 2.3, with the limit point L given by the argument to* `taylor`*, i.e., $x \to a$ in the first command, $x \to 0$ in the second and $h \to 0$ for the remaining commands.*

2.5.3 Condition Number

DEFINITION 2.4. (CONDITION NUMBER) *The condition number κ of a problem $\mathcal{P} : \mathbb{R}^n \to \mathbb{R}^m$ is the smallest number such that*

$$\frac{|\hat{x}_i - x_i|}{|x_i|} \le \varepsilon \quad for \ \ 1 \le i \le n \quad \implies \quad \frac{\|\mathcal{P}(\hat{x}) - \mathcal{P}(x)\|}{\|\mathcal{P}(x)\|} \le \kappa\varepsilon + o(\varepsilon), \quad (2.8)$$

where $o(\varepsilon)$ represents terms that are asymptotically smaller than ε.

A problem is *well conditioned* if κ is not too large; otherwise the problem is *ill conditioned*. Well-conditioned means that the solution of the problem with slightly perturbed data does not differ much from the solution of the problem with the original data. Ill-conditioned problems are problems for which the solution is very sensitive to small changes in the data.

EXAMPLE 2.3. *We consider the problem of multiplying two real numbers, $\mathcal{P}(x_1, x_2) := x_1 x_2$. If we perturb the data slightly, say*

$$\hat{x}_1 := x_1(1 + \varepsilon_1), \quad \hat{x}_2 := x_2(1 + \varepsilon_2), \qquad |\varepsilon_i| \le \varepsilon, \ i = 1, 2,$$

we obtain

$$\frac{\hat{x}_1 \hat{x}_2 - x_1 x_2}{x_1 x_2} = (1 + \varepsilon_1)(1 + \varepsilon_2) - 1 = \varepsilon_1 + \varepsilon_2 + \varepsilon_1 \varepsilon_2,$$

and since we assumed that the perturbations are small, $\varepsilon \ll 1$, we can neglect the product $\varepsilon_1 \varepsilon_2$ compared to the sum $\varepsilon_1 + \varepsilon_2$, and we obtain

$$\frac{|\hat{x}_1 \hat{x}_2 - x_1 x_2|}{|x_1 x_2|} \le 2\varepsilon.$$

Hence the condition number of multiplication is $\kappa = 2$, and the problem of multiplying two real numbers is well conditioned.

EXAMPLE 2.4. *Let A be a fixed, non-singular $n \times n$ matrix. Consider the problem of evaluating the matrix-vector product $\mathcal{P}(x) = Ax$ for $x = (x_1, \ldots, x_n)^\top$. Suppose the perturbed vector $\hat{x}$ satisfies $\hat{x}_i = x_i(1 + \varepsilon_i)$, $|\varepsilon_i| < \varepsilon$. Then considering the infinity norm, we have $\|\hat{x} - x\|_\infty \le \varepsilon\|x\|_\infty$, so that*

$$\frac{\|A\hat{x} - Ax\|_\infty}{\|Ax\|_\infty} \le \frac{\|A\|_\infty \|\hat{x} - x\|_\infty}{\|Ax\|_\infty} \le \frac{\varepsilon\|A\|_\infty \|x\|_\infty}{\|Ax\|_\infty}.$$

Since $\|\boldsymbol{x}\| = \|A^{-1}A\boldsymbol{x}\| \le \|A^{-1}\|\,\|A\boldsymbol{x}\|$, we in fact have

$$\frac{\|A\hat{\boldsymbol{x}} - A\boldsymbol{x}\|_\infty}{\|A\boldsymbol{x}\|_\infty} \le \varepsilon \|A\|_\infty \|A^{-1}\|_\infty,$$

so the condition number is $\kappa_\infty = \|A\|_\infty \|A^{-1}\|_\infty$. We will see in Chapter 3 that κ_∞ also plays an important role in the solution of linear systems of equations.

EXAMPLE 2.5. *Let us look at the condition number of subtracting two real numbers, $\mathcal{P}(x_1, x_2) := x_1 - x_2$. Perturbing again the data slightly as in the previous example, we obtain*

$$\frac{|(\hat{x}_1 - \hat{x}_2) - (x_1 - x_2)|}{|x_1 - x_2|} = \frac{|x_1\varepsilon_1 - x_2\varepsilon_2|}{|x_1 - x_2|} \le \frac{|x_1| + |x_2|}{|x_1 - x_2|}\varepsilon.$$

We see that if $\mathrm{sign}(x_1) = -\,\mathrm{sign}(x_2)$, which means the operation is an addition, then the condition number is $\kappa = 1$, meaning that the addition of two numbers is well conditioned. If, however, the signs are the same and $x_1 \approx x_2$, then $\kappa = \frac{|x_1| + |x_2|}{|x_1 - x_2|}$ becomes very large, and hence subtraction is ill conditioned in this case.

As a numerical example, taking $x_1 = \frac{1}{51}$ and $x_1 = \frac{1}{52}$, we obtain for the condition number $\kappa = \frac{\frac{1}{51} + \frac{1}{52}}{\frac{1}{51} - \frac{1}{52}} = 103$, and if we compute with three significant digits, we obtain $\hat{x}_1 = 0.196e{-}1$, $\hat{x}_2 = 0.192e{-}1$, and $\hat{x}_1 - \hat{x}_2 = 0.400e{-}3$, which is very different from the exact result $x_1 - x_2 = 0.377e{-}3$; thus, the large condition number reflects the fact that the solution is prone to large errors due to cancellation.

In our decimal example above, the loss of accuracy is easily noticeable, since the lost digits appear as zeros. Unfortunately, cancellation effects are rarely as obvious when working with most modern computers, which use binary arithmetic. As an example, let us first compute in MAPLE

```
> Digits:=30;
> a:=1/500000000000001;
```

$$\frac{1}{500000000000001}$$

```
> b:=1/500000000000002;
```

$$\frac{1}{500000000000002}$$

```
> ce:=a-b;
```

$$\frac{1}{250000000000001500000000000002}$$

```
> c:=evalf(a-b);
```

$$3.9999999999997600000000000011 \times 10^{-30} \tag{2.9}$$

This is the exact difference, rounded to thirty digits. We round the operands to 16-digit machine numbers:

```
> bm:=evalf[16](b);
```

$$1.999999999999992 \times 10^{-15}$$

```
> am:=evalf[16](a);
```

$$1.999999999999996 \times 10^{-15}$$

Now we calculate the difference to 30-digit precision

```
> c16:=am-bm;
```

$$4.0 \times 10^{-30}$$

and also the difference to 16-digit precision to emulate floating point operations:

```
> cf16:=evalf[16](am-bm);
```

$$4.0 \times 10^{-30}$$

which agrees with to the standard model $a\tilde{-}b = (a-b)(1+r)$ with $r = 0$. In MAPLE, *we see that the precision in this difference is reduced to two digits. Only two digits are displayed, which indicates a serious loss of precision due to cancellation.*

Performing the same operations in MATLAB, *we get:*

```
format long
>> a=1/500000000000001
a =
      1.99999999999996e-15
>> b=1/500000000000002
b =
      1.99999999999992e-15
>> c=a-b
c =
      3.944304526105059e-30
```

Comparing with the exact difference in Equation (2.9), we see again that only the first two digits are accurate. The apparently random digits appearing in the difference stem from binary-to-decimal conversion, and it would be hard to guess that they are meaningless. Had we looked at the same computation in hexadecimal format in MATLAB,

```
format hex
>> a=1/500000000000001
a =
   3ce203af9ee7560b
```

```
>> b=1/500000000000002
b =
   3ce203af9ee75601
>> c=a-b
c =
   39d4000000000000
```

we would have seen that MATLAB *also completes the result by zeros, albeit in binary.*

EXAMPLE 2.6.　*Consider once again the problem of evaluating the derivative of f numerically via the formula*

$$f'(x_0) \approx D_{x_0,h}(f) = \frac{f(x_0 + h) - f(x_0)}{h}.$$

Here, the function f acts as data to the problem. We consider perturbed data of the form

$$\hat{f}(x) = f(x)(1 + \varepsilon g(x)), \qquad |g(x)| \le 1,$$

which models the effect of roundoff error on a machine with precision ε. The condition number then becomes

$$\left| \frac{D_{x_0,h}(\hat{f}) - D_{x_0,h}(f)}{D_{x_0,h}(f)} \right| = \left| \frac{f(x_0+h)(1+\varepsilon\, g(x_0+h)) - f(x_0)(1 + \varepsilon\, g(x_0))}{f(x_0 + h) - f(x_0)} - 1 \right|$$

$$\le \frac{\varepsilon(|f(x_0)| + |f(x_0 + h)|)}{|f(x_0 + h) - f(x_0)|} \approx \frac{2\varepsilon|f(x_0)|}{|hf'(x_0)|}.$$

Thus, we have $\kappa \approx 2|f(x_0)|/|hf'(x_0)|$, meaning that the problem becomes more and more ill-conditioned as $h \to 0$. In other words, if h is too small relative to the perturbation ε, it is impossible to evaluate $D_{x_0,h}(f)$ accurately, no matter how the finite difference is implemented.

A related notion in mathematics is *well-* or *ill-posed problems*, introduced by Hadamard [60]. Let $A : X \to Y$ be a mapping of some space X to Y. The problem $Ax = y$ is well posed if

1. For each $y \in Y$ there exists a solution $x \in X$.

2. The solution x is unique.

3. The solution x is a continuous function of the the data y.

If one of the conditions is not met, then the problem is said to be *ill posed.* For example, the problem of calculating the derivative f' based on the values of f alone is an ill-posed problem in the continuous setting, as can be seen from the fact that $\kappa \to \infty$ as $h \to 0$ in Example 2.6. A sensible way of choosing h to obtain maximum accuracy is discussed in Chapter 5.

If a problem is ill-posed because condition 3 is violated, then it is also *ill conditioned*. But we can also speak of an ill-conditioned problem if the problem is well posed but if the solution is very sensitive with respect to small changes in the data. A good example of an ill-conditioned problem is finding the roots of the Wilkinson polynomial, see Chapter 5. It is impossible to "cure" an ill-conditioned problem by a good algorithm, but one should avoid transforming a well-conditioned problem into an ill-conditioned one by using a bad algorithm, e.g. one that includes ill-conditioned subtractions that are not strictly necessary.

2.6 Stable and Unstable Algorithms

An algorithm for solving a given problem $\mathcal{P} : \mathbb{R}^n \longrightarrow \mathbb{R}$ is a sequence of elementary operations,

$$\mathcal{P}(x) = f_n(f_{n-1}(\ldots f_2(f_1(x))\ldots)).$$

In general, there exist several different algorithms for a given problem.

2.6.1 Forward Stability

If the amplification of the error in the operation f_i is given by the corresponding condition number $\kappa(f_i)$, we naturally obtain

$$\kappa(\mathcal{P}) \leq \kappa(f_1) \cdot \kappa(f_2) \cdot \ldots \cdot \kappa(f_n).$$

DEFINITION 2.5. (FORWARD STABILITY) *A numerical algorithm for a given problem $\mathcal{P}$ is forward stable if*

$$\kappa(f_1) \cdot \kappa(f_2) \cdot \ldots \cdot \kappa(f_n) \leq C\kappa(\mathcal{P}), \tag{2.10}$$

where C is a constant which is not too large, for example $C = O(n)$.

EXAMPLE 2.7. *Consider the following two algorithms for the problem* $\mathcal{P}(x) := \frac{1}{x(1+x)}$:

1. $x \nearrow \begin{matrix} x \\ 1+x \end{matrix} \searrow x(1+x) \to \frac{1}{x(1+x)}$

2. $x \nearrow \begin{matrix} \frac{1}{x} \\ 1+x \to \frac{1}{1+x} \end{matrix} \searrow \frac{1}{x} - \frac{1}{1+x} \to \frac{1}{x(1+x)}$

In the first algorithm, all operations are well conditioned, and hence the algorithm is forward stable. In the second algorithm, however, the last operation is a potentially very ill-conditioned subtraction, and thus this second algorithm is not forward stable.

Roughly speaking, an algorithm executed in finite precision arithmetic is called *stable if the effect of rounding errors is bounded*; if, on the other hand, an algorithm increases the condition number of a problem by a large amount, then we classify it as *unstable*.

EXAMPLE 2.8. *As a second example, we consider the problem of calculating the values*

$$\cos(1), \cos(\frac{1}{2}), \cos(\frac{1}{4}), \ldots, \cos(2^{-12}),$$

or more generally,

$$z_k = \cos(2^{-k}), \quad k = 0, 1, \ldots, n.$$

By considering perturbations of the form $\hat{z}_k = \cos(2^{-k}(1+\varepsilon))$, we can calculate the condition number for the problem using Definition 2.4:

$$\left| \frac{\cos(2^{-k}(1+\varepsilon)) - \cos(2^{-k})}{\cos(2^{-k})} \right| \approx 2^{-k}\tan(2^{-k})\varepsilon \approx 4^{-k}\varepsilon \implies \kappa(\mathcal{P}) \approx 4^{-k}.$$

We consider two algorithms for recursively calculating z_k:

1. double angle: *we use the relation $\cos 2\alpha = 2\cos^2\alpha - 1$ to compute*

$$y_n = \cos(2^{-n}), \quad y_{k-1} = 2y_k^2 - 1, \quad k = n, n-1, \ldots, 1.$$

2. half angle: *we use $\cos\frac{\alpha}{2} = \sqrt{\frac{1+\cos\alpha}{2}}$ and compute*

$$x_0 = \cos(1), \quad x_{k+1} = \sqrt{\frac{1+x_k}{2}}, \quad k = 0, 1, \ldots, n-1.$$

The results are given in Table 2.3. We notice that the y_k computed by Algorithm 1 are significantly affected by rounding errors while the computations of the x_k with Algorithm 2 do not seem to be affected. Let us analyze the condition of one step of Algorithm 1 and Algorithm 2. For Algorithm 1, one step is $f_1(y) = 2y^2 - 1$, and for the condition of the step, we obtain from

$$\frac{f_1(y(1+\varepsilon)) - f_1(y)}{f_1(y)} = \frac{f_1(y) + f_1'(y)\cdot y\varepsilon + O(\varepsilon^2) - f_1(y)}{f_1(y)} = \frac{4y^2\varepsilon + O(\varepsilon^2)}{2y^2 - 1},$$

$$(2.11)$$

and from the fact that ε is small, that the condition number for one step is $\kappa_1 = \frac{4y^2}{|2y^2-1|}$. Since all the y_k in this iteration are close to one, we have $\kappa_1 \approx 4$. Now to obtain y_k, we must perform $n-k$ iterations, so the condition

2^{-k}	$y_k - z_k$	$x_k - z_k$
1	-0.0000000005209282	0.0000000000000000
5.000000e-01	-0.0000000001483986	0.0000000000000000
2.500000e-01	-0.0000000000382899	0.0000000000000001
1.250000e-01	-0.0000000000096477	0.0000000000000001
6.250000e-02	-0.0000000000024166	0.0000000000000000
3.125000e-02	-0.0000000000006045	0.0000000000000000
1.562500e-02	-0.0000000000001511	0.0000000000000001
7.812500e-03	-0.0000000000000377	0.0000000000000001
3.906250e-03	-0.0000000000000094	0.0000000000000001
1.953125e-03	-0.0000000000000023	0.0000000000000001
9.765625e-04	-0.0000000000000006	0.0000000000000001
4.882812e-04	-0.0000000000000001	0.0000000000000001
2.441406e-04	0.0000000000000000	0.0000000000000001

TABLE 2.3. *Stable and unstable recursions*

number becomes approximately 4^{n-k}. Thus, the constant C in (2.10) of the definition of forward stability can be estimated by

$$C \approx \frac{4^{n-k}}{\kappa(\mathcal{P})} \approx \frac{4^{n-k}}{4^{-k}} = 4^n.$$

This is clearly not a small constant, so the algorithm is not forward stable.

For Algorithm 2, one step is $f_2(x) = \sqrt{\frac{1+x}{2}}$. We calculate the one-step condition number similarly:

$$\frac{f_2(y+\varepsilon) - f_2(y)}{f_2(y)} = \frac{1}{2(1+x)}\varepsilon + O(\varepsilon^2).$$

Thus, for ε small, the condition number of one step is $\kappa_2 = \frac{1}{2|1+x|}$; since all x_k in this iteration are also close to one, we obtain $\kappa_2 \approx \frac{1}{4}$. To compute x_k, we need k iterations, so the overall condition number is 4^{-k}. Hence the stability constant C in (2.10) is approximately

$$C \approx \frac{4^{-k}}{\kappa(\mathcal{P})} \approx 1,$$

meaning Algorithm 2 is stable.

We note that while Algorithm 1 runs through the iteration in an unstable manner, Algorithm 2 performs the same iteration, but in reverse. Thus, if the approach in Algorithm 1 is unstable, inverting the iteration leads to the stable Algorithm 2. This is also reflected in the one-step condition number estimate (4 and $\frac{1}{4}$ respectively), which are the inverses of each other.

Finally, using for $n = 12$ a perturbation of the size of the machine precision $\varepsilon = 2.2204e-16$, we obtain for Algorithm 1 that $4^{12}\varepsilon = 3.7e-9$, which *is a good estimate of the error $5e{-}10$ of y_0 we measured in the numerical experiment.*

EXAMPLE 2.9. *An important example of an unstable algorithm, which motivated the careful study of condition and stability, is Gaussian elimination with no pivoting (see Example 3.5 in Chapter 3). When solving linear systems using Gaussian elimination, it might happen that we eliminate an unknown using a very small pivot on the diagonal. By dividing the other entries by the small pivot, we could be introducing artificially large coefficients in the transformed matrix, thereby increasing the condition number and transforming a possibly well-conditioned problem into an ill-conditioned one. Thus, choosing small pivots makes Gaussian elimination unstable — we need to apply a* pivoting strategy *to get a numerically satisfactory algorithm (cf. Section 3.2).*

Note, however, that if we solve linear equations using orthogonal transformations (Givens rotations, or Householder reflections, see Section 3.5), then the condition number of the transformed matrices remains constant. To see this, consider the transformation

$$A\boldsymbol{x} = \boldsymbol{b} \quad \Rightarrow \quad QA\boldsymbol{x} = Q\boldsymbol{b}$$

where $Q^{\top}Q = I$. Then the 2-norm condition number of QA (as defined in Theorem 3.5) satisfies $\kappa(QA) = \|(QA)^{-1}\|_2\,\|QA\|_2 = \|A^{-1}Q^{\top}\|_2\,\|QA\|_2 = \|A^{-1}\|_2\|A\|_2 = \kappa(A)$, since the 2-norm is invariant under multiplication with orthogonal matrices.

Unfortunately, as we can see in Example 2.8, it is generally difficult and laborious to verify whether an algorithm is forward stable, since the condition numbers required by (2.10) are often hard to obtain for a given problem and algorithm. A different notion of stability, based on perturbations in the initial data rather than in the results of the algorithm, will often be more convenient to use in practice.

2.6.2   Backward Stability

Because of the difficulties in verifying forward stability, Wilkinson introduced a different notion of stability, which is based on the *Wilkinson principle* we have already seen in Section 2.4:

> *The result of a numerical computation on the computer is the result of an exact computation with slightly perturbed initial data.*

DEFINITION 2.6. (BACKWARD STABILITY) *A numerical algorithm for a given problem $\mathcal{P}$ is backward stable if the result $\hat{y}$ obtained from the algorithm with data x can be interpreted as the exact result for slightly perturbed data $\hat{x}$, $\hat{y} = \mathcal{P}(\hat{x})$, with*

$$\frac{|\hat{x}_i - x_i|}{|x_i|} \leq C\,eps, \tag{2.12}$$

where C is a constant which is not too large, and eps is the precision of the machine.

Note that in order to study the backward stability of an algorithm, one does not need to calculate the condition number of the problem itself.

Also note that a backward stable algorithm does not guarantee that the error $\|\hat{\boldsymbol{y}} - \boldsymbol{y}\|$ is small. However, if the condition number κ of the problem is known, then the relative forward error can be bounded by

$$\frac{\|\hat{\boldsymbol{y}} - \boldsymbol{y}\|}{\|\boldsymbol{y}\|} = \frac{\|\mathcal{P}(\hat{\boldsymbol{x}}) - \mathcal{P}(\boldsymbol{x})\|}{\|\mathcal{P}(\boldsymbol{x})\|} \leq \kappa(\mathcal{P}) \max_i \frac{|\hat{x}_i - x_i|}{|x_i|} \leq \kappa(\mathcal{P}) \cdot C \, eps.$$

Thus, a backward stable algorithm is automatically forward stable, but not vice versa.

EXAMPLE 2.10. *We wish to investigate the backward stability of an algorithm for the scalar product $\boldsymbol{x}^\mathsf{T} \boldsymbol{y} := x_1 y_1 + x_2 y_2$. We propose for the algorithm the sequence of operations*

$$(x_1, x_2, y_1, y_2) \quad \begin{matrix} \nearrow & x_1 y_1 & \searrow \\ & & \\ \searrow & x_2 y_2 & \nearrow \end{matrix} \quad x_1 y_1 + x_2 y_2.$$

Using the fact that storing a real number x on the computer leads to a rounded quantity $x(1 + \varepsilon)$, $|\varepsilon| \leq eps$, and that each multiplication and addition again leads to a roundoff error of the size of eps, we find for the numerical result of this algorithm

$$(x_1(1+\varepsilon_1)y_1(1+\varepsilon_2)(1+\eta_1) + x_2(1+\varepsilon_3)y_2(1+\varepsilon_4)(1+\eta_2))(1+\eta_3) = \hat{x}_1\hat{y}_1 + \hat{x}_2\hat{y}_2,$$

where $|\varepsilon_i|, |\eta_i| \leq eps$, and

$$\hat{x}_1 = x_1(1 + \varepsilon_1)(1 + \eta_1), \qquad \hat{y}_1 = y_1(1 + \varepsilon_2)(1 + \eta_3),$$
$$\hat{x}_2 = x_2(1 + \varepsilon_3)(1 + \eta_2), \qquad \hat{y}_2 = y_2(1 + \varepsilon_4)(1 + \eta_3).$$

Hence the backward stability condition (2.12) is satisfied with the constant $C = 2$, and thus this algorithm is backward stable.

EXAMPLE 2.11. *As a second example, we consider the product of two upper triangular matrices A and B,*

$$A = \begin{pmatrix} a_{11} & a_{12} \\ 0 & a_{22} \end{pmatrix}, \qquad B = \begin{pmatrix} b_{11} & b_{12} \\ 0 & b_{22} \end{pmatrix}.$$

If we neglect for simplicity of presentation the roundoff error storing the entries of the matrix in finite precision arithmetic, we obtain using the standard algorithm for computing the product of A and B on the computer

$$\begin{pmatrix} a_{11}b_{11}(1+\varepsilon_1) & (a_{11}b_{12}(1+\varepsilon_2) + a_{12}b_{22}(1+\varepsilon_3))(1+\varepsilon_4) \\ 0 & a_{22}b_{22}(1+\varepsilon_5) \end{pmatrix},$$

where $|\varepsilon_j| \le eps$ for $j = 1, \ldots, 5$. If we define the slightly modified matrices

$$\hat{A} = \begin{pmatrix} a_{11} & a_{12}(1 + \varepsilon_3)(1 + \varepsilon_4) \\ 0 & a_{22}(1 + \varepsilon_5) \end{pmatrix},$$

$$\hat{B} = \begin{pmatrix} b_{11}(1 + \varepsilon_1) & b_{12}(1 + \varepsilon_2)(1 + \varepsilon_4) \\ 0 & b_{22} \end{pmatrix},$$

their product is exactly the matrix we obtained by computing AB numerically. Hence the computed product is the exact product of slightly perturbed A and B, and this algorithm is backward stable.

The notion of backward stability will prove to be extremely useful in Chapter 3, where we use the same type of analysis to show that Gaussian elimination is in fact stable when combined with a pivoting strategy, such as complete pivoting.

2.7   Calculating with Machine Numbers: Tips and Tricks

2.7.1   Associative Law

Consider the associative law for exact arithmetic:

$$(a + b) + c = a + (b + c).$$

This law does not hold in finite precision arithmetic. As an example, take the three numbers

$$a = 1.23456e{-}3, \quad b = 1.00000e0, \quad c = -b.$$

Then it is easy to see that, in decimal arithmetic, we obtain $(a + b) + c = 1.23000e{-}3$, but $a + (b + c) = a = 1.23456e{-}3$. It is therefore important to use parentheses wisely, and also to consider the order of operations.

Assume for example that we have to compute a sum $\sum_{i=1}^{N} a_i$, where the terms $a_i > 0$ are monotonically decreasing, i.e., $a_1 > a_2 > \cdots > a_n$. More concretely, consider the harmonic series

$$S = \sum_{i=1}^{N} \frac{1}{i}.$$

For $N = 10^6$, we compute an "exact" reference value using MAPLE with sufficient accuracy (`Digits:=20`):

```
Digits:=20;
s:=0;
for i from 1 to 1000000 do
  s:=s+1.0/i:
od:
s;
        14.392726722865723804
```

Using MATLAB with IEEE arithmetic, we get

```
N=1e6;
format long e
s1=0;
for i=1:N
  s1=s1+1/i;
end
s1
ans = 1.439272672286478e+01
```

We observe that the last three digits are different from the MAPLE result. If we sum again with MATLAB but in reverse order, we obtain

```
s2=0;
for i=N:-1:1
  s2=s2+1/i;
end
s2
ans = 1.439272672286575e+01
```

a much better result, since it differs only in the last digit from the MAPLE result! We have already seen this effect in the associative law example: if we add a small number to a large one, then the least significant bits of the smaller machine number are lost. Thus, it is better to start with the smallest elements in the sum and add the largest elements last. However, sorting the terms of the sum would mean more computational work than strictly required.

2.7.2 Summation Algorithm by W. Kahan

An accurate algorithm that does not require sorting was given by W. Kahan. The idea here is to keep as carry the lower part of the small term, which would have been lost when added to the partial sum. The carry is then added to the next term, which is small enough that it would not be lost.

ALGORITHM 2.3. *Kahan's Summation of* $\sum_{j=1}^{N} \frac{1}{j}$

```
s=0;                        % partial sum
c=0;                        % carry
for j=1:N
  y=1/j+c;
  t=s+y;                    % next partial sum, with roundoff
  c=(s-t)+y;                % recapture the error and store as carry
  s=t;
end
s=s+c
```

Doing so gives a remarkably good result, which agrees to the last digit with the MAPLE result:

```
s =   1.439272672286572e+01
```

2.7.3 Small Numbers

If $a + x = a$ holds in exact arithmetic, then we conclude that $x = 0$. This is no longer true in finite precision arithmetic. In IEEE arithmetic for instance, $1 + 1e{-}20 = 1$ holds; in fact, we have $1 + w = 1$ not only for $1e{-}20$, but for all positive machine numbers w with $w < eps$, where eps is the machine precision.

2.7.4 Monotonicity

Assume we are given a function f which is strictly monotonically increasing in $[a, b]$. Then for $x_1 < x_2$ with $x_i \in [a, b]$ we have $f(x_1) < f(x_2)$. Take for example $f(x) = \sin(x)$ and $0 < x_1 < x_2 < \frac{\pi}{2}$. Can we be sure that in finite precision arithmetic $\sin(x_1) < \sin(x_2)$ also holds? The answer in general is no. For *standard functions* however, special care was taken when they were implemented in IEEE arithmetic, so that at least monotonicity is maintained, only strict monotonicity is not guaranteed. In the example, IEEE arithmetic guarantees that $\sin(x_1) \leq \sin(x_2)$ holds.

As an example, let us consider the polynomial

$$f(x) = x^3 - 3.000001x^2 + 3x - 0.999999.$$

This function is very close to $(x - 1)^3$; it has the 3 isolated roots at

$$0.998586, \quad 1.00000, \quad 1.001414.$$

Let us plot the function f:

```
figure(1)
a=-1; b=3; h=0.1;
x=a:h:b; y=x.^3-3.000001*x.^2+3*x-0.999999;
plot(x,y)
line([a,b],[0,0])
legend('x^3-3.000001*x^2+3*x-0.999999')

figure(2)
a=0.998; b=1.002; h=0.0001;
x=a:h:b; y=x.^3-3.000001*x.^2+3*x-0.999999;
plot(x,y)
line([a,b],[0,0])
legend('x^3-3.000001*x^2+3*x-0.999999')

figure(3)
a=0.999999993; b=1.000000007; h=0.000000000005;
x=a:h:b; y=x.^3-3.000001*x.^2+3*x-0.999999;
axis([a b -1e-13 1e-13])
plot(x,y)
line([a,b],[0,0])
legend('x^3-3.000001*x^2+3*x-0.999999')
```

```
figure(4)                                    % using   Horner's rule
a=0.999999993; b=1.000000007; h=0.000000000005;
x=a:h:b; y=((x-3.000001).*x+3).*x-0.999999;
axis([a b -1e-13 1e-13])
plot(x,y)
line([a,b],[0,0])
legend('((x-3.000001)*x+3)*x-0.999999')
```

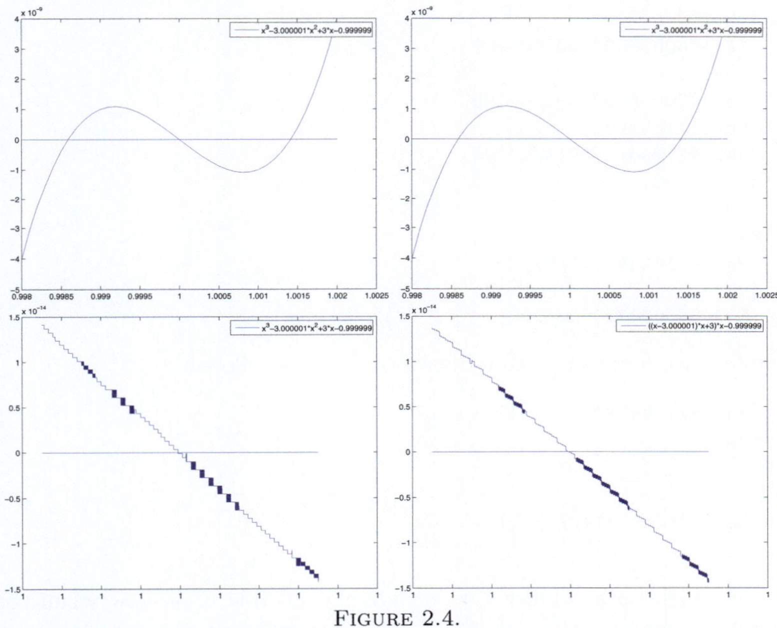

FIGURE 2.4.

Close to the middle root, monotonicity is lost in finite precision arithmetic. Using Horner's rule in the last graph, we see that the result is slightly better.

If we zoom in to the root at 1, we see in Figure 2.4 that f behaves like a step function and we cannot ensure monotonicity. The steps are less pronounced if we use for the evaluation Horner's rule, see Equation (5.62).

2.7.5 Avoiding Overflow

To avoid overflow, it is often necessary to modify the way quantities are computed. Assume for example that we wish to compute the polar coordinates of a given point (x, y) in the plane. To compute the radius $r > 0$, the textbook approach is to use

$$r = \sqrt{x^2 + y^2}.$$

However, if $|x|$ or $|y|$ is larger than $\sqrt{\texttt{realmax}}$, then x^2 or y^2 will overflow and produce the result $\texttt{Inf}$ and hence also $r = \texttt{Inf}$. Consider for example $x = 1.5e200$ and $y = 3.6e195$. Then

$$r^2 = 2.25e400 + 12.96e390 = 2.250000001296e400 > \texttt{realmax},$$

but $r = 1.500000000432e200$ would be well within the range of the machine numbers. To compute r without overflowing, one remedy is to factor out the large quantities:

```
>> x=1.5e200
x =      1.500000000000000e+200
>> y=3.6e195
y =      3.600000000000000e+195
>> if abs(x)>abs(y),
     r=abs(x)*sqrt(1+(y/x)^2)
   elseif y==0,
     r=0
   else
     r=abs(y)*sqrt((x/y)^2+1)
   end
r = 1.500000000432000e+200
```

A simpler program (with more operations) is the following:

```
m=max(abs(x),abs(y));
if m==0,
  r=0
else
  r=m*sqrt((x/m)^2+(y/m)^2)
end
```

Note that with both solutions we also avoid possible underflow when computing r.

2.7.6   Testing for Overflow

Assume we want to compute x^2 but we need to know if it overflows. With the IEEE standard, it is simple to detect this:

```
if x^2==Inf
```

Without IEEE, the computation might halt with an error message. A machine-independent test that works in almost all cases for normalized numbers is

```
if (1/x)/x==0     % then x^2 will overflow
```

In the case we want to avoid working with denormalized numbers, the test should be

```
if (eps/x)/x==0     % then x^2 will overflow
```

It is however difficult to guarantee that such a test catches overflow for all machine numbers.

In the IEEE standard, `realmin` and `realmax` are not quite symmetric, since the equation

$$\texttt{realmax} \times \texttt{realmin} = c \approx 4$$

holds with some constant c which depends on the processor and/or the version of MATLAB. In an ideal situation, we would have $c = 1$ in order to obtain perfect symmetry.

2.7.7 Avoiding Cancellation

We have already seen in Subsection 2.4.2 how to avoid cancellation when calculating the area of a circle. Consider as a second example for cancellation the computation of the exponential function using the Taylor series:

$$e^x = \sum_{j=0}^{\infty} \frac{x^j}{j!} = 1 + x + \frac{x^2}{2} + \frac{x^3}{6} + \frac{x^4}{24} + \ldots$$

It is well known that the series converges for any x. A naive approach is therefore (in preparation of the better version later, we write the computation in the loop already in a particular form):

ALGORITHM 2.4. *Computation of e^x, Naive Version*

```
function s=ExpUnstable(x,tol);
% EXPUNSTABLE computation of the exponential function
%   s=ExpUnstable(x,tol); computes an approximation s of exp(x)
%   up to a given tolerance tol.
%   WARNING: cancellation for large negative x.

s=1; term=1; k=1;
while abs(term)>tol*abs(s)
  so=s; term=term*x/k;
  s=so+term; k=k+1;
end
```

For positive x, and also small negative x, this program works quite well:

```
>> ExpUnstable(20,1e-8)
ans =    4.851651930670549e+08
>> exp(20)
ans =    4.851651954097903e+08
>> ExpUnstable(1,1e-8)
ans =    2.718281826198493e+00
>> exp(1)
ans =    2.718281828459045e+00
```

```
>> ExpUnstable(-1,1e-8)
ans =       3.678794413212817e-01
>> exp(-1)
ans =       3.678794411714423e-01
```

But for large negative x, e.g. for $x = -20$ and $x = -50$, we obtain

```
>> ExpUnstable(-20,1e-8)
ans =       5.621884467407823e-09
>> exp(-20)
ans =       2.061153622438558e-09
>> ExpUnstable(-50,1e-8)
ans =       1.107293340015503e+04
>> exp(-50)
ans =       1.928749847963918e-22
```

which are completely incorrect. The reason is that for $x = -20$, the terms in the series

$$1 - \frac{20}{1!} + \frac{20^2}{2!} - \cdots + \frac{20^{20}}{20!} - \frac{20^{21}}{21!} + \cdots$$

become large and have alternating signs. The largest terms are

$$\frac{20^{19}}{19!} = \frac{20^{20}}{20!} = 4.3e7.$$

The partial sums should converge to $e^{-20} = 2.06e-9$. But because of the growth of the terms, the partial sums become large as well and oscillate as shown in Figure 2.5. Table 2.4 shows that the largest partial sum has about the same size as the largest term. Since the large partial sums have to be

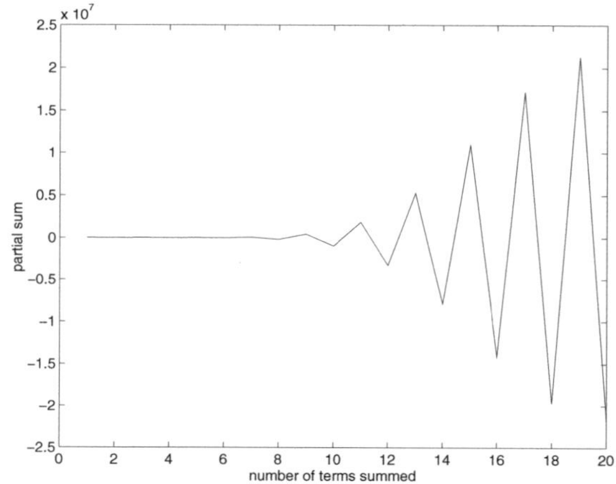

FIGURE 2.5. *Partial sum of the Taylor expansion of* e^{-20}

diminished by additions/subtractions of terms, this cannot happen without cancellation. Neither does it help to first sum up all positive and negative parts separately, because when the two sums are subtracted at the end, the result would again suffer from catastrophic cancellation. Indeed, since the result

$$e^{-20} \approx 10^{-17} \frac{20^{20}}{20!}$$

is about 17 orders of magnitude smaller than the largest intermediate partial sum and the IEEE Standard has only about 16 decimal digits of accuracy, we cannot expect to obtain even one correct digit!

number of terms summed	partial sum
20	$-2.182259377927747e + 07$
40	$-9.033771892137873e + 03$
60	$-1.042344520180466e - 04$
80	$6.138258384586164e - 09$
100	$6.138259738609464e - 09$
120	$6.138259738609464e - 09$
exact value	$2.061153622438558e\text{-}09$

TABLE 2.4. *Numerically Computed Partial Sums of* e^{-20}

2.7.8 Computation of Mean and Standard Deviation

A third example for cancellation is the recursive computation of the *mean* and the *standard deviation* of a sequence of numbers. Given the real numbers $x_1, x_2, \ldots, x_n$, the mean is

$$\mu_n = \frac{1}{n} \sum_{i=1}^{n} x_i. \tag{2.13}$$

One definition of the variance is

$$\text{var}(\boldsymbol{x}) = \frac{1}{n} \sum_{i=1}^{n} (x_i - \mu_n)^2. \tag{2.14}$$

The square-root of the variance is the standard deviation

$$\sigma_n = \sqrt{\text{var}(\boldsymbol{x})}. \tag{2.15}$$

Computing the variance using (2.14) requires two runs through the data x_i. By manipulating the variance formula as follows, we can obtain a new expression allowing us to compute both quantities with *only one run through the data*. By expanding the square bracket we obtain from (2.14)

$$\text{var}(\boldsymbol{x}) = \frac{1}{n} \sum_{i=1}^{n} (x_i^2 - 2\mu_n x_i + \mu_n^2) = \frac{1}{n} \sum_{i=1}^{n} x_i^2 - 2\mu_n \frac{1}{n} \sum_{i=1}^{n} x_i + \mu_n^2 \frac{1}{n} \sum_{i=1}^{n} 1,$$

which simplifies to

$$\sigma_n^2 = \frac{1}{n} \sum_{i=1}^{n} x_i^2 - \mu_n^2. \tag{2.16}$$

This relation leads to the classical recursive computation of mean, variance and standard deviation. In the following test, we use the values

```
x=100*ones(100,1)+1e-5*(rand(100,1)-0.5)
```

and compare the results with the MATLAB functions `mean`, `var` and `std`, which perform two runs through the data:

ALGORITHM 2.5.
*Mean, Standard Deviation – Classical Unstable
Computation*

```
format long
x=100*ones(100,1)+1e-5*(rand(100,1)-0.5);
s=0; sq=0; n=0;
while n<length(x),
  n=n+1;
  s=s+x(n);
  sq=sq+x(n)^2;
  mu=s/n;
end
means=[mu mean(x)]
sigma2=sq/n-mu^2;
variances=[sigma2 var(x,1)]
sigma=sqrt(sigma2);
standarddev=[sigma std(x,1)]
```

Each execution of these statements will be different since we use the function **rand** to generate the x_i. However, we typically get results like

```
means =
   1.0e+02 *
   1.00000000308131    1.00000000308131
variances =
   1.0e-11 *
   0.90949470177293    0.81380653750974
standarddev =
   1.0e-05 *
   0.30157829858478    0.28527294605513
```

which show that the classical formulas are numerically unstable. It may even happen that the standard deviation becomes complex because the variance becomes negative! Of course, this is a numerical effect due to severe cancellation, which can occur when using (2.16).

A better updating formula, which *avoids cancellation*, can be derived as follows:

$$
\begin{aligned}
n\sigma_n^2 &= \sum_{i=1}^{n}(x_i - \mu_n)^2 \\
&= \sum_{i=1}^{n-1}(x_i - \mu_n)^2 + (x_n - \mu_n)^2 \\
&= \sum_{i=1}^{n-1}\left((x_i - \mu_{n-1}) - (\mu_n - \mu_{n-1})\right)^2 + (x_n - \mu_n)^2 \\
&= \sum_{i=1}^{n-1}(x_i - \mu_{n-1})^2 - 2(\mu_n - \mu_{n-1})\sum_{i=1}^{n-1}(x_i - \mu_{n-1}) \\
&\qquad + (n-1)(\mu_n - \mu_{n-1})^2 + (x_n - \mu_n)^2 \\
&= (n-1)\sigma_{n-1}^2 + 0 + (n-1)(\mu_n - \mu_{n-1})^2 + (x_n - \mu_n)^2.
\end{aligned}
$$

For the mean we have the relation

$$
n\mu_n = (n-1)\mu_{n-1} + x_n,
$$

which implies

$$
\mu_{n-1} = \frac{n}{n-1}\mu_n - \frac{1}{n-1}x_n,
$$

and therefore

$$
(n-1)\left(\mu_n - \mu_{n-1}\right)^2 = (n-1)\left(\mu_n - \frac{n}{n-1}\mu_n + \frac{1}{n-1}x_n\right)^2 = \frac{(x_n - \mu_n)^2}{n-1}.
$$

Using this in the recursion for σ_n^2, we obtain

$$
n\sigma_n^2 = (n-1)\sigma_{n-1}^2 + \frac{n}{n-1}\left(x_n - \mu_n\right)^2,
$$

and finally

$$
\sigma_n^2 = \frac{n-1}{n}\sigma_{n-1}^2 + \frac{1}{n-1}\left(x_n - \mu_n\right)^2. \tag{2.17}
$$

This leads to the new algorithm

ALGORITHM 2.6.
Mean, Standard Deviation – Stable Computation

```
format long
x=100*ones(100,1)+1e-5*(rand(100,1)-0.5);
s=x(1);mu=s;sigma2=0;n=1;
while n<length(x),
```

```
    n=n+1;
    s=s+x(n);
    mu=s/n;
    sigma2=(n-1)*sigma2/n+(x(n)-mu)^2/(n-1);
end
means=[mu mean(x)]
variances=[sigma2 var(x,1)]
sigma=sqrt(sigma2);
standarddev=[sigma std(x,1)]
```

With this new algorithm, we now obtain significantly better results. A typical run gives

```
means =
    1.0e+02 *
    1.00000000308131    1.00000000308131
variances =
    1.0e-11 *
    0.81380653819342    0.81380653750974
standarddev =
    1.0e-05 *
    0.28527294617496    0.28527294605513
```

2.8 Stopping Criteria

An important problem when computing approximate solutions using a computer is to decide when the approximation is accurate enough. When computing in finite precision arithmetic, the properties discussed in the previous sections can often be exploited to design elegant algorithms that work *because of* (and not in spite of) rounding errors and the finiteness of the set of machine numbers.

2.8.1 Machine-independent Algorithms

Consider again as an example the computation of the exponential function using the Taylor series. We saw that we obtained good results for $x > 0$. Using the *Stirling Formula* $n! \sim \sqrt{2\pi} \left(\frac{n}{e}\right)^n$, we see that for a given x, the n-th term satisfies

$$t_n = \frac{x^n}{n!} \sim \frac{1}{\sqrt{2\pi}} \left(\frac{xe}{n}\right)^n \to 0, \quad n \to \infty.$$

The largest term in the expansion is therefore around $n \approx |x|$, as one can see by differentiation. For larger n, the terms decrease and converge to zero. Numerically, the term t_n becomes so small that in finite precision arithmetic we have

$$s_n + t_n = s_n, \quad \text{with} \quad s_n = \sum_{i=0}^{n} \frac{x^i}{i!}.$$

This is an elegant termination criterion which does not depend on the details of the floating point arithmetic but makes use of the finite number of digits in the mantissa. This way the algorithm is *machine-independent*; it would not work in exact arithmetic, however, since it would never terminate.

In order to avoid cancellation when $x < 0$, we use a property of the exponential function, namely $e^x = 1/e^{-x}$: we first compute $e^{|x|}$, and then $e^x = 1/e^{|x|}$. We thus get the following stable algorithm for computing the exponential function for all x:

ALGORITHM 2.7. *Stable Computation of e^x*

```
function s=Exp(x);
% EXP stable computation of the exponential function
%   s=Exp(x); computes an approximation s of exp(x) up to machine
%   precision.

if x<0, v=-1; x=abs(x); else v=1; end
so=0; s=1; term=1; k=1;
while s~=so
  so=s; term=term*x/k;
  s=so+term; k=k+1;
end
if v<0, s=1/s; end;
```

We now obtain very good results also for large negative x:

```
>> Exp(-20)
ans =      2.061153622438558e-09
>> exp(-20)
ans =      2.061153622438558e-09

>> Exp(-50)
ans =      1.928749847963917e-22
>> exp(-50)
ans =      1.928749847963918e-22
```

Note that we have to compute the terms recursively

$$t_k = t_{k-1}\frac{x}{k} \quad \text{and not explicitly} \quad t_k = \frac{x^k}{k!}$$

in order to avoid possible overflow in the numerator or denominator.

As a second example, consider the problem of designing an algorithm to compute the square root. Given $a > 0$, we wish to compute

$$x = \sqrt{a} \iff f(x) = x^2 - a = 0.$$

Applying *Newton's iteration* (see Section 5.2.5), we obtain

$$x - \frac{f(x)}{f'(x)} = x - \frac{x^2 - a}{2x} = \frac{1}{2}\left(x + \frac{a}{x}\right)$$

and the quadratically convergent iteration (also known as *Heron's formula*)

$$x_{k+1} = (x_k + a/x_k)/2. \tag{2.18}$$

When should we terminate the iteration? We could of course test to see if successive iterations are identical up to some relative tolerance. But here we can develop a much nicer termination criterion. The geometric interpretation of Newton's method shows us that if $\sqrt{a} < x_k$ then $\sqrt{a} < x_{k+1} < x_k$. Thus if we start the iteration with $\sqrt{a} < x_0$ then the sequence $\{x_k\}$ is *monotonically decreasing* toward $s = \sqrt{a}$. This monotonicity cannot hold forever on a machine with finite precision arithmetic. So when it is lost we have reached machine precision.

To use this criterion, we must ensure that $\sqrt{a} < x_0$. This is easily achieved, because one can see geometrically that after the first iteration starting with any positive number, the next iterate is always larger than $\sqrt{a}$. If we start for example with $x_0 = 1$, the next iterate is $(1 + a)/2 \geq \sqrt{a}$. Thus we obtain Algorithm 2.8.

ALGORITHM 2.8. *Computing $\sqrt{x}$ machine-independently*

```
function y=Sqrt(a);
% SQRT computes the square-root of a positive number
%    y=Sqrt(a); computes the square-root of the positive real
%    number a using Newton's method, up to machine precision.

xo=(1+a)/2; xn=(xo+a/xo)/2;
while xn<xo
  xo=xn; xn=(xo+a/xo)/2;
end
y=(xo+xn)/2;
```

Notice the elegance of Algorithm 2.8: there is no tolerance needed for the termination criterion. The algorithm computes the square root on any computer without knowing the machine precision by simply using the fact that there is always only a *finite set of machine numbers*. This algorithm would not work on a machine with exact arithmetic — it relies on finite precision arithmetic. Often these are the best algorithms one can design.

Another example of a *fool-proof and machine-independent algorithm* is given in Chapter 5. The bisection algorithm for finding a simple root makes use of the fact that there is only a finite set of machine numbers. Bisection is continued as long as there is a machine number in the interval (a, b). When the interval consists only of the endpoints then the iteration is terminated in a *machine-independent way*. See Algorithm 5.2 for details.

Machine-independent algorithms are not easy to find. We show in the next subsections two generic stopping criteria that are often used in practice when no machine-independent criterion is available.

2.8.2 Test Successive Approximations

If we are interested in the limit s of a convergent sequence x_k, a commonly used stopping criterion is to check the absolute or relative difference of two *successive approximations*

$$|x_{k+1} - x_k| < \texttt{tol} \quad \text{absolute or} \quad |x_{k+1} - x_k| < \texttt{tol}|x_{k+1}| \quad \text{relative ``error''}.$$

The test involves the absolute (or relative) difference of two successive iterates, to which one often refers somewhat sloppily as *absolute* or *relative error*. It is of course questionable whether the corresponding errors $|x_{k+1} - s|$ and $|x_{k+1} - s|/|s|$ are indeed small. This is certainly not the case if convergence is very slow (see Chapter 5 , Equation (5.101)), since we can be far away from the solution s and making very small steps toward it. In that case, the above stopping criterion will terminate the iteration prematurely.

Consider as an example the equation $xe^{10x} = 0.001$. A fixed point iteration is obtained by adding x an both sides and dividing by $1 + e^{10x}$,

$$x_{k+1} = \frac{0.001 + x_k}{1 + e^{10x_k}}. \tag{2.19}$$

If we start the iteration with $x_0 = -10$, we obtain the iterates

$$x_1 = -9.9990, \quad x_2 = -9.9980, \quad x_3 = -9.9970.$$

It would be incorrect to conclude that we are close to the solution $s \approx -9.99$, since the only solution of this equation is $s = 0.0009901473844$.

We will see in Chapter 5 that for fixed point iterations the Banach Fixed Point Theorem often allows us to derive a stopping criterion based on the difference of consecutive iterates, which guarantees asymptotically that the current approximation is within a given tolerance of the solution, see Equation (5.102).

2.8.3 Check the Residual

Another possibility to check whether an approximate solution is good enough is to insert this approximation into the equation to be solved, so that one can measure the amount by which the approximation fails to satisfy the equation. This discrepancy is called the *residual r*. For example, in case of the square root above, one might want to check if $r = x_k^2 - a$ is small in absolute value. In the case of a system of linear equations, $A\boldsymbol{x} = \boldsymbol{b}$, one checks how small the residual

$$\boldsymbol{r} = \boldsymbol{b} - A\boldsymbol{x}_k$$

becomes in some norm for an approximate solution $\boldsymbol{x}_k$.

Unfortunately, *a small residual does not guarantee that we are close to a solution either*! Take as an example the linear system

$$A\boldsymbol{x} = \boldsymbol{b}, \quad A = \begin{pmatrix} 0.4343 & 0.4340 \\ 0.4340 & 0.4337 \end{pmatrix} \quad \boldsymbol{b} = \begin{pmatrix} 1 \\ 0 \end{pmatrix}.$$

The exact solution is

$$x = \frac{1}{9}\begin{pmatrix} -43370000 \\ 43400000 \end{pmatrix} = \begin{pmatrix} -4.81888\ldots \\ 4.82222\ldots \end{pmatrix}10^6.$$

The entries of the matrix A are decimal numbers with 4 digits. The best 4-digit decimal approximation to the exact solution is

$$x_4 = \begin{pmatrix} -4819000 \\ 4822000 \end{pmatrix}.$$

Now if we compute the residual of that approximation we obtain the rather large residual

$$r_4 = b - Ax_4 = \begin{pmatrix} 144.7 \\ 144.6 \end{pmatrix};$$

We can easily guess solutions with smaller residuals, but which clearly are not better solutions. For example, we could have proposed as solution

$$x_1 = \begin{pmatrix} -1 \\ 1 \end{pmatrix} \quad \Rightarrow r_1 = b - Ax_1 = \begin{pmatrix} 1.0003 \\ 0.0003 \end{pmatrix},$$

which clearly has a smaller residual. In fact, the residual of $x = (0,0)^\top$ is $r = b = (1,0)^\top$, which is even smaller! *Thus, we cannot trust small residuals to always imply that we are close to a solution.*

2.9   Problems

PROBLEM 2.1. *Verify with some examples the standard model of arithmetic (2.2). For this purpose, write a* MAPLE *program and use extended arithmetic for the exact operations. Truncate the operands to "machine numbers" by using a statement like* **am:=evalf[7](a)** *to get a 7-digit number. Do the same for the machine operation.*

PROBLEM 2.2. *Single precision numbers can be defined in* MATLAB *with the function* **single**. *Run the statements*

```
>> format hex
>> y=single(6.5)
```

Convert the hexadecimal number to the corresponding binary number by expanding each hexadecimal digit into 4 binary digits (bits). Finally, interpret the 32 bits as a single precision floating point number. Verify that it really represents the decimal number 6.5.

PROBLEM 2.3. *Try to* compute *the parameters that define the finite arithmetic in* MATLAB. *Compare your results with the machine constants of the IEEE standard. Write* MATLAB *programs to compute*

1. *the machine precision eps. Hint: Use the definition that eps is the smallest positive machine number such that numerically* $1 + eps > 1$ *holds. Compare your result with the* MATLAB *constant* **eps**.

2. *the smallest positive normalized machine number* α. *Compare your result with* **realmin**.

3. *the smallest positive denormalized number. How can this number be computed by the IEEE constants?*

4. *the largest positive machine number* γ. *This is not easy to compute. An approximation is given by* $1/\alpha$. *Compare this value with the* MATLAB *constant* **realmax**.

PROBLEM 2.4. *Do the same as in the previous problem but for the finite precision arithmetic used in* MAPLE. *Use the standard value for the precision* **Digits:=10**.

1. *Explain first why the following* MAPLE *program to compute the machine precision*

```
eps:=1.0;
for i from 1 while 1.0+eps>1.0 do
  eps:=eps/2.0;
end do;
```

   does not work. Change the program and make it work! Recall that MAPLE *uses decimal rather than binary arithmetic.*

2. *Hint: To find the* MAPLE **realmin** *use in your loop the statement*

```
realmin:=realmin/1.0e100000000;
```

   otherwise you will wait too long! Then refine your guess iteratively by dividing with smaller factors. Convince yourself by an experiment that there are no denormalized numbers in MAPLE.

3. *Verify that* **realmax** $= 1/$**realmin** $= 1.0 \times 10^{2147483646}$.

PROBLEM 2.5. (IEEE QUADRUPLE PRECISION ARITHMETIC) *In problems where more precision is needed, the IEEE standard provides a specification for quadruple precision numbers, which occupy 128 bits of storage. Look up this specification on the web, such as the number of bits attributed to the mantissa and exponent, the machine precision eps, the range of representable numbers, etc. Give the sequence of bits representing the real number 11.25 in quadruple precision.*

PROBLEM 2.6. (MATRIX 2-NORM) *In this exercise, we will show that $\|A\|_2$ is the square root of the largest eigenvalue of $A^\top A$.*

1. *$A^\top A$ is clearly a square symmetric matrix. By the spectral theorem, there exists a complete set of orthonormal eigenvectors $\boldsymbol{v}_1, \ldots, \boldsymbol{v}_n$ with corresponding eigenvalues $\lambda_1, \ldots, \lambda_n$. Show that $\lambda_i \geq 0$ for all i.*

2. *Let $\boldsymbol{x} = \sum_{i=1}^{n} c_i \boldsymbol{v}_i$ be an arbitrary nonzero vector in $\mathbb{R}^n$, expressed in the basis of eigenvectors. Show that*

$$\|A\boldsymbol{x}\|_2^2 = \boldsymbol{x}^\top A^\top A \boldsymbol{x} = \sum_{i=1}^{n} c_i^2 \lambda_i. \tag{2.20}$$

3. *Assume without loss of generality that the eigenvalues are arranged in descending order, i.e., $\lambda_1 \geq \lambda_2 \geq \cdots \geq \lambda_n \geq 0$. Deduce from (2.20) that*

$$\|A\boldsymbol{x}\|_2^2 \leq \lambda_1 \sum_{i=1}^{n} c_i^2 = \lambda_1 \|\boldsymbol{x}\|_2^2.$$

Conclude that $\|A\|_2 \leq \sqrt{\lambda_1}$.

4. *Give an example of a vector $\boldsymbol{x}$ satisfying $\|A\boldsymbol{x}\|_2^2 = \lambda_1$. Conclude that $\|A\|_2 = \sqrt{\lambda_1}$.*

PROBLEM 2.7. (BIG- AND LITTLE-O NOTATION) *Prove the properties of $O(\cdot)$ and $o(\cdot)$ given in Lemma 2.1.* **Hint:** *For the first property, let $f_1 = O(g_1)$ and $f_2 = O(g_2)$ be two functions. Then use the definition of $O(\cdot)$ to show that $|f_1 + f_2| \leq C(|g_1| + |g_2|)$ for some constant C. Proceed similarly for the other properties.*

PROBLEM 2.8. *Determine the condition number of the elementary operations $/$ and $\sqrt{}$.*

PROBLEM 2.9. (CONDITION NUMBER AND DERIVATIVES) *Consider the problem $\mathcal{P} : x \mapsto f(x)$, in which $f : U \subset \mathbb{R} \to \mathbb{R}$ is a differentiable function on the open subset U. Give a formula for the condition number κ in terms of the function $f(x)$ and its derivative $f'(x)$.* **Hint:** *You may find (2.11) helpful.*

PROBLEM 2.10. *Consider the problem*

$$\mathcal{P} : D \subset \mathbb{R}^2 \to \mathbb{R}^2 : \begin{pmatrix} a \\ b \end{pmatrix} \mapsto \begin{pmatrix} x_1 \\ x_2 \end{pmatrix}, \tag{2.21}$$

where x_1 and x_2 are the roots of the second degree polynomial $p(x) = x^2 + 2ax + b$ and $D = \{(a,b) \in \mathbb{R}^2 \mid b < a^2\}$. Is this problem well conditioned? Use the infinity norm to calculate the condition number κ.

PROBLEM 2.11. *Study the forward and backward stability of the algorithm*

$$x_{1,2} = -a \pm \sqrt{a^2 - b}$$

for solving the problem (2.21). Also study the forward and backward stability of the improved algorithm in Problem 2.14 that uses Vieta's formula, and explain why the latter algorithm is preferable.

PROBLEM 2.12. (MONOTONICITY) *Suppose we are operating in a base-10 finite precision arithmetic with two significant digits in the mantissa. Let $\mathbb{M}$ be the set of machine numbers. Show that the function $f : \mathbb{M} \to \mathbb{M}$ defined by $x \mapsto x^2$ is not strictly monotonic by finding two machine numbers $a, b \in \mathbb{M}$ such that $a \neq b$, but $a \,\tilde{\times}\, a = b \,\tilde{\times}\, b$.*

PROBLEM 2.13. *Write a* MATLAB *function* `[r,phi] = topolar(x,y)` *to convert the Cartesian coordinates of a point (x, y) to polar coordinates (r, ϕ). In other words, the function should solve the equations*

$$\begin{aligned} x &= r \cos \phi \\ y &= r \sin \phi \end{aligned}$$

for r and ϕ. **Hint:** *study the* MATLAB *function* `atan2` *and avoid under- and overflow.*

PROBLEM 2.14. *Write a fool-proof* MATLAB *program to solve the quadratic equation*

$$x^2 + px + q = 0,$$

where p and q can be arbitrary machine numbers. Your program has to compute the solutions x_1 and x_2 if they lie in the range of machine numbers.
Hints: *consider the well-known formula*

$$x_{1,2} = -\frac{p}{2} \pm \sqrt{\left(\frac{p}{2}\right)^2 - q}.$$

If $|p|$ is large, you might get overflow when you square it. Furthermore, there is cancellation if $|p| \gg |q|$ for x_1.
Avoid overflow by rewriting of the formula appropriately. Avoid cancellation by using the relation: $x_1 x_2 = q$ (Vieta's formula).

PROBLEM 2.15. (LAW OF COSINES) *Given one angle γ and the two adjacent sides a and b of a triangle, the opposite side can be determined using the law of cosines:*

$$c = \sqrt{a^2 + b^2 - 2ab \cos \gamma}.$$

1. *Numerically we can expect problems for a small angle γ and if $a \simeq b \gg c$. The result c will be affected by cancellation in this case.*

   Change the formula to avoid cancellation by introducing $-2ab + 2ab$ in the square root and using the half angle formula

   $$\sin^2\left(\frac{\alpha}{2}\right) = \frac{1 - \cos\alpha}{2}.$$

   You should obtain the new and more reliable expression

   $$c = \sqrt{(a-b)^2 + 4ab\sin^2\left(\frac{\gamma}{2}\right)}$$

2. *Simulate on a pocket computer a 2-decimal-digit arithmetic by rounding the result after each operation to 2 decimal digits. Use the values $a = 5.6$, $b = 5.7$ and $\gamma = 5°$ and compute the side c with the two formulas for the Law of Cosines.*

PROBLEM 2.16. *The law of cosines can be used to compute the circumference of an ellipse.*

1. *Represent the ellipse in polar coordinates*

   $$r(\phi) = \frac{b}{\sqrt{1 - \epsilon^2\cos^2\phi}}, \quad \epsilon^2 = \frac{a^2 - b^2}{a^2}, \quad a \geq b.$$

2. *Consider now a partition $\phi_n = \frac{2\pi}{n}$ and the triangle with angle ϕ_n and the two adjacent sides $r(k\phi_n)$ and $r((k+1)\phi_n)$.*

   Compute the third side of this triangle (a chord on the ellipse) using the law of cosines. Sum up all the n chords to obtain this way an approximation of the ellipse circumference.

3. *Compare your approximation as $n \to \infty$ with the "exact" value, which you can obtain by numerically integrating the elliptic integral*

   $$U = a\int_0^{2\pi} \sqrt{1 - \epsilon^2\cos^2 t}\, dt \quad \epsilon^2 = \frac{a^2 - b^2}{a^2}, \quad a \geq b.$$

   Notice the difference that you obtain using the textbook formula and the stable formula for the law of cosines.

4. *Implement the* MATLAB *function*

```
function   [U,n]=Circumference(a,b)
% CIRCUMF computes the circumference of an ellipse
%    [U,n]=circumf(a,b) computes the circumference U of the ellipse
%    with semiaxes a and b and returns the number n of chords used to
%    approximate the circumference.
```

Use the stable formula and an elegant machine-independent termination criterion: start with $n = 4$ and double n in each step. The sequence of approximations should increase monotonically. Stop the iteration when the monotonicity is lost. Be careful to implement this algorithm efficiently — it takes a large number of operations. Avoid recomputations as much as possible.

PROBLEM 2.17. *The function $\ln(1+x)$ is evaluated inaccurately for small $|x|$.*

1. *Evaluate this function in* MATLAB *for $x = 0.1, 0.01, \ldots, 10^{-11}$.*

2. *Check the obtained values by computing the same in* MAPLE *using* `Digits := 20`.

3. *Program now in* MATLAB *and evaluate for the same arguments the function*

$$
\ln(1 + x) = \begin{cases} x & \text{if } 1 + x = 1 \text{ numerically,} \\ \dfrac{x \ln(1+x)}{(1+x) - 1} & \text{if } 1 + x \neq 1 \text{ numerically.} \end{cases}
$$

Comment on your results. Can you explain why this works? This transformation is another clever idea by W. Kahan.

PROBLEM 2.18. *We have seen that computing $f(x) = e^x$ using its Taylor series is not feasible for $x = -20$ because of catastrophic cancellation. However, the series can be used without problems for small $|x| < 1$.*

1. *Try therefore the following idea:*

$$
e^x = \left(\cdots \left(e^{\frac{x}{2^m}} \right)^2 \cdots \right)^2 .
$$

This means that we first compute a number m such that

$$
z = \frac{x}{2^m}, \quad \text{with} \quad |z| < 1.
$$

The we use the series to compute e^z and finally we get the result by squaring m times.

Write a MATLAB *function* `function y=es(x)` *that computes e^x this way and compare the results with the* MATLAB *function* `exp`.

2. *Perform an error analysis of this algorithm.*

PROBLEM 2.19. *The function*

$$f(x) = \ln((1 + x^4)^2 - 1)$$

is computed inexactly in IEEE arithmetic (the results may even be completely wrong) for small positive values of x. Already for $x \approx 10^{-3}$, we obtain only about 8 correct decimal digits. For $x \approx 10^{-4}$ we get in MATLAB *-Inf.*

Write a MATLAB *function* y=f(x) *which computes the correct function values for all* realmin $\leq x < 10^{-3}$.

PROBLEM 2.20. *When evaluating*

$$f(x) = \frac{e^x - 1 - x}{x^2}$$

on the computer, we observe a large relative error for values $x \approx 0$.

1. *Explain what happens.*

2. *Find a method to compute f for $|x| < 1$ to machine precision and write a* MATLAB *function for computing f.*

PROBLEM 2.21. *When evaluating*

$$f(x) = \frac{x^2}{(\cos(\sin x))^2 - 1}$$

on the computer, we observe a large relative error for values $x \approx 0$.

1. *Explain what happens.*

2. *Find a method to compute f for $|x| < 1$ to machine precision and write a* MATLAB *function for computing f.*

PROBLEM 2.22. (CHECKING SUCCESSIVE ITERATES AS STOPPING CRITERION)
Assume that the fixed-point iteration $x_{k+1} = F(x_k)$ yields a linearly convergent sequence $x_k \to s$ and that for the error $e_k = |x_k - s|$, the relation $e_{k+1} \leq ce_k$ holds with $0 < c < 1$. Investigate for what values of c we can conclude that if $|x_{k+1} - x_k| \leq \epsilon$ holds, then also $e_{k+1} < \epsilon$.

PROBLEM 2.23. *Write a* MATLAB *function to compute the sine function in a machine-independent way using its Taylor series. Since the series is alternating, cancellation will occur for large $|x|$.*

To avoid cancellation, reduce the argument x of $\sin(x)$ to the interval $[0, \frac{\pi}{2}]$. Then sum the Taylor series and stop the summation with the machine-independent criterion $s_n + t_n = s_n$, where s_n denotes the partial sum and t_n the next term. Compare the exact values for $[\boldsymbol{sin}(-10 + k/100)]_{k=0,\dots,2000}$

with the ones you obtain from your MATLAB *function and plot the relative error.*

PROBLEM 2.24. *The function*

$$f(x) = \frac{x^2}{1!} + 7\frac{x^4}{2!} + 17\frac{x^6}{3!} + \cdots = \sum_{n=1}^{\infty} (2n^2 - 1)\frac{x^{2n}}{n!}$$

should be evaluated to machine precision for x *in the range* $0 < x < 25$. *Write a* MATLAB *function for this purpose. Pay particular attention to*

a) *compute the result to machine precision with an elegant stopping criterion;*

b) *avoid any potential overflows.*

PROBLEM 2.25. *We would like to compute the integrals*

$$y_n = \int_0^1 \frac{x^n}{x + a}dx$$

for $n = 0, 1, 2, \ldots, 30$ *and* $a > 0$.

1. *Show that the following recurrence holds:*

$$y_n = \frac{1}{n} - ay_{n-1}, \quad y_0 = \log\left(\frac{1+a}{a}\right). \tag{2.22}$$

2. *Compute upper and lower bounds for the values* y_n *by choosing* $x = 0$, *respectively* $x = 1$, *in the denominator of the integrand.*

3. *Compute for* $a = 10$ *the sequence* $\{y_n\}$ *for* $n = 1, \ldots, 30$ *using (2.22) repeatedly. Print a table with the values and their bounds.*

4. *Solve (2.22) for* y_{n-1} *and compute again the sequence for* $a = 10$, *this time backwards starting from* $n = 30$. *Take as starting value the lower bound for* y_{30}.

5. *Finally, check your results by computing the integrals directly using the* MATLAB *function* **quad**.

PROBLEM 2.26. *Given an integer* n, *one often needs to compute the function values*

$$s_k = \sin(k\phi), \quad k = 1, \ldots, n \quad for \quad \phi = \frac{2\pi}{n}.$$

Instead of invoking the sine function n times to calculate s_k, one could also calculate them recursively using the trigonometric identities

$$\begin{pmatrix} \cos((k+1)\phi) \\ \sin((k+1)\phi) \end{pmatrix} = \begin{pmatrix} \cos\phi & -\sin\phi \\ \sin\phi & \cos\phi \end{pmatrix} \begin{pmatrix} \cos(k\phi) \\ \sin(k\phi) \end{pmatrix}. \tag{2.23}$$

Perform some experiments to compare the computing time versus the accuracy, then design a fast "mixed" algorithm.

PROBLEM 2.27. *Vector and matrix norms*

1. *Show that for vectors $\boldsymbol{x}$ the infinity norm $\|\boldsymbol{x}\|_\infty$ and the one norm $\|\boldsymbol{x}\|_1$ are norms according to Definition 2.1.*

2. *Show that for matrices A the Frobenius norm $\|A\|_F$, the infinity norm $\|A\|_\infty$ and the one norm $\|A\|_1$ are norms according to Definition 2.2. Show that each norm is submultiplicative.*

3. *Two vector norms $\|\cdot\|_A$ and $\|\cdot\|_B$ are called* equivalent *if there are positive constants C_1 and C_2 such that for all $\boldsymbol{x}$ we have*

$$C_1\|\boldsymbol{x}\|_A \le \|\boldsymbol{x}\|_B \le C_2\|\boldsymbol{x}\|_A.$$

Show that for $\boldsymbol{x} \in \mathbb{R}^n$,

$$\begin{aligned}
\|\boldsymbol{x}\|_2 &\le \|\boldsymbol{x}\|_1 &&\le \sqrt{n}\|\boldsymbol{x}\|_2, \\
\|\boldsymbol{x}\|_\infty &\le \|\boldsymbol{x}\|_2 &&\le \sqrt{n}\|\boldsymbol{x}\|_\infty, \\
\|\boldsymbol{x}\|_\infty &\le \|\boldsymbol{x}\|_1 &&\le n\|\boldsymbol{x}\|_\infty.
\end{aligned}$$

(Hint: Squaring the inequalities helps for the two norm.)

Chapter 3. Linear Systems of Equations

> *The conversion of a general system to triangular form via Gauss transformations is then presented, where the "language" of matrix factorizations is introduced.*
>
> *As we have just seen, triangular systems are "easy" to solve. The idea behind Gaussian elimination is to convert a given system $A\boldsymbol{x} = \boldsymbol{b}$ to an equivalent triangular system.*
>
> Golub and Van Loan, Matrix Computations, Third Edition, Johns Hopkins University Press, 1996

Prerequisites: Sections 2.2 (finite-precision arithmetic), 2.5 (conditioning) and 2.6 (stability) are required for this chapter.

Solving a system of linear equations is one of the most frequent tasks in numerical computing. The reason is twofold: historically, many phenomena in physics and engineering have been modeled by linear differential equations, since they are much easier to analyze than nonlinear ones. In addition, even when the model is nonlinear, the problem is often solved iteratively as a sequence of linear problems, e.g., by Newton's method (Chapter 5). Thus, it is important to be able to solve linear equations efficiently and robustly, and to understand how numerical artifacts affect the quality of the solution. We start with an introductory example, where we also mention Cramer's rule, a formula used by generations of mathematicians to write down explicit solutions of linear systems, but which is not at all suitable for computations. We then show in Section 3.2 the fundamental technique of Gaussian elimination with pivoting, which is the basis of LU decomposition, the workhorse present in all modern dense linear solvers. This decomposition approach to matrix computations, pioneered by Householder, represents a real paradigm shift in the solution of linear systems and is listed as one of the *top ten algorithms of the last century* [27] think factorization, not solution (see the quote above). In Section 3.3, we introduce the important concept of the condition number of a matrix, which is the essential quantity for understanding the condition of solving a linear system. We then use Wilkinson's Principle to show how the condition number influences the expected accuracy of the solution of the associated linear system. The special case of symmetric positive definite systems is discussed in Section 3.4, where the LU factorization can be expressed in the very special form $L = U^T$ due to symmetry, leading to the so-called

W. Gander et al., *Scientific Computing - An Introduction using Maple and MATLAB*, Texts in Computational Science and Engineering 11, DOI 10.1007/978-3-319-04325-8_3,

Cholesky factorization. An alternative for computing the solution of linear systems that does not require pivoting is shown in Section 3.5, where Givens rotations are introduced. We conclude this chapter with special factorization techniques for banded matrices in Section 3.6. The focus of the chapter is on direct methods; iterative methods for linear systems are discussed in Chapter 11. Moreover, we consider in this chapter only *square* linear systems (systems that have as many equations as unknowns) whose matrices are *non-singular*, i.e., systems that have a unique solution. In Chapter 6, we discuss how to solve problems with more equations than unknowns.

3.1   Introductory Example

As a simple example, we consider the geometric problem of finding the intersection point of three planes α, β and γ given in normal form:

$$
\begin{array}{rrrrrrrl}
\alpha & : & 4x_1 & + & x_2 & + & x_3 & = 2, \\
\beta & : & & & x_2 & + & 2x_3 & = 3, \\
\gamma & : & -5x_1 & + & & & 2x_3 & = 5.
\end{array}
\tag{3.1}
$$

If the normal vectors of the three planes are not *coplanar* (i.e., do not lie on the same plane themselves), then there is exactly one intersection point $\boldsymbol{x} = (x_1, x_2, x_3)^\top$ satisfying the three equations simultaneously. Equations (3.1) form a *system of linear equations*, with three equations in three unknowns. More generally, a system of n equations in n unknowns written component-wise is

$$
\begin{array}{ccccccc}
a_{11}x_1 & + & \cdots & + & a_{1n}x_n & = & b_1, \\
a_{21}x_1 & + & \cdots & + & a_{2n}x_n & = & b_2, \\
\vdots & & \vdots & & \vdots & & \vdots \\
a_{n1}x_1 & + & \cdots & + & a_{nn}x_n & = & b_n.
\end{array}
\tag{3.2}
$$

The constants a_{ij} are called the *coefficients* and the b_i form the *right-hand side*. The coefficients are collected in the matrix

$$
A = \begin{pmatrix}
a_{11} & a_{12} & \cdots & a_{1n} \\
a_{21} & a_{22} & \cdots & a_{2n} \\
\vdots & \vdots & \ddots & \vdots \\
a_{n1} & a_{n2} & \cdots & a_{nn}
\end{pmatrix}.
\tag{3.3}
$$

The right-hand side is the vector

$$
\boldsymbol{b} = \begin{pmatrix} b_1 \\ \vdots \\ b_n \end{pmatrix},
\tag{3.4}
$$

and we collect the x_i in the *vector of unknowns*

$$\boldsymbol{x} = \begin{pmatrix} x_1 \\ \vdots \\ x_n \end{pmatrix}. \tag{3.5}$$

In matrix-vector notation, the linear system (3.2) is therefore

$$A\boldsymbol{x} = \boldsymbol{b}. \tag{3.6}$$

For Example (3.1) we get

$$A = \begin{pmatrix} 4 & 1 & 1 \\ 0 & 1 & 2 \\ -5 & 0 & 2 \end{pmatrix} \quad \text{and} \quad \boldsymbol{b} = \begin{pmatrix} 2 \\ 3 \\ 5 \end{pmatrix}. \tag{3.7}$$

Often it is useful to consider the *columns of a matrix* A,

$$A = (\boldsymbol{a}_{:1}, \boldsymbol{a}_{:2}, \ldots, \boldsymbol{a}_{:n}), \tag{3.8}$$

where

$$\boldsymbol{a}_{:i} := \begin{pmatrix} a_{1i} \\ \vdots \\ a_{ni} \end{pmatrix} \tag{3.9}$$

denotes the ith column vector. In MATLAB we can address $\boldsymbol{a}_{:i}$ by the expression `A(:,i)`. Similarly we denote by

$$\boldsymbol{a}_{k:} = (a_{k1}, \ldots, a_{kn})$$

the kth *row vector* of A and thus

$$A = \begin{pmatrix} \boldsymbol{a}_{1:} \\ \vdots \\ \boldsymbol{a}_{n:} \end{pmatrix}. \tag{3.10}$$

The expression in MATLAB for $\boldsymbol{a}_{k:}$ is `A(k,:)`. Notice that in (3.1), the normal vectors to the planes are precisely the rows of A.

An common way to test whether the three normal vectors are coplanar uses *determinants*, which calculate the (signed) volume of the parallelepiped with edges given by three vectors. The determinant is, in fact, defined for a general $n \times n$ matrix A by the real number

$$\det(A) := \sum_{\boldsymbol{k}} (-1)^{\delta(\boldsymbol{k})} a_{1k_1} a_{2k_2} a_{3k_3} \ldots a_{nk_n}, \tag{3.11}$$

where the vector of indices $\boldsymbol{k} = \{k_1, \ldots, k_n\}$ takes all values of the permutations of the indices $\{1, 2, \ldots, n\}$. The sign is defined by $\delta(\boldsymbol{k})$, which equals 0

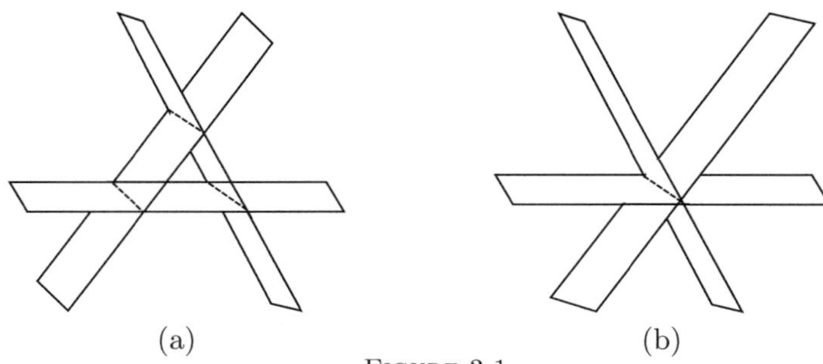

FIGURE 3.1.
Three planes with (a) no common intersection, (b)
infinitely many intersections along a straight line

or 1 depending on the permutation being even or odd. This formula is called the *Leibniz formula for determinants.*

EXAMPLE 3.1. *For $n = 2$ we have*

$$\det \begin{pmatrix} a_{11} & a_{12} \\ a_{21} & a_{22} \end{pmatrix} = a_{11}a_{22} - a_{12}a_{21}, \tag{3.12}$$

and for $n = 3$ we obtain

$$\det \begin{pmatrix} a_{11} & a_{12} & a_{13} \\ a_{21} & a_{22} & a_{23} \\ a_{31} & a_{32} & a_{33} \end{pmatrix} = \begin{array}{l} a_{11}a_{22}a_{33} - a_{11}a_{23}a_{32} + a_{12}a_{23}a_{31} \\ -a_{12}a_{21}a_{33} + a_{13}a_{21}a_{32} - a_{13}a_{22}a_{31}. \end{array} \tag{3.13}$$

In our geometric example, the rows of the coefficient matrix A are precisely the normal vectors to the planes, so $|\det(A)|$ is the volume of the parallelepiped generated by the three vectors. If this volume is zero, then the parallelepiped "collapses" onto the same plane, which implies our system either has no solution (no common intersection point, see Figure 3.1(a)), or has infinitely many solutions (intersection along a whole line or plane, see Figure 3.1(b)). However, if $\det(A) \neq 0$, then there is a unique intersection point, so the solution to (3.1) is unique.

Instead of using Definition (3.11), we can also compute the determinant using the *Laplace Expansion*. For each row i we have

$$\det(A) = \sum_{j=1}^{n} a_{ij}(-1)^{i+j} \det(M_{ij}), \tag{3.14}$$

where M_{ij} denotes the $(n-1) \times (n-1)$ submatrix obtained by deleting row i and column j of the matrix A. Instead of expanding the determinant as in (3.14) along a row, we can also use an expansion along a column.

The following recursive MATLAB program computes a determinant using the
Laplace Expansion for the first row:

ALGORITHM 3.1. *Determinant by Laplace Expansion*

```
function d=DetLaplace(A);
% DETLAPLACE determinant using Laplace expansion
%    d=DetLaplace(A); computes the determinant d of the matrix A
%    using the Laplace expansion for the first row.

n=length(A);
if n==1;
  d=A(1,1);
else
  d=0; v=1;
  for j=1:n
    M1j=[A(2:n,1:j-1) A(2:n,j+1:n)];
    d=d+v*A(1,j)*DetLaplace(M1j);
    v=-v;
  end
end
```

The following equation holds for determinants, see Problem 3.4:

$$\det(AB) = \det(A)\det(B). \tag{3.15}$$

This equation allows us to give an explicit expression for the solution of the
linear system (3.6). Consider replacing in the identity matrix

$$I = \begin{pmatrix} 1 & 0 & \cdots & 0 \\ 0 & 1 & \cdots & 0 \\ \vdots & & \ddots & \vdots \\ 0 & \cdots & 0 & 1 \end{pmatrix}$$

the i-th column vector $\boldsymbol{e}_i$ by $\boldsymbol{x}$ to obtain the matrix

$$E = (\boldsymbol{e}_1, \ldots, \boldsymbol{e}_{i-1}, \boldsymbol{x}, \boldsymbol{e}_{i+1}, \ldots, \boldsymbol{e}_n).$$

The determinant is simply

$$\det(E) = x_i, \tag{3.16}$$

as one can see immediately by expanding along the i-th row. Furthermore,

$$AE = (A\boldsymbol{e}_1, \ldots, A\boldsymbol{e}_{i-1}, A\boldsymbol{x}, A\boldsymbol{e}_{i+1}, \ldots, A\boldsymbol{e}_n),$$

and because $A\boldsymbol{x} = \boldsymbol{b}$ and $A\boldsymbol{e}_k = \boldsymbol{a}_{:k}$, we obtain

$$AE = (\boldsymbol{a}_{:1}, \ldots, \boldsymbol{a}_{:i-1}, \boldsymbol{b}, \boldsymbol{a}_{:i+1}, \ldots, \boldsymbol{a}_{:n}). \tag{3.17}$$

If we denote the matrix on the right hand side of (3.17) by A_i and if we compute the determinant on both sides we get

$$\det(A)\det(E) = \det(A_i).$$

Using Equation (3.16) we get

THEOREM 3.1. (CRAMER'S RULE) *For* $\det(A) \neq 0$, *the linear system* $A\boldsymbol{x} = \boldsymbol{b}$ *has the unique solution*

$$x_i = \frac{\det(A_i)}{\det(A)}, \quad i = 1, 2, \ldots, n, \tag{3.18}$$

where A_i *is the matrix obtained from* A *by replacing column* $\boldsymbol{a}_{:i}$ *by* $\boldsymbol{b}$.

The following MATLAB program computes the solution of a linear system with Cramer's rule:

ALGORITHM 3.2. *Cramer's Rule*

```
function x=Cramer(A,b);
% CRAMER solves a linear Sytem with Cramer's rule
%    x=Cramer(A,b); Solves the linear system Ax=b using Cramer's
%    rule. The determinants are computed using the function DetLaplace.

n=length(b);
detA=DetLaplace(A);
for i=1:n
  AI=[A(:,1:i-1), b, A(:,i+1:n)];
  x(i)=DetLaplace(AI)/detA;
end
x = x(:);
```

Cramer's rule looks simple and even elegant, but for computational purposes it is a disaster, as we show in Figure 3.2 for the Hilbert matrices, see Problem 3.3. The computational effort with Laplace expansion is $O(n!)$ (see Problem 3.5), while with Gaussian elimination (introduced in the next section), it is $O(n^3)$. Furthermore, the numerical accuracy due to finite precision arithmetic is very poor: it is very likely that cancellation occurs within the Laplace expansion, as one can see in Figure 3.2.

3.2 Gaussian Elimination

With *Gaussian Elimination*, one tries to reduce a given linear system to an equivalent system with a triangular matrix. As we will see, such systems are very easy to solve.

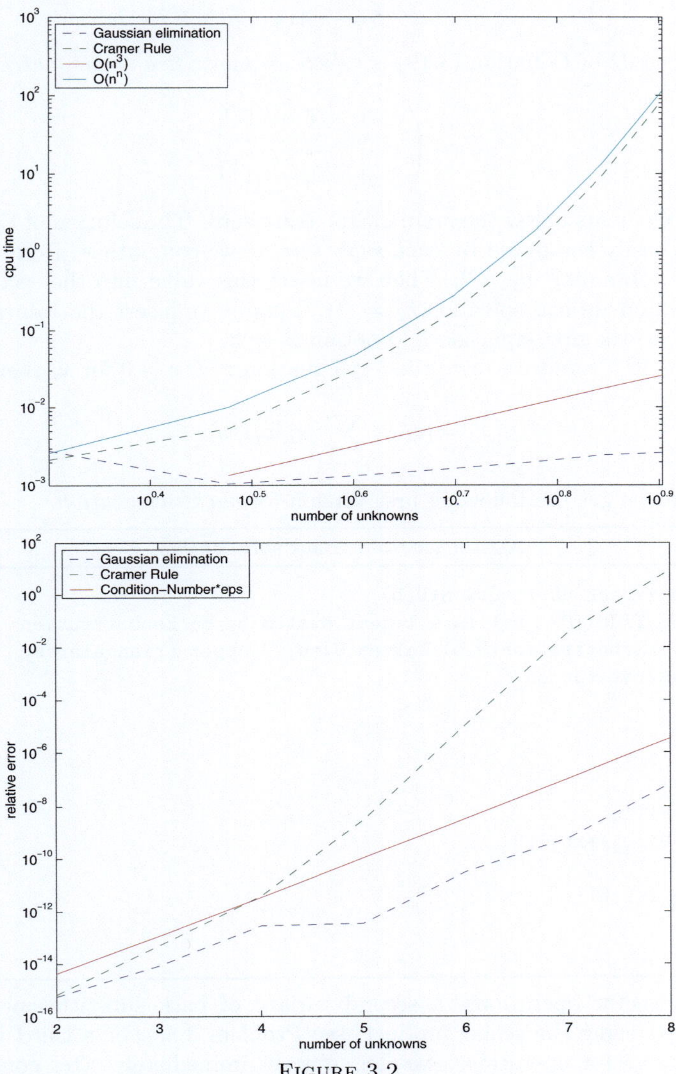

FIGURE 3.2.
*Comparison of speed (above) and accuracy (below) of
Cramer's rule with Gaussian Elimination*

EXAMPLE 3.2.

$$\begin{array}{rrrrr} 3x_1 & + & 5x_2 & - & x_3 & = & 2, \\ & & 2x_2 & - & 7x_3 & = & -16, \\ & & & - & 4x_3 & = & -8. \end{array} \qquad (3.19)$$

The matrix U of Equation (3.19) is called an *upper triangular matrix*,

$$U = \begin{pmatrix} 3 & 5 & -1 \\ 0 & 2 & -7 \\ 0 & 0 & -4 \end{pmatrix},$$

since all elements below the main diagonal are zero. The solution of Equation (3.19) is easily computed by *back substitution*: we compute x_3 from the last equation, obtaining $x_3 = 2$. Then we insert this value into the second last equation and we can solve for $x_2 = -1$. Finally we insert the values for x_2 and x_3 into the first equation and obtain $x_1 = 3$.

If $U \in \mathbb{R}^{n \times n}$ and we solve the i-th equation in $U\boldsymbol{x} = \boldsymbol{b}$ for x_i then

$$x_i = (b_i - \sum_{j=i+1}^{n} u_{ij}x_j)/u_{ii}.$$

Therefore we get the following first version for *back substitution*

ALGORITHM 3.3. *Back substitution*

```
function x=BackSubstitution(U,b)
% BACKSUBSTITUTION  solves a linear system by backsubstitution
%   x=BackSubstitution(U,b) solves Ux=b, U upper triangular by
%   backsubstitution

n=length(b);
for k=n:-1:1
  s=b(k);
  for j=k+1:n
    s=s-U(k,j)*x(j);
  end
  x(k)=s/U(k,k);
end
x=x(:);
```

With vector operations, a second variant of back substitution can be formulated using the scalar product, see Problem 3.8. For a third variant, also using vector operations, we can subtract immediately after computing x_i the i-th column of U multiplied by x_i from the right-hand side. This simplifies the process to the SAXPY variant[1] of back substitution:

---

[1] SAXPY, which stands for "scalar $a \cdot x$ plus y", is a basic linear algebra operation that overwrites a vector $\boldsymbol{y}$ with the result of $a\boldsymbol{x} + \boldsymbol{y}$, where a is a scalar. This operation is implemented efficiently in several libraries that can be tuned to the machine on which the code is executed.

ALGORITHM 3.4. *Back substitution, SAXPY-Variant*

```
function x=BackSubstitutionSAXPY(U,b)
% BACKSUBSTITUTIONSAXPY solves linear system by backsubstitution
%   x=BackSubstitutionSAXPY(U,b) solves Ux=b by backsubstitution by
%   modifying the right hand side (SAXPY variant)

n=length(b);
for i=n:-1:1
  x(i)=b(i)/U(i,i);
  b(1:i-1)=b(1:i-1)-x(i)*U(1:i-1,i);
end
x=x(:);
```

This algorithm costs n divisions and $(n-1)+(n-2)+\ldots+1 = \frac{1}{2}n^2 - \frac{1}{2}n$ additions and multiplications, and its complexity is thus $O(n^2)$.

We will now reduce in $n-1$ elimination steps the given linear system of equations

$$
\begin{array}{ccccccc}
a_{11}x_1 & + & a_{12}x_2 & +\ldots+ & a_{1n}x_n & = & b_1 \\
\vdots & & \vdots & & \vdots & & \vdots \\
a_{k1}x_1 & + & a_{k2}x_2 & +\ldots+ & a_{kn}x_n & = & b_k \\
\vdots & & \vdots & & \vdots & & \vdots \\
a_{n1}x_1 & + & a_{n2}x_2 & +\ldots+ & a_{nn}x_n & = & b_n
\end{array}
\tag{3.20}
$$

to an equivalent system with an upper triangular matrix. A linear system is transformed into an equivalent one by adding to one equation a multiple of another equation. An elimination step consists of adding a suitable multiple in such a way that one unknown is eliminated in the remaining equations.

To eliminate the unknown x_1 in equations #2 to #n, we perform the operations

```
for k=2:n
```
$$\{\text{new Eq. } \# \, k\} = \{\text{Eq. } \# \, k\} - \frac{a_{k1}}{a_{11}} \{\text{Eq.} \# \, 1\}$$
```
end
```

We obtain a reduced system with an $(n-1) \times (n-1)$ matrix which contains only the unknowns $x_2, \ldots, x_n$. This remaining system is reduced again by one unknown by freezing the second equation and eliminating x_2 in equations #3 to #n. We continue this way until only one equation with one unknown remains. This way we have reduced the original system to a system with an upper triangular matrix. The whole process is described by two nested loops:

```
for i=1:n-1
  for k=i+1:n
```

$$\{\text{new Eq. } \# \ k\} = \{\text{Eq. } \# \ k\} - \frac{a_{ki}}{a_{ii}} \{\text{Eq. } \# \ i\}$$

```
  end
end
```

The coefficients of the k-th new equation are computed as

$$a_{kj} := a_{kj} - \frac{a_{ki}}{a_{ii}} a_{ij} \quad \text{for} \quad j = i+1, \ldots, n, \tag{3.21}$$

and the right-hand side also changes,

$$b_k := b_k - \frac{a_{ki}}{a_{ii}} b_i.$$

Note that the k-th elimination step (3.21) is a *rank-one change* of the remaining matrix. Thus, if we append the right hand side to the matrix A by `A=[A, b]`, then the elimination becomes

```
for i=1:n-1
  A(i+1:n,i)=A(i+1:n,i)/A(i,i);
  A(i+1:n,i+1:n+1)=A(i+1:n,i+1:n+1)-A(i+1:n,i)*A(i,i+1:n+1);
end
```

where the inner loop over k has been subsumed by MATLAB's vector notation. Note that we did not compute the zeros in `A(i+1:n,i)`. Rather we used these matrix elements to store the factors necessary to eliminate the unknown x_i.

EXAMPLE 3.3. *We consider the linear system* $Ax = b$ *with* `A=invhilb(4)`,

$$A = \begin{pmatrix} 16 & -120 & 240 & -140 \\ -120 & 1200 & -2700 & 1680 \\ 240 & -2700 & 6480 & -4200 \\ -140 & 1680 & -4200 & 2800 \end{pmatrix} \qquad b = \begin{pmatrix} -4 \\ 6 \\ -180 \\ 140 \end{pmatrix}.$$

The right-hand side b *was chosen in such a way that the solution is* $x = (1, 1, 1, 1)^{\top}$. *If we apply the above elimination procedure to the* augmented *matrix* $[A, b]$, *then we obtain*

```
A =
   16.0000 -120.0000   240.0000 -140.0000    -4.0000
   -7.5000  300.0000  -900.0000  630.0000    30.0000
   15.0000   -3.0000   180.0000 -210.0000   -30.0000
   -8.7500    2.1000    -1.1667    7.0000     7.0000
>> U=triu(A)
U =
   16.0000 -120.0000   240.0000 -140.0000    -4.0000
         0  300.0000  -900.0000  630.0000    30.0000
         0         0   180.0000 -210.0000   -30.0000
         0         0         0    7.0000     7.0000
```

Thus the equivalent reduced system is

$$\begin{pmatrix} 16 & -120 & 240 & -140 \\ & 300 & -900 & 630 \\ & & 180 & -210 \\ & & & 7 \end{pmatrix} \boldsymbol{x} = \begin{pmatrix} -4 \\ 30 \\ -30 \\ 7 \end{pmatrix},$$

and has the same solution $\boldsymbol{x} = (1,1,1,1)^\top$.

There is unfortunately a glitch. Our elimination process may fail if, in step i, the i-th equation does not contain the unknown x_i, i.e., if the (i,i)-th coefficient is zero. Then we cannot use this equation to eliminate x_i in the remaining equations.

EXAMPLE 3.4.

$$\begin{array}{rcrcrcl} & & x_2 & + & 3x_3 & = & -6 \\ 2x_1 & - & x_2 & + & x_3 & = & 10 \\ -3x_1 & + & 5x_2 & - & 7x_3 & = & 10 \end{array}$$

Since the first equation does not contain x_1, *we cannot use it to eliminate* x_1 *in the second and third equation.*

Obviously, the solution of a linear system does not depend on the ordering of the equations. It is therefore very natural to reorder the equations in such a way that we obtain a *pivot-element* $a_{ii} \neq 0$. From the point of view of numerical stability, $a_{ii} \neq 0$ is not sufficient; we also need to reorder equations if the unknown x_i is only "weakly" contained. The following example illustrates this.

EXAMPLE 3.5. *Consider for* ε *small the linear system*

$$\begin{array}{rcrcl} \varepsilon x_1 & + & x_2 & = & 1, \\ x_1 & + & x_2 & = & 2. \end{array} \tag{3.22}$$

If we interpret the equations as lines in the plane, then their graphs show a clear intersection point near $x_1 = x_2 = 1$, *as one can see for* $\varepsilon = 10^{-7}$ *in Figure 3.3. The angle between the two lines is about* $45°$.

If we want to solve this system algebraically, then we might eliminate the first unknown in the second equation by replacing the second equation with the linear combination

$$\{Eq.\ \#2\} - \frac{1}{\varepsilon}\{Eq.\ \#1\}.$$

This leads to the new, mathematically equivalent system of equations

$$\begin{array}{rcrcl} \varepsilon x_1 & + & x_2 & = & 1, \\ & & \left(1 - \tfrac{1}{\varepsilon}\right) x_2 & = & 2 - \tfrac{1}{\varepsilon}. \end{array}$$

If we again interpret the two equations as lines in the plane, then we see that this time, the two lines are almost coinciding, as shown in Figure 3.3, where

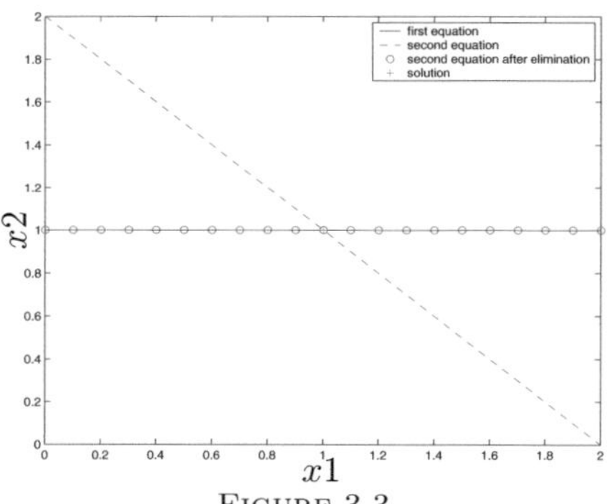

FIGURE 3.3.
*Example on how a well-conditioned problem can be
transformed into an ill-conditioned one when a small
pivot is used in Gaussian elimination*

*we used the circle symbol for the new second line in order to make it visible
on top of the first one. The intersection point is now very difficult to find —
the problem has become ill conditioned.*

*What went wrong? We eliminated the unknown using a very small pivot
on the diagonal. Doing this, we transformed a well-conditioned problem into
an ill-conditioned one. Choosing small pivots makes Gaussian elimination
unstable. If, however, we interchange the equations of (3.22) and eliminate
afterward, we get*

$$\begin{aligned} x_1 \quad + \quad x_2 &= 2 \\ (1 - \varepsilon)x_2 &= 1 - 2\varepsilon, \end{aligned} \qquad (3.23)$$

*a system for which the row vectors of the matrix are nicely linearly indepen-
dent, and so the intersection point is well defined and can be computed stably.*

This observation is the motivation for introducing a *pivoting strategy*.
We consider *partial pivoting*, which means that before each elimination step,
we look in the current column for the element with largest absolute value.
This element will then be chosen as the pivot. If we cannot find a nonzero
pivot element, this means that the corresponding unknown is absent from
the remaining equations, i.e., the linear system is singular. In finite precision
arithmetic, we cannot expect in the singular case that all possible pivot elements will be exactly zero, since rounding errors will produce rather small
(but nonzero) elements; these will have to be compared with the other matrix elements in order to decide if they should be treated as zeros. Therefore,
we will consider in the following program a pivot element to be zero if it is
smaller than $10^{-14}||A||_1$, a reasonable size in practice.

ALGORITHM 3.5.
Gaussian Elimination with Partial Pivoting

```
function x=Elimination(A,b)
% ELIMINATION solves a linear system by Gaussian elimination
%    x=Elimination(A,b) solves the linear system Ax=b using Gaussian
%    Elimination with partial pivoting. Uses the function
%    BackSubstitution

n=length(b);
norma=norm(A,1);
A=[A,b];                                 % augmented matrix
for i=1:n
  [maximum,kmax]=max(abs(A(i:n,i)));     % look for Pivot A(kmax,i)
  kmax=kmax+i-1;
  if maximum < 1e-14*norma;              % only small pivots
    error('matrix is singular')
  end
  if i ~= kmax                           % interchange rows
    h=A(kmax,:); A(kmax,:)=A(i,:); A(i,:)=h;
  end
  A(i+1:n,i)=A(i+1:n,i)/A(i,i);          % elimination step
  A(i+1:n,i+1:n+1)=A(i+1:n,i+1:n+1)-A(i+1:n,i)*A(i,i+1:n+1);
end
x=BackSubstitution(A,A(:,n+1));
```

Note that although we have to perform only $n-1$ elimination steps, the for-loop in `Elimination` goes up to n. This is necessary for testing whether a_{nn} becomes zero, which would indicate that the matrix is singular. The statements corresponding to the actual elimination have an empty index set, and thus have no effect for $i = n$.

EXAMPLE 3.6. *If we consider* `A=magic(4)` *and* $b = (1,0,0,0)^\top$, *then the call* `x=Elimination(A,b)` *will return the error message* `matrix is singular`. *This is correct, since the rank of A is 3.*

In MATLAB, a linear system $Ax = b$ is solved with the statement `x=A\b`. If A has full rank, the operator $\backslash$ solves the system using *partial pivoting*.

3.2.1 LU Factorization

Gaussian elimination becomes more transparent if we formulate it using matrix operations. Consider the matrix

$$
L_1 = \begin{pmatrix} 1 & 0 & 0 & \cdots & 0 \\ -l_{21} & 1 & 0 & \cdots & 0 \\ \vdots & \vdots & & \ddots & \vdots \\ -l_{n1} & 0 & 0 & \cdots & 1 \end{pmatrix}, \quad l_{j1} := \frac{a_{j1}}{a_{11}}, \; j = 2, \ldots, n. \tag{3.24}
$$

Multiplying the linear system $A\boldsymbol{x} = \boldsymbol{b}$ with L_1 from the left, we obtain

$$L_1 A\boldsymbol{x} = L_1\boldsymbol{b}, \tag{3.25}$$

and it is easy to see that this is the system that we get after the first elimination step: the first equation is unchanged and the remaining equations do not contain x_1 any more. Denoting the elements of the matrix $A^{(1)} := L_1 A$ by $a_{ik}^{(1)}$, we can eliminate in the same way the unknown x_2 with the matrix

$$L_2 = \begin{pmatrix} 1 & 0 & 0 & \cdots & 0 \\ 0 & 1 & 0 & \cdots & 0 \\ 0 & -l_{32} & 1 & \cdots & 0 \\ \vdots & \vdots & \vdots & \ddots & \vdots \\ 0 & -l_{n2} & 0 & \cdots & 1 \end{pmatrix}, \quad l_{j2} := \frac{a_{j2}^{(1)}}{a_{22}^{(1)}}, \; j = 3, \ldots, n, \tag{3.26}$$

from equations #3 to #n by multiplying the system (3.25) from the left by L_2,

$$L_2 L_1 A\boldsymbol{x} = L_2 L_1 \boldsymbol{b}, \tag{3.27}$$

and we obtain the new matrix $A^{(2)} := L_2 A^{(1)} = L_2 L_1 A$. Continuing this way, we obtain the matrices L_k and $A^{(k)}$ for $k = 1, \ldots, n-1$, and finally the system

$$A^{(n-1)}\boldsymbol{x} = L_{n-1} \cdots L_1 A\boldsymbol{x} = L_{n-1} \cdots L_1 \boldsymbol{b}, \tag{3.28}$$

where we have now obtained an upper triangular matrix U,

$$A^{(n-1)} = L_{n-1} \cdots L_1 A = U. \tag{3.29}$$

The matrices L_j are all lower triangular matrices. They are easy to invert, for instance

$$L_1^{-1} = \begin{pmatrix} 1 & 0 & 0 & \cdots & 0 \\ +l_{21} & 1 & 0 & \cdots & 0 \\ \vdots & \vdots & & \ddots & \vdots \\ +l_{n1} & 0 & 0 & \cdots & 1 \end{pmatrix}. \tag{3.30}$$

Thus we only have to invert the signs. Moving this way the L_i to the right hand side, we obtain

$$A = L_1^{-1} \ldots L_{n-1}^{-1} U. \tag{3.31}$$

The product of lower triangular matrices is again lower triangular, see Problem 3.6, and therefore

$$L := L_1^{-1} \ldots L_{n-1}^{-1} \tag{3.32}$$

is lower triangular. More specifically, we have

$$L = \begin{pmatrix} 1 & 0 & 0 & \cdots & 0 \\ l_{21} & 1 & 0 & \cdots & 0 \\ l_{31} & l_{32} & 1 & \cdots & 0 \\ \vdots & \vdots & \vdots & \ddots & \vdots \\ l_{n1} & l_{n2} & l_{n3} & \cdots & 1 \end{pmatrix}, \tag{3.33}$$

where l_{ij} are the multiplication factors which are used for the elimination. Equation (3.31) then reads

$$A = L \cdot U, \tag{3.34}$$

and we have obtained a decomposition of the matrix A into a product of two triangular matrices — the so-called *LU decomposition*. With partial pivoting, the equations (and hence the rows of A) are permuted, so we obtain not a *triangular decomposition* of the matrix A, but that of $\tilde{A} = PA$, where P is a *permutation matrix*, so that $\tilde{A}$ has the same row vectors as A, but in a different order.

THEOREM 3.2. (LU DECOMPOSITION) *Let $A \in \mathbb{R}^{n \times n}$ be a non-singular matrix. Then there exists a permutation matrix P such that*

$$PA = LU, \tag{3.35}$$

where L and U are the lower and upper triangular matrices obtained from Gaussian elimination.

PROOF. At the first step of the elimination, we look for the row containing the largest pivot, swap it with the first row, and then perform the elimination. Thus, we have

$$A^{(1)} = L_1 P_1 A,$$

where P_1 interchanges rows #1 and #k_1 with $k_1 > 1$. In the second step, we again swap rows #2 and #k_2 with $k_2 > 2$ before eliminating, i.e.,

$$A^{(2)} = L_2 P_2 A^{(1)} = L_2 P_2 L_1 P_1 A.$$

Continuing this way, we obtain

$$U = A^{(n-1)} = L_{n-1} P_{n-1} \cdots L_1 P_1 A, \tag{3.36}$$

where P_i interchanges rows #i and #k_i with $k_i > i$. Our goal is to make all the permutations appear together, instead of being scattered across the factorization. To do so, note that each L_i contains a single column of non-zero entries apart from the diagonal. Thus, it can be written as

$$L_i = I - \boldsymbol{v}_i \boldsymbol{e}_i^\top,$$

where $\boldsymbol{e}_i$ contains 1 at the i-th position and zeros everywhere else, and the first i entries of $\boldsymbol{v}_i$ are zero. By direct calculation, we see that

$$\begin{aligned}
P_{n-1} \cdots P_{i+1} L_i &= P_{n-1} \cdots P_{i+1} (I - \boldsymbol{v}_i \boldsymbol{e}_i^\top) \\
&= P_{n-1} \cdots P_{i+1} - \tilde{\boldsymbol{v}}_i \boldsymbol{e}_i^\top \\
&= \left[I - \tilde{\boldsymbol{v}}_i (P_{n-1} \cdots P_{i+1} \boldsymbol{e}_i)^\top \right] P_{n-1} \cdots P_{i+1},
\end{aligned}$$

where $\tilde{\boldsymbol{v}}_i = P_{n-1} \cdots P_{i+1} \boldsymbol{v}_i$ is a permuted version of $\boldsymbol{v}_i$. But the permutation $P_{n-1} \cdots P_{i+1}$ only permutes entries $i + 1$ to n in a vector; entries 1 to i

remain untouched. This means the first i entries of $\tilde{\boldsymbol{v}}_i$ are still zero, and $\boldsymbol{e}_i$ is unchanged by the permutation, i.e., $P_{n-1}\cdots P_{i+1}\boldsymbol{e}_i = \boldsymbol{e}_i$. So we in fact have

$$P_{n-1}\cdots P_{i+1}L_i = \tilde{L}_i P_{n-1}\cdots P_{i+1}$$

with $\tilde{L}_i = I - \tilde{\boldsymbol{v}}_i \boldsymbol{e}_i^\top$ still lower triangular. Now (3.36) implies

$$\begin{aligned}
U &= L_{n-1}P_{n-1}L_{n-2}P_{n-2}\cdots L_2 P_2 L_1 P_1 A \\
&= L_{n-1}\tilde{L}_{n-2}P_{n-1}P_{n-2}\cdots L_2 P_2 L_1 P_1 A \\
&= \cdots = (L_{n-1}\tilde{L}_{n-2}\cdots \tilde{L}_2 \tilde{L}_1)(P_{n-1}\cdots P_1)A.
\end{aligned}$$

Letting $P = P_{n-1}\cdots P_1$ and $L = \tilde{L}_1^{-1}\cdots \tilde{L}_{n-2}^{-1}L_{n-1}^{-1}$ completes the proof. $\square$

Note that the above proof shows that *we must swap entries in both L and U* when two rows are interchanged. It also means that there *exists* a row permutation such that all the pivots that appear during the elimination are the largest in their respective columns, so for analysis purposes we can assume that A has been "pre-permuted" this way. In practice, of course, the permutation is discovered during the elimination and not known in advance.

A linear system can be solved by the following steps:

1. *Triangular decomposition* of the coefficients matrix $PA = LU$.

2. Apply row changes to the right hand side, $\tilde{\boldsymbol{b}} = P\boldsymbol{b}$, and solve $L\boldsymbol{y} = \tilde{\boldsymbol{b}}$ by *forward substitution* (see Problem 3.9).

3. Solve $U\boldsymbol{x} = \boldsymbol{y}$ by *back substitution*.

The advantage of this arrangement is that for new right hand sides $\boldsymbol{b}$ we do not need to recompute the decomposition. It is sufficient to repeat steps 2 and 3. This leads to substantial computational savings, since the major cost lies in the factorization: to eliminate the first column, one needs $n-1$ divisions and $(n-1)^2$ multiplications and additions. For the second column one needs $n-2$ divisions and $(n-2)^2$ multiplications and additions until the last elimination, where one division and one addition and multiplication are needed. The total number of operations is therefore

$$\sum_{i=1}^{n-1} i + i^2 = \frac{n^3}{3} - \frac{n}{3},$$

which one can obtain from MAPLE using `sum(i+i^2,i=0..n-1)`. Hence the computation of the LU decomposition costs $O(n^3)$ operations, and is much more expensive than the forward and back substitution, for which the cost is $O(n^2)$.

The LU decomposition can also be used to compute *determinants* in a numerically sound way, since $\det(A) = \det(L)\det(U) = \det(U)$, see Problem 3.15, and is thus a decomposition useful in its own right. The implementation of the LU decomposition is left to the reader as an exercise in Problem 3.10.

3.2.2 Backward Stability

In order for the above algorithm to give meaningful results, we need to ensure that each of the three steps (LU factorization, forward and back substitution) is backward stable. For the factorization phase, we have seen in Section 3.2 that pivoting is essential for the numerical stability. One might wonder if partial pivoting is enough to guarantee that the algorithm is stable.

THEOREM 3.3. (WILKINSON) *Let A be an invertible matrix, and $\hat{L}$ and $\hat{U}$ be the numerically computed LU-factors using Gaussian elimination with partial pivoting, $|l_{ij}| \leq 1$ for all i, j. Then for the elements of $\hat{A} := \hat{L}\hat{U}$, we have*

$$|\hat{a}_{ij} - a_{ij}| \leq 2\alpha \min(i-1, j)\, eps + O(eps^2), \tag{3.37}$$

where $\alpha := \max_{ijk} |\hat{a}_{ij}^{(k)}|$.

PROOF. At step k of Gaussian elimination, we compute $\hat{a}_{ij}^{(k)}$ for $i > k$, $j > k$ using the update formula

$$\begin{aligned}
\hat{a}_{ij}^{(k)} &= (\hat{a}_{ij}^{(k-1)} - \hat{l}_{ik}\hat{a}_{kj}^{(k-1)}(1 + \varepsilon_{ijk}))(1 + \eta_{ijk}) \\
&= \hat{a}_{ij}^{(k-1)} - \hat{l}_{ik}\hat{a}_{kj}^{(k-1)} + \mu_{ijk},
\end{aligned} \tag{3.38}$$

where $|\mu_{ijk}|$ can be bounded using $|\varepsilon_{ijk}| \leq eps$ and $|\eta_{ijk}| \leq eps$:

$$\begin{aligned}
|\mu_{ijk}| &\leq \underbrace{|\hat{a}_{ij}^{(k-1)} - \hat{l}_{ik}\hat{a}_{kj}^{(k-1)}|}_{=\hat{a}_{ij}^{(k)} + O(eps)} |\eta_{ijk}| + |\hat{l}_{ik}||\hat{a}_{kj}^{(k-1)}||\varepsilon_{ijk}| + O(eps^2) \\
&\leq 2\alpha\, eps + O(eps^2).
\end{aligned}$$

In addition, for $i > j$, we have $\hat{a}_{ij}^{(j)} = 0$ and $\hat{l}_{ij} = \dfrac{\hat{a}_{ij}^{(j-1)}}{\hat{a}_{jj}^{(j-1)}}(1 + \varepsilon_{ijj})$, which implies

$$0 = \hat{a}_{ij}^{(j)} = \hat{a}_{ij}^{(j-1)} - \hat{l}_{ij}\hat{a}_{jj}^{(j-1)} + \mu_{ijj}, \quad \text{with} \quad |\mu_{ijj}| = |\hat{a}_{ij}^{(j-1)}\varepsilon_{ijj}| \leq \alpha\, eps.$$

Thus, (3.38) in fact holds whenever $i > j \geq k$ or $j \geq i > k$, with

$$|\mu_{ijk}| \leq 2\alpha\, eps + O(eps^2). \tag{3.39}$$

By the definition of $\hat{A} = \hat{L}\hat{U}$, we have

$$\hat{a}_{ij} = \sum_{k=1}^{\min(i,j)} \hat{l}_{ik}\hat{u}_{kj} = \sum_{k=1}^{\min(i,j)} \hat{l}_{ik}\hat{a}_{kj}^{(k-1)}. \tag{3.40}$$

For the case $i > j$, we obtain, using (3.38) with $i > j \geq k$, a telescopic sum for $\hat{a}_{ij}$:

$$\hat{a}_{ij} = \sum_{k=1}^{j}(\hat{a}_{ij}^{(k-1)} - \hat{a}_{ij}^{(k)} + \mu_{ijk}) = a_{ij} + \sum_{k=1}^{j}\mu_{ijk}, \tag{3.41}$$

since $a_{ij}^{(0)} = a_{ij}$ and $a_{ij}^{(j)} = 0$ for $i > j$. On the other hand, for $i \leq j$, we use (3.40) and (3.38) with $j \geq i > k$ to obtain

$$\hat{a}_{ij} = \sum_{k=1}^{i-1} (\hat{a}_{ij}^{(k-1)} - \hat{a}_{ij}^{(k)} + \mu_{ijk}) + \hat{l}_{ii}\hat{u}_{ij} = a_{ij} + \sum_{k=1}^{i-1} \mu_{ijk}, \tag{3.42}$$

where we used $\hat{l}_{ii} = 1$ and $\hat{u}_{ij} = \hat{a}_{ij}^{(i-1)}$. Combining (3.39), (3.41) and (3.42) yields the desired result. $\qquad\qquad\qquad\qquad\qquad\qquad\qquad\qquad\qquad\qquad\square$

This theorem shows that Gaussian elimination with partial pivoting is backward stable, if the *growth factor*

$$\rho := \frac{\alpha}{\max |a_{ij}|} \tag{3.43}$$

is not too large, which means that the elements $a_{ij}^{(k)}$ encountered during the elimination process are not growing too much.

Next, we show that the SAXPY variant of back substitution (Algorithm 3.4) is also backward stable. We show below the floating-point version of back substitution for solving $U\boldsymbol{x} = \boldsymbol{b}$, where the quantities ε_{jk} and η_{jk} all have moduli less than *eps*.

$\hat{b}_k^{(n)} := b_k$
for $k = n, n - 1, \ldots, 1$ do
$\qquad \hat{x}_k = \dfrac{\hat{b}_k^{(k)}}{u_{kk}}(1 + \varepsilon_{kk})$
$\qquad$ for $j = 1, \ldots, k - 1$ do
$\qquad\qquad \hat{b}_j^{(k-1)} = (\hat{b}_j^{(k)} - u_{jk}\hat{x}_k(1 + \varepsilon_{jk}))(1 + \eta_{jk})$
$\qquad$ end do
end do

THEOREM 3.4. (STABILITY OF BACK SUBSTITUTION) *Let $\hat{\boldsymbol{x}}$ be the numerical solution obtained when solving $U\boldsymbol{x} = \boldsymbol{b}$ using the SAXPY variant of back substitution. Then $\hat{\boldsymbol{x}}$ satisfies $\hat{U}\hat{\boldsymbol{x}} = b$, where*

$$|\hat{u}_{jk} - u_{jk}| \leq (n - k + 1)|u_{jk}|\, eps + O(eps^2).$$

PROOF. Define

$$\tilde{b}_j^{(k-1)} = \frac{\hat{b}_j^{(k-1)}}{(1 + \eta_{jk})(1 + \eta_{j,k+1}) \cdots (1 + \eta_{jn})}$$

for $k > j$. Then we can divide the update formula in the inner loop by $(1 + \eta_{jk}) \cdots (1 + \eta_{jn})$ to get

$$\tilde{b}_j^{(k-1)} = \tilde{b}_j^{(k)} - \hat{x}_k \cdot \frac{u_{jk}(1 + \varepsilon_{jk})}{(1 + \eta_{j,k+1}) \cdots (1 + \eta_{jn})}.$$

Moreover, the formula for calculating $\hat{x}_k$ in the outer loop implies

$$\tilde{b}_j^{(j)} = \frac{\hat{b}_j^{(j)}}{(1+\eta_{j,j+1})\cdots(1+\eta_{jn})} = \frac{u_{jj}\hat{x}_j}{(1+\varepsilon_{jj})(1+\eta_{j,j+1})\cdots(1+\eta_{jn})}.$$

Thus, using a telescoping sum, we get

$$b_j = \sum_{k=j+1}^{n} (\tilde{b}_j^{(k)} - \tilde{b}_j^{(k-1)}) + \tilde{b}_j^{(j)}$$

$$= \frac{u_{jj}\hat{x}_j}{(1+\varepsilon_{jj})(1+\eta_{j,j+1})\cdots(1+\eta_{jn})} + \sum_{k=j+1}^{n} \frac{u_{jk}\hat{x}_k(1+\varepsilon_{jk})}{(1+\eta_{j,k+1})\cdots(1+\eta_{jn})},$$

which shows that $\hat{U}\hat{x} = b$ with $|\hat{u}_{jk} - u_{jk}| \le (n-k+1)|u_{jk}|eps + O(eps^2)$, as required. $\qquad\square$

Since the $|u_{jk}|$ is bounded by $\alpha = \rho \cdot \max(a_{jk})$, we see that back substitution is also backward stable as long as the growth factor ρ is not too large. A similar argument shows that forward substitution is also backward stable.

3.2.3   Pivoting and Scaling

To achieve backward stability, one must choose a pivoting strategy that ensures that the growth factor ρ remains small. Unfortunately, there are matrices for which elements grow exponentially with partial pivoting during the elimination process, for instance the matrix

$$A = \begin{pmatrix} 1 & 0 & \cdots & & 0 & 1 \\ -1 & 1 & \ddots & & \vdots & \vdots \\ -1 & -1 & \ddots & & 0 & 1 \\ \vdots & \vdots & \ddots & & 1 & 1 \\ -1 & -1 & \cdots & & -1 & 1 \end{pmatrix}.$$

To start the elimination process in this matrix with partial pivoting, we have to add the first row to all the other rows. This leads to the value 2 in the last column of the matrix $A^{(1)}$, but none of the middle columns have changed. So now adding the second row of $A^{(1)}$ to all the following ones leads to the value 4 in the last column, and continuing like this, the last entry of $A^{(n-1)}$ will equal 2^{n-1}.

Partial pivoting is however used today almost universally when solving linear equations, despite the existence of matrices that grow exponentially during the elimination process. In fact, such matrices are rare[2]: a simple

---

[2] "...intolerable pivot-growth is a phenomenon that happens only to numerical analysts who are looking for that phenomenon", W. Kahan. Numerical linear algebra. *Canad. Math. Bull.*, *9:757–801, 1966.*

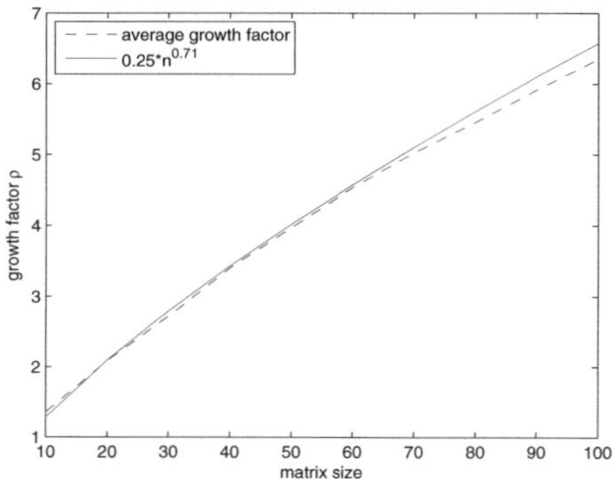

FIGURE 3.4.

Very slow growth of entries encountered during Gaussian elimination with partial pivoting applied to random matrices

MATLAB experiment with random matrices shows that Gaussian elimination with partial pivoting is a very stable process. We make use of the function LU in Problem 3.11.

```
N=500;
n=[10 20 30 40 50 60 70 80 90 100];
for j=1:length(n)
  m=0;
  for i=1:N
    A=rand(n(j));
    [L,U,P,rho]=LU(A);
    m=m+rho;
  end;
  g(j)=m/N
end;
plot(n,g,'--',n,0.25*n.^(0.71),'-');
legend('average growth factor','0.25*n^{0.71}','Location','NorthWest')
xlabel('matrix size'),ylabel('growth factor \rho')
```

In Figure 3.4, we show the results obtained for a typical run of this algorithm. In fact the elements grow sublinearly for random matrices, and thus, in this case, Theorem 3.3 shows that Gaussian elimination with partial pivoting is backward stable.

Partial pivoting may fail to choose the right pivot if the matrix is badly scaled. Consider the linear system $A\boldsymbol{x} = \boldsymbol{b}$

$$\begin{pmatrix} 10^{-12} & 1 & -1 \\ 3 & -4 & 5 \\ 40 & -60 & 0 \end{pmatrix} \boldsymbol{x} = \begin{pmatrix} 17 + 10^{-12} \\ -62 \\ -1160 \end{pmatrix}. \qquad (3.44)$$

The exact solution is $x = (1, 20, 3)^\top$. If we multiply the first equation with 10^{14} we get the system $By = c$,

$$\begin{pmatrix} 100 & 10^{14} & -10^{14} \\ 3 & -4 & 5 \\ 40 & -60 & 0 \end{pmatrix} y = \begin{pmatrix} 17 \cdot 10^{14} + 100 \\ -62 \\ -1160 \end{pmatrix} \qquad (3.45)$$

with of course the same solution.

With partial pivoting, for the system (3.44) the first pivot in the first elimination will be the element $a_{31} = 40$. For the second system (3.45), however, because of bad scaling, the element $a_{11} = 100$ will be chosen. This has the same bad effect as if we had chosen $a_{11} = 10^{-12}$ as pivot in system (3.44). Indeed, we obtain with MATLAB

$$x = \begin{pmatrix} 1 \\ 20 \\ 3 \end{pmatrix}, \quad y = \begin{pmatrix} 1.0025 \\ 20.003 \\ 3.0031 \end{pmatrix},$$

and the second solution is not very accurate, as expected.

Solving the system (3.45) by QR decomposition (Section 3.5) does not help either, as we can see below. But with *complete pivoting*, i.e., if we look for the largest element in modulus in the remaining matrix and interchange both rows (equations) and columns (reordering the unknowns), we get the correct solution (see Problem 3.13 for the function `EliminationCompletePivoting`). With the following MATLAB statements,

```
fak=1e14;
A=[ 100/fak 1 -1
         3 -4  5
        40 -60 0];
xe=[1 20 3 ]'; b=A*xe; x1=A\b;
B=[fak*A(1,:); A(2:3,:)]; c=b; c(1)=c(1)*fak; x2=B\c;
[Q,R]=qr(B); d=Q'*c; x3=R\d;
[x4,Xh,r,U,L,B,P,Q]=EliminationCompletePivoting(B,c,1e-15);
[xe x1 x2 x3 x4]
[norm(xe-x1) norm(xe-x2) norm(xe-x3) norm(xe-x4)]
```

we obtain the results

exact	A\b	B\c	QR	compl.piv.
xe	x1	x2	x3	x4
1.0000	1.0000	1.0025	1.0720	1.0000
20.0000	20.0000	20.0031	20.0456	20.0000
3.0000	3.0000	3.0031	3.0456	3.0000
	$\|$xe-x1$\|$	$\|$xe-x2$\|$	$\|$xe-x3$\|$	$\|$xe-x4$\|$
	1.3323e-15	5.0560e-03	9.6651e-02	2.9790e-15

which show that only Gaussian elimination with complete pivoting leads to a satisfactory solution in this case.

Wilkinson [149] proved that the *growth factors* ρ_n^c for complete pivoting are bound by

$$\rho_n^c \le n^{1/2} \left(2 \cdot 3^{1/2} \cdot \ldots \cdot n^{1/(n-1)} \sim n^{1/2} n^{\frac{1}{4}\log n} \right)$$

It was later conjectured that

$$g(n) := \sup_{A \in \mathbb{R}^{n \times n}} \rho_n^c(A) \le n.$$

However, this was proven false (see [71]). The limit $\lim_{n \to \infty} g(n)/n$ is an open problem. Though in practical problems the growth factors for complete pivoting turn out to be smaller than for partial pivoting, the latter is usually preferred because it is less expensive.

3.2.4   Sum of Rank-One Matrices

Gaussian elimination without pivoting, or the computation of the LU decomposition, may be interpreted as a sequence of *rank-one changes*. We consider for that purpose the matrix product as a sum of matrices of rank one:

$$A = LU = \sum_{k=1}^{n} \boldsymbol{l}_{:k}\boldsymbol{u}_{k:},$$

with columns of L and rows of U

$$L = [\boldsymbol{l}_{:1}, \boldsymbol{l}_{:2}, \ldots \boldsymbol{l}_{:n}], \quad U = \begin{pmatrix} \boldsymbol{u}_{1:} \\ \boldsymbol{u}_{2:} \\ \vdots \\ \boldsymbol{u}_{n:} \end{pmatrix}.$$

We define the matrices

$$A_j = \sum_{k=j}^{n} \boldsymbol{l}_{:k}\boldsymbol{u}_{k:}, \quad A_1 = A.$$

Because L and U are triangular, the first $j-1$ rows and columns of A_j are zero. Clearly

$$A_{j+1} = A_j - \boldsymbol{l}_{:j}\boldsymbol{u}_{j:}$$

holds. In order to eliminate row j and column j of A_j we can choose

$$\boldsymbol{u}_{j:} = (\ \underbrace{0, \ldots, 0}_{j-1 \text{ zeros}}\ , a_{j,j}^{(j)}, a_{j,j+1}^{(j)}, \ldots, a_{j,n}^{(j)})$$

and

$$l_{:j} = \begin{pmatrix} 0 \\ \vdots \\ 0 \\ 1 \\ a_{j+1,j}^{(j)}/a_{j,j}^{(j)} \\ \vdots \\ a_{n,j}^{(j)}/a_{j,j}^{(j)} \end{pmatrix}.$$

The choice for $u_{j:}$ and $l_{:j}$ is not unique – we could for example divide by the pivot the elements of $u_{j:}$ instead of $l_{:j}$. The result of these considerations is the MATLAB function

---

ALGORITHM 3.6.
Gaussian Elimination by Rank-one Modifications

```
function [L,U]=LUbyRank1(A);
% LUBYRANK1 computes the LU factorization
%    [L,U]=LUbyRank1(A); computes the LU-factorization of A with
%    diagonal pivoting as n rank-one modifications. The implementation
%    here is for didactic purposes only.

n=max(size(A));
L=[]; U=[];
for j=1:n
  u=[zeros(j-1,1);1; A(j+1:n,j)/A(j,j)];
  v=[zeros(j-1,1); A(j,j:n)'];
  A=A-u*v';
  L=[L u]; U=[U;v'];
end
```

EXAMPLE 3.7.

```
>> A=[ 17     24      1      8     15
        23      5      7     14     16
         4      6     13     20     22
        10     12     19     21      3
        11     18     25      2      9]
>> [L,U]=LUbyRank1(A)
L =
     1.0000         0         0         0         0
     1.3529    1.0000         0         0         0
     0.2353   -0.0128    1.0000         0         0
     0.5882    0.0771    1.4003    1.0000         0
     0.6471   -0.0899    1.9366    4.0578    1.0000
U =
    17.0000   24.0000    1.0000    8.0000   15.0000
```

```
       0   -27.4706      5.6471      3.1765     -4.2941
       0          0     12.8373     18.1585     18.4154
       0          0          0     -9.3786    -31.2802
       0          0          0          0      90.1734
>> norm(L*U-A)
ans =
   3.5527e-15
```

We saved the vectors in separate matrices L and U. This would of course not be necessary, since one could overwrite A with the decomposition. Furthermore, we did not introduce pivoting; this function is meant for didactic purposes and should not be used in real computations in the present form.

3.3 Condition of a System of Linear Equations

What can we say about the accuracy of the solution when we solve a system of linear equations numerically? A famous early investigation of this question was done by John von Neumann and H. H. Goldstine in 1947 [145]. They conclude their error analysis (over 70 pages long!) with quite pessimistic remarks on the size of linear systems that can be solved on a computer in finite precision arithmetic, see Table 3.1. The number of necessary multiplications they indicate are roughly n^3, which are the operations needed for inversion.

machine precision	10^{-8}	10^{-10}	10^{-12}
$n <$	15	50	150
# multiplications	3'500	120'000	3'500'000

TABLE 3.1.
Pessimistic Error analysis by
John von Neumann and H. H. Goldstine

It is the merit of Jim Wilkinson who discovered first by experiment that the bounds were too pessimistic and who developed the *backward error analysis* which explains much better what happens with calculations on the computer.

Consider the system of linear equations

$$A\boldsymbol{x} = \boldsymbol{b}, \quad \text{with } A \in \mathbb{R}^{n \times n} \text{ nonsingular.}$$

A perturbed system is $\hat{A}\hat{\boldsymbol{x}} = \hat{\boldsymbol{b}}$, and we assume that $\hat{a}_{ij} = a_{ij}(1 + \varepsilon_{ij})$ and $\hat{b}_i = b_i(1 + \varepsilon_i)$, with $|\varepsilon_{ij}| \leq \tilde{\varepsilon}_A$ and $|\varepsilon_i| \leq \tilde{\varepsilon}_b$. This perturbation could come from roundoff errors in finite precision arithmetic, or from measurements, or any other source; we are only interested in how much the solution $\hat{\boldsymbol{x}}$ of the perturbed system differs from the solution $\boldsymbol{x}$ of the original system. The element-wise perturbations imply

$$||\hat{A} - A|| \leq \varepsilon_A ||A||, \quad ||\hat{\boldsymbol{b}} - \boldsymbol{b}|| \leq \varepsilon_b ||\boldsymbol{b}||, \tag{3.46}$$

and if the norm is the 1-norm or infinity norm, we have $\varepsilon_A = \tilde{\varepsilon}_A$ and $\varepsilon_b = \tilde{\varepsilon}_b$, otherwise they differ just by a constant.

THEOREM 3.5. (CONDITIONING OF THE SOLUTION OF LINEAR SYSTEMS) *Consider two linear systems of equations* $A\boldsymbol{x} = \boldsymbol{b}$ *and* $\hat{A}\hat{\boldsymbol{x}} = \hat{\boldsymbol{b}}$ *satisfying (3.46), and assume that* A *is invertible. If* $\varepsilon_A \cdot \kappa(A) < 1$, *then we have*

$$\frac{||\hat{\boldsymbol{x}} - \boldsymbol{x}||}{||\boldsymbol{x}||} \le \frac{\kappa(A)}{1 - \varepsilon_A \kappa(A)}(\varepsilon_A + \varepsilon_B), \tag{3.47}$$

where $\kappa(A) := ||A|| \, ||A^{-1}||$ *is the* condition number *of the matrix* A.

PROOF. From

$$\hat{\boldsymbol{b}} - \boldsymbol{b} = \hat{A}\hat{\boldsymbol{x}} - A\boldsymbol{x} = (\hat{A} - A)\hat{\boldsymbol{x}} + A(\hat{\boldsymbol{x}} - \boldsymbol{x}),$$

we obtain

$$\hat{\boldsymbol{x}} - \boldsymbol{x} = A^{-1}(-(\hat{A} - A)\hat{\boldsymbol{x}} + \hat{\boldsymbol{b}} - \boldsymbol{b}),$$

and therefore, using Assumption (3.46),

$$||\hat{\boldsymbol{x}} - \boldsymbol{x}|| \le ||A^{-1}||(\varepsilon_A ||A|| \, ||\hat{\boldsymbol{x}}|| + \varepsilon_b ||\boldsymbol{b}||).$$

Now we can estimate $||\hat{\boldsymbol{x}}|| = ||\hat{\boldsymbol{x}} - \boldsymbol{x} + \boldsymbol{x}|| \le ||\hat{\boldsymbol{x}} - \boldsymbol{x}|| + ||\boldsymbol{x}||$, and $||\boldsymbol{b}|| = ||A\boldsymbol{x}|| \le ||A|| \, ||\boldsymbol{x}||$, which leads to

$$||\hat{\boldsymbol{x}} - \boldsymbol{x}|| \le ||A^{-1}|| \, ||A||(\varepsilon_A(||\boldsymbol{x}|| + ||\hat{\boldsymbol{x}} - \boldsymbol{x}||) + \varepsilon_b ||\boldsymbol{x}||),$$

and thus

$$||\hat{\boldsymbol{x}} - \boldsymbol{x}||(1 - \varepsilon_A \kappa(A)) \le \kappa(A)||\boldsymbol{x}||(\varepsilon_A + \varepsilon_b),$$

which concludes the proof. $\qquad\square$

This theorem shows that the condition of the problem of solving a linear system of equations is tightly connected to the condition number of the matrix A, regardless of whether the perturbation appears in the matrix or in the right hand side of the linear system. An alternative derivation of the perturbed solution when only the matrix is slightly changed is given in Section 6.4.

If $\kappa(A)$ is large then the solutions $\hat{\boldsymbol{x}}$ and $\boldsymbol{x}$ will differ significantly. According to Wilkinson's Principle (see Section 2.7), the result of a numerical computation is the exact result with slightly perturbed initial data. Thus $\hat{\boldsymbol{x}}$ will be the result of a linear system with perturbed data of order *eps*. *As a rule of thumb, we have to expect a relative error of* $eps \cdot \kappa(A)$ *in the solution.*

Computing the condition number is in general more expensive than solving the linear system. For example, if we use the spectral norm, the condition number can be computed by

$$\kappa_2(A) = ||A||_2 ||A^{-1}||_2 = \frac{\sigma_{\max}(A)}{\sigma_{\min}(A)} = \texttt{cond(A)} \text{ in MATLAB.}$$

EXAMPLE 3.8. *We consider the matrix*

```
>> A=[21.6257    51.2930     1.5724    93.4650
       5.2284    83.4314    37.6507    84.7163
      68.3400     3.6422     6.4801    52.5777
      67.7589     4.5447    42.3687     9.2995];
```

choose the exact solution and compute the right hand side

```
>> x=[1:4]'; b=A*x;
```

Now we solve the system

```
>> xa=A\b;
```

and compare the numerical solution to the exact one.

```
>> format long e
>> [xa x]
>> cond(A)
>> eps*cond(A)
ans =
   0.999999999974085    1.000000000000000
   1.999999999951056    2.000000000000000
   3.000000000039628    3.000000000000000
   4.000000000032189    4.000000000000000
ans =
      6.014285987206616e+05
ans =
      1.335439755935445e-10
```

We see that the numerically computed solution has about 5 incorrect decimal digits, which corresponds very well to the predicted error due to the large condition number of 6.014×10^5 of the matrix A. The factor eps $\kappa(A) = 1.33 \times 10^{-10}$ indicates well the error after the 10th decimal digit.

Examples of matrices with a very large condition number are the *Hilbert matrix H* with $h_{ij} = 1/(i+j-1)$, $i,j = 1,\ldots,n$ (H = hilb(n) in MATLAB) and *Vandermonde matrices*, which appear for example in interpolation and whose columns are powers of a vector v, that is $a_{ij} = v_i^{n-j}$ (A=vander(v) in MATLAB). Table 3.2 shows the condition numbers for $n = 3,\ldots,8$, where we have used v = [1:n] for the Vandermonde matrices.

Matrices with a small condition number are for example orthogonal matrices, U such that $U^\top U = I$, which gives for the spectral condition number $\kappa_2(U) = ||U||_2 ||U^{-1}||_2 = ||U||_2 ||U^\top||_2 = 1$, since orthogonal transformations preserve Euclidean length.

The *condition number $\kappa(A)$* satisfies several properties:

1. $\kappa(A) \geq 1$, provided the matrix norm satisfies the submultiplicative property, since $1 = ||I|| = ||A^{-1}A|| \leq ||A|| \, ||A^{-1}||$.

n	cond(hilb(n))	cond(vander([1:n]))
3	5.2406e+02	7.0923e+01
4	1.5514e+04	1.1710e+03
5	4.7661e+05	2.6170e+04
6	1.4951e+07	7.3120e+05
7	4.7537e+08	2.4459e+07
8	1.5258e+10	9.5211e+08

TABLE 3.2. *Matrices with large condition numbers*

2. $\kappa(\alpha A) = \alpha\kappa(A)$.

3. $\kappa(A) = \dfrac{\max_{\|\boldsymbol{y}\|=1}\|A\boldsymbol{y}\|}{\min_{\|\boldsymbol{z}\|=1}\|A\boldsymbol{z}\|}$, since

$$\|A^{-1}\| = \max_{\boldsymbol{x}\neq 0}\frac{\|A^{-1}\boldsymbol{x}\|}{\|\boldsymbol{x}\|} = \max_{\boldsymbol{z}\neq 0}\frac{\|\boldsymbol{z}\|}{\|A\boldsymbol{z}\|} = \min_{\boldsymbol{z}\neq 0}\left(\frac{\|A\boldsymbol{z}\|}{\|\boldsymbol{z}\|}\right)^{-1}.$$

Another useful property of the condition number is that it measures how far A is from a singular matrix, in a relative sense.

THEOREM 3.6. *Let* $E \in \mathbb{R}^{n\times n}$. *If* $\|E\| < 1$, *then* $I + E$ *is non-singular, and we have*

$$\frac{1}{1+\|E\|} \leq \|(I+E)^{-1}\| \leq \frac{1}{1-\|E\|}. \tag{3.48}$$

PROOF. To show that $I + E$ is non-singular, let us argue by contradiction by assuming that $I + E$ is singular. Then there exists a non-zero vector $\boldsymbol{x}$ such that $(I + E)\boldsymbol{x} = 0$. Then we have

$$\boldsymbol{x} = -E\boldsymbol{x} \implies \|\boldsymbol{x}\| = \|E\boldsymbol{x}\| \leq \|E\|\,\|\boldsymbol{x}\|.$$

Since $\boldsymbol{x} \neq 0$, we can divide both sides by $\|\boldsymbol{x}\|$ and conclude that $\|E\| \geq 1$, which contradicts the hypothesis that $\|E\| < 1$. Hence $I + E$ is non-singular. To show the first inequality in (3.48), we let $\boldsymbol{y} = (I + E)^{-1}\boldsymbol{x}$. Then

$$\boldsymbol{x} = (I + E)\boldsymbol{y} \implies \|\boldsymbol{x}\| \leq (1 + \|E\|)\|\boldsymbol{y}\|,$$

which implies

$$\frac{\|(I+E)^{-1}\boldsymbol{x}\|}{\|\boldsymbol{x}\|} = \frac{\|\boldsymbol{y}\|}{\|\boldsymbol{x}\|} \geq \frac{1}{1+\|E\|}.$$

Since this is true for all $\boldsymbol{x} \neq 0$, we can take the maximum of the left hand side over all $\boldsymbol{x} \neq 0$ and obtain

$$\|(I+E)^{-1}\| \geq \frac{1}{1+\|E\|},$$

as required. For the other inequality, note that

$$\|y\| \le \|x\| + \|y - x\| \le \|x\| + \| - Ey\| \le \|x\| + \|E\|\,\|y\|,$$

which implies

$$(1 - \|E\|)\|y\| \le \|x\|.$$

Since $1 - \|E\| > 0$, we can divide both sides by $(1 - \|E\|)\|x\|$ without changing the inequality sign. This yields

$$\frac{\|(I + E)^{-1}x\|}{\|x\|} = \frac{\|y\|}{\|x\|} \le \frac{1}{1 - \|E\|}.$$

Maximizing the left hand side over all $x \ne 0$ yields $\|(I+E)^{-1}\| \le 1/(1-\|E\|)$, as required. $\qquad\qquad\qquad\qquad\qquad\qquad\qquad\qquad\qquad\qquad\qquad\qquad\quad \Box$

COROLLARY 3.1. *Let $A \in \mathbb{R}^{n \times n}$ be a non-singular matrix, $E \in \mathbb{R}^{n \times n}$ and $r = \|E\|/\|A\|$. If $r\,\kappa(A) < 1$, then $A + E$ is non-singular and*

$$\frac{\|(A + E)^{-1}\|}{\|A^{-1}\|} \le \frac{1}{1 - r\,\kappa(A)}.$$

PROOF. By (3.48), we have

$$\|(A + E)^{-1}\| = \|(I + A^{-1}E)^{-1}A^{-1}\|$$

$$\le \|(I + A^{-1}E)^{-1}\|\,\|A^{-1}\| \le \frac{\|A^{-1}\|}{1 - \|A^{-1}E\|} \le \frac{\|A^{-1}\|}{1 - \|A^{-1}\|\,\|E\|}.$$

Substituting $\|E\| = r\|A\|$ and dividing by $\|A^{-1}\|$ gives the required result.
$\Box$

Note that the hypothesis implies that $A + E$ will be nonsingular provided that the perturbation E is small enough, in the relative sense, with respect to the condition number $\kappa(A)$. In other words, if $\kappa(A)$ is small, i.e., if A is well conditioned, then a fairly large perturbation is needed to reach singularity. On the other hand, there always exists a perturbation E with $\|E\|_2 = \|A\|_2/\kappa_2(A)$ such that $A + E$ is singular, cf. Problem 3.19.

3.4 Cholesky Decomposition

3.4.1 Symmetric Positive Definite Matrices

DEFINITION 3.1. (SYMMETRIC POSITIVE DEFINITE MATRICES) *A matrix is symmetric, $A = A^{\top}$, and positive definite if*

$$x^{\top}Ax > 0 \quad \textit{for all } x \ne 0. \tag{3.49}$$

Positive definite matrices occur frequently in applications, for example for discretizations of coercive partial differential equations, or in optimization, where the Hessian must be positive definite at a strict minimum of a function of several variables. Positive definite matrices have rather special properties, as shown in the following lemma.

LEMMA 3.1. *Let $A = A^\top \in \mathbb{R}^{n \times n}$ be positive definite. Then*

(a) *If $L \in \mathbb{R}^{m \times n}$ with $m \le n$ has rank m, then $LAL^\top$ is positive definite.*

(b) *Every principal submatrix of A is positive definite, i.e., for every non-empty subset $J \subset \{1, 2, \ldots, n\}$ with $|J| = m$, the $m \times m$ matrix $A(J, J)$ is positive definite.*

(c) *The diagonal entries of A are positive, i.e., $a_{ii} > 0$ for all i.*

(d) *The largest entry of A in absolute value must be on the diagonal, and hence positive.*

PROOF.

(a) Let $z \in \mathbb{R}^m$ be an arbitrary non-zero vector. Then the fact that L has rank m implies $y := L^\top z \neq 0$, since the rows of L would otherwise be linearly dependent, forcing L to have rank less than m. The positive definiteness of A, in turn, gives

$$z^\top LAL^\top z = y^\top Ay > 0.$$

(b) Let $J = \{j_1, \ldots, j_m\}$ be an arbitrary subset of $\{1, \ldots, n\}$ and $B = A(J, J)$. Define the $n \times m$ matrix R with entries

$$r_{jk} = \begin{cases} 1, & j = j_k \\ 0, & \text{otherwise.} \end{cases}$$

It is then easy to see that $B = R^\top AR$. It follows by letting $L = R^\top$ in (a) that B is positive definite.

(c) This follows from (b) by choosing J to be the one-element set $\{i\}$.

(d) We argue by contradiction. Suppose, on the contrary, that the largest entry in absolute value does not occur on the diagonal. Then there must be a pair (i, j) with $i \neq j$ such that $|a_{ij}|$ is larger than any entry on the diagonal; in particular, we have $|a_{ij}| > a_{ii}$ and $|a_{ij}| > a_{jj}$. From (b), we know that the principal submatrix

$$B = \begin{pmatrix} a_{ii} & a_{ij} \\ a_{ji} & a_{jj} \end{pmatrix}$$

must be positive definite. But taking $x^\top = (1, -\mathrm{sign}(a_{ij}))^\top$ gives

$$x^\top B x = a_{ii} + a_{jj} - 2|a_{ij}| < 0,$$

which contradicts the positive definiteness of B. Hence the largest entry in absolute value must occur on the diagonal (and hence be positive).

$\square$

This gives another characterization of symmetric positive definite matrices.

LEMMA 3.2. *Let $A = A^\top \in \mathbb{R}^{n \times n}$. Then A is positive definite if and only if all its eigenvalues are positive.*

PROOF. Suppose A is symmetric positive definite and λ an eigenvalue of A with eigenvector $v \neq 0$. Then

$$0 < v^\top A v = v^\top (\lambda v) = \lambda \|v\|_2^2.$$

We can now divide by $\|v\|_2^2 > 0$ to deduce that $\lambda > 0$.

Now suppose all eigenvalues of A are positive. Then by the spectral theorem, we can write $A = Q \Lambda Q^\top$, where Λ is a diagonal matrix containing the eigenvalues of A and $Q^\top Q = I$. But a diagonal matrix with positive diagonal entries is clearly positive definite, and Q is non-singular (i.e., it has full rank n). Hence, by Lemma 3.1(a), A is also positive definite. $\square$

THEOREM 3.7. *Let $A = A^\top \in \mathbb{R}^{n \times n}$ be symmetric positive definite. Then in exact arithmetic, Gaussian elimination with no pivoting does not break down when applied to the system $Ax = b$, and only positive pivots will be encountered. Furthermore, we have $U = DL^\top$ for a positive diagonal matrix D, i.e., we have the factorization $A = LDL^\top$.*

PROOF. The proof is by induction. As shown before, the diagonal element $a_{11} > 0$ and therefore can be used as pivot. We now show that after the first elimination step, the remaining matrix is again positive definite. For the first step we use the matrix

$$L_1 = \begin{pmatrix} 1 & 0 & 0 & \cdots & 0 \\ -\frac{a_{21}}{a_{11}} & 1 & 0 & \cdots & 0 \\ \vdots & \vdots & & \ddots & \vdots \\ -\frac{a_{n1}}{a_{11}} & 0 & 0 & \cdots & 1 \end{pmatrix}$$

and multiply from the left to obtain

$$L_1 A = \begin{pmatrix} a_{11} & b^\top \\ 0 & A_1 \end{pmatrix}, \quad b^\top = (a_{12}, \ldots a_{1n}).$$

Because of the symmetry of A, multiplying the last equation from the right with $L_1^\top$ we obtain

$$L_1 A L_1^\top = \begin{pmatrix} a_{11} & 0^\top \\ 0 & A_1 \end{pmatrix} \tag{3.50}$$

and the submatrix A_1 is not changed since the first column of $L_1 A$ is zero below the diagonal. This also shows that A_1 is symmetric. Furthermore, by Lemma 3.1(a), we know that $L_1 A L_1^\top$ is in fact positive definite, since L_1 has rank n. Since A_1 is a principal submatrix of $L_1 A L_1^\top$, it is also positive definite. Thus, the second elimination step can be performed again, and the process can be continued until we obtain the decomposition

$$L_{n-1} \cdots L_1 A L_1^\top \cdots L_{n-1}^\top = \begin{pmatrix} a_{11}^{(0)} & & 0 \\ & \ddots & \\ 0 & & a_{nn}^{(n-1)} \end{pmatrix} =: D,$$

where the diagonal entries $a_{kk}^{(k-1)}$ are all positive. By letting $L = L_1^{-1} \cdots L_{n-1}^{-1}$, we obtain the decomposition $A = LDL^\top$, as required. $\qquad\square$

Since the diagonal entries of D are positive, it is possible to define its square root

$$D^{1/2} = \mathrm{diag}(\sqrt{a_{11}^{(0)}}, \dots, \sqrt{a_{nn}^{(n-1)}}).$$

Then by letting $R = D^{1/2} L^\top$, we obtain the so-called *Cholesky Decomposition*

$$A = R^\top R.$$

We can compute R directly without passing through the LU decomposition as follows. Multiplying from the left with the unit vector $e_j^\top$, we obtain

$$\boldsymbol{a}_{j:} = e_j^\top R^\top R = \boldsymbol{r}_{:j}^\top R = \sum_{k=1}^{j} r_{kj}\, \boldsymbol{r}_{k:}. \tag{3.51}$$

If we assume that we already know the rows $\boldsymbol{r}_{1:}, \dots, \boldsymbol{r}_{j-1:}$, we can solve Equation (3.51) for $\boldsymbol{r}_{j:}$ and obtain

$$r_{jj}\, \boldsymbol{r}_{j:} = \boldsymbol{a}_{j:} - \sum_{k=1}^{j-1} r_{kj}\, \boldsymbol{r}_{k:} =: \boldsymbol{v}^\top.$$

The right hand side (the vector $\boldsymbol{v}$) is known. Thus multiplying from the right with $\boldsymbol{e}_j$ we get

$$r_{jj}\, \boldsymbol{r}_{j:}\boldsymbol{e}_j = r_{jj}^2 = v_j \Rightarrow r_{jj} = \sqrt{v_j} \quad \Rightarrow \boldsymbol{r}_{j:} = \frac{\boldsymbol{v}^\top}{\sqrt{v_j}}.$$

Thus we have computed the next row of R. We only need a loop over all rows and obtain the function `Cholesky`:

ALGORITHM 3.7. *Cholesky Decomposition*

```
function R=Cholesky(A)
```

```
% CHOLESKY computes the Cholesky decomposition of a matrix
%    R=Cholesky(A) computes the Cholesky decomposition A=R'R

n=length(A);
for j=1:n,
  v=A(j,j:n);
  if j>1,
    v=A(j,j:n)-R(1:j-1,j)'*R(1:j-1,j:n);
  end;
  if v(1)<=0
    error('Matrix is not positive definite')
  else
    h=1/sqrt(v(1));
  end
  R(j,j:n)=v*h;
end
```

To compute the Cholesky decomposition there is in MATLAB the built-in function `chol`. Using the Cholesky decomposition, we can write the quadratic form defined by the matrix A as a sum of squares:

$$\sum_{i=1}^{n}\sum_{j=1}^{n} a_{ij}^2 x_i x_j = \boldsymbol{x}^\top A \boldsymbol{x} = \boldsymbol{x}^\top R^\top R \boldsymbol{x} = ||R\boldsymbol{x}||^2 = \boldsymbol{y}^\top \boldsymbol{y} = \sum_{i=1}^{n} y_i^2, \quad \text{with } \boldsymbol{y} = R\boldsymbol{x}.$$

3.4.2 Stability and Pivoting

We have seen in Section 3.2.2 that for general non-symmetric matrices, pivoting is essential not only to ensure that the elimination process does not break down, but also to ensure stability. However, for symmetric positive definite matrices, we have shown that break down cannot occur in exact arithmetic, even when *no pivoting* is used. Thus, one might suspect that pivoting is also not needed for stability. This is indeed true, provided that the matrix is not too ill conditioned.

THEOREM 3.8. *Let* $A = A^\top \in \mathbb{R}^{n \times n}$ *be positive definite. If* $c_n \kappa_2(A)\, eps < 1$ *with* $c_n = 3n^2 + O(eps)$, *then Gaussian elimination (or Cholesky decomposition) does not break down. Moreover, if* $\hat{L}$ *and* $\hat{D}$ *are the numerically computed factors in the Cholesky decomposition and* $\hat{A} = \hat{L}\hat{D}\hat{L}^\top$ *is the numerically computed decomposition, then we have the estimate*

$$|\hat{a}_{ij} - a_{ij}| \le 3\alpha \min(i - 1, j)\, eps + O(eps^2), \tag{3.52}$$

where $\alpha = \max_{i,j} |a_{ij}|$.

Remark. The above result implies that if Cholesky breaks down, it is either because the matrix A is not positive definite, or that it is very ill conditioned. In practical computations, one rarely checks the condition

$c_n \kappa_2(A) \, eps < 1$ before the actual factorization, because the condition number $\kappa_2(A)$ is more expensive to calculate than the Cholesky factors itself.

If one suspects that A is so ill conditioned that the hypothesis may be violated, pivoting may help reduce the factorization errors due to rounding. For positive definite matrices, complete pivoting reduces to *diagonal pivoting*, since the largest entry always appears on the diagonal. We can then obtain the estimates in Theorem 3.3, provided each submatrix that appears after an elimination step remains positive definite.

PROOF. Let $\hat{A}^{(k)}$ be the $(n-k) \times (n-k)$ submatrix $\hat{A}^{(k)} = [\hat{a}_{ij}^{(k)}]_{k+1 \leq i,j \leq n}$ for $0 \leq k \leq n$. Just as in the proof of Theorem 3.3, we have the relation

$$\hat{a}_{ij}^{(k)} = (\hat{a}_{ij}^{(k-1)} - \hat{l}_{ik}\hat{a}_{kj}^{(k-1)}(1 + \varepsilon_{ijk}))(1 + \eta_{ijk}) = \underbrace{\hat{a}_{ij}^{(k-1)} - \frac{\hat{a}_{ik}^{(k-1)}\hat{a}_{kj}}{\hat{a}_{kk}^{(k-1)}}}_{b_{ij}^{(k)}} + \mu_{ijk},$$

where $\hat{l}_{ik} = \frac{\hat{a}_{ik}^{(k-1)}}{\hat{a}_{kk}^{(k-1)}}(1 + \delta_{ik})$, with $|\varepsilon_{ijk}|, |\eta_{ijk}|, |\delta_{ik}| < eps$. Here, $B^{(k)} = [b_{ij}^{(k)}]_{k+1 \leq i,j \leq n}$ is the submatrix we would have obtained if we had performed one step of Gaussian elimination on $\hat{A}^{(k-1)}$ in exact arithmetic, so that μ_{ijk} contains all the round-off errors associated with this step. Using the same manipulation as in the proof of Theorem 3.3, we can write

$$|\mu_{ijk}| \leq |b_{ij}^{(k)}| eps + \left| \frac{\hat{a}_{ik}^{(k-1)}\hat{a}_{kj}^{(k-1)}}{\hat{a}_{kk}^{(k-1)}} \right| \cdot 2eps + O(eps^2). \tag{3.53}$$

We now use the fact that $B^{(k)}$ is positive definite to bound the first two terms above. By Lemma 3.1(d), we have

$$\max_{ij} |b_{ij}^{(k)}| = \max_i b_{ii}^{(k)} = \max_i \left[\hat{a}_{ii}^{(k-1)} - \sum_{j=1}^{k} (\hat{a}_{ij}^{(k-1)})^2 \right] \leq \max_i \hat{a}_{ii}^{(k-1)}.$$

Moreover, we have

$$\max_{i,j} \left| \frac{\hat{a}_{ik}^{(k-1)}\hat{a}_{kj}^{(k-1)}}{\hat{a}_{kk}^{(k-1)}} \right| \stackrel{(*)}{=} \max_i \frac{(\hat{a}_{ik}^{(k-1)})^2}{\hat{a}_{kk}^{(k-1)}} = \max_i(\hat{a}_{ii}^{(k-1)} - b_{ii}^{(k)}) \stackrel{(\dagger)}{\leq} \max_i \hat{a}_{ii}^{(k-1)},$$

where the equality $(*)$ is due to the symmetry of $\hat{a}_{ij}^{(k-1)}$ and the inequality $(\dagger)$ is true because $b_{ii}^{(k)} > 0$. Thus, by letting $\alpha_k = \max_{i,j} |\hat{a}_{ij}^{(k)}| = \max_i \hat{a}_{ii}^{(k)}$, (3.53) becomes

$$|\mu_{ijk}| \leq 3\alpha_{k-1} \, eps + O(eps^2). \tag{3.54}$$

This implies $\hat{A}^{(k)} = B^{(k)} + E^{(k)}$ with

$$\|E^{(k)}\|_2 \leq 3(n-k) \, eps \|\hat{A}^{(k-1)}\|_2 + O(eps^2).$$

Thus, if we can show that $\hat{A}^{(k)}$ is positive definite with $c_k \kappa_2(A^{(k)}) eps < 1$, $c_k = 3(n-k)^2 + O(eps)$, then the induction is complete and we can conclude that the Cholesky factorization does not break down. To do so, we consider the matrix

$$\tilde{A}^{(k-1)} = \hat{A}^{(k-1)} + \begin{pmatrix} 0 & 0 \\ 0 & E^{(k)} \end{pmatrix} = \hat{A}^{(k-1)} + \tilde{E}^{(k)}.$$

In other words, if we perform one step of Gaussian elimination in exact arithmetic on $\tilde{A}^{(k-1)}$, the result would be $\hat{A}^{(k)}$. We argue that all the eigenvalues of $\tilde{A}^{(k-1)}$ are positive, which implies $\tilde{A}^{(k-1)}$ is positive definite. Assume the contrary, i.e., that the smallest eigenvalue of $\tilde{A}^{(k-1)}$ is zero or negative. Then since eigenvalues are a continuous function of the perturbation (see Chapter 7), there exists $0 < t \le 1$ such that $\hat{A}^{(k-1)} + t\tilde{E}^{(k)}$ has a zero eigenvalue, i.e., it is singular. But we have

$$\frac{\|t\tilde{E}^{(k)}\|_2 \kappa_2(\hat{A}^{(k-1)})}{\|\hat{A}^{(k-1)}\|_2} \le \frac{3(n-k)t\,eps + O(eps^2)}{3(n-k+1)^2\,eps + O(eps^2)} < 1,$$

so by Corollary 3.1, $\hat{A}^{(k-1)} + t E^{(k)}$ must be non-singular, a contradiction. Hence $\tilde{A}^{(k-1)}$ must be positive definite.

We now want to show that $\kappa_2(\hat{A}^{(k)}) \le \dfrac{1}{3(n-k)^2\,eps + O(eps^2)}$. We have

$$\begin{aligned}
\|\hat{A}^{(k)}\|_2 &= \max_{\|z\|_2=1} z^\top \hat{A}^{(k)} z \\
&= z^\top (\tilde{A}^{(k-1)}(k:n,k:n) - (\hat{a}_{kk}^{(k-1)})^{-1} \boldsymbol{a}_{:,k} \boldsymbol{a}_{:,k}^\top) z \\
&\le z^\top [\tilde{A}^{(k-1)}(k:n,k:n)] z \\
&= (0, z^\top) \tilde{A}^{(k-1)} \begin{pmatrix} 0 \\ z \end{pmatrix} \le \|\tilde{A}^{(k-1)}\|_2.
\end{aligned}$$

On the other hand, since $(\hat{A}^{(k)})^{-1}$ is a principal submatrix of $(\tilde{A}^{(k-1)})^{-1}$, we automatically have $\|(\hat{A}^{(k)})^{-1}\|_2 \le \|(\tilde{A}^{(k-1)})^{-1}\|_2$. Thus, we have $\kappa_2(\hat{A}^{(k)}) \le \kappa_2(\tilde{A}^{(k-1)})$. It now suffices to bound $\|\tilde{A}^{(k-1)}\|_2$ and $\|(\tilde{A}^{(k-1)})^{-1}\|_2$ individually:

$$\|\tilde{A}^{(k-1)}\|_2 \le \|\hat{A}^{(k-1)}\|_2 + \|E^{(k)}\|_2 \le \|\hat{A}^{(k-1)}\|_2(1 + 3(n-k)\,eps),$$

$$\begin{aligned}
\|(\tilde{A}^{(k-1)})^{-1}\|_2 &\le \frac{\|\hat{A}^{(k-1)}\|_2}{1 - 3(n-k)\kappa_2(\hat{A}^{(k-1)})\,eps + O(eps^2)} \\
&\le \frac{\|(\hat{A}^{(k-1)})^{-1}\|_2}{1 - \frac{3(n-k)}{c_{k-1}}} = \frac{c_{k-1}\|(\hat{A}^{(k-1)})^{-1}\|_2}{c_{k-1} - 3(n-k)}.
\end{aligned}$$

Multiplying the two quantities above gives

$$\kappa_2(\hat{A}^{(k)}) \leq \frac{\kappa_2(\hat{A}^{(k-1)})c_{k-1}(1+3(n-k)\,eps)}{c_{k-1}-3(n-k)}$$

$$\leq \frac{1}{eps}\frac{1+3(n-k)\,eps}{c_{k-1}-3(n-k)}$$

$$\leq \frac{1}{eps}\frac{1+3(n-k)\,eps}{3((n-k)^2+(n-k)+1)} \leq \frac{1}{3(n-k)^2\,eps},$$

so the induction step is complete. Finally, (3.54) implies that

$$\alpha_k \leq \alpha_{k-1}(1+3eps).$$

Thus, using the same analysis as in Theorem 3.3, we get the estimate (3.52), where

$$\alpha = \max_{0\leq k\leq n-1}\alpha_k \leq \alpha_0(1+3eps)^n = \max_{i,j}|a_{ij}| + O(eps).$$

$$\square$$

Note that the above theorem shows that the growth factor for the elimination process is essentially $\rho = 1$, i.e., no entries that appear during the elimination process can be larger than the largest entry in the original matrix A. Thus, Cholesky factorization is backward stable, even without pivoting.

3.5 Elimination with Givens Rotations

In this section, we discuss another method for elimination that requires more operations, but does not require pivoting and can also be used for solving least squares problems. We discuss here a straightforward implementation – a more sophisticated version is given in Section 6.5.3.

In the i-th step, we eliminate x_i in equations $i+1$ to n as follows: let

$$
\begin{aligned}
(i): \quad & a_{ii}x_i + \ldots + a_{in}x_n &&= b_i \\
& \quad\vdots && \vdots \\
(k): \quad & a_{ki}x_i + \ldots + a_{kn}x_n &&= b_k \\
& \quad\vdots && \vdots \\
(n): \quad & a_{ni}x_i + \ldots + a_{nn}x_n &&= b_n
\end{aligned}
\tag{3.55}
$$

be the reduced system. We wish to eliminate x_i from equation (k). In the case where $a_{ki} = 0$, nothing needs to be done because the unknown x_i is already eliminated. Otherwise, we multiply equation (i) with $-\sin\alpha$ and equation (k) with $\cos\alpha$ and replace equation (k) by the linear combination

$$(k)_{new} := -\sin\alpha \cdot (i) + \cos\alpha \cdot (k). \tag{3.56}$$

Therefore, we choose α so that

$$a_{ki}^{new} := -\sin\alpha \cdot a_{ii} + \cos\alpha \cdot a_{ki} = 0. \tag{3.57}$$

Since $a_{ki} \neq 0$, we compute from (3.57)

$$\cot \alpha = \frac{a_{ii}}{a_{ki}}, \tag{3.58}$$

and obtain, using well known trigonometric identities, the quantities co $=$ $\cos \alpha$ and si $= \sin \alpha$ by

$$\mathrm{cot} = a_{ii}/a_{ik}; \quad \mathrm{si} = 1/\sqrt{1 + \mathrm{cot}^2}; \quad \mathrm{co} = \mathrm{si} \times \mathrm{cot}; \tag{3.59}$$

In this elimination step, in addition to replacing equation (k), we also modify equation (i) with

$$(i)_{new} := \cos \alpha \cdot (i) + \sin \alpha \cdot (k). \tag{3.60}$$

This is done for stability purposes. Observe that pivoting now becomes unnecessary: for the situation when $a_{ii} = 0$ and $a_{ki} \neq 0$, we get $\cot \alpha = 0$, and therefore $\sin \alpha = 1$ and $\cos \alpha = 0$. The two assignments (3.56) and (3.60) simply exchange equations (k) and (i), as we would do with pivoting! Admittedly, however, the computational effort is doubled. This leads to the following program:

ALGORITHM 3.8.
Solving Linear Systems with Givens Rotations

```
function x=EliminationGivens(A,b);
% ELIMINATIONGIVENS solves a linear system using Givens-rotations
%   x=EliminationGivens(A,b) solves Ax=b using Givens-rotations. Uses
%   the function BackSubstitutionSAXPY.

n=length(A);
for i= 1:n
  for k=i+1:n
    if A(k,i)~=0
      cot=A(i,i)/A(k,i);                      % rotation angle
      si=1/sqrt(1+cot^2); co=si*cot;
      A(i,i)=A(i,i)*co+A(k,i)*si;             % rotate rows
      h=A(i,i+1:n)*co+A(k,i+1:n)*si;
      A(k,i+1:n)=-A(i,i+1:n)*si+A(k,i+1:n)*co;
      A(i,i+1:n)=h;
      h=b(i)*co+b(k)*si;                      % rotate right hand side
      b(k)=-b(i)*si+b(k)*co; b(i)=h;
    end
  end;
  if A(i,i)==0
    error('Matrix is singular');
  end;
end
x=BackSubstitutionSAXPY(A,b);
```

3.6 Banded matrices

A matrix is called *banded* if it contains nonzero elements only in a few diagonals next to the main diagonal. For example, consider the matrix

$$
A = \begin{pmatrix}
2 & 1 & -1 & 0 & 0 & 0 & 0 \\
-4 & 2 & 3 & 0 & 0 & 0 & 0 \\
0 & -12 & 3 & 1 & 2 & 0 & 0 \\
0 & 0 & -24 & 4 & -7 & 0 & 0 \\
0 & 0 & 0 & -40 & 5 & 1 & 4 \\
0 & 0 & 0 & 0 & -60 & 6 & -23 \\
0 & 0 & 0 & 0 & 0 & -84 & 7
\end{pmatrix}. \tag{3.61}
$$

A is a banded matrix with upper bandwidth $p = 2$ and lower bandwidth $q = 1$, thus in total there are $p + q + 1$ nonzero diagonals. It is easy to see that, when computing the LU decomposition without pivoting, the factors L and U occupy the same band, i.e., L is a lower banded matrix with q diagonals and U an upper banded matrix with p nonzero diagonals. For the matrix (3.61), the first Gaussian elimination step in which we eliminate x_1 will only modify the elements a_{21}, a_{22} and a_{23}. Performing the complete LU decomposition without pivoting we obtain the factors $A = LU$ with

$$
L = \begin{pmatrix}
1 & 0 & 0 & 0 & 0 & 0 & 0 \\
-2 & 1 & 0 & 0 & 0 & 0 & 0 \\
0 & -3 & 1 & 0 & 0 & 0 & 0 \\
0 & 0 & -4 & 1 & 0 & 0 & 0 \\
0 & 0 & 0 & -5 & 1 & 0 & 0 \\
0 & 0 & 0 & 0 & -6 & 1 & 0 \\
0 & 0 & 0 & 0 & 0 & -7 & 1
\end{pmatrix}
\quad
U = \begin{pmatrix}
2 & 1 & -1 & 0 & 0 & 0 & 0 \\
0 & 4 & 1 & 0 & 0 & 0 & 0 \\
0 & 0 & 6 & 1 & 2 & 0 & 0 \\
0 & 0 & 0 & 8 & 1 & 0 & 0 \\
0 & 0 & 0 & 0 & 10 & 1 & 4 \\
0 & 0 & 0 & 0 & 0 & 12 & 1 \\
0 & 0 & 0 & 0 & 0 & 0 & 14
\end{pmatrix}.
$$

Thus the band of matrix A might be overwritten by the LU decomposition if we, as usual, do not store the diagonal elements of L.

3.6.1 Storing Banded Matrices

Of course, we should not store a banded matrix as a full $n \times n$ matrix, but rather avoid storing zero elements. With lower bandwidth q and upper bandwidth p, a matrix A could be stored as a dense matrix B of size $n \times (p + q + 1)$. The nonzero diagonals of A become the columns of B (cf. Figure 3.5). The price we pay for saving memory in this fashion is that accessing the elements becomes more complicated, since the indexing requires more integer operations. Thus, there is always a trade off between saving memory versus saving operations. In addition, the speed of modern microprocessors is affected by complicated memory accesses, because of cache effects.

The mapping of the banded matrix A to the matrix B is computed by the following program

---

ALGORITHM 3.9. *Transformation for Banded Matrices*

```
function B=StoreBandMatrix(A,q,p)
% STOREBANDMATRIX stores the band of a matrix in a rectangular matrix
```

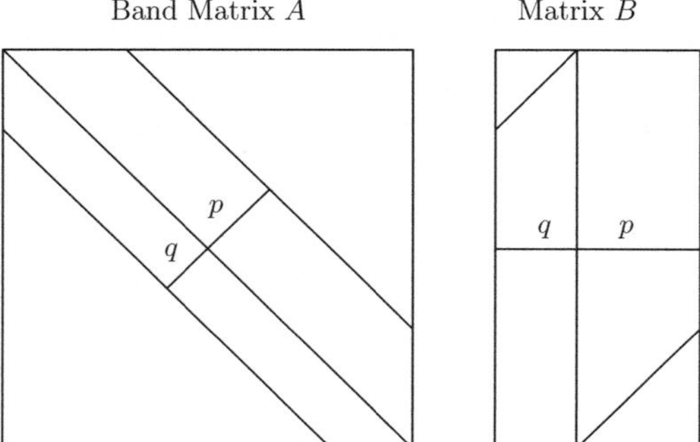

FIGURE 3.5. *Storing of banded matrices*

```
%    B=StoreBandMatrix(A) stores the band of A (with lower bandwidth p
%    and upper bandwidth q) in the rectangular matrix B of dimensions
%    n*p+q+1.

n=length(A);
B=zeros(n,p+q+1);                                    % reserve space
for i=1:n
  for j=max(1,i-q):min(n,i+p)
    B(i,j-i+q+1)=A(i,j);
  end
end
```

Our example matrix (3.61) is transformed with B=StoreBandMatrix(A,1,2) to

```
B =
        0     2     1    -1
       -4     2     3     0
      -12     3     1     2
      -24     4    -7     0
      -40     5     1     4
      -60     6   -23     0
      -84     7     0     0
```

Solving linear banded systems in this format with diagonal pivoting will be left as an exercise (see Problem 3.17).

3.6.2 Tridiagonal Systems

For *tridiagonal systems*, we will not use the above transformation, since it is easier to denote the three diagonals with the vectors c, d and e.

$$
A = \begin{pmatrix}
d_1 & e_1 \\
c_1 & d_2 & e_2 \\
& c_2 & d_3 & e_3 \\
& & \ddots & \ddots & \ddots \\
& & & c_{n-2} & d_{n-1} & e_{n-1} \\
& & & & c_{n-1} & d_n
\end{pmatrix}. \tag{3.62}
$$

Linear systems with a tridiagonal matrix can be solved in $O(n)$ operations. The LU decomposition with *no pivoting* generates two bidiagonal matrices

$$
L = \begin{pmatrix}
1 \\
l_1 & 1 \\
& l_2 & 1 \\
& & \ddots & \ddots \\
& & & l_{n-1} & 1
\end{pmatrix}
\quad \text{and} \quad
U = \begin{pmatrix}
u_1 & e_1 \\
& u_2 & e_2 \\
& & u_3 & \ddots \\
& & & \ddots & e_{n-1} \\
& & & & u_n
\end{pmatrix}.
$$

In order to compute L and U, we consider the elements c_k and d_{k+1} of the matrix A. Multiplying $L \cdot U$ and comparing elements we obtain the relations

$$
\begin{aligned}
l_k u_k &= c_k & \text{therefore} \quad l_k &= c_k/u_k, \\
l_k e_k + u_{k+1} &= d_{k+1} & \text{therefore} \quad u_{k+1} &= d_{k+1} - l_k e_k.
\end{aligned}
$$

The LU decomposition is thus computed by

```
u(1)=d(1);
for k=1:n-1
  l(k)=c(k)/u(k);
  u(k+1)=d(k+1)-l(k)*e(k);
end
```

Forward and back substitutions with L and U are straightforward. Note that we can overwrite the vectors c and d by l and u. Furthermore, the right hand side may also be overwritten with the solution. In the French literature, this algorithm is known as *Thomas' Algorithm*. We obtain the function

---

ALGORITHM 3.10.
Gaussian Elimination for Tridiagonal Systems: Thomas Algorithm

---

```
function [x,a,c]=Thomas(c,a,b,x);
% THOMAS Solves a tridiagonal linear system
%    [x,a,c]=Thomas(c,a,b,x) solves the linear system with a
```

```
%    tridiagonal matrix A=diag(c,-1)+diag(a)+diag(b,1). The right hand
%    side x is overwritten with the solution. The LU-decomposition is
%    computed with no pivoting resulting in L=eye+diag(c,-1),
%    U=diag(a)+diag(b,1).

n=length(a);
for k=1:n-1                          % LU-decomposition with no pivoting
  c(k)=c(k)/a(k);
  a(k+1)=a(k+1)-c(k)*b(k);
end
for k=2:n                            % forward substitution
  x(k)=x(k)-c(k-1)*x(k-1);
end
x(n)=x(n)/a(n);                      % backward substitution
for k=n-1:-1:1
  x(k)=(x(k)-b(k)*x(k+1))/a(k);
end
```

Tridiagonal systems occur, for example, when interpolating with splines (see Section 4.3.1) and also when solving one-dimensional boundary value problems.

3.6.3 Solving Banded Systems with Pivoting

With partial pivoting, rows are interchanged, destroying the band structure and introducing fill-in. If q is the lower bandwidth and p the upper bandwidth, then after interchanging rows, U will become an upper banded matrix with bandwidth $p + q$. In this section, we will develop a solver for banded systems which uses the matrix B where the nonzero diagonals of A are stored as columns.

We modify the core of the function **Elimination** to take advantage of the band structure and avoid operations with zero elements. It will not be possible to treat the right hand side as $(n + 1)$st column of A. Furthermore, we will "de-vectorize"– we cannot always use vector operations after the transformation $a_{i,j} = b_{i,j-i+q+1}$, and therefore rewrite the statements using loops.

The search for a pivot is limited to the lower bandwidth

```
maximum=0;
for k=i:min(i+q,n)
  if abs(B(k,i-k+q+1))>maximum,
    kmax=k; maximum=abs(B(k,i-k+q+1));
  end
end
```

The interchange of rows for pivoting was

```
h=A(kmax,:); A(kmax,:)=A(i,:); A(i,:)=h;
```

and now it becomes

```
for k=i:min(n,i+q+p)
  h=B(kmax,k-kmax+q+1);
  B(kmax,k-kmax+q+1)=B(i,k-i+q+1);
  B(i,k-i+q+1)=h;
end
h=b(kmax); b(kmax)=b(i); b(i)=h;
```

This could still be vectorized to

```
h=B(kmax,i-kmax+q+1:min(n,i+2*q+p-kmax+1));
B(kmax,i-kmax+q+1:min(n,i+2*q+p-kmax+1))=B(i,q+1:min(n,2*q+p+1));
B(i,q+1:min(n,2*q+p+1))=h;
h=b(kmax); b(kmax)=b(i); b(i)=h;
```

The elimination step

```
A(i+1:n,i)=A(i+1:n,i)/A(i,i);
A(i+1:n,i+1:n+1)=A(i+1:n,i+1:n+1)-A(i+1:n,i)*A(i,i+1:n+1);
```

is rewritten with for-loops

```
for k=i+1:n
  A(k,i)=A(k,i)/A(i,i);
end
for k=i+1:n
  for j=i+1:n+1
    A(k,j)=A(k,j)-A(k,i)*A(i,j);
  end
end
```

Now observing the upper and lower bandwidth and processing the right hand side separately, we get

```
for k=i+1:min(n,i+q)
   B(k,i-k+q+1)=B(k,i-k+q+1)/B(i,q+1);
end
for k=i+1:min(n,i+q)
  b(k)=b(k)-B(k,i-k+q+1)*b(i);
  for j=i+1:min(n,i+p+q)
    B(k,j-k+q+1)=B(k,j-k+q+1)-B(k,i-k+q+1)*B(i,j-i+q+1);
  end
end
```

Putting all together, we obtain the function `EliminationBandMatrix`:

ALGORITHM 3.11.
Gaussian Elimination with Partial Pivoting for Banded Matrices

---

```
function x=EliminationBandMatrix(p,q,B,b);
```

```
% ELIMINATIONBANDMATRIX solves a linear system with a banded matrix
%   x=EliminationBandMatrix(p,q,B,b); solves the banded linear system
%   Ax=b with partial pivoting. The columns of B contain the nonzero
%   diagonals of the matrix A. The first q columns of B contain the
%   lower diagonals of A (augmented by leading zeros) the remaining
%   columns of B contain the diagonal of A and the p upper diagonals,
%   augmented by trailing zeros. The vector b contains the right-hand
%   side.

n=length(B);
B=[B,zeros(n,q)];                           % augment B with q columns
normb=norm(B,1);
for i=1:n
  maximum=0;                                % search pivot
  for k=i:min(i+q,n)
    if abs(B(k,i-k+q+1))>maximum,
      kmax=k; maximum=abs(B(k,i-k+q+1));
    end
  end
  if maximum<1e-14*normb;                   % only small pivots
    error('matrix is singular')
  end
  if i~=kmax                                % interchange rows
    h=B(kmax,i-kmax+q+1:min(n,i+2*q+p-kmax+1));
    B(kmax,i-kmax+q+1:min(n,i+2*q+p-kmax+1))=B(i,q+1:min(n,2*q+p+1));
    B(i,q+1:min(n,2*q+p+1))=h;
    h=b(kmax); b(kmax)=b(i); b(i)=h;
  end
  for k=i+1:min(n,i+q)                      % elimination step
    B(k,i-k+q+1)=B(k,i-k+q+1)/B(i,q+1);
  end
  for k=i+1:min(n,i+q)
    b(k)=b(k)-B(k,i-k+q+1)*b(i);
    for j=i+1:min(n,i+p+q)
      B(k,j-k+q+1)=B(k,j-k+q+1)-B(k,i-k+q+1)*B(i,j-i+q+1);
    end
  end
end
for i=n:-1:1                                % back substitution
  s=b(i);
  for j=i+1:min(n,i+q+p)
    s=s-B(i,j-i+q+1)*x(j);
  end
  x(i)=s/B(i,q+1);
end
x=x(:);
```

3.6.4   Using Givens Rotations

We have seen in Section 3.5 that Givens rotations can be used as an alternative to LU decomposition to solve dense linear systems. This alternative is also available for banded systems: we show here how to proceed for *tridiagonal systems* with coefficient matrix A as shown in Equation (3.62). We use Givens rotation matrices $G^{(ik)}$, which differ in only four elements from the identity,

$$
\begin{aligned}
g_{ii} &= g_{kk} = c = \cos\alpha, \\
g_{ik} &= -g_{ki} = s = \sin\alpha.
\end{aligned}
$$

Multiplying the linear system from the left by $G^{(ik)}$ changes only two rows, $\boldsymbol{a}_{i:}$ and $\boldsymbol{a}_{k:}$:

$$
\begin{aligned}
\boldsymbol{a}_{i:}^{new} &:= \cos\alpha \cdot \boldsymbol{a}_{i:}^{old} + \sin\alpha \cdot \boldsymbol{a}_{k:}^{old}, \\
\boldsymbol{a}_{k:}^{new} &:= -\sin\alpha \cdot \boldsymbol{a}_{i:}^{old} + \cos\alpha \cdot \boldsymbol{a}_{k:}^{old}.
\end{aligned}
\tag{3.63}
$$

We can choose the angle α to zero elements in the matrix (see Section 3.5). We illustrate this for $n = 5$. In the first step we choose $G^{(12)}$ which combines the two first rows, and choose α such that $a_{21}^{new} = 0$:

$$
G^{(12)}
\begin{bmatrix}
x & x &   &   &   \\
x & x & x &   &   \\
  & x & x & x &   \\
  &   & x & x & x \\
  &   &   & x & x
\end{bmatrix}
=
\begin{bmatrix}
x & x & X &   &   \\
0 & x & x &   &   \\
  & x & x & x &   \\
  &   & x & x & x \\
  &   &   & x & x
\end{bmatrix}
$$

A fill-in element $a_{13} = X$ is generated. In the second step we take $G^{(23)}$ which changes the second and third row such that

$$
G^{(23)}
\begin{bmatrix}
x & x & X &   &   \\
0 & x & x &   &   \\
  & x & x & x &   \\
  &   & x & x & x \\
  &   &   & x & x
\end{bmatrix}
=
\begin{bmatrix}
x & x & X &   &   \\
0 & x & x & X &   \\
  & 0 & x & x &   \\
  &   & x & x & x \\
  &   &   & x & x
\end{bmatrix}
$$

zeroing $a_{32} = 0$, and generating the fill-in $a_{24} = X$. The next rotation with

$G^{(34)}$ yields

$$
G^{(34)}
\begin{array}{|ccccc|}
\hline
x & x & X & & \\
0 & x & x & X & \\
\hline
& 0 & x & x & \\
\hline
& & x & x & x \\
\hline
& & & x & x \\
\hline
\end{array}
=
\begin{array}{|ccccc|}
\hline
x & x & X & & \\
0 & x & x & X & \\
\hline
& 0 & x & x & X \\
\hline
& & 0 & x & x \\
\hline
& & & x & x \\
\hline
\end{array}
$$

Finally, we obtain A transformed to an upper triangular matrix R with $G^{(45)}$:

$$
G^{(45)}
\begin{array}{|ccccc|}
\hline
x & x & X & & \\
0 & x & x & X & \\
0 & x & x & X \\
\hline
& & 0 & x & x \\
\hline
& & & x & x \\
\hline
\end{array}
=
\begin{array}{|ccccc|}
\hline
x & x & X & & \\
0 & x & x & X & \\
& 0 & x & x & X \\
\hline
& & 0 & x & x \\
\hline
& & & 0 & x \\
\hline
\end{array}
$$

The solution is then obtained by back-substitution. For the transformation we need only the three diagonals of the matrix A. They will be overwritten with the elements of the upper banded matrix R.

$$
A = \begin{pmatrix}
d_1 & e_1 & & & \\
c_1 & d_2 & e_2 & & \\
& \ddots & \ddots & \ddots & \\
& & \ddots & \ddots & e_{n-1} \\
& & & c_{n-1} & d_n
\end{pmatrix}
\longmapsto
R = \begin{pmatrix}
d_1 & e_1 & c_1 & & \\
& \ddots & \ddots & \ddots & \\
& & \ddots & \ddots & c_{n-2} \\
& & & \ddots & e_{n-1} \\
& & & & d_n
\end{pmatrix}
$$

The following function `ThomasGivens` solves a tridiagonal system using Givens rotations. The right hand side is stored in b and overwritten with the solution.

ALGORITHM 3.12.

Solving Tridiagonal Systems with Givens rotations

```
function [b,d,e,c]=ThomasGivens(c,d,e,b);
% THOMASGIVENS solves a tridiagonal system of linear equations
%    [b,d,e,c]=ThomasGivens(c,d,e,b) solves a tridiagonal linear system
%    using Givens rotations. The coefficient matrix is
%    A=diag(c,-1)+diag(d)+diag(e,1), and the right hand side b is
%    overwritten with the solution. The R factor is also returned,
%    R=diag(d)+diag(e,1)+diag(c,2).

n=length(d);
e(n)=0;
for i=1: n-1                                    % elimination
   if c(i)~=0
     t=d(i)/c(i); si=1/sqrt(1+t*t); co=t*si;
```

```
    d(i)=d(i)*co+c(i)*si;  h=e(i);
    e(i)=h*co+d(i+1)*si;  d(i+1)=-h*si+d(i+1)*co;
    c(i)=e(i+1)*si;  e(i+1)=e(i+1)*co;
    h=b(i);  b(i)=h*co+b(i+1)*si;
    b(i+1)=-h*si+b(i+1)*co;
  end;
end;
b(n)=b(n)/d(n);                                      % backsubstitution
b(n-1)=(b(n-1)-e(n-1)*b(n))/d(n-1);
for i=n-2:-1:1,
  b(i)=(b(i)-e(i)*b(i+1)-c(i)*b(i+2))/d(i);
end;
```

3.7 Problems

PROBLEM 3.1. *Consider the linear system $A\boldsymbol{x} = \boldsymbol{b}$ with*

$$A = \begin{pmatrix} 1.2969 & 0.8648 \\ 0.2161 & 0.1441 \end{pmatrix} \quad \boldsymbol{b} = \begin{pmatrix} 0.8642 \\ 0.1440 \end{pmatrix} \quad \boldsymbol{x} = \begin{pmatrix} 2 \\ 2 \end{pmatrix}.$$

a) *Solve the linear system with* MATLAB *and compare the numerical solution to the exact one. Explain the difference.*

b) *Now consider the perturbed linear system which is obtained by changing the right hand side slightly:*

$$\boldsymbol{b} = \boldsymbol{b} + 10^{-9} \begin{pmatrix} -1 \\ 1 \end{pmatrix}.$$

Compute again the solution and discuss the results.

PROBLEM 3.2. *Consider the linear system $A\boldsymbol{x} = \boldsymbol{b}$ with*

$$A = \begin{pmatrix} 1 & 2 & -2 & -6 \\ -3 & -1 & -2 & \alpha \\ -4 & 3 & 9 & 16 \\ 5 & 7 & -6 & -15 \end{pmatrix} \quad \boldsymbol{b} = \begin{pmatrix} 1 \\ 1 \\ 1 \\ 1 \end{pmatrix}.$$

The element $a_{24} = \alpha$ has been lost. Assume, however, that before when α was available, the solution with MATLAB *turned out to be*

```
> x=A\b
x =
   1.0e+15 *

   0.7993
  -0.3997
   1.1990
  -0.3997
```

Can you determine with this information the missing integer matrix element $\alpha = a_{24}$?

PROBLEM 3.3. *Plot Figure 3.2. Generate for $2 \leq n \leq 8$ linear systems $A\boldsymbol{x} = \boldsymbol{b}$ using* A=hilb(n) *and* b=A*ones(n,1). *Then compute the solutions using Cramer's rule and also Gaussian elimination. Measure the solution time and compare the accuracy of the results. To compare the accuracy, plot the logarithm of the relative errors of both solutions as a function of n. Also include in your plot the quantity* log(cond(A)*eps) *and discuss the results.*

PROBLEM 3.4. *Prove that for two square matrices A and B, $\det(AB) = \det(A)\det(B)$. Hint: show first that the result holds for the elementary Gaussian elimination matrices L_j and that every matrix can be represented using products of such matrices.*

PROBLEM 3.5. (LAPLACE EXPANSION) *This problem shows that the Laplace expansion formula for calculating determinants requires $O(n!)$ operations, where n is the size of the matrix.*

1. *Let $T(n)$ be the number of operations required for a matrix of size n. Show that $T(n)$ satisfies the recurrence*

$$T(1) = 0, \qquad T(n) = 2n + nT(n-1).$$

2. *Show by induction that for $n \geq 2$,*

$$2 \cdot n! \leq T(n) \leq 6(n! - (n-1)!),$$

which implies $T(n)$ grows like $c \cdot n!$ with $2 \leq c \leq 6$.

PROBLEM 3.6. *Show that the product of two upper (respectively lower) triangular matrices is again an upper (respectively lower) triangular matrix.*

PROBLEM 3.7. *Determine for the linear system*

$$\begin{pmatrix} 2 & 1 & 3 & -1 \\ -6 & 0 & 1 & -1 \\ 4 & 2 & 0 & 1 \\ 2 & 2 & 2 & 0 \end{pmatrix} \begin{pmatrix} x_1 \\ x_2 \\ x_3 \\ x_4 \end{pmatrix} = \begin{pmatrix} 5 \\ -5 \\ 7 \\ 6 \end{pmatrix}$$

the three elimination matrices L_i and also the triangular decomposition of the matrix. The elimination may be performed without permutation of the rows.

PROBLEM 3.8. *Rewrite Algorithm 3.3* function BackSubstitution *using* MATLAB*'s scalar product notation for computing the x_i.*

PROBLEM 3.9. *Write a function* x=forwards(L,b) *to solve the system $L\boldsymbol{x} = \boldsymbol{b}$ with the lower diagonal matrix L by forward substitution using the SAXPY variant.*

PROBLEM 3.10. *In the* MATLAB *function* Elimination *(Algorithm 3.5) we store the factors used for the elimination in the transformed matrix A instead of the emerging zeros. Change this function and write a function* [L,U,P]=LU(A) *to compute the triangular decomposition* $PA = LU$. *Compare your results with the* MATLAB *built-in function* [L,U,P]=lu(X).

PROBLEM 3.11. *Modify function* [L,U,P]=LU(A) *from Problem 3.10 so that with* [L,U,P,alpha]=LU(A) *it also computes the largest element* $\alpha := \max_{ijk} |a_{ij}^{(k)}|$ *that occurs during the elimination process. This function is used to produce Figure 3.4.*

PROBLEM 3.12. Inverse iteration *is an algorithm to compute the smallest eigenvalue (in modulus) of a symmetric matrix A:*

$$
\begin{aligned}
&Choose\ \boldsymbol{x}_0 \\
&for\ k = 1, 2, \ldots, m\ (until\ convergence) \\
&\quad solve\ A\boldsymbol{x}_{k+1} = \boldsymbol{x}_k \\
&\quad normalize\ \boldsymbol{x}_{k+1} := \boldsymbol{x}_{k+1}/\|\boldsymbol{x}_{k+1}\| \\
&end
\end{aligned}
$$

Then $\lambda = \boldsymbol{x}_m^\top A \boldsymbol{x}_m / \boldsymbol{x}_m^\top \boldsymbol{x}_m$ *is an approximation for the smallest eigenvalue. A simple implementation of this algorithm is*

```
x=rand(n,1)
for k= 1:m
  x=A\x;
  x=x/norm(x);
end
lambda=x'*A*x
```

For large matrices, one can save operations if we compute the LU decomposition of the matrix A only once. The iteration is performed using the factors L and U. This way, each iteration needs only $O(n^2)$ *operations, instead of* $O(n^3)$ *with the program above. Use the programs* LU *from Problem 3.10,* BackSubstitution *from Problem 3.8, and* forwards *from Problem 3.9 to implement the inverse iteration. Experiment with a few matrices and compare your results with the correct eigenvalues obtained by* eig(A).

PROBLEM 3.13. *Solving a linear system. We are given a linear system* $A\boldsymbol{x} = \boldsymbol{b}$. *This time the matrix A is* $m \times n$ *with possibly* $m \neq n$ *and rank* $r \leq \min(m, n)$. *Eliminating variables will lead to a reduced system which will show if the system has solutions. Depending on the rank there might be infinitely many solutions, no solution or a unique solution. In order to determine the rank we need to reorder equations and unknowns. We look for pivot elements in the whole remaining matrix. This is called* complete pivoting.

Before an elimination step we search for the pivot with largest absolute value in the whole remaining matrix and move it to the diagonal by interchanging rows and columns. If the largest pivot is very small, say if with `norma=norm(A,1)` *we have* `abs(A(i,i)) < tol*norma` *then the elimination process should be terminated and the rank will be assumed to be* $r = i - 1$. *The reduced system will then have the form*

$$\begin{pmatrix} U_{r\times r} & B_{r\times n-r} \\ 0_{m-r\times r} & 0_{n-r\times n-r} \end{pmatrix} \begin{pmatrix} \tilde{x}_1 \\ \tilde{x}_2 \end{pmatrix} = \begin{pmatrix} c_1 \\ c_2 \end{pmatrix}. \tag{3.64}$$

We indicated the dimensions of the sub-matrices in Equation (3.64) by subscripts. The right hand side is partitioned in the same way.

If now $c_2 \neq 0$ *then the linear system has no solutions. If on the other hand* $c_2 = 0$ *then the variables* $\tilde{x}_2$ *can be chosen arbitrarily. For each choice of* $\tilde{x}_2$ *the fist part* $\tilde{x}_1$ *can be computed by back-substitution in*

$$U\tilde{x}_1 = c_1 - B\tilde{x}_2.$$

Thus the general solution of Equation (3.64) is

$$\begin{pmatrix} \tilde{x}_1 \\ \tilde{x}_2 \end{pmatrix} = \begin{pmatrix} U^{-1}c_1 \\ 0 \end{pmatrix} + \begin{pmatrix} -U^{-1}B \\ I \end{pmatrix} \tilde{x}_2,$$

and we obtain finally from this the general solution of $Ax = b$ *by*

$$x = P\begin{pmatrix} \tilde{x}_1 \\ \tilde{x}_2 \end{pmatrix},$$

where P *is the permutation matrix for the reordering of the columns (unknowns).*

Modify the function Elimination *by introducing complete pivoting and compute the general solution as described above. Your* MATLAB *function should follow the header*

```
function [x, Xh,r,U,L,B,P,Q]=EliminationCompletePivoting(A,b,tol)
% ELIMINATIONCOMPLETEPIVOTING Linear system solve with complete pivoting
%    [x,Xh,r,U,L,B,P,Q]=EliminationCompletePivoting(A,b,tol) computes a
%    solution x to the possibly non-square linear system Ax=b using
%    Gaussian elimination with complete pivoting, which produces the
%    decomposition A=Q'*L*[U,B]*P'.  Here Q and P are permutation
%    matrices, L is r x r lower unit triangular, U is n x r upper
%    triangular and B is r x (n-r), and r is the numerical rank: an
%    element A(i,j) of the remaining matrix A' during the elimination
%    process is considered to be zero if abs(A'(i,j))<tol*norm(A,1).  x
%    is a particular solution such that Ax=b.  Xh contains the
%    nullspace, linear independent solutions of Ax=0.
```

The computation should stop with an error message if it turns out that $A\boldsymbol{x} = \boldsymbol{b}$ has no solution. Check that you have computed the decomposition

$$A = Q^\top L[U, B]P^\top.$$

Check your function with the example `A=magic(10)`, `b=ones(10,1)` *and* `b=rand(10,1)`.

PROBLEM 3.14. (DIAGONALLY DOMINANT MATRICES) *A matrix is said to be* column diagonally dominant *if, for each column j, the absolute value of the diagonal entry is greater than the sum of the absolute values of the off-diagonal entries, i.e., if*

$$|a_{jj}| > \sum_{i \neq j} |a_{ij}| \qquad for\ j = 1, 2, \ldots, n.$$

Show that after one step of Gaussian elimination with no pivoting, the remaining $(n-1) \times (n-1)$ submatrix is also column diagonally dominant. Deduce that no row exchanges will occur throughout the elimination process, even when partial pivoting is used.

PROBLEM 3.15. *Modify the* MATLAB *function* `Elimination` *(Algorithm 3.5) to compute in a numerically stable way the determinant of a matrix. Observe that with Gaussian elimination we obtain*

$$PA = LU. \tag{3.65}$$

Taking the determinant we get

$$\det(P)\det(A) = \det(L)\det(U).$$

Now since L has a unit diagonal, it follows that $\det(L) = 1$. Furthermore $\det(P) = \pm 1$, depending on if the number of row changes is even or odd. Thus we obtain

$$\det(A) = (-1)^{\#\ row\ changes} \prod_{i=1}^{n} u_{ii}.$$

PROBLEM 3.16. *Write a* MATLAB-*function* `BandGivens` *which solves a banded linear system using Givens rotations. The coefficient matrix B contains the non-zero diagonals as columns (see Section 3.6.1). The header of your function should look:*

```
function x=BandGivens(p,q,B,b);
% BANDGIVENS solves a banded system of linear equations using
% Givens rotations. The diagonals (p upper, q lower) are
% stored as columns in B.
```

PROBLEM 3.17. *Write a MATLAB-function* B=luB(p,q,B) *which overwrites the given matrix B which has as columns the nonzero diagonals of a banded matrix (see Section 3.6.1) with the LU decomposition using diagonal pivoting.* **Hint:** *It might be simpler if you adapt first the elimination algorithm for an $n \times n$ matrix to the case of a banded matrix, and then use the transformation* StoreBandMatrix.m *(Algorithm 3.9).*

PROBLEM 3.18. *Compute the coefficients of a polynomial $P(t) = at^3 + bt^2 + ct + d$ such that $P(1) = 17$, $P(-1) = 3$, $P(0.5) = 7.125$ und $P(1.5) = 34.875$. Generate the linear system for the coefficients a, b, c and d and solve the system with MATLAB. What is the condition number of this system ?*

PROBLEM 3.19. *Suppose $A \in \mathbb{R}^{n \times n}$ is a nonsingular matrix. Recall that the 2-norm condition number $\kappa_2(A)$ is defined as*

$$\kappa_2(A) = \|A\|_2 \|A^{-1}\|_2.$$

Let $\boldsymbol{y}$ be a unit vector such that $\|A^{-1}\|_2 = \|A^{-1}\boldsymbol{y}\|_2$, and define $\boldsymbol{x} = \dfrac{A^{-1}\boldsymbol{y}}{\|A^{-1}\|_2}$. Finally, let $E = -A\boldsymbol{x}\boldsymbol{x}^T$.

1. *Show that $(A + E)\boldsymbol{x} = 0$. Conclude that $A + E$ is singular.*

2. *Show that $\|E\|_2 \leq 1/\|A^{-1}\|_2$. This implies the relative perturbation satisfies*

$$\frac{\|E\|_2}{\|A\|_2} \leq \frac{1}{\kappa_2(A)}.$$

PROBLEM 3.20. *(Sherman-Morrison-Woodbury formula) Let $A \in \mathbb{R}^{n \times n}$ be an invertible matrix, and $\boldsymbol{b}, \boldsymbol{u}, \boldsymbol{v} \in \mathbb{R}^n$.*

1. *Show that if $I + \boldsymbol{u}\boldsymbol{v}^{\top}$ is invertible, then there exists a σ such that*

$$(I + \boldsymbol{u}\boldsymbol{v}^{\top})^{-1} = I + \sigma\boldsymbol{u}\boldsymbol{v}^{\top}.$$

   What is a sufficient condition for $I + \boldsymbol{u}\boldsymbol{v}^{\top}$ to be invertible? Show that this is also a necessary condition.

2. *Suppose we know the LU decomposition of A, and also the solutions of the linear systems*

$$A\boldsymbol{y} = \boldsymbol{b} \quad and \quad A\boldsymbol{z} = \boldsymbol{u}. \tag{3.66}$$

   Find an efficient algorithm to solve

$$(A + \boldsymbol{u}\boldsymbol{v}^{\top})\boldsymbol{x} = \boldsymbol{b},$$

   which uses only the solutions of (3.66).

PROBLEM 3.21. *The* MATLAB *function* `hilb` *computes the Hilbert matrix. For instance*

```
>> A = hilb(4)
A =
    1.0000    0.5000    0.3333    0.2500
    0.5000    0.3333    0.2500    0.2000
    0.3333    0.2500    0.2000    0.1667
    0.2500    0.2000    0.1667    0.1429
```

The matrix elements are given by

$$a_{ij} = \int_0^1 t^{i-1} t^{j-1} dt = \frac{1}{i+j-1}. \tag{3.67}$$

Prove that for each n the matrix `A=hilb(n)` *is positive definite.* **Hint:** *consider the expression $\boldsymbol{x}^\top A \boldsymbol{x}$ and use Equation (3.67).*

PROBLEM 3.22. *The condition number of a rectangular matrix $A \in \mathbb{R}^{m \times n}$ can be defined by*

$$\kappa(A) := \frac{\max_{||\boldsymbol{x}||=1} ||A\boldsymbol{x}||}{\min_{||\boldsymbol{x}||=1} ||A\boldsymbol{x}||}.$$

Show for the Euclidean norm $|| \cdot ||_2$ that the equality

$$\kappa(A^\top A) = \kappa(A)^2$$

holds. Hint: Note that the symmetric matrix $A^\top A$ can be diagonalized, $A^\top A = Q^\top \Lambda Q$ with $\Lambda = diag(\lambda_1, \lambda_2, \ldots, \lambda_n)$, $\lambda_1 \geq \lambda_2 \geq \ldots \geq \lambda_n$, and show that $\max_{||\boldsymbol{x}||_2=1} ||A\boldsymbol{x}||_2^2 = \lambda_1$ and $\min_{||\boldsymbol{x}||_2=1} ||A\boldsymbol{x}||_2^2 = \lambda_n$.

PROBLEM 3.23. *Ill-conditioned systems of linear equations. To solve this problem, you will have to write a* MATLAB *program of about 12 lines using the functions* `rand`, `round`, `diag`, `eye`, `size`, `triu`, `tril`, `cond`. *The goal of this problem is to show that apparently harmless looking systems of linear equations may be very difficult to solve.*

a) *Generate an $n \times n$ matrix B with random integer elements in the range $b_{ij} \in [-10, 10]$. Choose for instance $n = 20$.*

b) *Remove the diagonal of B, save the upper triangular part in U and the lower triangular part in L, and put ones on the diagonals: $l_{ii} = u_{ii} = 1$.*

c) *Compute $A = L \cdot U$. What is the value of $\det(A)$ and why? Compute the determinant with* `det(A)` *and confirm your prediction. In case that you have doubts about the result, compute separately* `det(L)` *and* `det(U)`.

d) *Choose now an exact solution, for instance $\boldsymbol{x}_e = $* `ones(n,1)`, *and compute the corresponding right hand side $\boldsymbol{b} = A \boldsymbol{x}_e$.*

e) *Solve $A\,x = b$ using* MATLAB *and compare the solution with the exact x_e.*

f) *Explain the bad results by computing the condition number of A.*

Chapter 4. Interpolation

> *The question then arises as to how we can find the values of the function $\log_{10}(x)$ for values of the argument x which are intermediate between the tabulated values. The answer to this question is furnished by the theory of interpolation, which in its most elementary aspect may be described as the science of "reading between the lines of a mathematical table."*
>
> E. Whittaker and G. Robinson, The Calculus of Observations: a Treatise on Numerical Mathematics, 1924.
>
> *We wish to repeat that interpolation is only one way to approximate data. [...] For data with significant errors, the least squares approach is preferred.*
>
> D. Kahaner, C. Moler, S. Nash, Numerical Methods and Software, 1988.

Prerequisites: Chapters 2 and 3 are required.

Interpolation means inserting or blending in a missing value. It is the art of reading between the entries of a tabulated function (see first quote above). We start this chapter with several introductory examples in Section 4.1, through which we explain the interpolation principle. The most common interpolation technique is to use polynomials, and we show in Section 4.2 four classical techniques: using monomials, Lagrange polynomials, Newton polynomials, and orthogonal polynomials. The latter also leads naturally to a least squares approximation, which is more desirable if the data points are contaminated by errors (see second quote above, and Chapter 6). We then show that the representation in these different bases are related by the LU and QR factorizations of the corresponding matrices. We also explain the barycentric formula, give an estimate for the interpolation error, and discuss extrapolation, which is similar to interpolation, expect that the desired value lies outside the range of the given data. Section 4.3 is devoted to piecewise interpolation, which leads to the classical cubic splines. This section also contains the well-known Morrison-Woodbury formula. Section 4.4 addresses trigonometric interpolation and contains a detailed description of the fast Fourier transform.

W. Gander et al., *Scientific Computing - An Introduction using Maple and MATLAB*,
Texts in Computational Science and Engineering 11,
DOI 10.1007/978-3-319-04325-8_4,
© Springer International Publishing Switzerland 2014

4.1 Introductory Examples

Assume we know only some values $f(x_i)$ for $i = 0, \ldots, n$ of a function f,

$$\begin{array}{c|ccccccc} x & x_0, & x_1, & \cdots & x_i, & z & x_{i+1}, & \cdots, & x_n \\ \hline y = f(x) & y_0, & y_1, & \cdots & y_i, & ? & y_{i+1}, & \cdots, & y_n \end{array} \qquad (4.1)$$

Is there a way to compute or approximate the function value $f(z)$ for some given z without evaluating f? Why should we be interested in that problem?

1. An explicit formula for the function f *may not be known to us*, and the tabulated values (4.1) could be the only information we have. The values could have been obtained by some physical measurement, e.g., when we represent the temperature of the air outside a house during a day:

$$\begin{array}{c|ccccc} t & 8\,\text{am} & 9\,\text{am} & 11\,\text{am} & 1\,\text{pm} & 5\,\text{pm} \\ \hline T \text{ in } {}^\circ C & 12.1 & 13.6 & 15.9 & 18.5 & 16.1 \end{array}$$

   We might be interested in finding out what the temperature was at 10 am.

2. An important application of interpolation occurs in the processing of digital images: when a digital picture is enlarged, we have to *increase the number of pixels*. This has to be done by interpolating values for the additional pixels.

3. Another use of interpolation (maybe not so important anymore) is to compute intermediate values of a complicated tabulated function. Before the age of computers, tables of functions were very popular. Astronomers used for their computations a book with tables of the logarithms and trigonometric functions. Many of those tables have been replaced today by programs that compute the values of functions when needed.

   Consider for example the function

$$f(x) = \int_0^x e^{\sin t} dt.$$

   If we know the tabulated values

$$\begin{array}{c|cccccc} x & 0.4 & 0.5 & 0.6 & 0.7 & 0.8 & 0.9 \\ \hline y & 0.4904 & 0.6449 & 0.8136 & 0.9967 & 1.1944 & 1.4063 \end{array} \qquad (4.2)$$

   what is the value $f(0.66)$?

4. Finally, interpolation may also be used for data compression: in a large dense table, one can store only every tenth value and interpolate to obtain the deleted values when needed.

If the new value z is within the range of the interpolation points x_i then we speak of *interpolation*. If the desired value z is outside the range, we call the process *extrapolation*. Predictions are always extrapolations; we might, for instance, be interested in predicting what the temperature will be at 7 pm in the example above. Another nice example for extrapolation is the census demonstration in MATLAB. The task consists of estimating the population of the USA in the year 2010 based on census data for the years $1900, 1910, \ldots, 2000$. This example shows very well how sensitive the problem is. Depending on which model is used, one can obtain very different answers.

To interpolate the data given in Table (4.1) for some desired value z, we choose a model function $g(x)$. Typically this *interpolating function* should be easy to evaluate and have the following property:

$$g(x_k) = f(x_k) \text{ for some } x_k \text{ in the neighborhood of } z.$$

If g is known then $g(z)$ is taken as an approximation for $f(z)$. The hope and the aim is that the *interpolation error* $|g(z) - f(z)|$ will be small.

Note that if only the values of (4.1) are given and that we do not know anything more about the function f, then the problem of interpolation is ill posed. Any value can be chosen as "approximation" for $f(z)$. Consider the function

$$f(x) = \sin(x) + 10^{-3} \ln\left((x - 0.35415)^2\right).$$

If it is tabulated for $x = 0.33, 0.34, \ldots, 0.38$, one would not expect the singularity at $z = 0.35415$:

x	0.3300	0.3400	0.3500	0.3600	0.3700	0.3800
$f(x)$	0.3166	0.3250	0.3319	0.3420	0.3533	0.3636

Thus, if an interpolation procedure yields some value of $f(0.35415) \approx 0.3356$, one would probably be ready to accept this.

Often we assume that f is smooth. In that case more can be said about the interpolation error (see Section 4.2.2).

EXAMPLE 4.1. *We would like to interpolate $f(0.66)$ in table (4.2) and choose the function*

$$g(x) = \frac{a}{x - b}.$$

The two parameters a and b are determined by requiring that g interpolates both neighbor points of z, i.e.

$$g(0.6) = \frac{a}{0.6 - b} = 0.8136,$$

$$g(0.7) = \frac{a}{0.7 - b} = 0.9967.$$

With the MAPLE-*statement*

```
> solve({a/(0.6-b)=0.8136, a/(0.7-b)=0.9967},{a,b});
```

we obtain $a = -0.4429$ and $b = 1.1443$ thus

$$g(0.66) = -\frac{0.4429}{0.66 - 1.1443} = 0.9145 \approx f(0.66) = 0.9216.$$

By integrating $e^{\sin(t)}$ numerically one can check that the interpolation error is $|g(0.66) - f(0.66)| = 0.0071$, which is rather large. Indeed, the choice of the model function was not very clever; a better result could have been obtained with the linear function $g(x) = ax + b$, in which case we would get $g(0.66) = 0.9235$ and an error of 0.0019.

4.2   Polynomial Interpolation

A common choice of *model functions for interpolation are polynomials*, which are easy to evaluate and smooth, i.e., infinitely differentiable. Given the $n+1$ points in (4.1), we are looking for a polynomial $P(x)$ such that

$$P(x_i) = f(x_i), \quad i = 0, \ldots, n. \tag{4.3}$$

Since by (4.3) we have $n + 1$ constraints to satisfy, we need $n + 1$ degrees of freedom. Consider the n-th degree polynomial

$$P_n(x) = a_0 + a_1 x + \ldots + a_{n-1} x^{n-1} + a_n x^n. \tag{4.4}$$

The $n + 1$ coefficients $\{a_i\}$ have to be determined in such a way that (4.3) is satisfied. This leads to the *linear system* of equations

$$
\begin{array}{ccccccccc}
a_0 & + & a_1 x_0 & + & \ldots & + & a_{n-1} x_0^{n-1} & + & a_n x_0^n & = & f(x_0), \\
a_0 & + & a_1 x_1 & + & \ldots & + & a_{n-1} x_1^{n-1} & + & a_n x_1^n & = & f(x_1), \\
\vdots & & \vdots & & & & \vdots & & \vdots & = & \vdots \\
a_0 & + & a_1 x_n & + & \ldots & + & a_{n-1} x_n^{n-1} & + & a_n x_n^n & = & f(x_n).
\end{array}
$$

Written in matrix form, the system reads

$$
\underbrace{\begin{bmatrix}
1 & x_0 & \ldots & x_0^{n-1} & x_0^n \\
1 & x_1 & \ldots & x_1^{n-1} & x_1^n \\
\vdots & \vdots & & \vdots & \vdots \\
1 & x_n & \ldots & x_n^{n-1} & x_n^n
\end{bmatrix}}_{V}
\underbrace{\begin{pmatrix}
a_0 \\
\vdots \\
a_{n-1} \\
a_n
\end{pmatrix}}_{a}
=
\underbrace{\begin{pmatrix}
f(x_0) \\
f(x_1) \\
\vdots \\
f(x_n)
\end{pmatrix}}_{f},
\tag{4.5}
$$

The matrix V with the special structure containing the powers of the nodes is called a *Vandermonde Matrix*. We shall see in the next section that if all the nodes x_i are different, then the matrix V is non-singular and therefore there exists a unique solution to the linear system of equations.

On the other hand, Vandermonde matrices tend to be ill conditioned (see Table 3.2 in Chapter 3) and, as we will see, there are other ways to compute the interpolating polynomial by representing it in another basis of polynomials rather than with monomials.

4.2.1 Lagrange Polynomials

Instead of trying to directly interpolate the function values $f(x_i)$ at the nodes x_i with a polynomial of degree at most n, we can look for a representation of the interpolating polynomial $P_n(x)$ of the form

$$P_n(x) = \sum_{j=0}^{n} f(x_j)\, l_j(x), \qquad (4.6)$$

where the polynomials l_j should be of degree n as well and have to be determined so that $P_n(x)$ interpolates the function f at the nodes x_j. This is clearly the case if $l_i(x_i) = 1$ and $l_i(x_j) = 0$ for $i \neq j$, because then there is only one contribution to the sum from the i-th term if l_i is evaluated at x_i, namely

$$P_n(x_i) = \sum_{j=0}^{n} f(x_j)\, l_j(x_i) = f(x_i)\, l_i(x_i) = f(x_i).$$

Now to determine the $l_i(x)$ we have to find the polynomial which equals 1 at x_i and has n zeros at all the other nodes x_j, $i \neq j$. Such a polynomial can be written in factored form directly as

$$l_i(x) = \prod_{\substack{j=0 \\ j \neq i}}^{n} \frac{x - x_j}{x_i - x_j}, \quad i = 0, 1, \ldots, n. \qquad (4.7)$$

This polynomial $l_i(x)$ is clearly zero if evaluated at x_j for $j \neq i$, because one of the factors in the numerator vanishes there. On the other hand, when evaluated at x_i, the numerator and denominator become identical, so the polynomial equals one there, as required. These polynomials are called *Lagrange polynomials* and interpolation becomes very simple once the Lagrange polynomials are available, one simply forms the sum given in Equation (4.6).

Obviously the $n + 1$ Lagrange polynomials can only exist if the denominators in Equation (4.7) are nonzero. This implies that the nodes must be distinct, i.e., we must have $x_i \neq x_j$ for all $i \neq j$.

How do we know that the interpolation polynomial expanded in powers of x as in (4.4) and the polynomial constructed with the Lagrange basis functions (4.6) represent the same polynomial? One possibility is to expand (4.6) and reorder the terms and check that the expressions are indeed equal. There is, however, a simpler argument that shows that the polynomials are the same. Assume we have computed two interpolating polynomials $Q(x)$ and $P(x)$ each of degree n such that

$$Q(x_j) = f(x_j) = P(x_j), \quad j = 0, \ldots, n$$

holds. Then we can form the difference

$$d(x) = Q(x) - P(x).$$

d is certainly a polynomial of degree less or equal n. But because of the interpolation property of P and Q, we have

$$d(x_j) = Q(x_j) - P(x_j) = 0, \quad j = 0, \ldots, n.$$

A non-zero polynomial of degree less than or equal to n cannot have more than n zeros. But d has $n + 1$ distinct zeros; hence, it must be identically zero, meaning that $Q(x) \equiv P(x)$. Thus we have proved

THEOREM 4.1. (EXISTENCE AND UNIQUENESS OF THE INTERPOLATION POLYNOMIAL) *Given $n + 1$ distinct nodes x_j the Vandermonde matrix in (4.5) is non-singular and there exists a unique interpolating polynomial P_n of degree less or equal n with $P(x_j) = f(x_j)$ for $j = 0, \ldots, n$.*

EXAMPLE 4.2. *We interpolate again the value $f(0.66)$ for the function given in (4.2). We chose $n = 2$ and use the three points*

x	0.6	0.7	0.8
y	0.8136	0.9967	1.1944

to determine the interpolation polynomial (a quadratic function). We obtain

$$l_0(x) = \frac{x - 0.7}{0.6 - 0.7} \frac{x - 0.8}{0.6 - 0.8} \quad \Rightarrow \quad l_0(0.66) = 0.28$$

$$l_1(x) = \frac{x - 0.6}{0.7 - 0.6} \frac{x - 0.8}{0.7 - 0.8} \quad \Rightarrow \quad l_1(0.66) = 0.84$$

$$l_2(x) = \frac{x - 0.6}{0.8 - 0.6} \frac{x - 0.7}{0.8 - 0.7} \quad \Rightarrow \quad l_2(0.66) = -0.12$$

Thus we get

$$P_2(0.66) = 0.28 \cdot 0.8136 + 0.84 \cdot 0.9967 - 0.12 \cdot 1.1944 = 0.921708.$$

Since the exact value is $f(0.66) = 0.9216978$ the interpolation error is now $-1.01 \cdot 10^{-5}$ and the interpolated value has the same accuracy as the tabulated values.

The following MATLAB function interpolates with the Lagrange interpolation polynomial.

ALGORITHM 4.1. *Lagrange interpolation*

```
function yy=LagrangeInterpolation(x,y,xx)
% LAGRANGEINTERPOLATION interpolation using Lagrange polynomials
%    yy=LagrangeInterpolation(x,y,xx); uses the points (x,y) for the
%    Lagrange Form of the interpolating polynomial P and iterpolates
%    the values yy=P(xx)
```

```
n=length(x); nn=length(xx);
for i=1:nn,
  yy(i)=0;
  for k=1:n
    yy(i)=yy(i)+y(k)*prod((xx(i)-x([1:k-1,k+1:n])))...
                    /prod((x(k)-x([1:k-1,k+1:n]))));
  end;
end;
```

4.2.2 Interpolation Error

If the function f has continuous derivatives in the range of interpolation, then an error term can be derived for the interpolation formula.

THEOREM 4.2. (INTERPOLATION ERROR) *Let $f, f', \ldots, f^{(n+1)}$ be continuous in the interval $[x_0, x_n]$, where $x_0 < x_1 < \cdots < x_n$. If the polynomial P_n interpolates f in the nodes x_j then*

$$R_n(x) := f(x) - P_n(x) = \frac{(x - x_0)(x - x_1)\cdots(x - x_n)}{(n+1)!} f^{(n+1)}(\xi), \quad (4.8)$$

where ξ is some value between the nodes $x_0, x_1, \ldots, x_n$ and x.

PROOF. For $x = x_k$ the theorem trivially holds, since both sides vanish, so let us consider a fixed x such that $x \neq x_k$, $k = 0, \ldots, n$. Define

$$L(t) = \prod_{i=0}^{n} (t - x_i)$$

and consider the function

$$F(t) = f(t) - P_n(t) - cL(t), \quad \text{with} \quad c = \frac{f(x) - P_n(x)}{L(x)}.$$

Clearly
$$F(x_k) = 0, \quad k = 0, 1, \ldots, n,$$

but also $F(x) = 0$ because of the special choice of the constant c. Thus F has at least $n + 2$ distinct zeros. By Rolle's theorem, the continuity of F' implies that there is a zero of F' between any two zeros of F. Therefore F' has at least $n + 1$ distinct zeros. If we continue to take derivatives and count the zeros we finally find that

$$F^{(n+1)}(t) = f^{(n+1)}(t) - P_n^{(n+1)}(t) - cL^{(n+1)}(t) \quad (4.9)$$

has at least *one* zero ξ. Since P_n is of degree n we have

$$P_n^{(n+1)}(t) \equiv 0,$$

and, because L is a polynomial of degree $n+1$ with leading coefficient 1, we obtain

$$L^{(n+1)}(t) = (n+1)! \ .$$

If we insert this for $t = \xi$ in Equation (4.9) and if we solve the equation

$$F^{(n+1)}(\xi) = 0$$

for c then we get using the definition of c

$$c = \frac{f(x) - P_n(x)}{L(x)} = \frac{f^{(n+1)}(\xi)}{(n+1)!},$$

which is what we wanted to prove. $\qquad\square$

To estimate the interpolation error in Example 4.2 with the expression in (4.8), we need to compute the maximum of $|f'''|$ in the interval $[0.6, 0.8]$. Since

$$f'''(x) = (-\sin x + \cos^2 x)e^{\sin x},$$

and since f''' is monotonically decreasing in this interval with $f'''(0.6) = 0.2050$ and $f'''(0.8) = -0.4753$, we conclude

$$\max_{0.6 \leq x \leq 0.8} |f'''(x)| = |f'''(0.8)| = 0.4753.$$

Furthermore

$$\max_{0.6 \leq x \leq 0.8} |L(x)| = 3.849 \cdot 10^{-4},$$

and thus for all $x \in (0.6, 0.8)$ the error can be bounded by

$$|R_n(x)| \leq 3.049 \ 10^{-5}.$$

For $x = 0.66$ we have $|L(0.66)| = 3.36 \cdot 10^{-4}$ and thus we get the sharper bound

$$|R_n(0.66)| \leq 2.6617 \cdot 10^{-5}.$$

The estimate is about twice as large as the exact interpolation error.

Remarks:

- Normally the derivative $f^{(n+1)}$ is not available and so it is difficult to use the error term given above unless a bound on this derivative is known.

- From the product $L(x)$ in the error term, one can expect large interpolation errors towards the ends of the interval in which the interpolation is performed, since many of the terms in the product will be large. This is especially the case if many nodes are used. An impressive example was given by *Runge* and is reproduced in Figure 4.1. The function which is interpolated is

$$f(x) = \frac{1}{1 + x^2},$$

and the nodes are chosen to be equidistant on the interval $x \in [-5, 5]$. The polynomials indeed still interpolate the function values, but be-

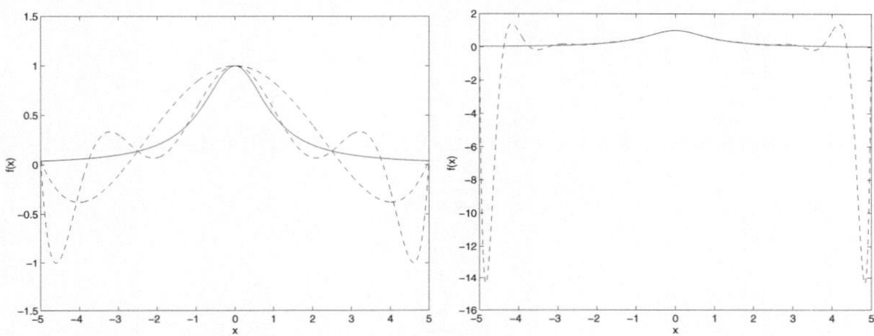

FIGURE 4.1.
Runges famous example, interpolated with equidistant nodes, on the left with a polynomial of degree four and eight, on the right with a polynomial of degree sixteen.

tween the nodes the interpolation error is unacceptably large for higher degree polynomials, especially near the boundary. A remedy is to use non-equidistant nodes which are more closely spaced at the ends of the interval, for example *Chebyshev nodes*, which are derived from the roots of the *Chebyshev polynomials*; see Problem 4.1. We discuss Chebyshev polynomials in detail in Section 11.5. Another possibility is to use piecewise polynomials, as we will see in the Section 4.3, which leads to spline interpolation in Section 4.3.1.

4.2.3 Barycentric Formula

Interpolating with the Lagrange formula (4.6) is not very efficient, since for every new location where we want to interpolate, we have to perform $O(n^2)$ operations. There exists a variant called the *Barycentric Formula* which requires only $O(n)$ operations per interpolation point. To derive it we consider

$$P_n(x) = \sum_{i=0}^{n} \prod_{\substack{j=0 \\ j \neq i}}^{n} \frac{x - x_j}{x_i - x_j} \, f(x_i).$$

We define the coefficients

$$\lambda_i = \frac{1}{(x_i - x_0) \cdots (x_i - x_{i-1})(x_i - x_{i+1}) \cdots (x_i - x_n)}, \quad i = 0, \ldots, n.$$

Now we write for $x \neq x_i$

$$P_n(x) = \sum_{i=0}^{n} \lambda_i \prod_{\substack{j=0 \\ j \neq i}}^{n} (x - x_j)\, f(x_i) = \prod_{j=0}^{n} (x - x_j) \sum_{i=0}^{n} \frac{\lambda_i}{x - x_i}\, f(x_i).$$

The interpolation formula is valid for any choice of the function values $f(x_i)$. For the constant function $f(x) = 1$ we obtain the relation

$$1 = \prod_{j=0}^{n} (x - x_j) \sum_{i=0}^{n} \frac{\lambda_i}{x - x_i} \quad \Rightarrow \quad \prod_{j=0}^{n} (x - x_j) = \frac{1}{\sum_{i=0}^{n} \frac{\lambda_i}{x - x_i}}.$$

Using this expression we finally obtain the *Barycentric Interpolation Formula*

$$P_n(x) = \frac{\displaystyle \sum_{i=0}^{n} \frac{\lambda_i}{x - x_i}\, f(x_i)}{\displaystyle \sum_{i=0}^{n} \frac{\lambda_i}{x - x_i}}. \tag{4.10}$$

To interpolate with the Barycentric Formula we compute first the coefficients λ_i in $O(n^2)$ operations with the following MATLAB function

---

ALGORITHM 4.2.
Coefficients for the Barycentric Representation

```
function lambda=BarycentricCoefficients(x)
% BARYCENTRICCOEFFICIENTS barycentric coefficients for interpolation
%    lambda=BarycentricCoefficients(x); computes the coefficients for
%    the barycentric representation of the interpolating polynomial
%    through the points (x,.)

n=length(x); x=x(:);
for k=1:n,
  lambda(k)=1/prod(x(k)-x([1:k-1,k+1:n]));
end;
```

For a new interpolation point z, we compute the weights $\mu_i = \lambda_i/(z - x_i)$ and evaluate then $P_n(z) = \sum_{i=0}^{n} \mu_i f(x_i) / \sum_{i=0}^{n} \mu_i$ with only $O(n)$ operations. We obtain the following MATLAB function:

ALGORITHM 4.3. *Barycentric Formula*

```
function yy=BarycentricInterpolation(x,y,lambda,xx)
% BARYCENTRICINTERPOLATION interpolate using the barycentric formula
%    yy=BarycentricInterpolation(x,y,lambda,xx) uses the points (x,y)
%    and the precalculated barycentric coefficients lambda for the
```

```
%    barycentric form of the interpolation polynomial P and iterpolates
%    the values yy=P(xx)

x=x(:); y=y(:); xx=xx(:);
nn=length(xx);
for i=1:nn,
  z=(xx(i)-x)+1e-30;  % prevents a division by zero
  mue=lambda'./z;
  yy(i)=mue'*y/sum(mue);
end;
```

4.2.4 Newton's Interpolation Formula

For *Newton's interpolation formula*, we use the *Newton polynomials* which
are defined as

$$\pi_k(x) = \prod_{j=0}^{k-1} (x - x_j).$$

The interpolating polynomial becomes

$$P_n(x) = d_0\pi_0(x) + d_1\pi_1(x) + \cdots d_n\pi_n(x). \tag{4.11}$$

The interpolation condition (4.3) leads to a system of linear equations for the
coefficients d_i:

$$\sum_{j=0}^{n} d_j\pi_j(x_i) = f(x_i), \quad i = 0,\ldots n. \tag{4.12}$$

The matrix of the system (4.12) is lower triangular

$$\begin{pmatrix} \pi_0(x_0) & \cdots & \pi_n(x_0) \\ \pi_0(x_1) & \cdots & \pi_n(x_1) \\ \vdots & \cdots & \vdots \\ \pi_0(x_n) & \cdots & \pi_n(x_n) \end{pmatrix} = \begin{pmatrix} 1 & & & & \\ 1 & x_1 - x_0 & & & \\ 1 & x_2 - x_0 & (x_2-x_0)(x_2-x_1) & & \\ \vdots & \vdots & \vdots & \ddots & \\ 1 & x_n - x_0 & (x_n-x_0)(x_n-x_1) & \cdots & \prod_{j=0}^{n-1}(x_n-x_j) \end{pmatrix}.$$

Thus the coefficients d_j could be computed by forward substitution:

$$d_0 = f(x_0), \quad d_i = \frac{f(x_i) - \sum_{j=0}^{i-1} d_j\pi(x_i)}{\pi_i(x_i)}, \quad i = 1,\ldots,n. \tag{4.13}$$

However, there is a simpler algorithm for computing these coefficients based
on *divided differences*. The definition is recursive:

$$\begin{aligned} f[x_i] &= f(x_i), \quad \text{zeroth divided difference,} \\ f[x_i, x_{i+1}, \ldots, x_{i+k}] &= \frac{f[x_{i+1}, x_{i+1}, \ldots, x_{i+k}] - f[x_i, x_{i+1}, \ldots, x_{i+k-1}]}{x_{i+k} - x_i}. \end{aligned}$$

For example, the first divided difference would evaluate to

$$f[x_0, x_1] = \frac{f[x_1] - f[x_0]}{x_1 - x_0} = \frac{f(x_1) - f(x_0)}{x_1 - x_0}.$$

The second divided difference would amount to

$$f[x_0, x_1, x_2] = \frac{f[x_1, x_2] - f[x_0, x_1]}{x_2 - x_0} = \frac{\frac{f(x_2)-f(x_1)}{x_2-x_1} - \frac{f(x_1)-f(x_0)}{x_1-x_0}}{x_2 - x_0}.$$

Divided differences are best arranged in a triangular array,

$$
\begin{array}{lllll}
x_0 & f(x_0) = f[x_0] \\
x_1 & f(x_1) = f[x_1] & f[x_0, x_1] \\
x_2 & f(x_2) = f[x_2] & f[x_1, x_2] & f[x_0, x_1, x_2] \\
\vdots & \vdots & \vdots & \vdots & \ddots \\
x_n & f(x_n) = f[x_n] & f[x_{n-1}, x_n] & f[x_{n-2}, x_{n-1}, x_n] & \cdots & f[x_0, \ldots, x_n].
\end{array}
$$

Assume now that the point $(x, f(x))$ is added at the top to the table of the divided differences:

$$
\begin{array}{llllll}
x & f(x) \\
x_0 & f(x_0) & f[x, x_0] \\
x_1 & f(x_1) & f[x_0, x_1] & \ddots \\
\vdots & \vdots & \vdots & \vdots & \ddots \\
x_{n-1} & f(x_{n-1}) & f[x_{n-2}, x_{n-1}] & \cdots & \cdots & f[x, x_0, \ldots, x_{n-1}] \\
x_n & f(x_n) & f[x_{n-1}, x_n] & \cdots & \cdots & f[x_0, \ldots, x_n] & f[x, x_0, \ldots, x_n].
\end{array}
$$

We then have

$$f[x, x_0] = \frac{f[x_0] - f[x]}{x_0 - x} \quad \Rightarrow \quad f(x) = f[x_0] + (x - x_0)f[x, x_0], \qquad (4.14)$$

and

$$f[x, x_0, x_1] = \frac{f[x_0, x_1] - f[x, x_0]}{x_1 - x} \quad \Rightarrow \quad f[x, x_0] = f[x_0, x_1] + (x - x_1)f[x, x_0, x_1].$$

Inserting this into (4.14) we find

$$
\begin{aligned}
f(x) &= f[x_0] + (x - x_0)\left(f[x_0, x_1] + (x - x_1)f[x, x_0, x_1]\right) \\
&= f(x_0) + (x - x_0)f[x_0, x_1] + (x - x_0)(x - x_1)f[x, x_0, x_1].
\end{aligned}
$$

If we continue eliminating the divided differences involving x in the same way, we obtain

$$f(x) = Q(x) + (x - x_0)(x - x_1) \cdots (x - x_n)f[x, x_0, x_1, \ldots, x_n], \qquad (4.15)$$

where

$$Q(x) = \sum_{k=0}^{n} f[x_0, x_1, \ldots, x_k] \prod_{q=0}^{k-1} (x - x_q) \qquad (4.16)$$

is a polynomial of degree n. Note that if we let $x \to x_i$ in (4.15) then we get $f(x_i) = Q(x_i)$ since the second term vanishes. Therefore we conclude that *Q must be the uniquely determined interpolation polynomial.* By comparing the expression in (4.16) with (4.11), we deduce that *the Newton coefficients are the divided differences along the diagonal of the divided difference table*:

$$d_i = f[x_0, x_1, \ldots, x_i], \quad i = 0, \ldots, n.$$

The following MATLAB function computes the coefficients using the divided differences scheme.

ALGORITHM 4.4. *Newton Coefficients*

```
function [d,D]=NewtonCoefficients(x,y)
% NEWTONCOEFFICIENTS divided differences for the Newton interpolation
%    [d,D]=NewtonCoefficients(x,y); computes the divided differences
%    needed for constructing the Newton form of the interpolating
%    polynomial through the points (x,y)

n=length(x)-1;                       % degree of interpolating polynomial
for i=1:n+1                          % divided differences
  D(i,1)=y(i);
  for j=1:i-1
    D(i,j+1)=(D(i,j)-D(i-1,j))/(x(i)-x(i-j));
  end
end
d=diag(D);
```

Note in the above algorithm that we first construct the entire divided difference table before extracting the diagonal. It is also possible to reduce the storage requirement of this algorithm to $O(n)$ by successively overwriting entries that are no longer needed, see Problem 4.2.

With the following MATLAB function, we evaluate the interpolation polynomial for the new arguments z in Horner form

$$
\begin{aligned}
P_n(z) \;=\;\; & f[x_0] + (z - x_0)\,(f[x_0, x_1] + (z - x_1)\,(f[x_0, x_1, x_2] + \cdots \\
& + (z - x_{n-1})\,(f[x_0, x_1, \ldots, x_n]) \cdots))
\end{aligned}
$$

by using the recurrence

$$
\begin{aligned}
p \;&:=\; f[x_0, x_1, \ldots, x_n] \\
p \;&:=\; (z - x_i)p + f[x_0, x_1, \ldots, x_i], \quad i = n-1, n-2, \ldots, 0.
\end{aligned}
$$

ALGORITHM 4.5. *Newton Interpolation*

```
function yy=NewtonInterpolation(x,d,xx)
```

```
% NEWTONINTERPOLATION interpolate using the Newton polynomial
%    yy=NewtonInterpolation(x,d,xx); uses the points (x,.)
%    and the precalculated divided difference coefficients d for the
%    Newton form of the interpolation polynomial P and iterpolates
%    the values yy=P(xx)

n=length(x)-1;
yy=d(n+1);
for i=n:-1:1
  yy=yy.*(xx-x(i))+d(i);
end;
```

The divided differences have many interesting properties. Comparing the expressions in (4.15) and the interpolation error (4.8)

$$f(x) = P_n(x) + \prod_{k=0}^{n} (x-x_k)\frac{f^{(n+1)}(\xi)}{(n+1)!} = P_n(x) + \prod_{k=0}^{n} (x-x_k)f[x,x_0,x_1,\ldots,x_n]$$

we see that the divided differences are approximations for the derivatives:

LEMMA 4.1. *For some ξ between the nodes $x, x_0, x_1, \ldots, x_n$ we have*

$$f[x,x_0,x_1,\ldots,x_n] = \frac{f^{(n+1)}(\xi)}{(n+1)!}.$$

The analogy of Newton's interpolation formula to Taylor expansions is now evident: the Taylor polynomial of f expanded about x_0,

$$T_n(x) = \sum_{k=0}^{n} \frac{f^{(k)}(x_0)}{k!}(x-x_0)^k,$$

looks very similar to the interpolation polynomial $Q_n(x)$ in (4.16).

A second useful property is *the symmetry* of the divided differences, e.g., $f[x_0,x_1,x_2] = f[x_2,x_1,x_0] = f[x_1,x_0,x_2]$.

THEOREM 4.3. (SYMMETRY OF DIVIDED DIFFERENCES) *The divided difference $f[x_i,x_{i+1},\ldots,x_{i+k}]$ is a symmetric function of its arguments and is given by*

$$f[x_i,x_{i+1},\ldots,x_{i+k}] = \sum_{j=i}^{i+k} \frac{f(x_j)}{\prod_{\substack{p=i\\p\neq j}}^{i+k}(x_j - x_p)}. \tag{4.17}$$

PROOF. This theorem can be proved by mathematical induction [112]. We will give here a simpler (though not quite complete) argument. The connection between the function values $f(x_i)$ and the divided differences is given by Equation (4.12) which is

$$\begin{pmatrix} 1 & & & & \\ 1 & x_1 - x_0 & & & \\ 1 & x_2 - x_0 & (x_2 - x_0)(x_2 - x_1) & & \\ \vdots & \vdots & \vdots & \ddots & \\ 1 & x_n - x_0 & (x_n - x_0)(x_n - x_1) & \cdots & \prod_{j=0}^{n-1}(x_n - x_j) \end{pmatrix} \begin{pmatrix} d_0 \\ d_1 \\ d_2 \\ \vdots \\ d_n \end{pmatrix} = \begin{pmatrix} f_0 \\ f_1 \\ f_2 \\ \vdots \\ f_n \end{pmatrix}.$$

Solving the first three of these equations by forward substitution for d_0, d_1 and d_2 gives us

$$\begin{pmatrix} d_0 \\ d_1 \\ d_2 \end{pmatrix} = \begin{pmatrix} 1 & & \\ \dfrac{1}{x_0 - x_1} & \dfrac{1}{x_1 - x_0} & \\ \dfrac{1}{(x_0 - x_1)(x_0 - x_2)} & \dfrac{1}{(x_1 - x_0)(x_1 - x_2)} & \dfrac{1}{(x_2 - x_0)(x_2 - x_1)} \end{pmatrix} \begin{pmatrix} f_0 \\ f_1 \\ f_2 \end{pmatrix}.$$

From these first elements of the inverse matrix the general pattern is evident, and we see that the symmetric expression (4.17) holds for the divided differences used for Newton's interpolation. $\square$

4.2.5 Interpolation Using Orthogonal Polynomials

In this section we consider yet another set of basis polynomials to represent the interpolation polynomial: *the set of orthogonal polynomials* belonging to the nodes x_i, $i = 0, \ldots, m$. Orthogonal polynomials are also used in quadrature (see Section 9.3.2).

In our case the scalar product is defined as

$$\langle p_j, p_k \rangle = \sum_{i=0}^{n} p_j(x_i) p_k(x_i). \tag{4.18}$$

DEFINITION 4.1. (ORTHOGONAL POLYNOMIALS) *A set of polynomials* $\{p_j\}_{j=0}^{N}$ *is said to be* orthogonal *with respect to the scalar product* $\langle \cdot, \cdot \rangle$ *if* p_j *has degree j for $1 \leq j \leq N$ and*

$$\langle p_j, p_k \rangle = 0, \quad j \neq k.$$

We define the norm *of a polynomial by* $\|p\|^2 = \langle p, p \rangle$.

THEOREM 4.4. *Let* $p_{-1}(x) \equiv 0$, $p_0(x) \equiv 1$, $\beta_0 = 0$ *and*

$$p_{k+1}(x) = (x - \alpha_{k+1})p_k(x) - \beta_k p_{k-1}(x), \quad k = 0, 1, 2, \ldots \tag{4.19}$$

with

$$\alpha_{k+1} = \frac{\langle x\, p_k, p_k \rangle}{\|p_k\|^2} \quad \text{and} \quad \beta_k = \frac{\|p_k\|^2}{\|p_{k-1}\|^2}.$$

Then the polynomials $p_j(x)$ are orthogonal.

PROOF. The proof is constructive by induction: it is in fact the so called *Lanczos Algorithm* (see Algorithm 9.4 in Section 9.3.2).

Base case: $\alpha_1 = \langle xp_0, p_0 \rangle / \|p_0\|^2$ and $p_1 = (x - \alpha_1)p_0$. Then

$$\langle p_0, p_1 \rangle = \langle p_0, (x - \alpha_1)p_0 \rangle = \underbrace{\langle p_0, x\, p_0 \rangle}_{=\langle x\, p_0, p_0 \rangle} - \alpha_1 \langle p_0, p_0 \rangle = 0$$

and thus p_1 is orthogonal to p_0 and the base case is established.

Induction hypothesis: Let $p_0, \ldots, p_k$ be orthogonal i.e. $\langle p_i, p_j \rangle = 0$ holds for $i, j \leq k$ and $i \neq j$.

Induction step: For $p_{k+1} = (x - \alpha_{k+1})p_k - \beta_k p_{k-1}$ the following holds

$$\begin{aligned}
\langle p_k, p_{k+1} \rangle &= \langle p_k, (x - \alpha_{k+1})p_k - \beta_k p_{k-1} \rangle \\
&= \underbrace{\langle p_k, x\, p_k \rangle}_{=\langle x\, p_k, p_k \rangle} - \alpha_{k+1} \langle p_k, p_k \rangle - \beta_k \underbrace{\langle p_k, p_{k-1} \rangle}_{=0} = 0.
\end{aligned}$$

Solving $p_k = (x - \alpha_k)p_{k-1} - \beta_{k-1}p_{k-2}$ for xp_{k-1} we get

$$xp_{k-1} = p_k + \alpha_k p_{k-1} + \beta_{k-1}p_{k-2}.$$

Multiplying with p_k we obtain

$$\langle p_k, xp_{k-1} \rangle = \langle p_{k-1}, xp_k \rangle = \langle p_k, p_k \rangle + \alpha_k \cdot 0 + \beta_{k1} \cdot 0 = \|p_k\|^2.$$

We use this in

$$\begin{aligned}
\langle p_{k-1}, p_{k+1} \rangle &= \langle p_{k-1}, (x - \alpha_{k+1})p_k - \beta_k p_{k-1} \rangle \\
&= \underbrace{\langle p_{k-1}, x\, p_k \rangle}_{=\|p_k\|^2} - \alpha_{k+1} \underbrace{\langle p_{k-1}, p_k \rangle}_{=0} - \beta_k \underbrace{\langle p_{k-1}, p_{k-1} \rangle}_{=\|p_k\|^2} = 0.
\end{aligned}$$

Finally, for $s < k - 1$,

$$\begin{aligned}
\langle p_s, p_{k+1} \rangle &= \langle p_s, (x - \alpha_{k+1})p_k - \beta_k p_{k-1} \rangle \\
&= \underbrace{\langle p_s, x\, p_k \rangle}_{=0,\ \text{for } s<k-1} - \alpha_{k+1} \underbrace{\langle p_s, p_k \rangle}_{=0,\ \text{for } s<k-1} - \beta_k \underbrace{\langle p_s, p_{k-1} \rangle}_{=0,\ \text{for } s<k-1} = 0,
\end{aligned}$$

where the first term vanishes because

$$\langle p_s, x\, p_k \rangle = \langle \underbrace{xp_s}_{\deg<k}, p_k \rangle = 0.$$

Hence, p_{k+1} is orthogonal to p_s for $s \leq k$.

$\square$

The coefficients α_k and β_k depend on the nodes x_i. In particular,

$$\alpha_1 = \frac{\langle x\,p_0, p_0\rangle}{\|p_0\|^2} = \frac{\langle x, 1\rangle}{n+1} = \frac{1}{n+1}\sum_{i=0}^{n} x_i,$$

thus α_1 is the average of the nodes. In general we get

$$\alpha_{k+1} = \frac{\displaystyle\sum_{i=0}^{n} x_i\,p_k(x_i)^2}{\displaystyle\sum_{i=0}^{n} p_k(x_i)^2} \quad \text{and} \quad \beta_k = \frac{\displaystyle\sum_{i=0}^{n} p_k(x_i)^2}{\displaystyle\sum_{i=0}^{n} p_{k-1}(x_i)^2}.$$

Consider now the *approximation problem* for $k \leq n$

$$b_0 p_0(x_j) + b_1 p_1(x_j) + \ldots + b_k p_k(x_j) \approx f(x_j), \quad j = 0, \ldots, n \qquad (4.20)$$

or in matrix notation $P\boldsymbol{b} \approx \boldsymbol{f}$

$$\begin{pmatrix} p_0(x_0) & p_1(x_0) & \cdots & p_k(x_0) \\ p_0(x_1) & p_1(x_1) & \cdots & p_k(x_1) \\ \vdots & \vdots & \cdots & \vdots \\ p_0(x_n) & p_1(x_n) & \cdots & p_k(x_n) \end{pmatrix} \begin{pmatrix} b_0 \\ \vdots \\ b_k \end{pmatrix} \approx \begin{pmatrix} f_0 \\ f_1 \\ \vdots \\ f_n \end{pmatrix}. \qquad (4.21)$$

If $k = n$, then we have $n+1$ equations for $n+1$ unknowns b_i. However, if $k < n$, then the system $P\boldsymbol{x} = \boldsymbol{b}$ cannot be satisfied exactly in general, so we must solve (4.21) as a *least squares problem* to obtain the best approximation. Since the matrix P is orthogonal due to the orthogonality of the polynomials the solution is easily obtained with the *normal equations* (see Equation (6.13) in Chapter 6):

$$P^{\top} P \boldsymbol{b} = P^{\top} \boldsymbol{f}.$$

Because $P^{\top} P$ is diagonal the solution is simply

$$b_j = \frac{\displaystyle\sum_{i=0}^{n} p_j(x_i) f_i}{\displaystyle\sum_{i=0}^{n} p_j(x_i)^2} = \frac{\boldsymbol{p}_j^{\top} \boldsymbol{f}}{\|\boldsymbol{p}_j\|^2}, \quad j = 0, \ldots, k.$$

The following MATLAB function `OrthogonalPolynomialCoefficients` computes the matrix P, the coefficients α_k and β_k and also the b_i.

ALGORITHM 4.6. *Coefficient of Orthogonal Polynomials*

```
function [alpha,beta,b,P]=OrthogonalPolynomialCoefficients(x,y)
% ORTHOGONALPOLYNOMIALCOEFFICIENTS 3 term recurrence coefficients
%    [alpha,beta,b,P]=OrthogonalPolynomialCoefficients(x,y); computes
```

```
%    the coefficients of the 3-term recurrence of the orthogonal
%    polynomials for the set of points (x,y). P is the orthogonal
%    matrix which contains as columns the values p_i(x) of the
%    orthogonal polynomials, and b contains the coefficients for the
%    expansion P(x)=b_1 p_1(x) + ... + b_m p_m(x).

m=length(x); x=x(:); y=y(:);
alpha(1)=sum(x)/m;
p1=ones(size(x)); p2=x-alpha(1);
P=[p1,p2];
b(1)=p1'*y/norm(p1)^2; b(2)=p2'*y/norm(p2)^2;
for k=1:m-2
  p0=p1; p1=p2;
  alpha(k+1)=p1'*(x.*p1)/norm(p1)^2;
  beta(k)=(norm(p1)/norm(p0))^2;
  p2=(x-alpha(k+1)).*p1-beta(k)*p0;
  P=[P,p2];
  b(k+2)=p2'*y/norm(p2)^2;
end
b=b(:);
```

If we wish to interpolate for new values xx we use the recurrence relation (4.19) to build up the matrix containing the values of the orthogonal polynomials at the points xx. A linear combination of those columns with k coefficients b_i gives us the best least squares polynomial of degree $k - 1$ to the given points (x, y). The matrix pp in the following MATLAB function OrthogonalInterpolation contains the values of the orthogonal polynomials for xx and the matrix yy holds in its columns the values of the best approximating polynomial for the argument xx.

ALGORITHM 4.7.
Interpolation with Orthogonal Polynomials

```
function [yy,pp]=OrthogonalInterpolation(alpha,beta,b,xx)
% ORTHOGONALINTERPOLATION interpolation using orthogonal polynomials
%    [yy,pp]=OrthogonalInterpolation(alpha,beta,b,xx) evaluates the
%    precomputed orthogonal polynomials defined by alpha, beta, b from
%    OrthogonalPolynomialCoefficients(x,y) for the values xx and
%    returns their values in pp and the approximations yy(:,i)=b_1
%    p_1(xx) + ... + b_i p_i(xx)

m=length(b);
p1=ones(size(xx)); p2=xx-alpha(1);
pp=[p1,p2];
yy=[p1*b(1) p1*b(1)+p2*b(2)];
for i=2:m-1
  p0=p1; p1=p2;
  p2=(xx-alpha(i)).*p1-beta(i-1)*p0;
```

```
  pp=[pp,p2];
  yy=[yy yy(:,i)+p2*b(i+1)];
end
```

EXAMPLE 4.3. *We use* 13 *equidistant points of the function* $f(x) = e^{\sin x}$:

```
x=[0:0.5:6]'; y=exp(sin(x));
plot(x,y,'o'); hold on
axis([1,6,-1,5]); axis('equal')
```

Now we compute the coefficients α_k, β_k *and the coefficients* b_i *for the approximating polynomials:*

```
[alpha,beta,b,P]=OrthogonalPolynomialCoefficients(x,y);
```

To plot the approximating polynomials we evaluate them for **xx**:

```
xx=[-1:0.1:7]';
yy=OrthogonalInterpolation(alpha,beta,b,xx);
```

Finally we plot some of the approximating polynomials (for degrees 1,4 and 7):

```
for k=2:3:8
  plot(xx,yy(:,k))
  residual=norm(y-P(:,1:k)*b(1:k))
end
```

and obtain Figure 4.2 and the residual values 2.1384 *for* $k = 2$, 0.5842 *for* $k = 5$ *and* 0.1032 *for* $k = 8$.

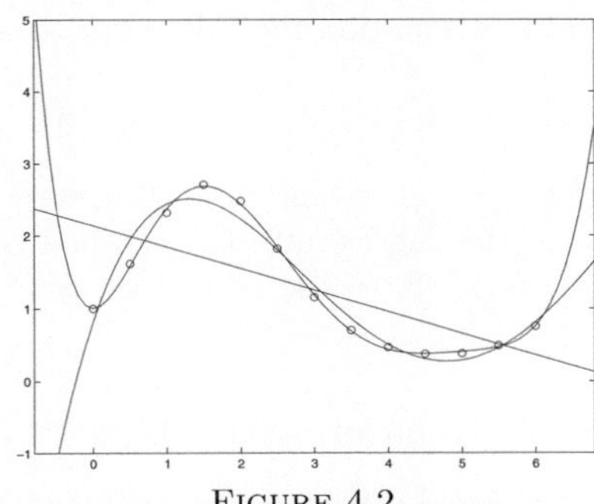

FIGURE 4.2.
Data Fitting with Polynomials of Degree 1,4 and 7

For *data fitting problems*, we often do not necessarily want to *interpolate* the data, since the data usually contains measurement or round-off errors that may be amplified by high-degree interpolation. Instead, we are usually satisfied with an *approximation*, e.g. by the polynomial of degree 7 shown in

Figure 4.2. Alternatively, we may know about a *measurement error* δ and wish to fit the data with the polynomial of lowest degree k for which the residual $\boldsymbol{r}_k$ satisfies $\|\boldsymbol{r}_k\| \le \delta$.

Let us denote by $P_k = [\boldsymbol{p}_0, \boldsymbol{p}_1, \ldots, \boldsymbol{p}_k]$ the matrix containing the first $k+1$ columns of P. The residual for the approximating polynomial of degree k can then be computed explicitly as

$$\boldsymbol{r}_k = \boldsymbol{y} - P_k \boldsymbol{b}_k, \quad \boldsymbol{b}_k^\top = (b_0, b_1, \ldots, b_k).$$

If we increase the degree by one, then

$$\boldsymbol{r}_{k+1} = \boldsymbol{y} - P_{k+1}\boldsymbol{b}_{k+1} = \boldsymbol{y} - P_k\boldsymbol{b}_k - b_{k+1}\boldsymbol{p}_{k+1} = \boldsymbol{r}_k - b_{k+1}\boldsymbol{p}_{k+1}.$$

If we now compute the length, then

$$\begin{aligned}
\|\boldsymbol{r}_{k+1}\|^2 &= (\boldsymbol{r}_k - b_{k+1}\boldsymbol{p}_{k+1})^\top (\boldsymbol{r}_k - b_{k+1}\boldsymbol{p}_{k+1}) \\
&= \|\boldsymbol{r}_k\|^2 - 2b_{k+1}\boldsymbol{p}_{k+1}^\top \boldsymbol{r}_k + b_{k+1}^2 \|\boldsymbol{p}_{k+1}\|^2.
\end{aligned}$$

But since $\boldsymbol{p}_{k+1}$ is orthogonal to P_k, we get

$$\boldsymbol{p}_{k+1}^\top \boldsymbol{r}_k = \boldsymbol{p}_{k+1}^\top (\boldsymbol{y} - P_k\boldsymbol{b}_k) = \boldsymbol{p}_{k+1}^\top \boldsymbol{y}.$$

Furthermore, the coefficients b_k are computed by

$$b_k = \frac{\boldsymbol{p}_k^\top \boldsymbol{y}}{\|\boldsymbol{p}_k\|^2}.$$

Thus, $\boldsymbol{p}_{k+1}^\top \boldsymbol{r}_k = \boldsymbol{p}_{k+1}^\top \boldsymbol{y} = b_{k+1}\|\boldsymbol{p}_{k+1}\|^2$. If we introduce this in the expression for $\|\boldsymbol{r}_{k+1}\|$, we obtain an equation that allows us to *compute the residuals recursively*:

$$\|\boldsymbol{r}_{k+1}\|^2 = \|\boldsymbol{r}_k\|^2 - b_{k+1}^2 \|\boldsymbol{p}_{k+1}\|^2. \tag{4.22}$$

We are now ready to write a program that computes best-approximating polynomials by increasing the degree until the residual becomes smaller than a certain tolerance *tol*. It is left to the reader as an exercise (see Problem 4.3).

4.2.6   Change of Basis, Relation with LU and QR

In the last sections, we discussed different representations of the interpolation polynomial. The first one used as *polynomial basis* the *monomials*

$$\boldsymbol{m}(x) = [1, x, x^2, \ldots, x^n]^\top.$$

The Lagrange interpolation formula uses the *Lagrange polynomial* basis

$$\boldsymbol{l}(x) = [l_0(x), l_1(x), \ldots, l_n(x)]^\top.$$

For the Newton interpolation, we use the basis of the *Newton polynomials*

$$\boldsymbol{\pi}(x) = [\pi_0(x), \pi_1(x), \ldots, \pi_n(x)]^\top \quad \text{with} \quad \pi_i(x) = \prod_{j=0}^{i-1}(x - x_j),$$

and for the approximation in the least squares sense, we introduced the basis of *orthogonal polynomials*, which also interpolates if $k = n$:

$$\boldsymbol{p}(x) = [p_0(x), p_1(x), \ldots, p_n(x)]^\top.$$

We introduce the vector of coefficients of the monomials

$$\boldsymbol{a} = [a_0, a_1, \ldots, a_n]^\top,$$

the vector of the function values

$$\boldsymbol{f} = [f_0, f_1, \ldots, f_n]^\top,$$

the vector of divided differences used for Newton's interpolation

$$\boldsymbol{d} = [d_0, d_1, \ldots, d_n]^\top,$$

and the vector of the coefficients of the orthogonal polynomials

$$\boldsymbol{b} = [b_0, b_1, \ldots, b_n]^\top.$$

With this notation, a scalar product describes the four representations of the interpolation polynomial:

$$P_n(x) = \boldsymbol{a}^\top \boldsymbol{m}(x) = \boldsymbol{f}^\top \boldsymbol{l}(x) = \boldsymbol{d}^\top \boldsymbol{\pi}(x) = \boldsymbol{b}^\top \boldsymbol{p}(x). \tag{4.23}$$

There must exist matrices which compute the *transformations between the polynomial bases*. The coefficients a_i used for the monomial basis are, as we discussed before, given by the solution of the linear system $V\boldsymbol{a} = \boldsymbol{f}$ with the Vandermonde matrix V. Now using Equation (4.23) we have

$$\boldsymbol{f}^\top \boldsymbol{l} = \boldsymbol{a}^\top \boldsymbol{m} = (V^{-1}\boldsymbol{f})^\top \boldsymbol{m} = \boldsymbol{f}^\top V^{-T}\boldsymbol{m}.$$

Since the last equation is valid for all possible choices of function values $\boldsymbol{f}$, we can conclude that

$$\boldsymbol{l} = V^{-T}\boldsymbol{m} \quad \Longrightarrow \quad V^\top \boldsymbol{l} = \boldsymbol{m}.$$

Thus, the transposed Vandermonde matrix maps the Lagrange polynomials to the monomials,

$$\begin{pmatrix} 1 & 1 & \cdots & 1 \\ x_0 & x_1 & \cdots & x_n \\ \vdots & \vdots & \vdots & \vdots \\ x_0^n & x_1^n & \cdots & x_n^n \end{pmatrix} \begin{pmatrix} l_0(x) \\ l_1(x) \\ \vdots \\ l_n(x) \end{pmatrix} = \begin{pmatrix} 1 \\ x \\ \vdots \\ x^n \end{pmatrix}. \tag{4.24}$$

Note that Equation (4.24) is just the *Lagrange interpolation representation of the monomials!*

There must be a similar relationship between the Lagrange and Newton polynomials. Let us denote by $U^\top \boldsymbol{d} = \boldsymbol{f}$ with

$$
U^\top = \begin{pmatrix}
1 & & & & \\
1 & x_1 - x_0 & & & \\
1 & x_2 - x_0 & (x_2 - x_0)(x_2 - x_1) & & \\
\vdots & \vdots & \vdots & \ddots & \\
1 & x_n - x_0 & (x_n - x_0)(x_n - x_1) & \cdots & \prod_{j=0}^{n-1}(x_n - x_j)
\end{pmatrix}
$$

the linear system (4.12) with which the divided differences can be computed from the function values. Using again Equation (4.23) we obtain

$$
\boldsymbol{d}^\top \boldsymbol{\pi} = \boldsymbol{f}^\top \boldsymbol{l} = \boldsymbol{d}^\top U \boldsymbol{l} \quad \Longrightarrow \quad U\boldsymbol{l}(x) = \boldsymbol{\pi}(x).
$$

Thus, the upper triangular matrix U maps the Lagrange polynomials to the Newton polynomials:

$$
\begin{pmatrix}
1 & 1 & 1 & \cdots & 1 \\
 & x_1 - x_0 & x_2 - x_0 & \cdots & x_n - x_0 \\
 & & (x_2 - x_0)(x_2 - x_1) & \cdots & (x_n - x_0)(x_n - x_1) \\
 & & & \ddots & \vdots \\
 & & & & \prod_{j=0}^{n-1}(x_n - x_j)
\end{pmatrix}
\begin{pmatrix}
l_0 \\ l_1 \\ l_2 \\ \vdots \\ l_n
\end{pmatrix}
=
\begin{pmatrix}
\pi_0 \\ \pi_1 \\ \pi_2 \\ \vdots \\ \pi_n
\end{pmatrix}.
$$

$$(4.25)$$

Note again that this result is obvious since (4.25) is the Lagrange representation of the Newton polynomials!

Next we would like to find a relation between the monomials and the Newton polynomials. We are looking for a lower triangular matrix L with $L\boldsymbol{\pi}(x) = \boldsymbol{m}(x)$. L has to be lower triangular, since the Newton polynomials and the monomials have the same degree. To determine L we need to know more about *divided differences.*

The divided differences may be interpreted as an elimination process. Let the function $H_p(x_0, \ldots, x_k)$ denote the sum of all homogeneous products of degree p of the variables $x_0, \ldots, x_k$. For example we have

$$
\begin{aligned}
H_p(x_0) &= x_0^p \\
H_1(x_0, \ldots, x_k) &= \sum_{j=0}^{k} x_j \\
H_p(x_0, x_1) &= \sum_{j=0}^{p} x_0^j x_1^{p-j} = \sum_{j=0}^{p} H_j(x_0) H_{p-j}(x_1).
\end{aligned}
$$

For these homogeneous products the following *Lemma of Miller* [91] holds:

LEMMA 4.2.

$$
H_p(x_0, \ldots, x_k) = \frac{H_{p+1}(x_0, \ldots, x_{k-1}) - H_{p+1}(x_1, \ldots, x_k)}{x_0 - x_k}
\qquad (4.26)
$$

PROOF. We rewrite the right-hand side of (4.26) as

$$\frac{H_{p+1}(x_0,\ldots,x_{k-1})-H_{p+1}(x_1,\ldots,x_k)}{x_0-x_k}=\frac{\sum_{s=1}^{p+1}(x_0^s-x_k^s)H_{p+1-s}(x_1,\ldots,x_{k-1})}{x_0-x_k}.$$

But

$$\frac{x_0^s-x_k^s}{x_0-x_k}=\sum_{q=0}^{s-1}x_0^q x_k^{q+1-s}=H_{s-1}(x_0,x_k).$$

Thus, the right-hand side of (4.26) is equal to

$$\sum_{s=0}^{p}H_s(x_0,x_k)H_{p-s}(x_1,\ldots,x_{k-1})=H_p(x_0,\ldots,x_k).$$

$\square$

Consider now the interpolation polynomial expanded in monomials:

$$P_n(x)=a_0+a_1x+\cdots a_nx^n,$$

and the system of linear equations (the interpolation condition) with the Vandermonde matrix

$$\sum_{j=0}^{n}a_jx_i^j=f(x_i),\quad i=0,\ldots n. \tag{4.27}$$

THEOREM 4.5. *The divided difference scheme eliminates the unknown coefficients of the interpolation polynomial in (4.27). More precisely,*

$$f[x_i,\ldots,x_{i+k}]=a_k+\sum_{j=k+1}^{n}a_jH_{j-k}(x_i,\ldots,x_{i+k}). \tag{4.28}$$

PROOF. The proof is by induction on the columns. For the base case, we have

$$f[x_i,x_{i+1}]=\frac{\sum_{j=0}^{n}a_jx_i^j-\sum_{j=0}^{n}a_jx_{i+1}^j}{x_i-x_{i+1}}=a_1+\sum_{j=2}^{n}a_jH_{j-1}(x_i,x_{i+1}).$$

Thus, the theorem is valid for the second column. Assume now it holds for the first $k-1$ columns. Then

$$f[x_i,\ldots,x_{i+k}]=\frac{f[x_i,\ldots,x_{i+k-1}]-f[x_{i+1},\ldots,x_{i+k}]}{x_i-x_{i+k}}$$

$$=\frac{\sum_{j=k}^{n}a_j\left(H_{j-k+1}(x_i,\ldots,x_{i+k-1})-H_{j-k+1}(x_{i+1},\ldots,x_{i+k})\right)}{x_i-x_{i+k}}.$$

By multiplying the denominator into the sum and by applying Lemma 4.2, we obtain

$$f[x_i, \ldots, x_{i+k}] = a_k + \sum_{j=k+1}^{n} a_j H_{j-k}(x_i, \ldots, x_{i+k}).$$

$\square$

As a conclusion, if the interpolated values $f(x_i)$ are values of a polynomial of degree n then

$$f[x_0, \ldots, x_n] = a_n = \frac{P_n^{(n)}(0)}{n!} = \text{ leading coefficient of } P_n \qquad (4.29)$$

and $f[x_0, \ldots, x_{n+k}] = 0$ for $k = 1, 2, \ldots$.

Furthermore, (4.28) gives us the relation between the divided differences d_i and the coefficients a_i. We have $L^\top a = d$ where

$$L = \begin{pmatrix} 1 & & & & \\ H_1(x_0) & 1 & & & \\ H_2(x_0) & H_1(x_0, x_1) & 1 & & \\ \vdots & \vdots & \ddots & \ddots & \\ H_n(x_0) & H_{n-1}(x_0, x_1) & \cdots & H_1(x_0, \ldots, x_{n-1}) & 1 \end{pmatrix}.$$

From Equation (4.23) we obtain

$$a^\top m = d^\top \pi = a^\top L \pi \quad \implies \quad L\pi(x) = m(x).$$

Thus we obtained the result: *the lower triangular matrix L maps the Newton polynomials to the monomials.*

From the relation $V^\top l = m$ between the Lagrange polynomials and the monomials and from $Ul = \pi$ relating the Lagrange polynomials to the Newton polynomials we can eliminate l and thus also obtain the connection between the Newton polynomials and the monomials

$$V^\top U^{-1} \pi = m \quad \implies \quad V^\top U^{-1} = L \quad \iff \quad V^\top = LU.$$

The last equation tells us that the transposed Vandermonde matrix has been factored into the product of a lower and an upper triangular matrix. Since the diagonal of L is one, the factorization is unique and nothing else than the factorization obtained by applying Gaussian elimination to $V^\top$.

THEOREM 4.6. *The LU decomposition of the Vandermonde matrix*

$$V = \begin{pmatrix} 1 & x_0 & \cdots & x_0^{n-1} & x_0^n \\ 1 & x_1 & \cdots & x_1^{n-1} & x_1^n \\ \vdots & \vdots & & \vdots & \vdots \\ 1 & x_n & \cdots & x_n^{n-1} & x_n^n \end{pmatrix}$$

is $V = U^{\top} L^{\top}$ respectively $V^{\top} = LU$ where

$$
L = \begin{pmatrix}
1 & & & & \\
H_1(x_0) & 1 & & & \\
H_2(x_0) & H_1(x_0, x_1) & 1 & & \\
\vdots & \vdots & \ddots & \ddots & \\
H_n(x_0) & H_{n-1}(x_0, x_1) & \cdots & H_1(x_0, \ldots, x_n) & 1
\end{pmatrix}
$$

and

$$
U = \begin{pmatrix}
1 & 1 & 1 & \cdots & 1 \\
& x_1 - x_0 & x_2 - x_0 & \cdots & x_n - x_0 \\
& & (x_2 - x_0)(x_2 - x_1) & \cdots & (x_n - x_0)(x_n - x_1) \\
& & & \ddots & \vdots \\
& & & & \prod_{j=0}^{n-1}(x_n - x_j)
\end{pmatrix}.
$$

The coefficients of the interpolating polynomial in the basis of monomials are the solution of $V\boldsymbol{a} = \boldsymbol{f}$. They may be computed using the LU decomposition $V = U^{\top} L^{\top}$ by

1. *solving $U^{\top}\boldsymbol{d} = \boldsymbol{f}$ by forward substitution, giving the intermediate vector $\boldsymbol{d}$ of the divided differences and*

2. *solving $L^{\top}\boldsymbol{a} = \boldsymbol{d}$ by back substitution.*

The polynomial basis can be computed by solving $V^{\top}\boldsymbol{l}(x) = \boldsymbol{m}(x)$ also using $V^{\top} = LU$:

1. *forward substitution for $L\boldsymbol{\pi}(x) = \boldsymbol{m}(x)$, yielding as intermediate vector the Newton polynomials.*

2. *backward substitution for $U\boldsymbol{l}(x) = \boldsymbol{\pi}(x)$, giving the Lagrange polynomials.*

Using MAPLE, we can confirm that the theorem indeed holds;

```
with(LinearAlgebra):
V:=VandermondeMatrix([x_0,x_1,x_2,x_3]);
VT:=Transpose(V);
(p,L,U):=LUDecomposition(VT);
U:=map(factor,U);
```

We obtain as predicted the decomposition

$$
L = \begin{bmatrix}
1 & 0 & 0 & 0 \\
x_0 & 1 & 0 & 0 \\
x_0^2 & x_0 + x_1 & 1 & 0 \\
x_0^3 & x_0^2 + x_1 x_0 + x_1^2 & x_0 + x_2 + x_1 & 1
\end{bmatrix}
$$

$$U = \begin{bmatrix} 1 & 1 & 1 & 1 \\ 0 & x_1 - x_0 & x_2 - x_0 & x_3 - x_0 \\ 0 & 0 & -(x_2 - x_1)(-x_2 + x_0) & (-x_3 + x_1)(-x_3 + x_0) \\ 0 & 0 & 0 & -(-x_3 + x_1)(-x_3 + x_2)(-x_3 + x_0) \end{bmatrix}.$$

For the orthogonal polynomial basis, the interpolation condition is

$$b_0 p_0(x_j) + b_1 p_1(x_j) + \cdots + b_n p_n(x_j) = f_j, \quad j = 0, \ldots, n.$$

The matrix of this linear system for the coefficients b_i,

$$P = \begin{pmatrix} p_0(x_0) & p_1(x_0) & \cdots & p_n(x_0) \\ p_0(x_1) & p_1(x_1) & \cdots & p_n(x_1) \\ \vdots & \vdots & \vdots & \vdots \\ p_0(x_n) & p_1(x_n) & \cdots & p_n(x_n) \end{pmatrix},$$

is orthogonal: $P^\top P = D^2$ where $D = \mathrm{diag}\{\|p_0\|, \|p_1\|, \ldots, \|p_n\|\}$. Thus, as we have seen before, the solution is easy to compute,

$$\boldsymbol{b} = D^{-2} P^\top \boldsymbol{f}.$$

Substituting this into (4.23), we get

$$P_n(x) = \boldsymbol{f}^\top \boldsymbol{l}(x) = \boldsymbol{b}^\top \boldsymbol{p}(x) = \boldsymbol{f}^\top P D^{-2} \boldsymbol{p}(x).$$

Since this is true for all $\boldsymbol{f}$, we conclude

$$\boldsymbol{l}(x) = P D^{-2} \boldsymbol{p}(x) \quad \Longleftrightarrow \quad P^\top \boldsymbol{l}(x) = \boldsymbol{p}(x).$$

There must exist a lower triangular matrix G which maps the orthogonal polynomials to the monomials

$$G\boldsymbol{p}(x) = \boldsymbol{m}(x).$$

If we write this equation for $x = x_0, x_1, \ldots, x_n$ then we get for G the matrix equation

$$GP^\top = V^\top \quad \Longleftrightarrow \quad V = PG^\top = \left(PD^{-1}\right)\left(DG^T\right).$$

Thus we have *decomposed the Vandermonde matrix V into a product of an orthonormal matrix $Q = PD^{-1}$ and an upper triangular matrix $R = DG^T$*, this is the QR decomposition, which we introduced on the one hand to solve in a very stable fashion linear systems, see Section 3.5, and also for the solution of least squares problems, see Subsection 6.5.1.

Therefore, to compute the matrix G, we can compute the QR factorization $V = QR$ in a standard way using e.g. Householder reflections and then compute $G = R^\top D^{-1}$. Alternatively, if we know V and P, we can solve for G in $GP^\top = V^\top$ to get $G = V^\top P D^{-2}$. Since the two representations for the interpolating polynomial $\boldsymbol{p}(x)^\top \boldsymbol{b} = \boldsymbol{m}(x)^\top \boldsymbol{a}$ both hold, we obtain by replacing $\boldsymbol{m}(x) = G\boldsymbol{p}(x)$ the relation

$$G^\top \boldsymbol{a} = \boldsymbol{b}.$$

Summarizing all the results we obtain the following theorem.

THEOREM 4.7. *Let $p_k(x)$ be the orthogonal polynomial of degree k, $\boldsymbol{p}(x) = (p_0(x), p_1(x), \ldots, p_n(x))^\top$ and $P^\top = [\boldsymbol{p}(x_0), \boldsymbol{p}(x_1), \ldots, \boldsymbol{p}(x_n)]$. Let D be the diagonal matrix $D = \mathrm{diag}\{\|p_0\|, \|p_1\|, \ldots, \|p_n\|\}$ and let $V = QR$ be the QR decomposition of the Vandermonde matrix. Then*

1. $Q = PD^{-1}$ *and* $R = D^{-1}P^\top V$.

2. *The basis transformation between the Lagrange polynomials $l_i(x)$ and the orthogonal polynomials $p_i(x)$ is $P^\top \boldsymbol{l}(x) = \boldsymbol{p}(x)$.*

3. *The basis transformation between the orthogonal polynomials $p_i(x)$ and the monomials $m_i(x) = x^i$ is $G\boldsymbol{p}(x) = \boldsymbol{m}(x)$ where $G = DR^\top = V^\top P D^{-2}$.*

4. *The coefficients of the monomials of the interpolating polynomial a_i and the coefficients of the orthogonal polynomials b_i are related by $G^\top \boldsymbol{a} = \boldsymbol{b}$.*

4.2.7 Aitken-Neville Interpolation

In this section we will discuss yet another algorithm for computing the interpolating polynomial. It is used to interpolate the function value $f(z)$ by computing the sequence $\{P_n(z)\}$ for $n = 0, 1, 2, \ldots$ which is obtained by using more and more interpolation points. The hope is that the sequence will converge to the function value $f(z)$.

Let $T_{ij}(x)$ be the polynomial of degree $\leq j$ that interpolates the data

$$\begin{array}{c|cccc} x & x_{i-j} & x_{i-j+1} & \cdots & x_i \\ \hline y & y_{i-j} & y_{i-j+1} & \cdots & y_i \end{array}.$$

We arrange these polynomials in a lower triangular matrix (the so called *Aitken-Neville scheme*):

$$
\begin{array}{c|ccccc}
x & y & & & & \\
\hline
x_0 & y_0 = T_{00} & & & & \\
x_1 & y_1 = T_{10} & T_{11} & & & \\
\vdots & \vdots & & \ddots & & \\
x_i & y_i = T_{i0} & T_{i1} & \cdots & T_{ii} & \\
\cdots & \cdots & \cdots & & \cdots & \ddots
\end{array}
\tag{4.30}
$$

THEOREM 4.8. *The polynomials T_{ij} can be computed with the recursion*

$$
\left.
\begin{aligned}
T_{i0} &= y_i \\
T_{ij} &= \frac{(x_i - x)T_{i-1,j-1} + (x - x_{i-j})T_{i,j-1}}{x_i - x_{i-j}} \\
j &= 1, 2, \ldots, i
\end{aligned}
\right\} \quad i = 0, 1, 2, \ldots \quad (4.31)
$$

PROOF. The proof is by induction. Base case: $T_{i,0}$ interpolates as required. Induction hypothesis: Let

$$T_{i,j-1} \text{ be the interpolation polynomial for } x_{i-j+1}, \ldots, x_i$$

and

$$T_{i-1,j-1} \text{ be the interpolation polynomial for } x_{i-j}, \ldots, x_{i-1}.$$

If we compute T_{ij} according to (4.31), then clearly T_{ij} is a polynomial of degree $\leq j$. If x_k is a common node of $T_{i-1,j-1}$ and $T_{i,j-1}$ then

$$T_{i-1,j-1}(x_k) = T_{i,j-1}(x_k) = y_k,$$

and hence also

$$T_{i,j}(x_k) = \frac{(x_i - x_k)y_k + (x_k - x_{i-j})y_k}{x_i - x_{i-j}} = y_k.$$

For the boundary x_i we obtain

$$T_{ij}(x_i) = \frac{0 \cdot T_{i-1,j-1}(x_i) + (x_i - x_{i-j})y_i}{x_i - x_{i-j}} = y_i$$

and similarly also $T_{ij}(x_{i-j}) = y_{i-j}$ holds. T_{ij} is therefore the interpolation polynomial as claimed in the theorem. $\qquad\square$

EXAMPLE 4.4. *Given the points*

$$
\begin{array}{c|cccc}
x & 0 & 5 & -1 & 2 \\
\hline
y & -5 & 235 & -9 & 19
\end{array},
$$

the recurrence (4.31) produces the scheme

$$
\begin{array}{cc|ccc}
x & y & & & \\
\hline
0 & -5 & & & \\
5 & 235 & 48x - 5 & & \\
-1 & -9 & \frac{122}{3}x + \frac{95}{3} & \frac{22}{3}x^2 + \frac{34}{3}x - 5 & \\
2 & 19 & \frac{28}{3}x + \frac{1}{3} & \frac{94}{9}x^2 - \frac{10}{9}x - \frac{185}{9} & \frac{14}{9}x^3 + \frac{10}{9}x^2 + \frac{32}{9}x - 5,
\end{array}
\qquad (4.32)
$$

which can be obtained by the MAPLE-*statements:*

```
with(LinearAlgebra):
```

```
x:=vector(4,[0, 5, -1, 2]);
y:=vector(4,[-5, 235, -9, 19]);
T:=Matrix(4,4);
for i from 1 to 4 do
  T[i,1]:=y[i];
  for j from 1 to i-1 do
    T[i,j+1]:=((x[i]-z)*T[i-1,j]+(z-x[i-j])*T[i,j])/(x[i]-x[i-j]);
  od;
od;
print(collect(T,z));
```

Usually the scheme is not computed for a variable x but for some fixed number x. The quantities T_{ik} are then simply numbers.

The following MATLAB function `AitkenInterpolation` computes the Aitken-Neville scheme (note that because the array indices have to start in MATLAB with 1 rather than 0, we had to adjust the loop variables):

ALGORITHM 4.8. *Aitken-Neville Scheme*

```
function T=AitkenInterpolation(x,y,z)
% AITKENINTERPOLATION interpolation using the Aitken Neville scheme
%    T=AitkenInterpolation(x,y,z) uses the ponts (x,y) to generate
%    the Aitken-Neville Scheme for the scalar interpolation value z.

n=length(x); T=zeros(n);
for i=1:n
  T(i,1)=y(i);
  for j=1:i-1
    T(i,j+1)=((x(i)-z)*T(i-1,j)+(z-x(i-j))*T(i,j))/(x(i)-x(i-j));
  end;
end;
```

EXAMPLE 4.5. *We compute the Aitken-Neville Scheme for the values of the function $f(x) = \int_0^x e^{\sin t} dt$ given in Table (4.2) for $x = 0.66$:*

x	y				
0.60	0.81360				
0.70	0.99670	0.92346			
0.80	1.19440	0.91762	0.92171		
0.50	0.64490	0.93797	0.92169	0.92172	
0.90	1.40630	0.94946	0.92188	0.92165	0.92171
0.40	0.49040	0.96667	0.92193	0.92189	0.92168	0.92171.

We see that we get $T_{22} = 0.9217\ldots$ (`T(3,3)` with MATLAB in Algorithm 4.8) the same value $P_2(0.66)$ as in Example 4.2 with Lagrange interpolation.

4.2.8 Extrapolation

Extrapolation is the same as interpolation, only the interpolation value z is outside the range of the given values x_j. Without loss of generality we can assume that $z = 0$ since we can always apply a shift of the independent variable $x' := x - z$.

Extrapolation is often used to *compute limits*. Let h be a discretization parameter and $T(h)$ an approximation of an unknown quantity a_0 with the following property:

$$\lim_{h \to 0} T(h) = a_0. \tag{4.33}$$

The usual assumption is that $T(0)$ is *difficult to compute* – maybe numerically unstable or requiring too many operations. If we compute some function values $T(h_i)$ for $h_i > 0$ and construct the interpolation polynomial P_n then $P_n(0)$ will be an approximation for a_0.

The sequence $\{P_n(0)\}$ for $n = 0, 1, 2, \ldots$ is given by the diagonal of the Aitken-Neville Scheme. Thus for extrapolation with polynomials the Aitken-Neville Scheme is the natural algorithm (in the problem section we discuss alternatives). The hope is that the scheme will converge to a_0. Indeed this is the case if there exists an asymptotic expansion for $T(h)$ of the form

$$T(h) = a_0 + a_1 h + \cdots + a_k h^k + R_k(h) \quad \text{with} \quad |R_k(h)| < C_k h^{k+1}, \tag{4.34}$$

and if the sequence $\{h_i\}$ is chosen such that

$$h_{i+1} < c h_i \quad \text{with} \quad 0 < c < 1,$$

i.e. if it converges sufficiently rapidly to zero. In this case, the diagonals of the Aitken-Neville scheme converge faster to a_0 than the columns [132].

Since $z = 0$, the recurrence (4.31) for computing the Aitken-Neville scheme simplifies to

$$T_{ij} = \frac{h_i T_{i-1,j-1} - h_{i-j} T_{i,j-1}}{h_i - h_{i-j}}. \tag{4.35}$$

Furthermore if we choose the special sequence

$$h_i = h_0 2^{-i}, \tag{4.36}$$

then the recurrence becomes

$$T_{ij} = \frac{2^{-j} T_{i-1,j-1} - T_{i,j-1}}{2^{-j} - 1}. \tag{4.37}$$

Often in the asymptotic expansion (4.34) the odd powers of h are missing, and

$$T(h) = a_0 + a_2 h^2 + a_4 h^4 + \cdots \tag{4.38}$$

holds. Such an example is the *asymptotic expansion of the trapezoidal rule*, see (9.30). In this case it is advantageous to extrapolate with a polynomial in the variable $x = h^2$. This way we obtain faster approximations of (4.38) of higher order. Instead of (4.35) we then use

$$T_{ij} = \frac{h_i^2 T_{i-1,j-1} - h_{i-j}^2 T_{i,j-1}}{h_i^2 - h_{i-j}^2}. \tag{4.39}$$

Moreover, if we use the sequence (4.36) for h_i, we obtain the recurrence

$$T_{ij} = \frac{4^{-j} T_{i-1,j-1} - T_{i,j-1}}{4^{-j} - 1}, \tag{4.40}$$

which is used in the algorithm of *Romberg* for integration (see (9.34) in Chapter 9).

Note that the Aitken-Neville extrapolation process is the same as *Richardson Extrapolation* (see Equation (9.35) in Chapter 9).

For the special choice of the sequence h_i according to (4.36) we obtain the following extrapolation algorithm:

ALGORITHM 4.9. *Extrapolation with Aitken-Neville*

```
function A=AitkenExtrapolation(T,h0,tol,factor);
% AITKENEXTRAPOLATION Aitken-Neville scheme for extrapolation
%    A=AitkenExtrapolation(T,h0,tol,factor); computes the
%    Aitken-Neville Scheme for the function T and h_i=h0/2^i until the
%    relative error of two diagonal elements is smaller than tol. The
%    parameter factor is 2 or 4, depending on the asymptotic expansion
%    of the function T.

h=h0; A(1,1)=T(h);
for i=2:15
  h=h/2; A(i,1)=T(h); vhj=1;
  for j=2:i
    vhj=vhj/factor;
    A(i,j)=(vhj*A(i-1,j-1)-A(i,j-1))/(vhj-1);
  end;
  if abs(A(i,i)-A(i-1,i-1))<tol*abs(A(i,i)), return
  end
end
warning(['limit of extrapolation steps reached. ', ...
         'Required tolerance may not be met.']);
```

EXAMPLE 4.6. *As shown in Chapter 2 in Figure 2.3 the difference quotient*

$$T(h) = \frac{f(x+h) - f(x)}{h} \tag{4.41}$$

is an approximation for $a_0 = f'(x)$ since $\lim_{h \to 0} T(h) = f'(x)$. If we expand the function f

$$f(x+h) = f(x) + \frac{f'(x)}{1!} h + \frac{f''(x)}{2!} h^2 + \cdots$$

and insert the series into (4.41), we obtain

$$T(h) = f'(x) + \frac{f''(x)}{2!}h + \frac{f'''(x)}{3!}h^2 + \cdots .$$

All the powers of h occur in this asymptotic expansion. If we use the sequence (4.36) for h_i, we need to extrapolate with the recurrence (4.37).

However, if we use the symmetric differences quotient

$$T(h) = \frac{f(x+h) - f(x-h)}{2h},$$

then we get

$$T(h) = f'(x) + \frac{f'''(x)}{3!}h^2 + \frac{f^{(5)}(x)}{5!}h^4 + \cdots \tag{4.42}$$

and we can extrapolate using the recurrence (4.40).

4.3 Piecewise Interpolation with Polynomials

As already shown with Runge's example (see Figure 4.1), the interpolation polynomial may not always produce the result that one would like. The following example demonstrates this clearly.

EXAMPLE 4.7. *The interpolating polynomial through the points*

$$
\begin{array}{c|ccccccc}
x & 1 & 2.5 & 3 & 5 & 13 & 18 & 20 \\
\hline
y & 2 & 3 & 4 & 5 & 7 & 6 & 3
\end{array}
\tag{4.43}
$$

has the graph shown in Figure 4.3.

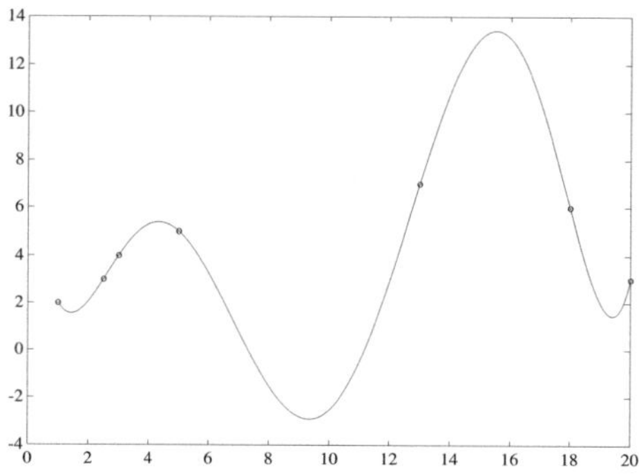

FIGURE 4.3. Undesirable Interpolation Result Leading to Negative Values

If the values in Table 4.43 were points of a function that must be positive by some physical argument, then it would not be desirable to approximate

the function by its interpolation polynomial, since the latter is negative in the interval $(7, 11)$. If we choose to interpolate piecewise with polynomials of lower degrees, e.g. by second degree polynomials for three consecutive points each, then we obtain Figure 4.4. This time the graph of the interpolation

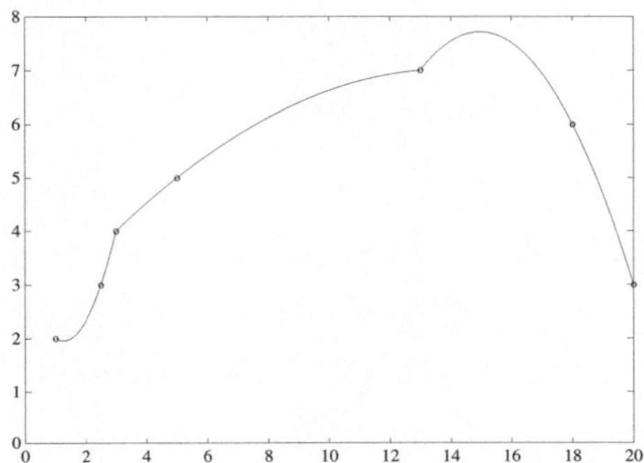

FIGURE 4.4. *Piecewise Interpolation with Parabolas*

function is all positive; however, the derivative is no longer continuous at the points where the polynomial pieces meet.

The idea of *spline interpolation* is to approximate piecewise with polynomials of lower degree in order to obtain a non-oscillating approximation that is also smooth (i.e., as many times differentiable as possible) at the knots where the pieces meet.

4.3.1 Classical Cubic Splines

To avoid discontinuous derivatives, we need to prescribe not only a common function value but also the same value for the derivative at the node where two piecewise polynomials meet. Assume the points

x	x_1	x_2	$\ldots$	x_n
y	y_1	y_2	$\ldots$	y_n

are given. The simplest possibility is to choose each x_i as node, and to interpolate in each interval

$$[x_i, x_{i+1}], \ i = 1, 2, \ldots, n - 1$$

by one polynomial P_i. The index i does not refer here to the degree of the polynomial but to the number of the interval. The polynomial P_i should satisfy the following conditions:

$$
\begin{array}{llll}
P_i(x_i) &= y_i, & P_i(x_{i+1}) &= y_{i+1} \qquad \text{interpolate} \\
P_i'(x_i) &= y_i', & P_i'(x_{i+1}) &= y_{i+1}' \qquad \text{continuous derivative}
\end{array}
\tag{4.44}
$$

Of course we still have to determine what our derivatives y_i' should be, since in general they are not given. The polynomial P_i has to satisfy four conditions – thus it is uniquely determined if we choose a degree of three. To simplify computations we make a change of variables

$$
t = \frac{x - x_i}{h_i} \quad \text{with} \quad h_i = x_{i+1} - x_i,
\tag{4.45}
$$

where t is a local variable in the i-th interval. Now

$$
Q_i(t) = P_i(x_i + th_i)
\tag{4.46}
$$

is a polynomial of degree three in t. The derivative is

$$
Q_i'(t) = h_i P_i'(x_i + th_i),
\tag{4.47}
$$

and thus the conditions (4.44) become

$$
\begin{array}{llll}
Q_i(0) &= y_i, & Q_i(1) &= y_{i+1}, \\
Q_i'(0) &= h_i y_i', & Q_i'(1) &= h_i y_{i+1}'.
\end{array}
\tag{4.48}
$$

A short computation (see Problem 4.16) yields

$$
\begin{aligned}
Q_i(t) =\ & y_i(1 - 3t^2 + 2t^3) + y_{i+1}(3t^2 - 2t^3) \\
& + h_i y_i'(t - 2t^2 + t^3) + h_i y_{i+1}'(-t^2 + t^3).
\end{aligned}
\tag{4.49}
$$

DEFINITION 4.2. (CUBIC HERMITE POLYNOMIALS, CARDINAL FORM)
The four polynomials

$$
\begin{array}{llll}
H_0^3(t) &= 1 - 3t^2 + 2t^3 & H_1^3(t) &= t - 2t^2 + t^3 \\
H_3^3(t) &= 3t^2 - 2t^3 & H_2^3(t) &= -t^2 + t^3
\end{array}
$$

are called cubic Hermite polynomials, and (4.49) is the cardinal form of the interpolating polynomial.

Q_i could be evaluated by the expression (4.49); however, a more efficient scheme is based on *Hermite interpolation*. For this, one forms differences in the following scheme by subtracting the value above from the one below:

$$
\begin{array}{ccccccc}
 & & h_i y_i' & & & & \\
 & & & \searrow & & & \\
a_0 = y_i & & & & a_2 & & \\
 & \searrow & & \nearrow & & \searrow & \\
 & & a_1 & & & & a_3 \qquad (4.50) \\
 & \nearrow & & \searrow & & \nearrow & \\
y_{i+1} & & & & b & & \\
 & & & \nearrow & & & \\
 & & h_i y_{i+1}' & & & &
\end{array}
$$

This way we obtain the coefficients a_0, a_1, a_2 and a_3 and we can compute

$$Q_i(t) = a_0 + (a_1 + (a_2 + a_3 t)(t - 1))t \tag{4.51}$$

using only 3 multiplications and 8 additions/subtractions. The verification that (4.51) yields the same polynomial as (4.49) is left as an exercise (see Problem 4.17).

In the next section, we will investigate several possibilities for choosing the derivatives y_i'. Assuming that these are known, and therefore also the polynomials Q_i for $i = 1, \ldots, n - 1$, the composed global function g is called a *cubic spline function*. To interpolate with g for a value $x = z$, we proceed in three steps:

1. Determine the interval which contains z, i.e. compute the index i for which $x_i \le z < x_{i+1}$.

2. Compute the local variable $t = (z - x_i)/(x_{i+1} - x_i)$.

3. Evaluate $g(z) = Q_i(t)$ with (4.50) and (4.51).

To find the interval which contains z, we can use a *binary search*. This is done in the following MATLAB function `SplineInterpolation`.

ALGORITHM 4.10. *Generic Cubic Spline*

```
function g=SplineInterpolation(x,y,ys,z);
% SPLINEINTERPOLATION generic interpolation with cubic splines
%   g=SplineInterpolation(x,y,ys,z); interpolates at the scalar
%   location z the data (x,y,ys) with a cubic spline function.
%   Here ys is the desired derivative at x.

n=length(x);
a=1; b=n; i=a;
while a+1~=b,
  i=floor((a+b)/2);
  if x(i)<z, a=i; else b=i; end
end
i=a; h=(x(i+1)-x(i));
t=(z-x(i))/h;
a0=y(i); a1=y(i+1)-a0; a2=a1-h*ys(i);
a3= h*ys(i+1)-a1; a3=a3-a2;
g=a0+(a1+(a2+a3*t)*(t-1))*t;
```

4.3.2 Derivatives for the Spline Function

As we saw in the previous section, we need derivatives at the nodes in order to construct a spline function. In principle, we could prescribe any value for them; however, it makes sense to estimate the derivatives from the given function values. A simple estimate for the derivative at point (x_i, y_i) is given by the slope of the straight line through the neighboring points (see Figure 4.5),

$$y_i' = \frac{y_{i+1} - y_{i-1}}{x_{i+1} - x_{i-1}} = \frac{y_{i+1} - y_{i-1}}{h_i + h_{i-1}} \quad i = 2, 3, \ldots, n - 1. \qquad (4.52)$$

Using (4.52), we can compute derivatives for all inner nodes. For the two

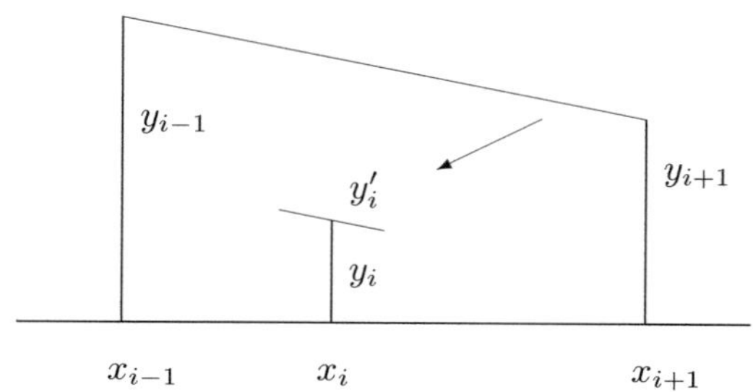

FIGURE 4.5. *Slope of the line through neighboring points*

boundary points, we have several possibilities:

1. Use the slope of the line through the first two (respectively the last two) points

$$y_1' = \frac{y_2 - y_1}{x_2 - x_1} \quad \text{respectively} \quad y_n' = \frac{y_n - y_{n-1}}{x_n - x_{n-1}}.$$

With MATLAB the derivatives `ys` for this case can be computed by

```
n=length(x);
ys=(y(3:n)-y(1:n-2))./(x(3:n)-x(1:n-2));
ys=[(y(2)-y(1))/(x(2)-x(1)); ys; (y(n)-y(n-1))/(x(n)-x(n-1))];
```

2. *Natural boundary conditions*: they are defined such that the second derivative vanishes. From the equations

$$Q_1''(0) = 0 \quad \text{and} \quad Q_{n-1}''(1) = 0$$

we obtain

$$y_1' = \frac{3}{2}\frac{y_2 - y_1}{h_1} - \frac{1}{2}y_2' \quad \text{respectively} \quad y_n' = \frac{3}{2}\frac{y_n - y_{n-1}}{h_{n-1}} - \frac{1}{2}y_{n-1}'.$$

3. An important special case are *periodic boundary conditions*. If the function values belong to a periodic function then $y_1 = y_n$ and we can choose

$$y_1' = y_n' = \frac{y_2 - y_{n-1}}{h_1 + h_{n-1}}.$$

This corresponds again to the slope of the line through neighboring points.

Spline functions which are computed with these constructed derivatives are called *defective spline functions*.

Let us now interpolate again the points given in Table 4.43. Using the slopes of the neighboring points and the natural boundary conditions we obtain the defective spline of Figure 4.6. We have also plotted the derivatives g' and g''. We see that g' (dotted line) is continuous while g'' (dashed line) is discontinuous.

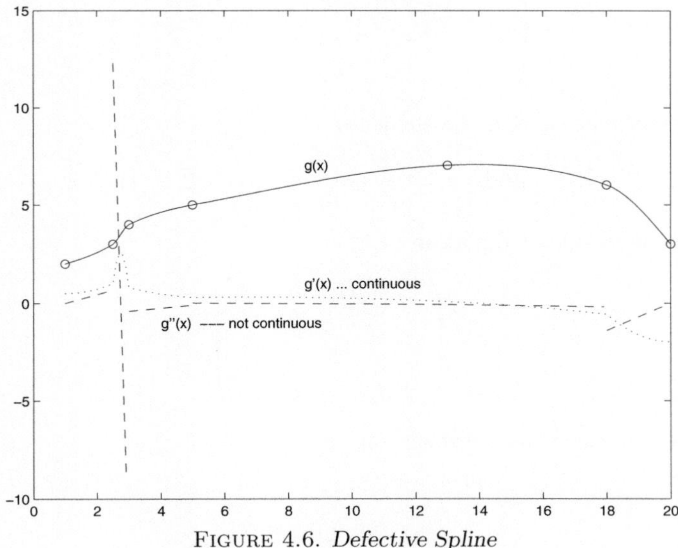

FIGURE 4.6. *Defective Spline*

The question is whether it is possible to choose the derivatives y_i' in such a way that the second derivative g'' will also be continuous. We would like to have

$$P_i''(x_{i+1}) = P_{i+1}''(x_{i+1}) \quad \text{for} \quad i = 1, 2, \ldots, n-2. \tag{4.53}$$

Written in the local variable t with $Q_i(t) = P_i(x_i + h_i t)$, the conditions (4.53) become

$$\frac{Q_i''(1)}{h_i^2} = \frac{Q_{i+1}''(0)}{h_{i+1}^2} \quad \text{for} \quad i = 1, 2, \ldots, n-2. \tag{4.54}$$

If we differentiate Q_i in (4.49) we get

$$Q_i''(t) = y_i(-6+12t) + y_{i+1}(6-12t) + h_i y_i'(-4+6t) + h_i y_{i+1}'(-2+6t), \quad (4.55)$$

and if we insert this into (4.54), we obtain

$$\frac{6}{h_i^2}(y_i - y_{i+1}) + \frac{2}{h_i}y_i' + \frac{4}{h_i}y_{i+1}' = \frac{6}{h_{i+1}^2}(y_{i+2} - y_{i+1}) - \frac{4}{h_{i+1}}y_{i+1}' - \frac{2}{h_{i+1}}y_{i+2}'. \quad (4.56)$$

Equation (4.56) is a linear equation for the unknown derivatives y_i', y_{i+1}' and y_{i+2}'. We can write such an equation for $i = 1, 2, \ldots, n-2$. In matrix notation, we get a linear system of $n-2$ equations with n unknowns,

$$A\mathbf{y}' = \mathbf{c}, \quad (4.57)$$

with $\mathbf{y}' = (y_1', y_2', \ldots, y_n')^\top$ being the vector of unknowns and

$$A = \begin{pmatrix} b_1 & a_1 & b_2 & & & \\ & b_2 & a_2 & b_3 & & \\ & & \ddots & \ddots & \ddots & \\ & & & b_{n-2} & a_{n-2} & b_{n-1} \end{pmatrix}$$

a *tridiagonal matrix* with the elements

$$b_i = \frac{1}{h_i} \quad \text{and} \quad a_i = \frac{2}{h_i} + \frac{2}{h_{i+1}} \quad i = 1, 2, \ldots, n-2.$$

The right hand side of Equation (4.57) is

$$\mathbf{c} = \begin{pmatrix} 3(d_2 + d_1) \\ 3(d_3 + d_2) \\ \vdots \\ 3(d_{n-1} + d_{n-2}) \end{pmatrix},$$

where we have used the abbreviation

$$d_i = \frac{y_{i+1} - y_i}{h_i^2} \quad i = 1, 2, \ldots, n-1. \quad (4.58)$$

Thus we have obtained $n-2$ linear equations for the unknown derivatives. We need two more equations to determine them uniquely. Just as for the defective splines, we can ask for further boundary conditions. We consider three possibilities:

1. *Natural boundary conditions:* $P_1''(x_1) = P_{n-1}''(x_n) = 0$. These conditions give two more equations:

$$\frac{2}{h_1}y_1' + \frac{1}{h_1}y_2' = 3d_1,$$

$$\frac{1}{h_{n-1}}y_{n-1}' + \frac{2}{h_{n-1}}y_n' = 3d_{n-1}, \quad (4.59)$$

which are obtained from $Q_1''(0) = Q_{n-1}''(1) = 0$ using (4.55). If we add them to the system of equations (4.57), then we can compute the derivatives y_i' by solving a linear system of equations with a *tridiagonal matrix*. Using the tridiagonal solver `Thomas` (see Algorithm 3.10 in Chapter 3), we obtain the derivatives in this case with the statements

```
h=x(2:n)-x(1:n-1);
a=2./h(1:n-2)+2./h(2:n-1);
b=1./h(1:n-1);
aa=[2/h(1); a; 2/h(n-1)]
bb=3*[d(1); d(2:n-1)+d(1:n-2); d(n-1)]
ys=Thomas(b,aa,b,bb)
```

The spline function $s : [x_1, x_n] \rightarrow \mathbb{R}$ that is determined piecewise by the $P_i(x)$, $i = 1, 2, \ldots, n-1$ this way is a simplified model of the shape of a thin wooden spline that passes through the given points, as the following theorem shows.

THEOREM 4.9. *For a given set of points (x_i, y_i), $i = 1, 2, \ldots, n$, let $s : [x_1, x_n] \rightarrow \mathbb{R}$ be the classical cubic spline function satisfying $s(x_i) = y_i$, $i = 1, 2, \ldots, n$. Then for any twice continuously differentiable function $f : [x_1, x_n] \rightarrow \mathbb{R}$ satisfying $f(x_i) = y_i$, $i = 1, 2, \ldots, n$ and*

$$s''(x_n)(f'(x_n) - s'(x_n)) = s''(x_1)(f'(x_1) - s'(x_1)), \qquad (4.60)$$

we have that

$$\int_{x_1}^{x_n} (s''(x))^2 dx \leq \int_{x_1}^{x_n} (f''(x))^2 dx, \qquad (4.61)$$

i.e. the spline function $s(x)$ minimizes the energy defined by the integral in (4.61)[1].

PROOF. Let $s(x)$ be the minimizer of $\int_{x_1}^{x_n} (s''(x))^2 dx$. Every twice continuously differentiable function passing through the points (x_i, y_i), $i = 1, 2, \ldots, n$ can be written in the form

$$f(x) := s(x) + \epsilon h(x),$$

where $\epsilon \in \mathbb{R}$ and $h(x)$ is a twice continuously differentiable function with zeros at x_i, $h(x_i) = 0$, $i = 1, 2, \ldots, n$. The minimality condition (4.61) becomes

$$\int_{x_1}^{x_n} (s''(x))^2 dx \leq \int_{x_1}^{x_n} (s''(x) + \epsilon h''(x))^2 dx$$

$$= \int_{x_1}^{x_n} (s''(x))^2 dx + 2\epsilon \int_{x_1}^{x_n} s''(x)h''(x)dx + \epsilon^2 \int_{x_1}^{x_n} h''(x)^2 dx.$$

---

[1] The correct energy integral of a thin wooden spline would actually be $\int_{x_1}^{x_n} s''(x)^2/(1+ s'(x)^2)^{5/2}dx$, but is too difficult to treat mathematically, see [25]: "Die Extremaleigenschaft des interpolierenden Splines wird häufig für die grosse praktische Nützlichkeit der Splines verantwortlich gemacht. Dies ist jedoch glatter 'Volksbetrug'."

For a fixed function $h(x)$, this condition is satisfied for all ϵ if and only if

$$\int_{x_1}^{x_n} s''(x)h''(x)dx = 0, \tag{4.62}$$

as one can see by differentiation. Integration by parts leads to

$$s''(x)h'(x)\big|_{x_0}^{x_n} - \int_{x_1}^{x_n} s'''(x)h'(x)dx = 0. \tag{4.63}$$

Now condition (4.60) implies that the boundary term in (4.63) vanishes. Since $s'''(x)$ is constant in each interval (x_{i-1}, x_i), $i = 2, 3, \ldots, n$, say equal to the constant C_i, the second term in (4.63) becomes

$$\int_{x_1}^{x_n} s'''(x)h'(x)dx = \sum_{i=2}^{n} C_i \int_{x_{i-1}}^{x_i} h'(x)dx = \sum_{i=2}^{n} C_i(h(x_i) - h(x_{i-1})) = 0,$$

since h vanishes at the nodes, and therefore (4.62) holds, which implies (4.61) and concludes the proof. $\square$

We see that in addition to the natural boundary conditions $s''(x_1) = s''(x_n) = 0$ from (4.60), the so called *clamped boundary conditions* $s'(x_1) = f'(x_1)$ and $s'(x_n) = f'(x_n)$ also lead to an energy minimizing spline among all interpolating functions f with these slopes at the endpoints. When interpolating a function f with a spline, it is better to use clamped boundary conditions, since with free boundary conditions, the approximation order is polluted from the boundary and drops from $O(h^4)$ to $O(h^2)$. As an alternative, one can use the approach described next, which does not require the knowledge of derivatives of f, but then looses the energy minimizing property toward the boundaries.

2. *Not-a-knot condition* of de Boor: Here we want the two polynomials in the first two and in the last two intervals to be the same:

$$P_1(x) \equiv P_2(x) \quad \text{and} \quad P_{n-2}(x) \equiv P_{n-1}(x). \tag{4.64}$$

We get here the two equations:

$$\frac{1}{h_1}y_1' + \left(\frac{1}{h_1} + \frac{1}{h_2}\right)y_2' = 2d_1 + \frac{h_1}{h_1 + h_2}(d_1 + d_2),$$

$$\left(\frac{1}{h_{n-2}} + \frac{1}{h_{n-1}}\right)y_{n-1}' + \frac{1}{h_{n-1}}y_n' = 2d_{n-1} + \frac{h_{n-1}}{h_{n-1} + h_{n-2}}(d_{n-1} + d_{n-2}). \tag{4.65}$$

At first sight condition (4.64) might look strange. The motivation is as follows. If the function values y_i are equidistant with step-size h and belong to a smooth differentiable function f, then according to the error estimate (4.8) we would expect an interpolation error $\sim h^4$. It can be shown in this case for a spline function with natural boundary conditions that the interpolation error is $\sim h^2$. The natural boundary condition is not at all *natural* from the viewpoint of approximating a function. In fact it is not clear why the function f that we wish to approximate should always have a vanishing second derivative at the endpoints. De Boor's Condition (4.64) yields an approximation $\sim h^4$ [24].

To obtain (4.65), it is sufficient to demand that the first two (respectively the last two) polynomials have the same third derivative. Since the third derivative of a cubic polynomial is constant, the two polynomials must be the same. For the first two polynomials, the equation

$$P_1'''(x_2) = P_2'''(x_2)$$

is equivalent to

$$\frac{Q_1'''(1)}{h_1^3} = \frac{Q_2'''(0)}{h_2^3},$$

which is

$$\frac{1}{h_1^2}y_1' + \left(\frac{1}{h_1^2} - \frac{1}{h_2^2}\right)y_2' - \frac{1}{h_2^2}y_3' = 2\left(\frac{d_1}{h_1} - \frac{d_2}{h_2}\right). \tag{4.66}$$

Equation (4.66) can be added to the system (4.57) as its first equation. Similarly, the second equation of (4.64) produces an equation that is added as the last equation to the system (4.57). Thus, we obtain once again n equations with n unknowns. Unfortunately, the matrix is no longer tridiagonal. If we wish to solve the system with a tridiagonal solver, we have to replace (4.66) by an equivalent one which contains only the unknowns y_1' and y_2'. Denote by (I) Equation (4.66) and by (II) the first equation of system (4.57). Both equations contain the unknowns y_1', y_2' and y_3'. With the linear combination

$$(I) + \frac{1}{h_2}(II)$$

we eliminate the unknown y_3' and obtain after the division with $\left(\frac{1}{h_1} + \frac{1}{h_2}\right)$ the equation

$$\frac{1}{h_1}y_1' + \left(\frac{1}{h_1} + \frac{1}{h_2}\right)y_2' = 2d_1 + \frac{h_1}{h_1 + h_2}(d_1 + d_2), \tag{4.67}$$

which we use now instead of (4.66) to preserve the tridiagonal structure. Similarly we obtain the second equation of (4.65).

To solve a linear system with a tridiagonal matrix we make again use of the function **Thomas** (see Algorithm 3.10).

Note that the MATLAB-function `YY=spline(X,Y,XX)` provides in `YY`, the values of the interpolating function at `XX`. The spline is ordinarily constructed using the not-a-knot end conditions.

3. *Periodic boundary conditions*: if y_i are function values of a periodic function, then $y_1 = y_n$. In addition, we require that the first and second derivatives be the same:

$$
\begin{aligned}
P_1'(x_1) &= P_{n-1}'(x_n), \\
P_1''(x_1) &= P_{n-1}''(x_n).
\end{aligned}
\tag{4.68}
$$

The first condition in (4.68) is equivalent to

$$
y_1' = y_n',
\tag{4.69}
$$

and the second, when expressed in Q_i, becomes

$$
\frac{Q_1''(0)}{h_1^2} = \frac{Q_{n-1}''(1)}{h_{n-1}^2},
$$

which yields the equation

$$
2\left(\frac{1}{h_1} + \frac{1}{h_{n-1}}\right) y_1' + \frac{1}{h_1} y_2' + \frac{1}{h_{n-1}} y_{n-1}' = 3(d_1 + d_{n-1}).
\tag{4.70}
$$

If we use (4.69) to eliminate the unknown y_n', we obtain a system of linear equations $B\boldsymbol{y}' = \boldsymbol{c}$ with $(n-1)$ unknowns and $(n-1)$ equations. The matrix has the form

$$
B =
\begin{pmatrix}
a_0 & b_1 & 0 & \cdots & & 0 & b_{n-1} \\
b_1 & a_1 & b_2 & 0 & \cdots & & 0 \\
0 & \ddots & \ddots & \ddots & & & \vdots \\
\vdots & & \ddots & \ddots & \ddots & & 0 \\
0 & & & \ddots & \ddots & & b_{n-2} \\
b_{n-1} & 0 & \cdots & & 0 & b_{n-2} & a_{n-2}
\end{pmatrix},
\tag{4.71}
$$

where we have again used the abbreviations

$$
b_i = \frac{1}{h_i}, \quad i = 1, 2, \ldots, n-1,
$$

and

$$
a_0 = \frac{2}{h_{n-1}} + \frac{2}{h_1},
$$

$$
a_i = \frac{2}{h_i} + \frac{2}{h_{i+1}}, \quad i = 1, 2, \ldots, n-2.
$$

The right hand side c of the system is computed by the coefficients d_i defined in Equation (4.58),

$$c = \begin{pmatrix} 3(d_1 + d_{n-1}) \\ 3(d_2 + d_1) \\ \vdots \\ 3(d_{n-1} + d_{n-2}) \end{pmatrix}. \tag{4.72}$$

Except for the two matrix elements at the bottom left and top right the matrix would be tridiagonal. In the next section we will see how to solve such a system efficiently.

4.3.3 Sherman–Morrison–Woodbury Formula

Let A be an $n \times n$ matrix and U, V be $n \times p$ with $p \leq n$ (often $p \ll n$). Then every solution x of the linear system

$$(A + UV^T)x = b \tag{4.73}$$

is also a solution of the *augmented system*

$$\begin{aligned} Ax &+ Uy &= b, \\ V^T x &- y &= 0. \end{aligned} \tag{4.74}$$

Now if we assume that the matrices in the following algebraic manipulations are invertible, then by solving the first equation for x, we obtain

$$x = A^{-1}b - A^{-1}Uy. \tag{4.75}$$

Introducing this in the second equation of (4.74), we get

$$y = (I + V^T A^{-1} U)^{-1} V^T A^{-1} b, \tag{4.76}$$

therefore

$$x = A^{-1}b - A^{-1}U(I + V^T A^{-1} U)^{-1} V^T A^{-1} b.$$

But from (4.73) and (4.74), we also have

$$x = (A + UV^T)^{-1}b$$

and

$$y = V^T(A + UV^T)^{-1}b.$$

Equating both expressions for x and y with the expressions (4.75) and (4.76) we get the matrix equations:

$$\begin{aligned} V^T(A + UV^T)^{-1} &= (I + V^T A^{-1} U)^{-1} V^T A^{-1}, \\ (A + UV^T)^{-1} &= A^{-1} - A^{-1}U(I + V^T A^{-1} U)^{-1} V^T A^{-1}. \end{aligned} \tag{4.77}$$

Equation (4.77) is called the *Shermann–Morrison–Woodbury Formula.* The (small) $p \times p$ matrix $I + V^T A^{-1} U$ is called *capacitance matrix.* The Sherman–Morrison–Woodbury formula is useful for computing the inverse of a *rank-p change* of the matrix A. Let us consider a few applications:

1. If A is sparse and/or $A\boldsymbol{v} = \boldsymbol{b}$ is easy to solve, then it pays to use the Sherman–Morrison–Woodbury formula to compute the solution of $(A + UV^T)\boldsymbol{x} = \boldsymbol{b}$. The algorithm is

   (a) Solve $A\boldsymbol{y} = \boldsymbol{b}$.

   (b) Compute the $n \times p$ matrix W by solving $AW = U$. This can be combined with the first step by simultaneously solving linear systems with the same coefficient matrix A with $p + 1$ right hand sides.

   (c) Form the capacitance matrix $C = I + V^T W \in \mathbb{R}^{p \times p}$ and solve the linear system $C\boldsymbol{z} = V^T \boldsymbol{y}$.

   (d) the solution is $\boldsymbol{x} = \boldsymbol{y} - W\boldsymbol{z}$.

2. Rank-1 change of the identity matrix. The Sherman-Morrison-Woodbury formula becomes

$$(I + \boldsymbol{u}\boldsymbol{v}^T)^{-1} = I - \frac{1}{1 - \boldsymbol{v}^T\boldsymbol{u}}\boldsymbol{u}\boldsymbol{v}^T.$$

If we have to solve a linear system $B\boldsymbol{x} = \boldsymbol{b}$ with $B = I + \boldsymbol{u}\boldsymbol{v}^T$, then the solution can be computed in $O(n)$ operations as a linear combination of the vectors $\boldsymbol{b}$ and $\boldsymbol{u}$:

$$\boldsymbol{x} = \boldsymbol{b} - \frac{\boldsymbol{v}^T\boldsymbol{b}}{1 - \boldsymbol{v}^T\boldsymbol{u}}\boldsymbol{u}.$$

3. Splines with periodic boundary conditions. The matrix B defined in Equation (4.71) is a rank-1 change of a tridiagonal matrix. With $\boldsymbol{e} = \boldsymbol{e}_1 + \boldsymbol{e}_{n-1} = (1, 0, \ldots, 0, 1)^T$ we have

$$B = A + \frac{1}{h_{n-1}}\boldsymbol{e}\boldsymbol{e}^T$$

with the tridiagonal matrix

$$A = \begin{pmatrix} \tilde{a}_0 & b_1 & & & & \\ b_1 & a_1 & b_2 & & & \\ & b_2 & \ddots & & \ddots & \\ & & & \ddots & a_{n-3} & b_{n-2} \\ & & & & b_{n-2} & \tilde{a}_{n-2} \end{pmatrix}.$$

The coefficients a_i and b_i are the same as given in Equation (4.71), but

$$\tilde{a}_0 = \frac{1}{h_{n-1}} + \frac{2}{h_1}, \quad \tilde{a}_{n-2} = \frac{2}{h_{n-2}} + \frac{1}{h_{n-1}}.$$

The solution of the linear system $B\boldsymbol{y}' = \boldsymbol{c}$ for the derivatives requires three steps. We make use of the function Thomas (Algorithm 3.10) to solve the linear systems with tridiagonal matrices:

(a) Solve $A\boldsymbol{u} = \boldsymbol{e}$ with Thomas.

(b) Solve $A\boldsymbol{v} = \boldsymbol{c}$ with Thomas.

(c) $\boldsymbol{y}' = \boldsymbol{v} - \dfrac{v_1 + v_{n-1}}{u_1 + u_{n-1} + h_{n-1}}\boldsymbol{u}.$

4.3.4 Spline Curves

Given n points in the plane (x_i, y_i) for $i = 1, 2, \ldots, n$, we would like to connect them by a curve. The numbering of the points is crucial; reordering them will give us another curve.

Plane curves are represented by parametric functions

$$(x(s), y(s)) \quad \text{with} \quad s_1 \le s \le s_n.$$

We can interpret the given points as function values for some (yet to be determined) parametrization

$$x(s_i) = x_i, \quad y(s_i) = y_i \quad i = 1, 2, \ldots, n.$$

The sequence $\{s_i\}$ of the parameter values can be chosen arbitrarily, we only have to pay attention to monotonicity that means the sequence must be strictly increasing $s_i < s_{i+1}$. Often the parameter is chosen to be the arc length, therefore it seems reasonable to parametrize using the distance between successive points

$$\begin{aligned} s_1 &= 0, \\ s_{i+1} &= s_i + \sqrt{(x_{i+1} - x_i)^2 + (y_{i+1} - y_i)^2}, \\ & \quad i = 1, 2, \ldots, n - 1. \end{aligned} \tag{4.78}$$

After computing the sequence $\{s_i\}$ according to (4.78) we have the data

s	s_1	s_2	$\ldots$	s_n
x	x_1	x_2	$\ldots$	x_n

for $x(s)$

and

s	s_1	s_2	$\ldots$	s_n
y	y_1	y_2	$\ldots$	y_n

for $y(s)$.

Both functions $x(s)$ and $y(s)$ can now be interpolated with any of the variants for spline interpolation and the curve can then be plotted. For closed curves, it is important to use the periodicity condition to avoid a cusp at the endpoints.

EXAMPLE 4.8. *For the points*

x	1.31	2.89	5.05	6.67	3.12	2.05	0.23	3.04	1.31
y	7.94	5.50	3.47	6.40	3.77	1.07	3.77	7.41	7.94

we obtain the curve of Figure 4.7 by using defective splines with periodic boundary conditions.

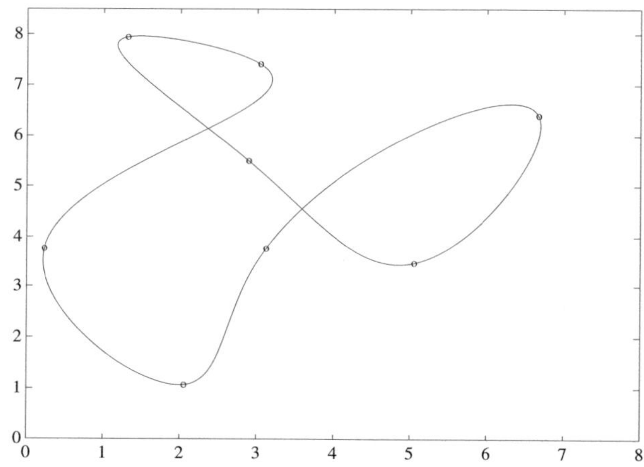

FIGURE 4.7. *Spline Curve*

4.4   Trigonometric Interpolation

In signal processing, it is often useful to study the different representations of a signal $f(t)$. For sound signals, it is usually the case that $f(t)$ is periodic (or nearly periodic) with period T, i.e., when $f(t + T) = f(t)$ for all $t \in \mathbb{R}$. In such cases, one often considers a *frequency* or *Fourier decomposition* of f:

$$f(t) \sim a_0 + \sum_{k=1}^{\infty} a_k \cos\left(\frac{2\pi kt}{T}\right) + \sum_{k=1}^{\infty} b_k \sin\left(\frac{2\pi kt}{T}\right), \qquad (4.79)$$

where $a_0 = \frac{1}{T} \int_0^T f(t)\, dt$ and

$$a_k = \frac{2}{T} \int_0^T f(t) \cos\left(\frac{2\pi kt}{T}\right) dt, \qquad b_k = \frac{2}{T} \int_0^T f(t) \sin\left(\frac{2\pi kt}{T}\right) dt$$

for $k \geq 1$ are the *real Fourier coefficients* of f. Through the change of variable $x = 2\pi t/T$, we can assume without loss of generality that $T = 2\pi$, i.e., we can consider 2π-periodic functions only. Note that in (4.79), we write $\sim$ instead of $=$ since it is a formal expansion, and we have not considered for a given t whether the infinite series converges or, when it does, whether it converges to $f(t)$ or to some other value.

From a practical point of view, the Fourier decomposition is useful because

(i) For sound signals, the functions $\sin(2\pi kt/T)$ and $\cos(2\pi kt/T)$ correspond to *harmonics* of the fundamental frequency $1/T$ and are naturally interpreted as pitch by the human ear. For other types of signals (e.g., electric signals), sinusoids of different frequencies also have natural interpretations.

(ii) The amplitude for higher frequencies tends to be much smaller than for low frequencies. Thus, it is possible to *compress the signal* by setting the high frequency coefficients to zero without dramatically affecting the quality of the signal.

REMARK. *The second point above actually has a theoretical basis: by the Riemann–Lebesgue lemma, for any integrable function $f : [a, b] \to \mathbb{R}$, we have*

$$\lim_{k\to\infty} \int_a^b f(t)\sin(kt)\,dt = 0, \qquad \lim_{k\to\infty} \int_a^b f(t)\cos(kt)\,dt = 0.$$

(For a proof, see for instance [108], p. 103). This implies the Fourier coefficients a_k and b_k both decay to zero for large k, although the decay rate depends on the smoothness properties of f. This is discussed in more detail in Section 4.4.3.

On digital computers, signals are usually represented as a discrete sequence of equidistant *samples*, i.e., instead of having $f(t)$ for all $t \in \mathbb{R}$, we instead have $y_j = f(t_j)$ for $t_j = jh$, where $h > 0$ is fixed and $nh = T$ for some integer value n. In other words, we have an n-periodic sequence $\{y_j\}$ that is supposed to represent $f(t)$ in some way. It is thus natural to ask the following questions:

1. How does one construct a 2π-periodic function $p(t)$ from the points $\{(t_j, y_j)\}_{j\in\mathbb{Z}}$?

2. Is the reconstructed function $p(t)$ identical to $f(t)$, and if not, how large can the error $|p(t) - f(t)|$ be?

3. How does one calculate the Fourier decomposition of $p(t)$ efficiently?

4. What is the relationship between the Fourier decomposition of $f(t)$ and $p(t)$?

4.4.1   Trigonometric Polynomials

Let $\{y_j\}_{j=0}^{2n-1}$ be an $2n$-periodic sequence. We define a *trigonometric polynomial of degree $2n$* by

$$
p_{2n}(x) = a_0 + \sum_{k=1}^{n} a_k \cos(kx) + \sum_{k=1}^{n-1} b_k \sin(kx).
$$

We would like to know whether there exists a trigonometric polynomial of degree $2n$ such that $p_{2n}(x_j) = y_j$ for $x_j = j\pi/n$, $j = 0, 1, \ldots, 2n - 1$, and whether such a polynomial is unique. To do so, we first rewrite the polynomial in *complex* form: for $-n \leq k \leq n$, we let

$$
c_k = \begin{cases}
\frac{1}{2}(a_k - ib_k), & 1 \leq k \leq n - 1, \\
\frac{1}{2}(a_k + ib_k), & -n + 1 \leq k \leq -1, \\
a_0, & k = 0, \\
a_n, & k = \pm n.
\end{cases}
$$

Using this definition, we calculate

$$
\begin{aligned}
p_{2n}(x) &= c_0 + \frac{1}{2}\sum_{k=1}^{n} a_k(e^{ikx} + e^{-ikx}) + \frac{1}{2i}\sum_{k=1}^{n-1} b_k(e^{ikx} - e^{-ikx}) \\
&= c_0 + \frac{1}{2}a_n(e^{inx} + e^{-inx}) + \sum_{k=1}^{n-1} \frac{a_k - ib_k}{2}e^{ikx} + \sum_{k=1}^{n-1} \frac{a_k + ib_k}{2}e^{-ikx} \\
&= \frac{1}{2}c_{-n}e^{-inx} + \sum_{k=-n+1}^{n-1} c_k e^{ikx} + \frac{1}{2}c_n e^{inx}.
\end{aligned}
$$

For later convenience, we introduce the shorthand notation

$$
p_{2n}(x) = \sideset{}{'}\sum_{k=-n}^{n} c_k e^{ikx},
$$

where the prime indicates that a weight of $\frac{1}{2}$ is needed for the first and last terms. Observe that we have an interpolation problem similar to the polynomial case: we would like to know whether, for any sequence $y_0, y_1, \ldots, y_{2n-1} \in \mathbb{C}$, there exists a unique set of coefficients $c_{-n}, c_{-n+1}, \ldots, c_n$ such that

$$
p_{2n}(x_j) = y_j = \sideset{}{'}\sum_{k=-n}^{n} c_k e^{ikx_j} = \sideset{}{'}\sum_{k=-n}^{n} c_k e^{\pi i j k/n}, \qquad j = 0, 1, \ldots, 2n - 1.
$$

To answer this question, first note that $e^{\pi i j k/n} = e^{\pi i(j+2n)k/n}$ is a $2n$-periodic sequence and that $c_n = c_{-n}$. Thus, if we extend the finite sequence

$c_{-n}, \ldots, c_{n-1}$ to a $2n$-periodic sequence, then we can write

$$y_j = \sum_{k=0}^{2n-1} c_k e^{\pi i j k/n},$$

since the terms for $k = \pm n$ are identical and can be combined. Thus, in matrix form, we have

$$\underbrace{\begin{pmatrix} 1 & 1 & \cdots & 1 \\ 1 & \omega_{2n} & \cdots & \omega_{2n}^{2n-1} \\ 1 & \omega_{2n}^2 & \cdots & \omega_{2n}^{2(2n-1)} \\ \vdots & \vdots & & \vdots \\ 1 & \omega_{2n}^{2n-1} & \cdots & \omega_{2n}^{(2n-1)^2} \end{pmatrix}}_{V} \underbrace{\begin{pmatrix} c_0 \\ c_1 \\ c_2 \\ \vdots \\ c_{2n-1} \end{pmatrix}}_{c} = \underbrace{\begin{pmatrix} y_0 \\ y_1 \\ y_2 \\ \vdots \\ y_{2n-1} \end{pmatrix}}_{y},$$

or $V\boldsymbol{c} = \boldsymbol{y}$, where $\omega_{2n} = e^{\pi i/n}$. In other words, a unique *trigonometric interpolating polynomial* of degree $2n$ exists if and only if V is non-singular.

THEOREM 4.10. *Let* $\{y_j\}_{j \in \mathbb{Z}}$ *be an* n*-periodic sequence, where* n *is even. Then the unique trigonometric polynomial of degree* n *that satisfies* $p_n(x_j) = y_j$ *for* $x_j = 2\pi j/n$ *is given by*

$$p_n(x) = \sum_{k=-n/2}^{n/2}{}' c_k e^{ikx},$$

where $\{c_k\}_{k \in \mathbb{Z}}$ *is the* n*-periodic sequence given by the* **discrete Fourier transform** *(DFT)*

$$c_k = (\mathcal{F}_n y)_k = \frac{1}{n} \sum_{j=0}^{n-1} y_j e^{-2\pi i j k/n}. \tag{4.80}$$

In addition, the y_j *can be obtained from the* c_k *via the* **inverse discrete Fourier transform** *(IDFT)*

$$y_j = (\mathcal{F}_n^{-1} c)_j = \sum_{k=0}^{n-1} c_k e^{2\pi i j k/n}. \tag{4.81}$$

PROOF. We show the existence and uniqueness of an interpolating trigonometric polynomial by explicitly producing an inverse of V. In fact, we show that $V^*V = nI$, where V^* is the conjugate transpose of V and I is the identity matrix. First, we note that the (j, k)th entry of V is $\omega_n^{(j-1)(k-1)}$, which means

$$[V^*V]_{kl} = \sum_{j=1}^{n} \omega_n^{-(k-1)(j-1)} \omega_n^{(j-1)(l-1)} = \sum_{j=0}^{n-1} \omega_n^{j(l-k)} \qquad \text{for } 1 \le k, l \le n.$$

If $k = l$, then each term in the sum is 1, so that $[V^*V]_{kk} = n$. If $k \neq l$, then $\omega_n^{k-l} \neq 1$ (since $1 \leq k, l \leq n$ and $\omega_n^m = 1$ if and only if m is divisible by n). Thus, we can write

$$\sum_{j=0}^{n-1} \omega_n^{j(k-l)} = \frac{1 - \omega_n^{n(k-l)}}{1 - \omega_n^{k-l}} = 0.$$

This shows that V^*V has n on the diagonal and 0 everywhere else, so $V^*V = nI$. Thus, $\boldsymbol{c} = \frac{1}{n}V^*\boldsymbol{y}$, so that

$$c_k = \frac{1}{n}\sum_{j=0}^{n-1} \omega_n^{-jk} y_j = \frac{1}{n}\sum_{j=0}^{n-1} y_j e^{-2\pi i j k/n},$$

as claimed by (4.80). Equation (4.81) is simply another way of writing $p_n(x_j)$ for $j = 0, 1, \ldots, n - 1$. $\qquad\square$

The above proof also shows that the matrix V, when divided by the constant $\sqrt{n}$, becomes an orthogonal matrix, so the problem of calculating the DFT and IDFT are well-conditioned (see Problem 4.20).

4.4.2 Fast Fourier Transform (FFT)

A straightforward implementation of the discrete Fourier transform using (4.80) requires $O(n)$ operations per element in the vector $\boldsymbol{c} = \mathcal{F}_n y$, making a total of $O(n^2)$ operations. However, when $n = 2^m$ is a power of 2, there are algorithms that can calculate $\mathcal{F}_n y$ in $O(n\log_2 n)$ operations only. Such algorithms are collectively known as the "fast" Fourier transform (FFT), and the FFT is listed as one of the *top ten algorithms of the last century* [27]; below we present a recursive version based on the original paper by Cooley and Tukey [18].

We will show only the algorithm for the inverse Fourier transform (4.81), the one for the forward transform being similar. First, let us split the sum in (4.81) into even and odd terms:

$$\begin{aligned}
(\mathcal{F}_n^{-1}y)_k &= \sum_{j=0}^{n-1} y_j e^{2\pi i j k/n} \\
&= \sum_{j=0}^{n/2-1} y_{2j} e^{2\pi i (2j)k/n} + \sum_{j=0}^{n/2-1} y_{2j+1} e^{2\pi i (2j+1)k/n} \\
&= \sum_{j=0}^{n/2-1} y_{2j} e^{\frac{2\pi i j k}{n/2}} + e^{2\pi i k/n} \sum_{j=0}^{n/2-1} y_{2j+1} e^{\frac{2\pi i j k}{n/2}}.
\end{aligned}$$

Notice that the two sums have exactly the same form as the inverse Fourier transform of length $n/2$. Thus, we can write

$$(\mathcal{F}_n^{-1}y)_k = (\mathcal{F}_{n/2}^{-1}y_e)_k + e^{2\pi i k/n}(\mathcal{F}_{n/2}^{-1}y_o)_k, \tag{4.82}$$

where y_e and y_o are the odd and even parts of the n-periodic sequence y respectively. This means we can recursively call our IFFT routine on the two smaller vectors and continue to subdivide the problem until we get a vector of length 1, for which the vector and its inverse Fourier transform are identical. We can then go back up the recursion tree and combine the odd and even elements according to (4.82) until we obtain the inverse Fourier transform of the original vector. The following MATLAB code implements this:

ALGORITHM 4.11. *Inverse Fast Fourier Transform*

```
function y=myIFFT(x)
% MYIFFT inverse fast Fourier transform
%   y=myIFFT(x); computes recursively the inverse Fourier tranform of
%   the vector x whose length must be a power of 2.

n=length(x);
if n==1,
  y=x;
else
  w=exp(2i*pi/n*(0:n/2-1)');
  ze=myIFFT(x(1:2:n-1));
  zo=myIFFT(x(2:2:n));
  y=[ze+w.*zo; ze-w.*zo];
end;
```

To compute the forward Fourier transform, one merely needs to replace $e^{2\pi i k/n}$ by $e^{-2\pi i k/n}$ and divide the end result by n (see Problem 4.18).

EXAMPLE 4.9. *To calculate the inverse Fourier transform of the vector* $[1, 2, 3, 4]$, *we have the following diagram:*

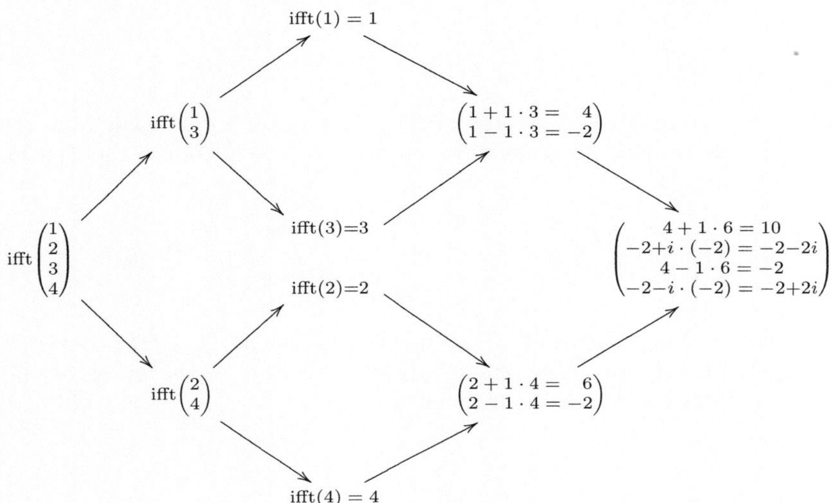

Note that when embedding the even and odd transforms into a larger vector, one needs to extend the shorter vectors periodically.

In the above example, the first half of the diagram contains no arithmetic operations, since the algorithm simply subdivides the problem until vectors of length one are obtained. The latter half of the diagram contains $\log_2 n$ columns; in each column, one must perform a total of n additions and n multiplications. Thus, the total cost of the algorithm is $O(n \log_2 n)$.

REMARK. *The* MATLAB *built-in functions* fft *and* ifft *implement the discrete Fourier transform and its inverse in compiled code. Note that unlike the definitions (4.80) and (4.81), the FFT in* MATLAB *puts the division by* n *in the inverse transform, i.e.,* MATLAB *uses*

$$(\tilde{\mathcal{F}}_n y)_k = \sum_{j=0}^{n-1} y_j e^{-2\pi i j k/n}, \qquad (\tilde{\mathcal{F}}_n^{-1} y)_j = \frac{1}{n} \sum_{k=0}^{n-1} e^{2\pi i j k/n}.$$

4.4.3 Trigonometric Interpolation Error

In the previous section, we have shown how to calculate the discrete Fourier transform of an n-periodic sequence $\{y_j\}_{j \in \mathbb{Z}}$, from which we can construct a trigonometric polynomial $p_n(x)$ such that

$$p_n\left(\frac{2\pi j}{n}\right) = y_j, \qquad j = 0, 1, \ldots, n-1,$$

whenever n is even. Now suppose the y_j are equidistant samples of a 2π-periodic function $f : \mathbb{R} \to \mathbb{C}$, i.e.,

$$p_n\left(\frac{2\pi j}{n}\right) = f\left(\frac{2\pi j}{n}\right) = y_j, \qquad j = 0, 1, \ldots, n-1.$$

What can one say about the interpolation error $p_n(x) - f(x)$ away from the interpolation points? To study this, we first need to introduce the *Fourier series* of $f(x)$

$$f(x) \sim \sum_{j=-\infty}^{\infty} \hat{f}(k) e^{ikx}, \qquad \hat{f}(k) = \frac{1}{2\pi} \int_0^{2\pi} f(x) e^{-ikx}\, dx,$$

which is the complex version of (4.79). We see that if $\sum_{k \in \mathbb{Z}} \hat{f}(k)$ converges absolutely, then by the Weierstrass M-test, the Fourier series converges uniformly and the "$\sim$" becomes an equality, i.e.,

$$f(x) = \sum_{j=-\infty}^{\infty} \hat{f}(k) e^{ikx}.$$

LEMMA 4.3. *Let $f : \mathbb{R} \to \mathbb{C}$ be 2π-periodic and absolutely integrable on $[0, 2\pi]$ and*

$$\hat{f}_n(k) = \frac{1}{n} \sum_{j=0}^{n-1} f\left(\frac{2\pi j}{n}\right) e^{-2\pi i j k/n}$$

be the kth element of the discrete Fourier transform of samples of f. If $\sum_{k \in \mathbb{Z}} \hat{f}(k)$ converges absolutely, then

$$\hat{f}_n(k) = \sum_{j \in \mathbb{Z}} \hat{f}(k + jn).$$

PROOF. If we write $x_l = 2\pi l/n$ and $\omega_n = e^{2\pi i/n}$, then

$$\hat{f}_n(k) = \frac{1}{n} \sum_{l=0}^{n-1} f(x_l) \omega_n^{-kl}$$

$$= \frac{1}{n} \sum_{l=0}^{n-1} \sum_{m=-\infty}^{\infty} \hat{f}(m) e^{imx_l} \omega_n^{-kl} = \sum_{m=-\infty}^{\infty} \hat{f}(m) \left[\frac{1}{n} \sum_{l=0}^{n-1} \omega_n^{(m-k)l} \right].$$

Now if $m - k$ is divisible by n, i.e., if $m = k + jn$ for some $j \in \mathbb{Z}$, then $\omega_n^{m-k} = 1$, so that the term inside square brackets is equal to 1. If, on the other hand, $m - k$ is not divisible by n, then $\omega_n^{m-k} \neq 1$, and an argument similar to the one in the proof in Theorem 4.10 shows that the inner sum vanishes. Thus, we have

$$\hat{f}_n(k) = \sum_{j=-\infty}^{\infty} \hat{f}(k + jn),$$

as required. $\qquad\qquad\qquad\qquad\qquad\qquad\qquad\qquad\qquad\qquad\qquad\square$

REMARK. *The discrete Fourier transform $\hat{f}_n(k)$ can be considered as an approximation of $\hat{f}(k)$ by the* trapezoidal rule *(see Chapter 9, Equation (9.14)): for a 2π-periodic function f, we have*

$$\hat{f}(k) = \frac{1}{2\pi} \int_0^{2\pi} f(x) e^{-ikx} \, dx \approx \frac{1}{2\pi} \cdot \frac{2\pi}{n} \sum_{j=0}^{n-1} \frac{f(x_j) e^{-ikx_j} + f(x_{j+1}) e^{-ikx_{j+1}}}{2}$$

$$= \frac{1}{2n} f(0) + \frac{1}{n} \sum_{j=1}^{n-1} f(x_j) e^{-2\pi i j k/n} + \frac{1}{2n} f(2\pi)$$

$$= \frac{1}{n} {\sum_{j=0}^{n}}' f(x_j) e^{-2\pi i j k/n} = \hat{f}_n(k).$$

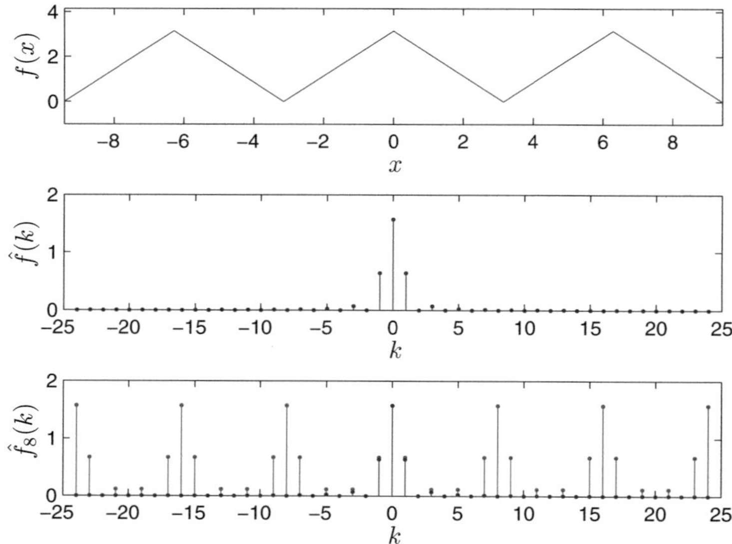

FIGURE 4.8.
Top: A 2π-periodic function $f(x)$. Middle: Fourier
coefficients $\hat{f}(k)$ of f. Bottom: A comparison between
$\hat{f}(k)$ (blue) and the discrete Fourier transform $\hat{f}_8(k)$
using 8 sample points (red). The difference between the
red and blue dots is due to aliasing.

REMARK. *The above lemma shows that the discrete Fourier coefficients
can be obtained by adding copies of the spectrum $\hat{f}(k)$ translated by a multiple
of n. Since $\hat{f}(k) \to 0$ as $|k| \to \infty$, one expects $\hat{f}_n(k)$ to be a poor approxima-
tion for large k, since $\hat{f}_n(k)$ is n-periodic. For low frequencies (small k), we
have*

$$\hat{f}_n(k) - \hat{f}(k) = \sum_{\substack{j \in \mathbb{Z} \\ j \neq 0}} \hat{f}(k + jn),$$

*so the error is a sum of coefficients belonging to high frequencies (above $n/2$
or below $-n/2$). Thus, if n is large enough that the range $[-n/2, n/2]$ contains all but the smallest coefficients, the error would be small; otherwise,
there would be a large error due to "contamination" by the high frequency
components. This phenomenon is known as* aliasing; *see Figure 4.8.*

THEOREM 4.11. (TRIGONOMETRIC INTERPOLATION ERROR) *Let f :
$\mathbb{R} \to \mathbb{C}$ be such that $\sum_{k \in \mathbb{Z}} \hat{f}(k)$ is absolutely convergent, and let $p_n(x) =$*

$\sum'_{|k|\leq n/2} \hat{f}_n(k)e^{ikx}$ *be its interpolating trigonometric polynomial of degree* n *(n even). Then*

$$|p_n(x) - f(x)| \leq 2 \sum'_{|k|\geq n/2} |\hat{f}(k)|.$$

PROOF. We have

$$|p_n(x) - f(x)| = \left| \sum'_{|k|\leq n/2} \hat{f}_n(k)e^{ikx} - \sum_{k\in\mathbb{Z}} \hat{f}(k)e^{ikx} \right|$$

$$= \left| \sum'_{|k|\leq n/2} (\hat{f}_n(k) - \hat{f}(k))e^{ikx} - \sum'_{|k|\geq n/2} \hat{f}(k)e^{ikx} \right|$$

$$\leq \sum'_{|k|\leq n/2} |\hat{f}_n(k) - \hat{f}(k)| + \sum'_{|k|\geq n/2} |\hat{f}(k)|$$

$$\leq \sum'_{|k|\leq n/2} \sum_{\substack{j\in\mathbb{Z}\\ j\neq 0}} |\hat{f}(k+jn)| + \sum'_{|k|\geq n/2} |\hat{f}(k)|.$$

The double sum above is in fact equal to $\sum'_{|m|\geq n/2} |\hat{f}(m)|$, since every m that is not an odd multiple of $n/2$ can be written as $k+jn$ with $|k| < n/2$ and $j \in \mathbb{Z}$ in a unique way. Moreover, there are exactly two ways of writing $k+jn$ when m is an odd multiple of $n/2$ (with $k = \pm n/2$), but the corresponding terms have a weight of $1/2$. Thus, each $\hat{f}(k+jn)$ corresponds to a different $\hat{f}(m)$ except for $k = \pm n/2$, when two terms double up to cancel with the half weights. The condition $j \neq 0$ excludes the range $m \in (-n/2, n/2)$ and leaves the half weights intact for $m = \pm n/2$. Hence, we conclude that

$$|p_n(x) - f(x)| \leq 2 \sum'_{|k|\geq n/2} |\hat{f}(k)|.$$

$\square$

Two immediate applications of the above theorem are as follows:

1. If f is *band-limited*, i.e., if there exists $M > 0$ such that $\hat{f}(k) = 0$ for all $|k| > M$, then $p_n(x) = f(x)$ whenever $n > 2M$. In other words, if we take enough samples of the function $f(x)$, then it is possible to reconstruct $f(x)$ perfectly by interpolation. This result is known as the *sampling theorem* [96].

2. If f is r times continuously differentiable, then by repeatedly integrating

by parts, we get

$$2\pi \hat{f}(k) = \int_0^{2\pi} f(x)e^{-ikx}\,dx = \frac{-1}{ik}\int_0^{2\pi} f'(x)e^{-ikx}\,dx$$

$$= \cdots = \left(\frac{-1}{ik}\right)^r \int_0^{2\pi} f^{(r)}(x)e^{-ikx}\,dx,$$

so that

$$|\hat{f}(k)| \le \frac{1}{|k|^r}\left(\frac{1}{2\pi}\int_0^{2\pi} |f^{(r)}(x)|\,dx\right) = \frac{C}{|k|^r}.$$

Hence,

$$|p_n(x) - f(x)| \le 2\sideset{}{'}\sum_{|k|\ge n/2} |\hat{f}(k)| \le 4C\int_{n/2}^{\infty} \frac{dy}{y^r} \le \frac{\tilde{C}}{n^{r-1}},$$

where $\tilde{C}$ is a constant that depends on the function f and on r, but is independent of k. Thus, the smoother the function f, the faster $p_n(x)$ converges to $f(x)$ as we increase the number of samples n.

REMARK. *For many r times differentiable functions f, one can show that the interpolation error in fact decays as $O(n^{-r})$, rather than the more pessimistic bound $O(n^{-r+1})$ above.*

4.4.4   Convolutions Using FFT

Let $\{y_j\}_{j\in\mathbb{Z}}$ and $\{z_j\}_{j\in\mathbb{Z}}$ be two n-periodic sequences. The *convolution* of $\boldsymbol{y}$ and $\boldsymbol{z}$ is also an n-periodic sequence whose kth element is defined as

$$(\boldsymbol{y} * \boldsymbol{z})_k = \sum_{j=0}^{n-1} y_{k-j}z_j. \tag{4.83}$$

The convolution has many important applications in signal processing. As an example, let $\{y_j\}_{j\in\mathbb{Z}}$ be a given n-periodic sound signal to which we would like to add an echo effect. It is easiest to see how an echo works by looking at how it acts on the *unit impulse* $\boldsymbol{\delta}$, i.e., when

$$\delta_k = \begin{cases} 1, & k \text{ divisible by } n, \\ 0 & \text{else.} \end{cases}$$

When one adds an echo to $\boldsymbol{\delta}$, the impulse is repeated after a fixed delay d, but with an attenuation factor α after each repetition. In other words, the transformed signal $\boldsymbol{z} = H\boldsymbol{\delta}$ looks like

$$z_k = \begin{cases} \alpha^m, & k = md,\ m = 0,1,\ldots,\lfloor\frac{n}{d}\rfloor, \\ 0 & 0 \le k \le n-1,\ k \ne md, \end{cases}$$

$$z_{n+k} = z_k.$$

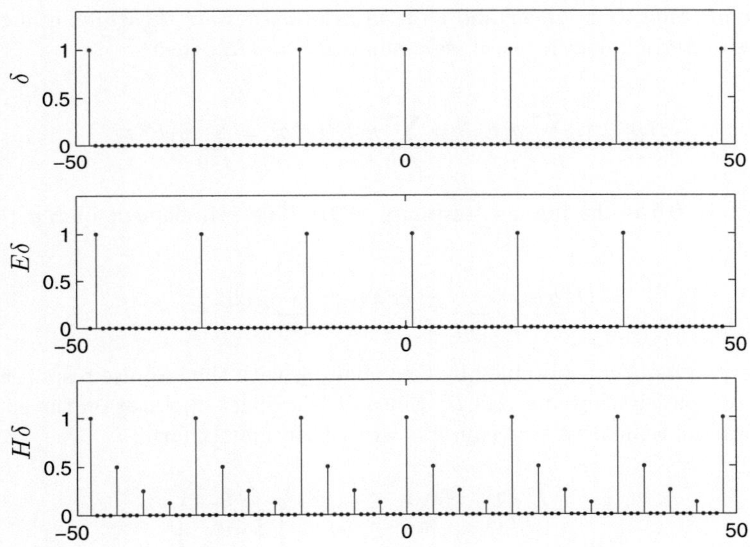

FIGURE 4.9.
*Top: Unit impulse δ for 16-periodic sequences. Middle:
Shifted impulse $E\delta$. Bottom: Impulse response of the
echo operator H, with delay $d = 4$ and attenuation factor
$\alpha = \frac{1}{2}$.*

The signal z is known as the *impulse response* of H, since it is obtained by applying H to the unit impulse δ, see Figure 4.9.

We now consider Hy for a general signal y. Let us introduce the shifting operator E satisfying $(Ez)_k = z_{k-1}$, i.e., it takes a sequence z and shifts it to the right by one position (see Figure 4.9). Then a general sequence y can be written as a sum of shifted impulses

$$y = \sum_{j=0}^{n-1} y_j E^j \delta.$$

Assuming that H is linear and shift invariant (or *time invariant* in signal processing terminology), i.e., if we assume $HE = EH$, then

$$Hy = H \sum_{j=0}^{n-1} y_j E^j \delta = \sum_{j=0}^{n-1} y_j E^j (H\delta) = \sum_{j=0}^{n-1} y_j E^j z,$$

where $z = H\delta$ is the impulse response of H. The kth element of Hy then satisfies

$$(Hy)_k = \sum_{j=0}^{n-1} y_j (E^j z)_k = \sum_{j=0}^{n-1} y_j z_{k-j}.$$

In other words, Hy is obtained by convolving y with the impulse response z, which completely characterizes H. Since H is a linear operator on the space of n-periodic sequences, one can also write H in matrix form:

$$Hy = \begin{pmatrix} z_0 & z_{n-1} & \cdots & z_1 \\ z_1 & z_0 & \cdots & z_2 \\ \vdots & & \ddots & \vdots \\ z_{n-1} & z_{n-2} & \cdots & z_0 \end{pmatrix} \begin{pmatrix} y_0 \\ y_1 \\ \vdots \\ y_{n-1} \end{pmatrix}.$$

The above matrix has constant values along each diagonal that "wraps around" when it hits the right edge; such matrices are called *circulant*. Every circulant matrix can be written as a *convolution*, and vice versa. The fundamental relationship between convolutions and discrete Fourier transforms is given by the following lemma.

LEMMA 4.4. *Let y and z be two periodic sequences. Then*

$$\mathcal{F}_n(y * z) = n(\mathcal{F}_n(y) \odot \mathcal{F}_n(z)),$$

where $\odot$ denotes pointwise multiplication.

Note that the lemma implies that the convolution operator is symmetric, i.e., $y * z = z * y$, despite the apparent asymmetry in the definition.

PROOF. We have

$$(\mathcal{F}_n(\boldsymbol{y} * \boldsymbol{z}))_k = \frac{1}{n} \sum_{j=0}^{n-1} \sum_{l=0}^{n-1} y_l z_{j-l} e^{-2\pi i jk/n}$$

$$= \frac{1}{n} \sum_{j=0}^{n-1} \sum_{l=0}^{n-1} y_l z_{j-l} e^{-2\pi i (j-l)k/n} e^{-2\pi i lk/n}$$

$$= \frac{1}{n} \left(\sum_{l=0}^{n-1} y_l e^{-2\pi i lk/n} \sum_{j=0}^{n-1} z_{j-l} e^{-2\pi i (j-l)k/n} \right)$$

$$= \frac{1}{n} \left(\sum_{l=0}^{n-1} y_l e^{-2\pi i lk/n} \sum_{j=0}^{n-1} z_j e^{-2\pi i jk/n} \right)$$

$$= n(\mathcal{F}_n\boldsymbol{y})_k(\mathcal{F}_n\boldsymbol{z})_k.$$

$\square$

A straightforward computation of the convolution using (4.83) requires $O(n^2)$ operations, but the above lemma suggests the following algorithm using the FFT:

ALGORITHM 4.12.
Computing the convolution of two n-periodic sequences
$\boldsymbol{y}$ *and* $\boldsymbol{z}$

---

1. Use the FFT to compute $\hat{\boldsymbol{y}} = \mathcal{F}_n\boldsymbol{y}$ and $\hat{\boldsymbol{z}} = \mathcal{F}_n\boldsymbol{z}$.

2. Compute $\hat{\boldsymbol{w}} = n(\hat{\boldsymbol{y}} \odot \hat{\boldsymbol{z}})$.

3. Use the IFFT to compute $\boldsymbol{w} = \mathcal{F}_n^{-1}\hat{\boldsymbol{w}}$.

Since steps 1 and 3 each cost $O(n \log n)$ and step 2 costs $O(n)$, the overall cost is $O(n \log n)$, which is much lower than $O(n^2)$ when n is large.

4.5 Problems

PROBLEM 4.1. *Interpolate Runge's function in the interval* $[-5, 5]$,

$$f(x) = \frac{1}{1 + x^2},$$

using Chebyshev nodes (see Equation (11.52)). Compare the resulting interpolation polynomial with the one with the equidistant nodes of Figure 4.1.

PROBLEM 4.2. *Rewrite Algorithm 4.4 to compute the diagonal of the divided difference table using only $O(n)$ storage.* Hint: *To see which entries can be overwritten at which stage, consider the diagram below:*

$$
\begin{array}{c|l}
x_0 & f[x_0] \\
 & \qquad\searrow \\
x_1 & f[x_1] \to f[x_0,x_1] \\
 & \qquad\searrow \qquad\qquad\searrow \\
x_2 & f[x_2] \to f[x_1,x_2] \to f[x_0,x_1,x_2] \\
 & \qquad\searrow \qquad\qquad\searrow \qquad\qquad\searrow \\
x_3 & f[x_3] \to f[x_2,x_3] \to f[x_1,x_2,x_3] \to f[x_0,\ldots,x_3]
\end{array}
$$

Here, $f[x_3]$ can be overwritten by $f[x_2,x_3]$, since it is no longer needed after the latter has been computed.

PROBLEM 4.3. *Given the "measured points"*

```
x=[0:0.2:7];                       % generate interpolation points
y=exp(cos(x))+0.1*rand(size(x));   % with some errors
```

write a MATLAB *function*

```
function [X,Y,n,rr]=fitpoly(x,y,delta)
% FITPOLY computes an approximating polynomial
%    [X,Y,k,rr]=fitpoly(x,y,delta) computes an approximating
%    polynomial of degree n to the points such that the norm of
%    the residual rr <= delta. (X,Y) are interpolated points for
%    plotting.
```

which computes the best polynomial in the least squares sense of lowest degree using the orthogonal basis such that the residual is $\|r\| \le \delta$.

Experiment with some values of `delta`. *To plot the polynomial, take 10 times more equally spaced interpolation points as the given points x to evaluate the approximating polynomial. Store the nodes of the interpolation points in X and compute the corresponding values of the best polynomial and store them in the vector Y. You then should be able to plot the points and the approximating polynomial by*

```
plot(x,y,'o');
hold on
plot(X,Y)
```

Compare your solution with the rather simple MATLAB *built-in function* `polyfit` *by using the degree n computed by* `fitpoly`.

PROBLEM 4.4. *Solving a nonlinear equation with* inverse interpolation *(see also Chapter 5).*

Consider the nonlinear scalar equation $f(x) = 0$. Starting with two function values $f(x_0)$ and $f(x_1)$ (preferably bracketing the solution) we compute *the following Aitken-Neville-Scheme for the interpolation value $z = 0$:*

$$
\begin{array}{c|ccccc}
f(x_1) & x_1 \\
f(x_2) & x_2 & x_3 := T_{22} \\
f(x_3) & x_3 & T_{32} & x_4 := T_{33} \\
f(x_4) & x_4 & T_{42} & T_{43} & x_5 := T_{44} \\
\cdots & \cdots & \cdots & \cdots & \cdots
\end{array}
$$

The extrapolated value in the diagonal $x_{i+1} := T_{ii}$ is written as new value $T_{i+1,1}$ in the first column of the scheme. Then we compute the function value $f(x_{i+1})$ and the new row i.e. the elements $T_{i+1,2}, \ldots, T_{i+1,i+1}$. If the scheme converges (use good starting values!) then the diagonal entries converge quadratically to a simple zero of f.

Write a program for inverse interpolation and solve the equations

$$
a)\ x - \cos x = 0 \qquad b)\ x = e^{\sqrt{\sin x}}.
$$

PROBLEM 4.5. *Assume you need to compute not only the function value $P_n(z)$ but also the derivative $P_n'(z)$ of an interpolation polynomial.*

Investigate what would be the best way to compute $P_n'(z)$. Consider the following representations of the interpolation polynomial: Lagrange, Barycentric, Newton, Orthogonal Polynomials, Aitken-Neville.

PROBLEM 4.6. *Use extrapolation to compute the derivative $f'(1)$ for*

$$
f(x) = x^2 \ln \left(\frac{\sqrt{x^3 + 1}\, e^x \left(x^3 + \sin x^2 + 1 \right)}{2 \left(\sin x + \cos^2 x + 3 \right) + \ln x} \right).
$$

PROBLEM 4.7. Extrapolation of π. *We will approximate the circumference of the unit circle by regular polygons. The circumference of a regular polygon with n corners on the unit circle is*

$$
U_n = 2n \sin \left(\frac{\pi}{n} \right). \tag{4.84}
$$

We introduce the variable

$$
h = \frac{1}{n}
$$

and the function

$$
T(h) = \frac{U_n}{2} = n \sin \left(\frac{\pi}{n} \right) = \frac{\sin(h\pi)}{h}. \tag{4.85}
$$

The Taylor series of $T(h)$ is

$$
T(h) = \pi - \frac{\pi^3}{3!} h^2 + \frac{\pi^5}{5!} h^4 \mp \cdots \tag{4.86}
$$

Because of $\lim_{h\to 0} T(h) = \pi$ *we can extrapolate* π *from the half circumfer-ences of some regular polygons. Only even powers of* h *occur, therefore we can extrapolate using (4.39).*

Write a program and extrapolate π *using the following table that contains the circumferences of polygons which can be computed by elementary mathe-matics:*

n	2	3	4	5	6	8	10
$\frac{U_n}{2}$	2	$\frac{3}{2}\sqrt{3}$	$2\sqrt{2}$	$\frac{5}{4}\sqrt{10 - 2\sqrt{5}}$	3	$4\sqrt{2 - \sqrt{2}}$	$\frac{5}{2}\left(\sqrt{5} - 1\right)$

PROBLEM 4.8. *(Euler–Mascheroni constant). The sequence*

$$s_n = 1 + \frac{1}{2} + \frac{1}{3} + \cdots + \frac{1}{n} - \ln n$$

converges. The limit has already been computed by Leonhard Euler and is denoted by γ. *Compute an approximation for* γ *by extrapolation.*

PROBLEM 4.9. *Compute the sum of the following series by extrapolation:*

$$a) \quad \sum_{k=1}^{\infty} \frac{1}{k^2} \qquad b) \quad \sum_{k=0}^{\infty} \frac{k^2 + k + 1}{3k^4 + 1}.$$

Hint: use $h = 1/n$ *and extrapolate the limit from the partial sums*

$$T(h) = s_n = \sum_{k=0}^{n} a_k.$$

Choose the sequence (4.36) for h_i.

PROBLEM 4.10. *Compute a table of the function*

$$f(x) = \prod_{n=1}^{\infty} \cos(\frac{x}{n})$$

for $x = 0, 0.1, \ldots, 1$. *Extrapolate each function value from partial products.*

PROBLEM 4.11. *The function* $f(x) = \sin x$ *is approximated by a poly-nomial of degree three* $P_3(x)$ *in such a way that the function values and derivatives match for* $x = 0$ *and for* $x = \pi$.

Compute the polynomial and determine the maximal interpolation error in the interval $(0, \pi)$.

PROBLEM 4.12. *Compute a polynomial of degree three that interpolates the following data:*

x	2	3
$f(x)$	1	2
$f'(x)$	0.5	−2

Compute the polynomial in two ways:

1. *Make an ansatz with unknown coefficients and solve the resulting linear system.*

2. *Use Equation (4.49), expand and order the powers so that the result can be compared with the coefficients above.*

PROBLEM 4.13. *The following table contains* equidistant *function values and derivatives of a function $f(x)$.*

x	h	$2h$	$\ldots$	nh
$f(x)$	y_1	y_2	$\cdots$	y_n
$f'(x)$	y_1'	y_2'	$\cdots$	y_n'

Compute an approximation of the integral

$$\int_{h}^{nh} f(x)\,dx$$

by interpolating the data with a cubic spline function and by integrating the spline function.
What quadrature rule is obtained?

PROBLEM 4.14. *Write a* MATLAB *program to interpolate with a defective spline. Use it to plot a spline through the points*

x	0.0	0.5	1.0	1.5	2.0	2.5	3.0	3.5	4.0
y	2.0	1.2	0.15	1.1	0.5	2.4	2.9	0.0	1.0

Program the three variants for the boundary values of the derivatives. For periodic boundaries set $y(n) = y(1)$.

PROBLEM 4.15. *Write a* MATLAB *function to compute the derivatives of a periodic defective spline function:*

```
% function ys=DerivativesPeriodicSpline(x,y)
% DERIVATIVESPERIODICSPLINE derivative of a periodic spline
%   ys=DerivativesPeriodicSpline(x,y) computes the
%   derivatives for the defective periodic spline passing
%   through the data (x,y). It is assumed that y(1)=y(n)
```

The equations

$$\begin{aligned} x &= \sin t \\ y &= \sin(2t - \tfrac{\pi}{4}) \end{aligned} \tag{4.87}$$

*define a closed curve. Compute with t=linspace(0,2*pi,8) $n = 8$ points of (4.87), interpolate them by a spline curve and compare the result with the exact curve.*

PROBLEM 4.16. *Verify the Equation (4.49). Hint: make the ansatz*

$$Q_i(t) = a + bt + ct^2 + dt^3$$

and determine the coefficients a, b, c and d by solving the Equations (4.48). MAPLE can help you.

PROBLEM 4.17. *Compute the difference scheme (4.50) algebraically and verify that the expression that is obtained with Equation (4.51) is the same as in Equation (4.49).*

PROBLEM 4.18. *Modify Algorithm 4.11 to compute the forward FFT for vectors of length 2^m. Suggestion: To implement the division by n cleanly, first create a driver program* myFFT *that calls the recursive algorithm* myFFT_rec, *then divide the result returned by* myFFT_rec *by n in the driver program.*

PROBLEM 4.19. *Generalize Algorithm 4.11 to handle vectors of length r^m for any integer $r \geq 2$. Hint: To obtain a recursive algorithm, split the input vector $\boldsymbol{v}$ into r subvectors $\boldsymbol{v_0}, \ldots, \boldsymbol{v_{r-1}}$ where $\boldsymbol{v_k}$ contains all components at positions equal to k mod r.*

PROBLEM 4.20. *For a set of samples $\boldsymbol{y} = (y_0, \ldots, y_{2n-1})^\top$, consider the problem of calculating the trigonometric interpolating polynomial $p_{2n}(x)$ with $p_{2n}(x_j) = y_j$, $x_j = j\pi/n$, $j = 0, \ldots, 2n-1$. Suppose the data is perturbed to $\tilde{\boldsymbol{y}} = (\tilde{y}_0, \ldots, \tilde{y}_{2n-1})^\top$, where $\tilde{y}_k = y_k(1 + \varepsilon_k)$ with $|\varepsilon_k| \leq \varepsilon$. The interpolating polynomial then becomes*

$$\tilde{p}_{2n}(x) = \sum_{|k| \leq n}' \tilde{c}_k e^{ikx},$$

so that $\tilde{p}_{2n}(x_j) = \tilde{y}_j$. The **condition number** *of the problem is defined to be the smallest constant $\kappa > 0$ satisfying*

$$\|\tilde{p}_{2n} - p_{2n}\|_{L^2(0,2\pi)} \leq \kappa \cdot \varepsilon \|\boldsymbol{y}\|_\infty,$$

where the L^2 norm is defined as

$$\|f\|^2_{L^2(0,2\pi)} = \int_0^{2\pi} |f(x)|^2 \, dx.$$

(a) We have seen that $\boldsymbol{y} = V\boldsymbol{c}$, where $V^*V = 2nI$. Conclude that

$$\|\tilde{\boldsymbol{c}} - \boldsymbol{c}\|_2 \leq \frac{1}{\sqrt{2n}} \|\tilde{\boldsymbol{y}} - \boldsymbol{y}\|_2 \leq \varepsilon \|\boldsymbol{y}\|_\infty.$$

(b) Using the orthogonality relation

$$\frac{1}{2\pi} \int_0^{2\pi} e^{ikx} e^{-ilx} \, dx = \delta_{kl},$$

deduce that

$$\|\tilde{p}_{2n} - p_{2n}\|_{L^2(0,2\pi)} \le \sqrt{2\pi}\|\tilde{c} - c\|_2 \le \varepsilon\sqrt{2\pi}\|\boldsymbol{y}\|_\infty.$$

What can one say about the conditioning of trigonometric interpolation?

PROBLEM 4.21. *Let $f(x)$ be the 2π-periodic function shown in Figure 4.8, which satisfies*

$$f(x) = \pi - |x| \qquad for\ |x| \le \pi.$$

(a) *Calculate its Fourier coefficients $\hat{f}(k)$ for all k. Hint: you should get*

$$\hat{f}(k) = \begin{cases} \frac{\pi}{2}, & k = 0, \\ \frac{2}{k^2\pi}, & k\ odd, \\ 0, & else. \end{cases}$$

(b) *Using either* myFFT *or the built-in* MATLAB *function* fft, *calculate $\hat{f}_n(k)$ for several n and k, and plot the difference $|\hat{f}_n(k) - \hat{f}(k)|$ as a function of n for $k = 1$. What is the decay rate?*

(c) *Verify Lemma 4.3 numerically for $n = 8$ and $k = 1$.*

(d) *By letting $x = 0$ in $f(x) = \sum_{k\in\mathbb{Z}} \hat{f}(k)e^{ikx}$, show that*

$$\sum_{k=0}^{\infty} \frac{1}{(2k+1)^2} = \frac{\pi^2}{8}.$$

PROBLEM 4.22. *Write a* MATLAB *function to evaluate the trigonometric interpolant $p_n(x)$ for a given set of samples $\boldsymbol{y}$:*

```
function yy=TrigonometricInterpolation(y,xx)
% TRIGONOMETRICINTERPOLIATION trigonometric interpolating polynomial
%   yy=TrigonometricInterpolation(y,xx) computes p(x), the trigonometric
%   interpolating polynomial through (x,y), x(j)=2*pi*(j-1)/length(y).
%   It returns yy, the values of p evaluated at xx.
```

To test your program, use

$$f(x) = 10\cos(x) - 3\sin(3x) + 5\cos(3x) - 15\cos(10x)$$

and plot the maximum error $\max |p_n(x) - f(x)|$ for $n = 4, 8, 16, 32, 64$. Verify that the maximum error is close to machine precision for $n = 32, 64$. What is the reason behind this?

PROBLEM 4.23. *Write a* MATLAB *function to add echoes to a given signal $\boldsymbol{y}$:*

```
function y=Echo(x, Fs, d, alpha)
% ECHO produces an echo effect
%    y=Echo(x,Fs,d,alpha) adds an echo to the sound vector x with a
%    delay of d seconds. 0 < alpha < 1 is the strength of the echo
%    and Fs is the sampling rate in Hz.
```

To test your program use one of the sound signals already available in MAT-LAB *(*chirp, gong, handel, laughter, splat *and* train*). For example, to load the* train *signal, use*

```
load train
```

which loads the variable y *containing the actual signal and* Fs*, the sampling rate. To play the signal, use*

```
sound(y,Fs)
```

Play the original as well as the transformed signals (with echoes) and compare.

PROBLEM 4.24. *One way of compressing a sound signal is to remove frequency components that have small coefficients, i.e., for a relative threshold τ, we transform a given signal y of length n into w, also of length n, whose discrete Fourier coefficients $\hat{w}_n$ satisfy*

$$\hat{w}_n(k) = \begin{cases} \hat{y}_n(k), & |\hat{y}_n(k)| > \tau \cdot \max_j |\hat{y}_n(j)|, \\ 0 & else. \end{cases}$$

Write a MATLAB *function that implements this compression scheme and returns a sparse version of the n-vector $\hat{w}_n$:*

```
function w=Compress(y,thres)
% COMPRESS removes small frequency components and compresses the signal
%    w=Compress(y,thres) removes all the frequencies whose amplitude is
%    less than thres times the maximum amplitude and compresses the
%    resulting sparse signal.
```

To play the compressed sound, use the command

```
sound(ifft(full(w)),Fs)
```

Compare the sound quality and the amount of storage required by the original and compressed sound signals for different values of τ. To compare the memory usage between the uncompressed and compressed signals, use the command whos *followed by a list of variable names.*

PROBLEM 4.25. *Let*

$$H = \begin{pmatrix} z_0 & z_{n-1} & \cdots & z_1 \\ z_1 & z_0 & \cdots & z_2 \\ \vdots & & \ddots & \vdots \\ z_{n-1} & z_{n-2} & \cdots & z_0 \end{pmatrix}$$

be a circulant matrix. Show that every vector of the form

$$\boldsymbol{v} = (1, e^{2\pi i k/n}, e^{4\pi i k/n}, \ldots, e^{2\pi i (n-1)k/n})^{\top}$$

is an eigenvector of H and compute its corresponding eigenvalue. What is the relationship between the eigenvalues of H and the discrete Fourier transform of $\boldsymbol{z} = (z_0, \ldots, z_{n-1})^{\top}$?

Chapter 5. Nonlinear Equations

> *Nonlinear equations are solved as part of almost all simulations of physical processes. Physical models that are expressed as nonlinear partial differential equations, for example, become large systems of nonlinear equations when discretized. Authors of simulation codes must either use a nonlinear solver as a tool or write one from scratch.*
>
> Tim Kelley, Solving Nonlinear Equations with Newton's Method, SIAM, 2003.

Prerequisites: This chapter requires Sections 2.5 (conditioning), 2.8 (stopping criteria), Chapter 3 (linear equations), as well as polynomial interpolation (§4.2) and extrapolation (§4.2.8).

Solving a *nonlinear equation* in one variable means: given a continuous function f on the interval $[a,b]$, we wish to find a value $s \in [a,b]$ such that $f(s) = 0$. Such a value s is called a *zero or root of the function* f or a *solution of the equation* $f(x) = 0$. For a multivariate function $f : \mathbb{R}^n \to \mathbb{R}^n$, solving the associated system of equations means finding a vector $s \in \mathbb{R}^n$ such that $f(s) = 0$. After an introductory example, we show in Section 5.2 the many techniques for finding a root of a scalar function: the fundamental bisection algorithm, fixed point iteration including convergence rates, and the general construction of one step formulas, where we naturally discover Newton's method[1], and also higher order variants. We also introduce Aitken acceleration and the ε-algorithm, and show how multiple zeros have an impact on the performance of root finding methods. Multistep iteration methods and how root finding algorithms can be interpreted as dynamical systems are also contained in this section. Section 5.3 is devoted to the special case of finding zeros of polynomials. In Section 5.4, we leave the scalar case and consider non-linear systems of equations, where fixed point iterations are the only realistic methods for finding a solution. The main workhorse for solving non-linear systems is then Newton's method[2], and variants thereof.

---

[1] Also called Newton-Raphson, since Newton wrote down the method only for a polynomial in 1669, while Raphson, a great admirer of Newton, wrote it as a fully iterative scheme in 1690

[2] First written for a system of 2 equations by Simpson in 1740

W. Gander et al., *Scientific Computing - An Introduction using Maple and MATLAB*, Texts in Computational Science and Engineering 11, DOI 10.1007/978-3-319-04325-8_5,

5.1   Introductory Example

We use *Kepler's Equation* as our motivating example: consider a *two-body problem* like a satellite orbiting the earth or a planet revolving around the sun. Kepler discovered that the orbit is an ellipse and the central body F (earth, sun) is in a focus of the ellipse. If the ellipse is eccentric (i.e., not a circle), then the speed of the satellite P is not uniform: near the earth it moves faster than far away. Figure 5.1 depicts the situation. Kepler also discovered

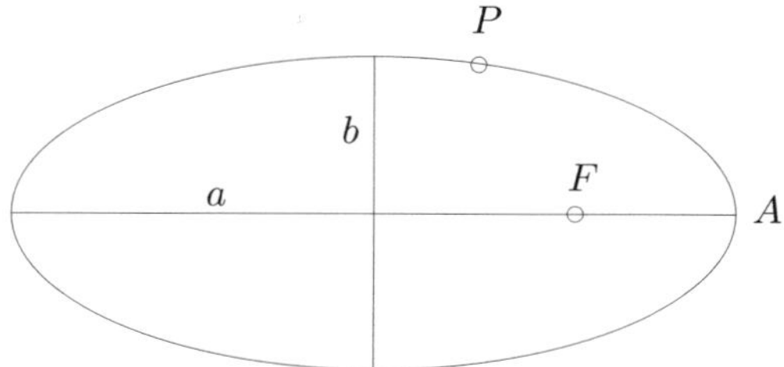

FIGURE 5.1. *Satellite P orbiting the earth F*

the law of this motion by carefully studying data from the observations by Tycho Brahe. It is called *Kepler's second law* and says that the travel time is proportional to the area swept by the radius vector measured from the focus where the central body is located, see Figure 5.2. We would like to use this

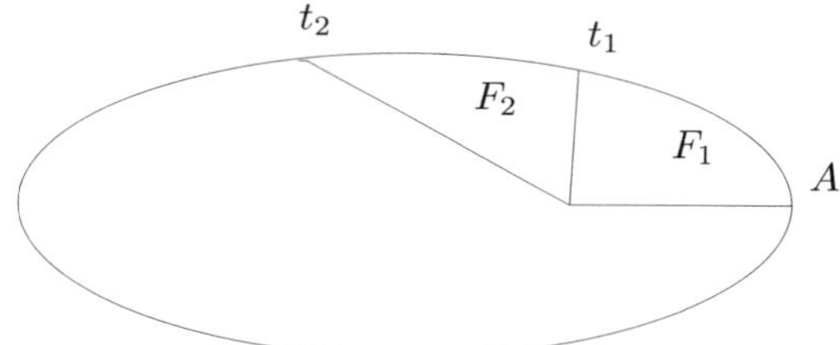

FIGURE 5.2.
Kepler's second law: if $F_1 = F_2$ then $t_2 = 2t_1$

law to predict where the satellite will be at a given time.

Assume that at $t = 0$ the satellite is at point A, the perihelion of the ellipse, nearest to the earth. Assume further that the time for completing a full orbit is T. The question is: where is the satellite at time t (for $t < T$)?

We need to compute the area $\triangle FAP$ that is swept by the radius vector as a function of the angle E (see Figure 5.3). E is called the *eccentric anomaly*. The equation of the ellipse with semi-axis a and b is

$$\begin{aligned} x(E) &= a \cos E, \\ y(E) &= b \sin E. \end{aligned}$$

To compute the infinitesimal area dI between two nearby radius vectors, we will use the cross product, see Figure 5.4. The infinitesimal vector of motion

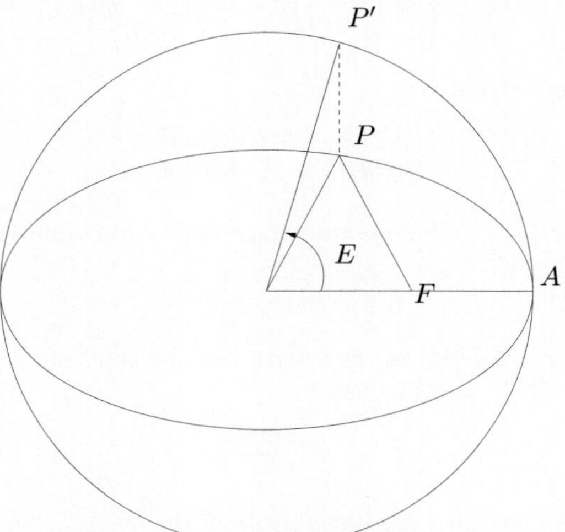

FIGURE 5.3. *Definition of E*

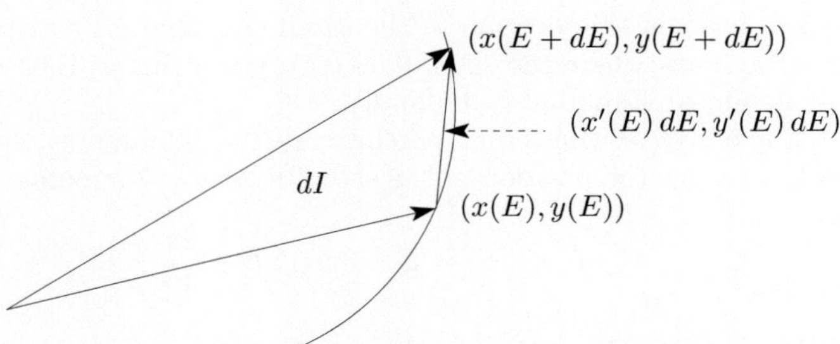

FIGURE 5.4. *Computing an infinitesimal area*

$(x'(E)\,dE, y'(E)\,dE)$ can be obtained by Taylor expansion from the difference of the two radius vectors. Taking the cross product, we get

$$dI = \frac{1}{2}\left|\left(\begin{array}{c} x(E) \\ y(E) \\ 0 \end{array}\right) \times \left(\begin{array}{c} x'(E) \\ y'(E) \\ 0 \end{array}\right)\right| dE = \frac{1}{2}(x(E)y'(E) - x'(E)y(E))dE$$

Inserting the derivatives

$$\begin{aligned} x'(E) &= -a\sin E \\ y'(E) &= b\cos E \end{aligned}$$

and integrating $I = \int_0^E dI$ we obtain the simple expression for the area

$$I = \Delta 0AP = \frac{1}{2}abE.$$

To obtain the area ΔFAP we now have to subtract from I the area of the triangle $\Delta 0FP$. This area is given by

$$\frac{aeb\sin E}{2},$$

where $e = \frac{\sqrt{a^2-b^2}}{a}$ is called the *eccentricity* of the ellipse. Thus we obtain the following function of $S(E)$ for the area swept by the radius:

$$S(E) = \frac{1}{2}ab(E - e\sin E).$$

Now according to Kepler's second law, $S(E)$ is proportional to the time t. Thus $E - e\sin E \sim t$. The proportionality factor must be $2\pi/T$, and therefore

$$E - e\sin E = \frac{2\pi}{T}t, \qquad \textit{Kepler's Equation.} \tag{5.1}$$

Kepler's Equation (5.1) defines the implicit function $E(t)$, i.e. gives the relation between the location of the satellite (angle E) and the time t. If we want to know where the satellite is for a given time t, then we have to solve the nonlinear Equation (5.1) for E.

Some typical values for a satellite are $T = 90$ minutes, and $e = 0.8$. If we want to know the position of the satellite for $t = 9$ minutes, then we have to solve

$$f(E) = E - 0.8\sin E - \frac{2\pi}{10} = 0. \tag{5.2}$$

5.2   Scalar Nonlinear Equations

Finding roots of scalar nonlinear equations is already a difficult task, and there are many numerical methods devoted to it. We show in this section some of the most popular ones, and not all of these methods can be generalized to higher dimensional problems. We start with the simplest but also most robust method called bisection.

5.2.1 Bisection

The first method for solving an equation $f(x) = 0$ which we will discuss in this chapter is called *bisection*. We assume that we know an interval $[a, b]$ for which $f(a) < 0$ and $f(b) > 0$. If f is continuous in $[a, b]$ then there must exist a zero $s \in [a, b]$. To find it, we compute the midpoint x of the interval and check the value of the function $f(x)$. Depending on the sign of $f(x)$, we can decide in which subinterval $[a, x]$ or $[x, b]$ the zero must lie. Then we continue this process of bisection in the corresponding subinterval until the size of the interval containing s becomes smaller than some given tolerance:

ALGORITHM 5.1. *Bisection – First Version*

```
while b-a>tol
  x=(a+b)/2
  if f(x)<0, a=x; else b=x; end
end
```

At each step of Algorithm 5.1, we compute a new interval (a_k, b_k) containing s. We have

$$b_k - a_k = \frac{1}{2}(b_{k-1} - a_{k-1}) = \frac{1}{2^k}(b - a).$$

Since the kth approximation of s is $x_k = (a_k + b_k)/2$, we obtain for the error

$$|x_k - s| \leq b_k - a_k = \frac{1}{2^k}(b - a) \to 0, \quad k \to \infty. \tag{5.3}$$

If we want the error to satisfy $|x_k - s| \leq tol$, then it suffices to have $(b-a)/2^k \leq tol$, so that

$$k > \ln\left(\frac{b - a}{tol}\right) / \ln 2. \tag{5.4}$$

Algorithm 5.1 can be improved. First, it does not work if $f(a) > 0$ and $f(b) < 0$. This can easily be fixed by multiplying the function f by -1. Second, the algorithm will also fail if the tolerance tol is too small: assume for example that the computer works with a mantissa of 12 decimal digits and that

$$\begin{aligned} a_k &= 5.34229982195 \\ b_k &= 5.34229982200 \end{aligned}$$

Then $a_k + b_k = 10.68459964395$ is the exact value and the rounded value to 12 digits is 10.6845996440. Now $x_k = (a_k + b_k)/2 = 5.34229982200 = b_k$. *Thus there is no machine number with 12 decimal digits between a_k and b_k and the midpoint is rounded to b_k.* Since $b_k - a_k = 5e{-}11$, a required tolerance of say $tol = 1e{-}15$ would be unreasonable and would produce an infinite loop with Algorithm 5.1.

However, it is easy to test if there is still at least one machine number in the interval. If for $x = (a + b)/2$ the condition `(a<x) & (x<b)` holds, then there exists such a number, otherwise we must terminate the iteration. Thus, we obtain the following MATLAB function (Algorithm 5.2):

ALGORITHM 5.2. *Bisection*

```
function [x,y]=Bisection(f,a,b,tol)
% BISECTION computes a root of a scalar equation
%    [x,y]=Bisection(f,a,b,tol) finds a root x of the scalar function
%    f in the interval [a,b] up to a tolerance tol. y is the
%    function value at the solution

fa=f(a); v=1; if fa>0, v=-1; end;
if fa*f(b)>0
   error('f(a) and f(b) have the same sign')
end
if (nargin<4), tol=0; end;
x=(a+b)/2;
while (b-a>tol) & ((a < x) & (x<b))
  if v*f(x)>0, b=x; else a=x; end;
  x=(a+b)/2;
end
if nargout==2, y=f(x); end;
```

Algorithm 5.2 is an example of a "fool-proof" and machine-independent algorithm (see Section 2.8.1): for any continuous function whose values have opposite signs at the two end points, the algorithm will find a zero of this function. With $tol = 0$, it will compute a zero of f to machine precision. The algorithm makes use of finite precision arithmetic and would not work on a computer with exact arithmetic.

EXAMPLE 5.1. *As a first example, we consider the function $f(x) = x + e^x$,*

```
>> [x,y]=Bisection(@(x) x+exp(x),-1,0)
x =
    -0.567143290409784
y =
    -1.110223024625157e-16
```

and we obtain the zero to machine precision. As a second example, we consider Kepler's Equation (5.2),

```
>> [E,f]=Bisection(@(E) E-0.8*sin(E)-2*pi/10,0,pi,1e-6)
E =
    1.419135586110581
f =
    -1.738227842773554e-07
```

where we asked for a precision of $1e-6$.

If the assumptions for bisection are not met, i.e., f has values with the same sign for a and b (see left figure in Figure 5.5) or f is not continuous in (a, b) (see right figure in Figure 5.5), then Algorithm 5.2 will fail.

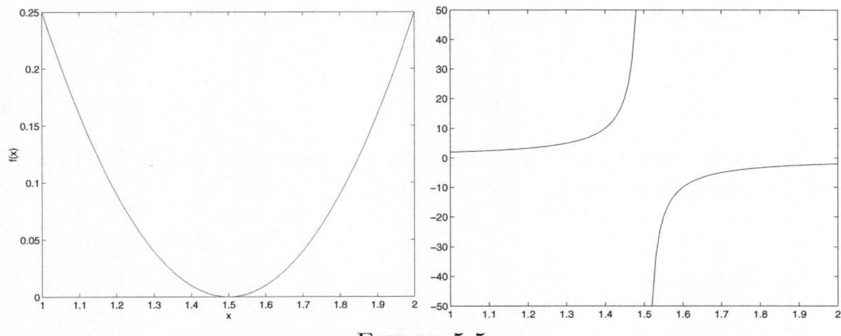

FIGURE 5.5.
Cases where the Bisection Algorithm 5.2 fails

5.2.2 Fixed Point Iteration

Consider the equation

$$f(x) = 0, \qquad (5.5)$$

where $f(x)$ is a real function defined and continuous in the interval $[a, b]$. Assume that $s \in [a, b]$ is a zero of $f(x)$. In order to compute s, we transform (5.5) algebraically into *fixed point form*,

$$x = F(x), \qquad (5.6)$$

where F is chosen so that $F(x) = x \iff f(x) = 0$. A simple way to do this is, for example, $x = x + f(x) =: F(x)$, but other choices are possible. Finding a zero of $f(x)$ in $[a, b]$ is then equivalent to finding a fixed point $x = F(x)$ in $[a, b]$. The fixed point form suggests the *fixed point iteration*

$$x_0 \text{ initial guess}, \quad x_{k+1} = F(x_k), \quad k = 0, 1, 2, \dots . \qquad (5.7)$$

The hope is that iteration (5.7) will produce a convergent sequence $x_k \to s$.

For example, consider

$$f(x) = xe^x - 1 = 0. \qquad (5.8)$$

When trying to solve this equation in MAPLE, we find

```
solve(x*exp(x)-1,x);
```

LambertW(1)

the well-known *LambertW function*, which we will encounter again later in this chapter, and also in Chapter 8, in the context of Differentiation. We will also encounter this function in Chapter 12 on Optimization when studying how to live as long as possible.

A first fixed point iteration for computing LambertW(1) is obtained by rearranging and dividing (5.8) by e^x,

$$x_{k+1} = e^{-x_k}. \tag{5.9}$$

With the initial guess $x_0 = 0.5$ we obtain the iterates shown in Table 5.1. Indeed x_k seems to converge to $s = 0.5671432...$

k	x_k	k	x_k	k	x_k
0	0.5000000000	10	0.5669072129	20	0.5671424776
1	0.6065306597	11	0.5672771960	21	0.5671437514
2	0.5452392119	12	0.5670673519	22	0.5671430290
3	0.5797030949	13	0.5671863601	23	0.5671434387
4	0.5600646279	14	0.5671188643	24	0.5671432063
5	0.5711721490	15	0.5671571437	25	0.5671433381
6	0.5648629470	16	0.5671354337	26	0.5671432634
7	0.5684380476	17	0.5671477463	27	0.5671433058
8	0.5664094527	18	0.5671407633	28	0.5671432817
9	0.5675596343	19	0.5671447237	29	0.5671432953

TABLE 5.1. *Iteration $x_{k+1} = \exp(-x_k)$*

A second fixed point form is obtained from $xe^x = 1$ by adding x on both sides to get $x+xe^x = 1+x$, factoring the left-hand side to get $x(1+e^x) = 1+x$, and dividing by $1 + e^x$, we obtain

$$x = F(x) = \frac{1+x}{1+e^x}. \tag{5.10}$$

This time the convergence is much faster — we need only three iterations to obtain a 10-digit approximation of s,

$$\begin{aligned}
x_0 &= 0.5000000000 \\
x_1 &= 0.5663110032 \\
x_2 &= 0.5671431650 \\
x_3 &= 0.5671432904.
\end{aligned}$$

Another possibility for a fixed point iteration is

$$x = x + 1 - xe^x. \tag{5.11}$$

This iteration function does not generate a convergent sequence. We observe here from Table 5.2 a chaotic behavior: no convergence but also no divergence to infinity.

k	x_k	k	x_k	k	x_k
0	0.5000000000	6	-0.6197642518	12	-0.1847958494
1	0.6756393646	7	0.7137130874	13	0.9688201302
2	0.3478126785	8	0.2566266491	14	-0.5584223793
3	0.8553214091	9	0.9249206769	15	0.7610571653
4	-0.1565059553	10	-0.4074224055	16	0.1319854380
5	0.9773264227	11	0.8636614202	17	0.9813779498

TABLE 5.2.

Chaotic iteration with $x_{k+1} = x_k + 1 - x_k e^{x_k}$

Finally we could also consider the fixed point form

$$x = x + xe^x - 1. \tag{5.12}$$

With this iteration function the iteration diverges to minus infinity:

$$
\begin{aligned}
x_0 &= 0.5000000000 \\
x_1 &= 0.3243606353 \\
x_2 &= -0.2270012400 \\
x_3 &= -1.4079030215 \\
x_4 &= -2.7523543838 \\
x_5 &= -3.9278929674 \\
x_6 &= -5.0052139570
\end{aligned}
$$

From these examples, we can see that there is an infinite number of possibilities in choosing an iteration function $F(x)$. Hence the question is, when does the iteration converge?

The fixed point iteration has a very nice geometric interpretation: we plot $y = F(x)$ and $y = x$ in the same coordinate system (see Figure 5.6). The intersection points of the two functions are the solutions of $x = F(x)$. The computation of the sequence $\{x_k\}$ with

$$
\begin{aligned}
x_0 &\quad &&\textit{choose initial value} \\
x_{k+1} &= F(x_k), &&k = 0, 1, 2, \ldots
\end{aligned}
$$

can be interpreted geometrically via sequences of lines parallel to the coordinate axes:

$$
\begin{aligned}
&x_0 &&\text{start with } x_0 \text{ on the } x\text{-axis} \\
&F(x_0) &&\text{go parallel to the } y\text{-axis to the graph of } F \\
&x_1 = F(x_0) &&\text{move parallel to the } x\text{-axis to the graph } y = x \\
&F(x_1) &&\text{go parallel to the } y\text{-axis to the graph of } F \\
&\text{etc.}
\end{aligned}
$$

One can distinguish four cases, two where $|F'(s)| < 1$ and the algorithm converges and two where $|F'(s)| > 1$ and the algorithm diverges. An example

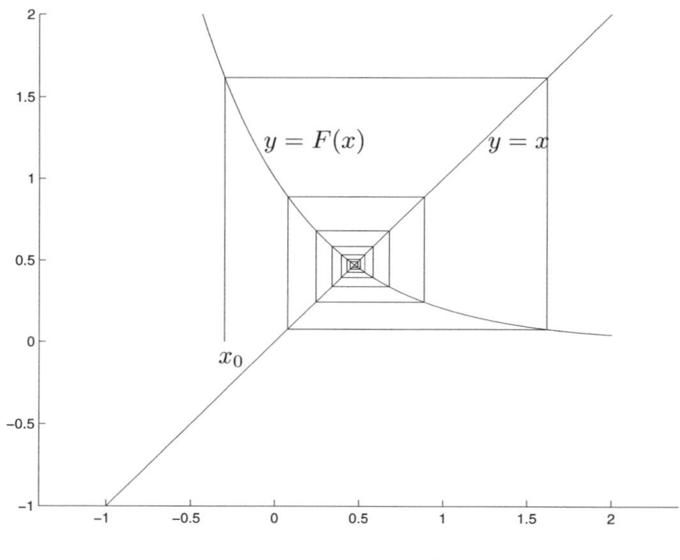

FIGURE 5.6. $x = F(x)$

for each case is given in Figure 5.7. A more general statement for convergence is the theorem of Banach, Theorem 5.5, which is explained in Section 5.4. For more information on the possibly chaotic behavior, as observed in Iteration (5.11), see Section 5.2.9.

5.2.3 Convergence Rates

In the previous section, we have seen geometrically that a fixed point iteration converges if $|F'(s)| < 1$. We have also observed that in case of convergence, some iterations (like Iteration (5.10)) converge much faster than others (like Iteration (5.9)). In this section, we would like to analyze the convergence speed. We observe already from Figure 5.7 that the smaller $|F'(s)|$, the faster the convergence.

DEFINITION 5.1. (ITERATION ERROR) *The error at iteration step k is defined by $e_k = x_k - s$.*

Subtracting the equation $s = F(s)$ from $x_{k+1} = F(x_k)$ and expanding in a Taylor series, we get

$$x_{k+1} - s = F(x_k) - F(s) = F'(s)(x_k - s) + \frac{F''(s)}{2!}(x_k - s)^2 + \cdots,$$

or expressed in terms of the error,

$$e_{k+1} = F'(s)e_k + \frac{F''(s)}{2!}e_k^2 + \frac{F'''(s)}{3!}e_k^3 + \cdots. \tag{5.13}$$

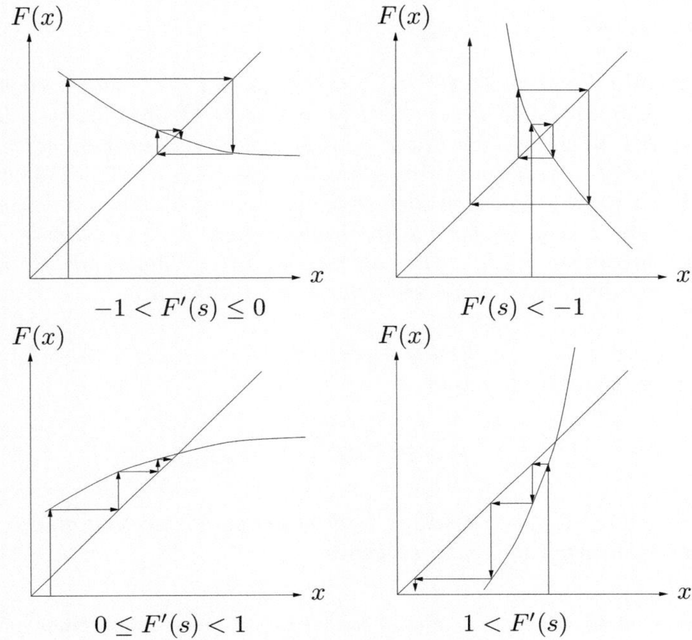

FIGURE 5.7.
Four different scenarios for the fixed point iteration

If $F'(s) \neq 0$, we conclude from (5.13), assuming that the error e_k goes to zero, that

$$\lim_{k \to \infty} \frac{e_{k+1}}{e_k} = F'(s).$$

This means that asymptotically $e_{k+1} \sim F'(s)e_k$. *Thus for large k the error is reduced in each iteration step by the factor* $|F'(s)|$. This is called *linear convergence* because the new error is a linear function of the previous one. Similarly we speak of *quadratic convergence* if $F'(s) = 0$ but $F''(s) \neq 0$, because $e_{k+1} \sim (F''(s)/2)\, e_k^2$. More generally we define:

DEFINITION 5.2. (CONVERGENCE RATE) *The rate of convergence of* $x_{k+1} = F(x_k)$ *is*

linear: *if $F'(s) \neq 0$ and $|F'(s)| < 1$,*

quadratic: *if $F'(s) = 0$ and $F''(s) \neq 0$,*

cubic: *if $F'(s) = 0$ and $F''(s) = 0$, but $F'''(s) \neq 0$,*

of order m**:** *if $F'(s) = F''(s) = \cdots = F^{(m-1)}(s) = 0$, but $F^{(m)}(s) \neq 0$.*

EXAMPLE 5.2.

1. *Consider Iteration (5.9): $x_{k+1} = F(x_k) = e^{-x_k}$. The fixed point is $s = 0.5671432904$. $F'(s) = -F(s) = -s = -0.5671432904$. Because $0 < |F'(s)| < 1$ we have linear convergence. With linear convergence the number of correct digits grows linearly. For $|F'(s)| = 0.5671432904$ the error is roughly halved in each step. So in order to obtain a new decimal digit one has to perform p iterations, where $(0.5671432904)^p = 0.1$. This gives us $p = 4.01$. Thus after about 4 iterations we obtain another decimal digit, as one can see by looking at Table 5.1.*

2. *If we want to solve Kepler's equation, $E - e\sin E = \frac{2\pi}{T}t$ for E, an obvious iteration function is*

$$E = F(E) = \frac{2\pi}{T}t + e\sin E.$$

Because $|F'(E)| = e|\cos E| < 1$, the fixed point iteration always generates a linearly convergent sequence.

3. *Iteration (5.10) $x_{k+1} = F(x_k)$ with $F(x) = \dfrac{1+x}{1+e^x}$ converges quadratically, because*

$$F'(x) = \frac{1 - xe^x}{(1 + e^x)^2} = -\frac{f(x)}{(1 + e^x)^2},$$

and since $f(s) = 0$ we have $F'(s) = 0$, and one can check that $F''(s) \neq 0$. With quadratic convergence, the number of correct digits doubles at each step asymptotically. If we have 3 correct digits at step k, $e_k = 10^{-3}$, then $e_{k+1} \approx e_k^2 = 10^{-6}$, and thus we have 6 correct digits in the next step $k + 1$.

The doubling of digits can be seen well when computing with MAPLE *with extended precision (we separated the correct digits with a $*$):*

```
> Digits:=59;
> x:=0.5;
> for i from 1 by 1 to 5 do x:=(1+x)/(1+exp(x)); od;

x:= .5
x:= .56*63110031972181530416491513817372818700809520366347554108
x:= .567143*1650348622127865120966596963665134313508187085567477
x:= .56714329040978*10286995766494153472061705578660439731056279
x:= .56714329040978387299999686622*088916713037266116513649733766
x:= .567143290409783872999996866221035554975381578718651250813513*
```

5.2.4 Aitken Acceleration and the ε-Algorithm

The ε-algorithm was invented by Peter Wynn [151] in order to accelerate the convergence of a sequence of real numbers. Wynn's algorithm is in turn an ingenious generalization of the Aitken[3] acceleration [2]: suppose we have a sequence of real numbers $\{x_n\}$ which converges very slowly toward a limit s, and we would like to accelerate its convergence. Often one observes that the sequence satisfies approximately the relation (linear convergence)

$$x_n - s \approx \rho(x_{n-1} - s) \approx \rho^n(x_0 - s), \tag{5.14}$$

or equivalently

$$x_n \approx s + C\rho^n, \tag{5.15}$$

for some unknown constants ρ, C, and s. We have already seen a typical example in Section 5.2.2, where the fixed point iteration

$$x_{n+1} = F(x_n)$$

was used to compute approximations to the fixed point $s = F(s)$. If F is differentiable, we have

$$x_n - s = F(x_{n-1}) - F(s) \approx F'(s)(x_{n-1} - s),$$

which is precisely of the form of Equation (5.14).

The idea of the *Aitken acceleration*, see [2], is to replace "$\approx$" in Equation (5.15) by "$=$", and then to determine ρ, C, and the limit s from three consecutive values x_n. Solving the small nonlinear system with MAPLE, we obtain

```
> solve({x[n-1]=s+C*rho^(n-1),x[n]=s+C*rho^n,x[n+1]
        =s+C*rho^(n+1)},{rho,C,s}):
> assign(%):
> simplify(s);
```

$$\frac{x_{n+1}x_{n-1} - x_n{}^2}{x_{n+1} - 2\,x_n + x_{n-1}}.$$

If Equation (5.14) is not satisfied exactly, the value obtained for s will depend on n. We thus obtain a new sequence $\{x'_n\}$ defined by

$$x'_{n-1} := \frac{x_{n+1}x_{n-1} - x_n{}^2}{x_{n+1} - 2\,x_n + x_{n-1}}, \tag{5.16}$$

which in general converges faster to the original limit s than the original sequence.

---

[3]Alexander Aitken (1895–1967) was one of the best mental calculators of all time, see for example M. L. Hunter: An exceptional talent for calculative thinking. British Journal of Psychology Bd.53, 1962, S. 243-258

DEFINITION 5.3. (SHANKS TRANSFORM) *For a given sequence* $\{x_n\}$, *the Shanks transform* $S(x_n)$ *is defined by*

$$S(x_n) := \frac{x_{n+1}x_{n-1} - x_n^{\,2}}{x_{n+1} - 2\,x_n + x_{n-1}}. \tag{5.17}$$

Thus, the Shanks transform represents none other than the Aitken acceleration of the sequence. We can now apply the Shanks transform again to the already accelerated sequence, and consider $S^2(x_n) = S(S(x_n))$, $S^3(x_n) = S(S(S(x_n)))$ etc., and further convergence accelerations may be obtained. As an example, we consider the sequence of partial sums

$$1 - \frac{1}{3} + \frac{1}{5} - \frac{1}{7} + \dots,$$

which converges to $\frac{\pi}{4}$, as one can see in MAPLE:

```
> sum((-1)^n/(2*n+1),n=0..infinity);
```

We first define the Shanks transform in MATLAB,

```
>> S=@(x) (x(1:end-2).*x(3:end)-x(2:end-1).^2)...
          ./(x(3:end)-2*x(2:end-1)+x(1:end-2));
```

and now apply it to the example sequence

```
>> n=20;
>> x=cumsum(-1./(1:2:2*n).*(-1).^(1:n));
>> x1=S(x);
>> x2=S(x1);
>> x3=S(x2);
>> 4*[x(end);x1(end);x2(end);x3(end)]
ans =
    3.091623806667840
    3.141556330284576
    3.141592555781680
    3.141592651495555
```

This shows that clearly multiple application of the Aitken acceleration further accelerates the convergence of this sequence.

An even better generalization is to assume a more complete model. We suppose that the initial sequence $\{x_n\}$ satisfies, instead of (5.15), approximately

$$x_n \approx s + C_1\rho_1^n + C_2\rho_2^n. \tag{5.18}$$

We then need to determine five parameters, and thus take the formula (5.18) for five consecutive iterates in order to obtain a nonlinear system to solve for ρ_1, ρ_2, C_1, C_2 and s. Shanks did this computation [124], and further generalized it.

DEFINITION 5.4. (GENERALIZED SHANKS TRANSFORM) *For a given sequence* $\{x_n\}$ *converging to* s *and satisfying approximately*

$$x_n \approx s + \sum_{i=1}^{\infty} a_i \rho_i^n,$$

the new sequence $\{s_{n,k}\}$ *obtained by solving the non-linear system of* $2k+1$ *equations*

$$x_{n+j} = s_{n,k} + \sum_{i=1}^{k} \tilde{a}_i \tilde{\rho}_i^{n+j}, \quad j = 0, 1, \ldots, 2k$$

for the $2k+1$ *unknowns* $s_{n,k}$, $\tilde{a}_i$ *and* $\tilde{\rho}_i$ *is called the* generalized Shanks transform $S_k(x_n)$ *of the sequence* $\{x_n\}$*. It is often also denoted by* s_{nk} *or* $e_k(x_n)$.

Unfortunately, the formulas obtained for the Shanks transform quickly become unwieldy for numerical calculations. In order to find a different characterization for the Shanks transform, let $P_k(x) = c_0 + c_1 x + \cdots + c_k x^k$ be the polynomial with zeros $\tilde{\rho}_1, \ldots, \tilde{\rho}_k$, normalized such that $\sum c_i = 1$, and consider the equations

$$c_0(x_n - s_{n,k}) = c_0 \sum_{i=1}^{k} \tilde{a}_i \tilde{\rho}_i^n$$

$$c_1(x_{n+1} - s_{n,k}) = c_1 \sum_{i=1}^{k} \tilde{a}_i \tilde{\rho}_i^{n+1}$$

$$\vdots \quad = \quad \vdots$$

$$c_k(x_{n+k} - s_{n,k}) = c_k \sum_{i=1}^{k} \tilde{a}_i \tilde{\rho}_i^{n+k}.$$

Adding all these equations, we obtain the sum

$$\sum_{j=0}^{k} c_j(x_{n+j} - s_{n,k}) = \sum_{i=1}^{k} \tilde{a}_i \tilde{\rho}_i^n \underbrace{\sum_{j=0}^{k} c_j \tilde{\rho}_i^j}_{P_k(\tilde{\rho}_i)=0},$$

and observing that $\sum c_i = 1$, the extrapolated value becomes

$$s_{n,k} = \sum_{j=0}^{k} c_j x_{n+j}. \tag{5.19}$$

If we knew the coefficients c_j of the polynomial with roots $\tilde{\rho}_j$, we could directly compute $s_{n,k}$ as a linear combination of successive iterates. .

P. Wynn established in 1956, see [151], the remarkable result that the quantities $s_{n,k}$ can be computed recursively. This procedure is called the ε-algorithm. Let $\varepsilon_{-1}^{(n)} = 0$ and $\varepsilon_0^{(n)} = x_n$ for $n = 0, 1, 2, \ldots$. From these values, the following table using the recurrence relation

$$\varepsilon_{k+1}^{(n)} = \varepsilon_{k-1}^{(n+1)} + \frac{1}{\varepsilon_k^{(n+1)} - \varepsilon_k^{(n)}} \tag{5.20}$$

is constructed:

$$
\begin{array}{ccccccc}
\varepsilon_{-1}^{(0)} \\
 & \varepsilon_0^{(0)} \\
\varepsilon_{-1}^{(1)} & & \varepsilon_1^{(0)} \\
 & \varepsilon_0^{(1)} & & \varepsilon_2^{(0)} \\
\varepsilon_{-1}^{(2)} & & \varepsilon_1^{(1)} & & \varepsilon_3^{(0)} \\
 & \varepsilon_0^{(2)} & & \varepsilon_2^{(1)} & & \cdots \\
\varepsilon_{-1}^{(3)} & & \varepsilon_1^{(2)} & & \cdots \\
 & \varepsilon_0^{(3)} & & \cdots \\
\varepsilon_{-1}^{(4)} & & \cdots
\end{array}
\tag{5.21}
$$

Wynn showed in 1956 that $\varepsilon_{2k}^{(n)} = s_{n,k}$ and $\varepsilon_{2k+1}^{(n)} = \frac{1}{S_k(\Delta x_n)}$, where $S_k(\Delta x_n)$ denotes the generalized Shanks transform of the sequence of the differences $\Delta x_n = x_{n+1} - x_n$. Thus every second column in the ε-table is in principle of interest. For the MATLAB implementation, we write the ε-table in the lower triangular part of the matrix E, and since the indices in MATLAB start at 1, we shift appropriately:

$$
\begin{aligned}
0 &= \varepsilon_{-1}^{(0)} = E_{11}, \\
0 &= \varepsilon_{-1}^{(1)} = E_{21} & x_1 &= \varepsilon_0^{(0)} = E_{22}, \\
0 &= \varepsilon_{-1}^{(2)} = E_{31} & x_2 &= \varepsilon_0^{(1)} = E_{32} & \varepsilon_1^{(0)} &= E_{33}, \\
0 &= \varepsilon_{-1}^{(3)} = E_{41} & x_3 &= \varepsilon_0^{(2)} = E_{42} & \varepsilon_1^{(1)} &= E_{43} & \varepsilon_2^{(0)} &= E_{44}.
\end{aligned}
\tag{5.22}
$$

We obtain the algorithm

ALGORITHM 5.3. *Scalar ε-Algorithm*

```
function [Er,E]=EpsilonAlgorithm(x,k);
% EPSILONALGORITHM epsilon algorithm of P. Wynn
%    [Er,E]=EpsilonAlgorithm(x,k) is an implementation of the epsilon
%    algorithm of P. Wynn. The first column of E is zero, the second
%    contains the elements of the sequence x whose convergence we want
%    to accelerate. 2k colums of the scheme are computed and returned
%    in E, while Er is a reduced version of E, containing only the
%    relevant columns (every second column).

n=2*k+1;
```

```
E=zeros(n+1,n+1);
for i=1:n
  E(i+1,2)=x(i);
end
for i=3:n+1
  for j=3:i
    E(i,j)=E(i-1,j-2)+1/(E(i,j-1)-E(i-1,j-1));
  end
end
Er=E(:,2:2:n+1);
```

EXAMPLE 5.3. *We will show the performance of the scalar ε-algorithm by accelerating the partial sums of the series*

$$1 - \frac{1}{2} + \frac{1}{3} - \frac{1}{4} \pm \cdots = \ln 2.$$

With the MATLAB-*statements we first compute partial sums and then apply the Epsilon Algorithm*

```
k=5;
v=1;
for j=1:2*k+1
  y(j)=v/j; v=-v;
end
x=cumsum(y)
E=EpsilonAlgorithm(x,k)
```

We obtain the result

```
x =
  Columns 1 through 7
    1.0000    0.5000    0.8333    0.5833    0.7833    0.6167    0.7595
  Columns 8 through 11
    0.6345    0.7456    0.6456    0.7365
E =
           0          0          0          0          0
  1.0000e+00          0          0          0          0
  5.0000e-01          0          0          0          0
  8.3333e-01  7.0000e-01          0          0          0
  5.8333e-01  6.9048e-01          0          0          0
  7.8333e-01  6.9444e-01  6.9333e-01          0          0
  6.1667e-01  6.9242e-01  6.9309e-01          0          0
  7.5952e-01  6.9359e-01  6.9317e-01  6.9315e-01          0
  6.3452e-01  6.9286e-01  6.9314e-01  6.9315e-01          0
  7.4563e-01  6.9335e-01  6.9315e-01  6.9315e-01  6.9315e-01
>> log(2)-E(10,k)
ans =  -1.5179e-07
```

It is quite remarkable how we can obtain a result with about 7 decimal digits of accuracy by extrapolation using only partial sums of the first 9 terms, especially since the last partial sum still has no correct digit!

It is possible to generalize the ε-algorithm to vector sequences, where the entries in the ε-scheme will be vectors, as encountered in Chapter 11. To do so, we need to simplify Algorithm 5.3 as follows: we will no longer store the entire table in a lower triangular matrix, but store only the diagonal elements and the last row of the matrix E of (5.22) in a vector e in reverse order. Thus, in step 3 for instance, we will have the corresponding variables

$$x_3 = \varepsilon_0^{(2)} = E_{42} = e_3, \quad \varepsilon_1^{(1)} = E_{43} = e_2, \quad \varepsilon_2^{(0)} = E_{44} = e_1.$$

The algorithm becomes:

ALGORITHM 5.4. *Scalar ε-Algorithm, low storage version*

```
function [W,e]=EpsilonAlgorithmLowStorage(x,k);
% EPSILONALGORITHMLOWSTORAGE epsilon algorithm with low storage
%   [w,e]=EpsilonAlgorithmLowStorage(x,k) is an implementation of the
%   epsilon algorithm of P. Wynn using only litte storage. It stores
%   only the diagonal of the epsilon table in W, and the last row of
%   the triangular array in e in reverse order, computing 2k steps.

e(1)=x(1);
for i=2:2*k+1
  v=0; e(i)=x(i);
  for j=i:-1:2
    d=e(j)-e(j-1); w=v+1/d;
    v=e(j-1); e(j-1)=w;
  end;
  W(i-1)=w;
end
```

Accelerating the same sequence as before, we obtain now with the call

```
>> [W,e]=EpsilonAlgorithmLowStorage(x,4)
```

the diagonal of E

```
W =
  Columns 1 through 5
  -2.0000e+00    7.0000e-01   -1.0200e+02    6.9333e-01   -3.8520e+03
  Columns 6 through 8
   6.9315e-01   -1.3698e+05    6.9315e-01
```

and the last row of E (in reverse order):

```
e =
  Columns 1 through 5
   6.9315e-01    4.9226e+05    6.9315e-01    6.5877e+04    6.9315e-01
  Columns 6 through 9
   2.0320e+03    6.9335e-01    9.0000e+00    7.4563e-01
```

5.2.5 Construction of One Step Iteration Methods

In this section we will show how to transform the equation $f(x) = 0$ *systematically* to a fixed point form $x = F(x)$ with a high convergence rate. We can construct these methods geometrically and algebraically.

Geometric Construction

The basic idea here is to approximate the function f in the neighborhood of a zero s by a simpler function h. The equation $f(x) = 0$ is then replaced by $h(x) = 0$ which should be easy to solve and one hopes that the zero of h is also an approximation of the zero of f. In general h will be so simple that we can solve $h(x) = 0$ analytically. If one must also solve $h(x) = 0$ iteratively, then we obtain a *method with inner and outer iterations*.

Let x_k be an approximation of a zero s of f. We choose for h a linear function that has for $x = x_k$ the same function value and derivative as f (i.e. the Taylor polynomial of degree one),

$$h(x) = f(x_k) + f'(x_k)(x - x_k). \tag{5.23}$$

The equation $h(x) = 0$ can be solved analytically,

$$h(x) = 0 \quad \Longleftrightarrow \quad x = x_k - \frac{f(x_k)}{f'(x_k)}.$$

Now the zero of h may indeed be a better approximation of s, as we can see from Figure 5.8. We have obtained the iteration

$$x_{k+1} = F(x_k) = x_k - \frac{f(x_k)}{f'(x_k)}, \tag{5.24}$$

which is called *Newton's Iteration* or the *Newton–Raphson Iteration*[4].

EXAMPLE 5.4. *We return to Kepler's Equation (5.2),*

$$f(E) = E - 0.8 \sin E - \frac{2\pi}{10} = 0.$$

Applying Newton's method, we obtain the iteration

$$E_{k+1} = E_k - \frac{f(E_k)}{f'(E_k)} = E_k - \frac{E_k - 0.8 \sin E_k - \frac{2\pi}{10}}{1 - 0.8 \cos E_K}.$$

Starting this iteration with $E = 1$ we obtain the values

```
1.531027719719945
1.424291078234385
1.419147688353853
1.419135783894323
1.419135783830583
1.419135783830583
```

---

[4]Raphson was a great admirer of Newton, and tried to understand and generalize the method Newton originally presented on the concrete case of one polynomial.

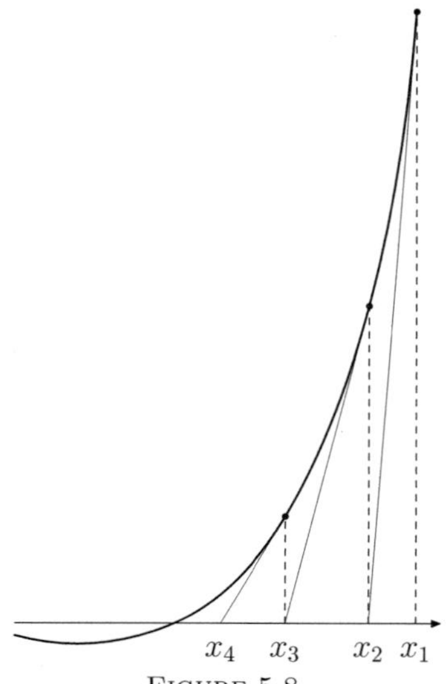

FIGURE 5.8.
Geometric Derivation of the Newton-Raphson Method

which clearly show quadratic convergence.

Instead of the Taylor polynomial of degree one, we can also consider the
function

$$h(x) = \frac{a}{x+b} + c. \tag{5.25}$$

We would like to determine the parameters a, b and c such that again h has
the same function value and the same first, and also second derivatives as f
at x_k:

$$
\begin{aligned}
f(x_k) &= h(x_k) &= \frac{a}{x_k+b} + c \\[2mm]
f'(x_k) &= h'(x_k) &= -\frac{a}{(x_k+b)^2} \\[2mm]
f''(x_k) &= h''(x_k) &= \frac{2a}{(x_k+b)^3}
\end{aligned}
\tag{5.26}
$$

We use MAPLE to define and solve the nonlinear system (5.26):

```
> h:=x->a/(x+b)+c;
> solve({h(xk)=f, D(h)(xk)=fp, D(D(h))(xk)=fpp, h(xn)=0}, {a,b,c,xn});
```

$$xn = \frac{fpp\,xk\,f - 2\,fp^2\,xk + 2\,f\,fp}{fpp\,f - 2\,fp^2}.$$

Rearranging, we obtain *Halley's Iteration*,

$$x_{k+1} = x_k - \frac{f(x_k)}{f'(x_k)} \frac{1}{1 - \dfrac{1}{2}\dfrac{f(x_k)f''(x_k)}{f'(x_k)^2}}. \tag{5.27}$$

Algebraic Construction

If $f(x) = 0$, adding x on both sides gives $x = F(x) = x + f(x)$. However, we cannot expect for arbitrary f that $|F'| = |1 + f'|$ is smaller than 1. But we can generalize this idea by premultiplying $f(x) = 0$ first with some function $h(x)$ which has yet to be determined. Thus, we consider the fixed point form

$$x = F(x) = x + h(x)f(x). \tag{5.28}$$

Let us try to choose $h(x)$ so to make $|F'(s)| < 1$ or even better $F'(s) = 0$. We have

$$F'(x) = 1 + h'(x)f(x) + h(x)f'(x).$$

For quadratic convergence, we need

$$F'(s) = 1 + h(s)f'(s) = 0. \tag{5.29}$$

To solve (5.29) for h, the following condition must hold

$$f'(s) \neq 0. \tag{5.30}$$

This gives us

$$h(s) = -\frac{1}{f'(s)}. \tag{5.31}$$

Condition (5.30) signifies that s must be a simple zero of f. Furthermore, the only condition on the choice of h is that its value at $x = s$ must be $-1/f'(s)$. Since s is unknown, the simplest choice for h is

$$h(x) = -\frac{1}{f'(x)}.$$

This choice leads to the iteration

$$x = F(x) = x - \frac{f(x)}{f'(x)}, \tag{5.32}$$

which is again *Newton's Iteration*. By the algebraic derivation, we have proved that *for simple zeros, (i.e., when $f'(s) \neq 0$) Newton's iteration generates a quadratically convergent sequence.*

Every fixed point iteration $x = F(x)$ can be regarded as a Newton iteration for some function g. This observation was already made in 1870 by

Schröder [120]. In order to determine g, we need to solve the differential equation

$$x - \frac{g(x)}{g'(x)} = F(x). \tag{5.33}$$

Rearranging (5.33), we obtain

$$\frac{g'(x)}{g(x)} = \frac{1}{x - F(x)},$$

which we can integrate to obtain

$$\ln|g(x)| = \int \frac{dx}{x - F(x)} + C,$$

and therefore

$$|g(x)| = C \exp\left(\int \frac{dx}{x - F(x)} \right). \tag{5.34}$$

EXAMPLE 5.5. *We consider the second fixed point form (5.10) of (5.8). Here we have*

$$F(x) = \frac{1 + x}{1 + e^x},$$

and using the approach above, we obtain

$$\int \frac{dx}{x - F(x)} = \int \frac{1 + e^x}{xe^x - 1} dx.$$

Dividing the numerator and denominator by e^x, and integrating gives

$$= \int \frac{e^{-x} + 1}{x - e^{-x}} dx = \ln|x - e^{-x}|.$$

Thus the fixed point form

$$x = \frac{1 + x}{1 + e^x}$$

is Newton's iteration for $f(x) = x - e^{-x} = 0$, where we dropped the absolute value, which is not important for root finding.

We can interpret Halley's Iteration as an "improved" Newton's iteration, since we can write

$$x = F(x) = x - \frac{f(x)}{f'(x)} G(t(x)), \tag{5.35}$$

with

$$G(t) = \frac{1}{1 - \frac{1}{2}t}, \tag{5.36}$$

and

$$t(x) = \frac{f(x)f''(x)}{f'(x)^2}. \tag{5.37}$$

If f has small curvature, i.e. $f''(x) \approx 0$, then $t(x) \approx 0$ which implies $G(t) \approx 1$. This shows that in the case of an f with small curvature, both methods are similar.

Which is the corresponding function g for which we can regard Halley's method (applied to $f(x) = 0$) as a Newton iteration? Surprisingly, the differential equation

$$x - \frac{g(x)}{g'(x)} = F(x) = x + \frac{2f'(x)f(x)}{f(x)f''(x) - 2f'(x)^2}$$

has a simple solution. We have

$$\frac{g'(x)}{g(x)} = -\frac{1}{2}\frac{f''(x)}{f'(x)} + \frac{f'(x)}{f(x)},$$

and integration yields

$$\ln|g(x)| = -\frac{1}{2}\ln|f'(x)| + \ln|f(x)| = \ln\left|\frac{f(x)}{\sqrt{f'(x)}}\right|.$$

We have therefore proved the following theorem:

THEOREM 5.1. *Halley's iteration (5.27) for $f(x) = 0$ is Newton's iteration (5.24) for the nonlinear equation*

$$\frac{f(x)}{\sqrt{f'(x)}} = 0.$$

Let us now analyze Halley-like iteration forms

$$x = F(x) = x - \frac{f(x)}{f'(x)}H(x) \tag{5.38}$$

and find conditions for the function $H(x)$ so that the iteration yields sequences that converge quadratically or even cubically to a simple zero of f. With the abbreviation

$$u(x) := \frac{f(x)}{f'(x)},$$

we obtain for the derivatives

$$\begin{aligned}
F &= x - uH, \\
F' &= 1 - u'H - uH', \\
F'' &= -u''H - 2u'H' - uH'',
\end{aligned}$$

and

$$\begin{aligned}
u' &= 1 - \frac{f f''}{f'^2}, \\
u'' &= -\frac{f''}{f'} + 2\frac{f f''^2}{f'^3} - \frac{f f'''}{f'^2}.
\end{aligned}$$

Because $f(s) = 0$ we have

$$u(s) = 0, \quad u'(s) = 1, \quad u''(s) = -\frac{f''(s)}{f'(s)}. \tag{5.39}$$

It follows for the derivatives of F that

$$F'(s) = 1 - H(s) \tag{5.40}$$

$$F''(s) = \frac{f''(s)}{f'(s)} H(s) - 2H'(s). \tag{5.41}$$

We conclude from (5.40) and (5.41) that for *quadratic convergence*, we must have

$$H(s) = 1,$$

and for cubic convergence, we need in addition

$$H'(s) = \frac{1}{2}\frac{f''(s)}{f'(s)}.$$

However, s is unknown; therefore, we need to choose H as a function of f and its derivatives:

$$H(x) = G(f(x), f'(x), \ldots).$$

For example, if we choose $H(x) = 1 + f(x)$, then because of $H(s) = 1$, we obtain an iteration with quadratic convergence:

$$x = x - \frac{f(x)}{f'(x)}(1 + f(x)).$$

If we choose

$$H(x) = G(t(x)), \tag{5.42}$$

with

$$t(x) = \frac{f(x)f''(x)}{f'(x)^2}, \tag{5.43}$$

then because

$$t(x) = 1 - u'(x),$$

we get

$$H'(x) = G'(t(x))t'(x) = -G'(t(x))u''(x).$$

This implies that

$$\begin{aligned} H(s) &= G(0), \\ H'(s) &= -G'(0)u''(s) = G'(0)\frac{f''(s)}{f'(s)}. \end{aligned} \tag{5.44}$$

We thus have derived a theorem

THEOREM 5.2. *(Gander [39]) Let s be a simple zero of f, and let G be any function with $G(0) = 1$, $G'(0) = \frac{1}{2}$ and $|G''(0)| < \infty$. Then the sequence generated by the fixed point form*

$$x = F(x) = x - \frac{f(x)}{f'(x)}\, G\left(\frac{f(x)f''(x)}{f'(x)^2}\right)$$

converges at least cubically to the simple zero s of f.

EXAMPLE 5.6. *Many well known iteration methods are special cases of Theorem 5.2. As we can see from the Taylor expansions of $G(t)$, they converge cubically:*

1. *Halley's method*

$$G(t) = \frac{1}{1 - \frac{1}{2}t} = 1 + \frac{1}{2}t + \frac{1}{4}t^2 + \frac{1}{8}t^3 + \cdots$$

2. *Euler's Iteration (see Problem 5.28)*

$$G(t) = \frac{2}{1 + \sqrt{1 - 2t}} = 1 + \frac{1}{2}t + \frac{1}{2}t^2 + \frac{5}{8}t^3 + \cdots$$

3. *Inverse quadratic interpolation (see Problem 5.30)*

$$G(t) = 1 + \frac{1}{2}t.$$

5.2.6 Multiple Zeros

The quadratic convergence of Newton's method was based on the assumption that s is a simple zero and that therefore $f'(s) \neq 0$. We will now investigate the convergence for zeros with multiplicity greater than one.

```
> F:=x->x-f(x)/D(f)(x);
> dF:=D(F)(x);
```

$$F := x \to x - \frac{\mathrm{f}(x)}{\mathrm{D}(f)(x)} \tag{5.45}$$

$$dF := \frac{\mathrm{f}(x)\, \mathrm{D}^{(2)}(f)(x)}{\mathrm{D}(f)(x)^2} \tag{5.46}$$

Let us now assume that $f(x)$ has a zero of multiplicity n at $x = s$. We therefore define $f(x)$ to be

```
> f:=x->(x-s)^n*g(x);
```

$$f := x \rightarrow (x - s)^n \, g(x)$$

where $g(s) \neq 0$. We inspect the first derivative of $F(x)$. If $F'(s) \neq 0$, then the iteration converges only linearly.

```
> dF;
```

$$(x - s)^n \, g(x) \left(\frac{(x - s)^n \, n^2 \, g(x)}{(x - s)^2} - \frac{(x - s)^n \, n \, g(x)}{(x - s)^2} + 2 \frac{(x - s)^n \, n \, \mathrm{D}(g)(x)}{x - s} \right.$$

$$\left. + (x - s)^n \, (\mathrm{D}^{(2)})(g)(x) \right) \bigg/ \left(\frac{(x - s)^n \, n \, g(x)}{x - s} + (x - s)^n \, \mathrm{D}(g)(x) \right)^2$$

Taking the limit of the above expression for $x \rightarrow s$ we obtain:

```
> limit(%, x=s);
```

$$\frac{n - 1}{n}$$

We have proved that *Newton's iteration converges only linearly with factor* $(n - 1)/n$ *if* $f(x)$ *has a zero of multiplicity* n. Thus, e.g., for a double root convergence is linear with factor $1/2$.

In order to find a remedy, we need to modify Newton's iteration to recover $F'(s) = 0$. We have seen that for multiple zeros, $F'(s) = \frac{n-1}{n}$. We therefore have for Newton's method

$$F'(s) = \left(s - \frac{f(s)}{f'(s)} \right)' = \frac{n - 1}{n} \implies \left(\frac{f(s)}{f'(s)} \right)' = 1 - \frac{n - 1}{n} = \frac{1}{n}.$$

Therefore, a possible remedy for reestablishing quadratic convergence for a double root, i.e. $F'(s) = 0$, is to take "double steps"

$$x_{k+1} = x_k - 2 \frac{f(x_k)}{f'(x_k)},$$

or, for a root of multiplicity n,

$$x_{k+1} = x_k - n \frac{f(x_k)}{f'(x_k)}. \tag{5.47}$$

However, one seldom knows the multiplicity in advance. So we should try to estimate n. An old proposal by Schröder of 1870 [121] is to use instead of the factor n

$$\frac{f'^2}{f'^2 - f f''}. \tag{5.48}$$

The resulting iteration looks almost like Halley's iteration (the factor $1/2$ is missing):

$$x_{k+1} = x_k - \frac{f(x_k)}{f'(x_k)} \frac{1}{1 - \dfrac{f(x_k) f''(x_k)}{f'(x_k)^2}}. \tag{5.49}$$

Interpreting again Iteration (5.49) as Newton's iteration by solving the differential equation

$$\frac{g(x)}{g'(x)} = \frac{f(x)}{f'(x)} \frac{1}{1 - \dfrac{f(x)f''(x)}{f'(x)^2}}$$

we obtain

$$\frac{g'(x)}{g(x)} = \frac{f' - \dfrac{f f''}{f'}}{f} = \frac{f'}{f} - \frac{f''}{f'}$$

and

$$\ln|g| = \ln|f| - \ln|f'| \quad \Longrightarrow \quad g(x) = \frac{f(x)}{f'(x)}.$$

Thus Schröder's method (5.49) is equivalent to applying Newton's method to $f(x)/f'(x) = 0$, *which cancels multiple roots automatically.*

5.2.7 Multi-Step Iteration Methods

In the previous sections we considered one step iteration methods: $x_{k+1} = F(x_k)$. Now we would like to make use of more information to compute the new iterate. We therefore consider iterations of the form

$$x_{k+1} = F(x_k, x_{k-1}, x_{k-2}, \ldots).$$

The idea is again to approximate f by a simpler function h and to compute the zero of h as an approximation of the zero s of f. Since more than one point is involved, it is natural to use *interpolation techniques*, see Chapter 4.

The simplest case is to use two points (x_0, f_0) and (x_1, f_1), where $f_0 = f(x_0)$ and $f_1 = f(x_1)$. If we approximate f by the linear function $h(x) = ax + b$ defined by the straight line through these two points, then the zero $x_2 = -b/a$ of h will be a new approximation to the zero s of f. We compute h and the zero x_2 by the following MAPLE statements:

```
> h:=x->a*x+b;
> solve({h(x0)=f0, h(x1)=f1, h(x2)=0}, {a,b,x2});
```

$$\left\{ a = \frac{-f1 + f0}{x0 - x1}, x2 = -\frac{-x1\,f0 + f1\,x0}{-f1 + f0}, b = \frac{-x1\,f0 + f1\,x0}{x0 - x1} \right\}$$

Thus we obtained the iteration

$$x_{k+1} = x_k - f_k \frac{x_k - x_{k-1}}{f_k - f_{k-1}} \qquad \textit{Secant Method.} \tag{5.50}$$

Notice that the secant method is obtained by approximating the derivative in Newton's method by a finite difference,

$$f'(x_k) \approx \frac{f(x_k) - f(x_{k-1})}{x_k - x_{k-1}}.$$

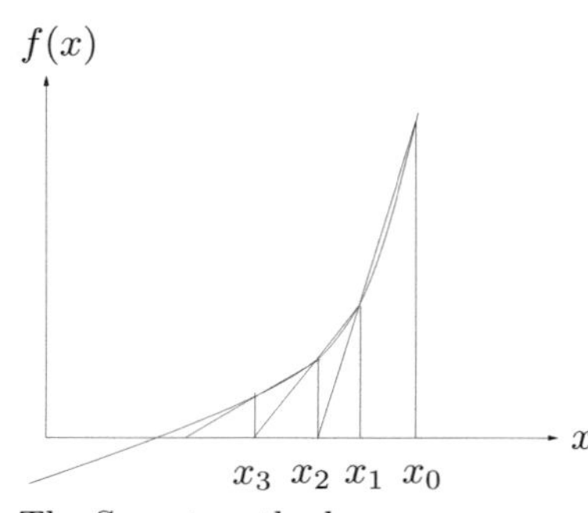

The Secant method

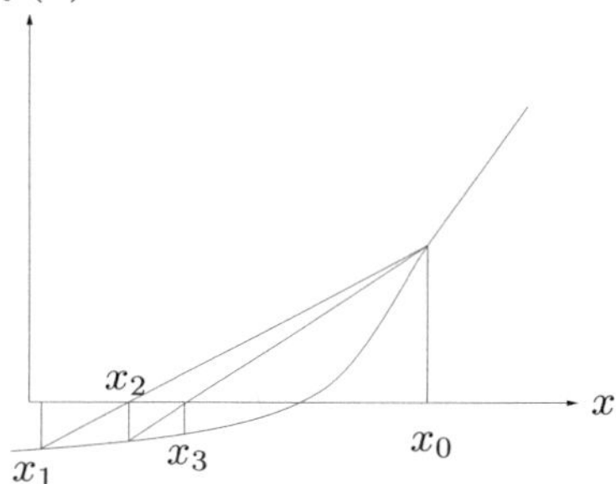

The Regula Falsi method

We determine the convergence rate of the secant method with the help of MAPLE:

```
> F:=(u,v)->u-f(u)*(u-v)/(f(u)-f(v));
```

$$F := (u,\, v) \rightarrow u - \frac{f(u)\,(u - v)}{f(u) - f(v)}$$

```
> x[k+1]=F(x[k],x[k-1]);
```

$$x_{k+1} = x_k - \frac{f(x_k)\,(x_k - x_{k-1})}{f(x_k) - f(x_{k-1})} \tag{5.51}$$

Using (5.51), we obtain the recurrence

$$e_{k+1} = F(s + e_k, s + e_{k-1}) - s$$

for the error $e_{k+1} = x_{k+1} - s$. The right-hand side can be expanded into a multivariate Taylor series about $e_k = 0$ and $e_{k+1} = 0$. We assume that s is a simple root ($f'(s) \neq 0$), and also that $f''(s) \neq 0$ holds. We set $f(s) = 0$ and compute the first term of the Taylor series expansion,

```
> f(s):=0:
> e2=normal(readlib(mtaylor)(F(s+e1,s+e0)-s,[e0,e1],4));
```

$$e2 = \frac{1}{2}\, \frac{e0\,(\mathrm{D}^{(2)})(f)(s)\,e1}{\mathrm{D}(f)(s)}$$

If we divide this leading coefficient by e_0 and e_1, we see that the limit of the quotient $e_2/(e_0\,e_1)$ is a constant different from zero. We make the ansatz that the convergence exponent is p and substitute $e_2 = K\,e_1{}^p$ and $e_1 = K\,e_0{}^p$, and divide by the constant K^p.

```
> %/e1/e0;
```

$$\frac{e2}{e1\,e0} = \frac{1}{2}\, \frac{(\mathrm{D}^{(2)})(f)(s)}{\mathrm{D}(f)(s)}$$

```
> simplify(subs(e2=K*e1^p,e1=K*e0^p,%/K^p),assume=positive);
```

$$e_0^{(p^2 - p - 1)} = \frac{1}{2} \frac{K^{-p}\,(\mathrm{D}^{(2)})(f)(s)}{\mathrm{D}(f)(s)}$$

This equation is valid for all errors e_0. Since the right hand side is constant, the left hand side must also be independent of e_0. This is the case only if the exponent of e_0 is zero. This condition is an equation for p (it is the *golden section equation* we will also encounter in Chapter 12), whose solution is

```
> solve(ln(lhs(%)),p);
```

$$\frac{1}{2}\sqrt{5} + \frac{1}{2},\ \frac{1}{2} - \frac{1}{2}\sqrt{5}$$

and gives the convergence exponent $p = (1 + \sqrt{5})/2$ for the secant method (note that the other solution is negative). *Thus we have shown that the secant method converges superlinearly*:

$$e_{k+1} \sim \frac{f''(s)}{2f'(s)}\, e_k^{1.618}.$$

A variant of the secant method is called *Regula Falsi*. The idea here is to start with two points in which the function values have different signs, as with bisection. Then compute the zero of h and *continue always with the new point and the one of the older points for which the function value has a different sign*. That way the zero is always bracketed between the values kept, as shown in Figure 5.2.7. Regula Falsi has the advantage versus the secant method that the zero is bracketed between the two points and the iteration will definitely converge, as with bisection. However, the convergence rate is no longer superlinear but only linear and can sometimes be slower than bisection.

We now consider three of the previous iteration values. In that case the function f can be approximated by an interpolating polynomial of degree two. Let (x_0, f_0), (x_1, f_1) and (x_2, f_2) be the three given previous iterations. Interpolating them by a quadratic polynomial

$$h(x) = ax^2 + bx + c,$$

and solving the quadratic equation $h(x) = 0$, we obtain the two solutions analytically. Choosing for x_3 the one that is closer to x_2, we compute $f_3 = f(x_3)$, and continue the process with (x_1, f_1), (x_2, f_2) and (x_3, f_3). This method is called *Müller's method* and it is shown in Figure 5.9. The convergence is again superlinear, see Problem 5.27.

Another possibility to approximate f with three points is by *inverse interpolation*. This time we interpolate the data by a polynomial of degree two in the variable y,

$$\begin{array}{c|ccc} y & f_0, & f_1, & f_2 \\ \hline x & x_0, & x_1, & x_2 \end{array}. \tag{5.52}$$

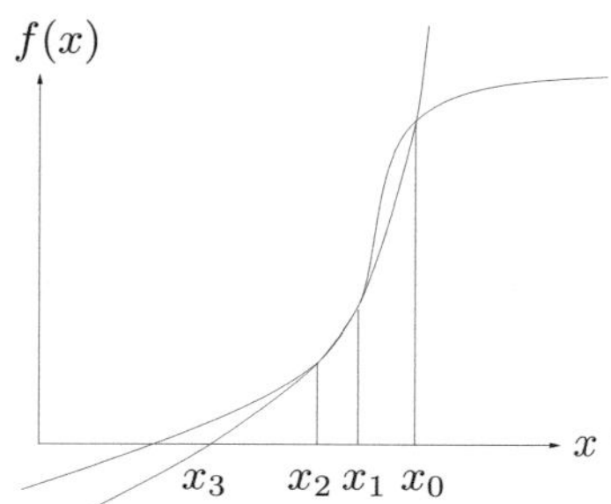

FIGURE 5.9. *Müller's method*

Thus $x = h(y)$ with $x_k = h(f_k)$, $k = 0, 1, 2$. We want to compute the new approximation $x_3 = h(0)$. This way we do not have to solve a quadratic equation, we merely need to evaluate a polynomial of degree two, see Problem 5.31. The convergence of this method is also superlinear. The method can be implemented using *Aitken-Neville interpolation*, see Section 4.2.7. Furthermore it can be generalized to include all computed points, see Problem 5.31.

5.2.8   A New Iteration Formula

Having considered the Newton and the secant methods to compute roots of a nonlinear equation, we now want to show how a new iteration method can be derived and analyzed, following an idea proposed in [46]. The new method is a combination of the Newton and the secant methods. It uses function values and first derivatives at two points. These four values define a (Hermite) interpolation polynomial (see Section 4.3.1) of degree three. A zero of this polynomial can be taken as the next approximation of the root. Unfortunately, the explicit expression for this zero is rather complex, so we propose to use inverse interpolation for the given data. We then need only to evaluate the resulting polynomial at $y = 0$ to obtain the new approximation for the root. In MAPLE, this is done with the following commands.

```
> p:=x->a*x^3+b*x^2+c*x+d:
> solve({p(f(x[0]))=x[0],D(p)(f(x[0]))=1/D(f)(x[0]),
>        p(f(x[1]))=x[1],D(p)(f(x[1]))=1/D(f)(x[1])},{a,b,c,d}):
> assign(%);
> p(0);
```

$$\Big(-f(x_1)^3\,\mathrm{D}(f)(x_1)\,f(x_0) + f(x_1)^3\,x_0\,\mathrm{D}(f)(x_0)\,\mathrm{D}(f)(x_1)$$
$$+ f(x_1)^2\,\mathrm{D}(f)(x_1)\,f(x_0)^2 - f(x_0)^3\,x_1\,\mathrm{D}(f)(x_0)\,\mathrm{D}(f)(x_1)$$
$$- f(x_1)^2\,\mathrm{D}(f)(x_0)\,f(x_0)^2 - 3\,f(x_1)^2\,x_0\,\mathrm{D}(f)(x_0)\,\mathrm{D}(f)(x_1)\,f(x_0)$$
$$+ f(x_1)\,f(x_0)^3\,\mathrm{D}(f)(x_0) + 3\,f(x_1)\,x_1\,\mathrm{D}(f)(x_0)\,\mathrm{D}(f)(x_1)\,f(x_0)^2 \Big) \Big/$$
$$\Big(\mathrm{D}(f)(x_0)\,\mathrm{D}(f)(x_1)\,(f(x_1) - f(x_0))\,(f(x_1)^2 - 2\,f(x_1)\,f(x_0) + f(x_0)^2) \Big)$$

The resulting expression is still not very simple. However, if the evaluation of f and f' is very expensive, it may still pay off since the convergence exponent is 2.73, as we will see.

For the convergence analysis we expand

$$e_{k+2} = F(s + e_k, s + e_{k-1}) - s.$$

into a multivariate Taylor series at $e_k = 0$ and $e_{k+1} = 0$, as we have done for the secant method:

```
> F:=unapply(%,x[0],x[1]):
> f(s):=0:
> e2=readlib(mtaylor)(F(s+e0,s+e1)-s,[e0,e1],8):
> eq:=normal(%);
```

$$eq := e2 = \frac{1}{24} e1^2\, e0^2 \Big(D(f)(s)^2\, (D^{(4)})(f)(s) + 15\,(D^{(2)})(f)(s)^3$$
$$- 10\,(D^{(2)})(f)(s)\,(D^{(3)})(f)(s)\,D(f)(s)\Big) \Big/ D(f)(s)^3$$

As with the Newton and Secant methods, we consider only simple roots, so that $D(f)(s) \neq 0$. If this condition holds, then the above equation tells us that in the limit,

```
> e2/e0^2/e1^2=const;
```

$$\frac{e2}{e0^2\, e1^2} = const$$

Let us again introduce the convergence exponent p and make the following substitutions:

```
> subs(e2=K*e1^p,e1=K*e0^p,%);
```

$$\frac{(K\, e0^p)^p}{K\, e0^2\, (e0^p)^2} = const$$

```
> simplify(%,assume=positive);
```

$$K^{(-1+p)}\, e0^{(p^2-2-2\,p)} = const$$

This equation must hold for all errors e_0. Since K, p and *const* are all constant, the exponent of e_0 must be zero.

```
> solve(p^2-2*p-2=0,p);
```

$$1 + \sqrt{3},\ 1 - \sqrt{3}$$

```
> evalf([%]);
```

$$[2.732050808, -.732050808]$$

Thus, the convergence exponent is $p = 1 + \sqrt{3}$ and we have *super-quadratic convergence*.

Let us use MAPLE to demonstrate the above convergence rate with an example. We use our algorithm to compute the zero of the function $f(x) = e^x + x$ starting with $x_0 = 0$ and $x_1 = 1$. For every iteration, we print the number of correct digits (first column) and its ratio to the number of correct digits in the previous step (second column). This ratio should converge to the convergence exponent $p = 2.73$.

```
> f:=x->exp(x)+x;
```

$$f := x \to e^x + x$$

```
> solve(f(x)=0,x);
```

$$-\mathrm{LambertW}(1)$$

```
> Digits:=500:
> x0:=0.0: x1:=1.0:
> for i to 6 do
>    x2:=evalf(F(x0,x1));
>    d2:=evalf(log[10](abs(x2+LambertW(1)))):
>    if i=1 then lprint(evalf(d2,20))
>    else lprint(evalf(d2,20),evalf(d2/d1,20))
>    fi;
>    x0:=x1: x1:=x2: d1:=d2:
> od:
```

```
 -2.7349672475721192576
 -7.8214175487946676893    2.8597847216407742880
 -23.118600850923949542    2.9558070140989569391
 -63.885801375143936026    2.7633939349141321183
 -176.01456902838525572    2.7551437915728687417
 -481.80650538330786806    2.7373103717659224742
```

The super-quadratic convergence rate with exponent $p = 1 + \sqrt{3}$ is now evident.

5.2.9   Dynamical Systems

We have so far shown a simplified view of fixed point iterations. A fixed point iteration of the form

$$x_{k+1} = F(x_k)$$

can do much more than just converge to a fixed point or diverge. The study of all the possible behaviors of such a fixed point iteration led to a new field in applied mathematics: *dynamical systems*. An interesting reference for dynamical systems in numerical analysis is [134]. To give a glimpse of

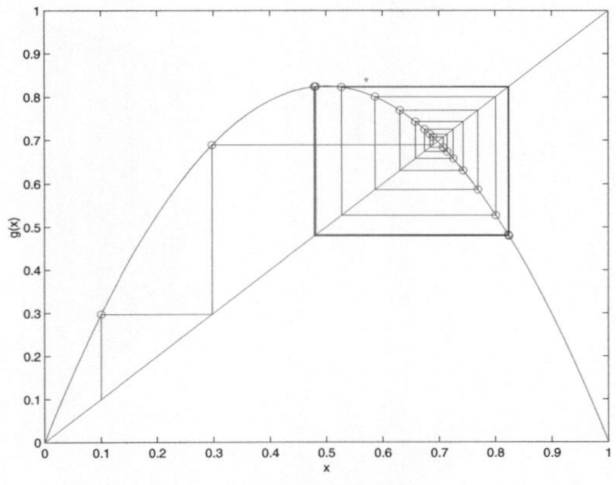

FIGURE 5.10.
Neither of the fixed points attracts the solution of the fixed point iteration. A two cycle is found.

how rich this field is, we will discuss the classical example of Mitchell Jay Feigenbaum. We consider the fixed point equation

$$x = F(x) = ax - ax^2, \quad a > 0$$

which could have come from an attempt to solve the nonlinear equation $f(x) = ax^2 - (a-1)x = 0$ using a fixed point iteration. We also assume that the parameter a in this problem is non-negative. The solutions are

$$s_1 = 0, \quad s_2 = \frac{a-1}{a}.$$

Which of those two does the fixed point iteration find ? Looking at the derivative

$$F'(x) = a - 2ax,$$

we find $F'(s_1) = a$ and $F'(s_2) = 2 - a$. Hence the fixed point iteration will converge to s_1 for $0 < a < 1$ and it will converge to s_2 for $1 < a < 3$. But what will happen if $a \geq 3$? Figure 5.10 shows for $a = 3.3$ that the trajectory is now neither attracted to s_1 nor to s_2. The sequence of iterates oscillates between two values (they form a stable periodic orbit with period two) around s_2. The two values are new fixed points of the mapping obtained when the original map is applied twice

$$F(F(x)) = a^2 x(1 - x)(1 - ax + ax^2).$$

The new fixed points representing period two solutions are

$$s_{3,4} = \frac{a + 1 \pm \sqrt{a^2 - 2a - 3}}{2a}.$$

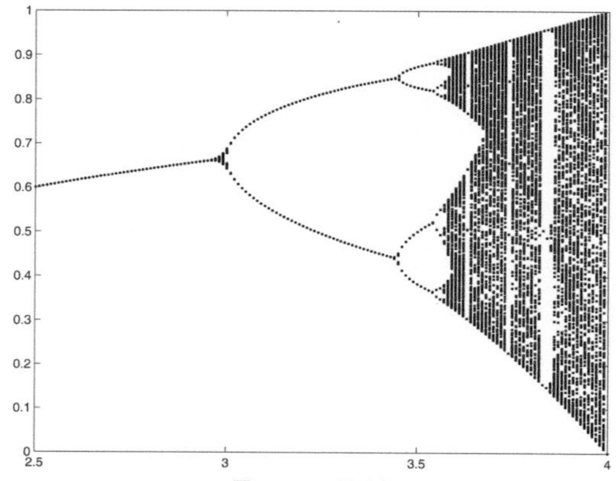

FIGURE 5.11.

*Period doubling and chaos in the simple fixed point
iteration $x_{k+1} = ax_k - a(x_k)^2$.*

For $a = 3.3$ we obtain $s_3 = 0.8236$ and $s_4 = 0.4794$. Evaluating the derivative
of $F(F(x))$ at the fixed point s_3, we find

$$\frac{d}{dx}F(F(s_3)) = -a^2 + 2a + 4.$$

The same is true for s_4. For the period two solution to be stable, we need
that this derivative is bounded in modulus by 1, which leads to $3 < a <
1 + \sqrt{6} = 3.4495$. Hence for $a > 3.4495$ the period two solution will not be
stable either and the trajectory will be rejected from the two fixed points
and the period two solution. What will happen next ? Figure 5.11 shows the
answer, computed with the MATLAB program

```
N1=100; N2=400;
a=(2.5:0.01:4)';
m=length(a);
x(1:m,1)=1/2;
for i=1:N2
  x(:,i+1)=a.*x(:,i)-a.*x(:,i).^2;
end;
plot(a,x(:,N1:N2),'.k'); xlabel('a');
```

A period 4 solution is formed, which becomes unstable as well for a bigger
than a certain value, whereupon a period 8 solution is formed. This process
is called *period doubling* and the ratio between consecutive intervals of period
doubling approaches a universal constant, the *Feigenbaum constant*. Table
5.3 shows this development for our example. The ratio converges to a limit

$$\lim_{n \to \infty} \frac{d_{n-1}}{d_n} = f = 4.6692016091 \quad \text{Feigenbaum constant.}$$

period	a_k	difference d_k	ratio d_k/d_{k+1}
	$a_1 = 3$		
2		$d_1 = 0.4495$	
	$a_2 = 4.4495$		$d_1/d_2 = 4.7515$
4		$d_2 = 0.0946$	
	$a_3 = 3.5441$		$d_2/d_3 = 4.6601$
8		$d_3 = 0.0203$	
	$a_4 = 3.5644$		

TABLE 5.3. *Approaching the Feigenbaum Constant*

The important observation (which we do not prove) is that the Feigenbaum constant is universal. If, instead of repeatedly iterating $F(x) = ax - ax^2$ we had used the function $F(x) = a \sin x$, we would have found the exact same constant describing the rate of bifurcation.

Because of this ratio of consecutive intervals, this period doubling will lead to an infinite period at a certain finite value of a. Using

$$d_n \approx \frac{d_{n-1}}{f}$$

we can compute an approximation for that finite value,

$$
\begin{aligned}
a_\infty &= a_4 + d_4 + d_5 + \cdots \\
&\approx a_4 + \frac{d_3}{f} + \frac{d_3}{f^2} + \cdots \\
&= a_4 + \frac{d_3}{f}\left(1 + \frac{1}{f} + \frac{1}{f^2} + \cdots\right) = a_4 + \frac{d_3}{f}\frac{1}{1 - 1/f} \\
a_\infty &\approx a_4 + \frac{d_3}{f-1} = 3.5699.
\end{aligned}
$$

As one can see in Figure 5.11 something interesting happens for $a > a_\infty$: the solution trajectory bounces back and forth without any apparent order for certain values of $a > a_\infty$. The system behaves *chaotically*. For other values of a, a new period doubling sequence is started as well, as one can see in Figure 5.11.

5.3   Zeros of Polynomials

Zeros of polynomials used to be important topic when calculations had to be performed by hand. Whenever possible, one tried to transform a given equation to a polynomial equation. For example if we wish to compute a solution of the goniomatric equation

$$\tan \alpha = \sin \alpha - \cos \alpha,$$

then using the change of variables

$$t = \tan \frac{\alpha}{2}, \quad \sin \alpha = \frac{2t}{1+t^2}, \quad \cos \alpha = \frac{1-t^2}{1+t^2}, \quad \tan \alpha = \frac{2t}{1-t^2},$$

we obtain

$$\frac{2t}{1-t^2} = \frac{2t}{1+t^2} - \frac{1-t^2}{1+t^2},$$

or equivalently, the following algebraic equation of degree 4.

$$t^4 + 4t^3 - 2t^2 + 1 = 0,$$

Surely this equation was easier to solve than the equivalent one involving trigonometric functions in the old days before computers were available.

The algebraic eigenvalue problem $A\boldsymbol{x} = \lambda \boldsymbol{x}$ can be reduced to computing the zeros of a polynomial, since the eigenvalues are the roots of the the *characteristic polynomial*:

$$P_n(\lambda) = \det(A - \lambda I).$$

This is however not an advisable numerical strategy for computing eigenvalues, as we will see soon, see also Chapter 7.

The Fundamental Theorem of Algebra states that every polynomial of degree $n \geq 1$,

$$P_n(x) = a_0 + a_1 x + \cdots + a_n x^n, \tag{5.53}$$

has at least one zero in $\mathbb{C}$. If s_1 is such a zero then we can factor P_n using Horner's scheme discussed in the next section to obtain

$$P_n(x) = P_{n-1}(x)(x - s_1). \tag{5.54}$$

The remaining zeros of P_n are also zeros of P_{n-1}. By continuing with the polynomial P_{n-1} we have *deflated* the zero s_1 from P_n. If s_2 is another zero of P_{n-1} then again we can deflate it and we obtain:

$$P_{n-1}(x) = P_{n-2}(x)(x - s_2) \Rightarrow P_n(x) = P_{n-2}(x)(x - s_1)(x - s_2).$$

Continuing in this way, we reach finally

$$P_n(x) = P_0(x)(x - s_1) \cdots (x - s_n).$$

Since $P_0(x) = \text{const}$, by comparing with (5.53) we must have that $P_0(x) = a_n$, and we have shown that P_n can be factored into n linear factors, and that a polynomial of degree n can have at most n zeros.

The following theorem estimates the region in the complex plane where the zeros of a polynomial can be.

THEOREM 5.3. *All the zeros of the polynomial* $P_n(x) = a_0 + a_1 x + \cdots + a_n x^n$ *with* $a_n \neq 0$ *lie in the disk around the origin in the complex plane with radius*

$$r = 2 \max_{1 \leq k \leq n} \sqrt[k]{\left| \frac{a_{n-k}}{a_n} \right|}.$$

PROOF. We want to show that if $|z| > r$, then $|P_n(z)| > 0$. Let $|z| > 2 \sqrt[k]{\left| \frac{a_{n-k}}{a_n} \right|}$ for all k. Then

$$2^{-k} |z|^k > \left| \frac{a_{n-k}}{a_n} \right| \iff 2^{-k} |a_n| |z|^n > |a_{n-k}| |z^{n-k}|. \tag{5.55}$$

Because of the triangle inequality $||x| - |y|| \leq |x + y| \leq |x| + |y|$, we conclude that

$$|P_n(z)| \geq |a_n z^n| - \sum_{k=1}^{n} |a_{n-k}| |z^{n-k}|,$$

and because of Equation (5.55) the right-hand side becomes

$$\geq |a_n| |z^n| - \sum_{k=1}^{n} 2^{-k} |a_n| |z^n| = |a_n| |z^n| (1 - \sum_{k=1}^{n} 2^{-k}) = |a_n| |z^n| (\tfrac{1}{2})^n > 0.$$

$\square$

In the work of Ruffini, Galois and Abel it was proved that, in general, explicit formulas for the zeros only exist for polynomials of degree $n \leq 4$. MAPLE knows these explicit formulas: for $n = 3$, they are known as the *Cardano formulas*, and the formulas for $n = 4$ were discovered by Ferrari and published in [16]. To solve a general polynomial equation of degree 4 algebraically, $x^4 + bx^3 + cx^2 + dx + e = 0$, the following commands are needed:

```
> solve(x^4+b*x^3+c*x^2+d*x+e=0,x);
> allvalues(%);
```

The resulting expressions are several pages long and maybe not very useful.

5.3.1 Condition of the Zeros

Zeros of polynomials became less popular when Jim Wilkinson discovered that they are often very ill conditioned. As part of a test suite for a new computer, Wilkinson constructed from the following polynomial of degree 20 (*Wilkinson's polynomial*) with roots $x_i = 1, 2, \ldots, 20$ by expanding the product:

$$P_{20}(x) = \prod_{i=1}^{20} (x - i) = x^{20} - 210 x^{19} + 20615 x^{18} - \cdots + 20!. \tag{5.56}$$

Then he used a numerical method to compute the zeros and he was astonished to observe that his program found some complex zeros, rather than reproducing the zeros $x_i = 1, 2, \ldots, 20$. After having checked very carefully that there was no programming error, and that the hardware was also working correctly, he then tried to understand the results by a backward error analysis. Let z be a simple zero of

$$P_n(x) = a_n x^n + a_{n-1} x^{n-1} + \cdots + a_0.$$

Let ε be a small parameter and let g be another polynomial of degree n. Consider the zero $z(\varepsilon)$ (corresponding to z, i.e., $z(0) = z$) of the perturbed polynomial h,

$$h(x) := P_n(x) + \varepsilon g(x).$$

We expand $z(\varepsilon)$ in a series

$$z(\varepsilon) = z + \sum_{k=1}^{\infty} p_k \varepsilon^k.$$

The coefficient $p_1 = z'(0)$ can be computed by differentiating

$$P_n(z(\varepsilon)) + \varepsilon g(z(\varepsilon)) = 0$$

with respect to ε. We obtain

$$P_n'(z(\varepsilon)) z'(\varepsilon) + g(z(\varepsilon)) + \varepsilon g'(z(\varepsilon)) z'(\varepsilon) = 0,$$

and therefore

$$z'(\varepsilon) = -\frac{g(z(\varepsilon))}{P_n'(z(\varepsilon)) + \varepsilon g'(z(\varepsilon))}.$$

Thus, for $\varepsilon = 0$,

$$z'(0) = -\frac{g(z)}{P_n'(z)}. \tag{5.57}$$

We now apply (5.57) to Wilkinson's polynomial (5.56). Wilkinson perturbed only the coefficient $a_{19} = -210$ by 2^{-23} (which was the machine precision for single precision on some early computers). This modification corresponds to the choice of $g(x) = x^{19}$ and $\varepsilon = -2^{-23}$, and we obtain for the zero $z_r = r$ the perturbation

$$\delta z_r \approx 2^{-23} \frac{r^{19}}{|P_{20}'(r)|} = 2^{-23} \frac{r^{19}}{(r-1)!\,(20-r)!}.$$

For $r = 16$, the perturbation is maximal and becomes

$$\delta z_{16} \approx 2^{-23} \frac{16^{19}}{15!\,4!} \approx 287.$$

Thus, the zeros are *ill conditioned*. Wilkinson computed the exact zeros using multiple precision and found e.g. the zeros $16.730 \pm 2.812\ i$. We can easily reconfirm this calculation using MAPLE:

```
> Digits:=50;
> p:=1:
> for i from 1 by 1 to 20 do
>    p:=p*(x-i)
> od:
> Z:=fsolve(p-2^(-23)*x^19,x,complex,maxsols=20);
> plot(map(z->[Re(z),Im(z)],[Z]),x=0..22,style=point,symbol=circle);
```

We do not actually need to work with 50 decimal digits; we obtain the same results even in MATLAB using standard IEEE arithmetic.

We can also use MAPLE to simulate the experiment of Wilkinson, using 7 decimal-digit arithmetic. For this simulation, it is important to represent the coefficients of the expanded polynomial as 7-digit numbers:

```
> Digits:=7;
> p:=1:
> for i from 1 by 1 to 20 do
>    p:=p*(x-i)
> od:
> PP:=expand(p);
> PPP:=evalf(PP);
> Z:=fsolve(PPP,x,complex,maxsols=20);
> plot(map(z->[Re(z),Im(z)],[Z]),x=0..28,style=point,symbol=circle);
```

Figure 5.12 shows how most of the zeros become complex numbers.

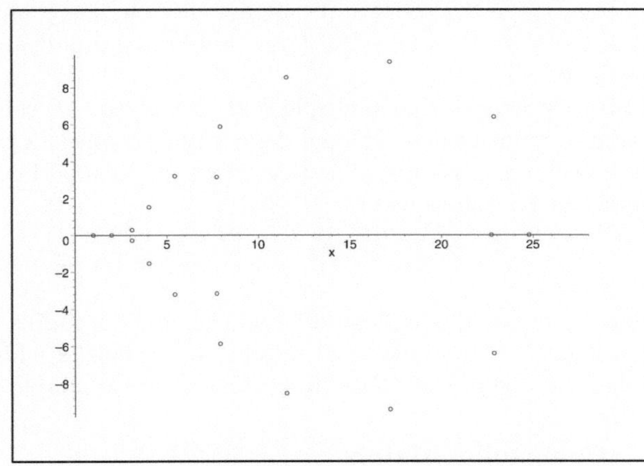

FIGURE 5.12. *Zeros of Wilkinson's Polynomial computed with 7 digits.*

In MATLAB, polynomials are represented by

$$P_n(x) = A(1)x^n + A(2)x^{n-1} + \cdots + A(n)x + A(n+1),$$

while in usual mathematical notation one prefers

$$P_n(x) = a_0 + a_1 x + a_2 x^2 + \cdots + a_n x^n.$$

In order to switch between the two representations, we must keep in mind the transformation

$$A(i) = a_{n+1-i}, \ i = 1, \ldots, n+1 \iff a_j = A(n+1-j), \ j = 0, \ldots, n.$$

In the following MATLAB script, we change the second coefficient of Wilkinson's polynomial by subtracting a small perturbation: $p_2 := p_2 - \lambda p_2$ where $\lambda = 0 : 1e-11 : 1e-8$. Note that the perturbations are cumulative; at the last iteration, the total perturbation gives $\tilde{p}_2 = p_2(1 - \mu)$, with $\mu \approx 5.0 \times 10^{-6}$.

ALGORITHM 5.5.
Experiment with Wilkinson's Polynomial

```
axis([-5 25 -10 10])
hold
P=poly(1:20)
for lamb=0:1e-11:1e-8
  P(2)=P(2)*(1-lamb);
  Z=roots(P);
  plot(real(Z),imag(Z),'o')
end
```

By computing the roots and plotting them in the complex plane, we can observe in Figure 5.13 how the larger roots are rapidly moving away from the real line.

We finally remark that multiple roots are always *ill conditioned*. To illustrate that, assume that we change the constant coefficient of $P_n(x) = (x-1)^n$ by the machine precision ε. The perturbed polynomial becomes $(x-1)^n - \varepsilon$ and its roots are solutions of

$$(x-1)^n = \varepsilon \quad \Rightarrow \quad x(\varepsilon) = 1 + \sqrt[n]{\varepsilon}.$$

The new roots are all simple and lie on the circle of radius $\sqrt[n]{\varepsilon}$ with center 1. For $\varepsilon = 2.2204e-16$ and $n = 10$ we get $\sqrt[10]{\varepsilon} = 0.0272$ which shows that the multiple roots "explode" quite dramatically into n simple ones.

5.3.2 Companion Matrix

As we have seen in the previous section, zeros of polynomials may be ill conditioned, so one had to find new methods for computing eigenvalues. New algorithms that work directly on the matrix, instead of forming the characteristic polynomial, have been successfully developed. Indeed, the QR algorithm (see Section 7.6) proved to be so successful that MATLAB has reversed

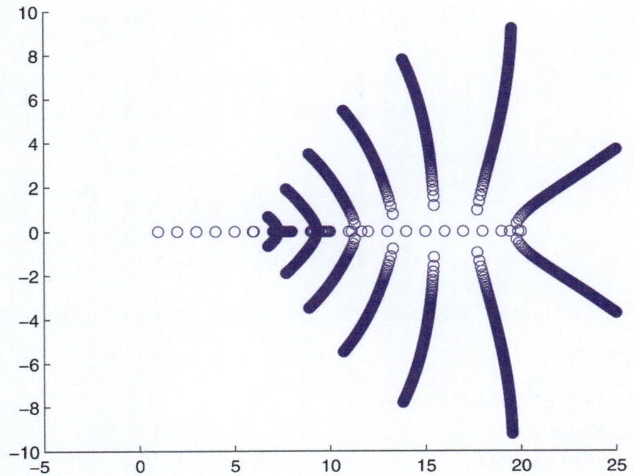

FIGURE 5.13. *Roots of the perturbed Wilkinson polynomial*

its thinking: to compute zeros of a polynomial, one should construct the *companion matrix*, which has the zeros of the polynomial as eigenvalues. Thus, instead of computing eigenvalues with the help of the characteristic polynomial, one computes zeros of polynomials with the QR eigenvalue algorithm! This is how the MATLAB function `roots` works.

DEFINITION 5.5. (COMPANION MATRIX) *The monic polynomial* $P_n(x) = x^n + a_{n-1}x^{n-1} + \cdots + a_0$ *has the* companion matrix

$$A = \begin{bmatrix} -a_{n-1} & -a_{n-2} & \cdots & -a_1 & -a_0 \\ 1 & 0 & \cdots & 0 & 0 \\ 0 & 1 & \ddots & 0 & 0 \\ \vdots & \ddots & \ddots & \ddots & \vdots \\ 0 & \cdots & 0 & 1 & 0 \end{bmatrix}.$$

By expanding the determinant $\det(\lambda I - A)$, we see that $P_n(\lambda)$ is the characteristic polynomial of A.

EXAMPLE 5.7. *Consider for $n = 4$ the matrix $C = \lambda I - A$ and compute its determinant by the expanding the first column. We get*

$$
\det \begin{pmatrix} \lambda + a_3 & a_2 & a_1 & a_0 \\ -1 & \lambda & 0 & 0 \\ 0 & -1 & \lambda & 0 \\ 0 & 0 & -1 & \lambda \end{pmatrix} = (\lambda + a_3) \det \begin{pmatrix} \lambda & 0 & 0 \\ 1 & \lambda & 0 \\ 0 & -1 & \lambda \end{pmatrix} + 1 \cdot \det \begin{pmatrix} a_2 & a_1 & a_0 \\ -1 & \lambda & 0 \\ 0 & -1 & \lambda \end{pmatrix}
$$

$$
= \lambda^4 + a_3 \lambda^3 + a_2 \lambda^2 + 1 \cdot \det \begin{pmatrix} a_1 & a_0 \\ -1 & \lambda \end{pmatrix}
$$

$$
= \lambda^4 + a_3 \lambda^3 + a_2 \lambda^2 + a_1 \lambda + a_0.
$$

The MATLAB function `compan(a)` computes this companion matrix. In MAPLE we get with

```
> p:=z^5+2*z^4+3*z^3+4*z^2+5*z+6;
> with(LinearAlgebra);
> A:=CompanionMatrix(p);
```

$$
\begin{bmatrix} 0 & 0 & 0 & 0 & -6 \\ 1 & 0 & 0 & 0 & -5 \\ 0 & 1 & 0 & 0 & -4 \\ 0 & 0 & 1 & 0 & -3 \\ 0 & 0 & 0 & 1 & -2 \end{bmatrix}
$$

a different definition of the companion matrix, which is often used in textbooks. Both definitions are equivalent, since by computing $\det(\lambda I - A)$ we obtain the characteristic polynomial of A.

5.3.3   Horner's Scheme

Often, one needs to divide a polynomial $P_n(x)$ by a linear factor $(x - z)$. We obtain

$$
\frac{P_n(x)}{x - z} = P_{n-1}(x) + \frac{r}{x - z}, \tag{5.58}
$$

where $P_{n-1}(x)$ is a polynomial of degree $n-1$ and r, a number, is the remainder. In secondary school, a method is taught for performing this calculation by hand, as is shown for the example of $P_3(x) = 3x^3 + x^2 - 5x + 1$ and $z = 2$:

$$
\begin{array}{l}
(3x^3 + x^2 - 5x + 1) : (x - 2) = \underbrace{3x^2 + 7x + 9}_{P_2(x)} \\[1ex]
\underline{-3x^3 + 6x^2} \\[1ex]
\qquad 7x^2 - 5x \\
\qquad \underline{- 7x^2 + 14x} \\[1ex]
\qquad\qquad 9x + 1 \\
\qquad\qquad \underline{- 9x + 18} \\[1ex]
\qquad\qquad\qquad 19 = r
\end{array} \tag{5.59}
$$

Thus Equation (5.58) for this example is

$$\frac{3x^3 + x^2 - 5x + 1}{x - 2} = 3x^2 + 7x + 9 + \frac{19}{x - 2}.$$

If we generalize this process, we get

$$
\begin{array}{l}
(a_n x^n + a_{n-1}x^{n-1} + a_{n-2}x^{n-2} + \cdots + a_0) \ :\ (x - z) = b_{n-1}x^{n-1} + \cdots + b_0 \\
\underline{-b_{n-1}x^n + b_{n-1}zx^{n-1}} \\
\qquad\quad b_{n-2}x^{n-1} + a_{n-2}x^{n-2} \\
\qquad\ \underline{-\ b_{n-2}x^{n-1} + b_{n-2}zx^{n-2}} \\
\qquad\qquad\quad b_{n-3}x^{n-2} + a_{n-3}x^{n-3} \\
\qquad\qquad\qquad \cdots \qquad\ \cdots \qquad \cdots \\
\qquad\qquad\qquad\quad b_0 x + \quad a_0 \\
\qquad\qquad\quad -\ \underline{\quad b_0 x + \quad b_0 z} \\
\qquad\qquad\qquad\qquad\qquad r
\end{array}
$$

$$(5.60)$$

From (5.60), we can see that the coefficients b_i of P_{n-1} are computed by the recurrence

$$
\begin{aligned}
b_{n-1} &= a_n \\
b_{i-1} &= zb_i + a_i, \quad i = n - 1, \cdots, 1,
\end{aligned}
\tag{5.61}
$$

and the remainder becomes $r = zb_0 + a_0$. For hand calculations, the recurrence relation (5.61) is written more conveniently as

$$
\begin{array}{cccccc}
a_n & a_{n-1} & a_{n-2} & \cdots & a_1 & a_0 \\
 & b_{n-1}z & b_{n-2}z & \cdots & b_1 z & b_0 z \\
\hline
x = z \quad b_{n-1} & b_{n-2} & b_{n-3} & \cdots & b_0\ | & r
\end{array}
\tag{5.62}
$$

The scheme (5.62) is called *simple Horner's scheme*.

The evaluation of Horner's scheme is computed in MATLAB by the following statements:

ALGORITHM 5.6. *Simple Horner's Scheme*

```
b(1)=a(1);
for i=2:n
  b(i)=b(i-1)*z+a(i)
end
r=b(n)*z+a(n+1)
```

For $n = 3$, $a = [3,\ 1,\ -5,\ 1]$ and $z = 2$ we obtain $b = [3,\ 7,\ 9]$ and $r = 19$.

What is the meaning of the remainder r? If we multiply Equation (5.58) by $(x - z)$, then we obtain

$$P_n(x) = P_{n-1}(x)(x - z) + r. \tag{5.63}$$

Inserting $x = z$ in Equation (5.63) we get $P_n(z) = r$, and thus

$$P_n(x) = P_{n-1}(x)(x - z) + P_n(z). \tag{5.64}$$

Since the remainder is the function value $P_n(z)$ *we can also use the division algorithm* (5.60) *to evaluate the polynomial!* This is even an efficient way to do it, as we will see. Straightforward evaluation of $P_n(z) = a_n z^n + a_{n-1} z^{n-1} + \cdots + a_1 z + a_0$ needs $\frac{n(n+1)}{2}$ multiplications and n additions. By factoring, we can reduce the number of multiplications to n:

$$P_n(z) = \Big(\cdots \Big(\big(\underbrace{a_n}_{b_{n-1}} z + a_{n-1} \big) z + a_{n-2} \Big) z + \cdots + a_1 \Big) z + a_0. \tag{5.65}$$

We recognize that the brackets are nothing but the coefficients of the polynomial P_{n-1}. Thus we have shown that Horner's scheme (5.62) can be used to

1. divide a polynomial by a linear factor,

2. evaluate a polynomial with n multiplications.

In MAPLE, we can convert a polynomial into *Horner's form* with the command:

```
> convert(3*x^3+x^2-5*x+1,horner,x);
```

and we obtain $1 + (-5 + (1 + 3x)x)x$.

If we just want to evaluate the polynomial, there is no need to save the coefficients b_i, so the MATLAB function becomes

---

ALGORITHM 5.7.
Evaluation of a Polynomial by Horner's Rule

```
function y=Horner(p,x);
% HORNER Evaluates a polynomial
%    y=Horner(p,x) evaluates the polynomial
%    y=p(1)*x^(n-1)+p(2)*x^(n-2)+...+p(n-1)*x+p(n)

n=length(p);
y=0;
for i=1:n
  y=y*x+p(i);
end
```

MATLAB provides a built-in function `polyval(p,x)` for polynomial evaluation. It works also when the parameter x is a matrix.

With the simple Horner's scheme, we compute the decomposition

$$P_n(x) = P_n(z) + (x - z)P_{n-1}(x). \tag{5.66}$$

Horner's scheme can be used repeatedly to decompose the polynomials $P_{n-1}(x)$, $P_{n-2}(x)$, ..., $P_1(x)$. Doing so leads to the equations:

$$
\begin{aligned}
P_n(x) &= P_n(z) &+ (x - z)P_{n-1}(x), \\
P_{n-1}(x) &= P_{n-1}(z) &+ (x - z)P_{n-2}(x), \\
&\;\;\vdots &\vdots \\
P_1(x) &= P_1(z) &+ (x - z)P_0(x).
\end{aligned}
\tag{5.67}
$$

If we stack the necessary Horner's schemes one above the other, we obtain the *complete Horner's scheme*. For example, for $P_3(x) = 3x^3 + x^2 - 5x + 1$ and $z = 2$, the scheme becomes

$$
\begin{array}{cccc|cccc}
3 & & 1 & & -5 & \mid & 1 & \\
& & 6 & & 14 & \mid & 18 & \\
\hline
3 & & 7 & & 9 & \mid & 19 & = P_3(z) \\
& & 6 & & 26 & \mid & & \\
\hline
3 & & 13 & \mid & 35 & & = P_2(z) & \\
& & 6 & \mid & & & & \\
\hline
3 & \mid & 19 & & = P_1(z), & & & \\
\hline
3 & & = P_0(z) & & & & &
\end{array}
\tag{5.68}
$$

We can read from the scheme that

$$
\begin{aligned}
P_2(x) &= 3x^2 + 7x + 9, \\
P_1(x) &= 3x + 13, \\
P_0(x) &= 3.
\end{aligned}
$$

Let us find an interpretation for the values $P_i(z)$. For this purpose, let us eliminate the quantities $P_{n-1}(x)$, $P_{n-2}(x)$, ..., $P_1(x)$ from top to bottom in (5.67). If we use only the first j equations, we obtain

$$P_n(x) = \sum_{k=0}^{j-1} (x - z)^k P_{n-k}(z) + (x - z)^j P_{n-j}(x), \tag{5.69}$$

or, for $j = n$,

$$P_n(x) = \sum_{k=0}^{n} P_{n-k}(z)(x - z)^k. \tag{5.70}$$

Since $P_0(x) = \text{const} = P_0(z)$. Equation (5.70) looks very much like the Taylor expansion of $P_n(x)$ at $x = z$,

$$P_n(x) = \sum_{k=0}^{n} \frac{P_n^{(k)}(z)}{k!}(x - z)^k. \tag{5.71}$$

Since the expansion (5.71) is unique, we must conclude by comparing with (5.70) that

$$P_{n-k}(z) = \frac{P_n^{(k)}(z)}{k!} \quad k = 0, 1, \ldots, n. \tag{5.72}$$

We have proved the following

THEOREM 5.4. (HORNER SCHEME) *The complete Horner's scheme is an algorithm to compute the Taylor expansion of a polynomial $P_n(x)$ for $x = z$.*

When computing the new expansion of $P_n(x)$ we may overwrite the coefficients a_i by the new coefficients b_i:

$$P_n(x) = \sum_{k=0}^{n} a_k x^k = \sum_{k=0}^{n} b_k (x - z)^k.$$

This is done in the following MATLAB function:

ALGORITHM 5.8.
Taylor Expansion by Complete Horner Scheme

```
function a=Taylor(a,z)
% TAYLOR re-expands the polynomial
%    a=Taylor(a,z) re-expands the polynomial
%    P(x)=a(1)*x^(n-1) +  ... + a(n-1)*x + a(n) to
%    a(1)*(x-z)^(n-1) +  ... + a(n-1)*(x-z) + a(n)
%    the coefficients are overwritten.

n=length(a);
for j=1:n-1
  for i=2:n-j+1
    a(i)=a(i-1)*z+a(i);
  end
end
```

Before step j in the function **Taylor**, we have (in mathematical notation):

$$
\begin{array}{ccccccc}
\cdot & \cdot & \cdot & \cdot & \cdot & \cdot & a_0 \\
\cdot & \cdot & \cdot & \cdot & \cdot & a_1 & \\
\cdot & \cdot & \cdot & \cdot & \cdot & & \\
a_n & a_{n-1} & \cdots & a_j & a_{j-1} & &
\end{array}
$$

and

$$P_{n-j}(x) = a_n x^{n-j} + a_{n-1} x^{n-j-1} + \cdots + a_j.$$

Because of (5.69) we obtain

$$P_n(x) = \sum_{k=0}^{j-1} a_k (x - z)^k + (x - z)^j \sum_{k=j}^{n} a_k x^{k-j}, \qquad (5.73)$$

and we see how the polynomial is transformed step-wise.

5.3.4 Number Conversions

Horner's rule can be used to *convert numbers to different base systems.* As an example, consider the binary number

$$u = 10011101101_2.$$

To compute the decimal value of u, we write

$$u = 1 \cdot 2^{10} + 0 \cdot 2^9 + 0 \cdot 2^8 + 1 \cdot 2^7 + 1 \cdot 2^6 + 1 \cdot 2^5 + 0 \cdot 2^4 + 1 \cdot 2^3 + 1 \cdot 2^2 + 0 \cdot 2^1 + 1.$$

Thus, u is the function value for $z = 2$ of a polynomial of degree 10 with the binary digits as coefficients,i.e., we have

$$u = P_{10}(2) = 1261_{10}.$$

More generally, a number in base b is represented as a string of characters. For instance, the string 2DFDC1C3E, interpreted as a hexadecimal number, is the polynomial

$$2 \cdot 16^8 + D \cdot 16^7 + F \cdot 16^6 + D \cdot 16^5 + C \cdot 16^4 + 1 \cdot 16^3 + C \cdot 16^2 + 3 \cdot 16 + E.$$

Assuming that the capital letters represent the hexadecimal digits $A = 10$, $B = 11$, $C = 12$, $D = 13$, $E = 14$ and $F = 15$, we have

$$\begin{aligned}
2DFDC1C3E_{16} &= 2 \cdot 16^8 + 13 \cdot 16^7 + 15 \cdot 16^6 + 13 \cdot 16^5 \\
&\quad + 12 \cdot 16^4 + 1 \cdot 16^3 + 12 \cdot 16^2 + 3 \cdot 16 + 14 \\
&= 12345678910_{10}
\end{aligned}$$

Thus, this conversion is simply an evaluation of a polynomial. This can be done using Horner's rule, provided we have converted the characters of the string to their numerical values.

In MATLAB, the characters are encoded in ASCII. Noting that zero has the *ASCII code* 48 and the the capital letter A has code 65, we can assign a numerical value x to an ASCII character with code c as follows:

$$x = \begin{cases} c - 48 & \text{if } c \text{ is a decimal digit} \\ c - 65 + 10 = c - 55 & \text{if } c \text{ is a capital letter} \end{cases}$$

This is implemented in the following function `BaseB2Decimal`:

ALGORITHM 5.9. *Number Conversion: Base b to Decimal*

```
function u=BaseB2Decimal(s,b)
% BASEBDECIMAL convert a number from and arbitrary base to decimal
%    u=BaseB2Decimal(s,b) converts a string s which represents an
%    integer in base b to the decimal number u using Horner's rule.

n=length(s); u=0;
for i=1:n
  if char(s(i))<=57,        % Assume decimal digit
    a=s(i)-48;              % subtract ASCII number of zero
  else                      % Assume capital letter
    a=s(i)-55;              % subtract ASCII number 65 of A and add 10
  end
  u=u*b+a;
end
```

EXAMPLE 5.8.

```
>> s='10110111111101110000011100001111110';
>> u=BaseB2Decimal(s,2)
u=1.234567891000000e+10

>> s= '2DFDC1C3E';
>> u=BaseB2Decimal(s,16)
u=1.234567891000000e+10

>> s='3E92HC76';
>> u=BaseB2Decimal(s,23)
u=1.234567891000000e+10
```

Note that the built-in MATLAB function `base2dec(s,b)` does the same as `BaseB2Decimal`. MATLAB also has the functions `bin2dec` and `hex2dec`.

If we want to convert a decimal number into a number in base b, we need to solve an inverse problem: Given the function value $u = P_n(z)$ and the argument $z = b$, we wish to find the coefficients of the polynomial $P_n(z)$. This problem does not have a unique solution unless we require that u, b, d_i must be integers and that $u \geq 0$, $b > d[i] \geq 0$.

Horner's scheme has to be computed in reverse direction, e.g. for $b = 2$ and $u = 57$, we get

$$
\begin{array}{ccccccc}
1 & & 1 & & 1 & & 0 & & 0 & & 1 \\
0 & & 2 & & 6 & & 14 & & 28 & & 56 \\
\uparrow & \nearrow & \uparrow & \nearrow & \uparrow & \nearrow & \uparrow & \nearrow & \uparrow & \nearrow & \uparrow \\
1 & & 3 & & 7 & & 14 & & 28 & & 57
\end{array}
\tag{5.74}
$$

and thus $57_{10} = 111001_2$. Starting with the decimal number 57, we divide the number by the base $b = 2$; the integer remainder is the new digit (i.e. the coefficient of the polynomial). Thus, we obtain the following algorithm `Decimal2BaseB`, which computes the digits d_i of a number u in the number system with base b. Notice that the digits with value > 9 are designed with capital letters $A, B, C, \ldots$. The resulting number is a string s, which is generated from the decimal representation of the digits stored in the vector d.

ALGORITHM 5.10.
Number Conversions: Decimal to Base b

```
function [s,d]=Decimal2BaseB(u,b)
% DECIMAL2BASEB converts an integer to another base representation
%    [s,d]=Decimal2BaseB(u,b) converts the integer u to base b. d
%    contains the digits as decimal numbers, s is a string containing
%    the digits of the base b represetation e.g. for b=16 (hexadecimal)
%    the digits are 1, 2, ... 9, A, B, C, D, E, F

i=1;
while u~=0
  d(i)=mod(u,b); u=floor(u/b); i=i+1;
end
n=i-1; d=d(n:-1:1); s=[];          % convert to string using char(65)=A
for i=1:n
  if d(i)<10
    s=[s num2str(d(i))];
  else
    s=[s char(65+d(i)-10)];
  end
end
```

EXAMPLE 5.9. *Let's convert the integer number $u = 12345678910$ to the systems with base $b = 2, 7, 16, 23$*

```
>> u=12345678910
u =
     1.234567891000000e+10

>> s=Decimal2BaseB(u,2)
s=1011011111110111000001110000111110

>> s=Decimal2BaseB(u,7)
s=614636352655

>> s=Decimal2BaseB(u,16)
s=2DFDC1C3E
```

```
>> s=Decimal2BaseB(u,23)
s=3E92HC76
```

Note that the built-in MATLAB function `dec2base(u,b)` does the same as `Decimal2BaseB`. There exist also the MATLAB functions `dec2bin` and `dec2hex`. Working with strings using only numbers and capital letters restricts the possible values for the base b; for example, `dec2base` imposes the restriction $2 \leq b \leq 36$. The same restriction applies in MAPLE: the function `convert(n, decimal, b)` converts a base b number (where b is an integer from 2 through 36) to a decimal number.

We return now to the computation of the zeros of polynomials. Since it is easy to compute the derivative of a polynomial using Horner's rule, it is straightforward to use Newton's method. We will consider three different implementations. The first one is straightforward, the second one will make use of various Taylor expansions and the third one is a variant for real zeros and will suppress rather than deflate zeros.

5.3.5 Newton's Method: Classical Version

To obtain the function value and the value of the derivative at z, we need to compute the first two rows of Horner's scheme. Since the zeros of a real polynomial might be complex, we have to start with a complex number for the iteration, otherwise the iteration sequence will never become complex. Our initial starting number will be $1 + i$, and after computing the first root, we will always take the last root as the initial value for the new root of the deflated polynomial. The implementation in MATLAB is easy, since MATLAB computes with complex numbers.

In the following MATLAB function, we stop the iteration when the function value of the polynomial becomes small. For multiple zeros, we have seen that Newton's iteration converges only linearly, see Section 5.2.6, and a test on successive iterates may not be very reliable. Near a zero x, we can expect that the function value drops to about (with machine precision ε)

$$|P_n(x)| \approx \varepsilon \sum_{i=0}^{n} |a_i x^i|. \tag{5.75}$$

We will use this equation with tolerance 10ε as a stopping criterion.

ALGORITHM 5.11.
Newton's Method for Zeros of Polynomials

```
function z=NewtonRoots(a)
% NEWTONROOTS computes zeros of a polynomial
%   z=NewtonRoots(a) computes zeros of the polynomial given by the
%   coefficients in the vector a using Newton's method
```

```
n=length(a); degree=n-1; z=[];
y=1+sqrt(-1);
for m=degree:-1:1
  p=1; h=0;                              % Newton iteration
  while abs(p)>10*eps*h
    x=y; p=0; ps=0; h=0;
    for k=1:m+1
      ps=p+ps*x; p=a(k)+p*x; h=abs(a(k))+h*abs(x);
    end
    y=x-p/ps;
  end
  b(1)=a(1);                             % deflation
  for k=2:m
    b(k)=b(k-1)*y+a(k);
  end
  z=[z; y];
  a=b;
end
```

For the Wilkinson polynomial, we obtain results that are comparable to the MATLAB built-in function **roots**:

```
> a=poly(1:20);
> x=NewtonRoots(a); y=roots(a);
> [norm(sort(x)-[1:20]')  norm(sort(y)-[1:20]')]
  ans=
        0.008482726803436   0.024155938346689
```

Of course we cannot expect `NewtonRoots` to work in all cases. Designing reliable software is much harder than teaching numerical methods!

5.3.6 Newton Method Using Taylor Expansions

Another implementation of Newton's method is based on expanding the polynomial at new points in the complex plane. Let x be an approximation of a zero, then

$$P_n(x + h) = a_0 + a_1 h + \cdots + a_n h^n.$$

Because $a_0 = P_n(x)$ and $a_1 = P_n'(x)$, the Newton step simply becomes

$$x := x - a_0/a_1.$$

We compute a new Taylor expansion of the polynomial at the new approximation and the effect is that the new coefficient a_0 decreases with each iteration. If finally x is a root, then $a_0 = 0$, and

$$P_n(x + h) = h(a_1 + a_2 h + \cdots + a_n h^{n-1}),$$

and thus the remaining roots are zeros of

$$P_{n-1}(x+h) = a_1 + a_2 h + \cdots + a_n h^{n-1}.$$

Therefore, deflation becomes here very simple. However, at each iteration, a complete Horner scheme must be computed, which is more work than for the Newton implementation presented earlier.

ALGORITHM 5.12.
Newton for Zeros of Polynomials – Second version

```
function z=NewtonTaylorRoots(a)
% NEWTONTAYLORROOTS computes the zeros of a polynomial
%    z=NewtonTaylorRoots(a) computes the zeros of the polynomial given
%    in the vector a using Newton's method by re-expanding the
%    polynomial

n=length(a); degree=n-1;
z=[]; x=0; h=1+sqrt(-1);
for m=degree:-1:1
  while abs(a(m+1))>norm(a)*eps               % Newton iteration
    x=x+h; a=Taylor(a,h); h=-a(m+1)/a(m);
  end
  a=a(1:m);                                    % deflation
  z=[z; x];
end
```

We again obtain a good result for Wilkinson's polynomial, for larger degrees however, `NewtonRoots` seems to work more reliably.

```
>> a=poly(1:20);
>> x=NewtonTaylorRoots(a);
>> norm(sort(x)-[1:20]')
ans=
      0.019332026554334
```

Deflation with an approximation of a root will not deliver accurate coefficients of the deflated polynomial. Wilkinson observed that it is numerically preferable to deflate smaller roots first.

5.3.7 Newton Method for Real Simple Zeros

The third Newton method we want to consider is useful when a polynomial is not represented by its coefficients and when it is known that the zeros are all real and simple. Two examples are *orthogonal polynomials*, as discussed in Section 4.2.5, or the leading principal minors of the matrix $\lambda I - T$, where T *symmetric and tridiagonal*. Both the orthogonal polynomials and the minor-polynomials are computed by a three-term recurrence relation.

Let T be symmetric and tridiagonal, and consider the submatrices

$$
T_i = \begin{pmatrix} \alpha_1 & \beta_1 & & \\ \beta_1 & \alpha_2 & \ddots & \\ & \ddots & \ddots & \beta_{i-1} \\ & & \beta_{i-1} & \alpha_i \end{pmatrix}, \quad i = 1, \ldots, n.
$$

Let $p_i(\lambda) = \det(\lambda I - T_i)$. By expanding the last row of the determinant, we obtain, using $p_{-1} \equiv 0$ and $p_0 \equiv 1$, a three-term recurrence for the polynomials,

$$
p_i(\lambda) = (\lambda - \alpha_i)p_{i-1}(\lambda) - \beta_{i-1}^2 p_{i-2}(\lambda), \qquad i = 1, \ldots, n. \tag{5.76}
$$

Thus to evaluate the determinant p_n for some value of λ we can use the recurrence (5.76). To compute zeros of p_n using Newton's method, we also need the derivative, which is obtained by differentiating (5.76),

$$
p_i'(\lambda) = p_{i-1}(\lambda) + (\lambda - \alpha_i)p_{i-1}'(\lambda) - \beta_{i-1}^2 p_{i-2}'(\lambda). \tag{5.77}
$$

Algorithm 5.13 below computes directly the Newton correction $\frac{p_n(x)}{p_n'(x)}$. In order to avoid possible over- or underflow, we scale both values if they become too large or too small. By this measure we do not compute the actual values of the functions p_n and p_n', which we do not need since we are only interested in the ratio.

ALGORITHM 5.13. *Newton Correction for* $\det(\lambda I - T)$

```
function r=NewtonCorrection(x,alpha,beta);
% NEWTONCORRECTION computes directly the Newton correction
%     r=NewtonCorrection(x,alpha,beta) evaluates the ratio of the
%     polynomial and its derivative given by the three-term
%     recurrence p_i(x)=(x-alpha_i)p_{i-1}(x)-beta_{i-1}^2
%     p_{i-2}(x) for the Newton correction

n=length(alpha);
p1s=0; p1=1; p2s=1; p2=x-alpha(1);
for k=1:n-1
  p0s=p1s; p0=p1; p1s=p2s; p1=p2;
  p2s=p1+(x-alpha(k+1))*p1s-beta(k)^2*p0s;
  p2=(x-alpha(k+1))*p1-beta(k)^2*p0;
  maxx=abs(p2)+abs(p2s);
  if maxx>1e20; d=1e-20;
  elseif maxx<1e-20; d=1e20;
  else d=1;
  end
  p1=p1*d; p2=p2*d; p1s=p1s*d; p2s=p2s*d;
end
r=p2/p2s;
```

Using Newton's method and the function `NewtonCorrection`, we now can compute a zero of p_n. However, deflation is not possible in the usual way by dividing the polynomial with the linear factor, because the polynomial is not represented by its coefficients. We have to use a new technique, called *suppression*, to avoid recomputing the same zero again. With suppression, we implicitly deflate a computed zero. If p_n is the polynomial and x_1 a first computed zero then we use

$$p_{n-1}(x) = \frac{p_n(x)}{x - x_1}$$

for the the next Newton iteration. We need again the derivative which is

$$p'_{n-1}(x) = \frac{p'_n(x)}{x - x_1} - \frac{p_n(x)}{(x - x_1)^2}.$$

Thus the Newton correction becomes

$$-\frac{p_{n-1}(x)}{p'_{n-1}(x)} = -\frac{p_n(x)}{p'_n(x) - \frac{p_n(x)}{x - x_1}}.$$

Suppose now that we have already computed the zeros $x_1, \ldots, x_k$. Then consider

$$p_{n-k}(x) = \frac{p_n(x)}{(x - x_1) \cdots (x - x_k)}.$$

Because of

$$\frac{d}{dx} \left(\prod_{i=1}^{k} (x - x_i)^{-1} \right) = - \left(\prod_{i=1}^{k} (x - x_i)^{-1} \right) \left(\sum_{i=1}^{k} (x - x_i)^{-1} \right),$$

the derivative becomes

$$p'_{n-k}(x) = \frac{p'_n(x)}{(x - x_1) \cdots (x - x_k)} - \frac{p_n(x)}{(x - x_1) \cdots (x - x_k)} \sum_{i=1}^{k} \frac{1}{x - x_i}.$$

Inserting this expression we obtain for the Newton step:

$$x_{new} = x - \frac{p_n(x)}{p'_n(x)} \frac{1}{1 - \frac{p_n(x)}{p'_n(x)} \sum_{i=1}^{k} \frac{1}{x - x_i}}. \tag{5.78}$$

Equation (5.78) defines the *Newton-Maehly* iteration, which suppresses already computed zeros. It can be used for any function, not only for polynomials.

In connection with the method of Newton-Maehly, Bulirsch and Stoer describe in [132] important details of the implementation of Newton-Maehly's method to compute zeros of polynomials with only real and simple zeros. They noticed that Newton's method for such polynomials has a bad global behavior. We have $p(x) = a_n x^n + \cdots + a_0$ and $p'(x) = n a_n x^{n-1} + \cdots + a_1$. Now if $a_n x^n$ is the dominant term, then the Newton step is approximately

$$x_{new} = x - \frac{p(x)}{p'(x)} \approx x - \frac{1}{n}x = x\left(1 - \frac{1}{n}\right).$$

Thus Bulirsch and Stoer propose to initially take *double steps*

$$x_{new} = x - 2\,\frac{p(x)}{p'(x)}$$

to improve global convergence. It is shown in [132] that the following holds (see Figure 5.14 for the notation): We denote by α the next local extremum

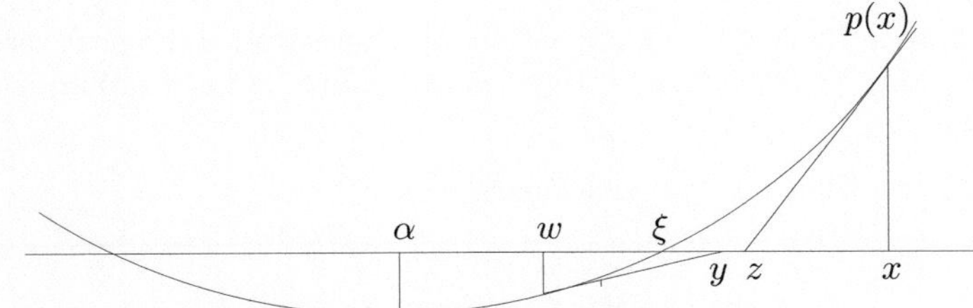

FIGURE 5.14. *Newton double step iteration*

of $p(x)$. For the usual Newton step starting from x we obtain z:

$$z = x - \frac{p(x)}{p'(x)}.$$

For the double Newton step starting from x we obtain w:

$$w = x - 2\frac{p(x)}{p'(x)}.$$

For the Newton step from w we get

$$y = w - \frac{p(w)}{p'(w)}.$$

The following then holds:

1. $\alpha < w$: with the double step iteration we arrive before the local extremum at $x = \alpha$.

2. $\xi \leq y \leq z$: the backward Newton step lies between the zero ξ and the simple Newton step z.

These considerations lead to the following algorithm:

1. Start with the Newton double step iteration *from the right of the first zero* and iterate

$$x = y; \quad y = x - 2\frac{p(x)}{p'(x)}$$

   until $y \geq x$. We are then in the situation described by Figure 5.14.

2. Continue to iterate with normal Newton steps

$$x = y; \quad y = x - \frac{p(x)}{p'(x)}$$

   until monotonicity is again lost and $y \geq x$. This time we have computed the zero $\xi = y$ to machine precision.

3. Suppress the new zero and continue the iteration for the next zero with the value $x < \xi$ found by the Newton double step iteration as initial value.

We obtain the following MATLAB function:

ALGORITHM 5.14.
Newton-Maehly for Zeros of Polynomials

```
function xi=NewtonMaehly(alpha,beta,K,x0);
% NEWTONMAEHLY computes zeros of a three term recurrence polynomial
%   xi=NewtonMaehly(alpha,beta,K,x0) computes by the Newton-Maehly
%   method K zeros of the polynomial defined by the three term
%   recurence with coefficients alpha and beta. The initial guess x0
%   must be bigger than all real zeros. Uses NewtonCorrection.m

xi=[];
for k=1:K,
  y=x0; x=x0*1.1;
  while y<x                                    % Newton double step
    x=y; r=NewtonCorrection(x,alpha,beta);
    s=sum(1./(x-xi(1:k-1)));                    % Maehly correction
    y=x-2*r/(1-r*s);
  end
  x0=x;
  y=x-r/(1-r*s); x=y+10;                        % single backward step
  while y<x                                     % Newton single steps
    x=y; r=NewtonCorrection(x,alpha,beta);
    y=x-r/(1-r*sum(1./(x-xi(1:k-1))));
  end
```

```
  xi(k)=y;
  k,y                                     % test output
end
xi=xi(:);
```

EXAMPLE 5.10. *We want to compute the zeros of the* Legendre *polynomial of degree $n = 20$, see Section 9.3. We define the coefficients of the corresponding tridiagonal matrix and call the function* **NewtonMaehly** *to compute the 10 positive zeros:*

```
n=20;
for i=1:n,
  alpha(i)=0;
  beta(i)=i/sqrt(4*i^2-1);
end
xi=NewtonMaehly(alpha,beta,10,1)
```

We obtain the values

$$
\begin{aligned}
&0.993128599185095\\
&0.963971927277914\\
&0.912234428251326\\
&0.839116971822219\\
&0.746331906460151\\
&0.636053680726515\\
&0.510867001950827\\
&0.373706088715420\\
&0.227785851141645\\
&0.076526521133497
\end{aligned}
$$

which are correct to the last digit.

5.3.8 Nickel's Method

A generalization of Newton's method is based on the following idea of Nickel. If we expand the polynomial at x

$$P_n(x + h) = a_0 + a_1 h + \cdots + a_n h^n,$$

and consider the equation

$$a_0 + a_1 h + \cdots + a_j h^j + \cdots + a_n h^n = 0, \tag{5.79}$$

then the Newton correction h is the solution of $a_0 + a_1 h = 0$ and will be only a good approximation of a solution of Equation (5.79) if the other terms $|a_2 h^2|, \ldots, |a_n h^n|$ are very small in comparison with $|a_1 h|$. Nickels idea is to replace Equation (5.79) by

$$a_0 + a_j h^j = 0, \tag{5.80}$$

where j is chosen in such a way that $|a_j h^j|$ is the largest term. This is the case if $a_j \neq 0$ and if the solution $|h|$ of (5.80),

$$h = \sqrt[j]{-\frac{a_0}{a_j}}, \tag{5.81}$$

is as small as possible. Thus we will choose j such that

$$|h| = \sqrt[j]{\left|\frac{a_0}{a_j}\right|} = \min_{\substack{1 \leq i \leq n \\ a_i \neq 0}} \sqrt[i]{\left|\frac{a_0}{a_i}\right|}. \tag{5.82}$$

Notice that the Newton correction is the special case $j = 1$ of the Nickel correction defined by (5.81) and (5.82). The Nickel correction prevents a division by zero if by accident $P_n'(x) = a_1 = 0$. For h we can choose any of the complex solutions of Equation (5.80). We will choose the principal value delivered by MATLAB.

ALGORITHM 5.15.
Nickel's Method for Zeros of Polynomials

```
function z=Nickel(a)
% NICKEL computes the zeros of the polynomial
%    z=Nickel(a)computes the zeros of the polynomial with coefficients
%    in the vector a using Nickel's method. Uses Taylor.m

n=length(a); degree=n-1;
z=[]; x=0; h=1+sqrt(-1);
for m=degree:-1:1
  while abs(a(m+1))>norm(a)*eps                    % Nickel iteration
    x=x+h; a=Taylor(a,h);
    if m==1,
      h=-a(2)/a(1);
    else
      [hh,j]=min(abs((-a(m+1)./a(m+1-[2:m])).^(1./[2:m])));
      h=(-a(m+1)/a(m+1-j))^(1/j);
    end
  end
  a=a(1:m);                                         % deflation
  z=[z;x];
end
```

The computation of the Nickel step is more expensive compared to Newton. The results are again comparable with the other methods if the degree of the polynomial is not too large.

```
> a=poly(1:20);
> x=Nickel(a);
> norm(sort(x)-[1:20]')
ans=  0.019332026554334
```

5.3.9 Laguerre's Method

The methods of Newton and Nickel generate quadratically convergent sequences of approximations for simple roots. We will now derive the *method of Laguerre*, which will converge cubically. The idea of this method is to approximate the polynomial $P(x)$ of degree n near a zero s by the special polynomial of the form $g(x) = a(x - x_1)(x - x_2)^{n-1}$. The three parameters a, x_1 an x_2 are chosen in such a way that P and g have, for some value x, the same function value and the same first and second derivatives:

$$\begin{aligned} P(x) &= g(x), \\ P'(x) &= g'(x), \\ P''(x) &= g''(x). \end{aligned} \tag{5.83}$$

Then we solve $g(x) = 0$ instead of $P_n(x) = 0$ by solving a quadratic equation, and hope that one of the solutions x_1 is a better approximation of the zero s. If we write Equations (5.83) explicitly, we obtain

$$I: \quad P_n(x) = a(x - x_1)(x - x_2)^{n-1},$$

$$II: \quad P_n'(x) = a(x - x_2)^{n-1} + a(x - x_1)(n - 1)(x - x_2)^{n-2},$$

$$III: \quad P_n''(x) = 2a(n - 1)(x - x_2)^{n-2} + a(n-1)(n-2)(x - x_1)(x - x_2)^{n-3}. \tag{5.84}$$

We can eliminate the unknown a by forming the quotients

$$\frac{II}{I} \quad : \quad \frac{P_n'(x)}{P_n(x)} = \frac{1}{x - x_1} + \frac{n - 1}{x - x_2},$$

$$\frac{III}{I} \quad : \quad \frac{P_n''(x)}{P_n(x)} = \frac{2(n - 1)}{(x - x_1)(x - x_2)} + \frac{(n - 1)(n - 2)}{(x - x_2)^2}.$$

Introducing the abbreviations

$$P = P_n(x), \quad P' = P_n'(x), \quad P'' = P_n''(x), \quad z = \frac{1}{x - x_1} \quad \text{and} \quad w = \frac{n - 1}{x - x_2},$$

we obtain

$$\begin{aligned} \frac{P'}{P} &= z + w, \\ \frac{P''}{P} &= 2zw + \frac{n - 2}{n - 1}w^2. \end{aligned} \tag{5.85}$$

If we solve the first equation in (5.85) for w and insert the result into the second, we obtain a quadratic equation for z,

$$z^2 - \frac{2P'}{nP}z + \left((1 - \frac{1}{n})PP'' - (1 - \frac{2}{n})P'^2 \right) \frac{1}{P^2} = 0. \tag{5.86}$$

The solution is

$$z = \frac{P'}{nP} \pm \sqrt{ \left(\frac{P'}{nP} \right)^2 - (1 - \frac{1}{n})\frac{PP''}{P^2} + (1 - \frac{2}{n}) \left(\frac{P'}{P} \right)^2 },$$

where by the root sign we mean the complex roots. Because of $x - x_1 = \frac{1}{z}$, we have

$$x - x_1 = \frac{nP}{P' \pm \sqrt{(n-1)^2 P'^2 - n(n-1)PP''}} \tag{5.87}$$

$$= \frac{P}{P'} \frac{n}{1 \pm \sqrt{(n-1)^2 - n(n-1)\frac{PP''}{P'^2}}}. \tag{5.88}$$

It makes sense to choose the sign in the denominator to be an addition and to choose the square root in (5.87) for which the real part is positive. This choice makes the absolute value of the denominator is larger, and thus the correction smaller. Doing so, we obtain *Laguerre's method*:

$$x_1 = x - \frac{P}{P'} \frac{n}{1 + \sqrt{(n-1)^2 - n(n-1)\frac{PP''}{P'^2}}}. \tag{5.89}$$

The implementation of Laguerre's method is left as Exercise 5.37. The proof that the method of Laguerre converges cubically is the Exercise 5.38.

5.4   Nonlinear Systems of Equations

Solving n *nonlinear equations* in n variables means that, for a given continuous function $\boldsymbol{f} : \mathbb{R}^n \longrightarrow \mathbb{R}^n$, we want to find a value $\boldsymbol{x} \in \Omega \subset \mathbb{R}^n$ such that $\boldsymbol{f}(\boldsymbol{x}) = 0$. We use the notation $\boldsymbol{x} = (x_1, x_2, \ldots, x_n)^\top$ and $\boldsymbol{f}(\boldsymbol{x}) = (f_1(\boldsymbol{x}), f_2(\boldsymbol{x}), \ldots, f_n(\boldsymbol{x}))^\top$.

In one dimension, the measure of distance between two numbers is the absolute value of the difference between those numbers. In several dimensions we have to use norms as measures of distance (see Chapter 2, Section 2.5.1). We also need the generalization of Taylor expansions from the one-dimensional case to *multivariate, vector-valued functions* $\boldsymbol{f} : \mathbb{R}^n \longrightarrow \mathbb{R}^m$. Recall that for the function $g : \mathbb{R} \longrightarrow \mathbb{R}$, a second-order Taylor expansion (with a third-order remainder term) is given by

$$g(t) = g(0) + g'(0)t + \frac{1}{2!}g''(0)t^2 + \frac{1}{3!}g'''(\tau)t^3, \tag{5.90}$$

where $0 \leq \tau \leq t$. In order to obtain the Taylor expansion for $\boldsymbol{f}$, we consider a perturbation in the direction $\boldsymbol{h} = (h_1, \ldots, h_n)^\top$ and consider each component f_i separately, i.e., we consider the scalar function

$$g(t) := f_i(\boldsymbol{x} + t\boldsymbol{h}).$$

In order to expand this scalar function using (5.90), we need the derivatives of g,

$$g'(0) \;=\; \sum_{j=1}^{n} \frac{\partial f_i}{\partial x_j}(\boldsymbol{x})h_j$$

$$g''(0) \;=\; \sum_{j=1}^{n}\sum_{k=1}^{n} \frac{\partial^2 f_i}{\partial x_j \partial x_k}(\boldsymbol{x})h_j h_k$$

$$g'''(\tau) \;=\; \sum_{j=1}^{n}\sum_{k=1}^{n}\sum_{l=1}^{n} \frac{\partial^3 f_i}{\partial x_j \partial x_k \partial x_l}(\boldsymbol{x}+\tau\boldsymbol{h})h_j h_k h_l.$$

Introducing these derivatives in the Taylor expansion (5.90) of g and evaluating at $t=1$, we naturally arrive at the Taylor expansion of each component f_i,

$$\begin{aligned}
f_i(\boldsymbol{x}+\boldsymbol{h}) \;=\;& f_i(\boldsymbol{x}) + \sum_{j=1}^{n}\frac{\partial f_i}{\partial x_j}(\boldsymbol{x})h_j + \frac{1}{2!}\sum_{j=1}^{n}\sum_{k=1}^{n}\frac{\partial^2 f_i}{\partial x_j \partial x_k}(\boldsymbol{x})h_j h_k \\
&+ \frac{1}{3!}\sum_{j=1}^{n}\sum_{k=1}^{n}\sum_{l=1}^{n}\frac{\partial^3 f_i}{\partial x_j \partial x_k \partial x_l}(\boldsymbol{x}+\tau_i\boldsymbol{h})h_j h_k h_l.
\end{aligned}$$

$$(5.91)$$

Since this explicit notation with sums is quite cumbersome, one often uses the notation of multilinear forms, and writes simultaneously for all components of $\boldsymbol{f}$

$$\boldsymbol{f}(\boldsymbol{x}+\boldsymbol{h}) = \boldsymbol{f}(\boldsymbol{x}) + \boldsymbol{f}'(\boldsymbol{x})(\boldsymbol{h}) + \frac{1}{2!}\boldsymbol{f}''(\boldsymbol{x})(\boldsymbol{h},\boldsymbol{h}) + \boldsymbol{R}(\boldsymbol{h}), \qquad (5.92)$$

where the remainder term can be estimated by $\boldsymbol{R}(\boldsymbol{h}) = O(\|\boldsymbol{h}\|^3)$, see Problem 5.41. Note that the first-order term can be written as a matrix-vector multiplication

$$\boldsymbol{f}'(\boldsymbol{x})(\boldsymbol{h}) = \begin{pmatrix} \frac{\partial f_1}{\partial x_1} & \frac{\partial f_1}{\partial x_2} & \cdots & \frac{\partial f_1}{\partial x_n} \\ \frac{\partial f_2}{\partial x_1} & \frac{\partial f_2}{\partial x_2} & \cdots & \frac{\partial f_2}{\partial x_n} \\ \vdots & \vdots & & \vdots \\ \frac{\partial f_m}{\partial x_1} & \frac{\partial f_m}{\partial x_2} & \cdots & \frac{\partial f_m}{\partial x_n} \end{pmatrix} \cdot \boldsymbol{h}.$$

The $m \times n$ matrix is often called the *Jacobian* matrix of $\boldsymbol{f}$, which we will denote by $J(\boldsymbol{x})$. In the special case of a scalar function with many arguments, $f : \mathbb{R}^n \longrightarrow \mathbb{R}$, which occurs often in optimization, we obtain

$$f(\boldsymbol{x}+\boldsymbol{h}) = f(\boldsymbol{x}) + f'(\boldsymbol{x})(\boldsymbol{h}) + \frac{1}{2!}f''(\boldsymbol{x})(\boldsymbol{h},\boldsymbol{h}) + O(\|\boldsymbol{h}\|^3), \qquad (5.93)$$

and now in the linear term, we see the transpose of the gradient appearing, $f'(\boldsymbol{x}) = (\nabla f(\boldsymbol{x}))^{\top}$. In the quadratic term, the bilinear form $f''(\boldsymbol{x})(\boldsymbol{h},\boldsymbol{h})$ can also be written in matrix notation as $f''(\boldsymbol{x})(\boldsymbol{h},\boldsymbol{h}) = \boldsymbol{h}^\top H(\boldsymbol{x})\boldsymbol{h}$, where $H(\boldsymbol{x})$ is a symmetric matrix with entries

$$(H(\boldsymbol{x}))_{ij} = \frac{\partial^2 f}{\partial x_i \partial x_j}(\boldsymbol{x}).$$

This matrix is known as the *Hessian matrix* of f. Thus, the Taylor expansion for the function $f : \mathbb{R}^n \longrightarrow \mathbb{R}$ can be written in matrix form as

$$f(\boldsymbol{x} + \boldsymbol{h}) = f(\boldsymbol{x}) + (\nabla f(\boldsymbol{x}))^\top \boldsymbol{h} + \frac{1}{2!}\boldsymbol{h}^\top H(\boldsymbol{x})\boldsymbol{h} + O(\|\boldsymbol{h}\|^3). \qquad (5.94)$$

5.4.1 Fixed Point Iteration

A fixed point iteration in several dimensions is very similar to one dimension. To find a zero of the function $\boldsymbol{f}(\boldsymbol{x})$ for $\boldsymbol{x} \in \Omega$, we need to define a continuous function $\boldsymbol{F}(\boldsymbol{x})$ on Ω such that

$$\boldsymbol{f}(\boldsymbol{x}) = 0 \quad \Longleftrightarrow \quad \boldsymbol{x} = \boldsymbol{F}(\boldsymbol{x}). \qquad (5.95)$$

For example, one could define

$$\boldsymbol{F}(\boldsymbol{x}) := \boldsymbol{x} - \boldsymbol{f}(\boldsymbol{x}),$$

but again there is an infinite number of possibilities for choosing a function $\boldsymbol{F}(\boldsymbol{x})$ which satisfies (5.95). To find a zero of $\boldsymbol{f}(\boldsymbol{x})$ in Ω is then equivalent to finding a fixed point of $\boldsymbol{F}(\boldsymbol{x})$ in Ω. For that purpose, the following fixed point iteration is used:

```
x1=initial guess;
x2=F(x1);
while norm(x2-x1)>tol*norm(x2)
  x1=x2;
  x2=F(x1);
end
```

Note that one can use any of the norms introduced in Section 2.5.1. The question of when the above iteration converges can be answered by Theorem 5.5 in the next section: $\boldsymbol{F}(\boldsymbol{x})$ has to be a contraction.

5.4.2 Theorem of Banach

We need some definitions to state the theorem. A *Banach space* $\mathcal{B}$ is a *complete normed vector space* over some number field $\mathcal{K}$ such as $\mathbb{R}$ or $\mathbb{C}$. "Normed" means that there exists a norm $\|\ \|$ with the following properties:

1. $\|\boldsymbol{x}\| \geq 0,\ \forall \boldsymbol{x} \in \mathcal{B}$, and $\|\boldsymbol{x}\| = 0 \iff \boldsymbol{x} = 0$

2. $\|\gamma \boldsymbol{x}\| = |\gamma|\|\boldsymbol{x}\|,\ \forall \gamma \in \mathcal{K}$ and $\forall \boldsymbol{x} \in \mathcal{B}$

3. $\|x + y\| \leq \|x\| + \|y\|$, $\forall x, y \in \mathcal{B}$ (triangle inequality)

"Complete" means that every Cauchy sequence converges in $\mathcal{B}$.

Let $A \subset \mathcal{B}$ be a closed subset (see [64] for a definition) and F a mapping $F : A \to A$. F is called *Lipschitz continuous* on A if there exists a constant $L < \infty$ such that $\|F(x) - F(y)\| \leq L\|x - y\|$ $\forall x, y \in A$. Furthermore, F is called a *contraction* if L can be chosen such that $L < 1$.

THEOREM 5.5. (BANACH FIXED POINT THEOREM) *Let A be a closed subset of a Banach space $\mathcal{B}$, and let F be a contraction $F : A \to A$. Then:*

a) *The contraction F has a unique fixed point s, which is the unique solution of the equation $x = F(x)$.*

b) *The sequence $x_{k+1} = F(x_k)$ converges to s for every initial guess $x_0 \in A$.*

c) *We have the a posteriori estimate :*

$$\|s - x_k\| \leq \frac{L^{k-l}}{1 - L}\|x_{l+1} - x_l\|, \quad for\ 0 \leq l \leq k. \tag{5.96}$$

PROOF. Because F is Lipschitz continuous, we have $\|x_{k+1} - x_k\| = \|F(x_k) - F(x_{k-1})\| \leq L\|x_k - x_{k-1}\|$ and therefore

$$\|x_{k+1} - x_k\| \leq L^{k-l}\|x_{l+1} - x_l\|, \quad 0 \leq l < k. \tag{5.97}$$

We claim that $\{x_k\}$ is a Cauchy sequence. We need to show that for a given $\varepsilon > 0$, there exists a number K such that $\forall m > 1$ and $\forall k \geq K$ the difference $\|x_{k+m} - x_k\| < \varepsilon$. This is the case since, using the triangle inequality, we obtain

$$\begin{aligned}
\|x_{k+m} - x_k\| &= \|x_{k+m} - x_{k+m-1} + x_{k+m-1} - x_{k+m-2} \pm \cdots - x_k\| \\
&\leq \sum_{i=k}^{k+m-1} \|x_{i+1} - x_i\| \\
&\leq (L^{m-1} + L^{m-2} + \cdots + L + 1)\|x_{k+1} - x_k\|.
\end{aligned}$$

Thus we have

$$\|x_{k+m} - x_k\| \leq \frac{1 - L^m}{1 - L}\|x_{k+1} - x_k\|. \tag{5.98}$$

Using (5.97), we get

$$\|x_{k+m} - x_k\| \leq \frac{1 - L^m}{1 - L}L^k\|x_1 - x_0\|, \tag{5.99}$$

and since $L < 1$, we can choose k so that the right hand side of (5.99) is smaller than ε. We have proved that $\{x_k\}$ is a Cauchy sequence and hence converges to some $s \in \mathcal{B}$; since A is a closed subset of $\mathcal{B}$, we deduce that $s \in A$.

We now show by contradiction that s is unique. If we had two fixed points $s_1 = F(s_1)$ and $s_2 = F(s_2)$ then

$$\|s_1 - s_2\| = \|F(s_1) - F(s_2)\| \le L\|s_1 - s_2\|,$$

thus $1 \le L$. But since we assumed $L < 1$ this is a contradiction and there cannot be two fixed points. Thus we have proved a) and b).

In order to prove c), we use (5.98) and then (5.97) to obtain

$$\|x_{k+m} - x_k\| \le \frac{1 - L^m}{1 - L} L^{k-l} \|x_{l+1} - x_l\|,$$

which holds for $0 \le l < k$ and for all m. If we let $m \to \infty$ we finally obtain

$$\|s - x_k\| \le \frac{1}{1 - L} L^{k-l} \|x_{l+1} - x_l\|.$$

$$\square$$

Consequences and remarks:

1. If we set $l = 0$ in Equation (5.96) we obtain an *a priori error estimate*:

$$\|s - x_k\| \le \frac{L^k}{1 - L} \|x_1 - x_0\|.$$

Using this error estimate and knowing L, we can predict how many iteration steps are necessary to obtain an error $\|s - x_k\| < \varepsilon$, namely

$$k > \frac{\ln\left(\frac{\varepsilon(1-L)}{\|x_1 - x_0\|} \right)}{\ln L}. \tag{5.100}$$

2. We can also obtain the convenient *a posteriori error estimate* by substituting $l = k - 1$ in Equation (5.96),

$$\|s - x_k\| \le \frac{L}{1 - L} \|x_k - x_{k-1}\|.$$

If $\|x_k - x_{k-1}\| < \varepsilon$, then

$$\|s - x_k\| \le \frac{L}{1 - L} \varepsilon. \tag{5.101}$$

Note that the bound on the right hand side of Equation (5.101) can be much larger than ε. In fact if e.g. $L = 0.9999$, then $\|s - x_k\| \le 9999\varepsilon$. Thus even if successive iterates agree to 6 decimal digits, the approximation may contain only two correct decimal digits.

Only when $L \le 0.5$ can we conclude from $\|x_k - x_{k-1}\| \le \varepsilon$ that $\|s - x_k\| \le \varepsilon$ (See Problem 2.22). In this case we may then terminate the *iteration by checking* successive iterates *and conclude that the error is also smaller than ε.*

A way to estimate L from the iteration at step $k+1$ is to use (5.97) for $k = l + 1$, which gives

$$L \approx \frac{\|\boldsymbol{x}_{k+1} - \boldsymbol{x}_k\|}{\|\boldsymbol{x}_k - \boldsymbol{x}_{k-1}\|}.$$

This, together with the *a posteriori* estimate, leads to the stopping criterion

$$\frac{\|\boldsymbol{x}_k - \boldsymbol{x}_{k-1}\| \|\boldsymbol{x}_{k+1} - \boldsymbol{x}_k\|}{\|\boldsymbol{x}_k - \boldsymbol{x}_{k-1}\| - \|\boldsymbol{x}_{k+1} - \boldsymbol{x}_k\|} < \texttt{tol}, \tag{5.102}$$

which guarantees asymptotically that the error is less than $\texttt{tol}$.

5.4.3 Newton's Method

We want to find $\boldsymbol{x}$ such that $\boldsymbol{f}(\boldsymbol{x}) = 0$. Expanding $\boldsymbol{f}$ at some approximation $\boldsymbol{x}_k$, we obtain

$$\boldsymbol{f}(\boldsymbol{x}) \approx \boldsymbol{f}(\boldsymbol{x}_k) + J(\boldsymbol{x}_k)\boldsymbol{h}, \quad \text{with} \quad \boldsymbol{h} = \boldsymbol{x} - \boldsymbol{x}_k,$$

where $J(\boldsymbol{x}_k)$ denotes the *Jacobian* evaluated at $\boldsymbol{x}_k$,

$$J = \begin{bmatrix} \frac{\partial f_1}{\partial x_1} & \frac{\partial f_1}{\partial x_2} & \cdots & \frac{\partial f_1}{\partial x_n} \\ \frac{\partial f_2}{\partial x_1} & \frac{\partial f_2}{\partial x_2} & \cdots & \frac{\partial f_2}{\partial x_n} \\ \vdots & \vdots & & \vdots \\ \frac{\partial f_n}{\partial x_1} & \frac{\partial f_n}{\partial x_2} & \cdots & \frac{\partial f_n}{\partial x_n} \end{bmatrix}.$$

Instead of solving $\boldsymbol{f}(\boldsymbol{x}) = 0$, we solve the linearized system

$$\boldsymbol{f}(\boldsymbol{x}_k) + J(\boldsymbol{x}_k)\boldsymbol{h} = 0$$

for the *Newton correction $\boldsymbol{h}$*, and obtain a new, hopefully better approximation

$$\boldsymbol{x}_{k+1} = \boldsymbol{x}_k + \boldsymbol{h} = \boldsymbol{x}_k - J(\boldsymbol{x}_k)^{-1}\boldsymbol{f}(\boldsymbol{x}_k). \tag{5.103}$$

Thus, Newton's method is the fixed point iteration

$$\boldsymbol{x}_{k+1} = \boldsymbol{F}(\boldsymbol{x}_k) = \boldsymbol{x}_k - J^{-1}(\boldsymbol{x}_k)\boldsymbol{f}(\boldsymbol{x}_k).$$

Note that this formula is a generalization of the formula derived in the one-dimensional case geometrically, see Subsection 5.2.5. It is important to note that the occurrence of the inverse of the Jacobian in the formula implies that one has to solve a linear system at each step of the Newton iteration.

THEOREM 5.6. (QUADRATIC CONVERGENCE OF NEWTON'S METHOD)
Suppose that $f : \mathbb{R}^n \to \mathbb{R}^n$ is three times continuously differentiable, and that Jacobian $f'(x)$ is invertible in a neighborhood of s, where $f(s) = 0$. Then, for x_k sufficiently close to s, the error $e_k := x_k - s$ in Newton's method satisfies

$$e_{k+1} = \frac{1}{2}(f'(x_k))^{-1} f''(x_k)(e_k, e_k) + O(\|e_k\|^3). \tag{5.104}$$

Hence Newton's method converges locally quadratically.

PROOF. We expand $f(x)$ in a Taylor series, see (5.92), and obtain

$$f(x) = f(x_k) + f'(x_k)(x - x_k) + \frac{1}{2}f''(x_k)(x - x_k, x - x_k) + O(\|x - x_k\|^3).$$

Setting $x := s$, where $f(s) = 0$, and subtracting the Newton iteration formula

$$0 = f(x_k) + f'(x_k)(x_{x+1} - x_k),$$

we obtain for the error $e_k = x_k - s$ the relation

$$0 = f'(x_k)(-e_{x+1}) + \frac{1}{2}f''(x_k)(e_k, e_k) + O(\|e_k\|^3),$$

which concludes the proof.   □

A short MATLAB implementation of Newton's method for an arbitrary system of nonlinear equations is given in Algorithm 5.16.

ALGORITHM 5.16.
Newton's method for a system of non-linear equations

```
function x=Newton(f,x0,tol,maxiter,fp);
% NEWTON solves a nonlinear system of equations
%    x=Newton(f,x0,tol,maxiter,fp); solves the nonlinear system of
%    equations f(x)=0 using Newtons methods, starting with the initial
%    guess x0 up to a tolerance tol, doing at most maxit iterations. An
%    analytical Jacobian can be given in the parameter fp

numJ=nargin<5;
if nargin<4, maxit=100; end;
if nargin<3, tol=1e6; end;

x=x0; i=0; dx=ones(size(x0));
while norm(f(x))>tol & norm(dx)>tol & i<maxiter
  if numJ, J=NumericalJacobian(f,x); else J=fp(x); end;
  dx=-J\f(x); x=x+dx; i=i+1;
end
if i>=maxiter
  error('Newton did not converge: maximum number of iterations exceeded');
end;
```

Note that in this implementation, one has the choice of giving the Jacobian matrix in analytic form, or letting the procedure compute an approximate *Jacobian matrix numerically*. For a numerical approximation, one usually uses a finite difference, i.e.

$$\frac{\partial f_i}{\partial x_j}(x_1, \ldots, x_n) \approx \frac{f_i(x_1, \ldots, x_j + h, \ldots, x_n) - f_i(x_1, \ldots, x_n)}{h}, \qquad (5.105)$$

for some discretization parameter h. A MATLAB implementation of this approach is given in Algorithm 5.17.

ALGORITHM 5.17.
Finite difference approximation for the Jacobian

```
function J=NumericalJacobian(f,x);
% NUMERICALJACOBIAN computes a Jacobian numerically
%    J=NumericalJacobian(f,x); computes numerically a Jacobian matrix
%    for the function f at x

for i=1:length(x)
  xd=x; h=sqrt(eps*(1+abs(xd(i)))); xd(i)=xd(i)+h;
  J(:,i)=(f(xd)-f(x))/h;
end;
```

In order to determine a numerically sensible choice for the discretization parameter h, one has to consider two approximation errors in the expression: first, the smaller one chooses h, the more accurate the finite difference approximation will be, since in the limit, it converges mathematically to the derivative. Numerically, however, there is a difference that needs to be computed in the numerator, and we have seen that differences of quantities that are very close in size suffer from cancellation, see Chapter 8, so h should not be too small. In order to get more insight, assume that f is a scalar function. Expanding using a Taylor series gives

$$\frac{f(x + h) - f(x)}{h} = f'(x) + \frac{h}{2}f''(x) + O(h^2), \qquad (5.106)$$

which shows clearly that h needs to be small in order to get a good approximation of the derivative $f'(x)$. On the other hand, let us study the roundoff error when computing the approximation,

$$\frac{f((x + h)(1 + \epsilon_1))(1 + \epsilon_2) - f(x)(1 + \epsilon_3)}{h}$$

$$\approx \frac{f((x + h) - f(x)}{h} + \frac{1}{h}(f'(x + h)(x + h)\epsilon_1 + f(x + h)\epsilon_2 - f(x)\epsilon_3)$$

where $|\epsilon_i| \leq eps$, the machine precision. A good choice is to balance the two sources of error, i.e.

$$\frac{h}{2}(\ldots) \approx \frac{1}{h}(\ldots)eps,$$

which indicates that a good choice for h is

$$h = \sqrt{eps}, \quad \text{or} \quad h = \sqrt{eps(1 + |x|)}.$$

EXAMPLE 5.11. *We would like to compute the intersection points of the circle of radius $r = 2$ centered at the origin with the ellipse with center $M = (3,1)$ and semi-axes (parallel to the coordinate axes) a and $b = 2$.*

We will solve this problem using the parametric representation of the circle and the ellipse. The circle is given by $(2\cos t, 2\sin t)$, $0 \leq t < 2\pi$. A point on the ellipse has the coordinates $(3 + a\cos s, 1 + 2\sin s)$, $0 \leq s < 2\pi$.

To find an intersection point we must solve the following nonlinear system of equations for s and t:

$$\begin{aligned} 2\cos t &= 3 + a\cos s, \\ 2\sin t &= 1 + 2\sin s, \end{aligned}$$

or

$$\boldsymbol{f}(\boldsymbol{x}) = \begin{pmatrix} 2\cos x_1 - 3 - a\cos x_2 \\ 2\sin x_1 - 1 - 2\sin x_2 \end{pmatrix} = \begin{pmatrix} 0 \\ 0 \end{pmatrix} \text{ with } \boldsymbol{x} = \begin{pmatrix} t \\ s \end{pmatrix}.$$

The Jacobian is

$$J = \begin{pmatrix} -2\sin x_1 & a\sin x_2 \\ 2\cos x_1 & -2\cos x_2 \end{pmatrix}.$$

We would like to compute the intersection points for $a = 1.3 : 0.5 : 7$. From the geometry it is clear that there are two points for a specific value of a in that range. In the following MATLAB script, we compute the solutions using Newton's method, with two different initial guesses for the unknown parameters $\boldsymbol{x} = (t, s)^{\top}$. We also plot the computed intersection points, the circle and the different ellipses (see Figure 5.15). To draw the ellipses we use the MATLAB function described in Section 6.7.2.

ALGORITHM 5.18. *Intersection of Circle and Ellipse*

```
PlotEllipse([0 0],2,2);
axis([-4 4, -3 5]); axis square; hold;
X1=[]; X2=[];
for a= 1.3:0.5:7
  f=@(x) [2*cos(x(1))-3-a*cos(x(2));2*sin(x(1))-1-2*sin(x(2))];
  fp=@(x) [-2*sin(x(1)) a*sin(x(2)); 2*cos(x(1)) -2*cos(x(2))];
  x=Newton(f,[0;4],1e-10,20);                    % first intersection
  X1=[X1; 2*cos(x(1)) 2*sin(x(1))];
  plot(2*cos(x(1)),2*sin(x(1)),'o');
```

```
  x=Newton(f,[1;3],1e-10,20);                    % second intersection
  X2=[X2; 2*cos(x(1)) 2*sin(x(1))];
  plot(2*cos(x(1)),2*sin(x(1)),'o');
  PlotEllipse([3 1],a,2);
  pause
end;
hold off
```

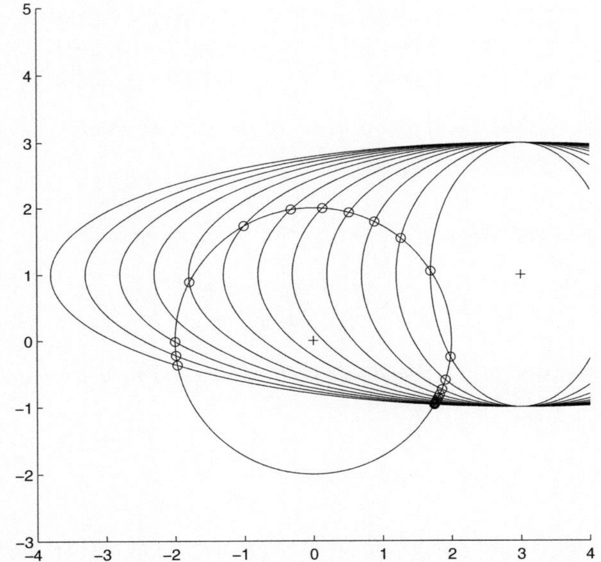

FIGURE 5.15. *Computing Intersection Points with Newton's Method*

Looking at Figure 5.15 and at the computed coordinates of the intersection points (see Table 5.4) we see that we missed one point! The convergence result in Theorem 5.6 is a local result, meaning Newton's method needs good starting values to converge. In this case, our initial guesses are not good enough to ensure that we do not converge to some other point we are not interested in.

Concerning the global convergence behavior of Newton's method, only very little is known. One can only completely analyze the behavior of Newton's method for certain simple examples; one of the best known is the following.

EXAMPLE 5.12. *We consider $f(z) = z^3 - 1$, with $z \in \mathbb{C}$. The equation $f(z) = 0$ has three complex roots on the unit circle. We could write this*

First points		Second points	
1.9845	−0.2486	1.7005	1.0528
1.9104	−0.5919	1.2685	1.5463
1.8594	−0.7367	0.8880	1.7920
1.8261	−0.8157	0.5217	1.9308
1.8038	−0.8640	0.1376	1.9953
1.7883	−0.8956	−0.3184	1.9745
1.7772	−0.9174	−1.0031	1.7302
1.7690	−0.9331	−1.7924	0.8873
1.7628	−0.9447 same!	1.7628	−0.9447
1.7580	−0.9536	−2.0000	−0.0136
1.7543	−0.9605	−1.9877	−0.2218
1.7512	−0.9660	−1.9662	−0.3662

TABLE 5.4. *Coordinates of Intersection Points*

function in real variables: if we set $z := x + iy$, we obtain

$$f(x,y) = \left(\begin{array}{c} x^3 - 3xy^2 - 1 \\ 3x^2 y - y^3 \end{array} \right).$$

It is however easier to use Newton's method directly in the complex formulation, which leads to the iteration

$$z_{k+1} = z_k - \frac{z_k^3 - 1}{3z_k^2}.$$

It is interesting to find out which of the complex roots will be found by Newton's method as we vary the starting values z_0; the set of initial values that lead to convergence to the same root is called the basin of attraction *of that root. The short* MATLAB *Algorithm 5.19 computes the basin of attraction for each root.*

ALGORITHM 5.19.
Newton's method applied to the complex equation
$$z^3 - 1 = 0$$

```
n=1000; m=30;
x=-1:2/n:1;
[X,Y]=meshgrid(x,x);
Z=X+1i*Y;
for i=1:m
  Z=Z-(Z.^3-1)./(3*Z.^2);
end;
image((round(imag(Z))+2)*10);          % transform roots to 10,20,30
```

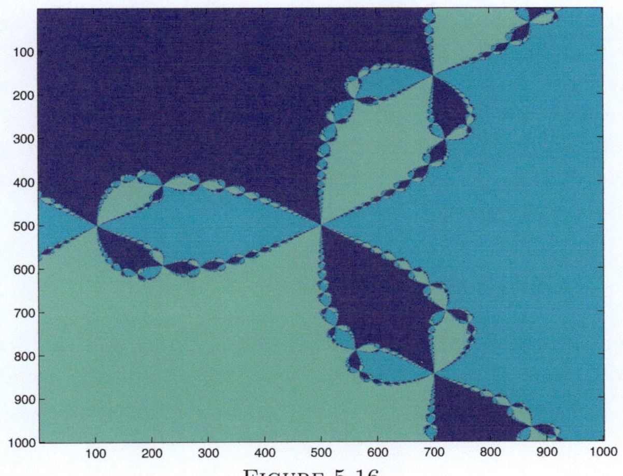

FIGURE 5.16.
*Solving the complex equation $z^3 - 1 = 0$ for many
starting values: a fractal appears*

*We show the result in Figure 5.16. We observe that Newton's method does
not always converge to the root of the equation that is closest to the initial
guess. On the contrary, the basins of attraction of the roots have a very
complicated structure, and similarly for their boundaries: they are fractal.
It is worthwhile playing with the parameters in the example Algorithm 5.19
to zoom into specific regions of the Figure, or even to reduce the number of
iterations.*

There are many variants of Newton's method to address convergence
problems or irregularities of the method, mostly developed in the context
of optimization, see Chapter 12. We show in the next subsection a different
approach that can be useful when solving very difficult nonlinear problems.

5.4.4 Continuation Methods

We have seen that Newton's method converges quadratically once it is close
enough to a root of the equation. In practice, however, it is often difficult to
find a good starting point for Newton's method to converge to the desired
solution. In such cases, a *Continuation method* can be of great help: suppose
we seek a solution to $\boldsymbol{f}(\boldsymbol{x}) = 0$ and Newton's method failed to converge to
the desired solution. A continuation method considers the problem

$$\boldsymbol{h}(\boldsymbol{x}, t) = t\boldsymbol{f}(\boldsymbol{x}) + (1 - t)\boldsymbol{g}(\boldsymbol{x}),$$

where $\boldsymbol{g}(\boldsymbol{x}) = 0$ has a known solution. Note that for $t = 0$, the solution to
$\boldsymbol{h}(\boldsymbol{x}, 0) = \boldsymbol{g}(\boldsymbol{x})$ is known, whereas for $t = 1$, the solution to $\boldsymbol{h}(\boldsymbol{x}, 1) = \boldsymbol{f}(\boldsymbol{x})$

is difficult to get. Thus, the idea is to take small steps Δt and consider neighboring problems for $t := t + \Delta t$; if Δt small enough, the neighboring problems are close, so we can hope to use the last solution as good starting values for the next problem.

More systematically, we note that $h(x, t)$ describes a curve $x(t)$ joining the known solution $x(0)$ to the hard-to-find solution $x(1)$. We can trace this curve by taking derivatives,

$$\frac{d}{dt} h(x(t), t) = h_x(x(t), t)\dot{x}(t) + h_t(x(t), t) = 0$$

where h_x denotes the Jacobian of h with respect to x and the dot denotes a derivative with respect to t. Hence the ordinary differential equation

$$\dot{x} = -(h_x(x, t))^{-1} h_t(x, t)$$

with the known initial value $x(0)$, describes the path from the known solution to $x(1)$, the solution that is unknown and hard to get. We will see in Chapter 10 how to track the path using a numerical solver for this system of ordinary differential equations.

5.5   Problems

PROBLEM 5.1. *Solve with bisection the equations*

$$a) \quad x^x = 50 \qquad b) \quad \ln(x) = \cos(x) \qquad c) \quad x + e^x = 0,$$

and describe your findings.

PROBLEM 5.2. *A goat is grazing on a circular meadow of radius r. The farmer leashes the goat and attaches it to a pole on the boundary of the meadow. Determine the length R of the rope so that the goat can only eat half the grass of the meadow by finding the equation for the ratio $x = R/r$ and solving it using bisection.*

PROBLEM 5.3. *Find x such that*

$$\int_0^x e^{-t^2}\, dt = 0.5.$$

Hint: *the integral cannot be evaluated analytically, so expand it in a series and integrate. Then find the zero by bisection of the resulting power series.*

PROBLEM 5.4. Binary search: *we are given an ordered sequence of numbers:*

$$x_1 \leq x_2 \leq \cdots \leq x_n$$

and a new number z. Write a program that computes an index value i such that either $x_{i-1} < z \le x_i$ or $i = 1$ or $i = n+1$ holds. The problem can be solved by considering the function

$$f(i) = x_i - z$$

and computing its "zero" by bisection.

PROBLEM 5.5. *Compute x where the following maximum is attained:*

$$\max_{0<x<\frac{\pi}{2}} \left(\frac{1}{4\sin x} + \frac{\sin x}{2x} - \frac{\cos x}{4x} \right).$$

PROBLEM 5.6. *Transform the following equations algebraically into a fixed point form and try to compute solutions with the fixed point iteration. In order to obtain an approximation for the fixed point, transform the equations first to $h(x) = g(x)$, where h are g well known and easy to sketch functions and read off approximations for the intersection points.*

a) $3x - \cos x = 0$

b) $2\sin x + e^x = 0$

c) $x\ln x - 1 = 0$

d) $x + \sqrt{x} = 1 + x^2$

e) $3x^2 + \tan x = 0$

PROBLEM 5.7. *Determine analytically the fixed point that is computed by the following iterations, and examine the rate of convergence:*

a) $x_0 = 1$, $x_{k+1} = 0.2(4x_k + \frac{a}{x_k})$

b) $x_0 = 1$, $x_{k+1} = 0.5(x_k + \frac{a}{x_k})$

c) $x_0 = 1$, $x_{k+1} = x_k(x_k^2 + 3a)/(3x_k^2 + a)$

PROBLEM 5.8. *What is the numerical value of the following expressions?*

$$a)\, 1 + \frac{1}{\sqrt{1 + \frac{1}{\sqrt{1+\cdots}}}} \qquad b)\, \sqrt[3]{1 + \sqrt[3]{1 + \sqrt[3]{1+\cdots}}}$$

PROBLEM 5.9. *Sketch the polynomial $P(x) = 2x^3 - 5.2x^2 - 4.1x + 3.1$ and determine approximations for the roots graphically. By solving the equation $P(x) = 0$ algebraically for the unknown in the terms with x, x^2 and x^3,* *respectively, one obtains three fixed point forms. Examine which of them can be used to compute which root and compute the roots to machine precision.*

PROBLEM 5.10. *The iteration*

$$x_0 = 1, \quad x_{k+1} = F(x_k) = \cos x_k, \quad k = 0, 1, \ldots$$

converges to $s = 0.7390851332\ldots$. How many steps would be necessary to compute s to 100 decimal digits? Predict the number and check it using multiple precision arithmetic with MAPLE.

PROBLEM 5.11. *The function $f(x) = xe^x - 1$ has only one zero at $x \approx 0.5$. Consider the fixed point form*

$$x = x + kf(x) =: F(x), \tag{5.107}$$

and determine k such that $F'(0.5) = 0$. Then compute the zero using Iteration (5.107).

PROBLEM 5.12. *Let $x = F(x)$ be a fixed point form for which the sequence $\{x_k\}$ diverges, because $|F'(s)| > 1$. Prove that then the fixed point form*

$$x = F^{[-1]}(x)$$

generates a locally convergent sequence to the fixed point s of $x = F(x)$.

As an application, compute the first positive solution of $x - \tan x = 0$. The fixed point form $x = \tan(x)$ does not converge, however

$$x = \arctan(x) + \pi$$

generates a convergent sequence.

PROBLEM 5.13. *The equation of Problem 5.3,*

$$\int_0^x e^{-t^2}\, dt = 0.5,$$

can be solved with the fixed point form

$$x = x + 0.5 - \int_0^x e^{-t^2}\, dt.$$

Why is it convergent?

PROBLEM 5.14. *The equation*

$$x^4 - 6x^3 + 12x^2 - 10x + 3 = 0$$

has a solution for $x = 1$. Analyze the convergence of Newton's iteration to that zero.

PROBLEM 5.15. *Consider the fixed point iteration $x_{k+1} = F(x_k)$, where*

$$F(x) = \alpha x + 1, \qquad 0 < \alpha < 1.$$

Using (5.34), determine the function g for which Newton's method leads to an equivalent iteration. Does this Newton's method not converge quadratically? Explain.

PROBLEM 5.16. *Divide the polynomial $P_4(x) = 2x^4 - 24x^3 + 100x^2 - 168x + 93$ by the linear factors $(x - z)$ for $z = 1, 2, 3, 4, 5$. Write down the calculations using Horner's scheme (Equation 5.62). What are the remainders r and the resulting coefficients for the polynomial $P_3(x)$ in the equation $P_4(x) = (x - z)P_3(x) + r$?*

PROBLEM 5.17. *Solve the following equations using Newton's and Halley's method in multiple precision (use `Digits := 1000`) arithmetic using* MAPLE. *Print consecutive iterates and observe the convergence.*

a) $x + e^x = 0$

b) $\ln(x) = \cos(x)$

PROBLEM 5.18. *We consider again Problem 5.3: find x such that*

$$f(x) = \int_0^x e^{-t^2}\, dt - 0.5 = 0.$$

Since a function evaluation is expensive (summation of the Taylor series) but the derivatives are cheap to compute, a higher order method is appropriate. Solve this equation with Newton's and Halley's methods.

PROBLEM 5.19. *In a triangle we know that the angle β is twice the angle α. Furthermore, the height $h_c = 5$ and the radius of the inscribed circle is $\rho = 2$ (see Figure 5.17). Determine the edges of the triangle.*

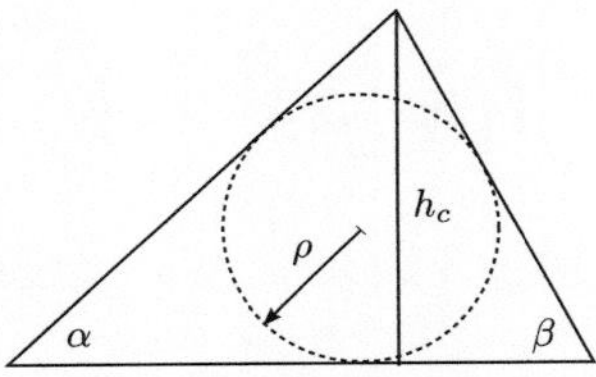

FIGURE 5.17. *Problem 5.19*

PROBLEM 5.20. *We are given the area $F = 12$ of a right-angled triangle and the section $p = 2$ of the hypotenuse (see Figure 5.18). Compute the edges of the triangle.*

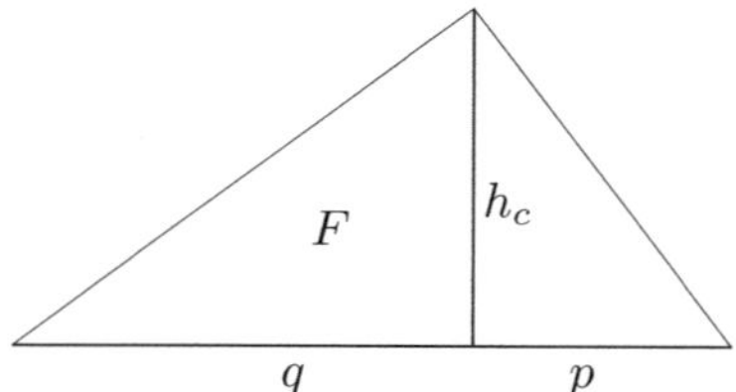

FIGURE 5.18. *Problem 5.20*

PROBLEM 5.21. *A truck is pulling an oil tank that has the shape of a cylinder with radius $r = 1.2m$ and length $l = 5m$. What is the height of the oil level when the tank is filled to one quarter?*

PROBLEM 5.22. *A pipe of radius $r = 4cm$ is suspended on a rope of length $L = 30cm$ (see Figure 5.19). What is the distance h of the pipe from the ceiling?*

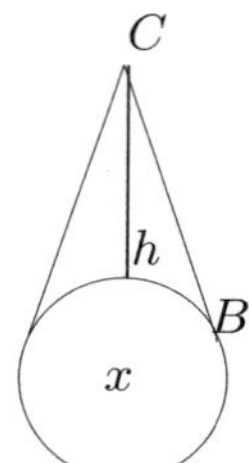

FIGURE 5.19. *Pipe problem*

PROBLEM 5.23. *Find a such that $\int_0^1 e^{at} dt = 2$.*

PROBLEM 5.24. *Let $p = 0.9$ and $\{a_j\} = \{0.1, 0.5, 1.0, 0.2, 5.0, 0.3, 0.8\}$. We are looking for $x > 0$ such that*

$$\prod_{j=1}^{n} (1 + xa_j) = 1 + p.$$

Hint: *by taking the logarithm, one gets the equation*

$$f(x) = \sum_{j=1}^{n} \ln(1 + xa_j) - \ln(1 + p) = 0,$$

which can be solved by Newton's method. Discuss $f''(x)$ and find an elegant termination criterion that makes use of monotonicity.

PROBLEM 5.25. *Use Halley's method and write a* MATLAB *function to compute the square root of a number. Invent an elegant termination criterion.*

PROBLEM 5.26. *The iteration (Newton with the wrong sign!):*

$$x_{k+1} = x_k + \frac{f(x_k)}{f'(x_k)}$$

strangely converges to a pole of the function f. Why?

PROBLEM 5.27. *Write a* MATLAB *function to compute a zero with Müller's method. Study empirically the convergence.*
Hint: *Use* MAPLE *to compute the coefficients of the quadratic equations. The naive approach will not be numerically stable! Consult [69] on page 198 for a good implementation. It is important to compute the* correction *for the next iterate and not directly the new iterate.*

PROBLEM 5.28. *Derive another iteration method (Euler's method [138]) by approximating $f(x)$ locally by a Taylor polynomial of order 2 (thus using the function values $f(x_k)$, $f'(x_k)$ and $f''(x_k)$). Write the resulting iteration in the form of Theorem 5.2 and, by doing, so prove that the sequence it generates converges cubically.*

PROBLEM 5.29. Ostrowski's method [99]*: Show that the fixed point iteration*

$$x_{k+1} = x_k - \frac{f(x_k)}{\sqrt{f'(x_k)^2 - f(x_k)f''(x_k)}}$$

also generates cubically converging sequences.

PROBLEM 5.30. *Derive yet another iteration method (quadratic inverse interpolation) by approximating the inverse function $f^{[-1]}$ by a Taylor polynomial of degree two in y. Use again the function values $f(x_k)$, $f'(x_k)$ and $f''(x_k)$) and remember how they are related to the derivatives of the inverse function.*

PROBLEM 5.31. *Computing zeros with inverse interpolation: We want to solve $f(x) = 0$. Let $f_i = f(x_i)$, $i = 0, 1, 2$. Interpolate the data*

$$\begin{array}{c|ccc} y & f_0, & f_1, & f_2 \\ \hline x & x_0, & x_1, & x_2 \end{array} \tag{5.108}$$

by a polynomial p of degree two in the variable y, $x = p(y)$ with $x_k = p(f_k)$, $k = 0, 1, 2$. The new iterate is $x_3 = p(0)$.
Hint: *The simplest algorithm here is to use the Aitken-Neviile scheme for the interpolation (see Section 4.2.7).*

The method can be generalized using not only three but all computed points, thus increasing the degree of the polynomial in each iteration. See problem 4.4 in Chapter 4.

PROBLEM 5.32. *Use bisection to create the following table:*

F	0	0.1π	0.2π	$\ldots$	π
h	0	?	?	$\ldots$	2

where the function $F(h)$ is given by

$$F(h) = \pi - 2\arccos\frac{h}{2} + h\sqrt{1 - \left(\frac{h}{2}\right)^2}.$$

PROBLEM 5.33. *(Newton's method in one dimension)*

1. *Implement Newton's method in* MATLAB.

2. *Use Newton's method to find a root of the function $f(x) = 1/x + \ln x - 2$, $x > 0$. List the correct number of digits at each iteration; notice the rate at which these correct digits increase. What kind of convergence is this (e.g. linear, quadratic, etc)?*

3. *Use Newton's method to find the zero of the function $f(x) = x^3$. What kind of convergence is this? Justify your answer.*

4. *Use Newton's method to find the zero of the function $f(x) = \arctan x$. Experiment with different starting values and notice the behavior of the algorithm. State what values yield convergence and divergence and try to determine a value which leads to oscillation. (Hint: Bisection can be used for the last part.)*

PROBLEM 5.34. Number conversions: *write a program to convert numbers given in base B_1 to numbers in base B_2. Hint: use the functions* `BaseB2Decimal` *and* `Decimal2BaseB`.

PROBLEM 5.35. *The equation $\cos(x) + \sin(2x) = -0.5$ can be transformed using $\sin(2x) = 2\sin x \cos x$ and $\cos x = \sqrt{1 - \sin^2 x}$ with the change of variable $z = \sin x$ to a polynomial equation of degree 4. Solve this equation with one of the methods discussed and compute all the solutions of the original equation.*

PROBLEM 5.36. *In a right-angled triangle ($\gamma = 90°$) the edge c measures 5 cm. The bisector of the angle α intersects the edge a in point P. The segment $\overline{PC}$ has length 1 cm. Compute the edge lengths a and b. Transform the equation to a polynomial equation using an appropriate substitution an solve this equation using* MATLAB*'s function* `root`.

PROBLEM 5.37. *Write a* MAPLE *program for Laguerre's method and compute a root of the polynomial*

$$-z^{47} + z^{45} + z^{44} + z^{43} + z^{42} + z^{41} - z^{40} + 1$$

to 250 digits. Start with $z_0 = 2$

PROBLEM 5.38. *Prove that Laguerre's method is a method that generates cubically convergent sequences for simple zeros.*

PROBLEM 5.39. *Write a* MATLAB *function to compute all the zeros of a polynomial using Laguerre's method.*

PROBLEM 5.40. *This problems shows what can be meant by* starting values in the neighborhood of a fixed point. *Compute with Newton's iteration the only zero of the function*

$$f(x) = \sqrt[11]{x^{11} - 1} + 0.5 + 0.05\sin(x/100).$$

Newton's iteration will only converge if you choose very good *starting values.*
   Hint: *Program the function as*

$$f(x) = \operatorname{sign}(x^{11} - 1)\,\sqrt[11]{|x^{11} - 1|} + 0.5 + 0.05\sin(x/100)$$

and compute the zero first by bisection. Notice that the derivative is non-negative.

PROBLEM 5.41. *For the general Taylor expansion of a function* $\boldsymbol{f}$: $\mathbb{R}^n \longrightarrow \mathbb{R}^m$, *estimate the remainder term* $R(\boldsymbol{h})$ *in (5.92) using Cauchy–Schwarz several times.*

PROBLEM 5.42. *Compute the intersection points of an ellipsoid with a sphere and a plane. The ellipsoid has the equation*

$$\left(\frac{x_1}{3}\right)^2 + \left(\frac{x_2}{4}\right)^2 + \left(\frac{x_3}{5}\right)^2 = 3.$$

The plane is given by $x_1 - 2x_2 + x_3 = 0$ *and the sphere has the equation* $x_1^2 + x_2^2 + x_3^2 = 49$

1. *How many solutions do you expect for this problem?*

2. *Solve the problem with the* `solve` *and* `fsolve` *commands from* MAPLE.

3. *Write a* MATLAB *script to solve the three equations with Newton's method. Vary the initial values so that you get all the solutions.*

Chapter 6. Least Squares Problems

> *A basic problem in science is to fit a model to observations subject to errors. It is clear that the more observations that are available, the more accurately will it be possible to calculate the parameters in the model. This gives rise to the problem of "solving" an overdetermined linear or nonlinear system of equations. It can be shown that the solution which minimizes a weighted sum of the squares of the residual is optimal in a certain sense.*
>
> Åke Björck, Numerical Methods for Least Squares Problems, SIAM, 1996.

Least squares problems appear very naturally when one would like to estimate values of parameters of a mathematical model from measured data, which are subject to errors (see quote above). They appear however also in other contexts, and form an important subclass of more general optimization problems, see Chapter 12. After several typical examples of least squares problems, we start in Section 6.2 with the linear least squares problem and the natural solution given by the normal equations. There were two fundamental contributions to the numerical solution of linear least squares problems in the last century: the first one was the development of the QR factorization by Golub in 1965, and the second one was the implicit QR algorithm for computing the singular value decomposition (SVD) by Golub and Reinsch (1970). We introduce the SVD, which is fundamental for the understanding of linear least squares problems, in Section 6.3. We postpone the description of the algorithm for its computation to Chapter 7, but use the SVD to study the condition of the linear least squares problem in Section 6.4. This will show why the normal equations are not necessarily a good approach for solving linear least squares problems, and motivates the use of orthogonal transformations and the QR decomposition in Section 6.5. Like in optimization, least squares problems can also have constraints. We treat the linear least squares problem with linear constraints in full detail in Section 6.6, and a special class with nonlinear constraints in Section 6.7. We then turn to nonlinear least squares problems in Section 6.8, which have to be solved by iteration. We show classical iterative methods for such problems, and like in the case of nonlinear equations, linear least squares problems arise naturally at each iteration. We conclude this chapter with an interesting example of least squares fitting with piecewise functions in Section 6.9. The currently best and most thorough reference for least squares methods is the book by Åke Björck [9].

W. Gander et al., *Scientific Computing - An Introduction using Maple and MATLAB*,
Texts in Computational Science and Engineering 11,
DOI 10.1007/978-3-319-04325-8_6,

6.1   Introductory Examples

We start this chapter with several typical examples leading to least squares problems.

EXAMPLE 6.1. *Measuring a road segment (Stiefel in his lectures at ETH[1]).*

$$A \quad\quad x_1 \quad\quad B \quad\quad x_2 \quad\quad C \quad\quad x_3 \quad\quad D$$

Assume that we have performed 5 measurements,

$$AD = 89m, \ AC = 67m, \ BD = 53m, \ AB = 35m \ und \ CD = 20m,$$

and we want to determine the length of the segments $x_1 = AB$, $x_2 = BC$ *und* $x_3 = CD$.

According to the observations we get a linear system with more equations than unknowns:

$$
\begin{aligned}
x_1 + x_2 + x_3 &= 89 \\
x_1 + x_2 &= 67 \\
x_2 + x_3 &= 53 \\
x_1 &= 35 \\
x_3 &= 20
\end{aligned}
\iff A\boldsymbol{x} = \boldsymbol{b}, \quad
A = \begin{pmatrix} 1 & 1 & 1 \\ 1 & 1 & 0 \\ 0 & 1 & 1 \\ 1 & 0 & 0 \\ 0 & 0 & 1 \end{pmatrix}, \quad
\boldsymbol{b} = \begin{pmatrix} 89 \\ 67 \\ 53 \\ 35 \\ 20 \end{pmatrix}.
$$

Notice that if we use the last three equations then we get the solution $x_1 = 35$, $x_2 = 33$ *and* $x_3 = 20$. *However, if we check the first two equations by inserting this solution we get*

$$
\begin{aligned}
x_1 + x_2 + x_3 - 89 &= -1, \\
x_1 + x_2 - 67 &= 1.
\end{aligned}
$$

So the equations contradict each other because of the measurement errors, and the over-determined system has no solution.

A remedy is to find an approximate solution that satisfies the equations as well as possible. For that purpose one introduces the residual vector

$$\boldsymbol{r} = \boldsymbol{b} - A\boldsymbol{x}.$$

One then looks for a vector $\boldsymbol{x}$ *that minimizes in some sense the residual vector.*

EXAMPLE 6.2. *The amount f of a component in a chemical reaction decreases with time t exponentially according to:*

$$f(t) = a_0 + a_1 e^{-bt}.$$

---

[1] Nehmen wir an, der Meister schickt seine zwei Lehrlinge aus, Strassenstücke zu vermessen. . .

If the material is weighed at different times, we obtain a table of measured values:

t	t_1	$\cdots$	t_m
y	y_1	$\cdots$	y_m

The problem now is to estimate the model parameters a_0, a_1 and b from these observations. Each measurement point (t_i, y_i) yields an equation:

$$f(t_i) = a_0 + a_1 e^{-bt_i} \approx y_i, \quad i = 1, \ldots m. \tag{6.1}$$

If there were no measurement errors, then we could replace the approximate symbol in (6.1) by an equality and use three equations from the set to determine the parameters. However, in practice, measurement errors are inevitable. Furthermore, the model equations are often not quite correct and only model the physical behavior approximately. *The equations will therefore in general contradict each other and we need some mechanism to balance the measurement errors, e.g. by requiring that (6.1) be satisfied as well as possible.*

EXAMPLE 6.3. *The next example comes from* coordinate metrology. *Here a coordinate measuring machine measures two sets of points on two orthogonal lines (see Figure 6.1). If we represent the line g_1 by the equations*

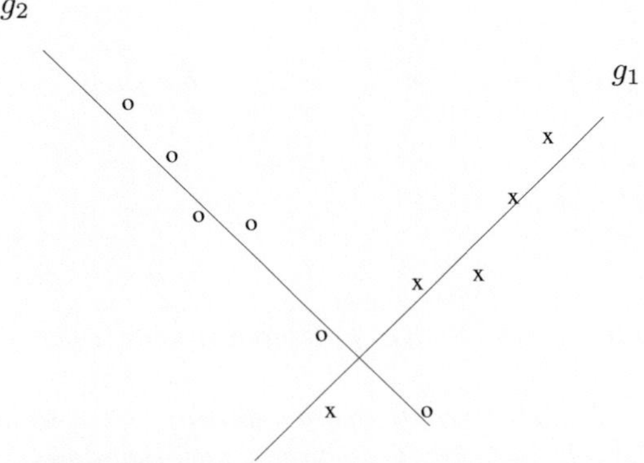

FIGURE 6.1. *Measured points on two orthogonal lines.*

$$g_1 : \quad c_1 + n_1 x + n_2 y = 0, \quad n_1^2 + n_2^2 = 1, \tag{6.2}$$

then $\boldsymbol{n} = (n_1, n_2)^{\top}$ is the normal vector on g_1. The normalizing equation $n_1^2 + n_2^2 = 1$ ensures the uniqueness of the parameters c, n_1 and n_2.

If we insert the coordinates of a measured point $P_i = (x_i, y_i)$ into Equation (6.2), we obtain the residual $r_i = c_1 + n_1 x_i + n_2 y_i$ and $d_i = |r_i|$ is the distance *of P_i from g_1. The equation of a line g_2 orthogonal to g_1 is*

$$g_2 : \quad c_2 - n_2 x + n_1 y = 0, \quad n_1^2 + n_2^2 = 1. \tag{6.3}$$

If we now insert the coordinates of q measured points Q_i into (6.3) and of p points P_i into Equation (6.2), we obtain the following system of equations for determining the parameters c_1, c_2, n_1 and n_2:

$$
\begin{pmatrix}
1 & 0 & x_{P_1} & y_{P_1} \\
1 & 0 & x_{P_2} & y_{P_2} \\
\vdots & \vdots & \vdots & \vdots \\
1 & 0 & x_{P_p} & y_{P_p} \\
0 & 1 & y_{Q_1} & -x_{Q_1} \\
0 & 1 & y_{Q_2} & -x_{Q_2} \\
\vdots & \vdots & \vdots & \vdots \\
0 & 1 & y_{Q_q} & -x_{Q_q}
\end{pmatrix}
\begin{pmatrix}
c_1 \\ c_2 \\ n_1 \\ n_2
\end{pmatrix}
\approx
\begin{pmatrix}
0 \\ 0 \\ \vdots \\ 0 \\ 0 \\ 0 \\ \vdots \\ 0
\end{pmatrix}
\qquad \text{subject to } n_1^2 + n_2^2 = 1.
$$

$$(6.4)$$

Note the difference between the least squares equations and the constraint: whereas equations of the type g_1 or g_2 only need to be satisfied approximately, the constraint $n_1^2 + n_2^2 = 1$ must be satisfied exactly *by the solution.*

EXAMPLE 6.4. *In control theory, one often considers a system like the one in Figure 6.2. The vectors $\boldsymbol{u}$ and $\boldsymbol{y}$ are the measured input and output signals at various points in time. Let $y_{t+i} = y(t + i\Delta t)$. A simple model assumes a linear relationship between the output and the input signal of the form*

$$y_{t+n} + a_{n-1} y_{t+n-1} + \cdots + a_0 y_t \approx b_{n-1} u_{t+n-1} + b_{n-2} u_{t+n-2} + \cdots + b_0 u_t. \quad (6.5)$$

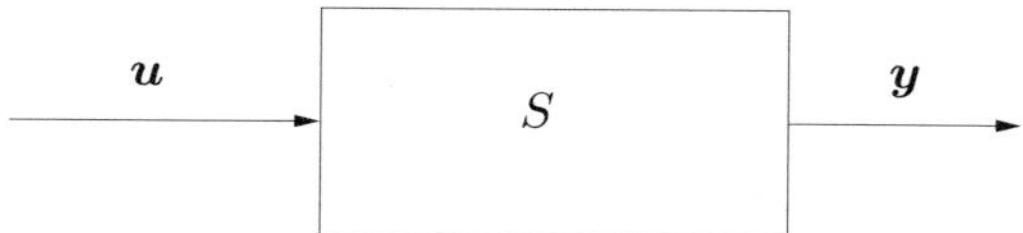

FIGURE 6.2. System with input $\boldsymbol{u}$ and output $\boldsymbol{y}$

The problem is to determine the parameters a_i and b_i from measurements of $\boldsymbol{u}$ and $\boldsymbol{y}$. For each time step we obtain a new equation of the form (6.5). If we write them all together, we get a system of linear equations:

$$
\begin{pmatrix}
y_{n-1} & y_{n-2} & \cdots & y_0 & -u_{n-1} & -u_{n-2} & \cdots & -u_0 \\
y_n & y_{n-1} & \cdots & y_1 & -u_n & -u_{n-1} & \cdots & -u_1 \\
y_{n+1} & y_n & \cdots & y_2 & -u_{n+1} & -u_n & \cdots & -u_2 \\
\vdots & \vdots & \vdots & \vdots & \vdots & \vdots & \vdots & \vdots
\end{pmatrix}
\begin{pmatrix}
a_{n-1} \\ a_{n-2} \\ \vdots \\ a_0 \\ b_{n-1} \\ b_{n-2} \\ \vdots \\ b_n
\end{pmatrix}
\approx
\begin{pmatrix}
-y_n \\ -y_{n+1} \\ -y_{n+2} \\ \vdots
\end{pmatrix}
$$

$$(6.6)$$

A matrix is said to be Toeplitz *if it has constant elements on the diagonals. Here in (6.6), the matrix is composed of two Toeplitz matrices. The number of equations is not fixed, since one can generate new equations simply by adding a new measurement.*

EXAMPLE 6.5. *In robotics and many other applications, one often encounters the* Procrustes *problem or one of its variants (see [45], Chapter 23). Consider a given body (e.g., a pyramid like in Figure 6.3) and a copy of the same body. Assume that we know the coordinates of m points $\boldsymbol{x}_i$ on*

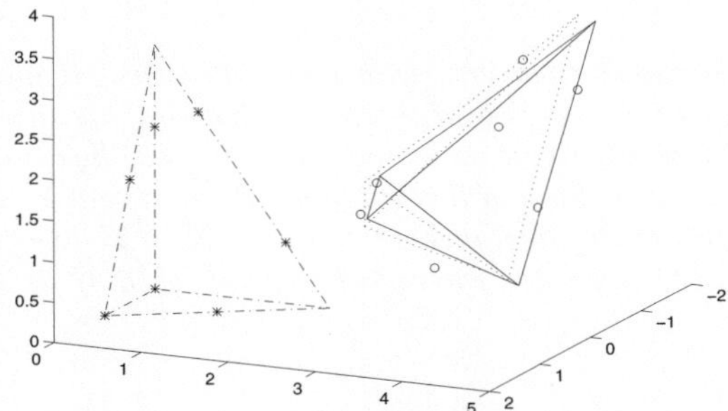

FIGURE 6.3. *Procrustes or registration problem*

the first body, and that the corresponding points $\boldsymbol{\xi}_i$ have been measured on the other body in another position in space. We would like to rotate and translate the second body so that it can be superimposed onto the first one as well as possible. In other words, we seek an orthogonal matrix Q (the product of three rotations) and a translation vector $\boldsymbol{t}$ such that $\boldsymbol{\xi}_i \approx Q\boldsymbol{x}_i + \boldsymbol{t}$ for $i = 1, \ldots, m$.

The above examples are illustrations of different classes of approximation problems. For instance, in Examples 6.1 and 6.4, the equations are linear. However, in Example 6.2 (chemical reactions), the system of equations (6.1) is nonlinear. In the metrology example 6.3, the equations are linear, but they are subject to the nonlinear constraint $n_1^2 + n_2^2 = 1$. Finally, we have also an nonlinear problem in Example 6.5, and it is not clear how to parametrize the unknown matrix Q. Nonetheless, in all the above examples, we would like to satisfy some equations as well as possible; this is indicated by the approximation symbol "$\approx$" and we have to define what we mean by that.

There are also least squares problems that are not connected with measurements like in the following example:

EXAMPLE 6.6. *We consider two straight lines g and h in space. Assume* *they are given by a point and a direction vector:*

$$g: \quad \mathbf{X} \;=\; \mathbf{P} + \lambda \mathbf{t}$$
$$h: \quad \mathbf{Y} \;=\; \mathbf{Q} + \mu \mathbf{s}$$

If they intersect each other, then there must exist a λ and a μ such that

$$\mathbf{P} + \lambda \mathbf{t} = \mathbf{Q} + \mu \mathbf{s}. \tag{6.7}$$

Rearranging (6.7) yields

$$\begin{pmatrix} t_1 & -s_1 \\ t_2 & -s_2 \\ t_3 & -s_3 \end{pmatrix} \begin{pmatrix} \lambda \\ \mu \end{pmatrix} = \begin{pmatrix} Q_1 - P_1 \\ Q_2 - P_2 \\ Q_3 - P_3 \end{pmatrix} \tag{6.8}$$

a system of three linear equations with two unknowns. If the equations are consistent, then we can use two of them to determine the intersection point. If, however, we have a pair of skew lines (i.e., if (6.8) has no solution) then we may be interested in finding the the point $\mathbf{X}$ on g and $\mathbf{Y}$ on h which are closest, i.e. for which the distance vector $\mathbf{r} = \mathbf{X} - \mathbf{Y}$ has minimal length $\|\mathbf{r}\|_2^2 \longrightarrow \min$. Thus, we are interested in solving (6.8) as a least squares problem.

6.2　Linear Least Squares Problem and the Normal Equations

Linear least squares problems occur when solving overdetermined linear systems, i.e. we are given more equations than unknowns. In general, such an overdetermined system has no solution, but we may find a meaningful approximate solution by minimizing some norm of the residual vector.

Given a matrix $A \in \mathbb{R}^{m \times n}$ with $m > n$ and a vector $\mathbf{b} \in \mathbb{R}^m$ we are looking for a vector $\mathbf{x} \in \mathbb{R}^n$ for which the norm of the residual $\mathbf{r}$ is minimized, i.e.

$$\|\mathbf{r}\| = \|\mathbf{b} - A\mathbf{x}\| \longrightarrow \min. \tag{6.9}$$

The calculations are simplest when we choose the 2-norm. Thus we will minimize the square of the length of the residual vector

$$\|\mathbf{r}\|_2^2 = r_1^2 + r_2^2 + \cdots + r_m^2 \longrightarrow \min. \tag{6.10}$$

To see that this minimum exists and is attained by some $\mathbf{x} \in \mathbb{R}^n$, note that $E = \{\mathbf{b} - A\mathbf{x} \mid \mathbf{x} \in \mathbb{R}^n\}$ is a non-empty, closed and convex subset of $\mathbb{R}^m$. Since $\mathbb{R}^m$ equipped with the Euclidean inner product is a Hilbert space, [108, Thm 4.10] asserts that E contains a unique element of smallest norm, so there exists an $\mathbf{x} \in \mathbb{R}^n$ (not necessarily unique) such that $\|\mathbf{b} - A\mathbf{x}\|_2$ is minimized.

The minimization problem (6.10) gave rise to the name *Least Squares Method*. The theory was developed independently by CARL FRIEDRICH

GAUSS in 1795 and ADRIEN-MARIE LEGENDRE who published it first in 1805. On January 1, 1801, using the least squares method, GAUSS made the best prediction of the orbital positions of the planetoid Ceres based on measurements of G. PIAZZI, and the method became famous because of this.

We characterize the least squares solution by the following theorem.

THEOREM 6.1. (LEAST SQUARES SOLUTION) *Let*

$$\mathcal{S} = \{\boldsymbol{x} \in \mathbb{R}^n \ with \ \|\boldsymbol{b} - A\boldsymbol{x}\|_2 \longrightarrow \min\}$$

be the set of solutions and let $\boldsymbol{r}_x = \boldsymbol{b} - A\boldsymbol{x}$ *denote the residual for a specific* $\boldsymbol{x}$*. Then*

$$\boldsymbol{x} \in \mathcal{S} \iff A^\top \boldsymbol{r}_x = 0 \iff \boldsymbol{r}_x \perp \mathcal{R}(A), \tag{6.11}$$

where $\mathcal{R}(A)$ *denotes the subspace spanned by the columns of* A*.*

PROOF. We prove the first equivalence, from which the second one follows easily.

"$\Leftarrow$": Let $A^\top \boldsymbol{r}_x = 0$ and $\boldsymbol{z} \in \mathbb{R}^n$ be an arbitrary vector. It follows that $\boldsymbol{r}_z = \boldsymbol{b} - A\boldsymbol{z} = \boldsymbol{b} - A\boldsymbol{x} + A(\boldsymbol{x} - \boldsymbol{z})$, thus $\boldsymbol{r}_z = \boldsymbol{r}_x + A(\boldsymbol{x} - \boldsymbol{z})$. Now

$$\|\boldsymbol{r}_z\|_2^2 = \|\boldsymbol{r}_x\|_2^2 + 2(\boldsymbol{x} - \boldsymbol{z})^\top A^\top \boldsymbol{r}_x + \|A(\boldsymbol{x} - \boldsymbol{z})\|_2^2.$$

But $A^\top \boldsymbol{r}_x = 0$ and therefore $\|\boldsymbol{r}_z\|_2 \geq \|\boldsymbol{r}_x\|_2$. Since this holds for every $\boldsymbol{z}$ then $\boldsymbol{x} \in \mathcal{S}$.

"$\Rightarrow$": We show this by contradiction: assume $A^\top \boldsymbol{r}_x = \boldsymbol{z} \neq 0$. We consider $\boldsymbol{u} = \boldsymbol{x} + \varepsilon \boldsymbol{z}$ with $\varepsilon > 0$:

$$\boldsymbol{r}_u = \boldsymbol{b} - A\boldsymbol{u} = \boldsymbol{b} - A\boldsymbol{x} - \varepsilon A\boldsymbol{z} = \boldsymbol{r}_x - \varepsilon A\boldsymbol{z}.$$

Now $\|\boldsymbol{r}_u\|_2^2 = \|\boldsymbol{r}_x\|_2^2 - 2\varepsilon \boldsymbol{z}^\top A^\top \boldsymbol{r}_x + \varepsilon^2 \|A\boldsymbol{z}\|_2^2$. Because $A^\top \boldsymbol{r}_x = \boldsymbol{z}$ we obtain

$$\|\boldsymbol{r}_u\|_2^2 = \|\boldsymbol{r}_x\|_2^2 - 2\varepsilon \|\boldsymbol{z}\|_2^2 + \varepsilon^2 \|A\boldsymbol{z}\|_2^2.$$

We conclude that, for sufficient small ε, we can obtain $\|\boldsymbol{r}_u\|_2^2 < \|\boldsymbol{r}_x\|_2^2$. This is a contradiction, since $\boldsymbol{x}$ cannot be in the set of solutions in this case. Thus the assumption was wrong, i.e., we must have $A^\top \boldsymbol{r}_x = 0$, which proves the first equivalence in (6.11). $\square$

The least squares solution has an important statistical property which is expressed in the following *Gauss-Markoff Theorem*. Let the vector $\boldsymbol{b}$ of observations be related to an unknown parameter vector $\boldsymbol{x}$ by the linear relation

$$A\boldsymbol{x} = \boldsymbol{b} + \boldsymbol{\epsilon}, \tag{6.12}$$

where $A \in \mathbb{R}^{m \times n}$ is a known matrix and $\boldsymbol{\epsilon}$ is a vector of random errors. In this *standard linear model* it is assumed that the random variables ϵ_j are uncorrelated and all have zero mean and the same variance.

THEOREM 6.2. (GAUSS-MARKOFF) *Consider the standard linear model (6.12). Then the best linear unbiased estimator of any linear function* $\boldsymbol{c}^\top \boldsymbol{x}$ *is the least square solution of* $\|A\boldsymbol{x} - \boldsymbol{b}\|_2^2 \longrightarrow \min.$

PROOF. Consult a statistics textbook, for example [94, p. 181].   □

Equation (6.11) can be used to determine the least square solution. From $A^\top r_x = 0$ it follows that $A^\top (b - Ax) = 0$, and we obtain the *Normal Equations* of Gauss:

$$A^\top A x = A^\top b. \tag{6.13}$$

EXAMPLE 6.7. *We return to Example 6.1 and solve it using the Normal Equations.*

$$A^\top A x = A^\top b \iff \begin{pmatrix} 3 & 2 & 1 \\ 2 & 3 & 2 \\ 1 & 2 & 3 \end{pmatrix} \begin{pmatrix} x_1 \\ x_2 \\ x_3 \end{pmatrix} = \begin{pmatrix} 191 \\ 209 \\ 162 \end{pmatrix}.$$

The solution of this 3×3 *system is*

$$x = \begin{pmatrix} 35.125 \\ 32.500 \\ 20.625 \end{pmatrix}.$$

The residual for this solution becomes

$$r = b - Ax = \begin{pmatrix} 0.7500 \\ -0.6250 \\ -0.1250 \\ -0.1250 \\ -0.6250 \end{pmatrix} \quad \textit{with } \|r\|_2 = 1.1726.$$

Notice that for the solution $x = (35, 33, 20)^\top$ *obtained by solving the last three equations we obtain a larger residual* $\|r\|_2 = \sqrt{2} = 1.4142.$

There is also a way to understand the normal equations geometrically from (6.11). We want to find a linear combination of columns of the matrix A to approximate the vector b. The space spanned by the columns of A is the range of A, $\mathcal{R}(A)$, which is a hyperplane in $\mathbb{R}^m$, and the vector b in general does not lie in this hyperplane, as shown in Figure 6.4. Thus, minimizing $\|b - Ax\|_2$ is equivalent to minimizing the length of the residual vector r, and thus the residual vector has to be orthogonal to $\mathcal{R}(A)$, as shown in Figure 6.4.

The normal equations (6.13) concentrate data since $B = A^\top A$ is a small $n \times n$ matrix, whereas A is $m \times n$. The matrix B is symmetric, and if $\mathrm{rank}(A) = n$, then it is also positive definite. Thus, the natural way to solve the normal equations is by means of the *Cholesky decomposition* (cf. Section 3.4.1):

1. Form $B = A^\top A$ (we need to compute only the upper triangle since B is symmetric) and compute $c = A^\top b$.

2. Decompose $B = R^\top R$ (Cholesky) where R is an upper triangular matrix.

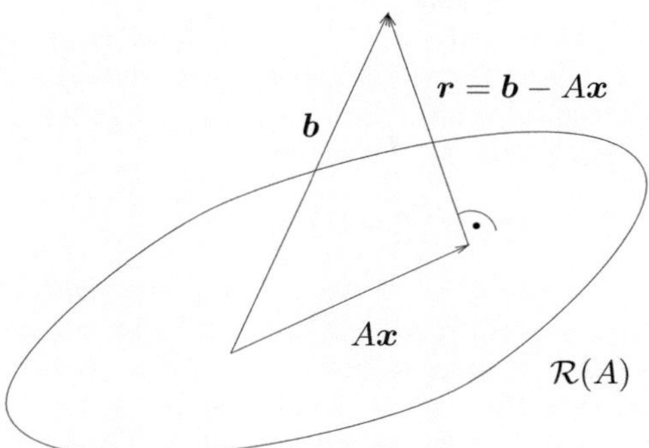

FIGURE 6.4. r *is orthogonal to* $\mathcal{R}(A)$

3. Compute the solution by forward- ($R^{\top}y = c$) and back-substitution ($Rx = y$).

We will see later on that there are numerically preferable methods for computing the least squares solution. They are all based on the use of *orthogonal matrices* (i.e. matrices B for which $B^{\top}B = I$).

Notice that when solving linear systems $Ax = b$ with n equations and n unknowns by Gaussian elimination, reducing the system to triangular form, we make use of the fact that *equivalent systems have the same solutions*:

$$Ax = b \iff BAx = Bb \quad \text{if } B \text{ is nonsingular.}$$

For a system of equations $Ax \approx b$ to be solved in the least squares sense, it no longer holds that multiplying by a nonsingular matrix B leads to an equivalent system. This is because the transformed residual Br may not have the same norm as r itself. However, if we restrict ourselves to the class of *orthogonal matrices*,

$$Ax \approx b \iff BAx \approx Bb \quad \text{if } B \text{ is orthogonal.}$$

then the least squares problems remain equivalent, since $r = b - Ax$ and $Br = Bb - BAx$ have the same length,

$$\|Br\|_2^2 = (Br)^{\top}(Br) = r^{\top}B^{\top}Br = r^{\top}r = \|r\|_2^2.$$

Orthogonal matrices and the matrix decompositions containing orthogonal factors therefore play an important role in algorithms for the solution of linear least squares problems. Often it is possible to simplify the equations by pre-multiplying the system by a suitable orthogonal matrix.

6.3 Singular Value Decomposition (SVD)

The singular value decomposition (SVD) of a matrix A is a very useful tool in the context of least squares problems. It is also very helpful for analyzing properties of a matrix. With the SVD one x-rays a matrix!

THEOREM 6.3. (SINGULAR VALUE DECOMPOSITION, SVD) *Let* $A \in \mathbb{R}^{m \times n}$ *with* $m \geq n$. *Then there exist orthogonal matrices* $U \in \mathbb{R}^{m \times m}$ *and* $V \in \mathbb{R}^{n \times n}$ *and a diagonal matrix* $\Sigma = \mathrm{diag}(\sigma_1, \ldots, \sigma_n) \in \mathbb{R}^{m \times n}$ *with* $\sigma_1 \geq \sigma_2 \geq \ldots \geq \sigma_n \geq 0$, *such that*

$$A = U \Sigma V^\top$$

holds. The column vectors of $U = [\boldsymbol{u}_1, \ldots, \boldsymbol{u}_m]$ *are called the* left singular vectors *and similarly* $V = [\boldsymbol{v}_1, \ldots, \boldsymbol{v}_n]$ *are the* right singular vectors. *The values* σ_i *are called the* singular values *of* A. *If* $\sigma_r > 0$ *is the smallest non-zero* singular value, *then the matrix* A *has rank* r.

PROOF. The 2-norm of A is defined by $\|A\|_2 = \max_{\|\boldsymbol{x}\|_2 = 1} \|A\boldsymbol{x}\|_2$. Thus there exists a vector $\boldsymbol{x}$ with $\|\boldsymbol{x}\|_2 = 1$ such that

$$\boldsymbol{z} = A\boldsymbol{x}, \quad \|\boldsymbol{z}\|_2 = \|A\|_2 =: \sigma.$$

Let $\boldsymbol{y} := \boldsymbol{z}/\|\boldsymbol{z}\|_2$. This yields $A\boldsymbol{x} = \sigma \boldsymbol{y}$ with $\|\boldsymbol{x}\|_2 = \|\boldsymbol{y}\|_2 = 1$.

Next we extend $\boldsymbol{x}$ into an orthonormal basis of $\mathbb{R}^n$. If $V \in \mathbb{R}^{n \times n}$ is the matrix containing the basis vectors as columns, then V is an orthogonal matrix that can be written as $V = [\boldsymbol{x}, V_1]$, where $V_1^\top \boldsymbol{x} = 0$. Similarly, we can construct an orthogonal matrix $U \in \mathbb{R}^{m \times m}$ satisfying $U = [\boldsymbol{y}, U_1]$, $U_1^\top \boldsymbol{y} = 0$. Now

$$A_1 = U^\top A V = \begin{bmatrix} \boldsymbol{y}^\top \\ U_1^\top \end{bmatrix} A\, [\boldsymbol{x}, V_1] = \begin{bmatrix} \boldsymbol{y}^\top A\boldsymbol{x} & \boldsymbol{y}^\top A V_1 \\ U_1^\top A\boldsymbol{x} & U_1^\top A V_1 \end{bmatrix} = \begin{bmatrix} \sigma & \boldsymbol{w}^\top \\ 0 & B \end{bmatrix},$$

because $\boldsymbol{y}^\top A\boldsymbol{x} = \boldsymbol{y}^\top \sigma \boldsymbol{y} = \sigma \boldsymbol{y}^\top \boldsymbol{y} = \sigma$ and $U_1^\top A\boldsymbol{x} = \sigma U_1^\top \boldsymbol{y} = 0$ since $U_1 \perp \boldsymbol{y}$.

We claim that $\boldsymbol{w}^\top := \boldsymbol{y}^\top A V_1 = 0$. In order to prove this, we compute

$$A_1 \begin{pmatrix} \sigma \\ \boldsymbol{w} \end{pmatrix} = \begin{pmatrix} \sigma^2 + \|\boldsymbol{w}\|_2^2 \\ B\boldsymbol{w} \end{pmatrix}$$

and conclude from that equation that

$$\left\| A_1 \begin{pmatrix} \sigma \\ \boldsymbol{w} \end{pmatrix} \right\|_2^2 = \left(\sigma^2 + \|\boldsymbol{w}\|_2^2 \right)^2 + \|B\boldsymbol{w}\|_2^2 \geq \left(\sigma^2 + \|\boldsymbol{w}\|_2^2 \right)^2.$$

Now since V and U are orthogonal, $\|A_1\|_2 = \|U^\top A V\|_2 = \|A\|_2 = \sigma$ holds and

$$\sigma^2 = \|A_1\|_2^2 = \max_{\|\boldsymbol{x}\|_2 \neq 0} \frac{\|A_1 \boldsymbol{x}\|_2^2}{\|\boldsymbol{x}\|_2^2} \geq \frac{\left\| A_1 \begin{pmatrix} \sigma \\ \boldsymbol{w} \end{pmatrix} \right\|_2^2}{\left\| \begin{pmatrix} \sigma \\ \boldsymbol{w} \end{pmatrix} \right\|_2^2} \geq \frac{\left(\sigma^2 + \|\boldsymbol{w}\|_2^2 \right)^2}{\sigma^2 + \|\boldsymbol{w}\|_2^2}.$$

The last equation reads

$$\sigma^2 \geq \sigma^2 + \|\boldsymbol{w}\|_2^2,$$

and we conclude that $\boldsymbol{w} = 0$. Thus we have obtained

$$A_1 = U^\top A V = \begin{bmatrix} \sigma & 0 \\ 0 & B \end{bmatrix}.$$

We can now apply the same construction to the sub-matrix B and thus finally end up with a diagonal matrix. $\qquad\square$

Although the proof is constructive, the singular value decomposition is not usually computed in this way. An efficient numerical algorithm was designed by Golub and Reinsch [148]. They first transform the matrix by orthogonal Householder transformations to bidiagonal form. Then the bidiagonal matrix is further diagonalized in a iterative process by a variant of the QR Algorithm. For details see Section 7.7 in Chapter 7 on eigenvalues.

If we write the equation $A = U\Sigma V^\top$ in partitioned form, in which Σ_r contains only the nonzero singular values, we get

$$A = [U_1, U_2] \begin{pmatrix} \Sigma_r & 0 \\ 0 & 0 \end{pmatrix} [V_1, V_2]^\top \qquad (6.14)$$

$$= U_1 \Sigma_r V_1^\top \qquad (6.15)$$

$$= \sum_{i=1}^{r} \sigma_i \, \boldsymbol{u}_i \boldsymbol{v}_i^\top . \qquad (6.16)$$

Equation (6.14) is the *full decomposition* with square matrices U and V. When making use of the zeros we obtain the *"economy"* or *"reduced"* version of the SVD given in (6.15). In MATLAB there are two variants to compute the SVD:

```
[U S V ]=svd(A)    % gives the full decomposition
[U S V ]=svd(A,0) % gives an m by n matrix U
```

The call `svd(A,0)` computes a version between full and economic with a non-square matrix $U \in \mathbb{R}^{m \times n}$. This form is sometimes referred to as the "thin SVD".

EXAMPLE 6.8. *The matrix A has rank one and its economy SVD is given by*

$$A = \begin{pmatrix} 1 & 1 & 1 \\ 1 & 1 & 1 \\ 1 & 1 & 1 \\ 1 & 1 & 1 \end{pmatrix} = \begin{pmatrix} \frac{1}{2} \\ \frac{1}{2} \\ \frac{1}{2} \\ \frac{1}{2} \end{pmatrix} (2\sqrt{3}) \begin{pmatrix} \frac{1}{\sqrt{3}} & \frac{1}{\sqrt{3}} & \frac{1}{\sqrt{3}} \end{pmatrix} .$$

With MATLAB *we get the thin version*

```
>> [U,S,V]=svd(ones(4,3),0)
U =
   -0.5000    0.8660   -0.0000
   -0.5000   -0.2887   -0.5774
   -0.5000   -0.2887    0.7887
   -0.5000   -0.2887   -0.2113
S =
    3.4641         0         0
         0    0.0000         0
```

```
           0            0            0
V =
      -0.5774       0.8165            0
      -0.5774      -0.4082      -0.7071
      -0.5774      -0.4082       0.7071
```

THEOREM 6.4. *If $A = U\Sigma V^\top$, then the column vectors of V are the eigenvectors of the matrix $A^\top A$ associated with the eigenvalues σ_i^2, $i = 1,\ldots,n$. The column vectors of U are the eigenvectors of the matrix $AA^\top$.*

PROOF.

$$A^\top A = (U\Sigma V^\top)^\top U\Sigma V^\top = VDV^\top, \quad D = \Sigma^\top \Sigma = \mathrm{diag}(\sigma_1^2,\ldots,\sigma_n^2). \quad (6.17)$$

Thus $A^\top AV = VD$ and σ_i^2 is an eigenvalue of $A^\top A$. Similarly

$$AA^\top = U\Sigma V^\top (U\Sigma V^\top)^\top = U\Sigma\Sigma^\top U^\top, \qquad (6.18)$$

where $\Sigma\Sigma^\top = \mathrm{diag}(\sigma_1^2,\ldots,\sigma_n^2,0,\ldots,0) \in \mathbb{R}^{m\times m}$. $\qquad\square$

THEOREM 6.5. *Let $A = U\Sigma V^\top$. Then*

$$\|A\|_2 = \sigma_1 \quad and \quad \|A\|_F = \sqrt{\sum_{i=1}^{n} \sigma_i^2}.$$

PROOF. Since U and V are orthogonal, we have $\|A\|_2 = \|U\Sigma V^\top\|_2 = \|\Sigma\|_2$. Now

$$\|\Sigma\|_2^2 = \max_{\|x\|_2=1} \|\Sigma x\|_2^2 = \max_{\|x\|_2=1} (\sigma_1^2 x_1^2 + \cdots + \sigma_n^2 x_n^2) \le \sigma_1^2 (x_1^2 + \cdots + x_n^2) = \sigma_1^2,$$

and since the maximum is attained for $x = e_1$ it follows that $\|A\|_2 = \sigma_1$. For the *Frobenius norm* we have

$$\|A\|_F = \sqrt{\sum_{i,j} a_{ij}^2} = \sqrt{\mathrm{tr}(A^\top A)} = \sqrt{\sum_{i=1}^{n} \sigma_i^2},$$

since the trace of a matrix equals the sum of its eigenvalues. $\qquad\square$

In (6.16), we have decomposed the matrix A as a sum of rank-one matrices of the form $u_i v_i^\top$. Now we have

$$\|u_i v_i^\top\|_2^2 = \max_{\|x\|_2=1} \|u_i v_i^\top x\|_2^2 = \max_{\|x\|_2=1} |v_i^\top x|^2 \|u_i\|_2^2 = \max_{\|x\|_2=1} |v_i^\top x|^2,$$

and since

$$\max_{\|x\|_2=1} |v_i^\top x|^2 = \max_{\|x\|_2=1} (\|v_i\|_2 \|x\|_2 \cos\alpha)^2 = \cos^2\alpha,$$

where α is the angle between the two vectors, we obtain

$$\max_{\|\boldsymbol{x}\|_2 = 1} \|\boldsymbol{v}_i^\top \boldsymbol{x}\|_2^2 = \|\boldsymbol{v}_i^\top \boldsymbol{v}_i\|_2^2 = 1.$$

We see from (6.16) that the matrix A is decomposed into a weighted sum of matrices which have all the same norm, and the singular values are the weights. The main contributions in the sum are the terms with the largest singular values. *Therefore we may approximate A by a lower rank matrix by dropping the smallest singular values, i.e., replacing their values by zero.* In fact we have the

THEOREM 6.6. *Let $A \in \mathbb{R}^{m \times n}$ have rank r and let $A = U\Sigma V^\top$. Let $\mathcal{M}$ denote the set of $m \times n$ matrices with rank $p < r$. The solution of*

$$\min_{X \in \mathcal{M}} \|A - X\|_2$$

is given by $A_p = \sum_{i=1}^{p} \sigma_i \boldsymbol{u}_i \boldsymbol{v}_i^\top$ and we have

$$\min_{X \in \mathcal{M}} \|A - X\|_2 = \|A - A_p\|_2 = \sigma_{p+1}.$$

PROOF. We have $U^\top A_p V = \mathrm{diag}(\sigma_1, \ldots, \sigma_p, 0, \ldots, 0)$ thus $A_p \in \mathcal{M}$ and

$$\|A - A_p\|_2 = \sigma_{p+1}.$$

Let $B \in \mathcal{M}$ and let the linear independent vectors $\boldsymbol{x}_1, \ldots, \boldsymbol{x}_{n-p}$ span the null space of B, so that $B\boldsymbol{x}_j = 0$. The two sets of vectors $\{\boldsymbol{x}_1, \ldots, \boldsymbol{x}_{n-p}\}$ and $\{\boldsymbol{v}_1, \ldots, \boldsymbol{v}_{p+1}\}$ contain altogether $n+1$ vectors. Hence, they must be linearly dependent, so we can write

$$\alpha_1 \boldsymbol{x}_1 + \alpha_2 \boldsymbol{x}_2 + \cdots + \alpha_{n-p} \boldsymbol{x}_{n-p} + \beta_1 \boldsymbol{v}_1 + \beta_2 \boldsymbol{v}_2 + \cdots + \beta_{p+1} \boldsymbol{v}_{p+1} = 0.$$

Moreover, not all $\alpha_i = 0$, otherwise the vectors $\boldsymbol{v}_i$ would be linearly dependent! Denote by

$$\boldsymbol{h} = -\alpha_1 \boldsymbol{x}_1 - \alpha_2 \boldsymbol{x}_2 - \cdots - \alpha_{n-p} \boldsymbol{x}_{n-p} = \beta_1 \boldsymbol{v}_1 + \beta_2 \boldsymbol{v}_2 + \cdots + \beta_{p+1} \boldsymbol{v}_{p+1} \neq 0,$$

and form the unit vector $\boldsymbol{z} = \boldsymbol{h}/\|\boldsymbol{h}\|_2 = \gamma_1 \boldsymbol{v}_1 + \gamma_2 \boldsymbol{v}_2 + \cdots + \gamma_{p+1} \boldsymbol{v}_{p+1}$. Then $B\boldsymbol{z} = 0$, $\boldsymbol{z}^\top \boldsymbol{z} = \gamma_1^2 + \cdots + \gamma_{p+1}^2 = 1$ and

$$A\boldsymbol{z} = U\Sigma V^\top \boldsymbol{z} = \sum_{i=1}^{p+1} \sigma_i \gamma_i \boldsymbol{u}_i.$$

It follows that

$$\|A - B\|_2^2 \geq \|(A - B)\boldsymbol{z}\|_2^2 = \|A\boldsymbol{z}\|_2^2 = \sum_{i=1}^{p+1} \sigma_i^2 \gamma_i^2$$

$$\geq \sigma_{p+1}^2 \sum_{i=1}^{p+1} \gamma_i^2 = \sigma_{p+1}^2 \|\boldsymbol{z}\|_2^2 = \sigma_{p+1}^2.$$

Thus, the distance from A to any other matrix in $\mathcal{M}$ is greater or equal to the distance to A_p. This proves the theorem. $\qquad\square$

6.3.1   Pseudoinverse

DEFINITION 6.1. (PSEUDOINVERSE) *Let $A = U\Sigma V^{\top}$ be the singular value decomposition with*

$$\Sigma = \begin{pmatrix} \Sigma_r \\ 0 \end{pmatrix} \in \mathbb{R}^{m \times n}, \quad \Sigma_r := \mathrm{diag}(\sigma_1, \ldots, \sigma_r, 0, \ldots, 0) \in \mathbb{R}^{n \times n}$$

with $\sigma_1 \geq \cdots \geq \sigma_r > 0$. Then the matrix $A^+ = V\Sigma^+ U^{\top}$ with

$$\Sigma^+ = (\Sigma_r^+ \;\; 0) \in \mathbb{R}^{n \times m}, \quad \Sigma_r^+ := \mathrm{diag}(\frac{1}{\sigma_1}, \ldots, \frac{1}{\sigma_r}, 0, \ldots, 0) \in \mathbb{R}^{n \times n} \quad (6.19)$$

is called the pseudoinverse *of A.*

We have discussed the SVD only for the case in which $A \in \mathbb{R}^{m \times n}$ with $m \geq n$. This was mainly for simplicity, since the SVD exists for any matrix: if $A = U\Sigma V^{\top}$, then $A^{\top} = V\Sigma^{\top} U^{\top}$ is the singular value decomposition of $A^{\top} \in \mathbb{R}^{n \times m}$. Usually the SVD is computed such that the singular values are ordered decreasingly. The representation $A^+ = V\Sigma^+ U^{\top}$ of the pseudoinverse is thus already a SVD, except that the singular values $\frac{1}{\sigma_1}, \cdots, \frac{1}{\sigma_r}$ are ordered increasingly. By simultaneously permuting rows and columns one can reorder the decomposition and bring it into standard form with decreasing elements in Σ^+.

THEOREM 6.7. (PENROSE EQUATIONS) *$Y = A^+$ is the only solution of the matrix equations*

$$\begin{array}{llll} (i) & AYA = A & (ii) & YAY = Y \\ (iii) & (AY)^{\top} = AY & (iv) & (YA)^{\top} = YA \end{array}$$

PROOF. It is simple to verify that A^+ is a solution: inserting the SVD e.g. into (i), we get

$$AA^+A = U\Sigma V^{\top} V\Sigma^+ U^{\top} U\Sigma V^{\top} = U\Sigma\Sigma^+\Sigma V^{\top} = U\Sigma V^{\top} = A.$$

More challenging is to prove uniqueness. To do this, assume that Y is any solution to (i)–(iv). Then

$$
\begin{aligned}
Y &= YAY \quad \text{because of (ii)} \\
  &= (YA)^\mathsf{T}Y = A^\mathsf{T}Y^\mathsf{T}Y \quad \text{because of (iv)} \\
  &= (AA^+A)^\mathsf{T}Y^\mathsf{T}Y = A^\mathsf{T}(A^+)^\mathsf{T}A^\mathsf{T}Y^\mathsf{T}Y \quad \text{because of (i)} \\
  &= A^\mathsf{T}(A^+)^\mathsf{T}YAY \quad \text{because of (iv)} \\
  &= A^\mathsf{T}(A^+)^\mathsf{T}Y = (A^+A)^\mathsf{T}Y \quad \text{because of (ii)} \\
  &= A^+AY \quad \text{because of (iv)} \\
  &= A^+AA^+AY \quad \text{because of (ii)} \\
  &= A^+(AA^+)^\mathsf{T}(AY)^\mathsf{T} = A^+(A^+)^\mathsf{T}A^\mathsf{T}Y^\mathsf{T}A^\mathsf{T} \quad \text{because of (iii)} \\
  &= A^+(A^+)^\mathsf{T}A^\mathsf{T} \quad \text{because of (i)} \\
  &= A^+(AA^+)^\mathsf{T} = A^+AA^+ \quad \text{because of (iii)} \\
Y &= A^+ \quad \text{because of (ii)}
\end{aligned}
$$

$\square$

6.3.2 Fundamental Subspaces

There are four *fundamental subspaces* associated with a matrix $A \in \mathbb{R}^{m \times n}$:

DEFINITION 6.2. (FUNDAMENTAL SUBSPACES OF A MATRIX)

1. $\mathcal{R}(A) = \{y \,|\, y = Ax, \quad x \in \mathbb{R}^n\} \subset \mathbb{R}^m$ *is the* range *or* column space.

2. $\mathcal{R}(A)^\perp$ *the orthogonal complement of* $\mathcal{R}(A)$.
   If $z \in \mathcal{R}(A)^\perp$ *then* $z^\mathsf{T}y = 0, \quad \forall y \in \mathcal{R}(A)$.

3. $\mathcal{R}(A^\mathsf{T}) = \{z \,|\, z = A^\mathsf{T}y, \quad y \in \mathbb{R}^m\} \subset \mathbb{R}^n$ *the* row space.

4. $\mathcal{N}(A) = \{x \,|\, Ax = 0\}$ *the* null space.

THEOREM 6.8. *The following relations hold:*

1. $\mathcal{R}(A)^\perp = \mathcal{N}(A^\mathsf{T})$. *Thus* $\mathbb{R}^m = \mathcal{R}(A) \oplus \mathcal{N}(A^\mathsf{T})$.

2. $\mathcal{R}(A^\mathsf{T})^\perp = \mathcal{N}(A)$. *Thus* $\mathbb{R}^n = \mathcal{R}(A^\mathsf{T}) \oplus \mathcal{N}(A)$.

In other words, $\mathbb{R}^m$ *can be written as a direct sum of the range of* A *and the null space of* A^T, *and an analogous result holds for* $\mathbb{R}^n$.

PROOF. Let $z \in \mathcal{R}(A)^\perp$. Then for any $x \in \mathbb{R}^n$, we have $Ax \in \mathcal{R}(A)$, so by definition we have

$$
0 = (Ax)^\mathsf{T}z = x^\mathsf{T}(A^\mathsf{T}z).
$$

Since this is true for all x, it follows that $A^\mathsf{T}z = 0$, which means $z \in \mathcal{N}(A^\mathsf{T})$ and therefore $\mathcal{R}(A)^\perp \subset \mathcal{N}(A^\mathsf{T})$.

On the other hand, let $\boldsymbol{y} \in \mathcal{R}(A)$ and $\boldsymbol{z} \in \mathcal{N}(A^\top)$. Then we have

$$\boldsymbol{y}^\top \boldsymbol{z} = (A\boldsymbol{x})^\top \boldsymbol{z} = \boldsymbol{x}^\top (A^\top \boldsymbol{z}) = \boldsymbol{x}^\top 0 = 0$$

which means that $\boldsymbol{z} \in \mathcal{R}(A)^\perp$. Thus also $\mathcal{N}(A^\top) \subset \mathcal{R}(A)^\perp$.

The second statement is verified in the same way. $\qquad\square$

One way to understand the problem of finding least squares approxima-tions is via projections onto the above subspaces. Recall that $P : \mathbb{R}^n \to \mathbb{R}^n$ is a *projector* onto a subspace $V \subset \mathbb{R}^n$ if $P^2 = P$ and $\mathcal{R}(P) = V$. Additionally, if P is symmetric, then it is an *orthogonal projector*. The following lemma shows a few properties of orthogonal projectors.

LEMMA 6.1. *Let $V \neq 0$ be a non-trivial subspace of $\mathbb{R}^n$. If P_1 is an orthogonal projector onto V, then $\|P_1\|_2 = 1$ and $P_1 \boldsymbol{v} = \boldsymbol{v}$ for all $\boldsymbol{v} \in V$. Moreover, if P_2 is another orthogonal projector onto V, then $P_1 = P_2$.*

PROOF. We first show that $P_1 \boldsymbol{v} = \boldsymbol{v}$ for all $\boldsymbol{v} \in V$. Let $0 \neq \boldsymbol{v} \in V$. Since $V = \mathcal{R}(P_1)$, there exists $\boldsymbol{x} \in \mathbb{R}^n$ such that $\boldsymbol{v} = P_1 \boldsymbol{x}$. Thus,

$$P_1 \boldsymbol{v} = P_1^2 \boldsymbol{x} = P_1 \boldsymbol{x} = \boldsymbol{v}.$$

Taking norms on both sides gives

$$\|P_1 \boldsymbol{v}\|_2 = \|\boldsymbol{v}\|_2,$$

which implies $\|P_1\|_2 \geq 1$. To show that $\|P_1\|_2 = 1$, let $\boldsymbol{y} \in \mathbb{R}^n$ be arbitrary. Then

$$\|P_1 \boldsymbol{y}\|_2^2 = \boldsymbol{y}^\top P_1^\top P_1 \boldsymbol{y} = \boldsymbol{y}^\top P_1^2 \boldsymbol{y} = \boldsymbol{y}^\top P_1 \boldsymbol{y}.$$

The Cauchy–Schwarz inequality now gives

$$\|P_1 \boldsymbol{y}\|_2^2 \leq \|\boldsymbol{y}\|_2 \|P_1 \boldsymbol{y}\|_2,$$

which shows, upon dividing both sides by $\|P_1 \boldsymbol{y}\|_2$, that $\|P_1 \boldsymbol{y}\|_2 \leq \|\boldsymbol{y}\|_2$. Hence, we conclude that $\|P_1\|_2 = 1$.

Now let P_2 be another orthogonal projector onto V. To show equality of the two projectors, we show that $(P_1 - P_2)\boldsymbol{y} = 0$ for all $\boldsymbol{y} \in \mathbb{R}^n$. Indeed, we have

$$\begin{aligned}
\|(P_1 - P_2)\boldsymbol{y}\|_2^2 &= \boldsymbol{y}^T (P_1 - P_2)^\top (P_1 - P_2)\boldsymbol{y} \\
&= \boldsymbol{y}^\top (P_1 - P_2)^2 \boldsymbol{y} \\
&= \boldsymbol{y}^\top (P_1 - P_1 P_2 - P_2 P_1 + P_2)\boldsymbol{y} \\
&= \boldsymbol{y}^\top (I - P_1)P_2 \boldsymbol{y} + \boldsymbol{y}^\top (I - P_2)P_1 \boldsymbol{y}. \qquad (6.20)
\end{aligned}$$

But for any $\boldsymbol{v} \in V$, we have

$$(I - P_1)\boldsymbol{v} = \boldsymbol{v} - P_1 \boldsymbol{v} = \boldsymbol{v} - \boldsymbol{v} = 0,$$

and similarly for $I - P_2$. Since $P_2 \boldsymbol{y} \in V$, we have $(I - P_1)P_2\boldsymbol{y} = 0$, so the first term in (6.20) vanishes. Exchanging the roles of P_2 and P_1 shows that the second term in (6.20) also vanishes, so $P_1 = P_2$. $\qquad\square$

Thanks to the above lemma, we see that the orthogonal projector onto a given V is in fact unique; we denote this projector by P_V. With the help of the pseudoinverse, we can describe *orthogonal projectors* onto the fundamental subspaces of A.

THEOREM 6.9. (PROJECTORS ONTO FUNDAMENTAL SUBSPACES)

$$
\begin{aligned}
&1. \quad P_{\mathcal{R}(A)} = AA^+ &\qquad &2. \quad P_{\mathcal{R}(A^\top)} = A^+A \\
&3. \quad P_{\mathcal{N}(A^\top)} = I - AA^+ &\qquad &4. \quad P_{\mathcal{N}(A)} = I - A^+A
\end{aligned}
$$

PROOF. We prove only the first relation; the other proofs are similar. Because of Relation (iii) in Theorem 6.7 we have $(AA^+)^\top = AA^+$. Thus $P_{\mathcal{R}(A)}$ is symmetric. Furthermore $(AA^+)(AA^+) = (AA^+A)A^+ = AA^+$ because of (i). Thus $P_{\mathcal{R}(A)}$ is symmetric and idempotent and is therefore an orthogonal projector. Now let $\boldsymbol{y} = A\boldsymbol{x} \in \mathcal{R}(A)$; then $P_{\mathcal{R}(A)}\boldsymbol{y} = AA^+\boldsymbol{y} = AA^+A\boldsymbol{x} = A\boldsymbol{x} = \boldsymbol{y}$. So elements in $\mathcal{R}(A)$ are projected onto themselves. Finally take $\boldsymbol{z} \perp \mathcal{R}(A) \iff A^\top \boldsymbol{z} = 0$ then $P_{\mathcal{R}(A)}\boldsymbol{z} = AA^+\boldsymbol{z} = (AA^+)^\top \boldsymbol{z} = (A^+)^\top A^\top \boldsymbol{z} = 0$. $\qquad\square$

Note that the projectors can be computed using the SVD. Let $U_1 \in \mathbb{R}^{m\times r}$, $U_2 \in \mathbb{R}^{m\times(n-r)}$, $V_1 \in \mathbb{R}^{n\times r}$, $V_2 \in \mathbb{R}^{n\times(n-r)}$ and $\Sigma_r \in \mathbb{R}^{r\times r}$ in the following SVD

$$
A = \begin{pmatrix} U_1 & U_2 \end{pmatrix} \begin{pmatrix} \Sigma_r & 0 \\ 0 & 0 \end{pmatrix} \begin{pmatrix} V_1^\top \\ V_2^\top \end{pmatrix}.
$$

Then inserting this decomposition into the expressions for the projectors of Theorem 6.9 we obtain:

$$
\begin{aligned}
&1. \quad P_{\mathcal{R}(A)} = U_1 U_1^\top &\qquad &2. \quad P_{\mathcal{R}(A^\top)} = V_1 V_1^\top \\
&3. \quad P_{\mathcal{N}(A^\top)} = U_2 U_2^\top &\qquad &4. \quad P_{\mathcal{N}(A)} = V_2 V_2^\top
\end{aligned}
$$

6.3.3 Solution of the Linear Least Squares Problem

We are now ready to describe the general solution for the linear least squares problem. We are given a system of equations with more equations than unknowns,

$$
A\boldsymbol{x} \approx \boldsymbol{b}.
$$

In general $\boldsymbol{b}$ will not be in $\mathcal{R}(A)$ and therefore the system will not have a solution. A consistent system can be obtained if we project $\boldsymbol{b}$ onto $\mathcal{R}(A)$:

$$
A\boldsymbol{x} = AA^+\boldsymbol{b} \iff A(\boldsymbol{x} - A^+\boldsymbol{b}) = 0.
$$

We conclude that $\boldsymbol{x} - A^+\boldsymbol{b} \in \mathcal{N}(A)$. That means

$$
\boldsymbol{x} - A^+\boldsymbol{b} = (I - A^+A)\boldsymbol{w}
$$

where we have generated an element in $\mathcal{N}(A)$ by projecting an arbitrary vector $\boldsymbol{w}$ onto it. Thus we have shown

THEOREM 6.10. (GENERAL LEAST SQUARES SOLUTION) *The general solution of the linear least squares problem* $A\boldsymbol{x} \approx \boldsymbol{b}$ *is*

$$\boldsymbol{x} = A^{+}\boldsymbol{b} + (I - A^{+}A)\boldsymbol{w}, \quad \boldsymbol{w} \text{ arbitrary.} \tag{6.21}$$

Using the expressions for projectors from the SVD we obtain for the general solution

$$\boldsymbol{x} = V_1 \Sigma_r^{-1} U_1^{\top} \boldsymbol{b} + V_2 \boldsymbol{c} \tag{6.22}$$

where we have introduced the arbitrary vector $\boldsymbol{c} := V_2^{\top} \boldsymbol{w}$. Notice that if we calculate $\|\boldsymbol{x}\|_2^2$ using e.g. (6.21), we obtain

$$\|\boldsymbol{x}\|_2^2 = \|A^{+}\boldsymbol{b}\|_2^2 + 2\boldsymbol{w}^{\top} \underbrace{(I - A^{+}A)^{\top} A^{+}}_{=0} \boldsymbol{b} + \|(I - A^{+}A)\boldsymbol{w}\|_2^2$$

$$= \|A^{+}\boldsymbol{b}\|_2^2 + \|(I - A^{+}A)\boldsymbol{w}\|_2^2 \geq \|A^{+}\boldsymbol{b}\|_2^2.$$

This calculation shows that any solution to the least squares problem must have norm greater than or equal to that of $A^{+}\boldsymbol{b}$; in other words, the pseudoinverse produces the *minimum-norm* solution to the least squares problem $A\boldsymbol{x} \approx b$. Thus, we have obtained an algorithm for computing both the general and the minimum norm solution of the linear least squares problem with (possibly) rank deficient coefficient matrix:

ALGORITHM 6.1.
General solution of the linear least squares problem
$$A\boldsymbol{x} \approx \boldsymbol{b}$$

---

1. Compute the SVD: `[U,S,V]=svd(A)`.

2. Make a rank decision, i.e. choose r such that $\sigma_r > 0$ and $\sigma_{r+1} = \cdots = \sigma_n = 0$. This decision is necessary because rounding errors will prevent the zero singular values from being exactly zero.

3. Set `V1=V(:,1:r)`, `V2= V(:,r+1:n)`, `Sr=S(1:r,1:r)`, `U1=U(:,1:r)`.

4. The solution with minimal norm is `xm=V1*(Sr\U1'*b)`.

5. The general solution is `x=xm+V2*c` with an arbitrary $\boldsymbol{c} \in \mathbb{R}^{n-r}$.

If A has full rank $(\text{rank}(A) = n)$ then the solution of the linear least squares problem is unique:

$$\boldsymbol{x} = A^{+}\boldsymbol{b} = V\Sigma^{+}U^{\top}\boldsymbol{b}.$$

The matrix A^+ is called pseudoinverse because in the full rank case the solution $Ax \approx b \Rightarrow x = A^+ b$ is the analogue of the solution $x = A^{-1}b$ of a linear system $Ax = b$ with nonsingular matrix $A \in \mathbb{R}^{n \times n}$.

The general least squares solution presented in Theorem 6.10 is also valid for a consistent system of equations $Ax = b$ where $m \leq n$, i.e. an underdetermined linear system with fewer equations than unknowns. In this case the $x = V_1 \Sigma_r^{-1} U_1^\top b$ solves the problem

$$\min \|x\|_2 \quad \text{subject to } Ax = b.$$

6.3.4 SVD and Rank

Consider the matrix

$$A = \begin{pmatrix}
-1.9781 & 4.4460 & -0.1610 & -3.8246 & 3.8137 \\
2.7237 & -2.3391 & 2.3753 & -0.0566 & -4.1472 \\
1.6934 & -0.1413 & -1.5614 & -1.5990 & 1.7343 \\
3.1700 & -7.1943 & -4.5438 & 6.5838 & -1.1887 \\
0.3931 & -3.1482 & 3.1500 & 3.6163 & -5.9936 \\
-7.7452 & 2.9673 & -0.1809 & 4.6952 & 1.7175 \\
-1.9305 & 8.9277 & 2.2533 & -10.1744 & 5.2708
\end{pmatrix} .$$

The singular values are computed by `svd(A)` as

$$\begin{aligned}
\sigma_1 &= 20.672908496836218 \\
\sigma_2 &= 10.575440102610981 \\
\sigma_3 &= 8.373932796689537 \\
\sigma_4 &= 0.000052201761324 \\
\sigma_5 &= 0.000036419750608
\end{aligned}$$

and we can observe a *gap* between σ_3 and σ_4. The two singular values σ_4 and σ_5 are about 10^5 times smaller than σ_3. Clearly the matrix A has rank 5. However, if the matrix elements comes from measured data and if the measurement uncertainty is $5 \cdot 10^{-5}$, one could reasonably suspect that A is in fact a perturbed representation of a rank-3 matrix, where the perturbation is due to measurement errors of the order of $5 \cdot 10^{-5}$. Indeed, reconstructing the matrix by setting $\sigma_4 = \sigma_5 = 0$ we get

```
[U,S,V]=svd(A);
S(4,4)=0; S(5,5)=0;
B=U*S*V'
```

We see no difference between A and B when using the MATLAB-format `short` because, according to Theorem 6.6, we have $\|A - B\|_2 = \sigma_4 = 5.2202 \, 10^{-5}$. Thus, in this case, one might as well declare that the matrix has *numerical rank* 3.

In general, if there is a distinct gap between the singular values one can define a threshold and remove small nonzero singular values which only occur because of the rounding effects of finite precision arithmetic or maybe because of measurement errors in the data. The default tolerance in MATLAB for the rank command is `tol=max(size(A))*eps(norm(A))`. Smaller singular values are considered to be zero.

There are full rank matrices whose singular values decrease to zero with no distinct gap. It is well known that the Hilbert matrix is positive definite (see Problem 7.1 in Chapter 7). The MATLAB statement `eig(hilb(14))` gives us as smallest eigenvalue (which is here equal to σ_{14}) the negative value $-3.4834\,10^{-17}$! With `svd(hilb(14))` we get $\sigma_{14} = 3.9007\,10^{-18}$ which is also not correct. Using the MAPLE commands

```
with(LinearAlgebra);
Digits:=40;
n:=14;
A:=Matrix(n,n);
for i from 1 to n do
  for j from 1 to n do
    A[i,j]:=1/(i+j-1);
  end do;
end do;
evalm(evalf(Eigenvalues(A)));
```

we obtain for $n = 14$ the singular values

$$
\begin{aligned}
&1.8306 \\
&4.1224 \cdot 10^{-01} \\
&5.3186 \cdot 10^{-02} \\
&4.9892 \cdot 10^{-03} \\
&3.6315 \cdot 10^{-04} \\
&2.0938 \cdot 10^{-05} \\
&9.6174 \cdot 10^{-07} \\
&3.5074 \cdot 10^{-08} \\
&1.0041 \cdot 10^{-09} \\
&2.2100 \cdot 10^{-11} \\
&3.6110 \cdot 10^{-13} \\
&4.1269 \cdot 10^{-15} \\
&2.9449 \cdot 10^{-17} \\
&9.8771 \cdot 10^{-20}
\end{aligned}
$$

For `A=hilb(14)` MATLAB computes `rank(A)=12` which is mathematically not correct but a reasonable numerical rank for such an ill-conditioned matrix. The SVD is the best tool to assign numerically a rank to a matrix.

6.4   Condition of the Linear Least Squares Problem

The principle of Wilkinson states that *the result of a numerical computation is the result of an exact computation for a slightly perturbed problem* (see Section 2.7). This result allows us to estimate the influence of finite precision arithmetic. A problem is said to be well conditioned if the results do not differ too much when solving a perturbed problem. For an ill-conditioned problem the solution of a perturbed problem may be very different.

Consider a system of linear equations $A\boldsymbol{x} = \boldsymbol{b}$ with $A \in \mathbb{R}^{n \times n}$ non-singular and a perturbed system $(A + \epsilon E)\boldsymbol{x}(\epsilon) = \boldsymbol{b}$, where ϵ is small, e.g. the machine precision. How do the solutions $\boldsymbol{x}(\epsilon)$ and $\boldsymbol{x} = \boldsymbol{x}(0)$ differ? We have already shown one way of estimating this difference in Chapter 3 (cf. Theorem 3.5), but here we illustrate another technique, which we will apply in the next section to the linear least squares problem. Let us consider the expansion

$$\boldsymbol{x}(\epsilon) = \boldsymbol{x}(0) + \dot{\boldsymbol{x}}(0)\epsilon + O(\epsilon^2).$$

The derivative $\dot{\boldsymbol{x}}(0)$ is obtained by differentiating:

$$(A + \epsilon E)\boldsymbol{x}(\epsilon) = \boldsymbol{b}$$
$$E\boldsymbol{x}(\epsilon) + (A + \epsilon E)\dot{\boldsymbol{x}}(\epsilon) = 0$$
$$\Rightarrow \dot{\boldsymbol{x}}(0) = -A^{-1}E\boldsymbol{x}(0).$$

Thus, we get

$$\boldsymbol{x}(\epsilon) = \boldsymbol{x}(0) - A^{-1}E\boldsymbol{x}(0)\,\epsilon + O(\epsilon^2).$$

Neglecting $O(\epsilon^2)$ and taking norms, we get

$$\|\boldsymbol{x}(\epsilon) - \boldsymbol{x}(0)\|_2 \leq \|A^{-1}\|_2\,\|\epsilon\,E\|_2\,\|\boldsymbol{x}(0)\|_2.$$

From the last equation, we conclude that the relative error satisfies

$$\frac{\|\boldsymbol{x}(\epsilon) - \boldsymbol{x}\|_2}{\|\boldsymbol{x}\|_2} \leq \underbrace{\|A^{-1}\|_2\,\|A\|_2}_{\kappa}\;\frac{\|\epsilon\,E\|_2}{\|A\|_2}. \qquad (6.23)$$
$$\text{condition number}$$

If we use the 2-norm as matrix norm,

$$\|A\|_2 := \max_{\boldsymbol{x} \neq 0} \frac{\|A\boldsymbol{x}\|_2}{\|\boldsymbol{x}\|_2} = \sigma_{\max}(A),$$

then the condition number is given by

$$\kappa = \frac{\sigma_{\max}(A)}{\sigma_{\min}(A)} = \texttt{cond(A)} \text{ in MATLAB}.$$

Thus, if $\|E\|_2 \approx \|A\|_2$, then according to the principle of Wilkinson we have to expect that the numerical solution may deviate by about κ units in the last digit from the exact solution.

We will now present an analogous result due to [56] for comparing the solutions of

$$\|\boldsymbol{b} - A\boldsymbol{x}\|_2 \longrightarrow \min \quad \text{and} \quad \|\boldsymbol{b} - (A + \epsilon E)\boldsymbol{x}(\epsilon)\|_2 \longrightarrow \min.$$

6.4.1   Differentiation of Pseudoinverses

Let A be a matrix whose elements are functions of k variables

$$A(\boldsymbol{\alpha}) \in \mathbb{R}^{m \times n}, \quad \boldsymbol{\alpha} = \begin{pmatrix} \alpha_1 \\ \vdots \\ \alpha_k \end{pmatrix}.$$

DEFINITION 6.3. (FRÉCHET DERIVATIVE) *The* Fréchet derivative *of the matrix $A(\boldsymbol{\alpha})$ is the 3-dimensional tensor*

$$\mathcal{D}A(\boldsymbol{\alpha}) = \left(\frac{\partial a_{ij}(\boldsymbol{\alpha})}{\partial \alpha_s} \right), \quad s = 1, \ldots, k$$

$\mathcal{D}A(\boldsymbol{\alpha})$ collects the gradients of all the matrix elements with respect to $\boldsymbol{\alpha}$.

We first derive some calculation rules for the operator $\mathcal{D}$:

1. If A is constant then $\mathcal{D}A = 0$.

2. Let $A \in \mathbb{R}^{m \times k}$ be constant. Then $\mathcal{D}[A\boldsymbol{\alpha}] = A$.

3. Product rule:

$$\mathcal{D}[A(\boldsymbol{\alpha})B(\boldsymbol{\alpha})] = \mathcal{D}A(\boldsymbol{\alpha})B(\boldsymbol{\alpha}) + A(\boldsymbol{\alpha})\mathcal{D}B(\boldsymbol{\alpha}).$$

Here, the product $\mathcal{D}A(\boldsymbol{\alpha})B(\boldsymbol{\alpha})$ means every layer of the tensor $\mathcal{D}A(\boldsymbol{\alpha})$ is multiplied by the matrix $B(\boldsymbol{\alpha})$. This rule is evident if we consider one element of the product which is differentiated with respect to α_s:

$$(\mathcal{D}\left[A(\boldsymbol{\alpha})B(\boldsymbol{\alpha})\right])_{i,j,s} = \frac{\partial}{\partial \alpha_s} \left(\sum_{t=1}^{n} a_{it}b_{tj} \right) = \sum_{t=1}^{n} \left(\frac{\partial a_{it}}{\partial \alpha_s}b_{tj} + a_{it}\frac{\partial b_{tj}}{\partial \alpha_s} \right).$$

4. Taylor expansion:

$$A(\boldsymbol{\alpha} + \boldsymbol{h}) = A(\boldsymbol{\alpha}) + \mathcal{D}A(\boldsymbol{\alpha})(\boldsymbol{h}) + O(\|\boldsymbol{h}\|_2^2).$$

Here, $\mathcal{D}A(\boldsymbol{\alpha})(\boldsymbol{h})$ is a matrix with the same size as $A(\boldsymbol{\alpha})$, where each matrix element is obtained by computing the scalar product between the gradient $\nabla_{\boldsymbol{\alpha}} a_{ij}$ (of length k) and $\boldsymbol{h}$, also of length k.

5. Derivative of the inverse:

$$\mathcal{D}\left[A^{-1}(\boldsymbol{\alpha})\right] = -A^{-1}(\boldsymbol{\alpha})\mathcal{D}A(\boldsymbol{\alpha})A^{-1}(\boldsymbol{\alpha}).$$

This results from differentiating $AA^{-1} = I$:

$$\mathcal{D}A(\boldsymbol{\alpha})A^{-1}(\boldsymbol{\alpha}) + A(\boldsymbol{\alpha})\mathcal{D}\left[A^{-1}(\boldsymbol{\alpha})\right] = 0.$$

THEOREM 6.11. (FRECHET PROJECTOR) *Let $\Omega \subset \mathbb{R}^k$ be an open set and let $A(\boldsymbol{\alpha}) \in \mathbb{R}^{m \times n}$ have constant rank for all $\boldsymbol{\alpha} \in \Omega$. Let $P_{\mathcal{R}(A)} = AA^+$ be the projector on the range of A. Then*

$$\mathcal{D}P_{\mathcal{R}(A)} = P_{\mathcal{N}(A^\top)}\mathcal{D}A\,A^+ + \left(P_{\mathcal{N}(A^\top)}\mathcal{D}A\,A^+\right)^\top, \quad \forall \boldsymbol{\alpha} \in \Omega. \tag{6.24}$$

PROOF.

$$P_{\mathcal{R}(A)}A = A \quad \text{because of Penrose Equation (i)}$$
$$\Rightarrow \mathcal{D}P_{\mathcal{R}(A)}A + P_{\mathcal{R}(A)}\mathcal{D}A = \mathcal{D}A$$
$$\mathcal{D}P_{\mathcal{R}(A)}A = (I - P_{\mathcal{R}(A)})\mathcal{D}A = P_{\mathcal{N}(A^\top)}\mathcal{D}A$$

Multiplying the last equation from the right with A^+ and observing that $P_{\mathcal{R}(A)} = AA^+$ we get

$$\mathcal{D}P_{\mathcal{R}(A)}P_{\mathcal{R}(A)} = P_{\mathcal{N}(A^\top)}\mathcal{D}A\,A^+. \tag{6.25}$$

A projector is idempotent, therefore

$$\mathcal{D}P_{\mathcal{R}(A)} = \mathcal{D}\left(P_{\mathcal{R}(A)}^2\right) = \mathcal{D}P_{\mathcal{R}(A)}P_{\mathcal{R}(A)} + P_{\mathcal{R}(A)}\mathcal{D}P_{\mathcal{R}(A)}. \tag{6.26}$$

Furthermore, $P_{\mathcal{R}(A)}$ is symmetric, which implies each layer of the tensor $\mathcal{D}P_{\mathcal{R}(A)}$ is also symmetric. Therefore

$$\left(\mathcal{D}P_{\mathcal{R}(A)}P_{\mathcal{R}(A)}\right)^\top = P_{\mathcal{R}(A)}\mathcal{D}P_{\mathcal{R}(A)}. \tag{6.27}$$

Substituting into (6.26) gives

$$\mathcal{D}P_{\mathcal{R}(A)} = \mathcal{D}P_{\mathcal{R}(A)}P_{\mathcal{R}(A)} + \left(\mathcal{D}P_{\mathcal{R}(A)}P_{\mathcal{R}(A)}\right)^\top,$$

which, together with (6.25), yields (6.24). $\qquad\square$

THEOREM 6.12. *Consider the projector $P_{\mathcal{R}(A^\top)} = A^+A$. With the same assumptions as in Theorem 6.11 we have*

$$\mathcal{D}P_{\mathcal{R}(A^\top)} = A^+\mathcal{D}A\,P_{\mathcal{N}(A)} + \left(A^+\mathcal{D}A\,P_{\mathcal{N}(A)}\right)^\top. \tag{6.28}$$

PROOF. The proof follows from the proof of Theorem 6.11 by exchanging $A \leftrightarrow A^+$ and $A^\top \leftrightarrow A$. $\qquad\square$

THEOREM 6.13. *Let $\Omega \subset \mathbb{R}^k$ be an open set and let the matrix $A(\boldsymbol{\alpha})$ be Fréchet differentiable for all $\boldsymbol{\alpha} \in \Omega$, with constant rank $r \leq \min(m, n)$. Then for every $\boldsymbol{\alpha} \in \Omega$ we have*

$$\mathcal{D}A^+(\boldsymbol{\alpha}) = -A^+\mathcal{D}A\,A^+ + A^+(A^+)^\top(\mathcal{D}A)^\top P_{\mathcal{N}(A^\top)} + P_{\mathcal{N}(A)}(\mathcal{D}A)^\top (A^+)^\top A^+. \tag{6.29}$$

PROOF. First, we have by Theorem 6.9

$$A^+ P_{\mathcal{N}(A^\top)} = A^+(I - AA^+) = A^+ - A^+ AA^+ = 0,$$

where the last equality follows from Penrose Equation (ii). Differentiating the above gives

$$\mathcal{D}A^+ P_{\mathcal{N}(A^\top)} + A^+ \mathcal{D}P_{\mathcal{N}(A^\top)} = 0$$

or

$$\mathcal{D}A^+ P_{\mathcal{N}(A^\top)} = -A^+ \mathcal{D}P_{\mathcal{N}(A^\top)}.$$

But $P_{\mathcal{N}(A^\top)} = I - P_{\mathcal{R}(A)}$, which implies

$$\mathcal{D}P_{\mathcal{N}(A^\top)} = -\mathcal{D}P_{\mathcal{R}(A)},$$

so by Theorem 6.11, we have

$$\begin{aligned}
\mathcal{D}A^+ P_{\mathcal{N}(A^\top)} &= A^+ \mathcal{D}P_{\mathcal{R}(A)} \\
&= \underbrace{A^+ P_{\mathcal{N}(A^\top)}}_{=0} \mathcal{D}AA^+ + A^+ (A^+)^\top (\mathcal{D}A^+)^\top P_{\mathcal{N}(A^\top)} \\
&= A^+ (A^+)^\top (\mathcal{D}A^+)^\top P_{\mathcal{N}(A^\top)}.
\end{aligned} \tag{6.30}$$

Similarly, using the relation $P_{\mathcal{N}(A)}A^+ = 0$ and Theorem 6.12, we derive

$$\begin{aligned}
P_{\mathcal{N}(A)}\mathcal{D}A^+ &= \mathcal{D}P_{\mathcal{R}(A^\top)}A^+ \\
&= A^+ \mathcal{D}A \underbrace{P_{\mathcal{N}(A)}A^+}_{=0} + P_{\mathcal{N}(A)}(\mathcal{D}A)^\top (A^+)^\top A^+ \\
&= P_{\mathcal{N}(A)}(\mathcal{D}A)^\top (A^+)^\top A^+
\end{aligned} \tag{6.31}$$

Finally, differentiating the relation $A^+ = A^+ AA^+$ gives

$$\begin{aligned}
\mathcal{D}A^+ &= (\mathcal{D}A^+)AA^+ + A^+(\mathcal{D}A)A^+ + A^+ A(\mathcal{D}A^+) \\
&= \mathcal{D}A^+(AA^+ - I) + A^+(\mathcal{D}A)A^+ + (A^+ A - I)\mathcal{D}A^+ + 2\mathcal{D}A^+ \\
&= 2\mathcal{D}A^+ + A^+(\mathcal{D}A)A^+ - \mathcal{D}A^+ P_{\mathcal{N}(A^\top)} - P_{\mathcal{N}(A)}\mathcal{D}A^+.
\end{aligned}$$

Thus, after rearranging we get

$$\mathcal{D}A^+ = -A^+ \mathcal{D}AA^+ + \mathcal{D}A^+ P_{\mathcal{N}(A^\top)} + P_{\mathcal{N}(A)}\mathcal{D}A^+. \tag{6.32}$$

Substituting (6.30) and (6.31) into (6.32) yields the desired result.   □

6.4.2 Sensitivity of the Linear Least Squares Problem

In this section, we want to compare the solution of the least squares problem $\|\boldsymbol{b} - A\boldsymbol{x}\|_2$ to that of the perturbed problem $\|\boldsymbol{b} - (A + \epsilon E)\boldsymbol{x}(\epsilon)\|_2$. Define $A(\epsilon) = A + \epsilon E$. The solution of the perturbed problem is

$$\boldsymbol{x}(\epsilon) = A(\epsilon)^+ \boldsymbol{b} = \left[A(0)^+ + \frac{dA^+}{d\epsilon}\epsilon + O(\epsilon^2) \right] \boldsymbol{b} = \boldsymbol{x} + \epsilon \frac{dA^+}{d\epsilon}\boldsymbol{b} + O(\epsilon^2). \quad (6.33)$$

Applying Theorem 6.13, we get

$$\frac{dA^+}{d\epsilon} = -A^+ \frac{dA}{d\epsilon} A^+ + A^+ A^{+\top} \frac{dA^\top}{d\epsilon} P_{\mathcal{N}(A^\top)} + P_{\mathcal{N}(A)} \frac{dA^\top}{d\epsilon} A^{+\top} A^+.$$

Multiplying from the right with $\boldsymbol{b}$ and using $dA/d\epsilon = E$, $A^+\boldsymbol{b} = \boldsymbol{x}$ and

$$P_{\mathcal{N}(A^\top)}\boldsymbol{b} = (I - AA^+)\boldsymbol{b} = \boldsymbol{b} - A\boldsymbol{x} = \boldsymbol{r},$$

we get

$$\frac{dA^+}{d\epsilon}\boldsymbol{b} = -A^+ E\boldsymbol{x} + A^+ A^{+\top} E^\top \boldsymbol{r} + P_{\mathcal{N}(A)} E^\top A^{+\top} \boldsymbol{x}.$$

Introducing this into (6.33), we obtain

$$\boldsymbol{x}(\epsilon) - \boldsymbol{x} = \epsilon \left(-A^+ E\boldsymbol{x} + P_{\mathcal{N}(A)} E^\top A^{+\top} \boldsymbol{x} + A^+ A^{+\top} E^\top \boldsymbol{r} \right) + O(\epsilon^2).$$

Neglecting the term $O(\epsilon^2)$ and taking norms, we get

$$\|\boldsymbol{x}(\epsilon) - \boldsymbol{x}\|_2 \leq |\epsilon| \Big(\|A^+\|_2 \|E\|_2 \|\boldsymbol{x}\|_2 + \|P_{\mathcal{N}(A)}\|_2 \|E^\top\|_2 \|A^{+\top}\|_2 \|\boldsymbol{x}\|_2$$
$$+ \|A^+\|_2 \|A^{+\top}\|_2 \|E^\top\|_2 \|\boldsymbol{r}\|_2 \Big).$$

Introducing the *condition number*

$$\kappa := \|A\|_2 \|A^+\|_2 = \frac{\sigma_1(A)}{\sigma_r(A)},$$

and observing that $\|P_{\mathcal{N}(A)}\|_2 = 1$ we get the the estimate:

THEOREM 6.14. (GOLUB-PEREYRA 1973)

$$\frac{\|\boldsymbol{x}(\epsilon) - \boldsymbol{x}\|_2}{\|\boldsymbol{x}\|_2} \leq \left(2\kappa + \kappa^2 \frac{\|\boldsymbol{r}\|_2}{\|A\|_2 \|\boldsymbol{x}\|_2} \right) \frac{\|\epsilon E\|_2}{\|A\|_2} + O(\epsilon^2). \quad (6.34)$$

Equation (6.34) tells us again what accuracy we can expect from the numerical solution. Here, we have to distinguish between good and bad models: when the model is good, the residual $\|\boldsymbol{r}\|_2$ must be small, since it is possible to find parameters that fit the data well. In this case, the error in the solution may deviate by about κ units in the last digit from the exact solution, just like for linear equations (6.23). However, when the model is bad, i.e. when $\|\boldsymbol{r}\|_2$ is large, the condition becomes much worse, so we must expect a larger error in the computed solution.

6.4.3 Normal Equations and Condition

If we want to solve $A\boldsymbol{x} \approx \boldsymbol{b}$ numerically using the normal equations, we have to expect worse numerical results than predicted by (6.34), even for good models. Indeed, if $A = U\Sigma V^{\top}$ has rank n then

$$\kappa(A^{\top}A) = \kappa(V\Sigma^{\top}U^{\top}U\Sigma V^{\top}) = \kappa(V\Sigma^{\top}\Sigma V^{\top}) = \frac{\sigma_1^2}{\sigma_n^2} = \kappa(A)^2. \qquad (6.35)$$

Thus, forming $A^{\top}A$ leads to a matrix with a squared condition number compared to the original matrix A. Intuitively, one also sees that forming $A^{\top}A$ may result in a loss of information, as shown by the following famous example by P. Läuchli:

$$A = \begin{pmatrix} 1 & 1 \\ \delta & 0 \\ 0 & \delta \end{pmatrix}, \quad A^{\top}A = \begin{pmatrix} 1+\delta^2 & 1 \\ 1 & 1+\delta^2 \end{pmatrix}.$$

If $\delta < \sqrt{\epsilon}$ (with ϵ = machine precision) then numerically $1 + \delta^2 = 1$ and the matrix of the normal equations becomes singular, even though A numerically has rank 2.

EXAMPLE 6.9. *We illustrate the theory with the following example. We generate a matrix A by taking the 6×5 segment of the inverse Hilbert-matrix divided by 35791. Next we construct a compatible right hand side $\boldsymbol{y}_1$ for the solution*

$$\boldsymbol{x} = \left(1, \frac{1}{2}, \frac{1}{3}, \frac{1}{4}, \frac{1}{5}\right)^{\top}.$$

Then we compare the numerical solution of $A\boldsymbol{x} \approx \boldsymbol{y}_1$ obtained by orthogonal transformations using MATLAB*'s* \ *operator with the solution of the the normal equations.*

In the second calculation we make the right hand side incompatible by adding a vector orthogonal to $\mathcal{R}(A)$. The amplification factor

$$2\kappa + \kappa^2 \frac{\|\boldsymbol{r}\|_2}{\|A\|_2 \, \|\boldsymbol{x}\|_2},$$

which is in the first case essentially equal to the condition number, grows in the second case because of the large residual and influences the accuracy of the numerical solution.

```
format compact, format long e
A=invhilb(6)/35791; A=A(:,1:5);              % generate matrix
y1=A*[1 1/2 1/3 1/4 1/5]';                   % consistent right hand side
x11=A\y1;
K=cond(A);
factor1=K*(2+K*norm(y1-A*x11)/norm(A)/norm(x11));
x21=(A'*A)\(A'*y1);                          % solve with normal equations
```

```
dy=[-4620 -3960 -3465 -3080 -2772 -2520]';
Check=A'*dy                           % dy is orthogonal to R(A)
y2=y1+dy/35791;                       % inconsistent right hand side
x12=A\y2;
factor2=K*(2+K*norm(y2-A*x12)/norm(A)/norm(x12));
x22=(A'*A)\(A'*y2);                   % solve with normal equations
O_solutions=[x11 x12]
factors=[factor1, factor2]
NE_solutions=[x21 x22]
SquaredCondition=K^2
```

We get the results

```
Check =
    -8.526512829121202e-14
    -9.094947017729282e-13
     1.455191522836685e-11
     1.455191522836685e-11
                          0
O_solutions =
     9.999999999129205e-01      1.000000008577035e+00
     4.999999999726268e-01      5.000000028639559e-01
     3.333333333220151e-01      3.333333345605716e-01
     2.499999999951682e-01      2.500000005367703e-01
     1.999999999983125e-01      2.000000001908185e-01
factors =
     9.393571042867742e+06      1.748388455005875e+10
NE_solutions =
     1.000039250650201e+00      1.000045308718515e+00
     5.000131762672949e-01      5.000152068771240e-01
     3.333389969395881e-01      3.333398690463054e-01
     2.500024822148246e-01      2.500028642353512e-01
     2.000008837044892e-01      2.000010196590261e-01
SquaredCondition =
     2.205979423052303e+13
```

We see that the first O_solution *has about 6 incorrect decimals which correspond well to the amplifying factor* $9.3\ 10^6$. *The second* O_solution *with incompatible right hand side has about 8 decimals incorrect, the factor in this case is* $1.7\ 10^{10}$ *and would even predict a worse result.*

The solutions with the normal equations, however, have about 12 incorrect decimal digits, regardless of whether the right hand side is compatible or not. This illustrates the influence of the squared condition number.

6.5 Algorithms Using Orthogonal Matrices

6.5.1 QR Decomposition

Consider a matrix $A \in \mathbb{R}^{m \times n}$ with $m \geq n$ and $\text{rank}(A) = n$. Then there exists the Cholesky decomposition of $A^\top A = R^\top R$ where R is an upper triangular matrix. Since R is non-singular we can write $R^{-\top} A^\top A R^{-1} = I$ or

$$(AR^{-1})^\top (AR^{-1}) = I.$$

This means that the matrix $Q_1 := AR^{-1}$ has *orthogonal columns*. Thus we have found the QR decomposition

$$A = Q_1 R. \tag{6.36}$$

Here, $Q_1 \in \mathbb{R}^{m\times n}$ and $R \in \mathbb{R}^{n\times n}$. We can always augment Q_1 to an $m \times m$ orthogonal matrix $Q := [Q_1, Q_2]$ and instead consider the decomposition

$$A = [Q_1, Q_2]\begin{pmatrix} R \\ 0 \end{pmatrix}. \tag{6.37}$$

The decomposition (6.37) is what MATLAB computes with the command `[Q,R] = qr(A)`. The decomposition (6.37) exists for any matrix A with full column rank.

We have shown in Section 6.2 that for an orthogonal matrix B the problems

$$A\boldsymbol{x} \approx \boldsymbol{b} \quad \text{and} \quad BA\boldsymbol{x} \approx B\boldsymbol{b}$$

are equivalent. Now if $A = Q\begin{pmatrix} R \\ 0 \end{pmatrix}$, then $B = Q^\top$ is orthogonal and $A\boldsymbol{x} \approx \boldsymbol{b}$ and $Q^\top A\boldsymbol{x} \approx Q^\top \boldsymbol{b}$ are equivalent. But

$$Q^\top A = \begin{pmatrix} R \\ 0 \end{pmatrix}$$

and the equivalent system becomes

$$\begin{pmatrix} R \\ 0 \end{pmatrix} \boldsymbol{x} \approx \begin{pmatrix} \boldsymbol{y}_1 \\ \boldsymbol{y}_2 \end{pmatrix}, \quad \text{with} \quad \begin{pmatrix} \boldsymbol{y}_1 \\ \boldsymbol{y}_2 \end{pmatrix} = Q^\top \boldsymbol{b}.$$

The square of the norm of the residual,

$$\|\boldsymbol{r}\|_2^2 = \|\boldsymbol{y}_1 - R\boldsymbol{x}\|_2^2 + \|\boldsymbol{y}_2\|_2^2,$$

is obviously minimal for $\hat{\boldsymbol{x}}$ where

$$R\hat{\boldsymbol{x}} = \boldsymbol{y}_1, \quad \hat{\boldsymbol{x}} = R^{-1}\boldsymbol{y}_1 \quad \text{and} \quad \min \|\boldsymbol{r}\|_2 = \|\boldsymbol{y}_2\|_2.$$

This approach is numerically preferable to the normal equations, since it does not change the condition number. This can be seen by noting that the singular values are not affected by orthogonal transformations: If $A = U\Sigma V^\top = Q\begin{pmatrix} R \\ 0 \end{pmatrix}$ then the singular value decomposition of $\begin{pmatrix} R \\ 0 \end{pmatrix}$ is

$$\begin{pmatrix} R \\ 0 \end{pmatrix} = (Q^\top U)\Sigma V^\top,$$

and thus R and A have the same singular values, which leads to

$$\kappa(A) = \kappa(R).$$

In the following section we will show how to compute the QR decomposition.

6.5.2 Method of Householder

DEFINITION 6.4. (ELEMENTARY HOUSEHOLDER MATRIX) *An elementary Householder matrix is a matrix of the form* $P = I - \boldsymbol{u}\boldsymbol{u}^{\top}$ *with* $\|\boldsymbol{u}\|_2 = \sqrt{2}$ *is*

Elementary Householder matrices have the following properties:

1. P is symmetric.

2. P is orthogonal, since
$$P^{\top}P = (I - \boldsymbol{u}\boldsymbol{u}^{\top})(I - \boldsymbol{u}\boldsymbol{u}^{\top}) = I - \boldsymbol{u}\boldsymbol{u}^{\top} - \boldsymbol{u}\boldsymbol{u}^{\top} + \boldsymbol{u}\underbrace{\boldsymbol{u}^{\top}\boldsymbol{u}}_{2}\boldsymbol{u}^{\top} = I.$$

3. $P\boldsymbol{u} = -\boldsymbol{u}$ and if $\boldsymbol{x} \perp \boldsymbol{u}$ then $P\boldsymbol{x} = \boldsymbol{x}$. If $\boldsymbol{y} = \alpha\boldsymbol{x} + \beta\boldsymbol{u}$ then $P\boldsymbol{y} = \alpha\boldsymbol{x} - \beta\boldsymbol{u}$. Thus P is a reflection across the hyperplane $\boldsymbol{u}^{\top}\boldsymbol{x} = 0$.

P will be used to solve the following basic problem: Given a vector $\boldsymbol{x}$, find an orthogonal matrix P such that

$$P\boldsymbol{x} = \begin{pmatrix} \sigma \\ 0 \\ \vdots \\ 0 \end{pmatrix} = \sigma\boldsymbol{e}_1.$$

Since P is orthogonal we have $\|P\boldsymbol{x}\|_2^2 = \|\boldsymbol{x}\|_2^2 = \sigma^2$ thus $\sigma = \pm\|\boldsymbol{x}\|_2$. Furthermore $P\boldsymbol{x} = (I - \boldsymbol{u}\boldsymbol{u}^{\top})\boldsymbol{x} = \boldsymbol{x} - \boldsymbol{u}(\boldsymbol{u}^{\top}\boldsymbol{x}) = \sigma\boldsymbol{e}_1$, thus $\boldsymbol{u}(\boldsymbol{u}^{\top}\boldsymbol{x}) = \boldsymbol{x} - \sigma\boldsymbol{e}_1$ and we obtain by normalizing

$$\boldsymbol{u} = \frac{\boldsymbol{x} - \sigma\boldsymbol{e}_1}{\|\boldsymbol{x} - \sigma\boldsymbol{e}_1\|_2}\sqrt{2}.$$

We can still choose the sign of σ, and we choose it such that no cancellation occurs in computing $\boldsymbol{x} - \sigma\boldsymbol{e}_1$,

$$\sigma = \begin{cases} \|\boldsymbol{x}\|_2, & x_1 < 0, \\ -\|\boldsymbol{x}\|_2, & x_1 \geq 0. \end{cases}$$

For the denominator we get $\|\boldsymbol{x} - \sigma\boldsymbol{e}_1\|_2^2 = (\boldsymbol{x} - \sigma\boldsymbol{e}_1)^{\top}(\boldsymbol{x} - \sigma\boldsymbol{e}_1) = \boldsymbol{x}^{\top}\boldsymbol{x} - 2\sigma\boldsymbol{e}_1^{\top}\boldsymbol{x} + \sigma^2$. Note that $-2\sigma\boldsymbol{e}_1^{\top} = 2|x_1|\|\boldsymbol{x}\|_2$, so the calculations simplify and we get

$$\boldsymbol{u} = \frac{\boldsymbol{x} - \sigma\boldsymbol{e}_1}{\sqrt{\|\boldsymbol{x}\|_2(|x_1| + \|\boldsymbol{x}\|_2)}}.$$

In order to apply this basic construction for the computation of the QR decomposition, we construct a sequence of n elementary matrices P_i:

$$P_i = \begin{pmatrix} I & 0 \\ 0 & I - \boldsymbol{u}_i\boldsymbol{u}_i^{\top} \end{pmatrix}.$$

We choose $\boldsymbol{u}_i \in \mathbb{R}^{m-i+1}$ such that zeros are introduced in the i-th column of A below the diagonal when multiplying $P_i A$. We obtain after n steps

$$P_n P_{n-1} \cdots P_1 A = \begin{pmatrix} R \\ 0 \end{pmatrix}$$

and, because each P_i is symmetric, Q becomes

$$Q = (P_n P_{n-1} \cdots P_1)^\top = P_1 P_2 \cdots P_n.$$

If we store the new diagonal elements (which are the diagonal of R) in a separate vector d we can store the Householder vectors $\boldsymbol{u}_i$ in the same location where we introduce zeros in A. This leads to the following *implicit QR factorization* algorithm:

ALGORITHM 6.2. *Householder QR Decomposition*

```
function [A,d]=HouseholderQR(A);
% HOUSEHOLDERQR computes the QR-decomposition of a matrix
%   [A,d]=HouseholderQR(A) computes an implicit QR-decomposition A=QR
%   using Householder transformations. The output matrix A contains
%   the Householder vectors u and the upper triangle of R. The
%   diagonal of R is stored in the vector d.

[m,n]=size(A);
for j=1:n,
  s=norm(A(j:m,j));
  if s==0, error('rank(A)<n'), end
  if A(j,j)>=0, d(j)=-s; else d(j)=s; end
  fak=sqrt(s*(s+abs(A(j,j))));
  A(j,j)=A(j,j)-d(j);
  A(j:m,j)=A(j:m,j)/fak;
  if j<n,        % transformation of the rest of the matrix G:=G-u*(u'*G)
    A(j:m,j+1:n)=A(j:m,j+1:n)-A(j:m,j)*(A(j:m,j)'*A(j:m,j+1:n));
  end
end
```

Algorithm `HouseholderQR` computes an upper triangular matrix R_h which is very similar to R_c obtained by the Cholesky decomposition $A^\top A = R_c^\top R_c$. The only difference is that R_h may have negative elements in the diagonal, whereas in R_c the diagonal entries are positive. Let D be a diagonal matrix with

$$d_{ii} = \begin{cases} 1 & r_{ii}^h > 0 \\ -1 & r_{ii}^h < 0 \end{cases}$$

then $R_c = DR_h$. The matrix Q is only implicitly available through the Householder vectors $\boldsymbol{u}_i$. This is not an issue because it is often unnecessary to compute and store the matrix Q explicitly; in many cases, Q is only needed as an *operator* that acts on vectors by multiplication. Using the implicit representation we can, for instance, compute the transformed right hand side of the least square equations $\boldsymbol{y} = Q^\top \boldsymbol{b}$ by applying the reflections $\boldsymbol{y} = P_n P_{n-1} \cdots P_1 \boldsymbol{b}$. This procedure is numerically preferable to forming the explicit matrix Q and then multiplying with $\boldsymbol{b}$.

ALGORITHM 6.3. *Transformation* $z = Q^\top y$

```
function z=HouseholderQTy(A,y);
% HOUSEHOLDERQTY applies Householder reflections transposed
%    z=HouseholderQTy(A,y); computes z=Q'y using the Householder
%    reflections Q stored as vectors in the matrix A by
%    A=HousholderQR(A)

[m,n]=size(A); z=y;
for j=1:n,
  z(j:m)=z(j:m)-A(j:m,j)*(A(j:m,j)'*z(j:m));
end;
```

If we wish to compute $\boldsymbol{z} = Q\boldsymbol{y}$ then because $Q = P_1 P_2 \cdots P_n$ it is sufficient to reverse the order in the for-loop:

ALGORITHM 6.4. *Transformation* $z = Qy$

```
function z=HouseholderQy(A,y);
% HOUSEHOLDERQY applies Householder reflections
%    z=HouseholderQy(A,y); computes z=Qy using the Householder
%    reflections Q stored as vectors in the matrix A by
%    A=HousholderQR(A)

[m,n]=size(A); z=y;
for j=n:-1:1,
  z(j:m)=z(j:m)-A(j:m,j)*(A(j:m,j)'*z(j:m));
end;
```

EXAMPLE 6.10. *We compute the QR decomposition of a section of the Hilbert matrix. We also compute the explicit matrix Q by applying the Householder transformations to the column vectors of the identity matrix:*

```
m=8;n=6;
H=hilb(m); A=H(:,1:n);
[AA,d]=HouseholderQR(A)
Q=[];
for i=eye(m),                              % compute Q explicit
  z=HouseholderQy(AA,i); Q=[Q z];
end
R=triu(AA);
```

```
for i=1:n, R(i,i)=d(i); end              % add diagonal to R
[norm(Q'*Q-eye(m)) norm(A-Q*R)]
[q,r]=qr(A);                             % compare with Matlab qr
[norm(q'*q-eye(m)) norm(A-q*r)]
```

The resulting matrix after the statement `[AA,d]=HouseholderQR(A)` contains the Householder vectors and the upper part of R. The diagonal of R is stored in the vector $\mathbf{d}$*:*

```
AA =
      1.3450    -0.7192    -0.5214    -0.4130    -0.3435    -0.2947
      0.3008     1.1852    -0.1665    -0.1612    -0.1512    -0.1407
      0.2005     0.3840    -1.0188     0.0192     0.0233     0.0254
      0.1504     0.3584     0.2452    -1.3185     0.0015     0.0023
      0.1203     0.3242     0.3945    -0.4332    -1.0779     0.0001
      0.1003     0.2925     0.4703    -0.2515     0.0111    -1.3197
      0.0859     0.2651     0.5059    -0.0801     0.3819    -0.5078
      0.0752     0.2417     0.5190     0.0641     0.8319     0.0235
d =
     -1.2359    -0.1499     0.0118     0.0007     0.0000     0.0000
ans =
     1.0e-15 *
     0.6834     0.9272
ans =
     1.0e-15 *
     0.5721     0.2461
```

We see that the results (orthogonality of Q and reproduction of A by QR) compare well with the MATLAB *function* **qr**.

6.5.3 Method of Givens

DEFINITION 6.5. (ELEMENTARY GIVENS ROTATION) *An elementary* Givens rotation *is the matrix*

$$
G := G_{i,k}(\alpha) =
\begin{bmatrix}
1 & & & & & & & & \\
 & \ddots & & & & & & & \\
 & & 1 & & & & & & \\
 & & & \cos\alpha & & & \sin\alpha & & \\
 & & & & 1 & & & & \\
 & & & & & \ddots & & & \\
 & & & & & & 1 & & \\
 & & & -\sin\alpha & & & \cos\alpha & & \\
 & & & & & & & 1 & \\
 & & & & & & & & \ddots \\
 & & & & & & & & & 1
\end{bmatrix}
\begin{matrix}
 \\ \\ \\ \leftarrow i \\ \\ \\ \\ \leftarrow k \\ \\ \\
\end{matrix}
$$

$G_{i,k}(\alpha)$ differs from the identity matrix only in the two columns and rows i and k with the elements $\cos\alpha$ and $\sin\alpha$. It is an orthogonal matrix: the columns have norm one and different columns are orthogonal to each other. It is called a rotation matrix since if we multiply G by a vector $\boldsymbol{x}$, the result $\boldsymbol{y} = G\boldsymbol{x}$ is the vector $\boldsymbol{x}$ rotated in the ik-plane by the angle α.

For least squares problems, it is convenient to work with a slight modification of these elementary matrices. This is because Givens rotation matrices are non-symmetric: thus, if $Q = G_3 G_2 G_1$ then $Q^\top = G_1^\top G_2^\top G_3^\top \neq G_1 G_2 G_3$.

DEFINITION 6.6. (ELEMENTARY GIVENS REFLECTION) *An elementary* Givens reflection *is the matrix*

$$
S := S_{i,k}(\alpha) =
\begin{pmatrix}
1 & & & & & & & & \\
& \ddots & & & & & & & \\
& & \cos\alpha & & & & \sin\alpha & & \\
& & & 1 & & & & & \\
& & & & \ddots & & & & \\
& & & & & 1 & & & \\
& & \sin\alpha & & & & -\cos\alpha & & \\
& & & & & & & 1 & \\
& & & & & & & & \ddots & \\
& & & & & & & & & 1
\end{pmatrix}
$$

$S_{i,k}(\alpha)$ is a modified Givens rotation matrix: row k has been multiplied by -1 and $S_{i,k}(\alpha)$ is now symmetric. It has the following properties:

1. S is orthogonal.

2. SA changes only two rows in A:

$$
\begin{aligned}
\boldsymbol{a}_{i:}^{\mathrm{new}} &= c\,\boldsymbol{a}_{i:} + s\,\boldsymbol{a}_{k:} \\
\boldsymbol{a}_{k:}^{\mathrm{new}} &= s\,\boldsymbol{a}_{i:} - c\,\boldsymbol{a}_{k:}
\end{aligned}
\quad, \quad \text{where } c = \cos\alpha \text{ and } s = \sin\alpha.
$$

3. Connection to Householder matrices: let

$$
\boldsymbol{u} = (0,\ldots,0,\sin\tfrac{\alpha}{2},0,\ldots,0,-\cos\tfrac{\alpha}{2},0,\ldots,0)^\top \sqrt{2},
$$

then $I - \boldsymbol{u}\boldsymbol{u}^\top = S_{i,k}(\alpha)$.

Givens transformations allow us to solve the following basic problem: given a vector $\boldsymbol{x}$, we wish to find an S that annihilates the k-th element, i.e., in $\boldsymbol{x}^{\mathrm{new}} := S\boldsymbol{x}$, we want to rotate x_k^{new} to zero.

Solution: Because $x_k^{\mathrm{new}} = s x_i - c x_k = 0$

$$
\Rightarrow \quad cot := \frac{x_i}{x_k}, \quad s := \frac{1}{\sqrt{1 + cot^2}}, \quad c := s * cot.
$$

Note that we do not need to compute the angle α explicitly to determine the matrix $S_{i,k}(\alpha)$.

To compute the QR decomposition of a matrix A, we now can apply Givens reflections to introduce zeros below the diagonal. The following pseudo-code computes the decomposition columnwise:

```
for i=1:n
    for k=i+1:m
        A=S(i,k,alpha)*A;                  % annihilate A(k,i)
    end;
end
```

Again we would like to store information about the rotation $S_{i,k}(\alpha)$ at the same place where we introduce the zero. The easiest would be to store the angle α, but we cannot since we do not compute it! Storing c and s separately would require space for two numbers instead of one. We could store only c and rederive s using $c^2 + s^2 = 1$, but this is numerically unstable if $|c| \approx 1$.

An elegant solution was proposed by G. W. Stewart [130]: we store the smaller of the two numbers c and s, and retrieve the other one in a numerically stable way from $c^2 + s^2 = 1$. In order to tell whether c or s has been stored, Stewart proposes to store s or $1/c$. More specifically, we have two possible formulas for computing $x_k^{\text{new}} = sx_i - cx_k = 0$:

1. $cot = c/s = x_i/x_k$, giving $s = 1/\sqrt{1 + cot^2}$ and $c = s * cot$

2. $tan = s/c = x_k/x_i$, giving $c = 1/\sqrt{1 + tan^2}$ and $s = c * tan$.

To avoid overflow, we will choose the one that gives an answer that is smaller than 1 for $|cot|$ or $|tan|$. We thus obtain the following pseudocode for computing and storing a reflection:

```
if x(k)==0                               % nothing to do, S is the identity
    c=1; s=0; Store 0
elseif abs(x(k))>=abs(x(i))
    h=x(i)/x(k); s=1/sqrt(1+h^2); c=s*h;
    if c~=0, Store 1/c else Store 1 end;
else
    h=x(k)/x(i); c=1/sqrt(1+h^2); s=c*h; Store s;
end
```

To reconstruct c and s from a stored number z we use the following pseudocode:

```
if z==1, c=0; s=1;
elseif abs(z)<1 then s=z; c=sqrt(1-s^2);
else c=1/z; s=sqrt(1-c^2)
end
```

We have now all the elements to write a MATLAB function to compute the QR decomposition using Givens reflections:

ALGORITHM 6.5. *Givens QR Decomposition*

```
function A=GivensQR(A);
% GIVENSQR Computes the QR-decomposition using Givens reflections
%    A=GivensQR(A); computes the QR-decomposition of the matrix A using
%    Givens reflections; after the decomposition, A contains the
%    implicit QR-decomposition of A. Instead of the zeros the
%    reflections are stored following a proposal of G. W. Stewart
%    (cf. srotg in BLAS). The decomposition is computed column wise.

[m n]=size(A);
for i=1:n,
  for k=i+1:m,
    if A(k,i)==0,                         % Compute  co and si
      co=1; si=0; z=0;
    else if abs(A(k,i))>=abs(A(i,i)),
      h=A(i,i)/A(k,i);                    % cot
      si=1/sqrt(1+h*h); co=si*h;
      if co~=0, z=1/co; else z=1; end
    else
       h=A(k,i)/A(i,i);                   % tan
       co=1/sqrt(1+h*h); si=co*h; z=si;
    end;
   end;
   A(i,i)=A(i,i)*co+A(k,i)*si;
   A(k,i)=z;                              % store co or si in A(k,i)
   if (si~=0)&(i<n),                      % Apply reflection
     S=[co,si;si,-co];
     A(i:k-i:k,i+1:n)=S*A(i:k-i:k,i+1:n);
   end;
  end
end
```

Note that in Algorithm 6.5, we make use of the MATLAB feature that allows us to select and overwrite several rows of a matrix at the same time. In a traditional programming language like Fortran, we would store the old rows in auxiliary vectors and then combine them to the new rows:

```
h1=A(i,i+1:n); h2=A(k,i+1:n);
A(i,i+1:n)=h1*co+h2*si;
A(k,i+1:n)=h1*si-h2*co;
```

In MATLAB, however, we can use the selection expression $A(i:k-i:k,i+1:n)$ to write the same computation in terms of the 2×2-matrix S as

```
S=[co,si;si,-co];
A(i:k-i:k,i+1:n)=S*A(i:k-i:k,i+1:n);
```

The function `GivensQR(A)` computes the QR decomposition of A. After the function call, we have `R=triu(A(1:n,1:n))` and Q is only given implicitly as an operator. If we wish to compute $\boldsymbol{y} = Q^{\top}\boldsymbol{x}$ then the following function can be used:

ALGORITHM 6.6. *Transformation* $\boldsymbol{y} := Q^{\top}\boldsymbol{x}$

```
function y=GivensQTy(A,y);
% GIVENSQTY applies Givens rotations transposed
%    y=GivensQTyqrgiv(A,y) computes  y=Q'*y; using the Givens
%    rotations Q stored in the matrix A computed by A=GivensQR(A).
%    For y=Q*y the for loops must be processed in reverse order.

[m,n]=size(A);
for i=1:n,                          % for i=n:-1:1,    for y=Q*y
  for k=i+1:m,                      % for k=m:-1:i+1, for y=Q*y
    if A(k,i)==1,                   % reconstruct co and si from A(k,i)
      co=0; si=1;
    else
      if abs(A(k,i))<1,
        si=A(k,i); co=sqrt(1-si*si);
      else
        co=1/A(k,i); si=sqrt(1-co*co);
      end;
    end;
    if si~=0,                       % Apply Givens reflector
      y(i:k-i:k)=[co,si;si,-co]*y(i:k-i:k);
    end
  end
end
```

EXAMPLE 6.11. *We compute the same example as before with House-holder:*

```
format compact
m=8;n=6;
H=hilb(m); A=H(:,1:n);
AA=GivensQR(A)
R=triu(AA(1:n,1:n));
Qt=[];
for y=eye(m)
  z=GivensQTy(AA,y); Qt=[Qt, z];
end
Q=Qt';
[norm(Q'*Q-eye(m)) norm(A-Q(:,1:n)*R)]
[q,r]=qr(A);
[norm(q'*q-eye(m)) norm(A-q*r)]
```

The orthogonality of Q and the representation of $A = QR$ are perfect and compare well with the MATLAB built-in function qr:

```
AA =
    1.2359    0.7192    0.5214    0.4130    0.3435    0.2947
    0.4472   -0.1499   -0.1665   -0.1612   -0.1512   -0.1407
    0.2857    0.6805    0.0118    0.0192    0.0233    0.0254
    0.2095    0.5135    1.5948   -0.0007   -0.0015   -0.0023
    0.1653    0.4116    0.6363    1.7994    0.0000    0.0001
    0.1365    0.3431    0.5371    1.4204    1.9844   -0.0000
    0.1162    0.2941    0.4643    0.6192    1.5366    2.1541
    0.1011    0.2572    0.4086    0.5485    0.6768    1.6456
ans =
    1.0e-15 *
    0.3719    0.2115
ans =
    1.0e-15 *
    0.5721    0.2461
```

The method of Givens needs about twice as many operations as the method of Householder. However, it can be efficient for sparse matrices, where only a few zeros have to be introduced to transform the matrix into upper triangular form. Furthermore, the QR decomposition may be computed row-wise, which may be advantageous for linear least squares problems where the number of equations is not fixed and additional equations are generated by new measurements.

EXAMPLE 6.12. *A matrix is said to be in upper Hessenberg form if it has zero entries below the first subdiagonal,*

$$H = \begin{pmatrix} * & * & * & * & * \\ * & * & * & * & * \\ 0 & * & * & * & * \\ 0 & 0 & * & * & * \\ 0 & 0 & 0 & * & * \\ 0 & 0 & 0 & 0 & * \end{pmatrix}.$$

Least squares problems of the type $Hy = r$ with an upper Hessenberg matrix H arise naturally in the GMRES algorithm for solving linear systems iteratively, see Section 11.7.6. To obtain the QR decomposition of such a matrix using Givens reflections, let us first eliminate the (2,1) entry; this is done by choosing a reflection $S_{1,2} := S_{1,2}(\alpha_1)$ that operates on the first two rows only, with the angle α_1 chosen so that the (2,1) entry is eliminated. The *transformed matrix then becomes*

$$H = \begin{pmatrix} * & * & * & * & * \\ 0 & * & * & * & * \\ 0 & * & * & * & * \\ 0 & 0 & * & * & * \\ 0 & 0 & 0 & * & * \\ 0 & 0 & 0 & 0 & * \end{pmatrix}.$$

But now the remaining matrix $H_{2:n,2:n}$ is also in upper Hessenberg form, so we can continue this process and choose $S_{23}, \ldots, S_{n,n+1}$ to transform H into an upper triangular matrix, i.e.,

$$\underbrace{S_{n,n+1} \cdots S_{12}}_{Q^\top} H = \begin{bmatrix} R \\ 0 \end{bmatrix}.$$

Note that the k-th step requires at most $2(n - k + 1)$ additions and the same number of multiplications, so the total cost is $O(n^2)$, as opposed to $O(mn^2)$ for a general dense matrix.

Givens rotations are also useful for updating QR decompositions, which we will see in Section 6.5.8. Another different use of upper Hessenberg matrices arise in the context of eigenvalue problems, see Section 7.5.2.

6.5.4   Fast Givens

In some architectures (particularly older ones), multiplication is a much more costly operation than addition, so reducing the number of multiplications can lead to faster running times. In the method of Givens, each rotation affects two rows of the matrix and transforms

$$\begin{array}{l} [x_1, x_2, \ldots, x_n] \\ [y_1, y_2, \ldots, y_n] \end{array} \quad \text{into} \quad \begin{array}{ll} cx_k & + & sy_k \\ sx_k & - & cy_k, \end{array} \quad k = 1, \ldots, n$$

which requires two multiplications per entry. The basic idea of Fast Givens is to consider the rows in factored form and delay some of the multiplications until the end of the algorithm, thus saving about half the multiplications in the process.

Let us apply a Givens reflection to two "scaled rows" in which the coefficients f_1, f_2 have been factored out:

$$\begin{pmatrix} c & s \\ s & -c \end{pmatrix} \quad \begin{array}{l} f_1 \, [x_1, x_2, \ldots, x_n] \\ f_2 \, [y_1, y_2, \ldots, y_n] \end{array}$$

After the multiplication these rows are changed to

$$\begin{array}{ll} cf_1 x_k & + & sf_2 y_k \\ sf_1 x_k & - & cf_2 y_k \end{array} \quad k = 1, \ldots, n$$

The goal now is to update the factored coefficient in such a way that one multiplication inside the row vectors is eliminated. We consider two possibilities, the first of which is

$$sf_2 \ (\beta_1 x_k + y_k) \qquad \text{with} \qquad \beta_1 = \frac{cf_1}{sf_2}, \quad \alpha_1 = -\frac{cf_2}{sf_1}.$$
$$sf_1 \ (x_k + \alpha_1 y_k)$$

Notice that y_k in the first row no longer has a coefficient multiplying it, and neither does x_k in the second row.

Now recall that the Givens reflection is chosen to annihilate one element in the matrix, say the first element y_1 of the second line. For this, c and s have to be chosen so that $sf_1 x_1 - cf_2 y_1 = 0$, which implies

$$c = \frac{f_1 x_1}{\sqrt{f_1^2 x_1^2 + f_2^2 y_1^2}}, \quad s = \frac{f_2 y_1}{\sqrt{f_1^2 x_1^2 + f_2^2 y_1^2}}.$$

Thus,

$$\alpha_1 = -\frac{c}{s}\frac{f_2}{f_1} = -\frac{f_1 x_1}{f_2 y_1}\frac{f_2}{f_1} = -\frac{x_1}{y_1}, \quad \beta_1 = \frac{c}{s}\frac{f_1}{f_2} = \frac{f_1 x_1}{f_2 y_1}\frac{f_1}{f_2} = -\alpha_1 \frac{f_1^2}{f_2^2}$$

and the new scaling factors become

$$f_1^{\text{new}} = sf_2, \quad f_2^{\text{new}} = sf_1.$$

For the effective computation of β_1, we only need the squares of the factors, so we will only store the quantities

$$d_i = \frac{1}{f_i^2}, \quad i = 1, 2.$$

Then using the expressions for s, α_1 and β_1

$$d_1^{\text{new}} = \frac{1}{s^2 f_2^2} = \frac{f_1^2 x_1^2 + f_2^2 y_1^2}{y_1^2 f_2^2 f_2^2} = d_2 \left(1 + \left(\frac{x_1}{y_1}\right)^2 \left(\frac{f_1}{f_2}\right)^2\right) = d_2\,(1 - \alpha_1 \beta_1).$$

Defining $\gamma_1 = -\alpha_1\beta_1 = (f_1 x_1)^2/(f_2 y_1)^2$ we obtain

$$d_1^{\text{new}} = d_2(1 + \gamma_1) \quad \text{and similarly} \quad d_2^{\text{new}} = \frac{1}{s^2 f_1^2} = d_1(1 + \gamma_1). \qquad (6.38)$$

The second choice of factors is

$$cf_1 \ (x_k + \beta_2 y_k) \qquad \text{with} \qquad \alpha_2 = \frac{sf_1}{cf_2} = -\frac{1}{\alpha_1}, \quad \beta_2 = \frac{sf_2}{cf_1} = \frac{1}{\beta_1}.$$
$$cf_2 \ (\alpha_2 x_k - y_k)$$

The equation $y_1^{\text{new}} = 0$ leads to the same expressions for c and s as before. The update of d_i becomes

$$d_1^{\text{new}} = \frac{1}{c^2 f_1^2} = d_1 \frac{f_1^2 x_1^2 + f_2^2 y_1^2}{f_1^2 x_1^2} = d_1(1 + \gamma_2) \quad \text{with} \quad \gamma_2 = \frac{1}{\gamma_1}. \qquad (6.39)$$

A similar computation yields

$$d_2^{\text{new}} = \frac{1}{c^2 f_2^2} = d_2(1 + \gamma_2).$$

Which of the two choices of factors should be used? We are working with scaled rows. Let $D = \text{diag}(d_1, \ldots, d_m)$ then

$$Q^\top A = D^{-1/2} \begin{pmatrix} \tilde{R} \\ 0 \end{pmatrix}, \quad \text{with} \quad D^{-1/2} = \begin{pmatrix} f_1 & 0 & \cdots & 0 \\ 0 & f_2 & \ddots & \vdots \\ \vdots & \ddots & \ddots & 0 \\ 0 & \cdots & 0 & f_m \end{pmatrix}.$$

Since $\gamma_i \geq 0$, the d_i grow after each reflection, meaning there is a danger of overflow. Because of $\gamma_1 \gamma_2 = 1$ it makes sense to choose the smaller $\gamma_i \leq 1$ to minimize the growth. However, even then it might still be necessary to monitor the growth and to multiply the equations with the factors in the diagonal matrix $D^{-1/2}$ from time to time. For simplicity, we choose not do this in the program below: we simply reduce the system $A\boldsymbol{x} \approx \boldsymbol{b}$ to

$$\begin{pmatrix} \tilde{R} \\ 0 \end{pmatrix} \boldsymbol{x} \approx \begin{pmatrix} \tilde{\boldsymbol{y}}_1 \\ \tilde{\boldsymbol{y}}_2 \end{pmatrix}$$

The scaling factors are given by the vector $\boldsymbol{d}$, the diagonal of the matrix D, and the solution is obtained by solving $\tilde{R}\boldsymbol{x} = \tilde{\boldsymbol{y}}_1$. The update of the diagonals of D according to (6.38) and (6.39) can be combined simply by swapping the diagonal elements in the first case.

ALGORITHM 6.7. *Fast Givens*

```
function [d,R,y]=FastGivens(A,b)
% FASTGIVENS reduce linear system to upper triangular form.
%   [d,R,y]=FastGivens(A,b) reduces Ax=B using fast Givens reflections
%   to Rx=y. d contains the scaling factors. If [q,r]=qr(A) then
%   abs(diag(d)^(-0.5)*R)=abs(r)

[m n]=size(A); d=ones(m,1);
for i=1:n,
  for k=i+1:m,
    if A(k,i)~=0,
      alpha=-A(i,i)/A(k,i); beta=-alpha*d(k)/d(i);
      gamma=-alpha*beta;
      if gamma<=1,
        A(i,i)=A(i,i)*beta+A(k,i);
        if i<n,
          h=A(i,i+1:n)*beta+A(k,i+1:n);
          A(k,i+1:n)=A(i,i+1:n)+A(k,i+1:n)*alpha;
```

```
        A(i,i+1:n)=h;
      end;
      h=b(i)*beta+b(k); b(k)=b(i)+b(k)*alpha; b(i)=h;
      h=d(i); d(i)=d(k); d(k)=h;                 % swap scaling factors
    else
      alpha=-1/alpha; beta=1/beta; gamma=1/gamma;
      A(i,i)=A(i,i)+beta*A(k,i);
      if i<n,
        h=A(i,i+1:n)+A(k,i+1:n)*beta;
        A(k,i+1:n)=A(i,i+1:n)*alpha -A(k,i+1:n);
        A(i,i+1:n)=h;
      end;
      h=b(i)+b(k)*beta; b(k)=b(i)*alpha-b(k); b(i)=h;
    end;
    d(i)=d(i)*(1+gamma); d(k)=d(k)*(1+gamma);  % update for both cases
  end;
  end
end
R=triu(A); y=b;
```

Note that in Algorithm 6.7, we did not write the row update in terms of multiplication by a 2×2 matrix, like we did in Algorithm 6.5. This is because want to avoid the multiplications with the factor 1. To compute one standard Givens reflection, we need about $4n$ multiplications and $2n$ additions. Fast Givens reduces this to $2n$ multiplications and $2n$ additions. Experiments show that the speedup is not by a factor of 2, but only about 1.4 to 1.6. This is due to overhead computations and also because multiplications are no longer much more expensive than additions on newer computer architectures.

6.5.5 Gram-Schmidt Orthogonalization

Yet another algorithm for calculating the QR decomposition is based on the idea of constructing an orthogonal basis in $\mathcal{R}(A)$. An overview on algorithms is given in [116]. For that purpose, the column vectors of $A = [\boldsymbol{a}_1, \ldots, \boldsymbol{a}_n]$ have to be orthogonalized. Assume that the orthonormal vectors $\boldsymbol{q}_1, \ldots, \boldsymbol{q}_{k-1}$ have already been computed and that they span the same space as $\boldsymbol{a}_1, \ldots, \boldsymbol{a}_{k-1}$. Now in order to construct the next basis vector $\boldsymbol{q}_k$, we take $\boldsymbol{a}_k$ and subtract its projections onto the previous basis vectors. If $\boldsymbol{a}_k$ was not linearly dependent from the previous vectors, the remainder will be nonzero and orthogonal to $\mathcal{R}(\boldsymbol{a}_1, \ldots, \boldsymbol{a}_{k-1})$, so it can be normalized to give the new basis vector. Thus we have to perform three steps:

1. Compute the projections $r_{ik}\boldsymbol{q}_i$ with $r_{ik} = \boldsymbol{q}_i^\top \boldsymbol{a}_k$.

2. Subtract the projections:

$$\boldsymbol{b}_k := \boldsymbol{a}_k - \sum_{i=1}^{k-1} r_{ik}\boldsymbol{q}_i. \tag{6.40}$$

3. Normalize: $r_{kk} := \|\boldsymbol{b}_k\|_2$ and $\boldsymbol{q}_k = \boldsymbol{b}_k/r_{kk}$

If we solve (6.40) for $\boldsymbol{a}_k$ then

$$\boldsymbol{a}_k = \sum_{i=1}^{k} r_{ik}\boldsymbol{q}_i, \quad k = 1,\ldots,n \iff A = QR.$$

Thus, we obtain again the QR decomposition of A. The elements of the upper triangular matrix R are the coefficients of the various projections. The following algorithm `ClassicalGramSchmidt` computes the QR decomposition using this *classical Gram-Schmidt orthogonalization*:

ALGORITHM 6.8. *Classical Gram-Schmidt*

```
function [Q,R]=ClassicalGramSchmidt(A)
% CLASSICALGRAMSCHMIDT classical Gram-Schmidt orthogonalization
%   [Q,R]=ClassicalGramSchmidt(A); computes the classical Gram-Schmidt
%   orthogonalization of the vectors in the columns of the matrix A

[m,n]=size(A); R=zeros(n);
Q=A;
for k=1:n,
  for i=1:k-1,
    R(i,k)=Q(:,i)'*Q(:,k);
  end                            % remove these two lines for
  for i=1:k-1,                   % modified-Gram-Schmidt
    Q(:,k)=Q(:,k)-R(i,k)*Q(:,i);
  end
  R(k,k)=norm(Q(:,k)); Q(:,k)=Q(:,k)/R(k,k);
end
```

Note that `ClassicalGramSchmidt` is numerically unstable: if we take the 15×10 section of the Hilbert matrix and try to compute an orthogonal basis then we notice that the vectors of Q are not orthogonal at all:

```
>> m=15; n=10; H=hilb(m); A=H(:,1:n);
>> [Q R]=ClassicalGramSchmidt(A);
>> norm(Q'*Q-eye(n))
ans = 2.9971
>> norm(Q*R-A)
ans = 2.9535e-17
```

However, the relation $A = QR$ is correct to machine precision. The reason for the numerical instability is as follows: when computing $\boldsymbol{b}_k$ using (6.40), cancellation can occur if the vectors are almost parallel. In that case, the result is a very small inaccurate $\boldsymbol{b}_k$ which, after normalization, is no longer orthogonal on the subspace spanned by the previous vectors because of the inaccuracies.

An interesting modification partly remedies that problem: if we compute the projections $r_{ik} = q_i^\top a_k$ and subtract the projection immediately from a_k

$$a_k := a_k - r_{ik}q_i$$

then this has no influence on the numerical value of subsequent projections because

$$r_{i+1,k} = q_{i+1}^\top(a_k - r_{ik}q_i) = q_{i+1}^\top a_k, \text{ since } q_{i+1}^\top q_i = 0.$$

Doing so, we reduce the norm of a_k with each projection and get the *Modified Gram-Schmidt* Algorithm, which is obtained by just eliminating the two lines

```
end                        % remove for
for i = 1:k-1,             % modified-Gram-Schmidt
```

in Algorithm `ClassicalGramSchmidt`:

ALGORITHM 6.9. *Modified Gram-Schmidt*

```
function [Q,R]=ModifiedGramSchmidt(A)
% MODIFIEDGRAMSCHMIDT modified Gram-Schmidt orthogonalization
%   [Q,R]=ModifiedGramSchmidt(A); computes the modified Gram-Schmidt
%   orthogonalization of the vectors in the columns of the matrix A

[m,n]=size(A); R=zeros(n);
Q=A;
for k=1:n,
  for i=1:k-1,
    R(i,k)=Q(:,i)'*Q(:,k);
    Q(:,k)=Q(:,k)-R(i,k)*Q(:,i);
  end
  R(k,k)=norm(Q(:,k)); Q(:,k)=Q(:,k)/R(k,k);
end
```

Now if we run our example again, we obtain better results than with classical Gram-Schmidt:

```
>> m=15; n=10; H=hilb(m); A=H(:,1:n);
>> [Q R]=ModifiedGramSchmidt(A);
>> norm(Q'*Q-eye(n))
ans = 1.7696e-05
>> norm(Q*R-A)
ans = 6.0510e-17
```

The approximate orthogonality of Q is now visible; however, the results are still not as good as with the algorithms of Householder or Givens.

Note that in Algorithm `ModifiedGramSchmidt`, we only process the k-th column of A in step k; the columns $k + 1, \ldots, n$ remain unchanged during this step. A mathematically and numerically identical version of modified Gram-Schmidt, popular in many textbooks, computes the projections for each new q_k and subtracts them immediately from all the column vectors of the remaining matrix:

ALGORITHM 6.10.
*Modified Gram-Schmidt, version updating whole
remaining matrix*

```
function [Q,R]=ModifiedGramSchmidt2(A);
% MODIFIEDGRAMSCHMIDT2 modified Gram-Schmidt orthogonalization version 2
%    [Q,R]=ModifiedGramSchmidt2(A); computes the modified Gram-Schmidt
%    orthogonalization of the vectors in the columns of the matrix A by
%    immediately updating all the remaining vectors as well during the
%    process

[m,n]=size(A); R=zeros(n);
for k=1:n
  R(k,k)=norm(A(:,k));
  Q(:,k)=A(:,k)/R(k,k);
  R(k,k+1:n)=Q(:,k)'*A(:,k+1:n);
  A(:,k+1:n)=A(:,k+1:n)-Q(:,k)*R(k,k+1:n);
end
```

In the k-th step of Algorithm `ModifiedGramSchmidt2` we compute the vector q_k from a_k. Then the k-th row of R is computed and the rest of A is updated by

$$A(:, k+1:n) = (I - q_k q_k^\top) A(:, k+1:n).$$

But $(I - q_k q_k^\top)$ is the orthogonal projector on the subspace orthogonal to q_k. This observation by Charles Sheffield in 1968 established a connection between modified Gram-Schmidt and the method of Householder. Consider the matrix

$$\tilde{A} = \begin{pmatrix} O \\ A \end{pmatrix}, \text{where } O \text{ is the } n \times n \text{ zero matrix.}$$

If we apply the first Householder transformation on $\tilde{A}$, i.e. we construct a vector u with $\|u\|_2 = \sqrt{2}$ such that

$$(I - uu^\top)\begin{pmatrix} O \\ a_1 \end{pmatrix} = \begin{pmatrix} \sigma \\ 0 \\ \vdots \\ 0 \end{pmatrix}$$

then

$$u = \frac{x - \sigma e_1}{\sqrt{\|x\|_2(x_1 + \|x\|_2)}}, \text{ with } x = \begin{pmatrix} 0 \\ \vdots \\ 0 \\ a_1 \end{pmatrix}.$$

Since $x_1 = 0$ (no cancellation in the first component) and $\sigma = \pm\|x\|_2 = \pm\|a_1\|_2$ we get

$$u = \left(\begin{array}{c} e_1 \\ \frac{a_1}{\|a_1\|_2} \end{array} \right) = \left(\begin{array}{c} e_1 \\ q_1 \end{array} \right)$$

and q_1 is the same vector as in modified Gram-Schmidt! The Householder transformation is

$$P_1 \begin{pmatrix} O \\ A \end{pmatrix} = \begin{pmatrix} O \\ A \end{pmatrix} - \begin{pmatrix} e_1 \\ q_1 \end{pmatrix} \begin{pmatrix} e_1^\top & q_1^\top \end{pmatrix} \begin{pmatrix} O \\ A \end{pmatrix} = \begin{pmatrix} -q_1^\top A \\ O_{n-1 \times n} \\ (I - q_1 q_1^\top)A \end{pmatrix}. \tag{6.41}$$

Thus we obtain the negative first row of R with $-q_1^\top A$ and the first transformation of the rest of the matrix $(I - q_1 q_1^\top)A$ as in modified Gram-Schmidt.

THEOREM 6.15. *The method of Householder applied to the matrix $\begin{pmatrix} O \\ A \end{pmatrix}$ is mathematically equivalent to modified Gram-Schmidt applied to A.*

After the call of `[AA,d]=HouseholderQR([zeros(n);A])` the matrix has been transformed to

$$AA = \begin{pmatrix} \tilde{R} \\ Q_h \end{pmatrix}, \quad \text{with } \operatorname{diag}(\tilde{R}) = I.$$

R_h is obtained by taking the negative strict upper part of $\tilde{R}$ and subtracting the diagonal elements stored in d. The orthogonal matrix Q_h is extracted from the Householder-vectors stored in AA. The matrices Q_h and R_h obtained this way by Householder transformations are the same matrices as computed by modified Gram-Schmidt. We illustrate this again with a submatrix of the Hilbert matrix:

```
format compact
m=7; n=4; H=hilb(m); A=H(:,1:n);
[Q1,R1]=ModifiedGramSchmidt2(A)
[AA,d]=HouseholderQR([zeros(n); A])
Q2=AA(n+1:n+m,:);
R2=-AA(1:n,1:n);
R2=R2-diag(diag(R2))-diag(d);
[norm(Q1-Q2) norm(R1-R2)]
```

We get the following results which illustrate well that both processes compute the same

```
Q1 =
      0.8133   -0.5438    0.1991   -0.0551
      0.4067    0.3033   -0.6886    0.4760
      0.2711    0.3939   -0.2071   -0.4901
      0.2033    0.3817    0.1124   -0.4396
      0.1627    0.3514    0.2915   -0.1123
      0.1356    0.3202    0.3892    0.2309
      0.1162    0.2921    0.4407    0.5206
```

```
R1 =
     1.2296      0.7116      0.5140      0.4059
          0      0.1449      0.1597      0.1536
          0           0      0.0108      0.0174
          0           0           0      0.0006
AA =
     1.0000     -0.7116     -0.5140     -0.4059
          0      1.0000     -0.1597     -0.1536
          0           0      1.0000     -0.0174
          0           0           0      1.0000
     0.8133     -0.5438      0.1991     -0.0551
     0.4067      0.3033     -0.6886      0.4760
     0.2711      0.3939     -0.2071     -0.4901
     0.2033      0.3817      0.1124     -0.4396
     0.1627      0.3514      0.2915     -0.1123
     0.1356      0.3202      0.3892      0.2309
     0.1162      0.2921      0.4407      0.5206
d =
    -1.2296     -0.1449     -0.0108     -0.0006
ans =
     1.0e-12 *
     0.1804      0.0001
```

6.5.6   Gram-Schmidt with Reorthogonalization

Our example indicated that modified Gram-Schmidt improves the orthogonality of Q in comparison with classical Gram-Schmidt a lot — however, it still cannot compete with Householder or Givens in that respect. In fact, for modified Gram-Schmidt the following estimate holds (with some constants c_1 and c_2, the condition number $\kappa = \kappa(A)$ and the machine precision ε) [9]:

$$\|I - Q^\top Q\|_2 \le \frac{c_1}{1 - c_2 \kappa \varepsilon} \kappa \varepsilon.$$

Thus, we must expect a loss of orthogonality if the condition of the matrix A is bad. A remedy is to reorthogonalize the vector $\boldsymbol{q}_k$ if it has been constructed from a small (an inaccurate) $\boldsymbol{b}_k$. If $\boldsymbol{q}_k$ is *reorthogonalized* with respect to $\boldsymbol{q}_i$, $i = 1, \ldots, k - 1$ then for full-rank matrices *one* reorthogonalization is sufficient. That "twice is enough" principle is analyzed in [50].

Note that when we reorthogonalize, we must also update R. Let us consider for that purpose the QR decomposition of the matrix

$$B = [\boldsymbol{q}_1, \ldots, \boldsymbol{q}_{k-1}, \boldsymbol{b}] = QR_1.$$

Then the following holds:

$$Q = [\boldsymbol{q}_1, \ldots, \boldsymbol{q}_{k-1}, \boldsymbol{q}_k], \quad R_1 = \begin{pmatrix} 1 & & & d_1 \\ & \ddots & & \vdots \\ & & 1 & d_{k-1} \\ & & & \|\boldsymbol{u}\|_2 \end{pmatrix}$$

with $d_i = \boldsymbol{q}_i^\top \boldsymbol{b}$, $i = 1, \ldots, k - 1$ and $\boldsymbol{u} = \boldsymbol{b} - \sum_{i=1}^{k-1} d_i \boldsymbol{q}_i$.

If we choose not to normalize the last column of Q, then the decomposition is $B = \bar{Q}\bar{R}_1$

with $\bar{Q} = [\boldsymbol{q}_1, \ldots, \boldsymbol{q}_{k-1}, \boldsymbol{u}]$ and

$$\bar{R}_1 = \begin{pmatrix} 1 & & d_1 \\ & \ddots & \vdots \\ & & 1 & d_{k-1} \\ & & & 1 \end{pmatrix} = I + \boldsymbol{d}\boldsymbol{e}_k^\top \quad \text{where } \boldsymbol{d} = \begin{pmatrix} d_1 \\ \vdots \\ d_{k-1} \\ 0 \end{pmatrix}.$$

Now assume the QR decomposition of $A = [\boldsymbol{a}_1, \ldots, \boldsymbol{a}_k]$ is $A = BR_2$ and we want to reorthogonalize the last column of B, i.e. we decompose $B = \bar{Q}\bar{R}_1$,

$$A = BR_2 = \bar{Q}\bar{R}_1 R_2.$$

Now

$$\bar{R}_1 R_2 = (I + \boldsymbol{d}\boldsymbol{e}_k^\top)R_2 = R_2 + \boldsymbol{d}\boldsymbol{e}_k^\top r_{kk}^{(2)}.$$

Again, if we do not normalize the last column of B then $r_{kk}^{(2)} = 1$ i.e. $\bar{R}_1 R_2 = R_2 + \boldsymbol{d}\boldsymbol{e}_k^\top$ and the update is simply

$$r_{ik} := r_{ik} + d_i, \quad i = 1, \ldots, k - 1.$$

The normalization is only performed after the reorthogonalization step. We shall now modify our function `ModifiedGramSchmidt` so that each vector is reorthogonalized. Doing so we need to compute the projections twice and add them to the new row of R. For this purpose we introduce the auxiliary variable `V`:

ALGORITHM 6.11.
Modified Gram-Schmidt with Reorthogonalization

```
function [Q,R]=ModifiedGramSchmidtTwice(A);
% MODIFIEDGRAMSCHMIDTTWICE modified Gram-Schmidt with reorthogonalization
%    [Q,R]=ModifiedGramSchmidtTwice(A); applies the modified
%    Gram-Schmidt procedure with reorthogonalization to the columns in
%    matrix A

[m,n]=size(A); R=zeros(n);
Q=A;
for k=1:n,
  for t=1:2                                   % reorthogonalize
    for i=1:k-1,
      V=Q(:,i)'*Q(:,k);                       % projections
      Q(:,k)=Q(:,k)-V*Q(:,i);
      R(i,k)=R(i,k)+V;
    end
  end
  R(k,k)=norm(Q(:,k)); Q(:,k)=Q(:,k)/R(k,k);
end
```

Using again our example we get this time

```
>> m=15; n=10; H=hilb(m); A=H(:,1:n);
>> [Q R]=ModifiedGramSchmidtTwice(A);
>> norm(Q'*Q-eye(n))
ans = 4.8199e-16
>> norm(Q*R-A)
ans = 6.7575e-17
```

a perfectly orthogonal matrix Q.

6.5.7 Partial Reorthogonalization

If we orthogonalize every vector twice, the number of computer operations is doubled. Therefore, it is natural to consider reorthogonalizing only when necessary. For this *partial reorthogonalization*, we need to decide when b_k is to be considered small.

In the following algorithm, which is a translation of the Algol program `ortho` by Heinz Rutishauser [111], a simple criterion for reorthogonalization is used. If b_k is 10 times smaller than a_k then we must expect to lose at least one decimal digit by cancellation, thus we will reorthogonalize in that case.

If the vectors are linearly dependent (the matrix is rank deficient) then cancellation will also occur when the vector is reorthogonalized. If with repeated reorthogonalizing the vector shrinks to rounding error level then it is assumed that the vector is linearly dependent and it is replaced by a zero vector.

We obtain thus the following algorithm:

ALGORITHM 6.12.
Gram-Schmidt with Reorthogonalization

```
function [Q,R,z]=GramSchmidt(Q);
% GRAMSCHMIDT modified Gram-Schmidt with partial reorthogonalization
%   [Q,R,z]=GramSchmidt(Q); computes the QR decomposition of the
%   matrix Q using modified Gram-Schmidt with partial
%   reorthogonalization; z is a vector that counts the
%   reorthogonalization steps per column.
%   Translation of the ALGOL procedure in H. Rutishauser: "Algol 60"

z=[]; [m,n]=size(Q); R=zeros(n);
for k=1:n,
  t=norm(Q(:,k));
  reorth=1;
  u=0;                               % count reorthogonalizations
  while reorth,
    u=u+1;
    for i=1:k-1,
```

```
      s= Q(:,i)'*Q(:,k);
      R(i,k)=R(i,k)+s;
      Q(:,k)=Q(:,k)-s*Q(:,i);
    end
    tt=norm(Q(:,k));
    if tt>10*eps*t & tt<t/10,        % if length short reorthogonalize
      reorth=1; t=tt;
    else
      reorth=0;
      if tt<10*eps*t, tt=0; end      % linearly dependent
    end
  end
  z=[z u];
  R(k,k)=tt;
  if tt*eps~=0, tt=1/tt; else tt=0; end
  Q(:,k)=Q(:,k)*tt;
end
```

Note that if the norm of b becomes very small, e.g., when `tt<10*eps*t`
in the implementation above, then `GramSchmidt` considers the new column
as linearly dependent and inserts a zero column in Q.

EXAMPLE 6.13. *The 7 column vectors of A are in $\mathbb{R}^5$. So they are linearly
dependent.*

```
A =[ 0      0      0      0      0      1      0
     83     1      0      0      0      1      973
     7      42     1      1      93     85     53
     9      65     42     70     91     76     26
     5      74     33     63     76     99     37]
[Q0,R0,z] = GramSchmidt(A)
[Q,R] = qr(A)
```

We obtain the results

```
Q0 =
        0         0         0         0         0    1.0000         0
   0.9889   -0.1388    0.0098    0.0515         0    0.0000         0
   0.0834    0.3841   -0.8231   -0.4098         0    0.0000         0
   0.1072    0.5977    0.5669   -0.5566         0   -0.0000         0
   0.0596    0.6899   -0.0309    0.7208         0   -0.0000         0
R0 =
  83.9285   15.8706    6.5532   11.3430   22.0426   22.1260  971.6480
        0  105.8968   48.2545   85.6870  142.5461  146.2353  -73.5970
        0         0   21.9672   36.9133  -27.3109  -29.9315  -20.4856
        0         0         0    6.0398  -33.9831   -5.7251   40.5828
        0         0         0         0         0         0         0
        0         0         0         0         0    1.0000   -0.0000
        0         0         0         0         0         0         0
z =
```

```
        1       1       1       2       1       2       1
Q =
          0  -0.0000  -0.0000   0.0000  -1.0000
    -0.9889  -0.1388   0.0098   0.0515  -0.0000
    -0.0834   0.3841  -0.8231  -0.4098  -0.0000
    -0.1072   0.5977   0.5669  -0.5566  -0.0000
    -0.0596   0.6899  -0.0309   0.7208   0.0000
R =
   -83.9285  -15.8706   -6.5532  -11.3430  -22.0426  -22.1260  -971.6480
         0  105.8968   48.2545   85.6870  142.5461  146.2353   -73.5970
         0        0   21.9672   36.9133  -27.3109  -29.9315   -20.4856
         0        0        0    6.0398  -33.9831   -5.7251    40.5828
         0        0        0        0   -0.0000   -1.0000    -0.0000
```

GramSchmidt discovered that the 5th vector was linearly dependent on the first four and replaced it by a zero vector. The 7th vector is also linearly dependent (as we would expect from the dimension of the column space) and is also replaced by a zero vector. Notice that the 4th and 6th vector had to be reorthogonalized. The standard MATLAB *function* **qr** *gives the same results without displaying the zero rows and columns in R and Q.*

$1 - \lambda$	G	M	Singular Values				
1.000e-09	3	3	8.593	2.091e-09	1.363e-09	1.543e-16	8.165e-18
1.000e-10	3	3	8.593	2.091e-10	1.363e-10	3.458e-17	2.122e-17
1.000e-11	3	3	8.593	2.091e-11	1.363e-11	5.955e-17	4.253e-17
1.000e-12	3	3	8.593	2.091e-12	1.363e-12	8.250e-17	2.081e-17
1.001e-13	3	3	8.593	2.095e-13	1.364e-13	7.927e-17	3.179e-17
1.010e-14	3	2	8.593	2.122e-14	1.383e-14	1.876e-16	4.088e-17
1.110e-15	2	1	8.593	2.398e-15	1.468e-15	1.150e-16	5.860e-17
2.220e-16	1	1	8.593	4.422e-16	3.197e-16	2.348e-16	1.697e-17
1.110e-16	1	1	8.593	5.609e-16	3.314e-16	1.055e-16	8.622e-18

TABLE 6.1. *Rank Computation with G (Gram-Schmidt) and M* (MATLAB *SVD*)

EXAMPLE 6.14. *As second example, we construct the following matrix $A \in \mathbb{R}^{8 \times 5}$ which has rank 3 if the parameter* **lam** $= \lambda < 1$. *However, for $\lambda = 1$ the matrix has rank 1. For $1 - \lambda \geq 10^{-14}$* GramSchmidt *finds the linear dependencies correctly and replaces the dependent columns by zeros. If we let $\lambda \to 1$ we can see in Table 6.1 how both functions* GramSchmidt *and the* MATLAB *function* **rank** *compute the rank.*

```
format short e
lam=0.99999999; e=0.00000001;
for i=1:9
  e=e/10; lam=lam+e*9;
  a=[1 lam 1 lam 1 1 1 lam]';
  b=[1 1 lam 1 1 1 lam 1]';
  c=[lam 1 1 lam 1 lam 1 1]';
```

```
A=[a+0.1*b-0.98*c, -0.321*a+0.07*b, 0.56*c, 0.3*a-3.1*b, b];
[q,r,z]=GramSchmidt(A)
rankgs=sum(any(q));
1-lam
[rankgs rank(A)]
svd(A)'
end
```

6.5.8   Updating and Downdating the QR Decomposition

Suppose we have computed the QR decomposition to find the model parameters that best fit the given data. It often happens that more data become available, data need to be thrown out, or the data have changed due to different operating conditions; in other words, the matrix has changed. Instead of recomputing the QR factorization from scratch, it is often more efficient to update the existing factorization. In this subsection, we show how to update the QR factors for the following types of changes in the matrix:

1. Rank-one update,

2. Deleting a column,

3. Adding a column,

4. Adding a row,

5. Deleting a row.

Rank-one update

Given a QR decomposition of the matrix $A = Q\binom{R}{0}$ with $A \in \mathbb{R}^{m \times n}$, $Q \in \mathbb{R}^{m \times m}$ with $m \geq n$ and two vectors $\boldsymbol{u} \in \mathbb{R}^m$ and $\boldsymbol{v} \in \mathbb{R}^n$, is there a simple way to compute the QR decomposition of the rank-1 modified matrix $A' = A + \boldsymbol{u}\boldsymbol{v}^\top = Q'R'$?

Multiplying with $Q^\top$ we get

$$Q^\top A' = Q^\top A + Q^\top \boldsymbol{u}\boldsymbol{v}^\top = \binom{R}{0} + \boldsymbol{w}\boldsymbol{v}^\top \quad \text{with} \quad \boldsymbol{w} = Q^\top \boldsymbol{u}.$$

The idea is to transform the vector $\boldsymbol{w}$ to a multiple of $\boldsymbol{e}_1$. Then the rank-1 correction $\boldsymbol{w}\boldsymbol{v}^\top$ can be added to the matrix R by just changing the first row.

The algorithm consists of three steps:

1. Choose Givens rotation G_k acting on row k and $k+1$ for $k = m, m-$

$1, \ldots, n+1$ such that the elements $w_{n+2}, \ldots, w_m$ are rotated to zero:

$$\underbrace{G_{n+1}^\top \ldots G_{m-2}^\top G_{m-1}^\top Q^\top}_{Q_1^\top} A' = \begin{pmatrix} R \\ 0 \end{pmatrix} + G_{n+1}^\top \ldots G_{m-2}^\top G_{m-1}^\top \boldsymbol{w}\boldsymbol{v}^\top$$

$$= \begin{pmatrix} R \\ 0 \end{pmatrix} + \begin{pmatrix} \boldsymbol{w}' \\ 0 \end{pmatrix} \boldsymbol{v}^\top.$$

This process generates

$$\begin{pmatrix} \boldsymbol{w}' \\ 0 \end{pmatrix} = G_{n+1}^\top \ldots G_{m-2}^\top G_{m-1}^\top \boldsymbol{w}$$

and $Q_1 = Q G_{m-1} G_{m-2} \ldots G_{n+1}$.

2. We now apply more Givens rotations to map $\boldsymbol{w}'$ to $\alpha \boldsymbol{e}_1$. These rotations also change Q_1 to Q_2 and the matrix R becomes an upper Hessenberg matrix H_1,

$$\underbrace{G_1^\top \ldots G_n^\top Q_1^\top}_{Q_2^\top} A' = \begin{pmatrix} H_1 \\ 0 \end{pmatrix} + \begin{pmatrix} \alpha \\ 0 \end{pmatrix} \boldsymbol{v}^\top = \begin{pmatrix} H \\ 0 \end{pmatrix}.$$

The transformed rank-1 correction is now added to the first row of the matrix H_1, thus the right hand side becomes a new Hessenberg matrix H.

3. In the last step we annihilate with further Givens rotations the subdiagonal of H and so we obtain the QR decomposition of the modified matrix,

$$\underbrace{G_n^\top \ldots G_1^\top Q_2^\top}_{Q'^\top} A' = G_n^\top \ldots G_1^\top \begin{pmatrix} H \\ 0 \end{pmatrix} = \begin{pmatrix} R' \\ 0 \end{pmatrix}.$$

In the the following algorithm we will make use of the MATLAB function `planerot`, which computes a Givens rotation matrix $G \in \mathbb{R}^{2\times 2}$. The call `[G,y]=planerot(x)` computes the matrix

$$G = \frac{1}{\|\boldsymbol{x}\|_2} \begin{pmatrix} x_1 & x_2 \\ -x_2 & x_1 \end{pmatrix} \quad \text{and vector} \quad \boldsymbol{y} = \begin{pmatrix} \|\boldsymbol{x}\|_2 \\ 0 \end{pmatrix}.$$

Since it makes use of the MATLAB built in function `norm`, it is safe to compute it this way. It would not be a good idea to replace the norm computation by $\sqrt{x_1^2 + x_2^2}$, see Section 2.7.5 and Problem 2.13.

ALGORITHM 6.13. *Rank-1 Update of QR decomposition*

```
function [Qs,Rs]=UpdateQR(Q,R,u,v)
```

```
% UPDATEQR Rank-1 update of the QR-Decomposition
%    [Qs,Rs]=UpdateQR(Q,R,u,v); If As=A+u v' and [Q,R]=qr(A) then we
%    compute Qs and Rs such that As=Qs Rs. Uses Matlab's PLANEROT.

[m,n]=size(R); Qs=Q; Rs=R; w=Q'*u;
for k=m:-1:n+2                          % annihilate w(n+2:m)
  G=planerot(w(k-1:k));
  w(k-1:k)=G*w(k-1:k);
  Qs(:,k-1:k)=Qs(:,k-1:k)*G';
end
for k=n+1:-1:2                          % annihilate w(2:n+1)
  G=planerot(w(k-1:k));
  w(k-1:k)=G*w(k-1:k);
  Rs(k-1:k,k-1:n)=G*Rs(k-1:k,k-1:n);
  Qs(:,k-1:k)=Qs(:,k-1:k)*G';
end
Rs(1,:)=Rs(1,:)+w(1)*v';               % Add rank-1 change to first row
for k=1:n                               % reduce Hessenberg matrix to Rs
  G=planerot(Rs(k:k+1,k));
  Rs(k:k+1,k:n)=G*Rs(k:k+1,k:n);
  Qs(:,k:k+1)=Qs(:,k:k+1)*G';
end
```

EXAMPLE 6.15. *We consider the 15×10 section of the Hilbert-matrix and the vectors $\boldsymbol{u} = (1, 2, \ldots, 15)^\top$ and $\boldsymbol{v} = (1, 2, \ldots, 10)^\top$. The test program*

```
m=15; n=10;
A=hilb(m); A=A(:,1:n);
[Q,R]=qr(A);
u=[1:m]'; v=[1:n]';
[Qs,Rs,]=UpdateQR(Q,R,u,v);
As=A+u*v';
disp('||As-Qs*Rs||'), norm(As-Qs*Rs)
disp('||Qs''*Qs-eye(m)||'), norm(Qs'*Qs-eye(m))
[Qk,Rk]=qr(As);
disp('||As-Qk*Rk||'), norm(As-Qk*Rk)
disp('||Qk''*Qk-eye(m)||'), norm(Qk'*Qk-eye(m))
disp('norm(Rs-Rk)'); norm(Rs-Rk)
disp('||abs(Rs)-abs(Rk)||'), norm(abs(Rs)-abs(Rk))
```

compares the updating technique with the explicit QR decomposition of the modified matrix, and gives the output

```
||As-Qs*Rs||
ans =
   3.8508e-13
||Qs'*Qs-eye(m)||
ans =
   1.3560e-15
```

```
||As-Qk*Rk||
ans =
   5.1102e-13
||Qk'*Qk-eye(m)||
ans =
   1.4212e-15
norm(Rs-Rk)
ans =
   1.3825e+03
||abs(Rs)-abs(Rk)||
ans =
   3.6950e-13
```

The results are the same.

Deleting a column

We first consider the partition of $A \in \mathbb{R}^{m \times n}$ into two matrices and the corresponding partitioning of the QR decomposition,

$$A = [A_1, A_2] = Q \begin{pmatrix} R_{11} & R_{12} \\ 0 & R_{22} \\ 0 & 0 \end{pmatrix}.$$

By multiplying the block columns, we conclude that

$$A_1 = Q \begin{pmatrix} R_{11} \\ 0 \end{pmatrix}$$

is the QR decomposition of A_1. So when removing the last column or some of the last columns the update is trivial.

Consider now the columns $A = [a_1, \ldots, a_k, \ldots, a_n]$ and the permutation matrix P which moves column k to the last column

$$\begin{aligned} AP &= [a_1, \ldots, a_{k-1}, a_{k+1} \ldots, a_n, a_k] \\ &= QRP = Q[r_1, \ldots, r_{k-1}, r_{k+1} \ldots, r_n, r_k]. \end{aligned}$$

The permuted matrix R becomes

$$RP = \begin{pmatrix} R_{11} & v & R_{13} \\ 0 & r_{kk} & w^\top \\ 0 & 0 & R_{33} \\ 0 & 0 & 0 \end{pmatrix} P = \begin{pmatrix} R_{11} & R_{13} & v \\ 0 & w^\top & r_{kk} \\ 0 & R_{33} & 0 \\ 0 & 0 & 0 \end{pmatrix}$$

a Hessenberg matrix. With Givens rotations G_k acting on rows k and $k + 1$, then $k + 1$ and $k + 2$, etc., we can transform the Hessenberg block again to upper triangular form,

$$G_{n-1}^\top \cdots G_k^\top \begin{pmatrix} w^\top & r_{kk} \\ R_{33} & 0 \end{pmatrix} = \bar{R}_{22}.$$

Thus we get the decomposition

$$AP = \bar{Q}\bar{R}, \quad \text{with} \quad \bar{Q} = Q\bar{G}_k \cdots \bar{G}_{n-1} \quad \text{and} \quad \bar{G}_i = \begin{pmatrix} I & 0 \\ 0 & G_i \end{pmatrix},$$

and the last column can simply be discarded.

ALGORITHM 6.14. *Remove a Column of QR*

```
function [Q,R]=RemoveColumnQR(Q,R,k);
% REMOVECOLUMNQR updating QR when a column is removed
%   [Q,R]=RemoveColumnQR(Q,R,k); finds a QR decomposition for the
%   matrix [A(:,1),...,A(:,k-1),A(:,k+1),..A(:,n)] where
%   [Q,R]=qr(A). Uses Matlab's function PLANEROT.

[m,n]=size(R);
R=R(:,[1:k-1,k+1:n]);
for j=k:n-1,
  G=planerot(R(j:j+1,j));
  R(j:j+1,j:n-1)=G*R(j:j+1,j:n-1);
  Q(:,j:j+1)=Q(:,j:j+1)*G';
end
```

Adding a column

Denote the new column by a_{n+1} and place it before column k forming the matrix $\bar{A} = [a_1, \ldots, a_{k-1}, a_{n+1}, a_k \ldots, a_n]$. Using an appropriate permutation matrix P we move the new column to the end $\bar{A} = [A, a_{n+1}]P$. Using the QR decomposition

$$A = Q\begin{pmatrix} R \\ 0 \end{pmatrix},$$

we get

$$Q^\top \bar{A} = \left[\begin{pmatrix} R \\ 0 \end{pmatrix}, Q^\top a_{n+1} \right] P.$$

Defining

$$Q^\top a_{n+1} = \begin{pmatrix} u \\ v \\ w \end{pmatrix}, \quad u \in \mathbb{R}^k, \ v \in \mathbb{R}^{n-k}, \ w \in \mathbb{R}^{m-n}$$

and reversing the permutation, we get

$$Q^\top \bar{A} = \begin{pmatrix} R_{11} & u & R_{12} \\ 0 & v & R_{22} \\ 0 & w & 0 \end{pmatrix}.$$

Now we apply Givens rotations G_j acting on rows j and $j + 1$ to annihilate elements of w and v up to v_1. When annihilating elements of v, the corresponding rows of R_{22} are also changed and we obtain again an upper triangular matrix.

ALGORITHM 6.15. *QR update by Adding a Column*

```
function [Q,R]=AddColumnQR(Q,R,k,w);
% ADDCOLUMNQR_update QR decomposition when a new column is added
%    [Q,R]=AddColumnQR(Q,R,k,w); finds the QR decomposition of
%    [A(:,1),A(:,2),... A(:,k-1),w,A(:,k),...A(:,n)] when [Q,R]=qr(A).
%    Uses Matlab PLANEROT

[m,n]=size(R);
R=[R(:,1:k-1),Q'*w,R(:,k:n)];
for j=m-1:-1:k
  G=planerot(R(j:j+1,k));
  R(j:j+1,k)=G*R(j:j+1,k);                  % annihilate new column
  if j<=n,                                  % transform also R
    R(j:j+1,j+1:n+1)=G*R(j:j+1,j+1:n+1);
  end
  Q(:,j:j+1)=Q(:,j:j+1)*G';                 % update Q
end
```

Adding a row

Let $A = QR$ be given with $A \in \mathbb{R}^{m \times n}$, $Q \in \mathbb{R}^{m \times p}$ and $R \in \mathbb{R}^{p \times n}$. We consider adding a $\boldsymbol{w}^\top$ as new last row of A,

$$\bar{A} = \begin{pmatrix} A \\ \boldsymbol{w}^\top \end{pmatrix}.$$

Then

$$\begin{pmatrix} Q^\top & 0 \\ 0 & 1 \end{pmatrix} \begin{pmatrix} A \\ \boldsymbol{w}^\top \end{pmatrix} = \begin{pmatrix} R \\ \boldsymbol{w}^\top \end{pmatrix}.$$

To eliminate the elements of vector $\boldsymbol{w}^\top$ we use again Givens rotations acting on $\boldsymbol{w}^\top$ and a row of R. For $p = n$ we get

$$G_n^\top \cdots G_2^\top G_1^\top \begin{pmatrix} R \\ \boldsymbol{w}^\top \end{pmatrix} = \begin{pmatrix} \bar{R} \\ 0 \end{pmatrix}$$

with $G_i^\top$ changing row i of R and $\boldsymbol{w}^\top$. By applying the Givens rotations to Q we get

$$\bar{Q} = \begin{pmatrix} Q & 0 \\ 0 & 1 \end{pmatrix} G_1 \cdots G_n$$

and thus $\bar{A} = \bar{Q} \binom{\bar{R}}{0}$.

Note that if the new row is not appended as last row but between the rows of A, then this means simply a permutation of the rows of $\bar{Q}$, since

$$\bar{Q}^\top P^\top P \bar{A} = (P\bar{Q})^\top \begin{pmatrix} \boldsymbol{a}_{1:} \\ \vdots \\ \boldsymbol{a}_{k-1:} \\ \boldsymbol{w}^\top \\ \boldsymbol{a}_{k:} \\ \vdots \\ \boldsymbol{a}_{m:} \end{pmatrix}$$

and $\bar{R}$ remains the same.

ALGORITHM 6.16. *QR update by Adding a Row*

```
function [Q,R]=AddRowQR(Q,R,k,w);
% ADDROWQR QR decomposition of modified matrix
%    [Q,R]=AddRowQR(Q,R,k,w); Given [Q,R]=qr(A),computes the QR
%    decomposition for [A(1,:);A(2,:), ...,A(k-1,:);w;A(k,:),...A(n,:)]

[p,n]=size(R); m=size(Q,1);            % Q is (m x p)
Q=[Q(1:k-1,:), zeros(k-1,1)            % add row and column to Q
   zeros(1,p), 1                       % and permute
   Q(k:m,:), zeros(m-k+1,1)];
R=[R; w];                              % augment R
k=p+1;
for j=1:min(n,p)
  G=planerot(R(j:k-j:k,j));
  R(j:k-j:k,j:n)=G*R(j:k-j:k,j:n);     % annihilate w
  Q(:,j:k-j:k)=Q(:,j:k-j:k)*G';        % update Q
end
```

EXAMPLE 6.16. *In the following example we consider a matrix $A \in \mathbb{R}^{5\times3}$ with rank 2:*

```
>> A=[17       24      -43
        23       5      244
         4       6      -14
        10      12       -2
        11      18      -55];
>> [m,n]=size(A);
>> [Q,R]=qr(A)
Q =
    -0.5234      0.5058      0.3869     -0.1275     -0.5517
    -0.7081     -0.6966      0.0492     -0.0642      0.0826
    -0.1231      0.1367     -0.6694     -0.7172     -0.0614
    -0.3079      0.1911     -0.6229      0.6816     -0.1271
```

```
    -0.3387      0.4514      0.1089    -0.0275      0.8179
R =
  -32.4808  -26.6311 -129.3073
         0   19.8943 -218.8370
         0         0    0.0000
         0         0         0
         0         0         0
>> p=rank(A)
p =
     2
>> q=Q(:,1:p)
q =
   -0.5234      0.5058
   -0.7081     -0.6966
   -0.1231      0.1367
   -0.3079      0.1911
   -0.3387      0.4514
>> r=R(1:p,:)
r =
  -32.4808   -26.6311  -129.3073
         0    19.8943  -218.8370
>> [norm(A-Q*R) norm(A-q*r)]
ans =
   1.0e-13 *
     0.1313      0.0775
```

We computed the QR decomposition with $Q \in \mathbb{R}^{5 \times 5}$ and, because A has rank 2, we also obtained the economic version with $q \in \mathbb{R}^{5 \times 2}$ and $r \in \mathbb{R}^{2 \times 3}$. Both products QR and qr represent the matrix A well. Next, we choose a new row $\mathbf{w}^\top$ and insert it as new third row:

```
>> k=3;
>> w=[1 2 3];
>> As=[A(1:k-1,:); w; A(k:m,:)]
As =
     17     24     -43
     23      5     244
      1      2       3
      4      6     -14
     10     12      -2
     11     18     -55
>> [Qs, Rs]=AddRowQR(Q,R,k,w)
Qs =
     0.5231      0.5039     -0.0460    -0.1275    -0.5517      0.3869
     0.7078     -0.6966      0.0195    -0.0642     0.0826      0.0492
     0.0308      0.0592      0.9978          0          0     -0.0000
     0.1231      0.1363     -0.0119    -0.7172    -0.0614     -0.6694
     0.3077      0.1902     -0.0208     0.6816    -0.1271     -0.6229
     0.3385      0.4500     -0.0371    -0.0275     0.8179      0.1089
Rs =
```

```
    32.4962    26.6801   129.3384
          0    19.9292  -218.5114
          0          0    11.9733
          0          0          0
          0          0          0
          0          0          0
>> [qs, rs]=AddRowQR(q,r,k,w)
qs =
     0.5231     0.5039     0.0460
     0.7078    -0.6966    -0.0195
     0.0308     0.0592    -0.9978
     0.1231     0.1363     0.0119
     0.3077     0.1902     0.0208
     0.3385     0.4500     0.0371
rs =
    32.4962    26.6801   129.3384
          0    19.9292  -218.5114
          0          0   -11.9733
>> [norm(As-Qs*Rs) norm(As-qs*rs)]
ans =
    1.0e-12 *
     0.1180     0.1165
```

*We see that in both cases the update procedure works well. The product of
the new matrices represents well the modified matrix.*

Deleting a row

Let $A = QR$ be the QR decomposition. If the k-th row should be removed,
then we perform first a permutation such that it becomes the first row:

$$A = QR, \quad PA = \begin{pmatrix} a_k^\top \\ \bar{A} \end{pmatrix} = PQR.$$

Our aim is to compute the QR decomposition of $\bar{A}$ which is A with row k
removed. Using Givens rotations, we transform the first row of PQ to the
first unit vector

$$PQG_{m-1}^\top \cdots G_1^\top = \begin{pmatrix} \pm 1 & 0 \\ x & \bar{Q} \end{pmatrix}$$

Note that by this operation the vector x must be zero! This because the
matrix is orthogonal. Thus we get

$$\begin{pmatrix} a_k^\top \\ \bar{A} \end{pmatrix} = (PQG_{m-1}^\top \cdots G_1^\top)(G_1 \cdots G_{m-1}R)$$

$$= \begin{pmatrix} \pm 1 & 0 \\ 0 & \bar{Q} \end{pmatrix} \begin{pmatrix} v^\top \\ \bar{R} \end{pmatrix}. \tag{6.42}$$

The multiplications with the Givens rotations $G_1 \cdots G_{m-1} R$ transform the matrix R into a Hessenberg matrix. Looking at (6.42), we can read off the solution $\bar{A} = \bar{Q}\bar{R}$ by discarding the first row.

ALGORITHM 6.17. *QR update by Removing a Row*

```
function [Q,R]=RemoveRowQR(Q,R,k);
% REMOVEROWQR update the QR decomposition when a row is removed
%    [Q,R]=RemoveRowQR(Q,R,k); computes the QR decomposition for
%    [A(1,:); A(2,:);...;A(k-1,:); A(k+1,:);...;A(n,:)] when
%    [Q,R]=qr(A).

[m,n]=size(R);
Q=[Q(k,:); Q(1:k-1,:); Q(k+1:m,:)];         % permute Q
for j=m-1:-1:1                               % map row to unit vector
  G=planerot(Q(1,j:j+1)');
  if j<=n
    R(j:j+1,j:n)=G*R(j:j+1,j:n);             % update R
  end;
  Q(:,j:j+1)=Q(:,j:j+1)*G';                  % update Q
end
Q=Q(2:m,2:m);R=R(2:m,:);
```

REMARKS.

1. Deleting a row is a special case of a rank-one modification: $\boldsymbol{u} = -\boldsymbol{e}_1$, $\boldsymbol{v} = \boldsymbol{a}_1$.

2. MATLAB offers two functions to add (respectively remove) rows and columns of a QR decomposition: `qrinsert`, `qrdelete`. They are essentially the same as our functions. There is also a MATLAB-built in function `qrupdate` which computes a rank-one update.

6.5.9   Covariance Matrix Computations Using QR

Let $A = Q\begin{pmatrix} R \\ 0 \end{pmatrix}$ be the QR decomposition. Then $A^\top A = R^\top R$ and thus the *covariance matrix* becomes

$$C = (A^\top A)^{-1} = R^{-1}R^{-\top}.$$

In this section, we will show how to compute elements of the covariance matrix using a minimal number of operations.

Computing individual elements

Since the covariance matrix C is symmetric, it is sufficient to compute elements in the upper triangle. Let $i \leq j$. Then

$$c_{ij} = e_i^\top R^{-1} R^{-\top} e_j = u_i^\top u_j \quad \text{where} \quad u_i = R^{-\top} e_i.$$

Thus u_i can be computed by forward substitution,

$$R^\top u_i = e_i.$$

Because the elements $u_i = 0$ for $i = 1, 2, \ldots, i - 1$ we need to start with forward substitution only at equation #i.

ALGORITHM 6.18. *Computes ith-Row of R^{-1}*

```
function u=ui(i,R)
% UI computes a row of the inverse of a triangular matrix
%    u=ui(i,R) computes the i-th row of the inverse of the upper
%    triangular matrix R

n=min(size(R)); u=zeros(1,i-1);
u(i)=1/R(i,i);
for j=i+1:n,
   u(j)=-u(i:j-1)*R(i:j-1,j)/R(j,j);
end
```

The element c_{ij} is then computed for $i < j$ by the statements

```
u1=ui(i,R); u2=ui(j,R); cij=u1(j:n)*u2(j:n)'
```

For a diagonal element this simplifies to

```
u=ui(i,R); cii=u(i:n)*u(i:n)'
```

Calculating the whole covariance matrix

To compute $C = R^{-1} R^{-\top}$ we need R^{-1}, that is the vectors $u_1, \ldots, u_n$. Then we multiply $R^{-1} R^{-\top}$ by using the triangular shape of R^{-1} and the symmetry and computing only the upper part of C.

ALGORITHM 6.19. *Covariance via R^{-1}*

```
function C=Covariance(R)
% COVARIANCE computes the covariance matrix
%    C=Covariance(R) computes the upper triangle of the covariance
%    matrix from the R factor of the QR-decomposition [Q,R]=qr(A)

n=min(size(R)); U=[];                        % U is R^(-1)
```

```
for i=1:n,
  u=ui(i,R); U=[U; u];
end;
for i=1:n,                                      % compute C=inv(R)*inv(R')
  for j=i:n,
    C(i,j)=U(i,j:n)*U(j,j:n)';
  end
end
```

Björck's Algorithm

Because $R^\top R = A^\top A$ it follows that $R^\top RC = I$ or

$$RC = R^{-\top}. \tag{6.43}$$

The diagonal elements on the right hand side of Equation (6.43) are $1/r_{ii}$. Surprisingly, this information is sufficient to compute by a recurrence the elements of the covariance matrix C.

Assume that the rows and columns $k+1, \dots, n$ of C are known, that is the elements $c_{ij} = c_{ji}$ for $j = k+1, \dots, n$ and $i \leq j$.

Consider the k-th diagonal element in (6.43),

$$r_k^\top c_k = \frac{1}{r_{kk}} \quad \Longleftrightarrow \quad r_{kk}c_{kk} + \sum_{j=k+1}^{n} r_{kj}c_{jk} = \frac{1}{r_{kk}}.$$

The elements after the summation sign are known, therefore

$$c_{kk} = \frac{1}{r_{kk}} \left(\frac{1}{r_{kk}} - \sum_{j=k+1}^{n} r_{kj}c_{jk} \right).$$

For an element on the k-th row with $i < k$ we have

$$r_i^\top c_k = 0 \quad \Longleftrightarrow \quad r_{ii}c_{ik} + \sum_{j=i+1}^{n} r_{ij}c_{jk} = 0.$$

Also here the elements after the summation symbol are known according to our assumption, thus

$$c_{ik} = -\frac{1}{r_{ii}} \sum_{j=i+1}^{n} r_{ij}c_{jk}, \quad i = k-1, \dots, 1.$$

Using these recurrence relations, the whole covariance matrix can be computed starting with the last column. We obtain

ALGORITHM 6.20. *Covariance Matrix, Björck Algorithm*

```
function C=CovarianceBjoerck(R)
% COVARIANCEBJOERCK computes the covariance matrix by Bjoerck's method
%   C=CovarianceBjoerck(R) computes the covariance matrix from the
%   R factor of A=QR by solving the system R C=R^-T

n=min(size(R));
C=zeros(n,n);
for k=n:-1:1,
  C(k,k)=(1/R(k,k)-R(k,k+1:n)*C(k,k+1:n)')/R(k,k);
  for i=k-1:-1:1,
    C(i,k)=-(R(i,i+1:k)*C(i+1:k,k)+R(i,k+1:n)*C(k,k+1:n)')/R(i,i);
  end
end
```

6.6   Linear Least Squares Problems with Linear Constraints

Given the matrices $A^{m \times n}$, $C^{p \times n}$ and the vectors $\boldsymbol{b}$ and $\boldsymbol{d}$, we are interested in finding a vector $\boldsymbol{x}$ such that

$$\|A\boldsymbol{x} - \boldsymbol{b}\|_2 \longrightarrow \min \quad \text{subject to} \quad C\boldsymbol{x} = \boldsymbol{d}. \tag{6.44}$$

We are interested in the case $p \le n \le m$. A solution exists only if the constraints are consistent i.e. if $\boldsymbol{d} \in \mathcal{R}(C)$.

A straightforward way is to eliminate the constraints and solve the reduced unconstrained problem. This can be done using the general solution of $C\boldsymbol{x} = \boldsymbol{d}$, see Section 6.3.3,

$$\boldsymbol{x} = C^+ \boldsymbol{d} + P_{\mathcal{N}(C)} \boldsymbol{y}, \quad \boldsymbol{y} \text{ arbitrary and} \quad P_{\mathcal{N}(C)} = I - C^+ C.$$

Now $\|A\boldsymbol{x} - \boldsymbol{b}\|_2^2 = \|AP_{\mathcal{N}(C)} \boldsymbol{y} - (\boldsymbol{b} - AC^+ \boldsymbol{d})\|_2^2 \longrightarrow \min$ is an unconstrained linear least squares problem in $\boldsymbol{y}$. The solution with minimal norm is

$$\tilde{\boldsymbol{y}} = (AP_{\mathcal{N}(C)})^+ (\boldsymbol{b} - AC^+ \boldsymbol{d}).$$

Thus $\boldsymbol{x} = C^+ \boldsymbol{d} + P_{\mathcal{N}(C)} (AP_{\mathcal{N}(C)})^+ (\boldsymbol{b} - AC^+ \boldsymbol{d})$. We can simplify this expression using the following lemma:

LEMMA 6.2. $P_{\mathcal{N}(C)} (AP_{\mathcal{N}(C)})^+ = (AP_{\mathcal{N}(C)})^+$.

PROOF. The matrix $(AP_{\mathcal{N}(C)})^+$ is the solution of the Penrose equations (see Theorem 6.7). We show that $Y = P_{\mathcal{N}(C)} (AP_{\mathcal{N}(C)})^+$ is also a solution, therefore by uniqueness they must be the same. In the proof below, we will write $P := P_{\mathcal{N}(C)}$ and use the relations $PP = P$ and $P^\top = P$, which hold because P is an orthogonal projector. We now substitute Y into the Penrose equations:

(i) $(AP)Y(AP) = A\underbrace{PP}_{P}(AP)^+ AP = AP(AP)^+ (AP) = AP.$

(ii) $Y(AP)Y = P(AP)^+ A \underbrace{PP}_{P} (AP)^+ = P((AP)^+ AP(AP)^+) = P(AP)^+ = Y.$

(iii) $(AP\,Y)^{\mathsf{T}} = (A \underbrace{PP}_{P} (AP)^+)^{\mathsf{T}} = A \underbrace{P}_{PP} (AP)^+ = AP\,Y.$

(iv) $(Y\,AP)^{\mathsf{T}} = (P(AP)^+ AP)^{\mathsf{T}} = [(AP)^+ AP]^{\mathsf{T}} P = (AP)^+ A \underbrace{PP}_{P}$

$$= (AP)^+ AP = ((AP)^+ AP)^{\mathsf{T}} = (AP)^{\mathsf{T}} [(AP)^+]^{\mathsf{T}} = \underbrace{P}_{PP} A^{\mathsf{T}} [(AP)^+]^{\mathsf{T}}$$

$$= P((AP)^+ AP)^{\mathsf{T}} = P((AP)^+ AP) = Y\,AP.$$

$\square$

As a consequence of the lemma, we have $P_{\mathcal{N}(C)} \tilde{\boldsymbol{y}} = \tilde{\boldsymbol{y}}$.

THEOREM 6.16. *Let* $\boldsymbol{d} \in \mathcal{R}(C)$ *and define* $\boldsymbol{x} = \tilde{\boldsymbol{x}} + \tilde{\boldsymbol{y}}$ *with*

$$\tilde{\boldsymbol{x}} = C^+ \boldsymbol{d}, \quad \tilde{\boldsymbol{y}} = (AP_{\mathcal{N}(C)})^+ (\boldsymbol{b} - A\tilde{\boldsymbol{x}}),$$

where $P_{\mathcal{N}(C)} = I - C^+ C$. *Then* $\boldsymbol{x}$ *is a solution to the problem (6.44); this solution is unique if and only if*

$$\mathrm{rank}\begin{pmatrix} A \\ C \end{pmatrix} = n.$$

Moreover, if (6.44) has more than one solution, then $\boldsymbol{x}$ *is the minimal norm solution.*

PROOF. We have already shown by construction that $\boldsymbol{x}$ is a solution. We note that the solution $\boldsymbol{x}$ cannot be unique if $\mathrm{rank}\begin{pmatrix} A \\ C \end{pmatrix} < n$, since there would exist a vector $\boldsymbol{w} \neq 0$ such that $\begin{pmatrix} A \\ C \end{pmatrix}\boldsymbol{w} = \begin{pmatrix} 0 \\ 0 \end{pmatrix}$ and $\boldsymbol{x} + \boldsymbol{w}$ is also a solution. Thus the condition is necessary; we will show in Section 6.6.2 that it is also sufficient.

We now show that $\boldsymbol{x}$ has minimal norm when the solution is not unique. Let $\boldsymbol{x}$ and $\hat{\boldsymbol{x}}$ be two solutions to (6.44). Since both must satisfy the constraint $C\boldsymbol{x} = d$, they can be written as

$$\boldsymbol{x} = C^+ \boldsymbol{d} + \tilde{\boldsymbol{y}} = \tilde{\boldsymbol{x}} + P\tilde{\boldsymbol{y}},$$
$$\hat{\boldsymbol{x}} = C^+ \boldsymbol{d} + P\boldsymbol{y} = \tilde{\boldsymbol{x}} + P\boldsymbol{y},$$

where we use again the abbreviation $P := P_{\mathcal{N}(C)}$ and the fact that $P\tilde{\boldsymbol{y}} = \tilde{\boldsymbol{y}}$. Since $P = P^{\mathsf{T}}$ and $PC^+ = (I - C^+ C)C^+ = 0$ (see Theorem 6.9), we see that $\tilde{\boldsymbol{x}} \perp P\tilde{\boldsymbol{y}}$, so we have

$$\|\boldsymbol{x}\|_2^2 = \|\tilde{\boldsymbol{x}}\|_2^2 + \|P\tilde{\boldsymbol{y}}\|_2^2 = \|\tilde{\boldsymbol{x}}\|_2^2 + \|\tilde{\boldsymbol{y}}\|_2^2.$$

Similarly, for $\hat{\boldsymbol{x}}$ we have

$$\|\hat{\boldsymbol{x}}\|_2^2 = \|\tilde{\boldsymbol{x}}\|_2^2 + \|P\boldsymbol{y}\|_2^2.$$

Now since $\tilde{\boldsymbol{y}}$ is the minimal norm solution, every other solution of the reduced unconstrained linear least squares problem has the form $\boldsymbol{y} = \tilde{\boldsymbol{y}} + \boldsymbol{w}$ with $\boldsymbol{w} \perp \tilde{\boldsymbol{y}}$. Also $\tilde{\boldsymbol{y}} \perp P\boldsymbol{w}$ holds since $\tilde{\boldsymbol{y}}^\top P\boldsymbol{w} = (P\tilde{\boldsymbol{y}})^\top \boldsymbol{w} = \tilde{\boldsymbol{y}}^\top \boldsymbol{w} = 0$. Thus

$$\|\hat{\boldsymbol{x}}\|_2^2 = \|\tilde{\boldsymbol{x}}\|_2^2 + \|P\boldsymbol{y}\|_2^2 = \|\tilde{\boldsymbol{x}}\|_2^2 + \|\tilde{\boldsymbol{y}}\|_2^2 + \|P\boldsymbol{w}\|_2^2 \geq \|\tilde{\boldsymbol{x}}\|_2^2 + \|\tilde{\boldsymbol{y}}\|_2^2 = \|\boldsymbol{x}\|_2^2.$$

Therefore $\boldsymbol{x} = \tilde{\boldsymbol{x}} + \tilde{\boldsymbol{y}}$ is the minimum norm solution. $\qquad\square$

6.6.1 Solution with SVD

Problem (6.44) can be solved in an elegant and even more general way using the SVD. We will assume that $A \in \mathbb{R}^{m \times n}$, $C \in \mathbb{R}^{p \times n}$ with $p < n < m$, that $\boldsymbol{b} \notin \mathcal{R}(A)$, that $\mathrm{rank}(C) = r_c < p$ and $\boldsymbol{d} \notin \mathcal{R}(C)$. This means that $C\boldsymbol{x} = \boldsymbol{d}$ is not consistent, but that we consider the more general problem

$$\|A\boldsymbol{x} - \boldsymbol{b}\|_2 \longrightarrow \min \quad \text{subject to} \quad \|C\boldsymbol{x} - \boldsymbol{d}\|_2 \longrightarrow \min. \qquad (6.45)$$

Since C has a nontrivial null space, the constraint $\|C\boldsymbol{x} - \boldsymbol{d}\|_2 \longrightarrow \min$ has many solutions and we want to minimize $\|A\boldsymbol{x} - \boldsymbol{b}\|_2$ over that set. The algorithm that we develop can then also be used for the special case of equality constraints when $C\boldsymbol{x} = \boldsymbol{d}$ is consistent.

In the first step we determine the general (least squares) solution of $C\boldsymbol{x} \approx \boldsymbol{d}$. Let $C^\top = TSR^\top$ be the SVD with $T \in \mathbb{R}^{n \times n}$ and $R \in \mathbb{R}^{p \times p}$. Since $\mathrm{rank}(C) = r_c < p$, we partition the matrices accordingly into the sub-matrices $T = [T_1, T_2]$ with $T_1 \in \mathbb{R}^{n \times r_c}$, $T_2 \in \mathbb{R}^{n \times (p - r_c)}$, $S_r = S(1{:}r_c, 1{:}r_c)$ and $R = [R_1, R_2]$ with $R_1 \in \mathbb{R}^{p \times r_c}$ and $R_2 \in \mathbb{R}^{p \times (p - r_c)}$. With this partition, $C = R_1 S_r T_1^\top$ and $C^+ = T_1 S_r^{-1} R_1^\top$. The general solution of $\|C\boldsymbol{x} - \boldsymbol{d}\|_2 \longrightarrow \min$ then becomes

$$\boldsymbol{x} = \boldsymbol{x}_m + T_2 \boldsymbol{z}, \text{ with } \boldsymbol{z} \in \mathbb{R}^{n - r_c} \text{ arbitrary and } \boldsymbol{x}_m = T_1 S_r^{-1} R_1^\top \boldsymbol{d}.$$

We now introduce this general solution into $\|A\boldsymbol{x} - \boldsymbol{b}\|_2$ and obtain

$$\|AT_2 \boldsymbol{z} - (\boldsymbol{b} - A\boldsymbol{x}_m)\|_2 \longrightarrow \min, \quad AT_2 \in \mathbb{R}^{m \times (n - r_c)}, \qquad (6.46)$$

an unconstrained least squares problem for $\boldsymbol{z}$. We compute the SVD of $AT_2 = U\Sigma V^\top$ and partition again the matrices according to the $\mathrm{rank}(AT_2) = r_a$, for which we will assume that $r_a < n - r_c$: $U = [U_1, U_2]$, $\Sigma_r = \Sigma(1 : r_a, 1 : r_a)$ and $V = [V_1, V_2]$. With this partition, the general solution of (6.46) is

$$\boldsymbol{z} = \boldsymbol{z}_m + V_2 \boldsymbol{w}, \text{ with } \boldsymbol{w} \in \mathbb{R}^{n - r_c - r_a} \text{ arbitrary, and } \boldsymbol{z}_m = V_1 \Sigma_r^{-1} U_1^\top (\boldsymbol{b} - A\boldsymbol{x}_m).$$

Thus the solution of Problem (6.45) is

$$\boldsymbol{x} = \boldsymbol{x}_m + T_2 \boldsymbol{z} = \boldsymbol{x}_m + T_2 V_1 \Sigma_r^{-1} U_1^\top (\boldsymbol{b} - A\boldsymbol{x}_m) + T_2 V_2 \boldsymbol{w}, \quad \boldsymbol{x}_m = T_1 S_r^{-1} R_1^\top \boldsymbol{d}$$
$$(6.47)$$

with $\boldsymbol{w} \in \mathbb{R}^{n - r_c - r_a}$ arbitrary.

EXAMPLE 6.17. *We consider* $A \in \mathbb{R}^{9 \times 6}$, $\mathrm{rank}(A) = 3$, $C \in \mathbb{R}^{3 \times 6}$, $\mathrm{rank}(C) = 2$ *and* $\mathrm{rank}\left(\begin{smallmatrix} A \\ C \end{smallmatrix}\right) = 5$. *Furthermore* $\boldsymbol{b} \notin \mathcal{R}(A)$ *and also* $\boldsymbol{d} \notin \mathcal{R}(C)$. *The solution (6.47) will not be unique, as we will see.*

```
>> A=[5 -1 -1  6  4  0
     -3  1  4 -7 -2 -3
      1  3 -4  5  4  7
      0  4 -1  1  4  5
      4  2  3  1  6 -1
      3 -3 -5  8  0  2
      0 -1 -4  4 -1  3
     -5  4 -3 -2 -1  7
      3  4 -3  6  7  7];
>> [m,n]=size(A);
>> b=[-4  1 -2  3  3  0 -1  3  1]';
>> ranksa=[rank(A) rank([A,b])]
ranksa=    3      4
>> C=[1 3 -2 3  8 0
     -3 0  0 1  9 4
     -2 3 -2 4 17 4];
>> [p n]=size(C);
>> d=[1 2 -3 ]';
>> ranksc=[rank(C) rank([C,d])]
ranksc=   2      3
```

Now we compute the minimal norm solution of $C\boldsymbol{x} \approx \boldsymbol{d}$, *its norm and the norm of the residual* $\boldsymbol{r} = \boldsymbol{d} - C\boldsymbol{x}_m$:

```
>> [T,S,R]=svd(C');
>> rc=rank(S); Sr=S(1:rc,1:rc);
>> T1=T(:,1:rc); T2=T(:,rc+1:n);
>> R1=R(:,1:rc); R2=R(:,rc+1:p);
>> xm=T1*(Sr\(R1'*d));
>> xm'
ans=-0.0783 -0.0778  0.0519 -0.0604 -0.0504  0.0698
>> [norm(xm) norm(d-C*xm)]
ans=    0.1611     3.4641
```

To obtain another solution $\boldsymbol{x}_g$ *of* $C\boldsymbol{x} \approx \boldsymbol{d}$, *we choose e.g.* $\boldsymbol{z} = (1,2,3,4)^\top$ *and obtain*

```
>> xg=xm+T2*[1 2 3 4]';
>> xg'
ans= 0.1391 0.6588 -2.7590 2.0783 -1.8586 3.7665
>> [norm(xg) norm(d-C*xg)]
ans=    5.4796     3.4641
```

We see that $\boldsymbol{x}_g$ *has the same residual, but* $\|\boldsymbol{x}_g\|_2 > \|\boldsymbol{x}_m\|_2$, *as we would expect. Note that the results for* $\boldsymbol{x}_g$ *may differ depending on the version of* MATLAB *used; this is because the matrix* T_2 *is not unique. We continue by eliminating the constraints and solve the unconstrained problem for* $\boldsymbol{z}$:

```
>> As=A*T2; c=b-A*xm;
```

```
>> nrc=n-rc
nrc=    4
>> [U Sig V]=svd(As);
>> ra=rank(Sig)
ra= 3
>> Sigr=Sig(1:ra,1:ra);
>> U1=U(:,1:ra); U2= U(:,ra+1:nrc);
>> V1=V(:,1:ra); V2= V(:,ra+1:nrc);
>> zm=V1*(Sigr\(U1'*c));
>> zm'
ans=  -0.0364 -0.1657 -0.2783 0.3797
```

*The matrix $A_s = AT_2$ is rank deficient, so the reduced unconstrained problem
has infinitely many solutions and the vector z_m is the minimal norm solution.
Thus the minimal norm solution of Problem (6.45) is $x_{min} = x_m + T_2 z_m$:*

```
>> xmin=xm+T2*zm;
>> xmin'
ans=0.1073 0.2230 0.2695 -0.1950 -0.0815 0.3126
```

We finally perform some checks; x_a is a solution of $Ax \approx b$ computed by
MATLAB.

```
>> xa=A\b;
Warning: Rank deficient, rank=3  tol=    3.0439e-14.
>> [norm(C*xm-d) norm(C*xg-d) norm(C*xmin-d) norm(C*xa-d)]
ans=    3.4641    3.4641    3.4641    6.1413
>> [norm(A*xm-b) norm(A*xg-b) norm(A*xmin-b) norm(A*xa-b)]
ans=    6.8668 93.3010 5.3106 5.3106
>> [norm(xm) norm(xg) norm(xmin) norm(xa)]
ans=    0.1611  5.4796  0.5256  0.5035
```

*Of course, x_a does not minimize $\|Cx - d\|_2$ and x_m does not minimize
$\|Ax - b\|_2$. However, x_{min} does minimize both norms as it should.*

*Another solution of Problem (6.45) is obtained by adding to x_{min} some
linear combination of the columns of the matrix $T_2 V_2$: $x = x_{min} + T_2 V_2 w$.
Since $T_2 V_2 \in \mathbb{R}^{6 \times 1}$ in our example, w is a scalar:*

```
T2*V2
ans =
    0.4656
   -0.2993
   -0.4989
   -0.6319
    0.1663
    0.1330
```

The general solution of Problem (6.45) for our example is therefore (with *arbitrary parameter λ)*

$$
x = \begin{pmatrix} 0.1073 \\ 0.2230 \\ 0.2695 \\ -0.1950 \\ -0.0815 \\ 0.3126 \end{pmatrix} + \lambda \begin{pmatrix} 0.4656 \\ -0.2993 \\ -0.4989 \\ -0.6319 \\ 0.1663 \\ 0.1330 \end{pmatrix}.
$$

We compute two other solutions and show that they also minimize the two norms but their own norm is larger than $\|x_{min}\|_2$:

```
>> x=xmin +T2*V2;
>> x'
ans= 0.5729 -0.0763 -0.2293 -0.8269  0.0848  0.4457
>> [norm(d-C*x) norm(b-A*x) norm(x)]
ans= 3.4641  5.3106  1.1297
>> x=xmin +5*T2*V2;
>> x'
ans= 2.4355 -1.2737 -2.2249 -3.3547  0.7500  0.9778
>> [norm(d-C*x) norm(b-A*x) norm(x)]
ans= 3.4641  5.3106  5.0276
```

Certainly, the SVD is the best method in such cases where rank decisions have to be made. By providing explicit expressions for the projectors and orthogonal bases for the subspaces, this method is well suited for rank deficient matrices and general solutions.

6.6.2   Classical Solution Using Lagrange Multipliers

Before the widespread use of the SVD as a computational tool, a classical approach for solving Problem (6.44) uses Lagrange multipliers (see Section 12.2.2 for details). The Lagrangian is

$$
L(x, \lambda) = \frac{1}{2}\|Ax - b\|_2^2 + \lambda^\top (Cx - d).
$$

Setting the partial derivatives to zero, we obtain

$$
\frac{\partial L}{\partial x} = A^\top (Ax - b) + C^\top \lambda = 0 \text{ and } \frac{\partial L}{\partial \lambda} = Cx - d = 0.
$$

Thus, we obtain the *Normal Equations*

$$
\begin{pmatrix} A^\top A & C^\top \\ C & 0 \end{pmatrix} \begin{pmatrix} x \\ \lambda \end{pmatrix} = \begin{pmatrix} A^\top b \\ d \end{pmatrix} \tag{6.48}
$$

The matrix of the Normal Equations (6.48) is symmetric, but not positive definite; for this reason, this system is also known as *saddle point equations,* since they correspond geometrically to saddle points, i.e., points that are neither local minima nor maxima. Because of the indefiniteness, we cannot use the Cholesky decomposition to solve (6.48); moreover, the matrix becomes singular in the case of rank deficient matrices, like the example in Section 6.6.1, or even when only C is rank deficient.

The Normal Equations (6.48) can be used to finish the proof of Theorem 6.16. We need to prove that if rank $\left(\begin{smallmatrix} A \\ C \end{smallmatrix}\right) = n$, then the solution of Problem (6.44) is unique. Assume that we have two solutions $\boldsymbol{x}_1$ and $\boldsymbol{x}_2$. Then both are solutions of the normal equations with some $\boldsymbol{\lambda}_1$ and $\boldsymbol{\lambda}_2$. If we take the difference of the normal equations, we obtain

$$\begin{pmatrix} A^\mathsf{T} A & C^\mathsf{T} \\ C & 0 \end{pmatrix} \begin{pmatrix} \boldsymbol{x}_1 - \boldsymbol{x}_2 \\ \boldsymbol{\lambda}_1 - \boldsymbol{\lambda}_2 \end{pmatrix} = \begin{pmatrix} 0 \\ 0 \end{pmatrix}. \tag{6.49}$$

Multiplying the first equation of (6.49) from the left with $(\boldsymbol{x}_1 - \boldsymbol{x}_2)^\mathsf{T}$, we get

$$\|A(\boldsymbol{x}_1 - \boldsymbol{x}_2)\|_2^2 + \underbrace{(C(\boldsymbol{x}_1 - \boldsymbol{x}_2))}_{=0}{}^\mathsf{T} (\boldsymbol{\lambda}_1 - \boldsymbol{\lambda}_2) = 0.$$

Thus $A(\boldsymbol{x}_1 - \boldsymbol{x}_2) = 0$ and also $C(\boldsymbol{x}_1 - \boldsymbol{x}_2) = 0$, which means, since rank $\left(\begin{smallmatrix} A \\ C \end{smallmatrix}\right) = n$, that $\boldsymbol{x}_1 = \boldsymbol{x}_2$, which is what we wanted to prove.

If both A and C have full rank, we may make use of the block structure of the matrix. Consider the ansatz

$$\begin{pmatrix} A^\mathsf{T} A & C^\mathsf{T} \\ C & 0 \end{pmatrix} = \begin{pmatrix} R^\mathsf{T} & 0 \\ G & -U^\mathsf{T} \end{pmatrix} \begin{pmatrix} R & G^\mathsf{T} \\ 0 & U \end{pmatrix}.$$

Multiplying the right hand side and equating terms we obtain

$$R^\mathsf{T} R = A^\mathsf{T} A \text{ thus } \texttt{R = chol(A'*A)}$$

$$R^\mathsf{T} G^\mathsf{T} = C^\mathsf{T} \text{ or } GR = C \text{ thus } G = CR^{-1}$$

and

$$GG^\mathsf{T} - U^\mathsf{T} U = 0 \text{ thus } \texttt{U = chol(G*G')}.$$

The whole algorithm becomes:

1. Compute the Cholesky decomposition $R^\mathsf{T} R = A^\mathsf{T} A$.

2. Solve for G^T by forward substituting $R^\mathsf{T} G^\mathsf{T} = C^\mathsf{T}$.

3. Compute the Cholesky decomposition $U^\mathsf{T} U = GG^\mathsf{T}$.

4. Solve for $\boldsymbol{y}_1$ and $\boldsymbol{y}_2$ by forward substitution,

$$\begin{pmatrix} R^\mathsf{T} & 0 \\ G & -U^\mathsf{T} \end{pmatrix} \begin{pmatrix} \boldsymbol{y}_1 \\ \boldsymbol{y}_2 \end{pmatrix} = \begin{pmatrix} A^\mathsf{T} \boldsymbol{b} \\ \boldsymbol{d} \end{pmatrix}.$$

5. Solve for $\boldsymbol{x}$ and $\boldsymbol{\lambda}$ by backward substitution,

$$\begin{pmatrix} R & G^{\mathsf{T}} \\ 0 & U \end{pmatrix} \begin{pmatrix} \boldsymbol{x} \\ \boldsymbol{\lambda} \end{pmatrix} = \begin{pmatrix} \boldsymbol{y}_1 \\ \boldsymbol{y}_2 \end{pmatrix}.$$

Compared to solving the Normal Equations (6.48) with Gaussian elimination, we save $(n+p)^3/6$ multiplications using this decomposition. This saving may be not significant for large m, because the dominant computational cost in this case is in forming the matrix $A^{\mathsf{T}}A$, which requires $mn^2/2$ multiplications. Just as in the unconstrained case, the above algorithm should not be used to solve Problem (6.44) numerically, since forming $A^{\mathsf{T}}A$ and GG^{T} is numerically not advisable.

6.6.3 Direct Elimination of the Constraints

We will assume that $C \in \mathbb{R}^{p \times n}$ has full rank and therefore $C\boldsymbol{x} = \boldsymbol{d}$ is consistent. Applying Gaussian elimination with *column pivoting*, we obtain the decomposition

$$CP = L[R, \ F]$$

with P a permutation matrix, L a unit lower triangular matrix, and R an upper triangular matrix. The constraints become

$$CPP^{\mathsf{T}}\boldsymbol{x} = L[R, \ F]\boldsymbol{y} = \boldsymbol{d} \text{ with } \boldsymbol{y} = P^{\mathsf{T}}\boldsymbol{x}.$$

If we partition $\boldsymbol{y} = [\boldsymbol{y}_1, \boldsymbol{y}_2]^{\mathsf{T}}$, we obtain

$$[R, \ F]\begin{pmatrix} \boldsymbol{y}_1 \\ \boldsymbol{y}_2 \end{pmatrix} = L^{-1}\boldsymbol{d} =: \boldsymbol{w}$$

with the solution $\boldsymbol{y}_2$ arbitrary and $\boldsymbol{y}_1 = R^{-1}(\boldsymbol{w} - F\boldsymbol{y}_2)$. Inserting $\boldsymbol{x} = P\boldsymbol{y}$ in $\|A\boldsymbol{x} - b\|_2$ and partitioning $AP = [A_1, A_2]$, we get the unconstrained problem

$$(A_2 - A_1 R^{-1}F)\boldsymbol{y}_2 \approx \boldsymbol{b} - A_1 R^{-1}\boldsymbol{w}.$$

A compact formulation of the algorithm is the following:

1. Form the combined system

$$\begin{pmatrix} C \\ A \end{pmatrix} \boldsymbol{x} \overset{=}{\approx} \begin{pmatrix} \boldsymbol{d} \\ \boldsymbol{b} \end{pmatrix}.$$

2. Eliminate p variables by Gaussian elimination with column pivoting,

$$\begin{pmatrix} R & F \\ 0 & \tilde{A} \end{pmatrix} \begin{pmatrix} \boldsymbol{y}_1 \\ \boldsymbol{y}_2 \end{pmatrix} \overset{=}{\approx} \begin{pmatrix} \boldsymbol{w} \\ \tilde{\boldsymbol{b}} \end{pmatrix}.$$

3. Continue elimination of $\tilde{A}\boldsymbol{y}_2 \approx \tilde{\boldsymbol{b}}$ with *orthogonal transformations* (Householder or Givens) to get

$$
\begin{pmatrix} R & F \\ 0 & R_2 \\ 0 & 0 \end{pmatrix} \begin{pmatrix} \boldsymbol{y}_1 \\ \boldsymbol{y}_2 \end{pmatrix} \approx \begin{pmatrix} \boldsymbol{w} \\ \boldsymbol{v}_1 \\ \boldsymbol{v}_2 \end{pmatrix}, \text{ with } \tilde{A} = Q\begin{pmatrix} R_2 \\ 0 \end{pmatrix} \text{ and } \begin{pmatrix} \boldsymbol{v}_1 \\ \boldsymbol{v}_2 \end{pmatrix} = Q^\top \tilde{\boldsymbol{b}}.
$$

4. Solve for $\boldsymbol{y}$ by back substitution

$$
\begin{pmatrix} R & F \\ 0 & R_2 \end{pmatrix} \boldsymbol{y} = \begin{pmatrix} \boldsymbol{w} \\ \boldsymbol{v}_1 \end{pmatrix} \text{ and } \boldsymbol{x} = P\boldsymbol{y}.
$$

The algorithm is implemented in the following MATLAB program

ALGORITHM 6.21. *Linearly Constrained Least Squares*

```
function x=LinearlyConstrainedLSQ(A,C,b,d)
% LINEARLYCONSTRAINEDLSQ solves linearly constrained least squares problem
%    x=LinearlyConstrainedLSQ(A,C,b,d); solves ||Ax-b||= min s.t. Cx=d
%    by direct Gaussian elimination. The reduced least squares problem
%    is solved using the standard Matlab \ operator.

[p,n]=size(C); [m,n]=size(A);
CA=[C,d;A,b];                            % augmented matrix
pp=[1:n];                                % permutation vector
for i=1:p,                               % eliminate p unknowns
  [h jmax]=max(abs(CA(i,i:n)));          % with Gaussian elimination
  jmax=i-1+jmax;
  if h==0, error('Matrix C is rank deficient'); end
  if jmax~=i                             % exchange columns
    h=CA(:,i); CA(:,i)=CA(:,jmax); CA(:,jmax)=h;
    zz=pp(i); pp(i)=pp(jmax); pp(jmax)=zz;
  end;
  CA(i+1:p+m,i)=CA(i+1:p+m,i)/CA(i,i);        % elimination
  CA(i+1:p+m,i+1:n+1)=CA(i+1:p+m,i+1:n+1) ...
                  -CA(i+1:p+m,i)*CA(i,i+1:n+1);
end;
y2=CA(p+1:m+p,p+1:n)\CA(p+1:m+p,n+1);         % solve lsq.-problem
y1=triu(CA(1:p,1:p))\(CA(1:p,n+1)-CA(1:p,p+1:n)*y2);
x(pp)=[y1;y2];                                % permute solution
x=x(:);
```

EXAMPLE 6.18. *If we interpolate the following 7 points by an interpolating polynomial of degree 6,*

```
x=[1; 2.5; 3; 5; 13; 18; 20];
y=[2;   3; 4; 5;  7;  6;  3];
```

```
plot(x,y,'o'); hold;
xx=1:0.1:20;
P=polyfit(x,y,6);                    % fit degree 6 polynomial
plot(xx, polyval(P,xx),':')
pause
```

*we obtain the dashed curve shown in Figure 6.5. The interpolation is really
not what one would like. We can obtain a smoother interpolation for example*

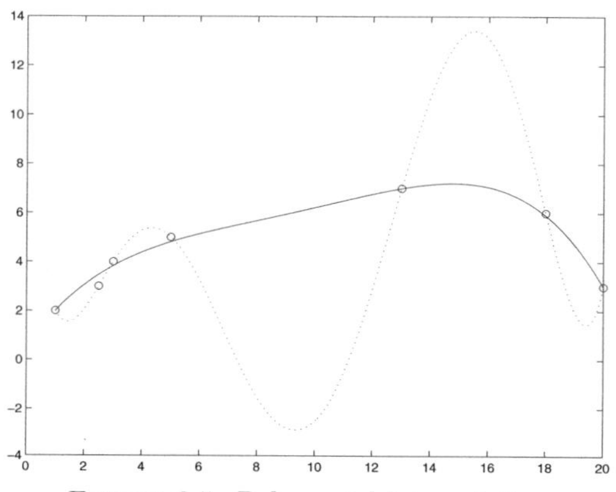

FIGURE 6.5. *Polynomial Interpolation*

*by giving up the interpolation condition, or maybe by demanding interpolation
only for a few points and using a least squares fit for the remaining ones.*

 In MATLAB, *a polynomial of degree d with coefficients* $\boldsymbol{p}$ *is represented as*

$$P_d(x) = p_1 x^d + p_2 x^{d-1} + \cdots + p_d x + p_{d+1}.$$

The interpolation and approximation conditions $P_d(x_i) = y_i$ *resp.* $P_d(x_i) \approx y_i$
lead to the constrained least squares problem

$$V\boldsymbol{p} \simeq \boldsymbol{y},$$

with the $m \times (d+1)$ *Vandermonde matrix* $V = (v_{ij})$ *with* $v_{ij} = x_i^{d-j+1}$. *We
now choose the degree* $d = 4$ *and interpolate* $p = 3$ *points, the first, the last
and the fifth:*

```
m=length(x); n=5;                    % choose degree 4
V=vander(x); V=V(:,m-n+1:m);
p=3;                                 % number of interpolating points
```

*We reorder the equations so that the first 3 are the ones with the interpolation
conditions:*

```
in=[1 5 7 2 3 4 6];                  % permute equations
```

```
Vp=V(in,:); yp=y(in);
C=Vp(1:p,:); A=Vp(p+1:m,:);
d=yp(1:p); b=yp(p+1:m);
P1=LinearlyConstrainedLSQ(A,C,b,d);
plot(xx, polyval(P1,xx))
```

As we can see from Figure 6.5, we obtain this time a much more satisfactory representation of the data. Comparing the function values VP_1 with $\boldsymbol{y}$, we spot the three interpolation points and see that the others are approximated in the least squares sense.

```
>> [V*P1 y]
ans =
    2.0000    2.0000
    3.4758    3.0000
    3.8313    4.0000
    4.8122    5.0000
    7.0000    7.0000
    5.9036    6.0000
    3.0000    3.0000
```

6.6.4   Null Space Method

There are several possibilities to avoid the Normal Equations (6.48) by direct elimination of the constraints. Since we assume that C has full rank p, we can express the general solution of $C\boldsymbol{x} = \boldsymbol{d}$ using the QR decomposition instead of the SVD as in Section 6.6.1.

We compute the QR decomposition of $C^{\top}$,

$$C^{\top} = [Q_1, Q_2]\begin{pmatrix} R \\ 0 \end{pmatrix}, \quad R \in \mathbb{R}^{p \times p}, Q_1 \in \mathbb{R}^{n \times p}.$$

Then the columns of Q_2 span the *null space* of $C^{\top}$: $\mathcal{N}(C) = \mathcal{R}(Q_2)$. With $Q = [Q_1, Q_2]$ and the new unknowns $\boldsymbol{y} = Q^{\top}\boldsymbol{x}$, the constraints become

$$C\boldsymbol{x} = CQ\boldsymbol{y} = [R^{\top}, 0]\boldsymbol{y} = R^{\top}\boldsymbol{y}_1 = \boldsymbol{d}, \quad \text{with} \quad \boldsymbol{y} = \begin{pmatrix} \boldsymbol{y}_1 \\ \boldsymbol{y}_2 \end{pmatrix}.$$

The general solution of the constraints is $\boldsymbol{y}_1 = R^{-\top}\boldsymbol{d}$ and $\boldsymbol{y}_2$ arbitrary. Introducing

$$A\boldsymbol{x} = AQQ^{\top}\boldsymbol{x} = AQ\begin{pmatrix} \boldsymbol{y}_1 \\ \boldsymbol{y}_2 \end{pmatrix} = A(Q_1\boldsymbol{y}_1 + Q_2\boldsymbol{y}_2)$$

into $\|A\boldsymbol{x} - \boldsymbol{b}\|_2$, we get an unconstrained least squares problem

$$\|AQ_2\boldsymbol{y}_2 - (\boldsymbol{b} - AQ_1\boldsymbol{y}_1)\|_2 \longrightarrow \min. \tag{6.50}$$

Thus we obtain the algorithm:

1. compute the QR decomposition $C^{\top} = [Q_1, Q_2]\begin{pmatrix} R \\ 0 \end{pmatrix}$.

2. Compute $\boldsymbol{y}_1$ by forward substitution $R^\top \boldsymbol{y}_1 = \boldsymbol{d}$ and $\boldsymbol{x}_1 = Q_1 \boldsymbol{y}_1$.

3. Form $\tilde{A} = AQ_2$ and $\tilde{\boldsymbol{b}} = \boldsymbol{b} - A\boldsymbol{x}_1$.

4. Solve $\tilde{A}\boldsymbol{y}_2 \approx \tilde{\boldsymbol{b}}$.

5. $\boldsymbol{x} = Q \begin{pmatrix} \boldsymbol{y}_1 \\ \boldsymbol{y}_2 \end{pmatrix} = \boldsymbol{x}_1 + Q_2 \boldsymbol{y}_2$.

ALGORITHM 6.22. *Null Space Method*

```
function x=NullSpaceMethod(A,C,b,d);
% NULLSPACEMETHOD solves a constrained least squares problem
%    x=NullSpaceMethod(A,C,b,d) solves the constrained least squares
%    problem ||Ax-b||=min s.t. Cx=d using the nullspace method.

[p n]=size(C);
[Q R]=qr(C');
y1=R(1:p,1:p)'\d;
x1=Q(:,1:p)*y1;
y2=(A*Q(:,p+1:n))\(b-A*x1);
x=x1+Q(:,p+1:n)*y2;
```

6.7 Special Linear Least Squares Problems with Quadratic Constraint

The SVD can be used very effectively to solve a very *particular least squares problem with a quadratic constraint,* as shown in the following theorem.

THEOREM 6.17. *Let* $A = U\Sigma V^\top$. *Then the problem*

$$\|A\boldsymbol{x}\|_2 \longrightarrow \min, \quad \text{subject to } \|\boldsymbol{x}\|_2 = 1 \tag{6.51}$$

has the solution $\boldsymbol{x} = \boldsymbol{v}_n$ *and the value of the minimum is* $\min_{\|\boldsymbol{x}\|_2=1} \|A\boldsymbol{x}\|_2 = \sigma_n$.

PROOF. We make use of the fact that for orthogonal V and $V^\top \boldsymbol{x} = \boldsymbol{y}$ we have $\|\boldsymbol{x}\|_2 = \|VV^\top \boldsymbol{x}\|_2 = \|V\boldsymbol{y}\|_2 = \|\boldsymbol{y}\|_2$:

$$\min_{\|\boldsymbol{x}\|_2=1} \|A\boldsymbol{x}\|_2^2 = \min_{\|\boldsymbol{x}\|_2=1} \|U\Sigma V^\top \boldsymbol{x}\|_2^2 = \min_{\|VV^\top \boldsymbol{x}\|_2=1} \|U\Sigma(V^\top \boldsymbol{x})\|_2^2$$

$$= \min_{\|\boldsymbol{y}\|_2=1} \|\Sigma \boldsymbol{y}\|_2^2 = \min_{\|\boldsymbol{y}\|_2=1} (\sigma_1^2 y_1^2 + \cdots + \sigma_n^2 y_n^2) \geq \sigma_n^2$$

The minimum is attained for $\boldsymbol{y} = \boldsymbol{e}_n$ thus for $\boldsymbol{x} = V\boldsymbol{y} = \boldsymbol{v}_n$. $\quad\square$

Such minimization problems appear naturally in a variety of geometric fitting problems, as we show in the following subsections.

6.7.1   Fitting Lines

We consider the problem of fitting lines by minimizing the sum of squares of the distances to given points (see Chapter 6 in [45]). In the plane we can represent a straight line uniquely by the equations

$$c + n_1 x + n_2 y = 0, \quad n_1^2 + n_2^2 = 1. \tag{6.52}$$

The unit vector (n_1, n_2) is the normal vector orthogonal to the line. A point is on the line if its coordinates (x, y) satisfy the first equation. On the other hand, if $P = (x_P, y_P)$ is some point not on the line and we compute

$$r = c + n_1 x_P + n_2 y_P,$$

then $|r|$ is its distance from the line. Therefore if we want to determine the line for which the sum of squares of the distances to given points is minimal, we have to solve the constrained least squares problem

$$\begin{pmatrix} 1 & x_{P_1} & y_{P_1} \\ 1 & x_{P_2} & y_{P_2} \\ \vdots & \vdots & \vdots \\ 1 & x_{P_m} & y_{P_m} \end{pmatrix} \begin{pmatrix} c \\ n_1 \\ n_2 \end{pmatrix} \approx \begin{pmatrix} 0 \\ 0 \\ \vdots \\ 0 \end{pmatrix} \quad \text{subject to } n_1^2 + n_2^2 = 1. \tag{6.53}$$

Let A be the matrix of the linear system (6.53). Using the QR decomposition $A = QR$, we can multiply by $Q^\top$ on the left and reduce the linear system to $R\mathbf{x} \approx 0$, i.e., the problem becomes

$$\begin{pmatrix} r_{11} & r_{12} & r_{13} \\ 0 & r_{22} & r_{23} \\ 0 & 0 & r_{33} \end{pmatrix} \begin{pmatrix} c \\ n_1 \\ n_2 \end{pmatrix} \approx \begin{pmatrix} 0 \\ 0 \\ 0 \end{pmatrix} \quad \text{subject to } n_1^2 + n_2^2 = 1. \tag{6.54}$$

Since the nonlinear constraint only involves two unknowns, we only have to solve

$$\begin{pmatrix} r_{22} & r_{23} \\ 0 & r_{33} \end{pmatrix} \begin{pmatrix} n_1 \\ n_2 \end{pmatrix} \approx \begin{pmatrix} 0 \\ 0 \end{pmatrix}, \quad \text{subject to} \quad n_1^2 + n_2^2 = 1. \tag{6.55}$$

The solution $\boldsymbol{n}$ is obtained using Theorem 6.17. We then obtain c by inserting the values into the first equation of (6.54).

    The quadratically constrained least squares problem

$$A\begin{pmatrix} \boldsymbol{c} \\ \boldsymbol{n} \end{pmatrix} \approx \begin{pmatrix} 0 \\ 0 \end{pmatrix}, \quad \text{subject to} \quad \|\boldsymbol{n}\|_2 = 1$$

is therefore solved by the following MATLAB function:

ALGORITHM 6.23.
Quadratically Constrained Linear Least Squares

```
function [c,n]=ConstrainedLSQ(A,dim);
% CONSTRAINEDLSQ solves a constraint least squares problem
%    [c,n]=ConstrainedLSQ(A,dim) solves the constrained least squares
%    problem A (c n)' ~ 0 subject to norm(n,2)=1, dim=length(n)

[m,p]=size(A);
if p<dim+1, error('not enough unknowns'); end;
if m<dim, error('not enough equations'); end;
m=min(m,p);
R=triu(qr(A));
[U,S,V]=svd(R(p-dim+1:m,p-dim+1:p));
n=V(:,dim);
c=-R(1:p-dim,1:p-dim)\R(1:p-dim,p-dim+1:p)*n;
```

Fitting two Parallel Lines

Suppose we wish to determine two parallel lines by measuring the coordinates
of points that lie on them. The measurements are given as two sets of points
$\{P_i\}, i = 1, \ldots, p$, and $\{Q_j\}, j = 1, \ldots, q$, and our goal is to find a pair of
parallel lines that best fit the two sets. Since the lines are parallel, their
normal vector must be the same. Thus the equations for the lines are

$$
\begin{aligned}
c_1 + n_1 x + n_2 y &= 0, \\
c_2 + n_1 x + n_2 y &= 0, \\
n_1^2 + n_2^2 &= 1.
\end{aligned}
$$

If we insert the coordinates of the two sets of points into these equations we
get the following constrained least squares problem for the four unknowns
n_1, n_2, c_1 and c_2:

$$
\begin{pmatrix}
1 & 0 & x_{P_1} & y_{P_1} \\
1 & 0 & x_{P_2} & y_{P_2} \\
\vdots & \vdots & \vdots & \vdots \\
1 & 0 & x_{P_p} & y_{P_p} \\
0 & 1 & x_{Q_1} & y_{Q_1} \\
0 & 1 & x_{Q_2} & y_{Q_2} \\
\vdots & \vdots & \vdots & \vdots \\
0 & 1 & x_{Q_q} & y_{Q_q}
\end{pmatrix}
\begin{pmatrix}
c_1 \\ c_2 \\ n_1 \\ n_2
\end{pmatrix}
\approx
\begin{pmatrix}
0 \\ 0 \\ \vdots \\ 0
\end{pmatrix}
\quad \text{subject to } n_1^2 + n_2^2 = 1.
$$

$$(6.56)$$

Again, we can use our function `ConstrainedLSQ` to solve this problem.

EXAMPLE 6.19. *In the following program some measured points are defined and two parallel lines are fitted and plotted.*

```
Px=[1:10]'
Py=[0.2 1.0 2.6 3.6 4.9 5.3 6.5 7.8 8.0 9.0]'
```

```
Qx=[1.5 2.6 3.0 4.3 5.0 6.4 7.6 8.5 9.9 ]'
Qy=[5.8 7.2 9.1 10.5 10.6 10.7 13.4 14.2 14.5]'
A=[ones(size(Px)) zeros(size(Px)) Px Py
  zeros(size(Qx)) ones(size(Qx)) Qx Qy]
[c,n]=ConstrainedLSQ(A,2)
clf; hold on;
axis([-1 11 -1 17])
PlotLine(Px,Py,'o',c(1),n,'-')
PlotLine(Qx,Qy,'+',c(2),n,'-')
hold off;
```

We have used the function `PlotLine` *to plot a line through given points.*

ALGORITHM 6.24. *Plot a line through points*

```
function PlotLine(x,y,s,c,n,t)
% PLOTLINE plots a line through a set of points
%   PlotLine(x,y,s,c,n,t) plots the set of points (x,y) using the
%   symbol s and plots the line c+n1*x+n2*y=0 using the line type
%   defined by t

plot(x,y,s)
xrange=[min(x) max(x)]; yrange=[min(y) max(y)];
if n(1)==0,                                % c+n2*y=0  => y=-c/n(2)
  x1=xrange(1); y1=-c/n(2);
  x2=xrange(2); y2=y1
elseif n(2)==0,                            % c+n1*x=0  => x=-c/n(1)
  y1=yrange(1); x1=-c/n(1);
  y2=yrange(2); x2=x1;
elseif xrange(2)-xrange(1)>yrange(2)-yrange(1),
  x1=xrange(1); y1=-(c+n(1)*x1)/n(2);
  x2=xrange(2); y2=-(c+n(1)*x2)/n(2);
else
  y1=yrange(1); x1=-(c+n(2)*y1)/n(1);
  y2=yrange(2); x2=-(c+n(2)*y2)/n(1);
end
plot([x1,x2],[y1,y2],t)
```

The results obtained by the program `mainparallel` *are the two lines*

$$0.5091 - 0.7146x + 0.6996y = 0,$$
$$-3.5877 - 0.7146x + 0.6996y = 0,$$

which are plotted in Figure 6.6.

6.7.2 Fitting Ellipses

We want to fit ellipses to measured points by minimizing the *algebraic distance* (see [146]). The solutions $\boldsymbol{x} = [x_1, x_2]$ of a quadratic equation

$$\boldsymbol{x}^\top A \boldsymbol{x} + \boldsymbol{b}^\top \boldsymbol{x} + c = 0 \tag{6.57}$$

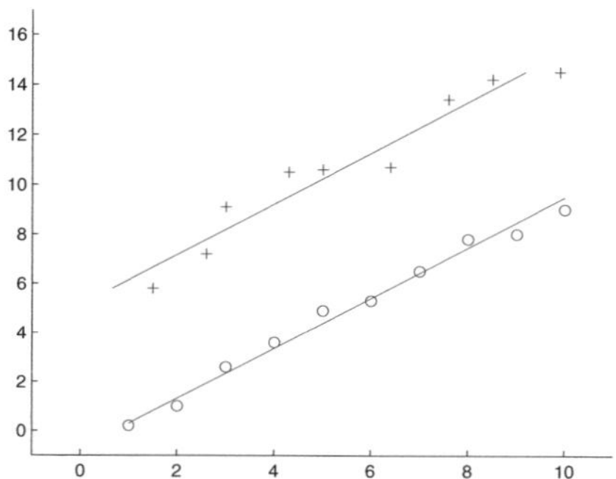

FIGURE 6.6. *Two parallel lines through measured points*

are points on an ellipse if A is symmetric and positive or negative definite (i.e. if $\det(A) = a_{11}a_{22} - a_{12}^2 > 0$). For each measured point, we substitute its coordinates $\boldsymbol{x}_i$ into (6.57) to obtain an equation for the unknown coefficients $\boldsymbol{u} = [a_{11}, a_{12}, a_{22}, b_1, b_2, c]$. Note that $a_{12} = a_{21}$ because of symmetry. Since (6.57) is homogeneous in the coefficients, we need some normalizing condition in order to make the solution unique. A possibility that can handle all cases is to normalize the coefficients by $\|\boldsymbol{u}\|_2 = 1$. We then obtain a quadratically constrained least squares problem

$$\|B\boldsymbol{u}\|_2 \longrightarrow \min \quad \text{subject to } \|\boldsymbol{u}\|_2 = 1.$$

If the rows of the matrix X contain the coordinates of the measured points then the matrix B is computed by the MATLAB statement

```
B=[X(:,1).^2  X(:,1).*X(:,2) X(:,2).^2 X(:,1) X(:,2) ones(size(X(:,1)))]
```

The solution $\boldsymbol{u} = [a_{11}, a_{12}, a_{22}, b_1, b_2, c]$ is obtained using Theorem 6.17. When the coefficients $\boldsymbol{u}$ are known, we can compute the geometric quantities, the axes and the center point by a *principle axis transformation*. To find a coordinate system in which the axes of the ellipse are parallel to the coordinate axes, we compute the eigenvalue decomposition

$$A = QDQ^\top, \quad \text{with } Q \text{ orthogonal and } D = \mathrm{diag}(\lambda_1, \lambda_2). \tag{6.58}$$

Introducing the new variable $\bar{\boldsymbol{x}} = Q^\top \boldsymbol{x}$, (6.57) becomes

$$\bar{\boldsymbol{x}}^\top Q^\top A Q \bar{\boldsymbol{x}} + (Q^\top \boldsymbol{b})^\top \bar{\boldsymbol{x}} + c = 0,$$

or, written in components (using $\bar{\boldsymbol{b}} = Q^\top \boldsymbol{b}$),

$$\lambda_1 \bar{x}_1^2 + \lambda_2 \bar{x}_2^2 + \bar{b}_1 \bar{x}_1 + \bar{b}_2 \bar{x}_2 + c = 0.$$

We transform this equation into its "normal form"

$$\frac{(\bar{x}_1 - \bar{z}_1)^2}{a^2} + \frac{(\bar{x}_2 - \bar{z}_2)^2}{b^2} = 1.$$

Here, $(\bar{z}_1, \bar{z}_2)$ is the center in the rotated coordinate system,

$$(\bar{z}_1, \bar{z}_2) = \left(-\frac{\bar{b}_1}{2\lambda_1}, -\frac{\bar{b}_2}{2\lambda_2}\right).$$

The axes are given by

$$a = \sqrt{\frac{\bar{b}_1^2}{4\lambda_1^2} + \frac{\bar{b}_2^2}{4\lambda_1\lambda_2} - \frac{c}{\lambda_1}}, \quad b = \sqrt{\frac{\bar{b}_1^2}{4\lambda_1\lambda_2} + \frac{\bar{b}_2^2}{4\lambda_2^2} - \frac{c}{\lambda_2}}.$$

To obtain the center in the unrotated system, we have to apply the change of coordinates $z = Q\bar{z}$. We now have all the elements to write a MATLAB function that fits an ellipse to measured points:

ALGORITHM 6.25. *Algebraic Ellipse Fit*

```
function [z,a,b,alpha]=AlgebraicEllipseFit(X);
% ALGEBRAICELLIPSEFIT ellipse fit, minimizing the algebraic distance
%   [z,a,b,alpha]=AlgebraicEllipseFit(X) fits an ellipse by minimizing
%   the algebraic distance to given points P_i=[X(i,1), X(i,2)] in the
%   least squares sense x'A x + bb'x + c=0. z is the center, a,b are
%   the main axes and alpha the angle between a and the x-axis.

[U S V]=svd([X(:,1).^2  X(:,1).*X(:,2) X(:,2).^2 ...
    X(:,1) X(:,2) ones(size(X(:,1)))]);
u=V(:,6); A=[u(1) u(2)/2; u(2)/2 u(3)];
bb=[u(4); u(5)]; c=u(6);
[Q,D]=eig(A);
alpha=atan2(Q(2,1),Q(1,1));
bs=Q'*bb; zs=-(2*D)\bs; z=Q*zs;
h=-bs'*zs/2-c; a=sqrt(h/D(1,1)); b=sqrt(h/D(2,2));
```

To plot an ellipse, it is best to use polar coordinates:

ALGORITHM 6.26. *Drawing an Ellipse*

```
function DrawEllipse(C,a,b,alpha)
% DRAWELLIPSE plots an ellipse
%   DrawEllipse(C,a,b,alpha) plots ellipse with center C, semiaxis a
%   and b and angle alpha between a and the x-axis

s=sin(alpha); c=cos(alpha);
Q =[c -s; s c]; theta=[0:0.02:2*pi];
u=diag(C)*ones(2,length(theta)) + Q*[a*cos(theta); b*sin(theta)];
plot(u(1,:),u(2,:));
plot(C(1),C(2),'+');
```

EXAMPLE 6.20. *We run the following program to fit an ellipse and generate Figure 6.7:*

```
X =[-2.8939    4.1521
    -2.0614    2.1684
    -0.1404    1.9764
     2.6772    3.0323
     5.1746    5.7199
     3.2535    8.1196
    -0.1724    6.8398 ]
axis([0 10 0 10]); axis('equal'); hold
plot(X(:,1),X(:,2),'o');
[z,a,b,alpha]=AlgebraicEllipseFit(X)
DrawEllipse(z,a,b,alpha)
```

We obtain the results

```
z = 1.2348
    4.9871
a = 2.3734
b = 4.6429
alpha = 2.0849
```

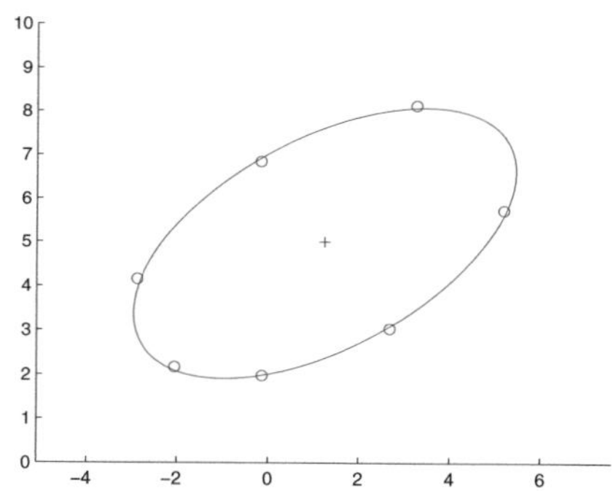

FIGURE 6.7. *Fitting an Ellipse to Measured Points*

6.7.3   Fitting Hyperplanes, Collinearity Test

The function `ConstrainedLSQ` can be used to fit an $(n-1)$-dimensional hyperplane in $\mathbb{R}^n$ to given points. Let the rows of the matrix $X = [\boldsymbol{x}_1, \boldsymbol{x}_2, \ldots, \boldsymbol{x}_m]^\top$ contain the coordinates of the given points, i.e. point P_i has the coordinates $\boldsymbol{x}_i = X(i,:)$, $i = 1, \ldots, m$. Then the call

```
[c, N] = ConstrainedLSQ ([ones(m,1) X], n);
```

determines the hyperplane in normal form $c + N_1 y_1 + N_2 y_2 + \ldots + N_n y_n = 0$.

In this section, we show how we can also compute best-fit hyperplanes of lower dimensions s, where $1 \leq s \leq n - 1$. We follow the theory developed in [127] and also published in [45]. An s-dimensional hyperplane α in $\mathbb{R}^n$ can be represented in parametric form,

$$\alpha: \quad y = p + a_1 t_1 + a_2 t_2 + \cdots + a_s t_s = p + At. \tag{6.59}$$

In this equation, p is a point on the plane and a_i are linearly independent direction vectors, thus the hyperplane is determined by the parameters p and $A = [a_1, \ldots, a_s]$.

Without loss of generality, we assume that A has orthonormal columns, i.e., $A^\top A = I_s$. If we now want to fit a hyperplane to the given set of points X, then we have to minimize the distance of the points to the plane. The distance d_i of a point $P_i = x_i$ to the hyperplane is given by

$$d_i = \min_t \|p - x_i + At\|_2.$$

To determine the minimum, we solve $\nabla d_i^2 = 2A^\top (p - x_i + At) = 0$ for t, and, since $A^\top A = I_s$, we obtain

$$t = A^\top (x_i - p). \tag{6.60}$$

Therefore the distance becomes

$$d_i^2 = \|p - x_i + AA^\top (x_i - p)\|_2^2 = \|P(x_i - p)\|_2^2,$$

where we denoted by $P = I - AA^\top$ the projector onto the orthogonal complement of the range of A, i.e. onto the null space of $A^\top$.

Our objective is to minimize the sum of squares of the distances of all points to the hyperplane, i.e., we want to minimize the function

$$F(p, A) = \sum_{i=1}^{m} \|P(x_i - p)\|_2^2. \tag{6.61}$$

A necessary condition is $\nabla F = 0$. We first consider the first part of the gradient, the partial derivative

$$\frac{\partial F}{\partial p} = -\sum_{i=1}^{m} 2P^\top P(x_i - p) = -2P(\sum_{i=1}^{m} x_i - mp) = 0,$$

where we made use of the property of an orthogonal projector $P^\top P = P^2 = P$. Since P projects the vector $\sum_{i=1}^{m} x_i - mp$ onto 0, this vector must be in the range of A, i.e.

$$p = \frac{1}{m} \sum_{i=1}^{m} x_i + A\tau. \tag{6.62}$$

Inserting this expression into (6.61) and noting that $PA = 0$, the objective function to be minimized simplifies to

$$G(A) = \sum_{i=1}^{m} \|P\hat{\boldsymbol{x}}_i\|_2^2 = \|P\hat{X}^\top\|_F^2, \qquad (6.63)$$

where we put

$$\hat{\boldsymbol{x}}_i = \boldsymbol{x}_i - \frac{1}{m}\sum_{i=1}^{m}\boldsymbol{x}_i,$$

and where we used the *Frobenius norm* of a matrix, $\|A\|_F^2 := \sum_{i,j} a_{ij}^2$, see Subsection 2.5.1. Now since P is symmetric, we may also write

$$G(A) = \|\hat{X}P\|_F^2 = \|\hat{X}(I - AA^\top)\|_F^2 = \|\hat{X} - \hat{X}AA^\top\|_F^2. \qquad (6.64)$$

If we define $Y := \hat{X}AA^\top$, which is a matrix of rank s, then we can consider the problem of minimizing

$$\|\hat{X} - Y\|_F^2 \longrightarrow \min, \quad \text{subject to} \quad \text{rank}(Y) = s. \qquad (6.65)$$

Problem (6.65) is similar to the problem solved in Theorem 6.6. The difference is that now we want to minimize the Frobenius norm and not the 2-norm.

THEOREM 6.18. *Let $A \in \mathbb{R}^{m \times n}$ have rank r and let $A = U\Sigma V^\top$. Let $\mathcal{M}$ denote the set of $m \times n$ matrices with rank $p < r$. A solution of*

$$\min_{X \in \mathcal{M}} \|A - X\|_F^2$$

is given by $A_p := \sum_{i=1}^{p} \sigma_i \boldsymbol{u}_i \boldsymbol{v}_i^\top$ and we have

$$\min_{X \in \mathcal{M}} \|A - X\|_F^2 = \|A - A_p\|_F^2 = \sigma_{p+1}^2 + \cdots + \sigma_n^2.$$

To prove the above theorem, we first need a lemma due to Weyl [147], which estimates the singular values of the sum of two matrices in terms of those of its summands.

LEMMA 6.3. (WEYL) *Let $A, B \in \mathbb{R}^{m \times n}$. Let $\sigma_i(A)$ denote the ith singular value of A in descending order, and similarly for B. Then for $i + j \leq \min\{m, n\} + 1$, we have*

$$\sigma_{i+j-1}(A + B) \leq \sigma_i(A) + \sigma_j(B).$$

PROOF. Let A_p denote the best rank-i approximation of A, i.e., if A has rank r and the SVD of A is $A = \sum_{k=1}^{r} \sigma_k \boldsymbol{u}_k \boldsymbol{v}_k^\top$, then $A_p = \sum_{k=1}^{p} \sigma_k \boldsymbol{u}_k \boldsymbol{v}_k^\top$. Then by Theorem 6.6, we have $\|A - A_i\|_2 = \sigma_{i+1}$. Let us now consider the matrix $R = A_{i-1} + B_{j-1}$, which has rank $\leq i + j - 2$. By the best approximation property in Theorem 6.6, we have

$$\sigma_{i+j-1}(A + B) = \|A + B - (A + B)_{i+j-2}\|_2 \leq \|A + B - R\|_2$$
$$\leq \|A - A_{i-1}\|_2 + \|B - B_{j-1}\|_2 = \sigma_i(A) + \sigma_j(B).$$

$\square$

PROOF. (Theorem 6.18) Let $\Sigma_p = \text{diag}(\sigma_1, \ldots, \sigma_p, 0, \ldots, 0)$, so that $A_p = U\Sigma_p V^\top$. Then the fact that the Frobenius norm is invariant under orthogonal transformations implies

$$\|A - A_p\|_F^2 = \|\Sigma - \Sigma_p\|_F^2 = \sigma_{p+1}^2 + \cdots + \sigma_n^2.$$

To prove that the choice $X = A_p$ minimizes $\|A - X\|_F^2$, let X be a matrix of rank $p < r$. Since $\sigma_{p+1}(X) = 0$, Lemma 6.3 implies for $i + p \leq n$

$$\sigma_{i+p}(A) = \sigma_{i+p}(A - X + X) \leq \sigma_i(A - X) + \sigma_{p+1}(X) = \sigma_i(A - X).$$

Thus, we have

$$\|A - X\|_F^2 = \sum_{i=1}^{n} \sigma_i^2(A - X) \geq \sum_{i=1}^{n-p} \sigma_i^2(A - X) \geq \sum_{i=p+1}^{n} \sigma_i^2(A).$$

In other words, no other choice of rank p matrix X can lead to a lower Frobenius norm error than A_p. $\square$

If $\hat{X} = U\Sigma V^\top$, then according to Theorem 6.18 the minimizing matrix of Problem (6.65) is given by $Y = U\Sigma_s V^\top$, where

$$\Sigma_s = \text{diag}(\sigma_1, \sigma_2 \ldots, \sigma_s, 0, \ldots, 0).$$

Now if $Y = U\Sigma_s V^\top$, we have to find a matrix A with orthonormal columns such that $\hat{X} A A^\top = Y$. It is easy to verify that if we choose $A = V_1$ where $V_1 = V(:, 1{:}s)$, then $\hat{X} A A^\top = U\Sigma_s V^\top$. Thus, the singular value decomposition of $\hat{X}$ gives us all the lower-dimensional hyperplanes that are best fits of the given set of points:

$$\boldsymbol{y} = \boldsymbol{p} + V_1 \boldsymbol{t}, \quad \text{with} \quad \boldsymbol{p} = \frac{1}{m} \sum_{i=1}^{m} \boldsymbol{x}_i.$$

Notice that $V_2 = V(:, s{+}1{:}n)$ also gives us the normal form of the hyperplane: here, the hyperplane is described as the solution of the linear equations

$$V_2^\top \boldsymbol{y} = V_2^\top \boldsymbol{p}.$$

A measure for the collinearity is the value of the minimum $\sigma_{s+1}^2 + \cdots + \sigma_n^2$.

In order to compute the hyperplanes, we therefore essentially have to compute one singular value decomposition. This is done in Algorithm 6.27.

ALGORITHM 6.27. *Computation of Hyperplanes.*

```
function [V,p]=HyperPlaneFit(X);
% HYPERPLANEFIT fits a hyperplane to a set of given points
%    [V,p]=HyperPlaneFit(X); fits a hyperplane of dimension s<n to a
%    set of given points X(i,:) in R^n. The hyperplane is given by
%    y=p+V(:,1:s)*tau (Parametric Form) or by the equations
%    V(:,s+1:n)'*(y-p)=0 (Normal Form)

m=max(size(X));
p=sum(X)'/m;
Xt=X-ones(size(X))*diag(p);
[U,S,V]=svd(Xt,0);
```

Note that the statement `[U,S,V]` = `svd(Qt,0)` computes the "economy size" singular value decomposition. If `Qt` is an m-by-n matrix with $m > n$, then only the first n columns of U are computed, and S is an n-by-n matrix.

6.7.4 Procrustes or Registration Problem

We consider a least squares problem in *coordinate metrology* (see [5], [15]): m points of a workpiece, called the *nominal points*, are given by their exact coordinates from construction plans when the workpiece is in its *nominal position* in a reference frame. We denote the coordinate vectors of the nominal points in this position by

$$\boldsymbol{x}_1, \ldots, \boldsymbol{x}_m, \quad \boldsymbol{x}_i \in \mathbb{R}^n, \quad 1 \leq n \leq 3.$$

Suppose now that a coordinate measuring machine gathers the same points of another workpiece. The machine records the coordinates of the *measured points*

$$\boldsymbol{\xi}_1, \ldots, \boldsymbol{\xi}_m, \quad \boldsymbol{\xi}_i \in \mathbb{R}^n, \quad 1 \leq n \leq 3,$$

which will be in a different frame than the frame of reference. The problem we want to solve is to find a frame transformation that maps the given nominal points onto the measured points. This problem is solved in [45].

We need to find a translation vector $\boldsymbol{t}$ and an orthogonal matrix Q with $\det(Q) = 1$ i.e., $Q^{\top}Q = I$ such that

$$\boldsymbol{\xi}_i = Q\boldsymbol{x}_i + \boldsymbol{t}, \quad \text{for } i = 1, \ldots, m. \tag{6.66}$$

For $m > 6$ in 3D-space, Equation (6.66) is an over-determined system of equations and is only consistent if the measurements have no errors. This is not the case for a real machine; therefore our problem is to determine the unknowns Q and $\boldsymbol{t}$ of the least squares problem

$$\boldsymbol{\xi}_i \approx Q\boldsymbol{x}_i + \boldsymbol{t}. \tag{6.67}$$

In the one-dimensional case, we are given two sets of points on the line. The matrix Q is just the constant 1 and we have to determine a scalar t such that

$$\xi_i \approx x_i + t, \quad i = 1, \ldots, m.$$

With the notation $A = (1, \ldots, 1)^\top$, $\boldsymbol{a} = (\xi_1, , \ldots, \xi_m)^\top$ and $\boldsymbol{b} = (x_1, \ldots, x_m)^\top$ the problem becomes

$$At \approx \boldsymbol{a} - \boldsymbol{b}. \tag{6.68}$$

Using the normal equations $A^\top A t = A^\top (\boldsymbol{a} - \boldsymbol{b})$ we obtain $mt = \sum_{i=1}^{m}(\xi_i - x_i)$ and therefore

$$t = \bar{\xi} - \bar{x}, \quad \text{with} \quad \bar{\xi} = \frac{1}{m}\sum_{i=1}^{m}\xi_i \quad \text{and} \quad \bar{x} = \frac{1}{m}\sum_{i=1}^{m}x_i. \tag{6.69}$$

We can generalize this result for $n > 1$. Consider

$$\boldsymbol{\xi}_i \approx \boldsymbol{x}_i + \boldsymbol{t}, \quad i = 1, \ldots, m.$$

In matrix notation, this least squares problem becomes (I is the $n \times n$ identity matrix):

$$\begin{pmatrix} I \\ I \\ \vdots \\ I \end{pmatrix} \boldsymbol{t} \approx \begin{pmatrix} \boldsymbol{\xi}_1 - \boldsymbol{x}_1 \\ \boldsymbol{\xi}_2 - \boldsymbol{x}_2 \\ \vdots \\ \boldsymbol{\xi}_m - \boldsymbol{x}_m \end{pmatrix}.$$

The normal equations are $m\boldsymbol{t} = \sum_{i=1}^{m}(\boldsymbol{\xi}_i - \boldsymbol{x}_i)$ an thus again

$$\boldsymbol{t} = \bar{\boldsymbol{\xi}} - \bar{\boldsymbol{x}}, \quad \text{with} \quad \bar{\boldsymbol{\xi}} = \frac{1}{m}\sum_{i=1}^{m}\boldsymbol{\xi}_i \quad \text{and} \quad \bar{\boldsymbol{x}} = \frac{1}{m}\sum_{i=1}^{m}\boldsymbol{x}_i.$$

Hence, we have shown that *the translation $\boldsymbol{t}$ is the vector connecting the two centers of gravity of the corresponding sets of points.*

Applying this result to the least squares problem for some fixed Q

$$\boldsymbol{\xi}_i \approx Q\boldsymbol{x}_i + \boldsymbol{t} \iff \sum_{i=1}^{m} \|Q\boldsymbol{x}_i + \boldsymbol{t} - \boldsymbol{\xi}_i\|_2^2 \longrightarrow \min, \tag{6.70}$$

we conclude that $\boldsymbol{t}$ is the vector connecting the two centers of gravity of the point sets $\boldsymbol{\xi}_i$ and $Q\boldsymbol{x}_i$, i.e.,

$$\boldsymbol{t} = \bar{\boldsymbol{\xi}} - Q\bar{\boldsymbol{x}}. \tag{6.71}$$

Using (6.71), we eliminate $\boldsymbol{t}$ in (6.70) and consider the problem

$$G(Q) = \sum_{i=1}^{m} \|Q(\boldsymbol{x}_i - \bar{\boldsymbol{x}}) - (\boldsymbol{\xi}_i - \bar{\boldsymbol{\xi}})\|_2^2 \longrightarrow \min. \tag{6.72}$$

Introducing the new coordinates

$$\boldsymbol{a}_i = \boldsymbol{x}_i - \bar{\boldsymbol{x}} \quad \text{and} \quad \boldsymbol{b}_i = \boldsymbol{\xi}_i - \bar{\boldsymbol{\xi}}$$

the problem is:

$$G(Q) = \sum_{i=1}^{m} \|Q\boldsymbol{a}_i - \boldsymbol{b}_i\|_2^2 \longrightarrow \min. \tag{6.73}$$

We can collect the vectors in matrices

$$A = (\boldsymbol{a}_1, \ldots, \boldsymbol{a}_m), \quad \text{and} \quad B = (\boldsymbol{b}_1, \ldots, \boldsymbol{b}_m),$$

where $A, B \in \mathbb{R}^{n \times m}$ and rewrite the function G using the *Frobenius norm*

$$G(Q) = \|QA - B\|_F^2.$$

Since the Frobenius norm of a matrix is the same for the transposed matrix, we finally obtain the *Procrustes problem*[51]: find an orthogonal matrix Q such that

$$\|B^\top - A^\top Q^\top\|_F^2 \longrightarrow \min.$$

The standard form of the Procrustes problem is formulated as follows: given two matrices C and D, both in $\mathbb{R}^{m \times n}$ with $m \geq n$, find an orthogonal matrix $P \in \mathbb{R}^{n \times n}$, such that

$$\|C - DP\|_F^2 \longrightarrow \min. \tag{6.74}$$

Thus, it suffices to let $C = B^\top$ and $D = A^\top$.

To solve the Procrustes problem (6.74), we need some properties of the *Frobenius norm*, which is given by

$$\|A\|_F^2 = \sum_{i=1}^{m}\sum_{j=1}^{n} a_{i,j}^2 = \sum_{j=1}^{n} \|\boldsymbol{a}_j\|_2^2, \quad \text{where } \boldsymbol{a}_j \text{ is the } j\text{th column of } A.$$

$$\tag{6.75}$$

Note that

$$\|A\|_F^2 = \operatorname{tr}(A^\top A) = \sum_{i=1}^{n} \lambda_i(A^\top A). \tag{6.76}$$

Equation (6.76) gives us some useful relations: if P is orthogonal, then $\|PA\|_F = \|A\|_F$. Additionally, since $\|A\|_F = \|A^\top\|_F$, we have $\|AP\|_F = \|A\|_F$.

$$\begin{aligned}
\|A + B\|_F^2 &= \operatorname{tr}((A + B)^\top (A + B)) \\
&= \operatorname{tr}(A^\top A + B^\top A + A^\top B + B^\top B) \\
&= \operatorname{tr}(A^\top A) + 2\operatorname{tr}(A^\top B) + \operatorname{tr}(B^\top B) \\
\|A + B\|_F^2 &= \|A\|_F^2 + \|B\|_F^2 + 2\operatorname{tr}(A^\top B) \tag{6.77}
\end{aligned}$$

We now apply (6.77) to the Procrustes problem:

$$\|C - DP\|_F^2 = \|C\|_F^2 + \|D\|_F^2 - 2\operatorname{tr}(P^\mathsf{T} D^\mathsf{T} C) \longrightarrow \min.$$

Computing the minimum is equivalent to maximizing

$$\operatorname{tr}(P^\mathsf{T} D^\mathsf{T} C) = \max.$$

Using the singular value decomposition $D^\mathsf{T} C = U\Sigma V^\mathsf{T}$, we obtain

$$\operatorname{tr}(P^\mathsf{T} D^\mathsf{T} C) = \operatorname{tr}(P^\mathsf{T} U\Sigma V^\mathsf{T}).$$

Since U, V are orthogonal, we may write the unknown matrix P in the following form

$$P = UZ^\mathsf{T} V^\mathsf{T}, \quad \text{with } Z \text{ orthogonal.}$$

Because similar matrices have the same trace, it follows that

$$\operatorname{tr}(P^\mathsf{T} D^\mathsf{T} C) = \operatorname{tr}(VZU^\mathsf{T} U\Sigma V^\mathsf{T}) = \operatorname{tr}(VZ\Sigma V^\mathsf{T}) = \operatorname{tr}(Z\Sigma)$$

$$= \sum_{i=1}^{n} z_{ii}\sigma_i \le \sum_{i=1}^{n} |z_{ii}|\sigma_i \le \sum_{i=1}^{n} \sigma_i,$$

where the inequality follows from $|z_{ii}| \le 1$ for any orthogonal matrix Z. Furthermore, the bound is attained for $Z = I$. Notice that if $D^\mathsf{T} C$ is rank deficient, the solution is not unique (cf. [65]). So we have proved the following theorem:

THEOREM 6.19. *The Procrustes problem* (6.74) *is solved by the orthogonal polar factor of* $D^\mathsf{T} C$, *i.e.* $P = UV^\mathsf{T}$ *where* $U\Sigma V^\mathsf{T}$ *is the singular value decomposition of* $D^\mathsf{T} C$.

The *polar decomposition* of a matrix is a generalization of the polar representation of complex numbers. The matrix is decomposed into the product of an orthogonal times a symmetric positive (semi-)definite matrix. The decomposition can be computed by the singular value decomposition or by other algorithms [40]. In our case we have

$$D^\mathsf{T} C = U\Sigma V^\mathsf{T} = \underbrace{UV^\mathsf{T}}_{\text{orthogonal}} \underbrace{V\Sigma V^\mathsf{T}}_{\substack{\text{positive} \\ \text{semidefinite}}}.$$

We are now ready to describe the algorithm to solve the Procrustes problem. Given measured points $\boldsymbol{\xi}_i$ and corresponding nominal points $\boldsymbol{x}_i$ for $i = 1, \ldots, m$. We want to determine $\boldsymbol{t}$ and Q orthogonal such that $\boldsymbol{\xi}_i \approx Q\boldsymbol{x}_i + \boldsymbol{t}$.

  1. Compute the centers of gravity:

$$\bar{\boldsymbol{\xi}} = \frac{1}{m} \sum_{i=1}^{m} \boldsymbol{\xi}_i \quad \text{and} \quad \bar{\boldsymbol{x}} = \frac{1}{m} \sum_{i=1}^{m} \boldsymbol{x}_i.$$

2. Compute the *relative coordinates*:

$$
\begin{aligned}
A &= [\boldsymbol{a}_1,\ldots,\boldsymbol{a}_m], & \boldsymbol{a}_i &= \boldsymbol{x}_i - \bar{\boldsymbol{x}} \\
B &= [\boldsymbol{b}_1,\ldots,\boldsymbol{b}_m], & \boldsymbol{b}_i &= \boldsymbol{\xi}_i - \bar{\boldsymbol{\xi}}
\end{aligned}
$$

3. The Procrustes problem becomes $\|C - DP\|_F^2 \longrightarrow \min$ with $C = B^\top$, $D = A^\top$ and $P = Q^\top$.

   Compute the singular value decomposition $AB^\top = U\Sigma V^\top$.

4. Then $Q^\top = UV^\top$ or $Q = VU^\top$ and $\boldsymbol{t} = \bar{\boldsymbol{\xi}} - Q\bar{\boldsymbol{x}}$.

For technical reasons, it may be important to decompose the orthogonal matrix Q into elementary rotations. The algorithm that we developed so far computes an orthogonal matrix, but there is no guarantee that Q can be represented as a product of rotations and that no reflection occurs. For Q to be representable as a product of rotations, it is necessary and sufficient to have $\det(Q) = 1$. Thus, if $\det(Q) = -1$, then a reflection is necessary and this may be of no practical use. In this case, one would like to find the best orthogonal matrix with $\det(Q) = 1$.

It is shown in [65] that the constrained Procrustes problem

$$
\|C - DP\|_F^2 \longrightarrow \min, \quad \text{subject to} \quad \det(P) = 1
$$

has the solution

$$
P = U \operatorname{diag}(1,\ldots,1,\mu)V^\top,
$$

where $D^\top C = U\Sigma V^\top$ is the singular value decomposition and $\mu = \det(UV^\top)$. The proof is based on the fact that for a real orthogonal $n \times n$ matrix Z with $\det(Z) < 1$, the trace is bounded by

$$
\operatorname{tr}(Z) \le n - 2 \quad \text{and} \quad \operatorname{tr}(Z) = n - 2 \iff \lambda_i(Z) = \{1,\ldots,1,-1\}.
$$

This can be seen by considering the real Schur form [51] of Z. The maximum of $\operatorname{tr}(Z\Sigma)$ is therefore $\sum_{i=1}^{n-1} \sigma_i - \sigma_n$ and is achieved for $Z = \operatorname{diag}(1,\ldots,1,-1)$.

Thus we obtain the MATLAB function `ProcrustesFit`

ALGORITHM 6.28.

```
function [t,Q]=ProcrustesFit(xi,x);
% PROCRUSTESFIT solves the Procrustes Problem
%    [t,Q]=ProcristesFit(xi,x) computes an orthogonal matrix Q and a
%    shift t such that xi=Qx+t

xiq=sum(xi')/length(xi); xiq=xiq';
xq=sum(x')/length(x); xq=xq';
A=x-xq*ones(1,length(x));
B=xi-xiq*ones(1,length(xi));
[U,sigma,V]=svd(A*B');
Q=V*diag([ones(1,size(V,1)-1) det(V*U')])*U';
t=xiq-Q*xq;
```

EXAMPLE 6.21. *As an example, we solve a Procrustes problem for $n = 2$. The following* MATLAB *program defines the nominal points $\boldsymbol{x}_k$ and the measured points $\boldsymbol{\xi}_k$. Then it computes and plots the fitted nominal points on the measurements.*

```
clf, clear
axis([0 10 0 10]), axis('equal')              % nominal points
hold, grid
x=[-4 -4 -2 -2 -2 -4 -4
    1  2  2  3  4  4  3];
plot(x(1,:),x(2,:),'-')                       % measured points
plot(x(1,:),x(2,:),'x')

xi=[-5.2572 -4.5528 -3.6564 -2.8239 -2.0555 -3.1761 -4.2007
     6.1206  6.6969  5.4162  6.0886  6.6329  8.2338  7.8175];
plot(xi(1,:),xi(2,:),'o')
pause
xiq=sum(xi,2)/length(xi);                      % centers of gravity
xq=sum(x,2)/length(x);
[t,Q]=ProcrustesFit(xi,x);
xx=Q*x+t*ones(1,length(x))                     % transform nominal points
plot(xx(1,:), xx(2,:),'-'), plot(xx(1,:), xx(2,:),'*')
```

The results are shown in Figure 6.8

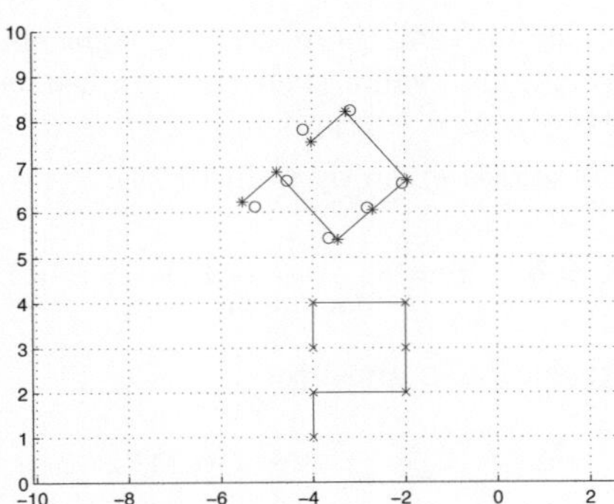

FIGURE 6.8. *Procrustes or Registration Problem*

6.7.5   Total Least Squares

The linear least squares problem $A\boldsymbol{x} \approx \boldsymbol{b}$ has so far been solved by projecting the vector $\boldsymbol{b}$ onto the range of A,

$$A\boldsymbol{x} = P_{\mathcal{R}(A)}\boldsymbol{b} = AA^{+}\boldsymbol{b}.$$

With "Total Least Squares", the system of equations is made consistent by changing both A and $\boldsymbol{b}$: we look for a matrix $\hat{A}$ and a vector $\hat{\boldsymbol{b}} \in \mathcal{R}(\hat{A})$ which differ as little as possible from the given data

$$\|[A, \boldsymbol{b}] - [\hat{A}, \hat{\boldsymbol{b}}]\|_F \longrightarrow \min, \quad \text{subject to} \quad \hat{\boldsymbol{b}} \in \mathcal{R}(\hat{A}).$$

The constraint states that $\hat{C} = [\hat{A}, \hat{\boldsymbol{b}}]$ must have rank n. Since in general $C = [A, \boldsymbol{b}]$ will have rank $n + 1$, our problem involves solving

$$\min_{\text{rank } \hat{C} = n} \|C - \hat{C}\|_F. \tag{6.78}$$

The solution of problem (6.78) is given by Theorem 6.17: let $[A, \boldsymbol{b}] = C = U\Sigma V^\top$ be the SVD. Then

$$[\hat{A}, \hat{\boldsymbol{b}}] = \hat{C} = \sum_{i=1}^{n} \sigma_i \boldsymbol{u}_i \boldsymbol{v}_i^\top = U\hat{\Sigma}V^\top, \quad \text{with} \quad \hat{\Sigma} = \text{diag}(\sigma_1, \ldots, \sigma_n, 0). \tag{6.79}$$

However, the constraint $\hat{\boldsymbol{b}} \in \mathcal{R}(\hat{A})$ is more than just a rank condition: if we define $\hat{C} = C + \Delta$ and write the condition $\hat{\boldsymbol{b}} \in \mathcal{R}(\hat{A})$ as $\hat{C}\boldsymbol{z} = 0$ with $\boldsymbol{z} = \left(\begin{smallmatrix} \boldsymbol{x} \\ -1 \end{smallmatrix}\right) \neq 0$, then the problem is equivalent to

$$\|\Delta\|_F^2 \longrightarrow \min \quad \text{subject to} \quad \exists\, \boldsymbol{x} \in \mathbb{R}^n : \quad (C + \Delta)\begin{pmatrix} \boldsymbol{x} \\ -1 \end{pmatrix} = 0. \tag{6.80}$$

This is equivalent to saying that there exists a right singular vector $\boldsymbol{v}$ corresponding to a zero singular value such that its last component does not vanish. Thus, if $\text{rank}(C) = n + 1$ and $v_{n+1,n+1} \neq 0$, then the total least squares solution exists and is given by $\hat{C}$ in (6.79):

$$\hat{A}\hat{\boldsymbol{x}} = \hat{\boldsymbol{b}}, \quad \hat{\boldsymbol{x}} = -\frac{1}{v_{n+1,n+1}}\boldsymbol{v}(1:n, n+1).$$

In this case, the perturbation is given by

$$\Delta = \hat{C} - C = -\sigma_{n+1}\boldsymbol{u}_{n+1}\boldsymbol{v}_{n+1}^\top.$$

This leads to the following MATLAB function:

ALGORITHM 6.29. *Total Least Squares*

```
function [x,v,At,bt]=TLS(A,b);
% TLS total least squares
%    [x,v,At,bt]=TLS(A,b) solves Ax~b by allowing changes in A and b.

C=[A,b];
[m,n]=size(C);
[U,Sigma,V]=svd(C,0);
```

```
v=V(:,n);
if v(n)==0
  disp('TLS Solution does not exist')
  x=[];
else
  x=-v(1:n-1)/v(n);
end
Ct=C-Sigma(n)*U(:,n)*V(:,n)';
At=Ct(:,1:n-1); bt=Ct(:,n);
```

THEOREM 6.20. (TOTAL LEAST SQUARES) *The total least squares solution satisfies the equation*

$$(A^\top A - \sigma_{n+1}^2 I)\hat{x} = A^\top b. \tag{6.81}$$

PROOF. Since $C = [A, b] = U\Sigma V^\top$ we have $Cv_{n+1} = \sigma_{n+1}u_{n+1}$ and therefore

$$C^\top C v_{n+1} = \sigma_{n+1}C^\top u_{n+1} = \sigma_{n+1}^2 v_{n+1}.$$

Dividing the last equation by $v_{n+1,n+1}$ and replacing $C = [A, b]$ we obtain

$$\left(\begin{array}{c|c} A^\top A & A^\top b \\ \hline b^\top A & b^\top b \end{array} \right) \begin{pmatrix} \hat{x} \\ -1 \end{pmatrix} = \sigma_{n+1}^2 \begin{pmatrix} \hat{x} \\ -1 \end{pmatrix},$$

and the first equation is our claim. $\square$

A variant of total least squares is given if some elements of the matrix A have no errors and therefore should remain unchanged. Let us consider $Ax \approx b$ with $A = [A_1, A_2] \in \mathbb{R}^{m \times n}$ where $A_1 \in \mathbb{R}^{m \times p}$ with $p < n$ has no error. The total least squares problem $\|E\|_F^2 + \|r\|_2^2 \longrightarrow \min$ subject to $(A + E)x = b + r$ becomes

$$\|\Delta\|_F^2 \longrightarrow \min \quad \text{subject to} \quad (C + \Delta)z = 0,$$

with $C = [A_1, A_2, b]$, $\Delta = [0, \Delta_2]$ and $\Delta_2 = [E_2, r]$. Using the QR decomposition of C, we can transform the constraint by pre-multiplying with $Q^\top$,

$$Q^\top(C + \Delta)z = \begin{pmatrix} R_{11} & R_{12} \\ 0 & R_{22} \end{pmatrix} z + \begin{pmatrix} 0 & \tilde{\Delta}_{12} \\ 0 & \tilde{\Delta}_{22} \end{pmatrix} z = 0, \tag{6.82}$$

and $R_{11} \in \mathbb{R}^{p \times p}$ is upper triangular. Now $\|\Delta\|_F^2 = \|Q^\top \Delta\|_F^2 = \|\tilde{\Delta}_{12}\|_F^2 + \|\tilde{\Delta}_{22}\|_F^2$ is minimized if we choose $\tilde{\Delta}_{12} = 0$ and minimize $\|\tilde{\Delta}_{22}\|_F$. So the algorithm for *constrained total least squares* is:

1. Compute $C = [A_1, A_2, b] = QR$ and reduce the constraint to (6.82).

2. Compute $\hat{v}$ the solution of the total least squares problem $(R_{22} + \tilde{\Delta}_{22})v = 0$.

3. Solve $R_{11}\boldsymbol{u} = -R_{12}\hat{\boldsymbol{v}}$ for $\boldsymbol{u}$.

4. $\boldsymbol{z} = \begin{pmatrix} \boldsymbol{u} \\ \hat{\boldsymbol{v}} \end{pmatrix} \in \mathbb{R}^{n+1}$ and $\boldsymbol{x} = -[z_1, \ldots, z_n]^\top / z_{n+1}$.

ALGORITHM 6.30. *Constrained Total Least Squares*

```
function  [z,x]=ConstrainedTLS(C,p);
% CONSTRAINEDTLS computes constrained total least squares solution
%    [z,x]=ConstrainedTLS(C,p); computes the constrained total least
%    squares solution with C=[A_1 A_2 b], where A1=C(:,1:p) is without
%    errors. One therefore solves [A_1 A_2]x ~b by modifying A2 and b
%    such that [A_1 A_2 b]z=0 or [A_1 A_2] x=b

[m,n]=size(C);
[Q,R]=qr(C);
R11=R(1:p,1:p); R12=R(1:p,p+1:n);
R22=R(p+1:n,p+1:n);
[m1,n1]=size(R22);
[x,v]=TLS(R22(:,1:n1-1),R22(:,n1));
u=-R11\R12*v;
z=[u;v]; x=-z(1:n-1)/z(n);
```

EXAMPLE 6.22. *We want to fit a 3-dimensional hyperplane in $\mathbb{R}^4$,*

$$z = c + n_1 x_1 + n_2 x_2 + n_3 x_3.$$

We shall simulate measured points, compute the coefficients c, n_1, n_2, n_3 and compare the results for our least squares methods.

The columns of the matrix $X^\top$ contain the coordinates of the "measurements" for x_1, x_2 and x_3. For simplicity we chose a grid of integer coordinates,

$$X^\top = \begin{pmatrix} 1 & 1 & 1 & 1 & 2 & 2 & 2 & 2 & 3 & 3 & 3 & 3 \\ 1 & 1 & 2 & 2 & 1 & 1 & 2 & 2 & 1 & 1 & 2 & 2 \\ 1 & 2 & 1 & 2 & 1 & 2 & 1 & 2 & 1 & 2 & 1 & 2 \end{pmatrix}.$$

We choose $c = 3$, $n_1 = -1$, $n_2 = 2$, $n_3 = 5$, add a first column of ones to X and compute the exact values

$$\boldsymbol{z} = \begin{pmatrix} \mathbf{1}, & X \end{pmatrix} \begin{pmatrix} c \\ n_1 \\ n_2 \\ n_3 \end{pmatrix}.$$

Then we add some random noise to the exact values and use the algorithms for least squares, total least squares and constrained total least squares to compute the coefficients from the data. With constrained total least squares *we compute 4 variants where we assume one to four columns of X to be without errors.*

```
clear, format compact
Xt=[ 1 1 1 1 2 2 2 2 3 3 3 3        % generate points
     1 1 2 2 1 1 2 2 1 1 2 2
     1 2 1 2 1 2 1 2 1 2 1 2];
X=Xt'; [m,n]=size(X);
cc=[3, -1, 2, 5]';                  % z=c+n_1*x_1+n_2*x_2+n_3*x_3
X=[ones(m,1) X];
ze=X*cc;                            % exact values for z
z=ze+(rand(m,1)-0.5);               % add noise
[ze,z]
c1=X\z;                             % Least squares fit
[c,N]=ConstrainedLSQ([X,z],4);      % Fit plane using ConstrainedLSQ
c2=-[c N(1) N(2) N(3)]'/N(4);
[c3,v,At,bt]=TLS(X,z);              % TLS
C=[X,z];                            % Constrained TLS
[zz,c4]=ConstrainedTLS(C,1);        % first column with no errors
[zz,c5]=ConstrainedTLS(C,2);        % 2 columns with no errors
[zz,c6]=ConstrainedTLS(C,3);        % 3 columns with no errors
[zz,c7]=ConstrainedTLS(C,4);        % 4 columns with no errors
[c1 c2 c3 c4 c5 c6 c7]
```

With this program we obtained the values

exact z_e	with noise z
9	9.2094
14	14.2547
11	10.7760
16	16.1797
8	8.1551
13	12.6626
10	9.6190
15	14.9984
7	7.4597
12	11.8404
9	9.0853
14	13.7238

and the computed coefficients are displayed in Table 6.2. Observe that the least squares solution (column 1) and the total least squares solution with all four columns without errors (last column) yield the same solution. Also minimizing the geometric distance using ConstrainedLSQ *(column 2) and CTLS 1 (column 4) compute the same, which shows that these two methods are mathematically equivalent.*

	LSQ	CLSQ	TLS	CTLS 1	CTLS 2	CTLS 3	CTLS 4
c	3.5358	3.4695	3.6021	3.4695	3.4612	3.4742	3.5358
n_1	-1.0388	-1.0416	-1.0543	-1.0416	-1.0388	-1.0388	-1.0388
n_2	1.8000	1.8129	1.7820	1.8129	1.8134	1.8000	1.8000
n_3	4.8925	4.9275	4.8900	4.9275	4.9289	4.9336	4.8925

TABLE 6.2. *Computed Coefficients*

6.8 Nonlinear Least Squares Problems

Let us consider some (nonlinear) function $f : \mathbb{R}^n \mapsto \mathbb{R}^m$ with $n \leq m$. We want to find a point $x \in \mathbb{R}^n$ such that $f(x) \approx 0$, or written in components,

$$f_i(x_1, x_2, \ldots, x_n) \approx 0, \quad i = 1, \ldots, m. \tag{6.83}$$

EXAMPLE 6.23. *In Example 6.2 we have* $x = (a_0, a_1, b)^\top \in \mathbb{R}^3$ *and* $f : \mathbb{R}^3 \mapsto \mathbb{R}^m$, *where*

$$f_i(x) = y_i - x_1 - x_2 e^{-x_3 t_i} \approx 0, \quad i = 1, \ldots, m.$$

Now by $f(x) \approx 0$, we mean we should make the 2-norm of f as small as possible:

$$\|f(x)\|_2^2 = \sum_{i=1}^{m} f_i(x)^2 \longrightarrow \min. \tag{6.84}$$

Just like for the linear least squares case, we associate to a given vector x the *residual vector* $r = f(x)$, whose 2-norm is the *residual sum of squares* that we seek to minimize.

6.8.1 Notations and Definitions

In order to develop algorithms to solve Problem (6.84), we need to recall some mathematical notations from calculus. Let $f : \mathbb{R}^n \mapsto \mathbb{R}$ be a continuous and differentiable function.

DEFINITION 6.7. (GRADIENT) *The* gradient *of f is the vector*

$$\operatorname{grad} f = \nabla f = \frac{\partial f(x)}{\partial x} = \left(\frac{\partial f(x)}{\partial x_1}, \frac{\partial f(x)}{\partial x_2}, \ldots, \frac{\partial f(x)}{\partial x_n} \right)^\top.$$

DEFINITION 6.8. (HESSIAN) *The* Hessian *of f is the $n \times n$ matrix*

$$\operatorname{hess} f = \nabla^2 f = \frac{\partial^2 f(x)}{\partial x^2} = H = (h_{ij}), \quad \text{with} \quad h_{ij} = \frac{\partial^2 f(x)}{\partial x_i \partial x_j}.$$

Note that the Hessian is a symmetric matrix (if the derivatives commute).

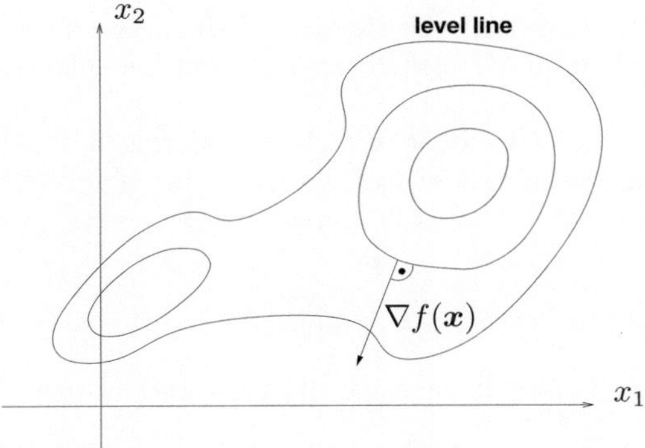

FIGURE 6.9. *Illustration of the gradient* $\nabla f(\boldsymbol{x})$.

For $n = 2$, we can draw the function f as a set of level curves (see Figure 6.9). The gradient at a given point $\boldsymbol{x}$ is a vector that points in the direction along which the function f increases most rapidly. It is also orthogonal to the level curve, as one can see using the Taylor expansion seen earlier in (5.94):

$$f(\boldsymbol{x} + \boldsymbol{h}) = f(\boldsymbol{x}) + \nabla f(\boldsymbol{x})^{\top} \boldsymbol{h} + \frac{1}{2}\boldsymbol{h}^{\top}\nabla^2 f(\boldsymbol{x})\boldsymbol{h} + O(\|\boldsymbol{h}\|_2^3). \qquad (6.85)$$

Since the a level curve is by definition a curve along which the value of f remains constant, if we choose the direction $\boldsymbol{h}$ to be along a level curve and only move a short distance away from $\boldsymbol{x}$, then we must have $f(\boldsymbol{x} + \boldsymbol{h}) = f(\boldsymbol{x}) + O(\|\boldsymbol{h}\|_2^2)$. Thus, we deduce that $\nabla f(\boldsymbol{x})^{\top}\boldsymbol{h} = 0$, i.e., the gradient is orthogonal to the level curve.

For a vector function $\boldsymbol{f} \in \mathbb{R}^m$, we expand each component and obtain

$$
\begin{aligned}
f_1(\boldsymbol{x} + \boldsymbol{h}) &= f_1(\boldsymbol{x}) + \nabla f_1(\boldsymbol{x})^{\top}\boldsymbol{h} + O(\|\boldsymbol{h}\|_2^2), \\
&\vdots \\
f_m(\boldsymbol{x} + \boldsymbol{h}) &= f_m(\boldsymbol{x}) + \nabla f_m(\boldsymbol{x})^{\top}\boldsymbol{h} + O(\|\boldsymbol{h}\|_2^2),
\end{aligned}
$$

or in matrix notation

$$\boldsymbol{f}(\boldsymbol{x} + \boldsymbol{h}) \approx \boldsymbol{f}(\boldsymbol{x}) + J(\boldsymbol{x})\boldsymbol{h},$$

where we have introduced the $m \times n$ *Jacobian* matrix

$$J(\boldsymbol{x}) = \begin{pmatrix} \nabla f_1^{\top} \\ \vdots \\ \nabla f_m^{\top} \end{pmatrix} = \begin{pmatrix} \nabla f_1, & \ldots, & \nabla f_m \end{pmatrix}^{\top} = \begin{pmatrix} \frac{\partial f_1}{\partial x_1} & \cdots & \frac{\partial f_1}{\partial x_n} \\ \vdots & \vdots & \vdots \\ \frac{\partial f_m}{\partial x_1} & \cdots & \frac{\partial f_m}{\partial x_n} \end{pmatrix}.$$

Notice the special case that the Hessian of the scalar function $f(\boldsymbol{x})$ is the same as the Jacobian of $\nabla f(\boldsymbol{x})$. In this case, the Jacobian is a square $n \times n$ matrix.

Let us now recall Newton's method for solving a system of nonlinear equations, which was presented in Chapter 5. Let $\boldsymbol{f}(\boldsymbol{x}) : \mathbb{R}^n \to \mathbb{R}^n$; we are looking for a solution of $\boldsymbol{f}(\boldsymbol{x}) = 0$. Expanding $\boldsymbol{f}$ at some approximation $\boldsymbol{x}_k$, we obtain

$$\boldsymbol{f}(\boldsymbol{x}) = \boldsymbol{f}(\boldsymbol{x}_k) + J(\boldsymbol{x}_k)\boldsymbol{h} + O(\|\boldsymbol{h}\|_2^2), \quad \text{with} \quad \boldsymbol{h} = \boldsymbol{x} - \boldsymbol{x}_k.$$

Instead of solving $\boldsymbol{f}(\boldsymbol{x}) = 0$, we solve the linearized system

$$\boldsymbol{f}(\boldsymbol{x}_k) + J(\boldsymbol{x}_k)\boldsymbol{h} = 0$$

for the *Newton correction* $\boldsymbol{h}$ and obtain a (hopefully better) approximation

$$\boldsymbol{x}_{k+1} = \boldsymbol{x}_k + \boldsymbol{h} = \boldsymbol{x}_k - J(\boldsymbol{x}_k)^{-1}\boldsymbol{f}(\boldsymbol{x}_k). \tag{6.86}$$

Note that computationally we would not invert the Jacobian; we would instead solve by Gaussian Elimination the linear system

$$J(\boldsymbol{x}_k)\boldsymbol{h} = -\boldsymbol{f}(\boldsymbol{x}_k)$$

for the correction $\boldsymbol{h}$.

6.8.2   Newton's Method

Given m nonlinear equations with n unknowns $(n \le m)$, we want to solve $\boldsymbol{f}(\boldsymbol{x}) \approx 0$ in the least squares sense, that is, we want

$$\Phi(\boldsymbol{x}) = \frac{1}{2}\|\boldsymbol{f}(\boldsymbol{x})\|_2^2 \longrightarrow \min. \tag{6.87}$$

A necessary condition for minimizing $\Phi(\boldsymbol{x})$ is $\nabla\Phi = 0$. We express this condition in terms of $\boldsymbol{f}$:

$$\frac{\partial\Phi(\boldsymbol{x})}{\partial x_i} = \sum_{l=1}^{m} f_l(\boldsymbol{x})\frac{\partial f_l}{\partial x_i}, \quad i = 1, \ldots, n, \tag{6.88}$$

or in matrix notation

$$\nabla\Phi(\boldsymbol{x}) = J(\boldsymbol{x})^\top \boldsymbol{f}(\boldsymbol{x}).$$

Thus we obtain as a necessary condition for minimizing $\Phi(\boldsymbol{x})$ a nonlinear system of n equations in n unknowns:

$$J(\boldsymbol{x})^\top \boldsymbol{f}(\boldsymbol{x}) = 0. \tag{6.89}$$

We can compute a solution of (6.89) using Newton's method (6.86). To do so, we need the Jacobian of $\nabla\Phi(\boldsymbol{x})$, which is the Hessian of $\Phi(\boldsymbol{x})$. If $\boldsymbol{x}_k$ is an approximation then we obtain the Newton correction by solving a linear system:

$$\nabla^2 \Phi(\boldsymbol{x}_k)\boldsymbol{h} = -J(\boldsymbol{x}_k)^\top \boldsymbol{f}(\boldsymbol{x}_k).$$

Let us express the Hessian in terms of the function $\boldsymbol{f}$. From (6.88), we compute the second derivatives,

$$\frac{\partial^2 \Phi(\boldsymbol{x})}{\partial x_i \partial x_j} = \sum_{l=1}^{m} \frac{\partial f_l}{\partial x_j} \frac{\partial f_l}{\partial x_i} + \sum_{l=1}^{m} f_l(\boldsymbol{x}) \frac{\partial^2 f_l}{\partial x_i \partial x_j}. \tag{6.90}$$

Now $\partial^2 f_l / \partial x_i \partial x_j$ is the (i,j) element of the Hessian of $f_l(\boldsymbol{x})$. Furthermore

$$\sum_{l=1}^{m} \frac{\partial f_l}{\partial x_j} \frac{\partial f_l}{\partial x_i} = \boldsymbol{J}_{:j}^\top \boldsymbol{J}_{:i}$$

is the scalar product of columns i and j of the Jacobian matrix. Therefore, we obtain in matrix notation

$$\nabla^2 \Phi(\boldsymbol{x}) = J^\top J + \sum_{l=1}^{m} f_l(\boldsymbol{x}) \nabla^2 f_l(\boldsymbol{x}).$$

The *Newton iteration for the nonlinear least squares problem* $\boldsymbol{f}(\boldsymbol{x}) \approx 0$ becomes

1. solve the linear system for the correction $\boldsymbol{h}$

$$\left(J(\boldsymbol{x}_k)^\top J(\boldsymbol{x}_k) + \sum_{l=1}^{m} f_l(\boldsymbol{x}_k) \nabla^2 f_l(\boldsymbol{x}_k) \right) \boldsymbol{h} = -J(\boldsymbol{x}_k)^\top \boldsymbol{f}(\boldsymbol{x}_k) \tag{6.91}$$

2. iterate: $\boldsymbol{x}_{k+1} = \boldsymbol{x}_k + \boldsymbol{h}$.

EXAMPLE 6.24. *Let us return to Example 6.2,*

$$f_l(\boldsymbol{x}) = x_1 + x_2 e^{-x_3 t_l} - y_l.$$

The gradient and the Hessian are readily computed using MAPLE:

```
with(LinearAlgebra):
with(VectorCalculus):
BasisFormat(false):
f:=x1+x2*exp(-x3*t)-y;
Gradient(f,[x1,x2,x3]);
Hessian(f,[x1,x2,x3]);
```

We obtain

$$\nabla f_l = \begin{pmatrix} 1 & e^{-x_3 t_l} & -x_2 t_l e^{-x_3 t_l} \end{pmatrix}^\top$$

and

$$\nabla^2 f_l = \begin{pmatrix} 0 & 0 & 0 \\ 0 & 0 & -t_l e^{-x_3 t_l} \\ 0 & -t_l e^{-x_3 t_l} & x_2 t_l^2 e^{-x_3 t_l} \end{pmatrix}$$

We can now write a program for the solution in MATLAB:

ALGORITHM 6.31.
Newton Method for Nonlinear Least Squares

```
function [x,fv]=NewtonLSQ(x,xi,eta)
% NEWTONLSQ Curve fiting with Newton's method
%   [x,fv]=NewtonLSQ(x,xi,eta) fits the function f(t)=x1+x2*exp(-x3*t)
%   to given points (xi,eta). x is the initial guess and is overwritten
%   with fitted parameters x1,x2 and x3. fv contains norm(f) for
%   each iteration

h=x; fv=[];
while norm(h)>100*eps*norm(x)
  ee=exp(-x(3)*xi);
  tee=xi.*ee;
  J=[ones(size(xi)),ee,-x(2)*tee ];        % Jacobian
  f=x(1)+x(2)*ee-eta;                       % residual
  s1=f'*tee; s2=x(2)*f'*(xi.*tee);
  A=J'*J+[0 0 0; 0 0 -s1; 0 -s1 s2];        % Hessian of Phi
  h=-A\(J'*f);
  x=x+h;
  xh=[x,h]
  res=norm(f)
  fv=[fv norm(f)];
end
```

In order to test the program we generate a test example:

```
format compact, format long
a0=1; a1=2; b=0.15;
xi=[1:0.3:7]'; etae=a0+a1*exp(-b*xi);       % compute exact values
rand('seed',0);                             % perturb to simulate
                                            % measurements
eta=etae+0.1*(rand(size(etae))-0.5);
[x1,fv1]= NewtonLSQ([1.8 1.8 0.1]',xi,eta); % first example
plot([1:14],fv1(1:14),'-')
pause                                       % second example using a
[x2,fv2]= NewtonLSQ([1.5 1.5 0.1]',xi,eta); % different initial guess
plot([1:14], fv1(1:14),'-',[1:14], fv2(1:14),':')
```

*The results are very interesting: the iterations converge to different so-
lutions. Because we print intermediate results (the matrix* **xh**, *showing the
current solution and the correction) we can observe quadratic convergence*

in both cases. In the first case with the starting values [1.8, 1.8, 0.1] we obtain the parameters $x1 = [2.1366, 0.0000, 0.0000]$, which means that the fitted function is the constant $f(t) = 2.1366$. In the second case we obtain $x2 = [1.1481, 1.8623, 0.1702]$, thus $f(t) = 1.1481 + 1.8623 e^{-0.1702\,t}$, a much better fit. Figure 6.10 shows the residual sum of squares for both cases. We see that in the first case (solid line) the iteration is trapped in a local minimum! After 6 iteration steps both sequences have reduced the residual sum of squares to about the final value, however, the first sequence does not converge there but moves away to a local minimum with a higher value of the residual sum of squares.

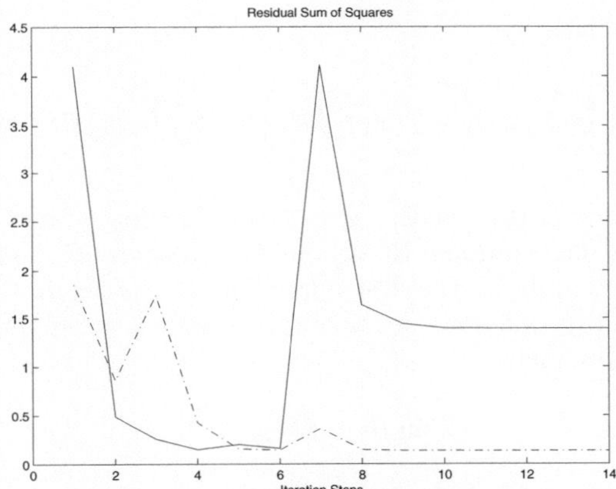

FIGURE 6.10. *Residual sum of squares*

There is another derivation of Newton's method which gives a different interpretation of the algorithm. Consider approximating Φ at a point $\boldsymbol{x}_k$ near the minimum with a quadratic form $Q(\boldsymbol{h})$ with $\boldsymbol{h} = \boldsymbol{x} - \boldsymbol{x}_k$ by expanding in a Taylor series :

$$\Phi(\boldsymbol{x}) = \frac{1}{2}\|\boldsymbol{f}(\boldsymbol{x})\|_2^2 \approx \Phi(\boldsymbol{x}_k) + \nabla\Phi(\boldsymbol{x}_k)^\top \boldsymbol{h} + \frac{1}{2}\boldsymbol{h}^\top \nabla^2\Phi(\boldsymbol{x}_k)\boldsymbol{h} =: Q(\boldsymbol{h}).$$

Then instead of minimizing Φ we minimize the quadratic form Q:

$$\nabla Q(\boldsymbol{h}) = 0 \iff \nabla^2\Phi(\boldsymbol{x}_k)\boldsymbol{h} + \nabla\Phi(\boldsymbol{x}_k) = 0.$$

But this is again Newton's method. Thus, we have shown that applying Newton's method to the (nonlinear) equations $\nabla\Phi(\boldsymbol{x}) = 0$ is equivalent to approximating Φ locally by a quadratic form Q and computing its minimum.

6.8.3   Gauss-Newton Method

The approximation of Φ by a quadratic form,

$$\Phi(\boldsymbol{x}_k + \boldsymbol{h}) \approx \Phi(\boldsymbol{x}_k) + \nabla\Phi(\boldsymbol{x}_k)^{\top}\boldsymbol{h} + \frac{1}{2}\boldsymbol{h}^{\top}\nabla^2\Phi(\boldsymbol{x}_k)\boldsymbol{h},$$

can be written in terms of $\boldsymbol{f}$,

$$\Phi(\boldsymbol{x}_k + \boldsymbol{h}) \approx \frac{1}{2}\boldsymbol{f}^{\top}\boldsymbol{f} + \boldsymbol{f}^{\top}J\boldsymbol{h} + \frac{1}{2}\boldsymbol{h}^{\top}J^{\top}J\boldsymbol{h} + \frac{1}{2}\boldsymbol{h}^{\top}\sum_{l=1}^{m}f_l(\boldsymbol{x}_k)\nabla^2 f_l(\boldsymbol{x}_k)\boldsymbol{h}.$$

Rearranging yields

$$\Phi(\boldsymbol{x}_k + \boldsymbol{h}) \approx \frac{1}{2}\|J(\boldsymbol{x}_k)\boldsymbol{h} + \boldsymbol{f}(\boldsymbol{x}_k)\|_2^2 + \frac{1}{2}\boldsymbol{h}^{\top}\sum_{l=1}^{m}f_l(\boldsymbol{x}_k)\nabla^2 f_l(\boldsymbol{x}_k)\boldsymbol{h}. \qquad (6.92)$$

The quadratic form approximating Φ consists therefore of two parts: the first involves only the Jacobian and the second the Hessians of the functions f_l. If we approximate Φ only by the first part $\Phi(\boldsymbol{x}) \approx \frac{1}{2}\|J(\boldsymbol{x}_k)\boldsymbol{h} + \boldsymbol{f}(\boldsymbol{x}_k)\|_2^2$, then the minimum of this quadratic form can be obtained by solving a *linear least squares problem*, namely

$$J(\boldsymbol{x}_k)\boldsymbol{h} + \boldsymbol{f}(\boldsymbol{x}_k) \approx 0. \qquad (6.93)$$

Computing the correction $\boldsymbol{h}$ by solving (6.93) is one step of the *Gauss-Newton method*. Another derivation of the Gauss-Newton method is obtained by linearizing $\boldsymbol{f}(\boldsymbol{x})$,

$$\boldsymbol{f}(\boldsymbol{x}) \approx \boldsymbol{f}(\boldsymbol{x}_k) + J(\boldsymbol{x}_k)\boldsymbol{h} \approx 0.$$

Summarizing, we obtain the algorithm *Gauss-Newton* for solving $\boldsymbol{f}(\boldsymbol{x}) \approx 0$:

1. Compute the Jacobian of $\boldsymbol{f}$ and solve the linear least squares problem for $\boldsymbol{h}$

$$J(\boldsymbol{x}_k)\boldsymbol{h} \approx -\boldsymbol{f}(\boldsymbol{x}_k).$$

2. Iterate $\boldsymbol{x}_{k+1} = \boldsymbol{x}_k + \boldsymbol{h}$.

Convergence is in general quadratic for Newton's method and only linear for Gauss-Newton. However, as we saw in Example 6.24, often the global behavior of the method is more important than local quadratic convergence. We will therefore consider in the next section a method that has been devised to prevent Newton's method from taking excessively large steps. The goal is to avoid convergence situations as shown in Example 6.24 where the residual sum of squares increases after step six.

6.8.4 Levenberg-Marquardt Algorithm

DEFINITION 6.9. (DESCENT DIRECTION) *The vector v is called a* descent *direction of $\Phi(x)$ at the point x if*

$$\nabla\Phi(x)^\top v < 0.$$

Let us explain this definition for $n = 2$: the gradient points into the direction of largest increase of $\Phi(x)$. If the scalar product of the gradient with a direction vector v is negative, then the angle α between the two vectors must be larger than $90°$. Thus, v must point downhill, as shown in Figure 6.11. An algebraic explanation is based on the Taylor expansion,

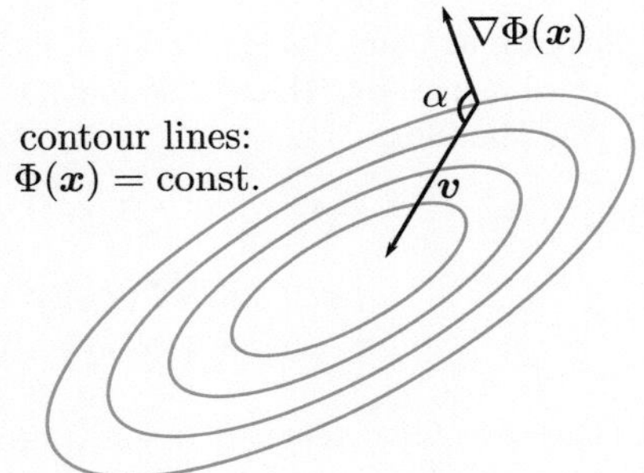

FIGURE 6.11. *Illustration of a descent direction v*

$$\Phi(x + \lambda v) = \Phi(x) + \underbrace{\nabla\Phi(x)^\top v}_{<0}\lambda + O(\lambda^2),$$

where we assumed that $\|v\|_2 = 1$. For sufficiently small $\lambda > 0$, we have $\Phi(x + \lambda v) < \Phi(x)$.

Let us now discuss whether the Newton and the Gauss-Newton corrections use descent directions:

1. Newton correction: $\nabla^2\Phi(x_k)h = -\nabla\Phi(x_k)$. If $\nabla^2\Phi(x_k)$ is nonsingular, i.e. the linear system has a solution, then

$$\nabla\Phi(x_k)^\top h = -\nabla\Phi(x_k)^\top(\nabla^2\Phi(x_k))^{-1}\nabla\Phi(x_k).$$

The matrix $\nabla^2\Phi$ is symmetric. If in addition it is positive definite, then the Newton correction is a descent direction.

Now the matrix $\nabla^2\Phi$ must be positive semidefinite in a neighborhood of a local minimum of Φ, as one can see as follows: if x^* is a local minimum, then we must have

(a) $\Phi(x^*) \le \Phi(x^* + h)$ for all $h \ne 0$ in a small neighborhood of x^*.

(b) $\nabla\Phi(\boldsymbol{x}^*) = 0$.

The Taylor expansion at $\boldsymbol{x}^*$ gives

$$\Phi(\boldsymbol{x}^* + \boldsymbol{h}) = \Phi(\boldsymbol{x}^*) + \underbrace{\nabla\Phi(\boldsymbol{x}^*)^{\top}\boldsymbol{h}}_{=0} + \frac{1}{2}\boldsymbol{h}^{\top}\nabla^2\Phi(\boldsymbol{x}^*)\boldsymbol{h} + O(\|\boldsymbol{h}\|_2^3).$$

Because $\Phi(\boldsymbol{x}^* + \boldsymbol{h}) \geq \Phi(\boldsymbol{x}^*)$, it follows that $\boldsymbol{h}^{\top}\nabla^2\Phi(\boldsymbol{x}^*)\boldsymbol{h} \geq 0$, which means $\nabla^2\Phi(\boldsymbol{x})$ must be positive semidefinite in a neighborhood of $\boldsymbol{x}^*$. Thus we have obtained

THEOREM 6.21. *If $\nabla^2\Phi(\boldsymbol{x}_k)$ is nonsingular for $\boldsymbol{x}_k$ in a neighborhood of a local minimum $\boldsymbol{x}^*$ then the Newton correction is a descent direction.*

2. Gauss-Newton correction: $J(\boldsymbol{x}_k)^{\top}J(\boldsymbol{x}_k)\boldsymbol{h} = -J(\boldsymbol{x}_k)^{\top}\boldsymbol{f}(\boldsymbol{x}_k) = -\nabla\Phi(\boldsymbol{x}_k)$. If $J(\boldsymbol{x}_k)^{\top}J(\boldsymbol{x}_k)$ is nonsingular, then this matrix and its inverse are positive definite and therefore

$$\nabla\Phi(\boldsymbol{x}_k)^{\top}\boldsymbol{h} = -\nabla\Phi(\boldsymbol{x}_k)^{\top}\left(J^{\top}J\right)^{-1}\nabla\Phi(\boldsymbol{x}_k) \tag{6.94}$$

$$= \begin{cases} < 0 & \text{if } \nabla\Phi(\boldsymbol{x}_k) \neq 0 \\ = 0 & \text{if } \nabla\Phi(\boldsymbol{x}_k) = 0 \end{cases} \tag{6.95}$$

If $\nabla\Phi(\boldsymbol{x}_k) = 0$, then we have reached a solution. To summarize, we have the result:

THEOREM 6.22. *If $J(\boldsymbol{x}_k)$ is not rank deficient in a neighborhood of a local minimum $\boldsymbol{x}^*$, then the Gauss-Newton correction is a descent direction.*

Locally the best descent direction is of course the negative gradient,

$$\boldsymbol{h} = -\nabla\Phi(\boldsymbol{x}_k) = -J(\boldsymbol{x}_k)^{\top}\boldsymbol{f}(\boldsymbol{x}_k).$$

However, often the locally optimal minimizing direction may not be the best for global optimization. The principal idea of Levenberg-Marquardt is a compromise between the negative gradient and the Gauss-Newton correction,

$$(J(\boldsymbol{x}_k)^{\top}J(\boldsymbol{x}_k) + \lambda I)\boldsymbol{h} = -J(\boldsymbol{x}_k)^{\top}\boldsymbol{f}(\boldsymbol{x}_k). \tag{6.96}$$

In (6.96), we have to choose a parameter λ. For $\lambda = 0$, we obtain the Gauss-Newton correction, whereas for $\lambda \gg 0$, the correction is along the direction of the negative gradient. There are several interpretations of the Levenberg-Marquardt correction:

- *Tikhonov-regularization*: If the matrix $J(\boldsymbol{x}_k)^{\top}J(\boldsymbol{x}_k)$ is singular, then for $\lambda > 0$ the matrix $J(\boldsymbol{x}_k)^{\top}J(\boldsymbol{x}_k) + \lambda I$ is again invertible and the correction can be computed.

- *Approximation of the Hessian*: We can regard λI as an approximation of

$$\sum_{l=1}^{m} f_l(\boldsymbol{x}_k)\nabla^2 f_l(\boldsymbol{x}_k).$$

Depending on the application on hand, one may choose to use a matrix other than λI, such as λD with D a diagonal matrix.

- *Limiting the norm of $\boldsymbol{h}$*: Consider the constrained minimization problem

$$\|J\boldsymbol{h} + \boldsymbol{f}\|_2^2 \longrightarrow \min \quad \text{subject to} \quad \|\boldsymbol{h}\|_2^2 \leq \alpha^2. \tag{6.97}$$

The unconstrained problem has a unique minimizer $\boldsymbol{h}_0$ given by the Gauss-Newton correction. We now have two cases:

1. if $\|\boldsymbol{h}_0\|_2 \leq \alpha$, then $\boldsymbol{h}_0$ also solves the constrained problem.
2. if $\|\boldsymbol{h}_0\|_2 > \alpha$, then the solution of the constrained problem must lie on the boundary, since there are no local minima inside the disk $\|\boldsymbol{h}\|_2 \leq \alpha$.

In the second case, the constrained problem can be solved using Lagrange multipliers as follows: consider the Lagrangian

$$L(\boldsymbol{h}, \lambda) = \frac{1}{2}\|J\boldsymbol{h} + \boldsymbol{f}\|_2^2 + \lambda(\|\boldsymbol{h}\|_2^2 - \alpha^2).$$

Setting the partial derivatives to zero, we obtain the equations

$$(J(\boldsymbol{x}_k)^\top J(\boldsymbol{x}_k) + \lambda I)\boldsymbol{h} = -J(\boldsymbol{x}_k)^\top \boldsymbol{f} \tag{6.98}$$
$$\|\boldsymbol{h}\|_2^2 = \alpha^2. \tag{6.99}$$

Solving (6.98) gives $\boldsymbol{h} = -R(\lambda)\boldsymbol{f}$, where

$$R(\lambda) = (J(\boldsymbol{x}_k)^\top J(\boldsymbol{x}_k) + \lambda I)^{-1} J(\boldsymbol{x}_k)^\top.$$

To calculate the value of α corresponding to this solution, let $J(\boldsymbol{x}_k) = U\Sigma V^\top$ be the SVD of $J(\boldsymbol{x}_k)$. Then

$$R(\lambda) = V(\Sigma^\top \Sigma + \lambda I)^{-1}\Sigma^\top U^\top,$$

so $R(\lambda)$ has singular values

$$\mu_i = \frac{\sigma_i(J)}{\sigma_i(J)^2 + \lambda},$$

and

$$\alpha^2 = \|\boldsymbol{h}\|_2^2 = \sum_{i=1}^{n} \mu_i^2 \tilde{f}_i^2,$$

where $\tilde{\boldsymbol{f}} = U^\top \boldsymbol{f}$. Thus, we see that $\alpha = \alpha(\lambda)$ is a strictly decreasing function of λ, with $\alpha(0) = \|\boldsymbol{h}_0\|_2$ and $\lim_{\lambda\to\infty} \alpha(\lambda) = 0$. This means for every $\lambda > 0$, there is an $\alpha(\lambda) < \|\boldsymbol{h}_0\|_2$ such that the Levenberg-Marquardt step $\boldsymbol{h}$ solves the constrained minimization problem (6.97) with $\alpha = \alpha(\lambda)$. In other words, the Levenberg-Marquardt method solves the minimization problem over a reduced set of admissible solutions, i.e., those that satisfy $\|\boldsymbol{h}\|_2 \leq \alpha(\lambda)$, effectively limiting the correction step to within a region near $\boldsymbol{x}_k$; this is known as a *trust region* in optimization, see Subsection 12.3.2.

Choosing a good λ is not easy. Besides trial and error, there are various propositions in the literature, such as the one by Brown and Dennis:

$$\lambda = c\|\boldsymbol{f}(\boldsymbol{x}_k)\|_2, \quad c = \begin{cases} 10 & \text{for } 10 \leq \|\boldsymbol{f}(\boldsymbol{x}_k)\|_2 \\ 1 & \text{for } 1 \leq \|\boldsymbol{f}(\boldsymbol{x}_k)\|_2 < 10 \\ 0.01 & \text{for } \|\boldsymbol{f}(\boldsymbol{x}_k)\|_2 < 1 \end{cases}$$

The computation of the Levenberg-Marquardt correction (6.96) is equivalent to solving the linear least squares problem

$$\begin{pmatrix} J \\ \sqrt{\lambda}I \end{pmatrix} \boldsymbol{h} \approx \begin{pmatrix} -\boldsymbol{f} \\ 0 \end{pmatrix}, \tag{6.100}$$

which is the numerically preferred way.

In MATLAB there exists the function `nlinfit` for nonlinear least squares data fitting by the Gauss-Newton method (in newer releases like R2011a it is in the Statistics Toolbox). The Jacobian is approximated by a finite difference and augmented by $0.01\ I$ (i.e. $\lambda = 10^{-4}$). Thus, the function in fact uses Levenberg-Marquardt-adjusted Gauss-Newton steps, which are computed using (6.100). The step is only accepted if the new residual norm is smaller than the previous one. Otherwise, the step is reduced by $\boldsymbol{h} = \boldsymbol{h}/\sqrt{10}$ until the condition is met. The function `nlinfit` therefore does not get trapped for Example 6.24 in the local minimum and is able also to deliver the correct answer for the starting values $[1.8,\ 1.8,\ 0.1]$. With the function

```
>> f=@(beta,x) beta(1)+beta(2)*exp(-beta(3)*x);
```

we obtain the same parameters as before with the Newton method:

```
>> beta=nlinfit(xi,eta,f,[1.8 1.8 0.1])
beta =
    1.148055903080751   1.862263964120192   0.170154093666413
```

6.9   Least Squares Fit with Piecewise Functions

We end the chapter with a good example of a constrained least squares problem with nonlinear constraints, which is solved in [45]. Consider the data set given in Table 6.3 and displayed in Figure 6.12. The problem is to find piecewise functions *with free knots* that best fit the data set. The choice of the knots is in fact part of the optimization problem, since the location of these knots may have a physical interpretation, such as phase changes. The constraint is that global function should be continuous in the knots that separate them; optionally, we may also require that the derivative be continuous.

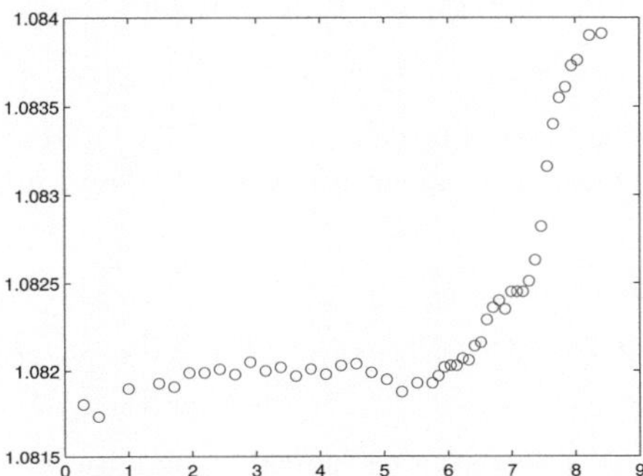

FIGURE 6.12. *Data Set*

TABLE 6.3. *Given Data Set*

x	y	x	y	x	y
0.288175	1.08181	4.562725	1.08204	6.794990	1.08240
0.525650	1.08174	4.800200	1.08199	6.889980	1.08235
1.000600	1.08190	5.037675	1.08195	6.984970	1.08245
1.475550	1.08193	5.275150	1.08188	7.079960	1.08245
1.713025	1.08191	5.512625	1.08193	7.174950	1.08245
1.950500	1.08199	5.750100	1.08193	7.269940	1.08251
2.187975	1.08199	5.845090	1.08197	7.364930	1.08263
2.425450	1.08201	5.940080	1.08202	7.459920	1.08282
2.662925	1.08198	6.035070	1.08203	7.554910	1.08316
2.900400	1.08205	6.130060	1.08203	7.649900	1.08340
3.137875	1.08200	6.225050	1.08207	7.744890	1.08355
3.375350	1.08202	6.320040	1.08206	7.839880	1.08361
3.612825	1.08197	6.415030	1.08214	7.934870	1.08373
3.850300	1.08201	6.510020	1.08216	8.029860	1.08376
4.087775	1.08198	6.605010	1.08229	8.219840	1.08390
4.325250	1.08203	6.700000	1.08236	8.409820	1.08391

Since the nature of the piecewise functions is not known, we will use polynomials of low degrees. The special case where all the polynomials have the same degree corresponds to a well-known problem called *least squares approximation by splines with free knots*. Many authors have worked on this problem; a good survey is given for example in [123].

Assume we are given some data points

$$
\begin{array}{c|cccc}
x & x_1, & x_2 & \cdots & x_N \\
\hline
y & y_1, & y_2, & \cdots & y_N
\end{array}.
\tag{6.101}
$$

where the x-coordinates are ordered: $a = x_1 < x_2 < \cdots < x_N = b$. We want to partition this set by two knots ξ and η such that

$$
x_1 < \cdots < x_{n_1} < \xi \le x_{n_1+1} < \cdots < x_{n_2} < \eta \le x_{n_2+1} < \cdots < x_N.
$$

These two knots define three intervals: (x_1, ξ), (ξ, η) and (η, x_N). We will fit in the least squares sense three given functions f, g and h in each interval to the data points,

$$
\begin{aligned}
f(\boldsymbol{a}, x_i) &\approx y_i^f & i &= 1, \ldots, n_1, \\
g(\boldsymbol{b}, x_i) &\approx y_i^g & i &= n_1 + 1, \ldots, n_2, \\
h(\boldsymbol{c}, x_i) &\approx y_i^h & i &= n_2 + 1, \ldots, N.
\end{aligned}
\tag{6.102}
$$

The superscript in the y-coordinates indicates that we have also partitioned the data points accordingly,

$$
\boldsymbol{x} = \begin{pmatrix} \boldsymbol{x}^f \\ \boldsymbol{x}^g \\ \boldsymbol{x}^h \end{pmatrix}, \quad
\boldsymbol{y} = \begin{pmatrix} \boldsymbol{y}^f \\ \boldsymbol{y}^g \\ \boldsymbol{y}^h \end{pmatrix}.
$$

The parameters of the functions f, g and h to be determined by the least squares fit are denoted by $\boldsymbol{a}$, $\boldsymbol{b}$ and $\boldsymbol{c}$.

To enforce continuity of the global piecewise function, we have to impose the constraints

$$
\begin{aligned}
f(\boldsymbol{a}, \xi) - g(\boldsymbol{b}, \xi) &= 0, \\
g(\boldsymbol{b}, \eta) - h(\boldsymbol{c}, \eta) &= 0.
\end{aligned}
\tag{6.103}
$$

If the first derivative is also required to be continuous at the knots, then the additional constraints

$$
\begin{aligned}
f'(\boldsymbol{a}, \xi) - g'(\boldsymbol{b}, \xi) &= 0, \\
g'(\boldsymbol{b}, \eta) - h'(\boldsymbol{c}, \eta) &= 0.
\end{aligned}
\tag{6.104}
$$

must also be considered.

Let us introduce the vector functions

$$
\boldsymbol{f}(\boldsymbol{a}) = \begin{pmatrix} f(\boldsymbol{a}, x_1) \\ \vdots \\ f(\boldsymbol{a}, x_{n_1}) \end{pmatrix}, \quad
\boldsymbol{g}(\boldsymbol{b}) = \begin{pmatrix} g(\boldsymbol{b}, x_{n_1+1}) \\ \vdots \\ g(\boldsymbol{b}, x_{n_2}) \end{pmatrix}, \quad
\boldsymbol{h}(\boldsymbol{c}) = \begin{pmatrix} h(\boldsymbol{c}, x_{n_2+1}) \\ \vdots \\ h(\boldsymbol{c}, x_N) \end{pmatrix},
$$

and the vector of unknowns

$$
\boldsymbol{z} = \begin{pmatrix} \boldsymbol{a} \\ \boldsymbol{b} \\ \boldsymbol{c} \\ \xi \\ \eta \end{pmatrix}, \quad
\boldsymbol{F}(\boldsymbol{z}) = \begin{pmatrix} \boldsymbol{f}(\boldsymbol{a}) \\ \boldsymbol{g}(\boldsymbol{b}) \\ \boldsymbol{h}(\boldsymbol{c}) \end{pmatrix} \quad \text{and} \quad
\boldsymbol{G}(\boldsymbol{z}) = \begin{pmatrix} f(\boldsymbol{a}, \xi) - g(\boldsymbol{b}, \xi) \\ g(\boldsymbol{b}, \eta) - h(\boldsymbol{c}, \eta) \end{pmatrix}.
$$

Then the problem of fitting a continuous global function becomes a *constrained nonlinear least squares problem*

$$\min_{\boldsymbol{z}} \| \boldsymbol{F}(\boldsymbol{z}) - \boldsymbol{y} \|_2 \quad \text{subject to} \quad \boldsymbol{G}(\boldsymbol{z}) = \boldsymbol{0}. \tag{6.105}$$

If we also require continuity of the derivative, then we have to replace $\boldsymbol{G}$ in (6.105) by

$$\boldsymbol{G}(\boldsymbol{z}) = \begin{pmatrix} f(\boldsymbol{a}, \xi) - g(\boldsymbol{b}, \xi) \\ g(\boldsymbol{b}, \eta) - h(\boldsymbol{c}, \eta) \\ f'(\boldsymbol{a}, \xi) - g'(\boldsymbol{b}, \xi) \\ g'(\boldsymbol{b}, \eta) - h'(\boldsymbol{c}, \eta) \end{pmatrix}.$$

We will solve Problem (6.105) by an alternating minimization procedure: in each iteration step, we will use the current values of ξ and η to allocate the points to the functions $\boldsymbol{f}$, $\boldsymbol{g}$ and $\boldsymbol{h}$. Then we will apply the *Gauss-Newton method* to solve the nonlinear least squares problem. For this, we linearize the functions $\boldsymbol{F}$ and $\boldsymbol{G}$ to obtain a linear least squares problem with linear constraints, from which we calculate the correction $\Delta \boldsymbol{z}$. Assume $\bar{\boldsymbol{z}}$ is an approximation of a solution of Problem (6.105). If we expand

$$\boldsymbol{F}(\bar{\boldsymbol{z}} + \Delta \boldsymbol{z}) \approx \boldsymbol{F}(\bar{\boldsymbol{z}}) + J_F \Delta \boldsymbol{z},$$

and do the same for $\boldsymbol{G}(\boldsymbol{z})$, we can replace Problem (6.105) by a constrained linear least squares problem with the Jacobian matrices for the correction

$$\begin{aligned} J_F\, \Delta \boldsymbol{z} &\approx \boldsymbol{y} - \boldsymbol{F}(\bar{\boldsymbol{z}}), \\ J_G\, \Delta \boldsymbol{z} &= -\boldsymbol{G}(\bar{\boldsymbol{z}}). \end{aligned} \tag{6.106}$$

We will use the `LinearlyConstrainedLSQ` Algorithm 6.21 for the solution.

6.9.1 Structure of the Linearized Problem

The linearly constrained least squares problem (6.106) is structured: the Jacobian matrix of $\boldsymbol{F}$ is block diagonal and contains the Jacobin matrices of the three functions f, g and h,

$$J_F = \begin{bmatrix} J_f & \boldsymbol{0} & \boldsymbol{0} & \boldsymbol{0} & \boldsymbol{0} \\ \boldsymbol{0} & J_g & \boldsymbol{0} & \vdots & \vdots \\ \boldsymbol{0} & \boldsymbol{0} & J_h & \boldsymbol{0} & \boldsymbol{0} \end{bmatrix}.$$

Since in our case the size of the matrix J_F is small, we will not treat it as a sparse matrix. The following function `DirectSum` comes in handy to construct such block-diagonal matrices :

ALGORITHM 6.32. *Generating Direct Sums*

```
function A=DirectSum(A,varargin)
```

```
% DIRECTSUM computes the direct sum of matrices
%    A=DirectSum(A1,A2,...,An) computes the direct sum of matrices of
%    arbitrary size (Peter Arbenz, May 30, 1997)

for k=1:length(varargin)
  [n,m]=size(A);
  [o,p]=size(varargin{k});
  A=[A zeros(n,p); zeros(o,m) varargin{k}];
end
```

For the continuity condition, the Jacobian of $\boldsymbol{G}$ is

$$J_G = \begin{bmatrix} \nabla f(\boldsymbol{a},\xi)^\top & -\nabla g(\boldsymbol{b},\xi)^\top & 0 & f'(\boldsymbol{a},\xi)-g'(\boldsymbol{b},\xi) & 0 \\ 0 & \nabla g(\boldsymbol{b},\eta)^\top & -\nabla h(\boldsymbol{c},\eta)^\top & 0 & g'(\boldsymbol{b},\eta)-h'(\boldsymbol{c},\eta) \end{bmatrix}.$$

Note that we denote with $\nabla f(\boldsymbol{a}, \xi)$ the gradient of f with respect to the parameters $\boldsymbol{a}$, while f' denotes the derivative of f with respect to the independent variable x.

If also the derivatives should be continuous then

$$J_G = \begin{bmatrix} \nabla f(\boldsymbol{a},\xi)^\top & -\nabla g(\boldsymbol{b},\xi)^\top & 0 & f'(\boldsymbol{a},\xi)-g'(\boldsymbol{b},\xi) & 0 \\ 0 & \nabla g(\boldsymbol{b},\eta)^\top & -\nabla h(\boldsymbol{c},\eta)^\top & 0 & g'(\boldsymbol{b},\eta)-h'(\boldsymbol{c},\eta) \\ \nabla f'(\boldsymbol{a},\xi)^\top & -\nabla g'(\boldsymbol{b},\xi)^\top & 0 & f''(\boldsymbol{a},\xi)-g''(\boldsymbol{b},\xi) & 0 \\ 0 & \nabla g'(\boldsymbol{b},\eta)^\top & -\nabla h'(\boldsymbol{c},\eta)^\top & 0 & g''(\boldsymbol{b},\eta)-h''(\boldsymbol{c},\eta) \end{bmatrix}. \tag{6.107}$$

6.9.2   Piecewise Polynomials

For polynomials, the Jacobians become Vandermonde matrices if we use the standard representation. If $f(\boldsymbol{a},x) = a_1 x^p + a_2 x^{p-1} + \cdots + a_p x + a_{p+1}$, then

$$J_f = \begin{pmatrix} x_1^p & x_1^{p-1} & \cdots & x_1 & 1 \\ \vdots & \vdots & \vdots & \vdots & \vdots \\ x_{n_1}^p & x_{n_1}^{p-1} & \cdots & x_{n_1} & 1 \end{pmatrix}.$$

If f, g and h are polynomials of degree p, q and r, then for the continuity of the global function, we need

$$J_G = \begin{pmatrix} \xi^p & \cdots & \xi & 1 & -\xi^q & \cdots & -\xi & -1 & 0 & \cdots & 0 & 0 & f'(\boldsymbol{a},\xi)-g'(\boldsymbol{b},\xi) & 0 \\ 0 & \cdots & 0 & 0 & \eta^q & \cdots & \eta & 1 & -\eta^r & \cdots & -\eta & -1 & 0 & g'(\boldsymbol{b},\eta)-h'(\boldsymbol{c},\eta) \end{pmatrix}.$$

For the continuity of the derivative we have

$$f'(\boldsymbol{a},\xi) = pa_1\xi^{p-1} + (p-1)a_2\xi^{p-2} + \cdots + a_p,$$

and thus we use in (6.107) the expression

$$\nabla f'(\boldsymbol{a},\xi)^\top = [p\xi^{p-1}, (p-1)\xi^{p-2}, \ldots, 2\xi, 1, 0].$$

For the following main program we need to first read the data, stored in a file `xy.m`. This file contains only the definition of the matrix `XY`:

```
XY = [0.288175     1.08181
      0.525650     1.08174
         ...          ...
      8.219840     1.08390
      8.409820     1.08391]
```

For the sake of brevity, we did not print all the data of Table 6.3, but we will print intermediate results and also plot the approximations. To plot the given points, we will use the function `PlotPoints`:

ALGORITHM 6.33. *Plot the Points*

```
function ax=PlotPoints(X)
% PLOTPOINTS plots the points X and returns the axis handle

clf; hold off;
plot(X(:,1),X(:,2),'o');
ax=axis; hold on;
```

During the iterations, the breakpoints ξ and η will change, so we need to repartition the data after each iteration. This is done with the function `Partition`:

ALGORITHM 6.34. *Partition of the Data*

```
function [n1,n2,xf,yf,xg,yg,xh,yh]=Partition(xi,eta,XY,N)
% PARTITION  Partitions the data XY into three data sets
%    [n1,n2,xf,yf,xg,yg,xh,yh]=Partition(xi,eta,XY,N) partition the
%    date set in XY into three subsets according to xi and eta.

n1=sum(XY(:,1)<xi); n2=sum(XY(:,1)<eta);
xf=XY(1:n1,1); yf=XY(1:n1,2);
xg=XY(n1+1:n2,1); yg=XY(n1+1:n2,2);
xh=XY(n2+1:N,1); yh=XY(n2+1:N,2);
```

The Jacobian matrices are Vandermonde matrices and are generated using the function `van`:

ALGORITHM 6.35. *Generate Jacobian*

```
function J=van(p,x)
% VAN construct a Vandermonde matrix
%   J=van(p,x) computes p+1 columns of the Vandermonde matrix of x

n=length(x); J=ones(n,1);
for j=1:p
  J=[x.^j J];
end
```

After each iteration, we will pause and plot the current approximation using the function `PlotFunctions`:

ALGORITHM 6.36. *Plot Functions*

```
function PlotFunctions(xi,eta,a,b,c,ax);
% PLOTFUNCTIONS  plots the three polynomials

xx=[0:0.1:xi]; plot(xx,polyval(a,xx),'r'), axis(ax)
xx=[xi:0.1:eta]; plot(xx,polyval(b,xx),'g'), axis(ax)
xx=[eta:0.1:9]; plot(xx,polyval(c,xx),'b'), axis(ax)
```

The initial approximations for coefficients of the polynomials are computed using the MATLAB function `polyfit` to fit each polynomial to the partitioned data sets in the least squares sense. For the evaluation of a polynomial, we use the MATLAB function `polyval`. Finally, to compute the coefficients of the derivative of a polynomial we use `dpoly` :

ALGORITHM 6.37. *Derivative of a Polynomial*

```
function DA=dpoly(n,A)
% DPOLY derivative of a polynomial
%   DA=dpoly(n,A) computes the coefficients DA of the derivative of
%   the polynomial of degree n given by the coefficients A.

DA=0;
for j=1:n, DA(j)=(n+1-j)*A(j); end
```

The following program `main1` computes the Gauss-Newton approximations. Of course the iterations may or may not converge. Even when the method converges, we cannot guarantee that the limit point is the global minimumm, since the problem has many local minima.

The degrees of the polynomials can be chosen by changing the statements for p, q and r. The variable `derivative` is used as a switch to decide whether the derivative should also be continuous at the break points (`derivative=1`) or not.

The iteration is stopped if no convergence has been reached after 40 iterations. It is also stopped if the break points switch their order, i.e. if $\eta < \xi$.

ALGORITHM 6.38. *Fitting Piecewise Polynomials*

```
% Piecewise Polynomial Fit: main1.m
xy; N=max(size(XY));                  % read data
ax=PlotPoints(XY)                     % plot given points
```

```matlab
p=3; q=3; r=3;                        % choose degrees and
derivatives=1                         % derivatives (1=continuos, 0=no)
xi0=6.4; eta0=8;                      % initialize break points
                                      % xi < eta

xi=xi0; eta=eta0;                     % first partition of data
[n1,n2,xf,yf,xg,yg,xh,yh]=Partition(xi,eta,XY,N);
Jf=van(p,xf);   Jg=van(q,xg);   Jh=van(r,xh);
                                      % initialize polynomial coef-
                                      % ficients by individual least
                                      % squares fit
a=polyfit(xf, yf,p); b=polyfit(xg,yg,q); c=polyfit(xh,yh,r);
                                      % plot initial approximation
                                      % functions
PlotFunctions(xi,eta,a,b,c, ax);
k=0;                                  % count interations
dp=1;                                 % initialize correction
while (norm(dp)>1e-5) & (k<40),
  k= k+1;
  JF=[DirectSum(Jf,Jg,Jh), zeros(N,2)];
  gradfxi=van(p,xi); gradgxi=van(q,xi);
  gradgeta=van(q,eta); gradheta=van(r,eta);
                                      % coeffs of derivatives
  da=dpoly(p,a); db=dpoly(q,b); dc=dpoly(r,c);
                                      % constraints for continuity
  JG=[gradfxi -gradgxi zeros(size(gradheta)) ...
      polyval(da,xi)-polyval(db,xi) 0
    zeros(size(gradfxi)) gradgeta -gradheta ...
      0 polyval(db,eta)-polyval(dc,eta)]
  if derivatives,                     % continuity of derivative
    gradfsxi =[[p:-1:1].*van(p-1,xi),0];
    gradgsxi=[[q:-1:1].*van(q-1,xi),0];
    gradgseta=[[q:-1:1].*van(q-1,eta),0];
    gradhseta=[[r:-1:1].*van(r-1,eta),0];

    dda=dpoly(p-1,da);                % coeffs of second derivative
    ddb=dpoly(q-1,db);
    ddc=dpoly(r-1,dc);
    JG=[JG
        gradfsxi -gradgsxi zeros(size(gradhseta))...
          polyval(dda,xi)-polyval(ddb,xi) 0
        zeros(size(gradfsxi)) gradgseta -gradhseta ...
          0 polyval(ddb,eta)-polyval(ddc,eta)]
  end
                                      % Right hand side for lsq
  z=[yf-polyval(a,xf);yg-polyval(b,xg);yh-polyval(c,xh)]
                                      % Right hand side for constraints
  mG=-[polyval(a,xi)-polyval(b,xi)
      polyval(b,eta)-polyval(c,eta)]
  if derivatives,                     % add constraints
```

```
      mG=-[-mG; polyval(da,xi)-polyval(db,xi)
              polyval(db,eta)-polyval(dc,eta)]
  end
                                      % solve for corrections
  dp=LinearlyConstrainedLSQ(JF,JG,z,mG);
  a=a+dp(1:p+1)';                     % update unknowns
  b=b+dp(p+2:p+q+2)'; c=c+dp(p+q+3:p+q+r+3)';
  xi=xi+dp(p+q+r+4); eta=eta+dp(p+q+r+5);
  if xi>eta, error('xi > eta'); end
  norm(dp)                            % print norm of correction
                                      % plot current approximation
  PlotFunctions(xi,eta,a,b,c, ax);
  pause
                                      % new partition of the data
  [n1,n2,xf,yf,xg,yg,xh,yh]=Partition(xi,eta,XY,N);
  Jf=van(p,xf); Jg=van(q,xg); Jh=van(r,xh);
end
                                      % plot final result
ax=PlotPoints(XY); PlotFunctions(xi,eta,a,b,c, ax);
                                      % compute residual and print
                                      % results
rf=norm(polyval(a,xf)-XY(1:n1,2));
rg=norm(polyval(b,xg)-XY(n1+1:n2,2));
rh=norm(polyval(c,xh)-XY(n2+1:N,2));
rr=norm([rf, rg, rh])               % residuals
dpnorm=norm(dp)                      % norm of last correction
disp('degrees of polynomials, derivative (yes=1), # of iterations')
[p q r derivatives k]
disp('breakpoints: initial, final')
[xi0 eta0 xi eta]
```

6.9.3 Examples

Figure 6.13 shows an approximation by classical smoothing splines of degree
3 with free knots. Using $\xi = 7.34$ and $\eta = 7.78$ as initial approximations for
the breakpoints, after 9 Gauss-Newton iterations the norm of the correction
vector drops to $4.3180e-06$ and the break points converge to $\xi = 7.3157$ and
$\eta = 7.7275$. The norm of the true residual for the three spline functions is
$3.4965e-04$.

However, the solution to which we converged may be only a local minimum
of the nonlinear least squares function. In fact, if we use other initial approximations for the break points, we may obtain no convergence or convergence
to other solutions. Table 6.4 shows some results. There are three different solutions with almost the same norm of the residual ($3.4965e-04$, $3.6819e-04$
and $3.6499e-04$). This shows that the problem is ill-conditioned.

In the next example, we choose linear functions and ask only for continuity
(see Figure 6.14). For the initial values $\xi = 7.34$ and $\eta = 7.78$ we obtain the

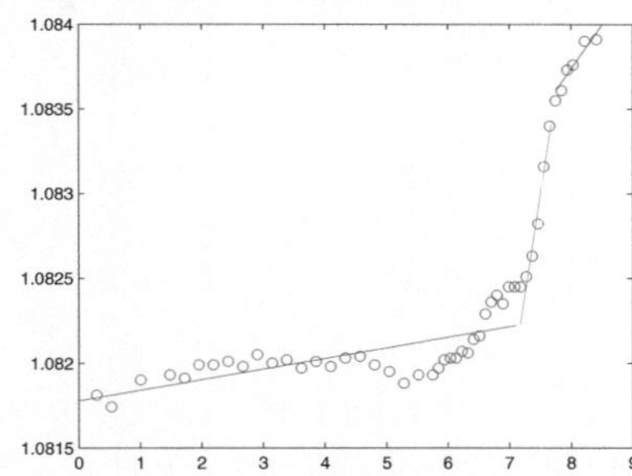

FIGURE 6.13.
*Classical Smoothing Splines
with Free Knots*

FIGURE 6.14.
*Piecewise Linear Functions
with Free Knots*

TABLE 6.4.
Different Solutions for Different Initial Breakpoints for Smoothing Splines

initial appr. final breakp.		norm of corr. **norm of res.**	# of inter.
7.34	7.78	4.3180e-06	9
7.31570	7.72746	**3.4965e-04**	
6.5	8	6.3398e-06	12
7.31570	7.72746	**3.4965e-04**	
6.4	8	5.3655e-06	21
6.03633	6.98520	**3.6819e-04**	
5	7	7.4487e-06	28
5.53910	7.04320	**3.6499e-04**	

break points $\xi = 7.1750$ and $\eta = 7.7623$.

Again by choosing different initial values for the break points, the Gauss-Newton method converges to different local minima (see Table 6.5). The ill conditioning here is even more pronounced.

Next we choose different degrees $p = 3$, $q = 1$ and $r = 2$ and fit a continuous global function, see Table 6.6. For the initial values $\xi = 7.34$ and $\eta = 7.78$ the break points become $\xi = 7.3552$ and $\eta = 7.6677$ and we obtain in 4 iterations Figure 6.15 with a residual norm of $3.5422e{-}04$. However, for the initial values $\xi = 6.5$ and $\eta = 8$ a better solution with break points 5.7587 and 7.3360 and a slightly reduced residual of 2.7396e-04 is obtained, see Figure 6.16. This example shows again how difficult it is to converge to a global minimum.

Finally, we increase the degrees to $p = 5$, $q = 3$ and $r = 2$ and ask for the continuity of the derivative as well, see Figure 6.17. Now the break points are $\xi = 7.3717$ and $\eta = 7.6494$.

TABLE 6.5. *Solutions for Piecewise Linear Functions*

initial appr.		norm of corr.	# of inter.
final breakp.		**norm of res.**	
7.34	7.78	7.8568e-14	4
7.17495	7.76231	**7.6797e-04**	
3	6	2.4061e-13	6
1.97596	6.57421	**8.1443e-04**	
5	7	4.5775e-13	7
6.12802	7.169537	**6.7463e-04**	

<table>
<tr><td>

FIGURE 6.15.

Polynomials with degrees $p = 3$, $q = 1$ and $r = 2$

</td><td>

FIGURE 6.16.

Polynomials with degrees $p = 3$, $q = 1$ and $r = 2$, second solution

</td></tr>
</table>

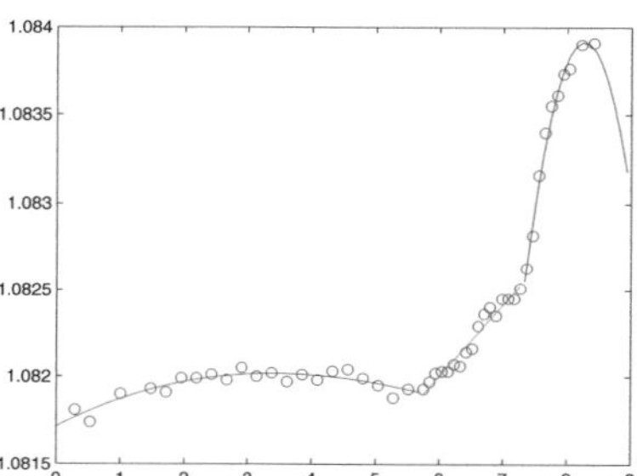

Again other starting values for the break points e.g. $\xi = 6$ and $\eta = 8$ lead to another solution, see Figure 6.18, with almost the same residual, see Table 6.7.

6.10 Problems

PROBLEM 6.1. *A person brings two packages to the post. One weighs 5 kg, the other 2 kg. However, at the post office both weigh together 8 kg. Adjust their weights using the least squares method.*

PROBLEM 6.2. *Let the matrix $A \in \mathbb{R}^{m \times n}$ with $m > n$ and $\mathrm{rank}(A) = n$. Prove that the matrix $A^\top A$ of the normal equations is positive definite.*

PROBLEM 6.3. *Let the matrix $A \in \mathbb{R}^{n \times n}$ be symmetric and positive definite. Prove that the singular values of A are the same as its eigenvalues. What is the relation between singular values and eigenvalues if A is symmetric but not positive definite?*

PROBLEM 6.4. *Let $A, B \in \mathbb{R}^{m \times n}$ and let $A = QBP^\top$ where Q and P are*

TABLE 6.6. *Solutions for $p = 3$, $q = 1$ and $r = 2$*

initial appr. final breakp.		norm of corr. **norm of res.**	# of inter.
7.34	7.78	3.6034e-06	4
7.3552	7.6677	**3.5422e-04**	
6.5	8	1.4420e-08	8
5.7587	7.3360	**2.7396e-04**	

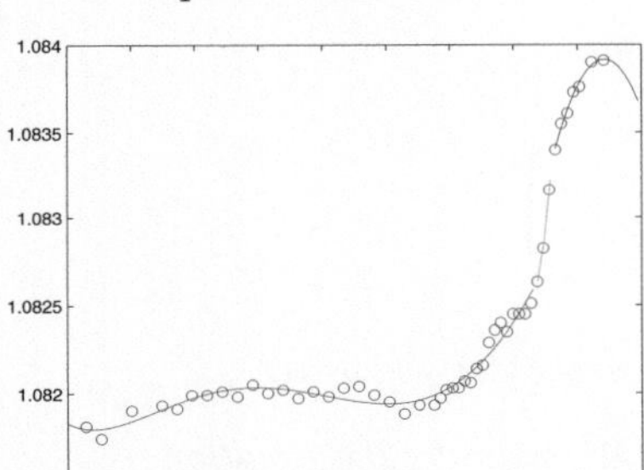

FIGURE 6.17.
Polynomials with degrees $p = 5$, $q = 3$ and $r = 2$

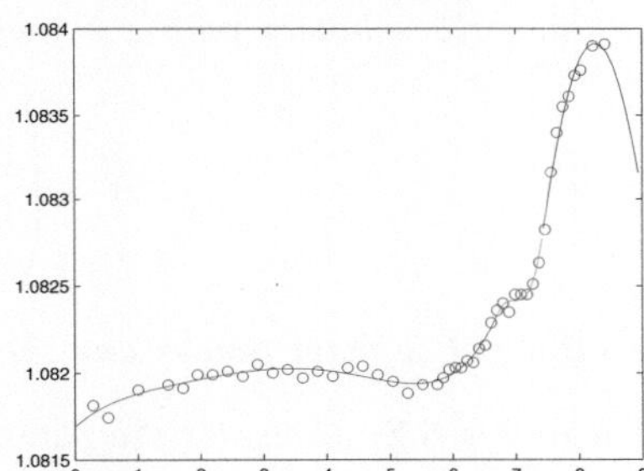

FIGURE 6.18.
Polynomials with degrees $p = 5$, $q = 3$ and $r = 2$, second solution

orthogonal matrices. Prove that

$$\|A\|_F = \|B\|_F.$$

PROBLEM 6.5. *Let $Q \in \mathbb{R}^{m \times n}$ with $m > n$ be an orthogonal matrix. Consider the matrix $P = QQ^{\top}$ Is P an orthogonal projector? If yes where does it projects? Justify your answer.*

TABLE 6.7. *Solutions for $p = 5$, $q = 3$ and $r = 2$, (continuous derivatives)*

initial appr. final breakp.		norm of corr. **norm of res.**	# of inter.
7.34	7.78	5.1228e-09	12
7.3717	7.6494	**2.8155e-04**	
6	8	8.9962e-06	15
6.7139	7.4377	**2.5334e-04**	

PROBLEM 6.6. *Consider the plane in $\mathbb{R}^3$ given by the equation*

$$x_1 + x_2 + x_3 = 0.$$

Construct a matrix P which projects a given point on this plane. Hint: consider first the orthogonal complement of the plane.

PROBLEM 6.7. *Given the matrix*

$$A = \begin{pmatrix} 1 & 2 & 6 \\ 1 & 3 & 7 \\ 1 & 4 & 8 \\ 1 & 5 & 9 \end{pmatrix}$$

Compute a Householder matrix P such that

$$PA = \begin{pmatrix} \sigma & x & x \\ 0 & x & x \\ 0 & x & x \\ 0 & x & x \end{pmatrix}$$

It is sufficient to determine σ and the Householder-vector $\boldsymbol{u}$.

PROBLEM 6.8. *Consider the plane in $\mathbb{R}^3$ given by the equation*

$$2x_1 - 2x_3 = 0.$$

Construct a matrix $P \in \mathbb{R}^{3\times3}$ which reflects a given point at this plane (computes the mirror image).

PROBLEM 6.9. *Let the measured points (t_k, y_k) for $i = k, \ldots, m$ be given. We want to fit the function $f(a,b) = ae^{bt}$ such that*

$$\sum_{k=1}^{m} \left(ae^{bt_k} - y_k \right)^2 \longrightarrow \min$$

using the Gauss-Newton method. Write up the system of equations for the first iteration step.

PROBLEM 6.10. *Let $\boldsymbol{b} = (b_1, b_2, b_3, b_4)$ be the measured lengths of the sides of a rectangle. Correct the measurements using least squares and determine side lengths $\boldsymbol{x}$ such that $\|\boldsymbol{x} - \boldsymbol{b}\|_2^2 \longrightarrow \min$ subject to the condition that two corresponding sides have the same length i.e. $x_1 = x_3$ and $x_2 = x_4$. Compute the corrected values.*

PROBLEM 6.11. *Let $A, B \in \mathbb{R}^{m\times n}$ and let $A = QBP^\top$ where Q and P are orthogonal matrices. Prove that*

$$\|A\|_2 = \|B\|_2.$$

PROBLEM 6.12. *Consider the matrix*

$$A = \begin{pmatrix} 1 \\ 2 \\ 3 \end{pmatrix}$$

1. *Compute and describe geometrically the 4 fundamental subspaces*

2. *Compute the 4 projectors on the fundamental subspaces.*

PROBLEM 6.13. *Consider the constrained least squares problem with given* $A \in \mathbb{R}^{m \times n}$ *and* $\boldsymbol{b} \in \mathbb{R}^m$ *and* $m > n$:

$$\min \|\boldsymbol{b} - \boldsymbol{y}\|_2^2 \quad \text{subject to } \boldsymbol{y} = A\boldsymbol{x}.$$

1. *Interpret the problem geometrically.*

2. *Assume A has full rank, give an explicit expression for the solution $\boldsymbol{y}$.*

3. *How can you compute $\boldsymbol{y}$ using the function* `ModifiedGramSchmidt` *(Algorithm 6.9)?*

PROBLEM 6.14. *A student wants to update the QR decomposition by removing a column using the function* `UpdateQR`. *He has defined* `ek` *to be the k-th unit vector.*

```
v=-A(:,k);
[Qs,Rs]=UpdateQR(Q,R,v,ek)
```

What is he effectively computing?

PROBLEM 6.15. *Minimizing the length of the residual vector $\boldsymbol{r} = \boldsymbol{b} - A\boldsymbol{x}$ is equivalent of minimizing the quadratic form*

$$Q(\boldsymbol{x}) = \boldsymbol{r}^\top \boldsymbol{r} = (\boldsymbol{b} - A\boldsymbol{x})^\top (\boldsymbol{b} - A\boldsymbol{x}) = \boldsymbol{b}^\top \boldsymbol{b} - 2\boldsymbol{x}^\top A^\top \boldsymbol{b} + \boldsymbol{x}^\top A^\top A\boldsymbol{x}.$$

By differentiating with respect to $\boldsymbol{x}$ and equating to zero show that the resulting equations are the normal equations.

PROBLEM 6.16. Savitzky-Golay Filter. *Noisy data often have to be filtered. One way to do this is to compute a least squares fit of a polynomial $P(x)$ of degree d through $2N + 1$ points left and right of the middle point x_i and to replace the function value y_i by the smoothed value $P(x_i)$.*

The smoothed value is an average of the neighboring points and thus the process is called a moving average *(see Chapter 9 in [45]).*

If the abscissas x_i are equidistant then the average is the same for all points and depends only on the degree of the polynomial and the number of points used for the average.

We consider therefore the data

x	$-N$	$\cdots$	-1	0	1	$\cdots$	N
y	y_{-N}	$\cdots$	y_{-1}	y_0	y_1	$\cdots$	y_N

We want to fit a polynomial

$$P(x) = b_d x^d + b_{d-1} x^{d-1} + \cdots + b_1 x + b_0$$

to this data. The coefficients b_i are obtained as solution of

$$P(i) \approx y_i, \quad for\ i = -N, \ldots, N$$

which is a linear least squares problem $Ab \approx y$. The smoothed value is $P(0) = b_0$. Since $b = A^+ y$ we obtain

$$b_0 = e_{p+1}^\top A^+ y = c^\top y$$

Write a MAPLE *script to compute the coefficient vector c.*

Hint: *in* MAPLE*'s* `CurveFitting` *package there is a function* `LeastSquares` *which is useful.*

PROBLEM 6.17. *A triangle has been measured, the measurements are as follows:*

$x_1 = \alpha$	$x_2 = \beta$	$x_3 = \gamma$	$x_4 = a$	$x_5 = b$	$x_6 = c$
$67°30'$	$52°$	$60°$	$172m$	$146m$	$165m$

The measurements x have errors. We would like to correct them so that the new values $\tilde{x} = x + h$ are consistent quantities of a triangle. They have to satisfy:

$$
\begin{aligned}
&\text{\textit{Sum of angles:}} && \tilde{x}_1 + \tilde{x}_2 + \tilde{x}_3 = 180° \\
&\text{\textit{Sine theorem:}} && \tilde{x}_4 \sin \tilde{x}_2 - \tilde{x}_5 \sin \tilde{x}_1 = 0 \\
& && \tilde{x}_5 \sin \tilde{x}_3 - \tilde{x}_6 \sin \tilde{x}_2 = 0
\end{aligned}
\qquad (6.108)
$$

Solve the constrained least squares problem $\|h\|_2 \longrightarrow$ min subject to the constraints (6.108). Replace the nonlinear constraints $f(\tilde{x}) = 0$ by the linearized equations $f(x) + Jh = 0$ where J is the Jacobian. Solve the linearized problem using e.g. `NullSpaceMethod`*. Iterate the process if necessary. Hint: Don't forget to work in radians!*

Check that for the new values also e.g. the cosine-theorem $c^2 = a^2 + b^2 - 2ab \cos(\gamma)$ holds.

You will notice that the corrections will be made mainly to the angles and much less to the lengths of the sides of the triangle. This is because the measurements have not the same absolute errors. While the error in last digit of the sides is about 1, the errors in radians of the angles are about 0.01.

Repeat your computation by taking in account with appropriate weighting the difference in measurement errors. Minimize not simply $\|\boldsymbol{h}\|_2$ but

$$\left\|\begin{pmatrix} 100h_1 \\ 100h_2 \\ 100h_3 \\ h_4 \\ h_5 \\ h_6 \end{pmatrix}\right\|_2$$

PROBLEM 6.18. *Prove that the following compact algorithm solves the linear least squares problem $A\boldsymbol{x} \approx \boldsymbol{b}$.*

1. *Form the augmented matrix*

$$\bar{A} = [A, \boldsymbol{b}] \in \mathbb{R}^{m \times n+1}.$$

2. *Compute the augmented normal equation matrix and decompose it using the Cholesky decomposition*

$$\bar{A}^{\mathsf{T}} \bar{A} = \bar{R}^{\mathsf{T}} \bar{R}.$$

3. *Partition*

$$\bar{R} = \begin{pmatrix} R & \boldsymbol{y} \\ 0 & \rho \end{pmatrix}. \tag{6.109}$$

4. *The least squares solution $\tilde{\boldsymbol{x}}$ is obtained by solving $R\boldsymbol{x} = \boldsymbol{y}$ with back-substitution.*

5. *The residual is $\|\boldsymbol{r}\|_2 = \|\boldsymbol{b} - A\tilde{\boldsymbol{x}}\|_2 = \rho$.*

PROBLEM 6.19. *Suppose you are given the decomposition of the matrix $A = LU$ where L is a lower and U an upper triangular matrix. Thus you can solve the system $A\boldsymbol{x} = \boldsymbol{b}$ in two steps*

1. *Solve $L\boldsymbol{y} = \boldsymbol{b}$ by forward substitution*

2. *Solve $U\boldsymbol{x} = \boldsymbol{y}$ by backward substitution.*

Example

$$L = \begin{pmatrix} 1 & & & & \\ 9 & 1 & & & \\ 1 & 5 & 1 & & \\ 9 & 10 & 5 & 1 & \\ 6 & 10 & 8 & 10 & 1 \end{pmatrix}, \quad U = \begin{pmatrix} 8 & 1 & 2 & 1 & 7 \\ & 3 & 10 & 4 & 0 \\ & & 10 & 9 & 8 \\ & & & 8 & 9 \\ & & & & 7 \end{pmatrix}, \quad \boldsymbol{b} = \begin{pmatrix} 15 \\ 138 \\ 39 \\ 211 \\ 209 \end{pmatrix}$$

Write two MATLAB *functions* `y = ForwardSubstitution(L,b)` *and* `x = BackSubstitution(U,y)` *so that they can be used to solve the linear system* $LU\boldsymbol{x} = \boldsymbol{b}$. *Both functions can be programmed using either scalar-product-operations or SAXPY-operations. A SAXPY is defined as a linear combination of two vectors:*

$$\boldsymbol{s} = \alpha\boldsymbol{x} + \boldsymbol{y}.$$

For modern processors the SAXPY variant is preferred, therefore program these versions.

PROBLEM 6.20. Fitting of circles. *We are given the measured points* (ξ_i, η_i):

ξ	0.7	3.3	5.6	7.5	6.4	4.4	0.3	-1.1
η	4.0	4.7	4.0	1.3	-1.1	-3.0	-2.5	1.3

Find the center (z_1, z_2) *and the radius* r *of a circle* $(x - z_1)^2 + (y - z_2)^2 = r^2$ *that approximate the points as well as possible. Consider the two cases*

1. Algebraic fit: *Rearrange the equation of the circle as*

$$2z_1 x + 2z_2 y + r^2 - z_1^2 - z_2^2 = x^2 + y^2. \tag{6.110}$$

With $c := r^2 - z_1^2 - z_2^2$, *we obtain with (6.110) for each measured point a linear equation for the unknowns* z_1, z_2 *and* c.

2. Geometric fit: *Use the Gauss-Newton method to minimize the sum of squares of the* distances

$$d_i = |\sqrt{(z_1 - \xi_i)^2 + (z_2 - \eta_i)^2} - r|.$$

Compute and plot in both cases the data points together with the circles. As a second example do the same with the data set

ξ	1	2	5	7	9	3
η	7	6	8	7	5	7

Hint: *A circle with center* (c_1, c_2) *and with radius* r *is best plotted using the parametric representation:*

$$x(t) = c_1 + r\cos t, \quad y(t) = c_2 + r\sin t \quad 0 \le t \le 2\pi.$$

PROBLEM 6.21. *The parametric form commonly used for the circle is given by*

$$\begin{aligned} x &= z_1 + r\cos\varphi \\ y &= z_2 + r\sin\varphi. \end{aligned}$$

The distance d_i of a point $P_i = (x_{i1}, x_{i2})$ may be expressed by

$$d_i^2 = \min_{\varphi_i} \left(x_{i1} - x(\varphi_i)\right)^2 + \left(x_{i2} - y(\varphi_i)\right)^2.$$

Now we want again compute a geometric fit, i.e. determine z_1, z_2 and r by minimizing

$$\sum_{i=1}^{m} d_i^2 \longrightarrow \min.$$

We can simultaneously minimize for z_1, z_2, r and $\{\varphi_i\}_{i=1\ldots m}$; i.e. find the minimum of the quadratic function

$$Q(\varphi_1, \varphi_2, \ldots, \varphi_m, z_1, z_2, r) = \sum_{i=1}^{m} \left(x_{i1} - x(\varphi_i)\right)^2 + \left(x_{i2} - y(\varphi_i)\right)^2.$$

This is equivalent to solving the nonlinear least squares problem

$$\begin{aligned} z_1 + r\cos\varphi_i - x_{i1} &\approx 0 \\ z_2 + r\sin\varphi_i - x_{i2} &\approx 0 \end{aligned} \qquad i = 1, 2, \ldots, m.$$

Solve this problem with the Gauss-Newton method. The Jacobian J is highly structured. Taking in account the structure when solving $J\boldsymbol{h} \approx -\boldsymbol{f}$ develop an effective algorithm.

PROBLEM 6.22. *Write a MATLAB programs to fit two orthogonal lines by minimizing the distance of measured points to the line. Assume that two different sets of points are given for the two lines involved.*

PROBLEM 6.23. *The function*

$$h(x) = 1 - \frac{3}{x}\frac{\sinh x - \sin x}{\cosh x - \cos x}$$

should be fitted to the following data:

x	0.100	0.800	1.500	2.200	2.900	3.600	4.300	5.000	5.700	6.400
y	0.049	0.051	0.153	0.368	0.485	0.615	0.712	0.717	0.799	0.790

Find the value of the parameter a such that $h(ax_i) \approx y_i$ for $i = 1, \ldots, 10$ in the least squares sense.

Solve the problem using the Gauss-Newton and also the Newton method. Compute the function and the derivatives using algorithmic differentiation (see Section 8.3).

PROBLEM 6.24. *We are given the coordinates of the points (ξ_i, η_i), $i = 1, \ldots, m$ in the plane. For a new point $P = (x_1, x_2)$ the distances $s_i = |P_i - P|$ from the given points have been measured:*

x_i	16	65	85	53	16	25
y_i	56	64	37	7	3	32
s_i	32	35	44	30	42	16

Compute the coordinates of P by solving the nonlinear least squares problem.

PROBLEM 6.25. *Fit the function $f(x) = k/(1 + be^{-ax})$ to the data*

x	0	2	3	4	5	6	7	8	9	10	11	12	13	14
y	.145	.19	.248	.29	.78	.78	1.16	1.4	1.94	2.3	2.5	2.8	3.12	3.32

using the MATLAB *function* **nlinfit**. *Plot the points and the fitted curve.*

PROBLEM 6.26. *Determine the parameters a and b such that the function $f(x) = ae^{bx}$ fits the following data*

x	30.0	64.5	74.5	86.7	94.5	98.9
y	4	18	29	51	73	90

Hint: If you fit $\log f(x)$ the problems becomes very easy!

PROBLEM 6.27. *The following program fits a polynomial of degree $n = 5$ to given points. The points are input with the mouse using the* MATLAB *function* **ginput**. *After each new point the coefficients are computed and the polynomial is plotted. The program is inefficient because it does not update the QR decomposition – the solution is fully recalculated for every new point.*

```
format compact
clf
axis([0 10 0 10])              % plot window
hold
degr=5;                        % degree of polynomial
n=degr+1;                      % number of coefficients
h=degr:-1:0                    % to generate matrix row
t=0:0.1:10;                    % to plot polynomial
[x,y]=ginput(1);
plot(x,y,'*r')                 % generate first row
A= x.^h;  b=y;  k=1;
while 1                        % stop with ctrl-w in plot window
  [x,y]=ginput(1);             % get new point
  plot(x,y,'*r')
  k=k+1;                       % count points
  A=[A; x.^h];                 % generate new row
  b=[b;y];                     % and right hand side
  if k>=n
    a=A\b;                     % solve for degrew coefficients
    a'                         % display coefficients
  end
  if k>degr
    p=polyval(a,t);            % evaluate polynomial
    plot(t,p)                  % and plot
  end
end
```

1. *Study the program and run a few examples with different degrees.*

2. *Replace the part between the comment signs so that the solution is updates with Givens rotations or -reflections. Each time when a new point is read the matrix R is updated with n Givens transformations. These Givens transformations annihilate the matrix-elements of the new equation and by back-substitution we obtain the new coefficients of the polynomial. Use the scalar product form for back-substitution.*

PROBLEM 6.28. *Assume you have decomposed a large matrix $A = QR$ and afterward you discover that the element a_{jk} is wrong. Use our update-techniques to fix the QR decomposition.*

Test your Algorithm for the small matrix `A= gallery(5); A=A(:,1:3)`

$$A = \begin{pmatrix} -9 & 11 & -21 \\ 70 & -69 & 141 \\ -575 & 575 & -1149 \\ 3891 & -3891 & 7782 \\ 1024 & -1024 & 2048 \end{pmatrix}$$

Change the element $a_{2,3} = 141$ to $a_{2,3} = 1414$ and compute the new decomposition.

PROBLEM 6.29. *Consider the augmented matrix $\bar{A} = [A, b]$. Show that the following "compact" algorithm solves the least squares problem $Ax \approx b$ and is equivalent with the method of normal equations:*

1. *decompose* $\bar{A}\bar{A}^\mathsf{T} = \bar{L}\bar{L}^\mathsf{T}$ *(Cholesky).*

2. *Set* $R = $ `L(1:n,1:n)`*,* $y = $ `L(n+1,1:n)'` *and* $\rho = $ `L(n+1,n+1)`*.*

3. *Solve* $Rx = y$ *by back-substitution.*

4. $\min \|b - Ax\|_2 = \rho$.

PROBLEM 6.30. *Derivation of modified Gram-Schmidt via matrix decomposition. Let A be a $m \times n$ matrix and consider the decomposition $A = QR$. If we set $L = R^T$ we can view the factorization as an outer product expansion*

$$A = QL^T = \sum_{i=1}^{n} q_i l_i^T$$

where $l_i^T = (0, \ldots, 0, r_{ii}, \ldots, r_{in})$ is the ith column vector of R. Observe that the first $i - 1$ columns of the rank one matrix $q_i l_i^T$ consist entirely of 0's. Define

$$A^{(k)} = A - \sum_{i=1}^{k-1} q_i l_i^T = \sum_{i=k}^{n} q_i l_i^T \quad k = 1, \ldots, n + 1. \tag{6.111}$$

Clearly the recursion holds

$$A^{(1)} = A, \quad A^{(k+1)} = A^{(k)} - \boldsymbol{q}_k \boldsymbol{l}_k^\top, \quad A^{n+1} = 0.$$

Assume now that $k-1$ columns of Q and $k-1$ rows of L are already computed. Multiply the recursion from the right be the unit vector e_k and from the left by $\boldsymbol{q}_k^\top$ to get expressions for the k-th column of Q and the k-th row of L. Write a MATLAB-program to compute this way the QR decomposition.

PROBLEM 6.31. *Consider the matrix A which is constructed by*

```
c=4.11;
m=13;
n=13;
condA_glob=c;
B=inv(pascal(m));
B=B(:,1:n);
[A,R]=qr(B,0);
C=inv(hilb(n));
[B,R]=qr(C,0);
A=A*diag([10.^(0:condA_glob/(n-1):condA_glob)])*B;
[m,n]=size(A);
```

Compute the QR decomposition

1. *with classical Gram-Schmidt*

2. *with modified Gram-Schmidt*

3. *via Cholesky decomposition of $A^\top A$*

4. *with MATLAB's function* $\boldsymbol{qr}$

In each case, test the departure from orthogonality using `norm(eye(13)-Q'*Q)`*. (This matrix was communicated by Alicja Smoktunowicz).*

PROBLEM 6.32. *We are given the following data concerning the growth of pigs. The weight of a pig has been measured over a period of 240 days. The values are given in the following table:*

t	0	10	20	30	40	50	60	70	80
y	1.64	2.68	5.81	7.45	9.98	12.51	15.34	19.07	23.24
t	90	100	110	120	130	140	150	160	170
y	28.90	35.60	42.90	51.39	61.07	69.71	79.54	88.03	95.18
t	180	190	200	210	220	230	240		
y	100.42	105.01	108.07	111.87	115.12	118.01	120.67		

The data suggest an exponential growth of the weight in a first phase followed by an exponential decrease of the growth to a final limit weight (which is not reached since the pigs are transformed to meat before that stage).

Thus it seems reasonable to approximate the data by two piecewise functions

$$F(x) = \begin{cases} f(\boldsymbol{a}, t) &= a_0 + a_1 \exp(a_3 t) & t < \xi \\ g(\boldsymbol{b}, t) &= b_0 + b_1 \exp(b_3 t) & t > \xi \end{cases}$$

We expect that by a least squares fit we will obtain exponents $a_3 > 0$ and $b_3 < 0$. The break point ξ is a free knot and the two functions should have the same value and the same derivatives for $t = \xi$.

Use the theory developed in Section 6.9 to determine the parameters by the Gauss-Newton method.

Depending on the initial values, you may have a hard time to converge to a solution. Use the Levenberg-Marquardt correction to avoid large correction steps.

Chapter 7. Eigenvalue Problems

> *The solution of the algebraic eigenvalue problem has for long had a particular fascination for me because it illustrates so well the difference between what might be termed classical mathematics and practical numerical analysis. The eigenvalue problem has a deceptively simple formulation and the background theory has been known for many years; yet the determination of accurate solutions presents a wide variety of challenging problems.*
>
> James Wilkinson, The Algebraic Eigenvalue Problem, Oxford University Press, 1988.

> *The interesting differences between various eigenvalue problems disappear when the formulation is made sufficiently abstract and the practical problems reduce to the single task of computing the eigenvalues of a square matrix with real or complex entries. Nevertheless there is one distinction worth maintaining: some matrices have real eigenvalues and others do not. The former usually come from so-called self-adjoint problems which are much nicer than the others.*
>
> Beresford Parlett, The Symmetric Eigenvalue Problem, SIAM Classics in Applied Mathematics, 1998.

Eigenvalue problems are ubiquitous in science and engineering; they occur whenever something is oscillating in a periodic motion. Historically, a very famous eigenvalue problem is related to the beautiful *Chladni figures*, which appear when fine grained sand or dust on a vibrating plate magically organizes itself to reveal the nodal lines of the vibrating plate [37]. In this same reference, the spectacular failure of the *Tacoma bridge* is shown also to be related to an eigenvalue problem. Eigenvalue problems in science and engineering are often formulated at the level of the differential equation, but the essential features of eigenvalue problems can be studied at the matrix level, see the second quote above. We start this chapter with an introductory example in Section 7.1, where a simple mass-spring problem modeled by a differential equation is reduced to a matrix eigenvalue problem. This section contains a nice MATLAB animation to illustrate such problems. In Section 7.2, we introduce the classical matrix eigenvalue problem, and review several important theoretical aspects of it. Since eigenvalues are the roots of the characteristic polynomial of a matrix, they can only be computed iteratively, as there are no closed form formulas for roots of polynomials of degree higher than four. Section 7.3 contains the oldest iterative algorithm

W. Gander et al., *Scientific Computing - An Introduction using Maple and MATLAB*,
Texts in Computational Science and Engineering 11,
DOI 10.1007/978-3-319-04325-8_7,

for computing eigenvalues, due to Jacobi, for the computation of planetary motion. We explain in detail a beautiful and sophisticated implementation due to Rutishauser. Today's most popular eigenvalue algorithms are, however, based on a different technique: the simple power method described in Section 7.4, together with its extension to orthogonal iterations. Modern algorithms also perform reductions to simpler form, described in Section 7.5, before starting the iteration. We then describe the QR Algorithm in Section 7.6, which is nowadays the most widely used algorithm for dense eigenvalue problems. We describe in detail the implicit version with shifts, and also convergence tests and deflation, which are essential to the success of the algorithm. The computation of the singular value decomposition (SVD) is the topic of Section 7.7. This decomposition remained for the longest time just a theoretical tool, until Golub and Reinsch found an efficient algorithm to compute it. We conclude this chapter with the description of an advanced version of the QD Algorithm, which is important whenever very high accuracy is required in the eigenvalues. The most important references for eigenvalue problems are Wilkinson [150], and Parlett for the symmetric eigenvalue problem [104]. For very recent results, see [80].

7.1   Introductory Example

We start this chapter by considering a mechanical system. Figure 7.1 shows three masses m connected with springs. We denote with $u_i(t)$ the displacements of the masses from the positions of rest.

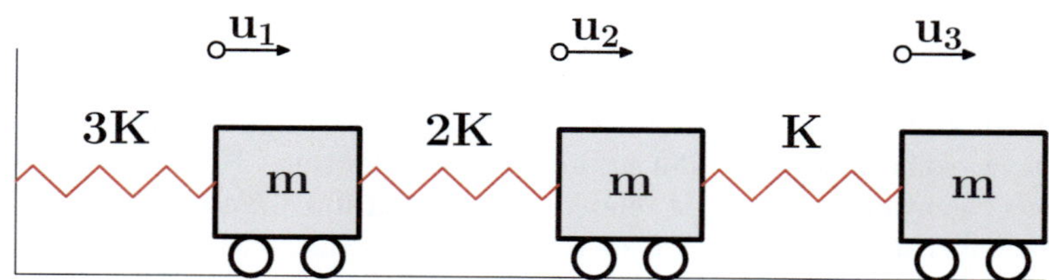

FIGURE 7.1. *Mass Spring System*

With the spring constants $3K$, $2K$ and K, the equations of motion according to Newton's law, $m\boldsymbol{u}'' = \boldsymbol{F}$, become

$$mu_1'' = -3Ku_1 - 2K(u_1 - u_2),$$
$$mu_2'' = -2K(u_2 - u_1) - K(u_2 - u_3),$$
$$mu_3'' = -K(u_3 - u_2).$$

Defining

$$\boldsymbol{u}(t) = \begin{pmatrix} u_1(t) \\ u_2(t) \\ u_3(t) \end{pmatrix}, \quad A = \begin{pmatrix} 5 & -2 & 0 \\ -2 & 3 & -1 \\ 0 & -1 & 1 \end{pmatrix},$$

we write these equations in matrix-vector notation

$$\mathcal{L}(\boldsymbol{u}) := -\frac{m}{K}\boldsymbol{u}'' - A\boldsymbol{u} = 0. \tag{7.1}$$

The operator $\mathcal{L}(\boldsymbol{u})$ is linear, which means that the solutions of (7.1) form a linear space — any linear combination of solutions is again a solution.

We are looking for special solutions using the ansatz

$$u_i(t) = v_i \cos(\omega t) \implies \boldsymbol{u}(t) = \boldsymbol{v}\cos(\omega t),$$

where v_i and ω are to be determined. Inserting the ansatz into (7.1), we obtain

$$-\frac{m}{K}\left(-\boldsymbol{v}\,\omega^2\cos(\omega t)\right) - A\boldsymbol{v}\cos(\omega t) = 0,$$

and dividing by $\cos(\omega t)$ leads to

$$A\boldsymbol{v} = \lambda\boldsymbol{v} \quad \text{with} \quad \lambda := \frac{m}{K}\omega^2. \tag{7.2}$$

Equation (7.2) defines an *eigenvalue problem*: we are looking for a vector $\boldsymbol{v} \neq 0$ and a scalar λ such that

$$A\boldsymbol{v} = \lambda\boldsymbol{v} \quad \Longleftrightarrow \quad (\lambda I - A)\boldsymbol{v} = 0.$$

A nontrivial solution $\boldsymbol{v} \neq 0$ can only exist if the matrix $\lambda I - A$ is singular, that is, if

$$P(\lambda) = \det(\lambda I - A) = 0. \tag{7.3}$$

The function $P(\lambda)$ is called the *characteristic polynomial* of the matrix A. A zero of $P(\lambda)$ is called an *eigenvalue* of A and a nontrivial solution $\boldsymbol{v}$ is a corresponding *eigenvector*.

In MATLAB, the command `poly` computes the coefficients of $P(\lambda)$

```
>> A=[5      -2       0
     -2       3      -1
      0      -1       1];
>> P=poly(A)
P =
    1.0000   -9.0000   18.0000   -6.0000
```

thus $P(\lambda) = \det(\lambda I - A) = \lambda^3 - 9\lambda^2 + 18\lambda - 6$. With `roots` we obtain three solutions, i.e., the three eigenvalues,

```
>> lambda=roots(P)
lambda =
    6.2899
    2.2943
    0.4158
```

To each eigenvalue, the associated eigenvector $\boldsymbol{v}^{(i)}$ is a solution of the homogeneous system

$$(A - \lambda_i I)\boldsymbol{v}^{(i)} = 0.$$

For instance for λ_1:

```
>> B=A-lambda(1)*eye(size(A))
B =
   -1.2899    -2.0000          0
   -2.0000    -3.2899    -1.0000
         0    -1.0000    -5.2899
```

and we have to find a nontrivial solution of

$$\begin{pmatrix} -1.2899 & -2.0000 & 0 \\ -2.0000 & -3.2899 & -1.0000 \\ 0 & -1.0000 & -5.2899 \end{pmatrix} \begin{pmatrix} v_1 \\ v_2 \\ v_3 \end{pmatrix} = \begin{pmatrix} 0 \\ 0 \\ 0 \end{pmatrix}. \tag{7.4}$$

Since $\boldsymbol{v}$ is only determined up to a constant factor, we may choose one component freely. Assuming $v_1 = 1$, we get the linear system (7.4)

$$\begin{pmatrix} -2.0000 & 0 \\ -3.2899 & -1.0000 \\ -1.0000 & -5.2899 \end{pmatrix} \begin{pmatrix} v_2 \\ v_3 \end{pmatrix} = \begin{pmatrix} 1.2899 \\ 2.0000 \\ 0 \end{pmatrix}. \tag{7.5}$$

which, when solved with MATLAB, gives

```
>> c=-B(1:2,2:3)\B(1:2,1)
c =
   -0.6450
    0.1219
>> c=[1;c]
c =
    1.0000
   -0.6450
    0.1219
>> Check=B*c
Check   =
   1.0e-13 *
          0
          0
     0.3764
```

Perfect! Continuing this way, our ansatz gives us three solutions:

$$\boldsymbol{u}^{(i)}(t) = \boldsymbol{v}^{(i)} \cos(\omega_i t) \quad \text{where} \quad \omega_i = \sqrt{\frac{K}{m}\lambda_i}.$$

A second ansatz $\boldsymbol{u}(t) = \boldsymbol{v}\sin(\omega t)$ yields three more solutions for the same eigenvalues and eigenvectors, thus all together we have 6 linearly independent solutions. The general solution is a combination of all of them. In matrix-vector notation, we get

$$\Omega = \operatorname{diag}(\omega_1, \omega_2, \omega_3) = \begin{pmatrix} \omega_1 & & \\ & \omega_2 & \\ & & \omega_3 \end{pmatrix},$$

$$\cos(\Omega) = \begin{pmatrix} \cos\omega_1 t & & \\ & \cos\omega_2 t & \\ & & \cos\omega_3 t \end{pmatrix},$$

and

$$\sin(\Omega) = \begin{pmatrix} \sin\omega_1 t & & \\ & \sin\omega_2 t & \\ & & \sin\omega_3 t \end{pmatrix}.$$

The general solution with arbitrary parameters

$$\boldsymbol{a} = \begin{pmatrix} \alpha_1 \\ \alpha_2 \\ \alpha_3 \end{pmatrix}, \quad \boldsymbol{b} = \begin{pmatrix} \beta_1 \\ \beta_2 \\ \beta_3 \end{pmatrix}$$

becomes

$$\boxed{\boldsymbol{u}(t) = V\left(\cos(\Omega t)\boldsymbol{a} + \sin(\Omega t)\boldsymbol{b}\right) \quad \text{with} \quad V = [\boldsymbol{v}^{(1)}, \boldsymbol{v}^{(2)}, \boldsymbol{v}^{(3)}].}$$

The following nice MATLAB program `Swing.m`, written by Oscar Chinellato, shows the three different eigenmodes and the motion of the masses as a linear combination of them.

ALGORITHM 7.1.

Simulation of a Mechanical System with Three Masses

```matlab
close all; m=[1 1 1]'; k=[3 2 1]';         % masses and spring constants
                                           % rest positions and initial
lo=[1 1 1]'; u0=[0.1, 0.1, 0.9]';          % displacements
A=[(-k(1)-k(2))/m(1), k(2)/m(1),           0
   k(2)/m(2),         (-k(2)-k(3))/m(2), k(3)/m(2)
   0,                 k(3)/m(3),         -k(3)/m(3)];
[U,l]=eig(A); l=diag(l); sl=sqrt(l); Ux=[U,u0];
w=1;                                       % w is the box width
for sp=1:4                                 % draw four plots:
  subplot(4,1,sp);                         % 1-3: eigensolutions
  axis([-1,10,0,3]);                       % 4: composite solution
  axis equal; axis off; hold on;
  o(1)=lo(1);                              % offsets for drawing
  o(2)=lo(1)+Ux(1,sp)+w+lo(2);
  o(3)=lo(1)+Ux(1,sp)+w+lo(2)+Ux(2,sp)+w+lo(3);
  bs(:,sp)=[                               % box plots
```

```
  plot([o(1)+Ux(1,sp),o(1)+Ux(1,sp)+w,o(1)+Ux(1,sp)+w,o(1)+ ...
    Ux(1,sp),o(1)+Ux(1,sp)],[0,0,w,w,0],'k-');
  plot([o(2)+Ux(2,sp),o(2)+Ux(2,sp)+w,o(2)+Ux(2,sp)+w,o(2)+ ...
    Ux(2,sp),o(2)+Ux(2,sp)],[0,0,w,w,0],'k-');
  plot([o(3)+Ux(3,sp),o(3)+Ux(3,sp)+w,o(3)+Ux(3,sp)+w,o(3)+ ...
    Ux(3,sp),o(3)+Ux(3,sp)],[0,0,w,w,0],'k-'); ];
  ss(:,sp)=[                              % spring plots
  plot([0:(o(1)+Ux(1,sp))/10:o(1)+Ux(1,sp)], ...
    [w/2+w/10,w/2+w/10*cos((1:9)*pi),w/2+w/10],'b-');
  plot([o(1)+Ux(1,sp)+w:(o(2)-o(1)+Ux(2,sp)-Ux(1,sp)-w)/10: ...
    o(2)+Ux(2,sp)],[w/2+w/10,w/2+w/10*cos((1:9)*pi),w/2+w/10],'b-');
  plot([o(2)+Ux(2,sp)+w:(o(3)-o(2)+Ux(3,sp)-Ux(2,sp)-w)/10: ...
    o(3)+Ux(3,sp)],[w/2+w/10,w/2+w/10*cos((1:9)*pi),w/2+w/10],'b-');
  ];
  tmp=plot([0,0,10],[2*w,0,0],'k-');       % draw axes
  set(tmp,'LineWidth',3);
end
drawnow;                                   % draw initial positions
t=0; dt=0.04;                              % perform simulation in time
while 1
  t=t+dt;                                  % compute the new position
  v=real(U*diag(exp(sl*t))*(U\Ux));
  for sp=1:4
    o(1)=10(1);                            % update the drawing
    o(2)=10(1)+v(1,sp)+w+10(2);
    o(3)=10(1)+v(1,sp)+w+10(2)+v(2,sp)+w+10(3);
    set(bs(1,sp),'XData',[o(1)+v(1,sp),o(1)+v(1,sp)+ ...
      w,o(1)+v(1,sp)+w,o(1)+v(1,sp),o(1)+v(1,sp)]);
    set(bs(2,sp),'XData',[o(2)+v(2,sp),o(2)+v(2,sp)+ ...
      w,o(2)+v(2,sp)+w,o(2)+v(2,sp),o(2)+v(2,sp)]);
    set(bs(3,sp),'XData',[o(3)+v(3,sp),o(3)+v(3,sp)+ ...
      w,o(3)+v(3,sp)+w,o(3)+v(3,sp),o(3)+v(3,sp)]);
    set(ss(1,sp),'XData',[0:(o(1)+v(1,sp))/10:o(1)+v(1,sp)]);
    set(ss(2,sp),'XData',[o(1)+v(1,sp)+w:(o(2)-o(1)+v(2,sp) ...
      -v(1,sp)-w)/10:o(2)+v(2,sp)]);
    set(ss(3,sp),'XData',[o(2)+v(2,sp)+w:(o(3)-o(2)+v(3,sp) ...
      -v(2,sp)-w)/10:o(3)+v(3,sp)]);
  end
  drawnow;                                 % draw updated positions
end
```

Figure 7.2 shows a snapshot of the animation.

7.2 A Brief Review of the Theory

7.2.1 Eigen-Decomposition of a Matrix

Let $\boldsymbol{v}_1$ be an eigenvector of $A \in \mathbb{R}^{n \times n}$ belonging to the eigenvalue λ_1. Then $A\boldsymbol{v}_1 = \boldsymbol{v}_1\lambda_1$. Assuming that A is *diagonalizable*, which means that there exist

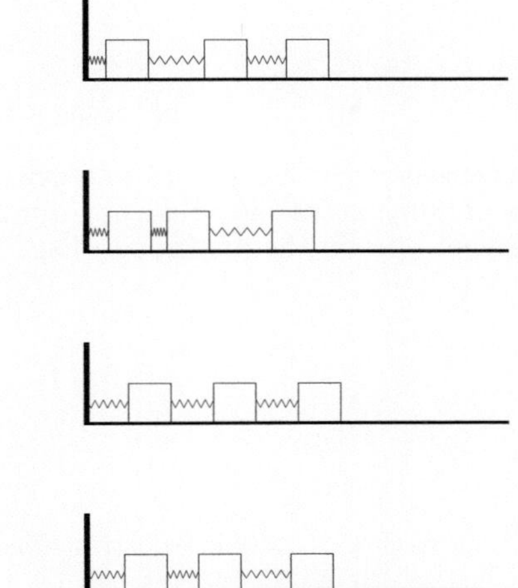

FIGURE 7.2. *A snapshot of the swinging masses*

n linearly independent eigenvectors $\boldsymbol{v}_j$, $j = 1, 2, \ldots, n$ with corresponding eigenvalues λ_j, we have analogously

$$A[\boldsymbol{v}_1, \boldsymbol{v}_2, \ldots, \boldsymbol{v}_n] = [\boldsymbol{v}_1\lambda_1, \boldsymbol{v}_2\lambda_2, \ldots, \boldsymbol{v}_n\lambda_n] = [\boldsymbol{v}_1, \boldsymbol{v}_2, \ldots, \boldsymbol{v}_n]\,\mathrm{diag}(\lambda_1, \ldots, \lambda_n).$$

Therefore we get

$$AV = VD \quad \text{with} \quad V = [\boldsymbol{v}_1, \boldsymbol{v}_2, \ldots, \boldsymbol{v}_n] \quad \text{and} \quad D = \begin{pmatrix} \lambda_1 & & & \\ & \lambda_2 & & \\ & & \ddots & \\ & & & \lambda_n \end{pmatrix},$$

which leads to the *eigen-decomposition* of the matrix A,

$$A = VDV^{-1}. \tag{7.6}$$

In MATLAB, this decomposition is computed with `[V,D]=eig(A)`. For our mass-spring example, the eigenmodes can be computed directly as follows:

```
>> A=[5 -2 0; -2 3 -1; 0 -1 1]
A =
     5    -2     0
    -2     3    -1
     0    -1     1
>> [V,D]=eig(A)
V =
    0.21493527624832   -0.50489606854382   -0.83599209744653
    0.49265588101116   -0.68305360433845    0.53919194773787
    0.84326330996266    0.52774779352380   -0.10192770232663
```

```
D =
    0.41577455678348                   0                   0
                   0    2.29428036027904                   0
                   0                   0    6.28994508293748
```

Unfortunately, the eigen-decomposition does not exist for all matrices, because not all matrices are diagonalizable. Take as example a *Jordan block* with all zero elements except ones in the first upper diagonal:

```
>> n=3;
>> J=diag(ones(n-1,1),1)
J =
     0     1     0
     0     0     1
     0     0     0
```

We now generate a well-conditioned matrix U and transform J to the similar matrix A

```
>> rand('state',0);          % to make the results reproducible
>> U=round(100*rand(n,n))
U =
    95    49    46
    23    89     2
    61    76    82
>> cond(U)
ans =
    4.8986
```

Let us form $A = UJU^{-1}$ and compute its eigen-decomposition:

```
>> A=U*J*inv(U)
>> [V,D]=eig(A)
A =
   -0.8224     0.6291     1.0435
   -0.8702    -0.6191     1.5886
   -0.9155    -0.0414     1.4414
V =
    0.8245               0.8245              0.8245
    0.1996               0.1996 + 0.0000i    0.1996 - 0.0000i
    0.5294               0.5294 + 0.0000i    0.5294 - 0.0000i
D =
    1.0e-05 *
    0.6238               0                   0
         0              -0.3119 + 0.5402i    0
         0               0                  -0.3119 - 0.5402i
```

We see that the eigenvector matrix V is a rank-one matrix, so the eigen-decomposition (7.6) does not exist. The column vectors of V are all the same, which means that there exists only *one* eigenvector. A should have the same eigenvalues as J, i.e., a triple eigenvalue zero. However, numerically the eigenvalue zero has "exploded" into three simple eigenvalues of modulus $\approx 10^{-5}$. In exact arithmetic, A is a *defective* matrix, since it has only one eigenvector for the multiple eigenvalue zero, just like the Jordan block.

Numerically, it can happen that a defective matrix is not detected. The eigen-decomposition can usually be computed because, in finite precision arithmetic, we do not compute the eigen-decomposition of A, but that of a perturbed matrix $A + E$ with $\|E\| \leq \varepsilon \|A\|$. An indication that the matrix may be defective is a high condition number for the eigenvector matrix V. An example is the matrix `gallery(5)`:

```
>> A=gallery(5)
A =
          -9          11         -21          63        -252
          70         -69         141        -421        1684
        -575         575       -1149        3451      -13801
        3891       -3891        7782      -23345       93365
        1024       -1024        2048       -6144       24572
>> rank(A)
ans =
     4
```

Thus, one eigenvalue must be zero. In fact,

```
>> [V,D]=eig(A);
>> V
V =
    0.0000  -0.0000+0.0000i  -0.0000-0.0000i   0.0000+0.0000i   0.0000-0.0000i
   -0.0206   0.0206+0.0001i   0.0206-0.0001i   0.0207+0.0001i   0.0207-0.0001i
    0.1398  -0.1397+0.0001i  -0.1397-0.0001i  -0.1397+0.0000i  -0.1397-0.0000i
   -0.9574   0.9574           0.9574           0.9574           0.9574
   -0.2519   0.2519-0.0000i   0.2519+0.0000i   0.2519-0.0000i   0.2519+0.0000i
>> diag(D)
ans =
   -0.0405
   -0.0118 + 0.0383i
   -0.0118 - 0.0383i
    0.0320 + 0.0228i
    0.0320 - 0.0228i
```

Again we see that A is defective (the columns of V look all the same, and the condition number is `cond(V)=1.0631e+11`), so there is only one eigenvector. A has an eigenvalue zero with multiplicity 5. If we compute the characteristic polynomial we get

```
>> poly(A)
ans =
    1.0000    0.0000   -0.0000    0.0000   -0.0000    0.0000
```

which looks very much like $P_5(\lambda) = \lambda^5$.

7.2.2　Characteristic Polynomial

We already defined the *characteristic polynomial*, which can be written also in factored form,

$$
\begin{aligned}
P_n(\lambda) &= \det(\lambda I - A) \\
&= \lambda^n + c_2\lambda^{n-1} + \cdots + c_n\lambda + c_{n+1} \\
&= (\lambda - \lambda_1)(\lambda - \lambda_2)\cdots(\lambda - \lambda_n).
\end{aligned}
$$

Expanding the factors and comparing coefficients, we get the well-known identities for the *trace* and the *determinant* of a matrix:

$$
c_2 = \sum_{i=1}^{n} a_{ii} = \sum_{i=1}^{n} \lambda_i = \text{trace of } A,
$$

and

$$
c_{n+1} = \det(A) = \lambda_1 \cdots \lambda_n.
$$

As seen already, the MATLAB command `c=poly(A)` computes the coefficients c_i. The corresponding command in MAPLE is

```
> with(LinearAlgebra);
> CharacteristicPolynomial(A,lambda);
```

To compute the roots of a polynomial

$$
P_n(\lambda) = c_1\lambda^n + c_2\lambda^{n-1} + \cdots + c_n\lambda + c_{n+1},
$$

MATLAB uses the function `roots(c)`. Thus we could compute the eigenvalues of a matrix A with `roots(poly(A))`. As shown in Chapter 5, this is not a good idea: small changes in the coefficients of a polynomial may change the roots dramatically, the problem is ill-conditioned. The MATLAB command `eig(A)` computes the eigenvalues of the matrix A directly using similarity transformations without computing the characteristic polynomial, and we will see the underlying algorithm in Section 7.6.

7.2.3　Similarity Transformations

Two matrices A and B are *similar* if an invertible matrix T exists such that

$$
A = TBT^{-1}.
$$

Similar matrices have the same characteristic polynomial, and hence the same eigenvalues, the same determinant and the same trace. The eigendecomposition (7.6) is also a similarity transformation of the matrix A: solving (7.6) for D, we get

$$
V^{-1}AV = D.
$$

Numerical methods like the QR Algorithm compute the eigenvalues using a sequence of *similarity transformations*. The matrix is transformed into a diagonal or triangular matrix, and its diagonal elements are the eigenvalues of the original matrix A.

Every real symmetric matrix A is similar to a symmetric tridiagonal matrix T. The transformation matrix can be constructed as a product of a finite sequence of elementary orthogonal matrices, see Subsection 7.5.3. A non-symmetric matrix can be reduced by a finite number of similarity transformations to *Hessenberg form* (see Subsection 7.5.2).

7.2.4 Diagonalizable Matrices

DEFINITION 7.1. (GEOMETRIC AND ALGEBRAIC MULTIPLICITY) *The* geometric multiplicity of an eigenvalue λ *is the dimension of its eigenspace, i.e., the dimension of the null space of $\lambda I - A$. The* algebraic multiplicity of an eigenvalue λ *is the multiplicity of λ as root of the characteristic polynomial.*

For example, the null space of a Jordan block `J = diag(ones(n-1,1),1)` consists only of the vector e_1, and thus is one dimensional. The geometric multiplicity is one. In contrast, the characteristic polynomial is $\det(\lambda I - J) = \lambda^n$ and therefore the algebraic multiplicity is n.

The algebraic and geometric multiplicities of a matrix A is invariant under similarity transformations. Thus, it follows that A is diagonalizable if and only if the algebraic multiplicity equals the geometric multiplicity for every eigenvalue.

Next, we state a few facts regarding the diagonalizability of certain classes of matrices. For a proof of these statements, consult a linear algebra textbook e.g. [85]:

- Let A^{H} denote the conjugate transpose of A. Then if A is normal, i.e., if $A^{\mathrm{H}}A = AA^{\mathrm{H}}$, then A is diagonalizable.

- Eigenvectors belonging to *different* eigenvalues are linearly independent.

  Therefore: if the matrix A has no multiple eigenvalues, then it is diagonalizable and the eigenvector matrix V is invertible: $A = VDV^{-1}$.

- Real symmetric matrices have only real eigenvalues and they are always diagonalizable. There exists an orthogonal basis of eigenvectors. Thus the eigen-decomposition is

$$AV = VD \quad \Longleftrightarrow \quad A = VDV^{\top}, \quad V^{\top}V = I \quad \text{orthogonal.}$$

7.2.5 Exponential of a Matrix

For a diagonalizable matrix $A = VDV^{-1}$, we have

$$A^2 = VDV^{-1}VDV^{-1} = VD^2V^{-1}.$$

Thus, if λ is an eigenvalue of A then its square, λ^2, is an eigenvalue of A^2 and the corresponding eigenvector is the same.

For the exponential function applied to a matrix, we can use the series expansion of the exponential to obtain

$$
\begin{aligned}
e^A &= I + \frac{A}{1!} + \frac{A^2}{2!} + \cdots + \frac{A^n}{n!} + \cdots \\
&= I + \frac{VDV^{-1}}{1!} + \frac{VD^2V^{-1}}{2!} + \cdots + \frac{VD^nV^{-1}}{n!} + \cdots \\
&= V\left(I + \frac{D}{1!} + \frac{D^2}{2!} + \cdots + \frac{D^n}{n!} + \cdots\right) V^{-1} \\
&= Ve^D V^{-1}.
\end{aligned}
$$

Notice that

$$
e^D = \begin{pmatrix} e^{\lambda_1} & & & \\ & e^{\lambda_2} & & \\ & & \ddots & \\ & & & e^{\lambda_n} \end{pmatrix},
$$

and therefore for quite general functions, e.g., those having a power series expansion like $f(x) = e^x$, the function applied to a matrix can be computed by

$$
f(A) = V \ \mathrm{diag}(f(\lambda_i)) \ V^{-1}.
$$

The exponential of the matrix e^A is computed in MATLAB[1] by `expm(A)`. The ending "m" after `exp` denotes a matrix function. Similarly, the command `W=sqrtm(A)` computes a matrix W such that $W^2 = A$. General *matrix functions* are computed in MATLAB with the function `funm`.

As an application of the exponential of a matrix, consider a *homogeneous linear system of differential equations* with constant coefficients,

$$
\boldsymbol{y}'(t) = A\,\boldsymbol{y}(t), \quad \boldsymbol{y}(0) = \boldsymbol{y}_0.
$$

The system has the solution

$$
\boldsymbol{y}(t) = e^{At}\boldsymbol{y}_0.
$$

7.2.6 Condition of Eigenvalues

As we have seen in Section 7.2.1, eigenvalues of non-symmetric matrices may be ill conditioned. In order to be more precise, let us consider, for a given matrix A, the family of matrices

$$
A(\epsilon) := A + \epsilon C, \tag{7.7}
$$

---

[1] See 'Nineteen Dubious Ways to Compute the Exponential of a Matrix' [93].

where the matrix C satisfies $\|C\| \leq \|A\|$. We first show that the eigenvalues $\lambda(\epsilon)$ of $A(\epsilon)$ are continuous functions of ϵ. Next, we show that if $\lambda(0)$ is a simple eigenvalue of A, then $\lambda(\epsilon)$ is also differentiable[2].

THEOREM 7.1. (CONTINUITY OF EIGENVALUES) *Let $P_A(\lambda)$ be the characteristic polynomial of A,*

$$P_A(\lambda) = \det(A - \lambda I) = (-1)^n \prod_{j=1}^{k} (\lambda - \lambda_j)^{m_j}, \qquad (7.8)$$

and let $\rho > 0$ be such that the discs $D_i := \{\lambda \in \mathbb{C}; |\lambda - \lambda_i| \leq \rho\}$ are disjoint. Then for $|\epsilon|$ sufficiently small, and for $i = 1, \ldots, k$, exactly m_i eigenvalues of $A(\epsilon) = A + \epsilon C$ (counted with their multiplicity) are in the disc D_i.

PROOF. To prove this result, we can use Rouché's Theorem, which states that if two functions $f(\lambda)$ and $g(\lambda)$ are analytic in the interior of the disc $D = \{\lambda; |\lambda - a| \leq \rho\}$, continuous on the boundary ∂D and if they satisfy $|f(\lambda) - g(\lambda)| < |f(\lambda)|$ on the boundary ∂D, then the two functions $f(\lambda)$ and $g(\lambda)$ have precisely the same number of zeros in the interior of D.

Setting $f(\lambda) := P_A(\lambda)$ and $g(\lambda) := P_{A+\epsilon C}(\lambda)$, we see from (7.8) that for $\lambda \in \partial D_i$, we have $|P_A(\lambda)| \geq \rho^n > 0$. We now show that the difference $P_{A+\epsilon C}(\lambda) - P_A(\lambda)$ contains the factor ϵ. Using the Leibniz formula for determinants (3.11) from Chapter 3,

$$\det(A) = \sum_{\boldsymbol{k}} (-1)^{\delta(\boldsymbol{k})} \prod_{j=1}^{n} a_{j,k_j},$$

we see that $\det(A)$ is a polynomial of its arguments, so that $\det(A - \lambda I + \epsilon C)$ is a polynomial in λ and ε, meaning it can be written as

$$\det(A - \lambda I + \varepsilon C) = p(\lambda) + \epsilon q(\lambda, \epsilon),$$

where $p(\lambda)$ and $q(\lambda, \epsilon)$ are polynomials. Moreover, by setting $\epsilon = 0$, we see that

$$p(\lambda) = \det(A - \lambda I) = P_A(\lambda).$$

thus, we have $\det(A - \lambda I + \epsilon C) - \det(A - \lambda I) = \epsilon q(\lambda, \epsilon)$. But the polynomial $q(\lambda, \epsilon)$ is bounded on the compact set $\partial D_i \times [-1, 1]$, say by $C_1 > 0$. Hence, for $|\epsilon| < \min(\rho^n/C_1, 1)$, we have

$$|P_{A+\epsilon C}(\lambda) - P_A(\lambda)| \leq |\epsilon| C_1 < \rho^n \leq |P_A(\lambda)|, \quad \text{for} \quad \lambda \in \partial D_i,$$

and Rouché's Theorem therefore implies that $P_{A+\epsilon C}(\lambda)$ and $P_A(\lambda)$ have the same number of zeros in D_i. $\qquad\square$

THEOREM 7.2. (DIFFERENTIABILITY OF EIGENVALUES) *Let λ_1 be a simple eigenvalue of A. Then, for $|\epsilon|$ sufficiently small, the matrix $A(\epsilon) := A + \epsilon C$*

---

[2]The formulation of these results here goes back to Ernst Hairer's numerical analysis course in Geneva.

has a unique eigenvalue $\lambda_1(\epsilon)$ close to λ_1. The function $\lambda_1(\epsilon)$ is analytic and has the expansion

$$\lambda_1(\epsilon) = \lambda_1 + \epsilon \frac{\boldsymbol{u}_1^{\mathrm{H}} C \boldsymbol{v}_1}{\boldsymbol{u}_1^{\mathrm{H}} \boldsymbol{v}_1} + O(\epsilon^2), \tag{7.9}$$

where $\boldsymbol{v}_1$ is the right eigenvector corresponding to λ_1, $A\boldsymbol{v}_1 = \lambda_1 \boldsymbol{v}_1$, and $\boldsymbol{u}_1$ is the left eigenvector corresponding to λ_1, $\boldsymbol{u}_1^{\mathrm{H}} A = \lambda_1 \boldsymbol{u}_1^{\mathrm{H}}$.

PROOF. With $p(\lambda, \epsilon) := P_{A+\epsilon C}(\lambda)$, we have

$$p(\lambda_1, 0) = 0, \quad \frac{\partial p}{\partial \lambda}(\lambda_1, 0) \neq 0,$$

since the eigenvalue is simple. Hence the implicit function theorem gives us a differentiable (even analytic) function $\lambda_1(\epsilon)$ in a neighborhood of $\epsilon = 0$, such that $\lambda_1(0) = \lambda_1$ and $p(\lambda_1(\epsilon), \epsilon) = 0$. Therefore there exists a vector $\boldsymbol{v}_1(\epsilon) \neq \boldsymbol{0}$ such that

$$(A(\epsilon) - \lambda_1(\epsilon)I)\boldsymbol{v}_1(\epsilon) = 0. \tag{7.10}$$

The matrix in (7.10) has rank $n - 1$, which means we can fix one component to equal 1, and then use Cramer's rule, see Theorem 3.1, to see that the other components of the vector $\boldsymbol{v}_1$ are rational functions of the elements in the matrix $A + \epsilon C - \lambda_1(\epsilon)C$, and hence differentiable.

In order to compute $\lambda_1'(0)$, we differentiate equation (7.10) with respect to ϵ to obtain, after setting ϵ to zero,

$$(A - \lambda_1 I)\boldsymbol{v}_1'(0) + (C - \lambda_1'(0)I)\boldsymbol{v}_1 = 0.$$

Now multiplying this relation with $\boldsymbol{u}_1^{\mathrm{H}}$, we obtain $\boldsymbol{u}_1^{\mathrm{H}}(C - \lambda_1'(0)I)\boldsymbol{v}_1 = 0$, which proves (7.9). $\qquad\square$

DEFINITION 7.2. (NORMAL MATRIX) *A complex square matrix A is a normal matrix if it commutes with its conjugate transpose: $A^{\mathrm{H}}A = AA^{\mathrm{H}}$.*

A normal matrix can be converted to a diagonal matrix by a unitary transformation, and every matrix that can be diagonalized by a unitary matrix is also normal.

For a normal matrix A, there exists a unitary matrix V such that $V^{\mathrm{H}}AV = \mathrm{diag}(\lambda_1, \ldots, \lambda_n)$, which means we have $\boldsymbol{u}_1 = \boldsymbol{v}_1$ in the notation of Theorem 7.2. Therefore (7.9) gives, up to $O(\epsilon^2)$,

$$|\lambda_1(\epsilon) - \lambda_1| \leq \epsilon \|C\|, \tag{7.11}$$

since $|\boldsymbol{v}_1^{\mathrm{H}} C \boldsymbol{v}_1| \leq \|\boldsymbol{v}_1\| \cdot \|C\| \cdot \|\boldsymbol{v}_1\|$. Hence, the computation of a simple eigenvalue of a normal matrix (for example a symmetric or skew-symmetric matrix) is well conditioned, using the definition of well-conditioning from Section 2.5.

On the other hand, if the matrix is non-normal, the computation of λ_1 can be very ill conditioned. Take for example the matrix

$$A = \begin{pmatrix} 1 & a \\ 0 & 2 \end{pmatrix}.$$

The eigenvalue $\lambda_1 = 1$ has the right and left normalized eigenvectors

$$\boldsymbol{v}_1 = \begin{pmatrix} 1 \\ 0 \end{pmatrix}, \quad \boldsymbol{u}_1 = \frac{1}{\sqrt{1+a^2}} \begin{pmatrix} 1 \\ -a \end{pmatrix},$$

and (7.9) gives us the estimate

$$\lambda_1(\epsilon) - \lambda_1 = \epsilon(c_{11} - ac_{21}) + O(\epsilon^2).$$

If we let $C = \begin{pmatrix} 0 & 0 \\ 1 & 0 \end{pmatrix}$, we obtain $\lambda_1(\epsilon) - \lambda_1 = -a\epsilon + O(\epsilon^2)$, which shows that for large values of a, the computation of $\lambda_1 = 1$ is ill conditioned.

A further typical example is the following: consider again a Jordan block

```
>> J=diag(ones(6,1),1)
J =
     0     1     0     0     0     0     0
     0     0     1     0     0     0     0
     0     0     0     1     0     0     0
     0     0     0     0     1     0     0
     0     0     0     0     0     1     0
     0     0     0     0     0     0     1
     0     0     0     0     0     0     0
```

The matrix J has an eigenvalue zero with multiplicity 7. Let us change the element $J_{7,1}$ slightly:

```
>> J(7,1)=1e-12
>> eig(J)
ans =
  -0.0174 + 0.0084i
  -0.0174 - 0.0084i
  -0.0043 + 0.0188i
  -0.0043 - 0.0188i
   0.0193
   0.0120 + 0.0151i
   0.0120 - 0.0151i
```

We obtain a huge change in the eigenvalues. In order to understand this, let $A = \lambda_1 I + J$ be a matrix with a single given eigenvalue λ_1. We compute the characteristic polynomial of the perturbed matrix $\lambda_1 I + J + \epsilon C$:

$$\det(\lambda_1 I + J + \epsilon C - \lambda I) = (\lambda_1 - \lambda)^n - (-1)^n \epsilon c_{n1} + O(\epsilon^2) + O(\epsilon|\lambda_1 - \lambda|).$$

If $c_{n1} \neq 0$, the terms $O(\epsilon^2)$ and $O(\epsilon|\lambda_1 - \lambda|)$ are negligible with respect to the term ϵc_{n1} and hence the eigenvalues of $A + \epsilon C$ are approximately given by the roots of

$$(\lambda_1 - \lambda)^n - (-1)^n \epsilon c_{n1} = 0 \quad \Longrightarrow \quad \lambda \approx \lambda_1 + (\epsilon c_{n1})^{\frac{1}{n}}, \tag{7.12}$$

the multiple complex roots of 1. The eigenvalues seem to explode, as we show in Figure 7.3 obtained with the MATLAB code

```
J=diag(ones(6,1),1);
C=rand(7);
for i=7:14
  plot(eig(eye(7)+J+10^(-i)*C),'o');
  hold on
end;
```

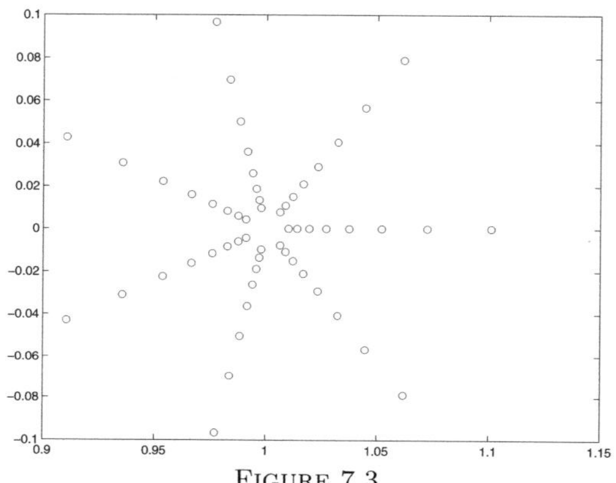

FIGURE 7.3.
Eigenvalues of a Jordan block are very ill conditioned

We see that for $\epsilon = 10^{-7}$, the error in the eigenvalue is about 0.1, and for $\epsilon = 10^{-14}$ the error is about 0.01, which confirms well the estimate (7.12) for $n = 7$. This is typical for non-symmetric matrices, whose eigenvalues may be very ill-conditioned.

Let us now consider the symmetric matrix $A = J + J^{\top}$:

```
>> A=J+J'
A =   0      1      0      0      0      0      0
      1      0      1      0      0      0      0
      0      1      0      1      0      0      0
      0      0      1      0      1      0      0
      0      0      0      1      0      1      0
      0      0      0      0      1      0      1
      0      0      0      0      0      1      0
>> eig(A)
ans =
  -1.84775906502257
  -1.41421356237310
  -0.76536686473018
  -0.00000000000000
   0.76536686473018
```

```
    1.41421356237309
    1.84775906502257
```

Now we make a similar small change of the elements $a_{1,7}$ and $a_{7,1}$ in order to preserve the symmetry:

```
>> A(7,1)=1e-12; A(1,7)=1e-12;
>> eig(A)
ans =
  -1.84775906502250
  -1.41421356237334
  -0.76536686472975
  -0.00000000000050
   0.76536686473061
   1.41421356237284
   1.84775906502265
```

We observe almost no changes in the eigenvalues! This illustrates how well conditioned the eigenvalues are for symmetric matrices. This is also the case for multiple eigenvalues, as we can see by making a small change in the identity matrix:

```
>> B=eye(7); B(7,1)=1e-12; B(1,7)=1e-12;
>> eig(B)
ans =
   0.99999999999900
   1.00000000000000
   1.00000000000000
   1.00000000000000
   1.00000000000000
   1.00000000000000
   1.00000000000100
```

Let us analyze this case of multiple eigenvalues. We first show a very useful result which indicates where eigenvalues of a matrix lie.

THEOREM 7.3. (GERSHGORIN) *Let $A \in \mathbb{C}^{n \times n}$. If λ is an eigenvalue of A, there exists an index i such that*

$$|\lambda - a_{ii}| \leq \sum_{\substack{j=1 \\ j \neq i}}^{n} |a_{ij}|, \tag{7.13}$$

or in other words, all the eigenvalues of A lie in the union of the discs

$$D_i := \{\lambda; |\lambda - a_{ii}| \leq \sum_{\substack{j=1 \\ j \neq i}}^{n} |a_{ij}|\}.$$

PROOF. Let $\boldsymbol{v} \neq 0$ be an eigenvector, and let us choose the index i such that $|v_i| \geq |v_j|$ for all $j = 1, \ldots, n$. Then the i-th equation of the eigenvalue equation $A\boldsymbol{v} = \lambda \boldsymbol{v}$ gives

$$\sum_{\substack{j=1 \\ j \neq i}}^{n} a_{ij} v_j = (\lambda - a_{ii}) v_i.$$

Now dividing by v_i and using the triangle inequality gives

$$|\lambda - a_{ii}| = \left| \sum_{\substack{j=1 \\ j \neq i}}^{n} a_{ij} \frac{v_j}{v_i} \right| \leq \sum_{\substack{j=1 \\ j \neq i}}^{n} |a_{ij}|,$$

which concludes the proof. $\qquad\qquad\qquad\qquad\qquad\qquad\qquad\qquad\qquad\square$

THEOREM 7.4. (SYMMETRIC MATRICES WITH MULTIPLE EIGENVALUES)
Let A and C be symmetric matrices, and let $\lambda_1, \ldots, \lambda_n$ (not necessarily distinct) be the eigenvalues of A. Then the eigenvalues $\lambda_j(\epsilon)$ of $A + \epsilon C$ satisfy

$$\lambda_j(\epsilon) = \lambda_j + \epsilon \boldsymbol{v}_j^\top C \boldsymbol{v}_j + O(\epsilon^2), \qquad\qquad (7.14)$$

where $\boldsymbol{v}_1, \ldots, \boldsymbol{v}_n$ is an orthonormal basis of $\mathbb{R}^n$ formed by the eigenvectors of A, $A\boldsymbol{v}_j = \lambda_j \boldsymbol{v}_j$, $\boldsymbol{v}_j^T \boldsymbol{v}_j = 1$, $\boldsymbol{v}_j^T \boldsymbol{v}_k = 0$ for $j \neq k$.

PROOF. We first diagonalize the matrix A such that identical eigenvalues are grouped together, i.e. we choose an orthogonal matrix W such that

$$W^T A W = \begin{pmatrix} \lambda_1 I & 0 \\ 0 & \Lambda \end{pmatrix}, \qquad W^T C W = \begin{pmatrix} C_{11} & C_{21}^T \\ C_{21} & C_{22} \end{pmatrix}, \qquad (7.15)$$

where Λ contains the remaining eigenvalues different from λ_1 on the diagonal, and C_{11} and C_{22} are symmetric. We now construct a matrix $B(\epsilon)$ such that the similarity transformation below (which preserves eigenvalues) leads to a block triangular matrix:

$$\begin{pmatrix} I & 0 \\ B(\epsilon) & I \end{pmatrix} \begin{pmatrix} \lambda_1 I + \epsilon C_{11} & \epsilon C_{21}^T \\ \epsilon C_{21} & \Lambda + \epsilon C_{22} \end{pmatrix} \begin{pmatrix} I & 0 \\ -B(\epsilon) & I \end{pmatrix} = \begin{pmatrix} D_1(\epsilon) & \epsilon C_{21}^T \\ 0 & D_2(\epsilon) \end{pmatrix}.$$
$$(7.16)$$

In order to do so, the matrix $B = B(\epsilon)$ needs to satisfy the equation

$$(\lambda_1 I - \Lambda)B + \epsilon B C_{11} + \epsilon C_{21} - \epsilon B C_{21}^T B - \epsilon C_{22} B = 0. \qquad (7.17)$$

Since $\lambda_1 I - \Lambda$ is invertible, the implicit function theorem implies that for ϵ small enough, (7.17) can be solved. We expand

$$B(\epsilon) = B(0) + \epsilon B'(0) + O(\epsilon^2).$$

We see from (7.17) that $B(0) = 0$, and by differentiating this equation, we get $B'(0) = (\Lambda - \lambda_1 I)^{-1} C_{21}$. Thus, we obtain the matrix valued function

$$B(\epsilon) = \epsilon (\Lambda - \lambda_1 I)^{-1} C_{21} + O(\epsilon^2). \qquad\qquad (7.18)$$

We can now compute $D_1(\epsilon)$ in (7.16),

$$D_1(\epsilon) = \lambda_1 I + \epsilon C_{11} + O(\epsilon^2). \tag{7.19}$$

Thus, the eigenvalues of $A(\epsilon)$ close to λ_1 are in fact the eigenvalues of $D_1(\epsilon)$. Furthermore, note that the matrix C_{11} is symmetric and can be diagonalized with an orthonormal matrix U, which leads to

$$\begin{pmatrix} U^T & 0 \\ & I \end{pmatrix} \begin{pmatrix} I & 0 \\ B(\epsilon) & I \end{pmatrix} W^T(A + \epsilon C)W \begin{pmatrix} I & 0 \\ -B(\epsilon) & I \end{pmatrix} \begin{pmatrix} U & 0 \\ & I \end{pmatrix}$$
$$= \begin{pmatrix} \widehat{D}_1(\epsilon) & \epsilon U^T C_{21}^T \\ 0 & D_2(\epsilon) \end{pmatrix},$$

$$\tag{7.20}$$

where $\widehat{D}_1(\epsilon) = \operatorname{diag}(\lambda_1 + \epsilon d_1, \ldots, \lambda_1 + \epsilon d_m) + O(\epsilon^2)$, with m being the multiplicity of the eigenvalue λ_1 and d_j being the eigenvalues of C_{11}. Using now Gershgorin's Theorem for the matrix $\widehat{D}_1(\epsilon)$ shows that the m eigenvalues of $A + \epsilon C$ which are close to λ_1 satisfy $\lambda_j(\epsilon) = \lambda_1 + \epsilon d_j + O(\epsilon^2)$.

Finally, we prove (7.14) by showing that the eigenvalues of C_{11} are given by $\boldsymbol{v}_j^\top C \boldsymbol{v}_j$, $j = 1, \ldots, m$, where $\boldsymbol{v}_j$ are eigenvectors of A associated with λ_1. To do so, recall that U was chosen so that

$$\operatorname{diag}(d_1, \ldots, d_m) = U^\top C_{11} U = (U^\top, 0) W^\top C W \begin{pmatrix} U \\ 0 \end{pmatrix}.$$

Thus, if we define

$$(\boldsymbol{v}_1, \boldsymbol{v}_2, \ldots) = V := W \begin{pmatrix} U & 0 \\ 0 & I \end{pmatrix},$$

then $d_j = \boldsymbol{v}_j^\top C \boldsymbol{v}_j$. Moreover,

$$V^\top A V = \begin{pmatrix} \lambda_1 U^\top U & 0 \\ 0 & \Lambda \end{pmatrix} = \begin{pmatrix} \lambda_1 I & 0 \\ 0 & \Lambda \end{pmatrix},$$

so the columns of V are indeed eigenvectors of A, which proves (7.14). $\qquad\square$

7.3   Method of Jacobi

The method we consider in this section was originally proposed for symmetric eigenvalue problems in 1846 by Jacobi [74], which is one of the earliest numerical analysis papers in the literature. In that paper, Jacobi was motivated by the problem of computing the motion of the planets in the solar system:

4.

Über ein leichtes Verfahren die in der Theorie der Säcularstörungen vorkommenden Gleichungen numerisch aufzulösen *).

(Von Herrn Professor Dr. *C. G. J. Jacobi.*)

1.

In der Theorie der Säcularstörungen und der kleinen Oscillationen wird man auf ein System linearer Gleichungen geführt, in welchem die Coëfficienten der verschiedenen Unbekannten in Bezug auf die Diagonale symmetrisch sind, die ganz constanten Glieder fehlen und zu allen in der Diagonale befindlichen Coëfficienten noch dieselbe Größe — x addirt ist. Durch Elimination der Unbekannten aus solchen linearen Gleichungen erhält man eine Bedingungsgleichung, welcher x genügen muß. Für jeden Werth von x, welcher diese Bedingungsgleichung erfüllt, hat man sodann aus den linearen Gleichungen die Verhältnisse der Unbekannten zu bestimmen. Ich werde hier zuerst die für ein solches System Gleichungen geltenden algebraischen Formeln ableiten, welche im Folgenden ihre Anwendung finden, und hierauf eine für die Rechnung sehr bequeme Methode mittheilen, wodurch man die numerischen Werthe der Größen x und der ihnen entsprechenden Systeme der Unbekannten mit Leichtigkeit und mit jeder beliebigen Schärfe erhält. Diese Methode überhebt der beschwerlichen Bildung und Auflösung der Gleichung, deren Wurzeln die Werthe von x sind,

Translated into English:

> The theory of secular perturbations of small oscillations leads to a symmetric system of linear equations in which the right hand side is missing and in which the quantity $-x$ is added to all diagonal elements. Applying the elimination process, we obtain an equation for x. For each solution x of this equation, we then have to determine the quotients of the unknowns. In the following I will first derive the algebraic expression for such a system of equations, which will be used in the application. Then I will present a very practical method for the computation, with which the numerical values of the quantity x and the corresponding system of unknowns can be easily computed and to any desired precision. The method avoids the tedious procedure of forming and solving the equation whose zeros are the values of x.

Heinz Rutishauser implemented this method in Algol for the Handbook [148][3]. The following listing is the original code (from punch-cards used in the computing center RZ-ETH), and is full of numerical jewels:

```
"PROCEDURE" JACOBI(N,EIVEC) TRANS:(A)RES:(D,V,ROT);
  "VALUE" N,EIVEC;
  "INTEGER" N,ROT; "BOOLEAN" EIVEC; "ARRAY" A,D,V;
```

---

[3] "Apart from this it is the algorithmic genius of H. Ruthishauser which has been my main source of inspiration" (Wilkinson, preface of the The Algebraic Eigenvalue Problem)

```
        "BEGIN"
          "REAL" SM,C,S,T,H,G,TAU,THETA,TRESH;
          "INTEGER" P,Q,I,J;
          "ARRAY" B,Z[1:N];
PROGRAM:
          "IF" EIVEC "THEN"
            "FOR" P:=1 "STEP" 1 "UNTIL" N "DO"
              "FOR" Q:=1 "STEP" 1 "UNTIL" N "DO"
                V[P,Q]:="IF" P=Q "THEN" 1.0 "ELSE" 0.0;
          "FOR" P:=1 "STEP" 1 "UNTIL" N "DO"
              "BEGIN" B[P]:=D[P]:=A[P,P];  Z[P]:=0 "END";
          ROT:=0;
          "FOR" I:=1 "STEP" 1 "UNTIL" 50 "DO"
SWP:
          "BEGIN"
            SM:=0;
            "FOR" P:=1 "STEP" 1 "UNTIL" N-1 "DO"
              "FOR" Q:=P+1 "STEP" 1 "UNTIL" N "DO"
                SM:=SM+ABS(A[P,Q]);
            "IF" SM=0 "THEN""GOTO" OUT;
            TRESH:="IF" I<4 "THEN" 0.2*SM/N^2 "ELSE" 0.0;
            "FOR" P:=1 "STEP" 1 "UNTIL" N-1 "DO"
              "FOR" Q:=P+1 "STEP" 1 "UNTIL" N "DO"
              "BEGIN"
                G:=100*ABS(A[P,Q]);
                "IF" I>4&ABS(D[P])+G=ABS(D[P])&
                ABS(D[Q])+G=ABS(D[Q]) "THEN" A[P,Q]:=0
                "ELSE"
                "IF" ABS(A[P,Q])>TRESH "THEN"
ROTATE:          "BEGIN"
                  H:=D[Q]-D[P];
                  "IF" ABS(H)+G=ABS(H) "THEN" T:=A[P,Q]/H
                  "ELSE"
                    "BEGIN"
                      THETA:=0.5*H/A[P,Q];
                      T:=1/(ABS(THETA)+SQRT(1+THETA^2));
                      "IF" THETA<0 "THEN" T:=-T
                    "END" COMPUTING TAN OF ROTATION ANGLE;
                  C:=1/SQRT(1+T^2);
                  S:=T*C;
                  TAU:=S/(1+C);
                  H:=T*A[P,Q];
                  Z[P]:=Z[P]-H;
                  Z[Q]:=Z[Q]+H;
                  D[P]:=D[P]-H;
                  D[Q]:=D[Q]+H;
                  A[P,Q]:=0;
                  "FOR" J:=1 "STEP" 1 "UNTIL" P-1 "DO"
                  "BEGIN"
                    G:=A[J,P];  H:=A[J,Q];
                    A[J,P]:=G-S*(H+G*TAU);
                    A[J,Q]:=H+S*(G-H*TAU)
                  "END" OF CASE 11J<P;
                  "FOR" J:=P+1 "STEP" 1 "UNTIL" Q-1 "DO"
                  "BEGIN"
                      G:=A[P,J]; H:=A[J,Q];
                      A[P,J]:=G-S*(H+G*TAU);
                      A[J,Q]:=H+S*(G-H*TAU)
                  "END" OF CASE P<J<Q;
                  "FOR" J:=Q+1 "STEP" 1 "UNTIL" N "DO"
                  "BEGIN"
                    G:=A[P,J];  H:=A[Q,J];
                    A[P,J]:=G-S*(H+G*TAU);
                    A[Q,J]:=H+S*(G-H*TAU)
                  "END" OF CASE Q<J1N;
                  "IF" EIVEC "THEN"
                    "FOR" J:=1 "STEP" 1 "UNTIL" N"DO"
```

```
        "BEGIN"
          G:=V[J,P];   H:=V[J,Q];
          V[J,P]:=G-S*(H+G*TAU);
          V[J,Q]:=H+S*(G-H*TAU)
        "END" OF CASE V;
      ROT:=ROT+1;
    "END" ROTATE;
  "END";
  "FOR" P:=1 "STEP" 1 "UNTIL" N "DO"
  "BEGIN"
  D[P]:=B[P]:=B[P]+Z[P];
  Z[P]:=0
  "END" P
"END" SWP;
OUT:
"END" JACOBI;
```

We will now explain the details of this implementation and also translate
the procedure to MATLAB.

If A is symmetric, $A = A^\top$, the eigenvalues λ_i are real and the eigenvectors
$\boldsymbol{v}_i$ can be chosen to be mutually orthogonal, thus we have the decomposition:

$$V^\top A V = \Lambda, \quad \Lambda = \mathrm{diag}(\lambda_1, \ldots, \lambda_n), \quad V^\top V = I.$$

The Jacobi algorithm iteratively applies *elementary orthogonal transforma-*
tions to the matrix $A_o := A$,

$$A_1 = U_0^\top A_0 U_0, \quad A_2 = U_1^\top A_1 U_1, \quad \ldots$$

such that the off-diagonal elements of A_k are reduced and so A_k converges
to the diagonal matrix Λ.

As measure for the deviation of A_k from a diagonal matrix, we introduce
the sum of the squares of the off-diagonal elements,

$$S_k = \mathrm{off}(A_k) := \sum_{i=1}^{n} \sum_{\substack{j=1 \\ j\neq i}}^{n} \left(a_{ij}^{(k)} \right)^2 = \|A_k\|_F^2 - \|\,\mathrm{diag}(A_k)\|_2^2.$$

The idea is to choose the transformations in such a way that the quantity
$\mathrm{off}(A_k)$ decreases: $S_0 > S_1 > S_2 > \cdots \geq 0$. The matrix A_k is related to A
through

$$A_k = U_{k-1}^\top U_{k-2}^\top \cdots U_0^\top A U_0 U_1 \cdots U_{k-1} = V_k^\top A V_k, \quad \text{with} \quad V_k = U_0 U_1 \cdots U_{k-1}.$$

Therefore if $\mathrm{off}(A_k) < \varepsilon$, then $\mathrm{diag}(A_k)$ and V_k will be approximations to the
eigenvalues and eigenvectors.

For the elementary transformations U_k, we use the *Givens rotations* introduced in Section 3.5; in the current context, they are also called *Jacobi*

rotations in honor of Jacobi, the inventor of this method:

$$
U_k = U(p,q,\phi) =
\begin{bmatrix}
1 & & & & & & & & \\
 & \ddots & & & & & & & \\
 & & 1 & & & & & & \\
 & & & \cos\phi & & & \sin\phi & & \\
 & & & & 1 & & & & \\
 & & & & & \ddots & & & \\
 & & & & & & 1 & & \\
 & & & -\sin\phi & & & \cos\phi & & \\
 & & & & & & & 1 & \\
 & & & & & & & & \ddots \\
 & & & & & & & & & 1
\end{bmatrix}
\begin{matrix} \\ \\ \\ \leftarrow p \\ \\ \\ \\ \leftarrow q \\ \\ \\ \end{matrix}
$$

$$(7.21)$$

The matrix U_k differs from the identity matrix only by 4 elements,

$$
u_{pp} = u_{qq} = \cos\phi, \quad u_{pq} = -u_{qp} = \sin\phi, \quad p < q.
$$

We now study the transformation $A'' = U_k^\top A U_k$ in two steps,

$$
A' = U_k^\top A \text{ and } A'' = A' U_k.
$$

The first transformation from the left replaces the rows p and q by the linear combinations:

$$
\begin{aligned}
A'(p,:) &= A(p,:)\cos\phi - A(q,:)\sin\phi, \\
A'(q,:) &= A(p,:)\sin\phi + A(q,:)\cos\phi.
\end{aligned}
\tag{7.22}
$$

The second transformation from the right changes the columns p and q:

$$
\begin{aligned}
A''(:,p) &= A'(:,p)\cos\phi - A'(:,q)\sin\phi, \\
A''(:,q) &= A'(:,p)\sin\phi + A'(:,q)\cos\phi.
\end{aligned}
\tag{7.23}
$$

We need the explicit transformation rule for the element in the pq-position. We use MAPLE to compute this:

```
A:=Matrix([[a_pp,a_pq],[a_pq,a_qq]]):
U:=Matrix([[c,s],[-s,c]]):
with(LinearAlgebra):
B:=Transpose(U).A.U:
expand(B[1,1]);
expand(B[2,2]);
expand(B[1,2]);
```

We obtain

$$
\begin{aligned}
& c^2\,a_{pp} - 2\,c\,s\,a_{pq} + s^2\,a_{qq}, \\
& s^2\,a_{pp} + 2\,c\,s\,a_{pq} + c^2\,a_{qq}, \\
& s\,c\,a_{pp} - s^2 a_{pq} + c^2 a_{pq} - c\,s\,a_{qq}.
\end{aligned}
\tag{7.24}
$$

In particular, for $B[1,2] = a''_{pq}$ we get, using the multi-angle trigonometric formula,

$$
\begin{aligned}
a''_{pq} &= (a_{pp} - a_{qq})\cos\phi\sin\phi + a_{pq}(\cos^2\phi - \sin^2\phi) \\
&= \tfrac{1}{2}(a_{pp} - a_{qq})\sin(2\phi) + a_{pq}\cos(2\phi).
\end{aligned}
$$

If we wish to have $a''_{pq} = 0$, then we must choose ϕ so that

$$
\cot(2\phi) = \frac{a_{qq} - a_{pp}}{2a_{pq}}.
$$

Defining θ to be this quantity,

$$
\theta := \cot(2\phi) = \frac{1}{\tan(2\phi)} = \frac{1 - \tan^2\phi}{2\tan\phi},
$$

we obtain first $t = \tan\phi$ by solving the quadratic equation

$$
t^2 + 2\theta t - 1 = 0. \tag{7.25}
$$

The above equation has two solutions, one corresponding to $|\phi| \leq \pi/4$ and the other to $|\phi| \geq \pi/4$. In order to understand which root to choose, consider the 2×2 near-diagonal case

$$
A = \begin{pmatrix} 1 & \epsilon \\ \epsilon & 2 \end{pmatrix}.
$$

Here $\theta = 1/2\epsilon$, so we have $t_1 \approx \epsilon$ and $t_2 \approx -1/\epsilon$, which corresponds to $\phi_1 \approx \epsilon$ and $\phi_2 \approx -\frac{\pi}{2} + \epsilon$. Substituting into (7.21), we see that ϕ_1 leads to a U_k that is close to the identity, whereas ϕ_2 gives one that interchanges the diagonal elements a_{11} and a_{22}. In other words, while both transformations reduce off(A_k), the larger angle leads to oscillatory behavior in the diagonal elements and hence should not be used.

Thus, we are interested in obtaining the smaller of the two solutions of (7.25); we can obtain this solution numerically stably without cancellation by the statements

```
theta=(A(q,q)-A(p,p))/2/A(p,q)
t=1/(abs(theta)+sqrt(1+theta^2))
if theta<0, t=-t end
```

Then from $t = \tan\phi$ we obtain $c = \cos\phi$ and $s = \sin\phi$ with

```
c=1/sqrt(1+t^2); s=c*t;
```

By choosing $\cos\phi$ and $\sin\phi$ this way, we annihilate two off-diagonal elements of A. What happens to the other elements? When multiplying from the left and right by the Jacobi rotation, all elements in the rows and columns p and

q are changed. Using the fact that the Frobenius norm is invariant under orthogonal transformations, we have

$$\|A''\|_F^2 = \|U_k^\top A U_k\|_F^2 = \|A\|_F^2.$$

But this implies that

$$\|A''\|_F^2 = \text{off}(A'') + \sum_{i=1}^n a_{ii}''^{\,2} = \text{off}(A) + \sum_{i=1}^n a_{ii}^2 = \|A\|_F^2$$

and thus

$$\text{off}(A'') = \text{off}(A) + a_{pp}^2 + a_{qq}^2 - (a_{pp}'')^2 - (a_{qq}'')^2.$$

The first transformation $A' = U_k^\top A$ changes the elements in rows p and q as follows

$$\begin{pmatrix} c & -s \\ s & c \end{pmatrix} \begin{pmatrix} a_{pi} \\ a_{qi} \end{pmatrix} = \begin{pmatrix} a_{pi}' \\ a_{qi}' \end{pmatrix} \quad i = 1, \ldots, n.$$

Since the matrix is orthogonal, the sum of squares is constant,

$$a_{pi}^2 + a_{qi}^2 = a'^2_{pi} + a'^2_{qi}.$$

The same holds for the second transformation for corresponding elements in the columns p and q. Therefore the sum of squares of the four elements in the intersections of rows and columns p and q remains the same,

$$a_{pp}^2 + a_{qq}^2 + 2a_{pq}^2 = a''^2_{pp} + a''^2_{qq} + 2a''^2_{pq}.$$

Since the rotations were chosen such that $a_{pq}'' = 0$ we obtain

$$a_{pp}^2 + a_{qq}^2 - (a'')_{pp}^2 - (a'')_{qq}^2 = -2a_{pq}^2,$$

and therefore

$$\text{off}(A'') = \text{off}(A) - 2a_{pq}^2,$$

so the matrix becomes more diagonal. The classical choice for p and q proposed by Jacobi[4] is to choose the largest element $|a_{pq}|$ in each step to maximize the decay of the off diagonal elements $S_k = \text{off}(A_k)$. With this strategy, because the largest element is larger than the average off-diagonal element

$$a_{pq}^2 \geq \frac{S_k}{n(n-1)},$$

we obtain

$$S_{k+1} \leq S_k \left(1 - \frac{2}{n(n-1)}\right) \leq S_0 \left(1 - \frac{2}{n(n-1)}\right)^{k+1},$$

---

[4] "Und so wurde auch bei allen folgenden Substitutionen im Allgemeinen die Regel befolgt, jedesmal den grössten von den ausserhalb der Diagonale vorhandnen Coëfficienten gleich Null zu machen" (And also in all the following substitutions we followed in general the rule of making the largest off-diagonal entry zero)

which implies at least *monotonic linear convergence*. One can show that asymptotically the convergence of the Jacobi algorithm is quadratic [67].

Finding the largest element requires an expensive search at each iteration, so sequentially annihilating each element is the preferred strategy for automated computations. The convergence of this variant was also analyzed by Henrici and Forsythe, but the proof of convergence is more difficult [33]. Newer variants treat several pairs p, q simultaneously and lead to naturally parallel algorithms.

Our discussion so far produces a first version of Jacobi's algorithm:

ALGORITHM 7.2. *Jacobi Algorithm Version 1*

```
function [d,V]=Jacobi1(A);
% JACOBI1 computes the eigen-decomposition of a symmetric matrix
%    [d,V]=Jacobi1(A) computes the eigenvalues d and the eigenvectors V
%    of the symmetric matrix A, a first version not using the symmetry
%    of the matrix in the algorithm

n=length(A); V=eye(size(A));
S=1;
while S~=0                                          % sweep
  for p=1:n-1
    for q=p+1:n
      theta=(A(q,q)-A(p,p))/2/A(p,q);               % rotate a_pq=0
      t=1/(abs(theta)+sqrt(1+theta^2));
      if theta<0, t=-t; end
      c=1/sqrt(1+t^2); s=c*t;
      h=A(p,:)*c-A(q,:)*s;                          % A=U^T*A
      A(q,:)=A(p,:)*s+A(q,:)*c;
      A(p,:)=h;
      h=A(:,p)*c-A(:,q)*s;                          % A=A*U
      A(:,q)=A(:,p)*s+A(:,q)*c;
      A(:,p)=h;
      h=V(:,p)*c-V(:,q)*s;                          % V=V*U
      V(:,q)=V(:,p)*s+V(:,q)*c;
      V(:,p)=h;
    end
  end
  A                                                 % test output
  S=norm(A,'fro')^2-norm(diag(A))^2
end
d=diag(A);
```

As an example, we compute the eigen-decomposition of a 4×4 matrix given in the book of Schwarz [122]:

```
>> A=[20    -7     3    -2
      -7     5     1     4
```

```
            3      1      3      1
           -2      4      1      2];
>> S=norm(A,'fro')^2-norm(diag(A))^2
S =
  160.0000
>> [d,V]=Jacobi1(A)
A =
   23.5231   -0.0091   -0.2385    0.1516
   -0.0091   -0.4376   -1.3977    0.9315
   -0.2385   -1.3977    6.1744   -0.0000
    0.1516    0.9315    0.0000    0.7401
S =
    5.8023
A =
   23.5274    0.0224   -0.0054    0.0106
    0.0224   -1.1601    0.0793    0.0025
   -0.0054    0.0793    6.4597   -0.0000
    0.0106    0.0025    0.0000    1.1731
S =
    0.0139
A =
   23.5274   -0.0000   -0.0000   -0.0000
   -0.0000   -1.1609   -0.0000    0.0000
   -0.0000   -0.0000    6.4605    0.0000
   -0.0000    0.0000    0.0000    1.1730
S =
    1.0940e-09
A =
   23.5274    0.0000   -0.0000    0.0000
    0.0000   -1.1609    0.0000   -0.0000
    0.0000   -0.0000    6.4605    0.0000
   -0.0000   -0.0000    0.0000    1.1730
S =
     0
d =
   23.5274
   -1.1609
    6.4605
    1.1730
V =
    0.9106    0.1729    0.2607    0.2699
   -0.3703    0.6750    0.5876    0.2492
    0.1078   -0.1168    0.5499   -0.8200
   -0.1484   -0.7077    0.5333    0.4390
```

We can clearly see fast convergence, which seems to be even more than quadratic in this example.

7.3.1   Reducing Cost by Using Symmetry

Note that a Jacobi rotation transforms the rows p and q using the same linear combination as for the columns p and q. Since the matrix is symmetric, we do not need to calculate the transformation on both the lower and upper triangular part; it suffices to work only on the upper triangular part. Assume that $1 \le p < q \le n$. Then

1. the elements in the columns $A(1 : p - 1, p)$ and $A(1 : p - 1, q)$ are changed by the multiplication from the right with $U_k^{\top}$.

2. the elements in the columns $A(p + 1 : q - 1, p)$ and $A(p + 1 : q - 1, q)$ would have to be linearly combined, and likewise for the elements in the rows $A(p, p+1 : q-1)$ and $A(q, p+1 : q-1)$. However, by symmetry we can replace the column $A(p + 1 : q - 1, p)$ by the same elements in the row $A(p, p + 1 : q - 1)$ and omit the updates for the elements in row q and column p. *Thus, we just combine the row elements $A(p, p+1 : q-1)$ with the column elements $A(p + 1 : q - 1, q)$ in this step.*

3. the elements in the rows $A(p, q + 1 : n)$ and $A(q, q + 1 : n)$ have to be linearly combined.

Doing so, we save about 50% of the work. In the following second version of our program we also skip the Jacobi rotation if the element $a_{pq} = 0$.

    Note that we have to explicitly compute the elements a_{pp}, a_{qq} and set $a_{pq} = 0$. Rutishauser recommends not to use the obvious formula

$$a_{pp}^{\text{new}} = \cos^2 \phi \, a_{pp} - 2 \cos \phi \sin \phi \, a_{pq} + \sin^2 \phi \, a_{qq}$$

and similarly for a_{qq}, but to compute the correction

$$
\begin{aligned}
a_{pp}^{\text{new}} - a_{pp} &= (a_{qq} - a_{pp}) \sin^2 \phi - 2 \cos \phi \sin \phi \, a_{pq} \\
&= \tan \phi \left((a_{qq} - a_{pp}) \cos \phi \sin \phi - 2 a_{pq} \cos^2 \phi \right) \\
&= -\tan \phi \left(\frac{1}{2}(a_{pp} - a_{qq}) \sin(2\phi) + a_{pq} \cos(2\phi) + a_{pq} \right) \\
&= -\tan \phi \left(a_{pq}^{\text{new}} + a_{pq} \right).
\end{aligned}
$$

Because we choose the angle to make $a_{pq}^{\text{new}} = 0$, we obtain with $t = \tan \phi$ the simpler updating formulas

$$a_{pp}^{\text{new}} = a_{pp} - t a_{pq}, \quad a_{qq}^{\text{new}} = a_{qq} + t a_{pq}. \tag{7.26}$$

Note that the correction to the diagonal elements must be the same with opposite sign, since the trace ($\sum_i a_{ii}$ which is also the sum of the eigenvalues $= \sum_i \lambda_i$) is invariant under Jacobi rotations.

Algorithm 7.3. *Jacobi Algorithm Version 2*

```
function [d,V]=Jacobi2(A);
% JACOBI2 computes the eigen-decomposition of a symmetric matrix
%   [d,V]=Jacobi2(A) computes the eigen-decomposition of the
%   symmetric matrix A as Jacobi1, but now using the symmetry

n=length(A); V=eye(size(A));
S=1;
while S~=0,                                             % sweep
  for p=1:n-1
    for q=p+1:n
      if A(p,q)~=0,                                     % rotate a_pq=0
        theta=(A(q,q)-A(p,p))/2/A(p,q);
        t=1/(abs(theta)+sqrt(1+theta^2));
        if theta<0, t=-t; end
        c=1/sqrt(1+t^2); s=c*t;
        h=A(1:p-1,p)*c-A(1:p-1,q)*s;                   % A=U^T*A
        A(1:p-1,q)=A(1:p-1,p)*s+A(1:p-1,q)*c;
        A(1:p-1,p)=h;
        h=A(p,p+1:q-1)*c-A(p+1:q-1,q)'*s;
        A(p+1:q-1,q)=A(p,p+1:q-1)'*s+A(p+1:q-1,q)*c;
        A(p,p+1:q-1)=h;
        h=A(p,q+1:n)*c-A(q,q+1:n)*s;                   % A=A*U
        A(q,q+1:n)=A(p,q+1:n)*s+A(q,q+1:n)*c;
        A(p,q+1:n)=h;
        h=A(p,q)*t;
        A(p,p)=A(p,p)-h;                               % update diagonal
        A(q,q)=A(q,q)+h;
        A(p,q)=0;
        h=V(:,p)*c-V(:,q)*s;                           % V=V*U
        V(:,q)=V(:,p)*s+V(:,q)*c;
        V(:,p)=h;
      end
    end
  end
  A                                                    % test output
  S=2*(norm(triu(A),'fro')^2-norm(diag(A))^2)
  Sneu=2*norm(triu(A,1),'fro')^2
end
d=diag(A);
```

The results for the Hilbert matrix for $n = 5$ are as follows:

```
>> A=hilb(5)
A =
    1.0000    0.5000    0.3333    0.2500    0.2000
    0.5000    0.3333    0.2500    0.2000    0.1667
    0.3333    0.2500    0.2000    0.1667    0.1429
    0.2500    0.2000    0.1667    0.1429    0.1250
    0.2000    0.1667    0.1429    0.1250    0.1111
>> S=2*(norm(triu(A),'fro')^2-norm(diag(A))^2)
S =
```

```
      1.3154
>> [d,V]=Jacobi2(A)
A =
      1.5651     0.0423     0.0003    -0.0302    -0.0017
      0.5000     0.2097     0.0018     0.0033    -0.0029
      0.3333     0.2500     0.0001     0.0002    -0.0002
      0.2500     0.2000     0.1667     0.0121          0
      0.2000     0.1667     0.1429     0.1250     0.0003
S =
      0.0055
Sneu =
      0.0055
A =
      1.5671    -0.0001    -0.0000     0.0000     0.0000
      0.5000     0.2085     0.0000    -0.0000    -0.0000
      0.3333     0.2500     0.0000     0.0000    -0.0000
      0.2500     0.2000     0.1667     0.0114          0
      0.2000     0.1667     0.1429     0.1250     0.0003
S =
      1.2527e-08
Sneu =
      1.2527e-08
A =
      1.5671    -0.0000     0.0000     0.0000    -0.0000
      0.5000     0.2085    -0.0000    -0.0000     0.0000
      0.3333     0.2500     0.0000    -0.0000    -0.0000
      0.2500     0.2000     0.1667     0.0114          0
      0.2000     0.1667     0.1429     0.1250     0.0003
S =
           0
Sneu =
      1.1682e-21
D =
      1.5671
      0.2085
      0.0000
      0.0114
      0.0003
V =
      0.7679    -0.6019     0.0062     0.2142    -0.0472
      0.4458     0.2759    -0.1167    -0.7241     0.4327
      0.3216     0.4249     0.5062    -0.1205    -0.6674
      0.2534     0.4439    -0.7672     0.3096    -0.2330
      0.2098     0.4290     0.3762     0.5652     0.5576
>> [Ve,De]=eig(A)
Ve =
     -0.0062     0.0472     0.2142    -0.6019     0.7679
      0.1167    -0.4327    -0.7241     0.2759     0.4458
     -0.5062     0.6674    -0.1205     0.4249     0.3216
      0.7672     0.2330     0.3096     0.4439     0.2534
     -0.3762    -0.5576     0.5652     0.4290     0.2098
De =
      0.0000          0          0          0          0
           0     0.0003          0          0          0
           0          0     0.0114          0          0
           0          0          0     0.2085          0
           0          0          0          0     1.5671
```

Note that we computed $S = \text{off}(A)$ here in two ways. We see that the first way leads to $S = 0$ in this example, whereas with the second, more efficient way $Sneu$, the iteration would not stop at the same time. Therefore, we must also improve the stopping criterion.

7.3.2 Stopping Criterion

A theorem by Henrici [67] states the following: let d_i be the diagonal elements of the matrix A_k in decreasing order, i.e., $d_1 \geq d_2 \geq \ldots \geq d_n$. Furthermore, let the eigenvalues of A be similarly ordered, i.e., $\lambda_1 \geq \lambda_2 \geq \ldots \geq \lambda_n$. Then

$$|d_j - \lambda_j| \leq \sqrt{S(A_k)}.$$

Thus, we know that if $S(A_k) < 10^{-10}$, then the diagonal elements of A_k will approximate the eigenvalues at least to about 5 decimal digits.

If a_{pq} is negligible compared to both diagonal elements, so that $a_{pp} + a_{pq} = a_{pp}$ and $a_{qq} + a_{pq} = a_{qq}$ hold numerically, then we just assign $a_{pq}^{\text{new}} = 0$. This is because such small numbers can no longer change the diagonal elements in view of (7.26), and can be safely neglected as a result.

7.3.3 Algorithm of Rutishauser

Here is now a translation to MATLAB of the original Algol procedure `Jacobi` as published in the Handbook [148]:

ALGORITHM 7.4. *Jacobi Algorithm*

```
function [d,rot,Z,V]=Jacobi(A)
% JACOBI computes eigenvalues of a symmetric matrix
%    [d,rot,Z,V]=Jacobi(A) computes the eigenvalues d and eigenvectors
%    V of the symmetric matrix A.  d=Jacobi(A) computes only the
%    eigenvalues without accumulating the vectors.  Only the upper
%    triangular triu(A,1) of the matrix A is used.  rot counts the
%    rotations and Z stores the sum of the absolute values of the off
%    diagonal elements after each sweep.  Jacobi is a translation of
%    the ALGOL 60 Procedure of H. Rutishauser from the Handbook
%    Wilkinson-Reinsch: Linear Algebra.

Z=[];
n=max(size(A));
eivec=1; if nargout<4, eivec=0; end
if eivec, V=eye(n); end
d=diag(A); b=d; z=zeros(size(d));
rot=0; i =0;
sm=sum(sum(abs(triu(A,1))));
while (sm ~=0  )& (i<50)
  i=i+1;                                 % sweep
  if  i<4, tresh=0.2*sm/n^2;
  else tresh=0;
  end
  for p=1:n-1
    for q=p+1:n
      g=100*abs(A(p,q));
```

```
          if (i>4)&(abs(d(p))+g==abs(d(p)))&(abs(d(q))+g==abs(d(q)))
            A(p,q)=0;
          else
            if abs(A(p,q))>tresh
              h=d(q)-d(p);                            % rotate
              if abs(h)+g==abs(h), t =A(p,q)/h;
              else
                theta=0.5*h/A(p,q);
                t=1/(abs(theta)+sqrt(1+theta^2));
                if theta < 0, t =-t; end
              end                                     % end of tan calculation
              c=1/sqrt(1+ t^2); s=t*c; tau=s/(1+c);
              h=t*A(p,q);
              z(p)=z(p)-h; z(q)=z(q)+h;
              d(p)=d(p)-h; d(q)=d(q)+h;
              A(p,q) =0;
              g=A(1:p-1,p); h=A(1:p-1,q);
              A(1:p-1,p)=g -s*(h+g*tau);
              A(1:p-1,q)=h +s*(g-h*tau);              % end of case 1<=j<p
              g=A(p,p+1:q-1); h=A(p+1:q-1,q);
              A(p,p+1:q-1)=g -s*(h'+g*tau);
              A(p+1:q-1,q)=h +s*(g'-h*tau);           % end of case p<j<q
              g=A(p,q+1:n); h=A(q,q+1:n);
              A(p,q+1:n)=g -s*(h+g*tau);
              A(q,q+1:n)=h +s*(g-h*tau);              % of cases q<j<=n
              if eivec
                g=V(:,p); h=V(:,q);
                V(:,p)=g -s*(h+g*tau);
                V(:,q)=h +s*(g-h*tau);
              end
              rot=rot + 1;
            end
          end
      end
  end
  b=b+z; d=b; z=zeros(size(d));
  sm=sum(sum(abs(triu(A,1))));
  Z=[Z,sm];
end
```

We compute again the eigenvalue decomposition for the Hilbert matrix
of order 5:

```
>> A=hilb(5);
>> [d,rot,Z,V]=Jacobi(A)
A =
    1.0000    0.0423   -0.0127   -0.0178   -0.0209
    0.5000    0.3333    0.0041    0.0023         0
    0.3333    0.2500    0.2000    0.0034    0.0035
```

```
    0.2500    0.2000    0.1667    0.1429    0.0047
    0.2000    0.1667    0.1429    0.1250    0.1111
sm =
    0.1117
A =
    1.0000   -0.0001   -0.0000   -0.0001    0.0001
    0.5000    0.3333    0.0000    0.0005    0.0006
    0.3333    0.2500    0.2000   -0.0001   -0.0001
    0.2500    0.2000    0.1667    0.1429         0
    0.2000    0.1667    0.1429    0.1250    0.1111
sm =
    0.0016
A =
    1.0000   -0.0000    0.0000    0.0000   -0.0000
    0.5000    0.3333   -0.0000    0.0000    0.0000
    0.3333    0.2500    0.2000   -0.0000         0
    0.2500    0.2000    0.1667    0.1429   -0.0000
    0.2000    0.1667    0.1429    0.1250    0.1111
sm =
    1.1035e-06
A =
    1.0000   -0.0000   -0.0000    0.0000    0.0000
    0.5000    0.3333   -0.0000   -0.0000   -0.0000
    0.3333    0.2500    0.2000    0.0000    0.0000
    0.2500    0.2000    0.1667    0.1429         0
    0.2000    0.1667    0.1429    0.1250    0.1111
sm =
    1.2645e-13
A =
    1.0000   -0.0000    0.0000   -0.0000   -0.0000
    0.5000    0.3333   -0.0000   -0.0000   -0.0000
    0.3333    0.2500    0.2000    0.0000         0
    0.2500    0.2000    0.1667    0.1429         0
    0.2000    0.1667    0.1429    0.1250    0.1111
sm =
    1.2053e-28
A =
    1.0000         0         0         0         0
    0.5000    0.3333         0         0         0
    0.3333    0.2500    0.2000         0         0
    0.2500    0.2000    0.1667    0.1429         0
    0.2000    0.1667    0.1429    0.1250    0.1111
sm =
     0
d =
    1.5671
    0.2085
    0.0000
    0.0114
```

```
    0.0003
rot =
    42
Z =
    0.1117      0.0016      0.0000      0.0000      0.0000           0
V =
    0.7679     -0.6019      0.0062      0.2142     -0.0472
    0.4458      0.2759     -0.1167     -0.7241      0.4327
    0.3216      0.4249      0.5062     -0.1205     -0.6674
    0.2534      0.4439     -0.7672      0.3096     -0.2330
    0.2098      0.4290      0.3762      0.5652      0.5576
```

The following program compares Jacobi with eig by generating 6 symmetric matrices $A = QDQ^T$ of order 10, 20, 40, 80, 160 and 320 with given eigenvalues D and diagonalizing them with Jacobi and the MATLAB built-in function eig:

```
nn=10*2.^(0:5);
for i=1:length(nn)
  n=nn(i)
  ew=1000*(rand(n,1)-0.5);
  Q=rand(n,n); Q=orth(Q);
  A=Q*diag(ew)*Q';
  d=Jacobi(A);
  [U,D]=eig(A);
  eigenv(i)=norm([sort(diag(D))-sort(ew)])/norm(ew);
  jac(i)=norm([sort(d)-sort(ew)])/norm(ew);
end;
```

A typical run of this program yields the results:

```
>> [eigenv' jac']
ans =
    1.0e-14 *
    0.0812      0.0688
    0.1021      0.0846
    0.1808      0.0690
    0.2479      0.1159
    0.2846      0.1470
    0.3589      0.1843
```

We observe that the relative error of the eigenvalues computed by Jacobi is smaller than with eig, often about half the size.

7.3.4   Remarks and Comments on Jacobi

We now explain the remaining numerical tricks in the implementation of the method of Jacobi by Rutishauser:

1. Instead of $S(A) = \text{off}(A)$, the sum sm of the absolute values of the off-diagonal elements is computed. This saves some computation time

because the elements need not to be squared. Note that computing S recursively by

$$S(A_{k+1}) = S(A_k) - \left(a_{pq}^{(k)}\right)^2$$

is not a good idea for numerical reasons: information is lost whenever S is large compared to an off-diagonal element.

2. The diagonal of the matrix is saved in the vector $\boldsymbol{d}$. By this the lower part of the matrix is untouched and A is not destroyed.

   Before each sweep, the diagonal is saved in vector $\boldsymbol{b}$: `b=d`. During the sweep, the diagonal $\boldsymbol{d}$ is used. However, the changes are also accumulated in vector $\boldsymbol{z}$. The sum of all changes after the sweep is then added as a whole to the diagonal

   ```
   b=b+z; d=b; z=zeros(size(d));
   ```

   This is numerically preferable and gives a more accurate diagonal than summing each (small) correction sequentially after each rotation to $\boldsymbol{d}$.

3. The termination criterion is $sm = 0$, safeguarded by a limitation of the total numbers of sweeps: $i < 50$.

4. During the first three sweeps, the program only annihilates elements above the threshold
$$\texttt{tresh} = 0.2 \times \frac{\texttt{sm}}{n^2}.$$

   Thus the first three sweeps are like a compromise between always searching for the largest element to be annihilated in the original variant of Jacobi, and systematically sweeping all elements.

5. At least three sweeps are done in all cases. In the fourth and consecutive sweeps, if an element is small compared to the diagonal elements, it is set to zero. More specifically,

   ```
   g=100*abs(A(p,q));
   if i>4 & abs(d(p))+g==abs(d(p)) & abs(d(q))+g==abs(d(q))
     A(p,q)=0;
   else ...
   ```

   It is an important feature of Rutishauser not to set such small elements to zero in the first four sweeps. This way, perturbed diagonal matrices are also diagonalized correctly.

6. If an element $g = 100 \times |a_{pq}|$ is small compared to the difference of the corresponding diagonal elements $h = d(q) - d(p)$,

   ```
   if abs(h)+g==abs(h), t=A(p,q)/h;
   ```

then $\tan\phi$ is computed as

$$t = \tan\phi \approx \frac{1}{2}\tan(2\phi) = \frac{a_{pq}}{d_q - d_p},$$

to save a few floating point operations.

7. Computing the transformed elements: Rutishauser again recommends working with corrections. Instead of using the obvious expression

$$a_{pj}^{\text{new}} = \cos\phi\, a_{pj} - \sin\phi\, a_{qj},$$

he considers

$$a_{pj}^{\text{new}} - a_{pj} = (\cos\phi - 1)\, a_{pj} - \sin\phi\, a_{qj} = -\sin\phi\left(a_{qj} + \frac{1 - \cos\phi}{\sin\phi}\, a_{pj}\right).$$

Now since

$$\texttt{tau} := \tan\frac{\phi}{2} = \frac{1 - \cos\phi}{\sin\phi} = \frac{1 - \cos^2\phi}{\sin\phi(1 + \cos\phi)} = \frac{\sin\phi}{1 + \cos\phi}$$

we obtain with $s = \sin\phi$ and $c = \cos\phi$ the expression

$$a_{pj}^{\text{new}} = a_{pj} - s \times (a_{qj} + \texttt{tau} \times a_{pj}),$$

and similarly

$$a_{qj}^{\text{new}} = a_{qj} - s \times (a_{pj} - \texttt{tau} \times a_{qj}).$$

Summary: `Jacobi` is one of the very beautiful, outstanding, *machine-independent* and foolproof working algorithms of Heinz Rutishauser. It has been neglected because the QR Algorithm needs fewer operations (though QR is not foolproof! [92]) to compute the eigenvalues of a matrix. However, with the fast hardware available today, `Jacobi` is again attractive as a very reliable algorithm for medium-size eigenvalue problems, and also for parallel architectures, since one can perform rotations in parallel for *disjoint* pairs of rows and columns (p, q) and (p', q'). For more details on parallel implementations, see Section 8.4.6 in [51] and the references therein.

7.4   Power Methods

We now show a very different idea for computing eigenvalues and eigenvectors of matrices based on the *power methods*. The basic method is designed to compute only one eigenvalue-eigenvector pair, but its generalization will later lead to the most widely used eigenvalue solver for dense matrices in Section 7.6.

7.4.1 Power Method

The power method for a matrix A is based on the simple iteration

$$\boldsymbol{x}_{k+1} = A\boldsymbol{x}_k, \tag{7.27}$$

where one starts with an arbitrary initial vector $\boldsymbol{x}_0$. We show in the next theorem that $\boldsymbol{x}_k := A^k \boldsymbol{x}_0$ approaches an eigenvector of A and the *Rayleigh quotient* $(\boldsymbol{x}_k^{\mathrm{H}} A \boldsymbol{x}_k)/(\boldsymbol{x}_k^{\mathrm{H}} \boldsymbol{x}_k)$ is an approximation of an eigenvalue of A.

THEOREM 7.5. (POWER METHOD CONVERGENCE) *Let* $A \in \mathbb{C}^{n \times n}$ *be a diagonalizable matrix with eigenvalues* $\lambda_1, \ldots, \lambda_n$ *and normalized eigenvectors* $\boldsymbol{v}_1, \ldots, \boldsymbol{v}_n$. *If* $|\lambda_1| > |\lambda_2| \geq |\lambda_3| \geq \ldots \geq |\lambda_n|$, *then the vectors* $\boldsymbol{x}_k$ *of the power iteration (7.27) satisfy*

$$\boldsymbol{x}_k = \lambda_1^k (a_1 \boldsymbol{v}_1 + O(|\lambda_2/\lambda_1|^k)), \tag{7.28}$$

where a_1 *is defined by the initial guess* $\boldsymbol{x}_0 = \sum_j a_j \boldsymbol{v}_j$. *If* $a_1 \neq 0$, *then the Rayleigh quotient satisfies*

$$\frac{\boldsymbol{x}_k^{\mathrm{H}} A \boldsymbol{x}_k}{\boldsymbol{x}_k^{\mathrm{H}} \boldsymbol{x}_k} = \lambda_1 + O(|\lambda_2/\lambda_1|^k). \tag{7.29}$$

If in addition, the matrix A *is normal, i.e. the eigenvectors are orthogonal, then the error in the eigenvalue (7.29) is* $O(|\lambda_2/\lambda_1|^{2k})$.

PROOF. We start by expanding the initial guess in the eigenvectors, $\boldsymbol{x}_0 = \sum_{j=1}^n a_j \boldsymbol{v}_j$. By induction, we get

$$\boldsymbol{x}_k = A^k \boldsymbol{x}_0 = \sum_{j=1}^n a_j \lambda_j^k \boldsymbol{v}_j = \lambda_1^k \left(a_1 \boldsymbol{v}_1 + \sum_{j=2}^n a_j \left(\frac{\lambda_j}{\lambda_1} \right)^k \boldsymbol{v}_j \right),$$

which proves (7.28). Inserting this expression into the numerator and denominator of the Rayleigh quotient, we obtain

$$\boldsymbol{x}_k^{\mathrm{H}} A \boldsymbol{x}_k = \boldsymbol{x}_k^{\mathrm{H}} A \boldsymbol{x}_{k+1} = \sum_{j=1}^n |a_j|^2 |\lambda_j|^{2k} \lambda_j + \sum_{j \neq l} \bar{a}_j a_l \bar{\lambda}_j^k \lambda_l^{k+1} \boldsymbol{v}_j^{\mathrm{H}} \boldsymbol{v}_l \tag{7.30}$$

$$\boldsymbol{x}_k^{\mathrm{H}} \boldsymbol{x}_k = \sum_{j=1}^n |a_j|^2 |\lambda_j|^{2k} + \sum_{j \neq l} \bar{a}_j a_l \bar{\lambda}_j^k \lambda_l^k \boldsymbol{v}_j^{\mathrm{H}} \boldsymbol{v}_l. \tag{7.31}$$

Now if $a_1 \neq 0$, we obtain for the Rayleigh quotient

$$\frac{\boldsymbol{x}_k^{\mathrm{H}} A \boldsymbol{x}_k}{\boldsymbol{x}_k^{\mathrm{H}} \boldsymbol{x}_k} = \frac{|a_1|^2 |\lambda_1|^{2k} \lambda_1 (1 + O(|\lambda_2/\lambda_1|^k)}{|a_1|^2 |\lambda_1|^{2k} (1 + O(|\lambda_2/\lambda_1|^k)}, \tag{7.32}$$

which proves (7.29). If the matrix A is normal, the second sum in the (7.30,7.31) vanishes because of orthogonality between $\boldsymbol{v}_j$ and $\boldsymbol{v}_l$, so the order term $O(|\lambda_2/\lambda_1|^k)$ can be replaced by $O(|\lambda_2/\lambda_1|^{2k})$ in (7.32), which proves (7.29). $\qquad\square$

The following MATLAB statements illustrate the power method:

```
A=[2 1 0;1 2 1;0 1 2];
x=[1 1 1]';
for i=1:4
  xn=A*x;
  (x'*xn)/(x'*x)
  x=xn;
end
ans =
    3.3333
ans =
    3.4118
ans =
    3.4141
ans =
    3.4142
>> eig(A)
ans =
    0.5858
    2.0000
    3.4142
>> x
x =
   116
   164
   116
```

Clearly the Rayleigh quotient converges to the largest eigenvalue, but one notices also that the iteration vector x_k grows during the iteration. It is therefore recommended to normalize the vector in each iteration, x=xn/norm(xn).

As we have seen in Theorem 7.5, the convergence of the power method can be very slow if the first two eigenvalues are very close, i.e., if $|\lambda_2/\lambda_1| \approx 1$. In addition, one can only compute the largest eigenvalue and associated eigenvector with the power method. This was however all that was needed for the Page Rank Algorithm of Google, and made it into a great success!

7.4.2 Inverse Power Method (Shift-and-Invert)

Suppose we know an approximation of the eigenvalue we would like to compute, say $\mu \approx \lambda_1$, where λ_1 now does not even have to be the largest eigenvalue of the matrix A. The idea of the inverse power method of Wielandt is to apply the power method to the matrix $(A - \mu I)^{-1}$. The eigenvalues of that matrix are $(\lambda_j - \mu)^{-1}$, and if μ is close to λ_1, then

$$\frac{1}{|\lambda_1 - \mu|} \gg \frac{1}{|\lambda_j - \mu|} \quad \text{for} \quad j \geq 2.$$

Hence the convergence of the method will be very fast. The inverse power iteration requires the solution of a linear system at each step, since

$$\boldsymbol{x}_{k+1} = (A - \mu I)^{-1}\boldsymbol{x}_k \quad \Longleftrightarrow \quad (A - \mu I)\boldsymbol{x}_{k+1} = \boldsymbol{x}_k.$$

Hence, one has to first compute an LU factorization of $A - \mu I$, see Section 3.2.1, and then each iteration of the inverse power method costs the same as the iteration of the original power method.

With the same example as before in MATLAB, we now use as a shift $\mu = 3.41$ and use the long format to show all digits:

```
format long;
A=[2 1 0;1 2 1;0 1 2];
mu=3.41;
[L,U]=lu(A-mu*eye(size(A)));
x=[1 1 1]';
for i=1:3
  xn=U\(L\x);
  1/(x'*xn)/(x'*x)+mu
  x=xn/norm(xn);
end
ans =
   3.410481976508708
ans =
   3.414213562649597
ans =
   3.414213562373096
>> eig(A)
ans =
   0.585786437626905
   2.000000000000000
   3.414213562373095
```

We see that the second iteration already has 9 digits of accuracy !

7.4.3 Orthogonal Iteration

We now generalize the power method in order to compute more than one eigenvalue-eigenvector pair. We consider a matrix A with eigenvalues satisfying

$$|\lambda_1| > |\lambda_2| > \ldots > |\lambda_n|. \tag{7.33}$$

With the power method, $\boldsymbol{x}_{k+1} = A\boldsymbol{x}_k$, we can compute an approximation to the largest eigenvalue, and associated eigenvector. In order to compute simultaneously an approximation to the second largest eigenvalue and eigenvector, we start with two linearly independent initial vectors, say $\boldsymbol{x}_0$ and $\boldsymbol{y}_0$ with $\boldsymbol{x}_0^{\mathrm{H}}\boldsymbol{y}_0 = 0$, and perform the iteration

$$\begin{aligned}
\boldsymbol{x}_{k+1} &= A\boldsymbol{x}_k, \\
\boldsymbol{y}_{k+1} &= A\boldsymbol{y}_k - \beta_{k+1}\boldsymbol{x}_{k+1},
\end{aligned} \tag{7.34}$$

where we choose β_{k+1} such that the new vector pair is again orthogonal, $\boldsymbol{x}_{k+1}^{\mathrm{H}}\boldsymbol{y}_{k+1} = 0$. By induction, we see that

$$
\begin{aligned}
\boldsymbol{x}_k &= A^k \boldsymbol{x}_0, \\
\boldsymbol{y}_k &= A^k \boldsymbol{y}_0 - \gamma_k \boldsymbol{x}_k,
\end{aligned}
$$

where γ_k is such that $\boldsymbol{x}_k^{\mathrm{H}}\boldsymbol{y}_k = 0$. This means that the first vector $\boldsymbol{x}_k$ is simply the vector of the power method applied to $\boldsymbol{x}_0$, and the second vector $\boldsymbol{y}_k$ is also a vector of the power method, applied to $\boldsymbol{y}_0$, but orthogonalized against the vector $\boldsymbol{x}_k$. Expanding the initial vectors in eigenvectors of A,

$$
\boldsymbol{x}_0 = \sum_{j=1}^{n} a_j \boldsymbol{v}_j, \qquad \boldsymbol{y}_0 = \sum_{j=1}^{n} b_j \boldsymbol{v}_j,
$$

the vectors of the iteration become

$$
\boldsymbol{x}_k = \sum_{j=1}^{n} a_j \lambda_j^k \boldsymbol{v}_j, \qquad \boldsymbol{y}_k = \sum_{j=1}^{n} (b_j - \gamma_k a_j) \lambda_j^k \boldsymbol{v}_j. \tag{7.35}
$$

While in the first vector $\boldsymbol{x}_k$ the term $a_1 \lambda_1^k \boldsymbol{v}_1$ will dominate as before, provided $a_1 \neq 0$, for the second vector $\boldsymbol{y}_k$ the orthogonality condition $\boldsymbol{x}_k^{\mathrm{H}}\boldsymbol{y}_k = 0$ implies

$$
\sum_{j=1}^{n}\sum_{l=1}^{n} \bar{a}_j (b_l - \gamma_k a_l) \bar{\lambda}_j^k \lambda_l^k \boldsymbol{v}_j^{\mathrm{H}} \boldsymbol{v}_l = 0, \tag{7.36}
$$

which defines γ_k. Since the term with $j = l = 1$ is again dominant, we see that $\gamma_k \approx b_1/a_1$. We suppose in what follows that $a_1 \neq 0$ and also that $a_1 b_2 - a_2 b_1 \neq 0$. Dividing (7.36) by $\bar{\lambda}_1^k$, we get

$$
\begin{aligned}
\bar{a}_1 (b_1 - \gamma_k a_1) \lambda_1^k &(1 + O(|\lambda_2/\lambda_1|^k)) \\
&= -\bar{a}_1 (b_2 - \gamma_k a_2) \lambda_2^k \left(\boldsymbol{v}_1^{\mathrm{H}} \boldsymbol{v}_2 + O(|\lambda_2/\lambda_1|^k) + O(|\lambda_3/\lambda_2|^k) \right).
\end{aligned}
$$

Inserting this result into the second equation in (7.35), we get

$$
\boldsymbol{y}_k = \lambda_2^k (b_2 - \gamma_k a_2) \left(\boldsymbol{v}_2 - \boldsymbol{v}_1^{\mathrm{H}} \boldsymbol{v}_2 \cdot \boldsymbol{v}_1 + O(|\lambda_2/\lambda_1|^k) + O(|\lambda_3/\lambda_2|^k) \right). \tag{7.37}
$$

Clearly, as $k \to \infty$, the second vector $\boldsymbol{y}_k$ approaches a multiple of the vector $\boldsymbol{v}_2 - \boldsymbol{v}_1^{\mathrm{H}} \boldsymbol{v}_2 \cdot \boldsymbol{v}_1$, which is the orthogonal projection of $\boldsymbol{v}_2$ on the hyperplane $\boldsymbol{v}_1^{\perp}$. For the eigenvalues, we have the following result:

THEOREM 7.6. *For the vectors* $\boldsymbol{x}_k$, $\boldsymbol{y}_k$ *given by (7.34), let*

$$
U_k := \left(\frac{\boldsymbol{x}_k}{\|\boldsymbol{x}_k\|_2}, \frac{\boldsymbol{y}_k}{\|\boldsymbol{y}_k\|_2} \right), \tag{7.38}
$$

which implies $U_k^{\mathrm{H}} U_k = I$. *If the condition (7.33) on the eigenvalues holds, then*

$$
U_k^{\mathrm{H}} A U_k \to \begin{pmatrix} \lambda_1 & * \\ 0 & \lambda_2 \end{pmatrix} \quad \textit{for} \quad k \to \infty. \tag{7.39}
$$

PROOF. The (1,1) element of the matrix $U_k^H A U_k$ is nothing else than the Rayleigh quotient of the power method for the first vector, and hence converges to λ_1. Using (7.37), we see that the (2,2) element satisfies

$$\frac{\boldsymbol{y}_k^H A \boldsymbol{y}_k}{\boldsymbol{y}_k^H \boldsymbol{y}_k} \to \frac{(\boldsymbol{v}_2 - \boldsymbol{v}_1^H \boldsymbol{v}_2 \cdot \boldsymbol{v}_1)^H (\lambda_2 \boldsymbol{v}_2 - \lambda_1 \boldsymbol{v}_1^H \boldsymbol{v}_2 \cdot \boldsymbol{v}_1)}{(\boldsymbol{v}_2 - \boldsymbol{v}_1^H \boldsymbol{v}_2 \cdot \boldsymbol{v}_1)^H (\boldsymbol{v}_2 - \boldsymbol{v}_1^H \boldsymbol{v}_2 \cdot \boldsymbol{v}_1)} = \frac{\lambda_2 (1 - |\boldsymbol{v}_1^H \boldsymbol{v}_2|^2)}{1 - |\boldsymbol{v}_1^H \boldsymbol{v}_2|^2} = \lambda_2.$$

Similarly, we obtain for the (2,1) element

$$\frac{\boldsymbol{y}_k^H A \boldsymbol{x}_k}{\|\boldsymbol{x}_k\|_2 \|\boldsymbol{y}_k\|_2} \to \frac{(\boldsymbol{v}_2 - \boldsymbol{v}_1^H \boldsymbol{v}_2 \cdot \boldsymbol{v}_1)^H \boldsymbol{v}_1 \lambda_1}{\|\boldsymbol{v}_2 - \boldsymbol{v}_1^H \boldsymbol{v}_2 \cdot \boldsymbol{v}_1\|_2 \|\boldsymbol{v}_1\|_2} = 0,$$

and finally for the (1,2) element

$$\frac{\boldsymbol{x}_k^H A \boldsymbol{y}_k}{\|\boldsymbol{x}_k\|_2 \|\boldsymbol{y}_k\|_2} \to \frac{\boldsymbol{v}_1^H (\lambda_2 \boldsymbol{v}_2 - \lambda_1 \boldsymbol{v}_1^H \boldsymbol{v}_2 \cdot \boldsymbol{v}_1)}{\|\boldsymbol{v}_1\|_2 \|\boldsymbol{v}_2 - \boldsymbol{v}_1^H \boldsymbol{v}_2 \cdot \boldsymbol{v}_1\|_2} = \frac{(\lambda_1 - \lambda_2) \boldsymbol{v}_1^H \boldsymbol{v}_2}{\sqrt{1 - |\boldsymbol{v}_1^H \boldsymbol{v}_2|^2}},$$

an expression which is in general non-zero. $\qquad\square$

Using the matrix U_k from (7.38), we can rewrite the iteration (7.34) in matrix form,

$$AU_k = U_{k+1} R_{k+1}, \tag{7.40}$$

where R_{k+1} is a 2×2 upper triangular matrix. We can now generalize this iteration with two vectors to the case of an arbitrary number of vectors, and even to n vectors: we choose an initial matrix U_0 with n orthonormal columns, which play the role of the initial vectors $\boldsymbol{x}_0, \boldsymbol{y}_0, \ldots$, and then perform Algorithm 7.5.

ALGORITHM 7.5. *Orthogonal Iteration*

```
for k=1,2,...
    Z_k = AU_{k-1};
    U_k R_k = Z_k;      % QR decomposition
end
```

If the condition (7.33) on the eigenvalues is satisfied and the matrix U_0 is appropriately chosen (i.e. $a_1 \neq 0$, $a_1 b_2 - a_2 b_1 \neq 0$, etc), a generalization of the previous theorem shows that the quantity

$$T_k := U_k^H A U_k \tag{7.41}$$

converges to a triangular matrix with the eigenvalues of the matrix A on its diagonal. We have thus transformed A into a triangular matrix using an orthogonal matrix, i.e. we have computed the *Schur decomposition* of A. Here is an example in MATLAB, where we compute the roots of the polynomial $(x-1)(x-2)(x-3) = x^3 - 6x^2 + 11x - 6$ using the companion matrix:

```
>> A=compan([1 -6 11 -6])
>> U=orth(hilb(3));
>> for k=1:8
     Z=A*U;
     [U,R]=qr(Z);
     U'*A*U
   end
A =
     6    -11     6
     1      0     0
     0      1     0
ans =
     2.4713    -4.3980   -11.2885
    -0.1939     2.2681     5.9455
     0.0233    -0.0017     1.2606
ans =
     2.7182    -4.6042   -11.6367
    -0.1024     2.0200     5.0316
     0.0119     0.0270     1.2618
ans =
     2.8427    -4.8734   -11.7033
    -0.0493     1.9628     4.5737
     0.0052     0.0307     1.1945
ans =
     2.9063    -5.1044   -11.6788
    -0.0245     1.9705     4.3541
     0.0021     0.0237     1.1233
ans =
     2.9412    -5.2619   -11.6429
    -0.0133     1.9879     4.2432
     0.0008     0.0152     1.0709
ans =
     2.9620    -5.3556   -11.6177
    -0.0078     1.9995     4.1829
     0.0003     0.0088     1.0385
ans =
     2.9750    -5.4071   -11.6038
    -0.0048     2.0048     4.1479
     0.0001     0.0047     1.0202
ans =
     2.9834    -5.4341   -11.5972
    -0.0031     2.0062     4.1266
     0.0000     0.0025     1.0104
```

We clearly see how the algorithm converges to the Schur decomposition, and the eigenvalues start to appear on the diagonal.

It is interesting to note that one can directly compute T_k in (7.41) from T_{k-1}: on the one hand, we obtain from (7.40) that

$$T_{k-1} = U_{k-1}^{\mathrm{H}} A U_{k-1} = (U_{k-1}^{\mathrm{H}} U_k) R_k. \tag{7.42}$$

On the other hand, we also have

$$T_k = U_k^{\mathrm{H}} A U_k = U_k^{\mathrm{H}} A U_{k-1} U_{k-1}^{\mathrm{H}} U_k = R_k (U_{k-1}^{\mathrm{H}} U_k). \tag{7.43}$$

One can therefore simply compute a QR factorization of the matrix T_{k-1} like in (7.42), and then multiply the two factors obtained in reverse order according to (7.43) to obtain T_k. This observation leads to the most widely used algorithm for computing eigenvalues of matrices, the *QR Algorithm*, which we will see in Section 7.6, although it has not been discovered this way historically. But first, we will introduce some useful reductions of matrices to simpler form in the next section.

7.5 Reduction to Simpler Form

In order to simplify computations, one often transforms matrices to a simpler form by introducing zeros. In the preceding section, we used Jacobi rotations on both sides of the matrix to annihilate an off-diagonal element. Sometimes, however, we only need a one-sided Givens rotation to annihilate one element.

7.5.1 Computing Givens Rotations

The standard task to annihilate an element is: given the vector $(x, y)^{\top}$, find an orthogonal matrix

$$G = \begin{pmatrix} c & s \\ -s & c \end{pmatrix}$$

with $c = \cos\phi$ and $s = \sin\phi$ such that

$$G^{\top} \begin{pmatrix} x \\ y \end{pmatrix} = \begin{pmatrix} r \\ 0 \end{pmatrix}. \tag{7.44}$$

Clearly $r = \pm\sqrt{x^2 + y^2}$, since the Givens matrix G is orthogonal and preserves the length of the vector. Using the orthogonality of G and (7.44), we have

$$G^{\top} G = \begin{pmatrix} c & -s \\ s & c \end{pmatrix} \begin{pmatrix} c & s \\ -s & c \end{pmatrix} = \begin{pmatrix} 1 & 0 \\ 0 & 1 \end{pmatrix} \quad \text{and} \quad \begin{pmatrix} c & -s \\ s & c \end{pmatrix} \begin{pmatrix} \frac{x}{r} \\ \frac{y}{r} \end{pmatrix} = \begin{pmatrix} 1 \\ 0 \end{pmatrix}.$$

Comparing columns above, we conclude that

$$c = \frac{x}{r} \quad \text{and} \quad s = -\frac{y}{r}. \tag{7.45}$$

If we do not need to fix the sign of r, the following function `GivensRotation` avoids possible overflow when computing r by factoring out x or y as necessary from the expression $\sqrt{x^2 + y^2}$.

ALGORITHM 7.6. *Givens Rotation*

```
function [c,s]=GivensRotation(x,y)
% GIVENSROTATION computes a Givens rotation matrix
%   [c,s]=GivensRotation(x,y) computes for the given numbers x and y
%   the Givens rotation matrix [c s;-s c] with s=sin(phi) and
%   c=cos(phi) such that [c s;-s c]'*[x;y]=[r;0]

if y==0
  c=1; s=0;
elseif abs(y)>=abs(x)
  h=-x/y; s=1/sqrt(1+h*h); c=s*h;
else
  h=-y/x; c=1/sqrt(1+h*h); s=c*h;
end;
```

Alternatively, one can use the built-in MATLAB function `planerot`: the statement `[G,y]=planerot(x)` computes a Givens matrix G such that the vector $\boldsymbol{x}$ is transformed to $\boldsymbol{y} = G\boldsymbol{x}$ with $y_2 = 0$. Computing G using (7.45) and the MATLAB function `norm` is also a good way numerically, since `norm` is taking care of any possible over- or underflow as described in Section 2.7.5.

7.5.2   Reduction to Hessenberg Form

A non-symmetric matrix A can be reduced by similarity transformations to Hessenberg form: an upper *Hessenberg matrix* is one that has zero entries below the first subdiagonal,

$$H = \begin{pmatrix} * & * & * & * & * \\ * & * & * & * & * \\ 0 & * & * & * & * \\ 0 & 0 & * & * & * \\ 0 & 0 & 0 & * & * \end{pmatrix}.$$

We will discuss here the similarity transformation of a matrix to Hessenberg form using Jacobi (also called Givens) rotations. In order to annihilate the elements $a_{31}, a_{41}, \ldots, a_{n1}$ we use Jacobi rotations in row p and column q with

$$p = 2, \quad q = 3, 4, \ldots, n.$$

The rotation angle, respectively $c = \cos\phi$ and $s = \sin\phi$, is chosen so that $a_{q1}^{\text{new}} = 0$. Similarly, we proceed with the other columns to annihilate the elements

$$a_{q,p-1}^{\text{new}} = 0, \quad q = p + 1, \ldots, n$$

using Jacobi rotations for rows and columns p and q. With each Jacobi rotation we transform two rows and two columns of A using (7.22) and (7.23). We obtain thus Algorithm 7.7 below.

ALGORITHM 7.7.
Similarity Transformation to Hessenberg Form

```
function [H,V]=Hessenberg1(A)
% HESSENBERG1 similarity transformation to Hessenberg form
%    [H,V]=Hessenberg1(A) transforms the given matrix A into
%    Hessenberg form H using the orthogonal transformations V,
%    such that H=V'*A*V

n=length(A); V=eye(n);
for p=2:n-1
  for q=p+1:n
    [c,s]=GivensRotation(A(p,p-1), A(q,p-1));  % rotate A(q,p-1)=0
    if s~=0
      h=A(p,:)*c-A(q,:)*s;                     % A=U^T*A
      A(q,:)=A(p,:)*s+A(q,:)*c;
      A(p,:)=h;
      h =A(:,p)*c-A(:,q)*s;                     % A=A*U
      A(:,q)= A(:,p)*s+A(:,q)*c;
      A(:,p)=h;
      h=V(:,p)*c-V(:,q)*s;                      % V=V*U
      V(:,q)=V(:,p)*s+V(:,q)*c;
      V(:,p)=h;
    end
  end
end
H=A;
```

As an example, we reduce for $n = 6$ the magic square matrix to Hessenberg form:

```
>> A=magic(6)
A =
    35     1     6    26    19    24
     3    32     7    21    23    25
    31     9     2    22    27    20
     8    28    33    17    10    15
    30     5    34    12    14    16
     4    36    29    13    18    11
>> [H,U]=Hessenberg1(A)
H =
    35.0000  -24.0722  -27.8661  -14.6821   -7.7773    4.2378
   -44.1588   55.7395   35.9545   -2.4720    2.5076   -4.6722
    -0.0000   50.9040   40.7834  -34.7565   -6.1114    5.6624
```

```
    -0.0000          0   -27.3916   -15.7856     7.2431   -14.7543
    -0.0000          0     0.0000    -4.6137     4.7703    -1.8187
    -0.0000          0    -0.0000          0     2.8029    -9.5076
U =
     1.0000          0          0          0          0          0
          0    -0.0679    -0.4910     0.6719     0.3346    -0.4369
          0    -0.7020     0.2549     0.4822    -0.1967     0.4136
          0    -0.1812    -0.5147    -0.1350    -0.7991    -0.2130
          0    -0.6794     0.0103    -0.5004     0.3421    -0.4135
          0    -0.0906    -0.6549    -0.2179     0.3061     0.6494
>> norm(U'*U-eye(6))
ans =
   3.8445e-16
>> norm(U'*A*U-H)
ans =
   1.4854e-14
```

The results in this example (up to signs) are the same as those obtained with
the MATLAB built-in function **hess**. Using the MATLAB function **planerot**
and the Givens matrix G, we can program the reduction to Hessenberg form
using matrix operations. Instead of computing the two linear combinations
of the rows p and q explicitly

```
h=A(p,:)*c-A(q,:)*s;                    % A=U^T*A
A(q,:)=A(p,:)*s+A(q,:)*c;
A(p,:)=h;
```

we could think of writing

```
[A(p,:);A(q,:)]=G*[A(p,:);A(q,:)]
```

However, this does not work in MATLAB. The right hand side would be ok,
but not the left hand side, since MATLAB does not allow multiple left-hand
sides written in this way. A simple trick for picking out 2 rows in the matrix
is to use the expression $A(p:q-p:q,:)$: the rows p to q are accessed with step
size $q - p$, thus there are only 2 rows addressed and we can use on the right
hand side the same expression. The statement becomes

```
A(p:q-p:q,:)=G*A(p:q-p:q,:);    % A=U^T*A
```

and we obtain the function **Hessenberg** for which the MATLAB program is
more compact:

ALGORITHM 7.8.
*Similarity Transformation to Hessenberg Form (Version
2)*

```
function [H,V]=Hessenberg(A)
% HESSENBERG similarity transformation to Hessenberg form
%   [H,V]=Hessenberg(A) transforms the given matrix A into
%   Hessenberg form H using the orthogonal transformations V,
```

```
%    such that H=V'*A*V

n=length(A); V=eye(n);
for p=2:n-1
  for q=p+1:n
    G=planerot(A(p:q-p:q,p-1));             % rotate A(q,p-1)=0
    if G(1,2)~=0
      A(p:q-p:q,:)=G*A(p:q-p:q,:);          % A=U^T*A
      A(:,p:q-p:q)=A(:,p:q-p:q)*G';          % A=A*U
      V(:,p:q-p:q)=V(:,p:q-p:q)*G';          % V=V*U
    end
  end
end
H=A;
```

Reducing again the matrix `magic(6)` we obtain this time:

```
>> [H2,U2]=Hessenberg(A)
H2 =
   35.0000   24.0722   27.8661  -14.6821    7.7773   -4.2378
   44.1588   55.7395   35.9545    2.4720    2.5076   -4.6722
    0.0000   50.9040   40.7834   34.7565   -6.1114    5.6624
   -0.0000        0   27.3916  -15.7856   -7.2431   14.7543
   -0.0000        0    0.0000    4.6137    4.7703   -1.8187
    0.0000        0    0.0000        0    2.8029   -9.5076
U2 =
    1.0000        0        0        0        0        0
        0    0.0679    0.4910    0.6719   -0.3346    0.4369
        0    0.7020   -0.2549    0.4822    0.1967   -0.4136
        0    0.1812    0.5147   -0.1350    0.7991    0.2130
        0    0.6794   -0.0103   -0.5004   -0.3421    0.4135
        0    0.0906    0.6549   -0.2179   -0.3061   -0.6494
>> norm(U2'*U2-eye(6))
ans =
   5.3139e-16
>> norm(U2'*A*U2-H2)
ans =
   1.7336e-14
```

The results are again the same up to a few different signs.

A matrix A can also be reduced to Hessenberg form by applying the iterative *Arnoldi Algorithm* 11.17, see Chapter 11. Though theoretically the Arnoldi Algorithm does the same reduction, numerically the Givens approach is preferable. The merits of the Arnoldi Algorithm lie in Krylov subspace approximations of very large matrices. By performing only a few iterations (relative to the dimension n), the Hessenberg matrix forms a basis of the Krylov subspace.

Another alternative is the reduction to Hessenberg form by means of Householder-Transformations, see Problem 7.4.

7.5.3 Reduction to Tridiagonal Form

If we apply Algorithm 7.7 to a symmetric matrix A, then the resulting Hessenberg matrix is also symmetric, since the symmetry is preserved with the Jacobi rotations. Thus, the resulting symmetric matrix H is tridiagonal.

Next, we would like to exploit the symmetry of the matrix. We will use only the elements below the diagonal. Assume that we have already introduced zeros in the columns $1, \dots, p-2$. The working array of our matrix becomes (the upper part is not displayed since it is untouched)

$$
\begin{array}{c}
\\ \\ \\ \\ \\
p \rightarrow \\ \\
q \rightarrow \\ \\ \\
\end{array}
\left(
\begin{array}{cccccccc}
a_{11} & & & & & & & \\
a_{21} & a_{22} & & & & & & \\
0 & \ddots & \ddots & & & & & \\
\vdots & & \ddots & & \ddots & & & \\
0 & \cdots & 0 & a_{p,p-1} & a_{p,p} & & & \\
\vdots & & \vdots & & | & \ddots & & \\
0 & \cdots & 0 & a_{q,p-1} & a_{qp} & - & a_{qq} & \\
\vdots & & \vdots & \vdots & | & & | & \ddots \\
0 & \cdots & 0 & a_{n,p-1} & a_{np} & & a_{nq} & a_{nn}
\end{array}
\right).
$$

In order to annihilate the element $a_{q,p-1}$, we perform a Jacobi rotation, which changes rows and columns p and q. We first consider the three elements in the cross points. From (7.24) we have, using the fact that $a_{pq} = a_{qp}$,

$$
\begin{aligned}
a''_{pp} &= c^2\, a_{pp} - 2\,c\,s\,a_{qp} + s^2\, a_{qq}, \\
a''_{qq} &= s^2\, a_{pp} + 2\,c\,s\,a_{qp} + c^2\, a_{qq}, \\
a''_{qp} &= s\,c\,a_{pp} - s^2 a_{qp} + c^2 a_{qp} - c\,s\,a_{qq}.
\end{aligned}
$$

These expressions can again be simplified by working with corrections. Define

$$
z = (a_{pp} - a_{qq})\,s + 2a_{qp}\,c.
$$

Then

$$
\begin{aligned}
a''_{pp} &= a_{pp} - z\,s, \\
a''_{qq} &= a_{qq} + z\,s, \\
a''_{qp} &= -a_{qp} + z\,c.
\end{aligned}
$$

On row p, there is only one element $a_{p,p-1}$ to be transformed, apart from the cross point element a_{pp}. The elements $a_{p+1,p}, \dots a_{q-1,p}$ in column p are transformed using the corresponding elements on row q. Similarly, the new elements on row q are transformed using the corresponding elements in column p. Finally, the remaining elements $q+1, \dots, n$ in columns p and q are transformed.

ALGORITHM 7.9.

Similarity Transformation to Tridiagonal Form

```
function [T,V]=Tridiagonalize(A)
% TRIDIAGONALIZE transforms a symmetric matrix to tridiagonal form
%    [T,V]=Tridiagonalize(A) transforms the symmetric matrix A into
%    tridiagonal form T using rotations such that T=V'*A*V

n=length(A); V=eye(n);
for p=2:n-1
  for q=p+1:n
    [c,s]=GivensRotation(A(p,p-1),A(q,p-1));
    if s~=0
      A(p,p-1)=A(p,p-1)*c-A(q,p-1)*s;          % one element on row p
      z=(A(p,p)-A(q,q))*s+2*A(q,p)*c;           % corner elements
      A(p,p)=A(p,p)-z*s; A(q,q)=A(q,q)+z*s;
      A(q,p)=-A(q,p)+z*c;
      h=A(p+1:q-1,p)*c-A(q,p+1:q-1)'*s;         % colum p and row q
      A(q,p+1:q-1)=A(p+1:q-1,p)*s+A(q,p+1:q-1)'*c;
      A(p+1:q-1,p)=h;
      h=A(q+1:n,p)*c-A(q+1:n,q)*s;              % rest of colums p and q
      A(q+1:n,q)=A(q+1:n,p)*s+A(q+1:n,q)*c;
      A(q+1:n,p)=h;
      h=V(:,p)*c-V(:,q)*s;                      % update V=V*U
      V(:,q)=V(:,p)*s+V(:,q)*c;
      V(:,p)=h;
    end
  end
end
c=diag(A,-1);
T=diag(diag(A))+diag(c,1)+diag(c,1)';
```

As an example, we reduce for $n = 6$ the Hilbert matrix to tridiagonal
form

```
>> A=hilb(6);
>> [T,V]=Tridiagonalize(A)
T =
    1.0000    0.7010         0         0         0         0
    0.7010    0.8162   -0.1167         0         0         0
         0   -0.1167    0.0602    0.0046         0         0
         0         0    0.0046    0.0018    0.0001         0
         0         0         0    0.0001    0.0000    0.0000
         0         0         0         0    0.0000    0.0000
V =
    1.0000         0         0         0         0         0
         0    0.7133    0.6231   -0.3076    0.0903   -0.0152
         0    0.4755   -0.1305    0.6780   -0.5177    0.1707
         0    0.3566   -0.3759    0.2638    0.5797   -0.5708
         0    0.2853   -0.4629   -0.2112    0.3447    0.7355
```

```
            0      0.2378    -0.4890    -0.5757    -0.5186    -0.3223
>> norm(V'*A*V-T)
ans =
   1.8936e-16
>> norm(V'*V-eye(6))
ans =
   4.2727e-16
>> [U,H]=hess(A)
U =
   -0.0074     0.0583     0.2793     0.7721    -0.5678          0
    0.1280    -0.4668    -0.7248    -0.0592    -0.4867          0
   -0.5234     0.6466    -0.1988    -0.2951    -0.4259          0
    0.7608     0.2687     0.2466    -0.3806    -0.3785          0
   -0.3616    -0.5371     0.5443    -0.4104    -0.3407          0
         0          0          0          0          0     1.0000
H =
    0.0000    -0.0000          0          0          0          0
   -0.0000     0.0003     0.0006          0          0          0
         0     0.0006     0.0138     0.0224          0          0
         0          0     0.0224     0.3198    -0.3764          0
         0          0          0    -0.3764     1.4534    -0.2935
         0          0          0          0    -0.2935     0.0909
>> norm(U*H*U'-A)
ans =
   6.3131e-16
>> norm(U'*U-eye(6))
ans =
   6.4322e-16
```

We observe here that the result of the MATLAB function **hess** is different
from the result of **Tridiagonalize**. Both are correct, since the reduction
to tridiagonal form is not unique; another possible reduction is the *Lanczos
Algorithm for tridiagonalization*, which is described in Section 11.7.3. Just
like the Arnoldi algorithm, the Lanczos algorithm is used to solve large symmetric eigenvalue problems. Note that both the algorithms of Arnoldi and
Lanczos are used in the constructive proof of Theorem 7.7 — the Implicit Q
Theorem.

7.6   QR Algorithm

The QR Algorithm is today the most widely used algorithm for computing
eigenvalues and eigenvectors of dense matrices. It allows us to compute the
roots of the characteristic polynomial of a matrix, a problem for which there
is no closed form solution in general for matrices of size bigger than 4×4, with
a complexity comparable to solving a linear system with the same matrix.

7.6.1 Some History

Heinz Rutishauser discovered some 50 years ago when testing a computer[5] that the iteration $A_0 = A$ with $A \in \mathbb{R}^{n \times n}$

$$
\begin{aligned}
A_i &= L_i R_i \quad \text{Gauss LU decomposition} \\
A_{i+1} &= R_i L_i \qquad\quad i = 1, 2, \ldots
\end{aligned}
\tag{7.46}
$$

produces a sequence of matrices A_i which, under certain conditions, converge to an upper triangular matrix which contains on the diagonal the eigenvalues of the original matrix A (see [113]).

J. G. F. Francis (USA) [34] and Vera Kublanovskaya (USSR) presented independently in 1961 a variant of this iteration, based on the QR decomposition instead of the Gauss LU decomposition (called by Rutishauser at that time the LR algorithm, because the Gauss triangular decomposition was known in German as "Zerlegung in eine **L**inks- und **R**echtsdreieckmatrix"). The QR Algorithm is listed as one of the *top ten algorithms of the last century* [27].

7.6.2 QR Iteration

The QR Algorithm uses as a basic ingredient the orthogonal iteration we have seen in Subsection 7.4.3 but, like in the inverse power method from Subsection 7.4.2, shifts are used in addition to obtain faster convergence. Given the matrix B_k and shift σ_k, one step of the QR Algorithm consists of

1. factoring $B_k - \sigma_k I = Q_k R_k$ (QR Decomposition),

2. forming $B_{k+1} = R_k Q_k + \sigma_k I$.

7.6.3 Basic Facts

The following basic relations are satisfied by matrices arising from the QR Algorithm. Their proof is left as an exercise, see Problem 7.5:

1. $B_{k+1} = Q_k^\top B_k Q_k$.

2. $B_{k+1} = P_k^\top B_1 P_k$ where $P_k := Q_1 \cdots Q_k$.

3. If $S_k = R_k \cdots R_1$ then the QR Decomposition of the matrix $\prod_{i=1}^{k}(B_1 - \sigma_i I)$ is $P_k S_k$.

---

[5] He wanted to test the sustainable floating point operations per second rate of the machine performing a clearly defined sequence of operations many times

7.6.4   Preservation of Form

If B_k is upper Hessenberg, then so is B_{k+1}, as one can see as follows: in the QR Decomposition of

$$B_k = Q_k R_k,$$

the orthogonal matrix Q_k is also upper Hessenberg. This is evident if we compute the decomposition with Gram-Schmidt, see Subsection 6.5.5. Hence the product

$$R_k Q_k = B_{k+1}$$

of an upper triangular matrix with an upper Hessenberg matrix is also an upper Hessenberg matrix.

In the special case when B_k is tridiagonal and symmetric, B_{k+1} is also tridiagonal and symmetric.

THEOREM 7.7. (IMPLICIT Q THEOREM) *For any square matrix B, if*

$$\hat{B} = Q^\top B Q$$

is an upper Hessenberg matrix with strictly positive sub-diagonal elements, then $\hat{B}$ and Q are completely determined by B and q_1, the first column of Q (or by B and $\boldsymbol{q}_n$).

PROOF. Take the columns of $BQ = Q\hat{B}$ in order:

$$Bq_1 = Q \begin{pmatrix} \hat{b}_{11} \\ \hat{b}_{21} \\ 0 \\ \vdots \\ 0 \end{pmatrix} = \boldsymbol{q}_1 \hat{b}_{11} + \boldsymbol{q}_2 \hat{b}_{21}.$$

By multiplying from the left with $\boldsymbol{q}_1^\top$, we obtain

$$\boldsymbol{q}_1^\top B \boldsymbol{q}_1 = \hat{b}_{11}.$$

Then $\|B\boldsymbol{q}_1 - \boldsymbol{q}_1 \hat{b}_{11}\| = \hat{b}_{21}$, thus $\boldsymbol{q}_2$ is also determined,

$$\boldsymbol{q}_2 = \frac{1}{\hat{b}_{21}} (B\boldsymbol{q}_1 - \boldsymbol{q}_1 \hat{b}_{11}).$$

The next step is to consider $B\boldsymbol{q}_2$, and the result follows by induction.   □

This constructive proof can be used as an algorithm to construct the orthonormal basis vectors $\boldsymbol{q}_2, \boldsymbol{q}_3$, etc. based on $\boldsymbol{q}_1$; this is in fact the *Algorithm of Arnoldi*. For a symmetric matrix, it becomes the *Lanczos Algorithm*.

7.6.5 Symmetric Tridiagonal Matrices

Let us consider the symmetric tridiagonal matrix

$$
B_k = \begin{pmatrix}
a_1 & b_1 & & & \\
b_1 & a_2 & \ddots & & \\
& \ddots & \ddots & b_{n-1} \\
& & b_{n-1} & a_n
\end{pmatrix}.
$$

The QR Algorithm with arbitrary shifts preserves this form. Using a QR step with explicit shift, we get

$$
B_k - \sigma I = \begin{pmatrix}
a_1 - \sigma & b_1 & & & \\
b_1 & a_2 - \sigma & \ddots & & \\
& \ddots & \ddots & b_{n-1} \\
& & b_{n-1} & a_n - \sigma
\end{pmatrix}.
$$

We now have to choose a Givens Rotation matrix G_1 such that

$$
\begin{pmatrix} c & -s \\ s & c \end{pmatrix} \begin{pmatrix} a_1 - \sigma \\ b_1 \end{pmatrix} = \begin{pmatrix} x \\ 0 \end{pmatrix}.
$$

Applying this Givens rotation, we obtain

$$
G_1^\top (B_k - \sigma I) = \begin{pmatrix}
x & x & \mathbf{x} & & \\
0 & x & x & & \\
& x & x & x & \\
& & x & x & x \\
& & & \ddots & \ddots & \ddots
\end{pmatrix},
$$

where $\mathbf{x}$ marks the fill-in. Continuing with Givens rotations to annihilate the sub-diagonal we get the matrix

$$
\underbrace{G_{n-1}^\top \cdots G_1^\top}_{Q_k^\top} (B_k - \sigma I) = \begin{pmatrix}
x & x & \mathbf{x} & & & \\
0 & x & x & \ddots & & \\
& \ddots & \ddots & \ddots & \mathbf{x} \\
& & 0 & x & x \\
& & & 0 & x
\end{pmatrix} = R_k.
$$

Now for the next iteration, we get

$$
B_{k+1} = R_k Q_k + \sigma I = R_k G_1 G_2 \cdots G_{n-1} + \sigma I.
$$

We illustrate this process with a numerical example. We first generate a 1-4-1 tridiagonal matrix:

```
>> n=7;
>> C=diag(ones(n-1,1),-1); B=4*diag(ones(n,1))+C+C';
```

Now we perform one step of QR with MATLAB and full matrix operations:

```
>> sigma=2                      % take a shift
>> [Q,R]=qr(B-sigma*eye(n))
>> B1=R*Q+sigma*eye(n)
Q =
    -0.8944     0.3586    -0.1952     0.1231    -0.0848     0.0620     0.0845
    -0.4472    -0.7171     0.3904    -0.2462     0.1696    -0.1240    -0.1690
          0    -0.5976    -0.5855     0.3693    -0.2544     0.1861     0.2535
          0          0    -0.6831    -0.4924     0.3392    -0.2481    -0.3381
          0          0          0    -0.7385    -0.4241     0.3101     0.4226
          0          0          0          0    -0.7774    -0.3721    -0.5071
          0          0          0          0          0    -0.8062     0.5916
R =
    -2.2361    -1.7889    -0.4472          0          0          0          0
          0    -1.6733    -1.9124    -0.5976          0          0          0
          0          0    -1.4639    -1.9518    -0.6831          0          0
          0          0          0    -1.3540    -1.9695    -0.7385          0
          0          0          0          0    -1.2863    -1.9789    -0.7774
          0          0          0          0          0    -1.2403    -1.9846
          0          0          0          0          0          0     0.6761
B1 =
     4.8000     0.7483    -0.0000     0.0000     0.0000     0.0000     0.0000
     0.7483     4.3429     0.8748     0.0000     0.0000    -0.0000     0.0000
          0     0.8748     4.1905     0.9250    -0.0000     0.0000     0.0000
          0          0     0.9250     4.1212     0.9500     0.0000     0.0000
          0          0          0     0.9500     4.0839     0.9643    -0.0000
          0          0          0          0     0.9643     4.0615    -0.5451
          0          0          0          0          0    -0.5451     2.4000
```

Next we compute the same with Givens rotations, using the short MATLAB function `DirectSum` we have seen in Algorithm 6.32:

```
>> [c,s]=GivensRotation(B(1,1)-sigma, B(2,1));
>> Gt=DirectSum([c,-s; s,c],eye(n-2));
>> Q1t=Gt;
>> R1=Gt*(B-sigma*eye(n));
>> for k=2:n-1
       [c,s]=GivensRotation(R1(k,k),R1(k+1,k));
       Gt=DirectSum(eye(k-1),[c,-s;s,c],eye(n-k-1));
       Q1t=Gt*Q1t;
       R1=Gt*R1;
   end
>> Q1=Q1t'
>> R1
>> B1g=R1*Q1+sigma*eye(n)
Q1 =
     0.8944    -0.3586     0.1952     0.1231    -0.0848     0.0620    -0.0845
     0.4472     0.7171    -0.3904    -0.2462     0.1696    -0.1240     0.1690
```

0	0.5976	0.5855	0.3693	-0.2544	0.1861	-0.2535
0	0	0.6831	-0.4924	0.3392	-0.2481	0.3381
0	0	0	-0.7385	-0.4241	0.3101	-0.4226
0	0	0	0	-0.7774	-0.3721	0.5071
0	0	0	0	0	-0.8062	-0.5916

```
R1 =
```

2.2361	1.7889	0.4472	0	0	0	0
0	1.6733	1.9124	0.5976	0	0	0
0	-0.0000	1.4639	1.9518	0.6831	0	0
0	-0.0000	0.0000	-1.3540	-1.9695	-0.7385	0
0	0.0000	-0.0000	0.0000	-1.2863	-1.9789	-0.7774
0	-0.0000	0.0000	-0.0000	-0.0000	-1.2403	-1.9846
0	0.0000	-0.0000	0.0000	0.0000	0.0000	-0.6761

```
B1g =
```

4.8000	0.7483	0.0000	0.0000	-0.0000	0.0000	-0.0000
0.7483	4.3429	0.8748	0.0000	-0.0000	0.0000	-0.0000
-0.0000	0.8748	4.1905	-0.9250	-0.0000	0.0000	-0.0000
-0.0000	-0.0000	-0.9250	4.1212	0.9500	0.0000	-0.0000
0.0000	0.0000	-0.0000	0.9500	4.0839	0.9643	-0.0000
-0.0000	-0.0000	0.0000	0.0000	0.9643	4.0615	0.5451
0.0000	0.0000	-0.0000	-0.0000	-0.0000	0.5451	2.4000

`B1g` is essentially the same matrix as `B1` – only the signs of the b_3 and b_{n-1} elements are different. We can also check the basic fact that $B_{k+1} = Q_k^\top B_k Q_k$:

```
>> B1a=Q1'*B*Q1
B1a =
```

4.8000	0.7483	-0.0000	-0.0000	-0.0000	-0.0000	-0.0000
0.7483	4.3429	0.8748	0.0000	-0.0000	0.0000	-0.0000
0	0.8748	4.1905	-0.9250	-0.0000	0.0000	-0.0000
0	0.0000	-0.9250	4.1212	0.9500	-0.0000	0.0000
-0.0000	-0.0000	-0.0000	0.9500	4.0839	0.9643	-0.0000
0.0000	0.0000	0.0000	0.0000	0.9643	4.0615	0.5451
-0.0000	-0.0000	-0.0000	0.0000	-0.0000	0.5451	2.4000

`B1g` and `B1a` are the same matrix.

Since

$$B_{k+1} = G_{n-1}^\top \cdots G_1^\top B_k G_1 G_2 \cdots G_{n-1},$$

we could also compute the transformation directly on B_k by applying the Givens rotations G_i and $G_i^\top$ immediately on the matrix. Note that

$$G_1^\top B_k G_1 = \begin{pmatrix} x & x & \mathbf{x} & & & \\ x & x & x & & & \\ \mathbf{x} & x & x & x & & \\ & x & x & x & & \\ & & & \ddots & \ddots & \ddots \end{pmatrix}.$$

The fill-in element $\mathbf{x}$ in the (1,3) and (3,1) position is called the *bulge*. With another Givens rotation $\tilde{G}_2 = G(2, 3, \phi)$ which annihilates $\mathbf{x}$ we can obtain

$$
\tilde{G}_2^\top G_1^\top B_k G_1 \tilde{G}_2 = \begin{pmatrix} x & x & 0 & & & \\ x & x & x & \mathbf{x} & & \\ 0 & x & x & x & & \\ \mathbf{x} & x & x & x & & \\ & & & \ddots & \ddots & \ddots \end{pmatrix},
$$

which means that we have *chased the bulge* down one row/column. If we continue in this fashion until we have restored the tridiagonal form of the matrix, we obtain a tridiagonal matrix $\tilde{B}_{k+1}$

$$
\tilde{B}_{k+1} = \tilde{G}_{n-1}^\top \cdots \tilde{G}_2^\top G_1^\top B_k G_1 \tilde{G}_2 \cdots \tilde{G}_{n-1}.
$$

In other words, $\tilde{B}_{k+1} = \tilde{Q}^\top B_k \tilde{Q}$ with $\tilde{Q} = G_1 \tilde{G}_2 \cdots \tilde{G}_{n-1}$. Compare this expression with $B_{k+1} = Q_k^\top B_k Q_k$, where $Q_k = G_1 G_2 \cdots G_{n-1}$. The first columns of the two orthogonal matrices $\tilde{Q}$ and Q_k are obviously the same, as we can see from

$$
\tilde{Q}\boldsymbol{e}_1 = G_1 \tilde{G}_2 \cdots \tilde{G}_{n-1}\boldsymbol{e}_1 = G_1 \boldsymbol{e}_1 \quad \text{since } \tilde{G}_i \boldsymbol{e}_1 = \boldsymbol{e}_1 \text{ for } i > 1,
$$

and

$$
Q_k \boldsymbol{e}_1 = G_1 G_2 \cdots G_{n-1} \boldsymbol{e}_1 = G_1 \boldsymbol{e}_1 \quad \text{since also } G_i \boldsymbol{e}_1 = \boldsymbol{e}_1 \text{ for } i > 1.
$$

Therefore, using the implicit Q Theorem, we conclude that $\tilde{Q}$ and Q_k must be essentially the same, so $\tilde{B}_{k+1}$ and B_{k+1} must also be the same matrix (up to possibly different signs in the off-diagonal elements), since both matrices are similar to B_k via an orthogonal matrix with the same first column. *The advantage of computing B_{k+1} with implicit shift this way is that we do not need to calculate R_k as intermediate result.*

Let us illustrate this with our example. This time, we chase the bulge and apply the necessary Givens rotations $\tilde{G}_i$ also to $B - \sigma I$. As we can see from the output, we obtain up to different signs the same upper triangular matrix! Thus, we can conclude that *the Givens rotations used to chase the bulge are essentially the same as those we need to compute the QR Decomposition of B_k.*

```
>> n=7;
>> C=diag(ones(n-1,1),-1); B=4*diag(ones(n,1))+C+C';
>> sigma=2;
>> [c,s]=GivensRotation(B(1,1)-sigma,B(2,1));
>> Gt=DirectSum([c,-s; s,c],eye(n-2));
>> tildeQt=Gt;
>> R1i=Gt*(B-sigma*eye(n));
>> B1i=Gt*B*Gt';
```

```
>> for k=2:n-1
     [c,s]=GivensRotation(B1i(k,k-1),B1i(k+1,k-1));
     Gt=DirectSum(eye(k-1),[c,-s;s,c],eye(n-k-1));
     tildeQt=Gt*tildeQt;
     R1i=Gt*R1i;
     B1i=Gt*B1i*Gt';
   end
>> tildeQ=tildeQt'
>> R1i
>> B1i
```

```
tildeQ =
    0.8944   -0.3586    0.1952    0.1231   -0.0848    0.0620   -0.0845
    0.4472    0.7171   -0.3904   -0.2462    0.1696   -0.1240    0.1690
         0    0.5976    0.5855    0.3693   -0.2544    0.1861   -0.2535
         0         0    0.6831   -0.4924    0.3392   -0.2481    0.3381
         0         0         0   -0.7385   -0.4241    0.3101   -0.4226
         0         0         0         0   -0.7774   -0.3721    0.5071
         0         0         0         0         0   -0.8062   -0.5916
R1i =
    2.2361    1.7889    0.4472         0         0         0         0
         0    1.6733    1.9124    0.5976         0         0         0
         0   -0.0000    1.4639    1.9518    0.6831         0         0
         0   -0.0000    0.0000   -1.3540   -1.9695   -0.7385         0
         0    0.0000   -0.0000    0.0000   -1.2863   -1.9789   -0.7774
         0   -0.0000    0.0000   -0.0000   -0.0000   -1.2403   -1.9846
         0    0.0000   -0.0000    0.0000    0.0000   -0.0000   -0.6761
B1i =
    4.8000    0.7483   -0.0000   -0.0000    0.0000   -0.0000    0.0000
    0.7483    4.3429    0.8748    0.0000   -0.0000    0.0000   -0.0000
   -0.0000    0.8748    4.1905   -0.9250    0.0000   -0.0000    0.0000
   -0.0000   -0.0000   -0.9250    4.1212    0.9500   -0.0000    0.0000
    0.0000    0.0000    0.0000    0.9500    4.0839    0.9643    0.0000
   -0.0000   -0.0000   -0.0000   -0.0000    0.9643    4.0615    0.5451
    0.0000    0.0000    0.0000    0.0000   -0.0000    0.5451    2.4000
```

7.6.6 Implicit QR Algorithm

We are now ready to program a QR step of the implicit QR Algorithm.
Suppose we know the matrix B and a first Givens transformation G_1. Let us
use MAPLE to compute the transformed matrix $B^{\text{new}} = G_1^{\mathsf{T}} B G_1$:

```
with(LinearAlgebra):
B:=Matrix([[a[1],b[1],0,0],[b[1],a[2],b[2],0],[0,b[2],a[3],b[3]],
          [0,0,b[3],a[4]]]):
G:=Matrix([[c,-s,0,0],[s,c,0,0],[ 0,0,1,0],[ 0,0,0,1]]):
Gs:=Transpose(G):
B1:=G.B.Gs;
```

We obtain:

$$
\begin{bmatrix}
(ca_1 - sb_1)\,c - (cb_1 - sa_2)\,s & (ca_1 - sb_1)\,s + (cb_1 - sa_2)\,c & -sb_2 & 0 \\
(sa_1 + cb_1)\,c - (sb_1 + ca_2)\,s & (sa_1 + cb_1)\,s + (sb_1 + ca_2)\,c & cb_2 & 0 \\
-sb_2 & cb_2 & a_3 & b_3 \\
0 & 0 & b_3 & a_4
\end{bmatrix}
$$

Thus the new elements become

$$
a_1^{\mathrm{new}} = (ca_1 - sb_1)\,c - (cb_1 - sa_2)\,s = c^2 a_1 - 2csb_1 + s^2 a_2 = a_1 - s(s(a_1 - a_2) + 2cb_1)
$$

$$
b_1^{\mathrm{new}} = (ca_1 - sb_1)\,s + (cb_1 - sa_2)\,c = cs(a_1 - a_2) + b_1(c^2 - s^2)
$$

$$
a_2^{\mathrm{new}} = (sa_1 + cb_1)\,s + (sb_1 + ca_2)\,c = s^2 a_1 + 2csb_1 + c^2 a_2 = a_2 + s(s(a_1 - a_2) + 2cb_1)
$$

$$
b_2^{\mathrm{new}} = cb_2
$$

$$
bulge = -sb_2.
$$

If we introduce the difference $d = a_1 - a_2$ and $z = s(s(a_1 - a_2) + 2cb_1) = s(sd + 2cb_1)$, then the computation of the first Givens transformation becomes in MATLAB

```
[c s]=GivensRotation(a(1)-sigma, b(1));   % first transformation
d=a(1)-a(2); z=s*(s*d+2*c*b(1));
a(1)=a(1)-z;
a(2)=a(2)+z;
b(1)=c*s*d+b(1)*(c^2-s^2);
bulge=-s*b(2);
b(2)=c*b(2);
```

Chasing the bulge is now done with $n - 2$ more Givens transformations for $k = 3, \ldots, n$. In the k-th step, c and s are chosen such that

$$
\begin{pmatrix} c & -s \\ s & c \end{pmatrix} \begin{pmatrix} b_{k-2} \\ bulge \end{pmatrix} = \begin{pmatrix} b_{k-2}^{\mathrm{new}} \\ 0 \end{pmatrix}.
$$

The second equation above tells us we must choose c and s so that

$$
s\, b_{k-2} + c\, bulge = 0.
$$

The transformed elements are then

$$
b_{k-2}^{\mathrm{new}} = c\, b_{k-2} - s\, bulge,
$$

$$
a_{k-1}^{\mathrm{new}} = c^2 a_{k-1} - 2csb_{k-1} + s^2 a_k = a_{k-1} - s(s(a_{k-1} - a_k) + 2cb_{k-1}),
$$

$$
b_{k-1}^{\mathrm{new}} = cs(a_{k-1} - a_k) + b_{k-1}(c^2 - s^2),
$$

$$
a_k^{\mathrm{new}} = s^2 a_{k-1} + 2csb_{k-1} + c^2 a_k = a_k + s(s(a_{k-1} - a_k) + 2cb_{k-1}),
$$

$$bulge = -sb_k \quad \text{the new bulge,}$$

$$b_k^{\text{new}} = cb_k.$$

```
for k=3:n                             % chasing the bulge
  [c s]=GivensRotation(b(k-2),bulge);
  d=a(k-1)-a(k); z=s*(s*d+2*c*b(k-1));
  a(k-1)=a(k-1)-z;
  a(k)=a(k)+z;
  b(k-2)=c*b(k-2)-s*bulge;
  b(k-1)=c*s*d+b(k-1)*(c^2-s^2);
  if k<n
    bulge=-s*b(k);
    b(k)=c*b(k);
  end
end
```

7.6.7 Convergence of the QR Algorithm

For the basic QR Iteration without shifts, if the matrix $A^{(1)} = A$ has distinct eigenvalues, then the super-diagonal elements $a_{ij}^{(k)}$ of $A^{(k)}$ behave asymptotically like $k_{ij}(\lambda_i/\lambda_j)^k$, where k_{ij} is a constant (see [148]). In the presence of p eigenvalues with equal modulus, the limiting matrix will not be diagonal, but contain a block of order p corresponding to the same modulus $|\lambda_i|$. To illustrate this fact, we consider the following example:

```
format short
Q=orth(hilb(6));                      % we construct a symmetric matrix
A=Q*diag([1 -1.01 3 4 5 6])*Q';
[Q,R]=qr(A);                          % watch the convergence
for k=1:200
  clc
  A=R*Q, [Q,R]=qr(A);
  pause(0.1)
end
```

Here, we constructed a matrix with two eigenvalues that are close in in modulus (1 and -1.01). After 200 iterations, we obtain the matrix

```
A =
    6.0000   -0.0000   -0.0000    0.0000   -0.0000   -0.0000
   -0.0000    5.0000   -0.0000   -0.0000    0.0000   -0.0000
   -0.0000   -0.0000    4.0000    0.0000    0.0000   -0.0000
   -0.0000   -0.0000   -0.0000    3.0000    0.0000    0.0000
   -0.0000   -0.0000   -0.0000   -0.0000   -0.8646    0.5207
    0.0000    0.0000    0.0000    0.0000    0.5207    0.8546
```

We see that some diagonal elements converged to the eigenvalues $3, 4, 5, 6$, but the lower 2×2 block has yet to converge to a diagonal matrix 1 and

-1.01. Nonetheless, if we perform 1000 iterations instead of only 200, then the 2×2 block becomes

```
   -1.0100      0.0002
    0.0002      1.0000
```

So we do indeed have convergence, but it is very slow. Since the speed of convergence depends on the quotient of two successive eigenvalues $|\lambda_i| > |\lambda_{i+1}|$,

$$\left| \frac{\lambda_{i+1}}{\lambda_i} \right|,$$

it is a good idea to accelerate convergence by shifting the spectrum, as we have already done in the inverse power method, see Subsection 7.4.2. Given some shift σ, if A has the eigenvalues $\lambda_1, \lambda_2, \ldots, \lambda_n$ then $B = A - \sigma I$ has the eigenvalues $\lambda_1 - \sigma, \lambda_2 - \sigma, \ldots, \lambda_n - \sigma$ and the convergence of the QR Algorithm applied to B depends now on

$$\left| \frac{\lambda_{i+1} - \sigma}{\lambda_i - \sigma} \right|.$$

So for $\lambda_5 = 1$ and $\lambda_6 = -1.01$ in our example, a shift of $\sigma = -0.9$ would give us

$$\left| \frac{\lambda_6 - \sigma}{\lambda_5 - \sigma} \right| = \frac{|-1.01 + 0.9|}{1 + 0.9} = 0.0579 \quad \text{instead of} \quad \left| \frac{\lambda_5}{\lambda_6} \right| = \frac{1}{1.01} = 0.9901,$$

i.e., much better convergence. In fact, the closer the shift is to an eigenvalue, the better the convergence will be. A simple estimate is to take as shift the last diagonal element

$$\sigma = a_{nn}.$$

The 2×2 block after 200 iterations displayed with more digits is:

```
   -0.864591608080753      0.520698825922572
    0.520698825922576      0.854591608080752
```

If we continue the QR steps with our 2×2 block, this time with a shift of $\sigma = a_{66}$, we obtain after three more steps

```
format long
A(5:6,5:6)
for k=1:3
  s =A(6,6); [Q,R]=qr(A-s*eye(6)); A=R*Q+s*eye(6) ;
  A(5:6,5:6)
  pause
end
```

```
-1.009047187652985    0.043752085277551
 0.043752085277547    0.999047187652984

-1.009999999785590    0.000020759740271
 0.000020759740275    0.999999999785589

-1.010000000000001    0.000000000000006
 0.000000000000002    1.000000000000000
```

Thus, we observe a fantastic convergence that is at least quadratic! An instructive example is the following: take the matrix

$$B_0 = A = \begin{pmatrix} 2 & a \\ \epsilon & 1 \end{pmatrix}, \tag{7.47}$$

where ϵ is a small number. Choosing for the shift $\sigma = 1$, we obtain

$$B_0 - \sigma I = \begin{pmatrix} 1 & a \\ \epsilon & 0 \end{pmatrix} = \begin{pmatrix} \frac{1}{\sqrt{1+\epsilon^2}} & \frac{-\epsilon}{\sqrt{1+\epsilon^2}} \\ \frac{\epsilon}{\sqrt{1+\epsilon^2}} & \frac{1}{\sqrt{1+\epsilon^2}} \end{pmatrix} \begin{pmatrix} \sqrt{1+\epsilon^2} & \frac{a}{\sqrt{1+\epsilon^2}} \\ 0 & \frac{-a\epsilon}{\sqrt{1+\epsilon^2}} \end{pmatrix} = Q_1 R_1,$$

and therefore for the next iteration

$$B_1 - \sigma I = R_1 Q_1 = \begin{pmatrix} x & x \\ \frac{-a\epsilon^2}{1+\epsilon^2} & x \end{pmatrix}.$$

We therefore see that when the matrix A is symmetric, i.e. $a = \epsilon$, the lower left element after the first operation is $O(\epsilon^3)$, which means that we have cubic convergence. This example also shows that for non-symmetric matrices, we can still expect quadratic convergence, since the lower left element is $O(\epsilon^2)$ after one iteration.

7.6.8 Wilkinson's Shift

An even better approximation the eigenvalue, which is known as *Wilkinson's shift*, is obtained by considering the last 2×2 block. Wilkinson's shift is defined to be the eigenvalue of the matrix

$$H = \begin{pmatrix} a_{n-1} & b_{n-1} \\ b_{n-1} & a_n \end{pmatrix}$$

that is closer to a_n. From $\det(H - \lambda I) = (a_{n-1} - \lambda)(a_n - \lambda) - b_{n-1}^2 = 0$, we get

$$\lambda_{1,2} = \frac{a_{n-1} + a_n}{2} \pm \sqrt{\delta^2 + b_{n-1}^2}, \quad \text{where } \delta = \frac{a_{n-1} - a_n}{2}.$$

Another way of writing $\lambda_{1,2}$ is

$$\lambda_{1,2} = a_n + \delta \pm \sqrt{\delta^2 + b_{n-1}^2} = a_n + x_{1,2},$$

where $x_{1,2}$ are the two solutions of the quadratic equation

$$x^2 - 2\delta x - b_{n-1}^2 = 0.$$

In order to obtain the shift λ closer to a_n, we need to compute the smaller of the two solutions $x_{1,2}$ in absolute value. Since the larger solution is given by

$$x_1 = \operatorname{sign}(\delta)(|\delta| + \sqrt{\delta^2 + b_{n-1}^2}) = \delta + \operatorname{sign}(\delta)\sqrt{\delta^2 + b_{n-1}^2}.$$

Vieta's theorem allows us to compute the smaller solution in a stable way by

$$x_2 = \frac{-b_{n-1}^2}{\delta + \operatorname{sign}(\delta)\sqrt{\delta^2 + b_{n-1}^2}},$$

giving the shift $\sigma = a_n + x_2$. In the special case of $\delta = 0$, we get

$$\lambda_{1,2} = a_n \pm |b_{n-1}|;$$

we recommend take the shift that is smaller in magnitude, i.e., $\sigma = a_n - |b_{n-1}|$, so that $H - \sigma I$ stays positive definite whenever H is. Thus, the Wilkinson shift is computed by

```
delta=(a(n)-a(n-1))/2;
if delta==0,
  sigma=a(n)-abs(b(n-1));
else
  sigma=a(n)-b(n-1)^2/(delta+sign(delta)*sqrt(delta^2+b(n-1)^2));
end
```

7.6.9 Test for Convergence and Deflation

With Wilkinson's shift, the element $b_{n-1}^k \to 0$ for $k \to \infty$. Convergence is more than cubic, as shown by B. Parlett [104]:

$$\frac{b_{n-1}^{(k+1)}}{\left(b_{n-1}^{(k)}\right)^3 \left(b_{n-2}^{(k)}\right)^2} \to \frac{1}{|\lambda_2 - \lambda_1|^3 |\lambda_3 - \lambda_1|} \qquad k \to \infty.$$

A straightforward test for convergence is to compare the off-diagonal element to the machine precision ε, multiplied by the arithmetic mean of the moduli of two diagonal elements:

$$|b_{n-1}| \leq \varepsilon \frac{|a_{n-1}| + |a_n|}{2} \qquad \Longrightarrow \qquad b_{n-1} := 0, \quad n := n - 1.$$

We propose here the equivalent *machine-independent criterion*

```
an=abs(a(n))+abs(a(n-1));
if an+b(n-1)==an, n=n-1; end
```

Reducing n by one means that we consider afterward only the remaining matrix with the last row and column removed, which is called *deflation*. Each deflation step reduces n until we finally obtain for $n = 2$ the matrix

$$
C = \begin{pmatrix}
x & \mathbf{x} & & & \\
\mathbf{x} & x & 0 & & \\
& 0 & \lambda_3 & \ddots & \\
& & \ddots & \ddots & 0 \\
& & & 0 & \lambda_n
\end{pmatrix}.
$$

Once we arrive at this stage, we can simply annihilate the $\mathbf{x}$-element in the 2×2 block explicitly using one Jacobi rotation, see (7.26).

7.6.10 Unreduced Matrices have Simple Eigenvalues

A symmetric tridiagonal matrix T is *unreduced* if all the off-diagonal elements are non-zero, i.e., $b_i \neq 0$. As shown in [104], such matrices have only simple eigenvalues, see also Problem 7.2. However, this result is only useful for theoretical considerations: Wilkinson gave a famous example of a tridiagonal matrix defined by

$$
a_i : [10, 9, \ldots, 1, 0, 1, \ldots, 10], \quad b_i = 1,
$$

for which the eigenvalues are simple but the two largest are very close,

```
10.746194182903322,
10.746194182903393.
```

If $b_k = 0$ for some k, then the eigenvalue problem can be *reduced* by splitting it into two independent problems. Let

$$
T = \begin{pmatrix} T_1 & \\ & T_2 \end{pmatrix}, \ T_1 = \begin{pmatrix}
a_1 & b_1 & & \\
b_1 & \ddots & \ddots & \\
& \ddots & \ddots & b_{k-1} \\
& & b_{k-1} & a_k
\end{pmatrix}, \ T_2 = \begin{pmatrix}
a_{k+1} & b_{k+1} & & \\
b_{k+1} & \ddots & \ddots & \\
& \ddots & \ddots & b_{n-1} \\
& & b_{n-1} & a_n
\end{pmatrix}.
$$

If $(\lambda, \boldsymbol{x})$ is an eigenpair of T_1 and $(\mu, \boldsymbol{y})$ an eigenpair of T_2, then

$$
\begin{pmatrix} T_1 & \\ & T_2 \end{pmatrix} \begin{pmatrix} \boldsymbol{x} \\ \boldsymbol{0} \end{pmatrix} = \lambda \begin{pmatrix} \boldsymbol{x} \\ \boldsymbol{0} \end{pmatrix} \quad \text{and} \quad \begin{pmatrix} T_1 & \\ & T_2 \end{pmatrix} \begin{pmatrix} \boldsymbol{0} \\ \boldsymbol{y} \end{pmatrix} = \mu \begin{pmatrix} \boldsymbol{0} \\ \boldsymbol{y} \end{pmatrix},
$$

so we have two eigenpairs of T.

The QR Algorithm described so far assumes that T is unreduced. In general, however, this will not be the case. If T arises from the transformation of full symmetric matrix into tridiagonal form using Givens rotations or Householder transformations, then multiple eigenvalues are likely. Furthermore, we may encounter a theoretically unreduced tridiagonal matrix which numerically has multiple eigenvalues, and hence numerically reducible. It is therefore necessary in each QR step to check if the matrix can be reduced. If the matrix can be reduced, e.g. if

$$T = \begin{pmatrix} T_1 & \\ & T_2 \end{pmatrix},$$

then a QR Iteration can be performed first on T_2 and then on T_1. Alternatively, in a parallel environment, one could diagonalize T_1 and T_2 simultaneously.

Putting all the parts together, we get the MATLAB function `ImplicitQR`:

ALGORITHM 7.10.
Implicit QR Algorithm for Tridiagonal Matrices

```
function d=ImplicitQR(a,b,test)
% IMPLICITQR computes eigevalues of a symmetric tridiagonal matrix
%    d=ImplicitQR(a,b) computes the eigenvalues d of the matrix
%    T=diag(a)+diag(b,1)+diag(b,-1) with the QR-algorithm with implicit
%    Wilkinson shift.
%    If test==1 we get testoutput for didactical reasons after each sweep

if nargin<3, test=0; end
n=length(a); I=1;
while n>1;
  for k=I:n-1                          % Check for small b_i and
    an=abs(a(k))+abs(a(k+1));          % possible deflation
    if an+b(k)==an, b(k)=0; I=k+1; end
  end
  if I==n;                             % deflation
    n=n-1; I=1;
  elseif I+1==n
    g=100*abs(b(I)); h=a(I+1)-a(I);
    if abs(h)+g==abs(h), t=b(I)/h;     % 2x2 block: annihilate b(I)
    else                               % explicitly by one rotation
      theta=0.5*h/b(I);
      t=1/(abs(theta)+sqrt(1+theta^2));
      if theta<0, t=-t; end
    end
    a(I)=a(I)-b(I)*t;
    a(I+1)=a(I+1)+b(I)*t;
    b(I)=0;
    n=n-2; I=1;                        % deflation
  else
    delta=(a(n)-a(n-1))/2;             % QR-step from I to n
    if delta==0,                       % with  Wilkinson shift
      sigma=a(n)-abs(b(n-1));
    else
```

```
    sigma=a(n)-b(n-1)^2/(delta+sign(delta)*sqrt(delta^2+b(n-1)^2));
  end
  [c s]=GivensRotation(a(I)-sigma, b(I));   % first transformation
  d=a(I)-a(I+1); z=s*(s*d+2*c*b(I));
  a(I)=a(I)-z;
  a(I+1)=a(I+1)+z;
  b(I)=c*s*d+b(I)*(c^2-s^2);
  bulge=-s*b(I+1);
  b(I+1)=c*b(I+1);
  for k= I+2:n                              % chasing the bulge
    [c s]=GivensRotation(b(k-2), bulge);
    balt= b(k-1);
    d=a(k-1)-a(k); z=s*(s*d+2*c*b(k-1));
    a(k-1)=a(k-1)-z;
    a(k)=a(k)+z;
    b(k-2)=c*b(k-2)-s*bulge;
    b(k-1)=c*s*d+b(k-1)*(c^2-s^2);
    if k<n
      bulge=-s*b(k);
      b(k)=c*b(k);
    end
  end
  if test                                   % test-output
    clc; ['working from I=' int2str(I) ' to n=' int2str(n)]
    T=diag(a)+diag(b,-1)+diag(b,1); T(1:n,1:n), pause
  end
  end
end
d=sort(a);
```

7.6.11 Specific Numerical Examples

Matrix B in [148, p. 238] : We test our algorithms on the matrix B and
its mirror reflection $\tilde{B}$:

$$
B = \begin{pmatrix}
1 & 10 & & & & & \\
10 & 10^2 & 10^3 & & & & \\
 & 10^3 & 10^4 & 10^5 & & & \\
 & & 10^5 & 10^6 & 10^7 & & \\
 & & & 10^7 & 10^8 & 10^9 & \\
 & & & & 10^9 & 10^{10} & 10^{11} \\
 & & & & & 10^{11} & 10^{12}
\end{pmatrix}, \quad
\tilde{B} = \begin{pmatrix}
10^{12} & 10^{11} & & & & & \\
10^{11} & 10^{10} & 10^9 & & & & \\
 & 10^9 & 10^8 & 10^7 & & & \\
 & & 10^7 & 10^6 & 10^5 & & \\
 & & & 10^5 & 10^4 & 10^3 & \\
 & & & & 10^3 & 10^2 & 10 \\
 & & & & & 10 & 1
\end{pmatrix}.
$$

Since the two matrices are related by a permutation $\tilde{B} = P^{\top} BP$, they
have the same eigenvalues.

The "exact" eigenvalues were computed with MAPLE using `Digits:=30;`
and copied into the MATLAB script:

```
disp('Matrix B Handbook p238')
exact=[-946347415.646935364911244813552, ...
```

```
        -946.34691970735030427802405636, ...
          .99989902019294252204158938608O, ...
         1046.33721478805627415639938322, ...
      1009899.03019971317042508022336, ...
   1046337712.68593889208601758446, ...
1010000009803.94060266114061589 ]';
a=[1 10^2 10^4 10^6 10^8 10^10 10^12];
b=[10 10^3 10^5 10^7 10^9 10^11];
B=diag(a)+diag(b,1)+diag(b,-1);
d=ImplicitQR(a,b); dj=Jacobi(B);
Eeig=norm(sort(eig(B))-exact),Ejac=norm(sort(dj)-exact),
Eqr=norm(d'-exact)
disp('Matrix B tilde=mirror of B at antidiagonal ')
n=length(a); aa=a(n:-1:1); bb=b(n-1:-1:1);
BB=diag(aa)+diag(bb,1)+diag(bb,-1);
dd=ImplicitQR(aa,bb); dj=Jacobi(BB);
Eeig=norm(sort(eig(BB))-exact),
Ejac=norm(sort(dj)-exact), Eqr=norm(dd'-exact)
```

The norm of the vector of the differences between the computed eigen-
values $\tilde{\lambda}_i$ and the exact ones, $\|\lambda - \tilde{\lambda}\|$, is displayed in the following table.
Our two functions Jacobi and ImplicitQR compare very well with eig
of MATLAB; in fact, Jacobi performs best with the matrix $\tilde{B}$.

Method	eig	Jacobi	ImplicitQR
B	1.2207e-04	1.2208e-04	3.7641e-06
$\tilde{B}$	1.2207e-04	6.0785e-07	1.2208e-04

Wilkinson Matrix: We consider the second matrix in [148, p. 238]. The
diagonal of this 21×21 matrix contains the elements

$$100, 90, 80, \ldots, 0, 10, 20, \ldots, 100,$$

and the sub-diagonal elements are $b_i = 1$. The exact eigenvalues (computed again with Digits:=30 in MAPLE) are:

```
 -.19709289103404678023365324112,
 9.90049425337547750582064624658,
10.09659543859792955659417818B2,
19.99950657441164459395750323B9,
20.00049662325265610078458873l5,
29.99999917290397223590507l3976,
30.00000828491903698796503586O,
39.99999999930925155775971387A8,
40.00000000069121141504537686A2,
49.99999999999965433730107396l9,
50.00000000000034600899698883B5,
60.00000000000034566269892603B1,
```

```
 60.000000000003458934099845950,
 70.00000000069074844224028612412,
 70.000000000690748442295235018,
 80.0000008270960277640949286021,
 80.0000008270960277640949384194,
 90.0004934255883554060424967604,
 90.0004934255883554060424967618,
100.0995057466245224941793753754,
100.0995057466245224941793753754
```

This matrix is unreduced, but nevertheless has very close pairs of eigenvalues. The results are given in the following table. Again our two
functions compare well with MATLAB's `eig`.

Method	eig	Jacobi	ImplicitQR
	6.5164e-14	2.2258e-14	1.5264e-13

Matrix 1-4-1: For $n = 20$, we consider the tridiagonal matrix with diagonal elements 4 and sub-diagonal elements 1. The results are again
comparable.

Method	eig	Jacobi	ImplicitQR
	3.468e-15	3.468e-15	6.296e-15

7.6.12 Computing the Eigenvectors

With the implicit QR algorithm, one can also compute the eigenvectors of
the given matrix by accumulating the Givens rotations to form the matrix of
the eigenvectors. This is left as an exercise, see Problem 7.9.

7.7 Computing the Singular Value Decomposition (SVD)

The singular value decomposition (SVD) of a matrix A is very useful in the
context of least squares problems, see Chapter 6.

Let $A \in \mathbb{R}^{m \times n}$ be a matrix with $m \geq n$. Then there exist orthog-
onal matrices $U \in \mathbb{R}^{m \times m}$ and $V \in \mathbb{R}^{n \times n}$ and a diagonal matrix $\Sigma = \mathrm{diag}(\sigma_1, \ldots, \sigma_n) \in \mathbb{R}^{m \times n}$ with $\sigma_1 \geq \sigma_2 \geq \ldots \geq \sigma_n \geq 0$, such that

$$A = U \Sigma V^{\top}$$

holds. If $\sigma_r > 0$ is the smallest *singular value* greater than zero, then the
matrix A has rank r. For a proof of the existence of this decomposition see
Chapter 6.

The column vectors of $U = [\boldsymbol{u}_1, \ldots, \boldsymbol{u}_m]$ are called the *left singular vectors*
and similarly $V = [\boldsymbol{v}_1, \ldots, \boldsymbol{v}_n]$ are the *right singular vectors*. The values σ_i
are called the *singular values* of A.

7.7.1 Transformations

If we already have a decomposition of A

$$A = PBQ^\top, \quad \text{with } P \text{ and } Q \text{ orthogonal,}$$

then it is sufficient to compute the singular value decomposition of B to obtain the SVD of A, since

$$B = U\Sigma V^\top \quad \Longleftrightarrow \quad A = (PU)\Sigma(QV)^\top.$$

Thus the singular values of A and B are the same, and the singular vectors of A are obtained by premultiplying U and V by with the orthogonal matrices P and Q. Two special transformations are useful in this context:

Chan Transformation: Compute the QR Decomposition of A,

$$A = Q\begin{pmatrix} R \\ 0 \end{pmatrix}.$$

Then compute the SVD of $R = U\Sigma V^\top$ to obtain $A = (QU)\Sigma V^\top$.

Bidiagonalization: Reduce A to *bidiagonal form* with orthogonal transformations

$$A = PBQ^\top, \quad B = \text{bidiagonal}$$

then compute the SVD of $B = U\Sigma V^\top$ to obtain $A = (PU)\Sigma(QV)^\top$.

In both cases, the main part of the computation — the expensive iterative part of the computation of the SVD — is executed on a smaller matrix, especially if $A \in \mathbb{R}^{m \times n}$ with $m \gg n$.

7.7.2 Householder-Rutishauser Bidiagonalization

For a given a matrix $A \in \mathbb{R}^{m \times n}$, we want to find two orthogonal matrices P and Q and a bidiagonal matrix B such that

$$A = PBQ^\top, \quad \text{with } B \text{ upper bidiagonal.}$$

To do so, we apply alternating Householder transformations P_i and Q_i to A from the left and right, see Subsection 6.5.2 for a definition of Householder transformations. The first step transforms the first column of A to a multiple of e_1,

$$P_1 A = \begin{pmatrix} x & x & \cdots & x \\ 0 & x & \cdots & x \\ \vdots & \vdots & \cdots & \vdots \\ 0 & x & \cdots & x \end{pmatrix}.$$

The second step annihilates elements of the first row:

$$
P_1 A Q_1 = \begin{pmatrix} x & x & 0 & \cdots & 0 \\ 0 & x & x & \cdots & x \\ \vdots & \vdots & \vdots & \cdots & \vdots \\ 0 & x & x & \cdots & x \end{pmatrix}.
$$

We continue in this manner until

$$
P_n \cdots P_1 A Q_1 \cdots Q_{n-1} = \begin{pmatrix} x & x & 0 & \cdots & 0 \\ 0 & x & x & \cdots & 0 \\ \vdots & \ddots & \ddots & \ddots & \vdots \\ 0 & 0 & \ddots & x & x \\ 0 & 0 & \ddots & 0 & x \\ 0 & 0 & 0 & \cdots & 0 \end{pmatrix}.
$$

Then, since $P_i = P_i^\top = P_i^{-1}$ holds, we have

$$
A = P B Q^\top, \quad \text{with} \quad P = P_1 \cdots P_n, \quad Q = Q_1 \cdots Q_n.
$$

For the details of the algorithm, we follow a derivation of Rutishauser: the matrix $P = I - \boldsymbol{u}\boldsymbol{u}^\top$ with $\|\boldsymbol{u}\| = \sqrt{2}$ is an *elementary Householder matrix*. These matrices have the following properties (see Subsection 6.5.2):

1. P is symmetric.

2. P is orthogonal.

3. $P\boldsymbol{u} = -\boldsymbol{u}$ and if $\boldsymbol{x} \perp \boldsymbol{u}$ then $P\boldsymbol{x} = \boldsymbol{x}$. Thus P is a reflection at the hyperplane $\boldsymbol{u}^\top \boldsymbol{x} = 0$.

P will be used to solve the following basic problem: Given a vector $\boldsymbol{x}$, find an orthogonal matrix P such that

$$
P\boldsymbol{x} = \begin{pmatrix} \sigma \\ 0 \\ \vdots \\ 0 \end{pmatrix} = \sigma \boldsymbol{e}_1.
$$

Since P is orthogonal we have $\|P\boldsymbol{x}\|^2 = \|\boldsymbol{x}\|^2 = \sigma^2$, and thus $\sigma = \pm\|\boldsymbol{x}\|$. Furthermore

$$
P\boldsymbol{x} = (I - \boldsymbol{u}\boldsymbol{u}^\top)\boldsymbol{x} = \boldsymbol{x} - \boldsymbol{u}(\boldsymbol{u}^\top \boldsymbol{x}) = \sigma \boldsymbol{e}_1,
$$

thus $\boldsymbol{u}(\boldsymbol{u}^\top \boldsymbol{x}) = \boldsymbol{x} - \sigma \boldsymbol{e}_1$ and we obtain by normalizing

$$
\boldsymbol{u} = \frac{\boldsymbol{x} - \sigma \boldsymbol{e}_1}{\|\boldsymbol{x} - \sigma \boldsymbol{e}_1\|} \sqrt{2}.
$$

We have the choice for the sign of σ. We choose the sign so that no cancellation occurs in computing $\boldsymbol{x} - \sigma \boldsymbol{e}_1$,

$$\sigma = \begin{cases} \|\boldsymbol{x}\|, & x_1 < 0, \\ -\|\boldsymbol{x}\|, & x_1 \geq 0. \end{cases}$$

So the calculations simplify to

$$\boldsymbol{u} = \frac{\boldsymbol{x} - \sigma \boldsymbol{e}_1}{\sqrt{\|\boldsymbol{x}\|(|x_1| + \|\boldsymbol{x}\|)}}.$$

In order to apply this basic construction for the bidiagonalization, we construct a sequence of n elementary matrices P_i,

$$P_i = \begin{pmatrix} I & 0 \\ 0 & I - \boldsymbol{u}_i \boldsymbol{u}_i^{\top} \end{pmatrix},$$

where we choose $\boldsymbol{u}_i \in \mathbb{R}^{m-i+1}$ such that zeros are introduced in the i-th column of A below the diagonal when we form the product $P_i A$. Analogously, Q_i will introduce zeros in row i. We obtain by this process

$$B = P_n P_{n-1} \cdots P_1 A Q_1 Q_2 \cdots Q_{n-1} \quad \Longleftrightarrow \quad A = PBQ^{\top},$$

and because of the symmetry of P_i and Q_i we have

$$P = (P_n P_{n-1} \cdots P_1)^{\top} = P_1 P_2 \cdots P_n \quad \text{and} \quad Q = Q_1 Q_2 \cdots Q_{n-1}.$$

If we store the new diagonal elements of B in vector $\boldsymbol{q}$ and the secondary diagonal in $\boldsymbol{e}$, then we can store the Householder vectors $\boldsymbol{u}_i$ at the same place where we introduce zeros in A. This leads to the following *implicit bidiagonalization* algorithm `Bidiagonalize.m`:

ALGORITHM 7.11. *Bidiagonalization*

```
function [q,e,A]=Bidiagonalize(A)
% BIDIAGONALIZE bidiagonalizes a matrix with Householder reflections
%   [q,e,A]=Bidiagonalize(A) computes B=diag(q)+diag(e(2:n),1) such
%   that A=P B Q' using Householder reflexions. A is overwritten with
%   the Householder-vectors.

[m,n]=size(A);
for i=1:n
  s=norm(A(i:m,i));                      % transform A(i:m,i) to
                                         % (q_i,0,...,0)
  if s==0, q(i)=0;
  else
    if A(i,i)>0, q(i)=-s; else q(i)=s; end
    fak=sqrt(s*(s+abs(A(i,i))));
    A(i,i)=A(i,i)-q(i);
```

```
   A(i:m,i)=A(i:m,i)/fak;
   A(i:m,i+1:n)=A(i:m,i+1:n)-A(i:m,i)*(A(i:m,i)'*A(i:m,i+1:n));
 end
 if i<n,
   s=norm(A(i,i+1:n));                    % tranformation A(i,i+1:n) to
                                          % (e_i,0...0)
   if s==0, e(i)=0;
   else
      if A(i,i+1)>0, e(i)=-s; else e(i)=s; end
      fak=sqrt(s*(s+abs(A(i,i+1))));
      A(i,i+1)=A(i,i+1)-e(i);
      A(i,i+1:n)=A(i,i+1:n)/fak;
      A(i+1:m,i+1:n)=A(i+1:m,i+1:n) - ...
                  (A(i+1:m,i+1:n)*A(i,i+1:n)')*A(i,i+1:n);
   end
 end
end
e=[0 e];                                  % insert 0 element in e (see
                                          % notation of bidiagonal matrix)
```

Note that the matrices $P = P_1 P_2 \cdots P_n$ and $Q = Q_1 Q_2 \cdots Q_{n-1}$ are given implicitly by the Householder vectors of P_k and Q_k. Thus, to form the product $P\boldsymbol{x}$, we need a procedure for applying the Householder transformations $P_k = I - \boldsymbol{u}_k \boldsymbol{u}_k^\top$ in reverse order. We can compute $\boldsymbol{y} = P\boldsymbol{x}$ using

$$\boldsymbol{y} = \boldsymbol{x}, \quad \boldsymbol{y} := (I - \boldsymbol{u}_k \boldsymbol{u}_k^T)\boldsymbol{y} = \boldsymbol{y} - \boldsymbol{u}_k(\boldsymbol{u}_k^T \boldsymbol{y}), \quad k = n, n-1, \ldots, 1.$$

See Problem 7.13.

7.7.3 Golub-Kahan-Lanczos Bidiagonalization

If $A = PBQ^\top$ with B upper bidiagonal and P, Q orthogonal, then we can compare columns in the two equations

$$AQ = PB, \quad \text{and} \quad A^\top P = QB^\top$$

and derive recursion formulas to compute the bidiagonal factorization and obtain thus the *Golub-Kahan-Lanczos Algorithm*, see Problem 7.14.

7.7.4 Eigenvalues and Singular Values

There is a connection between the singular values of A and the eigenvalues of $A^\top A$ and $AA^\top$. Let $m \geq n$, $A = U\Sigma V^\top$, and consider the symmetric matrices

$$A^\top A = V(\Sigma^\top \Sigma)V^\top, \quad AA^\top = U(\Sigma\Sigma^\top)U^\top.$$

We recognize the eigendecomposition: V contains the eigenvectors of the matrix $A^\top A$ and $\sigma_i^2, i = 1, \ldots, n$ are the corresponding eigenvalues. Similarly,

U are the eigenvectors of $AA^\top$ for the eigenvalues $\sigma_i^2, i = 1, \ldots, n$ plus the eigenvalue 0 with multiplicity $m - n$.

If $A \in \mathbb{R}^{n \times n}$ is symmetric, then $A^\top A = A^2$ and $\sigma_i^2 = \lambda_i^2$, therefore

$$\sigma_i = |\lambda_i|, \quad i = 1, \ldots, n.$$

Hence, for a symmetric positive definite matrix A, the eigenvalues and the singular values coincide.

One can also consider the following *augmented matrix*: let the SVD of $A \in \mathbb{R}^{m \times n}$ be $A = U\Sigma V^\top$, where $\Sigma = \mathrm{diag}(\Sigma_1, 0)$ with $\Sigma_1 \in \mathbb{R}^{r \times r}$ and $r = \mathrm{rank}(A)$. If we partition $U = (U_1, U_2)$ with $U_1 \in \mathbb{R}^{m \times r}$ and $V = (V_1, V_2)$ with $V_1 \in \mathbb{R}^{n \times r}$ accordingly, then we have the identity

$$C := \begin{pmatrix} 0 & A \\ A^\top & 0 \end{pmatrix} = P \begin{pmatrix} \Sigma_1 & 0 & 0 \\ 0 & -\Sigma_1 & 0 \\ 0 & 0 & 0 \end{pmatrix} P^\top$$

with

$$P = \frac{1}{\sqrt{2}} \begin{pmatrix} U_1 & U_1 & \sqrt{2}U_2 & 0_{m \times (n-r)} \\ V_1 & -V_1 & 0_{n \times (m-r)} & \sqrt{2}V_2 \end{pmatrix} \in \mathbb{R}^{(m+n) \times (m+n)}.$$

Therefore *the eigenvalues of C are $\pm\sigma_1, \pm\sigma_2, \ldots, \pm\sigma_r$, and zero with multiplicity $m + n - 2r$.* The augmented matrix is the basis for the first numerically reliable algorithm for computing the singular value decomposition, due to Golub and Kahan in 1965, see [55].

7.7.5   Algorithm of Golub-Reinsch

After the bidiagonalization of the matrix $A = PBQ^\top$, the singular values of

$$B = \begin{pmatrix} q_1 & e_2 & & & & \\ & q_2 & e_3 & & & \\ & & \ddots & \ddots & & \\ & & & \ddots & e_n \\ & & & & q_n \end{pmatrix}$$

are computed through a variant of the implicit shift QR Algorithm for the tridiagonal matrix

$$T = B^\top B = \begin{pmatrix} q_1^2 & q_1 e_2 & & & \\ q_1 e_2 & e_2^2 + q_2^2 & q_2 e_3 & & \\ & q_2 e_3 & \ddots & \ddots & \\ & & \ddots & e_{n-1}^2 + q_{n-1}^2 & q_{n-1} e_n \\ & & & q_{n-1} e_n & e_n^2 + q_n^2 \end{pmatrix}.$$

The main idea is to avoid forming the product $T = B^\top B$ and to work only with the bidiagonal matrix B. To derive this variant, let us consider the Implicit-shift QR Algorithm for T with shift σ. There, the first QR transformation is obtained by determining a Givens rotation G_1 such that

$$\begin{pmatrix} c & s \\ -s & c \end{pmatrix}^\top \begin{pmatrix} q_1^2 - \sigma \\ q_1 e_2 \end{pmatrix} = \begin{pmatrix} r \\ 0 \end{pmatrix}.$$

As seen before

$$r = \sqrt{(q_1^2 - \sigma)^2 + (q_1 e_2)^2}, \quad c = \frac{q_1^2 - \sigma}{r} \quad \text{and} \quad s = \frac{q_1 e_2}{r}.$$

We compute c and s numerically stably using the function `GivensRotation`, see Algorithm 7.6. Applying G_1 and its transpose to the matrix T yields

$$G_1^\top B^\top B G_1 = \begin{pmatrix} x & x & \mathbf{x} & & & \\ x & x & x & & & \\ \mathbf{x} & x & x & x & & \\ & x & x & x & & \\ & & \ddots & \ddots & \ddots & \end{pmatrix},$$

and the subsequent transformations would chase the bulge until we have restored the tridiagonal form.

To see how we can work only with the bidiagonal matrix B, consider the product BG_1:

$$BG_1 = \begin{pmatrix} x & x & & & \\ \mathbf{x} & x & x & & \\ & & x & x & \\ & & x & x & \\ & & & \ddots & \ddots \end{pmatrix}.$$

The bidiagonal form is destroyed by the bulge $\mathbf{x}$. We can chase the bulge by a Givens rotation P_1 from the left in the $(1, 2)$ plane:

$$P_1^\top BG_1 = \begin{pmatrix} x & x & \mathbf{x} & & \\ & x & x & & \\ & & x & x & \\ & & x & x & \\ & & & \ddots & \ddots \end{pmatrix},$$

followed by a rotation in the $(2, 3)$ plane from the right:

$$P_1^\top BG_1 G_2 = \begin{pmatrix} x & x & & & \\ & x & x & & \\ & \mathbf{x} & x & x & \\ & & x & x & \\ & & & \ddots & \ddots \end{pmatrix}.$$

Continuing this way, we restore the bidiagonal form

$$\tilde{B} = P_{n-1}^{\top} \cdots P_1^{\top} B G_1 \cdots G_{n-1} = P^{\top} B G.$$

But $\tilde{B}^{\top} \tilde{B} = G^{\top} B^{\top} B G$, and since the first column of G is the same as the first column of G_1, we obtain according to Theorem 7.7 the same result as if we had performed one QR Iteration on the matrix T. Thus, both algorithms are mathematically equivalent; nonetheless, the Golub-Reinsch iteration on B is numerically preferable because T is not formed and manipulated explicitly.

The following MATLAB Function `QRStep.m` computes one QR step on the submatrix

$$\begin{pmatrix} q_l & e_{l+1} & & \\ & q_{l+1} & \ddots & \\ & & \ddots & e_n \\ & & & q_n \end{pmatrix}.$$

ALGORITHM 7.12. *QR Step with Bidiagonal Matrix*

```
function [e,q]=QRStep(l,n,e,q)
% QRSTEP compute QR iteration step on a bidiagonal matrix
%    [e,q]=QRStep(l,n,e,q) computes one QR iteration step on the
%    bidiagonal matrix for the trailing pricipal submatrix (l:n,l:n)

an=e(n)^2+q(n)^2;                        % compute Wilkinson shift
an1=e(n-1)^2+q(n-1)^2;                    % with code from ImplicitQR.m
bn1=q(n-1)*e(n);
delta =(an-an1)/2;
if delta==0,
  sigma=an-abs(bn1);
else
  sigma=an-bn1^2/(delta+sign(delta)*sqrt(delta^2+bn1^2));
end
[c s]=GivensRotation(q(l)^2-sigma,q(l)*e(l+1));
h=q(l)*c-e(l+1)*s;                        % first transformation from the
e(l+1)=q(l)*s+e(l+1)*c; q(l)=h;           % right generates bulge
bulge=-q(l+1)*s; q(l+1)=q(l+1)*c;
for i=l:n-2                               % Chasing the bulge
  [c s]=GivensRotation(q(i),bulge);       % from left
  q(i)=q(i)*c-bulge*s; h=e(i+1)*c-q(i+1)*s;
  q(i+1)=e(i+1)*s+q(i+1)*c; e(i+1)=h;
  bulge=-e(i+2)*s; e(i+2)=e(i+2)*c;
  [c s]=GivensRotation(e(i+1),bulge);     % from right
  e(i+1)=e(i+1)*c-bulge*s;
  h=q(i+1)*c-e(i+2)*s; e(i+2)=q(i+1)*s+e(i+2)*c;
  q(i+1)=h;
  bulge=-q(i+2)*s; q(i+2)=q(i+2)*c;
```

```
end
[c s]=GivensRotation(q(n-1),bulge);      % last from left
q(n-1)=q(n-1)*c-bulge*s; h=e(n)*c-q(n)*s;
q(n)=e(n)*s+q(n)*c; e(n)=h;
```

We now address the issues of splitting, cancellation and convergence.

Splitting: If $e_i = 0$, then the bidiagonal matrix splits into two blocks whose singular values can be computed independently:

$$B = \begin{pmatrix} B_1 & 0 \\ 0 & B_2 \end{pmatrix}, \quad \mathrm{svd}(B) = \mathrm{svd}(B_1) \cup \mathrm{svd}(B_2).$$

If a split occurs, we work on the matrix B_2 first. In a parallel environment B_1 and B_2 could be processed independently.

If the split is for $i = n$ then $B_2 = q_n$ and q_n is a singular value (see convergence below).

Cancellation: If $q_i = 0$, then one of the singular values is zero. We can split the matrix B with a sequence of special Givens rotations applied from the left to zero out row i

$$G_{i,i+1}^{\mathsf{T}} \begin{pmatrix} q_1 & e_2 & & & & & & \\ & \ddots & \ddots & & & & & \\ & & q_{i-1} & e_i & & & & \\ & & & 0 & e_{i+1} & & & \\ & & & & q_{i+1} & \ddots & & \\ & & & & & & e_n & \\ & & & & & & & q_n \end{pmatrix} = \begin{pmatrix} q_1 & e_2 & & & & & & \\ & \ddots & \ddots & & & & & \\ & & q_{i-1} & e_i & & & & \\ & & & 0 & 0 & b & & \\ & & & & \tilde{q}_{i+1} & \ddots & & \\ & & & & & & e_n & \\ & & & & & & & q_n \end{pmatrix}.$$

The bulge b is removed with further Givens rotations $G_{i,k}$ for $k = i + 2, \ldots, n$. Then, because $e_{i+1} = 0$, the matrix splits again into two submatrices. Note that for these operations we need Givens rotations that annihilate the first and not the second component (`GivensRotation1`):

$$\begin{pmatrix} c & s \\ -s & c \end{pmatrix}^{\mathsf{T}} \begin{pmatrix} x \\ y \end{pmatrix} = \begin{pmatrix} 0 \\ r \end{pmatrix}.$$

The function `GivensRotation1` is left as an exercise (See Problem 7.15).

Test for negligibility: In finite precision arithmetic, we will of course not obtain exactly $e_i = 0$ or $q_i = 0$. Thus, we need a threshold to decide when these elements can be considered zero.

Golub and **Reinsch** [53] recommend

$$|e_{i+1}|, |q_i| \le \varepsilon \max_i(|q_i| + |e_i|) = \varepsilon \|B\|_1,$$

where $\varepsilon = eps$ is the machine precision.

Björck [9] suggests

$$|e_{i+1}| \le 0.5\,\varepsilon\,(|q_i| + |q_{i+1}|), \qquad |q_i| \le 0.5\,\varepsilon\,(|e_i| + |e_{i+1}|).$$

Linpack [26] uses Björck's approach, but omits the factor 0.5. However,

Stewart has effectively implemented the Linpack subroutine `SSVDC`, which can be downloaded from `http://www.netlib.org/`. He uses *machine-independent criteria* such as the one below for e_i:

```
          DO 390 LL = 1, M
            L = M - LL
            IF (L .EQ. 0) GO TO 400
            TEST = ABS(S(L)) + ABS(S(L+1))
            ZTEST = TEST + ABS(E(L))
            IF (ZTEST .NE. TEST) GO TO 380
              E(L) = 0.0E0
              GO TO 400
  380       CONTINUE
  390     CONTINUE
  400     CONTINUE
```

Convergence and deflation: When e_n is judged to be negligible, q_n can be accepted as singular value. The iteration then proceeds with the leading principal submatrix of order $n-1$, which contains the remaining singular values of B.

We are now ready to present the complete algorithm for computing the SVD:

ALGORITHM 7.13. *SVD Golub-Reinsch*

```
function q=SVDGolubReinsch(A)
% SVDGolubReinsch singular values by the Golub Reinsch algorithm
%   q=SVDGolubReinsch(A) computes the singular values of A using the
%   Golub-Reinsch Algorithm and a machine independent criterium
%   following G. W. Stewart.

[m,n]=size(A);
if n>m, A=A'; [m,n]=size(A);end
[q,e,A]=Bidiagonalize(A);
k=n;
while k~=1,
  splitting=false;
  l=k;
  while ~splitting & l>1,
    t=abs(q(l-1))+abs(q(l));
    if t+abs(e(l))==t;                      % splitting: e(l) is small
      splitting=true; e(l)=0;
    else
      t=abs(e(l))+abs(e(l-1));
```

```
     if t+abs(q(l-1))==t,               % q(l-1) is small, cancellation
       splitting=true; q(l-1)=0;
       bulge=e(l); e(l)=0;              % introduce zero row
       for kk=l:k-1
         [c s]=GivensRotation1(bulge,q(kk));
          q(kk)=s*bulge+c*q(kk);
          bulge=-s*e(kk+1);
          e(kk+1)=c*e(kk+1);
       end
       [c s]=GivensRotation1(bulge, q(k));
       q(k)=s*bulge+c*q(k);
     else
       l=l-1;                           % check elements on row before
     end
   end
 end
 if splitting & l==k,                   % convergence
   if q(k)<0, q(k)=-q(k); end           % adapt sign, singular values >= 0
   k=k-1;
 else
   [e,q]=QRStep(l,k,e,q);
 end
end
if q(1)<0, q(1)=-q(1); end
[q]=sort(q); q=q(n:-1:1)';              % sort singular values
```

EXAMPLE 7.1. *Consider $A \in \mathbb{R}^{n \times (n+1)}$ from page 150 of the Handbook [148]. For $n = 5$ we have*

$$A = \begin{pmatrix} 5 & -1 & -1 & -1 & -1 & -1 \\ 0 & 4 & -1 & -1 & -1 & -1 \\ 0 & 0 & 3 & -1 & -1 & -1 \\ 0 & 0 & 0 & 2 & -1 & -1 \\ 0 & 0 & 0 & 0 & 1 & -1 \end{pmatrix}$$

The exact singular values are $\sqrt{k(k+1)}$, $k = 0, \ldots n$.

```
n=200;
A=-ones(n,n+1);
A=A-tril(A);
A=A+[diag((n:-1:1)) zeros(n,1)];
k=n:-1:1;
exact=(sqrt(k.*(k+1)))';
[norm(svd(A)-exact) norm(SVDGolubReinsch(A')-exact)]
```

For $n = 200$, the norm of the difference between the exact and the computed eigenvalues becomes

```
ans =
```

```
1.0e-11 *
0.111027675811923    0.424274094705328
```

Both MATLAB*'s function **svd** and our program show comparable accuracy for this matrix.*

7.8   QD Algorithm

The *Quotient-Difference Algorithm*, or the *QD Algorithm*, has many applications. Its inventor, Heinz Rutishauser, first used it to compute the poles of a rational function and to transform a power series into a corresponding continued fraction [109, 110]. The algorithm can also be interpreted as the LR algorithm (7.46) for a tridiagonal matrix, so it can be used to calculate eigenvalues. Although the QR Algorithm is nowadays the standard method for solving eigenvalue problems, the QD Algorithm has been revived for that purpose thanks to a variant called the *differential QD Algorithm*, developed by K. Fernando and B. Parlett [31]. This variant proves to be very useful for computing singular values to a high degree accuracy. Much of the material in this section is based on lectures of Beresford Parlett given at ETH when he was visiting in 2003.

7.8.1   Progressive QD Algorithm

In order to understand the link between the QD and LR algorithms, we first need a few basic transformations involving tridiagonal matrices. Consider a non-symmetric unreduced tridiagonal matrix

$$
S = \begin{pmatrix}
a_1 & b_1 & & & \\
c_1 & a_2 & \ddots & & \\
& \ddots & \ddots & b_{n-1} \\
& & c_{n-1} & a_n
\end{pmatrix}, \tag{7.48}
$$

which has been obtained, for instance, from the reduction of a full symmetric matrix to tridiagonal form. A similarity transformation with the diagonal matrix D, where

$$
d_1 = 1, \quad d_i = \prod_{k=1}^{i-1} b_k, \quad i = 2, \ldots, n
$$

yields

$$
T = DSD^{-1} = \begin{pmatrix}
a_1 & 1 & & & \\
b_1 c_1 & a_2 & \ddots & & \\
& \ddots & \ddots & 1 \\
& & b_{n-1} c_{n-1} & a_n
\end{pmatrix}. \tag{7.49}
$$

Thus, the eigenvalues of a tridiagonal matrix essentially depend on $2n - 1$ parameters: the diagonal and the subdiagonal elements of the matrix T, i.e. the products of the off-diagonal elements of S. Moreover, the LU factorization of T has a special structure: we can write $T = LU$, where

$$L = \begin{pmatrix} 1 & & & \\ e_1 & 1 & & \\ & \ddots & \ddots & \\ & & e_{n-1} & 1 \end{pmatrix}, \quad U = \begin{pmatrix} q_1 & 1 & & \\ & q_2 & \ddots & \\ & & \ddots & 1 \\ & & & q_n \end{pmatrix}. \tag{7.50}$$

This gives us an equivalent parametrization of T by the vectors q and e:

$$T = \begin{pmatrix} q_1 & 1 & & \\ e_1 q_1 & q_2 + e_1 & \ddots & \\ & \ddots & \ddots & 1 \\ & & e_{n-1} q_{n-1} & q_n + e_{n-1} \end{pmatrix}. \tag{7.51}$$

By equating the two parametrizations of T, (7.49) and (7.51), we obtain Algorithm 7.14, which computes the vectors q and e of the matrix T in (7.51) from a given non-symmetric tridiagonal matrix S, (7.48).

ALGORITHM 7.14.
Compute QD Line from Tridiagonal Matrix

```
function [q,e]=QDLine(a,b,c)
% QDLINE compute qd-line from a tridiagonal matrix
%    [q,e]=QDLine(a,b,c) computes the elemente q(i) and e(i) of the
%    qd-line from the tridiagonal matrix given by T =(c(i-1),a(i),b(i))

n=length(a);
e(n)=0;                          % convenient for performing QDStep later
q(1)=a(1);
for i=1:n-1
  e(i)=b(i)*c(i)/q(i); q(i+1)=a(i+1)-e(i);
end
```

To get a more compact representation of T, we can collect the vectors q and e into a *QD line*, defined as

$$Z = \{q_1, e_1, q_2, e_2, \ldots, q_{n-1}, e_{n-1}, q_n, 0\}.$$

Let us now consider applying the LR algorithm (7.46) to T. The first step involves calculating $\hat{T} = UL$, where L and U are the LU factors of T, as shown in (7.50). Then the next LU decomposition $\hat{T} = \hat{L}\hat{U}$ can be obtained by equating the elements in $\hat{L}\hat{U} = UL$:

$$\hat{T} = \begin{pmatrix} \hat{q}_1 & 1 & & \\ \hat{e}_1\hat{q}_1 & \hat{q}_2+\hat{e}_1 & \ddots & \\ & \ddots & \ddots & 1 \\ & & \hat{e}_{n-1}\hat{q}_{n-1} & \hat{q}_n+\hat{e}_{n-1} \end{pmatrix} = \begin{pmatrix} q_1+e_1 & 1 & & \\ e_1q_2 & q_2+e_2 & \ddots & \\ & \ddots & \ddots & 1 \\ & & e_{n-1}q_n & q_n \end{pmatrix}. \tag{7.52}$$

Note that since $\hat{T} = UL = L^{-1}TL$, T and $\hat{T}$ must have the same eigenvalues. Thus, we obtain the spectrum-preserving transformation

$$Z = \{q_1, e_1, q_2, e_2, \ldots, q_{n-1}, e_{n-1}, q_n, 0\}$$
$$\longrightarrow \hat{Z} = \{\hat{q}_1, \hat{e}_1, \hat{q}_2, \hat{e}_2, \ldots, \hat{q}_{n-1}, \hat{e}_{n-1}, \hat{q}_n, 0\},$$

where

$$\hat{e}_0 = e_n = 0, \quad \hat{e}_{k-1} + \hat{q}_k = q_k + e_k, \quad \hat{e}_k\hat{q}_k = q_{k+1}e_k, \quad k = 1,\ldots,n. \tag{7.53}$$

Equations (7.53) are known as the *Rhombus Rules* of Stiefel. The name of these rules comes from the *QD scheme*, which is obtained if we iterate the LR steps. Denote by $e_i^{(0)} := e_i$, $q_i^{(0)} := q_i$ and by $e_i^{(1)} := \hat{e}_i$, $q_i^{(1)} := \hat{q}_i$. If we write the QD lines as diagonals, then we get the scheme

$$
\begin{array}{ccccccc}
q_1^{(0)} & & & & & & \\
& e_1^{(0)} & & & & & \\
q_1^{(1)} & & q_2^{(0)} & & & & \\
& e_1^{(1)} & & e_2^{(0)} & & & \\
q_1^{(2)} & & q_2^{(1)} & & q_3^{(0)} & & \\
\cdot & e_1^{(2)} & \cdot & e_2^{(1)} & \cdot & \cdot &
\end{array}
\tag{7.54}
$$

In each QD step, a new diagonal, or QD line, is computed by forming *quotients* and *differences*. Depending on whether the element is a q or an e, we apply the *Q-* or *E-Rhombus Rule*:

Quotient
E-Rule: $e_k^{(\nu+1)} = \dfrac{e_k^{(\nu)} q_{k+1}^{(\nu)}}{q_k^{(\nu+1)}}$

$$
\begin{array}{ccc}
 & e_k^{(\nu)} & \\
q_k^{(\nu+1)} \nearrow & & \searrow q_{k+1}^{(\nu)} \\
\searrow & & \nearrow \\
 & e_k^{(\nu+1)} &
\end{array}
$$

Difference
Q-Rule: $q_k^{(\nu+1)} = q_k^{(\nu)} + e_k^{(\nu)} - e_{k-1}^{(\nu+1)}$

$$
\begin{array}{ccc}
 & q_k^{(\nu)} & \\
e_{k-1}^{(\nu+1)} \nearrow & & \searrow e_k^{(\nu)} \\
\searrow & & \nearrow \\
 & q_k^{(\nu+1)} &
\end{array}
$$

These rules for computing the QD scheme (7.54) define the *QD Algorithm*, which was first described by Heinz Rutishauser [109, 110]. Rutishauser defines the *progressive QD step* for computing the next diagonal in the QD scheme by

ProgressiveQDStep

$$e_n = 0$$
$$\hat{q}_1 = q_1 + e_1$$
$$\text{for } k = 2 : n$$
$$\qquad \hat{e}_{k-1} = (e_{k-1}/\hat{q}_{k-1})q_k$$
$$\qquad \hat{q}_k = (q_k - \hat{e}_{k-1}) + e_k$$
$$\text{end}$$

In Parlett's variant, the step is defined by

ProgressiveQDStepParlett

$$\hat{e}_0 = 0$$
$$\text{for } k = 1 : n - 1$$
$$\qquad \hat{q}_k = (q_k - \hat{e}_{k-1}) + e_k$$
$$\qquad \hat{e}_k = e_k(q_{k+1}/\hat{q}_k)$$
$$\text{end for}$$
$$\hat{q}_n = q_n - \hat{e}_{n-1}$$

Note the different brackets in computing the e-element. Parlett's version avoids possible unnecessary underflow, but takes the risk that $q_{k+1}/\hat{q}_k$ might become large. Rutishauser's version is easier to implement in MATLAB (no zero index).

Although the QD algorithm bears no obvious resemblance to the matrix eigenvalue problem, it is in fact mathematically equivalent to an LR algorithm! This means it can be used to calculate the eigenvalues (but not the eigenvectors) of a matrix, as we show in the example below.

EXAMPLE 7.2. *(Progressive QD Algorithm without shifts) We apply the QD Algorithm to compute the eigenvalues of the matrix B*

$$
B = \begin{pmatrix}
4 & 1 & & & & & \\
1 & 4 & 1 & & & & \\
  & 1 & 4 & 1 & & & \\
  &   & 1 & 4 & 1 & & \\
  &   &   & 1 & 4 & 1 & \\
  &   &   &   & 1 & 4 & 1 \\
  &   &   &   &   & 1 & 4
\end{pmatrix}
$$

and perform 50 progressive QD steps. We omit the full output and print only the last three QD Lines together with the eigenvalues computed by `eig`.

```
n=7;
a=4* ones(1,n); b=ones(1,n-1)
```

```
B=diag(a)+diag(b,1)+diag(b,-1)
[q,e]=QDLineSymmetric(a,b)
for it=1:50
  q(1)=q(1)+e(1);                      % progressive qd-step without shift
  for k=2:n,
    e(k-1)=(e(k-1)/q(k-1))*q(k);
    q(k)=(q(k)-e(k-1))+e(k);
  end
  q, e
end
disp('q-line and Eigenvalues')
sort(q), eig(B)'
```

```
q =
   5.8136      5.4305      4.7762      4.0019      3.2349      2.5857      2.1524
e =
   0.0024      0.0019      0.0005      0.0001      0.0000      0.0000           0
q =
   5.8160      5.4301      4.7750      4.0016      3.2348      2.5857      2.1524
e =
   0.0022      0.0016      0.0004      0.0001      0.0000      0.0000           0
q =
   5.8182      5.4297      4.7739      4.0014      3.2348      2.5857      2.1524
e =
   0.0021      0.0014      0.0003      0.0001      0.0000      0.0000           0
q-line and Eigenvalues
ans =
   2.1524      2.5857      3.2348      4.0014      4.7739      5.4297      5.8182
ans =
   2.1522      2.5858      3.2346      4.0000      4.7654      5.4142      5.8478
```

*Comparing the last two lines of output, we notice that after 50 iterations we
have only about 2–4 correct decimal digits. To improve convergence, we need
to introduce shifts, just like in the QR Algorithm.*

7.8.2   Orthogonal LR-Cholesky Algorithm

Let us now consider the special case of a symmetric positive definite tridiagonal matrix

$$
A = \begin{pmatrix}
\alpha_1 & \beta_1 & & & \\
\beta_1 & \alpha_2 & \ddots & & \\
& \ddots & \ddots & \beta_{n-1} & \\
& & \beta_{n-1} & \alpha_n &
\end{pmatrix}.
$$

By Lemma 3.1(c), the diagonal elements satisfy $\alpha_i > 0$. Assume additionally
that A is *irreducible*, i.e., there exists no permutation matrix P such that
$P^{\top} A P$ is block upper triangular. Then we can also assume without loss of

generality that $\beta_i > 0$, since the matrix

$$\tilde{A} = E^{-1}AE, \qquad E = \mathrm{diag}(\sigma_1, \ldots, \sigma_n)$$

with $\sigma_i = \prod_{j=1}^{i} \mathrm{sign}(\beta_i)$ is similar to A and has positive subdiagonal entries. With these assumptions, we can conclude that the Cholesky decomposition $A = R^\top R$ exists and can make the following ansatz for R:

$$R = \begin{pmatrix} \sqrt{q_1} & \sqrt{e_1} & & \\ & \sqrt{q_2} & \ddots & \\ & & \ddots & \sqrt{e_{n-1}} \\ & & & \sqrt{q_n} \end{pmatrix}. \tag{7.55}$$

Now $R^\top R$ becomes

$$R^\top R = \begin{pmatrix} q_1 & \sqrt{e_1 q_1} & & \\ \sqrt{e_1 q_1} & q_2 + e_1 & \ddots & \\ & \ddots & \ddots & \sqrt{e_{n-1}q_{n-1}} \\ & & \sqrt{e_{n-1}q_{n-1}} & q_n + e_{n-1} \end{pmatrix}.$$

Comparing this expression with A, we obtain the Algorithm 7.15 to compute the quantities q_i and e_i from the α_i and β_i:

ALGORITHM 7.15.
Compute QD Line from Symmetric Tridiagonal Matrix

```
function [q,e]=QDLineSymmetric(a,b)
% QDLINESYMMETRIC qd-line from a s.p.d tridiagonal matrix
%    [q,e]=QDLineSymmetric(a,b) computes the elemente q(i) and e(i) from
%    the qd-line from the symmetric positive definite tridiagonal
%    matrix given by A=(b(i-1),a(i),b(i))

n=length(a);
e(n)=0;                          % convenient for performing QDStep later
q(1)=a(1);
for i=1:n-1
  e(i)=b(i)^2/q(i); q(i+1)=a(i+1)-e(i);
end
```

Note that $R^\top R$ is similar to the matrix T in (7.51): let $\Gamma = \mathrm{diag}(\gamma_i)$ with

$$\gamma_1 = 1, \quad \gamma_i = \prod_{k=1}^{i-1} \sqrt{e_i q_i}, \quad i = 2, \ldots, n.$$

Then $R^\top R = \Gamma^{-1} T \Gamma$. Thus, Algorithm 7.15 is actually a special case of Algorithm 7.14, since the QD line can also be computed by `[q,e]=QDLine(b,a,b)`.

Let us now consider the symmetric analogue of the LR algorithm. If A is symmetric and positive definite, then one step of the *LR-Cholesky Algorithm* becomes

$$A = R^\top R, \quad \hat{A} = RR^\top = \hat{R}^\top \hat{R}.$$

At first glance, both Cholesky decompositions are needed for the mapping from R to $\hat{R}$, but the following lemma shows that two decompositions are in fact related.

LEMMA 7.1. *If any two invertible matrices M_1 and M_2 satisfy $M_1^\top M_1 = M_2^\top M_2$, then $M_2 = QM_1$ for some orthogonal matrix Q.*

PROOF. Since M_1 and M_2 are invertible, there exists a matrix $F = M_1 M_2^{-1}$ such that $M_1 = FM_2$. Substituting this relation into $M_1^\top M_1 = M_2^\top M_2$ yields

$$M_1^\top M_1 = M_2^\top F^\top FM_2 = M_2^\top M_2 \iff F^\top F = I.$$

Thus F is orthogonal. $\square$

Applying this lemma to the relation $RR^\top = \hat{R}^\top \hat{R}$, we conclude that there exists an orthogonal matrix Q such that

$$R^\top = Q\hat{R}.$$

So the QR Decomposition of $R^\top$ is in fact $Q\hat{R}$!

EXAMPLE 7.3.

$$A = \begin{pmatrix} 4 & 1 & & & & & \\ 1 & 4 & 1 & & & & \\ & 1 & 4 & 1 & & & \\ & & 1 & 4 & 1 & & \\ & & & 1 & 4 & 1 & \\ & & & & 1 & 4 & 1 \\ & & & & & 1 & 4 \end{pmatrix}. \tag{7.56}$$

```
>> R=chol(A)
R =
    2.0000    0.5000         0         0         0         0         0
         0    1.9365    0.5164         0         0         0         0
         0         0    1.9322    0.5175         0         0         0
         0         0         0    1.9319    0.5176         0         0
         0         0         0         0    1.9319    0.5176         0
         0         0         0         0         0    1.9319    0.5176
         0         0         0         0         0         0    1.9319
>> tR=chol(R*R')
tR =
    2.0616    0.4697         0         0         0         0         0
         0    1.9484    0.5121         0         0         0         0
         0         0    1.9336    0.5171         0         0         0
```

0	0	0	1.9320	0.5176	0	0
0	0	0	0	1.9319	0.5176	0
0	0	0	0	0	1.9319	0.5176
0	0	0	0	0	0	1.8612

```
>> [Q, tR2]=qr(R')
Q =
```

-0.9701	0.2339	-0.0619	0.0166	-0.0044	0.0012	0.0003
-0.2425	-0.9354	0.2477	-0.0663	0.0178	-0.0048	-0.0013
0	-0.2650	-0.9291	0.2486	-0.0666	0.0179	0.0050
0	0	-0.2677	-0.9283	0.2487	-0.0666	-0.0185
0	0	0	-0.2679	-0.9282	0.2487	0.0692
0	0	0	0	-0.2679	-0.9282	-0.2582
0	0	0	0	0	-0.2679	0.9634

```
tR2 =
```

-2.0616	-0.4697	0	0	0	0	0
0	-1.9484	-0.5121	0	0	0	0
0	0	-1.9336	-0.5171	0	0	0
0	0	0	-1.9320	-0.5176	0	0
0	0	0	0	-1.9319	-0.5176	0
0	0	0	0	0	-1.9319	-0.5176
0	0	0	0	0	0	1.8612

We see that indeed **tR** *and* **tR2** *are the same (up to different signs).*

We are now ready to derive the recurrence relations for the orthogonal QD Algorithm. Let us compute the matrix Q for the transformation $R^{\top} \mapsto \hat{R}$ using Givens rotations,

$$
Q^{\top}\begin{pmatrix} a_1 & & & \\ b_1 & a_2 & & \\ & \ddots & \ddots & \\ & & b_{n-1} & a_n \end{pmatrix} = \begin{pmatrix} \hat{a}_1 & \hat{b}_1 & & \\ & \hat{a}_2 & \ddots & \\ & & \ddots & \hat{b}_{n-1} \\ & & & \hat{a}_n \end{pmatrix}.
$$

In the first step, we determine $c = \cos\alpha$ and $s = \sin\alpha$ such that

$$
\begin{pmatrix} c & -s \\ s & c \end{pmatrix}\begin{pmatrix} a_1 & 0 \\ b_1 & a_2 \end{pmatrix} = \begin{pmatrix} \hat{a}_1 & \hat{b}_1 \\ 0 & \tilde{a}_2 \end{pmatrix}.
$$

We will compute this transformation algebraically. Since $\hat{a}_1 = \sqrt{a_1^2 + b_1^2}$, using MAPLE we obtain

```
> solve({c*a[1]-s*b[1]=sqrt(a[1]^2+b[1]^2),s*a[1]+c*b[1]=0},{c,s});
```

$$
\left\{ c = \frac{a_1}{\sqrt{a_1^2 + b_1^2}}, s = -\frac{b_1}{\sqrt{a_1^2 + b_1^2}} \right\}
$$

```
> -s*a[2];
```

$$\frac{b_1 a_2}{\sqrt{a_1{}^2 + b_1{}^2}} \tag{7.57}$$

```
> c*a[2];
```

$$\frac{a_1 a_2}{\sqrt{a_1{}^2 + b_1{}^2}} \tag{7.58}$$

We can now obtain expressions for the three new elements $\hat{a}_1$, $\hat{b}_1$ and $\tilde{a}_2$ directly without invoking c and s:

$$\hat{a}_1 = \sqrt{a_1^2 + b_1^2}, \quad \hat{b}_1 = \frac{b_1 a_2}{\hat{a}_1}, \quad \tilde{a}_2 = \frac{a_1 a_2}{\hat{a}_1}.$$

Note that $\tilde{a}_2$ is an intermediate result, as it will become $\hat{a}_2$ after the next Givens rotation. We have thus shown that an *orthogonal QD step* (the transformation from $R \mapsto \hat{R}$ can be computed as follows:

OrthogonalQDStep [6]

$$
\begin{aligned}
&\tilde{a} = a_1; \\
&\text{for } k = 1 : n - 1 \\
&\qquad \hat{a}_k = \sqrt{\tilde{a}^2 + b_k^2}; \\
&\qquad \hat{b}_k = b_k(a_{k+1}/\hat{a}_k); \\
&\qquad \tilde{a} = \tilde{a}(a_{k+1}/\hat{a}_k); \\
&\text{end} \\
&\hat{a}_n = \tilde{a};
\end{aligned}
$$

Note that *there are no subtractions in this algorithm*: the elements are computed by means of multiplications and divisions. This is important for high accuracy.

7.8.3   Differential QD Algorithm

We wish to express the transformation $R \mapsto \hat{R}$ in the q-e-quantities. In order to do so, we need to identify the elements of R and $\hat{R}$ with the representation in (7.55),

$$a_i = \sqrt{q_i}, \quad b_i = \sqrt{e_i} \iff q_i = a_i^2, \quad e_i = b_i^2.$$

We can now derive the *differential QD step* by squaring the statements in `OrthogonalQDStep` (the new variable d is used for $\tilde{a}^2$):

DifferentialQDStep [7]

---

[6] in [31] denoted by **oqd**
[7] in [31] denoted by **dqd**

$$d = q_1;$$
$$\text{for } k = 1 : n - 1$$
$$\hat{q}_k = \tilde{a}^2 + b_k^2 = d + e_k;$$
$$\hat{e}_k = e_k(q_{k+1}/\hat{q}_k);$$
$$d = d(q_{k+1}/\hat{q}_k);$$
$$\text{end}$$
$$\hat{q}_n = d$$

This algorithm has already been formulated by Rutishauser in a manuscript that was published only posthumously [112].

An interesting link with Progressive QD becomes apparent if we eliminate the intermediate variable d: let $d = d_k$ at step k. Then from (7.58) and (7.57), we see that

$$d_{k+1} = c_k^2 q_{k+1} = q_{k+1} - s_k^2 q_{k+1} = q_{k+1} - \hat{e}_k.$$

With this formulation, we get back the original progressive QD Algorithm by Parlett (`ProgressiveQDStepParlett`).

EXAMPLE 7.4. *Algorithm* `DifferentialQDStep` *computes small eigenvalues to* high relative precision *while the precision of the smaller eigenvalues computed by* `ProgressiveQDStepParlett` *is relative to* $\sigma_1^2(B)$. *Fernando and Parlett show in [31] that for the matrix* $R \in \mathbb{R}^{64 \times 64}$ *defined by* $q_k = 1$, $k = 1, \ldots, 64$ *and* $e_k = 65536$, $k = 1, \ldots, 63$ *the differential QD Algorithm computes the smallest eigenvalue* 3.645449756934072e−304 *of* $B = R^\top R$ *to machine precision in two iterations; while* `ProgressiveQDStepParlett` *obtains the value 0 and* `eig(B)` *of* MATLAB *gives, even worse, a negative value* −5.420845440188925e−20 *as we can see by executing Algorithm 7.16.*

ALGORITHM 7.16. *Parlett Fernando Example*

```
format long e
n=64;
q=ones(1,n); e=65536*ones(1,n-1);
disp('Differential QD-Algorithm')
for p=1:3
  d=q(1);
  for k=1:n-1
    q(k)=d+e(k);
    e(k)=e(k)*(q(k+1)/q(k));
    d=d*(q(k+1)/q(k));
  end
  q(n)=d;
  p, min(q)
end

disp('Progressive QD-Algorithm')
q=ones(1,n); e=65536*ones(1,n-1);
for p=1:3
```

```
  e(n)=0;
  q(1)=q(1)+e(1);
  for k=2:n,
    e(k-1)=(e(k-1)/q(k-1))*q(k);
    q(k)=(q(k)-e(k-1)) + e(k);
  end
  p, min(q)
end

disp('Matlab eig')
q=ones(1,n); e=65536*ones(1,n-1);
R=diag(q)+diag(sqrt(e),1); B=R'*R;
min(eig(B))
```

7.8.4   Improving Convergence Using Shifts

Rutishauser points out in [112] that $e_k^{(j)} \to 0$ linearly for $j \to \infty$ with

$$e_k^{(j)} = O\left(\left[\frac{\lambda_{k+1}}{\lambda_k}\right]^j\right),$$

and therefore

$$q_k^{(j)} = \lambda_k + O(s^j), \quad \text{with } s = \max\left(\left|\frac{\lambda_{k+1}}{\lambda_k}\right|, \left|\frac{\lambda_k}{\lambda_{k-1}}\right|\right),$$

where we define $\lambda_0 = \infty$ and $\lambda_{n+1} = 0$. As a result, convergence can be very slow if two eigenvalues lie close together, just like with the unshifted QR algorithm. To accelerate convergence, let us consider the following modified progressive QD step, where a shift v has been introduced:

ProgressiveQDStepShifted [8]

$$
\begin{aligned}
&e_0' = 0 \\
&\text{for } k = 1 : n - 1 \\
&\qquad q_k' = ((q_k - e_{k-1}') - v) + e_k \\
&\qquad e_k' = e_k(q_{k+1}/q_k'); \\
&\text{end} \\
&q_n' = (q_n - e_{n-1}') - v
\end{aligned}
$$

The new QD elements are associated with the matrix

$$
T' = \begin{pmatrix}
q_1' & 1 & & & \\
e_1'q_1' & q_2' + e_1' & \ddots & & \\
& \ddots & \ddots & & 1 \\
& & e_{n-1}'q_{n-1}' & q_n' + e_{n-1}'
\end{pmatrix},
$$

---

[8]denoted **qds** in [31]

see (7.51). Expressing the diagonal of T' by $q'_k + e'_{k-1} = q_k + e_k - v$ and noting that $e'_k q'_k = e_k q_{k+1}$, we obtain after comparing with (7.52)

$$T' = \begin{pmatrix} q_1 + e_1 - v & 1 & & \\ e_1 q_2 & q_2 + e_2 - v & \ddots & \\ & \ddots & \ddots & \\ & & e_{n-1} q_n & q_n - v \end{pmatrix} = \hat{T} - vI.$$

Thus, the eigenvalues of matrix T' are those of matrix $\hat{T}$ (and also those of T) shifted by v.

We now consider the shifted version of the orthogonal QD step. Since A is assumed to be positive definite here, it makes sense to consider only positive shifts, i.e., $v = \tau^2$, which has the effect of shifting the spectrum of A closer to zero. In the orthogonal QD Algorithm, this means modifying the statements involving the diagonal elements $\hat{a}$ and $\tilde{a}$:

OrthogonalQDStepShifted[9]

$$\tilde{a} = a_1;$$
$$\text{for } k = 1 : n - 1$$
$$\hat{a}_k = \sqrt{\tilde{a}^2 + b_k^2 - \tau^2};$$
$$\hat{b}_k = b_k \, (a_{k+1}/\hat{a}_k);$$
$$\tilde{a} = \sqrt{\tilde{a}^2 - \tau^2} \, (a_{k+1}/\hat{a}_k);$$
$$\text{end}$$
$$\hat{a}_n = \sqrt{\tilde{a}^2 - \tau^2};$$

To keep $\hat{R}$ real, the shift must satisfy $\tau^2 \le \lambda_{min}(R^\top R)$ or $\tau \le \sigma_n(R)$. This constraint is not necessary for `DifferentialQDStep` and `ProgressiveQDStep` when introducing shifts. By defining $d = d_k = \tilde{a}_k^2 - \tau^2$, an addition can be saved in the algorithm

DifferentialQDStepShifted [10]

$$d = q_1 - \tau^2;$$
$$\text{for } k = 1 : n - 1$$
$$\hat{q}_k = d + e_k;$$
$$\hat{e}_k = e_k(q_{k+1}/\hat{q}_k);$$
$$d = d(q_{k+1}/\hat{q}_k) - \tau^2;$$
$$\text{end}$$
$$\hat{q}_n = d;$$

The choice of appropriate shifts is crucial and not easy. In order to preserve positive definiteness, the condition $v = \tau^2 \le \lambda_{min}(B^\top B)$ must be guaranteed, see [112].

---

[9] oqds in [31]
[10] dqds in [31]

EXAMPLE 7.5. *We perform here an experiment comparing the progressive QD Algorithm with shifts to the differential QD Algorithm with shifts. We choose as shifts $s = q_n$. It turns out that for the symmetric tridiagonal matrix with diagonal -2 and off-diagonal 1, we get convergence, with $q_n \to 0$. Deflation is possible if q_n is small compared to the sum of the shifts: if $q_n = 0$, then the sum of the shifts is an approximation of an eigenvalue.*

This algorithm is not stable since the QD lines do not remain positive and the eigenvalues are approximated by a sum containing positive and negative terms. Cancellation is very likely to occur.

First we perform progressive QD steps using Algorithm 7.17.

ALGORITHM 7.17. *Progressive QD step shifted*

```
function [q1,e1]=ProgressiveQDStepShifted(n,q,e,s)
% PROGERSSIVEQDSTEPSHIFTED progressive qd-step with shift
%   [q1,e1]=ProgressiveQDStepShifted(n,q,e,s) computes a progressive
%   qd-step with shift s for the given qd-line q, e. The new qd-line
%   is q1, e1.

e(n)=0;
q(1)=q(1)+e(1)-s;
for k=2:n,
  e(k-1)=(e(k-1)/q(k-1))*q(k);
  q(k)=(q(k)-e(k-1))+e(k) -s;
end
q1=q(1:n); e1=e(1:n);
```

In the following program, we define for $n = 200$ the $(1, -2, 1)$ tridiagonal matrix and compute the exact eigenvalues. Then we apply progressive QD steps to compute the eigenvalues:

ALGORITHM 7.18.
Experiment: Progressive QD versus Differential QD

```
n=200
a=-2*ones(1,n); b=ones(1,n-1);            % define tridiagonal symmetric
                                          % matrix
B=diag(a)+diag(b,1)+diag(b,-1);
exakt=sort(-4*sin([1:n]'*pi/2/(n+1)).^2); % exact eigenvalues
[q,e]=QDLineSymmetric(a,b);               % compute qd-line
lam=0; ew=[]; it=0; ss=[];
while n>0,
  s=q(n);
  ss=[ss s];
  lam=lam+s;
% [q,e]=ProgressiveQDStepShifted(n,q,e,s);% choose here a step
  [q,e]=DifferentialQDStepShifted(n,q,e,s);
```

```
  if lam+q(n)==lam                      % deflation
    ew=[ew;lam];                        % store eigenvalue
    n=n-1;
  end
  it=it+1;
end
it                                       % total iterations
norm(sort(ew)-exakt)/norm(exakt)         % relative error of computed
                                         % eigenvalues
[min(ss) max(ss)]                        % smallest and largest shift
```

We obtain the results

```
n =
   200
it =
   792
ans =
   1.2213e-10
ans =
   -1.0050    0.5488
```

In this example, `ProgressiveQDStepShifted` produces only about 10 correct decimal digits, using approximately 4 QD iterations per eigenvalue. The shifts were in the range $[-1.0050, 0.5488]$, which means that cancellation has occurred.

If we comment out the call to the progressive QD step with shift s in Algorithm 7.18 and replace it by the differential QD step with shifts (thus calling Algorithm 7.19),

---

ALGORITHM 7.19. *Differential QD step shifted*

```
function [q1,e1]=DifferentialQDStepShifted(n,q,e,s)
% DIFFERENTIALQDSTEPSHIFTED computes one differential qd-step with shift
%    [q1,e1]=DifferentialQDStepShifted(n,q,e,s) computes one
%    differential qd-step with shift s for the qd-line q, e. The new
%    qd-line is q1, e1.

d=q(1)-s;
for k=1:n-1
  q(k)=d+e(k);
  h=q(k+1)/q(k);
  e(k)=e(k)*h;
  d=d*h-s;
end
q(n)=d;
q1=q; e1=e;
```

we get better results (about 14 correct decimal digits), even though the shifts also have both signs:

```
n =
   200
it =
   799
ans =
   6.2047e-14
ans =
   -1.0050     0.8174
```

7.8.5   Connection to Orthogonal Decompositions

Fernando and Parlett observe in their paper [31] that an analogue of the orthogonal connection

$$R^\top = Q\hat{R}$$

is missing for the algorithms with shifts. They propose therefore to compute a $2n \times 2n$ orthogonal matrix Q such that

$$RR^\top = \hat{R}^\top \hat{R} + \tau^2 I \quad \Longleftrightarrow \quad Q^\top \begin{pmatrix} R^\top \\ 0 \end{pmatrix} = \begin{pmatrix} \hat{R} \\ \tau I \end{pmatrix}. \tag{7.59}$$

They add: "Moreover Q may be built up by well chosen plane rotations".

Urs von Matt describes in [144] how these "well chosen plane rotations" should be constructed. For this purpose, we need *generalized Givens transformations*, which are Givens rotations of the form

$$G = \begin{pmatrix} c & s \\ -s & c \end{pmatrix}, \quad s = \sin \alpha, \quad c = \cos \alpha,$$

where the rotation angle α is chosen in such a way that

$$G^\top \begin{pmatrix} x_1 \\ x_2 \end{pmatrix} = \begin{pmatrix} r \\ \sigma \end{pmatrix}$$

with given x_1, x_2 and σ (of course we must have $|\sigma| \leq \sqrt{x_1^2 + x_2^2}$, otherwise the transformation does not exist). Solving the linear equations

```
> solve({c*x[1]-s*x[2]=r, s*x[1]+c*x[2]=sigma},{s,c});
```

we get

$$\left\{ c = \frac{x_2\sigma + rx_1}{x_1{}^2 + x_2{}^2}, s = -\frac{-x_1\sigma + x_2 r}{x_1{}^2 + x_2{}^2} \right\}$$

Algorithm 7.20 avoids again overflow by appropriately factoring:

ALGORITHM 7.20. *Generalized Givens Rotation*

```
function [c,s,y]=GeneralizedGivensRotation(x,sig);
% GENERALIZEDGIVENSROTATION computes a generalized Givens rotation
%    [c,s,y]=GeneralizedGivensRotation(x,sig); computes for the given
%    number x(1), x(2) and sig a generalized Givens rotation to obtain
%    G'x = y = [r, sig]'

scale=max(abs(x));
if scale==0,
   c=1; s=0;
else
  x=x/scale; sig=sig/scale;
  norm2=norm(x);
  r=(norm2-sig)*(norm2+sig);
  if r<0, error('rotation does not exists'); end
  r=sqrt(r); norm2=norm2^2;
  s=([sig -r]*x)/norm2;
  c=([r sig]*x)/norm2;
  y=[scale*r; sig];
end
```

This generalized Givens rotation allows us to compute the LR step with shift
of (7.59). We now show how to construct step-by-step an orthogonal matrix
Q to transform

$$Q^\top \begin{pmatrix} R^\top \\ 0 \end{pmatrix} = \begin{pmatrix} \hat{R} \\ \sqrt{s}I \end{pmatrix}.$$

First, we use a generalized Givens rotation (acting on the two elements
marked in boldface) to introduce $\sigma = \sqrt{s}$ in the $(n+1,1)$-position. Next, we
annihilate the element b_1 with an ordinary Givens rotation, introducing $\hat{b}_1$
as fill-in and overwriting a_1':

$$
\begin{bmatrix}
\mathbf{a_1} & & & & \\
b_1 & a_2 & & & \\
 & b_2 & a_3 & & \\
 & & \ddots & \ddots & \\
\mathbf{0} & & & & \\
 & 0 & & & \\
 & & 0 & & \\
 & & & \ddots &
\end{bmatrix}
\mapsto
\begin{bmatrix}
\mathbf{a_1'} & & & & \\
\mathbf{b_1} & \tilde{a}_2 & & & \\
 & b_2 & a_3 & & \\
 & & \ddots & \ddots & \\
\sigma & & & & \\
 & 0 & & & \\
 & & 0 & & \\
 & & & \ddots &
\end{bmatrix}
\mapsto
\begin{bmatrix}
\hat{a}_1 & \hat{b}_1 & & & \\
0 & \mathbf{a_2} & & & \\
 & b_2 & a_3 & & \\
 & & \ddots & \ddots & \\
\sigma & & & & \\
 & \mathbf{0} & & & \\
 & & 0 & & \\
 & & & \ddots &
\end{bmatrix}.
$$

It is now clear how to repeat these two steps with the rest of the matrix. The
transformation is computed with Algorithm 7.21:

ALGORITHM 7.21. *Orthogonal LR Step with Shift*

```
function [aa,bb]=OrthogonalLRStepShifted(a,b,s)
% ORTHOGONALLRSTEPSHIFTED computes one orthogonal qd step with shift
%    [aa,bb]=OrthogonalLRStepShifted(a,b,s) computes the LR-step
%    R*R^T=Rt^T*Rt+s*I with R=diag(a)+diag(b,1) and shift s>0.  The
%    result is Rt=diag(aa)+diag(bb)

n=length(a);
sig=sqrt(s);
for k=1:n-1
  [si,co,y]=GeneralizedGivensRotation([a(k);0], sig);
  a(k)=y(1);
  [co,si]=GivensRotation(a(k),b(k));
  a(k)=co*a(k)-si*b(k);
  b(k)=-si*a(k+1);
  a(k+1)=co*a(k+1);
end
[si,co,y]=GeneralizedGivensRotation([a(n);0],sig);
a(n)=y(1);
aa=a; bb=b;
```

In the following example, we will compare the computation of one LR step
with the three methods: (1) direct matrix operations, (2) using von Matt's
generalized Givens rotations and (3) using Algorithm 7.22, the orthogonal
QD Algorithm with shift.

ALGORITHM 7.22. *Orthogonal QD step with shift*

```
function [aa,bb]=OrthogonalQDStepShifted(a,b,s)
% ORTHOGONALQDSTEPSHIFTED computes one orthogonal QD step with shift
%    [aa,bb]=OrthogonalQDStepShifted(a,b,s) computes the LR-step
%    R*R^T=Rt^T*Rt+s*I with R = diag(a)+diag(b,1) and shift s>0.  The
%    result is Rt=diag(aa)+diag(bb)

n=length(a);
sig=sqrt(s);
for k=1:n-1
  [si,co,y]=GeneralizedGivensRotation([a(k);0], sig);
  a(k)=y(1);
  [co,si]=GivensRotation(a(k),b(k));
  a(k)=co*a(k)-si*b(k);
  b(k)=-si*a(k+1);
  a(k+1)=co*a(k+1);
end
[si,co,y]=GeneralizedGivensRotation([a(n);0],sig);
a(n)=y(1);
aa=a; bb=b;
```

EXAMPLE 7.6. *We consider again the matrix A given in (7.56) and used in Example 7.3:*

ALGORITHM 7.23.
*Comparing Matrix QD step, Orthogonal LR Step and
Orthogonal QD step with Shifts*

```
format long
n=7;
A=diag(4*ones(1,n))+diag(ones(1,n-1),1)+diag(ones(1,n-1),-1)
s=2;
R=chol(A); a=diag(R); b=diag(R,1);          % qd-step with shift s
R1=chol(R*R'-s*eye(size(R)));               % as matrix operation
a1=diag(R1);
b1=diag(R1,1);
[a2,b2]=OrthogonalLRStepShifted(a,b,s);
[a3,b3]=OrthogonalQDStepShifted(a,b,s);
[a1,a2,a3]
[b1,b2,b3]
```

```
A =
     4     1     0     0     0     0     0
     1     4     1     0     0     0     0
     0     1     4     1     0     0     0
     0     0     1     4     1     0     0
     0     0     0     1     4     1     0
     0     0     0     0     1     4     1
     0     0     0     0     0     1     4
ans =
   1.500000000000000   1.500000000000000   1.500000000000000
   1.264911064067352   1.264911064067352   1.264911064067352
   1.174294790062638   1.174294790062637   1.174294790062637
   1.129219566043900   1.129219566043900   1.129219566043900
   1.102630728275509   1.102630728275509   1.102630728275510
   1.085124133798125   1.085124133798124   1.085124133798125
   0.939569046115558   0.939569046115557   0.939569046115557
ans =
   0.645497224367903   0.645497224367903   0.645497224367903
   0.788810637746615   0.788810637746615   0.788810637746615
   0.851439142005194   0.851439142005194   0.851439142005194
   0.885557232074404   0.885557232074404   0.885557232074404
   0.906921195399565   0.906921195399565   0.906921195399564
   0.921553497750770   0.921553497750770   0.921553497750769
```

We see that for this example the results are the same.

7.9   Problems

PROBLEM 7.1.

1. *The* Hilbert Matrix H *has as elements* $h_{ij} = \frac{1}{i+j-1}$, $i,j = 1,\ldots,n$. *Prove that the Hilbert Matrix is positive definite.*

   Hint: *Use the representation* $h_{ij} = \int\limits_0^1 t^{i-1}t^{j-1}\,dt$ *and the definition of positive definiteness!*

2. *The eigenvalues of symmetric positive definite matrices are all greater than zero.*

   Prove this by considering the eigen-decomposition $A = QDQ^\top$, *with* Q *orthogonal and* D *diagonal.*

3. *The* singular values of a matrix A *can be defined as the square-roots of the eigenvalues of* $A^\top A$:

$$\sigma_k(A) = \sqrt{\lambda_k(A^\top A)}.$$

   Prove that for symmetric positive definite matrices the singular values are equal to the eigenvalues.

4. *Finally check the results above by computing the eigenvalues and singular values on the computer.*

   (a) *use the function* `hilb` *to generate the Hilbert matrix of order* $n = 30$.

   (b) *compute, compare and comment the results you get when applying the* MATLAB *function* `eig` *and* `svd` *and our functions* `Jacobi` *and* `SVDGolubReinsch`. *Compute for comparison the exact eigenvalues using* MAPLE *and* `Digits:=60`.

PROBLEM 7.2. (UNREDUCED MATRICES HAVE SIMPLE EIGENVALUES)
   Let T be a symmetric tridiagonal matrix that is unreduced, *i.e.,* $t_{i,i-1} \neq 0$ *for $i = 2,\ldots,n$. Show that T has simple eigenvalues.*

Hint: *Let λ be an eigenvalue of A. Show that the first $n-1$ columns of the matrix $A - \lambda I$ are linearly independent, and conclude that λ has geometric multiplicity 1. Why is the symmetry assumption important?*

PROBLEM 7.3.
Consider the function `[V,D]=Jacobi1(A)` *(Algorithm 7.2), the first version of the algorithm of Jacobi.*

1. *Modify this function by choosing the largest element $|a_{pq}|$ of the matrix in each step to maximize the decay of the off-diagonal elements.*

2. *Compute S_k (the deviation of A_k from the diagonal matrix) with both functions for the following matrix A:*

$$\mathbf{A} = \begin{pmatrix} 0 & 1 & 2 & \dots & n \\ 1 & 0 & 1 & \dots & n-1 \\ 2 & 1 & 0 & \dots & n-2 \\ \vdots & \vdots & \vdots & \ddots & \vdots \\ n & n-1 & n-2 & \dots & 0 \end{pmatrix}.$$

For each $n = 5, 10, 20$, plot on the same figure the deviations S_k versus the number of steps (i.e., number of Jacobi rotations) k for both functions. Compare the performance of both methods.

Hints: *Use the functions* `max` *and* `semilogy`. *To generate the matrix A, use the function* `gallery('fiedler',n)`.

PROBLEM 7.4. *Program a reduction to Hessenberg form using Householder transformations.*

An elementary Householder matrix *is a matrix of the form*

$$P = \boldsymbol{I} - \boldsymbol{v}\boldsymbol{v}^\top$$

with $\|\boldsymbol{v}\| = \sqrt{2}$. The matrix P is symmetric, orthogonal and has the properties $P\boldsymbol{v} = -\boldsymbol{v}$ and $\boldsymbol{P}\boldsymbol{x} = \boldsymbol{x}$ for any $\boldsymbol{x} \perp \boldsymbol{v}$. Thus P is a reflection at the hyperplane $\boldsymbol{v}^\top \boldsymbol{x} = 0$.

Such matrices can be used for solving the following problem: Given a vector $\boldsymbol{x}$, find an orthogonal matrix P such that $P\boldsymbol{x} = \sigma\boldsymbol{e}_1$.

The Householder transformations $P^\top A P$ therefore seem to be a good method to zero out all the required elements and because they are orthogonal, they will preserve the eigenvalues of our matrix A.

Implement this process in MATLAB. *Your function should return the matrix H in the Hessenberg form and the orthogonal matrix Q such that*

$$Q^\top A Q = H.$$

Try to save as much time and space as possible.

Hints: *The original matrix can be overwritten during the process. Avoid matrix by matrix multiplication whenever possible.)*

PROBLEM 7.5. *Prove the 3 basic facts mentioned in Subsection 7.6.3:*

1. $B_{k+1} = Q_k^\top B_k Q_k$.

2. $B_{k+1} = P_k^\top B_1 P_k$ where $P_k := Q_1 \cdots Q_k$.

3. *If $S_k = R_k \cdots R_1$ then the QR Decomposition of the matrix $\prod_{i=1}^{k}(B_1 - \sigma_i I)$ is $P_k S_k$.*

PROBLEM 7.6. (VECTOR ITERATION/POWER METHOD) *Compute for* A =`magic(6)` *and* $\boldsymbol{v}_0 = \boldsymbol{e}_1$ *the sequences*

$$\boldsymbol{w} = A\boldsymbol{v}_{k-1}, \quad \boldsymbol{v}_k = \frac{\boldsymbol{w}}{\|\boldsymbol{w}\|}, \quad \lambda_{k-1} = (\boldsymbol{v}_{k-1})^{\top} A\boldsymbol{v}_{k-1}$$

Observe the convergence and compare the result with `[V,D] = eig(A)`.

PROBLEM 7.7. (INVERSE ITERATION/SHIFT-AND-INVERT) *Suppose we would like to compute the eigenvalue of* A =`magic(6)` *that is close to* $\mu = 10$. *To do so, apply vector iteration to the matrix* $B = (A - \mu I)^{-1}$.

What is the connection between the eigenvalues and eigenvectors of the matrices A *and* B? *Watch the convergence for the vector iteration with* B.

PROBLEM 7.8. (QR METHOD WITHOUT SHIFTS) *Perform QR iterations with the matrix* A =`magic(6)`:

$$A_i = Q_i R_i, \quad A_{i+1} = R_i Q_i.$$

Print after every iteration the matrices R_i *and* Q_i. *What do you see? Is the process convergent?*

PROBLEM 7.9. *Modify the* MATLAB *function* `d=ImplicitQR(a,b)` *and write a new function* `[d,V]=ImplicitQRVec(a,b,eivec)` *so that it computes for* `eivec=1` *also the eigenvectors of the symmetric tridiagonal matrix* T.

PROBLEM 7.10. *Write a main program and compute the eigenvalues* D *and vectors* V *of the Rosser matrix (see* `help rosser` *and* `A=rosser`*) and the matrix* `A=hadamard(8)` *(see also* `help hadamard`*). Compare the results of*

1. `eig`, *the built in function of* MATLAB,

2. `Jacobi`, *our translation of Rutishauser's ALGOL program, and*

3. *Reduction to tridiagonal form by* `Tridiagonalize` *followed by a call to* `ImplicitQRvec`.

Compute for each method the diffenrence $\|VDV^{\top} - A\|$ *and print the eigenvalues using* `format long e`.

PROBLEM 7.11. *The SVD is the best numerical tool to determine the rank of a matrix.*

1. *Compute the SVD of* $A = $ `magic(20)`.

   (a) *Find "the gap" and reconstruct* A *as matrix of lower rank* $\tilde{A}$.

   (b) *Check that* $\text{round}(\tilde{A})$ *should be equal to* A.

2. *Compute the rank of the matrix* `magic(21)`

PROBLEM 7.12. (RANK DETERMINATION BY THE DETERMINANT?)
There is a famous example by W. Kahan:

$$A_1 = \begin{pmatrix} 1 & -1 & \cdots & -1 \\ 0 & 1 & \ddots & \vdots \\ \vdots & \ddots & \ddots & -1 \\ 0 & \cdots & 0 & 1 \end{pmatrix}$$

1. *Compute the determinant, the singular values, the condition number, and the rank of A for $n = 50$ and interpret the results.*

2. *Change the minus signs in the definition of A_1 to plus signs and do the computations again with this new matrix B.*

3. *Compute low rank approximations of A and B by setting $\sigma_i = 0$ for $i = k, \ldots, n$.*

   Since the matrices have integer elements, you should round the low rank approximation. For which k is the matrix correctly reproduced?

PROBLEM 7.13. (COMPUTING WITH IMPLICIT BIDIAGONALIZATION)
With Algorithm 7.11, we bidiagonalize a matrix using Householder transformations,

$$P_n \cdots P_1 A Q_1 \cdots Q_{n-1} = B.$$

Since $P_i = P_i^\top = P_i^{-1}$ holds, we have

$$A = PBQ^\top, \quad \text{with} \quad P = P_1 \cdots P_n, \quad Q = Q_1 \cdots Q_{n-1}.$$

The matrices P and Q are not formed explicitly by `[q,e,A]=Bidiagonalize(A)`*; they are only given implicitly by the stored Householder vectors with which the matrix A has been overwritten.*

Given vectors $\boldsymbol{x} \in \mathbb{R}^n$ and $\boldsymbol{y} \in \mathbb{R}^m$, write 4 short MATLAB functions to compute after a call `[q,e,A]=Bidiagonalize(A)` *the products*

$$P\boldsymbol{y}, P^\top \boldsymbol{y}, Q\boldsymbol{x} \quad \text{and} \quad Q^\top \boldsymbol{x}.$$

For instance, with a `function z=Py(A,y)` *which computes $\boldsymbol{z} = P\boldsymbol{y}$, you could compute the matrix P explicitly by the statements:*

```
E=eye(m);                          % compute P
P=[];
for k=1:n
  y=E(:,k); z=Py(A,y); P=[P,z];
end
```


PROBLEM 7.14. (BIDIAGONALIZATION WITH GOLUB-KAHAN-LANCZOS ALGORITHM) *We would like to bidiagonalize the matrix $A \in \mathbb{R}^{m \times n}$ with*

$m \geq n$. In other words, we wish to find an orthogonal matrix $P \in \mathbb{R}^{m \times n}$, an upper bidiagonal matrix $B \in \mathbb{R}^{n \times n}$ and an orthogonal matrix $Q \in \mathbb{R}^{n \times n}$ such that $A = PBQ^\top$ holds.

    Consider for that the ansatz

$$AQ = PB, \quad A^\top P = QB^\top, \quad \text{with } Q = [\boldsymbol{q}_1, \ldots, \boldsymbol{q}_n] \text{ and } P = [\boldsymbol{p}_1, \ldots, \boldsymbol{p}_n].$$

Given some unit vector $\boldsymbol{q}_1$ we compute

Step 1: $AQe_1 = PBe_1 \iff A\boldsymbol{q}_1 = \boldsymbol{p}_1 a_1$

    Thus: $\boldsymbol{u} := A\boldsymbol{q}_1$, $a_1 := \|\boldsymbol{u}\|$, $\boldsymbol{p}_1 := \boldsymbol{u}/a_1$.

Step 2: $A^\top \boldsymbol{p}_1 = QB^\top e_1 = a_1 \boldsymbol{q}_1 + b_1 \boldsymbol{q}_2$

    Thus $\boldsymbol{u} := A^\top \boldsymbol{p}_1 - a_1 \boldsymbol{q}_1$, $b_1 := \|\boldsymbol{u}\|$, $\boldsymbol{q}_2 := \boldsymbol{u}/b_1$.

Step $k+1$: *Assume that we know*

$$Q_k = [\boldsymbol{q}_1, \ldots, \boldsymbol{q}_k] \, P_k = [\boldsymbol{p}_1, \ldots, \boldsymbol{p}_k] \text{ and } B_k = \begin{pmatrix} a_1 & b_1 & & \\ & \ddots & \ddots & \\ & & \ddots & b_{k-1} \\ & & & a_k \end{pmatrix}.$$

Then $A^\top \boldsymbol{p}_k = QB^\top e_k = a_k \boldsymbol{q}_k + b_k \boldsymbol{q}_{k+1}$ and

$$\boldsymbol{u} := A^\top \boldsymbol{p}_k - a_k \boldsymbol{q}_k, \quad b_k := \|\boldsymbol{u}\|, \quad \boldsymbol{q}_{k+1} := \boldsymbol{u}/b_k.$$

Furthermore $A\boldsymbol{q}_{k+1} = PBe_{k+1} = b_k \boldsymbol{p}_k + a_{k+1} \boldsymbol{p}_{k+1}$ thus

$$\boldsymbol{u} := A\boldsymbol{q}_{k+1} - b_k \boldsymbol{p}_k, \quad a_{k+1} := \|\boldsymbol{u}\|, \quad \boldsymbol{p}_{k+1} := \boldsymbol{u}/a_{k+1}.$$

We have computed the new quantities

$$b_k, \boldsymbol{q}_{k+1}, a_{k+1} \text{ and } \boldsymbol{p}_{k+1}.$$

Write a MATLAB `function [a,b,P,Q]=Lanczos(A)` *which bidiagonalizes the matrix A. Compare it with* `Bidiagonalize` *by computing the singular values of the resulting bidiagonal matrix. The bidiagonal matrices are not unique!*

PROBLEM 7.15.   *Implement a* MATLAB *program* `GivensRotation1.m` *which computes for a given vector of length two the Givens rotation which zeroes out the first element, see the end of Subsection 7.7.5.*

Chapter 8. Differentiation

Es bezeichne u den Wert der gesuchten Funktion in einem Netzpunkt und u_1, u_2, u_3, u_4 die Werte in den vier benachbarten Netzpunkten, und es sei ferner die Seite einer Masche gleich h, so wird die Differenzengleichung die Form annehmen[1]

$$u_1 + u_2 + u_3 + u_4 - 4u = h^2 \cdot C.$$

C. Runge, Über eine Methode die partielle Differentialgleichung $\Delta u = Constans$ numerisch zu integrieren, Zeitschrift für Mathematik und Physik, Vol. 56, page 226

Many algebraic computer languages now include facilities for the evaluation of functions of a complex variable. Such facilities can be effectively used for numerical differentiation. [...] Since this method is based on numerical quadrature, it does not show the sensitivity to roundoff error in the function evaluations that is characteristic of finite difference methods.

J. N. Lyness and C. B. Moler, Numerical Differentiation of Analytic Functions, SIAM J. Numer. Anal. Vol 4, No. 2, 1967.

Algorithmic, or automatic, differentiation (AD) is concerned with the accurate and efficient evaluation of derivatives for functions defined by computer programs. No truncation errors are incurred, and the resulting numerical derivative values can be used for all scientific computations that are based on linear, quadratic, or even higher order approximations to nonlinear scalar or vector functions.

Andreas Griewank, Evaluating Derivatives - Principles and Techniques of Algorithmic Differentiation, SIAM 2000.

Prerequisites: Notions from polynomial interpolation (§4.2) and extrapolation (§4.2.8) are required. For §8.3, we use the methods of bisection (§5.2.1) and Newton (§5.2.5), whereas for §8.3.4, Chapter 7 is needed.

In many applications, one needs to compute the derivative of some given function $f(x)$. If the function is an algebraic expression, then it is possible to evaluate $f'(x)$ analytically using differential calculus. Often, however, the function f and its derivative may be complicated or expensive to evaluate. For instance, the function may be given only as a table of values, so its derivative is not readily available. In other cases, the function f may be the output of a simulation, so that every function evaluation is in fact a

---

[1] Let u be the value of the solution sought in a gridpoint, and let u_1, u_2, u_3, u_4 be the values at the four neighboring gridpoints, and let h denote the mesh size, then the difference equation will be of the form

W. Gander et al., *Scientific Computing - An Introduction using Maple and MATLAB*,
Texts in Computational Science and Engineering 11,
DOI 10.1007/978-3-319-04325-8_8,
© Springer International Publishing Switzerland 2014

run of a larger program. Moreover, it is no longer obvious how one should calculate $f'(x)$. The most important application of derivative approximation is, however, the numerical solution of ordinary and partial differential equations. We start this chapter with the historically important example of the Brachystochrone in Section 8.1. There, we give a glimpse of variational calculus, and show how one can start from a concrete minimization problem in integral form and obtain a differential equation, which we then solve using approximations of the derivatives. In Section 8.2, we formally introduce finite difference approximations of derivatives, and show how one can easily generate such approximations using MAPLE, also for partial differential operators. A very different approach for obtaining derivatives, namely algorithmic differentiation, is explained in Section 8.3. We conclude this chapter with two interesting examples that use algorithmic differentiation: the first one is about circular billiards, whereas the second is about nonlinear eigenvalue problems.

8.1 Introductory Example

We start with the historical example of the *Brachystochrone problem*, which was posed in 1696 by Johann Bernoulli to his brother Jacob and the world: given two fixed points A and B in a vertical plane, find the curve from A to B such that a body gliding along it solely under the influence of gravity travels from A to B in the shortest possible time[2], see Figure 8.1. Galileo

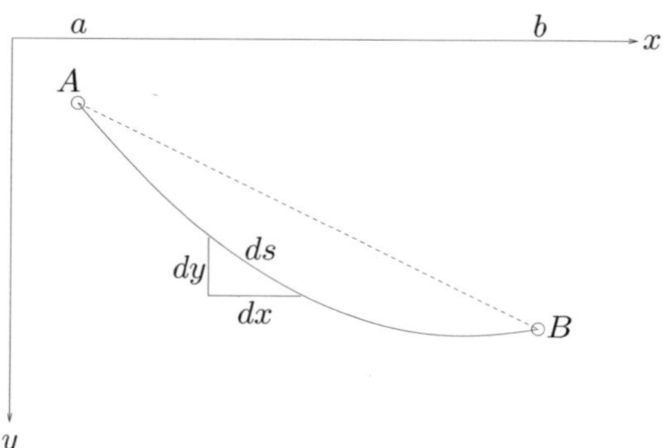

FIGURE 8.1. *The Brachystochrone Problem*

Galilei already knew in 1638 that the shortest path, namely the straight line between A and B, is not the fastest; his guess was that a circle would be the best solution, but this is incorrect.

We now present the historical formulation provided to de l'Hôpital, who admitted in a letter to Johann Bernoulli that he needed a pure mathematics

---

[2] In Latin: Datis in plano verticali duobus punctis A & B, assignare Mobili M viam AMB, per quam gravitate sua descendens, & moveri incipiens a puncto A, brevissimo tempore perveniat ad alterum punctum B.

formulation in order to be able to understand this problem. Since the velocity v of a free falling body satisfies $v = \sqrt{2gy}$ (from Galileo Galilei, where g is the constant of gravitation), the time for traveling along a small arc of length ds is $dJ = \frac{ds}{v} = \frac{ds}{\sqrt{2gy}}$. We therefore have to find a function $y(x)$ with $y(a) = A$, $y(b) = B$ such that the integral

$$J = \int_a^b \frac{\sqrt{dx^2 + dy^2}}{\sqrt{2gy}} = \int_a^b \frac{\sqrt{1 + p^2}}{\sqrt{2gy}}\, dx \qquad (8.1)$$

is minimized, where $p := \frac{dy}{dx}$ and we have used Pythagoras to compute ds, see Figure 8.1. The solution is therefore obtained by solving an optimization (minimization) problem, which could be solved by techniques from Chapter 12. Euler showed in 1744, however, that the solution of such a variational minimization problem also satisfies a differential equation. He obtained his result by discretization, but it was Lagrange, who provided in a letter to Euler in 1755, the formalism we are used to today: we add to $y(x)$ a fixed function $\delta y(x)$ with $\delta y(a) = \delta y(b) = 0$ (since the endpoints are fixed), multiplied by ϵ, and insert the result into (8.1).[3] This integral must be minimal for all functions $\delta y(x)$ at $\epsilon = 0$, i.e. setting $Z(x, y, p) := \frac{\sqrt{1+p^2}}{\sqrt{2gy}}$, the derivative of

$$J(\varepsilon) = \int_a^b Z(x, y + \varepsilon\delta y, p + \varepsilon\delta p)\, dx \qquad (8.2)$$

with respect to ϵ must be zero at $\epsilon = 0$. We can now simply compute the derivative. Denoting by $N(x, y, p) := \partial_y Z(x, y, p)$ and $P(x, y, p) := \partial_p Z(x, y, p)$ the partial derivatives that appear, we obtain

$$\frac{\partial J(\varepsilon)}{\partial \varepsilon}\Big|_{\varepsilon=0} = \int_a^b (N \cdot \delta y + P \cdot \delta p)\, dx = 0. \qquad (8.3)$$

Since δp is the derivative of δy, we can integrate by parts and obtain

$$\int_a^b (N - \partial_x P) \cdot \delta y \cdot dx = 0, \qquad (8.4)$$

where the end point contributions vanish because $\delta y(a) = \delta y(b) = 0$. Since δy is arbitrary, we conclude that the solution $y(x)$ of our problem must satisfy[4]

$$\partial_y Z(x, y, y') - \partial_x \partial_p Z(x, y, y') = 0, \quad y(a) = A,\ y(b) = B. \qquad (8.5)$$

We now derive the differential equation corresponding to the Brachystochrone problem using MAPLE:

---

[3] It took Euler 20 years to introduce this ϵ in order to explain the method to others.

[4] This last step, an immediate consequence for the young discoverer Lagrange, later caused the greatest difficulties and needs a certain care.

ALGORITHM 8.1.
Derivation of the Brachystochrone Equation

```
Z:=sqrt(1+p^2)/(sqrt(2*g*y));
N:=diff(Z,y);
P:=diff(Z,p);
P:=subs({p=p(x),y=y(x)},P);
N:=subs({p=p(x),y=y(x)},N);
de:=factor(N-diff(P,x));
de:=subs({p(x)=diff(y(x),x),diff(p(x),x)=diff(y(x),x,x)},de);
```

and we obtain the differential equation

$$de := -1/4 \, \frac{\sqrt{2}g \left(1 + \left(\frac{d}{dx}y\,(x)\right)^2 + 2\left(\frac{d^2}{dx^2}y\,(x)\right)y\,(x)\right)}{\left(gy\,(x)\right)^{3/2}\left(1 + \left(\frac{d}{dx}y\,(x)\right)^2\right)^{3/2}} = 0. \qquad (8.6)$$

MAPLE cannot currently solve this differential equation with the given boundary conditions using the command

```
> dsolve({de,y(a)=A,y(b)=B},y(x));
```

only an empty result is returned. Despite this, it is possible to obtain generic solution formulas without the boundary conditions using the command

```
> dsolve(de,y(x));
```

$$\left\{-\sqrt{-\left(y\,(x)\right)^2 + y\,(x)\,\_C1} + 1/2\,\_C1\,\arctan\left(\frac{y(x)-1/2\,\_C1}{\sqrt{-(y(x))^2+y(x)\_C1}}\right) - x - \_C2 = 0\right\},$$

$$\left\{\sqrt{-\left(y\,(x)\right)^2 + y\,(x)\,\_C1} - 1/2\,\_C1\,\arctan\left(\frac{y(x)-1/2\,\_C1}{\sqrt{-(y(x))^2+y(x)\_C1}}\right) - x - \_C2 = 0\right\}.$$

These implicit solutions have been known since the challenge of Johann Bernoulli.

Unfortunately, differential equations of the form (8.6) in general cannot be solved in closed form, especially if the problem is posed in more than one dimension, so one has to resort to numerical approximations. This is the most important area where numerical approximations for derivatives are needed: for the solution of partial differential equations and boundary value problems. There are many well-established methods, such as the finite difference method, the finite volume method, the finite element method and also spectral methods. Explaining all these approaches would fill more than another textbook: the finite element method, for example, goes back to the variational minimization problem, and there are several textbooks devoted entirely to it, see for instance [133, 73, 10]. In this chapter, we focus mainly on finite differences for approximating derivatives.

To solve the Brachystochrone problem approximately, we look for a discrete approximate solution $y_i \approx y(x_i)$ at the grid points $x_i = ih$, $i =$

$1, 2, \ldots, n$, where $h = (b - a)/(n + 1)$. Following the quote of Runge at the beginning of this chapter, we replace the derivatives in the numerator of the differential equation (8.6) by finite differences, see Section 8.2, and obtain for $i = 1, 2, \ldots, n$ the difference equation

$$1 + \left(\frac{y_{i+1} - y_{i-1}}{2h}\right)^2 + 2\frac{y_{i+1} - 2y_i + y_{i-1}}{h^2} y_i = 0, \quad y_0 = A, \ y_{n+1} = B. \quad (8.7)$$

This nonlinear system of equations can be solved using any of the methods described in Chapter 5. For example using a simple fixed point iteration, we obtain in MATLAB for $a = 0$, $b = 1$, and $y(a) = 0$

ALGORITHM 8.2.
Solution of the Brachystochrone Problem

```
n=20; h=1/(n+1); e=ones(n,1); x=(h:h:1-h)';
D=spdiags([-e e],[-1 1],n,n)/2/h;
D2=spdiags([e -2*e e],[-1 0 1],n,n)/h^2;
B=1/2; bc=zeros(n,1); bc(end)=B;
y=B*x;
w=0.001;
for i=1:1000
  r=e+(D*y+bc/2/h).^2+2*(D2*y+bc/h^2).*y;
  y=y+w*r;
  plot([0;x;1],-[0;y;B],'-o');
  drawnow
end;
```

The solution obtained is shown in Figure 8.2. As one can see, the fastest path in this example even dips below the target B, so that the body can pick up a lot of speed; the last part uphill is then traversed very rapidly.

The fixed point iteration we used is clearly not the best approach, as one can see from the number of iterations needed and from the relaxation parameter w, which needs to be very small for the method to converge. In Problem 8.1, we solve the equation using Newton's method; one can then determine for which B the Brachystochrone becomes a monotonically decreasing curve.

8.2 Finite Differences

The derivative of a function $f(x)$ is defined as the limit of a *finite difference*:

$$f'(x) = \lim_{h \to 0} \frac{f(x + h) - f(x)}{h}.$$

Therefore, to compute an approximation of f' for some argument z, we could use one-sided *finite differences*

$$f'(z) \approx \frac{f(z + h) - f(z)}{h} \quad \text{or} \quad f'(z) \approx \frac{f(z) - f(z - h)}{h}. \quad (8.8)$$

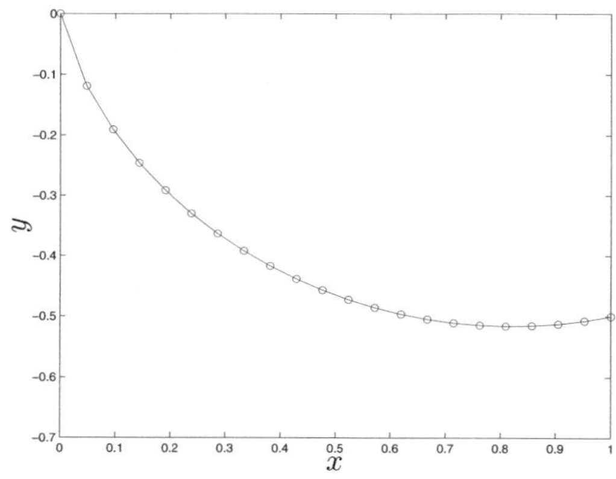

FIGURE 8.2.
*Finite difference approximation of the Brachysthochrone
solution*

These approximations compute the slope of the straight line through the two
points $(z, f(z))$ and $(z \pm h, f(z \pm h))$. Therefore we can only expect a good
approximation if the step size h is small. On the other hand, for small step
sizes of h, the result will be affected by cancellation because $f(z) \approx f(z \pm h)$,
as shown in Chapter 2. One should therefore not choose h smaller than
about $\sqrt{eps}$, where eps is the machine precision, see Figure 2.3. One way to
overcome this dilemma is to extrapolate the value (see Chapter 4, Example
4.6).

A different idea can be used when the function has an analytic continuation into the complex plane (see the quote of Moler and Lyness at the
beginning of this chapter): in the example of $f(z) = e^z$ used in Figure 2.3
in Chapter 2, this is the case. Let us expand the function in a Taylor series
along the direction ih, $i = \sqrt{-1}$,

$$f(z + ih) = f(z) + ihf'(z) - \frac{h^2}{2!}f''(z) - \frac{ih^2}{3!}f'''(z) + \dots . \qquad (8.9)$$

The derivative of f at z can thus also be approximated by

$$f'(z) \approx \operatorname{Im}(f(z + ih)/h),$$

and this approximation is even second order accurate. In addition, there are
no subtractions of numbers of similar size, and therefore no catastrophic cancellation can occur. We illustrate this with the following MATLAB example:

```
format long;
h=10.^((0:-1:-15)');
f=@(z) exp(z);
z=1;
fp=(f(z+h)-f(z))./h;
```

```
fp2=imag(f(z+1i*h)./h);
[exp(z)-fp exp(z)-fp2]
```

This results in the table of errors of the finite difference formula and the complex approximation formula

```
-1.952492442012559    0.430926541280203
-0.140560126414833    0.004528205018508
-0.013636827328035    0.000045304470618
-0.001359594073736    0.000000453046949
-0.000135918619438    0.000000004530470
-0.000013591453264    0.000000000045305
-0.000001358527480    0.000000000000453
-0.000000135505796    0.000000000000005
 0.000000051011672                    0
 0.000000228647356                    0
 0.000002893182615                    0
 0.000011774966813                    0
 0.000011774966812                    0
 0.004896756275163                    0
 0.053746569358670                    0
 0.053746569358670                    0
```

We clearly see that the finite difference formula has an error of $O(h)$ up to about $h = \sqrt{eps}$, and then the error increases again, while the complex approximation formula has an error of $O(h^2)$ until reaching machine precision, and then zero. More examples can be found in [128]. The underlying idea for the numerical differentiation of analytic functions goes back to [88], where arbitrary orders of derivatives are approximated *without the danger of catastrophic cancellation.*

We have seen in the above example that the finite difference formulas (8.8) are only first order accurate. An obvious improvement to one-sided formulas is to use the *symmetric difference*

$$\frac{f(z+h) - f(z-h)}{2h} = f'(z) + \frac{f^{(3)}(z)}{6}h^2 + \frac{f^{(5)}(z)}{120}h^4 + \cdots$$

which is the slope of the straight line through the neighboring points. This approximation of the derivative is also used when computing (defective) spline functions, see Figure 4.5 in Section 4.3.2.

An approximation for the second derivative $f''(z)$ can be obtained by first computing two values of the first derivative

$$f'(z - h/2) \approx \frac{f(z) - f(z-h)}{h} \quad \text{and} \quad f'(z + h/2) \approx \frac{f(z+h) - f(z)}{h}.$$

Now using the difference we get

$$f''(z) \approx \frac{f'(z + h/2) - f'(z - h/2)}{h} \approx \frac{\frac{f(z+h)-f(z)}{h} - \frac{f(z)-f(z-h)}{h}}{h},$$

the well-known approximation

$$f''(z) \approx \frac{f(z+h) - 2f(z) + f(z-h)}{h^2}.$$

(8.10)

It is however cumbersome to derive finite difference formulas manually this way, and we show a more convenient approach in the next subsection.

8.2.1 Generating Finite Difference Approximations

In this section, we want to develop formulas for *numerical differentiation* more systematically. Assume that we are given equidistant function values with step size $h = x_{k+1} - x_k$,

$$
\begin{array}{c|cccccc}
x & \cdots & -h & 0 & h & 2h & \cdots \\
\hline
f(x) & \cdots & f(-h) & f(0) & f(h) & f(2h) & \cdots
\end{array}.
$$

To compute approximations for the derivatives of f, we first compute an interpolating polynomial $P_n(z)$. Then we compute the derivatives of the polynomial and use them as approximations for the derivatives of the function f,

$$P_n^{(k)}(z) \approx f^{(k)}(z), \quad k = 1, 2, \ldots n.$$

We will compute the derivatives at some node $z = x_j = jh$. The change of variables $x' = x - jh$ is useful for computing the *discretization error* with the Taylor expansion for $x' = 0$. We obtain the following Maple procedure:

ALGORITHM 8.3.
Generating rules for numerical differentiation

```
FiniteDifferenceFormula:=proc(m,k,j)
# computes for m+1 equidistant points -jh, ..., -h, 0, h, ...,(m-j)h
# a finite difference approximation of the kth derivative evaluated
# at x=0
  local i, p, dp;
  p:=interp([seq(i*h,i=-j..(m-j))],[seq(f(i*h),i=-j..(m-j))],x);
  dp:=diff(p,x$k);
  simplify(eval(dp,x=0));
end:
```

The well-known symmetric difference formula for the second derivative (8.10) is computed by

```
> rule:=FiniteDifferenceFormula(2,2,1);
```

$$\text{rule} := \frac{f(-h) - 2\,f(0) + f(h)}{h^2}$$

To obtain the discretization error, we expand the difference

```
> err:=taylor((D@@2)(f)(0)-rule,h=0,8);
```

This yields

$$\text{err} := -1/12 \left(D^{(4)}\right)(f)(0)\,h^2 - \frac{1}{360}\left(D^{(6)}\right)(f)(0)\,h^4 + O\left(h^6\right).$$

Thus, we conclude

$$f''(0) = \frac{f(-h) - 2f(0) + f(h)}{h^2} - \frac{1}{12}f^{(4)}(0)h^2 + O(h^4).$$

It is now easy to reproduce the entire page 914 (*Coefficients for Differentiation*) given in the classic book of tables Abramowitz-Stegun [1] by this MAPLE procedure. To illustrate this, we compute for $k = 2$ (second derivative) and $m = 4$ (using 5 points) the coefficients for the approximation of the second derivative at the first, second and middle nodes and expand the error term

```
k:=2;
m:=4;
for j from 0 to m/2 do
  rule:=FiniteDifferenceFormula(m,k,j);
  err:=taylor(rule-(D@@k)(f)(0),h=0,m+3);
end do;
```

We obtain the following output, to which we have added the corresponding *stencil* (the double circle indicates where the second derivative is computed and below the grid points the weights are given for each node. They have to be divided by $12h^2$).

$$\text{rule} := -1/12\,\frac{-35\,f(0) + 104\,f(h) - 114\,f(2\,h) + 56\,f(3\,h) - 11\,f(4\,h)}{h^2}$$

$$\text{err} := \left(5/6\left(D^{(5)}\right)(f)(0)\,h^3 + \frac{119}{90}\left(D^{(6)}\right)(f)(0)\,h^4 + O\left(h^5\right)\right)$$

f''	35	-104	114	-56	11	$12h^2$

$$\text{rule} := 1/12\,\frac{11\,f(-h) - 20\,f(0) + 6\,f(h) + 4\,f(2\,h) - f(3\,h)}{h^2}$$

$$\text{err} := -1/12\left(D^{(5)}\right)(f)(0)\,h^3 - \frac{19}{360}\left(D^{(6)}\right)(f)(0)\,h^4 + O\left(h^5\right)$$

f''	11	-20	6	4	-1	$12h^2$

$$\text{rule} := 1/12 \, \frac{-f\left(-2\,h\right) + 16\,f\left(-h\right) - 30\,f\left(0\right) + 16\,f\left(h\right) - f\left(2\,h\right)}{h^2}$$

$$\text{err} := -\frac{1}{90}\left(D^{(6)}\right)(f)(0)\,h^4 + O\left(h^5\right)$$

$$f''\qquad -1\qquad 16\qquad -30\qquad 16\qquad -1\qquad \bigg|\qquad 12h^2$$

8.2.2　Discrete Operators for Partial Derivatives

The *Laplacian* operator in two space dimensions is defined as

$$\Delta u(x,y) = \frac{\partial^2 u}{\partial x^2} + \frac{\partial^2 u}{\partial y^2}.$$

For the discrete operator, we consider the function u on a grid with mesh size h,

$$u_{jk} := u(x_0 + jh, y_0 + kh).$$

If we approximate the second partial derivative with respect to x using (8.10), we get

$$\frac{\partial^2 u}{\partial x^2}(x_0, y_0) = \frac{1}{h^2}(u_{1,0} - 2u_{0,0} + u_{-1,0}) + O(h^2).$$

This centered approximation uses the point where the derivative is to be computed and the two neighboring points to the left and right of it. One often uses as graphical representation a *stencil* as shown in Figure 8.3.

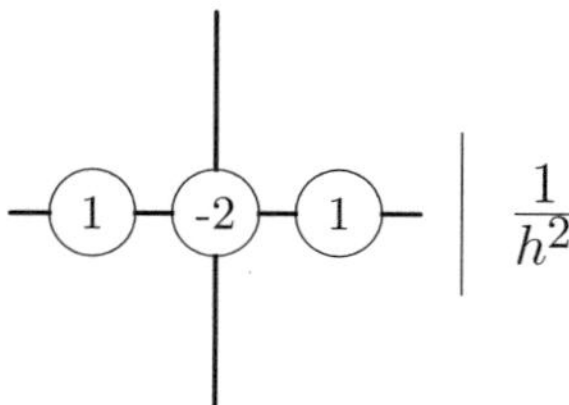

FIGURE 8.3. *Stencil for* $\frac{\partial^2 u}{\partial x^2}$

By discretizing in the same way the second partial derivative with respect to y, we obtain for the *Laplacian*

$$\Delta u(x_0, y_0) = \frac{\partial^2 u}{\partial x^2} + \frac{\partial^2 u}{\partial y^2} = \frac{1}{h^2}\left(u_{1,0} + u_{-1,0} + u_{0,1} + u_{0,-1} - 4u_{0,0}\right) + O(h^2).$$

$$(8.11)$$

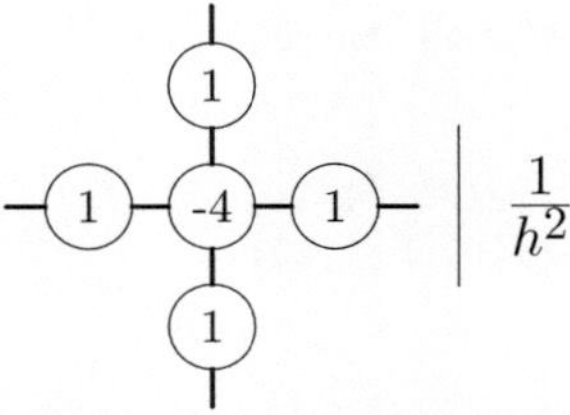

FIGURE 8.4. *Stencil for* $\Delta u(x,y) = \frac{\partial^2 u}{\partial x^2} + \frac{\partial^2 u}{\partial y^2}$

The corresponding stencil is displayed in Figure 8.4.

Next, we will discretize the operator with two second derivatives, $\frac{\partial^4 u}{\partial x^2 \partial y^2}$. In order to compute the second derivative with respect to y using the approximation (8.10), we need to know the partial second derivatives with respect to x at the points $(0,h), (0,0)$ and $(0,-h)$. We can approximate these values again using (8.10):

$$\frac{\partial^2 u}{\partial x^2}(0,h) = \frac{1}{h^2}(u_{-1,1} - 2u_{0,1} + u_{1,1}) + O(h^2), \tag{8.12}$$

$$\frac{\partial^2 u}{\partial x^2}(0,0) = \frac{1}{h^2}(u_{-1,0} - 2u_{0,0} + u_{1,0}) + O(h^2), \tag{8.13}$$

$$\frac{\partial^2 u}{\partial x^2}(0,-h) = \frac{1}{h^2}(u_{-1,-1} - 2u_{0,-1} + u_{1,-1}) + O(h^2). \tag{8.14}$$

Now, combining linearly the equations (8.12), (8.13) and (8.14) according to (8.10), we obtain

$$\frac{\partial^4 u}{\partial x^2 \partial y^2} = \frac{1}{h^4}(u_{1,1} + u_{-1,1} + u_{1,-1} + u_{-1,-1}$$
$$-2u_{1,0} - 2u_{-1,0} - 2u_{0,-1} - 2u_{0,1} + 4u_{0,0}) + O(h^2) \tag{8.15}$$

and the corresponding stencil is displayed in Figure 8.5.

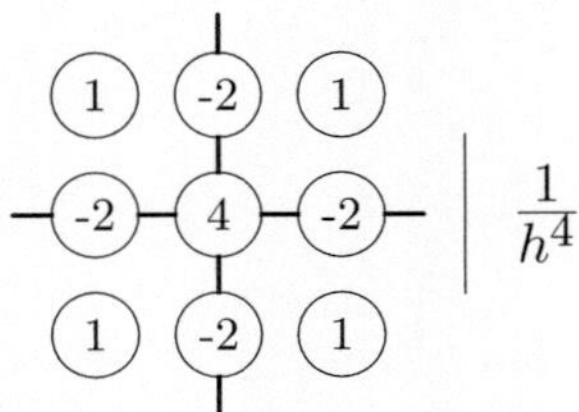

FIGURE 8.5. *Stencil for the operator* $\frac{\partial^4 u}{\partial x^2 \partial y^2}$

Note that the stencil matrix is computed by the outer product $\boldsymbol{v}\boldsymbol{v}^{\top}$ for

$$\boldsymbol{v} = \begin{pmatrix} 1 \\ -2 \\ 1 \end{pmatrix}.$$

For the operator $\frac{\partial^4 u}{\partial x^4}$ we will use 5 points. Using Algorithm 8.3, we get with the call

```
rule:=FiniteDifferenceFormula(5,4,2);
err:=taylor(rule-(D@@4)(f)(0),h=0,10);
```

the expressions

$$\text{rule:}= \frac{f\,(2\,h) - 4\,f\,(h) + 6\,f\,(0) + f\,(-2\,h) - 4\,f\,(-h)}{h^4},$$

$$\text{err:}= (1/6\,\left(D^{(6)}\right)(f)(0)\,h^2 + \tfrac{1}{80}\,\left(D^{(8)}\right)(f)(0)\,h^4 + O\left(h^6\right)),$$

which means

$$\frac{\partial^4 u}{\partial x^4}(0,0) = \frac{1}{h^4}\left(u_{-2,0} - 4u_{-1,0} + 6u_{0,0} - 4u_{1,0} + u_{2,0}\right) + O(h^2). \qquad (8.16)$$

We are now ready to compute the *biharmonic operator*

$$\Delta^2 u(x,y) = \frac{\partial^4 u}{\partial x^4} + 2\frac{\partial^4 u}{\partial x^2 \partial y^2} + \frac{\partial^4 u}{\partial y^4}$$

by combining (8.16), its analogue for y and (8.15). We obtain

$$\begin{aligned}
\frac{\partial^4 u}{\partial x^4} + 2\frac{\partial^4 u}{\partial x^2 \partial y^2} + \frac{\partial^4 u}{\partial y^4} \;=\; & \frac{1}{h^4}\left(20u_{0,0} - 8(u_{1,0} + u_{-1,0} + u_{0,1} + u_{0,-1})\right. \\
& + 2(u_{1,1} + u_{-1,1} + u_{1,-1} + u_{-1,-1}) \\
& \left. + u_{0,2} + u_{0,-2} + u_{2,0} + u_{-2,0}\right) + O(h^2).
\end{aligned}$$

$$(8.17)$$

The stencil for the biharmonic operator (8.17) is shown in Figure 8.6.

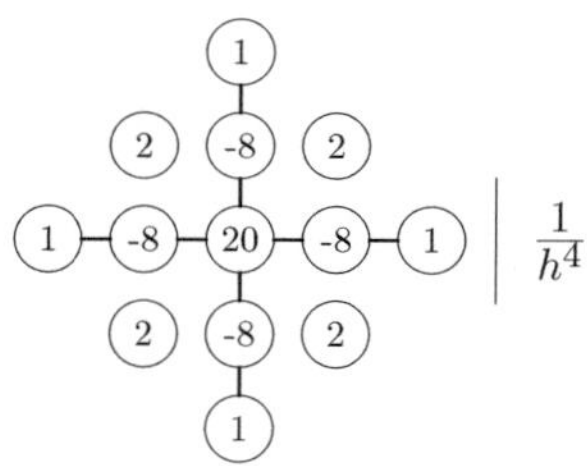

FIGURE 8.6.

Biharmonic operator: $\Delta^2 u(x,y) = \frac{\partial^4 u}{\partial x^4} + 2\frac{\partial^4 u}{\partial x^2 \partial y^2} + \frac{\partial^4 u}{\partial y^4}$

This stencil for the biharmonic operator can also be obtained differently, using the fact that the biharmonic operator is the Laplacian squared,

$$\Delta^2 u = \left(\frac{\partial^2}{\partial x^2} + \frac{\partial^2}{\partial y^2}\right)^2 u = \frac{\partial^4 u}{\partial x^4} + 2\frac{\partial^4 u}{\partial x^2 \partial y^2} + \frac{\partial^4 u}{\partial y^4}.$$

Thus, we can also compute the finite difference discretization of the biharmonic operator by computing the square of the finite difference stencil of the Laplacian, i.e. the product of the stencil with itself:

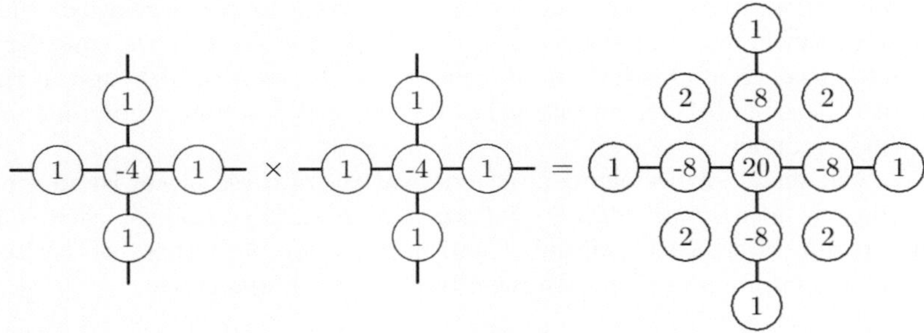

Proceeding similarly to the Kronecker product of matrices, one has to place one times the entire stencil to the left, the right, at the top and below, and subtract four times the stencil in the center, to obtain the stencil for the biharmonic operator.

8.3   Algorithmic Differentiation

Algorithmic differentiation or *automatic differentiation* is a technique for automatically augmenting computer programs, including very complex simulations, with statements for the computation of derivatives, also known as sensitivities. A good reference for this technique is the book by Griewank and Walther [58]. Software is available on the web-page `www-unix.mcs.anl.gov/autodiff/ADIFOR/`.

8.3.1   Idea Behind Algorithmic Differentiation

The key idea is that the results of a computer program that performs numerical computations are in fact functions of the input parameters. In order to simplify our exposition, we assume that the program begins by reading a value for the variable x and that, when the program terminates, the computed quantities y_1, y_2, ..., y_{10} are printed.

Assume for a moment that the computer can represent all real numbers in the range of computation $[-\text{realmax}, \text{realmax}]$. The results y_i are functions of x. Since the program P has to terminate in a finite number of steps and the computer can only perform the 4 basic operations $+, -, \times, /$, it is obvious that the results y_i must be piecewise rational functions of x:

$$y_i(x) = r_i(x), \quad i = 1, \ldots, 10.$$

One can differentiate these functions r_i piecewise, so the derivative exists almost everywhere. The situation is essentially the same if we consider a computer that computes only with a finite set of numbers, although in this case, it is possible to have an interval of the rational function containing only one single number.

We will show how we can compute derivatives of these piecewise rational functions by *algorithmic differentiation*. There exist compilers, as well as products for MATLAB in Matlab Central, that compute these derivatives without human intervention. We have to distinguish two cases:

1. The results y_i of program P are also in theory rational functions of x. In this case, algorithmic differentiation delivers the exact result.

2. The results y_i are theoretically not rational functions of x, but are only *approximated* (maybe very accurately) by rational functions. In this case, one has to analyze how well the derivative of the rational function approximates the correct derivative, and whether it can be used instead.

There are many problems for which the results are indeed rational functions of the input variables. Thus, it already pays to study algorithmic differentiation just for rational functions. In the second case, algorithmic differentiation is often superior to other approximations and therefore also useful. Admittedly, one has to be careful as we will show in the following example.

EXAMPLE 8.1. *We want to write a program that computes the exponential function $f(x) = e^x$ in the interval $[-0.1, 0.1]$. Using the partial sum*

$$g(x) = 1 + \frac{x}{1!} + \frac{x^2}{2!} + \frac{x^3}{3!}. \tag{8.18}$$

We have the approximation error

$$|g(x) - e^x| \leq 4.25 \cdot 10^{-6} \quad for \quad x \in [-0.1, 0.1],$$

which might be accurate enough for certain applications. If we compute the derivative of the rational function (8.18), which is just a polynomial in our example, we get

$$g'(x) = 1 + \frac{x}{1!} + \frac{x^2}{2!},$$

and hence for this derivative

$$|g'(x) - e^x| \leq 1.7 \cdot 10^{-4} \quad for \quad x \in [-0.1, 0.1],$$

which may or may not be accurate enough, since we might want to have the same precision as for the function value.

To better illustrate the relation between the exact derivative and the derivative of the rational function, we consider a program that computes the inverse function using a method for computing zeros. We are given the function f and a value x. We are looking for a value y such that $f(y) = x$, that is $y = f^{-1}(x)$. If we assume for simplicity that f is monotonically increasing in $[a, b]$ and if $f(a) < x$ and $f(b) > x$, then we can solve the equation with bisection (see Section 5.2.1).

EXAMPLE 8.2. *We compute the inverse function of $x = f(y) = y + e^y$ for tol = 0.1 using Algorithm 5.1. We get an implementation of the LambertW function:*

ALGORITHM 8.4.
LambertW function computed with bisection

```
function y=LambertW(x)
% LAMBERTW implementation of the LambertW function
%    y=LambertW(x) computes the solution of x=y+exp(y) using
%    bisection

f=@(y) y+exp(y);
tol=1e-1;
a=-2; b=2;
y=(a+b)/2;
while (b-a)>tol,
  if f(y)<x, a=y; else b=y; end;
  y=(a+b)/2;
end;
```

The bisection is terminated in Algorithm LambertW *when the interval becomes smaller than* tol= 0.1. *We have deliberately chosen a large value for* tol *in this example. When plotting the LambertW function, we notice that the approximation g of the inverse function is piecewise constant, see Figure 8.7. The plot was generated with the* MATLAB-*statements:*

```
x=[-2:0.01:2];
y=[];
for z=x
  y=[y LambertW(z)];
end
plot(x,y)
```

Choosing tol smaller makes the intervals smaller (this is also the case for the machine independent criterion) but the behavior remains the same. Because the rational function approximating the LambertW function is piecewise constant, algorithmic differentiation would give us here $g'(x) = 0$, which is completely incorrect. This is a situation where the piecewise rational function can approximate the desired function $f^{-1}(x)$ very well, but the derivative has nothing to do with the exact derivative.

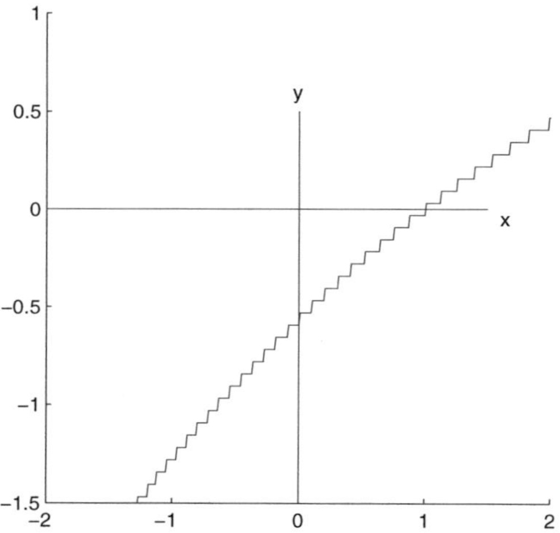

FIGURE 8.7. *LambertW function with bisection*

If we choose another algorithm for computing the zero, the resulting function g may approximate the derivative much better. Let us study the approximation we get when we apply Newton's method instead of bisection. Here we compute the sequence

$$y_{n+1} = y_n - \frac{f(y_n) - x}{f'(y_n)}, \quad n = 0, 1, 2, \ldots. \tag{8.19}$$

Every element y_n of the sequence is a function of x; in addition, we assume a constant initial guess, i.e., $y_0(x) = y_0 = \text{const}$. If we now differentiate the recurrence (8.19), we obtain

$$y'_{n+1} = y'_n - \frac{f'(y_n)\left(f'(y_n)y'_n - 1\right) - f''(y_n)y'_n\left(f(y_n) - x\right)}{f'(y_n)^2},$$

or, after some simplifications,

$$y'_{n+1} = \frac{1}{f'(y_n)} + \frac{f''(y_n)y'_n}{f'(y_n)^2}\left(f(y_n) - x\right). \tag{8.20}$$

Thus, y'_n satisfies another recurrence, with initial guess $y'_0 = 0$. Now, as $n \to \infty$, we have $y_n \to y$, where y is the solution of $f(y) = x$, so that $y = f^{-1}(x)$. Hence, (8.20) implies

$$\lim_{n \to \infty} y'_n = \frac{1}{f'(y)} = \frac{1}{f'\left(f^{-1}(x)\right)}. \tag{8.21}$$

The expression on the right hand side of (8.21) is indeed the derivative of the inverse function, as we can verify by differentiating the identity

$$f\left(f^{-1}(x)\right) = x.$$

We now consider the following function `LambertW2` to compute the sequence (8.19) for Example 8.2. We denote the derivative $f'(x) = 1 + e^x$ by `fs`.

ALGORITHM 8.5.
LambertW function computed with Newton's method

```
function y=LambertW2(x)
% LAMBERTW2 second implementation of the LambertW function
%    y=LambertW2(x) computes the solution of x=y+exp(y) using
%    Newton's method

f=@(y) y+exp(y); fs=@(y) 1+exp(y);
tol=1e-12;
ya=1; yn=0;
while abs(ya-yn)>=tol
  ya=yn;
  yn=ya-(f(ya)-x)/fs(ya);
end
y=yn;
```

To differentiate algorithmically the function `LambertW2` with respect to x, we have to introduce new variables, e.g. *yns* for the quantity *yn*. Furthermore, we need the function $f''(x) = e^x$, which we will denote by `fss`. The differentiated statement is inserted *before* the relating statement. Doing so we get the function `LambertW2P`:

ALGORITHM 8.6.
Algorithmic differentiated LambertW function

```
function [yn,yns]=LambertW2P(x)
% LAMBERTW2P compute LambertW function and its derivative
%    y=LambertW2P(x) computes the solution of x=y+exp(y) using
%    Newton's method, and also the derivative of the LambertW
%    function using algorithmic differentiation

f=@(y) y+exp(y); fs=@(y) 1+exp(y); fss=@(y) exp(y);
tol=1e-6;                          % enough because of extra iteration
ya=1; yns=0; yn=0;                 % where Newton doubles accuracy
while abs(ya-yn)>=tol
  yas=yns;ya=yn;
  yns=yas-(fs(ya)*(fs(ya)*yas-1)-fss(ya)*yas*(f(ya)-x))/fs(ya)^2;
  yn=ya-(f(ya)-x)/fs(ya);
```

```
end
yas=yns;ya=yn;                        % one extra iteration
yns=yas-(fs(ya)*(fs(ya)*yas-1)-fss(ya)*yas*(f(ya)-x))/fs(ya)^2;
yn=ya-(f(ya)-x)/fs(ya);
```

Note that when the termination criterion is met and thus `abs(ya-yn)<tol`
holds, we use `yn` as the "limit value". However, the value of the derivative
`yns` has been computed with the previous value of `yn` and, therefore, does
not have the same accuracy as `yn`. In order to achieve the same accuracy, it
is necessary to perform *one more iteration step* (*rule of Joss* [76]).

The following program computes the LambertW function and its derivative in two ways: by algorithmic differentiation and exactly by (8.21).

```
fs=@(x) 1+exp(x);
algdiff=[]; exact=[];
for x=-2:0.1:2
  [y,ys]=LambertW2P(x);
  algdiff=[algdiff; ys];
  exact=[exact; 1/fs(LambertW2(x))];
end
norm(algdiff-exact)

ans =
   3.2841e-16
```

We observe a perfect match of the values of the derivatives obtained by algorithmic differentiation with the "exact" ones. This example shows that the
derivative of the rational function that approximates a non-rational function
may indeed deliver reasonable and useful results.

A technique related to algorithmic differentiation, which has already been
used a long time ago, is to compute derivatives by differentiating recurrence
relations. This technique has been used, for instance, to compute derivatives
of orthogonal polynomials defined by three-term recurrence relations (see
Section 5.3.7).

8.3.2   Rules for Algorithmic Differentiation

The following rules can be used to introduce algorithmic differentiation systematically into any program that computes a function of the input variables:

1. Introduce new variables for the derivatives

2. Differentiate every statement according to the usual rules of calculus

$$
\begin{aligned}
(u \pm v)' &= u' \pm v', \\
(u \cdot v)' &= u'v + v'u, \\
\left(\frac{u}{v}\right)' &= \frac{u'v - v'u}{v^2}, \\
(u(v))' &= u'(v) \cdot v'.
\end{aligned}
$$

For the standard functions, use the usual rules like:

$$(\sin(u))' = \cos(u) \cdot u',$$
$$\text{if } u > 0 \text{ then } (abs(u))' = u \text{ else } (abs(u))' = -u.$$

3. Insert the differentiated statement *before* the statement to be differentiated.

EXAMPLE 8.3. *Evaluating a polynomial*

$$P(x) = p_1 x^{n-1} + p_2 x^{n-2} + \cdots p_{n-1}x + p_n$$

using Horner's scheme (see Section 5.3.3) is done using Algorithm 5.7:

```
y=0;
for i=1:n
  y=y*x+p(i);
end
```

With algorithmic differentiation we get

```
ys=0; y=0;
for i=1:n
  ys=ys*x+y;
  y=y*x+p(i);
end
```

We see that with the modified program we compute the first and second row of Horner's scheme. As shown in Section 5.3, the variable ys *is the exact derivative* $P'(x)$.

8.3.3 Example: Circular Billiard

We consider a *circular billiard table* and two balls located at the points P and Q, see Figure 8.8. In which direction must the ball at point P be hit, if it is to bounce off the boundary of the table exactly once and then hit the other ball located at Q? This problem has been discussed in [44], see also [28], where its solution is shown to be related to the caustic curves that appear inside partially illuminated coffee mugs.

The problem does not depend on the size of the circle. Therefore, without loss of generality, we may assume the radius of the table to be 1, i.e., we will consider the unit circle. Also, the problem remains the same if we rotate the table. Thus, we may assume that one ball (e.g. Q) is located on the x-axis.

The problem can now be stated as follows: In the unit circle, two arbitrary points $P = (p_x, p_y)$ and $Q = (a, 0)$ are given. We are looking for a reflection point $X = (\cos x, \sin x)$ (see Figure 8.9) on the circumference of the circle, such that a billiard ball traveling from P to X will hit Q after it bounces off

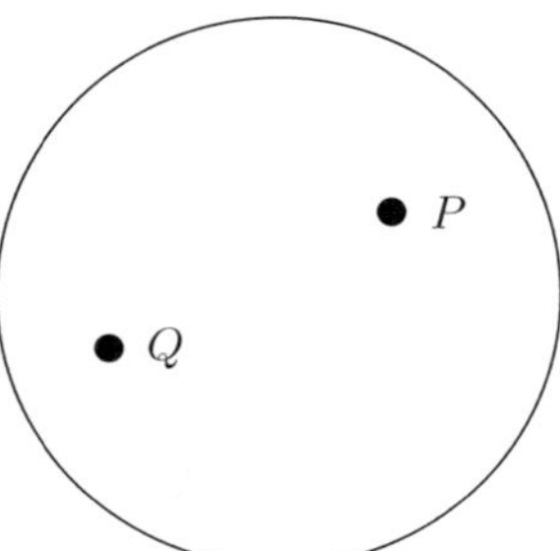

FIGURE 8.8. *Billiard table*

the edge. The problem is solved if we know the point X, which means that we are looking for the angle x.

The condition that must be satisfied is that the two reflection angles are equal, i.e., $\alpha_1 = \alpha_2$ in Figure 8.9. This is the case if the point X is the bisector of the angle QXP. Thus if, $\boldsymbol{e}_{XQ}$ is the unit vector in the direction XQ, and if $\boldsymbol{e}_{XP}$ is defined similarly, then the direction of the bisector is given by the sum $\boldsymbol{e}_{XQ} + \boldsymbol{e}_{XP}$. This vector must be orthogonal to the direction vector of the tangent g,

$$\boldsymbol{r} = \begin{pmatrix} \sin x \\ -\cos x \end{pmatrix}.$$

Therefore we obtain for the angle x the equation

$$f(x) = (\boldsymbol{e}_{XQ} + \boldsymbol{e}_{XP})^{\top} \boldsymbol{r} = 0. \tag{8.22}$$

Using MAPLE, we can give an explicit expression for the function f:

```
xp1:=px-cos(x);
xp2:=py-sin(x);
lp:=sqrt((xp1)^2+(xp2)^2);
ep1:=xp1/lp; ep2:=xp2/lp;              # unit vector XP

xq1:=a-cos(x);
xq2:=-sin(x);
lq:=sqrt((xq1)^2+(xq2)^2);
eq1:=xq1/lq; eq2:=xq2/lq;              # unit vector XQ

f:=(ep1+eq1)*sin(x)-(ep2+eq2)*cos(x);
```

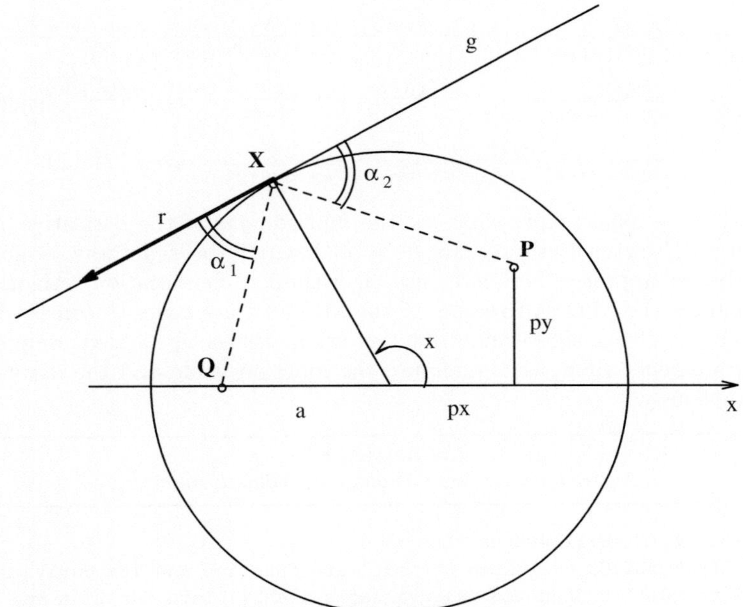

FIGURE 8.9. *Billiard problem*

Maple computes the function:

$$
f(x) := \left(\frac{px - \cos x}{\sqrt{(px - \cos x)^2 + (py - \sin x)^2}} + \frac{a - \cos x}{\sqrt{(a - \cos x)^2 + \sin x^2}} \right) \sin x
$$
$$
- \left(\frac{py - \sin x}{\sqrt{(px - \cos x)^2 + (py - \sin x)^2}} - \frac{\sin x}{\sqrt{(a - \cos x)^2 + \sin x^2}} \right) \cos x.
$$

$$(8.23)$$

We want to solve the equation $f(x) = 0$ for the unknown angle x. This can be done using Newton's method (see Equation (5.24) in Chapter 5). For this we need the derivative of f. The Maple statement `diff(f,x)` returns for f' the complicated expression

$$
\left(\frac{\sin(x)}{\sqrt{(px-\cos(x))^2+(py-\sin(x))^2}} - \frac{1}{2} \frac{(px-\cos(x))(2\,(px-\cos(x))\sin(x)-2\,(py-\sin(x))\cos(x))}{\left((px-\cos(x))^2+(py-\sin(x))^2\right)^{3/2}} + \right.
$$
$$
\left. \frac{\sin(x)}{\sqrt{(a-\cos(x))^2+(\sin(x))^2}} - \frac{1}{2} \frac{(a-\cos(x))(2\,(a-\cos(x))\sin(x)+2\,\sin(x)\cos(x))}{\left((a-\cos(x))^2+(\sin(x))^2\right)^{3/2}} \right) \sin(x)
$$
$$
+ \left(\frac{px-\cos(x)}{\sqrt{(px-\cos(x))^2+(py-\sin(x))^2}} + \frac{a-\cos(x)}{\sqrt{(a-\cos(x))^2+(\sin(x))^2}} \right) \cos(x)
$$

$$\left(\frac{\cos(x)}{\sqrt{(px-\cos(x))^2+(py-\sin(x))^2}} + \frac{1}{2} \frac{(py-\sin(x))(2\,(px-\cos(x))\sin(x)-2\,(py-\sin(x))\cos(x))}{\left((px-\cos(x))^2+(py-\sin(x))^2\right)^{3/2}} \right.$$

$$\left. + \frac{\cos(x)}{\sqrt{(a-\cos(x))^2+(\sin(x))^2}} - \frac{1}{2} \frac{\sin(x)(2\,(a-\cos(x))\sin(x)+2\,\sin(x)\cos(x))}{\left((a-\cos(x))^2+(\sin(x))^2\right)^{3/2}} \right) \cos(x)$$

$$+ \left(\frac{py-\sin(x)}{\sqrt{(px-\cos(x))^2+(py-\sin(x))^2}} - \frac{\sin(x)}{\sqrt{(a-\cos(x))^2+(\sin(x))^2}} \right) \sin(x).$$

Using these explicit expressions for the function f and the derivative f' to compute a Newton iteration step is probably not the right way to go. A much better approach here is to use algorithmic differentiation. Algorithm 8.7 contains the MATLAB version of the MAPLE statements to compute the function f. Using algorithmic differentiation, the newly added, differentiated statements allow us to compute the function value and the derivative simultaneously.

ALGORITHM 8.7.
Algorithmic differentiation of the Billiard function

```
function [f,fs]=BilliardFunction(x)
% BILLIARDFUNCTION evaluates the billiard function and its derivative
%    [f,fs]=BilliardFunction(x) evaluates the billiard function and
%    its derivative at x using algorithmic differentiation

global px py a
cs=-sin(x);    c=cos(x);
ss=cos(x);     s=sin(x);
xp1s=-cs;      xp1=px-c;
xp2s=-ss;      xp2=py-s;
xq1s=-cs;      xq1=a-c;
xq2s=-ss;      xq2=-s;
hs=(xp1*xp1s+xp2*xp2s)/sqrt(xp1^2+xp2^2);  h=sqrt(xp1^2+xp2^2);
ep1s=(h*xp1s-xp1*hs)/h^2;                  ep1=xp1/h;
ep2s=(h*xp2s-xp2*hs)/h^2;                  ep2=xp2/h;
hs=(xq1*xq1s+xq2*xq2s)/sqrt(xq1^2+xq2^2);  h=sqrt(xq1^2 + xq2^2);
eq1s=(h*xq1s-xq1*hs)/h^2;                  eq1=xq1/h;
eq2s=(h*xq2s-xq2*hs)/h^2;                  eq2=xq2/h;
fs=(ep1s+eq1s)*s+(ep1+eq1)*ss-(ep2s+eq2s)*c-(ep2+eq2)*cs;
f=(ep1+eq1)*s-(ep2+eq2)*c;
```

We can now use Newton's algorithm to compute a zero of the function f. With the ball positions $P = (0.5, 0.5)$ and $Q = (-0.6, 9)$ and the initial approximation for the angle $x = 0.8$, we obtain with the following program the reflection point $X = (0.5913, 0.8064)$:

```
global px py a
px=0.5; py=0.5; a=-0.6;                    % ball positions
tol=1e-9; h=2*tol; x=0.8;
while abs(h)>tol
```

```
  [y,ys]=BilliardFunction(x);
  h=-y/ys; x=x+h;                          % Newton step
end
angle=x
X=[cos(x) sin(x)]

angle =
    0.9381
X =
    0.5913     0.8064
```

Using other starting values x_0 for the angle, we get four solutions for the angle x and the reflection point $X = (X_1, X_2)$, which are displayed in Table 8.1. The fact that we obtain four solutions can be explained by a transformation

x_0	x	X_1	X_2
0.8	0.9381	0.5913	0.8064
1.6	2.2748	-0.6473	0.7623
3.0	2.7510	-0.9247	0.3808
5.0	5.0317	0.3139	-0.9494

TABLE 8.1. *Solutions for the billiard problem*

of the billiard equation into a fourth degree polynomial equation, see Problem 8.7. The number of real roots can then be related to the caustic curve one sees in a coffee mug, by imagining light rays entering the coffee mug from above, reflected like a billiard ball off the wall of the coffee mug, and arriving at the bottom of the mug. Moreover, the number of solutions are related to the light intensity, see [28].

The trajectories corresponding to the four solutions are displayed in Figure 8.10.

8.3.4   Example: Nonlinear Eigenvalue Problems

In this section, we will give another application of algorithmic differentiation. We follow the presentation given in [42]. Consider a matrix $C(\lambda)$ whose elements are functions of a parameter λ. In the case of an ordinary eigenvalue problem, we have $C(\lambda) = \lambda I - A$. A quadratic eigenvalue problem is defined by

$$\det(C(\lambda)) = 0, \quad \text{with} \quad C(\lambda) = \lambda^2 M + \lambda H + K.$$

If $\det(M) \neq 0$, it is possible to write an equivalent linear eigenvalue problem as follows: $\det(C(\lambda)) = 0$ means there exists a vector $\boldsymbol{v} \neq \boldsymbol{0}$ such that

$$C(\lambda)\boldsymbol{v} = \lambda^2 M\boldsymbol{v} + \lambda H\boldsymbol{v} + K\boldsymbol{v} = \boldsymbol{0}. \tag{8.24}$$

Introducing the vector $\boldsymbol{w} := \lambda\boldsymbol{v}$, we see that (8.24) has a non-zero solution if and only if the system

$$\begin{aligned} M\boldsymbol{w} &= \lambda M\boldsymbol{v}, \\ \lambda M\boldsymbol{w} &= -\lambda H\boldsymbol{v} - K\boldsymbol{v}, \end{aligned}$$

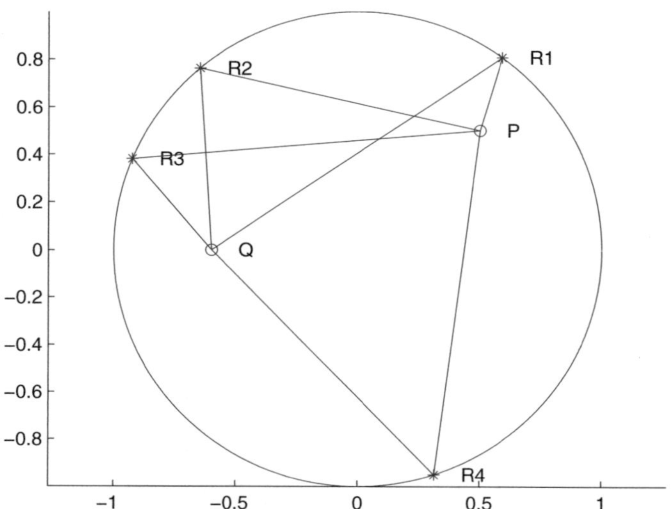

FIGURE 8.10. *Solutions for the billiard problem*

has a non-zero solution; this leads to the equivalent generalized eigenvalue problem with dimension $2n$

$$\det\left(\begin{bmatrix} M & 0 \\ 0 & K \end{bmatrix} - \lambda \begin{bmatrix} 0 & M \\ -M & -H \end{bmatrix}\right) = 0. \tag{8.25}$$

Instead of solving this equivalent linear eigenvalue problem, we will consider an alternative approach, which computes the zeros of $\det(C(\lambda))$ using a Newton iteration.

A good way to *compute the determinant* of a matrix is via its LU decomposition (see Exercise 3.15 in Chapter 3). The idea is to decompose the matrix C using Gaussian elimination, and then to compute the determinant of the factors:

$$C = PLU \quad \Longrightarrow \quad \det(C) = \det(P)\det(L)\det(U) = \pm 1 \prod_{k=1}^{n} u_{kk},$$

since the determinant of the permutation matrix P is ± 1 and the determinant of the unit lower triangular matrix L is 1. Algorithm 8.8 computes the determinant of a matrix:

ALGORITHM 8.8.
Computing a Determinant by Gaussian Elimination

```
function f=Determinant(C)
% DETERMINANT computes the determinant of a matrix
%   f=Determinant(C) computes the determinant of C via
%   LU-decomposition with partial pivoting

n=length(C);
f=1;
for i=1:n
  [cmax,kmax]=max(abs(C(i:n,i)));
    if cmax==0                                   % Matrix singular
      error('Matrix is singular')
    end
    kmax=kmax+i-1;
    if kmax~=i
      h=C(i,:); C(i,:)=C(kmax,:); C(kmax,:)=h;
      f=-f;
    end
    f=f*C(i,i);                                  % elimination step
    C(i+1:n,i)=C(i+1:n,i)/C(i,i);
    C(i+1:n,i+1:n)=C(i+1:n,i+1:n)-C(i+1:n,i)*C(i,i+1:n);
end
```

We would like to use Newton's method to compute the zeros of $P(\lambda) = \det(C(\lambda)) = 0$. For this, we need the derivative $P'(\lambda)$, which can be computed by *algorithmic differentiation* as discussed before, i.e., by differentiating each statement of the program that computes $P(\lambda)$. For instance, the statement that updates the determinant `f=f*C(i,i);` will be preceded by the statement for the derivative, i.e.

```
fs=fs*C(i,i)+f*Cs(i,i); f=f*C(i,i);
```

where we used the variable `Cs` for the matrix $C'(\lambda)$ and `fs` for the derivative of the determinant.

For larger matrices, however, there is the danger that the value of the determinant over- or underflows. To avoid this, notice that Newton's method does not require the values $f = \det(C(\lambda))$ and $fs = \frac{d}{d\lambda}\det(C(\lambda))$ individually; it is sufficient to compute the *ratio*

$$\frac{P(\lambda)}{P'(\lambda)} = \frac{f}{fs}.$$

Overflows can be reduced by working with the logarithm. Thus, instead of computing `f=f*C(i,i)` we can compute `lf=lf+log(C(i,i))`. It is even better to compute the derivative of the logarithm

$$lfs := \frac{d}{d\lambda}\log(f) = \frac{fs}{f},$$

which directly yields the *inverse Newton correction*.

Thus, instead of updating the logarithm $lf = lf + \log(c_{ii})$, we directly compute the derivative

$$lfs = lfs + \frac{(cs)_{ii}}{c_{ii}}.$$

These considerations lead to Algorithm 8.9 below.

ALGORITHM 8.9.
Computing the Determinant and the Derivative

```
function ffs=DeriveDeterminant(C,Cs)
% DERIVEDETERMINANT Newton correction for Determinant root finding
%    ffs=DeriveDeterminant(C,Cs) computes for the two given matrices
%    C(lambda) and its derivative C'(lambda), lambda being a real
%    parameter, the Newton correction ffs=f/fs for the determinant
%    function f(C(lambda))=0.

n=length(C); lfs=0;
for i=1:n
  [cmax,kmax]=max(abs(C(i:n,i)));
  if cmax==0                                   %  Matrix singular
    error('Matrix is singular')
  end
  kmax=kmax+i-1;
  if kmax~=i
    h=C(i,:); C(i,:)=C(kmax,:); C(kmax,:)=h;
    h=Cs(kmax,:); Cs(kmax,:)=Cs(i,:); Cs(i,:)=h;
  end
  lfs=lfs+Cs(i,i)/C(i,i);
  Cs(i+1:n,i)=(Cs(i+1:n,i)*C(i,i)-Cs(i,i)*C(i+1:n,i))/C(i,i)^2;
  C(i+1:n,i)=C(i+1:n,i)/C(i,i);
  Cs(i+1:n,i+1:n)=Cs(i+1:n,i+1:n)-Cs(i+1:n,i)*C(i,i+1:n)- ...
                  C(i+1:n,i)*Cs(i,i+1:n);
  C(i+1:n,i+1:n)=C(i+1:n,i+1:n)-C(i+1:n,i)*C(i,i+1:n);
end
ffs=1/lfs;
```

Consider the mass-spring system example from [137]. The matrix for the non-overdamped case is

$$C(\lambda) = \lambda^2 M + \lambda H + K \tag{8.26}$$

with

$$M = I, \ H = \tau\,\mathrm{tridiag}(-1,3,-1), \ K = \kappa\,\mathrm{tridiag}(-1,3,-1)$$

and with $\kappa = 5$, $\tau = 3$ and $n = 50$.

The following MATLAB script computes 3 eigenvalues. We use different initial values for Newton's iteration and converge to different eigenvalues. The results are given in Table 8.2.

```
n=50; tau=3; kappa=5;
e=ones(n-1,1);
H=diag(-e,-1)+diag(-e,1)+3*eye(n);
K=kappa*H; H=tau*H;
for lam=[1, -1.7-1.6*i, -2-i]
  ffs=1;
  while abs(ffs)>1e-14
    C=lam*(lam*eye(n)+H)+K; Cs=2*lam*eye(n)+H;
    ffs=DeriveDeterminant(C,Cs);
    lam=lam-ffs;
  end
  lam, C=lam*(lam*eye(n)+H)+K; det(C)
end
```

initial value	Eigenvalue λ	$f(\lambda) = \det(Q)$
1	-1.9101	$1.6609e{-}18$
$-1.7 - 1.6i$	$-1.5906 - 1.6649i$	$7.6365e21 - 1.3528e22i$
$-2 - i$	$-2.0142 - 1.7035e{-}38i$	$3.7877e{-}13 + 3.1863e{-}35i$

TABLE 8.2.
Three Eigenvalues of the Mass-Spring System

Notice the function value of $f(\lambda) = \det(Q)$ for the second (complex) eigenvalue: instead of being close to zero, it is huge! Did our computation of an approximate zero fail? Let us examine what happened by looking at function values of $f(\lambda)$ at neighboring points λ. We stop the program above after the second step, and then execute the MATLAB commands

```
L=[real(lam)*(1+10*eps)+1i*imag(lam)       % grid around lambda
real(lam)*(1-10*eps)+1i*imag(lam)
real(lam)+1i*imag(lam)*(1+10*eps)
real(lam)+1i*imag(lam)*(1-10*eps)
real(lam)*(1+10*eps)+1i*imag(lam)*(1+10*eps)
real(lam)*(1+10*eps)+1i*imag(lam)*(1-10*eps)
real(lam)*(1-10*eps)+1i*imag(lam)*(1+10*eps)
real(lam)*(1-10*eps)+1i*imag(lam)*(1-10*eps)
lam];
for k=1:9                                   % evaluate determinant
  Q=L(k)*(L(k)*eye(n)+H)+K;
  Qd(k)=det(Q)
end
QdIm=[Qd(7) Qd(3) Qd(5)                      % create image
      Qd(2) Qd(9) Qd(1)
      Qd(8) Qd(4) Qd(6)]
figure(1)
```

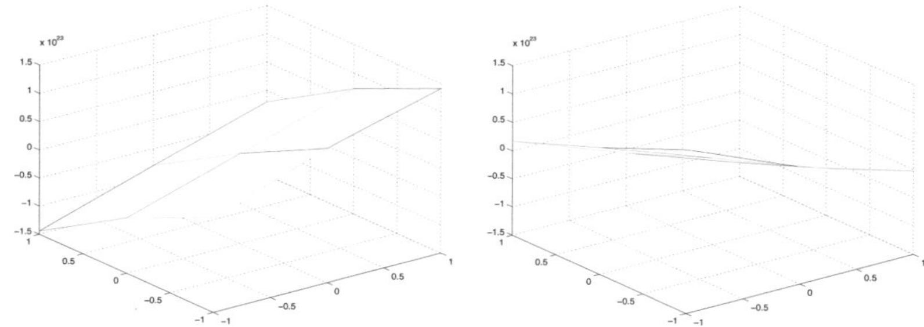

FIGURE 8.11.
Real part of the value of the determinant around
$\lambda = -1.5906 - 1.6649i$ *on the left, and imaginary part on
the right*

```
mesh(-1:1,-1:1,real(QdIm));
figure(2)
mesh(-1:1,-1:1,imag(QdIm));
```

We obtain the two figures shown in 8.11. We clearly see that there is a zero
in the area found by our program, but in the vicinity the function is huge.
This is because a polynomial of degree 100 grows very rapidly, so the machine
number next to the zero may very well have a function value of 10^{22}!

Note that the computational effort for solving the quadratic eigenvalue
problem by computing the determinant as described here is enormous. However, modern laptop computers are very powerful and the proposed way is numerically sound; consequently, this method can still be used to solve medium-
sized nonlinear eigenvalue problems, despite the large number of operations
(see Problem 8.8). One could also execute such a program in parallel, once
for each eigenvalue.

8.4   Problems

PROBLEM 8.1. *Solve the discretized Brachystochrone problem (8.7) using
Newton's method. Then find the value of B such that the solution of the
Brachystochrone problem is monotonically decreasing.*

PROBLEM 8.2. *When deriving the discrete operator (8.15) we first com-
puted approximations for $\frac{\partial^2 u}{\partial x^2}$ and then combined them for $\frac{\partial^4 u}{\partial x^2 \partial y^2}$. Convince
yourself that we get the same expression by interchanging the role of x and y.*

PROBLEM 8.3. *In [1], page 883, we also find formulas for numerical
differentiation where the derivative is not evaluated at an interpolation point
but at $x = ph$ with variable p. Change Algorithm 8.3 to a* MAPLE *function*

```
FiniteDifferenceFormulaP:=proc(m,k,j)
# computes for m+1 equidistant points -jh, ..., -h, 0, h, ...,(m-j)h
# a finite difference approximation of the kth derivative evaluated
# at x=p*h
```

and reproduce the formulas 25.3.4, 25.3.5 and 25.3.6 of [1].

PROBLEM 8.4. *Derive the 9-point Laplacian with error term.* **Hint:** *Use 5 points for the second derivatives.*

PROBLEM 8.5. *Derive the operator for the discrete biharmonic operator with error term $O(h^4)$.*

PROBLEM 8.6. *Write a* MATLAB *program and compute the solution to the billiard problem. Fix the position of the point P (e.g. choose $P = (0.5, 0.5)$) and let the ball Q move on the x-axis in steps h from -1 to 1 Compute the solutions for each position of Q. Display them as in Figure 8.10 and plot also the function f in another graphical window.*

PROBLEM 8.7. *Consider Equation (8.23) for the billiard problem. This is a goniometric equation (the unknown appears only as argument of trigono-metric functions). Such equations can be "rationalized" by introducing the new variable $t = \tan(x/2)$ and replacing*

$$\sin(x) = \frac{2t}{1+t^2}, \quad \cos(x) = \frac{1-t^2}{1+t^2}.$$

Make this change of variable in MAPLE *and generate the new equation for t. Although the new equations looks more complicated,* MAPLE *is able to solve it. Compute the solutions this way and interpret the result.*

PROBLEM 8.8. *Compute for $n = 50$ all the eigenvalues of the quadratic eigenvalue problem $\det(C(\lambda)) = 0$ with C defined by (8.26). Hints:*

1. *Use Algorithm 8.9 to compute the zeros of the determinant by Newton's method.*

2. *Start the iteration with a complex number, since some of the eigenvalues are complex.*

3. *When you have computed an eigenvalue λ_k, store it and* suppress *it to avoid recomputing it. Suppression leads to the Newton-Maehly iteration, see (5.78) in Section 5.3.*

4. *After suppressing λ_k, use some nearby value as starting point for the next Newton iteration, e.g. $\lambda_k(1 + 0.01i)$.*

5. *Finally, plot the eigenvalues as points in the complex plane.*

Compare your results with those obtained from `eig` *with the equivalent linear problem (8.25).*

Chapter 9. Quadrature

Prerequisites: Interpolation and extrapolation (§4.2, 4.2.8) are required. For
Gauss quadrature (§9.3), we also need the Newton–Maehly method (§5.3.7)
and eigenvalue solvers from Chapter 7.

In the context of scientific computing, quadrature means the approximate
evaluation of a *definite integral*

$$I = \int_a^b f(x)\, dx.$$

The word "quadrature" stems from a technique used by ancient Greeks to
measure areas in the plane by transforming them with straight edge and
compass into a square using area-preserving transformations. Finding an
area-preserving transformation that maps a circle into a square (*quadrature
of the circle*) became a famous unsolved problem until the 19th century, when
it was proved using Galois theory that the problem cannot be solved using

W. Gander et al., *Scientific Computing - An Introduction using Maple and MATLAB*,
Texts in Computational Science and Engineering 11,
DOI 10.1007/978-3-319-04325-8_9,
© Springer International Publishing Switzerland 2014

straight edge and compass. Since the definite integral measures the area between the graph of a function and the x-axis, it is natural to use the word "quadrature" to denote the approximation of this integral.

There are now very good algorithms to approximate definite integrals, and we will see two programs in this chapter that are used as a basis for the quadrature functions in MATLAB. But such algorithms cannot be completely foolproof, see the last quote above; we illustrate this with a well-known argument due to Kahan in Section 9.1. We also compare symbolic integration, which has made tremendous progress over the last decades, to numerical approximations of integrals. We then show in Section 9.2 the classical Newton-Cotes formulas, which one can easily derive nowadays with the help of MAPLE. We also present the Euler-Maclaurin summation formula and Romberg integration. Section 9.3 contains the major innovation of Gauss, which is to use the location of the quadrature nodes in addition to the weights to obtain more accurate quadrature formulas, which now bear his name. We explain the relation of these Gauss quadrature formulas to the theory of orthogonal polynomials, and give two highly efficient algorithms to compute such quadrature rules, the first using the more classical techniques of root finding, and the second due to Golub and Welsch. More information can be found in the standard reference on orthogonal polynomials and quadrature [48]. As for many numerical algorithms, adaptivity is a key ingredient, and this is the topic of Section 9.4, where we develop two locally adaptive quadrature algorithms. They are used as basis for the quadrature function `quad` and `quadl` in MATLAB and one of them is also used in GNU Octave.[1]

9.1 Computer Algebra and Numerical Approximations

If $F(x)$ is an antiderivative of $f(x)$, then

$$I = \int_a^b f(x)\, dx = F(b) - F(a).$$

Unlike computing derivatives, finding antiderivatives is often an art. One has to lookup tables or use various tricks and transformations like integration by parts

$$\int_a^b f'(x)g(x)\, dx = f(x)g(x)\Big|_a^b - \int_a^b f(x)g'(x)\, dx$$

to obtain a result. In addition, an explicit expression for the antiderivative often does not exist, e.g., for

$$F(x) = \int_0^x e^{-t^2}\, dt. \tag{9.1}$$

---

[1] http://www.gnu.org/software/octave/

Today, computer algebra systems help us a great deal in finding antiderivatives and have replaced integration tables as a tool for this purpose. Such systems use sophisticated algorithms (for instance the *Risch Algorithm*, see [12]) to find out if an antiderivative exists for certain classes of functions. In MAPLE, the function `int` is used for computing integrals. For the integral (9.1) we obtain the antiderivative expressed in terms of the error function:

```
int(exp(-x^2),x);
```

$$1/2\,\sqrt{\pi}\,erf(x),$$

and for the definite integral $\int_0^\infty e^{-x^2}\ln x\,dx$

```
int(exp(-x^2)*ln(x),x=0..infinity);
```

we get the expression

$$-1/4\,\sqrt{\pi}\,(\gamma + 2\,\ln(2)),$$

which may give us more insight than just the number $-.8700577268$.[2] On the other hand, even if an analytical expression for the antiderivative exists, it may be very complicated (several pages long) and thus be of little help for numerical calculations: try e.g.

```
int(exp(-x)*cos(6*x)^5*sin(5*x)^6,x);
```

From calculus, we know that a *Riemann sum* can be used to approximate a definite integral. To form a Riemann sum, we partition the interval $[a, b]$ into n subintervals $a = x_0 < x_1 < \cdots < x_n = b$ and approximate

$$I = \int_a^b f(x)\,dx \approx I_n := \sum_{i=1}^n w_i f(\xi_i), \tag{9.2}$$

with $w_i = x_i - x_{i-1}$ and $\xi_i \in [x_{i-1}, x_i]$. For example, we can choose $\xi_i = x_i$ as shown in Figure 9.1. By refining the subdivision of the interval of integration (or by letting $n \to \infty$) one can prove (e.g., for continuous functions) that $\lim_{n\to\infty} I_n = I$. Every rule for numerical quadrature has the form (9.2), but for practical computations, we can only use a finite number of function values to approximate the limit. The key problem in numerical quadrature is to determine the *nodes* ξ_i and the *weights* w_i to obtain the best possible approximation of an integral I with the least amount of work.

Because we cannot do better than invent clever "Riemann sums", we will never succeed in writing a program that will integrate every function correctly. W. Kahan gave the following argument: given a quadrature program, say `int(f,a,b)`, we first integrate the constant function $f(x) \equiv 1$ say on $[0, 1]$ and save the nodes ξ_k that `int` used for obtaining hopefully the result

---

[2] It appears that newer versions of MAPLE can no longer compute this integral. However, Wolfram Alpha computes the result when given the command `integrate exp(-x^2)*ln(x) dx from x=0 to infinity`.

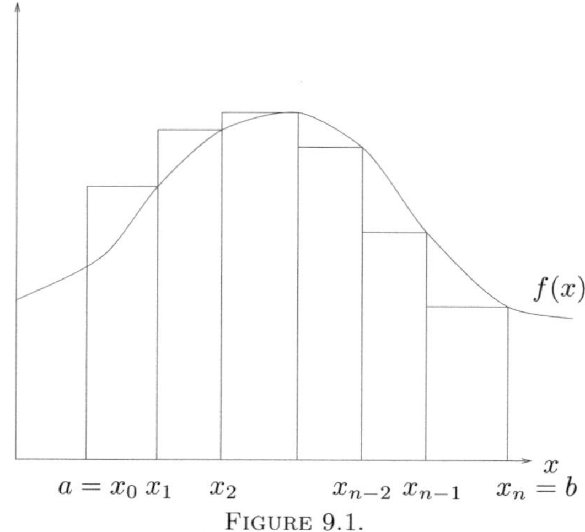

FIGURE 9.1.

The Riemann sum as an approximation to the integral

$I_n = 1$. Then we construct the function (with some appropriate constant $M > 0$)

$$g(x) = 1 + M \prod_{k=1}^{n} (x - \xi_k)^2.$$

and compute `int(g,0,1)`. Because $g(\xi_k) = 1$ we will obtain again the result $I_n = 1$, which will be wrong this time. This argument may not be valid if the program chooses different nodes ξ_k in every run. However, the point we want to make is that by computing a finite sum (9.2), we are sampling the function at a finite set of points. The resulting approximation can be wrong because the sampling may easily miss a large thin peak of the function that makes an essential contribution to the integral. Plot for instance the function

$$f(x) = \sin(x) + \frac{10^{-14}}{x - .54159265358979}$$

for the interval $[0, 1]$. It will look like $\sin(x)$, even though the integral $\int_0^1 f(x)\, dx$ does not exist.

The value for the integral computed by a Riemann sum can be interpreted as follows: we approximate the function f by *a piecewise constant* function

$$g(x) = f(\xi_i), \quad x_{i-1} \le x \le x_i, \quad i = 1, \ldots, n$$

and compute

$$\int_a^b g(x)\, dx \approx \int_a^b f(x)\, dx.$$

Any other approximation of f by a simpler function g which can be integrated analytically will lead to a quadrature rule.

9.2 Newton–Cotes Rules

One obvious way to approximate f is by polynomial interpolation. Here the nodes ξ_i, $i = 0, \ldots, n$ are supposed to be given. We only need to compute the weights. We construct the interpolation polynomial $P_n(x)$ for the data

$$
\begin{array}{c|ccc}
x & \xi_0, & \ldots, & \xi_n \\
\hline
f(x) & y_0, & \ldots, & y_n
\end{array}
$$

and by integration, obtain a quadrature rule:

$$
\int_a^b P_n(x)\, dx = \sum_{i=0}^n w_i y_i.
$$

For example, if we take $n = 1$ and as nodes the two end points $\xi_1 = a$ and $\xi_2 = b$, then the interpolating polynomial is of degree one,

$$
P_1(x) = \frac{y_2 - y_1}{b - a}(x - b) + y_2,
$$

and we obtain the *Trapezoidal Rule*

$$
\int_a^b P_1(x)\, dx = \frac{h}{2}(y_1 + y_2), \quad h = b - a. \tag{9.3}
$$

In general for the distinct nodes ξ_i, $i = 0, \ldots, n$ we can represent the interpolating polynomial using Lagrange interpolation (see Chapter 4)

$$
P_n(x) = \sum_{i=0}^n l_i(x) y_i, \quad l_i(x) = \prod_{\substack{j=0 \\ j \neq i}}^n \frac{x - \xi_j}{\xi_i - \xi_j}, \quad i = 0, \ldots, n.
$$

Integrating P_n we obtain *Newton–Cotes Quadrature Rules*

$$
\int_a^b P_n(x)\, dx = \sum_{i=0}^n w_i y_i, \quad \text{with } w_i = \int_a^b l_i(x)\, dx. \tag{9.4}
$$

The weights are the integrals of the Lagrange polynomials. In the literature, Newton–Cotes rules are always computed for equidistant nodes, but we do not need to restrict ourselves to the equidistant case. Thus we will refer to Newton–Cotes quadrature rules *when the nodes are prescribed* and the rule is obtained by integrating the interpolating polynomial.

With a computer algebra system like MAPLE, we can generate such formulas with just a few statements: there exists a function `interp` for computing the interpolating polynomial. Furthermore, integrals of polynomials can be computed exactly with every computer algebra system.

As an example, let us compute the classical *Newton–Cotes rules* for *equidistant* nodes of which two kinds are discussed in the literature: *closed* and *open* rules. If the interval (a, b) is divided into sub-intervals of length $h = (b - a)/n$, then the closed rules use the nodes $\xi_i = ih$ for $i = 0, 1, \ldots, n$. The open rules use only the interior points $\xi_i = ih$ for $i = 1, \ldots, n - 1$:

$$\int_0^{nh} f(x)\, dx \approx \begin{cases} \displaystyle\sum_{i=0}^{n} w_i\, f(ih) & \text{closed rule,} \\ \displaystyle\sum_{i=1}^{n-1} w_i\, f(ih) & \text{open rule.} \end{cases}$$

With the function

```
ClosedNewtonCotesRule:=n->factor(int(interp([seq(i*h,i=0..n)],
                   [seq(f(i*h),i=0..n)],z),z=0..n*h)):
```

we can thus generate the *Trapezoidal Rule* $(n = 1)$, *Simpson's Rule* $(n = 2)$, *Boole's Rule* $(n = 4)$ and the *Weddle Rule* $(n = 6)$:

```
TrapezoidalRule:=ClosedNewtonCotesRule(1);
```

$$\frac{h}{2}\left(f(0) + f(h)\right)$$

```
SimpsonsRule:=ClosedNewtonCotesRule(2);
```

$$\frac{h}{3}\left(f(0) + 4\,f(h) + f(2\,h)\right)$$

```
MilneRule:=ClosedNewtonCotesRule(4);
```

$$\frac{2}{45}\,h\left(7\,f(0) + 32\,f(h) + 12\,f(2\,h) + 32\,f(3\,h) + 7\,f(4\,h)\right)$$

```
WeddleRule:=ClosedNewtonCotesRule(6);
```

$$\frac{1}{140}\,h\,\big(41\,f(0) + 216\,f(h) + 27\,f(2\,h) + 272\,f(3\,h) \\ +\, 27\,f(4\,h) + 216\,f(5\,h) + 41\,f(6\,h)\big)$$

For $n = 8$ we obtain the following equidistant 9-point rule which is the first Newton–Cotes quadrature rule with negative weights:

```
ClosedNewtonCotesRule(8);
```

$$\frac{4}{14175}\, h\, (989\, f(0) + 5888\, f(h) - 928\, f(2\,h) + 10496\, f(3\,h)$$
$$-4540\, f(4\,h) + 10496\, f(5\,h) - 928\, f(6\,h) + 5888\, f(7\,h) + 989\, f(8\,h))$$

Negative weights are not favored in quadrature rules because they may lead to unnecessary loss of accuracy due to cancellation.

Open equidistant Newton–Cotes rules can be obtained similarly with the function:

```
OpenNewtonCotesRule:=n->factor(int(interp([seq(i*h,i=1..n-1)],
                      [seq(f(i*h),i=1..n-1)],z),z=0..n*h)):
```

For $n = 2, 3, 4$ and 5 we obtain the rules

$$\int_0^{2h} f(x)\, dx \;=\; 2hf(h) \qquad \text{mid-point rule,}$$

$$\int_0^{3h} f(x)\, dx \;=\; \frac{3h}{2}(f(h) + f(2h)),$$

$$\int_0^{4h} f(x)\, dx \;=\; \frac{4h}{3}(2f(h) - f(2h) + 2f(3h)),$$

$$\int_0^{5h} f(x)\, dx \;=\; \frac{5h}{24}(11f(h) + f(2h) + f(3h) + 11f(4h)).$$

We observe that the 3-point open rule already has a negative weight; the same is true for five or more points. Hence, these rules should be used with caution.

A second kind of open Newton–Cotes rules, which are not so popular in the literature, can also be derived easily. Such open rules are handy for integrating functions with a singularity at the end points, such as $\int_0^1 1/\sqrt{x}\, dx$.

Consider the same partition of the integration interval into n subintervals as for the closed rule: $x_i = ih$, $i = 0, \ldots, n$. This time, our nodes will be the *midpoints* of the intervals,

$$\xi_i = \frac{x_{i-1} + x_i}{2} = \left(i - \frac{1}{2}\right) h, \quad i = 1, \ldots, n. \tag{9.5}$$

Using the function

```
MidpointOpenNewtonCotesRule:=n->factor(int(interp([seq((i-1/2)*h,
             i=1..n)], [seq(f((i-1/2)*h),i=1..n)],z),z=0..n*h)):
```

we obtain this time open Newton–Cotes rules of higher order which have no negative coefficients. Notice that for $n = 1$ the resulting rule is the *midpoint rule*, which is well known in the literature:

```
MidpointOpenNewtonCotesRule(1);
```

$$f\left(1/2\,h\right)h$$

```
MidpointOpenNewtonCotesRule(2);
```

$$h\left(f\left(3/2\,h\right)+f\left(1/2\,h\right)\right)$$

```
MidpointOpenNewtonCotesRule(3);
```

$$\tfrac{3}{8}h\left(3\,f\left(\tfrac{1}{2}h\right)+2\,f\left(\tfrac{3}{2}h\right)+3\,f\left(\tfrac{5}{2}h\right)\right)$$

```
MidpointOpenNewtonCotesRule(4);
```

$$\tfrac{1}{12}h\left(13\,f\left(\tfrac{1}{2}h\right)+11\,f\left(\tfrac{3}{2}h\right)+11\,f\left(\tfrac{5}{2}h\right)+13\,f\left(\tfrac{7}{2}h\right)\right)$$

```
MidpointOpenNewtonCotesRule(5);
```

$$\tfrac{5}{1152}h\left(275\,f\left(\tfrac{1}{2}h\right)+100\,f\left(\tfrac{3}{2}h\right)+402\,f\left(\tfrac{5}{2}h\right)+100\,f\left(\tfrac{7}{2}h\right)+275\,f\left(\tfrac{9}{2}h\right)\right)$$

```
MidpointOpenNewtonCotesRule(6);
```

$$\frac{3}{640}h\left(247\,f_{1/2}+139\,f_{3/2}+254\,f_{5/2}+254\,f_{7/2}+139\,f_{9/2}+247\,f_{11/2}\right)$$

To save space, we abbreviate $f\left(\tfrac{1}{2}h\right)$ as $f_{1/2}$, and similarly for the other function values. (Note that this abbreviation is not appropriate for the computation of the error term, see next section). Finally,

```
MidpointOpenNewtonCotesRule(8);
```

$$\frac{1}{241920}h\left(295627\,f_{1/2}+71329\,f_{3/2}+471771\,f_{5/2}+128953\,f_{7/2}+128953\,f_{9/2}\right.$$

$$\left.+471771\,f_{11/2}+71329\,f_{13/2}+295627\,f_{15/2}\right)$$

Once again, the 7-point formula (as well as higher order rules) have negative coefficients.

Any other Newton–Cotes rule can be easily generated this way using MAPLE. In [131] we find one-sided formulas: for example, given four equidistant function values (three intervals), we can find an approximation for the integral over the third interval:

```
factor(int(interp([0,h,2*h,3*h],[f[0],f[1],f[2],f[3]],t),t=2*h..3*h));
```

$$\frac{h}{24}\left(f_0-5\,f_1+19\,f_2+9\,f_3\right).$$

9.2.1 Error of Newton–Cotes Rules

In [132], the authors cite an old result from Steffensen that gives an expression in the order p for the error of Newton–Cotes rules for equidistant function values: if the integrand f has sufficiently many continuous derivatives and P_n is the interpolating polynomial of degree n, then

$$\int_a^b P_n(x)\,dx - \int_a^b f(x)\,dx = h^{p+1} K f^{(p)}(\xi), \quad \xi \in (a,b). \tag{9.6}$$

Here, $h = (b-a)/n$ is the distance of the function values, and the constants p and K depend only on n and not on the integrand f. As an example, we will compute the error of the trapezoidal rule. We consider the function

$$E(z) = \int_{m-\frac{z}{2}}^{m+\frac{z}{2}} f(x)\,dx - \frac{z}{2}\left(f(m + \frac{z}{2}) + f(m - \frac{z}{2}) \right), \tag{9.7}$$

where $m = (a+b)/2$ is the midpoint of the integration interval. For $z = b-a$, the function value

$$E(b-a) = \int_a^b f(x)\,dx - \frac{b-a}{2}\left(f(b) + f(a) \right) \tag{9.8}$$

is the error which we are interested in. The first derivative of E is

$$E'(z) = -\frac{z}{4}\left(f'(m + \frac{z}{2}) - f'(m - \frac{z}{2}) \right). \tag{9.9}$$

According to the mean value theorem for derivatives, there exists a number ξ_z such that

$$f'(m + \frac{z}{2}) - f'(m - \frac{z}{2}) = z f''(m + \xi_z) \quad \text{with} \quad -\frac{z}{2} < \xi_z < \frac{z}{2}.$$

Thus

$$E'(z) = -\frac{z^2}{4} f''(m + \xi_z). \tag{9.10}$$

From (9.7), we see that $E(0) = 0$ and therefore

$$E(z) = \int_0^z E'(t)\,dt = -\frac{1}{4} \int_0^z t^2 f''(m + \xi_t)\,dt.$$

Applying now the mean value theorem for integrals, we get

$$E(z) = -\frac{1}{4} f''(\xi) \int_0^z t^2\,dt = -\frac{z^3}{12} f''(\xi), \tag{9.11}$$

with $a \leq m - \frac{z}{2} < \xi < m + \frac{z}{2} \leq b$. Letting $z = h = b - a$, we obtain the *error of the trapezoidal rule for one interval*:

$$E(b - a) = \int_a^b f(x)dx - \frac{b - a}{2}\left(f(b) + f(a)\right) = -\frac{h^3}{12}f''(\xi), \qquad (9.12)$$

with $a < \xi < b$. Thus, $p = 2$ and $K = \frac{1}{12}$.

DEFINITION 9.1. (ORDER OF A QUADRATURE FORMULA) *A numerical quadrature formula with nodes ξ_i and weights w_i is of order p, if it computes the exact integral value for all polynomials p_k up to degree $k \leq p - 1$, i.e.*

$$\int_a^b p_k(x)\, dx = \sum_{i=0}^{n} w_i p_k(\xi_i), \quad k = 0, 1, \ldots, p - 1. \qquad (9.13)$$

Looking at the error formula (9.12) we derived for the trapezoidal rule, we see that polynomials up to degree one, i.e. linear functions, are integrated exactly, since their second derivatives vanish, and thus trapezoidal rule is a quadrature formula of order $p = 2$ according to Definition 9.1. One might be wondering why the order is one lower than the error term in (9.12): this is because quadrature rules are usually applied not on a single interval, but on many, which leads to composite quadrature rules, and we will see later that their use over many intervals will lower the error term by one order.

Computing the Error by Taylor Expansion

It is much easier to compute the constants p and K by a Taylor expansion with MAPLE than by hand, like we did in (9.7)–(9.12). We start with the closed rules:

```
for i from 1 to 4 do
  rule:=ClosedNewtonCotesRule(i);
  err:=taylor(rule-int(f(x),x=0..i*h),h=0,i+4);
od;
```

We list below the first term of the Taylor series for each quadrature rule.

i	rule	error
1	Trapezoidal	$\frac{1}{12}f''(0)\,h^3$
2	Simpson	$\frac{1}{90}f^{(4)}(0)\,h^5$
3	$\frac{3}{8}$-Rule	$\frac{3}{80}f^{(4)}(0)\,h^5$
4	Boole	$\frac{8}{945}f^{(6)}(0)\,h^7$

Of course the error is not equal to the first terms of the Taylor series – we have to replace the value 0 with some ξ in the integration interval to obtain a correct expression for the error. However, the two constants p and K are computed correctly.

Similarly, we can compute the error for the open rules by

```
for i from 2 to 5 do
  rule:=OpenNewtonCotesRule(i);
  err:=taylor(rule-int(f(x),x=0..i*h),h=0,i+2);
od;
```

9.2.2 Composite Rules

Applying a quadrature rule to $\int_a^b f(x)\,dx$ may not yield an approximation with the desired accuracy. To increase the accuracy, one can partition the interval $[a, b]$ into subintervals and apply the quadrature rule to each subinterval. The resulting formula is known as a *composite rule*.

Trapezoidal Rule: Assume the interval $[a, b]$ is partitioned into n equidistant subintervals (x_i, x_{i+1}) of lengths $h = x_{i+1} - x_i = (b - a)/n$. If $y_i = f(x_i)$, then by applying the trapezoidal rule (9.3) to each subinterval we obtain the *composite trapezoidal rule*

$$T(h) = h \left(\frac{1}{2} y_0 + y_1 + \cdots + y_{n-1} + \frac{1}{2} y_n \right). \tag{9.14}$$

The integration error for this rule is the sum of the errors on each subinterval:

$$\sum_{i=0}^{n-1} \left(\int_{x_i}^{x_{i+1}} f(x) dx - T_i(h) \right) = -\frac{h^3}{12} \sum_{i=0}^{n-1} f''(\xi_i). \tag{9.15}$$

If we form the average

$$M = \frac{1}{n} \sum_{i=0}^{n-1} f''(\xi_i)$$

and if f'' is continuous, then there must again exist some $\xi \in [a, b]$ such that $M = f''(\xi)$. So using $n h = b - a$, (9.15) simplifies and we obtain for the *error of the composite trapezoidal rule*:

$$\int_a^b f(x) dx - T(h) = -\frac{h^3}{12} n f''(\xi) = -\frac{(b - a)h^2}{12} f''(\xi), \quad \xi \in [a, b].$$
$$\tag{9.16}$$

Basically, (9.16) tells us that if we halve the step size $h := h/2$ (and thus double the number of function evaluations), then the error will decrease by a factor of $(1/2)^2 = 1/4$. This now explains why the order of a quadrature rule is defined as in Definition 9.1: the error of the corresponding composite quadrature rule is of the order in the definition (here $p = 2$ for the trapezoidal rule), whereas the error over a single interval is one order higher, see (9.12) for the trapezoidal rule.

Suppose we wish to compute an integral to a certain accuracy given by a relative tolerance `tol`. Here, (9.16) is of little help, since it requires an estimate of the second derivative of the integrand. In this case, rather than estimating the error and predicting the necessary step size in advance, one should compute successive approximations $\{T(h_i)\}$ by halving the step size $h_{i+1} = h_i/2$, until two approximations match to the desired tolerance. As we refine the grid, it is essential to avoid re-computing function values, since the main cost of quadrature programs is the number of function evaluations they need for a certain accuracy.

To illustrate this, consider once again the composite trapezoidal rule. When we halve the step size h, the function values required are the ones on the coarser grid, plus the new function values located halfway between the old ones. For example, when $n = 4$, we have $h_2 = (b-a)/4$ and

$$h_3 = \frac{h_2}{2} = \frac{b-a}{8}.$$

If we denote the sum of the function values in (9.14) by s_i, then we can write

$$s_2 = \frac{1}{2}f(a) + f(a + h_2) + f(a + 2h_2) + f(a + 3h_2) + \frac{1}{2}f(b).$$

If we now halve the step size, the sum is updated by the new values

$$s_3 = s_2 + f(a + h_3) + f(a + 3h_3) + f(a + 5h_3) + f(a + 7h_3). \quad (9.17)$$

We see from (9.17) that the new sum of function values emerges from the old ones by adding the new function values. This leads to the following MATLAB function:

ALGORITHM 9.1. *Trapezoidal Rule*

```
function t=TrapezoidalRule(f,a,b,tol);
%TRAPEZOIDALRULE composite quadrature using the trapezoidal rule
%    t=TrapezoidalRule(f,a,b,tol); computes an approximation of
%    int_a^b f(x) dx to a relative tolerance tol using the
%    composite trapezoidal rule.

h=b-a; s=(f(a)+f(b))/2;
t=h*s; zh=1; told=2*t;
while abs(told-t)>tol*abs(t),
```

```
    told=t; zh=2*zh; h=h/2;
    s=s+sum(f(a+[1:2:zh]*h));
    t=h*s;
end;
```

EXAMPLE 9.1.

$$\int_0^1 \frac{xe^x}{(x+1)^2}\, dx = \frac{e-2}{2} = 0.3591409142\ldots$$

The table below shows the intermediate approximations t when calculated by `TrapezoidalRule(@(x) x.*exp(x)./(x+1).^2,0,1,1e-4)`:

i	h_i	$T(h_i)$	$\dfrac{T(h_i)-I}{T(h_{i-1})-I}$
0	1	0.339785228	
1	$\frac{1}{2}$	0.353083866	0.31293
2	$\frac{1}{4}$	0.357515195	0.26840
3	$\frac{1}{8}$	0.358726477	0.25492
4	$\frac{1}{16}$	0.359036783	0.25125
5	$\frac{1}{32}$	0.359114848	0.25031
6	$\frac{1}{64}$	0.359134395	0.25007

From the error law (9.16), we see that if f'' does not vary too much, then the error should decrease by the factor 0.25 at each step. By comparing the quotients of consecutive errors, we see that the results are in good agreement with the theory, and the order is indeed 2.

Simpson's Rule: For the composite Simpson Rule, we need to partition the interval $[a, b]$ into $2n$ subintervals. Let $h = \frac{b-a}{2n}$, $x_i = a + ih$ and $y_i = f(x_i)$ for $i = 0, 1, 2, \ldots, 2n$. Applying Simpson's Rule to two consecutive subintervals

$$\int_{x_{i-1}}^{x_{i+1}} f(x)\, dx \approx \frac{h}{3}(y_{i-1} + 4y_i + y_{i+1}), \tag{9.18}$$

we obtain the *Composite Simpson Rule*

$$S(h) = \frac{h}{3}\left(y_0 + 4y_1 + 2y_2 + 4y_3 + \cdots + 2y_{2n-2} + 4y_{2n-1} + y_{2n}\right).$$
$$\tag{9.19}$$

The *integration error* is

$$\left| \int_a^b f(x)dx - S(h) \right| = \frac{(b-a)h^4}{180} \left| f^{(4)}(\xi) \right|, \qquad (9.20)$$

with $a \leq \xi \leq b$. Equation (9.20) is obtained by summing the integration errors for two subintervals and again replacing the sum of the fourth derivatives by the mean value.

Again we would like to write a MATLAB function which will compute approximations to the integral by halving the integration step until a given tolerance is met. In Simpson's rule (9.19), the function values are summed with alternating weights 2 and 4 (and weight one for the endpoints). When we halve the step size, the new function values are evaluated between the old ones. All these new function values will be multiplied with weight 4. If we now introduce variables for sums of function values with the same weight — s_1, s_2 and s_4 — then we have the following update formula to compute the new Simpson approximation:

$$\begin{aligned}
S(h) &= \frac{h}{3}(s_1 + 4s_4 + 2s_2) \\
s_1^{new} &= s_1 \\
s_2^{new} &= s_2 + s_4 \\
s_4^{new} &= \text{sum of new function values} \\
S(h/2) &= \frac{h/2}{3}(s_1^{new} + 4s_4^{new} + 2s_2^{new})
\end{aligned}$$

This leads to the following MATLAB function:

ALGORITHM 9.2. *Simpson's Rule*

```
function s=SimpsonsRule(f,a,b,tol);
% SIMPSONSRULE composite quadrature using Simpson's rule
%   s=SimpsonsRule(f,a,b,tol); computes an approximation of
%   int_a^b f(x) dx to a relative tolerance tol using the
%   composite Simpson's rule.

h=(b-a)/2; s1=f(a)+f(b); s2=0;
s4=f(a+h); s=h*(s1+4*s4)/3;
zh=2; sold=2*s;
while abs(sold-s)>tol*abs(s),
   sold=s; zh=2*zh; h=h/2; s2=s2+s4;
   s4=sum(f(a+[1:2:zh]*h));
   s=h*(s1+2*s2+4*s4)/3;
end
```

EXAMPLE 9.2. *If we again compute the integral*

$$\int\limits_0^1 \frac{xe^x}{(x+1)^2}\,dx = \frac{e-2}{2} = 0.3591409142\ldots$$

with `SimpsonsRule(@(x) x.*exp(x)./(x+1).^2,0,1,1e-8)` *we obtain*

i	h_i	$S(h_i)$	$\dfrac{S(h_i)-I}{S(h_{i-1})-I}$
0	1	0.357516745	
1	$\frac{1}{2}$	0.358992305	0.09149
2	$\frac{1}{4}$	0.359130237	0.07184
3	$\frac{1}{8}$	0.359140219	0.06511
4	$\frac{1}{16}$	0.359140870	0.06317
5	$\frac{1}{32}$	0.359140911	0.06267
6	$\frac{1}{64}$	0.359140914	0.06254

From (9.20), we expect that if we halve the step size then the error should decrease by $(1/2)^4 = 0.0625$. We see that this is the case for this example, so Simpson's rule is indeed a fourth order method.

9.2.3 Euler–Maclaurin Summation Formula

The problem we would like to consider in this section is the summation of equidistant function values:

$$\sum_{i=\alpha}^{\beta} f(i).$$

Our goal is to find a "simple" expression for such sums. For instance, we know that

$$\sum_{i=1}^{n} i = \frac{n}{2}(n+1).$$

The idea comes from the *calculus of differences*, which was very popular before the age of computers. If we were given a function $s(x)$ with the property that

$$\Delta s(x) := s(x+1) - s(x) = f(x), \tag{9.21}$$

(s is called a *summation function*) then the summation problem would be solved since

$$\sum_{i=\alpha}^{\beta-1} f(i) = \sum_{i=\alpha}^{\beta-1} (s(i+1) - s(i)) = s(\beta) - s(\alpha).$$

The summation function is the analog of antiderivatives, but for sums instead of integrals.

EXAMPLE 9.3. *Consider $f(x) = x$ and the ansatz for the summation function $s(x) = ax^2 + bx + c$. The constant c is actually redundant and can be taken to be zero, since we use only differences of s. Now equating coefficients in*

$$\begin{aligned} s(x+1) - s(x) &= a((x+1)^2 - x^2) + b((x+1) - x) = 2ax + a + b \\ &= f(x) = x, \end{aligned}$$

we see that $2a = 1$ and $a + b = 0$ must hold. Thus we obtain

$$a = \frac{1}{2}, \ b = -\frac{1}{2} \ \text{ and } s(x) = \frac{x}{2}(x-1),$$

and

$$\sum_{i=1}^{n} i = s(n+1) - s(1) = \frac{n+1}{2}n$$

as before.

In the following, we will develop an expression for $\Delta f(x) = f(x+1) - f(x)$ using *formal power series*. Formal means that we are not concerned with whether or not the power series converge. Consider the Taylor series for $f(x+h)$ for $h = 1$,

$$f(x+1) = \sum_{k=0}^{\infty} f^{(k)}(x)\frac{1}{k!}.$$

Then

$$\Delta f(x) = f(x+1) - f(x) = \sum_{k=1}^{\infty} f^{(k)}(x)\frac{1}{k!}. \tag{9.22}$$

A similar relation holds for the derivatives of f:

$$\Delta f^{(i)}(x) = \sum_{k=1}^{\infty} f^{(k+i)}(x)\frac{1}{k!}. \tag{9.23}$$

The integral $F(x) = \int f(x)\,dx =: f^{(-1)}(x)$ can also be represented in this way:

$$\Delta F(x) = \int_{x}^{x+1} f(x)\,dx = \sum_{k=1}^{\infty} F^{(k)}(x)\frac{1}{k!} = \sum_{k=1}^{\infty} f^{(k-1)}(x)\frac{1}{k!}. \tag{9.24}$$

We now introduce the two infinite vectors

$$\boldsymbol{f} = \begin{pmatrix} f(x) \\ f'(x) \\ \vdots \\ f^{(m)}(x) \\ \vdots \end{pmatrix} \text{ and } \Delta\boldsymbol{f} = \begin{pmatrix} \Delta F(x) \\ \Delta f(x) \\ \vdots \\ \Delta f^{(m-1)}(x) \\ \vdots \end{pmatrix}.$$

Because of (9.22), (9.23) and (9.24), we have the relation

$$\Delta \boldsymbol{f} = A\boldsymbol{f} \tag{9.25}$$

with the upper triangular *Toeplitz matrix*

$$A = \begin{pmatrix} \frac{1}{1!} & \frac{1}{2!} & \frac{1}{3!} & \cdots \\ & \frac{1}{1!} & \frac{1}{2!} & \cdots \\ & & \frac{1}{1!} & \ddots \\ & & & \ddots \end{pmatrix}.$$

We would like to solve (9.25) for $\boldsymbol{f}$. The inverse of the matrix A is also a Toeplitz matrix of the same structure

$$A^{-1} = \begin{pmatrix} b_1 & b_2 & b_3 & \cdots \\ & b_1 & b_2 & \cdots \\ & & b_1 & \ddots \\ & & & \ddots \end{pmatrix},$$

and its entries are best computed using generating functions: letting $b(x) := b_1 + b_2 x + b_3 x^2 + \ldots$ and $a(x) := \frac{1}{1!} + \frac{1}{2!}x + \frac{1}{3!}x^2 + \ldots$, the relation $A^{-1}A = I$ is equivalent to saying that $b(x)a(x) = 1$, and therefore the coefficients b_i can be computed simply by dividing $1/a(x)$,

$$b_1 + b_2 x + b_3 x^2 + \cdots = \frac{1}{\frac{1}{1!} + \frac{1}{2!}x + \frac{1}{3!}x^2 + \cdots} = \frac{1}{\dfrac{e^x - 1}{x}} = \frac{x}{e^x - 1}. \tag{9.26}$$

The coefficients b_i are now obtained by expanding the function $x/(e^x - 1)$ in a Taylor series, which can be done with MAPLE,

```
series(x/(exp(x)-1),x=0,10);
```

$$1 - \frac{1}{2}x + \frac{1}{12}x^2 - \frac{1}{720}x^4 + \frac{1}{30240}x^6 - \frac{1}{1209600}x^8 + O\left(x^9\right).$$

Traditionally this series is written as

$$\frac{x}{e^x - 1} = \sum_{k=0}^{\infty} \frac{B_k}{k!} x^k,$$

where the B_k are the *Bernoulli numbers*:

k	0	1	2	3	4	5	6	7	8
B_k	1	$-\frac{1}{2}$	$\frac{1}{6}$	0	$-\frac{1}{30}$	0	$\frac{1}{42}$	0	$\frac{1}{30}$
$\frac{B_k}{k!}$	1	$-\frac{1}{2}$	$\frac{1}{12}$	0	$-\frac{1}{720}$	0	$\frac{1}{30240}$	0	$-\frac{1}{1209600}$

The Bernoulli numbers can be computed in MAPLE by the function `bernoulli(k)`. For $k \geq 3$ every Bernoulli number with odd index is zero. Thus we have found

$$
A^{-1} = \begin{pmatrix} \frac{B_0}{0!} & \frac{B_1}{1!} & \frac{B_2}{2!} & \cdots \\ & \frac{B_0}{0!} & \frac{B_1}{1!} & \cdots \\ & & \frac{B_0}{0!} & \ddots \\ & & & \ddots \end{pmatrix}.
$$

Now $\Delta \boldsymbol{f} = A\boldsymbol{f}$ can be solved by $\boldsymbol{f} = A^{-1}\Delta \boldsymbol{f} = \Delta(A^{-1}\boldsymbol{f})$ or, when written component-wise,

$$
f^{(p)}(x) = \Delta\left(\sum_{k=0}^{\infty} \frac{B_k}{k!}\, f^{(k-1+p)}(x) \right), \quad p = 0, 1, 2, \ldots .
$$

Specifically, for $p = 0$ we get an expression for the summation function:

$$
f(x) = \Delta s(x), \quad \text{with } s(x) = \sum_{k=0}^{\infty} \frac{B_k}{k!}\, f^{(k-1)}(x). \tag{9.27}
$$

Using the expression for $s(x)$ from (9.27), we obtain

$$
\sum_{i=\alpha}^{\beta-1} f(i) = s(\beta) - s(\alpha) = \frac{B_0}{0!}(f^{(-1)}(\beta) - f^{(-1)}(\alpha))
$$
$$
+ \frac{B_1}{1!}(f(\beta) - f(\alpha)) + \frac{B_2}{2!}(f'(\beta) - f'(\alpha)) + \cdots
$$

Adding $f(\beta)$ on both sides yields the *Euler–Maclaurin summation formula*:

$$
\sum_{i=\alpha}^{\beta} f(i) = \int_{\alpha}^{\beta} f(x)\, dx + \frac{f(\alpha) + f(\beta)}{2} + \sum_{j=1}^{\infty} \frac{B_{2j}}{(2j)!}\, (f^{(2j-1)}(\beta) - f^{(2j-1)}(\alpha)).
$$
$$\tag{9.28}$$

To use this result for quadrature, we need to generalize the Euler–Maclaurin summation formula for a sum of function values with a step size h instead of one. We choose first $\alpha = 0$ and $\beta = n$. Then we introduce the function g and the interval (a, b) with

$$
f(x) = g(a + xh), \quad h = \frac{b - a}{n}, \quad b = a + nh.
$$

For the derivatives and the integral, we get

$$
f^{(2j-1)}(x) = g^{(2j-1)}(a + xh)\, h^{2j-1}, \quad \int_0^n f(x)\, dx = \frac{1}{h} \int_a^b g(t)\, dt.
$$

After inserting g in (9.28), subtracting $(g(a) + g(b))/2$ on both sides and multiplying with h, we obtain an *asymptotic expansion for the trapezoidal rule*:

$$T(h) = \int_a^b g(t)\, dt + \sum_{j=1}^{\infty} \frac{B_{2j}}{(2j)!}\, h^{2j} \left(g^{(2j-1)}(b) - g^{(2j-1)}(a) \right). \qquad (9.29)$$

Recall that we have not taken convergence into account when deriving the Euler–Maclaurin summation formula. Indeed, the sums on the right-hand side in (9.28) and (9.29) often do not converge.

A rigorous derivation of the error term that does not assume g is infinitely differentiable is given in [132]. It has the form

$$T(h) = \int_a^b g(t)\, dt + \sum_{j=1}^{m} \frac{B_{2j}}{(2j)!}\, h^{2j} \left(g^{(2j-1)}(b) - g^{(2j-1)}(a) \right) + R_m, \qquad (9.30)$$

where the error term is given by

$$R_m = \frac{B_{2m+2}}{(2m+2)!}\, h^{2m+2}\, (b-a) g^{(2m+2)}(\xi), \ \text{ with } \xi \in (a,b).$$

Thus, for a fixed number of terms m, the error between $T(h)$ and the partial sum in (9.29) behaves like $O(h^{2m+2})$ as $h \to 0$.

EXAMPLE 9.4. *The* Riemann Zeta function *is defined for $Re(z) > 1$ by*

$$\zeta(z) = \sum_{k=1}^{\infty} \frac{1}{k^z},$$

and is extended to the rest of the complex plane (except for the point $z = 1$) by analytic continuation. The point $z = 1$ is a simple pole.

Some values like

$$\zeta(2) = \sum_{k=1}^{\infty} \frac{1}{k^2} = \frac{\pi^2}{6}, \quad \zeta(4) = \sum_{k=1}^{\infty} \frac{1}{k^4} = \frac{\pi^4}{90}$$

are known, and that MAPLE *knows them is revealed either with* Zeta(2) *or with*

```
sum('1/k^2','k'=1..infinity);
```

An open problem is to identify (not compute numerically!) the values for odd integers $\zeta(3)$, $\zeta(5)$ etc.

We will apply the Euler–Maclaurin summation formula to

$$\sum_{k=n}^{\infty} \frac{1}{k^2}.$$

We have $f(x) = 1/x^2$, and for the derivatives we obtain

$$f^{(2j-1)}(x) = -\frac{(2j)!}{x^{2j+1}}.$$

Thus

$$\sum_{k=n}^{\infty} \frac{1}{k^2} = \int_{k=n}^{\infty} \frac{dx}{x^2} + \frac{1}{2}\left(\frac{1}{n^2} + 0\right) + \sum_{j=1}^{\infty} \frac{B_{2j}}{(2j)!}\left(0 - \left(-\frac{(2j)!}{n^{2j+1}}\right)\right),$$

which simplifies to

$$\sum_{k=n}^{\infty} \frac{1}{k^2} = \frac{1}{n} + \frac{1}{2n^2} + \sum_{j=1}^{\infty} \frac{B_{2j}}{n^{2j+1}}.$$

We could have obtained this expansion also with the MAPLE *command*

```
assume(n>0);
series(sum('1/k^2','k'=n..infinity),n=infinity,11);
```

$$n^{-1} + 1/2\,n^{-2} + 1/6\,n^{-3} - 1/30\,n^{-5} + 1/42\,n^{-7} - 1/30\,n^{-9} + O(n^{-11}).$$

Using this expression in

$$\sum_{k=1}^{n-1} \frac{1}{k^2} = \frac{\pi^2}{6} - \sum_{j=n}^{\infty} \frac{1}{k^2},$$

we obtain after adding $1/n^2$ on both sides

$$\sum_{k=1}^{n} \frac{1}{k^2} = \frac{\pi^2}{6} - \frac{1}{n} + \frac{1}{2n^2} - \frac{1}{6n^3} + \cdots, \tag{9.31}$$

a summation formula for $f(k) = 1/k^2$. Assume we want to compute the sum for $n = 10^5$. Since $1/(6n^3) = 1.666e{-}16$ it is sufficient to compute

$$\frac{\pi^2}{6} - \frac{1}{n} + \frac{1}{2n^2} = 1.64492406689823$$

to obtain the correct value of $\sum_{k=1}^{100000} \frac{1}{k^2}$ to 15 digits.

For general $z \neq 1$, the expansion

$$\sum_{k=1}^{m-1} \frac{1}{k^z} + \frac{1}{2}\frac{1}{m^z} = \zeta(z) - \frac{1}{z-1}\sum_{j=0}^{\infty} \binom{1-z}{2j}\frac{B_{2j}}{m^{z-1+2j}} \tag{9.32}$$

holds. Equation (9.31) is the special case for $z = 2$.

9.2.4 Romberg Integration

The asymptotic expansion for the approximation of an integral by the trapezoidal rule given by the Euler–Maclaurin summation formula (9.30),

$$T(h) = \int_a^b g(t)\,dt + c_1 h^2 + c_2 h^4 + \cdots + c_m h^{2m} + R_m,$$

leads to another algorithm for quadrature. The idea here is to extrapolate to the limit $T(0)$ by interpolating the values $T(h_i)$ for some $h_i > 0$ by a polynomial. If the error term R_m is small, then the values $T(h_i)$ are close to the values of the polynomial

$$P_{2m}(h) = c_0 + c_1 h^2 + c_2 h^4 + \cdots + c_m h^{2m}, \text{ with } c_0 = \int_a^b g(t)\,dt.$$

To make use of the fact that only even powers of h occur, we consider a polynomial in the variable $x = h^2$. Thus, we wish to extrapolate to $x = 0$ the following data:

x	h_0^2	h_1^2	$\cdots$	h_i^2	$\cdots$
y	$T(h_0)$	$T(h_1)$	$\cdots$	$T(h_i)$	$\cdots$

We will use the *Aitken–Neville Scheme* for extrapolation (see Chapter 4). Since we do not know the value of m or the degree of the polynomial, we will keep adding new points by computing new trapezoidal rule approximations and hope that the Aitken–Neville scheme will converge to the desired integral value.

$$
\begin{aligned}
T(h_0) &= T_{00} \\
T(h_1) &= T_{10} \quad T_{11} \\
\;\vdots & \qquad\;\; \vdots \quad\;\; \vdots \quad\;\; \ddots \\
T(h_i) &= T_{i0} \quad T_{i1} \quad \cdots \quad T_{ii} \;.
\end{aligned}
\tag{9.33}
$$

An obvious choice for the sequence h_i is $h_i = h_{i-1}/2$, which was also used in the function `TrapezoidalRule`. By this choice, the amount of function evaluations (and thus the computational effort) doubles with each new row in the Aitken–Neville scheme. In [132], the authors discuss alternative choices for h_i, for which the number of function evaluations grows less rapidly.

If (x_i, y_i) are the given interpolation points and z is the argument at which the interpolation polynomial is to be evaluated, then the Aitken–Neville scheme is computed by the recursion

$$T_{ij} = \frac{(x_i - z)T_{i-1,j-1} + (z - x_{i-j})T_{i,j-1}}{x_i - x_{i-j}}$$

for $j = 1, 2, \ldots, i$ and $i = 1, 2, 3, \ldots$. In our case, $z = 0$ and $x_i = h_i^2$, and thus

$$T_{ij} = \frac{h_i^2 T_{i-1,j-1} - h_{i-j}^2 T_{i,j-1}}{h_i^2 - h_{i-j}^2}.$$

A further simplification occurs if we choose $h_i = h_{i-1}/2 = h_0/2^i$:

$$T_{ij} = \frac{\left(\frac{h_0}{2^i}\right)^2 T_{i-1,j-1} - \left(\frac{h_0}{2^{i-j}}\right)^2 T_{i,j-1}}{\left(\frac{h_0}{2^i}\right)^2 - \left(\frac{h_0}{2^{i-j}}\right)^2} = \frac{4^{-j} T_{i-1,j-1} - T_{i,j-1}}{4^{-j} - 1}.$$

This specialized scheme is called the *Romberg scheme*. The first column is computed by the trapezoidal rule $T_{i0} = T(h_i)$ for the step sizes $h_i = h_{i-1}/2$, and the other rows and columns of the triangular scheme are computed by

$$T_{ij} = \frac{4^{-j} T_{i-1,j-1} - T_{i,j-1}}{4^{-j} - 1}, \quad j = 1, 2, \ldots, i, \quad i = 0, 1, 2, \ldots \qquad (9.34)$$

The columns, rows and diagonals of the Romberg scheme all converge to the value of the integral; for smooth functions, the diagonal converges the fastest.

THEOREM 9.1. *The Romberg scheme computed by (9.34) contains in its first column the values of the trapezoidal rule, in its second column the values of Simpson's rule, and in its third column the approximations obtained by Boole's rule.*

The proof is simply a verification and is left as an exercise, see Problem 9.14. The further columns are no longer related to the Newton–Cotes rules.

Extrapolation with the Romberg scheme can also be interpreted as an algorithm for *elimination of lower order error terms* by taking the appropriate linear combinations. This process is called *Richardson Extrapolation* and is the same as Aitken–Neville Interpolation. Consider the asymptotic expansion of the trapezoidal rule,

$$\begin{aligned}
T(h) &= I + c_1 h^2 + c_2 h^4 + c_3 h^6 + \cdots, \\
T\left(\tfrac{h}{2}\right) &= I + c_1 \left(\tfrac{h}{2}\right)^2 + c_2 \left(\tfrac{h}{2}\right)^4 + c_3 \left(\tfrac{h}{2}\right)^6 + \cdots, \\
T\left(\tfrac{h}{4}\right) &= I + c_1 \left(\tfrac{h}{4}\right)^2 + c_2 \left(\tfrac{h}{4}\right)^4 + c_3 \left(\tfrac{h}{4}\right)^6 + \cdots.
\end{aligned} \qquad (9.35)$$

Forming the quantities

$$T_{11} = \frac{4T\left(\tfrac{h}{2}\right) - T(h)}{3} \quad \text{and} \quad T_{21} = \frac{4T\left(\tfrac{h}{4}\right) - T\left(\tfrac{h}{2}\right)}{3},$$

we obtain

$$\begin{aligned}
T_{11} &= I - \tfrac{1}{4} c_2 h^4 - \tfrac{5}{16} c_3 h^6 + \cdots, \\
T_{21} &= I - \tfrac{1}{64} c_2 h^4 - \tfrac{5}{1024} c_3 h^6 + \cdots.
\end{aligned}$$

Thus we have eliminated the term with h^2. Continuing with the linear combination

$$T_{22} = \frac{16T_{21} - T_{11}}{15} = I + \frac{1}{64}c_3 h^6 + \cdots$$

we eliminate the next term with h^4.

The following MATLAB function computes the Romberg scheme row by row until the relative difference of two consecutive diagonal elements is smaller than some given tolerance.

ALGORITHM 9.3. *Romberg Integration*

```
function [I,T]=Romberg(f,a,b,tol);
% ROMBERG quadrature using the Romberg scheme
%    [I,T]=Romberg(f,a,b,tol) computes an approximation of int_a^b f(x)
%    dx to a relative tolerance tol using the Romberg scheme.

h=b-a; intv=1; s=(f(a)+f(b))/2;
T(1,1)=s*h;
for i=2:15
  intv=2*intv; h=h/2;
  s=s+sum(f(a+[1:2:intv]*h));
  T(i,1)=s*h;
  vhj=1;
  for j=2:i
    vhj=vhj/4;
    T(i,j)=(vhj*T(i-1,j-1)-T(i,j-1))/(vhj-1);
  end;
  if abs(T(i,i)-T(i-1,i-1))<tol*abs(T(i,i)),
    I=T(i,i); return
  end
end
warning(['limit of extrapolation steps reached. ',...
        'Required tolerance may not be met.']);
I=T(i,i);
```

EXAMPLE 9.5. *If we use again as example the integral*

$$\int_0^1 \frac{xe^x}{(x+1)^2}\, dx = \frac{e-2}{2} = 0.35914091422952\ldots$$

we obtain with [I,T]=Romberg(@(x) x.*exp(x)./(x+1).^2,0,1,1e-8) *the* *following results:*

Romberg scheme columns *0 to 3*

```
0.33978522855738
0.35308386657870    0.35751674591915
0.35751519587192    0.35899230563633    0.35909067628414
0.35872647716421    0.35913023759497    0.35913943305888    0.35914020697594
0.35903678355577    0.35914021901962    0.35914088444793    0.35914090748585
0.35911484861929    0.35914087030714    0.35914091372630    0.35914091419104
0.35913439576246    0.35914091147685    0.35914091422149    0.35914091422935
```

columns *4 to 6*

```
0.35914091023295
0.35914091421734    0.35914091422123
0.35914091422950    0.35914091422951    0.35914091422952
```

If we compare the last value in the first column $T_{6,0} = 0.35913439576246$ with the exact value we see that only 4 decimal digits are correct. On the other hand the diagonal value $T_{6,6} = 0.35914091422952$ is correct to all 14 printed digits. Romberg extrapolation works very well for this example. Furthermore, if we look at the second column of the Romberg scheme, we observe that the values are the same as those obtained with the function `SimpsonsRule`*!*

EXAMPLE 9.6. *As a second example, we integrate $\int_0^1 \sqrt{x}\, dx = 2/3$ by the Romberg scheme. The call* `[I,T]=Romberg(@sqrt,0,1,1e-8)` *generates a warning message that 15 extrapolation steps were not sufficient to meet the tolerance of 10^{-8}. The values of the first column and the diagonal values are given in Table 9.1. We see from Table 9.1 that extrapolation does not work for this example. The extrapolated values are hardly better than the values obtained by trapezoidal rule. The reason is that the function $f(x) = \sqrt{x}$ is not differentiable for $x = 0$ and therefore the asymptotic expansion (9.30) does not hold.*

To compute the asymptotic expansion, we introduce $z = -1/2$ in Equation (9.32) and get

$$\sum_{k=1}^{m-1} \sqrt{k} + \frac{1}{2}\sqrt{m} = \zeta\left(-\frac{1}{2}\right) + \frac{2}{3}\sum_{j=0}^{\infty} \binom{3/2}{2j} \frac{B_{2j}}{m^{-3/2+2j}}. \tag{9.36}$$

If we integrate $\int_0^1 \sqrt{x}\, dx$ with the trapezoidal rule and choose $h = 1/m$ then

$$T(h) = h\left(\frac{1}{2}\sqrt{0} + \sqrt{h} + \sqrt{2h} + \cdots + \sqrt{(m-1)h} + \frac{1}{2}\sqrt{1}\right).$$

Thus

$$\frac{T(h)}{h^{3/2}} = \sqrt{1} + \sqrt{2} + \cdots + \sqrt{m-1} + \frac{1}{2}\sqrt{m}.$$

T(:,1)	diag(T)
0.50000000000000	0.50000000000000
0.60355339059327	0.63807118745770
0.64328304624275	0.65775660328156
0.65813022162445	0.66360756911229
0.66358119687723	0.66559286512947
0.66555893627894	0.66628769903384
0.66627081137851	0.66653274119989
0.66652565729683	0.66661932214828
0.66661654897653	0.66664992831868
0.66664888154995	0.66666074880826
0.66666036221898	0.66666457439141
0.66666443359297	0.66666592693598
0.66666587612718	0.66666640513240
0.66666638691157	0.66666657420034
0.66666656769401	0.66666663397489

TABLE 9.1. *First column and diagonal of the Romberg scheme for* $\int_0^1 \sqrt{x}\, dx$

Using Equation (9.36) we obtain after multiplying by $h^{3/2}$ *the asymptotic expansion*

$$T(h) = \zeta\left(-\frac{1}{2}\right) h^{3/2} + \frac{2}{3} \sum_{j=0}^{\infty} \binom{3/2}{2j} B_{2j}\, h^{2j}. \qquad (9.37)$$

The term $\zeta\left(-\frac{1}{2}\right) h^{3/2}$ *is disturbing the expansion in even powers of* h *and responsible for the bad convergence shown in Table 9.1.*

9.3 Gauss Quadrature

In the quadrature rules discussed so far, we have used a fixed location for the nodes ($\xi_i = a + ih$) at which the function was evaluated. The key idea of *Gauss quadrature rules* is to allow the nodes to vary as well; the additional freedom we gain can be used to increase the accuracy of the quadrature rule.

Without loss of generality, we consider functions which need to be integrated on the interval $[-1, 1]$. By the change of variables

$$x = \frac{b-a}{2}t + \frac{a+b}{2} \quad \Longleftrightarrow \quad t = \frac{2x - b - a}{b - a}, \qquad (9.38)$$

we can always transform a general finite interval $[a, b]$ to $[-1, 1]$, and thus

$$\int_a^b f(x)\, dx = \frac{b-a}{2} \int_{-1}^1 f\left(\frac{b-a}{2}t + \frac{a+b}{2}\right) dt.$$

For a given number of nodes n, we want to find the location of nodes ξ_k and the weights w_k in the quadrature rule

$$\int_{-1}^{1} f(x)\, dx \approx \sum_{k=1}^{n} w_k f(\xi_k), \tag{9.39}$$

which allows us to integrate exactly a polynomial of as high a degree as possible. Recall the order p of a quadrature rule in Definition 9.1, which stated that $p - 1$ equals the highest degree of polynomials that it integrates exactly.

THEOREM 9.2. (ORDER BOUND) *The order p of an n-point quadrature rule (9.39) is bounded by $2n$, i.e. an n-point quadrature rule can at most integrate a polynomial of degree $2n - 1$ exactly.*

PROOF. It is sufficient to give an example of a polynomial of degree $2n$ that cannot be integrated exactly by (9.39). Consider the polynomial which has the nodes as zeros,

$$Q(x) = \prod_{k=1}^{n} (x - \xi_k)^2.$$

The degree of Q is $2n$ and $\int_{-1}^{1} Q(x)\, dx > 0$ because $Q(x) \geq 0$. However, since the ξ_k are the nodes of the rule, the right hand side of (9.39) becomes 0 and the rule does not integrate Q exactly. $\qquad\square$

Very high order quadrature rules have in addition the very desirable property that their weights are positive:

THEOREM 9.3. (POSITIVITY OF WEIGHTS) *If an n-point quadrature rule (9.39) is of order $p \geq 2n-1$, then the weights w_k, $k = 1, 2, \ldots, n$, are positive.*

PROOF. Consider the Lagrange polynomials $l_i(x) := \prod_{k \neq i} \frac{x - \xi_k}{\xi_i - \xi_k}$ of degree $n - 1$, which satisfy $l_i(\xi_k) = 0$ for $i \neq k$ and $l_i(\xi_i) = 1$. Using this fact and that the quadrature rule is exact for polynomials of degree up to $2n - 2$, we get

$$w_i = \sum_{k=1}^{n} w_k l_i^2(\xi_k) = \int_{-1}^{1} l_i^2(\xi)\, d\xi > 0.$$

$$\square$$

Since every polynomial can be written as a sum of monomials,

$$p(x) = a_0 + a_1 x + a_2 x^2 + \ldots + a_m x^m,$$

and because the integral is a linear operator,

$$\int_{-1}^{1} p(x)\, dx = a_0 \int_{-1}^{1} dx + a_1 \int_{-1}^{1} x\, dx + \ldots + a_m \int_{-1}^{1} x^m\, dx,$$

it is sufficient to integrate all monomials exactly up to some degree m to obtain a rule that will integrate any polynomial exactly of degree less than or equal to m.

To derive conditions on the nodes and weights, consider the case $n = 2$. Then we have four degrees of freedom (two nodes and two weights) and thus we can satisfy 4 equations. We can therefore derive a quadrature rule which is exact for polynomials up to degree $2n - 1 = 3$ by solving the system of four equations obtained for the monomials 1, x, x^2 and x^3,

$$\int_{-1}^{1} 1\, dx \;=\; 2 = 1w_1 + 1w_2,$$

$$\int_{-1}^{1} x\, dx \;=\; 0 = \xi_1 w_1 + \xi_2 w_2,$$

$$\int_{-1}^{1} x^2\, dx \;=\; \frac{2}{3} = \xi_1^2 w_1 + \xi_2^2 w_2,$$

$$\int_{-1}^{1} x^3\, dx \;=\; 0 = \xi_1^3 w_1 + \xi_2^3 w_2.$$

Using MAPLE to solve the system of nonlinear equations, we find

```
eqns:={seq(int(x^k,x=-1..1)=w[1]*xi[1]^k+w[2]*xi[2]^k,k=0..3)};
sols:=solve(eqns,indets(eqns,name)):
convert(sols,radical);
```

$$\left\{ w_2 = 1, w_1 = 1, \xi_1 = 1/3\sqrt{3}, \xi_2 = -1/3\sqrt{3} \right\}.$$

We have thus computed the *Gauss–Legendre rule for $n = 2$*,

$$\int_{-1}^{1} f(x)\, dx \approx f\left(\frac{1}{\sqrt{3}}\right) + f\left(-\frac{1}{\sqrt{3}}\right). \tag{9.40}$$

Note that the node location is symmetric ($\xi_1 = -\xi_2 = \frac{1}{\sqrt{3}}$) and the weights are equal ($w_1 = w_2 = 1$).

Similarly, for $n = 3$ we obtain 6 equations for 6 unknowns which we can solve in the same way,

```
eqns:={seq(int(x^k,x=-1..1)=w[1]*xi[1]^k+w[2]*xi[2]^k+w[3]*xi[3]^k,
       k=0..5)};
sols:=solve(eqns,indets(eqns,name)):
convert(sols[1],radical);
```

We obtain

$$\left\{ w_1 = \frac{8}{9}, w_2 = 5/9, w_3 = 5/9, \xi_1 = 0, \xi_2 = 1/5\,\sqrt{5}\sqrt{3}, \xi_3 = -1/5\,\sqrt{5}\sqrt{3} \right\}.$$

(9.41)

Notice again the symmetry $\xi_2 = -\xi_3$ and $w_2 = w_3$. By exploiting this symmetry, we can halve the number of unknowns by making the ansatz for instance for $n = 2$

$$\int_{-1}^{1} f(x)\,dx \approx w_1(f(\xi_1) + f(-\xi_1))$$

with 2 parameters ξ_1, w_1 instead of $2n = 4$. For $n = 3$ we consider

$$\int_{-1}^{1} f(x)\,dx \approx w_0 f(0) + w_1(f(\xi_1) + f(-\xi_1))$$

with 3 parameters w_0, w_1, ξ_1 instead of $2n = 6$. These symmetric rules are always exact for monomials with odd degree. Therefore we only need to ask for exactness for the monomials with even exponents. For $n = 2$, e.g., we want the rule to be exact for 1 and x^2:

```
eqns:={int(1,x=-1..1)=2*w[1],int(x^2,x=-1..1)=2*xi[1]^2*w[1]};
sols:=solve(eqns,{xi[1],w[1]});
convert(sols, radical);
```

We obtain the same result as above:

$$\left\{ w_1 = 1, \xi_1 = 1/3\,\sqrt{3} \right\}.$$

For $n = 3$ we can ask the rule to be exact for the monomials 1, x^2 and x^4. It will be exact for polynomials of degree 5. The following MAPLE procedure **gauss** generates and solves the system of equations for *Gauss–Legendre quadrature rules* for a given n.

```
Gauss:=proc(n) local w,xi,i,j,firsteq,eqns,sols,m; global res;
  m:=trunc(n/2); w:=array(0..m); xi:=array(1..m);
  firsteq:=2*sum(w[i],i=1..m);
  if irem(n,2)=1 then firsteq:=firsteq+w[0]  fi;
  eqns:={2=firsteq};
  for j from 2 by 2 to 2*(n-1) do
    eqns:=eqns union{int(x^j,x=-1..1)=2*sum(w[i]*xi[i]^j,i=1..m)};
  od;
  if irem(n,2)=1 then sols:={w[0]}
    else sols:={}
  fi;
  for j from 1 to m do
```

```
   sols:=sols union {w[j],xi[j]};
 od;
 res:=solve(eqns,sols);
 evalf(res);
end:
```

The procedure `Gauss` works well for $n \leq 6$. For larger n, the system of equations becomes too difficult for MAPLE to solve. Also the analytical expressions grow enormously. The call

```
Digits:=20; Gauss(6);
```

computes the values

$$
\begin{aligned}
\xi_1 &= -0.23861918608319690851, & w_1 &= 0.46791393457269104747, \\
\xi_2 &= -0.66120938646626451369, & w_2 &= 0.36076157304813860749, \\
\xi_3 &= -0.93246951420315202781, & w_3 &= 0.17132449237917034509,
\end{aligned}
$$

and the 6 point Gauss quadrature rule (which by construction will be exact for polynomials up to degree 11) becomes

$$
\int\limits_{-1}^{1} f(x)\, dx \approx w_1(f(\xi_1) + f(-\xi_1)) + w_2(f(\xi_2) + f(-\xi_2)) + w_3(f(\xi_3) + f(-\xi_3)).
$$

With the procedure `Gauss(n)`, we generate n-point Gauss quadrature rules which are exact for polynomials of degree $2n-1$. This is the maximum degree for a rule with n points because of Theorem 9.2.

9.3.1 Characterization of Nodes and Weights

The brute force approach is only feasible for small values of n. For larger n, the system of nonlinear equations becomes too hard to solve for MAPLE, so one must resort to additional theory to compute the rules.

Our goal is to find nodes and weights to get an exact rule for polynomials of degree up to $2n - 1$,

$$
\int\limits_{-1}^{1} P_{2n-1}(x)\, dx = \sum_{i=1}^{n} w_i P_{2n-1}(\xi_i). \tag{9.42}
$$

We argue as follows (see [46]): consider the decomposition of the polynomial P_{2n-1} obtained by dividing by some other polynomial $Q_n(x)$ of degree n:

$$
P_{2n-1}(x) = H_{n-1}(x)Q_n(x) + R_{n-1}(x).
$$

Then $H_{n-1}(x)$ and the remainder $R_{n-1}(x)$ are polynomials of degree $n - 1$ and

$$
\int_{-1}^{1} P_{2n-1}(x)\, dx = \int\limits_{-1}^{1} H_{n-1}(x)Q_n(x)\, dx + \int_{-1}^{1} R_{n-1}(x)\, dx.
$$

Applying rule (9.42) on both sides for the integrals and subtracting both equations yields for the error the expression

$$
\int_{-1}^{1} P_{2n-1}(x)\,dx - \sum_{i=1}^{n} w_i P_{2n-1}(\xi_i) = \int_{-1}^{1} H_{n-1}(x)Q_n(x)\,dx
$$

$$
- \sum_{i=1}^{n} w_i H_{n-1}(\xi_i)Q_n(\xi_i) + \int_{-1}^{1} R_{n-1}(x)\,dx - \sum_{i=1}^{n} w_i R_{n-1}(\xi_i).
$$

Now we see that we can make the error zero by the following choices:

First, take $Q_n(x)$ to be the *orthogonal polynomial* (see Section 9.3.2) on the interval $[-1,1]$ corresponding to the scalar product

$$
(f,g) = \int_{-1}^{1} f(x)g(x)\,dx. \tag{9.43}
$$

By this choice and by the definition of orthogonal polynomials, the first term in the error vanishes: $\int_{-1}^{1} H_{n-1}(x)Q_n(x)\,dx = 0$. Q_n is a *Legendre Polynomial* (see the next section) and available in MAPLE as `orthopoly[P](n,x)`.

Second, choose as the nodes the (real) zeros of Q_n. Then the second term in the error will also vanish: $\sum_{i=1}^{n} w_i H_{n-1}(\xi_i)Q_n(\xi_i) = 0$.

Third, having fixed the nodes we can compute weights according to Newton–Cotes by integrating the interpolation polynomial. By construction, the rule will integrate every polynomial of degree $n - 1$ exactly, so that

$$
\int_{-1}^{1} R_{n-1}(x)\,dx = \sum_{i=1}^{n} w_i R_{n-1}(\xi_i)
$$

and the last two error terms cancel.

We summarize our considerations in the following theorem.

THEOREM 9.4. (GAUSS QUADRATURE) *The quadrature rule (9.39) has order $2n$ if the nodes ξ_i are the zeros of the orthogonal polynomial Q_n corresponding to the scalar product (9.43), and if the weights w_i are the integrals of the Lagrange polynomials of the interpolating polynomial. The rule of maximal order thus obtained is called the Gauss quadrature rule of order $2n$.*

So, we can compute a Gauss-Legendre quadrature rule, e.g. for $n = 20$, with the following MAPLE statements:

```
n:=20;
X:=sort([fsolve(orthopoly[P](n,x)=0,x)]);
Q:=int(interp(X,[seq(y[i],i=1..n)],z),z=-1..1);
```

We obtain the nodes

-0.9931285992	-0.9639719273	-0.9122344283	-0.8391169718
-0.7463319065	-0.6360536807	-0.5108670020	-0.3737060887
-0.2277858511	-0.07652652113	0.07652652113	0.2277858511
0.3737060887	0.5108670020	0.6360536807	0.7463319065
0.8391169718	0.9122344283	0.9639719273	0.9931285992

and the weights

$$
\begin{aligned}
Q := \; & 0.01640269573\,y_1 & +&0.01754850817\,y_{20} \\
+&0.03905195696\,y_2 & +&0.04058452479\,y_{19} \\
+&0.06184138764\,y_3 & +&0.06276883444\,y_{18} \\
+&0.08488117231\,y_4 & +&0.08217103415\,y_{17} \\
+&0.1133350868\,y_5 & +&0.1019128119\,y_{16} \\
+&0.1160385455\,y_6 & +&0.1170817403\,y_{15} \\
+&0.1314402339\,y_7 & +&0.1299214090\,y_{14} \\
+&0.1492030646\,y_8 & +&0.1419790993\,y_{13} \\
+&0.1602786570\,y_9 & +&0.1482908069\,y_{12} \\
+&0.1542562513\,y_{10} & +&0.1529135858\,y_{11}
\end{aligned}
$$

We note by comparing the two columns in this display that large numerical errors have occurred (the weights should be symmetric!). We are computing the rules in a notoriously unstable way. However, MAPLE offers us more precision by increasing the value of `Digits`. Another run of the above statements preceded by `Digits:=25` yields the weights

$$
\begin{aligned}
Q := \; & 0.017614007139152119874898044\,y_1 & +&0.0176140071391521183412206\,y_{20} \\
+&0.040601429800386943328637864\,y_2 & +&0.0406014298003869413271063\,y_{19} \\
+&0.062672048334109062281307366\,y_3 & +&0.0626720483341090627981819\,y_{18} \\
+&0.083276741576704739841578937\,y_4 & +&0.0832767415767047474924052\,y_{17} \\
+&0.101930119817240439673402856\,y_5 & +&0.1019301198172404347134974\,y_{16} \\
+&0.118194531961518411529158446\,y_6 & +&0.1181945319615184185321909\,y_{15} \\
+&0.131688638449176639263929757\,y_7 & +&0.1316886384491766275831639\,y_{14} \\
+&0.142096109318382038025646168\,y_8 & +&0.1420961093183820529433995\,y_{13} \\
+&0.149172986472603754276912149\,y_9 & +&0.1491729864726037461858007\,y_{12} \\
+&0.152753387130725847085646810\,y_{10} & +&0.1527533871307258476714246\,y_{11}
\end{aligned}
$$

Comparing the two columns again we see that they match to about 16 digits, sufficient for numerical calculations. Even if the rules were not symmetric, with two runs of the MAPLE statements with different precision we are able to obtain rules correct to the number of decimal digits we want.

9.3.2 Orthogonal Polynomials

We have seen in the last section that *orthogonal polynomials* play a key role in Gauss quadrature. Orthogonal polynomials are also used to approximate functions by interpolation (see Section 4.2.5). Here we use a different scalar product. Given the scalar product (9.43), the polynomials $p_k(x)$ of degree $k = 0, 1, 2, \ldots$ are said to be orthonormal on the interval $[-1, 1]$ if

$$(p_i, p_k) = \int_{-1}^{1} p_i(x)p_k(x)\, dx = \delta_{ik}.$$

Such polynomials can be constructed using the *Gram-Schmidt procedure* discussed for computing the QR decomposition in Chapter 6. One starts with the constant polynomial of degree zero

$$p_0(x) = \text{const.} = \frac{1}{\sqrt{2}},$$

whose value is determined by the normalizing condition

$$(p_0, p_0) = \int_{-1}^{1} p_0^2(x)\, dx = 1.$$

For the next polynomial, which is linear, we make the ansatz

$$p_1(x) = a_1 x + b_1$$

and determine the constants a and b by solving the two equations

$$\int_{-1}^{1} p_1^2(x)\, dx = 1 \text{ and } \int_{-1}^{1} p_1(x)p_0(x)\, dx = 0.$$

Using MAPLE, we find

```
p0:=1/sqrt(2): p1:=a1*x+b1:
sols:=solve({int(p1^2,x=-1..1)=1,int(p0*p1,x=-1..1)=0},{a1,b1});
convert(sols,radical);
```

$$\left\{ b1 = 0,\, a1 = 1/2\,\sqrt{3}\sqrt{2} \right\}$$

and get thus

$$p_1(x) = \sqrt{\frac{3}{2}}x.$$

For the next polynomial we make the ansatz

$$p_2(x) = a_2 x^2 + b_2 x + c_2$$

and determine the coefficients by solving the three equations

$$\int_{-1}^{1} p_2^2(x)\, dx = 1, \quad \int_{-1}^{1} p_2(x)p_1(x)\, dx = 0 \text{ and } \int_{-1}^{1} p_2(x)p_0(x)\, dx = 0.$$

Continuing with MAPLE to do the calculations, we find

```
p2:=a2*x^2+b2*x+c2:
sols:=solve({int(p2^2,x=-1..1)=1,int(p2*p1,x=-1..1)=0,
int(p2*p0,x=-1..1)=0},{a2,b2,c2});
convert(sols,radical);
```

$$\left\{ b2 = 0, c2 = 1/4\sqrt{5}\sqrt{2}, a2 = -3/4\sqrt{5}\sqrt{2} \right\}.$$

Taking the leading coefficient to be positive, we find

$$p_2(x) = \frac{3}{2}\sqrt{\frac{5}{2}}\,x^2 - \frac{1}{2}\sqrt{\frac{5}{2}}.$$

Continuing in this fashion, we can construct the orthogonal polynomials on the interval $[-1, 1]$, which are multiples of the *Legendre polynomials* (see Equation (9.49)). The Legendre polynomials P_i are normalized by $P_i(1) = 1$ while the polynomials p_i are orthonormal and normalized by $(p_i, p_i) = 1$.

There is, however, an easier way to compute orthogonal polynomials. As we will see, they satisfy a *three-term recurrence relation*,

$$xp_{k-1}(x) = \beta_{k-1}p_{k-2}(x) + \alpha_k p_{k-1}(x) + \beta_k p_k(x), \quad k = 1, 2, \ldots$$

with the initial conditions $p_{-1}(x) = 0$ and $\beta_0 = 0$.

To see why this is true, we first note that every polynomial q_k of degree k can be written as a linear combination of the orthogonal polynomials p_0, p_1 up to p_k,

$$q_k(x) = \sum_{j=0}^{k} c_j p_j(x). \tag{9.44}$$

Since the orthogonal polynomials form an orthonormal basis, the coefficients c_j are given by the scalar product

$$c_j = (p_j, q_k).$$

Now for the special choice $q_k(x) := xp_{k-1}(x)$ we get

$$c_j = (p_j, xp_{k-1}) = (xp_j, p_{k-1}) = 0, \quad \text{for } j \le k - 3,$$

since p_{k-1} is orthogonal to all polynomials of degree less than $k - 1$. Thus the sum (9.44) reduces to only three terms, namely

$$xp_{k-1}(x) = \sum_{j=0}^{k} c_j p_j(x) = c_{k-2}p_{k-2}(x) + c_{k-1}p_{k-1}(x) + c_k p_k(x), \tag{9.45}$$

and the coefficients are given by

$$\begin{aligned}
c_{k-2} &= (xp_{k-2}, p_{k-1}) &= \beta_{k-1}, \\
c_{k-1} &= (xp_{k-1}, p_{k-1}) &= \alpha_k, \\
c_k &= (xp_k, p_{k-1}) &= (p_k, xp_{k-1}) = \beta_k.
\end{aligned} \tag{9.46}$$

Equations (9.45) and (9.46) give us a direct iterative method, the *Lanczos Algorithm*, to compute the orthogonal polynomials without having to go through the Gram-Schmidt procedure. We summarize our discussion in the following theorem.

THEOREM 9.5. (THREE TERM RECURRENCE OF ORTHOGONAL POLYNOMIALS) *The orthogonal polynomials* $p_0(x), p_1(x), \ldots$ *satisfy the three-term recurrence relation*

$$xp_{k-1} = \beta_{k-1}p_{k-2} + \alpha_k p_{k-1} + \beta_k p_k, \qquad k = 1, 2, \ldots \tag{9.47}$$

with $\alpha_k = (xp_{k-1}, p_{k-1})$, $\beta_k = (xp_{k-1}, p_k)$ *and with the initialization* $\beta_0 := 0$, $p_{-1}(x) := 0$, *and* $p_0(x) := 1/\sqrt{2}$.

The following MAPLE procedure Lanczos computes the orthonormal polynomials p_k for $[-1, 1]$ using Theorem 9.5:

ALGORITHM 9.4.
Lanczos Algorithm to Generate Orthogonal Polynomials

```
Lanczos:=proc(p,alpha,beta,n)
  local k,q,x; p:=array(0..n); alpha:=array(1..n); beta:=array(1..n-1);
  p[0]:=1/sqrt(2);
  alpha[1]:=int(x*p[0]*p[0],x=-1..1);
  q:=(x-alpha[1])*p[0];
  for k from 2 to n do
    beta[k-1]:=sqrt(int(q*q,x=-1..1));
    p[k-1]:=expand(q/beta[k-1]);
    alpha[k]:=int(x*p[k-1]*p[k-1],x=-1..1);
    q:=(x-alpha[k])*p[k-1]-beta[k-1]*p[k-2];
  od;
  RETURN(NULL);      # results are returned in the variables passed in
end;
```

If we compute the first 10 polynomials and the coefficients of the recurrence relation (9.47)

```
N:=10; Lanczos(p,a,b,N);
for i from 0 to N-1 do print(simplify(p[i])); od;
for i from 1 to N-1 do print(b[i]); od;
for i from 1 to N do print(a[i]); od;
```

we obtain $\alpha_k = 0$ for all k and for β_k:

$$[\frac{\sqrt{3}}{3}, \frac{2\sqrt{15}}{15}, \frac{3\sqrt{35}}{35}, \frac{4\sqrt{7}}{21}, \frac{5\sqrt{11}}{33}, \frac{6\sqrt{143}}{143}, \frac{7\sqrt{195}}{195}, \frac{8\sqrt{255}}{255}, \frac{9\sqrt{323}}{323}].$$

A closer look reveals the rule for the sequence β_k:

$$\beta_0 = 0, \quad \beta_1 = \frac{1}{\sqrt{3}} = \frac{1}{\sqrt{4-1}}, \quad \beta_2 = \frac{2}{\sqrt{15}} = \frac{2}{\sqrt{16-1}}, \quad \beta_3 = \frac{3}{\sqrt{35}} = \frac{3}{\sqrt{36-1}}, \ldots$$

so that we can guess the general rule

$$\beta_k = \frac{k}{\sqrt{4k^2 - 1}}.$$

Thus the recursion for the orthonormal polynomials on $[-1, 1]$ simplifies (with $p_{-1} = 0$, $p_0 = 1/\sqrt{2}$) to

$$x\, p_{k-1}(x) = \frac{k-1}{\sqrt{4(k-1)^2 - 1}}\, p_{k-2}(x) + \frac{k}{\sqrt{4k^2 - 1}}\, p_k(x), \quad k = 1, 2, \ldots \tag{9.48}$$

As already said, the polynomial $p_k(x)$ is a multiple of the Legendre polynomial $P_k(x)$. For Legendre polynomials, the three-term recurrence (with $P_0(x) = 1$ and $P_1(x) = x$) is:

$$x\, P_k(x) = \frac{k}{2k+1} P_{k-1}(x) + \frac{k+1}{2k+1} P_{k+1}(x). \tag{9.49}$$

We have shown in this subsection how to compute the coefficients of the three-term recurrence for the orthonormal polynomials for the interval $[-1, 1]$. It is straightforward to generalize the MAPLE procedure **Lanczos** to compute orthogonal polynomials for an interval $[a, b]$ for the scalar product

$$(f, g) = \int_a^b w(x) f(x) w(x)\, dx \tag{9.50}$$

with some weight function $w(x) \geq 0$. For this and other generalizations, we refer to Chapter 18 written by Urs von Matt in [45].

If the coefficients α_k and β_k of the three-term recurrence (9.47) are known, then we can compute the polynomials and their zeros, which are then used as nodes for Gauss quadrature rules.

In principle, any of the root-finding methods discussed in Section 5.3 can be used to compute the zeros. However, it is not recommended to compute the coefficients of $p_n(x)$ explicitly in order to use a numerical method, since, as we have seen in Section 5.3.1, the zeros are very sensitive to changes in the coefficients of a polynomial. A much better way is to use the three-term recurrence (9.47) directly to compute the function value and derivatives of $p_n(x)$. As discussed in Section 5.3.7, the method of *Newton–Maehly* is designed for polynomials which have only simple and real zeros. This is the case for orthogonal polynomials:

THEOREM 9.6. (ZEROS OF ORTHOGONAL POLYNOMIALS) *The zeros of an orthogonal polynomial p_n with respect to the scalar product (9.50) on (a, b) are all real with multiplicity one and contained in (a, b).*

PROOF. Let $\xi_1, \ldots, \xi_m$ be the real and distinct zeros of p_n with odd multiplicities in $[a, b]$. Clearly $0 \leq m \leq n$ holds. We form the polynomial

$$q_m(x) = (x - \xi_1) \cdots (x - \xi_m), \quad (q_m = 1 \text{ if } m = 0),$$

and consider the integral

$$\int_a^b p_n(x)q_m(x)w(x)\,dx. \tag{9.51}$$

This integral is different from zero, since $p_n(x)q_m(x)$ does not change signs on the interval (a, b). On the other hand, since p_n is orthogonal to all polynomials of degree $\leq n - 1$, (9.51) can only be non-zero if $m \geq n$. Therefore we must have $m = n$ and thus all zeros must be simple and contained in (a, b). $\square$

9.3.3   Computing the Weights

If the nodes ξ_j are known, then the weights can be computed by integrating the Lagrange polynomials,

$$w_i = \int_a^b l_i(x)\,dx, \quad \text{with } l_i(x) = \prod_{\substack{j=1 \\ j\neq i}}^n \frac{x - \xi_j}{\xi_i - \xi_j}.$$

Since the nodes are known, a first simplification is obtained by factoring the orthogonal polynomial,

$$p_n(x) = a_n \prod_{j=1}^n (x - \xi_j).$$

Now, because of $p_n(\xi_i) = 0$,

$$p_n'(\xi_i) = \lim_{x\to\xi_i} \frac{p_n(x) - p_n(\xi_i)}{x - \xi_i} = \lim_{x\to\xi_i} \frac{a_n \prod_{j=1}^n (x - \xi_j)}{x - \xi_i} = a_n \prod_{\substack{j=1 \\ j\neq i}}^n (\xi_i - \xi_j).$$

Thus, $p_n'(\xi_i)$ is essentially the denominator of the Lagrange polynomial, so we can write

$$l_i(x) = \frac{p_n(x)}{p_n'(\xi_i)(x - \xi_i)}.$$

The weights then become

$$w_i = \frac{1}{p_n'(\xi_i)} \int_a^b \frac{p_n(x)}{(x - \xi_i)}\,dx.$$

THEOREM 9.7. (QUADRATURE WEIGHTS) *Define* $\Phi_0(x) = 0$ *and*

$$\Phi_i(x) = \int_a^b \frac{p_i(t) - p_i(x)}{t - x}\,dt, \qquad i = 1, 2, \ldots, n.$$

Then the functions $\Phi_i(x)$ satisfy the same three-term recurrence as the orthogonal polynomials

$$x\Phi_{i-1} = \beta_{i-1}\Phi_{i-2} + \alpha_i\Phi_{i-1} + \beta_i\Phi_i, \qquad i = 2, 3, \ldots, n,$$

and the weights are computed by

$$w_i = \frac{\Phi_n(\xi_i)}{p_n'(\xi_i)}, \qquad i = 1, 2, \ldots, n. \tag{9.52}$$

PROOF. Using the recurrence relation for the orthogonal polynomials

$$\beta_n p_n(x) = (x - \alpha_n)p_{n-1} - \beta_{n-1}p_{n-2}$$

in

$$\beta_n \Phi_n(x) = \int_a^b \frac{\beta_n p_n(t) - \beta_n p_n(x)}{t - x}\, dt,$$

we obtain

$$\beta_n \Phi_n(x) = \int_a^b \frac{(t - \alpha_n)p_{n-1}(t) - \beta_{n-1}p_{n-2}(t) - (x - \alpha_n)p_{n-1}(x) + \beta_{n-1}p_{n-2}(x)}{t - x}$$

$$= -\beta_{n-1}\Phi_{n-2}(x) + \int_a^b \frac{(t - \alpha_n)p_{n-1}(t) - (x - \alpha_n)p_{n-1}(x)}{t - x}\, dt.$$

Replacing $t - \alpha_n$ by $(t - x) + (x - \alpha_n)$ we obtain

$$= -\beta_{n-1}\Phi_{n-2}(x) + \underbrace{\int_a^b p_{n-1}(t)\, dt}_{=0} + (x - \alpha_n)\Phi_{n-1}(x).$$

$\square$

Using Theorem 9.7 and the Newton–Maehly method to compute the nodes (see Section 5.3.7), we can write a MATLAB function to compute Gauss–Legendre quadrature rules. We use the coefficients

$$\alpha_k = 0, \quad \beta_k = \frac{k}{\sqrt{4k^2 - 1}}, \quad k = 1, \ldots, n.$$

Furthermore $p_0(x) = 1/\sqrt{2}$ and

$$\Phi_1(x) = \int_a^b \frac{p_1(t) - p_1(x)}{t - x}\, dt = p_1(x)'(b - a) = \sqrt{\frac{3}{2}}\,(b - a).$$

The weights and nodes are adapted to the integration interval (a, b) using the change of variables (9.38).

ALGORITHM 9.5.

Gauss-Legendre Quadrature with Newton–Maehly

```
function [xi,w]=GaussByNewtonMaehly(a,b,n);
% GAUSSBYNEWTONMAEHLY Gauss nodes and weights using Newton Maehly
%    [xi,w]=GaussByNewtonMaehly(a,b,n) computes the n nodes and weights
%    for Gauss Legendre quadrature for the interval (a,b). The nodes
%    are computed with the Newton-Maehly algorithm, and the weights by
%    using the function Phi and the three term recurrence relation.
%    Since the rule is symmetric, only half of the nodes and weights
%    are computed.

xi=[];
for k= 1:n                          % coefficients for Gauss-Legendre
  beta(k)=k/sqrt(4*k^2-1);          % alpha(i)=0
end
phi=sqrt(3/2)*(b-a);                % =p_1'*int_a^b
p=1/sqrt(2);                        % =p_0
xm=(a+b)/2; xl=(b-a)/2;             % change of variables
anz=fix((n+1)/2); x0=1;
for k=1:anz,
  y=x0; finished=0;
  m=2;                             % Newton double step
  while ~finished,
    x=y;                           % evaluate polynom and derivative
    p1s=0; p1=p; p2s=sqrt(3/2); p2=sqrt(3/2)*x;
    for i=2:n
      p0s=p1s; p0=p1; p1s=p2s; p1=p2;
      p2s=(p1+x*p1s-beta(i-1)*p0s)/beta(i);
      p2=(x*p1-beta(i-1)*p0)/beta(i);
    end
    s=sum(1./(x-xi(1:k-1))+1./(x+xi(1:k-1)));
    y=x-m*p2/p2s/(1-p2/p2s*s);      % Newton-Maehly step
    if y>=x,
       if m==1, finished=1;
       end
       if ~finished;
         x0=x; m=1;                 % stop double step
         y=x-p2/p2s/(1-p2/p2s*s);   % Newton backstep
       end
    end
  end
  xi(k)=y;
  phi1=0; phi2=phi;
  for i=2:n
    phi0=phi1; phi1=phi2; phi2=(y*phi1-beta(i-1)*phi0)/beta(i);
```

```
   end
   w(k)=xl*phi2/p2s; w(n+1-k)=w(k);
end
for k=1:anz,                        % backchange variables
   y=xi(k); xi(k)=xm+xl*y; xi(n+1-k)=xm-xl*y;
end
xi=xi(:); w=w(:);
```

With the call [xi,w]=GaussByNewtonMaehly(-1,1,10), we obtain the values

ξ_i	w_i
0.97390652851717	0.06667134430869
0.86506336668898	0.14945134915058
0.67940956829902	0.21908636251598
0.43339539412925	0.26926671931000
0.14887433898163	0.29552422471475
-0.14887433898163	0.29552422471475
-0.43339539412925	0.26926671931000
-0.67940956829902	0.21908636251598
-0.86506336668898	0.14945134915058
-0.97390652851717	0.06667134430869

which are correct to all printed digits.

In the next section, we will describe another way to compute the weights and nodes for Gauss quadrature rules. This is the most elegant method for computing these quantities numerically, as we will see.

9.3.4 Golub–Welsch Algorithm

If we know the coefficients α_k and β_k, we can write the three-term recurrence relations (9.47) simultaneously in matrix form, namely

$$
x \begin{pmatrix} p_0(x) \\ p_1(x) \\ \vdots \\ p_{n-1}(x) \end{pmatrix} = \begin{pmatrix} \alpha_1 & \beta_1 & & & \\ \beta_1 & \alpha_2 & \beta_2 & & \\ & \beta_2 & \ddots & \ddots & \\ & & \ddots & \alpha_{n-1} & \beta_{n-1} \\ & & & \beta_{n-1} & \alpha_n \end{pmatrix} \begin{pmatrix} p_0(x) \\ p_1(x) \\ \vdots \\ p_{n-1}(x) \end{pmatrix} + \begin{pmatrix} 0 \\ \vdots \\ 0 \\ \beta_n p_n(x) \end{pmatrix},
$$

or, using vector notation,

$$
x\boldsymbol{p}(x) = T_n \boldsymbol{p}(x) + \boldsymbol{e}_n \beta_n p_n(x),
$$

where $\boldsymbol{e}_n$ denotes the canonical basis vector.

If ξ_i is a zero of $p_n(x)$, then

$$
\xi_i \boldsymbol{p}(\xi_i) = T_n \boldsymbol{p}(\xi_i).
$$

Hence, *the zeros ξ_i of the n-th orthogonal polynomial $p_n(x)$ are the eigenvalues of the matrix*

$$
T_n = \begin{bmatrix}
\alpha_1 & \beta_1 & & & & \\
\beta_1 & \alpha_2 & \beta_2 & & & \\
& \beta_2 & \ddots & & \ddots & \\
& & & \ddots & \alpha_{n-1} & \beta_{n-1} \\
& & & & \beta_{n-1} & \alpha_n
\end{bmatrix}.
$$

As a result, we can use an algorithm for computing eigenvalues of symmetric tridiagonal matrices in order to compute the nodes of a Gauss quadrature rule. Note that the eigenvectors

$$
\boldsymbol{p}(\xi_i) = \begin{pmatrix}
p_0(\xi_i) \\
p_1(\xi_i) \\
\vdots \\
p_{n-1}(\xi_i)
\end{pmatrix}
$$

are the function values of the orthogonal polynomials of lower degree evaluated at the zero ξ_i of p_n.

We have seen that the weights w_i are determined by integrating the Lagrange polynomials for the interpolating polynomial. Since the Gauss quadrature rule

$$
\int_{-1}^{1} f(x)\, dx \approx \sum_{k=1}^{n} w_k f(\xi_k)
$$

is exact for any polynomial of degree up to $2n - 1$, it will integrate the orthogonal polynomials $p_i(x)$ exactly for $i = 0, 1, \ldots, n - 1$:

$$
\begin{aligned}
\int_{-1}^{1} p_0(x)\, dx &= \sqrt{2} &= \sum_{i=1}^{n} p_0(\xi_i) w_i, \\
\int_{-1}^{1} p_1(x)\, dx &= 0 &= \sum_{i=1}^{n} p_1(\xi_i) w_i, \\
&\;\;\vdots & \vdots \\
\int_{-1}^{1} p_{n-1}(x)\, dx &= 0 &= \sum_{i=1}^{n} p_{n-1}(\xi_i) w_i,
\end{aligned}
$$

or, in matrix notation,

$$
\begin{bmatrix}
p_0(\xi_1) & p_0(\xi_2) & \cdots & p_0(\xi_n) \\
p_1(\xi_1) & p_1(\xi_2) & \cdots & p_1(\xi_n) \\
\vdots & \vdots & & \vdots \\
p_{n-1}(\xi_1) & p_{n-1}(\xi_2) & \cdots & p_{n-1}(\xi_n)
\end{bmatrix}
\begin{pmatrix}
w_1 \\
w_2 \\
\vdots \\
w_n
\end{pmatrix}
=
\begin{pmatrix}
\sqrt{2} \\
0 \\
\vdots \\
0
\end{pmatrix}.
$$

So to obtain the weights, we have to solve this linear system of equations, in matrix form

$$Pw = \sqrt{2}e_1.$$

But the columns in the matrix P are the eigenvectors of the symmetric matrix T_n, so they form an orthogonal matrix when scaled properly. To find this scaling, note that $p_0(x) = \text{const} = \frac{1}{\sqrt{2}}$. If we compute the eigen-decomposition

$$T_n = QDQ^\top,$$

using MATLAB `[Q,D]=eig(Tn)`, then the eigenvector matrix Q is orthogonal, i.e., $Q^\top Q = I$. In order to find P, we must scale the eigenvectors so that the first component of each eigenvector becomes $\frac{1}{\sqrt{2}}$.

If q is a column of Q, we have to scale it by dividing first by its first component and then multiplying by $\frac{1}{\sqrt{2}}$. In matrix notation, this is done by

$$P = QV$$

where V is a diagonal matrix containing the appropriate scaling factors

$$V = \text{diag}\left(\frac{1}{\sqrt{2}\,q_{1,1}}, \frac{1}{\sqrt{2}\,q_{1,2}}, \ldots, \frac{1}{\sqrt{2}\,q_{1,n}}\right).$$

Hence solving $Pw = \sqrt{2}e_1$ is equivalent to solving $QVw = \sqrt{2}e_1$ which is simply

$$w = \sqrt{2}V^{-1}Q^\top e_1 = 2\begin{pmatrix} q_{1,1}^2 \\ q_{1,2}^2 \\ \vdots \\ q_{1,n}^2 \end{pmatrix}.$$

Thus, all the information to compute the Gauss quadrature rules is contained in the matrix T_n. This result is the key idea of the *Golub–Welsch Algorithm* for computing Gauss quadrature rules [52]. *The nodes are the eigenvalues of the matrix T_n and the weights are the first components of the eigenvectors of T_n squared and multiplied by the factor 2.* These quantities can easily be obtained in MATLAB using brute force, which leads to the following short program for computing Gauss-Legendre quadrature rules using the Golub–Welsch idea, as shown in Trefethen and Bau [140]:

ALGORITHM 9.6.
Brute Force Implementation of the Golub–Welsch
approach by Trefethen and Bau

```
function [x,w]=GaussByGolubWelschBruteForce(n);
% GAUSSBYGOLUBWELSCHBRUTEFORCE Golub-Welsch brute force implementation
%    [x,w]=GaussByGolubWelschBruteForce(n); computes the n nodes and
```

```
%    weights for Gauss Legendre quadrature for the interval (a,b) using
%    the idea of Golub-Welsch, but in a brute force implementation.

beta=0.5./sqrt(1-(2*(1:n-1)).^(-2));
[Q,D]=eig(diag(beta,1)+diag(beta,-1));
[x,i]=sort(diag(D)); w=2*Q(1,i).^2';
```

The function `GaussByGolubWelschBruteForce` works well, but we have
to be aware that this short program *does not take into account the structure
of the problem* as it is proposed in the algorithm of Golub–Welsch [52]. *We
have only to solve a symmetric tridiagonal eigenvalue problem and from the
eigenvector matrix we only need the first row.*

With `GaussByGolubWelschBruteForce` the matrix is treated as full, and
the eigenvectors are computed in their entirety, not just their first components.

In Chapter 7, we have developed the QR Algorithm 7.10, which computes the eigenvalues of a symmetric tridiagonal matrix. The Golub-Welsch
approach uses this algorithm to compute the nodes and modifies it to compute the weights simultaneously. To do so, we need to apply the necessary
Jacobi rotations not only to the tridiagonal matrix, but also to a vector g that
is initialized to $g = e_1$. The accumulation of the rotations on g is equivalent
to computing

$$w = Q^\top e_1,$$

i.e., we obtain the first row of Q. To get the weights, the elements have to
be squared and multiplied by the factor 2. The following program results:

ALGORITHM 9.7.
Gauss-Quadrature: Golub-Welsch Algorithm

```
function [xi,w]=GaussByGolubWelsch(a,b)
% GAUSSBYGOLUBWELSCH compute Gauss nodes and weights using Golub-Welsch
%    [x,w]=GaussByGolubWelsch(n); computes the n nodes and weights for
%    Gauss Legendre quadrature for the interval (a,b) using the
%    algorithm of Golub-Welsch.

n=length(a); w=zeros(n-1,1); w=[1;w];
I=1;
while n>1;
  for k=I:n-1                            % Check for small b_i and
    an=abs(a(k))+abs(a(k+1));            % possible deflation
    if an+b(k)==an, b(k)=0; I=k+1; end
  end
  if I==n;                               % deflation
    n=n-1; I=1;
  elseif I+1==n
    g=100*abs(b(I)); h=a(I+1)-a(I);
```

```
    if abs(h)+g==abs(h),t=b(I)/h;          % 2x2 block: annihilate b(I)
    else                                    % explicitly by one rotation
      theta=0.5*h/b(I);
      t=1/(abs(theta)+sqrt(1+theta^2));
      if theta<0, t=-t; end
    end
    c=1/sqrt(1+t^2); s=t*c;                 % rotate weights vector
    hv=c*w(I)-s*w(I+1); w(I+1)=s*w(I)+c*w(I+1); w(I)=hv;
    a(I)=a(I)-b(I)*t; a(I+1)=a(I+1)+b(I)*t; b(I)=0;
    n=n-2; I=1;                             % deflation
  else
    delta =(a(n)-a(n-1))/2;                 % QR-step from I to n
    if delta==0,                            % with Wilkinson shift
      sigma=a(n)-abs(b(n-1));
    else
      sigma=a(n)-b(n-1)^2/(delta+sign(delta)*sqrt(delta^2+b(n-1)^2));
    end
    [c s]=GivensRotation(a(I)-sigma, b(I));% first transformation
    d=a(I)-a(I+1); z=s*(s*d+2*c*b(I));
    a(I)=a(I)-z; a(I+1)=a(I+1)+z;
    b(I)=c*s*d+b(I)*(c^2-s^2);
    bulge=-s*b(I+1); b(I+1)=c*b(I+1);
    hv=c*w(I)-s*w(I+1); w(I+1)=s*w(I)+c*w(I+1); w(I)=hv;
    for k= I+2:n                            % chasing the bulge
      [c s]=GivensRotation(b(k-2), bulge);
      balt= b(k-1);
      d=a(k-1)-a(k); z=s*(s*d+2*c*b(k-1));
      a(k-1)=a(k-1)-z; a(k)=a(k)+z;
      b(k-2)=c*b(k-2)-s*bulge;
      b(k-1)=c*s*d+b(k-1)*(c^2-s^2);
      if k<n
        bulge=-s*b(k); b(k)=c*b(k);
      end
      hv=c*w(k-1)-s*w(k); w(k)=s*w(k-1)+c*w(k); w(k-1)=hv;
    end
  end
end
[xi,i]=sort(a); xi=xi(:); w=2*w(i).^2;
```

With the statements

```
n=20
for k=1:n
  beta(k)=k/sqrt(4*k^2-1);
  alpha(k)=0;
end
[xi,w]=GaussByGolubWelsch(alpha,beta)
```

we obtain the weights and nodes:

w_i	ξ_i
0.01761400713915	-0.99312859918509
0.04060142980039	-0.96397192727791
0.06267204833411	-0.91223442825133
0.08327674157671	-0.83911697182222
0.10193011981724	-0.74633190646015
0.11819453196152	-0.63605368072652
0.13168863844918	-0.51086700195083
0.14209610931838	-0.37370608871542
0.14917298647260	-0.22778585114165
0.15275338713072	-0.07652652113350
0.15275338713073	0.07652652113350
0.14917298647260	0.22778585114165
0.14209610931838	0.37370608871542
0.13168863844918	0.51086700195083
0.11819453196152	0.63605368072651
0.10193011981724	0.74633190646015
0.08327674157670	0.83911697182222
0.06267204833411	0.91223442825133
0.04060142980039	0.96397192727791
0.01761400713915	0.99312859918509

We have presented three algorithms to compute the nodes and weights for
Gauss quadrature: `GaussByNewtonMaehly`, `GaussByGolubWelschBruteForce`
solving the full eigenvalue problem, and `GaussByGolubWelsch` solving the
tridiagonal eingenvalue problem and computing only the first row of the
eigenvector matrix. We cannot compare in a fair way the required computing time, since `eig` is a MATLAB *built-in function* and flop counts are no
longer reported in newer MATLAB releases. However, we can compare the accuracy of the results. For that purpose we need some exact reference values.
Therefore we first compute a Gauss quadrature rule for $n = 50$ using MAPLE
with extended precision

ALGORITHM 9.8.
Generate Reference Values for Gauss-Quadrature

```
Digits:=50;
n:=50;
X:=sort([fsolve(orthopoly[P](n,x)=0,x)]);
Q:=int(interp(X,[seq(y[i],i=1..n)],z),z=-1..1);
```

Now we compute the nodes and weights with the three MATLAB functions
and compare the results with the exact values by computing the difference

of the norm of the vectors. We obtain

MATLAB function	norm nodes	norm weights
GaussByNewtonMaehly	1.77×10^{-16}	9.25×10^{-16}
GaussByGolubWelschBruteForce	1.66×10^{-15}	2.83×10^{-15}
GaussByGolubWelsch	2.91×10^{-15}	4.85×10^{-15}

and we note that Newton–Maehly is very accurate.

9.4 Adaptive Quadrature

So far, to compute an integral to a certain accuracy, we have used a composite quadrature rule in which the integration step size is constant. We control the integration error by halving the step size until the relative difference of two approximations is smaller than some prescribed tolerance. By doing so, we generate equidistant subintervals and apply a quadrature rule (e.g. Simpson's rule) to each subinterval. For each subinterval, there will be a different integration error depending on the local behavior of the function. The global error will therefore depend mainly on the largest error in a subinterval.

To minimize the number of function evaluations, we should ideally work with a non-uniform partition of subintervals which is adapted to the local behavior of the function. For instance, larger integration steps can be taken where the function is flat, whereas smaller steps should be taken where the function has sharp transitions. The partition should be chosen in such a way that the error becomes roughly the same for each subinterval. This will generally not be the case for an equidistant partition, which tends to have too many points in the flat portions and not enough in the steep ones.

We will illustrate this by considering the integral

$$I = \int_0^1 \sqrt{x}\, dx = \frac{2}{3} = 0.666666666666\ldots.$$

If we compute I using the function `TrapezoidalRule(@sqrt,0,1,1e-5)` and print consecutive approximations, we get

h	$T(h)$
1	0.50000000000000
2^{-1}	0.60355339059327
2^{-2}	0.64328304624275
2^{-3}	0.65813022162445
2^{-4}	0.663358119687723
2^{-5}	0.66555893627894
2^{-6}	0.66627081137851
2^{-7}	0.66652565729683
2^{-8}	0.66661654897653
2^{-9}	0.66664888154995
2^{-10}	0.66666036221898
2^{-11}	0.66666443359297

Thus, for a result with 5 correct decimal digits, we need a step size of about $h = 2^{-10} = 9.765625e{-}4 \approx 10^{-3}$. We cannot estimate the error in the first subinterval $[0, h]$ with (9.12) because the derivative of $f(x) = \sqrt{x}$ has a singularity at $x = 0$ and the estimate is not valid. However, the estimate is applicable for the second subinterval $[h, 2h]$. Since $\max_{h \leq x \leq 2h} |f''(x)| = h^{-3/2}/4$, we obtain

$$|I - T(h)| = \frac{|f''(\xi)|}{12} h^3 \leq \frac{h^{3/2}}{48} = 6.357e{-}7 < 1e{-}5.$$

Already for the second subinterval, the step size is smaller than necessary for the tolerance! The disparity becomes even more pronounced for the last interval $[1 - h, 1]$, where the error estimate gives the bound $1.94e{-}11$. For this example, the step size h is dictated only by the first subinterval in which the function changes most.

How large could the step size be for the trapezoidal rule and `tol` $= 1e{-}5$ for the last subinterval? If we apply the error estimate (9.12) and use

$$\max_{1-h \leq x \leq 1} |f''(x)| = \frac{(1 - h)^{-\frac{3}{2}}}{4},$$

then we need to solve the equation

$$\frac{(1 - h)^{-\frac{3}{2}} h^3}{48} = 10^{-5}$$

to determine h. We obtain $h = 0.075$, which means that we could choose for the same error in the last subinterval a step size that is about 70 times larger.

The idea of adaptive quadrature is to try to choose a partition of subintervals automatically, such that the *error is about the same in all subintervals*.

All well-known software libraries (like IMSL, NAG, Netlib) offer general purpose adaptive quadrature routines. There are also such functions in MATLAB.

One way to achieve this automatic partition of subintervals is by the principle of *divide and conquer*. To compute

$$I = \int_a^b f(x)dx,$$

one proceeds as follows. First, one integrates f using two different numerical integration methods (or the same method with two different step sizes), thus obtaining the approximations I_1 and I_2. Typically, one, say I_1, is more accurate than the other. If the relative difference of the two approximations is smaller than some prescribed tolerance, one accepts I_1 as the value of the integral. Otherwise, the interval $[a, b]$ is *divided*, e.g., into two equal parts $[a, m]$ and $[m, b]$, where $m = (a + b)/2$, and the two respective integrals are computed (*conquered*) independently,

$$I = \int_a^m f(x)dx + \int_m^b f(x)dx.$$

Next, one recursively computes two approximations for each integral and, if necessary, continues to subdivide the smaller intervals. Using recursion is a very obvious way to implement an adaptive quadrature algorithm. In what follows, we will discuss one such algorithm, as described in [41] and [43].

9.4.1 Stopping Criterion

First, we need to decide when to stop the recursion. If I_1 and I_2 are two estimates for the integral, a conventional stopping criterion is

$$\text{if abs(i1-i2)<tol*abs(i1),} \tag{9.53}$$

where `tol` is some prescribed error tolerance. However, this criterion by itself is not sufficient. For our example, of $f(x) = \sqrt{x}$ integrated on $[0, 1]$ with I_1, defined by Simpson's rule using the step size h, and I_2 defined by Simpson's rule for the step size $h/2$, and $\text{tol} = 10^{-4}$ we obtain in MATLAB the error message:

```
???  Maximum recursion limit of 500 reached.  Use set(0,'RecursionLimit',N) to
change the limit.  Be aware that exceeding your available stack space can crash
MATLAB and/or your computer.
```

The reason is that, in this example, the two Simpson values never agree to 4 digits in the first interval containing 0.

A better criterion for terminating the recursion is based on the observation that it makes no sense to continue subdividing when the partial integral I_1

or I_2 becomes negligible compared to the whole integral. Therefore, we have to add the criterion

$$|I_1| < \eta \left| \int_a^b f(x)dx \right|, \qquad (9.54)$$

where η is another prescribed tolerance. Since the integral is not known in advance, we have to use an estimate instead. When both criteria (9.53) and (9.54) are used together with some reasonable choices of *tol* and η, a working algorithm can be obtained. For instance, if we use for the above example

$$\texttt{if abs(i1-i2)<1e-4*abs(i1) | abs(i1)<1e-4} \qquad (9.55)$$

as the stopping criterion, we obtain $I = 0.666617217$ in 41 function evaluations.

The stopping criterion (9.55), however, is still not satisfactory because the user has to choose `tol` and η, which depend on the machine and on the problem. To avoid possible bad choices by the user, one should eliminate such parameters whenever possible. To improve the criterion, we first need a rough estimate $is \approx \left| \int_a^b f(x)dx \right|$, with $is \neq 0$. The stopping criterion (9.54) would then be $|I_1| < \eta \cdot |is|$. But in order to eliminate η, we stop the recursion in a *machine-independent* way by (see also Subsection 2.8.1)

$$\texttt{if is+i1==is.} \qquad (9.56)$$

In the same spirit, we may as well replace the criterion (9.53) by

$$\texttt{if is+(i1-i2)==is.} \qquad (9.57)$$

Criterion (9.57) will in general be met before Criterion (9.56), and therefore we shall require only (9.57).

There are cases, e.g., $\int_0^1 \frac{1}{\sqrt{1-x^2}}\, dx$, where, when *ignoring the singularity* (this means artificially define the function value at the singularity to be equal to zero), the subdivision will continue until an interval contains no machine number other than the end points. In this case, we also need to terminate the recursion. Thus, our termination criterion is

$$\texttt{if (is+(i1-i2)==is) | (m<=a) | (m>=b),} \qquad (9.58)$$

where $\texttt{m} = (a + b)/2$. This, in particular, guarantees termination of the program.

Using the stopping criterion (9.58), we attempt to compute the integral to machine precision. If we wish to compute the integral with less accuracy, say within the tolerance `tol`, it suffices to magnify the estimated value `is` by

$$\texttt{is=is*tol/eps,}$$

where `eps` denotes the machine precision. Of course `is` will then no longer be an estimate of the integral.

9.4.2 Adaptive Simpson quadrature

As a first algorithm, we will use Simpson's method to develop two MATLAB functions for adaptive quadrature. The first `SimpsonAdaptive(f,a,b,tol)` will be used for initialization and it will call the second recursive function `SimpsonRecursion`.

For the first approximation I_1, we apply Simpson's rule to a single interval $[a, b]$. This requires the function values `fa` $= f(a)$, `fm` $= f(m)$ and `fb` $= f(b)$:

$$\texttt{i1} = \texttt{h}/1.5 * (\texttt{fa} + 4 * \texttt{fm} + \texttt{fb});$$

where `m` $= (a + b)/2$ and `h` $= (b - a)/4$. For the second approximation I_2, we apply Simpson's rule to the two subintervals $[a, m]$ and $[m, b]$, which requires two additional function values `fml` $= f(a + h)$ and `fmr` $= f(b - h)$:

$$\texttt{i2} = \texttt{h}/3 * (\texttt{fa} + 4 * (\texttt{fml} + \texttt{fmr}) + 2 * \texttt{fm} + \texttt{fb});$$

Now since we have two Simpson values, we can extrapolate a better value using the Romberg scheme and we overwrite the less accurate Simpson value I_1:

$$\texttt{i1} = (16 * \texttt{i2} - \texttt{i1})/15;$$

So we will use the Simpson value for the step size $h = (b - a)/4$ and the extrapolated value for the termination criterion (9.58). If we need to recur, then, in order to avoid recomputing the function values, we will pass to the next recursion level the function values `fa`, `fml` and `fm` for the interval $[a, m]$, and the values `fm`, `fmr` and `fb` for the second interval $[m, b]$ as parameters.

Our next concern is to compute the estimated value `is`. What we really need is just a rough estimate that indicates the order of magnitude of the integral. For this, we propose a Monte Carlo estimate of the integral, in which we also use the function values in the middle and at the end points of the interval (those values are used for Simpson's rule):

$$\texttt{is} = \frac{b - a}{8}\left(f(a) + f(m) + f(b) + \sum_{i=1}^{5} f(\xi_i)\right). \tag{9.59}$$

Here $m = (a + b)/2$ and $\xi = a + [.9501\ .2311\ .6068\ .4860\ .8913](b - a)$ is a vector of random numbers in (a, b). If by accident we get $is = 0$, then we use the value $\texttt{is} = b - a$, which means that for computing the estimate we replace the function f by the constant 1. With this choice of `is`, we adopt the stopping criterion (9.58).

We are now ready to present the function `SimpsonAdaptive`.

ALGORITHM 9.9.
Adaptive Quadrature using Simpson's Rule

```
function Q=SimpsonAdaptive(f,a,b,tol,trace,varargin)
```

```matlab
% SIMPSONADAPTIVE approximate integral using adaptive Simpson rule
%  Q=SimpsonAdaptive(f,a,b) approximates the integral of F(X) from a
%  to b to machine precision.  f is a function handle.  The function
%  f must return a vector of output values if given a vector of input
%  values.
%  If SimpsonAdaptive finds that the integrand f has a singularity,
%  i.e. encounters a function value NaN or Inf it replaces this value
%  by 0 and issues a warning.
%  Q=SimpsonAdaptive(f,a,b,tol) integrates to a relative error of
%  tol.  Q=SimpsonAdaptive(f,a,b,tol,trace) displays the left end
%  point of the current interval, the interval length, and the
%  partial integral.  Q=SimpsonAdaptive(f,a,b,tol,trace,p1,p2,...)
%  allows coefficients p1,... to be passed directly to the function
%  f: G =f(x,p1,p2,...).  To use default values for tol or trace, one
%  may pass the empty matrix ([]).
%  Walter Gander, 22.8.2000

global warn1 warn2
if (nargin<4),tol=[]; end;
if (nargin<5),trace=[]; end;
if (isempty(tol)),tol=eps; end;
if (isempty(trace)),trace=0; end;
if tol<=eps,tol=10*eps;   end
warn1=0; warn2=0;
x=[a (a+b)/2 b]; y=f(x,varargin{:});
for p= 1:length(y)
  if isinf(y(p)) | isnan(y(p)),
    y(p)=0; warn1=1;
  end
end
fa=y(1); fm=y(2); fb=y(3);
yy=f(a+[.9501 .2311 .6068 .4860 .8913]*(b-a),varargin{:});
for p= 1:length(yy)
  if isinf(yy(p)) | isnan(yy(p)),
    yy(p)=0; warn1=1;
  end
end
is=(b-a)/8*(sum(y)+sum(yy));
is=is*tol/eps; if is==0,is=b-a; end;
Q=SimpsonRecursive(f,a,b,fa,fm,fb,is,trace,varargin{:});
if warn1==1,
  warning(['Infinite or Not-a-Number function value encountered. ',...
   'Singularity likely. Required tolerance may not be met.']);
end
if warn2==1,
  warning(['Interval contains no more machine number. ',...
   'Singularity likely. Required tolerance may not be met.']);
end
```

```
function Q=SimpsonRecursive(f,a,b,fa,fm,fb,is,trace,varargin)
% SIMPSONRECURSIVE Recursive function used only by SIMPSONADAPTIVE

global warn1 warn2
m=(a+b)/2; h=(b-a)/4;
x=[a+h,b-h]; y=f(x,varargin{:});
for p=1:length(y)
  if isinf(y(p)) | isnan(y(p)),
    y(p)=0; warn1=1;
  end
end
fml=y(1); fmr=y(2); i1=h/1.5*(fa+4*fm+fb);
i2=h/3*(fa+4*(fml+fmr)+2*fm+fb); i1=(16*i2-i1)/15;
if (is+(i1-i2)==is) | (m<=a) | (b<=m),
  if ((m <= a) | (b<=m)), warn2=1; end;
  Q=i1;
  if (trace),disp([a b-a Q]),end;
else
  Q=SimpsonRecursive(f,a,m,fa,fml,fm,is,trace,varargin{:})+...
    SimpsonRecursive(f,m,b,fm,fmr,fb,is,trace,varargin{:});
end;
```

We go back to the integral that we considered at the beginning of this section,

$$I = \int_0^1 \sqrt{x}\,dx = \frac{2}{3} = 0.666666666666\ldots.$$

To compute the integral to a tolerance of $1e-5$, the composite trapezoidal rule required an integration step size of $h = 2^{-10}$, which implied about 1000 function evaluations. Using `SimpsonAdaptive(@sqrt,0,1,1e-5,1)` we get the result `0.66665999490706` with only 38 function evaluations. With `trace` turned on, the following intermediate values are printed:

i	x_i	$x_{i+1} - x_i$	$\int_{x_i}^{x_{i+1}} \sqrt{x}\,dx$
0	0.0000000	0.0078125	0.00045420327593
1	0.0078125	0.0078125	0.00084172670019
2	0.0156250	0.0156250	0.00238076263043
3	0.0312500	0.0312500	0.00673381360150
4	0.0625000	0.0625000	0.01904610104346
5	0.1250000	0.1250000	0.05387050881198
6	0.2500000	0.2500000	0.15236880834770
7	0.5000000	0.5000000	0.43096407049588

We notice that we needed only 8 subintervals. The smallest, the first one, has length 0.0078125 and the length of the last one is 0.5. We see that adaptive quadrature is remarkably efficient for this example.

Since `SimpsonAdaptive` works so well, we tried to make it as user-friendly and also as bullet-proof as possible. A tolerance of `eps` (the machine precision) makes `SimpsonAdaptive` work much harder than necessary, without yielding any noticeable gain in accuracy compared to, say, `tol = 10 · eps`. Thus we set for a required `tol ≤ eps` the tolerance to `tol = 10 · eps`.

For functions with a singularity, the integral may nevertheless exist. Consider as an example the integral

$$\int_0^1 \frac{\ln(1-x)}{x} \, dx = -\frac{\pi^2}{6} = -1.64493406684823.$$

For $x = 0$ and $x = 1$, the function has a singularity. One lazy way to deal with undefined function values and singularities is just to ignore them. This means, as we said before, to define artificially the function value to be equal to zero. Therefore we would have to program the integrand as follows:

```
function y=g(x)
y=[];
for z=x,
  if (z==0)
    y=[y 1];
  elseif z==1
    y=[y 0];
  else
    y=[y log(1-z)/z];
  end
end
```

However, we can detect and ignore such function values automatically in the function `SimpsonAdaptive` and make life for the user more comfortable (but also more dangerous!). In `SimpsonAdaptive`, we check after each function evaluation if `NaN` or `Inf` values occur, and replace them by zeros when they do. The user can now program the function simply as

```
g2=@(x) log(1-x)./x;
```

and with `SimpsonAdaptive(g2,0,1,1e-10)` we obtain -1.64493406583834 which is correct to 10 digits using 642 function evaluations. If `NaN` or `Inf` values occur `SimpsonAdaptive` will warn the user with

```
    Warning: Infinite or Not-a-Number function value encountered.
    Singularity likely. Required tolerance may not be met.
```

The danger of such an automated approach is that users of `SimpsonAdaptive` may "solve" unreasonable problems. For instance the integral

$$I = \int_0^1 \frac{dx}{3x - 1}$$

does not exist. The call `SimpsonAdaptive(@(x) 1./(3*x-1),0,1,1e-5)` delivers warning messages but also a wrong integral value of 0.13072984773219 using 1406 function evaluations.

9.4.3 Adaptive Lobatto quadrature

Gauss quadrature rules achieve a high order of approximation with a minimal number of points. However, if we refine the integration step size it is not possible to reuse the function values of the previous level. The goal in this section is therefore to compromise. When we consider Gauss quadrature rules in which some nodes are prescribed, then we speak of Gauss–Lobatto or, more generally, of Gauss–Kronrod rules. We will use *Gauss–Lobatto rules* (where the endpoints of the integration interval are used) and develop *Kronrod extensions* for some given nodes (see [43]).

To develop the theory, we will consider the integration interval $[-1, 1]$ and generalize later to an interval $[a, b]$ using a change of variables. The basic quadrature rule, which we first determine, will use the endpoints -1 and 1 plus two interior points, which by symmetry must be $-\xi_1$ and ξ_1. Thus the formula will be:

$$a\left(f(-1) + f(1)\right) + b\left(f(-\xi_1) + f(\xi_1)\right). \tag{9.60}$$

We want the basic formula to integrate polynomials of degrees as high as possible. Because of symmetry the monomials with odd degree are integrated exactly. There are three free parameters a, b and ξ_1 to determine the rule (9.60). We can require that it be exact for $f(x) = 1, x^2$ and x^4. Using MAPLE we set up the equations and solve them:

```
u1:=2*a+2*b:
u2:=2*a+2*b*xi[1]^2:
u3:= 2*a+2*b*xi[1]^4:
solve({u1=2,u2=2/3,u3=2/5},{a,b,xi[1]});
allvalues(%);
```

The result is

$$\left\{ b = 5/6, a = 1/6, \xi_1 = 1/5\sqrt{5} \right\},$$

and the basic rule becomes

$$\int_{-1}^{1} f(x)\, dx \approx I_2 = \frac{1}{6}\left(f(-1) + f(1)\right) + \frac{5}{6}\left(f(-\frac{1}{\sqrt{5}}) + f(\frac{1}{\sqrt{5}})\right). \tag{9.61}$$

We now need a second approximation in order to compare the integration error. We will add three more points (by symmetry one must be 0) and consider the Kronrod extension

$$A(f(-1) + f(1)) + B(f(-\xi_1) + f(\xi_1)) + C\left(f(-\frac{1}{\sqrt{5}}) + f(\frac{1}{\sqrt{5}})\right) + Df(0).$$

This time we have five parameters to determine (A, B, C, D and ξ_1), so we can require exactness for $f(x) = 1$, x^2, x^4, x^6 and x^8:

```
xi[2]:=1/sqrt(5);
u1:=2*A+2*B+2*C+D=2;
u2:=2*A+2*B*xi[1]^2+2*C*xi[2]^2=2/3;
u3:=2*A+2*B*xi[1]^4+2*C*xi[2]^4=2/5;
u4:=2*A+2*B*xi[1]^6+2*C*xi[2]^6=2/7;
u5:=2*A+2*B*xi[1]^8+2*C*xi[2]^8=2/9;
solve({u1,u2,u3,u4,u5},{A,B,C,D,xi[1]});
allvalues(%)[1];
```

The result is

$$\left\{ B = \frac{72}{245}, A = \frac{11}{210}, D = \frac{16}{35}, C = \frac{125}{294}, \xi_1 = 1/3\sqrt{6} \right\},$$

and the rule becomes

$$\int_{-1}^{1} f(x)\,dx \;\approx\; I_1 = \tfrac{11}{210}\left[f(-1) + f(1)\right] + \tfrac{72}{245}\left[f(-\sqrt{\tfrac{2}{3}}) + f(\sqrt{\tfrac{2}{3}})\right]$$

$$+ \tfrac{125}{294}\left[f(-\tfrac{1}{\sqrt{5}}) + f(\tfrac{1}{\sqrt{5}})\right] + \tfrac{16}{35}f(0). \tag{9.62}$$

The quadrature rule I_1 (9.62) will be more accurate than I_2 (9.61). In order to estimate how much more accurate (9.62) is compared to (9.61), we construct yet another Kronrod extension of (9.62) of the form

$$\texttt{is} \;=\; A\left[f(-1) + f(1)\right] + B\left[f(-\xi_1) + f(\xi_1)\right] + C\left[f(-\sqrt{\tfrac{2}{3}}) + f(\sqrt{\tfrac{2}{3}})\right]$$

$$+ D\left[f(-\xi_2) + f(\xi_2)\right] + E\left[f(-\tfrac{1}{\sqrt{5}}) + f(\tfrac{1}{\sqrt{5}})\right] + F\left[f(-\xi_3) + f(\xi_3)\right] + Gf(0) \tag{9.63}$$

by adding six more points. Using the ansatz (9.63) and requiring that it be exact for the monomials $1, x^2, x^4, \ldots x^{18}$, we obtain 10 nonlinear equations in 10 unknowns:

```
v1:=sqrt(2/3); v2:=sqrt(1/5);
u1:=2*A+2*B+2*C+2*D+2*E+2*F+G=2;
u2:=2*A+2*B*xi[1]^2+2*C*v1^2+2*D*xi[2]^2+2*E*v2^2+2*F*xi[3]^2=2/3;
u3:=2*A+2*B*xi[1]^4+2*C*v1^4+2*D*xi[2]^4+2*E*v2^4+2*F*xi[3]^4=2/5;
u4:=2*A+2*B*xi[1]^6+2*C*v1^6+2*D*xi[2]^6+2*E*v2^6+2*F*xi[3]^6=2/7;
u5:=2*A+2*B*xi[1]^8+2*C*v1^8+2*D*xi[2]^8+2*E*v2^8+2*F*xi[3]^8=2/9;
u6:=2*A+2*B*xi[1]^10+2*C*v1^10+2*D*xi[2]^10+2*E*v2^10+2*F*xi[3]^10=2/11;
u7:=2*A+2*B*xi[1]^12+2*C*v1^12+2*D*xi[2]^12+2*E*v2^12+2*F*xi[3]^12=2/13;
u8:=2*A+2*B*xi[1]^14+2*C*v1^14+2*D*xi[2]^14+2*E*v2^14+2*F*xi[3]^14=2/15;
u9:=2*A+2*B*xi[1]^16+2*C*v1^16+2*D*xi[2]^16+2*E*v2^16+2*F*xi[3]^16=2/17;
u10:=2*A+2*B*xi[1]^18+2*C*v1^18+2*D*xi[2]^18+2*E*v2^18+2*F*xi[3]^18=2/19;
sols:=solve({u1,u2,u3,u4,u5,u6,u7,u8,u9,u10},
          {A,B,C,D,E,F,G,xi[1],xi[2],xi[3]});
```

MAPLE is able to solve this system and gives a solution containing very complicated expressions (several pages long). However, evaluating the expressions as floating point numbers

```
Digits:=16;  evalf(sols);
```

yields the numerical values:

$$
\begin{aligned}
A &= 0.01582719197348004 & \xi_1 &= 0.9428824156954797 \\
B &= 0.09427384478891794 & \xi_2 &= 0.6418533423457821 \\
C &= 0.1550719873365840 & \xi_3 &= 0.2363831996621484 \\
D &= 0.1888215739147961 \\
E &= 0.1997734052268568 \\
F &= 0.2249264653333389 \\
G &= 0.2426110719014055
\end{aligned}
$$

with fortunately positive weights for the rule (9.63). The value obtained from this rule, which is exact for polynomials of degree 19, will be used for the estimated value `is` instead of the Monte Carlo sum, and also to determine the error between the two rules (9.61) and (9.62).

For an arbitrary interval $[a, b]$, the formulas (9.61) and (9.62) can be written respectively as

$$
I_2 = \frac{h}{6} \left\{ f(a) + f(b) + 5 \left[f(m - \beta h) + f(m + \beta h) \right] \right\} \tag{9.64}
$$

and

$$
\begin{aligned}
I_1 = \frac{h}{1470} \Big\{ &77 \left[f(a) + f(b) \right] + 432 \left[f(m - \alpha h) + f(m + \alpha h) \right] \\
&+ 625 \left[f(m - \beta h) + f(m + \beta h) \right] + 672 \, f(m) \Big\},
\end{aligned} \tag{9.65}
$$

where

$$
h = \frac{1}{2}(b - a), \quad m = \frac{1}{2}(a + b), \quad \alpha = \sqrt{\frac{2}{3}}, \quad \beta = \frac{1}{\sqrt{5}}.
$$

A similar reformulation holds for (9.63).

For the initial interval $[a, b]$ we will compute the ratio

$$
R = \frac{|I_1 - \mathtt{is}|}{|I_2 - \mathtt{is}|}.
$$

Since `is` is such a high order rule, it will deliver the "exact" value for smooth functions when compared with I_1 and I_2. We can assume that we are in that situation if it turns out that $R < 1$, and R tells us how much more accurate I_1 is compared to I_2. Since we will use the value I_1 in the recursive algorithm as final value we can also relax the tolerance in this case to

$$
\mathtt{tol} = \mathtt{tol}/R.
$$

At each recursive level, the current interval $[a, b]$ is subdivided into six subintervals when the error tolerance is not met, namely the intervals $[a, m - \alpha h]$, $[m - \alpha h, m - \beta h]$, $[m - \beta h, m]$, $[m, m + \beta h]$, $[m + \beta h, m + \alpha h]$, $[m + \alpha h, b]$ determined by (9.65):

$$x = a + (b - a) * [0.0918 \quad 0.2764 \quad 0.5000 \quad 0.7236 \quad 0.9082]$$

In order to reuse function values within the recursion, the function values of the end points of these six subintervals are passed to the six new calls of the quadrature function `LobattoAdaptive` for the next recursion level.

The termination criterion is essentially the same as for `SimpsonAdaptive`, with the modification that we terminate the recursion if *the smaller new interval no longer contains a machine number*. Putting it all together, we obtain the following MATLAB function:

ALGORITHM 9.10. *Adaptive Gauss–Lobatto Quadrature*

```
function Q=LobattoAdaptive(f,a,b,tol,trace,varargin)
% LOBTTOADAPTIVE approximate integral using adaptive Lobatto rule
%   Q=LobattoAdaptive(f,a,b) approximates the integral of f(x) from a
%   to b to machine precision.  f is a function handle.  The function
%   f must return a vector of output values if given a vector of input
%   values.
%   Q=LobattoAdaptive(f,a,b,tol) integrates to a relative error of
%   tol.  Q=LobattoAdaptive(f,a,b,tol,trace) displays the left end
%   point of the current interval, the interval length, and the
%   partial integral.  Q=LobattoAdaptive(f,a,b,tol,trace,p1,p2,...)
%   allows coefficients p1,... to be passed directly to the function
%   f: G=f(x,p1,p2,...).  To use default values for tol or trace, one
%   may pass the empty matrix ([]).
%   Reference:   Walter Gautschi, 08/03/98

global warn1 warn2
if(nargin<4), tol=[]; end;
if(nargin<5), trace=[]; end;
if(isempty(tol)), tol=eps; end;
if(isempty(trace)), trace=0; end;
if tol <= eps, tol=10*eps;   end
warn1=0; warn2=0;
m=(a+b)/2; h=(b-a)/2; alpha=sqrt(2/3); beta=1/sqrt(5);
x1=.942882415695480; x2=.641853342345781; x3=.236383199662150;
x=[a,m-x1*h,m-alpha*h,m-x2*h,m-beta*h,m-x3*h,m,m+x3*h,...
   m+beta*h,m+x2*h,m+alpha*h,m+x1*h,b];
y=f(x,varargin{:});
for p=1:length(y)
  if isinf(y(p)) | isnan(y(p)),
    y(p)=0; warn1=1;
  end
```

```matlab
end
fa=y(1); fb=y(13);
i2=(h/6)*(y(1)+y(13)+5*(y(5)+y(9)));
i1=(h/1470)*(77*(y(1)+y(13))+432*(y(3)+y(11))+ ...
      625*(y(5)+y(9))+672*y(7));
is=h*(.0158271919734802*(y(1)+y(13))+.0942738402188500 ...
    *(y(2)+y(12))+.155071987336585*(y(3)+y(11))+ ...
    .188821573960182*(y(4)+y(10))+.199773405226859 ...
    *(y(5)+y(9))+.224926465333340*(y(6)+y(8))+.242611071901408*y(7));
erri1=abs(i1-is); erri2=abs(i2-is);
R=1; if(erri2~=0), R=erri1/erri2; end;
if(R>0 & R<1), tol=tol/R; end;
is=is*tol/eps; if(is==0), is=b-a, end;
Q=LobattoRecursive(f,a,b,fa,fb,is,trace,varargin{:});
if warn1==1,
  warning(['Infinite or Not-a-Number function value  encountered. ',...
    'Singularity likely. Required tolerance may not be met.']);
end
if warn2==1,
  warning(['Interval contains no more machine number. ',...
    'Singularity likely. Required tolerance may not be met.']);
end

function Q=LobattoRecursive(f,a,b,fa,fb,is,trace,varargin)
% LOBATTORECURSIVE Recursive function used only by LobattoAdaptive

global warn1 warn2
h=(b-a)/2; m=(a+b)/2; alpha=sqrt(2/3); beta=1/sqrt(5);
mll=m-alpha*h; ml=m-beta*h; mr=m+beta*h; mrr=m+alpha*h;
x=[mll,ml,m,mr,mrr];
y=f(x,varargin{:});
for p=1:length(y)
   if isinf(y(p)) | isnan(y(p)),
      y(p)=0; warn1=1;
   end
end
fmll=y(1); fml=y(2); fm=y(3); fmr=y(4); fmrr=y(5);
i2=(h/6)*(fa+fb+5*(fml+fmr));
i1=(h/1470)*(77*(fa+fb)+432*(fmll+fmrr)+625*(fml+fmr)+672*fm);
m1=(a+mll)/2; m2=(b+mrr)/2;
if(is+(i1-i2)==is) | (m1<=a) | (b<=m2),
  if (m1<=a) | (b<=m2),  warn2 =1; end;
  Q=i1;
  if(trace), disp([a b-a Q]), end;
else
  Q=LobattoRecursive(f,a,mll,fa,fmll,is,trace,varargin{:})+...
    LobattoRecursive(f,mll,ml,fmll,fml,is,trace,varargin{:})+...
    LobattoRecursive(f,ml,m,fml,fm,is,trace,varargin{:})+...
    LobattoRecursive(f,m,mr,fm,fmr,is,trace,varargin{:})+...
```

```
    LobattoRecursive(f,mr,mrr,fmr,fmrr,is,trace,varargin{:})+...
    LobattoRecursive(f,mrr,b,fmrr,fb,is,trace,varargin{:});
end;
```

EXAMPLE 9.7. *We consider function No. 21 from the collection of test functions by Kahaner [78]:*

$$f(x) = \frac{1}{\cosh^2(10\,x - 2)} + \frac{1}{\cosh^4(100\,x - 40)} + \frac{1}{\cosh^6(1000\,x - 600)}.$$

This function has three maxima for $x = 0.2$, 0.4 and 0.6 in the interval $[0,1]$

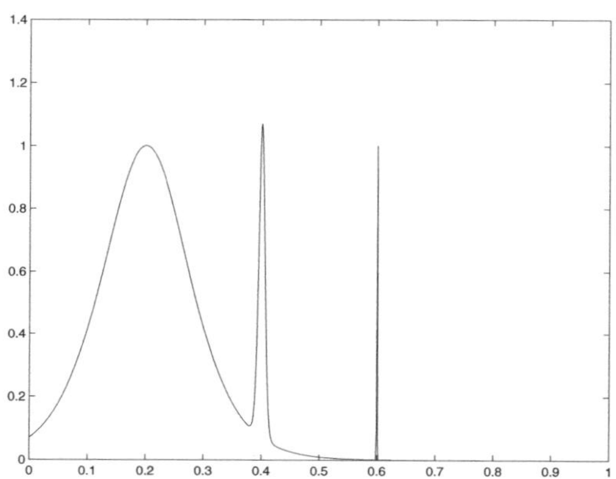

FIGURE 9.2. *Kahaner function No. 21*

as we can see from Figure 9.2. The peak for $x = 0.6$ is very sharp so that it can easily be missed (in fact MATLAB*'s command* fplot('g',[0,1]) *does not plot this peak). With* Digits:=30 *and* int(f(x),x=0..1) *we obtain with* MAPLE *the exact value of*

$$\int_0^1 f(x)\,dx = 0.210802735500549277375643255709.$$

We want to see how this function is integrated by our adaptive quadrature functions. With "trace on" in SimpsonAdaptive(@kahaner21,0,1,1e-5,1) *we obtain the values of Table 9.2. We see from this table,* SimpsonAdaptive *takes smaller steps for integrating the second peak at $x = 0.4$. However, the peak at $x = 0.6$ is missed, the step size is 0.125 for the interval $[0.5, 0.625]$. If we use a higher tolerance, we obtain*

tol	SimpsonAdaptive	# fct. evals.	LobattoAdaptive	# fct. evals.
10^{-5}	**0.20**973780201344	94	**0.20**973606921941	198
10^{-6}	**0.20**973611750548	134	**0.2108027**7613987	378
10^{-7}	**0.2108027**4315572	306	**0.2108027355**3093	498

x_i	$x_{i+1} - x_i$	$\displaystyle\int_{x_i}^{x_{i+1}} f(x)\,dx$
0.00000000	0.06250000	0.00842006439906
0.06250000	0.03125000	0.00932078843060
0.09375000	0.03125000	0.01514698663471
0.12500000	0.06250000	0.05107978398885
0.18750000	0.03125000	0.03096861623502
0.21875000	0.03125000	0.02767839620679
0.25000000	0.03125000	0.02088499381103
0.28125000	0.03125000	0.01383339993785
0.31250000	0.06250000	0.01321087362317
0.37500000	0.00781250	0.00086289185926
0.38281250	0.00781250	0.00131912330043
0.39062500	0.00390625	0.00177092854196
0.39453125	0.00390625	0.00332750783071
0.39843750	0.00390625	0.00407256527252
0.40234375	0.00390625	0.00297034509995
0.40625000	0.00781250	0.00201952110269
0.41406250	0.00781250	0.00050732087269
0.42187500	0.01562500	0.00062813537791
0.43750000	0.06250000	0.00122099740233
0.50000000	0.12500000	0.00045388158284
0.62500000	0.12500000	0.00003734896245
0.75000000	0.25000000	0.00000333154061

$$\text{ans} = 0.20973780201344$$

TABLE 9.2. Result of SimpsonAdaptive for Kahaner 21

We see that for `tol` $= 1e-5$ *both* `SimpsonAdaptive` *and* `LobattoAdaptive` *miss the peak at* $x = 0.6$. *For* `tol` $= 1e-6$ `SimpsonAdaptive` *still misses it but not* `LobattoAdaptive`. *And for for* `tol` $= 1e-7$ *both functions notice the peak and the results are much more accurate as we can see from the correct digits printed in boldface. Notice that Romberg extrapolation does not work well in this example. The reason is that the trapezoidal values also miss the peak until the step size is small enough.*

For `tol` $= 10^{-7}$ *the smallest step size used by* `LobattoAdaptive` *is* $h_{\min} = 1.564e-04$ *while the largest step size is* $h_{\max} = 0.184641$. *Thus the largest step* $h_{\max} \approx 1180\, h_{\min}$, *clearly a good example for adaptive quadrature.*

EXAMPLE 9.8. *We consider as a second example an integral with an infinite integration interval*

$$\int_0^{\infty} \frac{e^{-0.4x}\cos 2x}{x^{0.7}}\,dx.$$

The integrand has a singularity for $x = 0$. *The exact value can be computed using extended precision with* MAPLE. *With the statements*

```
Digits:=30; int(exp(-0.4*x)*cos(2*x)/x^0.7,x=0..infinity);
```

we obtain 2.21349827627298029505612097423.

Since the value of the integrand f is decaying exponentially and $f(100) =$ 8.24e−20 *we can obtain a good approximation by integrating over the finite interval* $[0, 100]$. *The call* LobattoAdaptive(@f,0,100,1e-14) *gives the result* 2.21349827627297 *correct to all but the last printed digits using 20'238 function evaluations.*

With SimpsonAdaptive(@f,0,100,1e-14) *we obtain* 2.21349827627298, *a similar result, however, using 147'214 function evaluations. The higher order approximation of* LobattoAdaptive *reduces the number of function evaluations quite impressively for this example.*

We can transform the integral to obtain a finite integration interval. The change of variables

$$t = e^{-x}, \quad x = -\ln t, \quad dx = -\frac{dt}{t}$$

(with renaming of the integration variable again as x) leads to

$$\int_0^\infty \frac{e^{-0.4x}\cos 2x}{x^{0.7}}\,dx = \int_0^1 \frac{x^{-0.6}\cos(2\ln x)}{(-\ln x)^{0.7}}\,dx.$$

By this transformation, we have introduced a second singularity at $x = 1$. *Now* $\lim_{x\to 0} f(x) = \lim_{x\to 1} f(x) = \infty$. *The plot of the function is shown in Figure 9.3. Integrating the transformed integral by ignoring the singularities,*

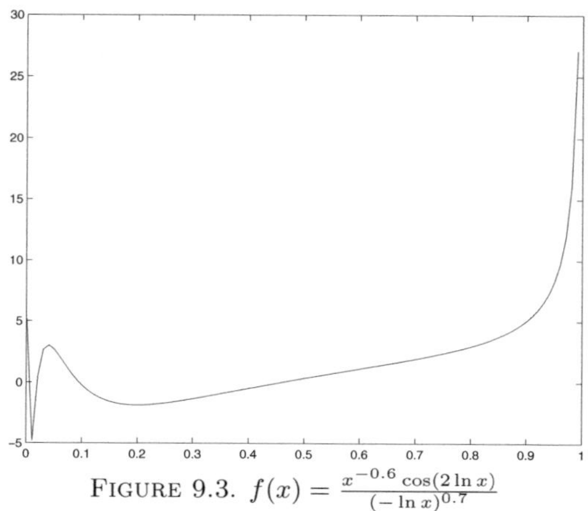

FIGURE 9.3. $f(x) = \dfrac{x^{-0.6}\cos(2\ln x)}{(-\ln x)^{0.7}}$

we obtain the following results

tol	LobattoAdaptive	# fct. evals	SimpsonAdaptive	# fct. evals
10^{-4}	**2.2**1331986623106	618	**2.2**1062338557919	198
10^{-6}	**2.2134**5478856698	1158	**2.2134**3307090916	522
10^{-8}	**2.2134**5412884202	2178	**2.2134**5143522493	1158
10^{-9}	**2.2134540**8916441	3048	**2.2134514**1188071	2062
10^{-10}	**2.2134540**8490537	5388	**2.2134514**1957675	3082
10^{-12}	**2.2134540820**4195	18318	**2.21345143**455330	8214
10^{-14}	**2.2134540820**8425	130668	**2.21345143**731996	18602
eps	**2.2134540820**8246	238848	**2.21345143**771824	24542

We observe that, as we increase the required accuracy, the values obtained by LobattoAdaptive *and* SimpsonAdaptive *seem to converge to different limits. However, both are wrong! The values are correct only to about 5 decimal digits. Here, we clearly have an example where ignoring the singularity is not the right way to go for high precision results.*

9.5 Problems

PROBLEM 9.1. *Derive* Kepler's Barrel Rule. *One would like to estimate the content of a wooden wine barrel by measuring the circumference of the barrel at the bottom* U_1 *the middle* U_2 *and the top* U_3. *The answer should be a simple formula containing the height* H *of the barrel and the three circumferences.*

PROBLEM 9.2. *Use the function* MidpointOpenNewtonCotesRule *(see Equation (9.5)) to generate the first 8 rules and compute the error of each rule.*

PROBLEM 9.3. *Construct a function* $f(x)$ *that is integrated incorrectly by the function* SimpsonAdaptive *but correctly by* LobattoAdaptive.

PROBLEM 9.4. *Write a* MAPLE *function that generates Newton–Cotes quadrature rules using the zeros*

$$z_k = \cos\left(\frac{\pi}{2n} + k\frac{\pi}{n}\right), \quad k = 0, 1, \ldots, n-1$$

or the extremal points (augmented by the boundary points)

$$x_k = \cos\left(\frac{k\pi}{n}\right), \quad k = 0, 1, \ldots, n$$

of the Chebyshev polynomials.

As described in Waldvogel [97], we distinguish three rules:

Fejér's first rule *uses as nodes the zeros* z_k, $k = 0, 1, \ldots, n-1$.

Clenshaw-Curtis *uses the extremal points x_k, $k = 0, 1, \ldots, n$.*

Fejér's second rule *uses the extremal points x_k for $k = 1, \ldots, n-1$ without the boundary points.*

a) Write three MAPLE *functions* `Fejer1`, `Fejer2` *and* `ClenshawCurtis` *that generate the rules. You will have to compute the weights using* `evalf` *and accept rounding errors, since the analytical expressions become complicated. However, the symmetry of the rules and raising* `Digits` *should help to control the errors.*

b) Generate for $n = 10$ the rules and compute the integral

$$\int_0^{10} \sqrt{x^5 + 1}\, e^{-0.03x^2\sqrt{x}}\, dx.$$

Hint: *you need to transform the integral to the interval $[-1, 1]$. Use the function* `LobattoAdaptive` *to compute the exact value for comparison. Also compute the integral with the rule* `ClosedNewtonCotesRule(10)` *and compare the results.*

PROBLEM 9.5. *How small do we have to choose the integration step h so that the integral $\int_2^5 x \ln x\, dx$ is integrated with the composite trapezoidal rule to an error of 10^{-8}? Estimate the step size using (9.16) and check your result with the* MATLAB *function* `TrapezoidalRule`.

PROBLEM 9.6. *Periodic functions that are integrated over a whole period (such integrals occur when computing Fourier coefficients) are integrated best with the trapezoidal rule.*

To understand this, write down the asymptotic expansion with error term for the trapezoidal rule, see (9.30).

To show this fact, compute $\int_0^T \sqrt{1 + \cos^2 x}\, dx$ for

a) $T = \pi$ *(full period) and*

b) $T = 4$

with `Romberg`. *Print the whole Romberg scheme and study its convergence.*

PROBLEM 9.7. *Compute the integral*

$$I = \int_0^2 (x^3 + 2x^2 - x + 1)\, dx$$

(i) analytically,   (ii) using Simpson's rule with $h = 1$. What is the integration error for Simpson's rule and why?

PROBLEM 9.8. *How small do we have to choose the integration step h to compute*

$$\int_0^{100} \frac{dx}{(1 + x)^2}$$

using the composite Simpson rule to an absolute error smaller than 10^{-6}? Estimate h using (9.20) and check your estimate by integrating with the function `SimpsonsRule`.

PROBLEM 9.9. *Extend the function* `SimpsonsRule` *so that parameters can be passed. Use it then to compute the elliptic integral*

$$A(k, \phi) = \int\limits_0^\phi \frac{dx}{\sqrt{1 - k^2 \sin^2 x}}$$

for $k = 0.4$ and $\phi = \frac{\pi}{2}$.

PROBLEM 9.10. *Consider the* mid-point rule

$$\int\limits_0^h f(x)\, dx \approx h f(h/2).$$

Derive the composite mid-point rule with error term. Write a MATLAB *function for the composite mid-point rule that computes approximations by halving the step size until a given tolerance is met.*

PROBLEM 9.11. *Use the* `MidpointOpenNewtonCotesRule` *rule (see Equation (9.5))*

$$\frac{b-a}{8}\left(3\,y_{1/2} + 2\,y_{3/2} + 3\,y_{5/2}\right) \quad with \quad h = \frac{b-a}{3}, \quad and$$

$$y_{1/2} = f(a + h/2), \quad y_{3/2} = f(a + 3h/2), \quad y_{5/2} = f(a + 5h/2)$$

to develop a MATLAB *function for a composite rule that computes approximations by reducing the step size until a given tolerance is met. Avoid recomputing function values.*

Hint: Use the function `SimpsonsRule` *as a model. Instead of halving the step size, you will have to divide the step size by three in order to reuse function values.*

PROBLEM 9.12. *Use again the* `MidpointOpenNewtonCotesRule` *rule (see Equation (9.5))*

$$\frac{b-a}{8}\left(3\,y_{1/2} + 2\,y_{3/2} + 3\,y_{5/2}\right)$$

to develop a MATLAB *function* `MidpointOpenNewtonCotesAdaptive` *for adaptive quadrature. Use* `SimpsonAdaptive` *as a model. Avoid recomputing function values.*

Hint: Instead of halving the step size, you will have to divide the step size by three in order to reuse function values. The extrapolation has to be done such that in

$$M(h) = I + c_1 h^4 + c_2 h^6 + \cdots$$

the term with $c_1 h^4$ is eliminated by a linear combination of the values $M(h)$ and $M(h/3)$.

PROBLEM 9.13. *Write a* MAPLE *script to compute*

$$\sum_{i=1}^{n} P_m(i),$$

where P_m is a polynomial of degree m.

Hint: *make an ansatz with a polynomial of degree $m+1$ as summation function or use the Euler–Maclaurin summation formula.*

As an example your program should compute the formula:

$$\sum_{i=1}^{n} i^3 = \frac{n^2}{4}\,(n+1)^2.$$

PROBLEM 9.14. *Show that in Romberg's scheme the second column contains the approximations of Simpson's rule and the third column contains the values of Boole's rule.*

PROBLEM 9.15. *We have seen that Romberg integration applied to $\int_0^1 \sqrt{x}$ does not work well. The reason is that the trapezoidal rule has the expansion*

$$T(h) = \int_a^b g(t)\,dt + \zeta\left(-\frac{1}{2}\right)\,h^{3/2} + c_1 h^2 + c_2 h^4 + \cdots + c_m h^{2m} + R_m$$

and the term $\zeta\left(-\frac{1}{2}\right)\,h^{3/2}$ is disturbing the convergence of the Romberg scheme.

1. *Compute the value $\zeta\left(-\frac{1}{2}\right)$ with the help of* MAPLE.

2. *Use this value in the modified trapezoidal rule $\tilde{T}(h)=T(h)-\zeta\left(-\frac{1}{2}\right)\,h^{3/2}$ for computing the Romberg scheme (modify the* MATLAB *function* Romberg*). Does the scheme now converge faster?*

PROBLEM 9.16. *Consider the interval of integration $[a,b] = [-1,1]$. Suppose we have an N-point quadrature rule such that the nodes are symmetric, i.e., we have $\xi_j = \xi_{N-j+1}$ for $j = 1,\ldots,N$. As usual, the weights are obtained by integrating the corresponding Lagrange polynomials.*

1. *Show that the weights are also symmetric, i.e., we have $w_j = w_{N-j+1}$ for $j = 1,\ldots,N$.*

2. *Show that the order p of this rule must be even.* **Hint:** *Suppose the quadrature rule has order $\geq 2k-1$, i.e., it is exact with respect to all*

polynomials of degree $2k - 2$. Using the fact that

$$\int_{-1}^{1} t^{2k-1}\, dt = 0,$$

show that the rule is also exact for all polynomials of degree $2k - 1$.
Hence, the rule has order $\geq 2k$.

PROBLEM 9.17. *Integrate with the Romberg method*

$$I = \int_{0}^{\pi} \frac{dx}{1 + \sin^2 x}.$$

Print the Romberg scheme and observe that the values in the first column
(trapezoidal rule) converge fastest. Why is this so?

PROBLEM 9.18. *Solve the equation*

$$f(x) = \int_{0}^{1} e^{xt^2}\, dt - 2 = 0$$

using Newton's method. Compute f and f' with SimpsonAdaptive.

PROBLEM 9.19. *Evaluate the integrals*

$$\int_{0}^{\pi} \sqrt{1 + \cos^2 x}\, dx$$

and

$$\int_{0}^{3} |\cos x|\, dx$$

with TrapezoidalRule, Romberg, SimpsonAdaptive and LobattoAdaptive
using tol$= 10^{-8}$. Count the function evaluations and explain why some
methods are best for these problems.

PROBLEM 9.20. *To compute integrals over an infinite interval, one technique is to make a change of variables to obtain a finite integration interval.
For instance one can use*

$$t = \frac{1}{x}, \quad t = \frac{1}{x+1} \quad or \quad t = e^{-x}.$$

With such a transformation we usually introduce singularities in the integrands. Ignore them and compute the following integrals:

$$a) \quad \int_0^\infty \frac{x}{e^x + 1}\, dx \qquad\qquad b) \quad \int_0^\infty \frac{\arctan x}{(x+1)^2}\, dx$$

$$c) \quad \int_0^\infty \frac{x^3 + 1}{1 + x^2 + x^5}\, dx \qquad d) \quad \int_1^\infty \frac{e^{-\frac{x}{2}}}{x}\, dx$$

Chapter 10. Numerical Ordinary Differential Equations

Numerical methods for ordinary differential equations fall naturally into two classes: those which use one starting value at each step ("one-step methods"), and those which are based on several values of the solution ("multistep methods" or "multi-value methods")

E. Hairer, S. P. Nørsett and G. Wanner, Solving Ordinary Differential Equations I, Nonstiff Problems, 1993.

It is convenient to represent a Runge-Kutta method by a partitioned tableau, of the form

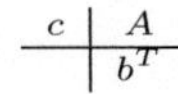

[...] As the order being sought increases, the algebraic conditions on the coefficients of the method become increasingly complicated.

J. C. Butcher, Numerical Methods for Ordinary Differential Equations, Wiley, second edition, 2008.

The numerical approximation of solutions to differential equations is of paramount importance in science and engineering, and there is a subclass of problems, namely ordinary differential equations, for which the field has reached a certain maturity. This is much less so for partial differential equations, for which the theory about existence of solutions is also much less complete. The focus of this chapter is therefore only on the numerical solution of ordinary differential equations. We start in Section 10.1 with several historical examples of ordinary differential equations to illustrate both how useful differential equations are for modeling, and how one might go about finding approximate solutions, since most differential equations cannot be solved in closed form. We then give an introduction to the theory of ordinary differential equations in Section 10.2, and present two basic solution methods, the power series expansion and the method of Euler. Next, we consider in Sections 10.3 and 10.4 the two main classes of numerical methods for ordinary differential equations (see the first quote above), namely Runge-Kutta methods and linear multistep methods. For both types of methods, we derive order conditions using MAPLE and prove convergence of the approximate solutions to the continuous ones. We also discuss the important concept of zero stability, which is necessary for the convergence of linear multistep methods.

W. Gander et al., *Scientific Computing - An Introduction using Maple and MATLAB*,
Texts in Computational Science and Engineering 11,
DOI 10.1007/978-3-319-04325-8_10,

Many problems of practical interest can only be solved adequately by special numerical methods, whose properties are related to the underlying dynamics of the problem. We will consider one such class of problems, known as *stiff problems*, in Section 10.5, where the concept of A-stability becomes important. Another class of problems is the long term integration of systems that contain invariant physical quantities (momentum, energy, etc.). Numerical methods that are able to preserve such quantities in the discrete solution are called *geometric integrators* and are the subject of discussion in Section 10.6. We briefly treat delay differential equations in Section 10.7. The best current reference on the numerical solution of ordinary differential equations, including stiff problems and geometric integration, are the three monographs by Hairer, Nørsett and Wanner [62], Hairer and Wanner [63] and Hairer, Lubich and Wanner [61]. The book by Butcher [14] also contains a section on general linear methods, which combine Runge-Kutta and linear multistep ideas.

10.1   Introductory Examples

Differential equations arose naturally after the invention of differential calculus by the two giants Newton and Leibniz. Newton considered in 1671 one of the first ordinary differential equations [95],

$$y' = 1 - 3x + y + x^2 + xy. \tag{10.1}$$

To find a function $y(x)$ satisfying this equation, Newton assumed that both x and y are small[1], and thus deduced from (10.1) that

$$y' \approx 1 \quad \text{for } x, y \text{ small.}$$

Assuming equality, the above approximation can be integrated to obtain

$$y \approx x.$$

Inserting this approximation back into (10.1) and keeping one more term, one finds a new approximation for the derivative,

$$y' \approx 1 - 3x + x = 1 - 2x,$$

which in turn after integration, leads to

$$y \approx x - x^2.$$

Inserting this again into (10.1) and keeping one more term leads to

$$y' \approx 1 - 2x + x^2,$$

and after integration to

$$y = x - x^2 + \frac{1}{3}x^3.$$

---

[1] In modern notation, this corresponds to the initial condition $y(0) = 0$.

This process can be continued in the same fashion to obtain a series approximation of the solution $y(x)$ of (10.1); this procedure is known today as the Taylor series approach, see Section 10.2.3.

Leibniz in turn worked on the famous "silver watch problem", which was first posed by Claude Perrault to Leibniz during the latter's stay in Paris between 1672 and 1676: suppose a silver watch with a chain attached to it is put on a table, as shown in Figure 10.1. What trajectory will the silver

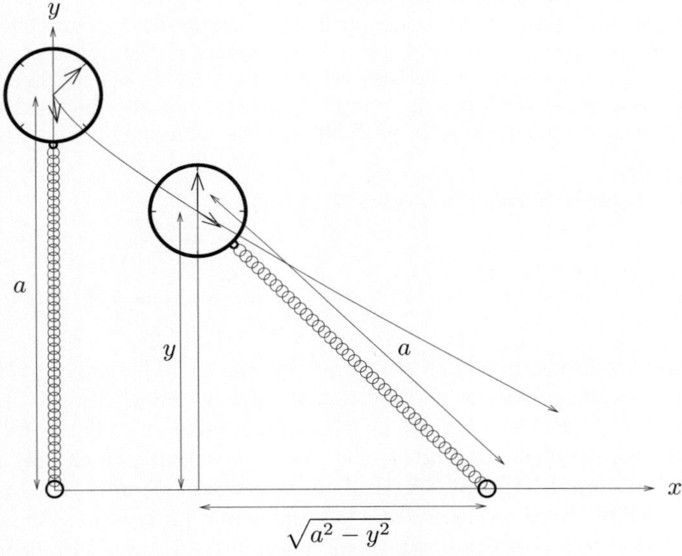

FIGURE 10.1.
Silver watch problem, solved by Leibniz and Bernoulli.

watch follow if one pulls on the chain along the x-axis? Since the watch always moves in the direction of the chain, we obtain the differential equation

$$y' = -\frac{y}{c} = -\frac{y}{\sqrt{a^2 - y^2}}.$$

To solve this problem, both Joh. Bernoulli and Leibniz invented independently the *method of separation of variables*: the equation is equivalent to

$$-\frac{\sqrt{a^2 - y^2}}{y}\,dy = dx,$$

which leads after integration to the implicit solution

$$x = \int \frac{\sqrt{a^2 - y^2}}{y}\,dy = \sqrt{a^2 - y^2} - a\operatorname{arctanh}\left(\frac{a}{\sqrt{a^2 - y^2}}\right) + C, \quad (10.2)$$

which can be obtained by MAPLE using the int command,

```
assume(a>0);
int(sqrt(a^2-y^2)/y,y);
```

$$\sqrt{a{\sim}^2 - y^2} - a{\sim}\operatorname{arctanh}\left(\frac{a{\sim}}{\sqrt{a{\sim}^2 - y^2}}\right)$$

In the MAPLE worksheet, a tilde appears after the symbol a, which shows that assumptions have been made on a with the assume command. If we do not use the assume command, then MAPLE does not simplify expressions containing square roots of a^2 and instead gives a different, but mathematically equivalent result, as long as a is positive. MAPLE can also be used to solve the differential equation directly with the dsolve command

```
assume(a>0);
dsolve(D(y)(x)=-y(x)/sqrt(a^2-y(x)^2),y(x));
```

$$x + \sqrt{a{\sim}^2 - (y(x))^2} - a{\sim}\operatorname{arctanh}\left(\frac{a{\sim}}{\sqrt{a{\sim}^2 - (y(x))^2}}\right) + \_C1 = 0$$

This simple example already shows that solutions to ODEs cannot always be obtained in explicit form, as the solution implicitly defined by (10.2) cannot be solved for y. Euler collected in a monumental effort all the ODEs for which he had managed to find a closed-form solution, which can now be found in Volumes XXII and XXIII of Euler's *Opera Omnia*. He suspected that many ODEs have no closed-form solutions, and proposed in [29] a simple approximation to the solution, which we now call the Forward Euler method, see Subsection 10.2.5. It took however almost a century before Liouville gave a first proof that the ODE

$$y' = x^2 + y^2$$

cannot be solved in terms of elementary functions [87]. It is instructive to ask Maple for a solution of this equation,

```
dsolve(D(y)(x)=y(x)^2+x^2,y(x));
```

$$y(x) = -\frac{x\left(\_C1\,\operatorname{BesselJ}\left(-3/4, 1/2\,x^2\right) + \operatorname{BesselY}\left(-3/4, 1/2\,x^2\right)\right)}{\_C1\,\operatorname{BesselJ}\left(1/4, 1/2\,x^2\right) + \operatorname{BesselY}\left(1/4, 1/2\,x^2\right)}.$$

This shows that more functions are considered to be elementary nowadays than at the time of Liouville. While neither Newton nor Euler proved that their respective approximate solutions were converging to a solution of the corresponding ODE, Cauchy proved rigorously in 1820 that Euler's method converges for general, nonlinear ODEs, and in 1835 that Newton's series expansion was also convergent. For more information on the fascinating history on the development of ordinary differential equations and their numerical solution, see [62, 63].

10.2 Basic Notation and Solution Techniques

After the introduction of some basic notation and terminology for differential equations, we show in this section how solutions to differential equations can be approximated by a series or Euler's historical method.

10.2.1 Notation, Existence of Solutions

An *ordinary differential equation* (abbreviated ODE) is an equation in which the unknown is a function of a single variable (often denoted t); it contains the function and its derivatives of various orders. ODEs are important for modeling time dependent processes in many disciplines, e.g. in engineering, physics, chemistry and economics. For instance, in classical mechanics, if the position, velocity, acceleration and various forces acting on a body are given, Newton's second law of motion allows us to express these variables as a differential equation and, by solving it, to compute the position of the body as a function of time.

A *first-order* differential equation for an unknown function $y(t)$ has the form

$$y' = f(t, y), \tag{10.3}$$

where the function $f(t, y)$ is given. A function $y(t)$ is a solution of (10.3) if

$$y'(t) = f(t, y(t))$$

is valid for all t.

EXAMPLE 10.1. *The differential equation*

$$y' = f(t, y) = y \tag{10.4}$$

has $y(t) = e^t$ as a solution.

In general, (10.3) will have infinitely many solutions: for instance, the functions $y(t) = C\, e^t$ with any $C \in \mathbb{R}$ are also solutions of (10.4). To obtain a unique solution, one has to prescribe an initial condition, *for instance by requiring that $y(0) = 1$.*

A differential equation of *order p* has the form

$$F(t, y, y', \ldots, y^{(p)}) = 0. \tag{10.5}$$

The *general solution* of (10.5) will contain in general p *constants of integration.* By requiring p *initial or boundary conditions,* a *particular solution* is selected from the family of solutions.

We have an *initial value problem* if all p conditions are given for the same value $t = t_0$. However, the p conditions can also be spread over different points t. A frequent case is to have the conditions at two values of t; in this case, we have a *two-point boundary value problem,* see also Section 8.1. A classical case stems from military applications: shooting a canon ball firing a cannon towards a target is described by a second-order differential equation with two boundary conditions

$$y'' = f(t, y, y'), \quad y(t_0) = y_0, \quad y(t_1) = y_1.$$

From a theoretical point of view, the distinction between initial and boundary value problems is not important. The conditions simply determine a particular solution. For the numerical methods, however, there is a big difference: initial value problems are easier to solve.

THEOREM 10.1. (EXISTENCE AND UNIQUENESS) *Let $f(t, y)$ be defined and continuous for $-\infty < a \leq t \leq b < \infty$ and $-\infty < y < \infty$ and suppose there exists a constant $L > 0$ such that for all $t \in [a, b]$ and all $y, \tilde{y} \in \mathbb{R}$, the Lipschitz condition*

$$|f(t, y) - f(t, \tilde{y})| \leq L|y - \tilde{y}| \tag{10.6}$$

holds. Then for any given $\eta \in \mathbb{R}$, there exists a unique function $y(t)$ with

1. y is continuously differentiable for $t \in [a, b]$.

2. $y' = f(t, y) \; \forall t \in [a, b]$.

3. $y(a) = \eta$.

PROOF. See the classical books of Henrici [68] or Hairer, Nørsett and Wanner [62]. $\qquad\qquad\square$

Equation (10.6) holds if the partial derivative $\partial f / \partial y$ exists, but this condition is not necessary.

EXAMPLE 10.2. *The function $f(t, y) = |y|$ meets the conditions of the Theorem 10.1.*

$$|f(t, y) - f(t, \tilde{y})| = ||y| - |\tilde{y}|| \leq |y - \tilde{y}|, \quad L = 1.$$

Thus we have a unique solution, namely

$$y(a) = \eta > 0 \implies y(t) = \eta e^{t-a}, \quad y' = \eta e^{t-a} = y = |y|,$$
$$y(a) = \eta < 0 \implies y(t) = \eta e^{a-t}, \quad y' = -\eta e^{a-t} = |y|.$$

As a second example, consider $f(t, y) = \sqrt{|y|}$. This time f is still continuous but does not satisfy the Lipschitz condition (10.6). For $y, \tilde{y} > 0$, let $\tilde{y} \to y$. Then

$$\frac{\left|\sqrt{|y|} - \sqrt{|\tilde{y}|}\right|}{|y - \tilde{y}|} \to \frac{1}{2\sqrt{|y|}}$$

and the expression is not bounded for $y \to 0$. To compute the solutions, we distinguish three cases:

a) *if $y > 0$, we can use the separation of variables technique from Section 10.1 to obtain*

$$y' = \sqrt{y} \implies 2\frac{dy}{2\sqrt{y}} = dt \implies \sqrt{y} = \frac{t}{2} + C \implies y = \left(\frac{t}{2} + C\right)^2.$$

b) *if $y < 0$, we can proceed similarly and get*

$$y' = \sqrt{|y|} = \sqrt{-y} \implies -2\frac{dy}{2\sqrt{-y}} = -dt \implies \sqrt{-y} = -\frac{t}{2} + C$$

Thus $y = -\dfrac{(t-a)^2}{4}$ with $a = 2C$.

c) *Also $y(t) = 0$ is a particular solution.*

We now construct for the initial condition $y(2) = -1$ the solution

$$y(t) = \begin{cases} -\dfrac{(t-4)^2}{4} & 2 \le t < 4, \\ 0 & 4 \le t \le a, \quad (a > 0) \\ \dfrac{(t-a)^2}{4} & t > a. \end{cases}$$

This solution is continuously differentiable and not unique, since it is a family of solutions for every a.

10.2.2 Analytical and Numerical Solutions

MAPLE can solve differential equations analytically. For instance, for $y' = \sqrt{|y|}$, $y(2) = -1$ we get

```
g(t):=rhs(dsolve({diff(y(t),t)=sqrt(abs(y(t))),y(2)=1},y(t)));
factor(g(t));
```

the function

$$g(t) = -\frac{(t-4)^2}{4},$$

so just the part of the solution in the neighborhood of $t = 2$. One can verify that this $g(t)$ no longer satisfies the ODE for $t > 4$.

As with antiderivatives, it is often impossible to solve a differential equation analytically, that is, by algebraic manipulations only. The equation

$$y' = t^2 + y^2$$

fulfills the condition of Theorem 10.1 and thus has a unique solution for $y(0) = 1$. However, as we have seen in Section 10.1, it took almost a century before Liouville gave a first proof that this differential equation cannot be solved in terms of elementary functions.

Even when an analytical solution is available, it may be of questionable usefulness. The linear differential equation

$$y'' = t^2 y + t + 1 - \frac{1}{t}$$

is solved in MAPLE with

```
dsolve(diff(y(t),t$2)=t^2*y(t)+t+1-1/t,y(t));
```

The result is a long complicated expression that may not be very informative.
The two independent solutions of the homogeneous equation are constructed
with a *BesselI* and a *BesselK* function, and the particular solution also contains includes *StruveL* and *hypergeometric functions*, in addition to the above
mentioned *BesselI* and *BesselK* functions. (Note: with MAPLE 15, it might
be necessary to write

```
dsolve(diff(y(t),t$2)=t^2*y(t)+t+1-1/t);
```

to get the result!)

Another example is the linear differential equation

$$y'' + 5y' + 4y = 1 - e^t, \quad y(0) = y'(0) = 0.$$

It has the solution

```
de:=diff(y(t),t$2)+5*diff(y(t),t)+4*y(t)=1-exp(t);
dsolve({de,y(0)=0,D(y)(0)=0},y(t));
f:=unapply(rhs(%),t);
```

$$f := t \mapsto -\frac{1}{6}\,e^{-t} + \frac{1}{60}\,e^{-4t} + 1/4 - \frac{1}{10}\,e^t.$$

We can solve this equation also using the `series` option

```
dsolve({de,y(0)=0, D(y)(0)=0},y(t),series,order=8);
g:=unapply(convert(rhs(%),polynom),t);
```

and get

$$g := t \mapsto -\frac{1}{6}t^3 + \frac{1}{6}t^4 - \frac{17}{120}t^5 + \frac{17}{180}t^6 - \frac{13}{240}t^7.$$

If we now want to evaluate the solution for small values of t, we obtain with
MAPLE the following table.

```
for t from 1e-3 by 1e-3 to 1e-2 do
  [t,f(t),g(t)]
od;
```

t	f analytical	g series
0.001	-1.0×10^{-10}	$-1.665001416 \times 10^{-10}$
0.002	-1.3×10^{-9}	$-1.330671193 \times 10^{-9}$
0.003	-4.5×10^{-9}	$-4.486534356 \times 10^{-9}$
0.004	-1.07×10^{-8}	$-1.062414468 \times 10^{-8}$
0.005	-2.07×10^{-8}	$-2.072960789 \times 10^{-8}$
0.006	-3.58×10^{-8}	$-3.578509721 \times 10^{-8}$
0.007	-5.67×10^{-8}	$-5.676886992 \times 10^{-8}$
0.008	-8.47×10^{-8}	$-8.465528415 \times 10^{-8}$
0.009	-1.205×10^{-7}	$-1.204148154 \times 10^{-7}$
0.010	-1.650×10^{-7}	$-1.650140728 \times 10^{-7}$

As we can see, the analytical solution suffers from cancellation, while the
approximate series solution is correct to 8 decimal digits. Thus, we should
be aware that when an analytical solution is evaluated numerically, the result
is not necessarily more accurate than an approximate solution obtained by a
numerical method.

10.2.3 Solution by Taylor Expansions

Following Newton's footsteps from Section 10.1, we now discuss methods for finding the solution of a differential equation in series form. The basic idea is to expand both sides of the differential equation $y' = f(t, y)$ in a Taylor series and compare coefficients. Let (t_0, y_0) be the expansion point and consider the ansatz

$$y(t_0 + h) = \sum_{k=0}^{\infty} a_k h^k \quad \text{with} \quad a_k = \frac{y^{(k)}(t_0)}{k!}. \tag{10.7}$$

Let us consider two methods for computing the coefficients of the series and demonstrate the procedure for the model problem

$$y' = t^2 + y^2, \quad y(0) = 1. \tag{10.8}$$

The first method computes derivatives by differentiating the differential equation. With the initial condition $y(t_0) = y_0$, we get

$$y' = f(t_0, y_0) \implies y'(t_0) \quad \text{is known,}$$
$$y'' = f_t + y' f_y \implies y''(t_0) \quad \text{is known,}$$
$$y''' = f_{tt} + 2y' f_{ty} + (y')^2 f_{yy} + f_y y'' \implies y'''(t_0) \quad \text{is known.}$$

The expressions soon get complicated in the general case.

For a specific differential equation, the computations are usually simpler. By differentiating (10.8), we get

$$\begin{aligned}
y' &= t^2 + y^2, \\
y'' &= 2t + 2yy', \\
y''' &= 2 + 2y'^2 + 2yy'', \\
y^{(4)} &= 6y'y'' + 2yy'''.
\end{aligned} \tag{10.9}$$

Using the initial condition $t_0 = 0$ and $y(t_0) = 1$ in (10.9), we obtain

$$y'(0) = 1, \ y''(0) = 2, \ y'''(0) = 8 \text{ and } y^{(4)}(0) = 28$$

and therefore

$$y(t) = 1 + t + \frac{2}{2!}t^2 + \frac{8}{3!}t^3 + \frac{28}{4!}t^4 + \cdots = 1 + t + t^2 + \frac{4}{3}t^3 + \frac{7}{6}t^4 + \cdots$$

The same result we obtain with MAPLE using the series option

```
dsolve({diff(y(t),t)=t^2+y(t)^2,y(0)=1},y(t),series);
```

$$y(t) = 1 + t + t^2 + \frac{4}{3}t^3 + \frac{7}{6}t^4 + \frac{6}{5}t^5 + O\left(t^6\right).$$

The idea is now to expand the solution locally by choosing a small step size Δt and then to advance to a next point of the solution,

$$t_{n+1} = t_n + \Delta t$$

$$y_{n+1} = y(t_n + \Delta t) = y(t_n) + \frac{y'(t_n)}{1!}\Delta t + \frac{y''(t_n)}{2!}\Delta t^2 + \cdots$$

We now solve our model problem (10.8) using (10.9) and compare the results with the MATLAB integrator `ode45`

ALGORITHM 10.1.
Solving $y' = t^2 + y^2$, $y(0) = 1$ by Taylorseries

```
t=0; y=1; dt=0.01; T=[t,y];
while t<0.8
   y1=t^2+y^2;
   y2=2*t+2*y*y1;
   y3=2+2*y1^2+2*y*y2;
   y4=6*y1*y2+2*y*y3;
   y=y+y1*dt+y2/2*dt^2+y3/6*dt^3+y4/24*dt^4;
   t=t+dt;
   T=[T;t,y];
end;
[t,Y]=ode45(@(t,y)t^2+y^2,[0:dt:0.8],1);
norm(T(:,2)-Y)
```

When we execute this program, we see that the norm of the difference of the function values is `4.7643e-04`, which shows good agreement between the two methods.

The second method is a power series approach: we use here the ansatz

$$y(t + \Delta t) = \sum_{i=0}^{\infty} a_{i+1} \Delta t^i. \tag{10.10}$$

We have shifted the index of the coefficients by one because MATLAB does not allow zero indices. Introducing this series into the differential equation

$$y' = t^2 + y^2,$$

we obtain with

$$y'(t + \Delta t) = \sum_{i=1}^{\infty} i a_{i+1} \Delta t^{i-1} = \sum_{k=0}^{\infty} (k+1) a_{k+2} \Delta t^k$$

and

$$y(t + \Delta t)^2 = \sum_{i=0}^{\infty} a_{i+1} \Delta t^i \sum_{j=0}^{\infty} a_{j+1} \Delta t^j = \sum_{k=0}^{\infty} \left(\sum_{j=1}^{k+1} a_j a_{k+2-j} \right) \Delta t^k$$

the equation

$$\sum_{k=0}^{\infty} (k+1) a_{k+2} \Delta t^k = (t + \Delta t)^2 + \sum_{k=0}^{\infty} \left(\sum_{j=1}^{k+1} a_j a_{k+2-j} \right) \Delta t^k.$$

Comparing the coefficients on both sides, we get with $a_1 = y(t_0)$ the equations:

$$
\begin{aligned}
k = 0: \qquad & a_2 = t^2 + a_1^2 \\
k = 1: \qquad & 2a_3 = 2t + a_1 a_2 + a_2 a_1 = 2t + 2a_1 a_2 \\
k = 2: \quad & 3a_4 = 1 + a_1 a_3 + a_2 a_2 + a_3 a_1 = 1 + 2a_1 a_3 + a_2^2 \\
k > 2: \qquad & (k+1)a_{k+2} = \sum_{j=1}^{k+1} a_j a_{k+2-j}.
\end{aligned}
$$

Thus, we can solve each equation sequentially for a new coefficient. The MATLAB function below computes the coefficients of (10.10)

ALGORITHM 10.2.
Taylorseries for $y' = t^2 + y^2$ using ansatz for $y(t + t)$

```
function a=DEQseries(t,y,n);
% DEQSERIES taylor series for a particular differential equation
%   a=DEQseries(t,y,n); computes n coefficients of the Taylor series
%   y(t+dt)=sum_{k=0} a_{k+1} dt^k of the solution of the differential
%   equation y'=t^2+y^2 using the power series ansatz

a=zeros(1,n);
a(1)=y;
a(2)=t^2+a(1)^2;
a(3)=t+a(1)*a(2);
a(4)=(1+2*a(1)*a(3)+a(2)^2)/3;
for k=4:n-1
   a(k+1)=a(1:k)*a(k:-1:1)'/(k);
end
```

We can choose the order n of the expansion, e.g. $n = 7$:

```
>> a=DEQseries(0,1,7)
a =
    1.0000    1.0000    1.0000    1.3333    1.1667    1.2000    1.2333
```

For $n = 5$, we obtain the same coefficients as with MAPLE using the series option.

10.2.4 Computing with Power Series

We need to manipulate power series when inserting an ansatz of the form (10.10) into a differential equation. We consider some basic operations. Let

$$
r(t) = \sum_{i=0}^{n} r_i t^i, \quad p(t) = \sum_{i=0}^{n} p_i t^i, \quad q(t) = \sum_{i=0}^{n} q_i t^i.
$$

a) Addition: $r(t) = p(t) \pm q(t), \quad r_i = p_i \pm q_i$

b) Multiplication: $r(t) = p(t)q(t), \quad r_i = \sum_{j=0}^{i} p_j q_{i-j}$

c) Division: $r(t) = p(t)/q(t) \iff r(t)q(t) = p(t)$ thus $\sum_{j=0}^{i} r_j q_{i-j} = p_i$.
If we solve the equation for r_i we get

$$r_i = \frac{1}{q_0}\left(p_i - \sum_{j=0}^{i-1} r_j q_{i-j} \right) \tag{10.11}$$

This formula reminds us very much of back substitution! Indeed, if we connect a power series

$$p(t) = p_0 + p_1 t + \cdots$$

to its *Toeplitz* matrix

$$P = \begin{pmatrix} p_0 & p_1 & p_2 & p_3 & \cdots \\ & p_0 & p_1 & p_2 & \cdots \\ & & p_0 & p_1 & \cdots \\ & & & p_0 & \cdots \\ & & & & \ddots \end{pmatrix},$$

then the series operations correspond to the matrix operations.

$$\begin{array}{rclcrcl} r & = & p \pm q & \iff & R & = & P \pm Q \\ r & = & pq & \iff & R & = & PQ \\ r & = & p/q & \iff & R & = & Q^{-1}P \iff QR = P. \end{array}$$

Now let P, Q and R be $n \times n$ Toeplitz matrices. Consider only the last column of R in the equation $QR = P$. We multiply by e_n and obtain $QRe_n = Pe_n$:

$$\begin{pmatrix} q_0 & q_1 & \cdots & q_n \\ & q_0 & \cdots & q_{n-1} \\ & & \ddots & \vdots \\ & & & q_0 \end{pmatrix} \begin{pmatrix} r_n \\ r_{n-1} \\ \vdots \\ r_0 \end{pmatrix} = \begin{pmatrix} p_n \\ p_{n-1} \\ \vdots \\ p_0 \end{pmatrix}.$$

It is now evident that the formula (10.11) is equivalent to solving for $r_0, \ldots, r_n$ by back substitution.

d) Exponential function: The trick here is to solve the differential equation for the exponential by differentiating $r(t) = e^{p(t)}$:

$$r'(t) = p'(t)e^{p(t)} = p'(t)r(t). \tag{10.12}$$

Inserting the series expansion into (10.12), we get

$$\sum_{i=0}^{n-1}(i+1)r_{i+1}t^i = \sum_{i=0}^{n-1}(i+1)p_{i+1}t^i \sum_{k=0}^{n} r_k t^k = \sum_{i=0}^{n}\left(\sum_{j=0}^{i}(j+1)p_{j+1}r_{i-j} \right) t^i,$$

so comparing coefficients gives

$$(i+1)r_{i+1} = \sum_{j=0}^{i}(j+1)p_{j+1}r_{i-j}.$$

By letting $i := i-1$ and performing the change of variables $k = i-j-1$, we obtain *Miller's formula*

$$r_0 = e^{p_0}, \quad r_i = \frac{1}{i}\sum_{k=0}^{i-1}(i-k)p_{i-k}r_k, \quad i = 1, 2, \ldots, n. \qquad (10.13)$$

EXAMPLE 10.3. *We consider* $p(t) = -1 + t + 2t^2 + 4t^3 - 2t^4 + t^5$, *compute* $e^{p(t)}$ *using Miller's formula (10.13) and compare the result we get with the corresponding Toeplitz matrix using* MATLAB*'s matrix function* **expm(P)**.

```
p=[-1 1 2 4 -2 1]
n=length(p)-1;                          % Miller for exponential
r=exp(p(1));
for i=1:n
  s=0;
  for k=0:i-1
    s=s+(i-k)*p(i-k+1)*r(k+1);
  end
  r=[r s/i];
end
P=triu(toeplitz(p));
R=expm(P);                              % exponential matrix
[R(:,n+1) r(end:-1:1)']                 % compare results


p =
    -1     1     2     4    -2     1
ans =
      4.1724      4.1724
      1.8547      1.8547
      2.2686      2.2686
      0.9197      0.9197
      0.3679      0.3679
      0.3679      0.3679
```

We get with both methods the same results

$$r(t) = e^{p(t)} = 0.3679 + 0.3679t + 0.9197t^2 + 2.2686t^3 + 1.8547t^4 + 4.1724t^5.$$

Using MAPLE *for the same computation,*

```
p:=-1+t+2*t^2+4*t^3-2*t^4+t^5;
series(exp(p),t=0);
evalf(%);
```

we obtain the same

$$.3678794412 + .3678794412\, t + .9196986030\, t^2 +$$
$$2.268589888\, t^3 + 1.854725516\, t^4 + 4.172365997\, t^5 + O(t^6).$$

e) Trigonometric functions $c(t) = \cos(p(t))$ **and** $s(t) = \sin(p(t))$**:** Here, we use the differential equations simultaneously for both series. For $t = 0$, we have the initial coefficients $c_0 = \cos(p_0)$ and $s_0 = \sin(p_0)$. Applying Miller's formula (10.13), we get

$$c_i = -\frac{1}{i} \sum_{k=0}^{i-1} (i - k) p_{i-k} s_k,$$
$$s_i = \frac{1}{i} \sum_{k=0}^{i-1} (i - k) p_{i-k} c_k.$$

The coefficients of the series for $\cos(p(t))$ and $\sin(p(t))$ can thus be computed by running these two recurrences simultaneously.

f) Composition of series: Given the two series

$$p(t) = p_0 + p_1 t + p_2 t^2 + \cdots,$$
$$q(t) = q_0 + q_1 t + q_2 t^2 + \cdots,$$

we want to compute the series for $p(q(t))$. If we simply insert the series we get

$$p(q(t)) = p_0 + p_1 (q_0 + q_1 t + q_2 t^2 + \cdots)^1$$
$$+ p_2 (q_0 + q_1 t + q_2 t^2 + \cdots)^2$$
$$+ p_3 (q_0 + q_1 t + q_2 t^2 + \cdots)^3$$
$$+ \cdots.$$

Assume we have expanded $q(t)^n$ and computed the coefficients

$$q(t)^n = q_0^{(n)} + q_1^{(n)} t + q_2^{(n)} t^2 + \cdots,$$

then we would have

$$p(q(t)) = p_0 + p_1 q_0^{(1)} + p_2 q_0^{(2)} + p_3 q_0^{(3)} + \cdots$$
$$+ (p_1 q_1^{(1)} + p_2 q_1^{(2)} + \cdots) t$$
$$+ (p_1 q_2^{(1)} + p_2 q_2^{(2)} + \cdots) t^2$$
$$+ \cdots.$$

The new coefficients above are all infinite sums that may or may not be convergent, so the coefficients may not actually exist! Therefore, the composition of series is only possible for series $q(t)$ with $q_0 = 0$ (these are *non-unit series*). In this case, the powers become

$$q(t)^n = q_n^{(n)} t^n + q_{n+1}^{(n)} t^{n+1} + q_{n+2}^{(n)} t^{n+2} + \cdots$$

and all the new coefficients are finite sums:

$$\begin{aligned} p(q(t)) =& p_0 + p_1 q_1^{(1)} t + (p_1 q_2^{(1)} + p_2 q_2^{(2)}) t^2 \\ &+ (p_1 q_3^{(1)} + p_2 q_3^{(2)} + p_3 q_3^{(3)}) t^3 \\ &+ \cdots. \end{aligned}$$

EXAMPLE 10.4.

$$e^{(e^t)} = e^{1 + t + \frac{t^2}{2!} + \frac{t^3}{3!} + \cdots} = e^1 e^{t + \frac{t^2}{2!} + \frac{t^3}{3!} + \cdots}.$$

By separating the first term, we obtain a non-unit series and thus

$$e^{(e^t)} = e \left(1 + t + t^2 + \frac{5}{6} t^3 + \frac{5}{8} t^4 + \frac{13}{30} t^5 + \cdots \right).$$

10.2.5 Euler's Method

Differential equations are a powerful tool for modeling, as we have already seen in the silver watch example. We now present another example that shows how easily one can model complicated problems using ODEs, and how quickly one can run into problems that cannot be solved analytically.

Suppose you are jogging along a given path described by your position in time, $(\xi(t), \eta(t))$, and suddenly a dog in a neighbor's garden sees you and starts chasing you at maximal speed w, as shown in Figure 10.2. What is the trajectory of the dog if we assume it is always running directly toward you?

Since the speed of the dog is constant, we have

$$\dot{x}^2 + \dot{y}^2 = w^2, \tag{10.14}$$

where we denote the time derivative by a dot above the variable. Since the dog is always running toward you, its direction is proportional to $(\xi - x, \eta - y)^T$, which implies

$$\begin{pmatrix} \dot{x}(t) \\ \dot{y}(t) \end{pmatrix} = \lambda(t) \begin{pmatrix} \xi(t) - x(t) \\ \eta(t) - y(t) \end{pmatrix}, \tag{10.15}$$

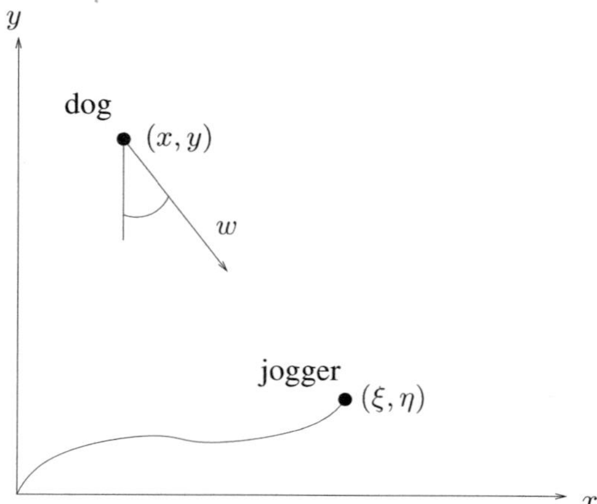

FIGURE 10.2. *Dog chasing a jogger*

where λ is a constant of proportionality. To find this constant, we substitute (10.15) into (10.14) to obtain

$$\lambda^2(t) = \frac{w^2}{(\xi(t) - x(t))^2 + (\eta(t) - y(t))^2}.$$

The trajectory of the dog is therefore described by the *system of ordinary differential equations*

$$\begin{pmatrix} \dot{x}(t) \\ \dot{y}(t) \end{pmatrix} = \frac{w}{\sqrt{(\xi(t) - x(t))^2 + (\eta(t) - y(t))^2}} \begin{pmatrix} \xi(t) - x(t) \\ \eta(t) - y(t) \end{pmatrix} \quad (10.16)$$

with initial conditions given by the initial position of the dog, $(x(0), y(0)) = (x_0, y_0)$. To find the trajectory of the dog, it remains to solve this initial value problem. While for the special case of the silver watch problem, an implicit, closed-form solution exists, there is no hope of finding a closed-form solution for a general jogging path $(\xi(t), \eta(t))$, so we need to resort to a numerical method.

A very natural and simple approach to obtain a numerical solution is the Forward Euler method, which was invented when Euler realized that probably not all ODEs can be solved in closed form. Instead of trying to solve the differential equation (10.16) directly, we use it to determine the direction the dog will choose initially, at position (x_0, y_0). This direction is given by the differential equation to be

$$\begin{pmatrix} \dot{x}(0) \\ \dot{y}(0) \end{pmatrix} = \frac{w}{\sqrt{(\xi_0 - x_0)^2 + (\eta_0 - y_0)^2}} \begin{pmatrix} \xi_0 - x_0 \\ \eta_0 - y_0 \end{pmatrix},$$

where we denote the initial position of the jogger by $(\xi_0, \eta_0) := (\xi(0), \eta(0))$. Note that all quantities on the right hand side are known. Euler's idea was

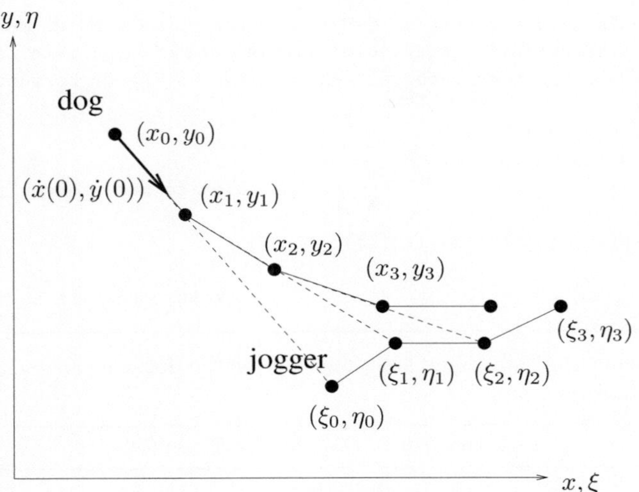

FIGURE 10.3.
*The Forward Euler (FE) method applied to the
dog-jogger problem.*

to advance in this direction for a short time Δt_1, as shown in Figure 10.3.
Then the new position (x_1, y_1) at time $t_1 = \Delta t_1$ is an approximation of the
exact position $(x(t_1), y(t_1))$, and one can repeat the same procedure starting
with position (x_1, y_1) and going for a short time Δt_1 along the direction

$$
\begin{pmatrix} \dot{x}(t_1) \\ \dot{y}(t_1) \end{pmatrix} = \frac{w}{\sqrt{(\xi(t_1) - x_1)^2 + (\eta(t_1) - y_1)^2}} \begin{pmatrix} \xi(t_1) - x_1 \\ \eta(t_1) - y_1 \end{pmatrix},
$$

where $(\xi_1, \eta_1) := (\xi(t_1), \eta(t_1))$ in Figure 10.3. Again the right hand side
is known, so one can advance again over a short time interval Δt_2 to find
a new position (x_2, y_2), which approximates the exact position of the dog
$(x(t_2), y(t_2))$ at $t_2 = t_1 + \Delta t_2$, and so on.

If one applies this procedure to a general system of ordinary differential
equations,

$$
\boldsymbol{y}' = \boldsymbol{f}(t, \boldsymbol{y}), \quad \boldsymbol{y}(t_0) = \boldsymbol{y}_0, \tag{10.17}
$$

one obtains the *Forward Euler method*

$$
\boldsymbol{y}_{k+1} = \boldsymbol{y}_k + \Delta t \boldsymbol{f}(t_k, \boldsymbol{y}_k). \tag{10.18}
$$

A MATLAB implementation of this method for a general system of ordinary
differential equations is

ALGORITHM 10.3. *Forward Euler ODE Solver*

```
function [t,y]=ForwardEuler(f,tspan,y0,n);
```

```
% FORWARDEULER solves system of ODEs using the Forward Euler method
%    [t,y]=ForwardEuler(f,tspan,y0,n) solves dy/dt=f(t,y) with initial
%    value y0 on the time interval tspan doing n steps of Forward Euler

dt=(tspan(2)-tspan(1))/n;
t=tspan(1):dt:tspan(2);
y(:,1)=y0(:);                            % colon to make column vector
for i=1:n,
  y(:,i+1)=y(:,i)+dt*f(t(i),y(:,i));
end;
t=t';y=y';                               % to return results in columns
```

If we implement the right hand side function for the dog-and-jogger problem
in the MATLAB function

ALGORITHM 10.4. *ODE for Dog Trajectory*

```
function xp=Dog(t,x);
% DOG ODE right hand side modeling the dog hunts jogger problem
%    xp=Dog(t,x); computes the right hand side of the ODE modeling the
%    dog hunts the jogger problem at time t and position x.

w=10;
xi=[8*t; 0];
xp=w/norm(xi-x)*(xi-x);
```

we obtain with the command

```
[t,x]=ForwardEuler(@Dog,[0,12],[60; 70],50);
```

an approximate solution path in the two columns of the variable x at time
values in the vector t. To display the results, one can use the MATLAB
commands

```
plot(x(:,1),x(:,2),'-',8*t, 0*t,'--');
hold on
plot(x(end,1),x(end,2),'k+','MarkerSize',20,'Linewidth',3);
hold off
axis([0 100 -10 70]);
legend('dog','jogger');
xlabel('x');
ylabel('y');
```

We show the results of this first experiment in Figure 10.4 on the top left.
On the top right, we show using similar commands in MATLAB a jogger that
notices the dog and tries to run back, on the bottom left a jogger running
on a circular track, and on the right the same situation, but with an old and
slow dog.

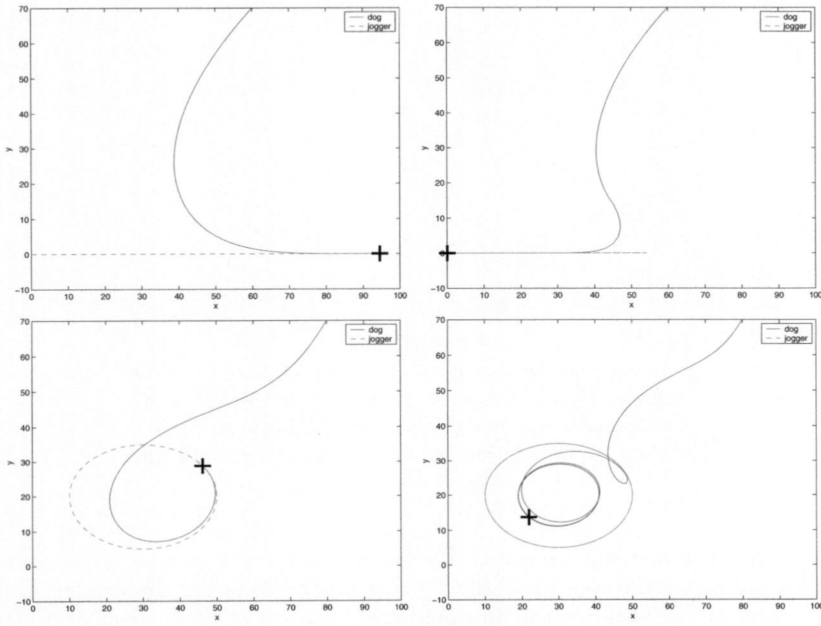

FIGURE 10.4.
Several jogging path, and the numerically computed
trajectory of the dog chasing the jogger.

As we have seen in the example with the dog and the jogger, it is easy to obtain approximate solutions, but how accurate are they when compared with the exact solution? Intuitively, making the time steps smaller should improve the accuracy of the approximation in Euler's method. Let us perform the following experiment in MATLAB to investigate this issue, using the programs developed above:

```
N=[5 10 20 40 80];
for i=1:length(N)
  [t,x]=ForwardEuler(@Dog,[0,10],[60; 70],N(i));
  plot(x(:,1),x(:,2),'-');
  hold on
end;
plot(8*t, 0*t,'--');
hold off
axis([0 100 -10 70]);
legend('5 Euler steps','10 Euler steps','20 Euler steps', ...
  '40 Euler steps','80 Euler steps','jogger');
xlabel('x');
ylabel('y');
```

The result is shown in the graph on the left of Figure 10.5. One can see that

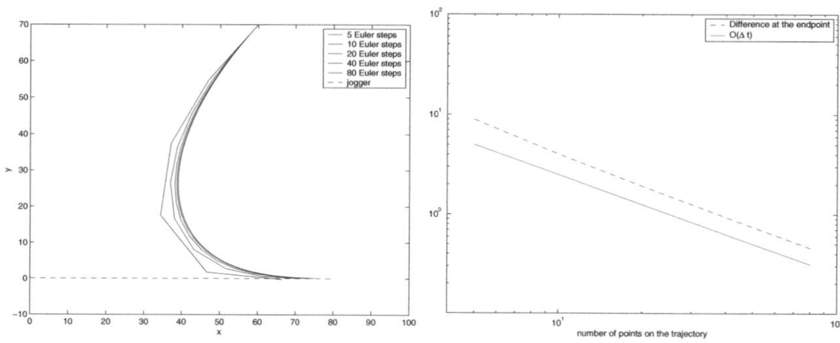

FIGURE 10.5.

*Computing the trajectory of the dog with Forward Euler
using 5, 10, 20, 40 and 80 time steps on the left, and
difference at the end of the dog path between these
different approximations and one with a very small time
step on the right.*

the more one refines the time step, the more accurate the trajectory becomes,
and it seems to converge to a limiting trajectory. To better understand this
convergence, we calculate the limiting trajectory using a highly refined Euler
approximation with 8000 steps, and compute the difference between the end
point of this trajectory and those of the various approximations. The result is
shown on the right of Figure 10.5. From this plot, we can see that apparently
the Euler approximation of the solution of the ODE is of order Δt. Let us look
at how accurate Euler's method is over one step. To do so, we compare one
step of Euler's method (10.18) to the Taylor expansion of the exact solution
of the system of ODEs (10.17) about the initial point t_0. We obtain

$$\boldsymbol{y}(t_0 + \Delta t) - \boldsymbol{y}_1 = \boldsymbol{y}(t_0) + \Delta t \boldsymbol{y}'(t_0) + O(\Delta t^2) - \boldsymbol{y}_0 - \Delta t \boldsymbol{f}(t_0, \boldsymbol{y}_0) = O(\Delta t^2),$$

$$(10.19)$$

where we have used the differential equation (10.17), which implies $\boldsymbol{y}'(t_0) =
\boldsymbol{f}(t_0, \boldsymbol{y}_0)$, and the fact that the initial condition states $\boldsymbol{y}(t_0) = \boldsymbol{y}_0$. Hence the
difference between the Euler approximation and the exact solution is $O(\Delta t^2)$
over one step, whereas at the end after many steps, the approximation is only
of $O(\Delta t)$, as we measured in Figure 10.5. In this figure, we can also see that
using 80 Euler steps leads to an error of about $\frac{1}{2}$, and we can estimate that
for a standard precision of $1e - 6$, one would need to double the number of
steps n times, where n satisfies the equation

$$\frac{1}{2}\left(\frac{1}{2}\right)^n = 1e - 6 \implies n = \frac{\log(2e - 6)}{\log(0.5)} \approx 18.93,$$

which gives approximately $80 \times 2^{18.93} \approx 40\,000\,000$ time steps. Euler's method
is therefore of limited practical interest, and one needs higher order methods,
which we will develop later in this chapter.

10.2.6 Autonomous ODE, Reduction to First Order System

To simplify the Taylor expansions in the error analysis, we may assume without loss of generality that the right hand side in the system of ordinary differential equations (10.17) is *autonomous*, i.e. $\boldsymbol{f}(t,\boldsymbol{y}) = \boldsymbol{f}(\boldsymbol{y})$. If not, the explicit dependence on the time variable t can be removed by introducing an additional variable: letting $z = t$, we obtain

$$\tilde{\boldsymbol{y}}(t) = \begin{pmatrix} \boldsymbol{y} \\ z \end{pmatrix}, \quad \tilde{\boldsymbol{y}}(t_0) = \begin{pmatrix} \boldsymbol{y}_0 \\ t_0 \end{pmatrix},$$

and the differential system $\boldsymbol{y}' = \boldsymbol{f}(t,\boldsymbol{y})$ becomes autonomous,

$$\tilde{\boldsymbol{y}}' = \tilde{\boldsymbol{f}}(\tilde{\boldsymbol{y}}) = \begin{pmatrix} \boldsymbol{f}(z,\boldsymbol{y}) \\ 1 \end{pmatrix}.$$

Numerical algorithms require as input a *first-order system*

$$\boldsymbol{y}' = \boldsymbol{f}(t,\boldsymbol{y}) \quad \text{with initial condition} \quad \boldsymbol{y}(t_0) = \boldsymbol{y}_0,$$

where

$$\boldsymbol{y} = \begin{pmatrix} y_1(t) \\ \vdots \\ y_n(t) \end{pmatrix}, \quad \boldsymbol{f}(t,\boldsymbol{y}) = \begin{pmatrix} f_1(t,\boldsymbol{y}) \\ \vdots \\ f_n(t,\boldsymbol{y}) \end{pmatrix}.$$

Differential equations of higher order are transformed into a first-order system by introducing additional unknowns, as we now show with the following example. Consider

$$y''' + 5y'' + 8y' + 6y = 10e^{-t}, \quad y(0) = 2, y'(0) = y''(0) = 0,$$

a linear inhomogeneous third-order differential equation with constant coefficients. To transform this equation to a first order system, we perform the following steps:

1. Solve the differential equation for the highest derivative,

$$y''' = -5y'' - 8y' - 6y + 10e^{-t}.$$

2. Introduce new functions

$$z_1 = y, \quad z_2 = y', \quad z_3 = y''.$$

3. Differentiate the new functions, replace the y functions and write the new system

$$\begin{aligned} z_1' &= y' = z_2 \\ z_2' &= y'' = z_3 \\ z_3' &= y''' = -5z_3 - 8z_2 - 6z_1 + 10e^{-t}. \end{aligned}$$

Since the differential equation is linear, we can use matrix-vector notation

$$z' = Az + \begin{pmatrix} 0 \\ 0 \\ 10e^{-t} \end{pmatrix} \text{ with } A = \begin{pmatrix} 0 & 1 & 0 \\ 0 & 0 & 1 \\ -6 & -8 & -5 \end{pmatrix} \text{ and } z(0) = \begin{pmatrix} 2 \\ 0 \\ 0 \end{pmatrix}$$

Note that MAPLE can often solve such linear differential equations analytically without requiring the user to transform it into a first-order system. Here, the statement

```
dsolve(diff(y(t),t$3)+5*diff(y(t),t$2)+8*diff(y(t),t)+6*y(t)=10*exp(-t),y(t));
```

gives the explicit general solution

$$y(t) = 5 \left(e^t\right)^{-1} + \_C1\, e^{-3\,t} + \_C2\, e^{-t} \cos(t) + \_C3\, e^{-t} \sin(t).$$

10.3   Runge-Kutta Methods

The Forward Euler method presented in Subsection 10.2.5 can be interpreted as an approximation of an integral. Integrating the differential equation (10.17) in time, we obtain

$$y(t_0 + \Delta t) - y(t_0) = \int_{t_0}^{t_0+\Delta t} f(\tau, y(\tau))d\tau,$$

and Forward Euler is a simple approximation of the area under the curve by a rectangle, like in a Riemann sum with one term,

$$\int_{t_0}^{t_0+\Delta t} f(\tau, y(\tau))d\tau \approx \Delta t f(t_0, y_0),$$

as shown in Figure 10.6. Runge-Kutta methods are based on the same idea, except one uses another (and generally better) approximation of this area.

10.3.1   Explicit Runge-Kutta Methods

Clearly, a better approximation would be to use the midpoint rule to approximate the integral, also indicated in Figure 10.6,

$$\int_{t_0}^{t_0+\Delta t} f(\tau, y(\tau))d\tau \approx \Delta t f(t_0 + \frac{\Delta t}{2}, y(t_0 + \frac{\Delta t}{2})).$$

Unfortunately, the midpoint value $y(t_0 + \frac{\Delta t}{2})$ is unknown; nonetheless, it can be approximated by an Euler step over half the interval,

$$y(t_0 + \frac{\Delta t}{2}) \approx y_0 + \frac{\Delta t}{2} f(t_0, y_0),$$

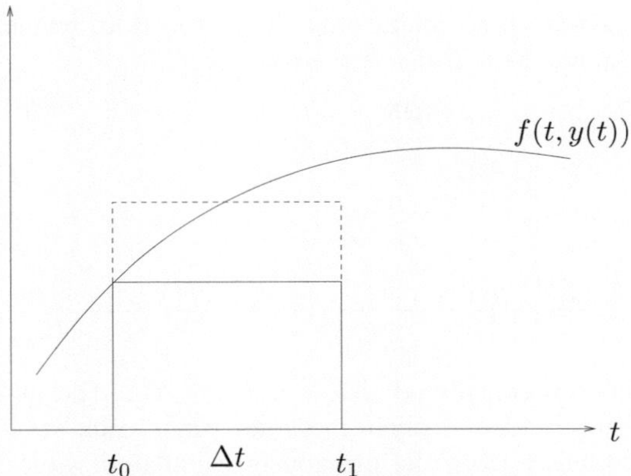

FIGURE 10.6. *Approximation of the integral.*

which leads to the new method

$$y_1 = y_0 + \Delta t f\left(t_0 + \frac{\Delta t}{2}, y_0 + \frac{\Delta t}{2} f(t_0, y_0)\right).$$

This is known as *Runge's method.* Like Euler's method, one step is performed using only data from the previous step, but now two function evaluations are needed per step. In general, a *one-step method* is of the form

$$y_{k+1} = y_k + \Delta t_{k+1} \phi_f(t_k, y_k, \Delta t_{k+1}) \tag{10.20}$$

together with an initial condition y_0 at time t_0, for example

$$
\begin{aligned}
\phi_f(t_0, \mathbf{y}_0, \Delta t) &= f(t_0, y_0) && \implies \text{Euler's method} \\
\phi_f(t_0, \mathbf{y}_0, \Delta t) &= f(t_0 + \tfrac{\Delta t}{2}, y_0 + \tfrac{\Delta t}{2} f(t_0, y_0)) && \implies \text{Runge's method}
\end{aligned}
$$

A systematic generalization of Runge's second order method yields the Runge-Kutta methods: An *explicit s-stage Runge-Kutta method* is given by

$$
\begin{aligned}
k_1 &= f(t_0, y_0) \\
k_2 &= f(t_0 + \Delta t c_2, y_0 + \Delta t a_{21} k_1) \\
k_3 &= f(t_0 + \Delta t c_3, y_0 + \Delta t(a_{31} k_1 + a_{32} k_2)) \\
&\ \vdots \\
k_s &= f(t_0 + \Delta t c_s, y_0 + \Delta t(a_{s1} k_1 + a_{s2} k_2 + \ldots + a_{s,s-1} k_{s-1})) \\
y_1 &= y_0 + \Delta t(b_1 k_1 + b_2 k_2 + \ldots + b_s k_s),
\end{aligned}
$$

$$\tag{10.21}$$

where k_j are called the *stage values*, and the coefficients c_i, a_{ij} and b_j are chosen to obtain a good method. J. C. Butcher, during his investigation into higher order Runge-Kutta methods, introduced the following compact notation to represent these coefficients. This notation, which first appeared in [13], is now known as a *Butcher tableau*:

$$
\begin{array}{c|ccccc}
0 & & & & & \\
c_2 & a_{21} & & & & \\
c_3 & a_{31} & a_{32} & & & \\
\vdots & \vdots & & \ddots & & \\
c_s & a_{s1} & a_{s2} & \cdots & a_{s,s-1} & \\
\hline
& b_1 & b_2 & \cdots & b_{s-1} & b_s
\end{array}
$$

So far we have only considered $a_{ij} = 0$ for $j \geq i$; the associated Runge Kutta methods are called explicit methods, since each stage value k_j can be evaluated explicitly from the previously calculated values. The Butcher tableau, however, suggests that one could also consider $a_{ij} \neq 0$ for $j \geq i$, and then the stage values k_j would be coupled by a system of equations. Such Runge-Kutta methods are called implicit methods and are useful for stiff problems, as we will show in Section 10.5.

The Forward Euler method is a one-stage explicit Runge-Kutta method,

$$
\begin{aligned}
\boldsymbol{k}_1 &= \boldsymbol{f}(t_0, \boldsymbol{y}_0) \\
\boldsymbol{y}_1 &= \boldsymbol{y}_0 + \Delta t \boldsymbol{k}_1
\end{aligned}
\qquad
\begin{array}{c|c}
0 & \\
\hline
& 1
\end{array}
$$

and Runge's second order method is a two-stage explicit Runge-Kutta method,

$$
\begin{aligned}
\boldsymbol{k}_1 &= \boldsymbol{f}(t_0, \boldsymbol{y}_0) \\
\boldsymbol{k}_2 &= \boldsymbol{f}(t_0 + \tfrac{1}{2}\Delta t, \boldsymbol{y}_0 + \tfrac{1}{2}\Delta t \boldsymbol{k}_1) \\
\boldsymbol{y}_1 &= \boldsymbol{y}_0 + \Delta t \boldsymbol{k}_2
\end{aligned}
\qquad
\begin{array}{c|cc}
0 & & \\
\tfrac{1}{2} & \tfrac{1}{2} & \\
\hline
& 0 & 1
\end{array}
$$

10.3.2   Local Truncation Error

In order to compare these new methods, we have to analyze their approximation qualities.

DEFINITION 10.1. (LOCAL TRUNCATION ERROR) *The local truncation error $\boldsymbol{\tau}$ for a one-step method of the form (10.20) approximating the system of ODEs (10.17) is defined by the difference between the exact solution and the numerical approximation after one step, i.e.*

$$
\boldsymbol{\tau} := \boldsymbol{y}(t_0 + \Delta t) - (\boldsymbol{y}_0 + \Delta t \phi_f(\boldsymbol{y}_0, \Delta t)).
$$

We have already seen in (10.19) that the local truncation error for Euler's method is $O(\Delta t^2)$. To find the local truncation error for Runge's method, we also use Taylor expansions, but this time both for the exact solution and the numerical method,

$$\boldsymbol{y}(t_0 + \Delta t) = \boldsymbol{y}(t_0) + \Delta t \boldsymbol{y}'(t_0) + \frac{\Delta t^2}{2}\boldsymbol{y}''(t_0) + O(\Delta t^3)$$

$$\boldsymbol{y}_1 = \boldsymbol{y}_0 + \Delta t \boldsymbol{f}(t_0, \boldsymbol{y}_0) + \frac{\Delta t^2}{2}\left(\boldsymbol{f}_t(t_0, \boldsymbol{y}_0) + \boldsymbol{f}_y(t_0, \boldsymbol{y}_0)\boldsymbol{y}'(t_0)\right) + O(\Delta t^3).$$

As for Euler's method, we can identify the first two terms in the two expansions because of the initial condition $\boldsymbol{y}(t_0) = \boldsymbol{y}_0$ and the differential equation $\boldsymbol{y}'(t_0) = \boldsymbol{f}(t_0, \boldsymbol{y}_0)$. For the third term, we need to differentiate the differential equation (10.17) to obtain

$$\boldsymbol{y}''(t_0) = \frac{d}{dt}\boldsymbol{f}(t_0, \boldsymbol{y}_0) = \boldsymbol{f}_t(t_0, \boldsymbol{y}_0) + \boldsymbol{f}_y(t_0, \boldsymbol{y}_0)\boldsymbol{y}'(t_0),$$

and hence the third term is also identical in both expansions. Therefore, the local truncation error of Runge's method is $\boldsymbol{\tau} = \boldsymbol{y}(t_0 + \Delta t) - \boldsymbol{y}_1 = O(\Delta t^3)$, which is better than for Euler's method.

Here is an implementation of Runge's method in MATLAB:

ALGORITHM 10.5. *Runge's Method for ODE*

```
function [t,y]=Runge(f,tspan,y0,n);
% RUNGE solves system of ODEs using Runge's method
%   [t,y]=Runge(f,tspan,y0,n) solves dy/dt=f(t,y) with initial
%   value y0 on the time interval tspan doing n steps of Runge's method.

dt=(tspan(2)-tspan(1))/n;
t=tspan(1):dt:tspan(2);
y(:,1)=y0(:);                             % colon to make column vector
for i=1:n,
  y(:,i+1)=y(:,i)+dt*f(t(i)+dt/2,y(:,i)+dt/2*f(t(i),y(:,i)));
end;
t=t';y=y';                                % to return results in columns
```

Applying Runge's method to the dog-and-jogger problem, also using 5, 10, 20, 40 and 80 steps, we obtain the results in Figure 10.7. One can see on the left that the trajectories converge more rapidly than for Euler's method, and on the right that the error at the end of the dog trajectory is $O(\Delta t^2)$. This method is hence known as Runge's second order method. With 80 steps, the approximation error is now only about $\frac{1}{100}$, and to reach a standard precision of $1e - 6$, one would need to double the number of steps n times, where n satisfies the equation

$$\frac{1}{100}\left(\frac{1}{4}\right)^n = 1e - 6 \implies n = \frac{\log(100e - 6)}{\log(0.25)} \approx 6.64,$$

which gives approximately $80 \times 2^{6.64} = 8\,000$ time steps. Even though each step of Runge's second order method is about twice as costly as an Euler step,

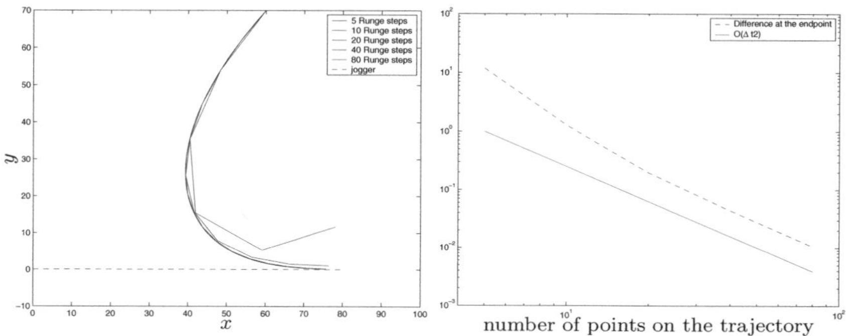

FIGURE 10.7.

Computing the trajectory of the dog with Runge's method using 5, 10, 20, 40 and 80 time steps on the left, and difference at the end of the dog path between these different approximations and one with a very small time step on the right.

since one needs to evaluate the function twice, this number of steps compares very favorably with the 40 000 000 steps needed by Euler's method for the same precision. It is therefore of great interest to construct even higher order methods.

10.3.3　Order Conditions

To find more Runge-Kutta methods, we need to introduce the concept of order for a one-step method.

DEFINITION 10.2. (ORDER) *The one-step method (10.20) approximating the system of ODEs (10.17) is of order p if, for all sufficiently differentiable functions $\boldsymbol{f}$ in (10.17), the local truncation error satisfies*

$$\boldsymbol{\tau} = O(\Delta t^{p+1}), \quad as \ \Delta t \longrightarrow 0.$$

Before justifying this definition with a convergence analysis, we investigate whether a different choice of coefficients in the general two-stage explicit Runge-Kutta method can yield a better method than Runge's. We assume in what follows that

$$c_1 = 0, \quad c_i = \sum_{j=1}^{i-1} a_{ij}, \quad i = 2,\ldots,s, \tag{10.22}$$

which implies that all stage values $\boldsymbol{k}_i$ are at least second order approximations of the time derivative of the solution at the corresponding time point $t_0 + c_i \Delta t$.

This can be seen from

$$
\begin{aligned}
\boldsymbol{k}_i &= \boldsymbol{f}(t_0 + c_i \Delta t, \boldsymbol{y}_0 + \Delta t \sum_{j=1}^{i-1} a_{ij}\boldsymbol{k}_j) \\[2mm]
&= \boldsymbol{f}(t_0, \boldsymbol{y}_0) + f_t(t_0, \boldsymbol{y}_0)c_i \Delta t + f_y(t_0, \boldsymbol{y}_0) \sum_{j=1}^{i-1} a_{ij}\boldsymbol{k}_j \Delta t \\[2mm]
&= \boldsymbol{f}(t_0, \boldsymbol{y}_0) + \left(f_t(t_0, \boldsymbol{y}_0)c_i + f_y(t_0, \boldsymbol{y}_0) \sum_{j=1}^{i-1} a_{ij}(\boldsymbol{y}'(t_0) + O(\Delta t)) \right) \Delta t \\[2mm]
&= \boldsymbol{y}'(t_0) + c_i \Delta t \boldsymbol{y}''(t_0, \boldsymbol{y}_0) + O(\Delta t^2) \; = \; \boldsymbol{y}'(t_0 + c_i \Delta t) + O(\Delta t^2).
\end{aligned}
$$

The conditions (10.22) are not necessary, but they greatly simplify the derivation of higher order methods, and they do not have a significant influence on the quality of the method, see Problem 10.6. The Butcher tableau of a two-stage explicit Runge-Kutta method then becomes

$$
\begin{array}{c|cc}
0 & & \\
a_{21} & a_{21} & \\
\hline
 & b_1 & b_2
\end{array}
$$

To simplify the Taylor expansions in the analysis of the local truncation error, we use MAPLE:

```
k[1]:=f(t0,y(t0));
k[2]:=taylor(f(t0+c[1]*dt,y(t0)+dt*a[2,1]*k[1]),dt,4);
y1:=taylor(y(t0)+dt*sum(b[j]*k[j],j=1..2),dt,4);
tau:=taylor(y(t0+dt),dt,4)-y1;
```

which gives as a result

$$
\begin{aligned}
\tau := & \; y(t0) + \mathrm{D}(y)(t0)\,dt + \frac{1}{2}\left(D^{(2)}\right)(y)(t0)\,dt^2 + \frac{1}{6}\left(D^{(3)}\right)(y)(t0)\,dt^3 + O(dt^4) \\[2mm]
& - (y(t0)(b_1 f(t0,y(t0)) + b_2 f(t0,y(t0)))\,dt \\[2mm]
& - b_2 \left(D_1(f)(t0,y(t0))c_1 + D_2(f)(t0,y(t0))a_{2,1}f(t0,y(t0))\right) dt^2 \\[2mm]
& - b_2 \left(\frac{1}{2}(D_{1,1})(f)(t0,y(t0))c_1{}^2 + c_1(D_{1,2})(f)(t0,y(t0))a_{2,1}f(t0,y(t0)) \right. \\[2mm]
& \left. + \frac{1}{2}a_{2,1}{}^2(f(t0,y(t0)))^2(D_{2,2})(f)(t0,y(t0)) \right) dt^3 + O(dt^4) \Bigg).
\end{aligned}
$$

This is however less convenient to read for the human eye, both because of the generic D operator notation for the derivative, and the fact that MAPLE does not know to substitute the derivatives of y with respect to t using the differential equation. If we tell MAPLE that D applied to y is a function that maps t into $f(t, y(t))$ with the command

```
D(y):=t->f(t,y(t));
```

and also to substitute the lengthy notation for the derivatives of f by the more compact notation

```
alias(f=f(t0,y(t0)),f[t]=D[1](f)(t0,y(t0)),f[y]=D[2](f)(t0,y(t0)));
alias(f[tt]=D[1,1](f)(t0,y(t0)),f[ty]=D[1,2](f)(t0,y(t0)));
alias(f[yy]=D[2,2](f)(t0,y(t0)));
```

then looking at the result τ again in MAPLE with

```
p:=collect(convert(tau,polynom),dt,factor);
```

gives

$$
\begin{aligned}
(\frac{1}{6}f_{tt} + \frac{1}{3}f_{ty}f + \frac{1}{6}f_{yy}f^2 + \frac{1}{6}f_y f_t + \frac{1}{6}f_y^2 f - \frac{1}{2}b_2 f_{tt}c_1^2 - b_2 c_1 f_{ty}a_{2,1}f \\
- \frac{1}{2}b_2 a_{2,1}^2 f^2 f_{yy})dt^3 + (\frac{1}{2}f_t + \frac{1}{2}f_y f - b_2 f_t c_1 - b_2 f_y a_{2,1}f)dt^2 \\
- f(-1 + b_1 + b_2)dt
\end{aligned}
$$

which is now very readable. From this expansion, we see that by choosing $b_1 + b_2 = 1$, the local truncation error becomes $O(\Delta t^2)$. How can we make the next coefficient zero as well ? Using MAPLE to simplify,

```
collect(op(2,%),f,factor);
```

we obtain

$$
-\frac{1}{2}f_y(-1 + 2b_2 a_{2,1})dt^2 f - \frac{1}{2}f_t(-1 + 2b_2 c_1)dt^2,
$$

and we now see the use of the simplifying assumption $c_1 = a_{2,1}$, as the choice $b_2 a_{21} = 1/2$ now leads to a local truncation error a local truncation error of $O(\Delta t^3)$. We also see that the coefficient of the $O(\Delta t^3)$ term cannot be made zero for a general function $\boldsymbol{f}$, so the highest order one can achieve with an explicit two-stage Runge-Kutta method is two. This calculation, however, reveals a free parameter: for example, we can choose b_1 as we wish and still get a local truncation error of order three. Three different choices have historically been used, leading to the methods

$$
\begin{aligned}
b_1 = \tfrac{1}{2} &\implies b_2 = \tfrac{1}{2} \quad, \quad a_{21} = 1, \\
b_1 = 0 &\implies b_2 = 1 \quad, \quad a_{21} = \tfrac{1}{2}, \\
b_1 = \tfrac{1}{4} &\implies b_2 = \tfrac{3}{4} \quad, \quad a_{21} = \tfrac{2}{3}.
\end{aligned}
$$

Unfortunately the terminology used in the literature is not uniform for these methods. The names *Heun, improved Euler* and *modified Euler* are used for different Butcher tableaus. Here is a summary:

$$
\begin{array}{c|cc}
0 & & \\
1 & 1 & \\
\hline
1 & \frac{1}{2} & \frac{1}{2}
\end{array}
\qquad
\begin{matrix}
\textit{Heun or}\\
\textit{modified Euler or}\\
\textit{improved Euler}
\end{matrix}
$$

$$
\begin{array}{c|cc}
0 & & \\
\frac{1}{2} & \frac{1}{2} & \\
\hline
1 & 0 & 1
\end{array}
\qquad
\begin{matrix}
\textit{Runge or}\\
\textit{improved Euler}
\end{matrix}
$$

$$
\begin{array}{c|cc}
0 & & \\
\frac{2}{3} & \frac{2}{3} & \\
\hline
1 & \frac{1}{4} & \frac{3}{4}
\end{array}
\qquad
\textit{Heun}
$$

The general difficulty in finding Runge-Kutta methods is that one must solve a nonlinear system of equations, which may or may not have a solution. For the loert order cases, these systems can still be studied using MAPLE: one has to extract the so-called *order conditions*, which are the coefficients in the Taylor expansion, and then solve the associated nonlinear system of equations. Dominik Gruntz gave in [45] a MAPLE program to compute Runge-Kutta formulas.

ALGORITHM 10.6.
Deriving Runge-Kutta Methods with MAPLE

```
RK:=proc(s,p)
  local TaylorPhi,RungeKuttaPhi,d,vars,eqns,k,i,j,dt;
  global a,b,c;
  D(y):=t->f(t,y(t)):                        # Taylor series
TaylorPhi:=convert(taylor(y(t+dt),dt=0,p+1),polynom):
TaylorPhi:=normal((TaylorPhi-y(t))/dt);
c[1]:=0:                                     # RK-ansatz
for i from 1 to s do
 k[i]:=taylor(f(t+c[i]*dt,y(t)+sum(a[i,j]*k[j],j=1..i-1)*dt),dt=0,p):
od:
RungeKuttaPhi:=sum(b[j]*k[j],j=1..s):
RungeKuttaPhi:=series (RungeKuttaPhi,dt,p):
RungeKuttaPhi:=convert(RungeKuttaPhi,polynom);
d:=expand(TaylorPhi-RungeKuttaPhi):
vars:={seq(c[i],i=2..s), seq(b[i],i=1..s),
        seq((seq(a[i,j],j=1..i-1)),i = 2..s)};
eqns:={coeffs(d,indets(d) minus vars)};
```

```
                                          # simplifying assumptions:
  eqns:=eqns union {seq(sum(a[i,'j'],'j'=1..i-1)-c[i],i=2..s)};
  solve(eqns,vars);
end;
```

For $s = 3$ stages, we do not even need the simplifying equations, as MAPLE manages to solve the nonlinear system without them. So we comment them out:

```
# eqns:=eqns union {seq(sum(a[i,'j'],'j'=1..i-1)-c[i],i=2..s)};
```

Using RK, we compute the general three-stage Runge-Kutta methods:

```
RK3:=RK(3,3);
```

$$\left\{ a_{2,1} = 1/6\,\frac{1}{b_3 a_{3,2}}, a_{3,1} = 1/2\,\frac{RootOf\left(1 - 4\,b_3 a_{3,2} - \_Z + 3\,\_Z^2 a_{3,2}\right) - 2\,b_3 a_{3,2}}{b_3}, \right.$$

$$a_{3,2} = a_{3,2}, b_1 = 1 - b_3 + 3\,b_3 a_{3,2} RootOf\left(1 - 4\,b_3 a_{3,2} - \_Z + 3\,\_Z^2 a_{3,2}\right) - 3\,b_3 a_{3,2},$$

$$b_2 = -3\,b_3 a_{3,2} RootOf\left(1 - 4\,b_3 a_{3,2} - \_Z + 3\,\_Z^2 a_{3,2}\right) + 3\,b_3 a_{3,2}, b_3 = b_3,$$

$$\left. c_2 = 1/6\,\frac{1}{b_3 a_{3,2}}, c_3 = 1/2\,\frac{RootOf\left(1 - 4\,b_3 a_{3,2} - \_Z + 3\,\_Z^2 a_{3,2}\right)}{b_3} \right\}$$

The solutions contains two free parameters, $a_{3,2}$ and b_3. A historically well-known method can be obtained by setting $a_{3,2} = \frac{2}{3}$ and $b_3 = \frac{3}{4}$. To evaluate the RootOf expressions, which represent in each case the same root of a second degree polynomial, one can use the MAPLE commands

```
assign(%);
a[3,2]:=2/3;
b[3]:=3/4;
b1:=allvalues(b[1]);
b2:=allvalues(b[2]);
a31:= allvalues(a[3,1]);
a22:=a[2,2];
```

We obtain two solutions: the first is the well-known method of Heun of order three, given on the left in Table 10.1, and on the right we show a second possible solution, which is further investigated in Problem 10.7.

    Runge-Kutta methods are very easy to implement: here is Heun's method in MATLAB:

ALGORITHM 10.7. *Heun's Order 3 ODE solver*

```
function [t,y]=Heun(f,tspan,y0,n);
% HEUN solves system of ODEs using Heun's method
%    [t,y]=Heun(f,tspan,y0,n) solves dy/dt=f(t,y) with initial
%    value y0 on the time interval tspan doing n steps of Heun's
%    method.
```

$$
\begin{array}{c|ccc}
0 & & & \\
\frac{1}{3} & \frac{1}{3} & & \\
\frac{2}{3} & 0 & \frac{2}{3} & \\
\hline
 & \frac{1}{4} & 0 & \frac{3}{4}
\end{array}
\qquad
\begin{array}{c|ccc}
0 & & & \\
\frac{1}{3} & \frac{1}{3} & & \\
-\frac{1}{3} & -1 & \frac{2}{3} & \\
\hline
 & -2 & \frac{9}{4} & \frac{3}{4}
\end{array}
$$

TABLE 10.1.
*Heun's Method (order 3) on the left, and an unknown
third order three stage explicit Runge Kutta method on
the right.*

```
dt=(tspan(2)-tspan(1))/n;
t=tspan(1):dt:tspan(2);
y(:,1)=y0(:);                      % colon to make column vector
for i=1:n,
  k1=f(t(i),y(:,i));
  k2=f(t(i)+dt/3,y(:,i)+dt/3*k1);
  k3=f(t(i)+2*dt/3,y(:,i)+2*dt/3*k2);
  y(:,i+1)=y(:,i)+dt*(1/4*k1+3/4*k3);
end;
t=t';y=y';                         % to return results in columns
```

Without the simplifying assumptions, MAPLE cannot solve the Runge-Kutta equations for $s = 4$. The computation is interrupted with the error message

```
[Length of output exceeds limit of 1000000].
```

However, using the simplifying assumptions, we obtain with

```
> RK4:=RK(4,4);
```

a set with four solutions. Choosing the third solution

```
> RK4[3];
```

we get

$$
\Big\{ a_{2,1} = \frac{1}{2}, a_{3,1} = -a_{3,2} + \frac{1}{2}, a_{3,2} = a_{3,2}, a_{4,1} = 0, a_{4,2} = \frac{1}{2}\frac{2\,a_{3,2}-1}{a_{3,2}}, a_{4,3} = \frac{1}{2a_{3,2}},
$$
$$
b_1 = \frac{1}{6}, b_2 = \frac{1}{6}\frac{-1+4\,a_{3,2}}{a_{3,2}}, b_3 = \frac{1}{6a_{3,2}}, b_4 = \frac{1}{6}, c_2 = \frac{1}{2}, c_3 = \frac{1}{2}, c_4 = 1 \Big\}
$$

With $\lambda = 2a_{3,2}$, this solution represents the tableau

$$
\begin{array}{c|cccc}
0 \\
\dfrac{1}{2} & \dfrac{1}{2} \\
\dfrac{1}{2} & \dfrac{1-\lambda}{2} & \dfrac{\lambda}{2} \\
1 & 0 & \dfrac{\lambda-1}{\lambda} & \dfrac{1}{\lambda} \\
\hline
& \dfrac{1}{6} & \dfrac{2\lambda-1}{3\lambda} & \dfrac{1}{3\lambda} & \dfrac{1}{6}
\end{array}
$$

$$
\begin{array}{c|cccc}
0 \\
\dfrac{1}{2} & \dfrac{1}{2} \\
\dfrac{1}{2} & 0 & \dfrac{1}{2} \\
1 & 0 & 0 & 1 \\
\hline
& \dfrac{1}{6} & \dfrac{2}{6} & \dfrac{2}{6} & \dfrac{1}{6}
\end{array}
$$

TABLE 10.2.

The classical Runge-Kutta method is a four stage fourth order method.

Note that for $\lambda = 1$, we obtain the classical Runge-Kutta algorithm given in Table 10.2. A MATLAB implementation of this classical Runge-Kutta method is

---

ALGORITHM 10.8. *Classical Runge-Kutta Method*

```
function [t,y]=RK4(f,tspan,y0,n);
% RK4 solves system of ODEs using classical 4th order Runge Kutta
%   [t,y]=RK4(f,tspan,y0,dt) solves dy/dt=f(t,y) with initial
%   value y0 on the time interval tspan doing n steps of the
%   classical 4th order Runge Kutta method.

dt=(tspan(2)-tspan(1))/n;
t=tspan(1):dt:tspan(2);
y(:,1)=y0(:);                        % colon to make column vector
for i=1:n,
  k1=f(t(i),y(:,i));
  k2=f(t(i)+0.5*dt,y(:,i)+0.5*dt*k1);
  k3=f(t(i)+0.5*dt,y(:,i)+0.5*dt*k2);
  k4=f(t(i)+dt,y(:,i)+dt*k3);
  y(:,i+1)=y(:,i)+dt*((k1+k4)/6+(k2+k3)/3);
end;
t=t';y=y';                           % to return results in columns
```

We show in Figure 10.8 how the four explicit Runge-Kutta methods we considered in this section compare when applied to the dog-and-jogger problem. These results clearly show the superiority of high order methods.

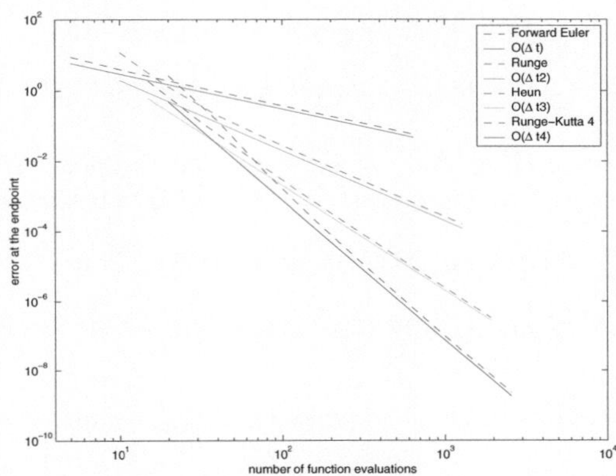

FIGURE 10.8.
*Error at the end point of the approximate trajectory of
the dog computed with various Runge-Kutta methods as
a function of the number of right hand side function
evaluations f needed to compute the approximate
trajectory. The higher order methods are clearly superior.*

Butcher investigated the best possible local truncation error one can get with a given number of function evaluations per step, see Table 10.3. This led to the analysis of Runge-Kutta methods using B-series; for a complete treatment, see [62].

f-evals	2	3	4, 5	6	7, 8	9, 10
Best τ_k	$O(\Delta t^2)$	$O(\Delta t^3)$	$O(\Delta t^4)$	$O(\Delta t^5)$	$O(\Delta t^6)$	$O(\Delta t^7)$

TABLE 10.3.
*Butcher's results about the best obtainable local
truncation error with a given number of function
evaluations in a Runge Kutta method*

10.3.4 Convergence

We now turn our attention to the fact that the *global truncation error* is one order lower than the local truncation error, which we have observed in all our numerical experiments. This is no coincidence, as the following theorem shows.

THEOREM 10.2. (CONVERGENCE OF ONE STEP METHODS) *Let $\boldsymbol{y}(t)$ be the solution of (10.17) on the interval $[t_0, t_n]$, and let $\boldsymbol{y}_j$, $j = 0, 1, \ldots, n$ be the numerical approximation of $\boldsymbol{y}(t)$ by the general one-step method (10.20) at time $t_0, t_1, \ldots, t_n$, $\Delta t_j := t_j - \Delta t_{j-1}$, $j = 1, 2, \ldots n$. If*

1. *the local truncation error of the one-step method (10.20) satisfies for all $t \in [t_0, t_{n-1}]$ and $\Delta t \le \Delta t_{\max} := \max_{1 \le j \le n} \Delta t_j$*

$$\|\boldsymbol{y}(t + \Delta t) - \boldsymbol{y}(t) - \Delta t \phi_f(t, \boldsymbol{y}(t), \Delta t)\| \le C \Delta t^{p+1}, \tag{10.23}$$

2. *the function $\phi_f(t, \boldsymbol{y}, \Delta t)$ of the one-step method (10.20) is locally Lipschitz in $\boldsymbol{y}$, i.e., there exists a constant $L > 0$ such that*

$$\|\phi_f(t, \boldsymbol{y}, \Delta t) - \phi_f(t, \boldsymbol{z}, \Delta t)\| \le L \|\boldsymbol{y} - \boldsymbol{z}\| \tag{10.24}$$

whenever $\Delta t \le \Delta t_{\max}$ and $(t, \boldsymbol{y})$, $(t, \boldsymbol{z})$ are in the neighborhood of the solution,

then the global truncation error satisfies for $\Delta t_{\max}$ small enough

$$\|\boldsymbol{y}(t_n) - \boldsymbol{y}_n\| \le \Delta t_{\max}^p \frac{C}{L} (e^{L(t_n - t_0)} - 1). \tag{10.25}$$

PROOF. Adding and subtracting $\boldsymbol{y}(t_{n-1})$ and $\Delta t_n \phi_f(t, \boldsymbol{y}(t_{n-1}), \Delta t_n)$, we obtain

$$
\begin{aligned}
\|\boldsymbol{y}(t_n) - \boldsymbol{y}_n\| \le\ & \|\boldsymbol{y}(t_n) - \boldsymbol{y}(t_{n-1}) - \Delta t_n \phi_f(t, \boldsymbol{y}(t_{n-1}), \Delta t_n)\| \\
& + \|\boldsymbol{y}(t_{n-1}) - \boldsymbol{y}_{n-1}\| \\
& + \|\Delta t_n \phi_f(t, \boldsymbol{y}(t_{n-1}), \Delta t_n) - \Delta t_n \phi_f(t, \boldsymbol{y}_{n-1}, \Delta t_n)\| \\
\le\ & C \Delta t_n^{p+1} + (1 + L \Delta t_n) \|\boldsymbol{y}(t_{n-1}) - \boldsymbol{y}_{n-1}\| \\
\le\ & C \Delta t_n^{p+1} + e^{\Delta t_n L} \|\boldsymbol{y}(t_{n-1}) - \boldsymbol{y}_{n-1}\|.
\end{aligned}
$$

Now we can apply this inequality again at step $n - 1$ to obtain

$$\|\boldsymbol{y}(t_n) - \boldsymbol{y}_n\| \le C \Delta t_n^{p+1} + e^{\Delta t_n L} C \Delta t_{n-1}^{p+1} + e^{(t_n - t_{n-2})L} \|\boldsymbol{y}(t_{n-2}) - \boldsymbol{y}_{n-2}\|.$$

We thus obtain by induction

$$\|\boldsymbol{y}(t_n) - \boldsymbol{y}_n\| \le C \sum_{j=0}^{n-1} \Delta t_{j+1}^{p+1} e^{(t_n - t_j)L} \le C \Delta t_{\max}^p \sum_{j=0}^{n-1} \Delta t_{j+1} e^{(t_n - t_j)L}.$$

Now note that the right hand side is in fact the Riemann sum approximating the integral of $e^{(t_n - t)L}$ from below on the interval $[t_0, t_n]$. Thus, the sum can be bounded above by

$$\|\boldsymbol{y}(t_n) - \boldsymbol{y}_n\| \le C \Delta t_{\max}^p \int_{t_0}^{t_n} e^{(t_n - t)L} dt = \Delta t_{\max}^p \frac{C}{L} (e^{(t_n - t_0)L} - 1),$$

which concludes the proof. $\qquad\square$

To apply this convergence result to Runge-Kutta methods, one has to show that the one-step method

$$\phi_f(t, \boldsymbol{y}, \Delta t) := \sum_{i=1}^{s} b_i k_i(\boldsymbol{y}) \tag{10.26}$$

defined by the Runge-Kutta coefficients is Lipschitz. This is true whenever the right hand side function $\boldsymbol{f}$ of (10.17) is itself Lipschitz: suppose $\boldsymbol{f}$ is Lipschitz with constant $\widetilde{L}$ in the neighborhood of the solution,

$$\|\boldsymbol{f}(t, \boldsymbol{y}) - \boldsymbol{f}(t, \boldsymbol{z})\| \le \widetilde{L} \|\boldsymbol{y} - \boldsymbol{z}\|.$$

Then we find that the first stage $\boldsymbol{k}_1(\boldsymbol{y})$ is also Lipschitz with the same constant, since $\boldsymbol{k}_1(\boldsymbol{y}) = \boldsymbol{f}(t, \boldsymbol{y})$. For the second stage, we obtain

$$\begin{aligned}
\|\boldsymbol{k}_2(\boldsymbol{y}) - \boldsymbol{k}_2(t, \boldsymbol{z})\| &\le \widetilde{L} \|\boldsymbol{y} + \Delta t a_{21} \boldsymbol{k}_1(\boldsymbol{y}) - (\boldsymbol{z} + \Delta t a_{21} \boldsymbol{k}_1(\boldsymbol{z}))\| \\
&\le \widetilde{L}(1 + \Delta t |a_{21}| \widetilde{L}) \|\boldsymbol{y} - \boldsymbol{z})\|.
\end{aligned}$$

Proceeding in the same fashion, we find for the third stage

$$\|\boldsymbol{k}_3(\boldsymbol{y}) - \boldsymbol{k}_3(t, \boldsymbol{z})\| \le \widetilde{L}(1 + \Delta t \widetilde{L}(|a_{31}| + |a_{32}|) + \Delta t^2 \widetilde{L}^2 |a_{32} a_{21}|) \|\boldsymbol{y} - \boldsymbol{z})\|.$$

Inserting these estimates into (10.26), we deduce that ϕ_f is indeed Lipschitz for Runge-Kutta methods,

$$\|\phi_f(t, \boldsymbol{y}, \Delta t) - \phi_f(t, \boldsymbol{y}, \Delta t)\| \le L \|\boldsymbol{y} - \boldsymbol{z}\|,$$

where L is given by

$$L = \tilde{L} \left(\sum_i |b_i| + \Delta t \widetilde{L} \sum_{ij} |b_i a_{ij}| + (\Delta t \widetilde{L})^2 \sum_{ijk} |b_i a_{ij} a_{jk}| + \cdots \right).$$

10.3.5 Adaptive Integration

So far we have not addressed the question of how to choose the time step Δt. To illustrate the paramount importance of this, we consider a problem from celestial mechanics, namely to compute the Arenstorf orbit of the *Restricted Three Body Problem*. In this problem, a small body (with negligible mass) is moving in the gravitational fields of two large bodies. As an example, imagine a satellite between the earth and the moon. We make the following assumptions:

1. The Earth, moon and satellite move in the same plane.

2. The Earth and the moon revolve around the common center of gravity, which is essentially that of the earth, since the mass M of the earth is 80.45 times larger than the mass m of the the moon.

3. The satellite has no influence on the earth and the moon, since its mass is too small.

To solve this problem, we choose a rotating coordinate system in which the two large bodies are fixed at the origin (earth) and at the point $(1,0)$ (moon). The length unit is the distance of the moon from the earth, and the time unit is chosen such that 2π corresponds to one month.

With $r = M/m = 80.45$ as the mass ratio between the earth and the moon, we define

$$a = \frac{1}{1+r} = 0.012277471 \quad \text{and} \quad b = 1 - a = 0.987722529.$$

The equations describing the position $(x(t), y(t))$ of the satellite under the gravitation forces are derived in [135]. They are

$$\begin{aligned}
x'' &= x + 2y' - b\frac{x+a}{D_1(x,y)} - a\frac{x-b}{D_2(x,y)}, \\
y'' &= y - 2x' - b\frac{y}{D_1(x,y)} - a\frac{y}{D_2(x,y)},
\end{aligned} \tag{10.27}$$

where $D_1(x,y) = \left((x+a)^2 + y^2\right)^{\frac{3}{2}}$ and $D_2(x,y) = \left((x-b)^2 + y^2\right)^{\frac{3}{2}}$. With the initial conditions

$$x(0) = 0.994,\ x'(0) = 0,\ y(0) = 0 \text{ and } y'(0) = -2.00158510637908,$$

the solution is a periodic orbit with period $t = 17.06521656015796$ (corresponding to 2.71 months), known as the *Arenstorf* orbit.

To apply the integrators we have developed so far, we need to convert the governing equations (10.27) into a *system of first order equations*, which is easily done by introducing two new variables $\tilde{x} := x'$ and $\tilde{y} := y'$, see Subsection 10.2.6. This leads to the right hand side function in MATLAB

ALGORITHM 10.9. *ODE for Arenstorf orbit*

```
function yp=Arenstorf(t,y);
% ARENSTORF ODE right hand side for the Arenstorf orbit problem
%    yp=Arenstorf(t,y); describes a system of ODEs which model the
%    flight of a light object between the earth and the moon.

a=0.012277471; b=1-a;
D1=((y(1)+a)^2+y(2)^2)^(3/2);
D2=((y(1)-b)^2+y(2)^2)^(3/2);
yp(1,1)=y(3);
yp(2,1)=y(4);
yp(3,1)=y(1)+2*y(4)-b*(y(1)+a)/D1-a*(y(1)-b)/D2;
yp(4,1)=y(2)-2*y(3)-b*y(2)/D1-a*y(2)/D2;
yp=yp(:);                              % make a column vector
```

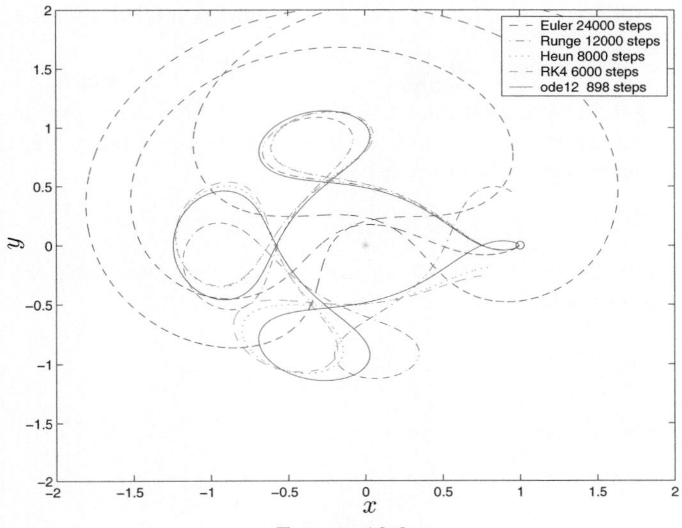

FIGURE 10.9.
Approximations of an Arenstorf orbit computed with fixed step Runge-Kutta methods and a new low order adaptive method.

Using the methods we have developed so far for this problem, we obtain the results in Figure 10.9, where care was taken to choose the number of steps such that the number of function evaluations for each method is the same. While higher order methods perform better than low order methods, none of them is able to find the closed orbit. However, a new method called *ode12* finds the closed orbit with excellent accuracy, using only two function evaluations per step and thus more than thirteen times cheaper than the other methods in terms of function evaluations. This is possible because the new method locally adapts its time step using an error estimator, so it uses a fine time step only where necessary to achieve the desired accuracy.

In order to determine a suitable time step Δt for the current integration step, we need to estimate the truncation error that will be made for this time step choice. Since the analytic formulas for the local truncation error τ are not very practical, one estimates the local truncation error using two approximations of different order (this can even be done taking into account components of the vector valued solution individually, see [62]),

$$
\begin{aligned}
\tilde{\boldsymbol{y}}_1 \quad &\text{such that} \quad \|\boldsymbol{\tau}\| = \|\boldsymbol{y}(\Delta t) - \tilde{\boldsymbol{y}}_1\| = C\Delta t^{p+1} + O(\Delta t^{p+2}) \\
\boldsymbol{y}_1 \quad &\text{such that} \quad \|\tilde{\boldsymbol{\tau}}\| = \|\boldsymbol{y}(\Delta t) - \boldsymbol{y}_1\| = O(\Delta t^{p+2}).
\end{aligned}
\tag{10.28}
$$

Taking the difference of the two computed values $\boldsymbol{y}_1$ and $\tilde{\boldsymbol{y}}_1$, we find

$$\|\boldsymbol{y}_1 - \tilde{\boldsymbol{y}}_1\| = C\Delta t^{p+1} + O(\Delta t^{p+2}), \tag{10.29}$$

which is a good approximation of the local truncation error $\|\boldsymbol{\tau}\|$ in $\tilde{\boldsymbol{y}}_1$. To obtain a local truncation error of the size of a given tolerance tol, the best time step choice, Δt_{opt} should satisfy

$$\mathrm{tol} = C\Delta t_{\mathrm{opt}}^{p+1} + O(\Delta t_{\mathrm{opt}}^{p+2}). \tag{10.30}$$

Dividing (10.29) by (10.30) and neglecting higher order terms, we find an equation for the optimal time step,

$$\frac{\|\boldsymbol{y}_1 - \tilde{\boldsymbol{y}}_1\|}{\mathrm{tol}} = \left(\frac{\Delta t}{\Delta t_{\mathrm{opt}}}\right)^{p+1},$$

which upon solving for Δt_{opt} gives

$$\Delta t_{\mathrm{opt}} = \Delta t \left(\frac{\mathrm{tol}}{\|\boldsymbol{y}_1 - \tilde{\boldsymbol{y}}_1\|}\right)^{\frac{1}{p+1}}. \tag{10.31}$$

Thus, using the two estimates $\boldsymbol{y}_1$ and $\tilde{\boldsymbol{y}}_1$ of different orders, it is possible to estimate not only the truncation error, but also the optimal time step one should have used for the prescribed tolerance. An adaptive integration method computes the two estimates using the current time step Δt, and then checks whether the local truncation error is small enough using (10.29): if so, the integration step is accepted and the method advances, otherwise the step is rejected. In both cases, a new Δt_{opt} is computed using (10.31) to continue the process, which means either recomputing the rejected step, or computing the next one. In the latter case, one assumes that the time step estimate from the previous integration step is not too far from the optimal choice for the new step.

To compute the two approximations $\boldsymbol{y}_1$ and $\tilde{\boldsymbol{y}}_1$ simultaneously, *embedded Runge-Kutta methods* are ideally suited. They contain a second set of parameters $\tilde{b}_j$, which give an approximation of a different order using the same stage values, so that no additional function evaluation are required.

0					
c_2	$a_{2,1}$				
c_3	$a_{3,1}$	$a_{3,2}$			
$\vdots$	$\vdots$	$\ddots$	$\ddots$		
c_s	$a_{s,1}$	$a_{s,2}$	$\cdots$	$a_{s,s-1}$	
1	b_1	b_2	$\cdots$	b_{s-1}	b_s
	$\tilde{b}_1$	$\tilde{b}_2$	$\cdots$	$\tilde{b}_{s-1}$	$\tilde{b}_s$

$y_1 = y_0 + h\sum_{i=1}^{s} b_i k_i$ with order p

$\tilde{y}_1 = y_0 + h\sum_{i=1}^{s} \tilde{b}_i k_i$ with order q

Usually we have $q = p \pm 1$.

As a first example, we consider Runge's method derived earlier,

$$
\begin{array}{c|cc}
0 & & \\
\frac{1}{2} & \frac{1}{2} & \\
\hline
\boldsymbol{y}_1 & 0 & 1 \\
\hline
\tilde{\boldsymbol{y}}_1 & \tilde{b}_1 & \tilde{b}_2
\end{array}
$$

Since we only need $\tilde{b}_1 + \tilde{b}_2 = 1$ for a first-order method, we can choose any pair of values satisfying this equation except $\tilde{b}_1 = 0$ and $\tilde{b}_2 = 1$, since then the methods would be identical and no error estimation would be possible. A simple choice is $\tilde{b}_1 = 1$ and $\tilde{b}_2 = 0$, which gives Forward Euler for the lower order method.

In a real code, one also adds some safeguards. First, the optimal time step derived in (10.31) is multiplied by a factor less than one, $\alpha_s \in [0.8, 0.95]$. To prevent the time step from decreasing or increasing too much, one also puts bounds on the maximum possible increase with a factor $\alpha_{\max} \in [1.5, 5]$ and the maximum possible decrease with a factor $\alpha_{\min} \in [0, 0.5]$. This leads to our first adaptive method in MATLAB:

ALGORITHM 10.10. *Simple Adaptive ODE Solver*

```
function [t,y,H]=ode12(f,tspan,y0,tol);
% ode12 solves system of ODEs using adaptive integration
%    [t,y]=ode12(f,tspan,y0,tol) solves dy/dt=f(t,y) with initial
%    value y0 on the time interval tspan using adaptive integration
%    with Euler and Runge's method and error tolerance tol.
%    For didactical reasons we store the used stepsizes in H.

dt=(tspan(2)-tspan(1))/10; H=dt;
y(:,1)=y0;
t(1)=tspan(1);
as=0.8; amin=0.2; amax=5;                 % heuristic parameters
i=1;
while t(i)<tspan(2),
  k1=f(t(i),y(:,i));
  k2=f(t(i)+dt/2,y(:,i)+dt/2*k1);
  yn=y(:,i)+dt*k2;                         % Runge step
  yt=y(:,i)+dt*k1;                         % Euler step
  toli=max([norm(yn,inf) norm(y(:,i),inf)]);
  toli=tol*(1+toli);                       % abs. and relative tol
  errest=norm(yn-yt);
  if errest<toli,                          % accept the step
    i=i+1;
    y(:,i)=yn;
    t(i)=t(i-1)+dt;
  end;
```

```
    dtopt=as*(dt*sqrt(toli/errest));       % estimate the new time step
    if dtopt<amin*dt,
      dt=amin*dt;
    elseif dtopt>amax*dt,
      dt=amax*dt;
    else
      dt=dtopt;
    end;
    if t(i)+dt>tspan(2),
      dt=tspan(2)-t(i);
    end;
    H=[H;dt];
  end;
t=t';
y=y';
```

The program returns $\boldsymbol{y}_1$, the higher order approximation, as the result, even though the error was estimated for the lower order approximation $\tilde{\boldsymbol{y}}_1$, since this higher order approximation is available for free; this is usually called *local extrapolation*. The program starts with a rather arbitrary first time step Δt equal to one tenth of the time interval over which the integration is performed. If this time step is too big, it is quickly reduced by the error estimator. For better estimates on the initial time step, see [62]. The last time step is adapted so that the integration does not overrun the end of the time interval.

With the following program we integrate the system requiring two different tolerances $\tau_1 = 0.005$ and $\tau_2 = \tau_1/10 = 0.0005$.

ALGORITHM 10.11. *Computing Arenstorf Orbit*

```
tol1=0.005;tol2=tol1/10;
tspan=[0,17.06521656015796];
y0=[0.994;0;0;-2.00158510637908];
figure(1),clf(1)
[t1,y1,h1]=ode12(@Arenstorf,tspan,y0,tol1);
plot(y1(:,1),y1(:,2));                       % plot orbit
axis([-1.5 1.5 -1.5 1.5]); hold
plot(0,0,'o')                                % position of earth
plot(1,0,'o')                                % position of moon
figure(2),clf(2)
axis([0,length(h1)+10,0,0.3]); hold
plot(h1)                                      % plot stepsizes
NumberOfSteps=length(t1)
RepeatedSteps=length(h1)-length(t1)
figure(3),clf(3)
[t2,y2,h2]=ode12(@Arenstorf,tspan,y0,tol2);
plot(y2(:,1),y2(:,2));
```

```
axis([-1.5 1.5 -1.5 1.5]); hold
plot(0,0,'o'), plot(1,0,'o')
figure(4),clf(4)
axis([0,length(h2)+10,0,0.1]); hold
plot(h2)
NumberOfSteps=length(t2)
RepeatedSteps=length(h2)-length(t2)
```

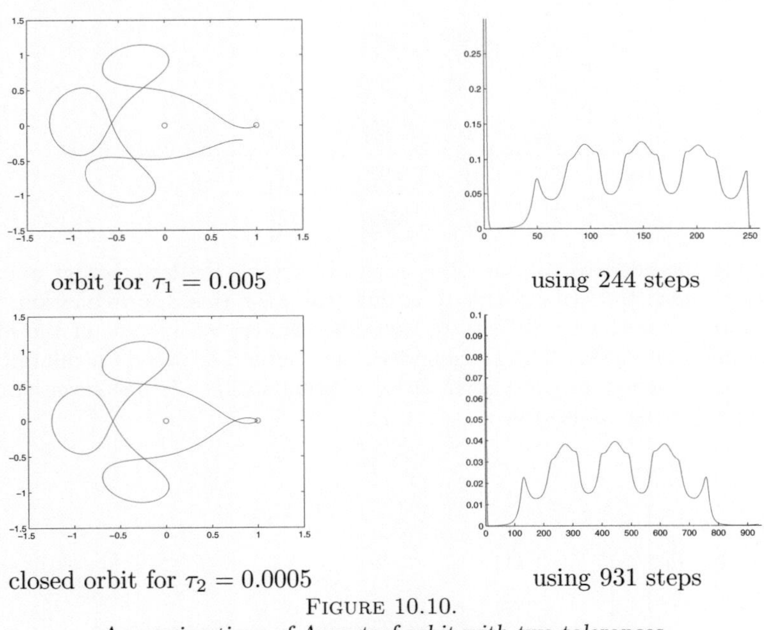

orbit for $\tau_1 = 0.005$ using 244 steps

closed orbit for $\tau_2 = 0.0005$ using 931 steps

FIGURE 10.10.
Approximations of Arenstorf orbit with two tolerances

The results of the two computations are shown in Figure 10.10. We notice that for $\tau_1 = 0.005$, we do not get the periodic orbit; we need to integrate with more precision, as we can see from the second plot of Figure 10.10. For τ_1, the integrator performed 244 integration steps, and 5 steps had to be repeated. For the lower tolerance $\tau_2 = 0.0005$, the integrator needed 931 steps and 6 times steps had to be repeated.

In fact, these repetitions occurred at the beginning of the integration. The statement `dt=(tspan(2)-tspan(1))/10;` for the first step in Algorithm 10.10 computes $\Delta t = 1.7065$, which is much too large for the required tolerance. For the tolerance τ_1, the first step has to be reduced five times to reach 0.00054609. After that, the error estimates works fine without repetitions. Note in Figure 10.10 that the step sizes become larger when the small body is far away.

Two commonly used embedded Runge-Kutta methods of higher order are the *Runge-Kutta-Fehlberg method,* and the *Dormand-Prince method.* Both are based on an embedding of a fourth order method inside a fifth order method. The Runge-Kutta-Fehlberg method has six stages, and the corresponding Butcher tableau is

$$
\begin{array}{c|cccccc}
0 \\
\frac{1}{4} & \frac{1}{4} \\
\frac{3}{8} & \frac{3}{32} & \frac{9}{32} \\
\frac{12}{13} & \frac{1932}{2197} & -\frac{7200}{2197} & \frac{7296}{2197} \\
1 & \frac{439}{216} & -8 & \frac{3680}{513} & -\frac{845}{4104} \\
\frac{1}{2} & -\frac{8}{27} & 2 & -\frac{3544}{2565} & \frac{1859}{4104} & -\frac{11}{40} \\
\hline
y_1 & \frac{16}{135} & 0 & \frac{6656}{12825} & \frac{28561}{56430} & -\frac{9}{50} & \frac{2}{55} \\
\tilde{y}_1 & \frac{25}{216} & 0 & \frac{1408}{2565} & \frac{2197}{4104} & -\frac{1}{5} & 0
\end{array}
$$

Fehlberg intended to continue the integration with the lower order approximation, since the error estimate is only valid for this approximation, and therefore chose the coefficients to minimize the error constant of the lower order approximation. The Dormand-Prince method is based on minimizing the error constant of the higher order approximation. It has seven stages, and the Butcher tableau is

$$
\begin{array}{c|ccccccc}
0 \\
\frac{1}{5} & \frac{1}{5} \\
\frac{3}{10} & \frac{3}{40} & \frac{9}{40} \\
\frac{4}{5} & \frac{44}{45} & -\frac{56}{15} & \frac{32}{9} \\
\frac{8}{9} & \frac{19372}{6561} & -\frac{25360}{2187} & \frac{64448}{6561} & -\frac{212}{729} \\
1 & \frac{9017}{3168} & -\frac{355}{33} & \frac{46732}{5247} & \frac{49}{176} & -\frac{5103}{18656} \\
1 & \frac{35}{384} & 0 & \frac{500}{1113} & \frac{125}{192} & -\frac{2187}{6784} & \frac{11}{84} \\
\hline
y_1 & \frac{35}{384} & 0 & \frac{500}{1113} & \frac{125}{192} & -\frac{2187}{6784} & \frac{11}{84} \\
\tilde{y}_1 & \frac{5179}{57600} & 0 & \frac{7571}{16695} & \frac{393}{640} & -\frac{92097}{339200} & \frac{187}{2100} & \frac{1}{40}
\end{array}
$$

As one can see from the table, Dormand and Prince used the lower order approximation also as the seventh stage of the Runge-Kutta method, a technique called FSAL (First Same As Last), which has the advantage that for accepted steps, there is no extra work, since the function has to be evaluated anyway for the following step. Implementations of these two methods are very similar to the program `ode12`, and are left as an exercise, see Problems 10.8 and 10.9.

10.3.6 Implicit Runge-Kutta Methods

Implicit Runge-Kutta (IRK) methods are obtained by allowing a complete Butcher tableau, including non-zero entries on and above the diagonal, i.e.

$$
\begin{array}{c|ccccc}
c_1 & a_{11} & a_{12} & \cdots & a_{1,s-1} & a_{1,s} \\
c_2 & a_{21} & a_{22} & \cdots & a_{2,s-1} & a_{2,s} \\
c_3 & a_{31} & a_{32} & & \vdots & \vdots \\
\vdots & \vdots & & \ddots & \vdots & \vdots \\
c_s & a_{s1} & a_{s2} & \cdots & a_{s,s-1} & a_{s,s} \\
\hline
& b_1 & b_2 & \cdots & b_{s-1} & b_s
\end{array}
$$

If only the diagonal entries are non-zero, one calls the method a diagonally implicit Runge-Kutta method (DIRK), and if all the the elements on the diagonal are the same, it is a singly diagonally implicit Runge-Kutta method (SDIRK). The numerical method associated with the complete Butcher tableau is

$$
\begin{aligned}
\boldsymbol{k}_i &= \boldsymbol{f}(t_0 + \Delta t c_i, \boldsymbol{y}_0 + \Delta t \textstyle\sum_{j=1}^{s} a_{i,j}\boldsymbol{k}_j), \quad i = 1, 2, \ldots, s \\
\boldsymbol{y}_1 &= \boldsymbol{y}_0 + \Delta t \textstyle\sum_{j=1}^{s} b_j \boldsymbol{k}_j.
\end{aligned}
\tag{10.32}
$$

This formulation immediately reveals a difficulty that was absent in the explicit methods: one cannot compute the stage value $\boldsymbol{k}_1$ without knowing all the other stage values, and the same is true for the other stage values. Hence, at each time step, an implicit system of nonlinear equations needs to be solved. The following theorem shows that for Δt small enough, there is always a solution of this nonlinear system, and the solution can be computed using a simple fixed point iteration.

THEOREM 10.3. *Let* $\boldsymbol{f} : \mathbb{R} \times \mathbb{R}^n \longrightarrow \mathbb{R}^n$ *be continuous and satisfy for all* t *the Lipschitz condition*

$$
\|\boldsymbol{f}(t, \boldsymbol{y}_1) - \boldsymbol{f}(t, \boldsymbol{y}_2)\| \le L\|\boldsymbol{y}_1 - \boldsymbol{y}_2\|.
\tag{10.33}
$$

If the time step Δt *satisfies*

$$
\Delta t < \frac{1}{L \max_i \sum_{j=1}^{s} |a_{ij}|},
\tag{10.34}
$$

then there exists a unique solution of (10.32), which can be obtained by a fixed-point iteration.

PROOF. We prove existence using the fixed-point iteration

$$
\boldsymbol{k}_i^{m+1} = \boldsymbol{f}(t_0 + \Delta t c_i, \boldsymbol{y}_0 + \Delta t \sum_{j=1}^{s} a_{i,j}\boldsymbol{k}_j^{m}), \quad i = 1, 2, \ldots, s.
$$

Collecting the stage vectors $\boldsymbol{k}_i$ into the big vector $\boldsymbol{K} := (\boldsymbol{k}_1^T, \ldots, \boldsymbol{k}_s^T)^T$, the fixed-point iteration becomes

$$\boldsymbol{K}^{m+1} = \boldsymbol{F}(\boldsymbol{K}^m), \qquad \boldsymbol{F}_i(\boldsymbol{K}^m) = \boldsymbol{f}\left(t_0 + \Delta t c_i, \boldsymbol{y}_0 + \Delta t \sum_{j=1}^{s} a_{ij} \boldsymbol{k}_j^m\right).$$

Now using the norm $\|\boldsymbol{K}\| := \max_i(\|\boldsymbol{k}_i\|)$ and the Lipschitz condition (10.33), repeated application of the triangle inequality leads to

$$\|\boldsymbol{F}(\boldsymbol{K}_1) - \boldsymbol{F}(\boldsymbol{K}_2)\| \leq \Delta t L \max_i \sum_{j=1}^{s} |a_{ij}| \|\boldsymbol{K}_1 - \boldsymbol{K}_2\|,$$

and thus, with (10.34), the fixed-point iteration is a contraction, and hence converges to a unique fixed point. $\qquad\qquad\qquad\qquad\qquad\qquad\qquad\qquad\qquad\quad\square$

Implicit methods are of great interest for stiff problems, as we will show in Section 10.5. The simplest implicit Runge Kutta method is the Backward Euler method, which one obtains by simply approximating the integral in the integrated differential equation (10.17),

$$\boldsymbol{y}(t_1) - \boldsymbol{y}(t_0) = \int_{t_0}^{t_1} \boldsymbol{f}(\tau, y(\tau)) d\tau,$$

by a Riemann sum with one term,

$$\int_{t_0}^{t_1} \boldsymbol{f}(\tau, y(\tau)) d\tau \approx \Delta t f(t_1, y(t_1)),$$

as shown in Figure 10.11. This leads to the Backward Euler method

$$\boldsymbol{y}_1 = \boldsymbol{y}_0 + \Delta t \boldsymbol{f}(t_1, \boldsymbol{y}_1). \tag{10.35}$$

If we want to write Backward Euler as an implicit Runge-Kutta method, the formula for advancing one step must be of the form

$$\boldsymbol{y}_1 = \boldsymbol{y}_0 + \Delta t \boldsymbol{k}_1 \tag{10.36}$$

and $\boldsymbol{k}_1 = \boldsymbol{f}(t_1, \boldsymbol{y}_1)$, which together with (10.36) gives

$$\boldsymbol{k}_1 = \boldsymbol{f}(t_1, \boldsymbol{y}_0 + \Delta t \boldsymbol{k}_1).$$

The Backward Euler method in the general Runge-Kutta notation is thus given by

$$
\begin{array}{rl}
\boldsymbol{k}_1 &= \boldsymbol{f}(t_1, \boldsymbol{y}_0 + \Delta t \boldsymbol{k}_1) \\
\boldsymbol{y}_1 &= \boldsymbol{y}_0 + \Delta t \boldsymbol{k}_1
\end{array}
\qquad
\begin{array}{c|c}
1 & 1 \\ \hline
 & 1
\end{array}
\tag{10.37}
$$

Using Definition 10.1, the local truncation error for the Backward Euler method is

$$\boldsymbol{\tau} = \boldsymbol{y}(t_1) - \boldsymbol{y}_0 - \Delta t \boldsymbol{f}(t_1, \boldsymbol{y}_1). \tag{10.38}$$

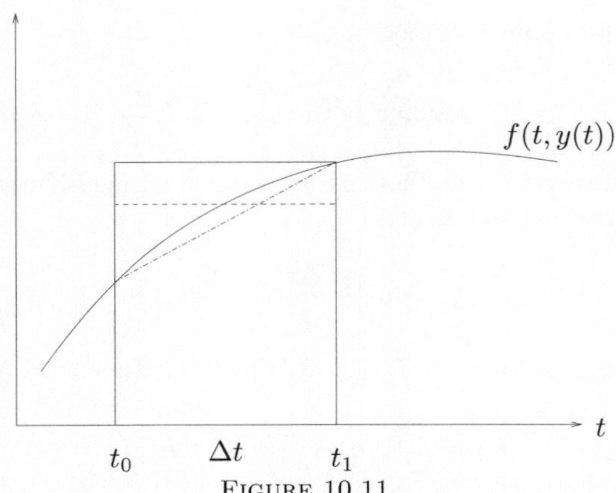

FIGURE 10.11.
*Approximation of the integral by an area which requires
knowledge of the function at the right end point.*

The first term on the right can be expanded in a Taylor series as in the case
of explicit methods,

$$\boldsymbol{y}(t_1) = \boldsymbol{y}(t_0) + \Delta t \boldsymbol{y}'(t_0) + O(\Delta t^2),$$

but the second term contains now $\boldsymbol{y}_1$, because the method is implicit. To
proceed, the key observation is that in the definition of Backward Euler
in (10.35), $\boldsymbol{y}_1$ on the left is defined implicitly by a function on the right
depending on $\boldsymbol{y}_1$, but the term depending on $\boldsymbol{y}_1$ on the right is multiplied by
Δt. We can therefore substitute $\boldsymbol{y}_1$ on the right in (10.38) using (10.35), and
obtain

$$\boldsymbol{\tau} = \boldsymbol{y}(t_1) - \boldsymbol{y}_0 - \Delta t \boldsymbol{f}(t_1, \boldsymbol{y}_0 + \Delta t \boldsymbol{f}(t_1, \boldsymbol{y}_1)). \qquad (10.39)$$

Expanding now for Δt small, the implicit dependence on $\boldsymbol{y}_1$ will show up
only in higher order terms, and if we repeat this substitution recursively, we
can always reach a level at which the implicit dependence no longer affects
the calculation of the truncation error. For Backward Euler, we obtain when
expanding (10.39)

$$\boldsymbol{\tau} = \boldsymbol{y}(t_0) + \Delta t \boldsymbol{y}'(t_0) - \boldsymbol{y}_0 - \Delta t \boldsymbol{f}(t_1, \boldsymbol{y}_0) + O(\Delta t^2) = O(\Delta t^2),$$

where we used $\boldsymbol{y}(0) = \boldsymbol{y}_0$ and the fact that the exact solution satisfies the
differential equation. Thus, the local truncation error is of order two and, according to Theorem 10.2, the Backward Euler method is a first-order method,
just like Forward Euler.

A better method can be obtained if we use the trapezoidal rule to approximate the integral,

$$\int_{t_0}^{t_1} \boldsymbol{f}(\tau, y(\tau)) d\tau \approx \frac{\Delta t}{2} (\boldsymbol{f}(t_0, \boldsymbol{y}(t_0)) + \boldsymbol{f}(t_1, \boldsymbol{y}(t_1))).$$

This leads to the implicit numerical method

$$y_1 = y_0 + \frac{\Delta t}{2}(f(t_0, y_0) + f(t_1, y_1)).$$

To write the trapezoidal method as an implicit Runge-Kutta method, the formula for advancing one step is

$$y_1 = y_0 + \frac{\Delta t}{2}(k_1 + k_2), \tag{10.40}$$

where $k_1 = f(t_0, y_0)$ an $k_2 = f(t_1, y_1)$, which together with (10.40) gives

$$k_2 = f(t_1, y_0 + \frac{\Delta t}{2}(k_1 + k_2)).$$

The *Implicit Trapezoidal Method* in the general Runge-Kutta notation is thus given by

$$
\begin{array}{rcl}
k_1 &=& f(t_0, y_0) \\
k_2 &=& f(t_1, y_0 + \frac{\Delta t}{2}(k_1 + k_2)) \\
y_1 &=& y_0 + \frac{\Delta t}{2}(k_1 + k_2)
\end{array}
\qquad
\begin{array}{c|cc}
0 & 0 & 0 \\
1 & \frac{1}{2} & \frac{1}{2} \\
\hline
 & \frac{1}{2} & \frac{1}{2}
\end{array}
\tag{10.41}
$$

and this is a second-order method, see Exercise 10.13.

Another implicit Runge-Kutta method can be obtained by using the midpoint rule to approximate the integral,

$$\int_{t_0}^{t_0+\Delta t} f(\tau, y(\tau))d\tau \approx \Delta t f(t_0 + \frac{\Delta t}{2}, y(t_0 + \frac{\Delta t}{2})).$$

Since the value $y(t_0 + \frac{\Delta t}{2})$ is not known, we approximate it by an average of y_0 and y_1. The associated numerical method becomes

$$y_1 = y_0 + \Delta t f(t_0 + \frac{\Delta t}{2}, \frac{y_0 + y_1}{2}).$$

Thus, the formula can be written as $y_1 = y_0 + \Delta t k_1$, where $k_1 = f(t_0 + \frac{\Delta t}{2}, \frac{y_0+y_1}{2}) = f(t_0 + \frac{\Delta t}{2}, y_0 + \frac{\Delta t}{2} k_1)$, which leads in the Runge-Kutta notation to

$$
\begin{array}{rcl}
k_1 &=& f(t_0 + \frac{\Delta t}{2}, y_0 + \frac{\Delta t}{2} k_1) \\
y_1 &=& y_0 + \Delta t k_1
\end{array}
\qquad
\begin{array}{c|c}
\frac{1}{2} & \frac{1}{2} \\
\hline
 & 1
\end{array}
\tag{10.42}
$$

and this is also a second-order method, see Exercise 10.14, even though it uses only one stage, compared to the implicit trapezoidal method (10.41).

This raises the question, as in the case of explicit Runge-Kutta methods, as to what the highest attainable order is for an implicit Runge-Kutta method with s stages. As in the case of explicit Runge-Kutta methods, we use Taylor expansions, but the implicit dependence of the k_i on themselves means we cannot expand them in a straightforward way for small Δt. Instead, we need to substitute the k_i recursively, as in the Backward Euler example, until the dependence enters only in the higher order terms, which can then be neglected for the truncation error analysis.

Let us see how this works for a general one-step implicit Runge-Kutta method in MAPLE. For simplicity, we work with an autonomous ODE. A first expansion is obtained with

```
k[1]:=f(y(t0)+dt*a[1,1]*K[1]);
D(y):=t->f(y(t));
alias(f=f(y(t0)),f[y]=D(f)(y(t0)),f[yy]=(D@@2)(f)(y(t0)));
y1:=taylor(y(t0)+dt*sum(b[j]*k[j],j=1..1),dt,3);
```

which leads to

$$y1 := \mathrm{y}(t0) + b_1\, f\, dt + b_1\, f_y\, a_{1,1}\, K_1\, dt^2 + \mathrm{O}(dt^3),$$

We see that the stage value k_1 is still present in the expansion, in the term of order Δt^2. If we substitute once more and expand,

```
k[1]:=subs(K[1]=k[1],k[1]);
y1:=taylor(y(t0)+dt*sum(b[j]*k[j],j=1..1),dt,3);
```

we obtain

$$y1 := \mathrm{y}(t0) + b_1\, f\, dt + b_1\, f_y\, a_{1,1}\, f\, dt^2 + \mathrm{O}(dt^3),$$

and the dependency on the stage value has disappeared from the expansion. We can now proceed as in the case of explicit Runge-Kutta methods,

```
tau:=taylor(y(t0+dt),dt,3)-y1;
p:=collect(convert(series(tau,dt,3),polynom),dt);
eqns:={coeffs(expand(p),[dt,f,f[y]])};
vars:=indets(eqns);
solve(eqns,vars);
```

to obtain the unique solution

$$\{b_1 = 1,\ a_{1,1} = \frac{1}{2}\},$$

which gives the *implicit midpoint method*, and we just showed that this is the only one-stage implicit Runge-Kutta method with local truncation error $O(\Delta t^3)$.

Similarly, we can derive the best two-stage implicit Runge-Kutta method using the following MAPLE program (to obtain the same level of recursive substitution, we use the intermediate variables kn)

ALGORITHM 10.12. *Generating Two Stage RK-Method*

```
for i from 1 to 2 do
  k[i]:=f(y(t0)+h*sum(a[i,j]*K[j],j=1..2));
od;
for i from 1 to 2 do
  kn[i]:=subs({K[1]=k[1],K[2]=k[2]},k[i]);
od;
for i from 1 to 2 do
  k[i]:=subs({K[1]=kn[1],K[2]=kn[2]},kn[i]);
od;
D(y):=t->f(y(t));
y1:=y(t0)+h*sum(b[j]*k[j],j=1..2);;
p:=collect(convert(taylor(y(t0+h)-y1,h,5),polynom),h);
vars:={a[1,1],a[1,2],a[2,1],a[2,2],b[1],b[2]};
eqns:={coeffs(expand(p),indets(p) minus vars)};
sols:=solve(eqns,vars);
allvalues(sols);
```

We obtain the results

$$\left\{ a_{1,1} = \frac{1}{4}, a_{1,2} = \frac{1}{4} + \frac{\sqrt{3}}{6}, a_{2,1} = \frac{1}{4} - \frac{\sqrt{3}}{6}, a_{2,2} = \frac{1}{4}, b_1 = \frac{1}{2}, b_2 = \frac{1}{2} \right\}$$

$$\left\{ a_{1,1} = \frac{1}{4}, a_{1,2} = \frac{1}{4} - \frac{\sqrt{3}}{6}, a_{2,1} = \frac{1}{4} + \frac{\sqrt{3}}{6}, a_{2,2} = \frac{1}{4}, b_1 = \frac{1}{2}, b_2 = \frac{1}{2} \right\}$$

This is the implicit two-stage Runge-Kutta method of order four invented by *Hammer and Hollingsworth*, which is known today as the *Gauss-Legendre Runge-Kutta Method of order four*. Its Butcher tableau is

$$
\begin{array}{c|cc}
\frac{1}{2} \pm \frac{\sqrt{3}}{6} & \frac{1}{4} & \frac{1}{4} \pm \frac{\sqrt{3}}{6} \\
\frac{1}{2} \mp \frac{\sqrt{3}}{6} & \frac{1}{4} \mp \frac{\sqrt{3}}{6} & \frac{1}{4} \\
\hline
 & \frac{1}{2} & \frac{1}{2}
\end{array}
$$

The nodes $\tau_{1,2} = \frac{1}{2} \pm \frac{\sqrt{3}}{6}$ are precisely the Gauss Quadrature nodes for the interval $(0,1)$. Transforming these two nodes to the interval $(-1,1)$, we obtain

$$\xi_{1,2} = 2\tau_{1,2} - 1 = \pm\frac{\sqrt{3}}{3}$$

which are the nodes of the Gauss-Legendre Quadrature rule for $n = 2$ (see Section 9.3).

A three-stage implicit Runge-Kutta method of order six found by *Kuntzmann and Butcher* is

$$
\begin{array}{c|ccc}
\frac{1}{2} \mp \frac{\sqrt{15}}{10} & \frac{5}{36} & \frac{2}{9} \mp \frac{\sqrt{15}}{15} & \frac{5}{36} \mp \frac{\sqrt{15}}{30} \\[2ex]
\frac{1}{2} & \frac{5}{36} \pm \frac{\sqrt{15}}{24} & \frac{2}{9} & \frac{5}{36} \mp \frac{\sqrt{15}}{24} \\[2ex]
\frac{1}{2} \pm \frac{\sqrt{15}}{10} & \frac{5}{36} - \frac{\sqrt{15}}{30} & \frac{2}{9} \pm \frac{\sqrt{15}}{15} & \frac{5}{36} \\[1ex]
\hline
& \frac{5}{18} & \frac{4}{9} & \frac{5}{18}
\end{array}
$$

The computation of the coefficients of the sixth order method currently represents the limit of feasibility in MAPLE. However, by noticing the link between the coefficients and Gauss quadrature in these methods, Kuntzmann and Butcher managed to systematically derive s-stage implicit Runge-Kutta methods of order $2s$, see [62].

10.4 Linear Multistep Methods

Linear multistep methods are fundamentally different from Runge-Kutta methods, since they use solution values from previous time steps to compute the value of the next step. There are two simple ways to derive special classes of linear multistep methods, using either the approximations of derivatives in Chapter 8, or the approximations of integrals in Chapter 9. The combination of both approaches then leads to general linear multistep methods.

We start by approximating the derivative in the ordinary differential equation

$$y' = f(y), \quad y(0) = y_0. \tag{10.43}$$

Applying a forward finite difference,

$$\frac{y(\Delta t) - y(0)}{\Delta t} \approx y'(0),$$

we obtain the numerical method

$$\frac{y_1 - y_0}{\Delta t} = f(y_0), \quad \text{or} \quad y_1 = y_0 + \Delta t f(y_0),$$

which is again the familiar *Forward Euler method*.

Applying a backward finite difference, we get

$$\frac{y_1 - y_0}{\Delta t} = f(y_1) \quad \text{or} \quad y_1 = y_0 + \Delta t f(y_1),$$

which is the *Backward Euler method*, i.e., the first order implicit Runge-Kutta method we had seen already earlier.

To obtain a higher order method, one can use a centered finite difference to approximate the derivative in (10.43),

$$\frac{y(2\Delta t) - y(0)}{2\Delta t} \approx y' = f(y),$$

which leads to the numerical method

$$\frac{\boldsymbol{y}_2 - \boldsymbol{y}_0}{2\Delta t} = \boldsymbol{f}(\boldsymbol{y}_1). \tag{10.44}$$

This method, known as the *Explicit Midpoint Rule*, is very different from all the methods we have seen so far, since it links approximations over two steps instead of one $\boldsymbol{y}_2$, $\boldsymbol{y}_1$ and $\boldsymbol{y}_0$. This methods uses two previous values to compute the next one, so it also needs two values to get started, $\boldsymbol{y}_0$ and $\boldsymbol{y}_1$. While $\boldsymbol{y}_0$ is given by the initial condition, $\boldsymbol{y}_1$ needs to be computed separately, for example using a Runge-Kutta method.

Numerical methods for ordinary differential equations based on approximating the derivative are called *backward differentiation formulas* (BDFs), which are a special case of Algorithm 8.3 in Section 8.2.1. They were introduced and studied by Gear, and can be computed with the simple MAPLE program

---

ALGORITHM 10.13.
Generating Backward Differentiation Formulas

```
BDF:=proc(k,q)
local p,i,t;
  p:=interp([seq(i*dt,i=0..k)],[seq(y[i],i=0..k)],t);
  simplify(subs(t=q*dt,diff(p,t)))=f(y[q]);
end:
```

The MAPLE command

```
BDF(2,1);
```

gives the result

$$-\frac{1}{2}\frac{-y_2 + y_0}{dt} = f(y_1)$$

the centered finite difference formula we have introduced above, and

```
BDF(2,2);
BDF(3,2);
BDF(3,3);
```

give the results

$$\frac{1}{2}\frac{3\,y_2 - 4\,y_1 + y_0}{dt} = f(y_2)$$

$$\frac{1}{6}\frac{2\,y_3 + 3\,y_2 - 6\,y_1 + y_0}{dt} = f(y_2)$$

$$-\frac{1}{6}\frac{-11\,y_3 + 18\,y_2 - 9\,y_1 + 2\,y_0}{dt} = f(y_3)$$

In general, by approximating the derivative with a finite difference, one finds a method of the form

$$\sum_{i=0}^{k} a_i \boldsymbol{y}_i = \Delta t \boldsymbol{f}(t_q, \boldsymbol{y}_q), \quad 0 \le q \le k. \tag{10.45}$$

Such a method is implicit if $q = k$, otherwise it is explicit.

Instead of approximating the derivative in the differential equation, one can integrate it and then use a numerical quadrature formula to approximate the integral, as we did already for Runge-Kutta methods. Integrating (10.43) in time, we obtain

$$\boldsymbol{y}(t_1) - \boldsymbol{y}(t_0) = \int_{t_0}^{t_1} \boldsymbol{f}(\boldsymbol{y}(t))dt,$$

and the two simplest approximations of the integral by a rectangle lead to the Forward and Backward Euler method. Using the trapezoidal rule leads to the *implicit trapezoidal method* we have seen in (10.41). This formulation, however, can also lead to new methods. For instance, one could use known values from the previous integration step to construct an approximate tangent for the approximate integral evaluation, as shown in Figure 10.12. This leads

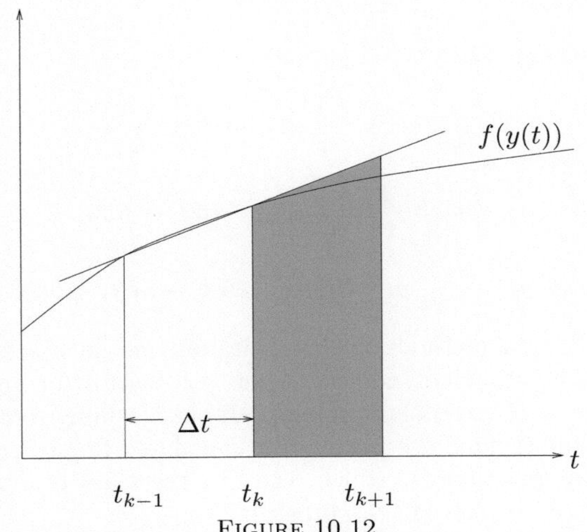

FIGURE 10.12.
Using information from previous function evaluations to construct an approximate tangent for the integral evaluation

to the numerical method

$$
\begin{aligned}
\boldsymbol{y}_{k+1} - \boldsymbol{y}_k &= \Delta t \left(\frac{\boldsymbol{f}(\boldsymbol{y}_{k-1}) + \boldsymbol{f}(\boldsymbol{y}_k)}{2} + \boldsymbol{f}(\boldsymbol{y}_k) - \boldsymbol{f}(\boldsymbol{y}_{k-1}) \right) \\
&= \frac{\Delta t}{2}(3\boldsymbol{f}(\boldsymbol{y}_k) - \boldsymbol{f}(\boldsymbol{y}_{k-1})),
\end{aligned}
$$

which is known as the *Adams–Bashforth two-step method.*

Using not only the previous point t_{k-1}, but also $t_{k-2}, t_{k-3}, \ldots$, and using a higher order interpolation polynomial, one can obtain an entire family of *Adams-Bashforth methods*, which can be computed with the MAPLE program

---

ALGORITHM 10.14.
Generating Adams-Bashforth Methods

```
AdamsBashforth:=proc(k)
local p,i,dt,t;
  p:=interp([seq(i*dt,i=0..k)],[seq(f[i],i=0..k)],t);
  y[k+1]-y[k]=factor(int(p,t=k*dt..(k+1)*dt));
end;
```

Using the MAPLE commands

```
AdamsBashforth(1);
AdamsBashforth(2);
AdamsBashforth(3);
```

we obtain the Adams-Bashforth methods

$$y_2 - y_1 = -\frac{1}{2}\, dt\, (-3\, f_1 + f_0)$$

$$y_3 - y_2 = \frac{1}{12}\, dt\, (23\, f_2 - 16\, f_1 + 5\, f_0)$$

$$y_4 - y_3 = -\frac{1}{24}\, dt\, (-55\, f_3 + 59\, f_2 - 37\, f_1 + 9\, f_0)$$

All Adams-Bashforth methods are explicit, but we have seen in Chapter 4 that the interpolation polynomial is not very useful for approximations outside the interval of interpolation points. If one is willing to use an implicit method, one could include the point to be calculated for the construction of the interpolation polynomial, which leads to the class of *Adams-Moulton methods*, computed by the MAPLE program

---

ALGORITHM 10.15.
Generating Adams-Moulton Methods

```
AdamsMoulton:=proc(k)
local p,i,dt,t;
  p:=interp([seq(i*dt,i=0..k)],[seq(f[i],i=0..k)],t);
  y[k]-y[k-1]=factor(int(p,t=(k-1)*dt..k*dt));
end:
```

Using the MAPLE commands

```
AdamsMoulton(1);
AdamsMoulton(2);
AdamsMoulton(3);
```

we obtain the Adams-Moulton methods

$$y_1 - y_0 = \frac{1}{2}\, dt\ (f_1 + f_0)$$

$$y_2 - y_1 = -\frac{1}{12}\, dt\ (-5\, f_2 - 8\, f_1 + f_0)$$

$$y_3 - y_2 = \frac{1}{24}\, dt\ (9\, f_3 + 19\, f_2 - 5\, f_1 + f_0)$$

Note that for $k = 0$, we obtain with `AdamsMoulton(0);` the result $y_0 - y_{-1} = dt\, f_0$, which is the *Backward Euler* Method, except it is not formulated as $y_1 - y_0 = dt\, f_1$, as one would like to have it.

In general, when we approximate the integral using a quadrature formula involving past integration steps, we get a numerical method of the form

$$\boldsymbol{y}_k - \boldsymbol{y}_{k-1} = \Delta t \sum_{i=0}^{k} b_i \boldsymbol{f}(\boldsymbol{y}_i), \tag{10.46}$$

and the method is explicit if $b_k = 0$, otherwise it is implicit.

Comparing methods based on approximating the derivative (10.45) and those based on approximating the integral (10.46), it is natural to combine the two, which leads to the *general linear multistep method*

$$\sum_{i=0}^{k} a_i \boldsymbol{y}_i = \Delta t \sum_{i=0}^{k} b_i \boldsymbol{f}(\boldsymbol{y}_i). \tag{10.47}$$

If $b_k = 0$, we have an explicit method, otherwise the method is implicit. The linear multistep method (10.47) can be scaled without changing it, so one often fixes $a_k = 1$.

10.4.1   Local Truncation Error

The fact that method (10.47) for $k > 1$ goes over more than one step makes it impossible to use the Definition 10.1 for the local truncation error of the method. We therefore need to generalize that definition.

DEFINITION 10.3. (LOCAL TRUNCATION ERROR FOR LINEAR MULTI-STEP METHODS) *Let $\boldsymbol{y}(t)$ be the solution of the system of ordinary differential equations (10.43), and let $\boldsymbol{y}_n$ be the approximation obtained by (10.47) using the exact solution for the previous values, $\boldsymbol{y}_i = \boldsymbol{y}(t_i)$, $i = 0, 1, \ldots, k-1$. Then the local truncation error $\boldsymbol{\tau}$ for the linear multistep method (10.47) is defined by*

$$\boldsymbol{\tau} := \boldsymbol{y}(t_k) - \boldsymbol{y}_k.$$

Definition 10.3 coincides with the earlier Definition 10.1 in the case of one-step methods.

We start by computing the local truncation error of the explicit centered finite difference method (explicit Midpoint Rule) (10.44),

$$\frac{\boldsymbol{y}_2 - \boldsymbol{y}_0}{2\Delta t} = \boldsymbol{f}(\boldsymbol{y}_1).$$

We obtain

$$
\begin{aligned}
\boldsymbol{\tau} &= \boldsymbol{y}(t_2) - (\boldsymbol{y}(t_0) + 2\Delta t \boldsymbol{f}(\boldsymbol{y}(t_1))) \\
&= \boldsymbol{y}(t_1) + \Delta t \boldsymbol{y}'(t_1) + \tfrac{\Delta t^2}{2}\boldsymbol{y}''(t_1) + O(\Delta t^3) \\
&\quad -(\boldsymbol{y}(t_1) - \Delta t \boldsymbol{y}'(t_1) + \tfrac{\Delta t^2}{2}\boldsymbol{y}''(t_1) + O(\Delta t^3) + 2\Delta t \boldsymbol{f}(\boldsymbol{y}(t_1)) \\
&= O(\Delta t^3),
\end{aligned}
$$

$$(10.48)$$

where we used the differential equation in the last step. This shows that the local truncation error of the explicit centered finite difference method is of order three.

DEFINITION 10.4. (CONSISTENCY) *A linear multistep method (10.47) is consistent of order p if the local truncation error $\tau = O(\Delta t^{p+1})$.*

10.4.2 Order Conditions

A great advantage of linear multistep methods, already indicated by our constructions of Adams and Gear methods, is that one can easily construct methods of high order, in contrast to the Runge-Kutta methods, where the order conditions lead to nonlinear systems that are difficult to solve. The order conditions linear multistep methods are given in the following theorem:

THEOREM 10.4. (ORDER CONDITIONS OF LINEAR MULTISTEP METHODS) *The linear multistep method (10.47) is consistent of order p if and only if the coefficients a_i and b_i, $i = 0, 1, \ldots k$ satisfy*

$$\sum_{i=0}^{k} a_i = 0, \qquad \sum_{i=0}^{k} a_i i^j = j \sum_{i=0}^{k} b_i i^{j-1} \quad for \quad j = 1, 2, \ldots, p. \qquad (10.49)$$

PROOF. We first compute the defect $\boldsymbol{d}$ defined by inserting the exact solution into the linear multistep method (10.47),

$$\boldsymbol{d} := \sum_{i=0}^{k} a_i \boldsymbol{y}(t_i) - \Delta t \sum_{i=0}^{k} b_i \boldsymbol{f}(\boldsymbol{y}(t_i)). \qquad (10.50)$$

Expanding $\boldsymbol{d}$ in a Taylor series for Δt small, we obtain

$$\boldsymbol{d} = \sum_{i=0}^{k} a_i \boldsymbol{y}(t_0 + i\Delta t) - \Delta t \sum_{i=0}^{k} b_i \boldsymbol{y}'(t_0 + i\Delta t)) = \sum_{j=0}^{\infty} d_j \frac{\Delta t^j}{j!} \boldsymbol{y}^{(j)}(t_0), \qquad (10.51)$$

where the d_j are given by

$$d_0 = \sum_{i=0}^{k} a_i, \quad d_j = \sum_{i=0}^{k} a_i i^j - j \sum_{i=0}^{k} b_i i^{j-1}.$$

Now in Definition 10.3 of the local truncation error, the exact solution is used for the previously computed steps, $\boldsymbol{y}_i = \boldsymbol{y}(t_i)$, $i = 0, 1, \ldots, k-1$, and hence $\boldsymbol{y}_k$ satisfies the equation

$$a_k \boldsymbol{y}_k + \sum_{i=0}^{k-1} a_i \boldsymbol{y}(t_i) - \Delta t b_k \boldsymbol{f}(\boldsymbol{y}_k) - \Delta t \sum_{i=0}^{k-1} b_i \boldsymbol{y}'(t_i) = 0.$$

Subtracting this from (10.51) leads to

$$a_k (\boldsymbol{y}(t_k) - \boldsymbol{y}_k) - \Delta t b_k (\boldsymbol{f}(\boldsymbol{y}(t_k)) - \boldsymbol{f}(\boldsymbol{y}_k)) = \sum_{j=0}^{\infty} d_j \frac{\Delta t^j}{j!} \boldsymbol{y}^{(j)}(t_0),$$

and hence $\boldsymbol{\tau} = \boldsymbol{y}(t_k) - \boldsymbol{y}_k = O(\Delta t^{p+1})$ if and only if $d_j = 0$ for $j = 0, 1, \ldots, p$, which concludes the proof. $\qquad\square$

Thus, we see that the order conditions for linear multistep methods are linear in a_i and b_i, so we can readily solve for them to obtain methods of maximal order. Using the small MAPLE program

---

ALGORITHM 10.16.
Generating Implicit Linear Multistep Methods

```
ImplicitLMM:=proc(k)
local i,j,a,b,eqns;
  a[k]:=1;
  eqns:={sum(a[i],i=0..k),seq(sum(a[i]*i^j,i=0..k)=
        j*sum(b[i]*i^(j-1),i=0..k),j=1..2*k)};
  solve(eqns,indets(eqns)); assign(%);
  sum(a[k-j]*y[k-j],j=0..k)= dt*sum(b[k-j]*f[k-j],j=0..k);
end:
```

---

the command

```
for i from 1 to 4 do ImplicitLMM(i) od;
```

gives the first four implicit linear multistep methods of maximal order,

$$y_1 - y_0 = dt \left(\frac{1}{2} f_1 + \frac{1}{2} f_0 \right)$$

$$y_2 - y_0 = dt \left(1/3\, f_2 + 4/3\, f_1 + 1/3\, f_0 \right)$$

$$y_3 + \frac{27}{11}\,y_2 - \frac{27}{11}\,y_1 - y_0 = dt\left(3/11\,f_3 + \frac{27}{11}\,f_2 + \frac{27}{11}\,f_1 + 3/11\,f_0\right)$$

$$y_4 + \frac{32}{5}\,y_3 - \frac{32}{5}\,y_1 - y_0 = dt\left(\frac{6}{25}\,f_4 + \frac{96}{25}\,f_3 + \frac{216}{25}\,f_2 + \frac{96}{25}\,f_1 + \frac{6}{25}\,f_0\right)$$

Similarly, the MAPLE program

ALGORITHM 10.17.
Generating Explicit Linear Multistep Methods

```
ExplicitLMM:=proc(k)
local i,j,a,b,eqns;
  a[k]:=1;b[k]:=0;
  eqns:={sum(a[i],i=0..k),seq(sum(a[i]*i^j,i=0..k)=
      j*sum(b[i]*i^(j-1),i=0..k),j=1..2*k-1)};
  solve(eqns,indets(eqns)); assign(%);
  sum(a[k-j]*y[k-j],j=0..k)= dt*sum(b[k-j]*f[k-j],j=0..k);
end:
```

---

leads after the MAPLE command

```
for i from 1 to 4 do ExplicitLMM(i) od;
```

to the first four explicit linear multistep methods of maximal order,

$$y_1 - y_0 = dt\,f_0$$

$$y_2 + 4\,y_1 - 5\,y_0 = dt\,(4\,f_1 + 2\,f_0)$$

$$y_3 + 18\,y_2 - 9\,y_1 - 10\,y_0 = dt\,(9\,f_2 + 18\,f_1 + 3\,f_0)$$

$$y_4 + \frac{128}{3}\,y_3 + 36\,y_2 - 64\,y_1 - \frac{47}{3}\,y_0 = dt\,(16\,f_3 + 72\,f_2 + 48\,f_1 + 4\,f_0)$$

Despite the ease of deriving high order methods, linear multistep methods can have serious stability problems, which unfortunately render many of them useless, as we will see in the next section.

10.4.3   Zero Stability

Even though possible stability problems in linear multistep methods were already pointed out by Rutishauser in [114], it was Dahlquist [21] who first systematically studied such phenomena. We use his historical example, namely the second of the explicit linear multistep methods with maximal order we derived in the previous paragraph,

$$y_{n+2} + 4y_{n+1} - 5y_n = \Delta t(4f(y_{n+1}) + 2f(y_n)), \tag{10.52}$$

and test how it performs on a simple test problem,

$$y' = y, \qquad y(0) = 1 \Longrightarrow y(t) = e^t. \tag{10.53}$$

The MAPLE program

ALGORITHM 10.18.
Solving $y' = f(y)$ with the Linear Multistep Method
(10.52)

```
ExplicitLMM2:=proc(f,y0,y1,dt,n)
local y,Y,i;
  y[0]:=y0; y[1]:=y1;
  Y:=[[0,y[0]],[dt,y[1]]];
  for i from 1 to n-1 do
    y[i+1]:=-4*y[i]+5*y[i-1]+dt*(4*f(y[i])+2*f(y[i-1]));
    Y:=[op(Y),[dt*(i+1),y[i+1]]];
  od;
  return Y;
end;
```

---

leads, after the MAPLE commands

```
f:=y->y;
y1:=ExplicitLMM2(f,1,exp(1/10),1/10,10):
y2:=ExplicitLMM2(f,1,exp(1/20),1/20,14):
y3:=ExplicitLMM2(f,1,exp(1/40),1/40,16):
y4:=ExplicitLMM2(f,1,exp(1/80),1/80,18):
y5:=ExplicitLMM2(f,1,exp(1/160),1/160,20):
y:=[seq([t/20,exp(t/20)],t=0..20)]:
plot({y,y1,y2,y3,y4,y5},t=0..1,0..3,color=black,axes=boxed);
```

to the results shown in Figure 10.13. Even though this method has a very small local truncation error, the approximation to the solution is not good, and it gets worse when one refines the discretization: the *method is unstable*. Note that this instability is not due to roundoff errors here, since we started with the exact values $y(0) = 1$ and $y(\Delta t) = e^{\Delta t}$, and the computations in MAPLE were performed in exact arithmetic, as one can verify with the MAPLE command

```
print(y1);
```

which shows the results of the MAPLE computation to be

$$[[0, 1], [\tfrac{1}{10}, e^{(1/10)}], [\tfrac{1}{5}, -\tfrac{18}{5} e^{(1/10)} + \tfrac{26}{5}], [\tfrac{3}{10}, \tfrac{454}{25} e^{(1/10)} - \tfrac{468}{25}],$$

$$[\tfrac{2}{5}, -\tfrac{10512}{125} e^{(1/10)} + \tfrac{11804}{125}], [\tfrac{1}{2}, \tfrac{248236}{625} e^{(1/10)} - \tfrac{273312}{625}],$$

$$[\tfrac{3}{5}, -\tfrac{5834808}{3125} e^{(1/10)} + \tfrac{6454136}{3125}], [\tfrac{7}{10}, \tfrac{137297224}{15625} e^{(1/10)} - \tfrac{151705008}{15625}],$$

$$[\tfrac{4}{5}, -\tfrac{3229875072}{78125} e^{(1/10)} + \tfrac{3569727824}{78125}], [\tfrac{9}{10}, \tfrac{75986390416}{390625} e^{(1/10)} - \tfrac{83976751872}{390625}],$$

$$[1, -\tfrac{1787638786848}{1953125} e^{(1/10)} + \tfrac{1975646150816}{1953125}]]$$

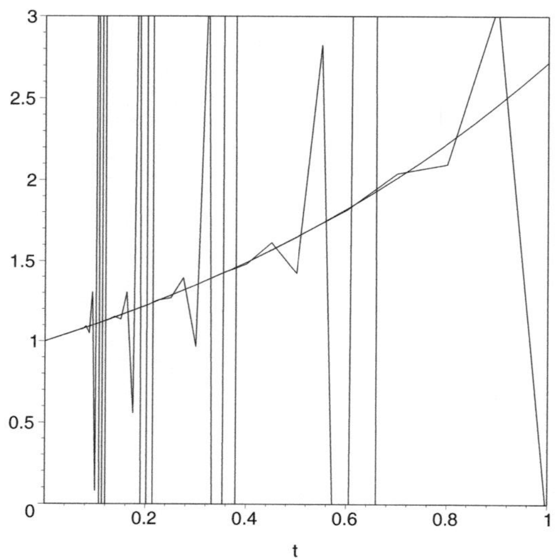

FIGURE 10.13.
*Instability of the second explicit linear multistep methods
with maximal order for the model problem $y' = y$.*

In order to understand what has gone wrong in this computation, we analyze
the recurrence relation obtained from (10.52) when applied to the simple
model problem (10.53),

$$y_{n+2} + 4(1 - \Delta t)y_{n+1} - (5 + 2\Delta t)y_n = 0.$$

Inserting the trial solution $y_n = \lambda^n$, we obtain for λ the *characteristic equation*

$$\lambda^2 + 4(1 - \Delta t)\lambda - (5 + 2\Delta t) = 0,$$

with the two roots

$$\lambda_1 = 2\Delta t - 2 + \sqrt{9 - 6\Delta t + 4\Delta t^2}$$
$$\lambda_2 = 2\Delta t - 2 - \sqrt{9 - 6\Delta t + 4\Delta t^2}.$$

The general solution of the recurrence relation is

$$y_n = C_1\lambda_1^n + C_2\lambda_2^n,$$

where the constants are determined by the initial conditions

$$y_0 = 1 = C_1 + C_2$$
$$y(\Delta t) = e^{\Delta t} = C_1\lambda_1 + C_2\lambda_2.$$

```
> solve({exp(dt)=C[1]*lambda[1]+C[2]*lambda[2],1=C[1]+C[2]},{C[1],C[2]});
> assign(%);
```

$$C_1 = \frac{\mathrm{e}^{dt} - \lambda_2}{\lambda_1 - \lambda_2}, C_2 = -\frac{\mathrm{e}^{dt} - \lambda_1}{\lambda_1 - \lambda_2}$$

We expand the two roots in the Taylor series

```
> R:=solve(lambda^2+4*(1-dt)*lambda-(5+2*dt)=0,lambda):
> lambda[1]:=R[1];lambda[2]:=R[2];
> series(lambda[1],dt=0);
> series(lambda[2],dt=0)
```

$$1 + dt + \frac{1}{2}dt^2 + \frac{1}{6}dt^3 + \frac{1}{72}dt^4 - \frac{5}{216}dt^5 + O\left(dt^6\right)$$

```
> series(lambda[2],dt=0);
```

$$-5 + 3\,dt - \frac{1}{2}dt^2 - \frac{1}{6}dt^3 - \frac{1}{72}dt^4 + \frac{5}{216}dt^5 + O\left(dt^6\right)$$

If C_2 does not vanish, the solution y_n will be dominated by λ_n^n for large n, since λ_2 is much larger in modulus than λ_1; thus, the solution will exhibit oscillations like $(-5)^n$. We have started the linear multistep method with the exact values $y_0 = 1$ and $y_1 = e^{\Delta t}$, which gives for the constants the values

```
> series(C[1],dt=0);
```

$$1 + \frac{1}{216}dt^4 + \frac{11}{1620}dt^5 + O\left(dt^6\right)$$

```
> series(C[2],dt=0);
```

$$-\frac{1}{216}dt^4 - \frac{11}{1620}dt^5 + O\left(dt^6\right)$$

and since $C_2 \neq 0$, the oscillations eventually dominate and destroy the approximation. Notice that the first term $C_1\lambda_1$

```
> series(C[1]*lambda[1],h=0);
```

$$1 + dt + \frac{1}{2}dt^2 + \frac{1}{6}dt^3 + \frac{1}{54}dt^4 - \frac{19}{1620}dt^5 + O\left(dt^6\right)$$

approximates $e^{\Delta t}$ to $O(\Delta t^4)$, while the second term

```
> series(C[2]*lambda[2],h=0);
```

$$\frac{5}{216}dt^4 + \frac{13}{648}dt^5 + O\left(dt^6\right)$$

is zero to $O(\Delta t^4)$.

In theory, one could have chosen the second initial value $y_1 = \lambda_2$ for this simple problem (which makes $C_2 = 0$) and obtain a useful approximation in exact arithmetic, as shown in Figure 10.14 on the left, which was obtained with the MAPLE commands

```
dt:=1/10;
y1:=ExplicitLMM2(f,1,2*dt-2+(9-6*dt+4*dt^2)^(1/2),dt,10):
y:=[seq([t/10,exp(t/10)],t=0..20)]:
plot({y,y1},t=0..2,0..7,color=black,axes=boxed);
```

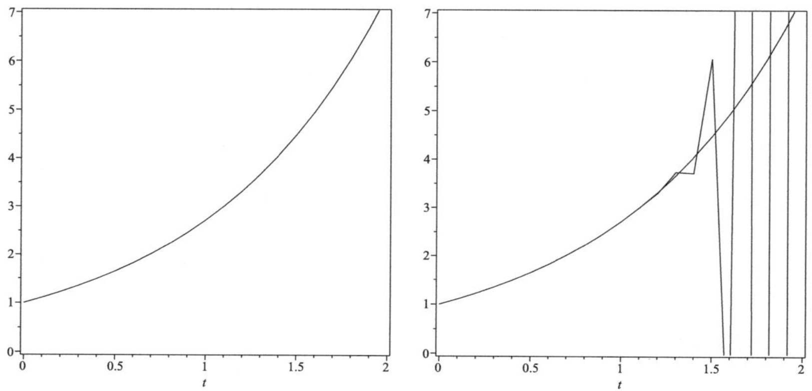

FIGURE 10.14.
*Exact arithmetic on the left, with appropriate starting
steps to cancel the oscillatory mode, and on the right the
same experiment in floating point arithmetic: the
method fails again!*

But when the solution is computed in floating point, roundoff errors suffice
to excite the *parasitic solution,* as shown in Figure 10.14 on the right; that
solution can be computed by adding an `evalf` command in `ExplicitLMM2`
for each floating point assignment. We see that despite its very small local
truncation error, this linear multistep method is completely useless! A useful
method would at least be able to solve the *test equation*

$$y' = 0, \quad y(0) = 1, \tag{10.54}$$

which has the constant solution $y(t) = 1$.

DEFINITION 10.5. (ZERO-STABILITY) *A method is defined to be zero-
stable if, when applied to the test equation (10.54), all the roots of the char-
acteristic equation are inside the unit circle, $|\lambda_i| \le 1$ and if $|\lambda_i| = 1$, λ_i must
be a simple root.*

Looking at the Forward Euler method, we find

$$y_{k+1} - y_k = 0$$

with the characteristic equation

$$\lambda - 1 = 0,$$

and thus the only root is $\lambda = 1$. The Forward Euler method is thus zero-
stable. The same is true for Backward Euler, since the characteristic equation
is the same. Adams-Bashforth and Adams-Moulton methods are also zero-
stable. For BDF(6), we get the characteristic equation

$$\frac{49}{20}\lambda^6 - 6\lambda^5 + \frac{15}{2}\lambda^4 - \frac{20}{3}\lambda^3 + \frac{15}{4}\lambda^2 - \frac{6}{5}\lambda + \frac{1}{6} = 0$$

which has the roots

```
>> p=[49/20, -6, 15/2,-20/3, 15/4, -6/5, 1/6]
>> roots(p)
ans =
   0.1453 + 0.8511i
   0.1453 - 0.8511i
   1.0000
   0.3762 + 0.2885i
   0.3762 - 0.2885i
   0.4061
>> abs(ans)
ans =
   0.8634
   0.8634
   1.0000
   0.4740
   0.4740
   0.4061
```

and thus is also zero-stable.

If we take however the linear multistep method (10.52), we obtain the characteristic equation

$$\lambda^2 + 4\lambda - 5 = 0$$

with the two roots $\lambda_1 = 1$ and $\lambda_2 = -5$ and clearly this method is not zero-stable. We see that the simple test equation (10.54) seems to be precisely the right tool for detecting whether a method has potential stability problems.

10.4.4 Convergence

The observation in the previous subsection leads to one of the most fundamental results in the numerical solution of differential equations, namely that stability and consistency imply convergence (in fact, this is an equivalence), and is proved for the case of ODEs in the following theorem.

THEOREM 10.5. (STABILILTY AND CONSISTENCY IMPLY CONVERGENCE)

A linear multistep method

$$y_{n+k} + \sum_{j=0}^{k-1} a_j y_{n+j} = \Delta t \sum_{j=0}^{k} b_j f(y_{n+j}), \quad n = 0, 1, 2, \ldots \qquad (10.55)$$

with starting values satisfying

$$\|e_{k-1}\| = \left\| \begin{pmatrix} y(t_{k-1}) - y_{k-1} \\ \vdots \\ y(t_0) - y_0 \end{pmatrix} \right\| \leq C\Delta t^p \qquad (10.56)$$

is convergent of order p if it is zero-stable and consistent of order p.

PROOF. From the definition of the linear multistep method (10.47), we obtain an implicit equation for y_{n+k}, namely

$$y_{n+k} = -\sum_{j=0}^{k-1} a_j\, y_{n+j} + \Delta t \sum_{j=0}^{k} b_j\, f(y_{n+j}), \quad n = 0, 1, \cdots . \tag{10.57}$$

Define the vector $\boldsymbol{z}_n := (y_{n+k-1}, y_{n+k-2}, \ldots, y_n)^\top$, and denote the solution of (10.57) by $y_{n+k} = \Phi(\boldsymbol{z}_n)$. We then obtain

$$y_{n+k} = -\sum_{j=0}^{k-1} a_j\, y_{n+j} + \Delta t \underbrace{\left(\sum_{j=0}^{k-1} b_j f(y_{n+j}) + b_k f(\Phi(\boldsymbol{z}_n)) \right)}_{=:g(\boldsymbol{z}_n)} . \tag{10.58}$$

We can therefore write the linear multistep method in matrix form,

$$\boldsymbol{z}_{n+1} = \begin{pmatrix} y_{n+k} \\ y_{n+k-1} \\ \vdots \\ y_{n+1} \end{pmatrix} = A \begin{pmatrix} y_{n+k-1} \\ y_{n+k-2} \\ \vdots \\ y_n \end{pmatrix} + \Delta t \begin{pmatrix} g(\boldsymbol{z}_n) \\ 0 \\ \vdots \\ 0 \end{pmatrix}, \tag{10.59}$$

with

$$A = \begin{pmatrix} -a_{k-1} & -a_{k-2} & \cdots & -a_0 \\ 1 & 0 & & \\ & & \ddots & \ddots & \\ & & & 1 & 0 \end{pmatrix}$$

The method (10.59) is now a one-step method of the form $\boldsymbol{z}_{n+1} = A\boldsymbol{z}_n + \Delta t \boldsymbol{g}(\boldsymbol{z}_n)$, $n = 0, 1, \cdots$. With $\boldsymbol{z}(t_{n+1}) = (y(t_{n+k}), y(t_{n+k-1}), \ldots, y(t_{n+1}))^\top$ we define the local error as for one-step methods to be $\boldsymbol{e}_{n+1} = \boldsymbol{z}(t_{n+1}) - \boldsymbol{z}_{n+1}$ and get by adding and subtracting

$$\boldsymbol{e}_{n+1} = \boldsymbol{z}(t_{n+1}) - A\boldsymbol{z}_n - \Delta t \boldsymbol{g}(\boldsymbol{z}_n) + A\boldsymbol{z}(t_n) + \Delta t \boldsymbol{g}(\boldsymbol{z}(t_n)) - A\boldsymbol{z}(t_n) - \Delta t \boldsymbol{g}(\boldsymbol{z}(t_n))$$

$$= A\left(\boldsymbol{z}(t_n) - \boldsymbol{z}_n\right) + \Delta t \left(\boldsymbol{g}(\boldsymbol{z}(t_n)) - \boldsymbol{g}(\boldsymbol{z}_n)\right) + \boldsymbol{z}(t_{n+1}) - A\boldsymbol{z}(t_n) - \Delta t \boldsymbol{g}(\boldsymbol{z}(t_n)).$$

Now we have

$$\boldsymbol{z}(t_{n+1}) - A\boldsymbol{z}(t_n) - \Delta t \boldsymbol{g}(\boldsymbol{z}(t_n))$$

$$= \begin{pmatrix} y(t_{n+k}) - \sum_{j=0}^{k-1} a_j y(t_{n+j}) - \Delta t \boldsymbol{g}(\boldsymbol{z}(t_n)) \\ 0 \\ \vdots \\ 0 \end{pmatrix} = \begin{pmatrix} y(t_{n+k}) - y_{n+k} \\ 0 \\ \vdots \\ 0 \end{pmatrix}$$

so consistency implies

$$\|\boldsymbol{z}(t_{n+1}) - A\boldsymbol{z}(t_n) - \Delta t \boldsymbol{g}(\boldsymbol{z}(t_n))\| = \tau \leq C_\tau \Delta t^{p+1}.$$

Since g is a finite sum of function values of f, it is Lipschitz whenever f is, and with the same constant L. Thus we get the estimate

$$\|e_{n+1}\| \leq \tau + \|A\|\,\|z(t_n) - z_n\| + \Delta t L\,\|z(t_n) - z_n\| = (\|A\| + \Delta t L)\,\|e_n\| + \tau.$$

By induction on n, we can sum the geometric series and obtain

$$\|e_n\| \leq (\|A\| + \Delta t L)^{n-k+1}\,\|e_{k-1}\| + \frac{(\|A\| + \Delta t L)^{n-k+1} - 1}{\|A\| + \Delta t L - 1}\,\tau. \qquad (10.60)$$

We now need to estimate the norm of A. If λ is a root of the characteristic equation $\lambda^k + a_{k-1}\lambda^{k-1} + a_{k-2}\lambda^{k-2} + \ldots + a_0 = 0$, then

$$A\begin{pmatrix} \lambda^{k-1} \\ \lambda^{k-2} \\ \vdots \\ \lambda \\ 1 \end{pmatrix} = \begin{pmatrix} -a_{k-1}\lambda^{k-1} - a_{k-2}\lambda^{k-2} - \ldots - a_0 \\ \lambda^{k-1} \\ \vdots \\ \lambda^2 \\ \lambda \end{pmatrix} = \lambda\begin{pmatrix} \lambda^{k-1} \\ \lambda^{k-2} \\ \vdots \\ \lambda \\ 1 \end{pmatrix}. \qquad (10.61)$$

Thus, we obtain an eigenvalue-eigenvector pair of A. Because we assumed that the method is zero-stable, we know that all eigenvalues of A are within the unit disk, and the ones on the boundary are simple. Let $\lambda_1, \ldots, \lambda_r$ be the eigenvalues of modulus one, which have to be simple by assumption, and denote by $T^{-1}AT = J$ a Jordan decomposition of the matrix A, with J of the form (see [51])

$$J = \begin{pmatrix} \lambda_1 & & & & & & \\ & \ddots & & & & & \\ & & \lambda_r & & & & \\ & & & \lambda_{r+1} & \varepsilon_{r+1} & & \\ & & & & \ddots & \ddots & \\ & & & & & \ddots & \varepsilon_{k-1} \\ & & & & & & \lambda_k \end{pmatrix}, \qquad (10.62)$$

with ε_j such that $|\lambda_j| + |\varepsilon_j| \leq 1$ for $j = r+1, \ldots, k$. Such a Jordan decomposition can always be obtained by diagonal scaling. We then choose as the norm in our estimate (recall that all norms in this finite dimensional setting are equivalent) $\|x\| := \|T^{-1}x\|_\infty$

$$\implies \|A\| = \sup_{x \neq 0} \frac{\|Ax\|}{\|x\|} = \sup_{x \neq 0} \frac{\|T^{-1}ATT^{-1}x\|_\infty}{\|T^{-1}x\|_\infty} = \sup_{y \neq 0} \frac{\|Jy\|_\infty}{\|y\|_\infty} = 1.$$

In this norm, we therefore have $\|A\| = 1$, and thus obtain from (10.60) using (10.56)

$$\|e_n\| \leq (1 + \Delta t L)^{n-k+1} C \Delta t^p + \frac{(1 + \Delta t L)^{n-k+1} - 1}{\Delta t L} C_\tau \Delta t^{p+1}. \qquad (10.63)$$

Now for a given time interval $(0, T)$ and time step Δt, we have $n \leq N := \frac{T}{\Delta t}$, and therefore $(1 + \Delta t L)^{n-k+1} \leq e^{L \Delta t (n-k+1)} \leq e^{L \Delta t (N-k+1)} \leq e^{LT}$, since $k \geq 1$, which gives for $n = k, k+1, \ldots, N$ the final estimate

$$\|e_n\| \leq e^{LT} (C + \frac{C_\tau}{L}) \Delta t^p \tag{10.64}$$

and hence the method is convergent of order p. $\square$

REMARK. *All consistent one-step methods are automatically zero-stable. This is easily seen for Runge-Kutta methods, for example from (10.32): since $f = 0$ when such a method is applied to the test equation (10.54), the only root of the characteristic equation is 1, and hence the method is zero-stable. The concept of zero-stability was therefore not needed in the convergence Theorem 10.2 for one-step methods.*

A very important result for linear multistep methods is *the first Dahlquist barrier* [21]: for p even, the best order attainable by a zero-stable linear p-step method is $q = p + 2$. For p odd, it is $q = p + 1$. If the linear multistep method is explicit, the best order attainable is only $q = p$. As we have seen, it is very easy to construct high order linear multistep methods, since one just has to satisfy the order conditions in Theorem 10.4. However, many of them are not convergent, since they are not zero-stable.

It is also much more difficult to construct adaptive linear multistep methods, since changing the step size locally for a new step means that equidistant function values are no longer available for the next time step. But it is possible to design such methods, see [63] ("Des war a harter Brockn, des ...").

10.5   Stiff Problems

It is difficult (and not necessarily helpful) to give a precise definition for when a problem is stiff. However, a good working definition is that a *problem is stiff* if the time step used by an explicit method in order to achieve a certain accuracy is much smaller than the time step one would need to use to resolve the shape of the solution[2]. Let us look at a simple example to illustrate this: we consider the model problem

$$y' = -10y, \quad y(0) = 1,$$

which has the closed form solution

$$y(t) = e^{-10t}.$$

This solution decays rapidly to zero. Suppose we approximate the solution with the backward differentiation formula,

$$\frac{y_{k+1} - y_{k-1}}{2 \Delta t} = -10 y_k,$$

---

[2] "Stiff equations are problems for which explicit methods don't work." [63]

for which we have seen that the local truncation error is $O(\Delta t^3)$, see (10.48). The method is also zero-stable, as one readily verifies, see Problem 10.10. The method is therefore convergent, according to Theorem 10.5. However, as Figure 10.15 shows, the method does not converge well to the solution. Although we used a very fine time step of $\Delta t = 1/100$ (the figure was pro-

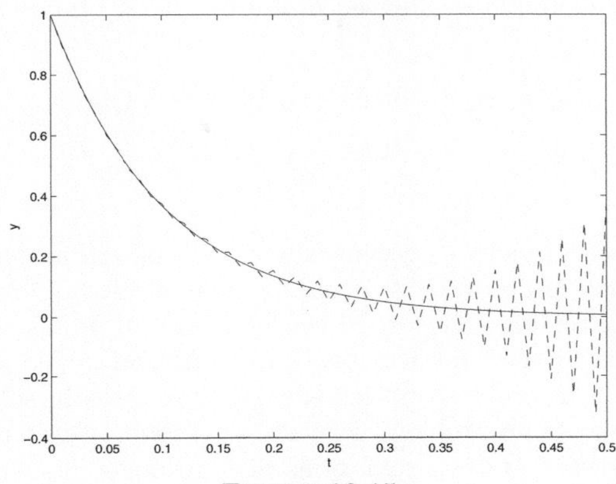

FIGURE 10.15.
*Oscillation of a backward differentiation formula
approximation for a simple model problem. Clearly this
oscillating approximation does not faithfully reproduce
the monotonically decaying physical solution.*

duced using the starting values $y_0 = 1$ and the Euler approximation for $y_1 = y_0 - 10\Delta t y_0$), the solution starts to oscillate about the true solution and fails to reproduce the physically correct result, which is very smooth; only a few points would suffice to represent it accurately. This is a typical case of a problem due to stiffness.

In order to understand this phenomenon, we consider, as in the case of zero stability, a test equation, namely

$$y' = ay, \quad y(0) = 1, \quad a \in \mathbb{C}, \tag{10.65}$$

which has the solution $y(t) = e^{at}$. Note that if the real part of a is negative, $\mathrm{Re}(a) < 0$, then $y(t)$ goes to zero as t goes to infinity. This property holds for many physical systems, for example for all diffusive problems, such as heat diffusion.

Applying the Explicit Midpoint Method to the test equation (10.65), we find

$$y_{k+1} - y_{k-1} = 2\Delta t a y_k.$$

To find the discrete solutions y_k, we solve again the characteristic equation

$$\lambda^2 - 2\Delta t a \lambda - 1 = 0,$$

which has the roots

$$\lambda_{1,2} = \Delta t a \pm \sqrt{\Delta t^2 a^2 + 1}.$$

Thus the general solution for y_k is

$$y_k = A\lambda_1^k + B\lambda_2^k.$$

In the numerical example we showed in Figure 10.15, we had $a = -10$, so we find

$$
\begin{aligned}
|\lambda_1| &= |-10\Delta t + \sqrt{100\Delta t^2 + 1}| < 1, \\
|\lambda_2| &= |-10\Delta t - \sqrt{100\Delta t^2 + 1}| > 1,
\end{aligned}
$$

Since $\Delta t > 0$, λ_2 will yield a growing numerical approximation, even though the exact solution $y(t) = e^{-10t}$ decays exponentially quickly to zero. It is precisely this growing solution which leads to the unfavorable oscillation properties of this method for stiff problems, and forces the time step to be unnaturally small in order to obtain a good approximate solution. In fact, if we integrate the same problem with $\Delta t = 1/1000$, then we obtain Figure 10.16, which shows that the oscillations are no longer visible. However, to

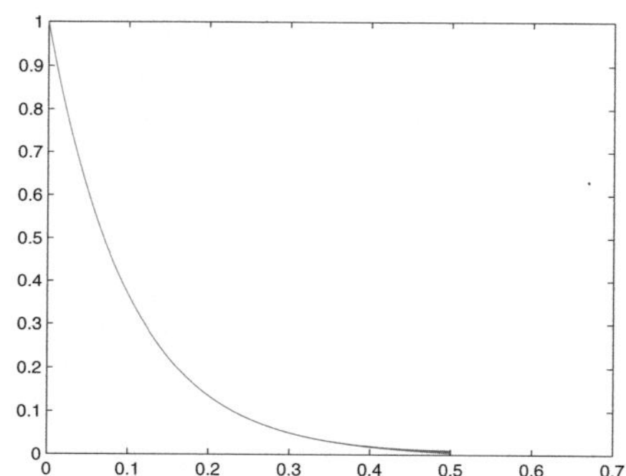

FIGURE 10.16.
*With a smaller step size the oscillations have
disappeared.*

achieve this, we would need 500 integration steps for such a smooth function, which shows that the method is not suited for integrating stiff equations.

Let us look at a second example which, when implemented in MATLAB, even seems to contradict Theorem 10.5. Consider the ODE

$$y'' + 101y' + 100y = 0, \quad y(0) = 1, y'(0) = -1. \tag{10.66}$$

The solution with the given initial conditions is $y(t) = e^{-t}$. We want to solve this equation using again the Explicit Midpoint Rule. For this, we first transform the ODE to a first order system by introducing the variables $z_1 = y$ and $z_2 = y'$, see Subsection 10.2.6,

$$z' = Az, \quad A = \begin{pmatrix} 0 & 1 \\ -100 & -101 \end{pmatrix}, \quad z(0) = \begin{pmatrix} 1 \\ -1 \end{pmatrix}$$

and run the following MATLAB program

ALGORITHM 10.19.
Solving Equation (10.66) with the Explicit Midpoint Rule

```
Steps=[0.04,0.02,0.01,0.005];
for dt=Steps
  tend=0.6;
  Y=[];
  t=0; y0=[1;-1];                    % exact initial conditions
  Y=[t,y0'];                         % save results
  t=dt; y1=[exp(-dt);-exp(-dt)];
  Y=[Y;t,y1'];
  while t<tend
    y=y0+2*dt*[0 1;-100 -101]*y1;
    t=t+dt; Y=[Y;t y'];
    y0=y1;y1=y;
  end
  T=Y(:,1);
  steps=max(size(Y))                 % number of integration steps
  error=norm(exp(-T)-Y(:,2))         % compare with exact solution
  plot(T,[Y(:,2),exp(-T)])
  pause
end
```

The result for the four integration step sizes are displayed in the following table

dt	# steps	error
0.04	16	2.2673e-04
0.02	31	0.2924
0.01	61	9.9013e+03
0.005	121	6.2028e+04

and in Figure 10.17. In this example, the solution gets worse when the step size is refined! This behavior seems to indicate an instability, which historically was called "weak instability" and started Dahlquist's career. However, according to Theorem 10.5 this method is convergent, but only if the step size is made *much, much smaller*, see Problem 10.17.

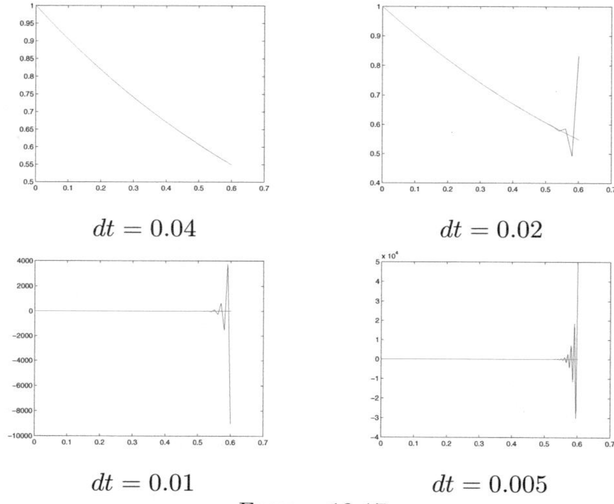

$$dt = 0.04 \qquad\qquad dt = 0.02$$

$$dt = 0.01 \qquad\qquad dt = 0.005$$

FIGURE 10.17.
*Solving Equation (10.66) using the Explicit Mid Point
Rule*

10.5.1   A-Stability

In order to find good methods for such types of problems, we need to introduce the concepts of a *region of absolute stability* and *A-stability*. As we noticed in the previous paragraph, the roots of the characteristic equation obtained when applying the method to the test equation (10.65) depend on the parameter $\mu := a\Delta t$, $\lambda_i = \lambda_i(a\Delta t) = \lambda_i(\mu)$. This observation leads to the following definition.

DEFINITION 10.6. (REGION OF ABSOLUTE STABILITY) *The* region of absolute stability R *for a numerical method for solving an ODE is defined by*

$$R := \left\{ \mu : |\lambda_i(\mu)| < 1 \ \forall \lambda_i(\mu) \right\},$$

where $\lambda_i(\mu)$, $\mu = a\Delta t$ are the roots of the characteristic equation when the method is applied to the test equation $y' = ay$, $a \in \mathbb{C}$.

We see that if Δt for a given method is chosen such that $a\Delta t$ is in the region of absolute stability, then the numerical solution decays to zero.

DEFINITION 10.7. (A-STABILITY) *A method is called A-stable if its region of absolute stability R contains the left half of the complex plane $\{z \in \mathbb{C} : \mathrm{Re}(z) < 0\}$.*

A-stability is a desirable property for a method, because if the method is A-stable, then the numerical approximations obtained from the method decays to zero whenever the exact solution of the test equation (10.65) does

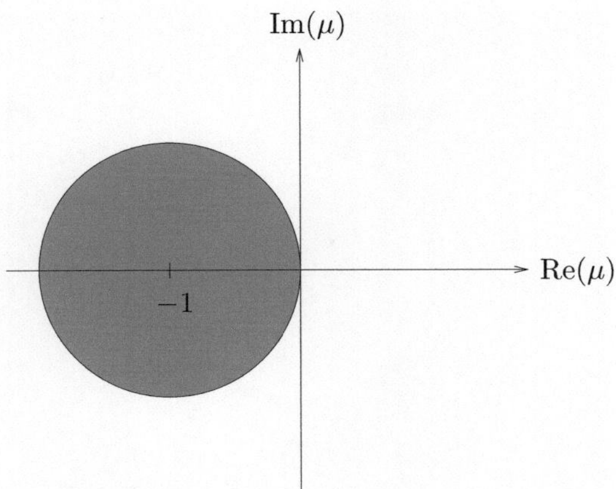

FIGURE 10.18.
*Region of absolute stability for the Forward Euler
method*

as well. Hence, one obtains physically meaningful solutions independent of the step size used.

It is best to look at some examples to understand the concept of A-stability. We start with the Forward Euler method, which gives, when applied to the test equation (10.65), the method

$$y_{k+1} = y_k + \Delta t\, a y_k = (1 + \Delta t\, a)y_k.$$

The only root of its characteristic equation is therefore

$$\lambda_1 = 1 + \Delta t\, a =: 1 + \mu.$$

The region of absolute stability for the Forward Euler method is therefore

$$|\lambda_1(\mu)| = |1 + \mu| < 1$$

as shown in Figure 10.18. It becomes clear now why this region is so important: as soon as Δt is small enough for the Forward Euler method, $a\Delta t$ for $\mathrm{Re}(a) < 0$ will lie inside its region of absolute stability, and the numerical approximation to the solution will decay to zero, as does the exact solution.

For the Backward Euler method applied to the test equation (10.65), we find

$$\frac{y_{k+1} - y_k}{\Delta t} = a y_{k+1},$$

and thus the characteristic equation has the only root

$$\lambda_1 = \frac{1}{1 - a\Delta t} =: \frac{1}{1 - \mu}.$$

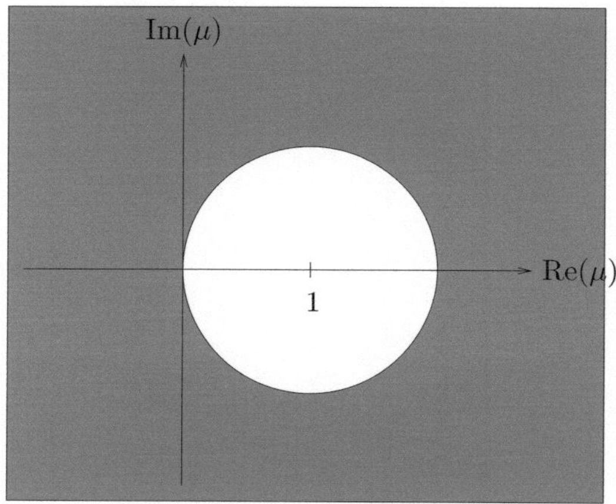

FIGURE 10.19.
*Region of absolute stability for the Backward Euler
method*

The values of μ for which

$$|\lambda_1(\mu)| = \frac{1}{|1 - \mu|} < 1$$

are thus the values for which

$$|1 - \mu| > 1.$$

Hence the region of absolute stability for Backward Euler lies outside the circle centered at 1 with unit radius, as shown in Figure 10.19. Since the region of absolute stability contains the left half of the complex plane, Backward Euler is the first method we encounter that is A-stable. When using the Backward Euler method for the test equation (10.65), numerical approximations will decay whenever the exact solution also decays, i.e. whenever $\text{Re}(a) < 0$, regardless of the time step Δt chosen. In other words, in order for Backward Euler to produce a physically correct decaying solution, the time step Δt does not have to be small; thus, unlike for Forward Euler, the size of Δt is only determined by the desired accuracy, and not by the artificial requirement that the solution should be physically correct.

One also sees that Backward Euler is in fact *overstable*, since there are values for a with positive real part for which the Backward Euler approximation decays to zero, even though the exact solutions e^{at} grow exponentially. This is however less of a problem in applications, since few physically relevant problems have exponentially growing solutions.

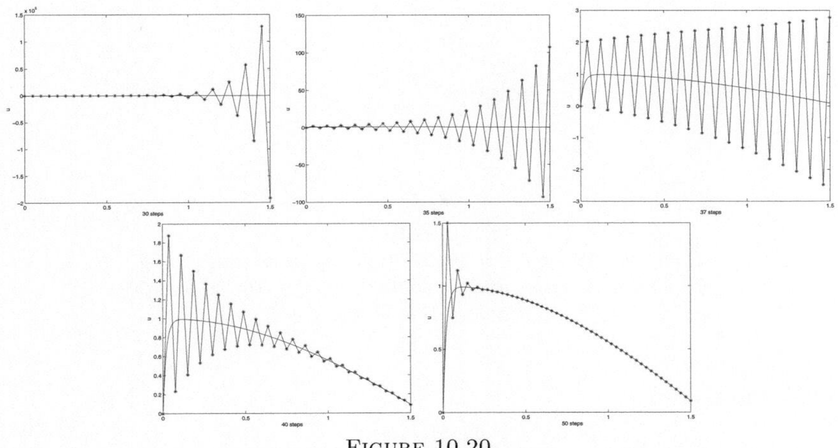

FIGURE 10.20.
*Forward Euler approximation of a stiff model problem by
Curtiss and Hirschfelder, using 30, 35, 37, 40 and 50
integration steps.*

10.5.2 A Nonlinear Example

We have so far only looked at the linear test equation (10.65). We show now
a second, stiff and nonlinear example from the historical paper by Curtiss
and Hirschfelder [20], written as a system

$$\begin{aligned}
\dot{x} &= -50(x - \cos(y)) \\
\dot{y} &= 1
\end{aligned} \tag{10.67}$$

in order to explain why the concept of A-stability is still relevant. We show
in Figure 10.20 the x-component approximation obtained from Forward Euler for various values of Δt. We clearly see that the approximate solutions
obtained by Forward Euler are very inaccurate, even for a small Δt that requires a large number of time steps to integrate; the approximations at the
beginning of the time interval are especially inaccurate.

Similarly, we show in Figure 10.21 the x-component approximation obtained from Backward Euler for various values of Δt. Even for only 10 time
steps, the physical behavior of the solution is already correct, and for 30
time steps, shown in the second panel of Figure 10.21, the approximation is
quite accurate. This is in contrast to Forward Euler with 30 time steps, the
first panel in Figure 10.20, which gives a completely useless approximation.
Clearly, these two first-order methods perform very differently on this model
problem.

To understand why A-stability is still the important concept for this nonlinear example, we linearize the general nonlinear system of ordinary differ-

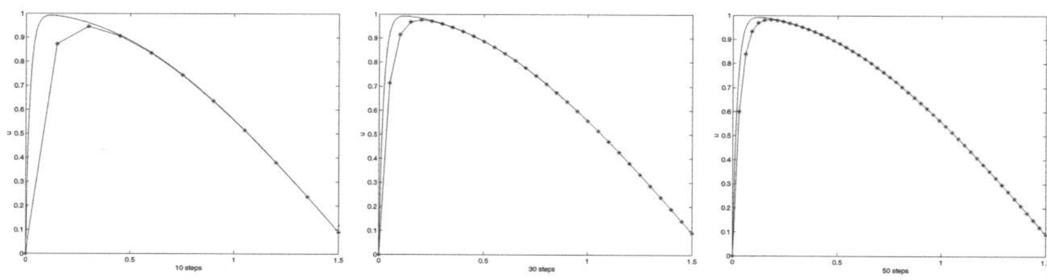

FIGURE 10.21.
*Backward Euler approximation of the same stiff model
problem by Curtiss and Hirschfelder, using 10, 30 and 50
integration steps.*

ential equations

$$y' = f(y), \tag{10.68}$$

and look for a nearby solution. We expand the right hand side

$$\begin{aligned}(y + \Delta y)' &= f(y + \Delta y) \\ &= f(y) + \nabla f(y)\Delta y + O(\|\Delta y\|^2),\end{aligned} \tag{10.69}$$

where $\nabla f(y)$ denotes the Jacobian of f at y. If the solution is close by, we can neglect the term $O(\|\Delta y\|^2)$, and taking the difference between the original equation (10.68) and the expanded one (10.69), we find the difference to be

$$(\Delta y)' = \nabla f(y)\Delta y, \tag{10.70}$$

which is now a linear differential equation. If, furthermore, the Jacobian matrix ∇f is diagonalizable, $\nabla f = S \Lambda S^{-1}$, $\Lambda = \mathrm{diag}(\mu_1, \mu_2, \ldots)$, then we can also diagonalize (10.70) (otherwise one can continue the argument with a Jordan decomposition). Denoting by $z := S^{-1}y$, we get for z_i the scalar differential equations

$$\dot{z}_i = \mu_i z_i, \quad i = 1, 2, \ldots$$

which are now of the form of the simple test equation (10.65) used to determine the region of absolute stability of a numerical method (although the μ_i in general still depend on time). For our nonlinear example, we obtain

$$\nabla f(y) = \begin{pmatrix} -50 & 0 \\ 0 & 0 \end{pmatrix},$$

and thus the eigenvalues are $a_1 = 0$ and $a_2 = -50$. Since Backward Euler is A-stable, it will only produce decaying solutions, and the eigenvalue $a_2 = -50$ will not affect the quality of the approximation. In contrast, the Forward Euler method needs to satisfy the condition $|1 + a_i \Delta t| < 1$, $i = 1, 2$ to be in the region of absolute stability, which corresponds for $a_2 = -50$ to $\Delta t \leq \frac{2}{50}$, or 38 or more time steps in the example in Figure 10.20. The comparison with the two approximate solutions for Forward Euler with 37 and 40 time steps

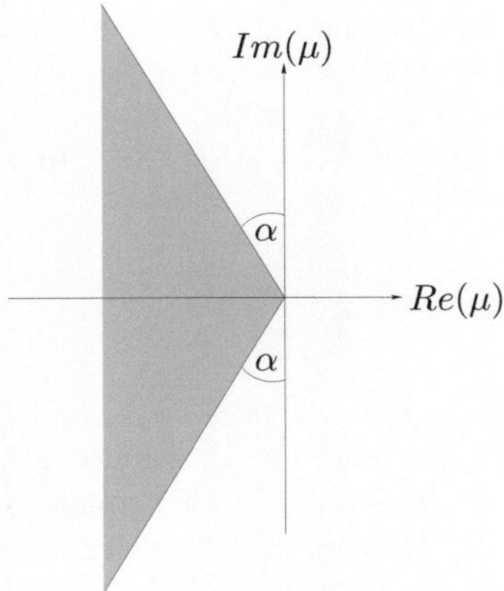

FIGURE 10.22.
*Region that must be included in the region of absolute
stability of a method in order for the method to be
A(α)-stable*

in Figure 10.20 illustrates why the concept of A-stability is very important
for stiff problems.

A very important general result is the *second Dahlquist barrier* [22]: an
A-stable linear multistep method cannot have an order higher than two. To
circumvent this barrier, the slightly weaker condition of *A(α)-stability* has
been introduced: a method is $A(\alpha)$-stable if its region of absolute stability
includes the region depicted in Figure 10.22.

10.5.3 Differential Algebraic Equations

Stiff equations are often obtained from *singular perturbation problems*. As an
example, we consider the system of ordinary differential equations

$$\begin{aligned}
\dot{y} &= f(y, z), \\
\varepsilon\dot{z} &= g(y, z),
\end{aligned} \tag{10.71}$$

Such problems often occur in chemical reaction simulations, where there are
slow and fast reactions, and also in circuit simulation and fluid dynamics.
When ε becomes small, these equations become very stiff: computing the
Jacobian of the right hand side function in (10.71) after division by ε yields

$$J = \begin{pmatrix} f_y & f_z \\ \frac{1}{\varepsilon}g_y & \frac{1}{\varepsilon}g_z \end{pmatrix}.$$

The eigenvalues of J are

$$\lambda_{1,2} = \frac{f_y + \frac{1}{\varepsilon}g_z \pm \sqrt{(f_y + \frac{1}{\varepsilon}g_z)^2 - \frac{4}{\varepsilon}(f_y g_z - f_z g_y)}}{2}.$$

The two eigenvalues behave very differently when ε becomes small: one of the eigenvalues will behave like $\frac{1}{\varepsilon}$, whereas the other will be of order one; as a result, the system becomes very stiff.

What happens in the limit as $\varepsilon \longrightarrow 0$? In that case, the singularly perturbed system (10.71) becomes

$$\begin{aligned} \dot{y} &= f(y,z), \\ 0 &= g(y,z), \end{aligned} \qquad (10.72)$$

which is called a *differential algebraic equation (DAE)*. Solutions of such equations must satisfy the algebraic constraint $g(y(t), z(t)) = 0$, i.e., they have to lie on the manifold defined by this equation.

In order to solve (10.71) for small values of ε, one is interested in methods that have robust convergence estimates independent of ε. Such methods can then also be used to solve (10.72) directly. There is a large body of literature available that considers such problems, see [63] and the references therein; specialized methods for (10.72) can for example be obtained by applying a Runge-Kutta method or a linear multistep method to the original problem (10.71), and then letting ε go to zero in the definition of the method. Alternatively, one can differentiate the algebraic constraint equation $g(y,z) = 0$ in order to obtain

$$g_y \dot{y} + g_z \dot{z} = 0 \quad \Longrightarrow \quad \dot{z} = -g_z^{-1} g_y f,$$

which is again a differential equation and can be solved together with the first differential equation by any method we have seen. DAEs for which this is possible are called index-1 DAEs. Not all DAEs have index 1, as the following example shows:

$$\begin{aligned} \dot{y} &= f(y,z), \\ 0 &= g(y). \end{aligned} \qquad (10.73)$$

Computing a derivative of the constraint equation $g(y) = 0$, we obtain $g_y(y) f(y,z) = 0$, which is still an algebraic constraint. However, differentiating once more works, which is why (10.73) is called an index-2 DAE. In general, methods are developed for index-1 DAEs, just as most methods were developed for first order ODEs; higher index DAEs must be reduced to index-1 before standard methods can be applied. A comprehensive treatment of numerical methods for such problems can be found in [63].

10.6   Geometric Integration

Geometric integration has established itself as a mature field of research over the last decade, and comprehensive monographs are available [61, 84]. A geometric integrator is a numerical method that preserves geometric properties of the exact flow of a differential equation. We illustrate two important concepts of geometric integration on the non-trivial example of the *Lotka-Volterra Equations* [143]. We consider a predator species y and its prey x. The Lotka-Volterra system of differential equations describes the evolution of x and y by

$$\begin{aligned} \dot{x} &= x - xy \quad;\quad x(0) = \hat{x}, \\ \dot{y} &= -y + xy \quad;\quad y(0) = \hat{y}, \end{aligned} \tag{10.74}$$

where we denote the time derivative on the left by a dot. The growth rate of the prey population $\dot{x}$ is proportional to the current population x minus the number of prey-predator encounters, proportional to xy. This is quite intuitive, since if there is more prey, they will produce more offspring, but prey-predator encounters are potentially fatal for the prey. In the other equation, the growth rate of the predator population $\dot{y}$ is proportional to the predator-prey encounters xy minus the current population y. This is also quite natural, since predator-prey encounters are potentially beneficial for the predator, and a large predator population will rapidly be reduced when there is little prey. In order to simplify the exposition, we have not introduced the biologically important constants of proportionality here.

It is well known that the solution to (10.74) is cyclic for all initial values $\hat{x}$, $\hat{y}$ in the first quadrant of the xy plane. The cycles are around the equilibrium point $\bar{x} = 1$, $\bar{y} = 1$, which is obtained by setting the time derivatives on the left hand side of (10.74) to zero. An explicit solution can be obtained in parametric form [129], see also Problem 10.18, namely

$$\begin{aligned} x &= \tfrac{1}{2}\tau \pm \tfrac{1}{2}\sqrt{\tau^2 - 4Ce^\tau}, \\ y &= \tfrac{1}{2}\tau \mp \tfrac{1}{2}\sqrt{\tau^2 - 4Ce^\tau}, \end{aligned} \tag{10.75}$$

where the constant C is given by

$$C = \hat{x}\hat{y}e^{\hat{x}+\hat{y}}.$$

Figure 10.23 on the left shows the solution for initial values $\hat{x} = 0.5$ and $\hat{y} = 0.5$.

We now apply a very simple numerical method to solve the Lotka-Volterra equations approximately, namely the Forward Euler method. This leads to the discrete iteration formula with approximations at time t_n,

$$\begin{aligned} \frac{x_{n+1} - x_n}{\Delta t} &= x_n - x_n y_n \quad;\quad x_0 = \hat{x}, \\ \frac{y_{n+1} - y_n}{\Delta t} &= -y_n + x_n y_n \quad;\quad y_0 = \hat{y}. \end{aligned} \tag{10.76}$$

This is an explicit iteration formula. The solution obtained with a time step $\Delta t = 0.1$ and initial values $\hat{x} = 0.5$, $\hat{y} = 0.5$ is shown in Figure 10.23 on the right, together with the exact solution. The numerical solution obtained by Forward Euler spirals outwards, which is physically incorrect. In fact, most classical methods we have seen exhibit spiraling: explicit methods typically spiral outward, and implicit methods inward, see Problems 10.19 and 10.20.

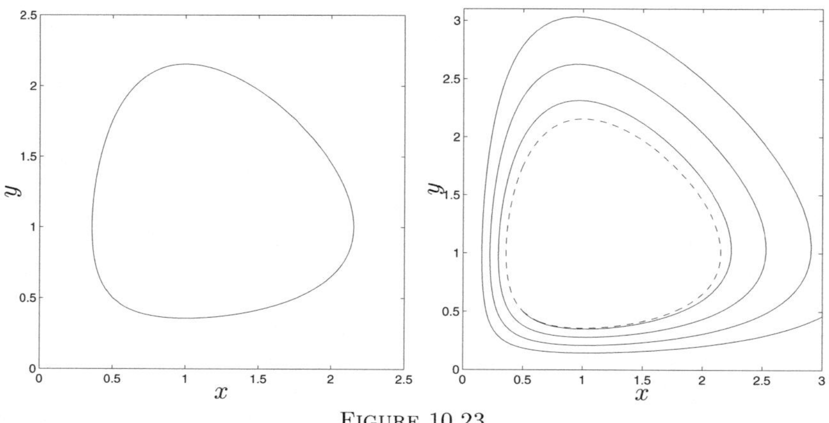

FIGURE 10.23.
*Exact solution of the Lotka-Volterra System on the left,
and outward spiraling approximate solution computed
with Forward Euler on the right*

10.6.1 Symplectic Methods

We now show how a simple modification of the Forward Euler method leads
to a numerical approximation that does not exhibit any spiraling: We replace
x_n in the product term of the second equation of (10.76) by its newest value
x_{n+1} to get

$$
\begin{aligned}
\frac{x_{n+1} - x_n}{\Delta t} &= x_n - x_n y_n \quad ; \quad x_0 = \hat{x}, \\
\frac{y_{n+1} - y_n}{\Delta t} &= -y_n + x_{n+1} y_n \quad ; \quad y_0 = \hat{y}.
\end{aligned}
\tag{10.77}
$$

This modification has an analogue in numerical linear algebra, where one
calls the iteration in (10.76) the Jacobi iteration, see Section 11.3.2, and the
modified one in (10.77) the Gauss-Seidel iteration, see Section 11.3.3. Figure
10.24 shows the solution obtained with the modified Forward Euler method
with time step $\Delta t = 0.1$ and initial values $\hat{x} = 0.5$, $\hat{y} = 0.5$. The exact
solution is plotted as a dashed curve. Note that the modified Forward Euler
method is still explicit.

We now explain why this modification in Euler's method removes the
physically incorrect behavior of the original Forward Euler method. We first
rewrite the Lotka-Volterra equations (10.74) as

$$
\begin{aligned}
\dot{x} &= -xy\frac{\partial H}{\partial y}; \quad x(0) = \hat{x}, \\
\dot{y} &= xy\frac{\partial H}{\partial x}; \quad y(0) = \hat{y},
\end{aligned}
\tag{10.78}
$$

with the so-called Hamiltionian function

$$
H(x,y) = x + y - \ln x - \ln y.
$$

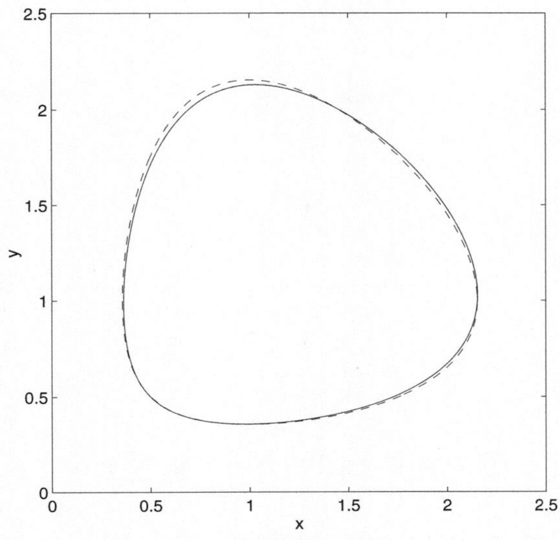

FIGURE 10.24.
Cyclic solution obtained with the modified Euler method

A system of the form (10.78) is known as a Poisson system; (if the factor xy were not present in (10.78), it would be a Hamiltonian system; See [61] for a comprehensive treatment of the numerical approximation of Hamiltonian and Poisson systems). Hamiltonian systems are symplectic, which means area-preserving in our two-dimensional example. In other words, the flow preserves the infinitesimal quantity defined by the wedge product $dx \wedge dy$. It is therefore desirable for a numerical approximation to be symplectic as well, i.e. to have a flow which also preserves that same quantity $dx \wedge dy$. In our case of a Poisson system, a similar quantity is preserved, as we show in the following theorem.

THEOREM 10.6. *The system (10.78) preserves the weighted area* $(dx \wedge dy)/xy$.

PROOF. Let Ω_0 be a subset of $\mathbb{R}^2$ at time t_0 and Ω_1 the set into which Ω_0 is mapped by (10.78) at time t_1, as shown in Figure 10.25. Preservation of $(dx \wedge dy)/xy$ is equivalent to

$$\int_{\Omega_0} \frac{1}{xy}\, dxdy = \int_{\Omega_1} \frac{1}{xy}\, dxdy.$$

We now look at the domain D in x, y, t space with the boundary ∂D given by Ω_0 at t_0, Ω_1 at t_1 and the set of trajectories emerging from the boundary

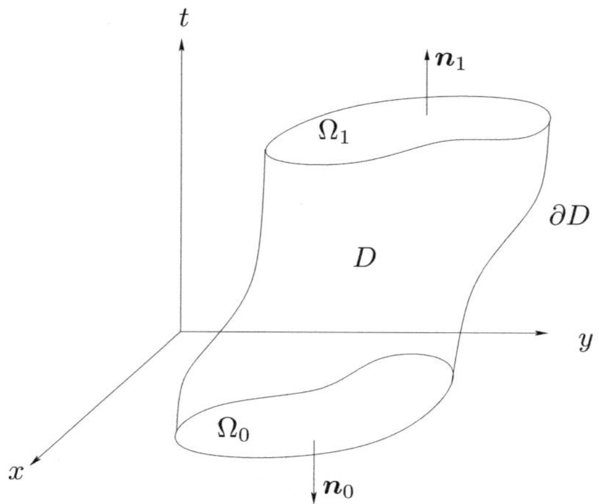

FIGURE 10.25. *Area preserving mapping*

of Ω_0 and ending on the boundary of Ω_1. Consider the vector field

$$\boldsymbol{v} := \frac{1}{xy} \begin{pmatrix} \dot{x} \\ \dot{y} \\ 1 \end{pmatrix}$$

in x, y, t space. Integrating this vector field over the boundary ∂D of D, we obtain

$$\int_{\partial D} \boldsymbol{v} \cdot \boldsymbol{n} = \int_{\Omega_0} \boldsymbol{v} \cdot \boldsymbol{n_0} + \int_{\Omega_1} \boldsymbol{v} \cdot \boldsymbol{n_1} = \int_{\Omega_1} \frac{1}{xy}\, dxdy - \int_{\Omega_0} \frac{1}{xy}\, dxdy \;,$$

where $\boldsymbol{n_0} = (0,0,-1)^T$ and $\boldsymbol{n_1} = (0,0,1)^T$ denote the outward unit normal of Ω_0 and Ω_1. There is no other contribution to the surface integral, because the vector field $\boldsymbol{v}$ is by construction parallel to the trajectories, which form the rest of the boundary ∂D. Applying the divergence theorem to the left hand side of the same equation, we get

$$\int_{\partial D} \boldsymbol{v} \cdot \boldsymbol{n} = \int_{D} \nabla \cdot \boldsymbol{v} = \int_{D} -\frac{\partial H^2}{\partial x \partial y} + \frac{\partial H^2}{\partial x \partial y} + 0 = 0 \;,$$

which concludes the proof. $\qquad\square$

We now show that the modified Euler method also preserves the same quantity.

THEOREM 10.7. *The modified Euler scheme (10.77) preserves the weighted area $(dx \wedge dy)/xy$.*

PROOF. To simplify notation, we rewrite one step of (10.77) as

$$\begin{aligned} X &= \Delta tx + x - \Delta txy, \\ Y &= -\Delta ty + y + \Delta tXy, \end{aligned} \tag{10.79}$$

where we have set $X := x_{n+1}$, $Y := y_{n+1}$, $x := x_n$ and $y := y_n$, and solved for the unknowns X and Y. Taking derivatives of both sides, we get

$$\begin{aligned} dX &= \Delta t dx + dx - \Delta t dx\, y - \Delta t x dy, \\ dY &= -\Delta t dy + dy + \Delta t dX\, y + \Delta t X dy. \end{aligned}$$

Now we can compute the wedge product $dX \wedge dY$. We obtain, after some manipulations and using the fact that $dX \wedge dX = 0$ and $dY \wedge dY = 0$,

$$\begin{aligned} dX \wedge dY &= dx \wedge dy \Big(\Delta t^3 (x - 2xy + xy^2) + \Delta t^2(-1 + 2x + y - 2xy) \\ &\quad + \Delta t(x - y) + 1 \Big). \end{aligned}$$

But the product of X and Y is

$$\begin{aligned} XY &= xy \Big(\Delta t^3(x - 2xy + xy^2) + \Delta t^2(-1 + 2x + y - 2xy) \\ &\quad + \Delta t(x - y) + 1 \Big), \end{aligned}$$

and therefore, we have

$$\frac{1}{XY} dX \wedge dY = \frac{1}{xy} dx \wedge dy,$$

and our scheme preserves the weighted area. $\qquad\square$

By KAM theory [119], a numerical method does not spiral if it preserves (weighted) area, and therefore the modified Euler method will produce physically correct "closed" trajectories. For more information, see the comprehensive treatment in [61].

10.6.2 Energy Preserving Methods

By looking at the Lotka-Volterra system written in the form (10.78), one notices that the direction of the derivatives is orthogonal to the gradient of H. Therefore the solution curve is identical to the level set of H that passes through the initial data. In other words, the Hamiltonian H is preserved and does not vary over time. For a numerical method not to spiral, we can require the method to preserve H as well. A simple calculation shows that for the modified Euler Method (10.79), the Hamiltonian H is not preserved in general:

$$\begin{aligned} H(X,Y) - H(x,y) &= -\ln(x + \Delta t(-x + xy)) - \ln(y + \Delta t(y - xy)) \\ &\quad + x + \Delta t(-x + xy) + y + \Delta t(y - xy) \\ &\quad + ln(x) + ln(y) - x - y \\ &= -\ln(1 - \Delta t + \Delta ty) - \ln(1 + \Delta t - \Delta tx) \\ &\quad - \Delta tx + \Delta ty \\ &\neq 0. \end{aligned}$$

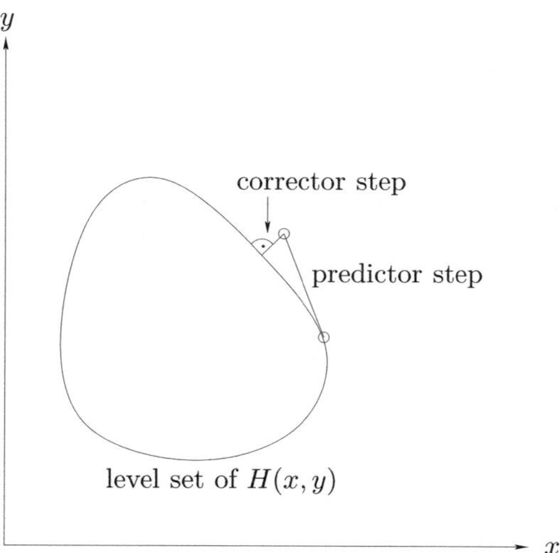

FIGURE 10.26.
Corrector step to get back onto the level set of H

However, if we add a small correction to the Euler step, we can make it H-preserving. We take Euler's Method as a predictor

$$\begin{aligned}
X_p &= \Delta tx + x - \Delta txy, \\
Y_p &= -\Delta ty + y + \Delta tX_py
\end{aligned}$$

and we do a corrector step to get back on the level set, as shown in Figure 10.26. A good direction for the correction is the gradient of H, which we use to project the predicted value back onto the level set of H. This leads to the corrector step

$$\begin{aligned}
X &= X_p + \alpha\frac{\partial H}{\partial x}(X,Y). \\
Y &= Y_p + \alpha\frac{\partial H}{\partial y}(X,Y),
\end{aligned}$$

where α has to be determined such that H is preserved. Note that these equations are implicit in X and Y, but we can replace the values on the right hand side by the predicted values X_p and Y_p, which only perturbs the projection direction slightly,

$$\begin{aligned}
X &= X_p + \alpha\frac{\partial H}{\partial x}(X_p, Y_p), \\
Y &= Y_p + \alpha\frac{\partial H}{\partial y}(X_p, Y_p).
\end{aligned} \tag{10.80}$$

To determine α, we have to solve the nonlinear equation

$$H(X(x,y,\Delta t,\alpha), Y(x,y,\Delta t,\alpha)) - H(x,y) = 0.$$

We are interested in a root close to zero, because we expect the error of the predicted values to be small. In most cases, searching within the interval $[-\Delta t, \Delta t]$ is a good choice. It remains to prove that the order of accuracy of the predictor method is not changed by the corrector step.

THEOREM 10.8. *The order of accuracy of the corrected H-preserving method is $O(\Delta t)$.*

PROOF. The local truncation error $\tau(x,y,t)$ is defined by the equation

$$\frac{X-x}{\Delta t} - \frac{\alpha}{\Delta t}\left(1 - \frac{1}{x+\Delta t}(-x+xy)\right) = \dot{x} + \tau(x,y,t).$$

Since we know that Euler has a local truncation error of order Δt^2, it suffices to show that the correction is also of order Δt^2, or equivalently that $\alpha = \alpha(\Delta t)$ is of order $O(\Delta t^2)$. Define

$$
\begin{aligned}
F(x,y,\alpha,\Delta t) \quad &:= \quad H(X,Y) - H(x,y) \\
&= \quad -\ln\left(XP + \alpha\left(1 - \tfrac{1}{XP}\right)\right) - \ln\left(YP + \alpha\left(1 - \tfrac{1}{YP}\right)\right) \\
&\quad + XP + \alpha\left(1 - \tfrac{1}{XP}\right) + YP + \alpha\left(1 - \tfrac{1}{YP}\right) \\
&\quad + \ln x + \ln y - x - y,
\end{aligned}
$$

$$(10.81)$$

where X and Y are given by (10.80). We expand $\alpha(x,y,\Delta t)$ in a Taylor series about $\Delta t = 0$,

$$\alpha(x,y,\Delta t) = \alpha(x,y,0) + \Delta t\frac{\partial \alpha}{\partial \Delta t}(x,y,0) + O(\Delta t^2).$$

Setting Δt in (10.81) to zero gives $\alpha(x,y,0) = 0$. To compute the first derivative, we can apply implicit differentiation

$$\frac{\partial \alpha}{\partial \Delta t} = -\frac{F_{\Delta t}}{F_\alpha}.$$

After a longer computation, we find $\frac{\partial \alpha}{\partial \Delta t}(x,y,0) = 0$, and we have shown that

$$\alpha(x,y,\Delta t) = O(\Delta t^2),$$

which concludes our proof. $\qquad\square$

Figure (10.27) shows the solution obtained with the H-preserving scheme. The trajectory is identical with the trajectory of the exact solution, and the only error is a time shift in this simple two-dimensional example.

The simple example of the Lotka-Volterra equations only gives a glimpse of the rich area of geometric numerical integration. For a complete treatment of Hamiltonian and Poisson systems and symplectic integrators, integration methods on manifolds, geometric properties of symmetric integration methods and reversibility, and highly oscillatory problems, see the reference book [61].

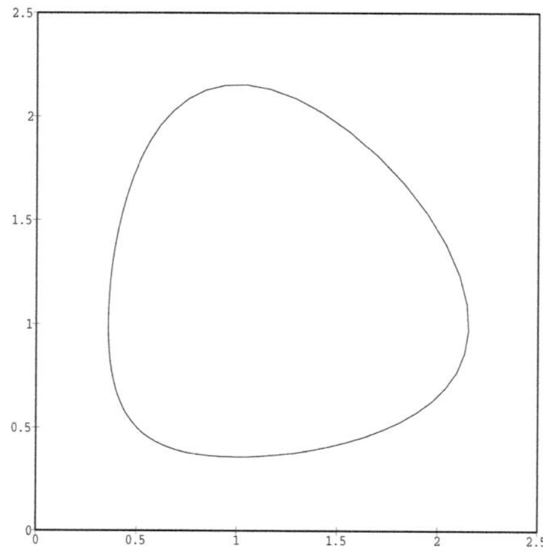

FIGURE 10.27.
Numerical result of an H preserving numerical method

10.7 Delay Differential Equations

Differential equations are obtained when one assumes that the rate of change of a physical system only depends on its present state and not on the past. This is often only an approximation; in some instances, a more realistic equation would also take into account the value of the solution with a certain delay, i.e.

$$\dot{y}(t) = f(y(t), y(t - \tau)), \quad \tau \geq 0. \tag{10.82}$$

Such a problem does not only need an initial condition, $y(0) = y_0$, but an entire initial function $y(t) = y_0(t)$ for $t \in [-\tau, 0]$. Typical examples of such equations occur in models with feedback, or in biology, because the size of a new generation depends on the size of the past generation, and procreation is rarely possible instantaneously. A well-studied example is the blow fly equation, which models the density of blow flies:

$$\dot{u} = -\delta u + \nu \Delta u + \rho u(t - \tau)e^{-\alpha u(t-\tau)}.$$

Here, $\delta > 0$ is the death rate of the blow fly, $\nu > 0$ is the diffusion rate at which the blow fly migrates randomly, and $\alpha, \rho > 0$ are growth rates. This equation contains partial derivatives and is too complex for illustrating the essential properties of delay differential equations, so we instead consider the much simpler model problem from [62],

$$\dot{y}(t) = -y(t - 1), \quad y(t) = 1 \text{ for } -1 \leq t \leq 0.$$

This means that $\dot{y} = -1$ for $0 \leq t \leq 1$, and we can simply integrate to obtain $y(t) = -t + C$ with C some constant. (If the right hand side depended on

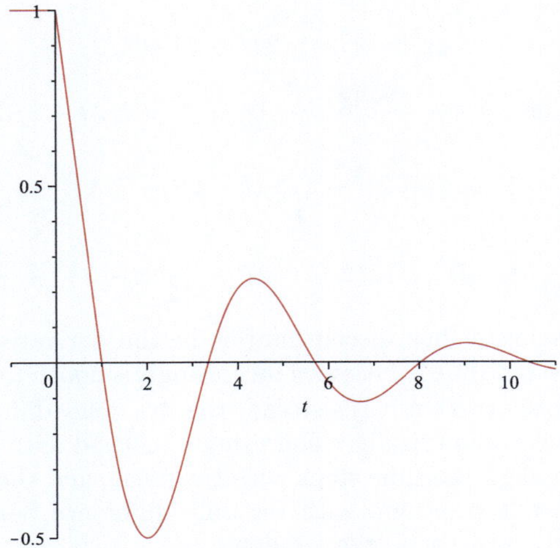

FIGURE 10.28.
Analytical solution of a simple delay equation

$y(t)$ as well, one would have to solve the corresponding ODE.) Using the initial condition $y(0) = 1$, we find $C = 1$, and therefore $y(t) = 1 - t$ for $0 \leq t \leq 1$. We can now proceed the same way for $1 \leq t \leq 2$, and obtain $y(t) = \frac{1}{2}t^2 - 2t + \frac{3}{2}$ for $1 \leq t \leq 2$. We see that solving a delay differential equation with a delay bounded away from zero amounts to solving differential equations on time intervals defined by the delay. We show in Figure 10.28 the solution obtained with the MAPLE commands

ALGORITHM 10.20.
Analytical solution of a simple delay equation

```
N:=10;
y[-1]:=t->1;
for i from 0 to N do
  c:='c';r:=int(-y[i-1](t-1),t)+c;
  rr:=unapply(r,t);
  c:=solve(rr(i)=y[i-1](i),c);
  y[i]:=unapply(r,t);
od;
for i from -1 to N do
  P[i]:=plot(y[i](t),t=i..i+1);
od;
plots[display](seq(P[j],j=-1..N));
```

The first few lines of the result obtained are

$$y_0 := t \mapsto -t + 1$$

$$y_1 := t \mapsto 1/2\,t^2 - 2\,t + 3/2$$

$$y_2 := t \mapsto -1/6\,(t-1)^3 + t^2 - 7/2\,t + 8/3$$

$$y_3 := t \mapsto 1/24\,(t-2)^4 - 1/3\,(t-1)^3 + 7/4\,t^2 - \frac{37}{6}\,t + \frac{125}{24}$$

$$y_4 := t \mapsto -\frac{1}{120}\,(t-3)^5 + 1/12\,(t-2)^4 - \frac{7}{12}\,(t-1)^3 + \frac{37}{12}\,t^2 - \frac{91}{8}\,t + \frac{54}{5}$$

Note that the solution has discontinuities in the derivatives at the points where it is connected, but it becomes increasingly smooth as time progresses.

This example shows that it is also easy to solve delay differential equations with a fixed delay τ numerically when using a method with a fixed time step Δt: it suffices to align the time steps with the delay, and then to proceed like in the above MAPLE program, with the only difference being that one now solves numerically over each time window.

Things get substantially more complicated when the time delay is variable, $\tau = \tau(t)$, or if one wants to use a variable step size Δt. In that case, one also needs approximations of the function at earlier time points where the solution has not necessarily been computed by the numerical method. For a fully general method, one needs a global approximation of the solution, such as one provided by linear multistep methods or by modern Runge-Kutta methods with dense output, see [62].

10.8   Problems

PROBLEM 10.1. *Solve the differential equation by the Taylor expansion method:*

$$y''' - x^2 y = \frac{2}{x+1}, \quad y(0) = 2, y'(0) = 0, y''(0) = 1.$$

Compute the first 6 coefficients of $y(x) = \sum_k a_k x^k$. *Check your result using* MAPLE.

PROBLEM 10.2. *Transform the following system of differential equations into a first order system.*

$$\begin{aligned} r'' - r\Theta^2 &= -\frac{2}{r^2} \\ r\Theta'' + 2r'\Theta' &= 0 \end{aligned}$$

PROBLEM 10.3. *A dog would like to cross a river of width b. He starts at point* $(b, 0)$ *with the goal to swim to* $(0, 0)$ *where he has found a sausage. He swims at a constant speed* v_D *and his nose always points to the sausage. The* *river flows north in direction of the y-axis at a velocity v_R that is constant everywhere.*

a) *Derive the differential equation describing the trajectory $z(t) = (x(t), y(t))^{\top}$ of the dog.*

b) *Program a MATLAB function $zp=dog(t,z)$ describing the differential equation. The velocities v_D and v_R may be declared as global variables.*

c) *Use the built-in MATLAB function `quiver` and plot the slope field for $b = 1$, $v_R = 1$ and for the different swimming speeds $v_D = 0.8$, 1.0 and 1.5.*

 Note: `quiver(X,Y,Xp,Yp)` needs 4 matrices. X and Y contain the coordinates of the points and Xp and Yp the two components of the velocity at that point. To compute these you can use the function dog e.g.

```
z=dog(0,[X(k,j),Y(k,j)]); Xp(k,j)=z(1); Yp(k,j)=z(2);
```

d) *Develop a MATLAB integrator for the method of Heun of order 2*

```
function y=OdeHeun(f,y0,tend,n)
% ODEHEUN solve ODE using Heun's method
%    y=OdeHeun(f,y0,tend,n) integrates y'=f(t,y),
%    y(0)=y0 with Heun from t=0 to tend using a fixed
%    step size h=tend/n
```

 which integrates a given system of differential equations $\boldsymbol{y'} = \boldsymbol{f}(t, \boldsymbol{y})$ and stores the results in the matrix Y. The i-th row of the matrix Y contains the values
$$[t_i, y_1(t_i), \ldots, y_n(t_i)].$$

 Compute and plot the trajectories for the three swimming speeds. You may want to stop the integration before executing all n steps when the dog arrives close to the origin or, in the case of $v_D < v_R$, when the dog is near the y-axis.

PROBLEM 10.4. *Apply the method*

$$
\begin{array}{c|ccc}
0 & & & \\
1 & 1 & & \\
\frac{1}{2} & \frac{1}{4} & \frac{1}{4} & \\
\hline
 & \frac{1}{6} & \frac{1}{6} & \frac{4}{6}
\end{array}
$$

to the differential equation

$$\boldsymbol{y'}(t) = A\boldsymbol{y}(t), \quad \boldsymbol{y}(0) = \boldsymbol{y}_0$$

Compute the expression $F(Ah)$ one obtains when integrating one step with step size h:

$$\boldsymbol{y}_1 = F(Ah)\boldsymbol{y}_0.$$

PROBLEM 10.5. *The differential equation*

$$y'(t) = g(t), \quad y(t_0) = 0 \tag{10.83}$$

has the solution

$$y(t) = \int_{t_0}^{t} g(t)\, dt.$$

Which quadrature rules do you get when integrating (10.83) by the methods of Euler, Heun (order2) and classical Runge Kutta? Perform one step with step size h with the three methods.

PROBLEM 10.6. *Determine the conditions for an explicit Runge Kutta method to be second order without the assumption (10.22) on the coefficients c_1 and c_2. Discuss the utility of this extra degree of freedom.*

PROBLEM 10.7. *Implement Heun's method as well as the second possibility found in Table 10.1, and compare their performance on the Arenstorf problem in Section 10.3.5. Compute on finer and finer meshes and plot the error in loglog scale. Which method is better? Are there other advantages/disadvantages for the two methods?*

PROBLEM 10.8. *Implement the Runge-Kutta-Fehlberg adaptive integration method from Section 10.3.5, and test it on the Arenstorf orbit model problem. How many function evaluations do you need to reach the same accuracy as with* ode12*?*

PROBLEM 10.9. *Implement the Dormand-Prince adaptive integration method from Section 10.3.5, and test it on the Arenstorf orbit model problem. Is this method more accurate than the Runge-Kutta-Fehlberg method?*

PROBLEM 10.10. *Show that for the second backward differentiation formula*

$$\frac{y_{k+1} - y_{k-1}}{2\Delta t} = -10 y_k,$$

the local truncation error is $O(\Delta t^3)$, and that the method is zero-stable.

PROBLEM 10.11. *Let A be a matrix with the Jordan decomposition $T^{-1}AT = J$,*

$$J = \begin{pmatrix} \lambda & 1 & \\ & \lambda & 1 \\ & & \lambda \end{pmatrix}.$$

Show that for any given $\varepsilon > 0$, there exists a diagonal scaling matrix D such that $(TD)^{-1}ATD = \tilde{J}$ is a new Jordan decomposition of A with

$$\tilde{J} = \begin{pmatrix} \lambda & \varepsilon & \\ & \lambda & \varepsilon \\ & & \lambda \end{pmatrix}.$$

PROBLEM 10.12. *Prove that the second and third Taylor methods are convergent of second and third order respectively.*

PROBLEM 10.13. *Prove that the the implicit trapezoidal method (10.41) is second order accurate.*

PROBLEM 10.14. *Prove that the the implicit midpoint method (10.42) is second order accurate.*

PROBLEM 10.15. *For the Backward Euler method, the Trapezoidal method and the Adams-Moulton two step method do the following:*

1. *Compute the local truncation error.*

2. *Check if the method is zero-stable.*

3. *Compute the region of absolute stability (Note: For the Adams-Moulton two-step method, it is hard to determine the region of absolute stability analytically from the formula for the roots of the characteristic polynomial. Instead, to find the region of absolute stability numerically, you can use the following program, which tries points in the complex plane and then gives you a picture of the area where all the roots given by the function f have magnitude less than one:*

```
function A=RAS(f,X,Y,n);
% RAS computes the region of absolute stability
%    y=RAS(f,X,Y,n); computes the region of absolute stability
%    where f is a vector valued function giving the values of
%    all the roots. The area [X]x[Y] is scanned with nxn points
%    and the point is ploted gray if all the roots for that
%    value are less than 1 in magnitude.

dx=(X(2)-X(1))/n;
dy=(Y(2)-Y(1))/n;
A=zeros(n+1,n+1);
for j=1:n+1;
  for k=1:n+1;
    mu=X(1)+(j-1)*dx+i*(Y(1)+(k-1)*dy); % i=sqrt(-1)
    la=feval(f,mu);
    if max(abs(la))<1,
       A(k,j)=1;
```

```
      else
        A(k,j)=0;
      end;
   end;
 end;
 x=(X(1):dx:X(2));
 y=(Y(1):dy:Y(2));
 colormap([0 0 0;0.5 0.5 0.5]);
 image(x,y,A*2);
```

You may also use this program for the other two methods to check if your analytic reasoning is correct.

4. *Is the method A-stable?*

PROBLEM 10.16.

Implement the following methods to solve a system of ordinary differential equations:

- *Forward Euler. Use as header*

```
function [t,u]=Euler(f,T0,Tfinal,u0,dt);
% EULER solves ordinary differential equation using Forward Euler
%   [t,u]=Euler(f,T0,Tfinal,u0,dt) solves the differential
%   equation du/dt=f(u), u(T0)=u0 with the Forward Euler method
%   up to time Tfinal using the time step dt
```

- *Adams-Bashforth two step method. To get the second starting value, use one Euler step.*

- *Heun's method*

1. *Apply the above methods to the problem of the jogger and the dog to find the trajectory of the dog. The differential equation for this problem is:*

$$\dot{x} = \frac{w}{\sqrt{(\xi_0 + tv - x)^2 + y^2}}(\xi_0 + tv - x)$$

$$\dot{y} = \frac{w}{\sqrt{(\xi_0 + tv - x)^2 + y^2}}(-y).$$

Use as initial condition for the dog $x_0 = 60$ and $y_0 = 70$ and as speed of the dog $w = 10$. The jogger starts running at $\xi_0 = 0$ at constant speed $v = 8$.

2. *Compare your results for $\Delta t = 1$ and $\Delta t = 0.1$ with the highly accurate solution obtained from the built-in solver for ordinary differential equations ode45 (type help ode45) in* MATLAB. *Which method is closest to the* MATLAB *solution ?*

3. *Why would it be much harder to implement Backward Euler and use it to solve the above problem ?*

PROBLEM 10.17. *Discretize the system of differential equations*

$$z' = Az, \quad A = \begin{pmatrix} 0 & 1 \\ -100 & -101 \end{pmatrix}, \quad z(0) = \begin{pmatrix} 1 \\ -1 \end{pmatrix}$$

using the explicit midpoint rule. Solve the recurrence relation you obtain analytically, and determine the step size needed in order for the method to have an error smaller than 0.01 at $z(0.6)$. Could you have run the method in MATLAB *with such a small step size ?*

PROBLEM 10.18. *Solve the Lotka-Volterra system of ordinary differential equations (10.74) in parametric form to obtain (10.75). Hint: use the approach shown in [129].*

PROBLEM 10.19. *Apply the Runge and Heun methods to approximately solve the Lotka-Volterra system of ordinary differential equations (10.74). Plot the solution in phase space. What do you observe?*

PROBLEM 10.20. *Apply the Backward Euler method to approximately solve the Lotka-Volterra system of ordinary differential equations (10.74). Plot the solution in phase space. What do you observe?*

Chapter 11. Iterative Methods for Linear Systems

> *Schwerlich werden Sie je wieder direct eliminieren, wenigstens nicht, wenn Sie mehr als 2 Unbekannte haben. Das indirecte Verfahren lässt sich halb im Schlaf ausführen, oder man kann während desselben an andere Dinge denken[1].*
>
> C. F. Gauss, in a letter to his friend Gerling, 1823.
>
> *With the growth in speed and complexity of modern digital computers has come an increase in the use of computers by those who wish to find or approximate the solutions of partial differential equations in several variables. [...] The author was fortunate to have been associated with the Mathematics Group of the Bettis Atomic Power Laboratory where very large matrix problems (of order 20'000 in two dimensions!) are solved on fast computers in the design of nuclear reactors.*
>
> R. Varga, Matrix Iterative Analysis, Prentice-Hall, 1962.
>
> *Until recently, direct solution methods were often preferred to iterative methods in real applications because of their robustness and predictable behavior. However, a number of efficient iterative solvers were discovered and the increased need for solving very large linear systems triggered a noticeable and rapid shift toward iterative techniques in many applications. [...] It was found that the combination of preconditioning and Krylov subspace iterations could provide efficient and simple general- purpose procedures that could compete with direct solvers.*
>
> Y. Saad, Iterative Methods for Sparse Linear Systems, SIAM, 2003.
>
> *Quite frequently in life, the most elegant solution to a problem also turns out to be the most efficient one for practical purposes. This heuristical observation certainly applies to Krylov subspace methods.*
>
> J. Liesen and Z. Strakos, Krylov Subspace Methods, Principles and Analysis, Oxford University Press, 2012.

Prerequisites: Chapter 3 is required, in particular the section on symmetric positive definite matrices (§3.4). Some properties of eigenvalues (§7.2) are also needed.

The solution of sparse linear systems by iterative methods has become one of the core application and research areas of high performance computing. The size of systems that are solved routinely has increased tremendously over time, see the quotes by Gauss and Varga above. This is because the discretization of partial differential equations can lead to systems that are

---

[1] You will in the future hardly ever eliminate directly anymore, at least not when you have more than two unknowns. The indirect procedure can be done while one is half asleep, or one can think about other things while doing it.

W. Gander et al., *Scientific Computing - An Introduction using Maple and MATLAB*,
Texts in Computational Science and Engineering 11,
DOI 10.1007/978-3-319-04325-8_11,

arbitrarily large. Iterative methods for solving linear systems can be divided into the older class of stationary iterative methods, and the more modern class of non-stationary methods. The class of stationary iterative methods, the first of which was invented by Gauss, are nowadays rarely used as standalone solvers, although they remain core components of more sophisticated algorithms, such as multi-grid or domain decomposition methods. They are also important for preconditioning the newer class of non-stationary iterative methods, in particular Krylov methods. We start in Section 11.1 with an introductory example on the propagation of heat in an enclosed space. Using approximation techniques for derivatives from Chapter 8, we obtain a large sparse linear system that one has to solve at each time step, and such systems are best solved by iterative methods. We then show in Section 12.2 how matrix splittings can be used to derive fixed point iterations of the same type as those seen in Section 5.2.2 for nonlinear problems. We introduce the basic concepts of residual, error and difference of iterates, and present the classical convergence criteria for such methods. In Section 11.3 we introduce the reader to the general theory of regular splittings, and present the classical methods of Jacobi, Gauss-Seidel, SOR and Richardson. This method of Richardson is actually much simpler than the one originally proposed by Richardson, which is a non-stationary iterative method; we will see this in Section 11.4, along with two other non-stationary methods, those of Conjugate Residuals and Steepest Descent. Each of the above methods tries to do the locally optimal thing at each step; while this seems to be a good tactic, it is generally not the best strategy[2]. Instead, a better strategy would be to optimize convergence globally over many steps, based on information known a priori or gathered during the iterations. A global optimization strategy based on a priori spectral information uses Chebyshev polynomials; the resulting Chebyshev semi-iterative method, which we show in Section 11.5, realizes the dream Richardson had when he presented his method, unaware of the results of Chebyshev. In general, global strategies can be formulated using extrapolation techniques, which we show in Section 11.6 following the approach of Brezinski [11]. This leads naturally to the Krylov subspace methods in Section 11.7, which are the best iterative methods for linear systems known today. We briefly mention the fundamental idea of preconditioning in Section 11.8; a complete treatment of this important topic would easily fill an entire textbook by itself.

The authoritative book on stationary iterative methods is still the book 'Matrix Iterative Analysis' by Varga [142]. Both stationary and non-stationary iterative methods are treated in the comprehensive book by Saad [118], and an excellent, recent research monograph on Krylov subspace methods is the book by Strakos and Liesen [86].

---

[2] According to Eduard Stiefel, who was professor at ETH but also a colonel in the Swiss Army.

11.1 Introductory Example

The solution of partial differential equations by *discretization* often leads to large and sparse linear systems of equations. To illustrate this, let us consider the propagation of heat in an enclosed space, e.g., a room in a building. The temperature is a function of time t and space $\boldsymbol{x}$, which we denote by $u(t, \boldsymbol{x})$. Heat evolves from some initial distribution to an equilibrium for $t \to \infty$. This evolution is described by the *heat equation*,

$$u_t = c\Delta u \quad \text{in some domain } \Omega,$$

together with appropriate initial and boundary conditions, for example $u(0, \boldsymbol{x}) = u_0(\boldsymbol{x})$, a given initial temperature, and $u = g$ on the boundary $\partial\Omega$.

If we are only interested in the stationary equilibrium, then for $t \to \infty$ we have $u_t = 0$ and the equation simplifies to

$$\Delta u = 0 \quad \text{in } \Omega, \tag{11.1}$$

together with the boundary condition $u = g$ on $\partial\Omega$.

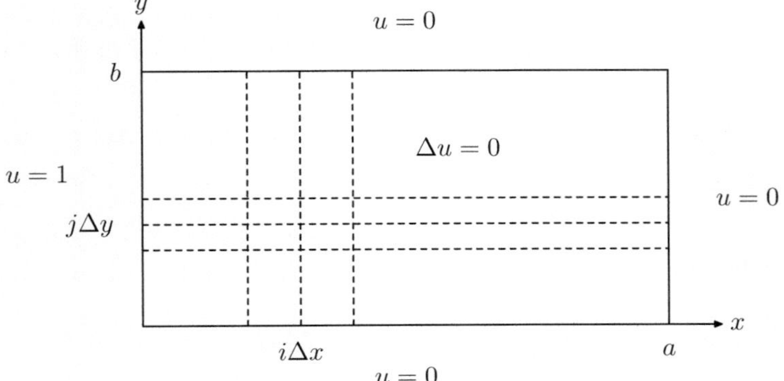

FIGURE 11.1. *Heat equation*

Let us consider a simple two dimensional example, where the domain Ω is a rectangle with sides a and b and the stationary heat distribution is a function of two variables $u(x, y)$, see Figure 11.1, where the boundary conditions are also displayed. We discretize by choosing the mesh sizes

$$\Delta x = \frac{a}{m+1}, \quad \Delta y = \frac{b}{n+1}$$

and search for an approximation $u_{ij} \approx u(i\Delta x, j\Delta y)$. The *Laplace operator* in two dimensions,

$$\Delta u = \frac{\partial^2 u}{\partial x^2} + \frac{\partial^2 u}{\partial y^2},$$

is discretized using finite differences, see Chapter 8, Equation (8.11):

$$\frac{\partial^2 u}{\partial x^2} \approx \frac{u(x + \Delta x, y) - 2u(x, y) + u(x - \Delta x, y)}{\Delta x^2},$$

and similarly

$$\frac{\partial^2 u}{\partial y^2} \approx \frac{u(x, y + \Delta y) - 2u(x, y) + u(x, y - \Delta y)}{\Delta y^2}.$$

Introducing these approximations into (11.1) and multiplying with $\Delta x^2 \Delta y^2$, we obtain for the approximation at point $(i\Delta x, j\Delta y)$ the equation

$$u_{i-1,j}\Delta y^2 - 2\Delta y^2 u_{i,j} + \Delta y^2 u_{i+1,j} + u_{i,j-1}\Delta x^2 - 2\Delta x^2 u_{i,j} + \Delta x^2 u_{i,j+1} = 0.$$
$$(11.2)$$

We now introduce the vectors

$$\boldsymbol{u}_j = \begin{pmatrix} u_{1,j} \\ \vdots \\ u_{m,j} \end{pmatrix},$$

which contain the values $u_{i,j}$ for $i = 1, \ldots, m$ on a horizontal line corresponding to j. Writing the equation (11.2) for each point i on the line, we obtain

$$i = 1 : -2(\Delta x^2 + \Delta y^2)u_{1,j} + u_{2,j}\Delta y^2 + u_{1,j-1}\Delta x^2 + u_{1,j+1}\Delta x^2 = -u_{0,j}\Delta y^2$$
$$i = 2 : u_{1,j}\Delta y^2 - 2(\Delta x^2 + \Delta y^2)u_{2,j} + \Delta y^2 u_{3,j} + u_{2,j-1}\Delta x^2 + u_{2,j+1}\Delta x^2 = 0$$
$$\cdots$$

Dividing by Δx^2 and writing this for $i = 1, \ldots, m$ as a vector equation, we get

$$\boldsymbol{u}_{j-1} + T\boldsymbol{u}_j + \boldsymbol{u}_{j+1} = -\left(\frac{\Delta y}{\Delta x}\right)^2 \boldsymbol{e}_1, \qquad (11.3)$$

where we have the unit vector $\boldsymbol{e}_1$ because of the left boundary condition, where $u = 1$, and the tridiagonal matrix T is given by

$$T = \begin{pmatrix} -2\gamma & \delta & & \\ \delta & -2\gamma & \ddots & \\ & \ddots & \ddots & \delta \\ & & \delta & -2\gamma \end{pmatrix} \quad \text{with} \quad \delta = \left(\frac{\Delta y}{\Delta x}\right)^2 \quad \text{and} \quad \gamma = 1 + \delta.$$

If we now introduce the vector of all unknowns

$$\boldsymbol{u} = \begin{pmatrix} \boldsymbol{u}_1 \\ \vdots \\ \boldsymbol{u}_n \end{pmatrix},$$

and write the equations (11.3) all stacked on top of one another, then we obtain a large sparse system of linear equations with $m \times n$ equations for $m \times n$ unknowns,

$$A\boldsymbol{u} = \boldsymbol{f},$$

with the block tridiagonal matrix

$$A = \begin{bmatrix} T & I & & \\ I & T & \ddots & \\ & \ddots & \ddots & I \\ & & I & T \end{bmatrix}, \tag{11.4}$$

and the right hand side vector $\boldsymbol{f}$ containing the boundary conditions. The matrix A is sparse, with only 5 nonzero elements per row! Moreover, A has a regular structure. If we assume e.g. that $m = 100$ and $n = 50$ then we have 5'000 unknowns and 25 million matrix entries, of which only 25'000 (1%) are nonzero.

The basic idea of an iterative method is to generate a sequence of iterates

$$\boldsymbol{u}^{(k+1)} = F_k(\boldsymbol{u}^{(k)}), \qquad k = 0, 1, 2, \ldots \tag{11.5}$$

such that $\boldsymbol{u}^{(k)} \to \boldsymbol{u}$, the solution of $A\boldsymbol{u} = \boldsymbol{f}$. In contrast to the direct methods we have seen in Chapter 3, iterative methods do not need to manipulate the matrix A explicitly; it is sufficient to have a rule that computes the product $\boldsymbol{y} = A\boldsymbol{x}$ for any given vector $\boldsymbol{x}$. Thus, the user of an iterative method is not required to store the matrix A in the memory of the computer; instead, he only needs to *provide a program for the operator* `y = multiply(A,x)`, which returns the result for a given vector $\boldsymbol{x}$ after transformation by A.

11.2 Solution by Iteration

Like the fixed point iteration we encountered in the solution of non-linear equations in Section 5.2.2, linear systems can be solved by fixed point iteration. To do so, one has to transform the system of linear equations into fixed point form, which is achieved in general by splitting the matrix into two parts.

11.2.1 Matrix Splittings

One way to derive an iterative method for solving the linear system of equations

$$A\boldsymbol{x} = \boldsymbol{b}, \quad A \in \mathbb{R}^{n \times n}, \ \boldsymbol{b} \in \mathbb{R}^n,$$

is to *split the matrix* into $A = M - N$. Assuming that M is invertible, this splitting induces an iterative method as follows. Starting with some initial approximation $\boldsymbol{x}_0$, we iterate by solving for $\boldsymbol{x}_{k+1}$

$$M\boldsymbol{x}_{k+1} = N\boldsymbol{x}_k + \boldsymbol{b}, \quad k = 0, 1, 2, \ldots, \tag{11.6}$$

and hope that $\boldsymbol{x}_k$ will converge to the desired solution. This is equivalent to letting $F_k(\boldsymbol{x}) = M^{-1}N\boldsymbol{x} + M^{-1}\boldsymbol{b}$ in (11.5) for all k. An iterative method is called *stationary* if it can be expressed in the simple form (11.6), where neither M nor N depends on the iteration count k. Iteration (11.6) is a *single-stage iteration*, since $\boldsymbol{x}_{k+1}$ depends only on *one* former iterate $\boldsymbol{x}_k$.

Of course, to obtain an efficient method, the *matrix splitting* must be such that solving linear systems with the matrix M requires fewer operations than for the original system. On the other hand, we would like to minimize the number of iterations. In the extreme case where we chose $M = A$, and thus $N = 0$, we converge in one iteration to the solution. Therefore, the choice of the splitting should be such that

- M is a good approximation of A,

- $M\boldsymbol{x} = \boldsymbol{y}$ is easy and cheap to solve.

The fact that we have two generally conflicting criteria means that a compromise is usually required.

11.2.2 Residual, Error and the Difference of Iterates

We will write iterative methods based on the splitting $A = M - N$ in the *standard form*

$$M\boldsymbol{x}_{k+1} = N\boldsymbol{x}_k + \boldsymbol{b} \iff \boldsymbol{x}_{k+1} = M^{-1}N\boldsymbol{x}_k + M^{-1}\boldsymbol{b}. \tag{11.7}$$

DEFINITION 11.1. (ITERATION MATRIX) *For any matrix splitting $A = M - N$ with M invertible, the matrix $G := M^{-1}N$ is called the* iteration matrix.

Since

$$M^{-1}N\boldsymbol{x}_k + M^{-1}\boldsymbol{b} = M^{-1}(M - A)\boldsymbol{x}_k + M^{-1}\boldsymbol{b} = \boldsymbol{x}_k + M^{-1}(\boldsymbol{b} - A\boldsymbol{x}_k),$$

we can also write the iteration in the *correction form*

$$\boldsymbol{x}_{k+1} = \boldsymbol{x}_k + M^{-1}\boldsymbol{r}_k, \tag{11.8}$$

where $\boldsymbol{r}_k = \boldsymbol{b} - A\boldsymbol{x}_k$ is called the *residual*. The residual is a measure of how good the approximation $\boldsymbol{x}_k$ is, see Subsection 2.8.3. The matrix M is called a *preconditioner*, since by iterating with (11.8), if $\boldsymbol{x}_k \to \boldsymbol{x}_\infty$, then $\boldsymbol{x}_\infty$ is the solution of the *preconditioned system*

$$M^{-1}A\boldsymbol{x} = M^{-1}\boldsymbol{b}.$$

Subtracting the iteration from the split system,

$$\left.\begin{array}{rl} M\boldsymbol{x} &= N\boldsymbol{x} + \boldsymbol{b} \\ M\boldsymbol{x}_{k+1} &= N\boldsymbol{x}_k + \boldsymbol{b} \end{array}\right\} \Rightarrow M(\boldsymbol{x} - \boldsymbol{x}_{k+1}) = N(\boldsymbol{x} - \boldsymbol{x}_k),$$

and introducing the *error* $\boldsymbol{e}_k := \boldsymbol{x} - \boldsymbol{x}_k$, we obtain a *recurrence for the error*,

$$M\boldsymbol{e}_{k+1} = N\boldsymbol{e}_k \iff \boldsymbol{e}_{k+1} = M^{-1}N\boldsymbol{e}_k. \tag{11.9}$$

From (11.8), it follows that

$$\underbrace{\boldsymbol{b} - A\boldsymbol{x}_{k+1}}_{\boldsymbol{r}_{k+1}} = \underbrace{\boldsymbol{b} - A\boldsymbol{x}_k}_{\boldsymbol{r}_k} - AM^{-1}\boldsymbol{r}_k,$$

and we get a *recurrence for the residual vectors*,

$$\boldsymbol{r}_{k+1} = (I - AM^{-1})\boldsymbol{r}_k = (I - AM^{-1})^k \boldsymbol{r}_0. \tag{11.10}$$

Note that $M - A = N$, and therefore

$$I - AM^{-1} = NM^{-1} = M(M^{-1}N)M^{-1},$$

which shows that the iteration matrices $I - AM^{-1}$ in (11.10) and $M^{-1}N$ in (11.9) are similar, and thus have the same eigenvalues.

Consider the *difference of consecutive iterates*,

$$\boldsymbol{u}_k = \boldsymbol{x}_{k+1} - \boldsymbol{x}_k.$$

Using the iteration (11.7), we get

$$\begin{aligned}
\boldsymbol{u}_k &= \boldsymbol{x}_{k+1} - \boldsymbol{x}_k = M^{-1}N\boldsymbol{x}_k + M^{-1}\boldsymbol{b} - M^{-1}N\boldsymbol{x}_{k-1} - M^{-1}\boldsymbol{b} \\
&= M^{-1}N(\boldsymbol{x}_k - \boldsymbol{x}_{k-1}) \\
&= M^{-1}N\boldsymbol{u}_{k-1}.
\end{aligned}$$

Hence the differences between consecutive iterates obey the same recurrence as the error. Furthermore, from

$$\begin{aligned}
\boldsymbol{u}_k &= M^{-1}N(\boldsymbol{x}_k - \boldsymbol{x} + \boldsymbol{x} - \boldsymbol{x}_{k-1}) \\
&= M^{-1}N(-\boldsymbol{e}_k + \boldsymbol{e}_{k-1}) = -M^{-1}N\boldsymbol{e}_k + \boldsymbol{e}_k \\
&= (I - M^{-1}N)\boldsymbol{e}_k,
\end{aligned}$$

we see that the difference between consecutive iterates and the true error are related by

$$\boldsymbol{u}_k = M^{-1}A\boldsymbol{e}_k.$$

Finally, from (11.8) we have $M\boldsymbol{u}_k = \boldsymbol{r}_k$, a connection between the difference of consecutive iterates and the residual. We summarize these results in the following theorem.

THEOREM 11.1. (ERROR, RESIDUAL AND DIFFERENCE OF ITERATES) *For a non-singular matrix $A \in \mathbb{R}^{n \times n}$ and $\boldsymbol{b} \in \mathbb{R}^n$, let $\boldsymbol{x} \in \mathbb{R}^n$ be the solution of $A\boldsymbol{x} = \boldsymbol{b}$, and let $A = M - N$ be a splitting with M non-singular. Choose $\boldsymbol{x}_0$ and compute the sequence of iterates $\boldsymbol{x}_{k+1} = M^{-1}N\boldsymbol{x}_k + M^{-1}\boldsymbol{b}$. Let*

$e_k := x - x_k$ be the error and $r_k := b - Ax_k$ be the residual at step k, and let $u_k := x_{k+1} - x_k$ be the difference of two consecutive iterates. Then these vectors can be computed by the recurrences

$$e_{k+1} \;=\; M^{-1}Ne_k, \tag{11.11}$$

$$u_{k+1} \;=\; M^{-1}Nu_k, \tag{11.12}$$

$$r_{k+1} \;=\; (I - AM^{-1})r_k = NM^{-1}r_k, \tag{11.13}$$

and we have the relation

$$Mu_k = r_k = Ae_k. \tag{11.14}$$

Equation (11.14) has a simple interpretation. The solution of $Ax = b$ can be obtained by solving $Ae_k = r_k$ with $x = x_k + e_k$. However, we cannot solve systems with the matrix A directly because we assume that A is too large to be stored and factored. Therefore we replace the problem by an easier one and solve $Mu_k = r_k$, and then iterate using $x_{k+1} = x_k + u_k$.

11.2.3   Convergence Criteria

We turn now to the question of when an iteration of the form (11.7) or (11.8) converges. Assume that $Ax = b$ has a unique solution and consider the splitting $A = M - N$ with an invertible matrix M. Then for any vector norm $\| \cdot \| : \mathbb{R}^n \to \mathbb{R}^+$ and the corresponding *induced matrix norm* $\|A\| := \sup_{\|x\|=1} \|Ax\|$ (see Section 2.5.1), we conclude by taking norms for the error recurrence (11.9) that

$$\|e_{k+1}\| \le \|M^{-1}N\| \, \|e_k\| \le \ldots \le \|M^{-1}N\|^{k+1} \|e_0\|.$$

Thus, we have convergence if we have $\|M^{-1}N\| < 1$ for the chosen norm. This is a sufficient, but not a necessary, condition for convergence: a triangular matrix R with zero diagonal may have a norm $\|R\| > 1$, but since R is nilpotent, we have $R^k \to 0$ for $k \to \infty$. Therefore we need another property that describes convergence.

DEFINITION 11.2. (SPECTRAL RADIUS) *The* spectral radius *of a matrix $A \in \mathbb{R}^{n \times n}$ is*

$$\rho(A) := \max_{j=1,\ldots,n} |\lambda_j(A)|,$$

where $\lambda_j(A)$ denotes the j-th eigenvalue of A.

LEMMA 11.1. *For all induced matrix norms (see Section 2.5.1, Equation (2.4)), $\rho(A) \le \|A\|$ holds.*

PROOF. Let (λ, x) be an eigenpair of A. From the eigenvector-eigenvalue relation $Ax = \lambda x$, we conclude that

$$|\lambda|\|x\| = \|\lambda x\| = \|Ax\| \le \|A\| \, \|x\|.$$

Now since $\|\boldsymbol{x}\| \neq 0$, we obtain $|\lambda| \leq \|A\|$, and since this holds for any eigenvalue λ, we also get $\rho(A) := \max_{j=1\ldots n} |\lambda_j(A)| \leq \|A\|$. $\qquad\square$

THEOREM 11.2. (CONVERGENCE OF STATIONARY ITERATIVE METHODS) *Let $A \in \mathbb{R}^{n \times n}$ be non-singular, $A = M - N$ with M non-singular and $\boldsymbol{b} \in \mathbb{R}^n$. The stationary iterative method*

$$M\boldsymbol{x}_{k+1} = N\boldsymbol{x}_k + \boldsymbol{b}$$

converges for any initial vector $\boldsymbol{x}_0$ to the solution $\boldsymbol{x}$ of the linear system $A\boldsymbol{x} = \boldsymbol{b}$ if and only if $\rho(M^{-1}N) < 1$.

PROOF. We first show the 'only if' part with a proof by contrapositive and assume that $|\lambda_m| = \rho(M^{-1}N) \geq 1$. Choosing $\boldsymbol{x}_0$ such that $\boldsymbol{e}_0 = \boldsymbol{x} - \boldsymbol{x}_0$ is a corresponding eigenvector, and applying the error recurrence (11.9), we get

$$\boldsymbol{e}_{k+1} = M^{-1}N\boldsymbol{e}_k = \cdots = (M^{-1}N)^{k+1}\boldsymbol{e}_0 = \lambda_m^{k+1}\boldsymbol{e}_0.$$

Thus, if $|\lambda_m| > 1$, then $|\lambda_m^{k+1}| \to \infty$, so the error cannot converge to zero. If $|\lambda_m| = 1$, then we also have no convergence since the error does not decrease.

For the 'if' part, we assume that $\rho(M^{-1}N) < 1$. We then consider the *Jordan decomposition* (see [51], page 317)

$$M^{-1}N = VJV^{-1}, \quad \text{with } V, J \in \mathbb{C}^{n \times n} \text{ and } V \text{ nonsingular.}$$

The matrix J is block-diagonal

$$J = \begin{bmatrix} J_{m_1}(\lambda_1) & 0 & 0 & \cdots & 0 \\ 0 & J_{m_2}(\lambda_2) & 0 & \cdots & 0 \\ \vdots & & \ddots & \ddots & \vdots \\ 0 & \cdots & 0 & J_{m_{s-1}}(\lambda_{s-1}) & 0 \\ 0 & \cdots & \cdots & 0 & J_{m_s}(\lambda_s) \end{bmatrix}$$

with

$$J_{m_i}(\lambda_i) = \begin{bmatrix} \lambda_i & 1 & 0 & \cdots & 0 \\ 0 & \lambda_i & 1 & \cdots & 0 \\ \vdots & \vdots & \ddots & \ddots & \vdots \\ 0 & 0 & \cdots & \lambda_i & 1 \\ 0 & 0 & \cdots & 0 & \lambda_i \end{bmatrix} \in \mathbb{C}^{m_i \times m_i}, \quad i = 1, \ldots, s.$$

Now $(M^{-1}N)^k = VJ^kV^{-1}$, and since J is block diagonal, we get

$$
J^k = \begin{bmatrix}
J^k_{m_1}(\lambda_1) & 0 & 0 & \cdots & & 0 \\
0 & J^k_{m_2}(\lambda_2) & 0 & \cdots & & 0 \\
\vdots & \ddots & \ddots & \ddots & & \vdots \\
0 & \cdots & 0 & J^k_{m_{s-1}}(\lambda_{s-1}) & & 0 \\
0 & \cdots & \cdots & & 0 & J^k_{m_s}(\lambda_s)
\end{bmatrix}.
$$

It is easy to verify with the following MAPLE script

```
with(LinearAlgebra):
J:=Matrix([[lambda,1,0,0,0],
[0,lambda,1,0,0],
[0,0,lambda,1,0],
[0,0,0,lambda,1],
[0,0,0,0,lambda]]);
J.J;
J.J.J;
J.J.J.J;
J.J.J.J.J;
```

the well-known expression for the powers of a Jordan block

$$
J^k_{m_i}(\lambda_i) = \begin{bmatrix}
\lambda_i^k & \binom{k}{1}\lambda_i^{k-1} & \binom{k}{2}\lambda_i^{k-2} & \cdots & \binom{k}{m_i-1}\lambda_i^{k-m_i+1} \\
0 & \lambda_i^k & \binom{k}{1}\lambda_i^{k-1} & \cdots & \binom{k}{m_i-2}\lambda_i^{k-m_i+2} \\
\vdots & \vdots & \ddots & \ddots & \vdots \\
0 & 0 & \cdots & \lambda_i^k & \binom{k}{1}\lambda_i^{k-1} \\
0 & 0 & \cdots & 0 & \lambda_i^k
\end{bmatrix}. \tag{11.15}
$$

Therefore, if $\rho(M^{-1}N) < 1$, then $|\lambda_i| < 1$ for all i, so that

$$
\lim_{k\to\infty} J^k_{m_i}(\lambda_i) = 0.
$$

for all Jordan blocks. It follows that $\lim_{k\to\infty} J^k = 0$. But this implies

$$
\lim_{k\to\infty} (M^{-1}N)^k = \lim_{k\to\infty} VJ^kV^{-1} = V(\lim_{k\to\infty} J^k)V^{-1} = 0.
$$

$\square$

REMARK. *One can wonder whether there is a relation between the spectral radius and the norm of a matrix. For symmetric matrices, the two concepts are definitely related:*

LEMMA 11.2. *For symmetric matrices $A \in \mathbb{R}^{n \times n}$, the spectral radius equals the 2-norm, $\rho(A) = \|A\|_2$.*

PROOF. *Using the definition of the 2-norm in Section 2.5.1, we obtain*

$$\|A\|_2^2 = \lambda_{\max}(A^\top A) = \lambda_{\max}(A^2) = \max |\lambda(A)|^2 = \rho(A)^2.$$

$\square$

However, in general, the spectral radius is not a norm! For norms, it follows from $\|A\| = 0$ that $A = 0$. Not so for the spectral radius: for an upper triangular matrix R with zero diagonal (thus all eigenvalues are zero) we have $\rho(R) = 0$ but $R \neq 0$. Furthermore, the triangle inequality

$$\rho(A + B) \leq \rho(A) + \rho(B)$$

does not hold: take for example

$$A = \begin{pmatrix} 0 & 1 \\ 0 & 0 \end{pmatrix} \quad B = \begin{pmatrix} 0 & 0 \\ 1 & 0 \end{pmatrix}.$$

Then $\rho(A + B) = 1$ but $\rho(A) = \rho(B) = 0$. Nonetheless, the following result holds:

THEOREM 11.3. *Let $A \in \mathbb{R}^{n \times n}$. Then for any given $\varepsilon > 0$, there exists a norm $\| \cdot \|$, which depends on A and ε, such that*

$$\|A\| \leq \rho(A) + \varepsilon.$$

For a proof, see Problem 11.4.

11.2.4 Singular Systems

Iterative methods can often be used to solve the system $A\boldsymbol{x} = \boldsymbol{b}$ even when A is singular, as long as a solution exists. For instance, the matrix A could come from the normal equations

$$C^\top C \boldsymbol{x} = C^\top \boldsymbol{b}, \qquad C \in \mathbb{R}^{m \times n},$$

which always have a solution. If $\operatorname{rank}(C) < n$, then $A = C^\top C$ is singular, so the system (which is consistent) has infinitely many solutions. In that case, for any splitting $A = M - N$ with M non-singular, the iteration matrix $M^{-1}N$ must necessarily have 1 as an eigenvalue, since $A\boldsymbol{v} = 0$, $\boldsymbol{v} \neq 0$ implies $M^{-1}N\boldsymbol{v} = \boldsymbol{v}$. This necessarily means $\rho(M^{-1}N) \geq 1$.

What can we say about the convergence of the method? Let m be such that $|\lambda_m| = \rho(M^{-1}N)$. We have seen that if $|\lambda_m| = 1$, then the error $\boldsymbol{e}_k$ does not decrease. However, this does not imply non-convergence in the singular case, since the solution $\boldsymbol{x}$ is not unique; it is very well possible that the iteration converges to some $\tilde{\boldsymbol{x}}$ with $A\tilde{\boldsymbol{x}} = \boldsymbol{b}$, but $\tilde{\boldsymbol{x}} \neq \boldsymbol{x}$, so $\boldsymbol{e}_k \not\to 0$. Let us consider once again the recurrence

$$\boldsymbol{e}_k = (M^{-1}N)^k \boldsymbol{e}_0 \tag{11.16}$$

and the Jordan decomposition $M^{-1}N = VJV^{-1}$. Multiplying (11.16) by V^{-1} from the left on both sides, we get

$$V^{-1}\boldsymbol{e}_k = J^k V^{-1}\boldsymbol{e}_0.$$

From (11.15), we see that $J^k_{m_i}(\lambda_i)$ does not converge if either $|\lambda_i| > 1$ or $|\lambda_i| = 1$ but $\lambda_i \neq 1$. For $\lambda_i = 1$, we also have divergence if $m_i > 1$, i.e., if the Jordan block is non-trivial. However, if

1. all eigenvalues other than $\lambda = 1$ have modulus strictly less than 1, and

2. if $\lambda = 1$ has as many linearly independent eigenvectors as its algebraic multiplicity,

then

$$\lim_{k\to\infty} J^k = \begin{bmatrix} I & 0 \\ 0 & 0 \end{bmatrix} =: P,$$

with the size of I equal to the algebraic multiplicity of $\lambda = 1$. This means

$$\lim_{k\to\infty} \boldsymbol{e}_k = \lim_{k\to\infty} V J^k V^{-1}\boldsymbol{e}_0 = VPV^{-1}\boldsymbol{e_0} \neq 0,$$

so $\tilde{\boldsymbol{x}} = \boldsymbol{x} - \lim_{k\to\infty} \boldsymbol{e}_k$ exists. Moreover, we have

$$A(\boldsymbol{x} - \tilde{\boldsymbol{x}}) = M(I - M^{-1}N)VPV^{-1}\boldsymbol{e}_0 = MV(I - J)PV^{-1}\boldsymbol{e}_0.$$

But

$$I - J = \begin{bmatrix} 0 & 0 \\ 0 & \tilde{J} \end{bmatrix}, \qquad P = \begin{bmatrix} I & 0 \\ 0 & 0 \end{bmatrix},$$

so $(I - J)P = 0$, which implies

$$A(\boldsymbol{x} - \tilde{\boldsymbol{x}}) = 0 \implies A\tilde{\boldsymbol{x}} = A\boldsymbol{x} = \boldsymbol{b}.$$

Thus, the iteration *converges* to another solution $\tilde{\boldsymbol{x}}$, which depends on the initial error $\boldsymbol{e}_0$!

11.2.5   Convergence Factor and Convergence Rate

Let us return to the case where A is non-singular and consider the error recurrence (11.11). By induction, we get

$$\boldsymbol{e}_k = (M^{-1}N)^k \boldsymbol{e}_0,$$

and taking norms yields a bound for the error reduction over k steps,

$$\frac{\|e_k\|}{\|e_0\|} \le \|(M^{-1}N)^k\|.$$

We are interested in knowing how many iterations are needed for the error reduction to reach some given tolerance ε,

$$\frac{\|e_k\|}{\|e_0\|} \le \|(M^{-1}N)^k\| < \varepsilon,$$

so let us write the right-hand side as

$$\|(M^{-1}N)^k\| = \left(\|(M^{-1}N)^k\|^{\frac{1}{k}}\right)^k < \varepsilon.$$

Taking the logarithm, we get

$$k > \frac{\ln(\varepsilon)}{\ln\left(\|(M^{-1}N)^k\|^{\frac{1}{k}}\right)}. \tag{11.17}$$

This equation does not seem very useful at first glance, since the number of necessary iterations k also appears on the right hand side. However, for large k, we can get a good estimate using the following lemma:

LEMMA 11.3. *For any matrix* $G \in \mathbb{R}^{n \times n}$ *with spectral radius* $\rho(G)$ *and any induced matrix norm, we have*

$$\lim_{k \to \infty} \|G^k\|^{\frac{1}{k}} = \rho(G).$$

PROOF. Let x and λ be an eigenpair of G. Then

$$Gx = \lambda x \quad \implies \quad G^k x = \lambda^k x$$

and using $\|G^k x\| \le \|G^k\|\,\|x\|$, we find

$$|\lambda|^k \|x\| \le \|G^k\|\,\|x\| \quad \implies \quad |\lambda| \le \|G^k\|^{\frac{1}{k}},$$

which implies that $\rho(G) \le \|G^k\|^{\frac{1}{k}}$.

On the other hand, let $\varepsilon > 0$ be given and consider the matrix

$$G(\varepsilon) = \frac{1}{\rho(G) + \varepsilon} G.$$

If λ is an eigenvalue of G, then $G(\varepsilon)$ has the eigenvalue $\lambda/(\rho(G) + \varepsilon)$, and therefore

$$\rho(G(\varepsilon)) = \frac{\rho(G)}{\rho(G) + \varepsilon} < 1.$$

Hence $\lim_{k\to\infty}\|G(\varepsilon)^k\| = 0$, which means that for all fixed $\varepsilon > 0$, we have for k sufficiently large

$$\|G(\varepsilon)^k\| = \frac{\|G^k\|}{(\rho(G)+\varepsilon)^k} < 1 \iff \|G^k\|^{\frac{1}{k}} < \rho(G)+\varepsilon.$$

Combining both bounds, we get any $\varepsilon > 0$, we have

$$\rho(G) \le \|G^k\|^{\frac{1}{k}} < \rho(G)+\varepsilon$$

for large enough k, which completes the proof. $\square$

DEFINITION 11.3. (CONVERGENCE FACTOR) *The* mean convergence factor *of an iteration matrix G over k steps is the number*

$$\rho_k(G) = \|G^k\|^{\frac{1}{k}}. \tag{11.18}$$

The asymptotic convergence factor *is the spectral radius*

$$\rho(G) = \lim_{k\to\infty} \rho_k(G). \tag{11.19}$$

DEFINITION 11.4. (CONVERGENCE RATE) *The* mean convergence rate *of an iteration matrix G over k steps is the number*

$$R_k(G) = -\ln\left(\|G^k\|^{\frac{1}{k}}\right) = -\ln(\rho_k(G)). \tag{11.20}$$

The asymptotic convergence rate *is*

$$R_\infty(G) = -\ln(\rho(G)). \tag{11.21}$$

We now return to the question of how many iteration steps k are necessary until the error reduction reaches a given tolerance ε, or until the error decreases by a factor δ, with $\varepsilon = 1/\delta$. The answer is

$$k \simeq -\frac{\ln \varepsilon}{R_\infty(M^{-1}N)} = \frac{\ln \delta}{R_\infty(M^{-1}N)}.$$

As an example, suppose $\rho(M^{-1}N) = 0.8$. Then, to obtain another decimal digit of the solution, we need to take $\delta = 10$ or $\varepsilon = 0.1$ to get

$$\frac{\ln 10}{-\ln 0.8} = 10.31.$$

Thus, we need about 10 iterations.

11.3 Classical Stationary Iterative Methods

We have seen in Subsection 11.2 that the iterative solution of linear systems can be achieved by splitting the associated matrix into two parts and iterating. We will see in this section the classical approaches for splitting a matrix, and study the convergence behavior of the associated methods.

11.3.1 Regular Splittings and M-Matrices

In the 1960s, substantial research was devoted to finding general criteria for matrices and their associated splittings that lead to convergent stationary iterative methods. We follow in this subsection the pioneering work of Varga [142].

DEFINITION 11.5. (NON-NEGATIVE MATRIX) *A matrix $A \in \mathbb{R}^{n \times n}$ is said to be* non-negative *(non-positive) if $a_{ij} \geq 0$ (respectively $a_{ij} \leq 0$) for $i, j = 1, \ldots, n$.*

Non-negative matrices have a very interesting property, which was discovered by Perron and Frobenius.

THEOREM 11.4. (PERRON-FROBENIUS (1907/1912)) *Let $A \in \mathbb{R}^{n \times n}$ be a non-negative matrix. Then A has a non-negative real eigenvalue λ which equals the spectral radius of A, $\lambda = \rho(A)$, and a corresponding eigenvector which is non-negative.*

PROOF. See [142], Theorem 2.7. $\qquad\square$

The original result of Perron and Frobenius is slightly stronger: it requires also that the matrix A be *irreducible* (i.e. there exists no permutation matrix P such that $P^{\top} A P$ is block upper triangular), and then 'non-negative' can be replaced by 'positive' in Theorem 11.4.

We now introduce a particular class of splittings for matrices, for which one can obtain very general convergence results.

DEFINITION 11.6. (REGULAR SPLITTING) *A splitting $A = M - N$ is said to be* regular *if M is invertible and if both M^{-1} and N are non-negative.*

THEOREM 11.5. *Let $A \in \mathbb{R}^{n \times n}$ be a given matrix, and $A = M - N$ be a regular splitting. Then*

$$\rho(M^{-1}N) < 1 \quad \Longleftrightarrow \quad A \text{ is invertible and } A^{-1} \text{ is non-negative.}$$

PROOF. If $\rho(M^{-1}N) < 1$, then $A = M - N = M(I - M^{-1}N)$ is invertible. Furthermore, using the Neumann series, we obtain

$$A^{-1} = (I - M^{-1}N)^{-1}M^{-1} = \sum_{j=0}^{\infty}(M^{-1}N)^j M^{-1},$$

which shows that A^{-1} is non-negative.

On the other hand, suppose A is invertible and A^{-1} is non-negative. Since M is invertible, we obtain from $A = M - N = M(I - M^{-1}N)$ that $(I - M^{-1}N)$ is also invertible. Thus

$$A^{-1}N = (M(I - M^{-1}N))^{-1}N = (I - M^{-1}N)^{-1}M^{-1}N. \qquad (11.22)$$

By assumption, both matrices M^{-1} and N are non-negative, and thus their product is also non-negative. Using Theorem 11.4, there exists a non-negative vector $\boldsymbol{x}$ such that

$$M^{-1}N\boldsymbol{x} = \lambda\boldsymbol{x}, \quad \lambda = \rho(M^{-1}N). \qquad (11.23)$$

Using (11.22), we obtain

$$A^{-1}N\boldsymbol{x} = (I - M^{-1}N)^{-1}M^{-1}N\boldsymbol{x} = \frac{\rho(M^{-1}N)}{1 - \rho(M^{-1}N)}\boldsymbol{x}.$$

Now because $A^{-1}N$ and $\boldsymbol{x}$ are non-negative, we get

$$\frac{\rho(M^{-1}N)}{1 - \rho(M^{-1}N)} \geq 0 \quad \implies \quad 1 - \rho(M^{-1}N) \geq 0 \quad \implies \quad 0 \leq \rho(M^{-1}N) \leq 1.$$

Finally, we need to exclude the possibility that $\rho(M^{-1}N) = 1$. If the latter were to hold, then by (11.23), we would have

$$M^{-1}N\boldsymbol{x} = \rho(M^{-1}N)\boldsymbol{x} = \boldsymbol{x} \quad \implies \quad (I - M^{-1}N)\boldsymbol{x} = 0, \quad \boldsymbol{x} \neq 0,$$

which would in turn contradict the invertibility of $I - M^{-1}N$. Thus, we must have $\rho(M^{-1}N) < 1$, as required. $\qquad \square$

DEFINITION 11.7. (M-MATRIX) *The matrix $A \in \mathbb{R}^{n \times n}$ is an M-matrix if each of the following conditions holds:*

1. $a_{ii} > 0$ for $i = 1, \ldots, n$,

2. $a_{ij} \leq 0$ for $i \neq j$, $i, j = 1, \ldots, n$,

3. A is invertible,

4. $A^{-1} \geq 0$.

EXAMPLE 11.1. *Consider the boundary value problem*

$$-u''(x) = f(x), \quad u(0) = u(1) = 0.$$

After discretizing with step size $h = 1/(n+1)$, $x_k = kh$, $u_k = u(x_k)$ and approximating

$$u''(x_k) \approx \frac{u_{k-1} - 2u_k + u_{k+1}}{h^2}$$

we obtain the linear system

$$\begin{pmatrix} 2 & -1 & & \\ -1 & 2 & \ddots & \\ & \ddots & \ddots & -1 \\ & & -1 & 2 \end{pmatrix} \begin{pmatrix} u_1 \\ u_2 \\ \vdots \\ u_n \end{pmatrix} = h^2 \begin{pmatrix} f_1 \\ f_2 \\ \vdots \\ f_n \end{pmatrix}.$$

Since the matrix of the system A has the the positive inverse $B = A^{-1}$, where B is computed explicitly with the MATLAB *statements*

```
B=diag([n:-1:1])*tril(ones(n)*diag([1:n]));
B=(B+tril(B,-1)')/(n+1);
```

we conclude that A is a M-matrix. This can also be proved analytically, see Problem 11.6.

As a consequence of Theorem 11.5, we obtain the following general convergence result.

COROLLARY 11.1. *Let $A \in \mathbb{R}^{n \times n}$ be an M-matrix and $A = M - N$ be a regular splitting. Then the stationary iteration $M\boldsymbol{x}_{k+1} = N\boldsymbol{x}_k + \boldsymbol{b}$ converges to the solution of $A\boldsymbol{x} = \boldsymbol{b}$.*

A further result, which is attributed to Householder [72] and John [75], is the following theorem (see also [8]).

THEOREM 11.6. (HOUSEHOLDER-JOHN (1955/1956)) *Let $A \in \mathbb{R}^{n \times n}$ be symmetric and non-singular. Let $A = M - N$ be a splitting with a real, non-singular matrix M and assume that $N + M^{\top}$ is positive definite. Then*

$$\rho(M^{-1}N) < 1 \iff A \text{ is positive definite.}$$

PROOF. We first prove that if A is positive definite, then $\rho(M^{-1}N) < 1$. Let $(\lambda, \boldsymbol{x})$ be a possibly complex eigenpair of the matrix $M^{-1}N$, $M^{-1}N\boldsymbol{x} = \lambda\boldsymbol{x}$. Then

$$\begin{aligned} A &= M(I - M^{-1}N) \\ \implies \quad A\boldsymbol{x} &= (1 - \lambda)M\boldsymbol{x} \\ \implies \quad \boldsymbol{x}^{\mathrm{H}}A\boldsymbol{x} &= (1 - \lambda)\boldsymbol{x}^{\mathrm{H}}M\boldsymbol{x}, \end{aligned}$$

which shows that $\lambda \neq 1$, since A is positive definite. If we take the conjugate transpose of the last equation, we get

$$(\boldsymbol{x}^{\mathrm{H}}A\boldsymbol{x})^{\mathrm{H}} = \boldsymbol{x}^{\mathrm{H}}A^{\mathrm{H}}\boldsymbol{x} = \boldsymbol{x}^{\mathrm{H}}A\boldsymbol{x} = (1 - \bar{\lambda})\boldsymbol{x}^{\mathrm{H}}M^{\mathrm{H}}\boldsymbol{x}.$$

Since M is real, we have $M^{\mathrm{H}} = M^{\top}$. Let $Q = N + M^{\top} = M + M^{\top} - A$, which is by assumption positive definite. Dividing both equations by the factor inside the parentheses (which cannot vanish, since $\lambda \neq 1$) and adding them, we get

$$\underbrace{\left(\frac{1}{1-\lambda} + \frac{1}{1-\bar{\lambda}}\right)}_{2\,\mathrm{Re}\,\frac{1}{1-\lambda}} \boldsymbol{x}^{\mathrm{H}} A \boldsymbol{x} = \boldsymbol{x}^{\mathrm{H}} \underbrace{(M + M^{\mathrm{H}})}_{Q+A} \boldsymbol{x}$$

$$\implies \quad 2\,\mathrm{Re}\,\frac{1}{1-\lambda} = \frac{\boldsymbol{x}^{\mathrm{H}} Q \boldsymbol{x} + \boldsymbol{x}^{\mathrm{H}} A \boldsymbol{x}}{\boldsymbol{x}^{\mathrm{H}} A \boldsymbol{x}} = 1 + \frac{\boldsymbol{x}^{\mathrm{H}} Q \boldsymbol{x}}{\boldsymbol{x}^{\mathrm{H}} A \boldsymbol{x}} > 1$$

since both Q and A are positive definite. We therefore have

$$2\,\mathrm{Re}\,\frac{1}{1-\lambda} > 1,$$

which, by letting $\lambda = \alpha + i\beta$ with $\alpha, \beta \in \mathbb{R}$, gives

$$2\,\mathrm{Re}\,\frac{1}{1-\lambda} = \frac{2(1-\alpha)}{(1-\alpha)^2 + \beta^2} > 1 \quad \Longleftrightarrow \quad |\lambda|^2 = \alpha^2 + \beta^2 < 1 \implies \rho(M^{-1}N) < 1,$$

which concludes the proof in this direction.

To prove the other direction, namely that if $\rho(M^{-1}N) < 1$ then A is positive definite, we first need the following lemma:

LEMMA 11.4. *Under the same conditions as in Theorem 11.6, the following identity holds*

$$A - (M^{-1}N)^{\top} A (M^{-1}N) = (I - M^{-1}N)^{\top}(M^{\top} + N)(I - M^{-1}N).$$

To prove this lemma, we replace $A = M - N$ and expand the left and the right hand sides and show that they are the same. The expansion of the right hand side gives

$$\begin{aligned} &= (I - N^{\top} M^{-\top})(M^{\top} + N)(I - M^{-1}N) \\ &= (M^{\top} + N - N^{\top} - N^{\top} M^{-\top} N)(I - M^{-1}N) \end{aligned}$$

Because of the symmetry $A^{\top} = A \iff M^{\top} - N^{\top} = M - N$, we get

$$\begin{aligned} &= (M - N^{\top} M^{-\top} N)(I - M^{-1}N) \\ &= M - N^{\top} M^{-\top} N - N + N^{\top} M^{-\top} N M^{-1} N. \end{aligned}$$

Expanding the left hand side, we obtain

$$\begin{aligned} &= M - N - N^{\top} M^{-\top}(M - N) M^{-1} N \\ &= M - N - (N^{\top} M^{-\top} M - N^{\top} M^{-\top} N) M^{-1} N \\ &= M - N - (N^{\top} M^{-\top} N - N^{\top} M^{-\top} N M^{-1} N) \end{aligned}$$

the same expression as for the right hand side, which concludes the proof of the lemma.

Continuing with the proof of Theorem 11.6, we now prove the second part by showing the contrapositive, i.e., we suppose that A is not positive definite and show that this implies $\rho(M^{-1}N) \geq 1$. Let $x_0 \in \mathbb{R}^n$ be given. We consider the sequence of vectors

$$x_{k+1} = M^{-1}Nx_k.$$

Applying the lemma, we obtain

$$Ax_k - (M^{-1}N)^\top A \underbrace{(M^{-1}N)x_k}_{x_{k+1}} = (I - M^{-1}N)^\top \underbrace{(M^\top + N)}_{Q} \underbrace{(I - M^{-1}N)x_k}_{x_k - x_{k+1}}.$$

Multiplying the last equation from the left with $x_k^\top$ yields

$$x_k^\top Ax_k - x_{k+1}^\top Ax_{k+1} = (x_k - x_{k+1})^\top Q(x_k - x_{k+1}) \geq 0,$$

since Q is positive definite. Therefore the sequence $x_k^\top Ax_k$ satisfies

$$x_k^\top Ax_k \geq x_{k+1}^\top Ax_{k+1},$$

and is thus non increasing. Since A is assumed to be non-singular but not positive definite, we can find an initial vector x_0 such that

$$0 > x_0^\top Ax_0 \geq x_1^\top Ax_1 \geq \cdots .$$

This means that x_k cannot converge to 0. Thus $\rho(M^{-1}N) \geq 1$. $\qquad\square$

In the following sections, we present some classical stationary iterative methods. These classical methods are defined in terms of the matrices obtained by splitting A into the strictly lower triangular part L, the diagonal part $D = \mathrm{diag}(A)$, and the strictly upper triangular part U,

$$A = L + D + U.$$

11.3.2 Jacobi

The *Jacobi method* (or Point-Jacobi method) is based on solving for every variable locally, with the other variables frozen at their old values. One iteration (or sweep) corresponds to solving for every variable once:

ALGORITHM 11.1. *Point-Jacobi Iteration Step*

```
for i=1:n
  tmp(i)=(b(i)-A(i,[1:i-1 i+1:n])*x([1:i-1 i+1:n]))/A(i,i);
end
x=tmp(:);
```

This corresponds to the splitting $A = M - N$ with

$$M = D, \quad N = -L - U \quad \Longrightarrow \quad D\boldsymbol{x}_{k+1} = -(L + U)\boldsymbol{x}_k + \boldsymbol{b}. \qquad (11.24)$$

A classical convergence result for the Jacobi method is the following:

THEOREM 11.7. (CONVERGENCE OF JACOBI) *If the matrix $A \in \mathbb{R}^{n \times n}$ is strictly diagonally dominant , i.e.*

$$|a_{ii}| > \sum_{j \neq i} |a_{ij}| \quad for \ i = 1, \dots, n, \qquad (11.25)$$

then the Jacobi iteration (11.24) converges.

PROOF. The iteration matrix of the Jacobi method is $G_{\mathrm{J}} = -D^{-1}(L+U)$, and with Condition (11.25), we obtain that

$$\|G_{\mathrm{J}}\|_{\infty} = \max_{i=\{1,\dots,n\}} \frac{1}{|a_{ii}|} \sum_{j \neq i} |a_{ij}| < 1.$$

Now using Lemma 11.1, we obtain

$$\rho(G_{\mathrm{J}}) \leq \|G_{\mathrm{J}}\|_{\infty} < 1,$$

which, together with Theorem 11.2, concludes the proof. $\square$

We now apply Theorem 11.6 to the method of Jacobi.

THEOREM 11.8. *Let $A \in R^{n \times n}$ be symmetric and non-singular, and consider the splitting $A = D + L + L^{\top}$ with L strictly lower triangular and D diagonal and positive definite, $d_{ii} > 0$. Then the Jacobi iteration converges if and only if A and $2D - A$ are positive definite.*

PROOF. If A and $Q := 2D - A$ are positive definite, then according to Theorem 11.6 Jacobi converges: it suffices to set $M = D$ and $N = -(L + L^{\top})$, which implies that $N + M^{\top} = 2D - A$.

For the converse, we assume now that Jacobi is convergent, and we want to show that A and Q are positive definite. Consider an eigenpair of the iteration matrix

$$M^{-1}N\boldsymbol{x} = \underbrace{-D^{-1}(L + L^{\top})}_{G_{\mathrm{J}}} \boldsymbol{x} = \lambda \boldsymbol{x}. \qquad (11.26)$$

Since D is positive definite, we can take its square-root, $D = D^{\frac{1}{2}}D^{\frac{1}{2}}$. The matrix G_{J} is then similar to the symmetric matrix $D^{\frac{1}{2}}G_{\mathrm{J}}D^{-\frac{1}{2}}$, therefore the eigenvalues of the iteration matrix are real, and since we suppose that Jacobi is convergent, we have $|\lambda| < 1$. Furthermore, from the eigenvalue equation (11.26), we have

$$(L + L^{\top})\boldsymbol{x} = -\lambda D\boldsymbol{x}, \qquad (11.27)$$

and adding $D\boldsymbol{x}$ on both sides gives

$$
\begin{aligned}
A\boldsymbol{x} &= (1-\lambda)D\boldsymbol{x} \\
\Longleftrightarrow \quad \underbrace{D^{-\frac{1}{2}}AD^{-\frac{1}{2}}}_{\tilde{A}}\underbrace{D^{\frac{1}{2}}\boldsymbol{x}}_{\boldsymbol{y}} &= (1-\lambda)\underbrace{D^{\frac{1}{2}}\boldsymbol{x}}_{\boldsymbol{y}}.
\end{aligned}
$$

This means that $1-\lambda > 0$ is an eigenvalue of $\tilde{A}$ with the corresponding eigenvector $\boldsymbol{y}$. Thus $\tilde{A}$ is positive definite, which means that $0 < \boldsymbol{y}^\top \tilde{A}\boldsymbol{y} = \boldsymbol{x}^\top A\boldsymbol{x}$ for all $\boldsymbol{y} \neq 0$, and hence also for all $\boldsymbol{x} = D^{-\frac{1}{2}}\boldsymbol{y} \neq 0$. Therefore, A is also positive definite.

A similar argument holds for Q: Multiplying (11.27) by -1 and adding $D\boldsymbol{x}$ on both sides yields

$$
\underbrace{(D - (L + L^\top))}_{Q}\boldsymbol{x} = (1+\lambda)D\boldsymbol{x}.
$$

As before,

$$
D^{-\frac{1}{2}}QD^{-\frac{1}{2}}\boldsymbol{y} = (1+\lambda)\boldsymbol{y},
$$

and since $1 + \lambda > 0$, we conclude that Q is also positive definite. $\qquad\square$

For the block tridiagonal matrix A given in the introduction in (11.4),

$$
A = \begin{bmatrix}
T & I & & \\
I & T & \ddots & \\
& \ddots & \ddots & I \\
& & I & T
\end{bmatrix},
$$

it is natural to use a block splitting $A = D + L + L^\top$, where

$$
D = \begin{bmatrix}
T & & & \\
& T & & \\
& & \ddots & \\
& & & T
\end{bmatrix}
\quad \text{and} \quad
L = \begin{bmatrix}
0 & & & \\
I & 0 & & \\
& \ddots & \ddots & \\
& & I & 0
\end{bmatrix}.
$$

We then get a *Block Jacobi Iteration*, which is sometimes also called the *Line Jacobi Iteration*, because each block corresponds to a line of unknowns in the grid.

By modifying Theorem 11.8 slightly, we can show that Block Jacobi converges for this matrix. Indeed, we know that $-A$ is a positive definite matrix (see Problem 11.14), so according to Theorem 11.8, it suffices to show that

$$
A - 2D = \begin{bmatrix}
-T & I & & \\
I & -T & \ddots & \\
& \ddots & \ddots & I \\
& & I & -T
\end{bmatrix}
$$

is also positive definite. But $-A$ and $A - 2D$ are related by

$$A - 2D = -SAS^\top, \qquad S = \operatorname{diag}(I, -I, I, -I, \ldots).$$

Hence $A - 2D$ is positive definite as well. Now Theorem 11.8 implies that the block Jacobi splitting $M = -D$, $N = (L + L^\top)$ leads to a convergent method.

11.3.3   Gauss-Seidel

The *Gauss-Seidel method* is similar to the Jacobi method, except that it uses updated values as soon as they are available. Thus, in Algorithm 11.1, we just replace the variable `tmp` by the column vector `x`.

ALGORITHM 11.2. *Gauss-Seidel Iteration Step*

```
for i=1:n
  x(i)=(b(i)-A(i,[1:i-1 i+1:n])*x([1:i-1 i+1:n]))/A(i,i);
end
```

This corresponds to the splitting

$$M = D + L, \quad N = -U \quad \Rightarrow (D + L)\boldsymbol{x}_{k+1} = -U\boldsymbol{x}_k + \boldsymbol{b}. \qquad (11.28)$$

Since Gauss-Seidel uses the updated values as soon as they are available, one can expect that convergence will be faster than with Jacobi. This is often the case, e..g. for the model problem $\Delta u = 0$, one can show that only half of the iteration steps are needed for the same accuracy. However, here is an example where Jacobi converges and Gauss-Seidel does not. Consider the matrix

$$A = \begin{pmatrix} -1 & 0 & -1 \\ -1 & 1 & 0 \\ 1 & 2 & -3 \end{pmatrix}.$$

The Jacobi iteration matrix is

$$G_{\mathrm{J}} = -D^{-1}(L + U) = \begin{pmatrix} 0 & 0 & -1 \\ 1 & 0 & 0 \\ \frac{1}{3} & \frac{2}{3} & 0 \end{pmatrix},$$

and has the eigenvalues $0.37 \pm 0.86i$ and -0.74. Thus $\rho(G_{\mathrm{J}}) = 0.944$ and the iteration converges.

On the other hand, the iteration matrix of Gauss-Seidel is

$$G_{\mathrm{GS}} = -(D + L)^{-1}U = \begin{pmatrix} 0 & 0 & -1 \\ 0 & 0 & -1 \\ 0 & 0 & -1 \end{pmatrix},$$

which has the eigenvalues $0, 0, -1$ with $\rho(G_{\mathrm{GS}}) = 1$. The iteration therefore does not converge in general.

Instead of analyzing the convergence of the Gauss-Seidel method, we now show an important generalization with improved convergence behavior. The convergence analysis of the generalization then also contains Gauss-Seidel as a special case.

11.3.4 Successive Over-relaxation (SOR)

The *successive over-relaxation method* (SOR) is derived from the Gauss-Seidel method by introducing an "extrapolation" parameter ω. The component x_i is computed as for Gauss-Seidel but then averaged with its previous value.

ALGORITHM 11.3. *SOR Iteration Step*

```
for i=1:n
  x(i)=omega*(b(i)-A(i,[1:i-1 i+1:n])*x([1:i-1 i+1:n]))/...
       A(i,i)+(1-omega)*x(i);
end
```

SOR can also be derived by considering the Gauss-Seidel iteration form

$$(D + L)\boldsymbol{x} = -U\boldsymbol{x} + \boldsymbol{b}. \tag{11.29}$$

Multiplying (11.29) by ω and adding on both sides the expression $(1-\omega)D\boldsymbol{x}$, we obtain the SOR iteration

$$(D + \omega L)\boldsymbol{x}_{k+1} = (-\omega U + (1-\omega)D)\boldsymbol{x}_k + \omega\boldsymbol{b}. \tag{11.30}$$

SOR is therefore based on the splitting $A = M - N$ with

$$M = \frac{1}{\omega}D + L \text{ and } N = -U + \left(\frac{1}{\omega} - 1\right)D, \tag{11.31}$$

where we divided (11.30) by ω in order to remove the factor ω in front of the right hand side term $\boldsymbol{b}$ and obtain the stationary iterative method in standard form (11.7). Notice that for $\omega = 1$, we get as a special case the Gauss-Seidel method.

We first show that one cannot choose the relaxation parameter arbitrarily if one wants to obtain a convergent method. This general, very elegant result is due to Kahan from his PhD thesis [77].

THEOREM 11.9. (KAHAN (1958)) *Let $A \in \mathbb{R}^{n \times n}$ and $A = L + D + U$ with D invertible. If*

$$G_{\text{SOR}} = (D + \omega L)^{-1}(-\omega U + (1-\omega)D) \tag{11.32}$$

is the SOR iteration matrix, then the inequality

$$\rho(G_{\text{SOR}}) \geq |\omega - 1| \tag{11.33}$$

holds for all $\omega \in \mathbb{R}$.

PROOF. The key idea is to insert DD^{-1} between the factors of G_{SOR},

$$\begin{aligned} G_{\text{SOR}} &= (D + \omega L)^{-1}DD^{-1}(-\omega U + (1-\omega)D) \\ &= (I + \omega D^{-1}L)^{-1}(-\omega D^{-1}U + (1-\omega)I). \end{aligned}$$

Now the determinant of $(I + \omega D^{-1} L)$ equals 1, since this matrix is lower triangular with unit diagonal, which implies that the determinant of its inverse also equals 1. Therefore

$$\det(G_{\mathrm{SOR}}) = \det(-\omega D^{-1} U + (1 - \omega)I) = (1 - \omega)^n,$$

since this second factor is upper triangular with $1 - \omega$ on the diagonal. The determinant of a matrix is equal to the product of its eigenvalues, which in our case yields

$$\prod_{j=1}^{n} \lambda_j(G_{\mathrm{SOR}}) = (1-\omega)^n \quad \Longrightarrow \quad |1-\omega|^n \leq \left(\max_j |\lambda_j(G_{\mathrm{SOR}})| \right)^n = \rho(G_{\mathrm{SOR}})^n,$$

and this implies the result after taking the n-th root. $\qquad\square$

From this elegant result, we can conclude that

$$\rho(G_{\mathrm{SOR}}) < 1 \quad \Longrightarrow \quad 0 < \omega < 2,$$

and thus for convergence of SOR it is necessary to choose $0 < \omega < 2$.

The next result is a general convergence result for SOR, due to Ostrowski [98] and Reich [105]. It is for a restricted class of matrices, and does not answer the question yet on how to choose ω in order to obtain a fast method.

THEOREM 11.10. (OSTROWSKI-REICH (1954/1949)) *Let $A \in \mathbb{R}^{n \times n}$ be symmetric and invertible, with positive diagonal elements, $D > 0$. Then SOR converges for all $0 < \omega < 2$ if and only if A is positive definite.*

PROOF. Since A is symmetric, the SOR iteration is

$$\underbrace{\left(\frac{1}{\omega} D + L \right)}_{M} x_{k+1} = \underbrace{\left(\frac{1 - \omega}{\omega} D - L^\top \right)}_{N} x_k + b.$$

M is non-singular, since $D > 0$. Consider

$$Q = N + M^\top = \frac{2 - \omega}{\omega} D.$$

Q is positive definite because $D > 0$ and $0 < \omega < 2$. Now we can apply Theorem 11.6 to conclude the proof. $\qquad\square$

An important question, which remained unanswered so far, is how to choose the parameter ω in SOR in order to obtain a fast method. The pioneer in this area was David Young, who answered this question for a large class of matrices in his PhD thesis [153] [3].

DEFINITION 11.8. (PROPERTY A) *A matrix $A \in \mathbb{R}^{n \times n}$ has* Property A *if there exists a permutation matrix P such that*

$$P^\top A P = \begin{pmatrix} D_1 & F \\ E & D_2 \end{pmatrix}, \quad \text{with } D_1 \text{ and } D_2 \text{ diagonal.}$$

---

[3] An electronic version is available at `http://www.ma.utexas.edu/CNA/DMY/`

EXAMPLE 11.2. *If we discretize the one-dimensional Laplacian $\frac{\partial^2}{\partial x^2}$ by finite differences on a regular grid, the discrete operator becomes*

$$A = \frac{1}{h^2} \begin{bmatrix} -2 & 1 & & & & \\ 1 & -2 & 1 & & & \\ & 1 & \ddots & \ddots & & \\ & & \ddots & \ddots & 1 \\ & & & 1 & -2 \end{bmatrix} \in \mathbb{R}^{n \times n}.$$

If n is even, then A can be permuted as required for Property A,

$$P^{\top}AP = \left[\begin{array}{cccc|cccc} -2 & & & & 1 & & & \\ & \ddots & & & 1 & \ddots & & \\ & & \ddots & & & \ddots & \ddots & \\ & & & -2 & & & 1 & 1 \\ \hline 1 & 1 & & & -2 & & & \\ & \ddots & \ddots & & & \ddots & & \\ & & \ddots & 1 & & & \ddots & \\ & & & 1 & & & & -2 \end{array} \right],$$

with P constructed from the unit vectors $\boldsymbol{e}_j$ as follows:

$$P = [\boldsymbol{e}_1, \boldsymbol{e}_3, \ldots \boldsymbol{e}_{n-1}, \boldsymbol{e}_2, \boldsymbol{e}_4, \ldots, \boldsymbol{e}_n].$$

For matrices with Property A, a useful relation exists between the eigenvalues of the associated Jacobi and SOR iteration matrices. We first need

LEMMA 11.5. *Let B have a zero diagonal and the block structure*

$$B = \begin{bmatrix} 0 & F \\ E & 0 \end{bmatrix} = L + U,$$

where L and U are the strictly lower and upper triangular parts. If μ is an eigenvalue of B, then so is $-\mu$. Furthermore, B and $\alpha L + \frac{1}{\alpha}U$ are similar, and thus have the same eigenvalues for all $\alpha \neq 0$.

PROOF. Consider the diagonal matrix

$$S = \begin{bmatrix} I & 0 \\ 0 & \alpha I \end{bmatrix} \implies SBS^{-1} = \begin{bmatrix} 0 & \frac{1}{\alpha}F \\ \alpha E & 0 \end{bmatrix} = \alpha L + \frac{1}{\alpha}U.$$

Furthermore, if $\binom{u}{v}$ is an eigenvector of B,

$$B\binom{u}{v} = \mu\binom{u}{v},$$

then $\binom{u}{-v}$ is an eigenvector with eigenvalue $-\mu$, because

$$B\binom{u}{-v} = \binom{-Fv}{Eu} = \binom{-\mu u}{\mu v} = -\mu\binom{u}{-v}.$$

$\square$

THEOREM 11.11. *Let* $A \in \mathbb{R}^{n\times n}$ *have Property A, and let*

$$\tilde{A} = P^{\top} A P = \begin{bmatrix} D_1 & F \\ E & D_2 \end{bmatrix} = L + D + U,$$

with all diagonal elements of D_1 *and* D_2 *nonzero. Let*

$$G_{\mathrm{SOR}} = (D + \omega L)^{-1}(-\omega U + (1 - \omega)D)$$

be the SOR iteration matrix for $\tilde{A}$ *and*

$$G_{\mathrm{J}} = -D^{-1}(L + U)$$

the Jacobi iteration matrix. Then for $\omega \neq 0$, *we have that* λ *is a nonzero eigenvalue of* G_{SOR} *if and only if* μ, *the solution of*

$$(\lambda + \omega - 1)^2 = \lambda\omega^2\mu^2,$$

is an eigenvalue of G_{J}.

PROOF. As before, we have

$$
\begin{aligned}
G_{\mathrm{SOR}} &= (D + \omega L)^{-1}DD^{-1}(-\omega U + (1 - \omega)D) \\
&= (I + \omega D^{-1}L)^{-1}(-\omega D^{-1}U + (1 - \omega)I),
\end{aligned}
$$

therefore λ is an eigenvalue if and only if

$$
\begin{aligned}
\det((I + \omega D^{-1}L)^{-1}(-\omega D^{-1}U + (1 - \omega)I) - \lambda I) &= 0 \\
\Longleftrightarrow \det(-\omega D^{-1}U + (1 - \omega)I - \lambda(I + \omega D^{-1}L)) &= 0 \\
\Longleftrightarrow \det\big((\lambda + \omega - 1)I + \omega D^{-1}(\lambda L + U)\big) &= 0.
\end{aligned}
$$

Now factoring out $\omega\sqrt{\lambda}$ (which is non-zero by assumption) in the determinant, we get

$$\Longleftrightarrow \det\left(\frac{\lambda + \omega - 1}{\omega\sqrt{\lambda}}I + D^{-1}(\sqrt{\lambda}L + \frac{1}{\sqrt{\lambda}}U)\right) = 0. \qquad (11.34)$$

This equation means that $\frac{\lambda+\omega-1}{\omega\sqrt{\lambda}}$ is an eigenvalue of $-D^{-1}(\sqrt{\lambda}L + \frac{1}{\sqrt{\lambda}}U)$. Using Lemma 11.5, we obtain

$$-D^{-1}\left(\sqrt{\lambda}L + \frac{1}{\sqrt{\lambda}}U\right) = -\left(\sqrt{\lambda}D^{-1}L + \frac{1}{\sqrt{\lambda}}D^{-1}U\right) \text{ is similar to } G_{\mathrm{J}} = -D^{-1}(L+U),$$

therefore they have the same eigenvalues. Since μ and $-\mu$ are eigenvalues of G_{J}, we get

$$\pm\mu = \frac{\lambda+\omega-1}{\omega\sqrt{\lambda}} \qquad \Longleftrightarrow \qquad (\lambda+\omega-1)^2 = \lambda\omega^2\mu^2. \tag{11.35}$$

$\square$

We are now ready to prove the most important result for SOR methods: the optimal choice of the relaxation parameter ω, given by Young in his thesis [153].

THEOREM 11.12. (OPTIMAL SOR PARAMETER (YOUNG 1950)) *Let A, $\tilde{A}$ and G_{J} be defined as in Theorem 11.11. If the eigenvalues $\mu(G_{\mathrm{J}})$ are real and $\rho(G_{\mathrm{J}}) < 1$, then the optimal SOR parameter ω for $\tilde{A}$ is*

$$\omega_{\mathrm{opt}} = \frac{2}{1 + \sqrt{1 - \rho(G_{\mathrm{J}})^2}}.$$

PROOF. From (11.35), we obtain

$$\pm\mu\sqrt{\lambda} = \frac{\lambda+\omega-1}{\omega}. \tag{11.36}$$

For μ fixed, the left hand side of (11.36) is the equation of a parabola, whereas the right hand side is that of a straight line passing through the point $(1,1)$ with slope $1/\omega$. Figure 11.2 shows the roots of this equation. We see that the roots can be real or complex, depending on ω. If the roots are real, λ_1 is always bigger in modulus than λ_2, and λ_1 is increasing in μ, and decreasing in ω. Solving for λ, we obtain the quadratic equation

$$\lambda^2 + (2(\omega-1) - \omega^2\mu^2)\lambda + (\omega-1)^2 = 0 \tag{11.37}$$

with the solutions

$$\lambda_{1,2} = \frac{1}{2}\left(\omega^2\mu^2 - 2(\omega-1) \pm \sqrt{\omega^2\mu^2(\omega^2\mu^2 - 4(\omega-1))}\right).$$

With the discriminant $d(\omega,\mu) := \omega^2\mu^2 - 4(\omega-1)$, we have that $d(0,\mu) = 4$ and $d(2,\mu) = -4 + 4\mu^2 < 0$, and d becomes zero if

$$4 - 4\omega + \omega^2\mu^2 = 0 \iff \omega_{1,2} = \frac{2 \pm 2\sqrt{1-\mu^2}}{\mu^2}.$$

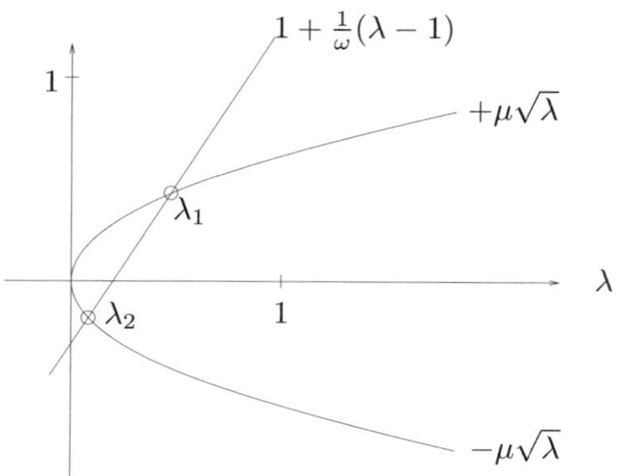

FIGURE 11.2.
Relation between the roots of Jacobi and SOR

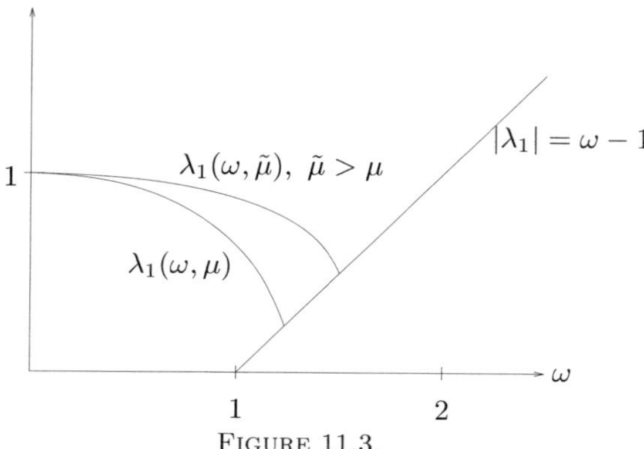

FIGURE 11.3.
Larger eigenvalue in modulus as functions of ω

Of the two possible values for ω, we only have to consider the smaller one,

$$\omega_1(\mu) = \frac{2 - 2\sqrt{1 - \mu^2}}{\mu^2} = \frac{2}{1 + \sqrt{1 - \mu^2}},$$

since the second value $\omega_2 > 2$ cannot lead to a convergent method, see Theorem 11.9. Thus for $\omega \in (0, \omega_1)$, the eigenvalues are real and λ_1 is the larger one in modulus.

For larger values $\omega \in (\omega_1, 2)$, the discriminant is negative and $\lambda_{1,2}$ are complex conjugates. From (11.37), we see that $\lambda_1 \lambda_2 = (\omega - 1)^2$, so that in the complex case we have $|\lambda_1| = |\omega - 1|$. We thus obtain for each eigenvalue μ of the Jacobi method the corresponding curve shown in Figure 11.3, and since in the real case λ_1 is increasing with μ, we obtain

$$
\rho(G_{\mathrm{SOR}}) = \begin{cases} \frac{1}{2}\left(\omega^2\rho(G_J)^2 - 2(\omega-1) + \sqrt{\omega^2\rho(G_J)^2(\omega^2\rho(G_J)^2 - 4(\omega-1))}\right), \\ \qquad\qquad \text{for } \omega \in (0,\omega_1), \text{ decreasing,} \\[2mm] |\omega - 1|, \qquad \text{for } \omega \in (\omega_1, 2), \text{ linearly increasing.} \end{cases}
$$

The minimum of $\rho(G_{\mathrm{SOR}})$ is thus reached for $\omega_{\mathrm{opt}} = \omega_1(\rho(G_J))$. $\qquad\square$

For ω_{opt}, the optimized convergence factor of SOR becomes

$$
\rho(G_{\mathrm{SOR}})_{\min} = \omega_{\mathrm{opt}} - 1 = \left(\frac{\rho(G_J)}{1 + \sqrt{1 - \rho(G_J)^2}}\right)^2 = \frac{1 - \sqrt{1 - \rho(G_J)^2}}{1 + \sqrt{1 - \rho(G_J)^2}}.
$$
$$(11.38)$$

In general, we do not know in advance the spectral radius $\rho(G_J)$, so we cannot compute ω_{opt} for SOR and have to rely on estimates. The rule of thumb is to try to overestimate ω, because of the steeper slope on the left, see Figure 11.4. To illustrate this, we consider the function

---

ALGORITHM 11.4. *SOR Spectral Radius*

```
function r=Rho(omega,mu);
% RHO computes spectral radius for SOR
%   r=Rho(omega,mu) computes the spectral radius of the iteration
%   matrix for SOR, given the spectral radius mu of the Jacobi
%   iteration matrix and the relaxation parameter omega

if omega<2/(1+sqrt(1-mu^2))
  r=((omega*mu)^2-2*(omega-1)+sqrt((omega*mu)^2*...
    ((omega*mu)^2-4*(omega-1)))))/2;
else
  r=abs(omega-1);
end;
```

---

and plot it for various values of μ using the commands

```
axis([0,2,0,1])
hold on
xx=[0:0.01:2];
for mu=0.1:0.1:0.9
  yy=[];
  for x=xx
    yy=[yy rho(x,mu)];
  end
  plot(xx,yy)
end
```

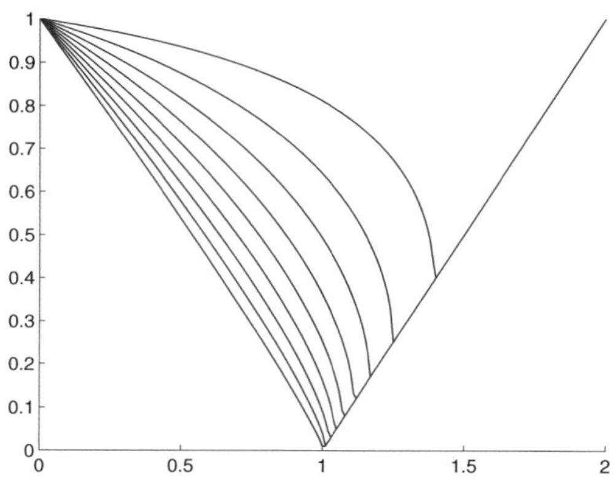

FIGURE 11.4.

$\rho(G_{\mathrm{SOR}})$ *as function of* ω *for* $\mu = 0.1 : 0.1 : 0.9$

We can see that for $\mu = 0.9$, the choice of $\omega \approx 1.4$ yields a convergence factor of about 0.4. This is a drastic improvement over Jacobi, leading to about 8 to 9 times fewer iterations, since $\mu^{8.5} \approx 0.4$.

The optimal choice of the relaxation parameter ω also improves the convergence factor asymptotically: one gains a square root, as one can see when setting $\rho(G_J) = 1 - \varepsilon$, and then expanding the corresponding optimized $\rho(G_{SOR})$ for small ε. Using the MAPLE commands

```
rho:=mu^2/(1+sqrt(1-mu^2))^2;
mu:=1-epsilon;
series(rho,epsilon,1);
```

we obtain $\rho(G_{SOR}) = 1 - 2\sqrt{2\varepsilon} + O(\varepsilon)$, which shows that indeed the improvement is a square root.

11.3.5   Richardson

In order to explain the Richardson iteration, we consider the correction form (11.8) of the stationary iterative method,

$$\boldsymbol{x}_{k+1} = \boldsymbol{x}_k + M^{-1}\boldsymbol{r}_k.$$

If we choose $M^{-1} = \alpha I$, which corresponds to the splitting $M = \frac{1}{\alpha}I$ and $N = \frac{1}{\alpha}I - A$, we obtain the *Method of Richardson*,[4]

$$\boldsymbol{x}_{k+1} = \boldsymbol{x}_k + \alpha(\boldsymbol{b} - A\boldsymbol{x}_k) = (I - \alpha A)\boldsymbol{x}_k + \alpha\boldsymbol{b}. \tag{11.39}$$

---

[4]It is important to state here that in the original paper of Richardson [106], a different value of α was used at each iteration step, $\alpha = \alpha_k$, which leads to a much more sophisticated method, as we will see later in Section 11.4.

THEOREM 11.13. (CONVERGENCE OF RICHARDSON) *Let $A \in \mathbb{R}^{n \times n}$ be symmetric and positive definite. Then*

a) *Richardson converges if and only if $0 < \alpha < \dfrac{2}{\rho(A)}$.*

b) *Convergence is optimal for $\alpha_{\mathrm{opt}} = \dfrac{2}{\lambda_{\max}(A) + \lambda_{\min}(A)}$.*

c) *The asymptotic convergence factor is*

$$\rho(I - \alpha_{\mathrm{opt}} A) = \frac{\kappa(A) - 1}{\kappa(A) + 1}, \ where \ \kappa(A) := \frac{\lambda_{\max}(A)}{\lambda_{\min}(A)}.$$

PROOF. Since A is positive definite, we have

$$0 < \lambda_{\min}(A) \le \lambda_j(A) \le \lambda_{\max}(A) = \rho(A).$$

The iteration matrix $M^{-1}N = I - \alpha A$ has the eigenvalues $1 - \alpha\lambda_j(A)$, so

$$\rho(I - \alpha A) < 1 \quad \Longleftrightarrow \quad 1 - \alpha\lambda_{\min}(A) < 1 \ \text{ and } \ 1 - \alpha\lambda_{\max}(A) > -1.$$

Thus we have convergence for

$$0 < \alpha < \frac{2}{\lambda_{\max}(A)}.$$

To minimize $\rho(I - \alpha A)$, we consider Figure 11.5, which shows the curves $|1 - \alpha\lambda_{\max}|$ and $|1 - \alpha\lambda_{\min}|$. We see that the optimal α is determined by the

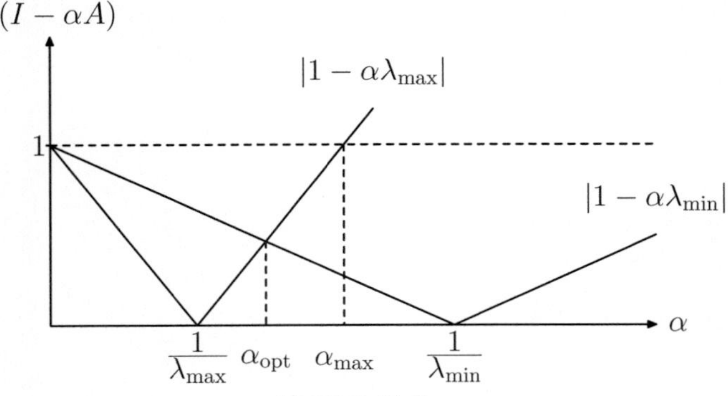

FIGURE 11.5.
Determining an optimal α for the Richardson iteration

intersection point between the two lines, since the curves $|1 - \alpha\lambda_j|$ for other j lie in between these two lines. This means

$$\alpha\lambda_{\max}(A) - 1 = 1 - \alpha\lambda_{\min}(A) \quad \Longrightarrow \quad \alpha_{\mathrm{opt}} = \frac{2}{\lambda_{\max} + \lambda_{\min}}.$$

For the optimal α, we get

$$\rho_{\text{opt}} = \rho(I - \alpha_{\text{opt}} A) = \frac{\lambda_{\max} - \lambda_{\min}}{\lambda_{\max} + \lambda_{\min}} = \frac{\kappa(A) - 1}{\kappa(A) + 1} \quad \text{with} \quad \kappa(A) = \frac{\lambda_{\max}}{\lambda_{\min}}.$$

$$(11.40)$$

$\square$

11.4 Local Minimization by Nonstationary Iterative Methods

In contrast to the stationary iterative methods, the increment $\boldsymbol{x}_{k+1} - \boldsymbol{x}_k$ chosen by non-stationary iterative methods varies from one iteration to the next. We assume throughout this section that the matrix $A \in \mathbb{R}^{n \times n}$ symmetric is positive definite unless otherwise stated. To start, we consider the method of Richardson in its original form in [106], where the parameter α is allowed to change in each iteration,

$$\boldsymbol{x}_{k+1} = \boldsymbol{x}_k + \alpha_k(\boldsymbol{b} - A\boldsymbol{x}_k). \tag{11.41}$$

We consider for $\mu \in \mathbb{R}$ the norm (it is a norm since A is positive definite)

$$\|\boldsymbol{r}\|_{A^{-\mu}}^2 = \boldsymbol{r}^\top A^{-\mu} \boldsymbol{r}, \tag{11.42}$$

and, following Ritz's idea[5], we want to determine the value of α_k that minimizes

$$Q(\alpha_k) := \|\boldsymbol{r}_{k+1}\|_{A^{-\mu}}^2. \tag{11.43}$$

Differentiating with respect to α_k and noting that iteration (11.41) can be written as a recurrence for residuals,

$$\boldsymbol{r}_{k+1} = (I - \alpha_k A)\boldsymbol{r}_k, \tag{11.44}$$

we get

$$\frac{dQ}{d\alpha_k} = 0 = 2\boldsymbol{r}_{k+1}^\top A^{-\mu} \frac{d}{d\alpha_k} \boldsymbol{r}_{k+1} = 2((I - \alpha_k A)\boldsymbol{r}_k)^\top A^{-\mu}(-A\boldsymbol{r}_k).$$

Solving for α_k, we obtain

$$\alpha_k = \frac{\boldsymbol{r}_k^\top A^{1-\mu} \boldsymbol{r}_k}{\boldsymbol{r}_k^\top A^{2-\mu} \boldsymbol{r}_k}. \tag{11.45}$$

Useful choices for μ are 0 and 1, since otherwise we would have to solve equations with the matrix A, or perform more than one multiplication with A in order to determine α_k.

---

[5] It was Ritz who first proposed in 1908 to find approximate solutions to minimization problems by searching in a small, well-chosen subspace.

11.4.1 Conjugate Residuals

For $\mu = 0$, (11.45) gives

$$\alpha_k = \frac{r_k^\top A r_k}{\|A r_k\|_2^2},\qquad (11.46)$$

so in this case we are minimizing locally

$$\|r_{k+1}\|_2^2 = \|A e_{k+1}\|_2^2 = \|e_{k+1}\|_{A^2}^2 \longrightarrow \min .$$

Using (11.44) for r_{k+1} and (11.46) for α_k, a short calculation shows that

$$r_{k+1}^\top A r_k = 0,$$

which means that the *residuals are conjugate*, and therefore this method is called the *conjugate residuals algorithm*.

ALGORITHM 11.5. *Conjugate Residuals Algorithm*

$$
\begin{aligned}
&k = 0;\ x = x_0\ ;\\
&r = b - Ax;\\
&\text{while not converged}\\
&\quad k = k + 1;\\
&\quad a_r = Ar;\\
&\quad \alpha = \frac{r^\top a_r}{\|a_r\|^2};\\
&\quad x = x + \alpha r;\\
&\quad r = r - \alpha a_r;\\
&\text{end}
\end{aligned}
$$

11.4.2 Steepest Descent

Another method, called the *Steepest Descent method*, can be derived by choosing $\mu = 1$; then (11.45) gives

$$\alpha_k = \frac{\|r_k\|^2}{r_k^\top A r_k}.\qquad (11.47)$$

THEOREM 11.14. (STEEPEST DESCENT PROPERTIES) *Let $A \in \mathbb{R}^{n \times n}$ be symmetric and positive definite. With the choice of α_k given by (11.47), the following functionals are being minimized by the method (11.41):*

a) $\|r_{k+1}\|_{A^{-1}}^2 = r_{k+1}^\top A^{-1} r_{k+1}.$

b) $Q(x_k + \alpha r_k)$ *as a function of α, where $Q(x) = \frac{1}{2} x^\top A x - b^\top x.$*

c) $\|e_{k+1}\|_A^2 = e_{k+1}^\top A e_{k+1}$.

PROOF. Statement a) holds by construction. Statement c) follows from Equation (11.14), which states that $A e_{k+1} = r_{k+1}$.

For statement b), $Q(x)$ is a quadratic form whenever A is symmetric and positive definite, and it attains its unique minimum at $x = A^{-1}b$. The gradient of Q is

$$\nabla Q = Ax - b = -r,$$

which means that the residual points into the direction of steepest descent. Minimizing $Q(x + \alpha r)$ with respect to α for fixed x and r means to descend in the direction of steepest descent. In order to determine α, we expand Q,

$$Q(x + \alpha r) = \frac{1}{2}(x + \alpha r)^\top A(x + \alpha r) - b^\top (x + \alpha r),$$

and compute the derivative with respect to α and set it to zero,

$$\frac{dQ}{d\alpha} = r^\top Ax + \alpha r^\top Ar - b^\top r = 0,$$

which leads with $r^\top Ax - r^\top b = -r^\top r$ to $\alpha = \frac{r^\top r}{r^\top Ar}$, as required by (11.47). $\square$

The pairs of consecutive residuals obtained by the this *method of steepest descent* are not conjugate but *orthogonal,* as one can see by a short calculation evaluating the scalar product $r_{k+1}^\top r_k$.

ALGORITHM 11.6. *Steepest Descent Algorithm*

$k = 0;\ x = x_0\ ;$
$r = b - Ax;$
while not converged
    $k = k + 1;$
    $a_r = Ar;$
    $\alpha = \dfrac{r^\top r}{r^\top a_r};$
    $x = x + \alpha r;$
    $r = r - \alpha a_r;$
end

THEOREM 11.15. (STEEPEST DESCENT CONVERGENCE) *Let $A \in \mathbb{R}^{n \times n}$ be symmetric and positive definite. Then the method of steepest descent converges.*

PROOF. The idea is to show that the sequence of $\|\boldsymbol{r}_k\|_{A^{-1}}$ goes to zero as k goes to infinity. We compute

$$\begin{aligned}
\|\boldsymbol{r}_{k+1}\|^2_{A^{-1}} &= \boldsymbol{r}^\top_{k+1} A^{-1} \boldsymbol{r}_{k+1} \\
&= (\boldsymbol{r}_k - \alpha_k A \boldsymbol{r}_k)^\top A^{-1} \boldsymbol{r}_{k+1} \\
&= \boldsymbol{r}^\top_k A^{-1} \boldsymbol{r}_{k+1} - \alpha_k \boldsymbol{r}^\top_k A^\top A^{-1} \boldsymbol{r}_{k+1} \\
&= \boldsymbol{r}^\top_k A^{-1} \boldsymbol{r}_{k+1},
\end{aligned}$$

where we have used the fact that $A^\top A^{-1} = I$ (A is symmetric) and the consecutive residuals are orthogonal. Replacing $\boldsymbol{r}_{k+1}$ in the last expression, we obtain

$$\|\boldsymbol{r}_{k+1}\|^2_{A^{-1}} = \boldsymbol{r}^\top_k A^{-1}(\boldsymbol{r}_k - \alpha_k A \boldsymbol{r}_k) = \|\boldsymbol{r}_k\|^2_{A^{-1}} - \alpha_k \|\boldsymbol{r}_k\|^2_2.$$

Inserting now the value of α_k and dividing by $\|\boldsymbol{r}_k\|^2_{A^{-1}}$, we get

$$\frac{\|\boldsymbol{r}_{k+1}\|^2_{A^{-1}}}{\|\boldsymbol{r}_k\|^2_{A^{-1}}} = 1 - \frac{\|\boldsymbol{r}_k\|^4_2}{(\boldsymbol{r}^\top_k A \boldsymbol{r}_k)(\boldsymbol{r}^\top_k A^{-1} \boldsymbol{r}_k)}.$$

By the Kantorovitch inequality, see Theorem 11.16 below, the right hand side can be bounded using $\kappa := \kappa(A)$, the condition number of A, and we obtain

$$\frac{\|\boldsymbol{r}_{k+1}\|^2_{A^{-1}}}{\|\boldsymbol{r}_k\|^2_{A^{-1}}} \le 1 - \frac{4}{(\sqrt{\kappa} + \frac{1}{\sqrt{\kappa}})^2} = \left(\frac{\kappa - 1}{\kappa + 1}\right)^2 < 1.$$

Therefore $\|\boldsymbol{r}_k\|_{A^{-1}} \to 0$ for $k \to \infty$. $\qquad\square$

THEOREM 11.16. (KANTOROVITCH INEQUALITY) *Let $A \in \mathbb{R}^{n \times n}$ be symmetric and positive definite. Then for all $\boldsymbol{x} \in \mathbb{R}^n$, $\boldsymbol{x} \ne 0$, we have*

$$1 \le \frac{\boldsymbol{x}^\top A \boldsymbol{x} \cdot \boldsymbol{x}^\top A^{-1} \boldsymbol{x}}{(\boldsymbol{x}^\top \boldsymbol{x})^2} \le \frac{\left(\sqrt{\kappa(A)} + \left(\sqrt{\kappa(A)}\right)^{-1}\right)^2}{4}, \tag{11.48}$$

where $\kappa(A) := \frac{\lambda_{\max}(A)}{\lambda_{\min}(A)}$ is the condition number of A.

PROOF. Let $A = Q \Lambda Q^\top$ be the eigen-decomposition of A, with Q orthogonal and $\Lambda = \mathrm{diag}(\lambda_1, \ldots, \lambda_n)$, where $0 < \lambda_1 \le \cdots \le \lambda_n$ because A is positive definite. Then defining $\Lambda^{1/2} = \mathrm{diag}(\sqrt{\lambda_1}, \ldots, \sqrt{\lambda_n})$, we can write

$$\begin{aligned}
\boldsymbol{x}^\top \boldsymbol{x} &= \boldsymbol{x}^\top Q Q^\top \boldsymbol{x} && \text{(Q is orthogonal)} \\
&= \boldsymbol{x}^\top Q \Lambda^{1/2} \Lambda^{-1/2} Q^\top \boldsymbol{x} \\
&\le \|\Lambda^{1/2} Q^\top \boldsymbol{x}\|_2 \|\Lambda^{-1/2} Q^\top \boldsymbol{x}\|_2 && \text{(Cauchy--Schwarz)} \\
&= (\boldsymbol{x}^\top Q \Lambda Q^\top \boldsymbol{x})^{1/2} (\boldsymbol{x}^\top Q \Lambda^{-1} Q^\top \boldsymbol{x})^{1/2} \\
&= (\boldsymbol{x}^\top A \boldsymbol{x})^{1/2} (\boldsymbol{x}^\top A^{-1} \boldsymbol{x})^{1/2}.
\end{aligned}$$

Squaring and dividing by $(\boldsymbol{x}^\top \boldsymbol{x})^2$ then yields the first inequality.

For the second inequality, let $c > 0$ be a constant (to be chosen later) and let $\tilde{A} := cA$. Then

$$
\begin{aligned}
(\boldsymbol{x}^\top A\boldsymbol{x})^{1/2}(\boldsymbol{x}^\top A^{-1}\boldsymbol{x})^{1/2} &= (\boldsymbol{x}^\top (cA)\boldsymbol{x})^{1/2}(\boldsymbol{x}^T (cA)^{-1}\boldsymbol{x})^{1/2} \\
&\leq \frac{1}{2}(\boldsymbol{x}^\top \tilde{A}\boldsymbol{x} + \boldsymbol{x}^\top \tilde{A}^{-1}\boldsymbol{x}) \\
&= \frac{1}{2}\boldsymbol{x}^\top Q(\tilde{\Lambda} + \tilde{\Lambda}^{-1})Q^\top \boldsymbol{x},
\end{aligned}
$$

where we have used the arithmetic-geometric mean inequality $\sqrt{ab} \leq \frac{1}{2}(a+b)$ for $a, b > 0$. If we now define the function $f(\lambda) = \lambda + \frac{1}{\lambda}$, then the inequality above becomes

$$
\begin{aligned}
(\boldsymbol{x}^\top A\boldsymbol{x})^{1/2}(\boldsymbol{x}^\top A^{-1}\boldsymbol{x})^{1/2} &\leq \frac{1}{2}(Q^\top \boldsymbol{x})^\top f(\tilde{\Lambda})(Q^\top \boldsymbol{x}) \\
&\leq \frac{1}{2}\|f(\tilde{\Lambda})\|_2 \|Q^\top \boldsymbol{x}\|_2^2 = \frac{1}{2}\max_j |f(\tilde{\lambda}_j)|\boldsymbol{x}^\top \boldsymbol{x}.
\end{aligned}
$$

But $f(\lambda) > 0$ and $f''(\lambda) = 2/\lambda^3 > 0$ for $\lambda > 0$, so f is a positive convex function over the interval $[\tilde{\lambda}_1, \tilde{\lambda}_n]$. Thus, we have

$$
\max_j |f(\tilde{\lambda}_j)| = \max\{f(\tilde{\lambda}_1), f(\tilde{\lambda}_n)\},
$$

so picking $c = 1/\sqrt{\lambda_1 \lambda_n}$ gives

$$
\tilde{\lambda}_1 = \sqrt{\frac{\lambda_1}{\lambda_n}}, \qquad \tilde{\lambda}_n = \sqrt{\frac{\lambda_n}{\lambda_1}},
$$

which in turn yields

$$
\max_j |f(\tilde{\lambda}_j)| = f(\tilde{\lambda}_1) = f(\tilde{\lambda}_n) = \sqrt{\frac{\lambda_1}{\lambda_n}} + \sqrt{\frac{\lambda_n}{\lambda_1}}.
$$

Finally, noting that $\kappa(A) = \lambda_n/\lambda_1$, we obtain

$$
(\boldsymbol{x}^\top A\boldsymbol{x})^{1/2}(\boldsymbol{x}^\top A^{-1}\boldsymbol{x})^{1/2} \leq \frac{1}{2}\left(\sqrt{\kappa(A)} + \frac{1}{\sqrt{\kappa(A)}}\right)\boldsymbol{x}^\top \boldsymbol{x},
$$

which, upon squaring and dividing by $(\boldsymbol{x}^\top \boldsymbol{x})^2$, gives the second inequality. $\square$

11.5 Global Minimization with Chebyshev Polynomials

The choice of α_k in the previous section such that $\|\boldsymbol{r}_{k+1}\|_{A^{-\mu}}^2 \longrightarrow$ min means that we want to do the best for the current iteration. This is the best *tactic*, but perhaps not the best *overall strategy*. To illustrate this, we consider the

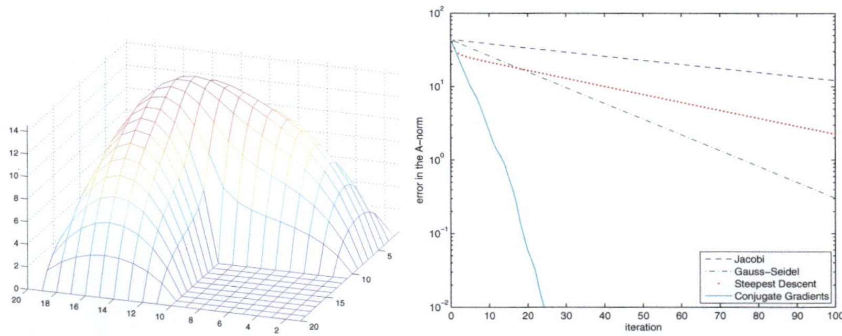

FIGURE 11.6.
*Solution of the Poisson equation on an L-shaped domain
on the left, and reduction of the error, measured in the
A-norm, by Jacobi, Gauss-Seidel, Steepest Descent and
Conjugate Gradients on the right, starting with a zero
initial guess*

numerical solution of a Poisson equation on an L-shaped domain shown in
Figure 11.6 on the left. To obtain the system matrix and exact solution for
this problem, we use the MATLAB commands

```
n=20;
G=numgrid('L',n);
A=delsq(G);
b=ones(size(A,1),1);
u=A\b;
U=G;
U(G>0)=u(G(G>0));
mesh(U);
axis([1 n 1 n 0 max(u)])
view([200,30])
```

When solving this discretized problem using the methods of Jacobi (§11.3.2),
Gauss-Seidel (§11.3.3) and Steepest Descent (§11.4.2), starting with an initial
guess of zero, we obtain the convergence curves shown in Figure 11.6 on the
right. We observe that Gauss-Seidel converges faster than Jacobi, about
twice as fast. Steepest descent converges even faster initially, but it then
seems to slow down, eventually becoming even slower than Gauss-Seidel.
This slow convergence behavior of stationary iterative methods and locally
optimal methods is typical. In Figure 11.7, we show for each method the
approximation obtained after the first, fifth, tenth and twentieth iterations.
Clearly, doing the best locally, as in steepest descent, is not beneficial for the
overall performance. The performance of the Conjugate Gradient method,
which we will explain later in Subsection 11.7.1, is also shown in Figure 11.7
on the right, and in the last row of Figure 11.7. This method performs much
better than all the others. In order to get some more insight, we show in

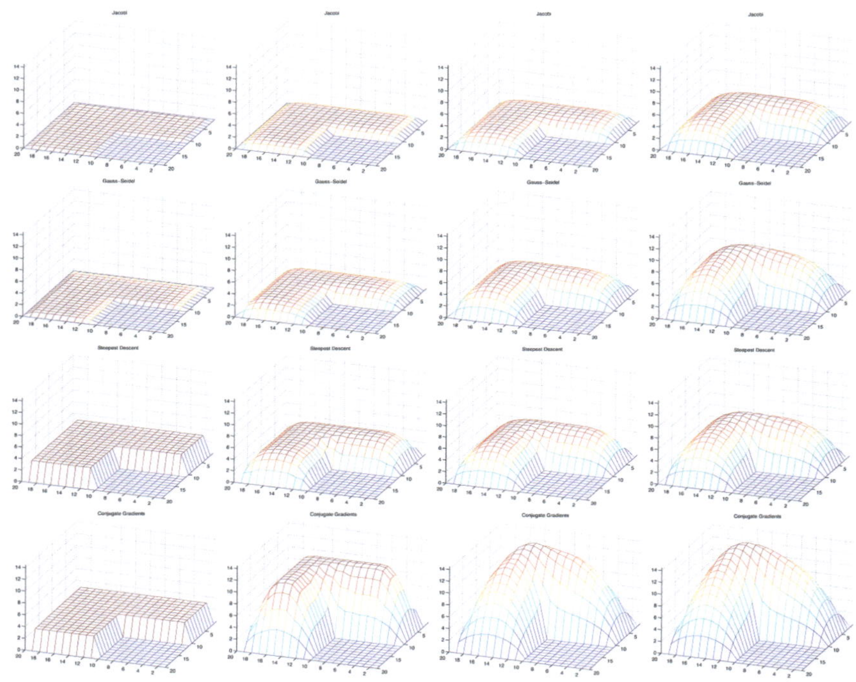

FIGURE 11.7.
Iteration 1, 5, 10, and 20 of Jacobi (top row),
Gauss-Seidel, Steepest Descent and Conjugate gradients
(bottom row) to approximate the solution shown in
Figure 11.6 on the left

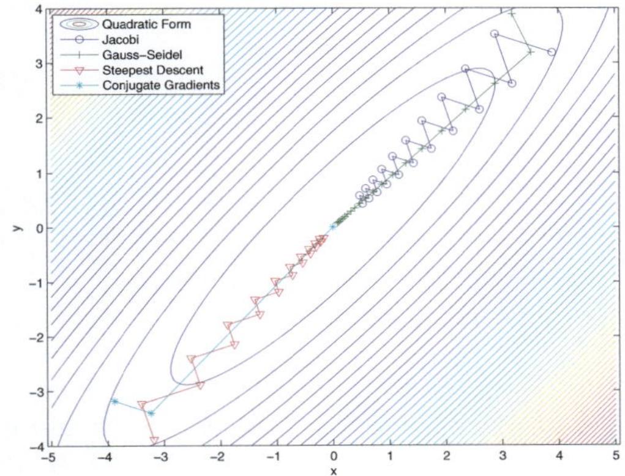

FIGURE 11.8.
Convergence paths of Jacobi, Gauss-Seidel, Steepest Descent and Conjugate Gradients for a two dimensional model problem

Figure 11.8 the results obtained for a much simpler problem, namely the matrix A obtained from the MATLAB commands

```
V=[1  1
   1 -1]/sqrt(2);
E=[1 0
   0 20];
A=V'*E*V;
```

As we can see in Figure 11.8, we try in this example to compute the zero solution. Each method starts from a different initial guess, but because of symmetry, these initial guesses are equivalent, and are only chosen to be different for the sake of better visibility of the approximations computed by the different iterative methods. We clearly see that Jacobi, Gauss-Seidel and Steepest Descent search over and over again into the same directions, whereas Conjugate Gradients search each direction only once, and thus finds the solution in two steps in this two-dimensional example. Conjugate Gradients is trying to do the best globally at each step, in contrast to Steepest Descent, which tries to do the best locally at each step. The remainder of this chapter focuses on how to obtain such globally best methods and approximations thereof.

Let us consider again the method of Richardson with variable parameters α_k. The recurrence for the residuals is

$$\boldsymbol{r}_{k+1} = (I - \alpha_k A)\boldsymbol{r}_k.$$

Because of $\boldsymbol{r}_k = A\boldsymbol{e}_k$, we have the same recurrence also for the error

$$\boldsymbol{e}_{k+1} = (I - \alpha_k A)\boldsymbol{e}_k.$$

Thus, the error after k steps is

$$e_k = (I - \alpha_{k-1}A)(I - \alpha_{k-2}A)\cdots(I - \alpha_0 A)e_0 = P_k(A)e_0. \qquad (11.49)$$

DEFINITION 11.9. (RESIDUAL POLYNOMIAL) *The polynomial P_k in (11.49), which satisfies $P_k(0) = 1$, is called the* residual polynomial.

The residual polynomial is determined by the choice of α_k; its zeros are $z_j = 1/\alpha_j$. The parameters α_k should be chosen so that the *global error* is minimized.

If we could choose $\alpha_k = 1/\lambda_k$, where λ_k is the k-th eigenvalue of A, then, because of the Cayley-Hamilton Theorem, we would have $P_n(A) = 0$ after at most n steps, since every matrix is annihilated by its characteristic polynomial. In fact, we might even reach the solution earlier, since it is sufficient to iterate until $P_k(A)e_0 = 0$.

DEFINITION 11.10. (MINIMAL POLYNOMIAL) *The* minimal polynomial *of a matrix $A \in \mathbb{R}^{n \times n}$ is the polynomial $\tilde{P}_m(x)$ with smallest degree m such that $\tilde{P}_m(A) = 0$.*

By the Cayley-Hamilton Theorem, we know that the minimal polynomial is a divisor of the characteristic polynomial, so $m \leq n$.

DEFINITION 11.11. (MINIMAL POLYNOMIAL WITH RESPECT TO A VECTOR) *The* minimal polynomial *of a matrix $A \in \mathbb{R}^{n \times n}$ with respect to a vector $\boldsymbol{v}$ is the polynomial $P_m(x)$ with smallest degree m such that $P_m(A)\boldsymbol{v} = \boldsymbol{0}$.*

Note that the degree of the minimal polynomial with respect to a vector might be lower than the minimal polynomial of the matrix. For instance, if $\boldsymbol{v}$ is the linear combination of two eigenvectors with distinct eigenvalues λ_1 and λ_2, then $P_m(x) = (1 - x/\lambda_1)(1 - x/\lambda_2)$ satisfies $P_m(A)\boldsymbol{v} = 0$, so the minimal polynomial with respect to $\boldsymbol{v}$ has degree two.

For optimal convergence, we would ideally like to use the *minimal polynomial with respect to e_0*, but unfortunately this is not known in general. However, since the inequality

$$\|e_k\| = \|P_k(A)e_0\| \leq \|P_k(A)\|\|e_0\|$$

holds for any polynomial P_k, one can try to choose P_k in order to minimize the upper bound $\|P_k(A)\|$. For symmetric positive definite matrices, the eigen-decomposition $A = Q\Lambda Q^\top$ gives

$$\|P_k(A)\| = \|QP_k(\Lambda)Q^\top\| \leq \|Q\|\|Q^\top\|\|P_k(\Lambda)\| = \kappa(Q)\|P_k(\Lambda)\|,$$

where $\|P_k(\Lambda)\|$ is a diagonal matrix with elements $P_k(\lambda_j)$. Moreover, if the spectral norm $\|\cdot\|_2$ is used, then $\kappa(Q) = 1$, since the spectral norm is invariant under orthogonal transformations. Thus, to minimize the bound, we would like to choose the polynomial P_k so that it is small on the spectrum of A. In general, the spectrum of A will not be known, but it may be possible to estimate the interval $[a, b]$ within which the eigenvalues lie in the case where A has only real eigenvalues:

$$a \leq \lambda_1 \leq \lambda_2 \leq \cdots \leq \lambda_n \leq b.$$

In this case we have

$$\max_i |P_k(\lambda_i)| \leq \max_{a \leq x \leq b} |P_k(x)|.$$

Therefore we consider the following minimization problem: given a degree k, we want to determine a polynomial P_k with $P_k(0) = 1$ such that

$$\max_{a \leq x \leq b} |P_k(x)| \longrightarrow \min.$$

For indefinite A, i.e., when the interval $[a, b]$ contains zero, the minimum is necessarily greater than or equal to one because of the constraint $P_k(0) = 1$; however, for positive definite A, i.e., when $a > 0$, a good choice of P_k will lead to a minimum strictly less than 1, and hence will give rise to useful algorithms. As we will see, the polynomials that are closest to zero in a given interval are given by the *Chebyshev polynomials*.

DEFINITION 11.12. (CHEBYSHEV POLYNOMIALS) *The* Chebyshev polynomials *are defined by*

$$C_k(t) := \cos(k \arccos t), \quad -1 \leq t \leq 1, \quad k = 0, 1, \ldots \tag{11.50}$$

We show in Figure 11.9 the first seven Chebyshev polynomials, which we plotted using the MAPLE commands

```
> with(orthopoly);
> plot([seq(T(i,t),i=0..6)],t=-1..1);
```

Why is C_k a polynomial? Let $t = \cos \alpha$, so that

$$C_k(\cos \alpha) = \cos(k\alpha).$$

The function $\cos(k\alpha)$ for multiple angles can be represented as a polynomial in $\cos \alpha$, for example

$$
\begin{array}{llll}
k = 0: & C_0(\cos \alpha) = \cos(0) = 1 & \implies C_0(t) = 1, \\
k = 1: & C_1(\cos \alpha) = \cos(\alpha) & \implies C_1(t) = t, \\
k = 2: & C_2(\cos \alpha) = \cos(2\alpha) = 2\cos(\alpha)^2 - 1 & \implies C_2(t) = 2t^2 - 1, \\
k = 3: & C_3(\cos \alpha) = \cos(3\alpha) = 4\cos(\alpha)^3 - 3\cos(\alpha) & \implies C_3(t) = 4t^3 - 3t.
\end{array}
$$

We now state some properties of the Chebyshev polynomials:

1. Summing the trigonometric identities

$$\cos(k+1)\alpha = \cos(k\alpha + \alpha) = \cos k\alpha \cos \alpha - \sin k\alpha \sin \alpha,$$

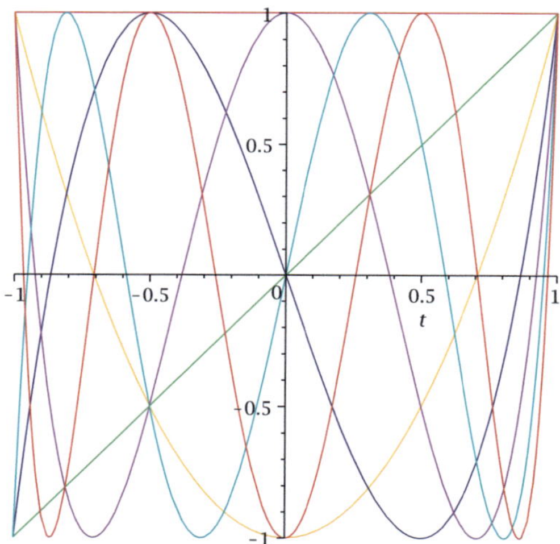

FIGURE 11.9. *The first seven Chebyshev polynomials*

$$\cos(k-1)\alpha = \cos(k\alpha - \alpha) = \cos k\alpha \cos \alpha + \sin k\alpha \sin \alpha,$$

we get

$$\cos(k+1)\alpha + \cos(k-1)\alpha = 2\cos k\alpha \cos \alpha,$$

which translates into the well-known *three-term recurrence relation* for Chebyshev polynomials,

$$C_0 = 1, \quad C_1 = t, \quad C_{k+1}(t) = 2tC_k(t) - C_{k-1}(t), \quad k = 1, 2, \ldots \tag{11.51}$$

2. From the recurrence relation, we see that C_k is a polynomial of degree k with integer coefficients and leading coefficient 2^{k-1},

$$C_k(t) = 2^{k-1}t^k + \ldots$$

3. For k even, C_k is an even function, i.e., $C_k(t) = C_k(-t)$ for all t. This can be proved by induction from the recurrence relation. Similarly, for k odd, we have $C_k(t) = -C_k(-t)$. These properties can also be observed in Figure 11.9.

4. From Definition 11.12, we see that $|C_k(t)| \leq 1$ for $|t| \leq 1$, see also Figure 11.9.

5. The trigonometric identity for $k > l$,

$$\cos(k+l)\alpha + \cos(k-l)\alpha = 2\cos l\alpha \cos k\alpha,$$

translates into the *abbreviated recurrence relation*

$$C_{k+l} = 2C_l C_k - C_{k-l}.$$

6. The zeros t_j of C_k are the solutions of $C_k(\cos\alpha) = \cos k\alpha = 0$, that is, $k\alpha = -\pi/2 + j\pi$ for $j = 1, 2\ldots, k$, and thus

$$t_j = \cos\left(-\frac{\pi}{2k} + \frac{j\pi}{k}\right), \quad j = 1, 2\ldots, k. \tag{11.52}$$

7. The extrema ξ_j of C_k are solutions of $|\cos k\alpha| = 1$, that is $k\alpha = j\pi$ for $j = 0, \ldots, k$ (including the boundary values for $t = 1, -1$), and thus

$$\xi_j = \cos\frac{j\pi}{k}, \quad j = 0, \ldots, k.$$

Moreover, we have a maximum for even j and a minimum for odd j, i.e.,

$$C_k(\xi_j) = (-1)^j.$$

In particular, $C_k(1) = 1$ and $C_k(-1) = (-1)^k$, see also Figure 11.9.

8. The Chebyshev polynomials are *orthogonal polynomials* for the scalar product

$$(f, g) := \int_{-1}^{1} \frac{f(t)g(t)}{\sqrt{1-t^2}}\, dt.$$

PROOF. With the variable transformation $t = \cos\alpha$, we get

$$
\begin{aligned}
\int_{-1}^{1} \frac{C_i(t)C_j(t)}{\sqrt{1-t^2}}\, dt &= \int_0^{\pi} \cos i\alpha \cos j\alpha\, d\alpha \\
&= \int_0^{\pi} \frac{1}{2}\left(\cos(i+j)\alpha + \cos(i-j)\alpha\right)\, d\alpha \\
&= \begin{cases} \pi, & i = j = 0, \\ \frac{\pi}{2}, & i = j \neq 0, \\ 0, & i \neq j. \end{cases}
\end{aligned}
$$

$\square$

9. C_k has the smallest deviation from zero:

THEOREM 11.17. (SMALLEST DEVIATION FROM ZERO OF CHEBYSHEV POLYNOMIALS) *The Chebyshev polynomials have in the interval $[-1, 1]$ the smallest deviation from zero among all polynomials of the same degree k with leading coefficient 2^{k-1}.*

PROOF. Let Q_k be a polynomial with leading coefficient 2^{k-1}, and suppose that

$$\max_{-1\leq t\leq 1} |Q_k(t)| \leq \max_{-1\leq t\leq 1} |C_k(t)| = 1. \tag{11.53}$$

We now use the extremal points $1 = \xi_0, \ldots, \xi_k$ where $|C_k(\xi_i)| = 1$. The difference function

$$R_{k-1}(t) = C_k(t) - Q_k(t)$$

is a polynomial of degree $k - 1$, since the leading coefficient of C_k and Q_k is the same. Using (11.53), we conclude that

$$
\begin{aligned}
R_{k-1}(\xi_0) &\geq 0, \\
R_{k-1}(\xi_1) &\leq 0, \\
&\ \ \vdots \\
R_{k-1}(\xi_k)(-1)^k &\geq 0.
\end{aligned}
$$

It follows that the polynomial $R_{k-1}(t)$ has at least one zero in each closed interval $[\xi_0, \xi_1]$, $[\xi_1, \xi_2]$, $\ldots [\xi_{k-1}, \xi_k]$. If a zero t is on the boundary of an interval, it has to be counted twice, since at such a point we have $C_k'(t) = 0$ and $Q_k'(t) = 0$, and hence also the derivative $R_{k-1}'(t)$ vanishes, which indicates at least a double root. In total, the number of zeros is therefore greater than or equal to k. But since the degree of $R_{k-1}(t)$ is $k - 1$, this can only happen if $R_{k-1}(t) \equiv 0$, and therefore $Q_k \equiv C_k$. $\qquad\qquad\square$

10. Scaling and normalization: by a change of variables, we can define the Chebyshev polynomials for the interval $[a, b]$ as follows. If $-1 \leq t \leq 1$, we consider $a \leq x \leq b$ with

$$x = \frac{a+b}{2} + \frac{b-a}{2}t \iff t = -1 + 2\frac{x-a}{b-a},$$

and

$$\widetilde{C}_k(x) := C_k\left(-1 + 2\frac{x-a}{b-a}\right).$$

We want to normalize the polynomials so that the function value for $x = 0$ is 1. The normalized Chebyshev polynomial is then

$$Q_k(x) = \frac{C_k\left(-1 + 2\frac{x-a}{b-a}\right)}{C_k\left(\frac{a+b}{a-b}\right)}. \tag{11.54}$$

We can now translate the result of Theorem 11.17 to the interval $[a, b]$.

THEOREM 11.18. (MINIMIZING PROPERTY OF CHEBYSHEV POLYNOMIALS) *Let* $\Pi_k^1 := \{Q_k \text{ polynomial of degree } k \text{ with } Q_k(0) = 1\}$. *Then*

$$\min_{R_k \in \Pi_k^1} \max_{a \leq x \leq b} |R_k(x)| = \max_{a \leq x \leq b} \left| \frac{C_k\left(-1 + 2\frac{x-a}{b-a}\right)}{C_k\left(\frac{a+b}{a-b}\right)} \right| = \frac{1}{\left| C_k\left(\frac{a+b}{a-b}\right) \right|}.$$

Often, instead of $Q_k(0) = 1$, the normalization at $x = 1$ is used, which leads to the polynomials

$$P_k(x) = \frac{C_k\left(-1 + 2\frac{x-a}{b-a}\right)}{C_k\left(\frac{2-b-a}{b-a}\right)}, \quad \text{with } P_k(1) = 1.$$

11. Bound for $C_k\left(\frac{a+b}{a-b}\right)$: the Chebyshev polynomials have been defined for $|t| = |\cos\alpha| \leq 1$ by the formula $C_k(\cos\alpha) = \cos k\alpha$, and we have seen in 2. that this indeed defines a polynomial, whose domain is not restricted to $-1 \leq t \leq 1$. The following lemma gives a convenient formula for evaluating Chebyshev polynomials when $|t| > 1$.

LEMMA 11.6. *For $t \in \mathbb{R}$, the Chebyshev polynomial C_k is given by*

$$C_k(t) = \frac{1}{2}\left(\left(t + \sqrt{t^2 - 1}\right)^k + \left(t - \sqrt{t^2 - 1}\right)^k\right). \qquad (11.55)$$

PROOF. For $t = \cos\alpha$ with $|t| \leq 1$, the Chebyshev polynomial is by definition (11.50)

$$C_k(t) = \cos k\alpha = \frac{1}{2}\left(e^{ik\alpha} + e^{-ik\alpha}\right) = \frac{1}{2}\left(\left(e^{i\alpha}\right)^k + \left(e^{-i\alpha}\right)^k\right)$$

$$= \frac{1}{2}\left((\cos\alpha + i\sin\alpha)^k + (\cos\alpha - i\sin\alpha)^k\right).$$

Now using that $\sin\alpha = \sqrt{1 - t^2}$, we obtain the polynomial as a function of t,

$$C_k(t) = \frac{1}{2}\left(\left(t + i\sqrt{1 - t^2}\right)^k + \left(t - i\sqrt{1 - t^2}\right)^k\right)$$

$$= \frac{1}{2}\left(\left(t + \sqrt{t^2 - 1}\right)^k + \left(t - \sqrt{t^2 - 1}\right)^k\right),$$

and hence the last expression defines the same polynomial. $\qquad\square$

Note that when expanding the parentheses in (11.55) using the binomial theorem, we obtain

$$C_k(t) = \frac{1}{2}\sum_{j=0}^{k}\binom{k}{j}\left(t^{k-j}(t^2 - 1)^{j/2} + (-1)^j t^{k-j}(t^2 - 1)^{j/2}\right)$$

$$= \sum_{j=0}^{\lfloor k/2 \rfloor}\binom{k}{2j}t^{k-2j}(t^2 - 1)^j,$$

so the expression indeed represents a polynomial. Moreover, when $t > 1$, then each term in the sum is positive, so we have the simple bound $C_k(t) \geq t^k$ for $t > 1$. The next theorem gives a more refined bound for the case $t = \frac{a+b}{a-b}$.

THEOREM 11.19. (GROWTH ESTIMATE OF CHEBYSHEV POLYNOMI-ALS) *The maximum of the shifted and scaled Chebyshev polynomial Q_k in (11.54) can be bounded by*

$$\frac{1}{\left| C_k\left(\frac{a+b}{a-b}\right)\right|} \leq 2 \left| \frac{\sqrt{b} - \sqrt{a}}{\sqrt{b} + \sqrt{a}} \right|^k . \tag{11.56}$$

PROOF. Applying Lemma 11.6 with $t = \frac{a+b}{a-b}$ yields

$$t \pm \sqrt{t^2 - 1} = \frac{a+b}{a-b} \pm \sqrt{\frac{b^2 + 2ab + a^2 - b^2 + 2ab - a^2}{(a-b)^2}}$$

$$= \frac{a + b \pm 2\sqrt{ab}}{a-b} = \frac{(\sqrt{a} \pm \sqrt{b})^2}{(\sqrt{a} + \sqrt{b})(\sqrt{a} - \sqrt{b})} .$$

We therefore find the value of the Chebyshev polynomial to be

$$C_k\left(\frac{a+b}{a-b}\right) = \frac{1}{2}\left[\left(\frac{\sqrt{a} + \sqrt{b}}{\sqrt{a} - \sqrt{b}}\right)^k + \left(\frac{\sqrt{a} - \sqrt{b}}{\sqrt{a} + \sqrt{b}}\right)^k \right] .$$

Moreover, we have inside the square brackets the sum of an expression and its inverse, which have the same sign. We can therefore estimate the modulus of C_k from below by the larger of the two expressions, namely

$$\left| C_k\left(\frac{a+b}{a-b}\right) \right| \geq \frac{1}{2}\left| \frac{\sqrt{a} + \sqrt{b}}{\sqrt{a} - \sqrt{b}} \right|^k ,$$

which leads to the desired estimate.   □

Combining Theorems 11.18 and 11.19, the global strategy of minimizing the error over all iterations yields an error of

$$\boldsymbol{e}_k = Q_k(A)\boldsymbol{e}_0 \quad \text{with} \quad Q_k(x) = \frac{C_k\left(-1 + 2\frac{x-a}{b-a}\right)}{C_k\left(\frac{a+b}{a-b}\right)} ,$$

and if the bounds on the spectrum are tight, i.e., if $a = \lambda_{\min}(A)$ and $b = \lambda_{\max}(A)$, then the estimate can be related to the condition number $\kappa(A)$ of the matrix A,

$$\|e_k\| \leq 2 \left(\frac{\sqrt{\frac{b}{a}} - 1}{\sqrt{\frac{b}{a}} + 1} \right)^k \|e_0\| = 2 \left(\frac{\sqrt{\kappa(A)} - 1}{\sqrt{\kappa(A)} + 1} \right)^k \|e_0\|.$$

The algorithm that results from these considerations is to choose first the degree k of the polynomial, and then apply Richardson's method with variable parameters α_j according to the zeros of the Chebyshev polynomial,

$$\frac{1}{\alpha_j} = \frac{a+b}{2} + \frac{b-a}{2} \cos\left(\frac{2j-1}{k} \frac{\pi}{2} \right), \quad j = 1, 2, \ldots, k$$

in a cyclic fashion. Unfortunately, this algorithm can have numerical instabilities if the α_j are not used in appropriate order, see for example [3], and references therein.

11.5.1 Chebyshev Semi-Iterative Method

In this section, a beautiful algorithm from the PhD thesis of Gene Golub [54] is described. The convergence rate of this algorithm is the same as the one obtained with the strategy described in the previous section, but the method is for a general splitting and has an elegant, compact implementation. We consider again a splitting $A = M - N$ and the basic stationary iteration

$$M\boldsymbol{x}_{i+1} = N\boldsymbol{x}_i + \boldsymbol{b}, \quad i = 1, \ldots, k.$$

Note that M and N need not be symmetric; for instance, the Gauss–Seidel splitting can be used. We would like to accelerate convergence by computing a linear combination of the iterates,

$$\boldsymbol{y}_k = \sum_{j=0}^{k} \gamma_j \boldsymbol{x}_j.$$

Clearly if all $\boldsymbol{x}_j$ are equal to the exact solution $\boldsymbol{x}$, then we would like to get $\boldsymbol{y} = \boldsymbol{x}$ also. Therefore, we need the normalization

$$\sum_{j=0}^{k} \gamma_j = 1.$$

Our goal will be to choose the coefficients γ_j such that $\boldsymbol{y}_k$ converges to $\boldsymbol{x}$ as fast as possible.

With $\boldsymbol{e}_j = \boldsymbol{x} - \boldsymbol{x}_j = (M^{-1}N)^j \boldsymbol{e}_0$ and $\boldsymbol{x} = \sum \gamma_j \boldsymbol{x}$, the error becomes

$$\boldsymbol{x} - \boldsymbol{y}_k = \sum_{j=0}^{k} \gamma_j (\boldsymbol{x} - \boldsymbol{x}_j) = \sum_{j=0}^{k} \gamma_j (M^{-1}N)^j \boldsymbol{e}_0 = P_k(M^{-1}N)\boldsymbol{e}_0.$$

Therefore the error $\boldsymbol{x} - \boldsymbol{y}_k$ becomes zero if the γ_j are the coefficients of the minimal polynomial of the iteration matrix $G = M^{-1}N$ with respect to $\boldsymbol{e}_0$, with the normalization $P_k(1) = \sum_{j=0}^{k} \gamma_j = 1$ and $\gamma_k \neq 0$.

We are again looking for polynomials which are small, but this time on the spectrum of the iteration matrix $M^{-1}N$. For the special case of a symmetric iteration matrix $M^{-1}N$, the method is convergent if and only if the eigenvalues $\lambda_i(M^{-1}N)$ satisfy

$$-1 < a = \lambda_1 \leq \lambda_2 \leq \cdots \leq \lambda_n = b < 1.$$

Therefore

$$\|P_k(M^{-1}N)\|_2 = \max_{\lambda_i}|P_k(\lambda_i)| \leq \max_{a \leq \lambda \leq b}|P_k(\lambda)|,$$

and the bound is minimized, as shown in the previous section, for

$$P_k(x) = \frac{C_k\left(-1 + 2\frac{x-a}{b-a}\right)}{C_k\left(\frac{2-b-a}{b-a}\right)}. \tag{11.57}$$

In the resulting algorithm, as we will see, it is not necessary to save the vectors $\boldsymbol{x}_i$ to compute the $\boldsymbol{y}_k$. The recurrence relation for Chebyshev polynomials will allow us to compute the sequence $\boldsymbol{y}_k$ without even computing the $\boldsymbol{x}_i$!

We use the polynomials P_k from (11.57), which are normalized to satisfy $P_k(1) = 1$. Let

$$t = -1 + 2\frac{x-a}{b-a} \quad \text{and} \quad \mu = \frac{2-b-a}{b-a}.$$

Then, using the Chebyshev recurrence, (11.57) becomes

$$C_k(\mu)P_k(x) = C_k(t) \tag{11.58}$$

$$= 2tC_{k-1}(t) - C_{k-2}(t). \tag{11.59}$$

Using (11.58) for $k-1$ and $k-2$ in (11.59), we obtain

$$C_k(\mu)P_k(x) = 2tC_{k-1}(\mu)P_{k-1}(x) - C_{k-2}(\mu)P_{k-2}(x),$$

and thus

$$P_k(x) = 2t\frac{C_{k-1}(\mu)}{C_k(\mu)}P_{k-1}(x) - \frac{C_{k-2}(\mu)}{C_k(\mu)}P_{k-2}(x). \tag{11.60}$$

The quotients of $C_k(\mu)$ can be simplified using the Chebyshev recurrence:

$$C_k(\mu) = 2\mu C_{k-1}(\mu) - C_{k-2}(\mu)$$

$$\implies \quad 1 = 2\mu\frac{C_{k-1}(\mu)}{C_k(\mu)} - \frac{C_{k-2}(\mu)}{C_k(\mu)}$$

$$\implies \quad \frac{C_{k-2}(\mu)}{C_k(\mu)} = 2\mu\underbrace{\frac{C_{k-1}(\mu)}{C_k(\mu)}}_{=:\,\omega_k} - 1,$$

and the recurrence (11.60) simplifies to

$$P_k(x) = t\frac{\omega_k}{\mu}P_{k-1}(x) + (1 - \omega_k)P_{k-2}(x).$$

Writing

$$t\frac{\omega_k}{\mu} = \frac{2x - a - b}{b - a}\frac{b - a}{2 - a - b}\omega_k = \left(\frac{2}{2 - a - b}x - \frac{a + b}{2 - a - b}\right)\omega_k$$

$$= (\gamma x + 1 - \gamma)\omega_k, \quad \text{with} \quad \gamma := \frac{2}{2 - a - b},$$

we can simplify the recurrence even more,

$$P_k(x) = (\gamma x + 1 - \gamma)\omega_k P_{k-1}(x) + (1 - \omega_k)P_{k-2}(x). \tag{11.61}$$

Finally, it is even possible to compute the coefficients ω_k recursively. Using the Chebyshev recurrence again, we get

$$\omega_k = 2\mu\frac{C_{k-1}(\mu)}{C_k(\mu)} = 2\mu\frac{C_{k-1}(\mu)}{2\mu C_{k-1}(\mu) - C_{k-2}(\mu)}$$

$$= \frac{2\mu}{2\mu - \dfrac{C_{k-2}(\mu)}{C_{k-1}(\mu)}} = \frac{4\mu^2}{4\mu^2 - \underbrace{2\mu\dfrac{C_{k-2}(\mu)}{C_{k-1}(\mu)}}_{\omega_{k-1}}}$$

and the recurrence for the ω_k is therefore for $k = 2, 3, \ldots$

$$\omega_k = \frac{4\mu^2}{4\mu^2 - \omega_{k-1}}, \quad \omega_1 = 2. \tag{11.62}$$

Let us show that the sequence ω_k converges and determine its limit. First, we show that $\mu > 1$. In fact, if $-1 < a < b < 1$, then

$$\begin{array}{rcll}
b - a & < & 1 - a & \mid \quad +1 \\
1 + b - a & < & 2 - a & \mid \quad -b \\
b - a < 1 - a & < & 2 - a - b & \mid \quad : b - a \\[2mm]
1 < \dfrac{1 - a}{b - a} & < & \dfrac{2 - a - b}{b - a} = \mu.
\end{array}$$

The fixed points s of the iteration (11.62), see also Section 5.2.2, are solutions of the equation

$$s = F(s) = \frac{4\mu^2}{4\mu^2 - s}, \quad s_{1,2} = 2\mu^2 \pm 2\mu\sqrt{\mu^2 - 1}.$$

To determine which fixed point is attractive, let us compute the interval over which $|F'(s)| < 1$. We have

$$F'(s) = \frac{4\mu^2}{(4\mu^2 - s)^2} > 0,$$

so $|F'(s)| < 1$ is satisfied if and only if

$$4\mu^2 < (4\mu^2 - s)^2 \iff (4\mu^2 - 2\mu - s)(4\mu^2 + 2\mu - s) > 0.$$

So $F'(s) < 1$ if $s \in (-\infty, 4\mu^2 - 2\mu) =: I_1$ or $s \in (4\mu^2 + 2\mu, \infty) =: I_2$. Now since

$$F'(s_1) = \frac{1}{(\mu - \sqrt{\mu^2 - 1})^2} = (\mu + \sqrt{\mu^2 - 1})^2 > 1,$$

we see that $s_1 \notin I_1 \cup I_2$, so this fixed point is not attractive. But

$$s_2 = 2\mu^2 - 2\mu\sqrt{\mu^2 - 1} = 4\mu^2 - 2\mu \underbrace{(\mu + \sqrt{\mu^2 - 1})}_{>1} < 4\mu^2 - 2\mu,$$

so $s_2 \in I_1$. Moreover, for any $s \in I_1$, we have

$$F(s) = \frac{4\mu^2}{4\mu^2 - s} < \frac{4\mu^2}{2\mu} = 2\mu < 2\mu \underbrace{(2\mu - 1)}_{>1} = 4\mu^2 - 2\mu,$$

which means F maps I_1 into I_1. Thus, by Banach's fixed point theorem (Theorem 5.5), the sequence ω_k with starting point $\omega_1 = 2 \in I_1$ converges to the unique fixed point $s_2 \in I_1$, i.e.,

$$\omega_k \;\to\; s_2 = 2\mu^2 - 2\mu\sqrt{\mu^2 - 1} = \frac{2}{1 + \sqrt{1 - \frac{1}{\mu^2}}}. \tag{11.63}$$

Now we go back to the convergence acceleration. As we have seen, the error is

$$\boldsymbol{x} - \boldsymbol{y}_k = P_k(M^{-1}N)\boldsymbol{e}_0. \tag{11.64}$$

We use the recurrence relation (11.61) to replace $P_k(M^{-1}N)$,

$$\boldsymbol{x} - \boldsymbol{y}_k = (\gamma M^{-1}N + (1-\gamma)I)\omega_k \underbrace{P_{k-1}(M^{-1}N)\boldsymbol{e}_0}_{\boldsymbol{x}-\boldsymbol{y}_{k-1}} + (1-\omega_k)\underbrace{P_{k-2}(M^{-1}N)\boldsymbol{e}_0}_{\boldsymbol{x}-\boldsymbol{y}_{k-2}}.$$

We multiply with -1 and expand,

$$\begin{aligned}
\boldsymbol{y}_k - \boldsymbol{x} = {}& \gamma\omega_k M^{-1}N\boldsymbol{y}_{k-1} + (1-\gamma)\omega_k\boldsymbol{y}_{k-1} + (1-\omega_k)\boldsymbol{y}_{k-2} \\
& - \gamma\omega_k M^{-1}N\boldsymbol{x} - (1-\gamma)\omega_k\boldsymbol{x} - (1-\omega_k)\boldsymbol{x}.
\end{aligned}$$

If we replace $M^{-1}N\boldsymbol{x}$ by $\boldsymbol{x} - M^{-1}\boldsymbol{b}$, which holds because of $M\boldsymbol{x} = N\boldsymbol{x} + \boldsymbol{b}$, we see that all the terms involving $\boldsymbol{x}$ cancel, and we are left with

$$\begin{aligned}
\boldsymbol{y}_k &= \gamma\omega_k M^{-1}N\boldsymbol{y}_{k-1} + (1-\gamma)\omega_k\boldsymbol{y}_{k-1} + (1-\omega_k)\boldsymbol{y}_{k-2} + \gamma\omega_k M^{-1}\boldsymbol{b} \\
&= \omega_k\gamma\left[(M^{-1}N - I)\boldsymbol{y}_{k-1} + M^{-1}\boldsymbol{b}\right] + \omega_k\boldsymbol{y}_{k-1} + (1-\omega_k)\boldsymbol{y}_{k-2}.
\end{aligned}$$

Notice that the expression in brackets can be simplified,

$$(M^{-1}N - I)\boldsymbol{y}_{k-1} + M^{-1}\boldsymbol{b} = -M^{-1}A\boldsymbol{y}_{k-1} + M^{-1}\boldsymbol{b} = M^{-1}(\boldsymbol{b} - A\boldsymbol{y}_{k-1}) =: \boldsymbol{z}_{k-1},$$

where $\boldsymbol{z}_{k-1}$ is the preconditioned residual.

We are now ready for the first variant of the algorithm. This algorithm accelerates the basic stationary iteration $M\boldsymbol{x}_{k+1} = N\boldsymbol{x}_k + \boldsymbol{b}$. The interval $[a, b] \subset (-1, 1)$ contains the spectrum of the iteration matrix $M^{-1}N$, and at step k the error is bounded by

$$\|\boldsymbol{x} - \boldsymbol{y}_k\| \leq \frac{1}{|C_k(\mu)|}\|\boldsymbol{x} - \boldsymbol{y}_0\|.$$

ALGORITHM 11.7. *Chebyshev-Semi-iterative Method I*

$w = 2$; $\mu = \frac{2-a-b}{b-a}$; $\gamma = \frac{2}{2-a-b}$;
choose $\boldsymbol{y}_0$;
$\boldsymbol{r} = \boldsymbol{b} - A\boldsymbol{y}_0$;
Solve $M\boldsymbol{z} = \boldsymbol{r}$;
$\boldsymbol{y}_1 = \boldsymbol{y}_0 + \gamma\boldsymbol{z}$;
$k = 1$; $c = 4\mu^2$;
while not converged
$\quad w = c/(c - w)$;
$\quad \boldsymbol{r} = \boldsymbol{b} - A\boldsymbol{y}_k$;
$\quad$ Solve $M\boldsymbol{z} = \boldsymbol{r}$;
$\quad k = k + 1$;
$\quad \boldsymbol{y}_k = w(\gamma\boldsymbol{z} + \boldsymbol{y}_{k-1} - \boldsymbol{y}_{k-2}) + \boldsymbol{y}_{k-2}$;
end

Remarks

1. The larger $\mu = \frac{2-a-b}{b-a} = 1 + 2\frac{1-b}{b-a}$ is, the smaller $1/|C_k(\mu)|$ becomes, and the better the iteration converges.

2. In the special case of $b = \rho(M^{-1}N) = -a < 1$, we have

$$\gamma = 1, \quad \mu = \frac{1}{\rho} \quad \text{and the error is bounded by} \quad \frac{1}{|C_k(\frac{1}{\rho})|} \leq \rho^k.$$

   Thus, if the basic iteration converges rapidly, then the accelerated iteration converges even faster.

3. A drawback of the method is that we need to know the parameters a and b or ρ.

4. In deriving the method, we have assumed that the iteration matrix $M^{-1}N$ is symmetric. However, for our arguments to hold, it is sufficient for the iteration matrix to have only real eigenvalues smaller than one in modulus. For non-symmetric matrices, this is unfortunately rarely the case: for example if we choose SOR as the basic iteration,

$$G_{\text{SOR}} = (D + \omega L)^{-1}((1 - \omega)D - \omega U),$$

the matrix is non-symmetric and we will in general have complex eigenvalues. A symmetrized version is therefore of interest, as we will see shortly.

We now rewrite the semi-iterative method in slightly different form, when the iteration matrix is only known as G,

$$\boldsymbol{x}_{k+1} = M^{-1}N\boldsymbol{x}_k + M^{-1}\boldsymbol{b} =: G\boldsymbol{x}_k + \boldsymbol{d}.$$

With M and N, the method was given by

$$\boldsymbol{y}_k = \omega_k\gamma\left[(M^{-1}N - I)\boldsymbol{y}_{k-1} + M^{-1}\boldsymbol{b}\right] + \omega_k\boldsymbol{y}_{k-1} + (1 - \omega_k)\boldsymbol{y}_{k-2},$$

which becomes, when using G and $\boldsymbol{d}$,

$$\boldsymbol{y}_k = \omega_k\left[\gamma(G\boldsymbol{y}_{k-1} + \boldsymbol{d}) + (1 - \gamma)\boldsymbol{y}_{k-1}\right] + (1 - \omega_k)\boldsymbol{y}_{k-2}. \tag{11.65}$$

The following algorithm accelerates the basic stationary iteration $\boldsymbol{x}_{k+1} = G\boldsymbol{x}_k + \boldsymbol{d}$, when the interval $[a, b] \subset (-1, 1)$ contains the spectrum of the iteration matrix G.

ALGORITHM 11.8. *Chebyshev-Semi-iterative Method II*

$w = 2$; $\mu = \frac{2-a-b}{b-a}$; $\gamma = \frac{2}{2-a-b}$;
choose $\boldsymbol{y}_0$;
$\boldsymbol{y}_1 = G\boldsymbol{y}_0 + \boldsymbol{d}$;
$k = 1$; $c = 4\mu^2$;
while not converged
    $w = c/(c - w)$;
    $z = G\boldsymbol{y}_k + \boldsymbol{d}$;
    $k = k + 1$;
    $\boldsymbol{y}_k = w(\gamma z + (1 - \gamma)\boldsymbol{y}_{k-1}) + (1 - w)\boldsymbol{y}_{k-2}$;
end

11.5.2   Acceleration of SSOR

As mentioned before, the iteration matrix of SOR is in general non-symmetric and may have complex eigenvalues. It is however possible to symmetrize SOR by performing an additional backward step. By doing so we obtain the *symmetric successive over relaxation iteration*, or SSOR. One iteration step consist of two parts:

Forward half step: $\quad (D + \omega L)\boldsymbol{x}_{k+\frac{1}{2}} = \omega \boldsymbol{b} + (1 - \omega)D\boldsymbol{x}_k - \omega U\boldsymbol{x}_k,$

Backward half step: $\quad (D + \omega U)\boldsymbol{x}_{k+1} = \omega \boldsymbol{b} + (1 - \omega)D\boldsymbol{x}_{k+\frac{1}{2}} - \omega L\boldsymbol{x}_{k+\frac{1}{2}}.$

THEOREM 11.20. *Let A be symmetric with positive diagonal $D > 0$. Then the eigenvalues of the iteration matrix of SSOR are real and non-negative.*

PROOF. Without loss of generality, we can assume that $D = I$, since otherwise we can transform the system $A\boldsymbol{x} = \boldsymbol{b}$ and consider the equivalent system $D^{-\frac{1}{2}}AD^{-\frac{1}{2}}(D^{\frac{1}{2}}\boldsymbol{x}) = D^{-\frac{1}{2}}\boldsymbol{b}$, whose matrix is symmetric with unit diagonal.

Let $U = L^{\top}$ and $A = I + L + L^{\top}$; the iteration matrix is then

$$G_{\text{SSOR}} = (I + \omega U)^{-1}\left((1 - \omega)I - \omega L\right)(I + \omega L)^{-1}\left((1 - \omega)I - \omega U\right).$$

Factoring out ω, we get

$$G_{\text{SSOR}} = \left(\frac{1}{\omega}I + U\right)^{-1}\left(\frac{1 - \omega}{\omega}I - L\right)\left(\frac{1}{\omega}I + L\right)^{-1}\left(\frac{1 - \omega}{\omega}I - U\right).$$

Now observe that the matrices $\left(\frac{1-\omega}{\omega}I - L\right)\left(\frac{1}{\omega}I + L\right)^{-1}$ commute, therefore

$$\left(\frac{1}{\omega}I + U\right) G_{\text{SSOR}} \left(\frac{1}{\omega}I + U\right)^{-1} = H(\omega)H(\omega)^{\top}$$

with

$$H(\omega) = \left(\frac{1}{\omega}I + L\right)^{-1}\left(\frac{1 - \omega}{\omega}I - L\right).$$

The matrix G_{SSOR} is thus similar to the symmetric positive (semi-)definite matrix $HH^{\top}$, and therefore the eigenvalues are real and non-negative. $\qquad\square$

The following algorithm accelerates SSOR using the Chebyshev Semi-Iterative method, when the eigenvalues of the iteration operator G_{SSOR} satisfy $0 \le a \le \lambda(G_{\text{SSOR}}) \le b < 1$. We denote the overrelaxation parameter in SSOR by ω_{SSOR}, since the Chebyshev semi-iterative method uses the variable ω already.

ALGORITHM 11.9. *Chebyshev Acceleration of SSOR*

$w = 2;\ \mu = \frac{2-a-b}{b-a};\ \gamma = \frac{2}{2-a-b};$
choose $\boldsymbol{y}_0;$
$\boldsymbol{y}_1 = \text{SSOR}(\boldsymbol{y}_0, \omega_{\text{SSOR}});\ k = 1;\ c = 4\mu^2;$
while not converged

$\quad w = c/(c - w);$

$$z = \mathrm{SSOR}(\boldsymbol{y}_k, \omega_{\mathrm{SSOR}});$$
$$k = k + 1;$$
$$\boldsymbol{y}_k = w(\gamma \boldsymbol{z} + (1 - w)\boldsymbol{y}_{k-1}) + (1 - w)\boldsymbol{y}_{k-2};$$
end

11.6 Global Minimization by Extrapolation

This section is based on the theory developed by Sidi [126, 125] and Brezinski [11]. We consider the linear system $A\boldsymbol{x} = \boldsymbol{b}$ with $A \in \mathbb{R}^{n \times n}$ and $\boldsymbol{b} \in \mathbb{R}^n$, a splitting $A = M - N$, and the basic stationary iteration in the two forms

$$M\boldsymbol{x}_{i+1} = N\boldsymbol{x}_i + \boldsymbol{b} \tag{11.66}$$

$$\boldsymbol{x}_{i+1} = G\boldsymbol{x}_i + \boldsymbol{d}, \quad G := M^{-1}N, \quad \boldsymbol{d} := M^{-1}\boldsymbol{b}. \tag{11.67}$$

Here, we do not assume that A is symmetric, but we do require A to be non-singular. This is equivalent to assuming that 1 is not an eigenvalue of G, i.e., that the system

$$(I - G)\boldsymbol{x} = \boldsymbol{d},$$

has a unique solution.

The idea of extrapolation methods is the same as with the acceleration by Chebyshev polynomials: one tries to form a clever linear combination of the iterates $\boldsymbol{x}_i$ in order to converge more quickly to the desired solution. We denote as before the weights of the linear combination by γ_i for $i = 0, 1, \ldots, k$, with the normalization condition $\sum_{i=0}^{k} \gamma_i = 1$. We have

$$\boldsymbol{x}_i = \boldsymbol{x} - \boldsymbol{e}_i \quad \Longrightarrow \quad \underbrace{\sum_{i=0}^{k} \gamma_i \boldsymbol{x}_i}_{\boldsymbol{y}_k} = \boldsymbol{x} \underbrace{\sum_{i=0}^{k} \gamma_i}_{1} - \sum_{i=0}^{k} \gamma_i \boldsymbol{e}_i,$$

and therefore

$$\boldsymbol{y}_k = \boldsymbol{x} - \sum_{i=0}^{k} \gamma_i \boldsymbol{e}_i. \tag{11.68}$$

For $\boldsymbol{y}_k$ to be a good approximation, we would like to have

$$\sum_{i=0}^{k} \gamma_i \boldsymbol{e}_i \approx 0. \tag{11.69}$$

Because of the error recurrence (11.9), we have $\boldsymbol{e}_i = G\boldsymbol{e}_{i-1}$, which means

$$\sum_{i=0}^{k} \gamma_i G^i \boldsymbol{e}_0 = P_k(G)\boldsymbol{e}_0 \approx 0. \tag{11.70}$$

Again, as in the previous section, we conclude that $\boldsymbol{y}_k = \boldsymbol{x}$ if the polynomial P_k is the minimal polynomial of G with respect to the vector $\boldsymbol{e}_0$.

LEMMA 11.7. *If $P_k(G)e_0 = 0$, then $P_k(G)e_m = 0$ and $P_k(G)u_m = 0$ for $m \geq 0$, where $u_m := x_{m+1} - x_m$.*

PROOF. Using Equations (11.12) and (11.14), we get

$$u_m = Gu_{m-1}, \quad \text{and} \quad u_m = (I - G)e_m, \tag{11.71}$$

and since polynomials in G commute, we obtain from $P_k(G)e_0 = 0$ that

$$0 = G^m P_k(G)e_0 = P_k(G)G^m e_0 = P_k(G)e_m,$$

and similarly

$$0 = (I - G)P_k(G)e_m = P_k(G)(I - G)e_m = P_k(G)u_m.$$

$\square$

As a consequence of this lemma, we could try to obtain $P_k(G)u_0 \approx 0$. Written in terms of the coefficients of the polynomial, this expression becomes

$$P_k(G)u_0 = \gamma_0 u_0 + \gamma_1 \underbrace{Gu_0}_{u_1} + \cdots + \gamma_k \underbrace{G^k u_0}_{u_k} = [u_0, u_1, \ldots, u_k]\gamma. \tag{11.72}$$

Introducing the matrix $U_k \in \mathbb{R}^{n \times (k+1)}$ defined by

$$U_k := [u_0, u_1, \ldots, u_k], \tag{11.73}$$

we would like to determine the coefficients γ_i in the vector $\gamma \in \mathbb{R}^{k+1}$ such that

$$U_k \gamma \approx 0, \quad \text{subject to the constraint} \quad \sum_{i=0}^{k} \gamma_i = 1. \tag{11.74}$$

In order to remove the constraint, we parametrize the γ_i by k parameters ξ_i,

$$
\begin{aligned}
\gamma_0 &= 1 - \xi_0 \\
\gamma_1 &= \xi_0 - \xi_1 \\
&\cdots \\
\gamma_i &= \xi_{i-1} - \xi_i \\
&\cdots \\
\gamma_k &= \xi_{k-1},
\end{aligned}
$$

which leads to[6] $\gamma = S\xi + \mathbf{e}_1$, $\xi \in \mathbb{R}^k$, with the matrix

$$S = \begin{bmatrix} -1 & & & \\ 1 & -1 & & \\ & 1 & \ddots & \\ & & \ddots & -1 \\ & & & 1 \end{bmatrix}, \quad S \in \mathbb{R}^{(k+1) \times k}. \tag{11.75}$$

---

[6] In this chapter, we use $\mathbf{e}_i$ to denote the i-th standard basis vector, in order to distinguish it from e_i, the i-th error vector.

With this new parametrization, Equation (11.74) becomes

$$U_k \boldsymbol{\gamma} = U_k S \boldsymbol{\xi} + U_k \mathbf{e}_1 \approx 0.$$

The matrix $U_k S$ consists of the columns

$$U_k S = [-\boldsymbol{u}_0 + \boldsymbol{u}_1, -\boldsymbol{u}_1 + \boldsymbol{u}_2, \ldots, -\boldsymbol{u}_{k-1} + \boldsymbol{u}_k].$$

We set $\boldsymbol{w}_j = \boldsymbol{u}_{j+1} - \boldsymbol{u}_j$ and define

$$W_k := [\boldsymbol{w}_0, \ldots, \boldsymbol{w}_k] \in \mathbb{R}^{n \times (k+1)}.$$

With this notation, Equation (11.74) becomes

$$W_{k-1} \boldsymbol{\xi} \approx -\boldsymbol{u}_0. \tag{11.76}$$

We also introduce the matrix

$$X_k = [\boldsymbol{x}_0, \ldots, \boldsymbol{x}_k] \in \mathbb{R}^{n \times (k+1)}.$$

Then once the coefficients γ_i or ξ_i have been computed, the extrapolated vector is obtained from the formula

$$\boldsymbol{y}_k = X_k \boldsymbol{\gamma} = X_k S \boldsymbol{\xi} + X_k \mathbf{e}_1 = U_{k-1} \boldsymbol{\xi} + \boldsymbol{x}_0.$$

We therefore obtain two forms for the *extrapolation algorithm*:

ALGORITHM 11.10. *Generic Extrapolation Algorithms*

First Form: Choose k and solve

$$U_k \boldsymbol{\gamma} \approx 0 \quad \text{s.t.} \quad \sum_{j=0}^{k} \gamma_j = 1, \tag{11.77}$$

$$\boldsymbol{y}_k = X_k \boldsymbol{\gamma}. \tag{11.78}$$

Second Form: Choose k and solve

$$W_{k-1} \boldsymbol{\xi} + \boldsymbol{u}_0 \approx 0, \tag{11.79}$$

$$\boldsymbol{y}_k = \boldsymbol{x}_0 + U_{k-1} \boldsymbol{\xi}. \tag{11.80}$$

DEFINITION 11.13. (KRYLOV SPACE) *Let $A \in \mathbb{R}^{n \times n}$ and $\boldsymbol{r} \in \mathbb{R}^n$. The associated* Krylov space *of dimension k is*

$$\mathcal{K}_k(A, \boldsymbol{r}) = \text{span}\{\boldsymbol{r}, A\boldsymbol{r}, A^2 \boldsymbol{r}, \ldots, A^{k-1} \boldsymbol{r}\}. \tag{11.81}$$

From (11.80), it follows that

$$\boldsymbol{y}_k - \boldsymbol{x}_0 = [\boldsymbol{u}_0, \boldsymbol{u}_1, \ldots, \boldsymbol{u}_{k-1}]\boldsymbol{\xi} = [\boldsymbol{u}_0, G\boldsymbol{u}_0, \ldots, G^{k-1}\boldsymbol{u}_0]\boldsymbol{\xi},$$

and therefore this difference lies in a Krylov space,

$$\boldsymbol{y}_k - \boldsymbol{x}_0 \in \mathcal{K}_k(G, \boldsymbol{u}_0). \tag{11.82}$$

If we choose $\boldsymbol{x}_0 = 0$, we have $\boldsymbol{x}_1 = G\boldsymbol{x}_0 + \boldsymbol{d} = \boldsymbol{d}$, and thus $\boldsymbol{u}_0 = \boldsymbol{x}_1 - \boldsymbol{x}_0 = \boldsymbol{d}$, which implies

$$\boldsymbol{y}_k \in \mathcal{K}_k(G, \boldsymbol{d}). \tag{11.83}$$

Notice that in the case $\boldsymbol{x}_0 = 0$ we also have

$$\boldsymbol{x}_k \in \mathcal{K}_k(G, \boldsymbol{d}) = \mathcal{K}_k(M^{-1}N, M^{-1}\boldsymbol{b}), \tag{11.84}$$

since

$$\begin{aligned}
\boldsymbol{x}_1 &= \boldsymbol{d} \\
\boldsymbol{x}_2 &= G\boldsymbol{x}_1 + \boldsymbol{d} \\
&\cdots \\
\boldsymbol{x}_k &= \sum_{j=0}^{k-1} G^j \boldsymbol{d}.
\end{aligned}$$

11.6.1 Minimal Polynomial Extrapolation (MPE)

We consider the first form of the generic extrapolation algorithm,

$$U_k \boldsymbol{\gamma} \approx 0 \quad \text{s.t.} \quad \sum_{j=0}^{k} \gamma_j = 1,$$

and we now have to define what we mean by solving $U_k\boldsymbol{\gamma} \approx 0$ approximately. The idea of *Minimal Polynomial Extrapolation (MPE)* is to fix the last coefficient to equal one, and then to solve

$$U_{k-1}\boldsymbol{c} \approx -\boldsymbol{u}_k \tag{11.85}$$

as a least squares problem, see Section 6.2. This leads to the polynomial

$$P_k(G) = c_0 + c_1 G + \cdots + c_{k-1} G^{k-1} + G^k, \tag{11.86}$$

which is an approximation of the minimal polynomial. With $c_k = 1$, we then obtain the scaled coefficients

$$\gamma_j = \frac{c_j}{\sum_{i=0}^{k} c_i}, \quad j = 0, \ldots, k. \tag{11.87}$$

However, these coefficients can only be computed if $\sum_{i=0}^{k} c_i \neq 0$. Otherwise, the extrapolated value $\boldsymbol{y}_k$ does not exist.

If we solve the least squares problem (11.85) using the normal equations, see (6.13) in Chapter 6, we obtain

$$U_{k-1}^\top U_{k-1} \boldsymbol{c} = -U_{k-1}^\top \boldsymbol{u}_k.$$

With $c_k = 1$, moving the right hand side to the left and dividing by $\sum_{j=0}^k c_j$, we get

$$U_{k-1}^\top U_k \boldsymbol{\gamma} = 0, \quad \text{s.t.} \quad \sum_{j=0}^k \gamma_j = 1. \tag{11.88}$$

THEOREM 11.21. *Let the coefficients $\boldsymbol{\gamma}$ be determined by (11.88). Let $\boldsymbol{y}_k = X_k \boldsymbol{\gamma}$ be the extrapolated approximation and consider the preconditioned residual*

$$\boldsymbol{r}_k = \boldsymbol{d} - (I - G)\boldsymbol{y}_k = M^{-1}(\boldsymbol{b} - A\boldsymbol{y}_k).$$

Then

$$\boldsymbol{r}_k = U_k \boldsymbol{\gamma} \quad and \quad \boldsymbol{r}_k \perp \mathcal{K}_k(G, \boldsymbol{u}_0).$$

PROOF. Since the coefficients γ_i sum up to one, we have

$$\boldsymbol{d} = [\boldsymbol{d}, \boldsymbol{d}, \dots, \boldsymbol{d}]\boldsymbol{\gamma} =: D\boldsymbol{\gamma}.$$

Using this relation, we obtain for the residual

$$\boldsymbol{r}_k = \boldsymbol{d} - (I-G)\boldsymbol{y}_k = \boldsymbol{d} - (I-G)X_k\boldsymbol{\gamma} = D\boldsymbol{\gamma} - (I-G)X_k\boldsymbol{\gamma} = (D+GX_k - X_k)\boldsymbol{\gamma}. \tag{11.89}$$

Now since $X_k = [\boldsymbol{x}_0, \boldsymbol{x}_1, \dots, \boldsymbol{x}_k]$ and $\boldsymbol{x}_{k+1} = G\boldsymbol{x}_k + \boldsymbol{d}$, we have $D + GX_k = [\boldsymbol{x}_1, \boldsymbol{x}_2, \dots, \boldsymbol{x}_{k+1}]$ and from (11.89) we obtain

$$\boldsymbol{r}_k = ([\boldsymbol{x}_1, \dots, \boldsymbol{x}_{k+1}] - [\boldsymbol{x}_0, \dots, \boldsymbol{x}_k])\boldsymbol{\gamma} = U_k\boldsymbol{\gamma}.$$

For the second result, by (11.88), we have $U_{k-1}^\top \boldsymbol{r}_k = 0$, which shows that $\boldsymbol{r}_k \perp \mathcal{K}_k(G, \boldsymbol{u}_0)$. $\square$

The following MATLAB function MPE is meant to illustrate the theory presented. It can be optimized for the purpose of just solving linear equations, in which case a few operations can be saved, see [125].

ALGORITHM 11.11. *MPE*

```
function [Y,X,U,Gamma]=MPE(G,d,x0,n)
% MPE Minimal Polynomial Extrapolation
%    [Y,X,U,Gamma]=MPE(G,d,x0,n) performs n number of iteration steps
%    x(j+1)=G*x(j)+d starting with x0 and computes an extrapolated
%    vector y at each iteration, stored in the matrix Y. The basic
%    iteration steps and their differences are stored in X and U
%    respectively, and Gamma contains the coefficients of the
```

```
%    polynomial at each step.

n=length(d);
Y=[]; U=[]; Gamma=zeros(n+1,n+1);
xk=x0; X=xk;
for k=0:n
  xk1=G*xk+d; X=[X,xk1];
  uk=xk1-xk; U=[U,uk];
  c=[-U(:,1:k)\uk;1];
  y=X(:,1:k+1)*c/sum(c);
  Gamma(:,k+1)=[c/sum(c); zeros(n-k,1)];
  Y=[Y,y];
  xk=xk1;
end;
```

EXAMPLE 11.3. *We consider the linear system with the non-symmetric matrix*

$$A = \begin{pmatrix} 0 & -4 & -8 & -2 \\ -4 & -7 & -7 & -8 \\ -9 & -5 & -4 & -5 \\ 0 & -5 & -9 & -6 \end{pmatrix}. \tag{11.90}$$

and with the right hand side b such that the solution vector x is $x=[1, 2, \ldots, n]$:

```
>> A=[ 0     -4      -8      -2
      -4     -7      -7      -8
      -9     -5      -4      -5
       0     -5      -9      -6   ];
>> n=max(size(A)); x=(1:n)'; b=A*x;
>> x0=x-5 ;                          % Starting vector
>> G=eye(size(A))-A;
>> [Y,X,U,Gamma]=MPE(G,b,x0,4);
```

We first note that the basic iteration $x_{i+1} = Gx_i + b$ diverges:

```
>> X
X =
       -4       -74     -1784    -38804     -873794    -19498334
       -3      -133     -3058    -68863    -1539508    -34438633
       -2      -117     -2472    -56187    -1249542    -27984777
       -1      -101     -2486    -55001    -1235066    -27588941
```

The extrapolated solutions, however, converge:

```
>> Y
Y =
   -4.0000   -0.7314    0.2662    1.0638    1.0000
   -3.0000    3.0703    3.1914    1.6232    2.0000
   -2.0000    3.3698    1.8713    3.1228    3.0000
   -1.0000    3.6694    4.6950    4.1728    4.0000
```

According to Theorem 11.21, the residual is given by $\boldsymbol{r}_k = U_k\boldsymbol{\gamma}$. We obtain for the residuals of all approximations

```
>> U*Gamma
ans =
   -70.0000     6.5787    -2.8739    -0.1797          0
  -130.0000     0.5106     3.0638    -0.1411          0
  -115.0000   -10.4047    -1.6869     0.0451     0.0000
  -100.0000     6.6965    -0.0313     0.2573    -0.0000
```

This should be compared to the explicitly computed true residuals for verification: for the norm of the difference, we indeed get

```
>> R=b*ones(1,5)-A*Y;
>> norm(R-U*Gamma)
ans =
   8.6075e-12
```

We can also verify that the new residual is orthogonal to the Krylov space, $\boldsymbol{r}_k \perp \mathcal{K}_k(G, \boldsymbol{u}_0)$:

```
>> u0=U(:,1);
>> K=[u0, G*u0, G^2*u0, G^3*u0]
K =
         -70        -1710       -37020       -834990
        -130        -2925       -65805      -1470645
        -115        -2355       -53715      -1193355
        -100        -2385       -52515      -1180065
>> K'*R
ans =
   4.5025e+04   -1.6485e-11    2.9985e-11   -2.0119e-10    1.8503e-10
   1.0093e+06   -4.2113e+03    7.0391e-10   -4.5384e-09    4.8380e-09
   2.2575e+07   -6.9924e+04   -2.9673e+03   -1.0101e-07    1.0288e-07
   5.0488e+08   -1.7299e+06   -5.6105e+04    3.8879e+01    2.3325e-06
```

The result is a lower triangular matrix. The upper triangle is essentially zero, which shows the orthogonality $\boldsymbol{r}_k \perp \boldsymbol{u}_j$ for $j = 0, 1, \ldots, k-1$ (up to the accuracy possible, given the size of the elements in the matrix K, which is naturally not well conditioned, $\kappa(K) \approx 5.8294e+09$).

Finally, the coefficients $\boldsymbol{c}$ are approximations of the coefficients of the minimal polynomial P_k, see Equation (11.86). We compare the coefficients obtained in the last step with the coefficients of the characteristic polynomial of G:

```
>> poly(G)
ans =
    1.0000   -21.0000   -57.0000   637.0000  -984.0000
>> coeff=Gamma(:,5)'/Gamma(5,5);
>> coeff(5:-1:1)-poly(G)
ans =
   1.0e-08 *
        0     0.0007    -0.0139    -0.0733     0.3246
```

and we observe a good match.

11.6.2 Reduced Rank Extrapolation (RRE)

We consider now the second form of the generic extrapolation algorithm (see Algorithm 11.10). In this case we solve approximately

$$W_{k-1}\boldsymbol{\xi} \approx -\boldsymbol{u}_0 \tag{11.91}$$

as a least squares problem. The extrapolated vector then becomes

$$\boldsymbol{y}_k = \boldsymbol{x}_0 + U_{k-1}\boldsymbol{\xi}. \tag{11.92}$$

The normal equations for (11.91) are

$$W_{k-1}^{\top}W_{k-1}\boldsymbol{\xi} = -W_{k-1}^{\top}\boldsymbol{u}_0. \tag{11.93}$$

Now since $W_{k-1} = U_k S$ and $\boldsymbol{u}_0 = U_k \boldsymbol{e}_1$ it follows that

$$0 = W_{k-1}^{\top}U_k S\boldsymbol{\xi} + W_{k-1}^{\top}\boldsymbol{u}_0 = W_{k-1}^{\top}U_k(S\boldsymbol{\xi} + \boldsymbol{e}_1) = W_{k-1}^{\top}U_k\boldsymbol{\gamma}.$$

This gives us an alternative formulation for determining the coefficients for RRE,

$$W_{k-1}^{\top}U_k\boldsymbol{\gamma} = 0, \quad \text{s.t.} \sum_{j=0}^{k}\gamma_j = 1. \tag{11.94}$$

On the other hand, let $\boldsymbol{e} = (1,1,\ldots,1)^{\top}$ be the vector of all ones, and consider the least squares problem with a linear constraint,

$$U_k\boldsymbol{\gamma} \approx 0, \quad \text{s.t.} \sum_{j=0}^{k}\gamma_j = \boldsymbol{e}^{\top}\boldsymbol{\gamma} = 1.$$

Writing the normal equations for the constrained minimization problem (see Equation (6.48)), we obtain

$$\begin{aligned} U_k^{\top}U_k\boldsymbol{\gamma} &= \lambda\boldsymbol{e} \\ \boldsymbol{e}^{\top}\boldsymbol{\gamma} &= 1. \end{aligned} \tag{11.95}$$

By subtracting consecutive equations, we can eliminate the unknown λ. This is done by multiplying the matrix $S^{\top}$ from (11.75) from the left. The reduced system becomes

$$S^{\top}U_k^{\top}U_k\boldsymbol{\gamma} = W_{k-1}^{\top}U_k\boldsymbol{\gamma} = 0, \quad \boldsymbol{e}^{\top}\boldsymbol{\gamma} = 1. \tag{11.96}$$

So we obtain the same equation as (11.94). These results show again an orthogonality relation, which is summarized in the following theorem.

THEOREM 11.22. *Let the coefficient vector $\boldsymbol{\gamma}$ be determined by (11.94), let $\boldsymbol{y}_k = X_k\boldsymbol{\gamma}$ be the extrapolated approximation, and consider the preconditioned residual*

$$\boldsymbol{r}_k = \boldsymbol{d} - (I - G)\boldsymbol{y}_k = M^{-1}(\boldsymbol{b} - A\boldsymbol{y}_k) = U_k\boldsymbol{\gamma}.$$

Then this residual is orthogonal to a Krylov space,

$$\boldsymbol{r}_k \perp \mathcal{K}_k(G, \boldsymbol{w}_0).$$

Notice that for MPE and RRE, the preconditioned residuals for the extrapolated values are orthogonal to *different* Krylov spaces, since the initial vector is different. However, Krylov spaces are invariant with respect to

$$\begin{aligned}\textbf{Scaling:} \quad & \mathcal{K}_k(A, \boldsymbol{r}) = \mathcal{K}_k(\sigma A, \tau \boldsymbol{r}), \\ \textbf{Translation:} \quad & \mathcal{K}_k(A, \boldsymbol{r}) = \mathcal{K}_k(A - \sigma I, \boldsymbol{r})\end{aligned} \tag{11.97}$$

for any matrix A and vector $\boldsymbol{r}$. In our case, since $\boldsymbol{w}_0 = (G - I)\boldsymbol{u}_0 = -M^{-1}A\boldsymbol{u}_0$ and $M^{-1}N = I - M^{-1}A$, it follows that

$$\begin{aligned}\mathcal{K}_k(G, \boldsymbol{w}_0) &= \mathcal{K}_k(M^{-1}N, \boldsymbol{w}_0) = \mathcal{K}_k(M^{-1}A, M^{-1}A\boldsymbol{u}_0) \\ &= M^{-1}A\mathcal{K}_k(M^{-1}A, \boldsymbol{u}_0),\end{aligned}$$

and thus we have for

MPE: $\boldsymbol{r}_k \perp \mathcal{K}_k(M^{-1}A, \boldsymbol{u}_0)$

RRE: $\boldsymbol{r}_k \perp M^{-1}A\,\mathcal{K}_k(M^{-1}A, \boldsymbol{u}_0)$.

11.6.3   Modified Minimal Polynomial Extrapolation (MMPE)

We now discuss a variant of MPE. To approximate the coefficients γ_i of the minimal polynomial, we wish to solve approximately

$$U_k \boldsymbol{\gamma} \approx 0 \quad \text{s.t.} \quad \sum_{j=0}^{k} \gamma_j = 1.$$

Instead of solving this as a least squares problem, one can also use a Galerkin approach. Let $Q_{k-1} = [\boldsymbol{q}_0, \boldsymbol{q}_1, \dots, \boldsymbol{q}_{k-1}] \in \mathbb{R}^{n \times k}$ be any matrix with full rank k. MMPE determines the coefficients by solving

$$\begin{aligned}Q_{k-1}^{\top} U_k \boldsymbol{\gamma} &= 0, \\ \sum_{j=0}^{k} \gamma_j &= 1.\end{aligned} \tag{11.98}$$

Now $\boldsymbol{r}_k$ is orthogonal to $\mathcal{S} = \operatorname{span}\{\boldsymbol{q}_0, \boldsymbol{q}_1, \dots, \boldsymbol{q}_{k-1}\}$. The space $\mathcal{S}$ is so far arbitrary, but one can find a matrix $\tilde{G}$ for which it is a Krylov space. To this end, we augment Q_{k-1} to a full $n \times n$ matrix by adding $n - k$ linearly independent columns,

$$\tilde{Q} = [\boldsymbol{q}_0, \boldsymbol{q}_1, \dots, \boldsymbol{q}_{k-1}, \tilde{\boldsymbol{q}}_k, \tilde{\boldsymbol{q}}_{k+1}, \dots, \tilde{\boldsymbol{q}}_{n-1}],$$

and denote by $\tilde{q}_i := q_i$ the first columns $i = 0, \ldots, k-1$. We want to determine a matrix $\tilde{G}$ such that

$$\tilde{q}_{i+1} = \tilde{G}\tilde{q}_i$$

holds. This means that, with some vector x to be chosen, the matrix equation

$$[\tilde{q}_1, \ldots, \tilde{q}_{n-1}, x] = \tilde{G}\tilde{Q}$$

must hold, and thus

$$\tilde{G} = [\tilde{q}_1, \ldots, \tilde{q}_{n-1}, x]\tilde{Q}^{-1}.$$

How should one choose the vectors q_i? A simple choice is to take randomly selected linearly independent unit vectors e_j. Doing so means to choose k equations randomly from the linear system

$$U_k\gamma = 0,$$

and to solve them together with the normalization condition $\sum_{j=0}^{k}\gamma_j = 1$. There are however many other choices for q_i; one particular choice is discussed next.

11.6.4 Topological ε-Algorithm (TEA)

This extrapolation variant is the same as MMPE but with a specific matrix $Q_{k-1} \in \mathbb{R}^{n \times k}$. We choose some vector $q \in \mathbb{R}^n$ and define

$$Q_{k-1} := [q, G^\top q, \ldots, (G^\top)^{k-1}q],$$

where $G = M^{-1}N$. With this matrix, we solve as before the linear system

$$\begin{aligned} \sum_{j=0}^{k}\gamma_j &= 1, \\ Q_{k-1}^\top U_k\gamma &= 0. \end{aligned} \tag{11.99}$$

Note that using the recurrence $u_{k+1} = Gu_k$, the product matrix $Q_{k-1}^\top U_k$ becomes

$$Q_{k-1}^\top U_k = \begin{pmatrix} q^\top u_0 & q^\top u_1 & \cdots & q^\top u_k \\ q^\top u_1 & q^\top u_2 & \cdots & q^\top u_{k+1} \\ \vdots & \vdots & \cdots & \vdots \\ q^\top u_{k-1} & q^\top u_k & \cdots & q^\top u_{2k-1} \end{pmatrix}. \tag{11.100}$$

As a consequence, we do not need to compute Q_{k-1} explicitly; however, more basic iteration steps are necessary to compute the entire matrix, since we need the vectors u_i up to $i = 2k - 1$.

In the following program, we compute approximations γ for $k = 1, 2, \ldots, n$ by updating the matrix of the linear system (11.99).

ALGORITHM 11.12. TEA

```
function [Y,X,U]=TEA(G,d,x0,q,n)
% TEA Topological Epsilon Algorithm
%    [Y,X,U]=mpe(G,d,x0,n) performs n number of iteration
%    steps x(j+1)=G * x(j) + d starting with x0 and computes an
%    extrapolated vector y at each iteration, stored in the matrix
%    Y. The basic iteration steps and their differences are stored
%    in X and U respectively.

X=[x0]; Y=[x0]; U=[]; xk=x0; QU=1;
for k=1:n
   m=k+1;                        % size of linear system
   xk1=G*xk+d; X=[X,xk1];        % compute two basic iterations
   xk2=G*xk1+d; X=[X,xk2];
   uk1=xk1-xk; uk2=xk2-xk1; % form the new differences
   U=[U,uk1, uk2];
   QU(1,m)=1;                    % update the matrix for gamma
   for j=2:m-2                   % copy elements in new column m
     QU(j,m)=QU(j+1,m-1);
   end
   for j=1:m-2                   % copy elements for new row m
     QU(m,j)=QU(m-1,j+1);
   end
   QU(m,m-1)=q'*uk1;             % update elements in corner
   if m>2,
     QU(m-1,m)=QU(m,m-1);
   end
   QU(m,m)=q'*uk2;
   ee=zeros(m,1); ee(1)=1;   % right hand side
   gamma=QU\ee;
   y=X(:,1:k+1)*gamma;
   Y=[Y,y];
   xk=xk2;
end;
```

EXAMPLE 11.4. *We consider the linear system $A\boldsymbol{x} = \boldsymbol{b}$ with the non-symmetric matrix (11.90). As basic iteration we use $\boldsymbol{x}_{k+1} = (I - A)\boldsymbol{x}_k + \boldsymbol{b}$. As vector $\boldsymbol{q}$ we choose the starting vector $\boldsymbol{q} = \boldsymbol{x}_0$.*

```
A=[0   -4   -8   -2
   -4   -7   -7   -8
   -9   -5   -4   -5
    0   -5   -9   -6];
n=max(size(A)); x=(1:n)'; b=A*x;
x0=x-5;                                  % Starting vector
q=x0;G=eye(size(A))-A;
```

```
[Y,X]=TEA(G,b,x0,x0,n);
Y, format short e, X
```

We obtain the following results: the basic iteration diverges,

```
X =
  Columns 1 through 6
  -4.0000e+00   -7.4000e+01   -1.7840e+03   -3.8804e+04   -8.7379e+05   -1.9498e+07
  -3.0000e+00   -1.3300e+02   -3.0580e+03   -6.8863e+04   -1.5395e+06   -3.4439e+07
  -2.0000e+00   -1.1700e+02   -2.4720e+03   -5.6187e+04   -1.2495e+06   -2.7985e+07
  -1.0000e+00   -1.0100e+02   -2.4860e+03   -5.5001e+04   -1.2351e+06   -2.7589e+07
  Columns 7 through 9
  -4.3631e+08   -9.7555e+09   -2.1817e+11
  -7.7011e+08   -1.7222e+10   -3.8514e+11
  -6.2555e+08   -1.3991e+10   -3.1287e+11
  -6.1718e+08   -1.3801e+10   -3.0864e+11
```

However, the extrapolated values converge:

```
Y =
   -4.0000   -0.7757    0.1403    3.2917    1.0000
   -3.0000    2.9880    3.1933   -2.3914    2.0000
   -2.0000    3.2971    2.0914    5.9430    3.0000
   -1.0000    3.6062    4.5696    3.0331    4.0000
```

11.6.5   Recursive Topological ε-Algorithm

Given a sequence of vectors, one could attempt to apply the scalar ε-algorithm seen in Section 5.2.4 to each component. This turns out not to be very successful.

In order to generalize the ε-scheme for a vector sequence $\{\boldsymbol{x}_j\}$, we have to give a meaning to the "inverse of a vector" used in the recursion (5.20). One possibility proposed by Wynn [152] is to use the *pseudo-inverse* (also called Samelson inverse in the case of a vector), defined by

$$\boldsymbol{y}^{-1} := \frac{1}{\|\boldsymbol{y}\|^2}\boldsymbol{y}^{\top}. \tag{11.101}$$

It has been conjectured by Wynn and proved by McLeod in [90] that if the sequence $\{\boldsymbol{x}_j\}$ with limit $\boldsymbol{x}$ has the property that for some $\beta_i \in \mathbb{R}$,

$$\sum_{i=0}^{k} \beta_i(\boldsymbol{x}_{m+i} - \boldsymbol{x}) = 0, \quad m = 0, 1, \ldots$$

then $\varepsilon_{2k}^{(m)}$ defined by the ε-Algorithm satisfy $\varepsilon_{2k}^{(m)} = \boldsymbol{x}$ if $\sum_{i=0}^{k} \beta_i \neq 0$. This is however all that one knows about this method, called the *vector epsilon algorithm* (VEA).

A second generalization for vector sequences is based on a different interpretation of "inverse" [11]. Brezinski defines the inverse of an ordered pair $(\boldsymbol{a}, \boldsymbol{b})$ of vectors satisfying $\boldsymbol{a}^\top \boldsymbol{b} \neq 0$ to be the ordered pair $(\boldsymbol{b}^{-1}, \boldsymbol{a}^{-1})$, where

$$\boldsymbol{b}^{-1} = \frac{\boldsymbol{a}}{\boldsymbol{b}^\top \boldsymbol{a}}, \quad \boldsymbol{a}^{-1} = \frac{\boldsymbol{b}}{\boldsymbol{b}^\top \boldsymbol{a}}.$$

For real sequences $\{\boldsymbol{x}_j\}$, two algorithms result with this interpretation [136]. The first one is TEA1 (note that the difference in $\Delta \varepsilon_m^{(n)}$ is with respect to n):

ALGORITHM 11.13. *Topological ε-Algorithm TEA1*

Choose an arbitrary vector $\boldsymbol{q}$ and set

$$\varepsilon_{-1}^{(n)} = \boldsymbol{0}, \quad \varepsilon_0^{(n)} = \boldsymbol{x}_n, \quad n = 0, 1, 2, \ldots$$

$$\varepsilon_{2m+1}^{(n)} = \varepsilon_{2m-1}^{(n+1)} + \frac{\boldsymbol{q}}{\boldsymbol{q}^\top \Delta \varepsilon_{2m}^{(n)}}$$
$$\varepsilon_{2m+2}^{(n)} = \varepsilon_{2m}^{(n+1)} + \frac{\Delta \varepsilon_{2m}^{(n)}}{(\Delta \varepsilon_{2m+1}^{(n)})^\top \Delta \varepsilon_{2m}^{(n)}} \qquad m, n = 0, 1 \ldots$$

To update $\varepsilon_{2m+1}^{(n)}$, we use the inverse of $\Delta \varepsilon_{2m}^{(n)}$ with respect to $\boldsymbol{q}$, whereas to update $\varepsilon_{2m+2}^{(n)}$, the inverse of $\varepsilon_{2m+1}^{(n)}$ with respect to $\Delta \varepsilon_{2m}^{(n)}$ is used. A second variant, TEA2, uses the difference in the numerator of the last line of the algorithm at step $n + 1$, see Problem 11.33.

For the MATLAB implementation, we rewrite Algorithm 5.4 to work for vectors. As stated in Algorithm 11.13, we have to distinguish between even and odd columns.

ALGORITHM 11.14. MATLAB *implementation of TEA1*

```
function [x,W,E]=TEA1(A,b,x0,n);
% TEA1 topological epsilon algorithm, first variant
%   [x,W,E]=TEA1(A,b,x0,n); topological epsilon algorithm to
%   accelerate the convergence of the sequence x_{j+1}=(I-A)*x_j+b
%   starting with x0. Returns the last result in x, the accelerated
%   vectors on the diagonal of the epsilon table in W, and the vectors
%   in the last row of the epsilon table in reverse order in E.

W=[]; r0=b-A*x0;
y=r0/norm(r0)^2; E(:,1)=x0;
for i=2:2*n+1
  v=zeros(size(x0));
  E(:,i)=(eye(size(A))-A)*E(:,i-1)+b;
```

```
  for j=i:-1:2
    if rem(i+j,2) == 0,                             % i+j even
      de(:,j-1) = E(:,j)-E(:,j-1);
      w=v+y/(y'*de(:,j-1));
    else                                            % i+j odd
      w=v+de(:,j-1)/(de(:,j-1)'*(E(:,j)-E(:,j-1)));
    end;
    v=E(:,j-1);
    E(:,j-1)=w;
  end;
  W=[W w];
end
x=E(:,1);
```

It is shown in [11] that TEA1 computes a generalized Shanks transform and yields the same extrapolated values as those obtained from the "normal equations" with the TEA algorithm described in Section 11.6.4. This is the case if we choose

$$q = r_0 = b - Ax_0.$$

We will verify this in the following example.

EXAMPLE 11.5.

```
A=[0   -4   -8   -2
  -4   -7   -7   -8
  -9   -5   -4   -5
   0   -5   -9   -6];
n=length(A); x=(1:n)'; b=A*x;
x0=x-5;                                    % Starting vector
[s,W]=TEA1(A,b,x0,4); te=W(:,2:2:8)
q=b-A*x0; G=eye(size(A))-A;
[Y,X]=TEA(G,b,x0,q,n); Y
```

We indeed get the same extrapolated vectors:

```
te =
    -0.7314   -26.5624     0.8810     1.0000
     3.0703    -0.0670     1.6982     2.0000
     3.3698    42.1720     3.2542     3.0000
     3.6694   -22.8856     4.0633     4.0000
Y =
    -4.0000    -0.7314   -26.5624     0.8810     1.0000
    -3.0000     3.0703    -0.0670     1.6982     2.0000
    -2.0000     3.3698    42.1720     3.2542     3.0000
    -1.0000     3.6694   -22.8856     4.0633     4.0000
```

11.7 Krylov Subspace Methods

Krylov subspace methods are non-stationary methods that search for an approximate solution of $Ax = b$ in some lower dimensional Krylov subspace.

11.7.1 The Conjugate Gradient Method

The *Conjugate Gradient Method* (CG) is a method for solving linear systems of equations with symmetric and positive definite coefficient matrices. The preconditioned conjugate gradient method is used nowadays as a standard method in many software libraries, and CG is the starting point of Krylov methods, which are listed among the *top ten algorithms of the last century* [27].

Let $A \in \mathbb{R}^{n \times n}$ be a symmetric positive definite matrix. Then solving the linear system $A\boldsymbol{x} = \boldsymbol{b}$ is equivalent to finding the minimum of the quadratic form

$$Q(\boldsymbol{x}) = \frac{1}{2}\boldsymbol{x}^\top A\boldsymbol{x} - \boldsymbol{x}^\top \boldsymbol{b}, \tag{11.102}$$

as one can see by differentiation, see Problem 11.20. The conjugate gradient method belongs to the class of *descent methods*, see Section 11.4, and also Section 12.3.1 in the chapter on optimization. Such methods compute the minimum of the quadratic form (11.102) iteratively: at step k,

1. a direction vector $\boldsymbol{p}_k$ is chosen, and

2. the new vector $\boldsymbol{x}_{k+1} = \boldsymbol{x}_k + \alpha_k \boldsymbol{p}_k$ is computed, where α_k is chosen such that

$$Q(\boldsymbol{x}_k + \alpha_k \boldsymbol{p}_k) \longrightarrow \min .$$

An interesting class of methods is obtained if we choose *conjugate directions*, that is if

$$\boldsymbol{p}_i^\top A\boldsymbol{p}_j = 0, \quad \text{for } i \neq j. \tag{11.103}$$

Conjugate directions have the following properties:

THEOREM 11.23. *Let $A \in \mathbb{R}^{n \times n}$ be a symmetric positive definite matrix, let $\boldsymbol{x}_0 \in \mathbb{R}^n$ be given, and let the direction vectors $\boldsymbol{p}_i \neq 0$, $i = 0, 1, \ldots, n-1$ be conjugate. Then the sequence $\boldsymbol{x}_k$ defined by*

$$\boldsymbol{x}_{k+1} = \boldsymbol{x}_k + \alpha_k \boldsymbol{p}_k, \quad \text{with} \quad \alpha_k = \frac{\boldsymbol{p}_k^\top \boldsymbol{r}_k}{\boldsymbol{p}_k^\top A\boldsymbol{p}_k} \quad \text{and} \quad \boldsymbol{r}_k = \boldsymbol{b} - A\boldsymbol{x}_k$$

converges in at most n steps to the solution of $A\boldsymbol{x} = \boldsymbol{b}$.

PROOF. We will give two proofs for this theorem: the first proof will illustrate the meaning of the parameter α. Let $\boldsymbol{x}_{k+1} = \boldsymbol{x}_k + \alpha \boldsymbol{p}$. Then minimizing

$$Q(\boldsymbol{x}_{k+1}) = \frac{1}{2}\boldsymbol{x}_{k+1}^\top A\boldsymbol{x}_{k+1} - \boldsymbol{b}^\top \boldsymbol{x}_{k+1}$$

with respect to α yields the equation

$$\frac{\partial Q}{\partial \alpha} = \boldsymbol{p}^\top A\boldsymbol{x}_{k+1} - \boldsymbol{b}^\top \boldsymbol{p} = \boldsymbol{p}^\top A(\boldsymbol{x}_k + \alpha \boldsymbol{p}) - \boldsymbol{b}^\top \boldsymbol{p} = 0,$$

and we get the solution

$$\alpha = \frac{\boldsymbol{p}^\top \boldsymbol{r}_k}{\boldsymbol{p}^\top A \boldsymbol{p}}.$$

Thus, by this choice of α, we are minimizing the quadratic form Q along the line $\boldsymbol{x}_k + \alpha \boldsymbol{p}$.

Before continuing the proof for the general case, we first look at the case where the matrix A is diagonal, which will be helpful later. In that case we have

$$Q(\boldsymbol{x}) = \frac{1}{2} \sum_i a_{ii} x_i^2 - \boldsymbol{b}^\top \boldsymbol{x},$$

and as conjugate direction we can take the unit vectors $\boldsymbol{e}_i$, since $\boldsymbol{e}_i^\top A \boldsymbol{e}_j = a_{ij} = 0$ for $i \neq j$. In this special case, Q is minimal for $x_i = b_i / a_{ii}$. Now let $\boldsymbol{x}_0$ be some initial vector. Then

$$\boldsymbol{x}_1 = \boldsymbol{x}_0 + \alpha_0 \boldsymbol{e}_1 \quad \text{with} \quad \alpha_0 = \frac{\boldsymbol{e}_1^\top (\boldsymbol{b} - A\boldsymbol{x}_0)}{\boldsymbol{e}_1^\top A \boldsymbol{e}_1} = \frac{b_1 - a_{11} x_{0,1}}{a_{11}}.$$

Thus, in the first iteration step, only the first element of the initial vector is changed, and we obtain

$$\boldsymbol{x}_1 = \begin{pmatrix} \frac{b_1}{a_{11}} \\ x_{0,2} \\ \vdots \\ x_{0,n} \end{pmatrix}.$$

In the second iteration step, we obtain

$$\boldsymbol{x}_2 = \begin{pmatrix} \frac{b_1}{a_{11}} \\ \frac{b_2}{a_{22}} \\ x_{0,3} \\ \vdots \\ x_{0,n} \end{pmatrix},$$

and so on, so after n steps we do get the exact solution $\boldsymbol{x}_n = \boldsymbol{x}$.

Now let A be a general symmetric positive definite matrix, and $P = [\boldsymbol{p}_0, \dots, \boldsymbol{p}_{n-1}]$ be the conjugate directions. Then $P^\top A P = D$ with D diagonal and $d_{ii} = \boldsymbol{p}_i^\top A \boldsymbol{p}_i > 0$ (note this is not the eigenvalue decomposition, since P is not orthogonal!). With the change of variables $\boldsymbol{x} = P\boldsymbol{y}$, we get

$$Q(\boldsymbol{x}) = \frac{1}{2}(P\boldsymbol{y})^\top A P \boldsymbol{y} - \boldsymbol{b}^\top P \boldsymbol{y} = \frac{1}{2} \boldsymbol{y}^\top D \boldsymbol{y} - (P^\top \boldsymbol{b})^\top \boldsymbol{y} =: R(\boldsymbol{y}).$$

Minimizing this quadratic form with respect to $\boldsymbol{y}$ is equivalent to solving the diagonal linear system $D\boldsymbol{y} = P^\top \boldsymbol{b}$. Using the unit vectors $\boldsymbol{e}_i$ as conjugate directions, we see that the coefficients are

$$\alpha_k = \frac{\boldsymbol{e}_k^\top (P^\top \boldsymbol{b} - D \boldsymbol{y}_k)}{\boldsymbol{e}_k^\top D \boldsymbol{e}_k} = \frac{\boldsymbol{p}_k^\top (\boldsymbol{b} - A \boldsymbol{x}_k)}{\boldsymbol{p}_k^\top A \boldsymbol{p}_k},$$

and thus they are the same as for the original variable $\boldsymbol{x}$. As shown before in the special case, the iteration

$$\boldsymbol{y}_{k+1} = \boldsymbol{y}_k + \alpha_k \boldsymbol{e}_k \qquad (11.104)$$

converges in n steps. Multiplying (11.104) from the left by P, we see that

$$\boldsymbol{x}_{k+1} = \boldsymbol{x}_k + \alpha_k \boldsymbol{p}_k$$

also converges in n steps, which concludes the first proof.

The second proof shows and uses the orthogonality between the residuals and the direction vectors. Using $\boldsymbol{x}_{k+1} = \boldsymbol{x}_k + \alpha \boldsymbol{p}_k$ and observing that the $\boldsymbol{p}_j$ are conjugate, we get

$$(\boldsymbol{b} - A\boldsymbol{x}_{k+1})^{\top}\boldsymbol{p}_j = (\boldsymbol{b} - A\boldsymbol{x}_k - \alpha_k A\boldsymbol{p}_k)^{\top}\boldsymbol{p}_j = (\boldsymbol{b} - A\boldsymbol{x}_k)^{\top}\boldsymbol{p}_j, \quad j < k.$$

Since this relation holds for all $k > j$, we see by induction that

$$(\boldsymbol{b} - A\boldsymbol{x}_{k+1})^{\top}\boldsymbol{p}_j = (\boldsymbol{b} - A\boldsymbol{x}_k)^{\top}\boldsymbol{p}_j = (\boldsymbol{b} - A\boldsymbol{x}_{k-1})^{\top}\boldsymbol{p}_j = \cdots = (\boldsymbol{b} - A\boldsymbol{x}_{j+1})^{\top}\boldsymbol{p}_j,$$

i.e. all these scalar products have the same value. The last product is

$$(\boldsymbol{b} - A\boldsymbol{x}_{j+1})^{\top}\boldsymbol{p}_j = (\boldsymbol{b} - A\boldsymbol{x}_j - \alpha_j A\boldsymbol{p}_j)^{\top}\boldsymbol{p}_j = (\boldsymbol{b} - A\boldsymbol{x}_j)^{\top}\boldsymbol{p}_j - \frac{\boldsymbol{p}_j^{\top}\boldsymbol{r}_j}{\boldsymbol{p}_j^{\top}A\boldsymbol{p}_j}\boldsymbol{p}_j^{\top}A\boldsymbol{p}_j = 0,$$

So this means that all products are zero, and that the residuals for $k > j$ are orthogonal to the direction vectors

$$\boldsymbol{p}_j \perp \boldsymbol{r}_k, \quad k > j. \qquad (11.105)$$

If we consider in particular $\boldsymbol{r}_n$, then

$$\boldsymbol{r}_n \perp P = [\boldsymbol{p}_0, \boldsymbol{p}_1, \ldots, \boldsymbol{p}_{n-1}].$$

Now $P^{\top}AP = D$ implies $\det(P)^2 \det(A) = \det(D)$, and thus $\operatorname{rank}(P) = n$, which implies that $\boldsymbol{r}_n = 0$. $\qquad\qquad\square$

The previous theorem shows that conjugate directions can be used to construct an iterative method that converges in a finite number of steps. As a result, we are interested in finding conjugate directions for a given matrix A. One possibility is to use the eigenvectors of A: if

$$AQ = QD, \quad Q^{\top}Q = I, \quad D = \operatorname{diag}(\lambda_1, \ldots, \lambda_n)$$

is the eigen-decomposition of A, then the eigenvectors Q are conjugate:

$$Q^{\top}AQ = D.$$

However, computing the eigen-decomposition is too expensive for our purpose, as it is more expensive than solving the linear system, see Chapter 7.

Stiefel and Hestenes [70] found a much more efficient algorithm to construct such directions, now known as *the conjugate gradient algorithm*:

ALGORITHM 11.15.
Conjugate Gradient Algorithm: CGHS

---

choose $\boldsymbol{x}_0$, $\boldsymbol{p}_0 = \boldsymbol{r}_0 := \boldsymbol{b} - A\boldsymbol{x}_0$; ;
for $k = 0 : n - 1$

$$\alpha_k = \frac{\boldsymbol{r}_k^\top \boldsymbol{p}_k}{\boldsymbol{p}_k^\top A \boldsymbol{p}_k}; \tag{11.106}$$

$$\boldsymbol{x}_{k+1} = \boldsymbol{x}_k + \alpha_k \boldsymbol{p}_k; \tag{11.107}$$

$$\boldsymbol{r}_{k+1} = \boldsymbol{r}_k - \alpha_k A \boldsymbol{p}_k; \tag{11.108}$$

$$\beta_k = -\frac{\boldsymbol{p}_k^\top A \boldsymbol{r}_{k+1}}{\boldsymbol{p}_k^\top A \boldsymbol{p}_k}; \tag{11.109}$$

$$\boldsymbol{p}_{k+1} = \boldsymbol{r}_{k+1} + \beta_k \boldsymbol{p}_k; \tag{11.110}$$

end

THEOREM 11.24. *Let $A \in \mathbb{R}^{n \times n}$ be symmetric and positive definite. Then the direction vectors $\boldsymbol{p}_j$ in Algorithm 11.15 are conjugate, and $\boldsymbol{p}_j \neq 0$ if $\boldsymbol{x}_j$ is not the exact solution. Furthermore the residual vectors $\boldsymbol{r}_j$ are orthogonal, $\boldsymbol{r}_j^\top \boldsymbol{r}_i = 0$ for $i \neq j$, and $\boldsymbol{r}_i \perp \boldsymbol{p}_j, i > j$.*

PROOF. The proof is the same as with BiCG, see the proof of Theorem 11.39. $\qquad\square$

In Algorithm 11.15, each iteration seemingly requires two matrix-vector multiplications, $A\boldsymbol{p}_k$ and $A\boldsymbol{r}_{k+1}$. However, it is possible to avoid the second one as follows. Using Statements (11.108) and (11.110), we get, because of the conjugacy and the orthogonality,

$$\boldsymbol{r}_{k+1}^\top \boldsymbol{p}_{k+1} = (\boldsymbol{r}_k - \alpha_k A\boldsymbol{p}_k)^\top \boldsymbol{p}_{k+1} = \boldsymbol{r}_k^\top \boldsymbol{p}_{k+1} = \boldsymbol{r}_k^\top (\boldsymbol{r}_{k+1} + \beta_k \boldsymbol{p}_k) = \beta_k \boldsymbol{r}_k^\top \boldsymbol{p}_k.$$

Solving for β_k, and using again Statement (11.110) and that the residuals are orthogonal to the direction vectors, Equation (11.105), we get

$$\beta_k = \frac{\boldsymbol{r}_{k+1}^\top \boldsymbol{p}_{k+1}}{\boldsymbol{r}_k^\top \boldsymbol{p}_k} = \frac{\boldsymbol{r}_{k+1}^\top (\boldsymbol{r}_{k+1} + \beta_k \boldsymbol{p}_k)}{\boldsymbol{r}_k^\top (\boldsymbol{r}_k + \beta_{k-1} \boldsymbol{p}_{k-1})} = \frac{\|\boldsymbol{r}_{k+1}\|^2}{\|\boldsymbol{r}_k\|^2}.$$

Because of $\boldsymbol{r}_k^\top \boldsymbol{p}_k = \|\boldsymbol{r}_k\|^2$, the computation of α_k simplifies also to

$$\alpha_k = \frac{\|\boldsymbol{r}_k\|^2}{\boldsymbol{p}_k^\top A \boldsymbol{p}_k}. \tag{11.111}$$

With these observations, we obtain a version of CG which uses only one matrix-vector multiplication per iteration step:

ALGORITHM 11.16. *Conjugate Gradients*

```
function [X,R,P,alpha,beta]=CG(A,b,x0,m);
% CG conjugate gradient method
%    [X,R,P,alpha,beta]=CG(A,b,x0,m) computes an approximation for the
%    solution of the linear system Ax=b performing m steps of the
%    conjugate gradient method starting with x0, and returns in the
%    matrix X the iterates, in the matrix R the residuals, in P the
%    conjugate directions and the coefficients alpha and beta.

x=x0; r=b-A*x; p=r;
R =r; P=p; X=x;
oldrho=r'*r;
for k=1:m
  Ap=A*p;
  alpha(k)=oldrho/(p'*Ap);
  x=x+alpha(k)*p; r=r-alpha(k)*Ap;
  rho=r'*r;
  beta(k)=rho/oldrho; oldrho=rho;
  p=r+beta(k)*p;
  X=[X,x]; R=[R,r]; P=[P,p];
end
```

THEOREM 11.25. *For the vectors generated by the Conjugate Gradient Algorithm 11.15, we have for* $i = 0, 1, 2, \ldots$

$$A\boldsymbol{p}_i \in \mathrm{span}\{\boldsymbol{p}_0, \ldots, \boldsymbol{p}_{i+1}\}, \tag{11.112}$$

$$\boldsymbol{r}_i \in \mathrm{span}\{\boldsymbol{p}_0, \ldots, \boldsymbol{p}_i\} = \mathcal{K}_{i+1}(A, \boldsymbol{p}_0) = \mathcal{K}_{i+1}(A, \boldsymbol{r}_0). \tag{11.113}$$

PROOF. The proof is by induction, and we start by showing the first two statements, without the Krylov spaces. Using (11.110), (11.108) and $\boldsymbol{r}_0 = \boldsymbol{p}_0$, we have

$$\boldsymbol{p}_1 = \boldsymbol{r}_1 + \beta_0 \boldsymbol{p}_0 = \boldsymbol{p}_0 - \alpha_0 A\boldsymbol{p}_0 + \beta_0 \boldsymbol{p}_0,$$

and solving for $A\boldsymbol{p}_0$, we obtain

$$A\boldsymbol{p}_0 = \frac{1}{\alpha_0}(1 + \beta_0)\boldsymbol{p}_0 - \frac{1}{\alpha_0}\boldsymbol{p}_1 \in \mathrm{span}\{\boldsymbol{p}_0, \boldsymbol{p}_1\},$$

and with $\boldsymbol{r}_0 = \boldsymbol{p}_0 \in \mathrm{span}\{\boldsymbol{p}_0\}$, the statement holds for $i = 0$. We now assume that the statement holds for $i = 0, \ldots, k$. Then using (11.108), we obtain

$$\boldsymbol{r}_{k+1} = \boldsymbol{r}_k - \alpha_k A\boldsymbol{p}_k = \sum_{j=0}^{k} \nu_j \boldsymbol{p}_j - \alpha_k \sum_{j=0}^{k+1} \eta_j \boldsymbol{p}_j,$$

and thus $r_{k+1} \in \mathrm{span}\{p_0, \ldots, p_{k+1}\}$. Furthermore, using again (11.110) and (11.108), we get

$$p_{k+2} = r_{k+2} + \beta_{k+1}p_{k+1} = r_{k+1} - \alpha_{k+1}Ap_{k+1} + \beta_{k+1}p_{k+1},$$

and therefore

$$Ap_{k+1} = \frac{1}{\alpha_{k+1}}(r_{k+1} + \beta_{k+1}p_{k+1} - p_{k+2}) \in \mathrm{span}\{p_0, \ldots, p_{k+2}\},$$

which concludes the induction step. It remains to prove that the vectors are members of Krylov subspaces, which we also do by induction. For $i = 0$, there is nothing to prove, so assume that

$$\mathrm{span}\{p_0, \ldots, p_k\} = \mathcal{K}_{k+1}\{A, p_0\} \quad \text{for } k < n-1.$$

This implies that $A^k p_0 \in \mathrm{span}\{p_0, \ldots, p_k\}$, and using (11.112) we get

$$A^{k+1}p_0 = A\left(\sum_{j=0}^{k} \nu_j p_j\right) = \sum_{j=0}^{k} \nu_j Ap_j \in \mathrm{span}\{p_0, \ldots, p_{k+1}\}. \qquad (11.114)$$

In order to show that $\mathcal{K}_{k+2}(A, p_0) \subset \mathrm{span}\{p_0, \ldots, p_{k+1}\}$, we take an element from the Krylov space, and using (11.114) and the induction hypothesis, we obtain

$$\sum_{j=0}^{k+1} \eta_j A^j p_0 = \underbrace{\eta_{k+1} A^{k+1} p_0}_{\in \mathrm{span}\{p_0, \ldots, p_{k+1}\}} + \underbrace{\sum_{j=0}^{k} \eta_j A^j p_0}_{\in \mathrm{span}\{p_0, \ldots, p_k\}},$$

and thus $\mathcal{K}_{k+2}(A, p_0) \subset \mathrm{span}\{p_0, \ldots, p_{k+1}\}$.

To show the converse, $\mathrm{span}\{p_0, \ldots, p_{k+1}\} \subset \mathcal{K}_{k+2}(A, p_0)$, we consider

$$\sum_{j=0}^{k+1} \eta_j p_j = \underbrace{\eta_{k+1} p_{k+1}}_{I} + \underbrace{\sum_{j=0}^{k} \eta_j p_j}_{II}.$$

The second term II is in $\mathcal{K}_{k+1}(A, p_0) \subset \mathcal{K}_{k+2}(A, p_0)$ by induction hypothesis. For the first term I, using the statements (11.110) and (11.108), we get

$$\eta_{k+1} p_{k+1} = \eta_{k+1}(r_{k+1} + \beta_k p_k) = \eta_{k+1}(r_k - \alpha_k Ap_k + \beta_k p_k).$$

Now again by induction hypothesis $r_k, p_k \in \mathcal{K}_{k+1}(A, p_0)$, and since

$$Ap_k = A\left(\sum_{j=0}^{k} \nu_j A^j p_0\right) \in \mathcal{K}_{k+2}(A, p_0),$$

also the first term I is in $\mathcal{K}_{k+2}(A, \boldsymbol{p}_0)$, which concludes the proof. $\qquad\square$

The convergence analysis of CG is most naturally performed in a norm associated to the system matrix: let $A \in \mathbb{R}^{n \times n}$ be a symmetric ($A^\top = A$) and positive definite ($\boldsymbol{x}^\top A \boldsymbol{x} > 0, \forall \boldsymbol{x} \neq \boldsymbol{0}$) matrix. Then one can define a *vector norm associated with A*, the so called *A-norm* or *energy norm*,

$$\|\boldsymbol{x}\|_A := \sqrt{\boldsymbol{x}^\top A \boldsymbol{x}}.$$

THEOREM 11.26. (OPTIMALITY OF CG) *Let $\boldsymbol{x}$ be the solution of $A\boldsymbol{x} = \boldsymbol{b}$, with $A \in \mathbb{R}^{n \times n}$ symmetric and positive definite. Then, for each k, the approximation $\boldsymbol{x}_{k+1}$ obtained by CG minimizes the A-norm of the error,*

$$\|\boldsymbol{x} - \boldsymbol{x}_{k+1}\|_A = \min_{\tilde{\boldsymbol{x}} \in \mathcal{S}_k} \|\boldsymbol{x} - \tilde{\boldsymbol{x}}\|_A,$$

where the affine Krylov subspace $\mathcal{S}_k$ is defined by

$$\mathcal{S}_k = \boldsymbol{x}_0 + \mathcal{K}_{k+1}(A, \boldsymbol{r}_0).$$

PROOF. We compute the minimum of the norm of the error over the space $\mathcal{S}_k$ and show that we obtain the same vector with CG. Each vector $\tilde{\boldsymbol{x}} \in \mathcal{S}_k$ can be written as

$$\tilde{\boldsymbol{x}} = \boldsymbol{x}_0 + \sum_{j=0}^{k} \nu_j \boldsymbol{p}_j.$$

Because $\boldsymbol{r}_0 = \boldsymbol{b} - A\boldsymbol{x}_0 = A\boldsymbol{e}_0 = A(\boldsymbol{x} - \boldsymbol{x}_0)$, it follows that

$$\boldsymbol{x} - \tilde{\boldsymbol{x}} = \underbrace{\boldsymbol{x} - \boldsymbol{x}_0}_{A^{-1}\boldsymbol{r}_0} - \sum_{j=0}^{k} \nu_j \boldsymbol{p}_j,$$

and therefore we obtain

$$\|\boldsymbol{x} - \tilde{\boldsymbol{x}}\|_A^2 = \left(A^{-1}\boldsymbol{r}_0 - \sum_{j=0}^{k} \nu_j \boldsymbol{p}_j\right)^\top A\left(A^{-1}\boldsymbol{r}_0 - \sum_{j=0}^{k} \nu_j \boldsymbol{p}_j\right)$$

$$= \boldsymbol{r}_0^\top A^{-1}\boldsymbol{r}_0 - 2\boldsymbol{r}_0^\top \sum_{j=0}^{k} \nu_j \boldsymbol{p}_j + \sum_{j=0}^{k} \nu_j^2 \boldsymbol{p}_j^\top A \boldsymbol{p}_j, \qquad (11.115)$$

where all other products are zero because of the conjugacy of the direction vectors $\boldsymbol{p}_j$. Equation (11.115) is a quadratic form in the ν_j with a diagonal matrix containing $\boldsymbol{p}_j^\top A \boldsymbol{p}_j$ on the diagonal, so the minimum is easy to compute,

$$\frac{\partial \|\boldsymbol{x} - \tilde{\boldsymbol{x}}\|_A^2}{\partial \nu_j} = -2\boldsymbol{r}_0^\top \boldsymbol{p}_j + 2\nu_j \boldsymbol{p}_j^\top A \boldsymbol{p}_j = 0 \quad \Longrightarrow \quad \nu_j = \frac{\boldsymbol{r}_0^\top \boldsymbol{p}_j}{\boldsymbol{p}_j^\top A \boldsymbol{p}_j}, \quad j = 0, \ldots, k.$$

On the other hand, by Statement (11.108), we have

$$\begin{aligned}
\boldsymbol{r}_j &= \boldsymbol{r}_{j-1} - \alpha_{j-1} A \boldsymbol{p}_{j-1} \\
&= \boldsymbol{r}_{j-2} - \alpha_{j-2} A \boldsymbol{p}_{j-2} - \alpha_{j-1} A \boldsymbol{p}_{j-1} \\
&\;\;\vdots \\
\boldsymbol{r}_j &= \boldsymbol{r}_0 - \sum_{i=0}^{j-1} \alpha_i A \boldsymbol{p}_i,
\end{aligned}$$

and it follows that $\boldsymbol{p}_j^\top \boldsymbol{r}_j = \boldsymbol{p}_j^\top \boldsymbol{r}_0$. Using this relation in the result for ν_j, we get

$$\nu_j = \frac{\boldsymbol{r}_0^\top \boldsymbol{p}_j}{\boldsymbol{p}_j^\top A \boldsymbol{p}_j} = \frac{\boldsymbol{r}_j^\top \boldsymbol{p}_j}{\boldsymbol{p}_j^\top A \boldsymbol{p}_j} = \alpha_j.$$

Now using Statement (11.107), we see that

$$\begin{aligned}
\boldsymbol{x}_{k+1} &= \boldsymbol{x}_k + \alpha_k \boldsymbol{p}_k \\
&= \boldsymbol{x}_{k-1} + \alpha_{k-1} \boldsymbol{p}_{k-1} + \alpha_k \boldsymbol{p}_k \\
&\;\;\vdots \\
&= \boldsymbol{x}_0 + \sum_{i=0}^{k} \alpha_i \boldsymbol{p}_i = \boldsymbol{x}_0 + \sum_{i=0}^{k} \nu_i \boldsymbol{p}_i,
\end{aligned}$$

which shows that $\boldsymbol{x}_{k+1}$ indeed minimizes $\|\boldsymbol{x} - \tilde{\boldsymbol{x}}\|_A$ for all $\tilde{\boldsymbol{x}} \in \mathcal{S}_k$. $\qquad \square$

We are now ready to give a different interpretation of Theorem 11.26. Since we chose $\boldsymbol{p}_0 = \boldsymbol{r}_0$, the approximation $\boldsymbol{x}_k \in \mathcal{S}_{k-1}$ is

$$\boldsymbol{x}_k = \boldsymbol{x}_0 + \sum_{j=0}^{k-1} c_j A^j \boldsymbol{r}_0.$$

Hence it follows that

$$\boldsymbol{r}_k = \boldsymbol{b} - A \boldsymbol{x}_k = \boldsymbol{r}_0 - \sum_{j=0}^{k-1} c_j A^{j+1} \boldsymbol{r}_0 = R_k(A) \boldsymbol{r}_0$$

where the polynomial R_k with $R_k(0) = 1$ is the *residual polynomial*. Using Equation (11.14), $A^{-1} \boldsymbol{r}_k = \boldsymbol{x} - \boldsymbol{x}_k$, we can write

$$\|\boldsymbol{x} - \boldsymbol{x}_k\|_A^2 = \boldsymbol{r}_k^\top A^{-1} A A^{-1} \boldsymbol{r}_k = \boldsymbol{r}_k^\top A^{-1} \boldsymbol{r}_k = \|R_k(A) \boldsymbol{r}_0\|_{A^{-1}}^2.$$

Note that since A is symmetric and positive definite, A^{-1} is also symmetric and positive definite, so it can be used to define a norm as well. Theorem 11.26 now becomes

THEOREM 11.27. *Let $\boldsymbol{x}_k$ be the approximation obtained by CG at the k-th step. Let R_k be a polynomial of degree k with $R_k(0) = 1$. Then*

$$\|\boldsymbol{e}_k\|_A = \|\boldsymbol{x} - \boldsymbol{x}_k\|_A = \min_{R_k(0)=1} \|R_k(A)\boldsymbol{r}_0\|_{A^{-1}}.$$

As an immediate corollary, we have

COROLLARY 11.2. *If A has only $m \leq n$ different eigenvalues, then CG converges in at most m steps.*

PROOF. This theorem follows immediately from Theorem 11.27, since the minimal polynomial for A has degree m. $\qquad\square$

We notice also a relation to the semi-iterative method described in Section 11.5.1, see (11.64), which states that the error for the simple splitting $A = M - N$ with $M = I$ and $N = I - A$ satisfies the relation

$$\boldsymbol{x} - \boldsymbol{y}_k = P_k(I - A)\boldsymbol{e}_0, \quad P_k(1) = 1,$$

and the semi-iterative method minimizes a bound of the the 2-norm $\|P_k(I - A)\|_2$ over the spectrum of the iteration matrix $I - A$. The minimization problem solved implicitly by the conjugate gradient method is very similar — the minimization is for a polynomial in A, which explains why the polynomials have a different normalization. In addition, it is the A^{-1}-norm rather than the 2-norm that we seek to minimize. The advantage of CG is that *no parameters have to be estimated*. In addition, CG finds the best polynomial for the given spectrum of A, which explains why in practice the CG algorithm converges better if the eigenvalues are clustered.

Let us derive a general error estimate. Here, it is impossible to take into account all the various spectra possible, so we again assume that the spectrum of A is contained in an interval and use Chebyshev polynomials. Using the eigen-decomposition

$$A = QDQ^\top, \quad Q^\top Q = I, \quad D = \operatorname{diag}(\lambda_1, \ldots, \lambda_n),$$

we obtain

$$\|R_k(A)\boldsymbol{r}_0\|_{A^{-1}}^2 = \boldsymbol{r}_0^\top Q R_k(D)^2 D^{-1} Q^\top \boldsymbol{r}_0.$$

Let $\boldsymbol{t} := Q^\top \boldsymbol{r}_0$, then

$$\|R_k(A)\boldsymbol{r}_0\|_{A^{-1}}^2 = \boldsymbol{t}^\top R_k(D)^2 D^{-1} \boldsymbol{t} = \sum_{i=1}^{n} t_i^2 \frac{R_k(\lambda_i)^2}{\lambda_i}$$

$$\leq \max_i R_k(\lambda_i)^2 \sum_{i=1}^{n} \frac{t_i^2}{\lambda_i}.$$

Now with $\boldsymbol{r}_0 = A(\boldsymbol{x} - \boldsymbol{x}_0)$, we observe that

$$\sum_{i=1}^{n} \frac{t_i^2}{\lambda_i} = \boldsymbol{r}_0^\top Q^\top D^{-1} Q \boldsymbol{r}_0 = \boldsymbol{r}_0^\top A^{-1} \boldsymbol{r}_0$$

$$= (\boldsymbol{x} - \boldsymbol{x}_0)^\top A (\boldsymbol{x} - \boldsymbol{x}_0)$$

$$= \|\boldsymbol{x} - \boldsymbol{x}_0\|_A^2 = \|e_0\|_A^2.$$

Assume that $0 < a = \lambda_1 \leq \lambda_2 \leq \cdots \leq \lambda_n = b$, then

$$\max_i R_k(\lambda_i)^2 \leq \max_{a \leq \lambda \leq b} |R_k(\lambda)|^2.$$

Using the results of Theorem 11.18 and 11.19,

$$\min_{R_k(0)=1} \max_{a \leq \lambda \leq b} |R_k(\lambda)| = \frac{1}{\left| C_k \left(\frac{a+b}{a-b} \right) \right|} \leq 2 \left(\frac{\sqrt{\kappa} - 1}{\sqrt{\kappa} + 1} \right)^k,$$

together with Theorem 11.27, we get:

THEOREM 11.28. (CONVERGENCE ESTIMATE FOR CG) *Let $\boldsymbol{x}_k$ be the approximation at the k-th step of the CG algorithm. Then the error $e_k := \boldsymbol{x} - \boldsymbol{x}_k$ satisfies the estimate*

$$\|e_k\|_A \leq 2 \left(\frac{\sqrt{\kappa} - 1}{\sqrt{\kappa} + 1} \right)^k \|e_0\|_A,$$

where $\kappa = \lambda_n / \lambda_1$ is the condition number of the matrix A.

The *Rayleigh quotient* is bounded by the extremal eigenvalues, so that

$$\lambda_1 \leq \frac{\boldsymbol{x}^\top A \boldsymbol{x}}{\boldsymbol{x}^\top \boldsymbol{x}} = \frac{\|\boldsymbol{x}\|_A^2}{\|\boldsymbol{x}\|_2^2} \leq \lambda_n \quad \Longrightarrow \quad \sqrt{\lambda_1} \|\boldsymbol{x}\|_2 \leq \|\boldsymbol{x}\|_A \leq \sqrt{\lambda_n} \|\boldsymbol{x}\|_2.$$

From this inequality, we get the error estimate of Theorem 11.28 in the 2-norm,

$$\|e_k\|_2 \leq 2 \left(\frac{\sqrt{\kappa} - 1}{\sqrt{\kappa} + 1} \right)^k \sqrt{\kappa} \|e_0\|_2. \tag{11.116}$$

This error estimate does not necessarily imply that the error decreases monotonically in each step, since it is just an upper bound. However, we also have the following result:

THEOREM 11.29. (MONOTONICITY OF CG) *With each CG step, the error decreases. More specifically, if $\boldsymbol{x}_{k-1} \neq \boldsymbol{x}$, then*

$$\|\boldsymbol{x} - \boldsymbol{x}_k\|_2 < \|\boldsymbol{x} - \boldsymbol{x}_{k-1}\|_2.$$

PROOF. Taking the difference on the right, and adding and subtracting $\boldsymbol{x}_k$, we obtain the estimate

$$\boldsymbol{x} - \boldsymbol{x}_{k-1} = (\boldsymbol{x} - \boldsymbol{x}_k) + (\boldsymbol{x}_k - \boldsymbol{x}_{k-1})$$

$$\implies \|\boldsymbol{x} - \boldsymbol{x}_{k-1}\|_2^2 = \|\boldsymbol{x} - \boldsymbol{x}_k\|_2^2 + \|\boldsymbol{x}_k - \boldsymbol{x}_{k-1}\|_2^2 + 2(\boldsymbol{x}_k - \boldsymbol{x}_{k-1})^\top (\boldsymbol{x} - \boldsymbol{x}_k).$$

Thus the theorem holds if we can show that $2(\boldsymbol{x}_k - \boldsymbol{x}_{k-1})^\top (\boldsymbol{x} - \boldsymbol{x}_k) > 0$. Assuming that $\boldsymbol{x} = \boldsymbol{x}_m$ with $m \leq n$, we obtain from Statement (11.107) that

$$\boldsymbol{x}_m = \boldsymbol{x}_{m-1} + \alpha_{m-1}\boldsymbol{p}_{m-1}$$

$$= \boldsymbol{x}_{m-2} + \alpha_{m-2}\boldsymbol{p}_{m-2} + \alpha_{m-1}\boldsymbol{p}_{m-1} \quad \text{etc.}$$

$$\implies \quad \boldsymbol{x}_m = \boldsymbol{x}_k + \alpha_k\boldsymbol{p}_k + \cdots + \alpha_{m-1}\boldsymbol{p}_{m-1}.$$

It therefore follows that

$$\boldsymbol{x} - \boldsymbol{x}_k = \boldsymbol{x}_m - \boldsymbol{x}_k = \alpha_k\boldsymbol{p}_k + \cdots + \alpha_{m-1}\boldsymbol{p}_{m-1}.$$

With $\boldsymbol{x}_k - \boldsymbol{x}_{k-1} = \alpha_{k-1}\boldsymbol{p}_{k-1}$, we obtain

$$(\boldsymbol{x}_k - \boldsymbol{x}_{k-1})^\top (\boldsymbol{x} - \boldsymbol{x}_k) = \alpha_{k-1}(\alpha_k\boldsymbol{p}_{k-1}^\top\boldsymbol{p}_k + \cdots + \alpha_{m-1}\boldsymbol{p}_{k-1}^\top\boldsymbol{p}_{m-1}).$$

Now all $\alpha_j > 0$ because of (11.111) and the positive definiteness of A. Thus, the result holds if we can show that

$$\boldsymbol{p}_{k-1}^\top\boldsymbol{p}_j > 0, \quad j \geq k.$$

By Statement (11.110), we have

$$\boldsymbol{p}_j = \boldsymbol{r}_j + \beta_{j-1}\boldsymbol{p}_{j-1}$$

$$= \boldsymbol{r}_j + \beta_{j-1}(\boldsymbol{r}_{j-1} + \beta_{j-2}\boldsymbol{p}_{j-2}) \quad \text{etc.}$$

$$\implies \quad \boldsymbol{p}_j = \boldsymbol{r}_j + \beta_{j-1}\boldsymbol{r}_{j-1} + \cdots + (\beta_{j-1}\cdots\beta_k)\boldsymbol{r}_k + (\beta_{j-1}\cdots\beta_{k-1})\boldsymbol{p}_{k-1}.$$

Using (11.105) that $\boldsymbol{p}_{k-1}^\top\boldsymbol{r}_j = 0$ for $j \geq k$, we get

$$\boldsymbol{p}_{k-1}^\top\boldsymbol{p}_j = (\beta_{j-1}\cdots\beta_{k-1})\boldsymbol{p}_{k-1}^\top\boldsymbol{p}_{k-1} = (\beta_{j-1}\cdots\beta_{k-1})\|\boldsymbol{p}_{k-1}\|^2 > 0,$$

since $\beta_i = \|\boldsymbol{r}_{i+1}\|^2/\|\boldsymbol{r}_i\|^2 > 0$. $\qquad\qquad\square$

Often CG is considered as a *convergence accelerator* for a stationary iterative method. To see this, we first rewrite the conjugate gradient iteration as a *three-term recurrence* by eliminating the direction vectors $\boldsymbol{p}_k$. We have

$$\boldsymbol{x}_{k+1} = \boldsymbol{x}_k + \alpha_k\boldsymbol{p}_k = \boldsymbol{x}_k + \alpha_k(\boldsymbol{r}_k + \beta_{k-1}\boldsymbol{p}_{k-1}).$$

If we replace $\boldsymbol{p}_{k-1}$ by using $\boldsymbol{x}_k = \boldsymbol{x}_{k-1} + \alpha_{k-1}\boldsymbol{p}_{k-1}$, we obtain

$$\boldsymbol{x}_{k+1} = \boldsymbol{x}_k + \alpha_k\left(\boldsymbol{r}_k + \frac{\beta_{k-1}}{\alpha_{k-1}}(\boldsymbol{x}_k - \boldsymbol{x}_{k-1})\right)$$

$$= \underbrace{\left(1 + \frac{\alpha_k\beta_{k-1}}{\alpha_{k-1}}\right)}_{=:\rho_{k+1}}\boldsymbol{x}_k + \alpha_k\boldsymbol{r}_k - \underbrace{\frac{\alpha_k\beta_{k-1}}{\alpha_{k-1}}}_{1-\rho_{k+1}}\boldsymbol{x}_{k-1}$$

$$\boldsymbol{x}_{k+1} = \rho_{k+1}\boldsymbol{x}_k + \alpha_k\boldsymbol{r}_k + (1 - \rho_{k+1})\boldsymbol{x}_{k-1} \qquad\qquad (11.117)$$

With $\delta_{k+1} := \alpha_k/\rho_{k+1}$ and substituting $\boldsymbol{b} - A\boldsymbol{x}_k$ for $\boldsymbol{r}_k$, we get

$$\boldsymbol{x}_{k+1} = \rho_{k+1} \underbrace{(\boldsymbol{x}_k + \delta_{k+1}(\boldsymbol{b} - A\boldsymbol{x}_k))}_{\delta_{k+1}((I-A)\boldsymbol{x}_k + \boldsymbol{b}) + (1-\delta_{k+1})\boldsymbol{x}_k} + (1 - \rho_{k+1})\boldsymbol{x}_{k-1},$$

and finally

$$\boldsymbol{x}_{k+1} = \rho_{k+1}\left[\delta_{k+1}\left((I - A)\boldsymbol{x}_k + \boldsymbol{b}\right) + (1 - \delta_{k+1})\boldsymbol{x}_k\right] + (1 - \rho_{k+1})\boldsymbol{x}_{k-1}. \tag{11.118}$$

Notice the similarity with the second variant (the fixed point version) of the semi-iterative method (11.65) applied to the simple splitting $A = M - N$, $M = I$ and $N = I - A$,

$$\boldsymbol{y}_{k+1} = \omega_{k+1}\left[\gamma((I - A)\boldsymbol{y}_k + \boldsymbol{b}) + (1 - \gamma)\boldsymbol{y}_k\right] + (1 - \omega_{k+1})\boldsymbol{y}_{k-1}.$$

We see the corresponding quantities

$$\rho_{k+1} \,\widehat{=}\, \omega_{k+1}, \quad \delta_{k+1} \,\widehat{=}\, \gamma = \mathrm{const},$$

which shows that CG also accelerates the basic iteration

$$\boldsymbol{x}_{k+1} = (I - A)\boldsymbol{x}_k + \boldsymbol{b}.$$

In the semi-iterative method, we defined

$$\gamma = \frac{2}{2 - a - b},$$

where $[-1, 1] \supset [a, b]$ contained the spectrum of the iteration matrix $I - A$. With CG, we do not need to estimate the parameters a and b, since they are computed by the algorithm.

We now show that CG also computes a *reduction of the original matrix to a tridiagonal matrix*, with interesting spectral properties. We will see this reduction again in form of the symmetric Lanczos algorithm in Section 11.7.3. By multiplying Equation (11.117) with $-A$, and adding $\boldsymbol{b} = \rho_{k+1}\boldsymbol{b} + (1 - \rho_{k+1})\boldsymbol{b}$ on both sides, we obtain a recurrence for the residual vectors,

$$\boldsymbol{r}_{k+1} = (\rho_{k+1}I - \alpha_k A)\boldsymbol{r}_k + (1 - \rho_{k+1})\boldsymbol{r}_{k-1}. \tag{11.119}$$

Because of $\boldsymbol{r}_k = A\boldsymbol{e}_k$, this three term recurrence is also valid for the error. Solving for $A\boldsymbol{r}_k$, we obtain

$$A\boldsymbol{r}_k = \frac{1 - \rho_{k+1}}{\alpha_k}\boldsymbol{r}_{k-1} + \frac{\rho_{k+1}}{\alpha_k}\boldsymbol{r}_k - \frac{1}{\alpha_k}\boldsymbol{r}_{k+1} \tag{11.120}$$

for $k = 1, 2, \ldots, n - 1$, with $\boldsymbol{r}_n = 0$. For $k = 0$, using Equation (11.108), we obtain

$$\boldsymbol{r}_1 = \boldsymbol{r}_0 - \alpha_0 A\boldsymbol{p}_0,$$

and since $\boldsymbol{p}_0 = \boldsymbol{r}_0$ in CG, we get

$$A\boldsymbol{r}_0 = \frac{1}{\alpha_0}\boldsymbol{r}_0 - \frac{1}{\alpha_0}\boldsymbol{r}_1.$$

Defining the coefficients for $k = 1, \ldots, n-1$

$$c_k := \frac{1 - \rho_{k+1}}{\alpha_k} = -\frac{\beta_{k-1}}{\alpha_{k-1}}$$

$$a_{k+1} := \frac{\rho_{k+1}}{\alpha_k} = \frac{1}{\alpha_k} + \frac{\beta_{k-1}}{\alpha_{k-1}}, \quad a_1 = \frac{1}{\alpha_0} \tag{11.121}$$

$$b_k := -\frac{1}{\alpha_{k-1}}$$

and the matrix of the residuals

$$R_k := [\boldsymbol{r}_0, \ldots, \boldsymbol{r}_{k-1}],$$

Equation (11.120) becomes

$$AR_n = R_n T_n, \quad T_n = \begin{bmatrix} a_1 & c_1 & & & \\ b_1 & a_1 & \ddots & & \\ & \ddots & \ddots & c_{n-1} \\ & & b_{n-1} & a_n \end{bmatrix}, \tag{11.122}$$

and we obtain the following theorem:

THEOREM 11.30. *CG transforms the matrix A to the similar tridiagonal matrix T_n (11.122). The coefficients are given by Equations (11.121).*

Note that the transformation matrix R_n has orthogonal (but not orthonormal) columns,

$$R_n^\top R_n = D^2, \quad D_n = \mathrm{diag}(d_0, d_1, \ldots, d_{n-1}), \quad d_i = \|\boldsymbol{r}_i\|.$$

THEOREM 11.31. *Let $A \in \mathbb{R}^{n \times n}$, $R \in \mathbb{R}^{n \times p}$ with full rank p and $F \in \mathbb{R}^{p \times p}$ with $p < n$. Assume that the matrix decomposition $AR = RF$ holds. Then each eigenvalue of F is also an eigenvalue of A.*

PROOF. Let λ and $\boldsymbol{y}$ be an eigenpair of F: $F\boldsymbol{y} = \lambda\boldsymbol{y}$. Then $\boldsymbol{x} = R\boldsymbol{y} \neq 0$, since R has full rank, and

$$A\boldsymbol{x} = AR\boldsymbol{y} = RF\boldsymbol{y} = R\lambda\boldsymbol{y} = \lambda\boldsymbol{x},$$

and thus λ and $\boldsymbol{x}$ is an eigenpair of A. $\qquad\square$

If we perform fewer than n CG steps, we get the decomposition

$$AR_k = R_k T_k + b_k \boldsymbol{r}_k \mathbf{e}_k^\top. \tag{11.123}$$

Note that because

$$\mathbf{c}_k^\top = \frac{1}{d_{k-1}^2} \mathbf{r}_{k-1}^\top R_k$$

the correction can also be written as

$$AR_k = R_k T_k + \frac{b_k}{d_{k-1}^2} \mathbf{r}_k \mathbf{r}_{k-1}^\top R_k$$

or

$$\left(A - \frac{b_k}{d_{k-1}^2} \mathbf{r}_k \mathbf{r}_{k-1}^\top \right) R_k = R_k T_k. \tag{11.124}$$

COROLLARY 11.3. *If we perform $k < n$ CG steps and obtain the matrix decomposition (11.123), then the eigenvalues of T_k are eigenvalues of the rank-one modified matrix*

$$A - \frac{b_k}{d_{k-1}^2} \mathbf{r}_k \mathbf{r}_{k-1}^\top.$$

EXAMPLE 11.6. *We want to demonstrate the properties of the CG algorithm with a numerical example. We construct a positive definite matrix with given eigenvalues, choose an exact solution and compute the corresponding right hand side.*

```
>> n=6
>> d=[1 2 3 4 5 6 ];          % construct positive
>> Q=orth(hilb(n));           % definite matrix
>> A=Q'*diag(d)*Q;
>> xexact=(1:n)';             % choose exact solution
>> b=A*xexact;                % compute right hand side
>> x0=xexact-5;               % choose starting vector
```

We now perform n CG steps and construct the tridiagonal matrix T_n

```
>> k=n;                       % number of iterations
>> [X,P,R,alpha,beta]=CG(A,b,x0,k )
>> c=-beta(1:k-1)./alpha(1:k-1);   % construct
>> a=[1/alpha(1) 1./alpha(2:k)-c]; % tridiagonal matrix
>> b=-1./alpha(1:k-1);
>> T=diag(a)+diag(c,1)+diag(b,-1)
```

```
T =
    4.8074   -0.0590        0        0        0        0
   -4.8074    3.4322   -0.3802        0        0        0
        0    -3.3732    3.2222   -1.0030        0        0
        0         0    -2.8420    3.9367   -0.7171        0
        0         0         0    -2.9337    3.1342   -0.2641
        0         0         0         0    -2.4170    2.4673
```

For the the decomposition we obtain a good match

```
>> RR=R(:,1:k);
>> norm(A*RR-RR*T)
ans =
   3.2710e-14
```

And because the matrices A and T_n are similar, i.e., $R_n^{-1}AR_n = T_n$, they have the same eigenvalues:

```
>> [eig(A) eig(T)]
ans =
      1.0000    6.0000
      2.0000    5.0000
      6.0000    1.0000
      3.0000    4.0000
      4.0000    3.0000
      5.0000    2.0000
```

Next, we illustrate the statements of Theorem 11.24: we show by computing $R_k^\top R_k$ that the residual vectors are orthogonal, and by computing $P_k^\top A P_k$ that the direction vectors are conjugate. Furthermore, as shown in (11.105), we see that $\boldsymbol{p}_j \perp \boldsymbol{r}_k$ for $k > j$ because the matrix $R_k^\top P_k$ is upper triangular:

```
>> disp('Residuals are orthogonal')
>> format short e
>> RR'*RR
>> disp('and directions conjugate')
>> PP=P(:,1:k);
>> PP'*A*PP
>> disp('p_j orth r_k for k>j')
>> RR'*PP
Residuals are orthogonal
ans =
     3.1826e+03   -2.8422e-14            0   -7.1054e-15   -1.4211e-14    9.5479e-15
    -2.8422e-14    3.9090e+01   4.4409e-15   -1.1935e-15   -1.2212e-15    1.0270e-15
              0    4.4409e-15   4.4059e+00    9.5757e-16    2.2204e-16   -1.8041e-16
    -7.1054e-15   -1.1935e-15   9.5757e-16    1.5549e+00    3.2439e-16   -1.4311e-16
    -1.4211e-14   -1.2212e-15   2.2204e-16    3.2439e-16    3.8009e-01    1.2143e-16
     9.5479e-15    1.0270e-15  -1.8041e-16   -1.4311e-16    1.2143e-16    4.1531e-02
and directions conjugate
ans =
     1.5300e+04   -1.7053e-13            0   -1.0658e-14   -7.1054e-14    3.5527e-14
    -5.6843e-14    1.3186e+02   1.7764e-14   -1.3323e-15   -6.2172e-15    3.4417e-15
    -4.2633e-14    1.4211e-14   1.2522e+01    2.6645e-15    1.1102e-16   -8.3267e-17
     2.4869e-14   -8.8818e-16   1.4433e-15    4.5615e+00    9.4369e-16   -6.4185e-16
    -7.1054e-14   -6.6613e-15   1.1102e-16    1.3600e-15    9.1868e-01    3.0878e-16
     3.4639e-14    3.4417e-15  -1.2490e-16   -6.2103e-16    3.0531e-16    9.1501e-02
p_j orth r_k for k>j
ans =
     3.1826e+03    3.9090e+01   4.4059e+00    1.5549e+00    3.8009e-01    4.1531e-02
    -2.8422e-14    3.9090e+01   4.4059e+00    1.5549e+00    3.8009e-01    4.1531e-02
              0    4.4409e-15   4.4059e+00    1.5549e+00    3.8009e-01    4.1531e-02
    -7.1054e-15   -1.6653e-15   8.8818e-16    1.5549e+00    3.8009e-01    4.1531e-02
    -1.4211e-14   -1.4433e-15   5.5511e-17    3.5388e-16    3.8009e-01    4.1531e-02
     9.5479e-15    1.1657e-15  -4.1633e-17   -1.5786e-16    8.3267e-17    4.1531e-02
```

Looking at the iterates stored in the columns of X, we note as stated in Corollary 11.2 that we reach the exact solution in n steps. Furthermore, the error decreases at each step, as stated in Theorem 11.29:

```
>> format short
>> disp('reach exact solution in n steps')
>> Solution=X
>> disp('norm of error is monotonically decreasing')
>> Errors=X-xexact*ones(1,k+1);
>> ErrorNorms=diag(Errors'*Errors)'
reach exact solution in n steps
Solution =
    -4.0000   -0.3176    0.3744    0.5204    0.7403    0.9183    1.0000
    -3.0000    2.5718    2.9425    2.4399    2.1744    2.0063    2.0000
    -2.0000    1.9671    2.5428    2.6852    3.0159    3.0464    3.0000
    -1.0000    2.4707    3.6503    4.1187    4.0319    4.0181    4.0000
          0    5.8627    5.1656    4.9925    4.8873    5.0184    5.0000
     1.0000    6.5633    5.7554    6.0088    6.0782    5.9834    6.0000
norm of error is monotonically decreasing
ErrorNorms =
   150.0000    6.5302    1.6982    0.5369    0.1179    0.0098    0.0000
```

As second example, we use the same matrix but we perform only $k < n$ iteration steps.

EXAMPLE 11.7.

```
>> n=6
>> d=[1 2 3 4 5 6 ];          % construct positive
>> Q=orth(hilb(n));           % definite matrix
>> A=Q'*diag(d)*Q;
>> xexact=(1:n)';             % choose exact solution
>> b=A*xexact;                % compute right hand side
>> x0=xexact-5;               % choose starting vector
>> k=n-2                      % number of iterations
>> [X,R,P,alpha,beta]=CG(A,b,x0,k );
>> c=-beta(1:k-1)./alpha(1:k-1);    % construct
>> a=[1/alpha(1) 1./alpha(2:k)-c];  % tridiagonal matrix
>> b=-1./alpha(1:k-1);
>> T=diag(a)+diag(c,1)+diag(b,-1)

n =
     6
k =
     4
T =
     4.8074   -0.0590         0         0
    -4.8074    3.4322   -0.3802         0
          0   -3.3732    3.2222   -1.0030
          0         0   -2.8420    3.9367
```

```
>> RR=R(:,1:k);
>> disp('A*R-R*T')
>> A*RR-RR*T
>> disp('b(k)*R(:,k+1)*e')
>> b(k)=-1/alpha(k);
>> e=zeros(1,k); e(k)=1;
>> b(k)*R(:,k+1)*e

A*R-R*T
ans =
          0   -0.0000    0.0000   -0.7992
    -0.0000   -0.0000    0.0000    0.6336
    -0.0000   -0.0000    0.0000    0.4801
    -0.0000   -0.0000    0.0000   -0.0846
          0         0   -0.0000   -1.1510
          0    0.0000         0    0.8177
b(k)*R(:,k+1)*e
ans =
          0         0         0   -0.7992
          0         0         0    0.6336
          0         0         0    0.4801
          0         0         0   -0.0846
          0         0         0   -1.1510
          0         0         0    0.8177
```

We see here with only $k = 4$ iterations that the difference $AR_4 - R_4 T_4$ is given as stated in Equation (11.123) by the matrix $b_k \boldsymbol{r}_k \boldsymbol{e}_k^\top$. As stated in Corollary 11.3, the eigenvalues of the small matrix T_k are eigenvalues of a rank-one modification of the matrix A.

```
>> EigOfA=sort(eig(A))
>> D2=diag(R'*R);
>> Rank1Matrix=b(k)/D2(k)*R(:,k+1)*R(:,k)';
>> EigOfRank1Mod=sort(eig(A-Rank1Matrix))
>> EigOfT=sort(eig(T))
EigOfA =
    1.0000
    2.0000
    3.0000
    4.0000
    5.0000
    6.0000
EigOfRank1Mod =
    1.4225
    1.9350
    3.4399
    3.6665
    4.9187
    5.6173
```

```
EigOfT =
    1.4225
    3.4399
    4.9187
    5.6173
```

We see that the eigenvalues of matrix T indeed match eigenvalues of the rank one modification of A.

Finally, we illustrate Corollary 11.2, which says that if the matrix A has only m distinct eigenvalues, then CG converges in m steps. We construct a matrix of order $n = 6$ with three double eigenvalues. After $m = 3$ iterations, the Conjugate Gradient method computes the exact solution. Furthermore, the eigenvalues of the matrix T are precisely the three diffenrent eigenvalues of A.

EXAMPLE 11.8.

```
>> n=6
>> d=[2 2 3 3 5 5 ];            % construct positive
>> Q=orth(hilb(n));             % definite matrix
>> A=Q'*diag(d)*Q;
>> xexact=(1:n)';              % choose exact solution
>> b=A*xexact;                 % compute right hand side
>> x0=xexact-5;                % choose starting vector
>> k=3                         % number of iterations
>> [X,R,P,alpha,beta] = CG(A,b,x0,k );
>> c=-beta(1:k-1)./alpha(1:k-1);    % construct
>> a=[1/alpha(1) 1./alpha(2:k)-c];  % tridiagonal matrix
>> b=-1./alpha(1:k-1);
>> T=diag(a)+diag(c,1)+diag(b,-1)
>> Solutions=X
>> R
>> EigOfT=sort(eig(T))
n =

     6

k =

     3

T =
    4.7524   -0.0984         0
   -4.7524    3.0497   -0.0590
         0   -2.9514    2.1978
Solutions =
   -4.0000   -0.1696    0.9336    1.0000
   -3.0000    2.0224    2.3796    2.0000
   -2.0000    2.0933    2.6702    3.0000
   -1.0000    1.9694    3.9136    4.0000
         0    5.3463    5.0693    5.0000
    1.0000    7.2356    5.9136    6.0000
R =
```

```
   18.2038      2.8792      0.0769      0.0000
   23.8686      0.5602     -0.8330      0.0000
   19.4531      1.2999      0.6714      0.0000
   14.1117      5.4462      0.0700     -0.0000
   25.4077     -1.3433     -0.1319      0.0000
   29.6341     -4.5149      0.2627      0.0000
EigOfT =
    2.0000
    3.0000
    5.0000
```

11.7.2   Arnoldi Process

The *Arnoldi process* constructs an orthogonal basis of the Krylov space

$$\mathcal{K}_k(A, \boldsymbol{r}_0) = \operatorname{span}\{\boldsymbol{r}_0, A\boldsymbol{r}_0, A^2\boldsymbol{r}_0, \ldots, A^{k-1}\boldsymbol{r}_0\}$$

using Gram-Schmidt orthogonalization. It computes the QR decomposition of the matrix $K_k \in \mathbb{R}^{n \times k}$,

$$K_k := [\boldsymbol{r}_0, A\boldsymbol{r}_0, \ldots, A^{k-1}\boldsymbol{r}_0] = Q_k R_k, \tag{11.125}$$

with $Q_k \in \mathbb{R}^{n \times k}$ orthogonal and $R_k \in \mathbb{R}^{k \times k}$ upper triangular. In the j-th step, the vector $A^j \boldsymbol{r}_0$ is orthonormalized to produce $\boldsymbol{q}_{j+1}$. This is done using the Gram-Schmidt process, by first computing the projections on the vectors $\boldsymbol{q}_1, \ldots, \boldsymbol{q}_j$ which have already been constructed,

$$r_{i,j+1} = \boldsymbol{q}_i^{\top} A^j \boldsymbol{r}_0, \quad i = 1, \ldots, j.$$

Then the corresponding components are subtracted,

$$\boldsymbol{u} = A^j \boldsymbol{r}_0 - \sum_{i=1}^{j} r_{i,j+1} \boldsymbol{q}_i, \tag{11.126}$$

and the result is normalized

$$r_{j+1,j+1} = \|\boldsymbol{u}\|, \quad \boldsymbol{q}_{j+1} = \frac{1}{r_{j+1,j+1}} \boldsymbol{u}. \tag{11.127}$$

Idea of Arnoldi: If we are only interested in the orthogonal basis defined by the matrix Q_k, then we can replace the vector $A^j \boldsymbol{r}_0$ by $A\boldsymbol{q}_j$. This is justified because as long as the Krylov matrix K_k has full rank, it follows from $K_k = Q_k R_k$ that $Q_k = K_k R_k^{-1}$, and thus with $\boldsymbol{\alpha} = R_k^{-1} \boldsymbol{e}_j$, we obtain

$$\boldsymbol{q}_j = \alpha_0 \boldsymbol{r}_0 + \alpha_1 A\boldsymbol{r}_0 + \cdots + \alpha_{j-1} A^{j-1}\boldsymbol{r}_0, \quad \text{with } \alpha_{j-1} = \frac{1}{r_{j,j}} \neq 0.$$

So $\boldsymbol{q}_j$ contains a component which points in the direction of $A^{j-1}\boldsymbol{r}_0$, and therefore $A\boldsymbol{q}_j$ can be used instead of $A^j\boldsymbol{r}_0$. With this modification, the orthogonalization step becomes

$$h_{j+1,j}\boldsymbol{q}_{j+1} = A\boldsymbol{q}_j - \sum_{i=1}^{j} h_{ij}\boldsymbol{q}_i, \tag{11.128}$$

where $h_{ij} = \boldsymbol{q}_i^\top A\boldsymbol{q}_j$ are the projections and $h_{j+1,j}$ is the normalizing factor.

Note that we could have obtained a similar relation by inserting the relation $\boldsymbol{q}_{j+1} = \boldsymbol{u}/r_{j+1,j+1}$ (cf. (11.127)) into (11.126):

$$r_{j+1,j+1}\boldsymbol{q}_{j+1} = A^j\boldsymbol{r}_0 - \sum_{i=1}^{j} r_{i,j+1}\boldsymbol{q}_i, \tag{11.129}$$

Comparing (11.128) with (11.129), we see that we have shifted the second index in h_{ij} by one relative to the r_{ij}. The reason is that we want to write (11.128) now in matrix form as

$$A\boldsymbol{q}_j = \sum_{i=1}^{j+1} h_{ij}\boldsymbol{q}_i, \quad j = 1,\ldots,k, \tag{11.130}$$

where the first vector $\boldsymbol{q}_1$ no longer appears on the left hand side. Using the matrices $Q_k = [\boldsymbol{q}_1,\ldots,\boldsymbol{q}_k] \in \mathbb{R}^{n\times k}$ and

$$\tilde{H}_k = \begin{bmatrix} h_{11} & h_{12} & \cdots & h_{1k} \\ h_{21} & h_{22} & \cdots & h_{2k} \\ & \ddots & \ddots & \vdots \\ & & \ddots & h_{k,k} \\ & & & h_{k+1,k} \end{bmatrix} \in \mathbb{R}^{(k+1)\times k},$$

(11.130) can now be written as

$$AQ_k = Q_{k+1}\tilde{H}_k.$$

If we define the upper *Hessenberg matrix* $H_k \in \mathbb{R}^{k\times k}$ to contain the first k rows of $\tilde{H}_k$, then, after $k < n$ steps of the Arnoldi process, we have the equation

$$AQ_k = Q_k H_k + h_{k+1,k}\boldsymbol{q}_{k+1}\boldsymbol{e}_k^\top. \tag{11.131}$$

Since the column vectors $\boldsymbol{q}_j$ of Q_k are orthonormal, it follows from Equation (11.131) that

$$Q_j^\top AQ_j = H_j, \quad j = 1,\ldots,k. \tag{11.132}$$

The matrices H_j are often called the *projections of the matrix A onto the Krylov subspace*. It can happen that for some k we get $h_{k+1,k}\boldsymbol{q}_{k+1} = 0$. In this case, $AQ_k = Q_k H_k$ holds, which means we have computed an invariant subspace spanned by the columns of Q_k; this event is therefore called a "lucky breakdown". In this case, if λ and x is an eigenpair of H_k, then λ and $y = Q_k x$ is an eigenpair of A.

In the special case of $k = n$, we have $q_{n+1} = 0$, and thus

$$AQ_n = Q_n H_n \quad \Longleftrightarrow \quad Q_n^\top A Q_n = H_n,$$

which means that A has been transformed by an orthogonal similarity transformation to *upper Hessenberg form*. This is usually done as shown in Chapter 7 by using elementary orthogonal matrices. The Arnoldi process is another way of computing this transformation, but it is less stable numerically because it uses a Gram-Schmidt process. Using the more stable modified Gram-Schmidt process (cf. Section 6.5.5), we get the algorithm

---

ALGORITHM 11.17.
Arnoldi algorithm to compute $A = QTQ^\top$

```
function [H,Q,v]=Arnoldi(A,r,k);
% ARNOLDI Arnoldi process
%    [H,Q,v]=Arnoldi(A,r,k) applies k<=n steps of the Arnoldi process to
%    the matrix A starting with the vector r. Computes Q orthogonal and
%    H upper Hessenberg such that AQ=QH+ve_k^T, with Q(:,1)=r/norm(r).

ninf=norm(A,inf);
Q(:,1)=r/norm(r);
for j=1:k
  v=A*Q(:,j);
  for i = 1:j
    H(i,j)=Q(:,i)'*v;
    v=v-H(i,j)*Q(:,i);
  end
  if j<k
    H(j+1,j)=norm(v);
    if abs(H(j+1,j))+ninf==ninf
      disp('lucky breakdown occured!');
      H=H(1:end-1,:);
      break;
    end
    Q(:,j+1)=v/H(j+1,j);
  end
end
```

As an example, we consider for $n = 4$ the magic square matrix from MATLAB:

```
>> A=magic(4);
>> r=(1:4)';
```

```
>> [H,Q,v]=Arnoldi(A,r,4)
H =
    28.3333    13.2567    -4.1306    -1.3734
    12.7550     3.7347    12.9021     3.9661
         0      3.8955    -2.2508     1.9173
         0           0     9.9142     4.1827
Q =
     0.1826     0.7539     0.4895     0.3984
     0.3651     0.4628    -0.2153    -0.7785
     0.5477     0.0573    -0.6830     0.4798
     0.7303    -0.4628     0.4975    -0.0702
v =
   1.0e-15 *
   -0.2220
    0.8882
    0.4441
   -0.1110
>> norm(A*Q-Q*H)
ans =
   2.0368e-15
```

If we start however with a vector of all ones, we obtain

```
>> r=ones(4,1);
>> [H,Q,v]=Arnoldi(A,r,3)
lucky breakdown occured!
H =
    34
Q =
    0.5000
    0.5000
    0.5000
    0.5000
v =
     0
     0
     0
     0
```

and Arnoldi found that the vector of all ones is an eigenspace, as expected
for a magic square.

11.7.3   The Symmetric Lanczos Algorithm

If we apply the Arnoldi process to a symmetric matrix, $A = A^{\top}$, then the
matrix $Q_k^{\top} A Q_k = H_k$ is also symmetric, which means it is tridiagonal,

$$
T_k = \begin{bmatrix} \alpha_1 & \beta_1 & & \\ \beta_1 & \alpha_2 & \ddots & \\ & \ddots & \ddots & \beta_{k-1} \\ & & \beta_{k-1} & \alpha_k \end{bmatrix}
$$

The recurrence relation (11.128) simplifies to the *symmetric Lanczos iteration*,

$$
\beta_j \boldsymbol{q}_{j+1} = A\boldsymbol{q}_j - \alpha_j \boldsymbol{q}_j - \beta_{j-1}\boldsymbol{q}_{j-1}. \tag{11.133}
$$

To obtain $\boldsymbol{q}_{j+1}$ in the next iteration step, we compute

1. $\boldsymbol{v} := A\boldsymbol{q}_j$.

2. To obtain α_j, we multiply (11.133) from the left by $\boldsymbol{q}_j^\top$ and get, using orthogonality of the $\boldsymbol{q}_i$, that $\alpha_j := \boldsymbol{q}_j^\top \boldsymbol{v}$.

3. The right hand side of (11.133) then becomes $\tilde{\boldsymbol{v}} := \boldsymbol{v} - \alpha_j \boldsymbol{q}_j - \beta_{j-1}\boldsymbol{q}_{j-1}$.

4. We now compute $\beta_j = \|\tilde{\boldsymbol{v}}\|$, and store the coefficients α_j and β_j in T. The new vector $\boldsymbol{q}_{j+1}$, a *Lanczos vector*, is then given by $\boldsymbol{q}_{j+1} := \tilde{\boldsymbol{v}}/\beta_j$.

In order to obtain an elegant algorithm, it is convenient to assume that $\beta_0 = 0$. Unfortunately, MATLAB arrays cannot be indexed with 0, so we need an if-statement for the first iteration. We refrain from testing if $\beta_j = 0$, since orthogonality of the Lanczos vectors $\boldsymbol{q}_j$ is lost more quickly than with Arnoldi. We thus obtain the Lanczos algorithm

---

ALGORITHM 11.18.
*Lanczos algorithm to compute $A = QTQ^\top$ for a
symmetric A*

```
function [alpha,beta,Q,v]=Lanczos(A,r,k)
% LANCZOS Symmetric Lanczos process
%    [alpha,beta,Q,r]=Lanczos(A,r,k) applies k<=n steps of the
%    symmetric Lanczos process to the symmetric matrix A starting with
%    the vector r. Computes Q orthogonal and a symmetric tridiagonal
%    matrix given by the diagonal in the vector alpha, and the super-
%    and subdiagonal in the vector beta.

Q(:,1)=r/norm(r);
if k==1, beta=[]; end;
for j=1:k
  v=A*Q(:,j);
  alpha(j)=Q(:,j)'*v;
  v=v-alpha(j)*Q(:,j);
  if j>1
    v=v-beta(j-1)*Q(:,j-1);
  end
```

```
  if j<k
    beta(j)=norm(v);
    Q(:,j+1)=v/beta(j) ;
  end
end
```

EXAMPLE 11.9. *We consider the symmetric matrix*

```
>> A=[ 8  6  8  2 11  2
       6  2 17 13 11  1
       8 17  6 10  8  1
       2 13 10  6 20  5
      11 11  8 20 16 15
       2  1  1  5 15 20];
```

and the initial vector

```
>> r=[1 2 3 4 5 6]';
```

and first check the relation (11.131) for $k = 3$,

$$AQ_k = Q_k H_k + h_{k+1,k} \boldsymbol{q}_{k+1} \boldsymbol{e}_k^\top.$$

```
>> [alpha,beta,Q,v2]=Lanczos(A,r,3)
alpha =
    50.7692    14.7761     1.3219
beta =
    14.7045    14.5237
Q =
     0.1048     0.4864     0.0702
     0.2097     0.5166    -0.2150
     0.3145    -0.0450     0.7933
     0.4193     0.0636     0.3545
     0.5241     0.3789    -0.2845
     0.6290    -0.5890    -0.3359
v2 =
    -4.6316
     7.4022
     2.2040
    -1.3569
    -2.8748
     0.5028
>> T=diag(alpha)+diag(beta,-1)+diag(beta,1);
>> A*Q-Q*T
ans =
         0          0    -4.6316
         0     0.0000     7.4022
         0     0.0000     2.2040
         0          0    -1.3569
         0     0.0000    -2.8748
         0          0     0.5028
```

The products AQ_k and Q_kT_k differ indeed by the vector $\mathbf{v2} = h_{k+1,k}\mathbf{q}_{k+1}$.

Next, we apply the Arnoldi process to the symmetric matrix A and compare the resulting decompositions and the orthogonality of the matrix Q with the decompositions obtained using Householder transformations, as implemented in the the MATLAB function **hess**.

```
>> [Q,T]=hess(A);
>> HES=[norm(A*Q-Q*T) norm(Q'*Q-eye(6))]
>> [T,Q,v3]=Arnoldi(A,r,6);
>> ARN=[norm(A*Q-Q*T) norm(Q'*Q-eye(6))]
>> [alpha,beta,Q,v2]=Lanczos(A,r,6);
>> T=diag(alpha)+diag(beta,-1)+diag(beta,1);
>> LAN=[norm(A*Q-Q*T) norm(Q'*Q-eye(6))]
```

The results are summarized in the following table, and show the discrepancy between AQ and QT as well as the loss of orthogonality of Q.

	$\|AQ - QT\|$	$\|Q^{\top}Q - I\|$
Householder	0.1214e−13	0.0048e−13
Arnoldi	0.9668e−13	0.0766e−13
Sym. Lanczos	0.4655e−9	0.0067e−9

We see already with this small example that Arnoldi is not quite as good as the reduction by elementary Householder matrices. But Arnoldi is much better than symmetric Lanczos, where we observe substantial loss of orthogonality.

It is interesting to study how well for $k < n$ the eigenvalues of the projected matrix

$$Q_k^{\top} AQ_k = H_k \quad (\text{or} \quad T_k)$$

approximate the eigenvalues of the original matrix A. With

```
for k=1:6
  [alpha,beta,Q,v2]=Lanczos(A,r,k);
  T=diag(alpha)+diag(beta,-1)+diag(beta,1);
  if k>1
    D(1:k,k)=sort(eig(T));
  else
    D(1,1)=alpha;
  end;
end
D
```

we get the results

exact	$k = 1$	$k = 2$	$k = 3$	$k = 4$	$k = 5$	$k = 6$
−13.9139	50.7692	9.5327	−9.0918	−13.8082	−13.9068	−13.9139
−13.1962		56.0127	19.4813	2.7841	1.4462	−13.1962
1.4648			56.4777	21.0970	5.9899	1.4648
6.0244				56.4927	21.1281	6.0244
21.1282					56.4927	21.1282
56.4927						56.4927

We see nicely how the eigenvalues of the matrices T_k approximate first the eigenvalues with largest modulus $|\lambda_i(A)|$.

After n Arnoldi or Lanczos steps, we would expect that $h_{n+1,k}\boldsymbol{q}_{n+1} = 0$. However, we get with

```
>> [alpha,beta,Q1,v1]=Lanczos(A,r,6);
>> T=diag(alpha)+diag(beta,-1)+diag(beta,1);
>> [T2,Q2,v2]=Arnoldi(A,r,6);
>> [v1 v2]
ans =
   1.0e-09 *
   -0.1256   -0.0000
   -0.1727   -0.0000
   -0.1651    0.0000
   -0.2067    0.0000
   -0.2751    0.0000
   -0.1595    0.0001
```

The vector $h_{k+1,k}\boldsymbol{q}_{n+1}$ is much smaller with Arnoldi than with Lanczos; the Lanczos process does not terminate and it is interesting to see what happens to the eigenvalues of the tridiagonal matrix when we continue with the iteration:

```
D=[];
for k=7:10
  [alpha,beta,Q,v2]=Lanczos(A,r,k);
  T=diag(alpha)+diag(beta,-1)+diag(beta,1);
  D(1:k,k-6)=sort(eig(T));
end
D
```

$k = 7$	$k = 8$	$k = 9$	$k = 10$
-13.9139	-13.9139	-13.9139	-13.9139
-13.1962	-13.1962	-13.1962	-13.1962
1.4648	1.4648	0.5184	-12.5049
6.0244	6.0244	1.4648	1.4648
21.1282	20.6959	6.0244	4.0302
56.4917	21.1282	21.0941	6.0244
56.4927	56.4927	21.1282	21.1265
	56.4927	56.4927	21.1282
		56.4927	56.4927
			56.4927

We observe the well-known effect [103], cf. [86], that multiple copies of the eigenvalues, also called ghost *eigenvalues, are produced.*

If we are only interested in the tridiagonal matrix, i.e. we wish to compute only the coefficients α_i and β_i, then, with a careful implementation, we need only two vectors for the Lanczos process. Two observations help for that purpose: first, we can compute new values for $\boldsymbol{v}$ and $\boldsymbol{w}$ in the same for-loop (see Statement (1) in Algorithm 11.19), and second, α_j can be computed (see Statement (2)) by

$$\alpha_j = \boldsymbol{q}_j^\top A \boldsymbol{q}_j = \boldsymbol{q}_j^\top (A\boldsymbol{q}_j - \beta_{j-1}\boldsymbol{q}_{j-1}).$$

This is possible because of $\boldsymbol{q}_j^\top \boldsymbol{q}_{j-1} = 0$.

ALGORITHM 11.19. *Economic Lanczos*

$[\boldsymbol{\alpha}, \boldsymbol{\beta}] = \text{EconomicLanczos}(A, \boldsymbol{r}, k)$

$\qquad \boldsymbol{w} = 0; \qquad\qquad\qquad\qquad$ % $\boldsymbol{w} = \boldsymbol{q}_0 = 0$

$\qquad \boldsymbol{v} = \boldsymbol{r}_0;$

$\qquad \beta_0 = \|\boldsymbol{r}_0\|;$

$\qquad \text{for } j = 1 : k$

(1) $\qquad \begin{pmatrix} \boldsymbol{w} \\ \boldsymbol{v} \end{pmatrix} := \begin{pmatrix} \boldsymbol{v}/\beta_{j-1} \\ \beta_{j-1}\boldsymbol{w} \end{pmatrix} \qquad$ % $\boldsymbol{w} = \boldsymbol{q}_j$

$\qquad\qquad\qquad\qquad\qquad\qquad\qquad$ % $\boldsymbol{v} = \beta_{j-1}\boldsymbol{q}_{j-1}$

$\qquad\qquad \boldsymbol{v} = A\boldsymbol{w} - \boldsymbol{v}; \qquad\qquad$ % $\boldsymbol{v} = A\boldsymbol{q}_j - \beta_{j-1}\boldsymbol{q}_{j-1}$

(2) $\qquad\qquad \alpha_j = \boldsymbol{w}^\top \boldsymbol{v} \qquad\qquad\quad$ % $\quad = \boldsymbol{q}_j^\top (A\boldsymbol{q}_j - \beta_{j-1}\boldsymbol{q}_{j-1})$

$\qquad\qquad \boldsymbol{v} = \boldsymbol{v} - \alpha_j \boldsymbol{w} \qquad\qquad$ % $\quad = A\boldsymbol{q}_j - \alpha_j\boldsymbol{q}_j - \beta_{j-1}\boldsymbol{q}_{j-1}$

$\qquad\qquad \beta_j = \|\boldsymbol{v}\|$

$\qquad \text{end}$

11.7.4   Solving Linear Equations with Arnoldi

We consider the linear system $A\boldsymbol{x} = \boldsymbol{b}$. Let $\boldsymbol{x}_0$ be an initial approximation to the solution and $\boldsymbol{r}_0 = \boldsymbol{b} - A\boldsymbol{x}_0$ the corresponding residual. After k Arnoldi steps, we have a matrix Q_k whose column vectors form an orthonormal basis of the Krylov subspace $\mathcal{K}_k(A, \boldsymbol{r}_0)$. For an approximate solution, we make the ansatz

$$\boldsymbol{x}_k = \boldsymbol{x}_0 + Q_k \boldsymbol{y}. \tag{11.134}$$

One way to determine the vector $\boldsymbol{y}$ is to impose a *Galerkin condition* and to require the residual $\boldsymbol{r}_k = \boldsymbol{b} - A\boldsymbol{x}_k$ to be orthogonal to the Krylov subspace. This method is known under the name of *Full Orthogonalization Method* or *FOM*. Requiring orthogonality means

$$\boldsymbol{r}_k \perp \mathcal{K}_k(A, \boldsymbol{r}_0)$$

$$\Longleftrightarrow \qquad 0 = Q_k^\top \boldsymbol{r}_k = Q_k^\top (\boldsymbol{b} - A\boldsymbol{x}_k)$$

$$= Q_k^\top (\boldsymbol{b} - A\boldsymbol{x}_0 - AQ_k\boldsymbol{y}) = Q_k^\top (\boldsymbol{r}_0 - AQ_k\boldsymbol{y})$$

$$\Longleftrightarrow \qquad Q_k^\top AQ_k \boldsymbol{y} = Q_k^\top \boldsymbol{r}_0.$$

Using (11.131) and $Q_k^\top \boldsymbol{r}_0 = \|\boldsymbol{r}_0\|\boldsymbol{e}_1$, which is true since $\boldsymbol{q}_1 = \frac{\boldsymbol{r}_0}{\|\boldsymbol{r}_0\|}$ and $\boldsymbol{q}_j^\top \boldsymbol{q}_1 = 0$, $j > 1$, we obtain

$$H_k \boldsymbol{y} = \|\boldsymbol{r}_0\|\boldsymbol{e}_1. \tag{11.135}$$

Thus, we have to solve this small $k \times k$ linear system with the Hessenberg matrix H_k to get the approximation $\boldsymbol{x}_k = \boldsymbol{x}_0 + Q_k\boldsymbol{y}$.

The approximation $\boldsymbol{x}_k$ may be computed after each Arnoldi step. But for a large system, one might want to do so only every once in a while. Note that the new residual is an element of the next Krylov space, since

$$\boldsymbol{x}_k = \boldsymbol{x}_0 + Q_k \boldsymbol{y} \quad \Longrightarrow \quad \boldsymbol{r}_k = \boldsymbol{b} - A\boldsymbol{x}_k = \boldsymbol{r}_0 \ - AQ_k \boldsymbol{y} \in \mathcal{K}_{k+1}(A, \boldsymbol{r}_0).$$
$$(11.136)$$

Furthermore,

$$\begin{aligned}
\boldsymbol{r}_k = \boldsymbol{r}_0 \ - AQ_k \boldsymbol{y} &= \boldsymbol{r}_0 \ - Q_{k+1}\tilde{H}_k \boldsymbol{y} \\
&= \boldsymbol{r}_0 \ - Q_k H_k \boldsymbol{y} - h_{k+1,k}\boldsymbol{q}_{k+1}y_k \\
&= Q_k \underbrace{\left(\|\boldsymbol{r}_0\| \boldsymbol{e}_1 - H_k \boldsymbol{y} \right)}_{0} - h_{k+1,k}\boldsymbol{q}_{k+1}y_k,
\end{aligned}$$

which shows that $\boldsymbol{r}_k$ and $\boldsymbol{q}_{k+1}$ are collinear.

In the following algorithm, we compute the new solution in each iteration step.

ALGORITHM 11.20. *FOM for Solving Linear Systems*

```
function [X,R,H,Q]=FOM(A,b,x0);
% FOM full orthogonalization method
%    [X,R,H,Q]=FOM(A,b,x0) computes the decomposition A=QHQ',
%    Q orthogonal and H upper Hessenberg, Q(:,1)=r/norm(r),
%    using Arnoldi in order to solve the system Ax=b with the
%    full orthogonalization method. X contains the iterates
%    and R the residuals

n=length(A); X=x0;
r=b-A*x0; R=r; r0norm=norm(r);
Q(:,1)=r/r0norm;
for k=1:n
  v =A*Q(:,k);
  for j=1:k
    H(j,k)=Q(:,j)'*v; v=v-H(j,k)*Q(:,j);
  end
  e0=zeros(k,1); e0(1)=r0norm;                    % solve system
  y=H\e0; x= x0+Q*y;
  X=[X x];
  R=[R b-A*x];
  if k<n
    H(k+1,k)=norm(v); Q(:,k+1)=v/H(k+1,k);
  end
end
```

We are now ready to state and prove a first equivalence theorem between a Krylov subspace method and an extrapolation method.

THEOREM 11.32. (EQUIVALENCE OF MPE AND FOM) *Let $A \in \mathbb{R}^{n \times n}$ be an invertible matrix, $A = M - N$ be a splitting of A with M invertible, and $b \in \mathbb{R}^n$. Then for any given starting vector x_0, applying FOM to the preconditioned system $M^{-1}Ax = M^{-1}b$ or applying MPE to the stationary iterative method $Mx_{k+1} = Nx_k + b$ lead to identical iterates.*

PROOF. As we have seen, FOM discussed above determines its iterates x_k^{FOM} such that

$$r_k^{\mathrm{FOM}} \perp \mathcal{K}_k(M^{-1}A, r_0),$$

where $r_0 := M^{-1}b - M^{-1}Ax_0$ is the preconditioned residual, and also r_k^{FOM} denotes preconditioned residuals. Theorem 11.21 shows that MPE determines its iterates y_k^{MPE} such that

$$r_k^{\mathrm{MPE}} \perp \mathcal{K}_k(M^{-1}N, u_0).$$

We first show that these two Krylov spaces are the same: from $M^{-1}N = M^{-1}(M - A) = I - M^{-1}A$, and using the translation and scaling invariance, see (11.97), we can replace $M^{-1}N$ by $M^{-1}A$ without changing the Krylov space. We now look at the initial vector: by the definition of u_0 (see statement of Lemma 11.7) and that of the stationary iterative method, we have

$$\begin{aligned}
u_0 &= x_1 - x_0 = M^{-1}Nx_0 + M^{-1}b - x_0 \\
&= M^{-1}b + (M^{-1}N - I)x_0 = M^{-1}b - M^{-1}Ax_0 = r_0.
\end{aligned} \tag{11.137}$$

Thus, the initial vector of the Krylov spaces is the same, which implies that the Krylov spaces are identical for the two methods.

We now look at the residuals. For FOM, (11.134) says that $x_k^{\mathrm{FOM}} - x_0$ can be written as a linear combination of the columns of Q_k. Since Q_k also contains a basis for the Krylov subspace $\mathcal{K}_k(M^{-1}A, r_0)$, we deduce that

$$x_k^{\mathrm{FOM}} = x_0 + \sum_{j=0}^{k-1} \delta_j (M^{-1}A)^j r_0$$

for some coefficients δ_j. The residual is therefore

$$r_k^{\mathrm{FOM}} = M^{-1}b - M^{-1}Ax_k^{\mathrm{FOM}} = r_0 - M^{-1}A \sum_{j=0}^{k-1} \delta_j (M^{-1}A)^j r_0$$

$$= P_k^{\mathrm{FOM}}(M^{-1}A)r_0,$$

where P_k^{FOM} is a polynomial of degree k with $P_k^{\mathrm{FOM}}(0) = 1$. For MPE, we have from Theorem 11.21 and $u_0 = r_0$ that

$$\begin{aligned}
r_k^{\mathrm{MPE}} &= U_k \gamma = [u_0, u_1, \ldots, u_k]\gamma \\
&= [u_0, M^{-1}Nu_0, \ldots, (M^{-1}N)^k u_0]\gamma \\
&= [r_0, M^{-1}Nr_0, \ldots, (M^{-1}N)^k r_0]\gamma \\
&= q_k(M^{-1}N)r_0
\end{aligned}$$

for some polynomial q_k of degree k with $q_k(1) = 1$, since the γ_j sum up to 1. This can be rewritten as

$$r_k^{\mathrm{MPE}} = q_k(M^{-1}N)r_0 = q_k(I - M^{-1}A)r_0 = P_k^{\mathrm{MPE}}(M^{-1}A)r_0,$$

where P_k^{MPE} is a polynomial of degree k with $P_k^{\mathrm{MPE}}(0) = 1$. Hence both residuals have the same polynomial form, and the coefficients are determined by the same orthogonality condition, which implies that the residuals are the same, $r_k^{\mathrm{FOM}} = r_k^{\mathrm{MPE}}$, and thus the iterates must coincide. $\qquad\square$

EXAMPLE 11.10. *We illustrate this equivalence with a small example:*

```
A=[ 0      -4      -8      -2
    -4      -7      -7      -8
    -9      -5      -4      -5
     0      -5      -9      -6  ];
n=length(A); x=(1:n)'; b=A*x;
x0=x-5;                                    % Starting vector
[X,R,H,Q]=FOM(A,b,x0);
disp('Solutions with FOM'); X
[Y,X,U,Gamma]=MPE(eye(size(A))-A,b,x0,n);
disp('Solutions with MPE'); Y


Solutions with FOM
X =
      -4.0000     -0.7314      0.2662      1.0638      1.0000
      -3.0000      3.0703      3.1914      1.6232      2.0000
      -2.0000      3.3698      1.8713      3.1228      3.0000
      -1.0000      3.6694      4.6950      4.1728      4.0000
Solutions with MPE
Y =
      -4.0000     -0.7314      0.2662      1.0638      1.0000
      -3.0000      3.0703      3.1914      1.6232      2.0000
      -2.0000      3.3698      1.8713      3.1228      3.0000
      -1.0000      3.6694      4.6950      4.1728      4.0000
```

11.7.5 Solving Linear Equations with Lanczos

With a symmetric matrix A, it is possible to compute $\boldsymbol{x}_k$ from $\boldsymbol{x}_{k-1}$ using the Lanczos algorithm and a short recurrence. The Lanczos vectors Q_k need not to be stored.

We first study the *LDU decomposition of the symmetric tridiagonal matrix* T_k,

$$T_k = \begin{bmatrix} \alpha_1 & \beta_1 & & \\ \beta_1 & \alpha_2 & \ddots & \\ & \ddots & \ddots & \beta_{k-1} \\ & & \beta_{k-1} & \alpha_k \end{bmatrix} = L_k D_k L_k^{\top}, \tag{11.138}$$

with $D_k = \mathrm{diag}(d_1, \ldots, d_k)$ and

$$
L_k = \begin{bmatrix} 1 & & & \\ \mu_1 & \ddots & & \\ & \ddots & \ddots & \\ & & \mu_{k-1} & 1 \end{bmatrix}.
$$

By comparing the matrix elements of the product

$$
L_k D_k L_k^\top = \begin{bmatrix} d_1 & d_1\mu_1 & & \\ d_1\mu_1 & d_1\mu_1^2 + d_2 & d_2\mu_2 & \\ & d_2\mu_2 & d_2\mu_2^2 + d_3 & \ddots \\ & & \ddots & \ddots \end{bmatrix}
$$

with T_k, we get the equations required to compute this decomposition,

$$
d_1 = \alpha_1, \quad \mu_1 = \frac{\beta_1}{d_1},
$$

$$
d_2 = \alpha_2 - d_1\mu_1^2 = \alpha_2 - \beta_1\mu_1, \quad \mu_2 = \frac{\beta_2}{d_2},
$$

and for $i \geq 2$

$$
d_i = \alpha_i - \beta_{i-1}\mu_{i-1}, \quad \mu_i = \frac{\beta_i}{d_i}. \tag{11.139}
$$

Notice that (11.139) can be used for $i = k$ to update the decomposition from T_k to T_{k+1}.

With the ansatz $\boldsymbol{x}_k = \boldsymbol{x}_0 + Q_k \boldsymbol{y}$, we need to solve (11.135), which now reads

$$
T_k \boldsymbol{y} = Q_k^\top \boldsymbol{r}_0 = \|\boldsymbol{r}_0\|\boldsymbol{e}_1 \quad \Rightarrow \boldsymbol{y} = T_k^{-1} Q_k^\top \boldsymbol{r}_0.
$$

Using this and the decomposition (11.138) in the expression for $\boldsymbol{x}_k$, we obtain

$$
\begin{aligned}
\boldsymbol{x}_k &= \boldsymbol{x}_0 + Q_k \boldsymbol{y} \\
&= \boldsymbol{x}_0 + Q_k T_k^{-1} Q_k^\top \boldsymbol{r}_0 \\
&= \boldsymbol{x}_0 + Q_k (L_k D_k L_k^\top)^{-1} Q_k^\top \boldsymbol{r}_0 \\
&= \boldsymbol{x}_0 + F_k \boldsymbol{z},
\end{aligned}
$$

where we defined

$$
F_k = Q_k L_k^{-\top}, \quad \text{and} \quad \boldsymbol{z} = D_k^{-1} L_k^{-1} Q_k^\top \boldsymbol{r}_0 \iff L_k D_k \boldsymbol{z} = Q_k^\top \boldsymbol{r}_0.
$$

It follows from the definition of F_k that $F_k L_k^\top = Q_k$, written more explicitly as

$$
F_k L_k^\top = [\boldsymbol{f}_1, \mu_1\boldsymbol{f}_1 + \boldsymbol{f}_2, \ldots, \mu_{k-1}\boldsymbol{f}_{k-1} + \boldsymbol{f}_k] = Q_k.
$$

Thus, comparing the last column leads to the relation

$$\boldsymbol{f}_k = \boldsymbol{q}_k - \mu_{k-1}\boldsymbol{f}_{k-1}.$$

Since the matrix $L_k D_k$ of the linear system $L_k D_k \boldsymbol{z} = Q_k^\top \boldsymbol{r}_0$, which defines $\boldsymbol{z}$, is lower bi-diagonal and the vector on the right hand side is zero except for the first element, we can update the last component z_k of $\boldsymbol{z}$ by

$$\mu_{k-1}d_{k-1}z_{k-1} + d_k z_k = 0 \quad \implies \quad z_k = -\frac{1}{d_k}\mu_{k-1}d_{k-1}z_{k-1}.$$

Putting all this together, we obtain

$$\begin{aligned}
\boldsymbol{x}_k &= \boldsymbol{x}_0 + F_k \boldsymbol{z} \\
&= \underbrace{\boldsymbol{x}_0 + F_{k-1}(z_1, \ldots, z_{k-1})^\top}_{\boldsymbol{x}_{k-1}} + \boldsymbol{f}_k z_k.
\end{aligned}$$

By inserting a few additional statements into the k-th Lanczos step, we obtain an algorithm which computes recursively at each iteration step a new approximation for the solution of the linear system. It turns out that the resulting algorithm computes the same approximations as the *conjugate gradient method*, see Section 11.7.1!

ALGORITHM 11.21.
Symmetric Lanczos: Conjugate Gradient Step

The k-th iteration step of Algorithm 11.18 is augmented by the statements to solve linear equations.

compute $\boldsymbol{q}_{k+1}$, β_k and α_{k+1} with Algorithm 11.18;

$$\mu_k = \frac{\beta_k}{d_k};$$

$$d_{k+1} = \alpha_{k+1} - \beta_k \mu_k;$$

$$\boldsymbol{f}_{k+1} = \boldsymbol{q}_{k+1} - \mu_k \boldsymbol{f}_k;$$

$$z_{k+1} = -\frac{1}{d_{k+1}}\mu_k d_k z_k;$$

$$\boldsymbol{x}_{k+1} = \boldsymbol{x}_k + \boldsymbol{f}_{k+1} z_{k+1};$$

EXAMPLE 11.11. *In this example, we solve a linear system using Algorithm 11.21 and compare the solution with the results obtained by Conjugate Gradient. The program is more complicated than necessary because MATLAB does not allow indices smaller than 1 for arrays. The example demonstrates the equivalence of Algorithm 11.21 with the Conjugate Gradient Method.*

```
disp('Solving linear Equations with sym Lanczos is the same as CG')
```

```matlab
clear, clc, format compact
A=[ 8   6   8   2 11   2
    6   2 17 13 11   1
    8 17   6 10   8   1
    2 13 10   6 20   5
   11 11   8 20 16 15
    2   1   1   5 15 20];
n=length(A); xexact=(1:n)'; b=A*xexact;
x0=xexact-5; x=x0; XX=x; r=b-A*x;
Q(:,1)=r/norm(r);
for k=1:n
  v=A*Q(:,k);
  alpha(k)=Q(:,k)'*v;
  if k==1,
    d(1)=alpha(1);
    F(:,1)=Q(:,1);
    z(1)=norm(r)/d(1);
    x=x+r/alpha(1);   XX=[XX, x];
    v=v-alpha(1)*Q(:,1);
  else
    d(k)=alpha(k)-beta(k-1)*mue(k-1);
    F(:,k)=Q(:,k)-mue(k-1)*F(:,k-1);
    z(k)=-mue(k-1)*d(k-1)*z(k-1)/d(k);
    x=x+z(k)*F(:,k); XX=[XX,x];
    v=v-alpha(k)*Q(:,k)-beta(k-1)*Q(:,k-1);
  end
  beta(k)=norm(v);
  Q(:,k+1)=v/beta(k) ;
  mue(k)=beta(k)/d(k);
end
end
XX
X=CG(A,b,x0,n)
disp('Check that the vectors in F are conjugate F''*A*F')
F'*A*F
```

```
XX =
    -4.0000   -0.7205   -0.7533    0.3889    1.3119    0.9069    1.0000
    -3.0000    1.4318    1.5185    1.7551    1.5332    1.8937    2.0000
    -2.0000    2.4318    1.2991    2.1021    2.2058    3.4478    3.0000
    -1.0000    3.9636    6.0616    5.1314    4.3583    3.9727    4.0000
          0    7.1795    5.8986    5.8365    5.7317    4.7367    5.0000
     1.0000    4.9000    5.4587    4.7593    5.3541    6.2149    6.0000
X =
    -4.0000   -0.7205   -0.7533    0.3889    1.3119    0.9069    1.0000
    -3.0000    1.4318    1.5185    1.7551    1.5332    1.8937    2.0000
    -2.0000    2.4318    1.2991    2.1021    2.2058    3.4478    3.0000
    -1.0000    3.9636    6.0616    5.1314    4.3583    3.9727    4.0000
          0    7.1795    5.8986    5.8365    5.7317    4.7367    5.0000
     1.0000    4.9000    5.4587    4.7593    5.3541    6.2149    6.0000
```

As we can see, with this implementation of the Symmetric Lanczos Algorithm, we obtain the same approximation vectors as those computed by the Conjugate Gradient Algorithm. The vectors $\boldsymbol{f}_k$ are direction vectors and indeed conjugate:

```
Check that the vectors in F are conjugate F'*A*F
ans =
    56.4103    -0.0000    -0.0000    -0.0000    -0.0000    -0.0000
    -0.0000    -9.9583    -0.0000     0.0000     0.0000     0.0000
    -0.0000    -0.0000    15.5937    -0.0000    -0.0000    -0.0000
    -0.0000     0.0000    -0.0000    10.1442     0.0000    -0.0000
    -0.0000     0.0000    -0.0000     0.0000     1.1705     0.0000
    -0.0000     0.0000    -0.0000    -0.0000     0.0000   -18.5937
```

11.7.6 Generalized Minimum Residual: GMRES

The *Generalized Minimum Residual method*, proposed by Saad and Schulz [117], constructs an approximation $\boldsymbol{x}_k \in \mathcal{K}_k(A, \boldsymbol{b})$ not by making the new residual orthogonal to the current Krylov subspace, but by requiring that $\|\boldsymbol{r}_k\| = \|\boldsymbol{b} - A\boldsymbol{x}_k\|$ be minimized.

We will assume in the following that we start the iteration with $\boldsymbol{x}_0 = 0$. Then $\boldsymbol{r}_0 = \boldsymbol{b}$ and $\boldsymbol{x}_k = Q_k \boldsymbol{y} \in \mathcal{K}_k(A, \boldsymbol{b})$.

After k Arnoldi steps, we have the decomposition

$$AQ_k = Q_{k+1}\tilde{H}_k.$$

With the ansatz $\boldsymbol{x}_k = Q_k\boldsymbol{y}$, we get for the residual

$$\boldsymbol{r}_k = \boldsymbol{b} - A\boldsymbol{x}_k = \boldsymbol{b} - AQ_k\boldsymbol{y} = \boldsymbol{b} - Q_{k+1}\tilde{H}_k\boldsymbol{y} = Q_{k+1}\left(\|\boldsymbol{b}\|\boldsymbol{e}_1 - \tilde{H}_k\boldsymbol{y}\right),$$
$$(11.140)$$

since $\boldsymbol{q}_1 = \boldsymbol{b}/\|\boldsymbol{b}\|$. Because Q_{k+1} is orthogonal, we have $\|Q_{k+1}\boldsymbol{z}\| = \|\boldsymbol{z}\|$, as we can see from

$$\|Q_{k+1}\boldsymbol{z}\|^2 = \boldsymbol{z}^\top Q_{k+1}^\top Q_{k+1}\boldsymbol{z} = \boldsymbol{z}^\top \boldsymbol{z} = \|\boldsymbol{z}\|^2, \quad \forall \boldsymbol{z}.$$

So the length of the residual (11.140) is minimized for the solution $\boldsymbol{y}$ of the least squares problem

$$\tilde{H}_k\boldsymbol{y} \approx \|\boldsymbol{b}\|\boldsymbol{e}_1. \qquad (11.141)$$

THEOREM 11.33. (FOUNDATION OF GMRES) *Let $A \in \mathbb{R}^{n \times n}$, $\boldsymbol{b} \in \mathbb{R}^n$ and let $AQ_k = Q_{k+1}\tilde{H}_k$ be the decomposition computed by the Arnoldi iteration at step k. Then the minimizer $\boldsymbol{x}_k$ of*

$$\min_{\boldsymbol{x} \in \mathcal{K}_k(A,\boldsymbol{b})} \|\boldsymbol{b} - A\boldsymbol{x}\|_2 = \|\boldsymbol{b} - A\boldsymbol{x}_k\|_2$$

is given by $\boldsymbol{x}_k = Q_k\boldsymbol{y}$, where $\boldsymbol{y} \in \mathbb{R}^k$ is the solution of the least squares problem (11.141).

The least squares problem (11.141) has a special structure: the matrix $\tilde{H}_k$ is upper Hessenberg and there are $(k+1)$ equations and k unknowns. Such problems are best solved by applying k Givens rotations to reduce the system to an equivalent system with an upper triangular matrix, see Section 3.5.

GMRES solves implicitly a polynomial approximation problem. Since $\boldsymbol{x}_k \in \mathcal{K}_k(A, \boldsymbol{b})$, it is a linear combination of the basis vectors $A^j \boldsymbol{b}$, $j = 1, \ldots, k$,

$$\boldsymbol{x}_k = \sum_{j=0}^{k-1} \alpha_j A^j \boldsymbol{b}.$$

For the residual $\boldsymbol{r}_k = \boldsymbol{b} - A\boldsymbol{x}_k$, we get

$$\boldsymbol{r}_k = \boldsymbol{b} - \sum_{j=0}^{k-1} \alpha_j A^{j+1} \boldsymbol{b} = \sum_{j=0}^{k} \beta_j A^j \boldsymbol{b}, \quad \beta_0 = 1, \ \beta_j = \alpha_{j-1}.$$

With the *residual polynomial*

$$p_k(A) = \sum_{j=0}^{k} \beta_j A^j, \quad p(0) = 1,$$

and denoting the set of all polynomials of degree k by $\mathcal{P}_k$, the minimization problem can also be stated as

$$\min_{\boldsymbol{x} \in \mathcal{K}_k(A,\boldsymbol{b})} \|\boldsymbol{b} - A\boldsymbol{x}\|_2 = \min_{\alpha_j \in \mathbb{R}} \|\boldsymbol{r}_k\|_2 = \min_{\substack{p \in \mathcal{P}_k \\ p(0)=1}} \|p(A)\boldsymbol{b}\|_2. \tag{11.142}$$

This polynomial approximation problem allows us to obtain a convergence bound for GMRES.

THEOREM 11.34. *Let $A \in \mathbb{R}^{n \times n}$ be diagonalizable, $A = S\Lambda S^{-1}$, with the diagonal matrix $\Lambda = \mathrm{diag}(\lambda_1, \ldots, \lambda_n)$ and $\boldsymbol{b} \in \mathbb{R}^n$. Then at the k-th step of the GMRES iteration, we have*

$$\frac{\|\boldsymbol{b} - A\boldsymbol{x}_k\|_2}{\|\boldsymbol{b}\|_2} \le \kappa(S) \min_{\substack{p \in \mathcal{P}_k \\ p(0)=1}} \max_i |p(\lambda_i)|.$$

PROOF. Since A is not assumed to be symmetric, the eigenvectors S and eigenvalues λ_j may be complex. The approximation problem (11.142) is

$$\min_{\boldsymbol{x} \in \mathcal{K}_k(A,\boldsymbol{b})} \|\boldsymbol{b} - A\boldsymbol{x}\|_2 = \min_{\substack{p \in \mathcal{P}_k \\ p(0)=1}} \|p(A)\boldsymbol{b}\|_2 = \min_{\substack{p \in \mathcal{P}_k \\ p(0)=1}} \|p(S\Lambda S^{-1})\boldsymbol{b}\|_2$$

$$= \min_{\substack{p \in \mathcal{P}_k \\ p(0)=1}} \|S\, p(\Lambda)\, S^{-1}\boldsymbol{b}\|_2$$

$$\le \underbrace{\|S\|_2 \|S^{-1}\|_2}_{\kappa(S)} \|\boldsymbol{b}\|_2 \min_{\substack{p \in \mathcal{P}_k \\ p(0)=1}} \max_i |p(\lambda_i)|.$$

$\square$

Therefore, in order to estimate the convergence rate of GMRES, we need to find polynomials that are small on a given set of complex numbers. The following lemma will be helpful in finding such polynomials:

LEMMA 11.8. *Let $C(0,r)$ denote the circle of radius r centered at the origin. Let $\gamma \in \mathbb{C}$ with $|\gamma| > r$, and denote by $\mathcal{P}_k$ the set of polynomials p_k of degree k. Then*

$$\min_{\substack{p \in \mathcal{P}_k \\ p(\gamma)=1}} \max_{z \in C(0,r)} |p(z)| = \left(\frac{r}{|\gamma|} \right)^k,$$

and the minimum is attained for $\hat{p}(z) = \left(\frac{z}{\gamma} \right)^k$.

PROOF. The polynomial $\hat{p}(z) = \left(\frac{z}{\gamma} \right)^k$ satisfies $\hat{p}(\gamma) = 1$ and

$$\max_{z \in C(0,r)} |\hat{p}(z)| = \left(\frac{r}{|\gamma|} \right)^k.$$

Thus, it is sufficient to show that no polynomial with the same normalization is smaller. Consider any polynomial p_k of degree k and its normalized version,

$$g_k(z) = \frac{p_k(z)}{p_k(\gamma)}.$$

Assume that

$$|g_k(z)| = \left| \frac{p_k(z)}{p_k(\gamma)} \right| < \left| \left(\frac{z}{\gamma} \right) \right|^k \qquad \forall z \in C(0,r)$$

holds. By the Theorem of Rouché, if for two analytic functions f, g we have $|g(z)| < |f(z)|$ for all $z \in C(0,r)$, then f and $f - g$ have the same number of zeros inside $C(0,r)$. Now $\left(\frac{z}{\gamma} \right)^k$ has a zero of multiplicity k at 0 and

$$\left(\frac{z}{\gamma} \right)^k - \frac{p_k(z)}{p_k(\gamma)} = (z - \gamma)q_{k-1}(z),$$

with some polynomial q_{k-1} of degree $k-1$ with at most $k-1$ zeros in the interior of $C(0,r)$. Since γ is outside of $C(0,r)$, we have a contradiction to the theorem of Rouché, and therefore no such polynomial exists. $\square$

Note that with a change of variables, we get

$$\min_{\substack{p \in \mathcal{P}_k \\ p(\gamma)=1}} \max_{z \in C(c,r)} |p(z)| = \left(\frac{r}{|\gamma - c|} \right)^k.$$

As a next ingredient in our search for small polynomials, we will need the conformal mapping

$$J(w) = \frac{w + w^{-1}}{2},$$

which is called the *Joukowski transformation*. It maps the circle $C(0, r)$ defined by $w = re^{i\Theta}$ to

$$
z = \frac{1}{2}\left(re^{i\Theta} + \frac{1}{r}e^{-i\Theta}\right) = \frac{1}{2}\left(r\cos\Theta + ir\sin\Theta + \frac{1}{r}\cos\Theta - \frac{i}{r}\sin\Theta\right)
$$
$$
= \frac{1}{2}\left(\left(r + \frac{1}{r}\right)\cos\Theta + i\left(r - \frac{1}{r}\right)\sin\Theta\right),
$$

an ellipse with foci at -1 and 1 and principal axes $a = \left(r + \frac{1}{r}\right)/2$ and $b = \left(r - \frac{1}{r}\right)/2$. Without loss of generality, we may assume $r > 1$, since r and $1/r$ produce the same ellipse. For $r = 1$ the ellipse is degenerate: the circle is mapped to the real interval $[-1, 1]$.

The Joukowski transformation can be used for expressing the Chebyshev polynomials, which are defined by the recurrence relation (11.51):

LEMMA 11.9. *Let $w \in \mathbb{C}$ and $z = (w + w^{-1})/2$. Then the Chebyshev polynomials are*

$$
C_k(z) = \frac{w^k + w^{-k}}{2}. \tag{11.143}
$$

PROOF. We verify by induction that the quantities $C_k(z)$ defined by (11.143) obey the three-term recurrence relation (11.51). For $k = 0$, we have $C_0(z) = (w^0 + w^{-0})/2 = 1$, and for $k = 1$, we obtain $C_1(z) = (w^1 + w^{-1})/2 = z$ as required. Assume now that (11.143) holds for $j \leq k$. Then

$$
C_{k+1}(z) = 2zC_k(z) - C_{k-1}(z)
$$
$$
= (w + w^{-1})\frac{w^k + w^{-k}}{2} - \frac{w^{k-1} + w^{-(k-1)}}{2}
$$
$$
= \frac{w^{k+1} + w^{-k+1} + w^{k-1} + w^{-k-1} - w^{k-1} - w^{-(k-1)}}{2}
$$
$$
= \frac{w^{k+1} + w^{-(k+1)}}{2},
$$

which concludes the proof. $\qquad\square$

Note that if we solve the equation

$$
z = (w + w^{-1})/2 \quad\Longleftrightarrow\quad w^2 - 2zw + 1 = 0
$$

for w, we get

$$
w_1 = z + \sqrt{z^2 - 1}, \quad w_2 = z - \sqrt{z^2 - 1} = \frac{1}{w_1}.
$$

Inserting w_1 or w_2 into (11.143), we find

$$
C_k(z) = \frac{\left(z + \sqrt{z^2 - 1}\right)^k + \left(z - \sqrt{z^2 - 1}\right)^k}{2},
$$

the same expression for the Chebyshev polynomials as in Lemma 11.6.

THEOREM 11.35. *Let E denote the ellipse obtained from the circle $C(0,r)$ using the Joukowski transformation $J(w)$, and let γ be a point outside E. Let w_γ be the solution with larger modulus of $J(w) = \gamma$. Then*

$$\frac{r^k}{|w_\gamma|^k} \le \min_{\substack{p \in \mathcal{P}_k \\ p(\gamma)=1}} \max_{z \in E} |p(z)| \le \frac{r^k + r^{-k}}{|w_\gamma^k + w_\gamma^{-k}|}. \tag{11.144}$$

PROOF. Every polynomial $p \in \mathcal{P}_k$ with $p(\gamma) = 1$ can be written as

$$p(z) = \frac{\sum_{j=0}^{k} \alpha_j z^j}{\sum_{j=0}^{k} \alpha_j \gamma^j}.$$

Using the Joukowski transformation $z = J(w)$, this becomes

$$p(z) = \frac{\sum_{j=0}^{k} \tilde{\alpha}_j (w^j + w^{-j})}{\sum_{j=0}^{k} \tilde{\alpha}_j (w_\gamma^j + w_\gamma^{-j})}. \tag{11.145}$$

In particular, for $\tilde{\alpha}_k = 1$ and $\tilde{\alpha}_j = 0$, $j = 0, 1, \ldots, k-1$, we obtain a normalized Chebyshev polynomial,

$$p^*(z) = \frac{w^k + w^{-k}}{w_\gamma^k + w_\gamma^{-k}} = \frac{C_k(z)}{C_k(\gamma)}.$$

Letting $w = re^{i\Theta}$, we get

$$\left| r^k e^{ik\Theta} + \frac{1}{r^k} e^{-ik\Theta} \right| \le \left| r^k e^{ik\Theta} \right| + \left| \frac{1}{r^k} e^{-ik\Theta} \right| = r^k + \frac{1}{r^k},$$

and the maximum is attained for $\Theta = 0$. Thus the upper bound in (11.144) is proved.

In order to prove the lower bound, we rewrite (11.145) as

$$p(z) = \frac{w^{-k}}{w_\gamma^{-k}} \frac{\sum_{j=0}^{k} \tilde{\alpha}_j (w^{k+j} + w^{k-j})}{\sum_{j=0}^{k} \tilde{\alpha}_j (w_\gamma^{k+j} + w_\gamma^{k-j})}.$$

The absolute value becomes

$$|p(z)| = \frac{r^{-k}}{|w_\gamma|^{-k}} |q_{2k}(w)|,$$

where $q_{2k}(w)$ is a polynomial of degree $2k$, with $q_{2k}(w_\gamma) = 1$. By Lemma 11.8, we have

$$|q_{2k}(w)| \ge \frac{r^{2k}}{|w_\gamma|^{2k}} \quad \Longrightarrow \quad \max_{z \in E} |p(z)| \ge \frac{r^{-k}}{|w_\gamma|^{-k}} \frac{r^{2k}}{|w_\gamma|^{2k}} = \frac{r^k}{|w_\gamma|^k},$$

which concludes the proof. $\square$

Since the difference between the two expressions

$$\frac{r^k}{|w_\gamma|^k} \quad \text{and} \quad \frac{r^k + r^{-k}}{|w_\gamma^k + w_\gamma^{-k}|}$$

converges to zero for $k \to \infty$, we conclude that for large k the complex Chebyshev polynomial

$$p^*(z) = \frac{w^k + w^{-k}}{w_\gamma^k + w_\gamma^{-k}} = \frac{C_k(z)}{C_k(\gamma)}$$

is optimal. But $p^*(z)$ is only optimal asymptotically for k large, which is different from the real case, where the Chebyshev polynomials are optimal for all k.

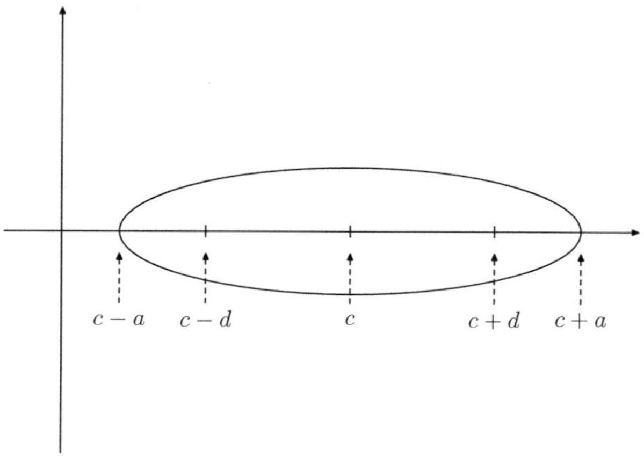

FIGURE 11.10. *Ellipse*

As in the real case, an ellipse $E(c, d, a)$ with center $c \in \mathbb{R}$, major axis a and distance of focal points d, see Figure 11.10, is obtained by a change of variables,

$$\hat{C}_k(z) = \frac{C_k\left(\frac{z-c}{d}\right)}{C_k\left(\frac{\gamma-c}{d}\right)}.$$

The maximum is attained at $z = c + a$, as we have seen in the proof of Theorem 11.35, and hence

$$\max_{z \in E(c,d,a)} |\hat{C}_k(z)| = \left|\frac{C_k\left(\frac{a}{d}\right)}{C_k\left(\frac{\gamma-c}{d}\right)}\right| = \frac{C_k\left(\frac{a}{d}\right)}{|C_k\left(\frac{\gamma-c}{d}\right)|},$$

where we removed the modulus in the numerator, since $a > d$ and thus $\frac{a}{d} > 1$, which means $C_k(\frac{a}{d}) > 0$.

THEOREM 11.36. (CONVERGENCE ESTIMATE FOR GMRES) *Let $A \in \mathbb{R}^{n \times n}$ be diagonalizable, $A = S\Lambda S^{-1}$ with $\Lambda = \mathrm{diag}(\lambda_1, \ldots, \lambda_n)$, and assume that all $\lambda_j \in E(c, d, a)$ and that the origin is not contained in the ellipse. Then at the k-th iteration of GMRES, we have*

$$\frac{\|\boldsymbol{b} - A\boldsymbol{x}_k\|_2}{\|\boldsymbol{b}\|_2} \leq \kappa(S) \frac{C_k\left(\frac{a}{d}\right)}{|C_k\left(\frac{c}{d}\right)|}.$$

PROOF. From Theorem 11.34 we know that

$$\frac{\|\boldsymbol{b} - A\boldsymbol{x}_k\|_2}{\|\boldsymbol{b}\|_2} \leq \kappa(S) \min_{\substack{p \in \mathcal{P}_k \\ p(0)=1}} \max_i |p(\lambda_i)|.$$

Then Theorem 11.35, the change of variables, and setting $\gamma = 0$ show that

$$\min_{\substack{p \in \mathcal{P}_k \\ p(0)=1}} \max_{z \in E(c,d,a)} |p(\lambda)| \leq \frac{C_k\left(\frac{a}{d}\right)}{|C_k\left(-\frac{c}{d}\right)|} = \frac{C_k\left(\frac{a}{d}\right)}{|C_k\left(\frac{c}{d}\right)|},$$

since $|C_k(-z)| = |C_k(z)|$. $\qquad\square$

For $c > a > d$, we get

$$\frac{C_k\left(\frac{a}{d}\right)}{|C_k\left(\frac{c}{d}\right)|} = \frac{\left(\frac{a}{d} + \sqrt{\left(\frac{a}{d}\right)^2 - 1}\right)^k + \left(\frac{a}{d} + \sqrt{\left(\frac{a}{d}\right)^2 - 1}\right)^{-k}}{\left|\left(\frac{c}{d} + \sqrt{\left(\frac{c}{d}\right)^2 - 1}\right)^k + \left(\frac{c}{d} + \sqrt{\left(\frac{c}{d}\right)^2 - 1}\right)^{-k}\right|}$$

$$\approx \left(\frac{\frac{a}{d} + \sqrt{\left(\frac{a}{d}\right)^2 - 1}}{\frac{c}{d} + \sqrt{\left(\frac{c}{d}\right)^2 - 1}}\right)^k = \left(\frac{a + \sqrt{a^2 - d^2}}{c + \sqrt{c^2 - d^2}}\right)^k \quad \text{for large } k. \qquad (11.146)$$

If the spectrum of GMRES does not lie in an ellipse excluding the origin, more refined estimates can be found in the research literature.

In exact arithmetic, we can make another statement about the convergence of GMRES.

THEOREM 11.37. *Let $A \in \mathbb{R}^{n \times n}$ be invertible, $\boldsymbol{b} \in \mathbb{R}^n$ and p_m be the minimal polynomial of A. Then GMRES applied to the linear system $A\boldsymbol{x} = \boldsymbol{b}$ converges to the exact solution in at most m iterations.*

PROOF. Since A is invertible, the minimal polynomial $p_m(A)$ has the constant coefficient $\alpha_0 \neq 0$. Thus the polynomial

$$p^*(A) = \frac{1}{\alpha_0} p_m(A)$$

satisfies $p^*(0) = 1$ and $p^*(A) = 0$. GMRES minimizes the residual, which is equivalent to the polynomial approximation problem

$$\min_{\boldsymbol{x}_k \in \mathcal{K}_k(A,\boldsymbol{b})} \|\boldsymbol{b} - A\boldsymbol{x}_k\|_2 = \min_{\substack{p \in \mathcal{P}_k \\ p(0)=1}} \|p(A)\boldsymbol{b}\|_2 = \|p^*(A)\boldsymbol{b}\|_2 = 0, \quad \text{when } k = m.$$

$\qquad\square$

We make three final remarks about GMRES:

1. The iterates of FOM and GMRES are related: in FOM, ones solves at each iteration the system of linear equations $H_k y_k = ||b||_2 e_1$, whereas GMRES solves the least squares problem $||\tilde{H}_k y_k - ||b||_2 e_1||_2 \longrightarrow \min$. Using Givens rotations both for the least squares problem and the linear system, we see that in FOM, only the last Givens rotation is not applied.

2. Each iteration generates a new vector, which has to be stored in memory. For large (but sparse) matrices, this can quickly use up the available memory and lead to problems. Therefore, one often uses GMRES with restarts: after m iterations, the vectors $q_1, q_2, \ldots, q_m$ are discarded, the best approximation x_m found so far is retained, and GMRES is restarted in order to solve the following system for the corrections $\tilde{x}$:

$$A\tilde{x} = b - Ax_m = r_m.$$

3. When A is symmetric, $A = A^\top$, GMRES, which uses the Arnoldi algorithm to construct the orthogonal basis of $\mathcal{K}_k(A, b)$, requires many more operations than necessary. It is more efficient to use the Lanczos iteration in this case. We obtain then the MINRES algorithm, proposed by Paige and Saunders in 1975, see [101].

11.7.7 Classical Lanczos for Non-Symmetric Matrices

The symmetric Lanczos algorithm, see Section 11.7.3, which is mathematically equivalent to the Arnoldi Algorithm when applied to a symmetric matrix $A \in \mathbb{R}^{n \times n}$, computes in n steps the decomposition

$$AQ_n = Q_n T_n,$$

with T_n tridiagonal and Q_n orthogonal. If A is non-symmetric, then it is no longer possible to have a tridiagonal matrix T_n together with an orthogonal Q_n. To extend the Lanczos algorithm to non-symmetric matrices, one possibility is to insist on reducing A to tridiagonal form, but using a similarity transformation with a non-singular matrix V instead of an orthogonal matrix,

$$V^{-1}AV = \begin{pmatrix} \alpha_1 & \gamma_1 & & \\ \beta_1 & \ddots & \ddots & \\ & \ddots & \ddots & \gamma_{n-1} \\ & & \beta_{n-1} & \alpha_n \end{pmatrix} := T. \tag{11.147}$$

With the definitions

$$V = [v_1, \ldots, v_n], \quad W = V^{-\top} = [w_1, \ldots, w_n],$$

the two matrices are said to be *biorthogonal*, since $W^\top V = I$ holds. Comparing coefficients of the equations $AV = VT$ and $A^\top W = WT^\top$ by multiplying from the right with $\mathbf{e}_j$, we find

$$AV\mathbf{e}_j = VT\mathbf{e}_j \tag{11.148}$$

$$\implies \quad A\mathbf{v}_j = \gamma_{j-1}\mathbf{v}_{j-1} + \alpha_j\mathbf{v}_j + \beta_j\mathbf{v}_{j+1}, \quad \text{with } \gamma_0\mathbf{v}_0 := 0, \tag{11.149}$$

and similarly

$$A^\top \mathbf{w}_j = \beta_{j-1}\mathbf{w}_{j-1} + \alpha_j\mathbf{w}_j + \gamma_j\mathbf{w}_{j+1}, \quad \text{with } \beta_0\mathbf{w}_0 := 0. \tag{11.150}$$

Multiplying (11.149) by $\mathbf{w}_j^\top$ and using bi-orthogonality, we have

$$\mathbf{w}_j^\top A\mathbf{v}_j = \alpha_j,$$

and thus the α_j are determined. For β_j and γ_j, we obtain from (11.149) and (11.150)

$$\beta_j\mathbf{v}_{j+1} = (A - \alpha_j I)\mathbf{v}_j - \gamma_{j-1}\mathbf{v}_{j-1} =: \mathbf{r}_j, \tag{11.151}$$

$$\gamma_j\mathbf{w}_{j+1} = (A^\top - \alpha_j I)\mathbf{w}_j - \beta_{j-1}\mathbf{w}_{j-1} =: \mathbf{p}_j. \tag{11.152}$$

Here the choice of symbol for $\mathbf{r}_j$ and $\mathbf{p}_j$ is not arbitrary; we will see later that $\mathbf{r}_j$ is related to the residual when using the non-symmetric Lanczos process for solving linear systems. Using again that $W^\top V = I$, we get

$$1 = \mathbf{v}_{j+1}^\top\mathbf{w}_{j+1} = \frac{1}{\beta_j}\frac{1}{\gamma_j}\mathbf{r}_j^\top\mathbf{p}_j \quad \Longleftrightarrow \quad \beta_j\gamma_j = \mathbf{r}_j^\top\mathbf{p}_j. \tag{11.153}$$

This is the only condition for β_j and γ_j, and we obtain the generic non-symmetric Lanczos process:

ALGORITHM 11.22. *Non-symmetric Lanczos Process*

```
choose v₁ and w₁ such that v₁ᵀw₁ = 1;
γ₀ = 0; β₀ = 0;
for j = 1 : n
    αⱼ = wⱼᵀAvⱼ;
    rⱼ = (A − αⱼI)vⱼ − γⱼ₋₁vⱼ₋₁;
    pⱼ = (Aᵀ − αⱼI)wⱼ − βⱼ₋₁wⱼ₋₁;
    choose βⱼ and γⱼ such that βⱼγⱼ = rⱼᵀpⱼ;
    vⱼ₊₁ = rⱼ/βⱼ;
    wⱼ₊₁ = pⱼ/γⱼ;
end
```

In the original publication of Lanczos [82], the choice $\beta_j = 1$ is made, which implies that $\gamma_j = \boldsymbol{r}_j^\top \boldsymbol{p}_j$, but Rutishauser [115] observed already[7] that this choice leads to vectors $\boldsymbol{v}_j$ and $\boldsymbol{w}_j$ with bigger and bigger elements. To avoid this growth, one can use β_j and γ_j to scale these vectors. A "canonical choice" according to Golub and van Loan [51] is to choose

$$\beta_j = \|\boldsymbol{r}_j\| \Rightarrow \boldsymbol{v}_{j+1} = \frac{\boldsymbol{r}_j}{\|\boldsymbol{r}_j\|}, \tag{11.154}$$

which then implies that

$$\gamma_j = \frac{1}{\beta_j} \boldsymbol{r}_j^\top \boldsymbol{p}_j = \boldsymbol{v}_{j+1}^\top \boldsymbol{p}_j.$$

A different choice for β_j and γ_j was proposed by Saad in [118], namely

$$\beta_j = \sqrt{|\boldsymbol{r}_j^\top \boldsymbol{p}_j|} \quad \Longrightarrow \quad \gamma_j = \beta_j \, \mathrm{sign}(\boldsymbol{r}_j^\top \boldsymbol{p}_j). \tag{11.155}$$

In this case, the matrix T has the property that $t_{j+1,j} = \pm t_{j,j+1}$, which means it is symmetric in modulus. A third variant analyzed by Gutknecht in [59] is given by

$$\beta_j = -\alpha_j - \gamma_{j-1}, \quad \gamma_j = \frac{1}{\beta_j} \boldsymbol{r}_j^\top \boldsymbol{p}_j, \tag{11.156}$$

which leads to an interesting property we will see later when solving linear systems.

If we perform only $j < n$ iteration steps of the non-symmetric Lanczos process, we obtain the matrices

$$V_j = [\boldsymbol{v}_1, \dots, \boldsymbol{v}_j], \quad W_j = [\boldsymbol{w}_1, \dots, \boldsymbol{w}_j], \quad T_j = \begin{pmatrix} \alpha_1 & \gamma_1 & & \\ \beta_1 & \ddots & & \ddots \\ & \ddots & \ddots & \gamma_{j-1} \\ & & \beta_{j-1} & \alpha_j \end{pmatrix},$$

and the relations

$$AV_j = V_j T_j + \beta_j \boldsymbol{v}_{j+1} \boldsymbol{e}_j^\top = V_j T_j + \boldsymbol{r}_j \boldsymbol{e}_j^\top,$$
$$A^\top W_j = W_j T_j^\top + \gamma_j \boldsymbol{w}_{j+1} \boldsymbol{e}_j^\top = W_j T_j^\top + \boldsymbol{p}_j \boldsymbol{e}_j^\top.$$

Because of bi-orthogonality, we have $W_j^\top V_j = I$ and $W_j^\top \boldsymbol{v}_{j+1} = 0$, which implies that

$$W_j^\top A V_j = T_j.$$

---

[7]Im allgemeinen wird man beobachten, dass die Grössenordnung der iterierten Vektoren x_k, y_k mit fortschreitendem k rasch unbequem wird.

ALGORITHM 11.23.
MATLAB *implementation: Non-Symmetric Lanczos
Process*

```
function [V,W,T,r,p] = NonSymmetricLanczos(A,v,n,variant)
% NONSYMMETRICLANCZOS Non-symmetric Lanczos process
%   [V,W,T,r,p]=NonSymmetricLanczos(A,v,n,variant) generates
%   biorthogonal vectors in the matrices V and W and a tridiagonal
%   matrix T with the starting vector v using the non-symmetric
%   Lanczos process, where the variant parameter can be either
%   'GolubVanLoan','Saad' or 'Gutknecht'. The auxiliary vectors r and
%   p are also returned.

V=v; W=v/norm(v)^2;
beta=0; gamma=0;
alpha=W(:,1)'*A*V(:,1); T(1,1)=alpha;
for j=1:n,
  if j==1,
    r=A*V(:,j)-alpha*V(:,j);
    p=A'*W(:,j)-alpha*W(:,j);
  else
    r=A*V(:,j)-alpha*V(:,j)-gamma*V(:,j-1);
    p=A'*W(:,j)-alpha*W(:,j)-beta*W(:,j-1);
  end
  switch variant
    case 'GolubVanLoan'
      beta=norm(r); gamma=r'*p/beta;
    case 'Saad'
      beta=sqrt(abs(r'*p));gamma=beta*sign(r'*p);
    case 'Gutknecht'
      beta=-alpha-gamma; gamma=r'*p/beta;
  end
  T(j+1,j)=beta; T(j,j+1)=gamma;
  V=[V,r/beta]; W=[W,p/gamma];
  alpha=W(:,j+1)'*A*V(:,j+1); T(j+1,j+1)=alpha;
end;
```

EXAMPLE 11.12. *We show here with a small example the results of the
three previously presented variants for the non-symmetric Lanczos process.
Consider again the linear system with the non-symmetric matrix (11.90).*

*In the program above, the three choices of β_j are obtained by choosing the
appropriate variant in the function NonSymmetricLanczos. For the choice of
β_j according to Golub/van Loan, we obtain*

```
>> A=[ 0    -4    -8    -2
      -4    -7    -7    -8
      -9    -5    -4    -5
```

```
        0    -5    -9    -6  ]
>> n=max(size(A)); b=A*[1:n]';
>> disp('Golub, van Loan')
>> [V,W,T,r,p]=NonSymmetricLanczos(A,b,4,'GolubVanLoan')
Golub, van Loan
V =
   -4.0000e+01  -1.6353e-01  -8.8595e-01  -5.2322e-02   3.5472e-01
   -7.1000e+01   1.8505e-02   4.5684e-01  -4.9383e-01   6.1695e-01
   -5.1000e+01   7.9721e-01   7.8243e-02  -2.6143e-01   4.9710e-01
   -6.1000e+01  -5.8083e-01  -1.6201e-02   8.2767e-01   4.9642e-01
W =
   -3.0857e-03  -2.4368e-01  -9.1984e-01  -4.1643e-01   3.6955e-01
   -5.4771e-03  -6.7756e-01   4.2378e-01  -8.7086e-01   4.8720e-01
   -3.9343e-03   1.1877e+00  -7.3598e-02   5.4206e-01   6.2768e-01
   -4.7057e-03  -4.4536e-02   1.7146e-01   8.3350e-01   5.1632e-01
T =
   -2.1107e+01   3.2058e-02            0            0            0
    1.8509e+02   3.0348e+00   7.2491e+00            0            0
             0   3.9635e+00   1.2989e+00  -2.7085e+00            0
             0            0   1.6194e+00  -2.2699e-01  -5.7562e-13
             0            0            0   2.3336e-12  -2.1361e+01
r =
    8.2778e-13
    1.4397e-12
    1.1600e-12
    1.1585e-12
p =
   -2.1272e-13
   -2.8044e-13
   -3.6131e-13
   -2.9721e-13
```

Note that after $n = 4$ steps we should get $r = 0$, and thus we expect $V(:, 5) = 0$. However, rounding errors prevent this from happening, and we obtain instead $V(:, 5) = r/\|r\| \neq 0$. Similarly, p should also be zero but is not. Checking bi-orthogonality and the tridiagonal matrix, we find

```
>> V=V(1:n,1:n); W=W(1:n,1:n); T=T(1:n,1:n);
>> disp('Check bi-orthogonality: norm(V''*W)=')
>> norm(V'*W-eye(n))
>> disp('Check decomposition with V: norm(A*V-V*T)=');
>> norm(A*V-V*T)
>> disp('and decomposition with W: norm(A''W-W*T'')=');
>> norm(A'*W-W*T')
Check bi-orthogonality: norm(V'*W)=
ans =
    3.0741e-12
Check decomposition with V: norm(A*V-V*T)=
ans =
    2.3336e-12
```

```
and decomposition with W: norm(A'W-W*T')=
ans =
   5.8546e-13
```

For the Saad variant, we get comparable results, with a bit more accuracy for p *but less for* r,

```
Saad
V =
  -4.0000e+01  -1.2425e+01  -4.9777e+01  -2.2731e+00   3.1378e+01
  -7.1000e+01   1.4061e+00   2.5668e+01  -2.1455e+01   5.4971e+01
  -5.1000e+01   6.0576e+01   4.3961e+00  -1.1358e+01   4.4348e+01
  -6.1000e+01  -4.4134e+01  -9.1027e-01   3.5959e+01   4.4071e+01
W =
  -3.0857e-03  -3.2070e-03  -1.6372e-02  -9.5852e-03   3.9859e-03
  -5.4771e-03  -8.9171e-03   7.5426e-03  -2.0045e-02   5.5094e-03
  -3.9343e-03   1.5630e-02  -1.3099e-03   1.2477e-02   7.1128e-03
  -4.7057e-03  -5.8612e-04   3.0516e-03   1.9185e-02   5.8233e-03
T =
  -2.1107e+01   2.4359e+00            0            0            0
   2.4359e+00   3.0348e+00   5.3602e+00            0            0
            0   5.3602e+00   1.2989e+00  -2.0943e+00            0
            0            0   2.0943e+00  -2.2699e-01  -1.2047e-12
            0            0            0   1.2047e-12  -2.1363e+01
r =
   3.7801e-11
   6.6223e-11
   5.3426e-11
   5.3092e-11
p =
  -4.8017e-15
  -6.6371e-15
  -8.5687e-15
  -7.0152e-15
Check bi-orthogonality: norm(V'*W)=
ans =
   7.2457e-14
Check decomposition with V: norm(A*V-V*T)=
ans =
   1.0718e-10
and decomposition with W: norm(A'W-W*T')=
ans =
   1.3773e-14
```

Note that the tridiagonal matrix is symmetric up to signs. Finally, for the Gutknecht variant, we obtain

```
Gutknecht
V =
  -4.0000e+01  -1.4340e+00   9.2866e+00   1.2058e-01  -2.4168e-12
```

```
   -7.1000e+01    1.6228e-01   -4.7887e+00    1.1381e+00   -4.2392e-12
   -5.1000e+01    6.9911e+00   -8.2016e-01    6.0247e-01   -3.4282e-12
   -6.1000e+01   -5.0936e+00    1.6982e-01   -1.9074e+00   -3.4025e-12
W =
   -3.0857e-03   -2.7788e-02    8.7754e-02    1.8070e-01   -5.3222e+10
   -5.4771e-03   -7.7264e-02   -4.0429e-02    3.7789e-01   -7.1131e+10
   -3.9343e-03    1.3543e-01    7.0213e-03   -2.3522e-01   -9.1392e+10
   -4.7057e-03   -5.0786e-03   -1.6357e-02   -3.6168e-01   -7.5392e+10
T =
   -2.1107e+01    2.8113e-01             0             0             0
    2.1107e+01    3.0348e+00   -8.6649e+00             0             0
             0   -3.3159e+00    1.2989e+00   -5.9546e-01             0
             0             0    7.3660e+00   -2.2699e-01   -1.5374e-24
             0             0             0    8.2245e-01   -2.1362e+01
r =
   -1.9877e-12
   -3.4865e-12
   -2.8195e-12
   -2.7984e-12
p =
    8.1823e-14
    1.0936e-13
    1.4051e-13
    1.1591e-13
Check bi-orthogonality: norm(V'*W)=
ans =
    1.1837e-12
Check decomposition with V: norm(A*V-V*T)=
ans =
    5.6469e-12
and decomposition with W: norm(A'W-W*T')=
ans =
    2.2766e-13
```

These results are again comparable to the previous ones.

Saad proposed in [118] to use the non-symmetric Lanczos process for solving linear systems. Let x_0 be a given approximation to the solution of $Ax = b$ and $r_0 = b - Ax_0$ be the corresponding residual. Choosing $v_1 = r_0$ and $w_1 = r_0/\|r_0\|^2$, we search for an approximate solution in the space spanned by V_j, $x_j := x_0 + V_j z$. This choice leads to the residual

$$r_j = b - Ax_j = r_0 - AV_j z.$$

In order to determine the approximate solution, we impose a Galerkin condition, but now on the space spanned by W_j,

$$0 = W_j^\top r_j = W_j^\top r_0 - W_j^\top AV_j z = \varepsilon_1 - T_j z. \qquad (11.157)$$

This idea of using two different spaces, one for approximating the solution and one for imposing orthogonality, is often called the *Petrov-Galerkin* method, due to its historical origins in finite element methods.

We see from (11.157) that in order to determine z, we must solve the tridiagonal system

$$T_j z = \mathbf{e}_1, \tag{11.158}$$

from which we obtain $x_j = x_0 + V_j z$. Note that it is not necessary to compute x_j in each step: we can compute the residual independently, since

$$
\begin{aligned}
r_j &= r_0 - AV_j z = r_0 - (V_j T_j + \beta_j v_{j+1} \mathbf{e}_j^\top) z \\
&= r_0 - V_j T_j z - \beta_j v_{j+1} \mathbf{e}_j^\top z \\
&= -\beta_j v_{j+1} \mathbf{e}_j^\top z,
\end{aligned}
$$

where we used on the second line that $V_j T_j z = V_j \mathbf{e}_1 = r_0$, from (11.158) and the definition of r_0. Therefore,

$$r_j = -\beta_j v_{j+1} z_j \quad \Longrightarrow \quad \|r_j\| = |\beta_j| |z_j| \|v_{j+1}\|. \tag{11.159}$$

We can thus compute the residual from the solution z in (11.158), which means we can decide to compute the approximation x_j only when the residual is sufficient small.

The following MATLAB function `Saad` computes a fixed number of Lanczos steps and stores the approximation vectors together with the norm of the corresponding residuals. It is meant as an implementation for educational purposes: for example, it is not necessary to store the W matrix if we only want to solve linear systems of equations.

ALGORITHM 11.24.
*Non-Symmetric Lanczos for Linear Systems: Saad
Variant*

```
function [X,res,V,W,T]=Saad(A,b,x0,n);
% SAAD Non-symmetric Lanczos for linear systems
%    [X,res,V,W,T]=Saad(A,b,x0,n) computes an approximate solution of
%    the linear system Ax=b performing n non-symmetric Lanczos
%    iterations starting with the initial vector x0. The matrix X
%    contains the approximate solutions and the vector res the norms of
%    the corresponding residuals. The method also returns the
%    biorthogonal vectors in V and W and the tridiagonal matrix T

X=x0; r0=b-A*x0; res=norm(r0);
V=r0; W=r0/norm(r0)^2;
beta=0; gamma=0; r=r0;
alpha=W(:,1)'*A*V(:,1);
T(1,1)=alpha;
for j=1:n,
  z=T\eye(j,1);
```

```
   x=x0+V*z;
   X=[X x];
   if j==1,
     r=A*V(:,j)-alpha*V(:,j);
     p=A'*W(:,j)-alpha*W(:,j);
   else
     r=A*V(:,j)-alpha*V(:,j)-gamma*V(:,j-1);
     p=A'*W(:,j)-alpha*W(:,j)-beta*W(:,j-1);
   end
   beta=sqrt(abs(r'*p)); gamma=beta*sign(r'*p);
   res=[res norm(r)*abs(z(j))];
   T(j+1,j)=beta; T(j,j+1)=gamma;
   V=[V, r/beta]; W=[W, p/gamma];
   alpha=(W(:,j+1)'*A*V(:,j+1));
   T(j+1,j+1)=alpha;
end;
```

EXAMPLE 11.13. *As a first example, we consider again the system with
the matrix A from (11.90):*

```
A=[ 0      -4      -8      -2
    -4      -7      -7      -8
    -9      -5      -4      -5
     0      -5      -9      -6];
x=[1:4]'; b=A*x;
x0=zeros(4,1);
[X,res]=Saad(A,b,x0,4)

X =

           0      1.8951      1.3020      0.7390      1.0000
           0      3.3639      3.1276      4.0556      2.0000
           0      2.4163      4.3198      2.1920      3.0000
           0      2.8901      1.1090      3.1811      4.0000
res =
    113.8552      8.7694     10.4821      2.3045      0.0000
```

*The norm of the residuals is not monotonically decreasing, as we can see
already in this small example, but we nonetheless obtain the exact solution
after $n = 4$ steps, as predicted by the theory. However, this will no longer be
the case for larger n because of rounding errors. To see this, run the following
MATLAB program, which solves randomly generated 50×50 linear systems.
Observe the plots of the norm of residuals:*

```
for k=1:30
  clf
  A=rand(50);
  condition=cond(A)
  b=A*[1:50]';
  x0=zeros(50,1);
```

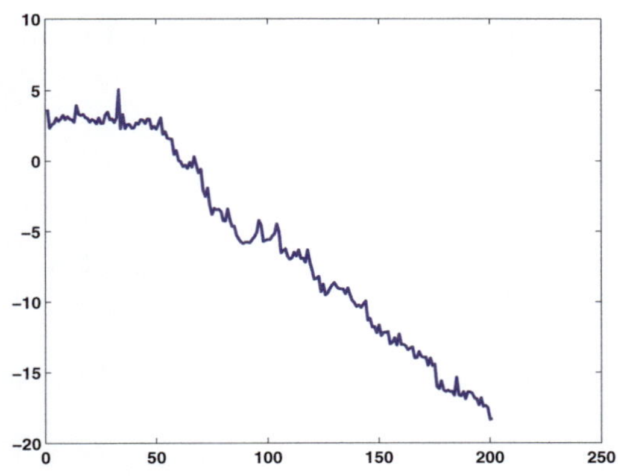

FIGURE 11.11.
Non-symmetric Lanczos for solving linear systems: norm of residuals

```
   [X,res]=Saad(A,b,x0,200);
   plot(log10(res));
   pause
end
```

A typical plot of the logarithms of the norm of the residuals is shown in Figure 11.11. We see with this larger example that we do not obtain the exact solution after $n = 50$ steps. Convergence does occur, but only after about $200 = 4n$ steps. Again convergence is not monotonic.

In order to avoid having to solve a small linear system at each iteration, one can proceed like in Subsection 11.7.5 to obtain an updating formula for the iterates x_k. In [59], Gutknecht chooses the starting vectors like in the Saad variant, $v_1 = r_0$ and $w_1 = r_0/\|r_0\|^2$, but uses the choice of coefficients (11.156), which is different from Saad. That choice leads to the updating formula

$$x_{j+1} = -(v_j + \alpha_j x_j + \gamma_{j-1} x_{j-1})/\beta_j. \tag{11.160}$$

In the special case of a symmetric matrix, the algorithm thus obtained becomes the conjugate gradient algorithm, see Section 11.7.1.

LEMMA 11.10. *Let $V = [v_1, \ldots, v_j]$ be the Lanczos vectors obtained from j steps of the non-symmetric Lanczos process with initial vectors $v_1 = r_0$, $w_1 = r_0/\|r_0\|^2$ and the choice of β_j according to (11.156). If the approximate solutions x_j are calculated using the recurrence (11.160), then the v_j are precisely the residual vectors with respect to x_j, i.e.,*

$$v_j = b - Ax_j.$$

PROOF. The proof is by induction. We first have $b - Ax_0 = r_0 = v_1$ because of the initialization. Assume now that the statement holds for $1, \ldots, j$. Then, using $b - v_j = Ax_j$, we get

$$
\begin{aligned}
b - Ax_{j+1} &= b + (Av_j + \alpha_j Ax_j + \gamma_{j-1} Ax_{j-1})/\beta_j \\
&= b + (Av_j + \alpha_j(b - v_j) + \gamma_{j-1}(b - v_{j-1}))/\beta_j \\
&= b + \underbrace{(Av_j - \alpha_j v_j - \gamma_{j-1} v_{j-1})}_{\beta_j v_{j+1},\text{ see (11.151)}} /\beta_j + b(\alpha_j + \gamma_{j-1})/\beta_j.
\end{aligned}
$$

Because of (11.156), we have $\beta_j = -\alpha_j - \gamma_{j-1}$, and the vector b cancels, which leads to

$$
b - Ax_{j+1} = v_{j+1}.
$$

$\square$

ALGORITHM 11.25.
Non-Symmetric Lanczos for Linear Systems: BiORES of
Gutknecht

```
function [X,res,V,W,T]=BiORES(A,b,x0,n);
% BIORES Non-symmetric Lanczos for linear equations
%   [X,res,V,W,T]=BiORES(A,b,x0,n) computes an approximate solution of
%   the linear system Ax=b performing n non-symmetric Lanczos
%   iterations starting with the initial vector x0. X contains the
%   successive approximations computed by the proposal of Gutknecht,
%   and the vector res contains the norms of the corresponding
%   residuals. The method also returns the biorthogonal vectors in V
%   and W and the tridiagonal matrix T

X=x0;
r0=b-A*x0; res=norm(r0);
V=r0; W=r0/norm(r0)^2;
beta=0; gamma=0; r=r0;
alpha=(W(:,1)'*A*V(:,1));
T(1,1)=alpha;
for j=1:n,
  if j==1
    r=A*V(:,j)-alpha*V(:,j);
    p=A'*W(:,j)-alpha*W(:,j);
  else
    r=A*V(:,j)-alpha*V(:,j)-gamma*V(:,j-1);
    p=A'*W(:,j)-alpha*W(:,j)-beta*W(:,j-1);
  end
  beta=-alpha-gamma;
  if j==1
    x=-(V(:,j)+alpha*X(:,j))/beta;
```

```
    else
      x=-(V(:,j)+alpha*X(:,j)+gamma*X(:,j-1))/beta;
    end
    X=[X x];
    gamma=r'*p/beta;
    res=[res norm(r/beta)];
    T(j+1,j)=beta; T(j,j+1)=gamma;
    V=[V,r/beta]; W=[W,p/gamma];
    alpha=(W(:,j+1)'*A*V(:,j+1));
    T(j+1,j+1)=alpha;
end
```

EXAMPLE 11.14. *With the matrix A from (11.90), we obtain the same results as with Saad's variant:*

```
>> A=[ 0      -4      -8      -2
       -4      -7      -7      -8
       -9      -5      -4      -5
        0      -5      -9      -6];
>> x=[1:4]'; b=A*x;
>> x0=zeros(4,1);
>> [X,res]=BiORES(A,b,s0,4)
X =
          0    1.8951    1.3020    0.7390    1.0000
          0    3.3639    3.1276    4.0556    2.0000
          0    2.4163    4.3198    2.1920    3.0000
          0    2.8901    1.1090    3.1811    4.0000
res =
    113.8552    8.7694   10.4821    2.3045    0.0000
```

Like in the case of MPE and FOM, we can now prove an equivalence result between TEA and Non-Symmetric Lanczos (NSL) for solving linear systems.

THEOREM 11.38. (EQUIVALENCE OF TEA AND NSL) *Let $A \in \mathbb{R}^{n \times n}$ be an invertible matrix, $A = M - N$ be a splitting of A with M invertible, and $\boldsymbol{b} \in \mathbb{R}^n$. Then, for any given starting vector $\boldsymbol{v}_0$, applying non-symmetric Lanczos (NSL) to the preconditioned system $M^{-1}A\boldsymbol{v} = M^{-1}\boldsymbol{b}$ and applying TEA to the stationary iterative method $M\boldsymbol{v}_{k+1} = N\boldsymbol{v}_k + \boldsymbol{b}$, with the particular choice in TEA of $\boldsymbol{q} := \boldsymbol{r}_0 = M^{-1}\boldsymbol{b} - M^{-1}A\boldsymbol{v}_0$, lead to identical iterates.*

PROOF. The proof is similar to the proof of Theorem 11.32. TEA is a special minimum polynomial extrapolation method, and as such determines its iterates $\boldsymbol{v}_k^{\mathrm{TEA}}$ such that $\boldsymbol{r}_k^{\mathrm{TEA}} = U_k \boldsymbol{\gamma}$, with the matrix U_k defined in (11.73),

$$U_k = [\boldsymbol{u}_0, (I - M^{-1}A)\boldsymbol{u}_0, \ldots, (I - M^{-1}A)^k \boldsymbol{u}_0].$$

Proceeding as in (11.137) in the proof of Theorem 11.32, we can show that $\boldsymbol{u}_0 = \boldsymbol{r}_0$, and hence

$$\boldsymbol{r}_k^{\mathrm{TEA}} \in \mathcal{K}_{k+1}(M^{-1}A, \boldsymbol{r}_0).$$

Non-symmetric Lanczos applied to the linear system $M^{-1}Av = M^{-1}b$ generates the subspace

$$\text{span}\{V_k\} = \mathcal{K}_k(M^{-1}A, r_0),$$

and the residual satisfies

$$
\begin{aligned}
r_k^{NSL} &= M^{-1}b - M^{-1}Av_k^{NSL} \\
&= M^{-1}b - M^{-1}A(v_0 + V_k z) \\
&= r_0 - M^{-1}AV_k z \in \mathcal{K}_{k+1}(M^{-1}A, r_0).
\end{aligned}
$$

Hence both residuals lie in the same Krylov space, and are determined by k coefficients, since the initial residual is the same. For TEA, the coefficients are determined by the orthogonality condition $Q_{k-1}^{\top} r_k^{\text{TEA}} = 0$, see (11.98), where the assumption $q = r_0$ implies

$$Q_{k-1} := [q, G^{\top} q, \ldots, (G^{\top})^{k-1} q] \quad \text{with} \quad G = I - M^{-1}A.$$

The coefficients for NSL are determined by $W_k^{\top} r_k^{\text{NSL}} = 0$ (the orthogonality condition) and

$$\text{span}\{W_k\} = \mathcal{K}_k((M^{-1}A)^{\top}, r_0).$$

Therefore the coefficients in both cases are determined by the same orthogonality condition, which means that the residuals, and thus the iterates must coincide. $\qquad\square$

EXAMPLE 11.15. *The following example illustrates the equivalence.*

```
A =[ 0      -4     -8     -2
     -4     -7     -7     -8
     -9     -5     -4     -5
      0     -5     -9     -6   ];

n=length(A); x=(1:n)'; b=A*x;
x0=x-5;                             % Starting vector
[X,res,V,W,T]=BiORES(A,b,x0,n);
[x,W,E]=TEA1(A,b,x0,4);
disp('Solutions with BiORES'); X
disp('Solutions with TEA1'); W(:,2:2:8)

Solutions with BiORES
X =
     -4.0000    -0.7314   -26.5624     0.8810     1.0000
     -3.0000     3.0703    -0.0670     1.6982     2.0000
     -2.0000     3.3698    42.1720     3.2542     3.0000
     -1.0000     3.6694   -22.8856     4.0633     4.0000
Solutions with TEA1
ans =
     -0.7314   -26.5624     0.8810     1.0000
      3.0703    -0.0670     1.6982     2.0000
      3.3698    42.1720     3.2542     3.0000
      3.6694   -22.8856     4.0633     4.0000
```

11.7.8 Biconjugate Gradient Method (BiCG)

Based on the non-symmetric Lanczos process, Lanczos already developed the *Biconjugate Gradient Method* in [83], see also Fletcher [32]. The following algorithm computes approximations $\boldsymbol{x}_j$ of the solution of the linear system $A\boldsymbol{x} = \boldsymbol{b}$. In total, five vectors are used: two *direction vectors* $\boldsymbol{p}_j, \tilde{\boldsymbol{p}}_j$, two *residual vectors* $\boldsymbol{r}_j, \tilde{\boldsymbol{r}}_j$ and the *approximation of the solution* $\boldsymbol{x}_j$. All these vectors are updated at each iteration step by two-term recurrences.

ALGORITHM 11.26. *Biconjugate Gradient*

$$\boldsymbol{r}_0 = \boldsymbol{b} - A\boldsymbol{x}_0; \quad \tilde{\boldsymbol{r}}_0 = \boldsymbol{r}_0;$$
$$\boldsymbol{p}_{-1} = \tilde{\boldsymbol{p}}_{-1} = 0;$$
$$\rho_{-1} = 1;$$
$$\text{for } j = 0 : n$$
$$\quad \rho_j = \tilde{\boldsymbol{r}}_j^\top \boldsymbol{r}_j; \quad \beta_j = \frac{\rho_j}{\rho_{j-1}};$$

$$\boldsymbol{p}_j = \boldsymbol{r}_j + \beta_j \boldsymbol{p}_{j-1}; \qquad (1)$$
$$\tilde{\boldsymbol{p}}_j = \tilde{\boldsymbol{r}}_j + \beta_j \tilde{\boldsymbol{p}}_{j-1}; \qquad (2)$$
$$\sigma_j = \tilde{\boldsymbol{p}}_j^\top A \boldsymbol{p}_j;$$

$$\alpha_j = \frac{\rho_j}{\sigma_j}$$

$$\boldsymbol{r}_{j+1} = \boldsymbol{r}_j - \alpha_j A \boldsymbol{p}_j; \qquad (3)$$
$$\tilde{\boldsymbol{r}}_{j+1} = \tilde{\boldsymbol{r}}_j - \alpha_j A^\top \tilde{\boldsymbol{p}}_j; \qquad (4)$$
$$\boldsymbol{x}_{j+1} = \boldsymbol{x}_j + \alpha_j \boldsymbol{p}_j; \qquad (5)$$
$$\text{end}$$

Note that

$$\alpha_j = \frac{\tilde{\boldsymbol{r}}_j^\top \boldsymbol{r}_j}{\tilde{\boldsymbol{p}}_j^\top A \boldsymbol{p}_j}, \qquad \beta_j = \frac{\tilde{\boldsymbol{r}}_j^\top \boldsymbol{r}_j}{\tilde{\boldsymbol{r}}_{j-1}^\top \boldsymbol{r}_{j-1}}.$$

THEOREM 11.39. (PROPERTIES OF BiCG) *The following relations hold for the vector sequences generated by the BiCG-algorithm:*

a) *Bi-orthogonality:* $\tilde{\boldsymbol{r}}_i^\top \boldsymbol{r}_j = \boldsymbol{r}_i^\top \tilde{\boldsymbol{r}}_j = 0$ *for* $j < i$.

b) *Bi-conjugacy:* $\tilde{\boldsymbol{p}}_i^\top A \boldsymbol{p}_j = \boldsymbol{p}_i^\top A^\top \tilde{\boldsymbol{p}}_j = 0$ *for* $j < i$.

c) $\tilde{\boldsymbol{r}}_i^\top \boldsymbol{p}_j = \boldsymbol{r}_i^\top \tilde{\boldsymbol{p}}_j = 0$ *for* $j < i$.

PROOF. The proof is by induction, and we need to prove a) and b) simultaneously. We start with the base case for a): Since $\boldsymbol{p}_0 = \boldsymbol{r}_0$, we have

$$\boldsymbol{r}_1 = \boldsymbol{r}_0 - \alpha_0 A \boldsymbol{p}_0 = \boldsymbol{r}_0 - \alpha_0 A \boldsymbol{r}_0,$$

and since $\tilde{\boldsymbol{r}}_0 = \tilde{\boldsymbol{p}}_0$ from (2), we have $\alpha_0 = \frac{\tilde{\boldsymbol{r}}_0^\top \boldsymbol{r}_0}{\tilde{\boldsymbol{r}}_0^\top A \boldsymbol{r}_0}$, and thus

$$\boldsymbol{r}_1^\top \tilde{\boldsymbol{r}}_0 = \tilde{\boldsymbol{r}}_0^\top \boldsymbol{r}_1 = \tilde{\boldsymbol{r}}_0^\top \boldsymbol{r}_0 - \alpha_0 \tilde{\boldsymbol{r}}_0^\top A \boldsymbol{r}_0 = \tilde{\boldsymbol{r}}_0^\top \boldsymbol{r}_0 - \frac{\tilde{\boldsymbol{r}}_0^\top \boldsymbol{r}_0}{\tilde{\boldsymbol{r}}_0^\top A \boldsymbol{r}_0} \tilde{\boldsymbol{r}}_0^\top A \boldsymbol{r}_0 = 0.$$

Similarly, we find that $\tilde{\boldsymbol{r}}_1^\top \boldsymbol{r}_0 = 0$. The base case for b) also holds, since $\boldsymbol{p}_0^\top A^\top \tilde{\boldsymbol{p}}_{-1} = 0$ and $\tilde{\boldsymbol{p}}_0^\top A \boldsymbol{p}_{-1} = 0$ by the initialization $\boldsymbol{p}_{-1} = \tilde{\boldsymbol{p}}_{-1} = 0$.

We assume now that a) and b) hold for $j < i$, from which we want to deduce that a) and b) also hold for $j < i + 1$. Using Statement (4) in BiCG, we get

$$\tilde{\boldsymbol{r}}_{i+1}^\top \boldsymbol{r}_j = (\tilde{\boldsymbol{r}}_i - \alpha_i A^\top \tilde{\boldsymbol{p}}_i)^\top \boldsymbol{r}_j = \tilde{\boldsymbol{r}}_i^\top \boldsymbol{r}_j - \alpha_i \tilde{\boldsymbol{p}}_i^\top A \boldsymbol{r}_j. \tag{11.161}$$

Now we use Statement (1) and solve for $\boldsymbol{r}_j$ in order to replace the second occurrence of $\boldsymbol{r}_j$ on the right in (11.161),

$$\tilde{\boldsymbol{r}}_{i+1}^\top \boldsymbol{r}_j = \tilde{\boldsymbol{r}}_i^\top \boldsymbol{r}_j - \alpha_i \tilde{\boldsymbol{p}}_i^\top A (\boldsymbol{p}_j - \beta_j \boldsymbol{p}_{j-1})$$
$$= \underbrace{\tilde{\boldsymbol{r}}_i^\top \boldsymbol{r}_j}_{I} - \underbrace{\alpha_i \tilde{\boldsymbol{p}}_i^\top A \boldsymbol{p}_j}_{II} + \underbrace{\alpha_i \beta_j \tilde{\boldsymbol{p}}_i^\top A \boldsymbol{p}_{j-1}}_{III}.$$

By the induction hypothesis, the term III is zero for $j < i + 1$, and for $j < i$, both terms I and II are also zero. For $j = i$, the terms I and II cancel, since

$$\alpha_i = \frac{\tilde{\boldsymbol{r}}_i^\top \boldsymbol{r}_i}{\tilde{\boldsymbol{p}}_i^\top A \boldsymbol{p}_i}.$$

Therefore $\tilde{\boldsymbol{r}}_{i+1}^\top \boldsymbol{r}_j = 0$ for $j < i + 1$. Similarly, one can also show that $\boldsymbol{r}_{i+1}^\top \tilde{\boldsymbol{r}}_j = 0$, which completes the induction step of a). To complete the induction step of b), we use Statement (1) of BiCG to get

$$\boldsymbol{p}_{i+1}^\top A^\top \tilde{\boldsymbol{p}}_j = (\boldsymbol{r}_{i+1} + \beta_{i+1} \boldsymbol{p}_i)^\top A^\top \tilde{\boldsymbol{p}}_j$$
$$= \boldsymbol{r}_{i+1}^\top A^\top \tilde{\boldsymbol{p}}_j + \beta_{i+1} \boldsymbol{p}_i^\top A^\top \tilde{\boldsymbol{p}}_j.$$

With Statement (4) from BiCG, this becomes

$$\boldsymbol{p}_{i+1}^\top A^\top \tilde{\boldsymbol{p}}_j = \frac{1}{\alpha_j} \left(\boldsymbol{r}_{i+1}^\top \tilde{\boldsymbol{r}}_j - \boldsymbol{r}_{i+1}^\top \tilde{\boldsymbol{r}}_{j+1} \right) + \beta_{j+1} \boldsymbol{p}_i^\top A^\top \tilde{\boldsymbol{p}}_j, \tag{11.162}$$

and for $j < i$ every term on the right is zero, since we have already shown that $\boldsymbol{r}_{i+1}^\top \tilde{\boldsymbol{r}}_j = 0$ for $j < i + 1$. Similarly, for $j = i$, we have already shown that $\boldsymbol{r}_{i+1}^\top \tilde{\boldsymbol{r}}_i = 0$, so replacing α_j in (11.162) gives

$$\boldsymbol{p}_{i+1}^\top A^\top \tilde{\boldsymbol{p}}_i = - \underbrace{\frac{\boldsymbol{r}_{i+1}^\top \tilde{\boldsymbol{r}}_{i+1}}{\tilde{\boldsymbol{r}}_i^\top \boldsymbol{r}_i}}_{\beta_{i+1}} \underbrace{\tilde{\boldsymbol{p}}_i^\top A \boldsymbol{p}_i}_{} + \beta_{i+1} \underbrace{\boldsymbol{p}_i^\top A^\top \tilde{\boldsymbol{p}}_i}_{\tilde{\boldsymbol{p}}_i^\top A \boldsymbol{p}_i} = 0,$$

since the two terms cancel. We can verify by a similar calculation that also $\tilde{\boldsymbol{p}}_{i+1}^{\top} A \boldsymbol{p}_j = 0$ and thus the induction step for b) is also proved, which completes the proof by induction for a) and b).

For c), using Statement (2) of BiCG to replace $\tilde{\boldsymbol{p}}_j$, we get

$$
\begin{aligned}
\boldsymbol{r}_i^{\top} \tilde{\boldsymbol{p}}_j &= \boldsymbol{r}_i^{\top} \left(\tilde{\boldsymbol{r}}_j + \beta_j \tilde{\boldsymbol{p}}_{j-1} \right) \\
&= \boldsymbol{r}_i^{\top} \left(\tilde{\boldsymbol{r}}_j + \beta_j (\tilde{\boldsymbol{r}}_{j-1} + \beta_{j-1} \tilde{\boldsymbol{p}}_{j-2}) \right) \\
&= \boldsymbol{r}_i^{\top} \left(\tilde{\boldsymbol{r}}_j + \beta_j \tilde{\boldsymbol{r}}_{j-1} + \beta_j \beta_{j-1} \tilde{\boldsymbol{r}}_{j-2} + \cdots \right) \\
&= 0 \quad \text{for } j < i.
\end{aligned}
$$

Similarly, one can also show that $\tilde{\boldsymbol{r}}_i^{\top} \boldsymbol{p}_j = 0$, which concludes the proof. $\qquad \square$

The following MATLAB implementation of BiCG is for educational purposes only: we store non-essential vectors in order to illustrate the relations stated in Theorem 11.39. For solving the linear system, it would be sufficient to return only the solution $\boldsymbol{x}$.

ALGORITHM 11.27. MATLAB *implementation of BiCG*

```
function [X,res,R,Rt,P,Pt]=BiCG(A,b,x0,n);
% BICG Bi-Conjugate Gradient Method
%    [X,res,R,Rt,P,Pt]=BiCG(A,b,x0,n); computes an approximate solution
%    of the linear system Ax=b performing n steps of the Bi-Cojugate
%    Gradient algorithm starting with the initial vector x0. The matrix
%    X contains the approximate solutions and the vector res the norms
%    of the corresponding residuals. The matrices R, Rt and P, Pt
%    contain the residuals and directions.

x=x0;r=b-A*x0; rt=r;
p=zeros(size(b)); pt=p;
rho=1; res=norm(r);
X=x0; P=[]; Pt=[]; R=r; Rt=rt;
for j=1:n
  rhoold=rho; rho=rt'*r;
  beta=rho/rhoold;
  p=r+beta*p; pt=rt+beta*pt;
  P=[P p]; Pt=[Pt pt];                        % not needed
  Ap=A*p;
  alpha=rho/(pt'*Ap);
  r=r-alpha*Ap; rt=rt-alpha*A'*pt;
  R=[R r]; Rt=[Rt rt];                         % not needed
  x=x+alpha*p;
  X=[X x];                                     % not needed
  res=[res norm(r)];
end;
```

For a symmetric matrix A, this algorithm results in the two-term recurrence version of CG called *Orthomin*. One can also show that BiCG is equivalent to BiORES based on the non-symmetric Lanczos algorithm, see Algorithm 11.25. By comparing the two algorithms, one sees that the only difference is the left starting vector, which is normalized in BiORES. We illustrate this now by the following example:

EXAMPLE 11.16. *We use again our matrix A (11.90).*

```
clc,format short e
format compact
A=[ 0     -4     -8     -2
    -4     -7     -7     -8
    -9     -5     -4     -5
     0     -5     -9     -6];
n=length(A); x=(1:n)'; b=A*x;
x0=x-5;                           % Starting vector
[X1,res,V,W,T]=BiORES(A,b,x0,4);
[X2,res,R,Rt,P,Pt]=BiCG(A,b, x0,4);
disp('Bi-orthogonality of residuals R''*Rt');
R(:,1:4)'*Rt(:,1:4)
disp('Bi-conjugacy of directions: P''*A''*Pt respectively Pt''*A*P')
P'*A'*Pt
format short
disp('Orthogonality of residuals and directions  R''*Pt')
disp('respectively Rt''*P for  i>j' )
R(:,1:4)'*Pt
Rt(:,1:4)'*P
pause
clc,format short e
disp('Tridiagonal matrix by BiCG')
P\A*P
% Pt\A'*Pt
disp('Tridiagonal matrix by Lanczos')
T(1:4,1:4)
disp('are not the same, but similar, the eigenvalues are')
[eig(P\A*P) eig(T(1:4,1:4))]
pause
clc
disp('BiORES approximations '); X1
disp('BiCG approximations'); X2
disp('Solutions are the same')
disp('right Lanczos vectors V'); V
disp('BiCG residual vectors R'); R
disp('component wise quotient of left Lanczos vectors W ')
disp('and BiCG vectors Rt')
W(:,1:4)./Rt(:,1:4)
```

We obtain the following results: we first check the orthogonality,

```
Bi-Orthogonality of residuals R'*Rt
```

```
ans =
    4.5025e+04    3.4106e-12   -3.0923e-10   -1.2506e-10
    9.0949e-13   -1.0686e+02   -2.9559e-12   -5.9686e-13
   -6.9122e-11    2.2737e-13   -2.2676e+04   -8.4583e-11
   -2.1743e-11    3.5705e-13   -1.0877e-10    1.4832e+01
Bi-conjugacy of directions: P'*A'*Pt respectively Pt'*A*P
ans =
   -9.6425e+05   -7.2760e-11   -1.2107e-08    2.6776e-09
   -1.8190e-11    2.7904e+01   -1.4552e-11   -2.1828e-11
   -2.7940e-09   -4.3656e-11   -1.3902e+06   -5.1805e-09
    4.7524e-10   -2.8411e-11   -6.6534e-09   -1.8344e+01
```

which confirms Statement a) and b) of Theorem 11.39. Next, we verify

```
Orthogonality of residuals and directions  R'*Pt
respectively Rt'*P for  i>j
ans =
   1.0e+04 *
    4.5025   -0.0107   -2.2676    0.0015
    0.0000   -0.0107   -2.2676    0.0015
   -0.0000   -0.0000   -2.2676    0.0015
   -0.0000    0.0000   -0.0000    0.0015
ans =
   1.0e+04 *
    4.5025   -0.0107   -2.2676    0.0015
    0.0000   -0.0107   -2.2676    0.0015
   -0.0000   -0.0000   -2.2676    0.0015
   -0.0000   -0.0000   -0.0000    0.0015
```

*and see that the two matrices are upper triangular, which confirms Statement
c) of Theorem 11.39, namely that $\tilde{r}_i^\top p_j = r_i^\top \tilde{p}_j = 0$ for $j < i$.*

The projected matrix $T = P^{-1}AP = \tilde{P}^{-1}A^\top\tilde{P}$ becomes

```
Tridiagonal matrix by BiCG
ans =
   -2.1365e+01   -6.1974e-04    2.8422e-14    1.0550e-14
    2.1416e+01   -5.5667e+01   -1.3009e+04   -6.8472e-12
   -2.6645e-15    2.6111e-01    6.1269e+01   -8.0895e-04
   -2.5580e-13   -2.6645e-13   -6.1309e+01   -1.2367e+00
Tridiagonal matrix by Lanczos
ans =
   -2.1416e+01   -5.0830e-02            0            0
    2.1416e+01   -2.1028e-01    5.5406e+01            0
            0    2.6111e-01    5.9030e+00    4.0103e-02
            0            0   -6.1309e+01   -1.2768e+00
```

*The two tridiagonal matrices are not the same, but they have the same eigen-
values*

```
ans =
```

```
    -2.1363e+01                -2.1363e+01
     7.4918e+00                 7.4918e+00
    -1.5644e+00 + 4.4941e-01i  -1.5644e+00 + 4.4941e-01i
    -1.5644e+00 - 4.4941e-01i  -1.5644e+00 - 4.4941e-01i
```

So these matrices are similar.

Finally, we compare the approximations for the solutions. With BiORES, we get

```
BiORES approximations
X1 =
    -4.0000e+00   -7.3140e-01   -2.6562e+01    8.8103e-01    1.0000e-00
    -3.0000e+00    3.0703e+00   -6.6986e-02    1.6982e+00    2.0000e+00
    -2.0000e+00    3.3698e+00    4.2172e+01    3.2542e+00    3.0000e+00
    -1.0000e+00    3.6694e+00   -2.2886e+01    4.0633e+00    4.0000e+00
```

and with BiCG, the approximations are the same,

```
BiCG approximations
X2 =
    -4.0000e+00   -7.3140e-01   -2.6562e+01    8.8103e-01    1.0000e+00
    -3.0000e+00    3.0703e+00   -6.6986e-02    1.6982e+00    2.0000e+00
    -2.0000e+00    3.3698e+00    4.2172e+01    3.2542e+00    3.0000e+00
    -1.0000e+00    3.6694e+00   -2.2886e+01    4.0633e+00    4.0000e+00
```

If we compare the right Lanczos vectors obtained by the non-symmetric Lanczos process

```
right  Lanczos vectors V
V =
    -7.0000e+01    6.5787e+00    2.5134e+02    9.5342e-01   -6.1963e-12
    -1.3000e+02    5.1063e-01   -6.5599e+01   -3.0242e-01   -1.0725e-11
    -1.1500e+02   -1.0405e+01   -2.3614e+02   -1.2463e+00   -8.6352e-12
    -1.0000e+02    6.6965e+00    1.8090e+02    1.1590e+00   -8.6963e-12
```

to the vectors r_j obtained by BiCG, we see that they are also the same,

```
BiCG residual vectors R
R =
    -7.0000e+01    6.5787e+00    2.5134e+02    9.5342e-01    7.3446e-12
    -1.3000e+02    5.1063e-01   -6.5599e+01   -3.0242e-01   -2.6455e-12
    -1.1500e+02   -1.0405e+01   -2.3614e+02   -1.2463e+00   -9.1305e-12
    -1.0000e+02    6.6965e+00    1.8090e+02    1.1590e+00    5.1730e-12
```

This is not the case for the left Lanczos vectors W, because the initial one was normalized in BiORES, in contrast to BiCG. However, the vectors $\tilde{r}_j$ of BiCG are just multiples of the left Lanczos vectors, as we can see by dividing component-wise both matrices:

```
component wise quotient of left Lanczos vectors and und BiCG vectors
ans =
     2.2210e-05   -9.3576e-03   -4.4100e-05    6.7420e-02
```

```
   2.2210e-05  -9.3576e-03  -4.4100e-05   6.7420e-02
   2.2210e-05  -9.3576e-03  -4.4100e-05   6.7420e-02
   2.2210e-05  -9.3576e-03  -4.4100e-05   6.7420e-02
```

As shown in Theorem 11.38, the non-symmetric Lanczos algorithm for solving linear systems and the Topological ε-Algorithm are equivalent, and we have also seen that they are equivalent to BiCG. In the following example, we show that TEA1 gives the same approximation solutions x_k as those shown above, up to differences due to finite precision arithmetic.

EXAMPLE 11.17. *We use again our matrix A (11.90) as in Example 11.16:*

```
A=[ 0     -4     -8     -2
   -4     -7     -7     -8
   -9     -5     -4     -5
    0     -5     -9     -6];
format long
clc
n=length(A); x=(1:n)'; b=A*x;
x0=x-5;                                  % Starting vector
[X,res,R,Rt,P,Pt]=BiCG(A,b,x0,n);
[x,W,E]=TEA1(A,b,x0,4);

disp('Solutions with BiCG')
X(:,2:n+1)
disp('Solutions with TEA1')
W(:,2:2:8)
disp('Eigenvalues of basic iteration matrix')
eig(eye(size(A))-A)
rho=max(abs(ans))

format short e
disp('last x_j used for extrapolation')
E(:,9)

disp('Basic iterations')
G=eye(size(A))-A;
x=x0; xx=[];
for k=1:8
  x=G*x+b;
  xx=[xx x];
end
xx
```

The results, the vectors of consecutive approximations, are as predicted the same:

```
Solutions with BiCG
ans =
  -0.731397459165154 -26.562405416410403   0.881029980440719   0.999999999998884
```

```
     3.070261861550428   -0.066986452894987    1.698228789336742    2.000000000000207
     3.369847031371532   42.171996268315795    3.254246071134837    3.000000000001025
     3.669432201192636  -22.885556136309262    4.063267002656218    3.999999999999152
Solutions with TEA1
ans =
    -0.731397459165152  -26.562405418626817    0.881029980719667    0.999999983283914
     3.070261861550421   -0.066986453163246    1.698228789299351    1.999999943830457
     3.369847031371535   42.171996271646179    3.254246070822271    3.000000040378477
     3.669432201192635  -22.885556138588385    4.063267002885153    4.000000016511125
```

We note that the values of BiCG are more precise. One reason is that the basic iteration for TEA1, $\boldsymbol{x}_{j+1} = (I - A)\boldsymbol{x}_j + \boldsymbol{b}$, diverges for this example. In fact, the spectral radius of $I - A$ is larger than one:

```
Eigenvalues of basic iteration matrix
ans =
 22.363100417299790
 -6.491823701736879
  2.564361642218540 + 0.449408785422842i
  2.564361642218540 - 0.449408785422842i
rho =
 22.363100417299790
```

so the "approximations" of the basic iterations do not converge, but grow quickly. We obtain $\boldsymbol{x}_8$ as final vector with a norm of $\approx 10^{11}$:

```
last x_j used for extrapolation
ans =
  -2.1817e+11
  -3.8514e+11
  -3.1287e+11
  -3.0864e+11
```

It is amazing that even from these divergent approximations, the ε-algorithm TEA1 manages to extract a result with about 8 correct digits. Experiments show, however, that even when the basic iteration converges, the results of TEA1 seem to be more affected by rounding errors than BiCG.

11.7.9 Further Krylov Methods

Any linear system with a non-singular matrix A can be reduced to an equivalent system with a symmetric and positive definite matrix.

DEFINITION 11.14. (GAUSS TRANSFORMATION) *Let $A\boldsymbol{x} = \boldsymbol{b}$ be a linear system with $A \in \mathbb{R}^{n \times n}$ non-singular. Then*

1. $A^{\top}A\boldsymbol{x} = A^{\top}\boldsymbol{b}$ is called the first Gauss transformation.

2. $AA^{\top}\boldsymbol{y} = \boldsymbol{b}$ with $\boldsymbol{x} = A^{\top}\boldsymbol{y}$ is called the second Gauss transformation.

The coefficient matrices $A^{\top}A$ and $AA^{\top}$ are symmetric and positive definite, so CG can be applied to the transformed systems. The only serious drawback is that the condition number is squared, and hence the convergence is slowed down and errors due to the finite precision arithmetic are increased.

Gauss transformations are useful for linear systems with rectangular matrices. If $A \in \mathbb{R}^{m \times n}$ with $m > n$, then we have more equations than unknowns and the system is overdetermined, which means it has no solution in general. Using the first Gauss transformation $A^\top A x = A^\top b$, we get the *normal equations*, and the solutions are the least squares solutions minimizing the residual $\|b - Ax\|_2$, see Chapter 6. For $A \in \mathbb{R}^{m \times n}$ with $m < n$ with full rank, we have more unknowns than equations, so there are in general many solutions. In this case, the second Gauss transformation $AA^\top y = b$ with $x = A^\top y$ computes the solution with minimal norm.

With the first Gauss transformation, followed by the application of CG, we obtain the *CGNR-Method (conjugate gradient on normal equations)*. This method minimizes the residuals, since

$$\|x - x_k\|_{A^\top A}^2 = (x - x_k)^\top A^\top A (x - x_k) = \|A(x - x_k)\|_2^2 = \|b - Ax_k\|_2^2. = \|r_k\|_2^2$$

Therefore, this algorithm is similar to CR, Algorithm 11.5, or GMRES, but the minimization is done over a different Krylov space $(x_0 + \mathcal{K}_k(A^\top A, r_0)$ rather than $x_0 + \mathcal{K}_k(A, r_0))$.

Applying the conjugate gradient method after the second Gauss transformation, we obtain the *CGNE-Method*, which is also called *Craig's Method*.

As we mentioned before, these two methods are only useful in connection with a good preconditioner, since the condition number is squared before the CG method is applied. One way to avoid squaring the condition number is to use the Biconjugate Gradient (BiCG) method, which has been discussed in the previous section. One drawback, however, is that the convergence of BiCG is often erratic, and the algorithm can suffer from breakdowns. A stabilized version, called BiCGStab, was developed by van der Vorst [141] and is currently widely used.

A different idea, proposed by Freund and Nachtigal [35], is to only approximately minimize the residual in the Krylov space in a least squares sense. This leads to the so-called *quasi minimum residual* (QMR) method.

There are many more variants than we have discussed in this chapter, for example special methods for symmetric but indefinite systems, like MINRES and SYMMLQ developed by Paige and Saunders in [101] and [102], or flexible Krylov methods, where the preconditioner can change from iteration to iteration (e.g. FGMRES). For an overview, see the Templates book [7], and for further reading, we recommend the excellent book by Saad [118].

11.8 Preconditioning

Historically, the CG method generated great interest as a method for computing the solution of a linear system of equations in at most n iteration steps. When it was realized that this is no longer true in finite precision arithmetic, interest quickly faded, until one realized that CG is much more interesting as an iterative method, because it minimizes at each step the error over the current Krylov subspace, as stated in Theorem 11.26. Theorem 11.28 shows that the rate of convergence depends on the condition number of the matrix — convergence is improved if the condition number is smaller. Also, by Corollary 11.2, convergence is faster if A has multiple eigenvalues.

The idea of *preconditioning* is to replace the linear system $A\boldsymbol{x} = \boldsymbol{b}$ by a new system $\hat{A}\hat{\boldsymbol{x}} = \hat{\boldsymbol{b}}$ with a smaller condition number or a better clustered spectrum. More specifically, we wish to find a non-singular matrix S such that

$$\hat{A}\hat{\boldsymbol{x}} = \hat{\boldsymbol{b}}, \quad \text{with} \quad \hat{A} = SAS^\top, \quad \hat{\boldsymbol{x}} = S^{-T}\boldsymbol{x}, \quad \hat{\boldsymbol{b}} = S\boldsymbol{b},$$

and determine S such that the condition number is reduced, i.e., $\kappa(\hat{A}) < \kappa(A)$.

The best choice for S would be $S = A^{-\frac{1}{2}}$, because then $\hat{A} = I$, but computing this S is a more difficult problem than solving the given system of equations. Obviously solving equations with the matrix S must be easy and the transformation $A \to \hat{A} = SAS^\top$ must be done implicitly, since we do not want to destroy any possible sparsity pattern of A. Therefore, we will reformulate the CG algorithm for the transformed system such that only the original variables A, $\boldsymbol{x}$, $\boldsymbol{b}$ and the new matrix S are used.

> CG algorithm for the transformed system:
> choose $\hat{\boldsymbol{x}}_0$, $\hat{\boldsymbol{p}}_0 = \hat{\boldsymbol{r}}_0 := \hat{\boldsymbol{b}} - \hat{A}\hat{\boldsymbol{x}}_0$;
> for $k = 0 : n - 1$
>
> $$\hat{\alpha}_k = \frac{\|\hat{\boldsymbol{r}}_k\|^2}{\hat{\boldsymbol{p}}_k^\top \hat{A}\hat{\boldsymbol{p}}_k}; \tag{11.163}$$
>
> $$\hat{\boldsymbol{x}}_{k+1} = \hat{\boldsymbol{x}}_k + \hat{\alpha}_k\hat{\boldsymbol{p}}_k; \tag{11.164}$$
>
> $$\hat{\boldsymbol{r}}_{k+1} = \hat{\boldsymbol{r}}_k - \hat{\alpha}_k\hat{A}\hat{\boldsymbol{p}}_k; \tag{11.165}$$
>
> $$\hat{\beta}_k = \frac{\|\hat{\boldsymbol{r}}_{k+1}\|^2}{\|\hat{\boldsymbol{r}}_k\|^2}; \tag{11.166}$$
>
> $$\hat{\boldsymbol{p}}_{k+1} = \hat{\boldsymbol{r}}_{k+1} + \hat{\beta}_k\hat{\boldsymbol{p}}_k; \tag{11.167}$$
> end

We start with a relation between the preconditioned and unpreconditioned residuals: for $k = 0, 1, \ldots$, we have

$$\hat{\boldsymbol{r}}_k = \hat{\boldsymbol{b}} - \hat{A}\hat{\boldsymbol{x}}_k = S\boldsymbol{b} - SAS^\top S^{-T}\boldsymbol{x}_k = S\boldsymbol{r}_k. \tag{11.168}$$

By definition, we also have $\boldsymbol{x}_k = S^\top \hat{\boldsymbol{x}}_k$, and we define two new vectors $\boldsymbol{p}_k := S^\top \hat{\boldsymbol{p}}_k$ and $\tilde{\boldsymbol{r}}_k := S^\top S\boldsymbol{r}_k$. Replacing $\hat{A} = SAS^\top$ and $\hat{\boldsymbol{r}}_k$ from (11.168) in Statement (11.163), we obtain

$$\hat{\alpha}_k = \frac{\tilde{\boldsymbol{r}}_k^\top \boldsymbol{r}_k}{\boldsymbol{p}_k^\top A\boldsymbol{p}_k},$$

and multiplying Statement (11.164) by $S^\top$, we get

$$x_{k+1} = S^\top \hat{x}_{k+1} = S^\top \hat{x}_k + \hat{\alpha}_k S^\top \hat{p}_k = x_k + \hat{\alpha}_k p_k.$$

We then multiply Statement (11.165) by S^{-1}, and replacing $\hat{A} = SAS^\top$ we get

$$r_{k+1} = r_k - \hat{\alpha}_k A p_k.$$

Substituting (11.168) into Statement (11.166) and using the definition of $\tilde{r}_k$, we get

$$\hat{\beta}_k = \frac{\tilde{r}_{k+1}^\top r_{k+1}}{\tilde{r}_k^\top r_k},$$

and finally, we transform the last statement by multiplying (11.167) by $S^\top$,

$$p_{k+1} = S^\top \hat{p}_{k+1} = S^\top \hat{r}_{k+1} + \hat{\beta}_k S^\top \hat{p}_k = S^\top S r_{k+1} + \hat{\beta}_k p_k = \tilde{r}_{k+1} + \hat{\beta}_k p_k.$$

Comparing these transformed statements with the original CG Algorithm 11.15, we see that very few modifications are necessary in order to obtain the preconditioned version: only three instances of r_k need to be replaced by $\tilde{r}_k$, and $\tilde{r}_k$ needs to be computed from $\tilde{r}_k = S^\top S r_k =: M^{-1} r_k$, where M is the symmetric and positive definite preconditioner, and the matrix S itself is not needed. This leads to the following MATLAB implementation, where we renamed $\hat{\alpha}$ and $\hat{\beta}$ again to α and β.

ALGORITHM 11.28.
Preconditioned Conjugate Gradient Algorithm

```
function [X,R,P,alpha,beta]=PCG(A,b,x0,M,m);
% CG preconditioned conjugate gradient method
%  [X,R,P,alpha,beta]=PCG(A,b,x0,M,m) computes an approximation for
%  the solution of the linear system Ax=b performing m steps of the
%  preconditioned conjugate gradient method with preconditioner M,
%  starting with x0, and returns in the matrix X the iterates, in the
%  matrix R the residuals, and in a and b the coefficients alpha and
%  beta.

x=x0; r=b-A*x;
rt=M\r; p=rt;
R=r; P=p; X=x;
oldrho=rt'*r;
for k=1:m
  Ap=A*p;
  alpha(k)=oldrho/(p'*Ap);
  x=x+alpha(k)*p;
  r=r-alpha(k)*Ap;
  rt=M\r;
  rho=rt'*r;
```

```
    beta(k)=rho/oldrho;
    oldrho=rho;
    p=rt+beta(k)*p;
    X=[X,x];
    R=[R,r];
    P=[P,p];
end;
```

How should we choose the preconditioning matrix $M = (S^\top S)^{-1}$? We have

$$\hat{A} = S A S^\top \iff S^\top \hat{A} S^{-T} = S^\top S A,$$

and thus $\hat{A}$ is similar to $(S^\top S)A = M^{-1}A$. Now $\hat{A}$ is well conditioned if

$$(S^\top S)A \approx I \iff M \approx A.$$

The preconditioning matrix M has to be symmetric positive definite and a good approximation to A, but it must be easy to solve systems of linear equations involving M, since we need to compute $\tilde{r}_k$ in every step. For $M = I$, we are back to the classical conjugate gradient algorithm.

The quest for effective preconditioners M is currently a major research area in numerical analysis. If one has an effective matrix M from a stationary iterative method with the splitting $A = M - N$, it is a good candidate for preconditioning CG, since in the symmetric positive definite case, $\rho(M^{-1}N) \ll 1$ implies that the real eigenvalues are clustered around zero, and thus the eigenvalues of the preconditioned system matrix $M^{-1}A = I - M^{-1}N$ are clustered around one, which implies a small condition number. As we have seen, a small condition number guarantees rapid convergence for CG.

Two rather different types of preconditioners are currently being developed and analyzed: the first type are algebraic preconditioners, which are based on the information contained in the matrix alone; examples include *incomplete LU decompositions (ILU), sparse approximate inverses (SPAI, AINV)* and *algebraic multigrid methods*. The second type are *geometric preconditioners*, which also use information from the underlying physical problem that gives rise to the discrete linear system; preconditioners of this kind include *domain decomposition methods* and *geometric multigrid methods*. Algebraic preconditioners are very easy to use, since they are available as black boxes, but geometric preconditioners are often more effective. An excellent introduction to this subject can be found in Saad [118].

11.9   Problems

PROBLEM 11.1.   *Verify the Theorem of Perron–Frobenius for the matrix* `A=magic(19)`.

PROBLEM 11.2.   *Equation (11.15) was verified with* MAPLE *for* $n = 5$. *Prove this in general.*

PROBLEM 11.3. *By applying the power series to a Jordan block, prove that*

$$
f\left(\begin{bmatrix} \lambda & 1 & 0 & \cdots & 0 \\ 0 & \lambda & 1 & \cdots & 0 \\ \vdots & \ddots & \ddots & \ddots & \vdots \\ 0 & \cdots & 0 & \lambda & 1 \\ 0 & \cdots & 0 & 0 & \lambda \end{bmatrix}\right) = \begin{bmatrix} \frac{f(\lambda)}{0!} & \frac{f'(\lambda)}{1!} & \frac{f''(\lambda)}{2!} & \cdots & \frac{f^{(m)}(\lambda)}{m!} \\ 0 & \frac{f(\lambda)}{0!} & \frac{f'(\lambda)}{1!} & \cdots & \frac{f^{(m-1)}(\lambda)}{(m-1)!} \\ \vdots & \ddots & \ddots & \ddots & \vdots \\ 0 & \cdots & 0 & \frac{f(\lambda)}{0!} & \frac{f'(\lambda)}{1!} \\ 0 & \cdots & 0 & 0 & \frac{f(\lambda)}{0!} \end{bmatrix}.
$$

PROBLEM 11.4. *For a given matrix A and a tolerance $\varepsilon > 0$, construct an induced matrix norm $\| \cdot \|$ such that $\|A\| \le \rho(A) + \varepsilon$.*

(a) Let $V \in \mathbb{R}^{n \times n}$ be non-singular, and define $\|x\| := \|Vx\|_\infty$. Show that $\| \cdot \|$ is a vector norm that induces the matrix norm

$$
\|A\| = \|VAV^{-1}\|_\infty.
$$

(b) Show that there exists a diagonal matrix D such that

$$
DJ_{m_i}(\lambda)D^{-1} = \begin{bmatrix} \lambda & \varepsilon & & \\ & \ddots & \ddots & \\ & & \lambda & \varepsilon \\ & & & \lambda \end{bmatrix}.
$$

Hint: *The diagonal entries of D can be chosen to be decreasing powers of ε.*

(c) Find an induced norm $\| \cdot \|$ such that $\|A\| \le \rho(A) + \varepsilon$.

PROBLEM 11.5. *Write a* MATLAB *function* y=multiply(A,x) *which computes efficiently the multiplication $y = Ax$ with the matrix A of Equation (11.4).*

Choose $a = 2$, $b = 1$, $m = 100$ and $n = 50$.

Multiply $y = Ax$ for some random vector x and measure the time with the MATLAB *commands* tic *and* toc.

PROBLEM 11.6. *It is well known that*

$$
B = \begin{pmatrix} 2 & -1 & & \\ -1 & 2 & \ddots & \\ & \ddots & \ddots & -1 \\ & & -1 & 2 \end{pmatrix} \in \mathbb{R}^{n \times n} \tag{11.169}
$$

has the eigenvalues λ_i and eigenvectors Q with

$$q_{ij} = \sqrt{\frac{2}{n+1}} \sin \frac{ij\pi}{n+1}, \quad \lambda_i = 2 - 2\cos\frac{i\pi}{n+1}. \tag{11.170}$$

(a) *Consider the Jacobi splitting $B = M - N$ and prove that the eigenvalues of the iteration matrix $M^{-1}N$ are real.*

(b) *Use (11.170) to compute the eigenvalues of the iteration matrix.*

(c) *Use Theorem 11.5 to conclude that B is an M-matrix.*

PROBLEM 11.7. *Consider the linear system of equations $A\boldsymbol{x} = \boldsymbol{b}$ with the matrix*

$$A = \begin{pmatrix} 2 & -1 & & \\ -1 & 2 & \ddots & \\ & \ddots & \ddots & -1 \\ & & -1 & 2 \end{pmatrix}$$

and choose $\boldsymbol{b}$ such that the exact solution is the vector `[1:n]`.

Write an experimental program for $n = 1000$ using first the Jacobi method. Perform 100 Jacobi iteration steps and estimate the spectral radius using $\boldsymbol{u}_{k+1} = M^{-1}N\boldsymbol{u}_k$ and $\rho \approx \|\boldsymbol{u}_{k+1}\|/\|\boldsymbol{u}_k\|$.

Compare the estimated spectral radius to the true one.

Then compute with the estimated spectral radius according to the theory of David Young the optimal over-relaxation parameter ω and continue for 100 more iteration steps with SOR.

Compute in each step the true error $\boldsymbol{e}_k = \boldsymbol{x} - \boldsymbol{x}_k$ and store its norm in a vector. Finally use `semilogy` and plot the error norm as a function of the iteration steps. You should notice a better convergence with SOR.

PROBLEM 11.8. *Following the same approach as in the introduction, derive a linear system which approximates the three dimensional Poisson equation*

$$\Delta u = f \quad in \ \Omega = (0,1)^3$$

with homogeneous boundary conditions, $u = 0$ on $\partial\Omega$.

PROBLEM 11.9. *Write the classical iteration methods Jacobi, Gauss-Seidel, SOR and Richardson in the correction form (11.8).*

PROBLEM 11.10. *Recall that for $\boldsymbol{x}$ in $\mathbb{R}^n$ and $1 \leq p < +\infty$ the following vector norms can be defined:*

$$\|\boldsymbol{u}\|_p = \left(\sum_{i=1}^n |u_i|^p \right)^{\frac{1}{p}}, \qquad \|\boldsymbol{u}\|_\infty = \max_{1 \mid eqi \leq n} \{|u_i|\}.$$

For a matrix A in $\mathbb{R}^{n \times m}$, we define the norms $\| \ \|_{pq}$ by

$$\|A\|_{pq} = \sup_{x \neq 0} \frac{\|Ax\|_p}{\|x\|_q},$$

for $1 \leq p, q \leq +\infty$. The norm $\| \ \|_{pp}$ is simply denoted by $\| \ \|_p$. We also define the Frobenius norm*:*

$$\|A\|_F^2 = \sum_{i=1}^{n} \sum_{j=1}^{m} a_{ij}{}^2.$$

1. *Show that*

$$\|A\|_1 = \max_{j=1,\ldots,m} \sum_{i=1}^{n} |a_{ij}|, \qquad \|A\|_\infty = \max_{i=1,\ldots,n} \sum_{j=1}^{m} |a_{ij}|,$$

$$\|A\|_2 = [\rho(A^{\mathrm{H}}A)]^{\frac{1}{2}}, \qquad\qquad \|A\|_F = [\mathrm{tr}(A^{\mathrm{H}}A)]^{\frac{1}{2}} = [\mathrm{tr}(AA^{\mathrm{H}})]^{\frac{1}{2}}.$$

2. *Show that the Frobenius norm has the* submultiplicative property

$$\|AB\|_F \leq \|A\|_F \|B\|_F.$$

PROBLEM 11.11. *Let $p \in \mathbb{N}^+$. Show that*

1. *if $0 < \rho < 1$, then $\displaystyle\lim_{k \to \infty} \binom{k}{p} \rho^k = 0$.*

2. $\displaystyle\lim_{k \to \infty} \binom{k}{p}^{\frac{1}{k}} = 1.$

PROBLEM 11.12. *Let $A \in \mathbb{C}^{n \times n}$ be a square matrix whose spectral radius $\rho(A)$ satisfies $\rho(A) < 1$. Show that $I - A$ is invertible and that*

$$(I - A)^{-1} = \sum_{j=0}^{\infty} A^j.$$

This series is a particular case of the Neumann series (generalization of the geometric series to matrices).

PROBLEM 11.13.

1. *Let $A \in \mathbb{R}^{n \times n}$ be a normal matrix, and suppose that the spectrum of A, denoted by $\lambda(A)$, is real. Show that $A = A^{\mathrm{H}}$.*

2. *Let $A = M - N$, and suppose that $M^{-1}N$ is normal. Show that the error of the stationary iterative method associated with this splitting satisfies*

$$\|e_k\|_2 \leq (\rho(M^{-1}N))^k \|e_0\|_2.$$

PROBLEM 11.14. *Let us consider the matrix A given in (11.4), which arises from the discretization of the 2D Laplacian:*

$$
A = \begin{bmatrix} T & I & & \\ I & T & \ddots & \\ & \ddots & \ddots & I \\ & & I & T \end{bmatrix}, \qquad
T = \begin{bmatrix} -2\gamma & \delta & & \\ \delta & -2\gamma & \ddots & \\ & \ddots & \ddots & \delta \\ & & \delta & -2\gamma \end{bmatrix},
$$

with $\delta = (\Delta y / \Delta x)^2$, $\gamma = 1 + \delta$. We wish to show that $-A$ is positive definite.

(a) *Find the coefficients α and β such that $T = \alpha I + \beta B$, where B is the tridiagonal matrix defined in (11.169). Calculate the eigenvalues of T explicitly to conclude that $-T$ is positive definite.*

(b) *Let $T = U \Lambda U^\top$ be the eigenvalue decomposition of T, where $\Lambda = \mathrm{diag}(\lambda_1, \ldots, \lambda_m)$ and $U^\top U = I$. Show that*

$$
\begin{bmatrix} U^\top & & & \\ & U^\top & & \\ & & \ddots & \\ & & & U^\top \end{bmatrix} A \begin{bmatrix} U & & & \\ & U & & \\ & & \ddots & \\ & & & U \end{bmatrix} = \begin{bmatrix} \Lambda & I & & \\ I & \Lambda & \ddots & \\ & \ddots & \ddots & I \\ & & I & \Lambda \end{bmatrix}.
$$

(c) *Show that there is a permutation P such that*

$$
P \begin{bmatrix} \Lambda & I & & \\ I & \Lambda & \ddots & \\ & \ddots & \ddots & I \\ & & I & \Lambda \end{bmatrix} P^\top = \begin{bmatrix} T_1 & & & \\ & T_2 & & \\ & & \ddots & \\ & & & T_m \end{bmatrix},
$$

where

$$
T_j = \begin{bmatrix} \lambda_j & 1 & & \\ 1 & \lambda_j & \ddots & \\ & \ddots & \ddots & 1 \\ & & 1 & \lambda_j \end{bmatrix}.
$$

Hint: *Instead of grouping the unknowns in the 2D domain horizontally, group them vertically.*

(d) *Diagonalize the T_j and show that their eigenvalues are all negative. Conclude that $-A$ is positive definite.*

PROBLEM 11.15. *For $n_x, n_y \in \mathbb{N}$ and $h \in \mathbb{R}^+$, with $n = n_x n_y$, we define the square matrix $A \in \mathbb{R}^{n \times n}$ representing the discretized Laplacian by (see* *also Section 11.1)*

$$
A = \frac{1}{h^2}
\begin{bmatrix}
-4 & 1 & & & 1 & & & & \\
1 & -4 & \ddots & & & 1 & & & \\
& \ddots & \ddots & 1 & & & \ddots & & \\
& & 1 & -4 & & & & 1 & \\
1 & & & & -4 & 1 & & & \ddots \\
& 1 & & & 1 & -4 & \ddots & & \\
& & \ddots & & & \ddots & \ddots & 1 & \\
& & & 1 & & & 1 & -4 & \\
& & & & \ddots & & & & \ddots
\end{bmatrix} .
$$

Note that the submatrices are of size $n_x \times n_x$. We say that a square matrix B has Property A (see Definition 11.8), if there exists a permutation matrix P such that

$$
P^{\mathsf{T}} B P = \begin{bmatrix} D_1 & F \\ E & D_2 \end{bmatrix},
$$

where D_1, D_2 are diagonal matrices. Show that the discrete Laplacian matrix A above has property A.

PROBLEM 11.16.

1. *Implement the methods of Jacobi and Gauss-Seidel in* MATLAB, *using as header*

```
[x,res]=Jacobi(A,b,x0,tol,maxiter)
% JACOBI Method of Jacobi to solve linear systems
%    [x,res]=Jacobi(A,b,x0,tol,maxiter) solves the system of
%    linear equations Ax=b using the method of Jacobi,
%    starting with the initial guess x0. The algorithm stops
%    when either maxiter number of iterations have been
%    performed, or the 2-norm of the relative residual
%    is smaller than tol. The result is returned in x, and
%    res contains the history of the relative residual norms.
```

and similarly for Gauss-Seidel and SOR.

2. *Apply your methods to the discretized Laplacian of Problem 11.15 with right hand side $\boldsymbol{b} = [1, 1, \ldots, 1]^{\mathsf{T}}$.*

3. *Test your codes on the discretized advection-diffusion operator B defined by*

$$
B = A + \alpha C,
$$

where $\alpha \in \mathbb{R}^+$ (for example $\alpha = 1$) and C is the discretized advection operator given by

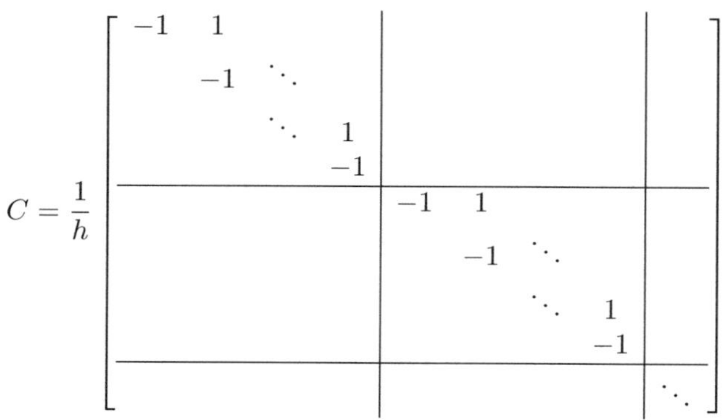

PROBLEM 11.17. *Show that if a matrix A has property A, then Gauss-Seidel converges two times as fast as Jacobi.* Hint: *Use from Theorem 11.11 the relation $(\lambda + \omega - 1)^2 = \lambda\omega^2\mu^2$, where λ is an eigenvalue of the iteration matrix of SOR, μ is an eigenvalue of the iteration matrix of Jacobi and ω is the relaxation parameter of SOR.*

PROBLEM 11.18.

1. *Show that the eigenvalues of the matrix A describing the discretized Laplacian in Problem 11.15 are*

$$\lambda_{i,j} = -\frac{4}{h^2}\left(\sin^2\left(\frac{i}{2(n_x+1)}\right)\pi\right) + \sin^2\left(\frac{j}{2(n_y+1)}\right)\pi\right)$$

   for the natural numbers $1 \le i \le n_x$, $1 \le j \le n_y$ when the matrix blocks are of size $n_x \times n_x$ and the whole matrix is of size $n_x n_y \times n_x n_y$ (n_x represents the number of lines in the grid and n_y the number of columns, and the nodes are enumerated column-wise).

2. *Compute the optimal parameter for Richardson's method applied to the matrix A.*

3. *Compute the optimal parameter for SOR applied to the matrix A.*

PROBLEM 11.19.

1. *Implement Richardson and SOR using the header*

```
function [x,res]=Richardson(A,b,alpha,x0,tol,maxiter)
% RICHARDSON Method of Richardson to solve linear systems
%    [x,res]=Richardson(A,b,x0,alpha,tol,maxiter) solves
```

```
%    the system of linear equations Ax=b using the method
%    of Richarson with relaxation parameter alpha, starting
%    with the initial guess x0. The algorithm stops when
%    either maxiter number of iterations have been performed,
$    or the 2-norm of the relative residual is smaller than
%    tol. The result is returned in x, and res contains the
%    history of the relative residual norms.
```

and similarly for SOR.

2. *Test Richardson and SOR on the discretized Laplacian using various parameters and the optimal ones.*

3. *Refine the mesh in order to obtain bigger and bigger matrices A and compare the convergence of Jacobi, Gauss-Seidel (see Problem 11.16) with Richardson and SOR using the optimal parameter. Represent the logarithm of the residual as a function of the logarithm of the mesh size.*

PROBLEM 11.20. *Show by differentiation that if $A \in \mathbb{R}^{n \times n}$ is a symmetric positive definite matrix, and $\boldsymbol{b} \in \mathbb{R}^n$, then solving the linear system $A\boldsymbol{x} = \boldsymbol{b}$ is equivalent to finding the minimum of the quadratic form $Q(\boldsymbol{x}) := \frac{1}{2}\boldsymbol{x}^\top A \boldsymbol{x} - \boldsymbol{x}^\top \boldsymbol{b}$.*

PROBLEM 11.21. *It is well known that*

$$
B = \begin{pmatrix} 2 & -1 & & \\ -1 & 2 & \ddots & \\ & \ddots & \ddots & -1 \\ & & -1 & 2 \end{pmatrix} \in \mathbb{R}^{m \times m}
$$

has the eigenvalues λ_i eigenvectors Q with

$$
q_{ij} = \sqrt{\frac{2}{m+1}} \sin \frac{ij\pi}{m+1}, \qquad \lambda_i = 2 - 2\cos\frac{i\pi}{m+1}. \tag{11.171}
$$

Generalize this result and relate the eigenvalues and eigenvectors of matrices of the form

$$
C = \begin{pmatrix} a & b & & \\ b & a & \ddots & \\ & \ddots & \ddots & b \\ & & b & a \end{pmatrix}
$$

to the eigenvalues and eigenvectors of the matrix B.

PROBLEM 11.22. *Solve the same equation as in Problem 11.7. You can reuse your experimental program for $n = 1000$. Perform again 100 Jacobi* *iteration steps and estimate the spectral radius ρ using $\boldsymbol{u}_{k+1} = M^{-1}N\boldsymbol{u}_k$ and $\rho \approx \|\boldsymbol{u}_{k+1}\|/\|\boldsymbol{u}_k\|$.*

Then switch to the semi-iterative method. Use the fact that the spectrum of the iteration matrix $M^{-1}N$ is real and symmetric so that you can use $\mu = 1/\rho$ and $\gamma = 1$ in Algorithm 2.3. Plot again the error history using semilogy.

Experiment with different values of ρ to get a feeling how sensitive the choice of this parameter is.

PROBLEM 11.23. *The linear system with the matrix*

$$A = \begin{pmatrix} 2 & -1 & & \\ -1 & 2 & \ddots & \\ & \ddots & \ddots & -1 \\ & & -1 & 2 \end{pmatrix} \in \mathbb{R}^{n \times n}$$

could be preconditioned with the matrix

$$M = \begin{pmatrix} 1 & -1 & & \\ -1 & 2 & \ddots & \\ & \ddots & \ddots & -1 \\ & & -1 & 2 \end{pmatrix} = FF^{\top} \quad with \quad F = \begin{pmatrix} 1 & & & \\ -1 & 1 & \ddots & \\ & \ddots & \ddots & \\ & & -1 & 1 \end{pmatrix}$$

Suppose you are solving the preconditioned system $M^{-1}A\boldsymbol{x} = M^{-1}\boldsymbol{b}$ with GMRES. How many iteration steps are necessary in the worst case? Justify your claim.

PROBLEM 11.24. *Change Algorithm 11.18 to*

```
[alpha,beta,v]=Lanczos22(A,v)
```

so that only two vectors for your computation are used, and do not store the matrix Q. Eliminate the if-statements in the loop to please modern processors!

PROBLEM 11.25. *Show that any matrix $A \in \mathbb{R}^{n \times n}$ can be transformed into upper Hessenberg form using a sequence of Householder reflections*

$$A = QHQ^{\top}, \qquad Q = P_1 \cdots P_n,$$

where $P_i = I - 2\boldsymbol{w}\boldsymbol{w}^{\top}$ with $\|\boldsymbol{w}\|_2 = 1$.

PROBLEM 11.26. *Let A be a real invertible $m \times m$ matrix, and let $\boldsymbol{b}$ be a real vector in $\mathbb{R}^m$. Suppose that for all $k \geq 1$ we have $\mathcal{K}_k := \operatorname{span}(\boldsymbol{b}, A\boldsymbol{b}, \ldots, A^{k-1}\boldsymbol{b})$. We define the algorithm of Arnoldi, see Subsection 11.7.2, by*

$$\boldsymbol{q}_1 = \boldsymbol{b}/\|\boldsymbol{b}\|$$

$$
\begin{aligned}
&\textit{for } n = 1, 2, 3, \ldots, \\
&\qquad \boldsymbol{v} = A\boldsymbol{q}_n \\
&\qquad \textit{for } j = 1, \ldots, n \\
&\qquad\qquad h_{j,n} = \boldsymbol{q}_j^{\mathrm{T}} \boldsymbol{v} \\
&\qquad\qquad \boldsymbol{v} = \boldsymbol{v} - h_{j,n}\boldsymbol{q}_j \\
&\qquad \textit{end;} \\
&\qquad h_{n+1,n} = \|\boldsymbol{v}\| \\
&\qquad \boldsymbol{q}_{n+1} = \boldsymbol{v}/h_{n+1,n} \\
&\textit{end;}
\end{aligned}
$$

We suppose there exists $n \leq m$ such that $h_{n+1,n} = 0$.

1. *Show that $A\boldsymbol{q_n}$ is in $\mathcal{K}_{n+1}$.*

2. *Deduce that $\mathcal{K}_{p+1} = \mathcal{K}_{n+1}$ for all integer $p \geq n$.*

3. *Using the concept of minimal polynomial of a matrix, show that the solution $\boldsymbol{x}$ of the system of linear equations $A\boldsymbol{x} = \boldsymbol{b}$ is in $\mathcal{K}_{n+1}$.*

PROBLEM 11.27.

1. *Construct an algorithm to compute the QR decomposition of an upper Hessenberg matrix H of size $k \times k$ using Givens rotations and only $O(k^2)$ operations.*

2. *Suppose that we know already the QR decomposition of the matrix given by the first $k - 1 \times k - 1$ block of H. How can this knowledge be used to compute the new QR decomposition of H in $O(k)$ operations?*

PROBLEM 11.28. *Implement Algorithm 11.19 and combine it with Algorithm 11.21, so that you get an implementation of the conjugate gradient algorithm to solve linear equations.*

PROBLEM 11.29. *GMRES searches for an approximate solution $\boldsymbol{x}_n$ of the linear system of equations $A\boldsymbol{x} = \boldsymbol{b}$ in the Krylov space $\mathcal{K}_n(A, \boldsymbol{b}) = \mathrm{span}(\boldsymbol{b}, \ldots, A^{n-1}\boldsymbol{b})$. We suppose that we know a good initial approximation $\boldsymbol{x}_0$ of the solution. How could one modify the problem $A\boldsymbol{x} = \boldsymbol{b}$ in order to incorporate this information in $\mathcal{K}_n$?*

PROBLEM 11.30. *We want to apply GMRES to linear systems of equations $A\boldsymbol{x} = \boldsymbol{b}$ with system matrices of the form*

$$
A = \begin{bmatrix} I & B \\ 0 & I \end{bmatrix}.
$$

In how many iterations at most will GMRES converge, independently of the choice of right hand side $\boldsymbol{b}$?

PROBLEM 11.31. *Implement the Arnoldi Algorithm 11.17 and the symmetric Lanczos Algorithm 11.18 in* MATLAB, *and test them on the matrix obtained from discretizing the Laplacian on an L-shaped region,*

```
>> G=numgrid('L',14);
>> A=delsq(G);
```

Compute the first few eigenvectors v, *which model the lowest few vibration modes of a drum, and visualize them using*

```
>> u=G;
>> u(G>0)=v(G(G>0));
```

and then the commands **surfc** *or* **mesh** *on* **u**. *Try also other region shapes, type* **help numgrid** *for more information.*

PROBLEM 11.32. *a) Implement GMRES in* MATLAB *using the header*

```
function x=GMRES(A,b,tol)
% GMRES approximate solutions of linear systems of equations
%   x=GMRES(A,b,tol) solves approximately the linear system
%   of equations Ax=b by computing approximations x in the
%   Krylov subspace spanned by the vectors (b,...,A^(k-1)b)
%   such that norm(b-A*x,2) is minimized. The algorithms
%   stops when this norm is smaller than tol, or after
%   size(A,1) iterations.
```

b) Test GMRES applied to the discretized Laplacian (see Problem 11.31) with right hand side $b = [1, \ldots, 1]$. *Compare the convergence speed with Jacobi, Gauss-Seidel and SOR from Problem 11.16 using a* **semilogy** *plot with the number of iterations on the x axis and the norm of the residuals on the y axis.*

c) Verify numerically your answer in Problem 11.23.

PROBLEM 11.33. *Implement the second variant of the topological ε-algorithm TEA2,*

ALGORITHM 11.29. *Topological ε-Algorithm TEA2*

Choose an arbitrary vector q *and set*

$$\varepsilon_{-1}^{(n)} = 0, \quad \varepsilon_0^{(n)} = x_n, \quad n = 0, 1, 2, \ldots$$

$$\varepsilon_{2m+1}^{(n)} = \varepsilon_{2m-1}^{(n+1)} + \frac{q}{q^\top \Delta \varepsilon_{2m}^{(n)}}$$

$$\varepsilon_{2m+2}^{(n)} = \varepsilon_{2m}^{(n+1)} + \frac{\Delta \varepsilon_{2m}^{(n+1)}}{(\Delta \varepsilon_{2m+1}^{(n)})^\top \Delta \varepsilon_{2m}^{(n+1)}} \qquad m, n = 0, 1 \ldots$$

---

using as a model the MATLAB *implementation of TEA1, see Algorithm 11.14.*

PROBLEM 11.34. *Show that the direction vector $\boldsymbol{p}_k$ determined by the conjugate gradient method is the residual vector of the least squares problem*

$$AP_{k-1}\boldsymbol{z} \approx \boldsymbol{r}_{k-1}.$$

Chapter 12. Optimization

Optimization problems are ubiquitous in science and engineering, and even in our daily lives, when we try to optimize our way to go to work, when we choose the line we stand in at the supermarket, or when we make decisions on the education of our children. As mentioned in the first quote above, men and nature love to optimize! In a mathematical formulation, optimization problems always consist of a scalar objective function, which should be minimized or maximized over a set of parameters, which might be constrained (see the second quote above). We begin this chapter in Section 12.1 with several simple examples, which show the breadth of problems that fall into the category of optimization problems, and give a classification of optimization problems, explaining which types of problems will be treated in this chapter, and which ones will not. We then present the classical mathematical treatment of optimization problems in Section 12.2 and give the necessary and sufficient conditions for a point to be a local optimum. Constraints are included using Lagrange multipliers, and we will also see the well-known Karush-Kuhn-Tucker conditions. Section 12.3 is then dedicated to the numerical solution of unconstrained optimization problems.

W. Gander et al., *Scientific Computing - An Introduction using Maple and MATLAB*,
Texts in Computational Science and Engineering 11,
DOI 10.1007/978-3-319-04325-8_12,
© Springer International Publishing Switzerland 2014

We introduce and present elementary convergence proofs for two widely-used methods, namely line search and trust region methods, both of which use derivative information from the objective function. We also explain the very popular direct method of Nelder–Mead, which does not use any derivative information, but instead uses a discrete set of rules to move a simplex along the function to be minimized. In Section 12.4, we present methods for constrained optimization. We start with the famous simplex algorithm for linear programming due to Dantzig in 1951, and then present for the general case methods based on penalty and barrier functions, interior point methods, and also sequential quadratic optimization. A substantial part of this chapter was greatly inspired by the excellent lecture notes of Gould and Leyffer from the Durham summer school 2002, see [57] and the first quote above. Another good standard reference is the book by Gill, Murray and Wright [49].

12.1 Introductory Examples

Optimization problems come in many different forms. We show in this section several examples, and also give a characterization of optimization problems we will treat in this chapter, and of problems that will be left out.

12.1.1 How much daily exercise is optimal ?

In his John von Neumann lecture at the annual SIAM meeting, Joe Keller asked the following question and proposed a very simple model to answer it: suppose at birth, every human being is given a fixed number of heartbeats, and once these heartbeats are used up, life ends. How should one optimally use these heartbeats to have as long a life as possible? A first immediate idea is to stay in bed and rest, so the heart rate stays low, and one uses the heartbeats as economically as possible. Another idea, however, comes from the fact that a well-trained heart beats much more slowly when the person is at rest than the heart of an untrained person. So exercise could increase the lifespan. Unfortunately, during exercise, the heart beats faster, so that one uses up the heartbeats faster, in the hope to gain them back during rest. So is there an optimum?

Suppose that the untrained heart beats 80 times a minute when a person is at rest, and that during exercise, it beats 120 times per minute. If a person spends a fraction x of his time exercising, then this person uses on average

$$f(x) := 120x + g(x)(1 - x)$$

heartbeats per minute, where the unknown function g should be close to 80 for x small, meaning the person hardly does any exercise, and probably around 50 for x approaching 1, when the person is extremely well trained. Since it is known that a little exercise every day decreases the resting heart rate considerably, a simple model for g would be exponential decay, i.e.

$$g(x) := 50 + 30e^{-100x},$$

where the choice -100 is quite arbitrary here, and should be much more carefully researched with the help of a medical doctor. Figure 12.1 shows the function f for this example, and there is clearly a minimum, which means

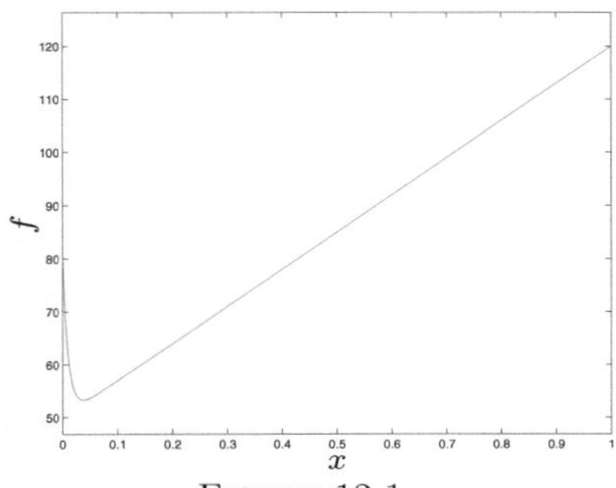

FIGURE 12.1.
*A simple model of the average heart beat of a person
which exercises the fraction x of time.*

there is an optimal choice of x that minimizes the average use of heartbeats, and which leads to the longest life possible. From calculus, we know that we need to set the derivative to zero to find the minimum, which with MAPLE is easily achieved by

```
> f:=120*x+(50+30*exp(-100*x))*(1-x);
> fp:=diff(f,x);
> solve(fp,x);
```

$$-\frac{1}{100}\,\mathrm{LambertW}\left(\frac{7}{3}\,e^{101}\right) + \frac{101}{100}$$

It is interesting to see that the closed-form solution of this problem involves the *LambertW function* we have already encountered in Chapter 5 on the numerical solution of nonlinear equations. To obtain a numerical value in MAPLE , we type

```
> evalf(%);
```

and MAPLE returns 0.0373019079, which means that one should exercise a bit over 50 minutes every day. If there had been no closed-form solution in MAPLE , one could have used any of the nonlinear equation solvers from Chapter 5 to find the solution, or the MAPLE command `fsolve`.

Is it also possible to find the minimum directly without knowing the derivative? For this one-dimensional problem, an algorithm like Algorithm 5.2 (Bisection) in Chapter 5 would be nice, but it is not possible to tell from the midpoint of an initial interval $[a, b]$ if the minimum lies in the left or right half of the interval. However, if one computes the function value for two distinct points x_1 and x_2 in $[a, b]$, $x_1 < x_2$, and if $f(x_1)$ is smaller than $f(x_2)$, as in the example in Figure 12.2, then a minimum must lie in the interval

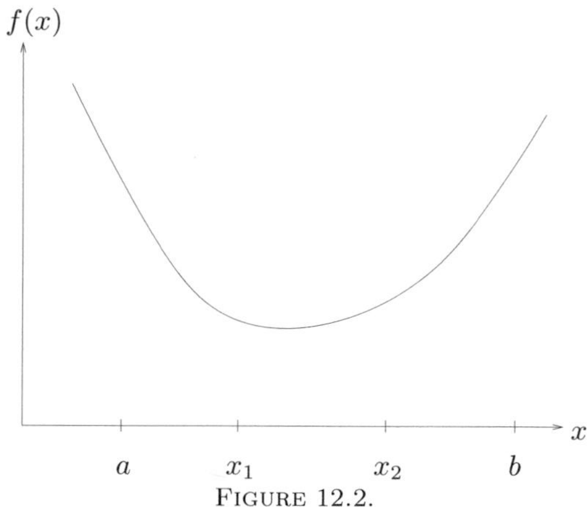

FIGURE 12.2.

Example of a minimum search algorithm similar to bisection, using two points x_1 and x_2 in the search interval $[a, b]$.

$[a, x_2]$. On the other hand, if $f(x_1)$ is bigger than $f(x_2)$, then a minimum must lie in the interval $[x_1, b]$, and hence one can continue to search for the minimum in a smaller interval, as in bisection. A simple choice for x_1 and x_2 is to choose one third and two thirds of the interval $[a, b]$, but this choice has the big disadvantage that one needs to evaluate f twice in each step of the algorithm, since in the new interval, neither x_1 in the former, nor x_2 in the latter case lies at one third or two thirds of the new interval. A better strategy is to set

$$x_1 = \lambda a + (1 - \lambda)b, \quad x_2 = (1 - \lambda)a + \lambda b, \tag{12.1}$$

and to choose λ that allows one to reuse function values that have already been computed. This means that in the former case, the old x_1 would need to become the new x_2 in the interval $[a, x_2]$, i.e.

$$x_1 = (1 - \lambda)a + \lambda x_2.$$

Substituting the values from equation (12.1) and solving for λ yields

$$\lambda = \frac{-1 + \sqrt{5}}{2} \approx 0.6180, \tag{12.2}$$

where we have chosen the positive root. In the latter case, the posibility of reuse implies

$$x_2 = \lambda x_1 + (1 - \lambda)b,$$

which leads after substitution from equation (12.1) to the same value of λ. This choice of λ corresponds to the *golden section* (see also Chapter 5), and leads to the following *golden section minimization* algorithm in MATLAB:

ALGORITHM 12.1. *Golden Section Minimization*

```
function x=Minimize(f,a,b)
% MINIMIZE finds a minimum of a scalar function
%   x=Minimize(f,a,b) searches for a minimum of the scalar function f
%   in the interval [a,b]

fa=f(a); fb=f(b);
l=(-1+sqrt(5))/2;
x1=l*a+(1-l)*b; x2=(1-l)*a+l*b;
fx1=f(x1); fx2=f(x2);
while a<x1 & x1<x2 & x2<b
  if fx1>fx2
    a=x1;
    x1=x2; fx1=fx2;
    x2=(1-l)*a+l*b; fx2=f(x2);
  else
    b=x2;
    x2=x1; fx2=fx1;
    x1=l*a+(1-l)*b; fx1=f(x1);
  end;
end;
x=x1;
```

Running the algorithm above on the daily exercise example leads to the same result as before

```
>> f=@(x) 120*x+(50+30*exp(-100*x))*(1-x);
>> x=Minimize(f,0,1)
x =
   0.037301908030998
```

12.1.2 Mobile Phone Networks

Mobile phone systems can be used almost everywhere because there is a dense network of base stations to which the mobile phone can connect. Each base station covers a certain neighborhood, and when the mobile phone moves from one neighborhood into another during a call, the call is automatically transferred from the old station to the new one. The dense covering of areas with base stations causes problems: if two base stations use the same frequency to connect to two different mobile phones simultaneously, the signals may interfere and the quality of the communication deteriorates. This is the case in particular if the two base stations are physically close to each other. Therefore, one would like to assign different frequencies to different base stations. However, in practice, there are fewer frequencies available than there are base stations, so the frequencies have to be reused. A natural question is how one should assign frequencies to base stations so that interference among simultaneous calls is minimized. This problem is known as the *frequency assignment problem*, see for example [36].

We derive a mathematical formulation of the frequency assignment problem: consider n base stations S_i, $i = 1, 2, \ldots, n$. We first assume that each

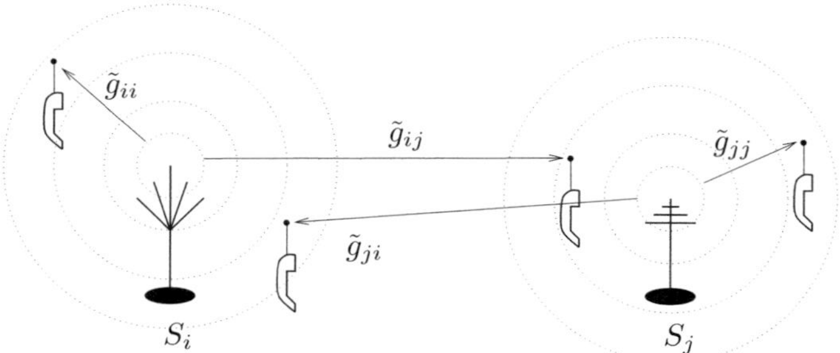

FIGURE 12.3.
*Two base stations and positions of worst reception and
interference*

station S_i sends signals with the same frequency and at a power level p_i. To describe how the power level affects interference, we consider pairs of base stations S_i and S_j, as shown in Figure 12.3. We define the entries of the link gain matrix $\tilde{G} = [\tilde{g}_{ij}]$ as follows:

- $\tilde{g}_{ii}p_i$ is the minimum signal power received by a phone when connected to the station S_i (sending with power p_i) within the area of S_i.

- $\tilde{g}_{ij}p_j$ is the maximum interference power received by a phone within the area of station S_i that is emitted by the station S_j (sending with power p_j).

For simplicity, we neglect natural noise from other sources. The worst possible signal-to-noise ratio q_i of a phone connected to station S_i can now be defined as the ratio between the minimum signal power from the station S_i and the sum of the interference powers $\tilde{g}_{ij}p_j$ of the other stations S_j, $j \neq i$,

$$q_i := \frac{\tilde{g}_{ii}p_i}{\sum_{j \neq i} \tilde{g}_{ij}p_j} = \frac{p_i}{\sum_{j \neq i} \frac{\tilde{g}_{ij}}{\tilde{g}_{ii}}p_j} = \frac{p_i}{\left(\sum_j \frac{\tilde{g}_{ij}}{\tilde{g}_{ii}}p_j\right) - p_i}. \tag{12.3}$$

Defining the normalized link gain matrix $G = [g_{ij}]$ by

$$g_{ij} := \begin{cases} \frac{\tilde{g}_{ij}}{\tilde{g}_{ii}} & i \neq j, \\ 0 & \text{otherwise,} \end{cases} \tag{12.4}$$

the worst signal-to-noise ratio from (12.3) becomes

$$q_i = \frac{p_i}{\sum_j g_{ij}p_j}. \tag{12.5}$$

In the simple example with one frequency, we can only choose the sending power p_i at each station S_i to maximize the smallest signal-to-noise ratio q_i over all stations. The optimum is obtained when the signal-to-noise ratio is the same for all stations, as one can see as follows: starting with all the q_i equal and decreasing the sending power p_j of station S_j, we necessarily decrease the signal-to-noise ratio q_j since $g_{jj} = 0$. On the other hand, increasing the power p_j of station S_j would necessarily decrease the signal-to-noise ratio of coupled neighboring stations, because $g_{jk} \geq 0$.

Since the matrix G is non-negative, we have by the *Perron–Frobenius Theorem* 11.4 that the eigenvalue with largest modulus of G, $\lambda_{\max}$, is real positive, and the corresponding eigenvector $\phi_{\max}$ is non-negative. Using this eigenvector as our power distribution, $\boldsymbol{p} := \phi_{\max}$, we obtain for the signal-to-noise ratio

$$q_i = \frac{p_i}{\sum_j g_{ij} p_j} = \frac{p_i}{\lambda_{\max} p_i} = \frac{1}{\lambda_{\max}}, \tag{12.6}$$

and thus every station can guarantee at least a signal to noise ratio of $\frac{1}{\lambda_{\max}}$, the best possible solution with one frequency. If several frequencies are available, the goal is to collect stations in groups so that for each group $\frac{1}{\lambda_{\max}}$ is maximized, or in other words, the largest eigenvalue of the link gain matrix G for each group is minimized.

Suppose we are given a normalized link gain matrix $G \in \mathbb{R}^{n \times n}$ which has zeros on the diagonal and non-negative entries otherwise,

$$G = \begin{pmatrix} 0 & g_{12} & g_{13} & \cdots & g_{1n} \\ g_{21} & 0 & g_{23} & \cdots & g_{2n} \\ \vdots & & \ddots & & \vdots \\ g_{n1} & g_{n2} & \cdots & \cdots & 0 \end{pmatrix}, \quad g_{ij} \geq 0, 1 \leq i, j \leq n,$$

and we have two frequencies to be assigned. In this case, we need to find a decomposition of a permutation Π of the matrix G into four subblocks,

$$\Pi^{\top} G \Pi = \begin{pmatrix} A & B \\ C & D \end{pmatrix}, \quad \begin{array}{ll} A \in \mathbb{R}^{p \times p}, & B \in \mathbb{R}^{p \times (n-p)}, \\ C \in \mathbb{R}^{(n-p) \times p}, & D \in \mathbb{R}^{(n-p) \times (n-p)}, \end{array}$$

such that the maximum of the largest eigenvalues of the submatrices on the diagonal is minimized,

$$\min_{\Pi, p} (\max(\lambda_{\max}(A), \lambda_{\max}(D))). \tag{12.7}$$

An exhaustive search to solve this minimax problem is performed by the MATLAB function

---

ALGORITHM 12.2.
Exhaustive Search for Best Partition of Link Gain
Matrix G

```
function s=BruteForcePartition(G);
% BRUTEFORCEPARTITION find best two-frequency partition
%   s=BruteForcePartition(G); finds for a given link gain matrix G the
%   best partition into four submatrices, such that the maximum of the
%   largest eigenvalues of the two diagonal blocks is minimized.  The
%   result is given by the binary index s for one of the submatrices.

n=size(G,1);
for k=1:2^(n-1)-1
  s=logical(double(dec2bin(k,n))-'0');
  r(k)=max([max(eig(G(s,s))) max(eig(G(~s,~s)))]);
end
[rmin,k]=min(r);
disp(rmin)
s=logical(double(dec2bin(k,n))-'0');
```

To test this function, we need various configurations of base stations. Since base stations are usually intelligently distributed, we use for testing purposes base stations on a rectangular grid with random perturbations, as in the MATLAB code

---

ALGORITHM 12.3. *Generate Link Gain Matrix*

```
function [A,x,y]=GenerateProblem(n,m,de);
% GENERATEPROBLEM generates organized random set of base stations
%   [A,x,y]=GenerateProblem(n,m,de); generates n x m base station
%   locations on a rectangular unitary grid, each at a random location
%   in a square of side 2*de centered at the corresponding gridpoint,
%   and computes the corresponding link gain matrix using the fact
%   that the signal decays like 1/r^3.  The output is the link gain
%   matrix A and the coordinates of the base stations (x,y).

[X,Y]=meshgrid(1:n,1:m);
X=X+de*(rand(size(X))-1/2);
Y=Y+de*(rand(size(Y))-1/2);
x=X(:);
y=Y(:);
for i=1:n*m-1,
  for j=i+1:n*m,
    A(i,j)=1/(sqrt((x(i)-x(j))^2+(y(i)-y(j))^2)^3);
    A(j,i)=A(i,j);
  end;
end;
plot(x,y,'o','MarkerSize',12,'LineWidth',2)
```

We show on the left of Figure 12.4 the result of the commands

```
>> [A,x,y]=GenerateProblem(4,3,0.3);
>> s=BruteForcePartition(A);
>> plot(x(s),y(s),'*',x(~s),y(~s),'o');
```

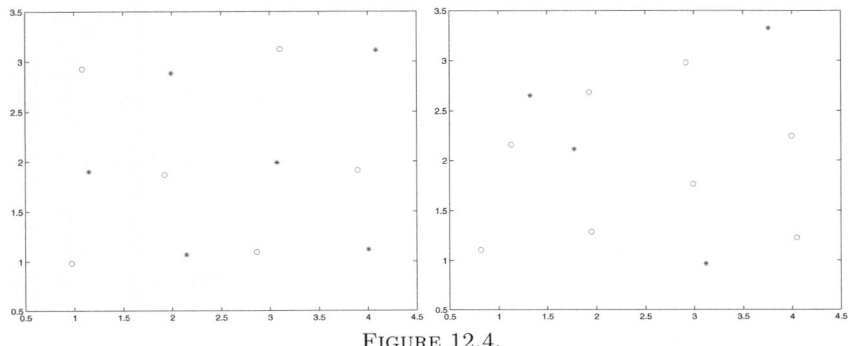

$$\text{FIGURE } 12.4.$$

*Two configurations of base stations, and the best
two-frequency assignment found by exhaustive search.*

The minimum of the largest eigenvalue was found to be 1.2026, and one can
see intuitively why this solution is optimal: frequencies are assigned in order
to separate the base stations' regions of influence. On the right of Figure
12.4, we show the result when the randomness parameter has been increased
to 0.7. In this case, the minimum of the largest eigenvalue was found to be
2.9017, and it is already less evident why this partition should be optimal.

The exhaustive search is currently feasible for matrices of size up to $20 \times$
20, for which one has to try 2^{19} configurations. Since the matrices arising
in radio communication problems are on the order of 100×100, a feasible
strategy to compute approximations to the best solution is needed.

To do so, we first relate (12.7) to a problem involving norms: suppose
the normalized link gain matrix G is symmetric, which would hold in many
situations. Then G is diagonalizable, $G = Q^\top \Lambda Q$, where Q is orthogonal and
Λ is diagonal with the eigenvalues of G on the diagonal. Thus

$$\|G\|_2 = \|Q^\top \Lambda Q\|_2 = \|\Lambda\|_2 = \lambda_{\max}(G)$$

and the min-max problem in (12.7) can be written equivalently using the
spectral norm,

$$\min_{\Pi,p}(\max(\|A\|_2, \|D\|_2)). \tag{12.8}$$

Often the spectral norm is ideal for minimization problems, but in this case
different norms are easier to handle. One could for example approximately
minimize the *one* or *infinity norm*,

$$\|A\|_1 = \max_j \sum_i |a_{ij}|, \qquad \|A\|_\infty = \max_i \sum_j |a_{ij}|,$$

or the *Frobenius norm,*

$$\|A\|_F = \sqrt{\sum_{i,j} |a_{ij}|^2}.$$

The physical intuition behind the use of different norms is the following: If we manage to keep only weak links (small link gain) in the sub-matrices A and D and put strong links (large link gain) into the off-diagonal blocks B and C, then the strong links lose their importance, because they link stations with different frequencies. So physically minimizing norms of A and D is still meaningful, even in the non-symmetric case.

Suppose we want to minimize the Frobenius norm of the two sub-blocks A and D,

$$\min_{\Pi,p}(\max(\|A\|_F^2, \|D\|_F^2)). \tag{12.9}$$

Let $\boldsymbol{p} \in \mathbb{R}^n$ be a partitioning vector, $p_i \in \{-1, 1\}$. Then the sum of the Frobenius norms of A and D can be written as

$$\begin{aligned}
\|A\|_F^2 + \|D\|_F^2 &= \frac{1}{4} \sum_{i,j} g_{ij}^2 (p_i + p_j)^2 \\
&= \frac{1}{4} \sum_{i,j} g_{ij}^2 (p_i^2 + p_j^2) + \frac{1}{2} \sum_{i,j} g_{ij}^2 p_i p_j \\
&= \frac{1}{2}(\|G\|_F^2 + \boldsymbol{p}^\top (G \odot G)\boldsymbol{p}),
\end{aligned}$$

where the *Hadamard product* $G \odot G$ denotes the component-wise product, i.e. in our case the matrix containing the elements of G squared (computed in MATLAB by G.^ 2). Since the term $\|G\|_F^2$ is independent of $\boldsymbol{p}$, minimizing the sum of the two Frobenius norms $\|A\|_F^2 + \|D\|_F^2$ is equivalent to minimizing

$$\min_{p_i \in \{-1,1\}} \boldsymbol{p}^\top (G \odot G)\boldsymbol{p}. \tag{12.10}$$

This formulation corresponds to the *spectral bisection problem* in *graph partitioning,* which has been shown to be NP-hard in the literature.

To find an approximation to the solution of (12.10), one can relax the constraint on $\boldsymbol{p}$ to contain only elements $+1$ or -1 and allows arbitrary $\boldsymbol{p}$. Thus the minimum is obtained for $\boldsymbol{p}$ being the eigenvector of $G \odot G$ associated with the smallest eigenvalue $\lambda_{min}(G \odot G)$. Such a search is performed by the MATLAB code

---

ALGORITHM 12.4.
Best Partition of Link Gain Matrix G using Eigenvector
of $G \odot G$

```
function s=SpectralPartition(G);
% SPECTRALPARTITION find approximate two frequency partition
```

```
%    s=SpectralPartition(G); finds for a given link gain matrix G an
%    approximation to the best partition into four submatrices, such
%    that the maximum of the largest eigenvalues of the two diagonal
%    blocks is minimized.  The result is given by the binary index s
%    for one of the submatrices.

n=size(G,1);
G2=G.^2;
[V,E]=eig(G2);
[dummy,k]=min(diag(E));
[v,id]=sort(V(:,k));
for p=1:n-1
  r(p)=max([max(eig(G(id(1:p),id(1:p))))...
       max(eig(G(id(p+1:n),id(p+1:n))))]);
end
[rmin,p]=min(r);
disp(rmin)
s=zeros(1,n);
s(id(1:p))=1;
s=logical(s);
```

It remains to assign frequencies to individual stations based on the values in the vector $\boldsymbol{p}$. A simple way is to assign the same frequency to all stations i for which p_i has the same sign. A more elaborate approach is used in the algorithm above: the link gain matrix G is first permuted symmetrically in increasing order of p_i. We then try the $n - 1$ possible partitions obtained by grouping stations $\{1, \ldots, k\}$ for $k = 1, \ldots, n - 1$. The partition that gives the smallest eigenvalues is chosen for the assignment. Similar techniques are employed for the graph coloring problem, see [4] and [6] for details.

Using the same configuration as in Figure 12.4, the spectral approximate optimization yields the results shown in Figure 12.5. On the left, with the randomness parameter equal to 0.3, the approximate minimum of the largest eigenvalue was found to be 1.2026, and is identical to the result of the exhaustive search. On the right in Figure 12.5, we show the results when the randomness parameter has been increased to 0.7, and now the approximate minimum of the largest eigenvalue was found to be 3.0173, which is not the optimal partition found by the exhaustive search in Figure 12.4 on the right. On the other hand, with the approximate optimization, we can find approximate solutions to much larger problems. The commands

```
>> [A,x,y]=GenerateProblem(6,6,0.3);
>> s=SpectralPartition(A);
>> plot(x(s),y(s),'*',x(~s),y(~s),'o');
```

lead to the result in Figure 12.6 on the left, and on the right with the increased randomness parameter 0.7.

There are many other techniques that can be applied to solve the frequency assignment problem approximately, see for example [66] for the use of genetic search algorithms.

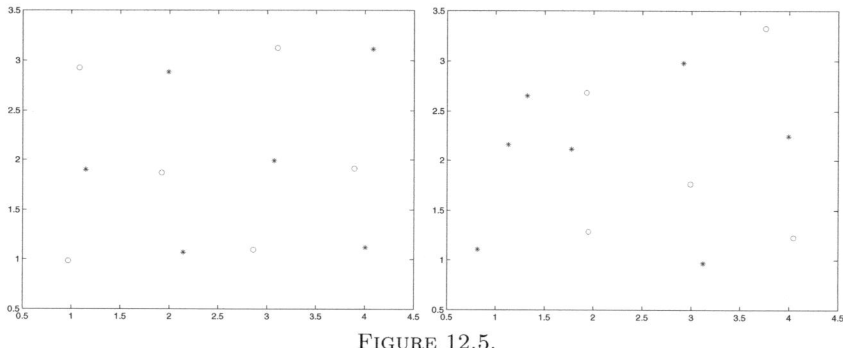

FIGURE 12.5.
*Two configurations of base stations, and the best two
frequency assignment found by approximate spectral
search.*

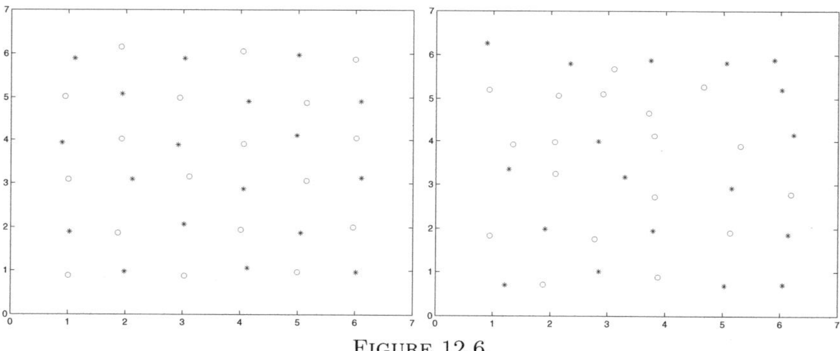

FIGURE 12.6.
*Two larger configurations of base stations, and the best
two frequency assignment found by approximate spectral
optimization.*

12.1.3 A Problem from Operations Research

A construction company specialized in transport has two offices at different locations, O_1 and O_2, with 8 trucks available at O_1 and 6 trucks available at O_2. The company is currently servicing two construction sites C_1 and C_2. The site C_1 needs 4 trucks to be operational, and the site C_2 needs 7 trucks. The distances from office O_1 to construction site C_1 is 8 kilometers, and to construction site C_2 it is 9 kilometers. The distances from office O_2 to construction site C_1 is 3 kilometers, and to construction site C_2 it is 5 kilometers, as illustrated in Figure 12.7. The construction company would like to minimize fuel cost and hence send trucks to the construction sites such that the overall distance the trucks have to travel is minimized. If we denote by x_1 the number of trucks sent from O_1 to C_1, by x_2 the number of trucks

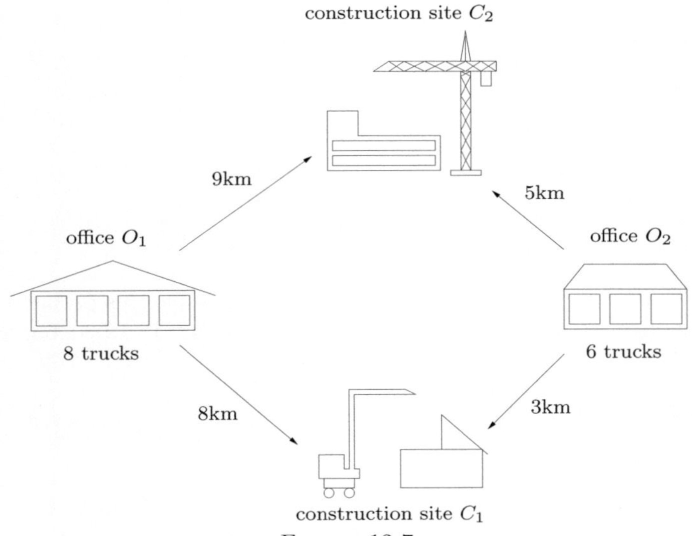

FIGURE 12.7.
A construction company specialized in transport wants
to optimize the use of their trucks.

sent from O_1 to C_2, by x_3 the number of trucks sent from O_2 to C_1, by x_4 the number of trucks sent from O_2 to C_2, the company needs to make a choice satisfying the following conditions:

$$
\begin{aligned}
x_1 + x_2 &\leq 8 && \text{no more than 8 trucks at } O_1, \\
x_3 + x_4 &\leq 6 && \text{no more than 6 trucks at } O_2, \\
x_1 + x_3 &= 4 && \text{need 4 trucks at } C_1, \\
x_2 + x_4 &= 7 && \text{need 7 trucks at } C_2, \\
x_i &\geq 0 && \text{number of trucks need to be non-negative.}
\end{aligned}
\tag{12.11}
$$

In addition, the choice of x_i, $i = 1, 2, 3, 4$ needs to be such that the fuel use is minimized, i.e. the objective function

$$8x_1 + 9x_2 + 3x_3 + 5x_4 \longrightarrow \min.$$

To simplify this problem, one can first eliminate the variables x_3 and x_4 using the equalities in (12.11),

$$x_3 = 4 - x_1, \qquad x_4 = 7 - x_2,$$

which leads to the simpler optimization problem

$$
\begin{aligned}
5x_1 + 4x_2 + 47 &\longrightarrow \min \\
x_1 + x_2 &\le 8 \\
-x_1 - x_2 &\le -5 \\
x_1 &\le 4 \\
x_2 &\le 7 \\
x_1 &\ge 0 \\
x_2 &\ge 0.
\end{aligned}
\tag{12.12}
$$

Since we have now only two variables left, one can solve this problem graphically, as shown in Figure 12.8. The inequalities in (12.12) determine the

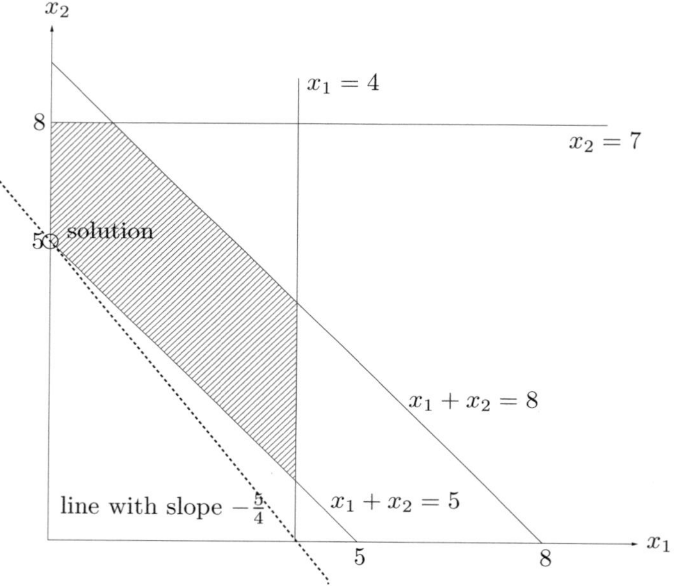

FIGURE 12.8.
Graphical solution of the operations research problem of the construction company.

region shown in Figure 12.8, which contains all admissible solutions of the optimization problem. Minimizing $5x_1 + 4x_2 + 47$ is equivalent to minimizing the function $f(x_1, x_2) = 5x_1 + 4x_2$ over the region of admissible solutions. The geometric interpretation of f is a plane that intersects the x_1-x_2 plane along the line $5x_1 + 4x_2 = 0$. Minimizing f means looking for the lowest point on the plane in the admissible region. This is equivalent to finding the point in the admissible region which has the smallest distance from the line $5x_1 + 4x_2 = 0$. By moving this line with slope $-\frac{5}{4}$ parallel till it hits the admissible region, we find the solution $x_1 = 0$ and $x_2 = 5$, which implies

$x_3 = 4$ and $x_4 = 2$. This solution leads to 4 trucks at construction site C_1 and 7 trucks at construction site C_2, as required, and the trucks drive the overall distance

$$8x_1 + 9x_2 + 3x_3 + 5x_4 = 67\text{km},$$

which is the smallest distance possible. The solution is interesting, since 5 trucks drive the longest distance in the network at the optimal solution which minimizes the overall distance, whereas an intuitive, less carefully chosen solution could have sent only one truck for the longest distance, which would have led to a worse overall solution of 71 km.

12.1.4 Classification of Optimization Problems

In general, an *optimization problem* at the mathematical level is given by an objective function $f : K \longrightarrow \mathbb{R}$, and one searches for $x^* \in K$, such that $f(x^*) \leq f(x)$ for all $x \in K$. Often, K is a subset of $\mathbb{R}^n$ defined by *constraints*, and the optimization problem is formulated as

$$
\begin{aligned}
f(x) &\longrightarrow & \min \quad & x \in \mathbb{R}^n, \\
c_j(x) &\geq & 0 \quad & j \in I, \\
c_j(x) &= & 0 \quad & j \in E,
\end{aligned}
\tag{12.13}
$$

where I is an index set denoting the inequality constraints, and E is an index set denoting the equality constraints of the problem, whose sizes we denote by $m_I = |I|$ and $m_E = |E|$, and $c_j : \mathbb{R}^n \longrightarrow \mathbb{R}$ are the constraints.

One can roughly divide *optimization problems* into the following categories:

1. Optimization problems without constraints: $m_I = m_E = 0$

    (a) Quadratic optimization problems: the objective function f is quadratic, i.e. $f(x) = x^\top A x - b^\top x$, where $A \in \mathbb{R}^{n \times n}$ is a symmetric matrix. Such problems require the solution of a linear system, as shown in Chapters 3, 11, and the conjugate gradient algorithm can be conveniently used for its solution.

    (b) General nonlinear optimization problems without constraints: f is neither quadratic nor linear.

2. Optimization problems with linear constraints: the functions c_j are affine.

    (a) Problems with equality constraints only, $m_I = 0$.

    (b) Problems with inequality constraints:

        i. Linear programming, if f is linear.

        ii. Quadratic-linear optimization problems, if f is quadratic.

        iii. Nonlinear optimization problems with linear constraints, if f is neither linear nor quadratic.

   (c) Nonlinear optimization problems:

      i. Nonlinear optimization problems with equality constraints.

      ii. General nonlinear optimization problems.

3. Optimal control problems, where x is a function of one or several parameters.

4. Combinatorial optimization problems, where the set $K \subset \mathbb{R}^n$ is discrete, or even finite.

We will not address the combinatorial optimization problems beyond the example in Section 12.1.2 on mobile phone networks. The techniques used for such problems are substantially different from the techniques used for all the other cases. Optimal control problems, however, can be handled by the techniques of this chapter after discretization.

12.2   Mathematical Optimization

DEFINITION 12.1. (LOCAL AND GLOBAL MINIMA) *A function* $f : K \longrightarrow \mathbb{R}$ *with* $K \subset \mathbb{R}^n$ *has a* local minimum *at the point* $x^* \in K$ *if there exists a neighborhood* U *of* x^* *such that*

$$f(x^*) \leq f(x), \quad \forall x \in K \cap U. \tag{12.14}$$

The function f *has a* global minimum *at the point* $x^* \in K$ *if*

$$f(x^*) \leq f(x), \quad \forall x \in K. \tag{12.15}$$

We say that the minimum is strict if we have $f(x^*) < f(x)$ *for all* $x \neq x^*$ *in (12.14) or (12.15).*

Note that to obtain results for maxima, it suffices to simply change the sign and replace f by $-f$.

Definitions (12.14) and (12.15) are not very useful in practice, so the topic of this section is to establish other, more easily verifiable conditions that are necessary or sufficient for the occurrence of local minima. Only in rare cases can one make a statement about global minima.

12.2.1   Local Minima

We start with a simple one-dimensional example: let $f : \mathbb{R} \longrightarrow \mathbb{R}$ be given by $f(x) = \frac{1}{2}x^5 - \frac{3}{40}x^4 - \frac{3}{5}x^3 + \frac{3}{10}$, a function shown in Figure 12.9, obtained by the MAPLE commands

```
> f:=1/2*x^5-3/40*x^4-3/5*x^3+3/10;
> plot(f,x=-1.1..1.2,axes=boxed);
```

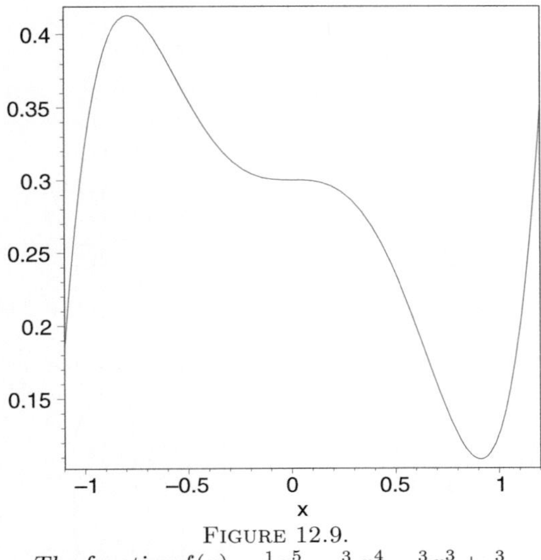

FIGURE 12.9.
The function $f(x) = \frac{1}{2}x^5 - \frac{3}{40}x^4 - \frac{3}{5}x^3 + \frac{3}{10}$.

Clearly the function has a maximum on the left, and a minimum on the right. We know from calculus that if f is twice continuously differentiable, then

$$\left\{ \begin{array}{rcl} f'(x^*) & = & 0 \\ f''(x^*) & > & 0 \end{array} \right\} \Longrightarrow \left\{ \begin{array}{c} f \text{ has a local} \\ \text{minimum at } x^* \end{array} \right\} \Longrightarrow \left\{ \begin{array}{rcl} f'(x^*) & = & 0 \\ f''(x^*) & \geq & 0 \end{array} \right\}.$$

$$(12.16)$$

The conditions on the left are sufficient conditions for f to have a minimum, and the conditions on the right are necessary conditions for f to have a minimum: if they are not satisfied, there cannot be a minimum. For the example, we can verify with the MAPLE commands

```
> fp:=diff(f,x);
> fpp:=diff(f,x,x);
> sols:=solve(fp,x);
```

$$sols := 0,\, 0,\, \frac{3}{50} + \frac{3\sqrt{201}}{50},\, \frac{3}{50} - \frac{3\sqrt{201}}{50}$$

```
> x:=sols[1];
> evalf([f,fp,fpp]);
```

$$[.3000000000,\, 0.,\, 0.]$$

```
> x:=sols[3];
> evalf([f,fp,fpp]);
```

$$[.1084415149,\, 0.,\, 3.527111813]$$

```
> x:=sols[4];
> evalf([f,fp,fpp]);
```

$$[.4127585747, 0., -2.658791812]$$

The example also shows that the vanishing derivative at $x = 0$ does not imply an extremum, since the second derivative vanishes there as well. The fact that a strict minimum does not imply a strictly positive second derivative is easily seen by looking at the function $f(x) = x^4$ for example.

To generalize the result (12.16) to n dimensions, we use Taylor's theorem with remainder: for $f : \mathbb{R}^n \longrightarrow \mathbb{R}$ twice continuously differentiable, we have

$$f(\boldsymbol{x}^* + \boldsymbol{x}) = f(\boldsymbol{x}^*) + f'(\boldsymbol{x}^*)\boldsymbol{x} + \frac{1}{2}f''(\boldsymbol{x}^*)(\boldsymbol{x}, \boldsymbol{x}) + r(\boldsymbol{x})\|\boldsymbol{x}\|^2, \qquad (12.17)$$

where $r(\boldsymbol{x}) \to 0$ as $\boldsymbol{x} \to \boldsymbol{0}$. In (12.17), f' is the transpose of the *gradient* of f

$$(f'(\boldsymbol{x}^*))^\top = \nabla f(\boldsymbol{x}^*) = \begin{pmatrix} \frac{\partial f}{\partial x_1}(\boldsymbol{x}^*) \\ \frac{\partial f}{\partial x_2}(\boldsymbol{x}^*) \\ \vdots \\ \frac{\partial f}{\partial x_n}(\boldsymbol{x}^*) \end{pmatrix},$$

and f'' is the bilinear form defined by the *Hessian* of f,

$$f''(\boldsymbol{x}^*)(\boldsymbol{x}, \boldsymbol{x}) = \boldsymbol{x}^\top \nabla^2 f(\boldsymbol{x}^*)\boldsymbol{x} = \boldsymbol{x}^\top H(\boldsymbol{x}^*)\boldsymbol{x},$$

where the symmetric Hessian matrix is given by

$$H(\boldsymbol{x}^*) = \begin{pmatrix} \frac{\partial^2 f}{\partial x_1^2}(\boldsymbol{x}^*) & \frac{\partial^2 f}{\partial x_1 \partial x_2}(\boldsymbol{x}^*) & \cdots & \frac{\partial^2 f}{\partial x_1 \partial x_n}(\boldsymbol{x}^*) \\ \frac{\partial^2 f}{\partial x_1 \partial x_2}(\boldsymbol{x}^*) & \frac{\partial^2 f}{\partial x_2^2}(\boldsymbol{x}^*) & \cdots & \frac{\partial^2 f}{\partial x_2 \partial x_n}(\boldsymbol{x}^*) \\ \vdots \ddots & & \vdots & \\ \frac{\partial^2 f}{\partial x_1 \partial x_n}(\boldsymbol{x}^*) & \frac{\partial^2 f}{\partial x_2 \partial x_n}(\boldsymbol{x}^*) & \cdots & \frac{\partial^2 f}{\partial x_n^2}(\boldsymbol{x}^*) \end{pmatrix}.$$

THEOREM 12.1. (UNCONSTRAINED SUFFICIENT AND NECESSARY OPTIMALITY CONDITIONS) *Let* $f : \mathbb{R}^n \longrightarrow \mathbb{R}$ *be twice continuously differentiable, and let* $\boldsymbol{x}^* \in \mathbb{R}^n$. *Then*

$$\left\{ \begin{array}{c} \nabla f(\boldsymbol{x}^*) = 0 \\ \boldsymbol{x}^\top H(\boldsymbol{x}^*)\boldsymbol{x} > 0 \\ \textit{for all } \boldsymbol{x} \in \mathbb{R}^n,\ \boldsymbol{x} \neq 0 \end{array} \right\} \Longrightarrow \left\{ \begin{array}{c} \textit{f has a} \\ \textit{local minimum} \\ \textit{at } \boldsymbol{x}^* \end{array} \right\} \Longrightarrow \left\{ \begin{array}{c} \nabla f(\boldsymbol{x}^*) = 0 \\ \boldsymbol{x}^\top H(\boldsymbol{x}^*)\boldsymbol{x} \geq 0 \\ \textit{for all } \boldsymbol{x} \in \mathbb{R}^n \end{array} \right\}.$$

$$(12.18)$$

PROOF. To show the sufficient condition, let $\lambda_{\min}$ be the smallest eigenvalue of $H(\boldsymbol{x}^*)$. Then

$$\boldsymbol{x}^\top H(\boldsymbol{x}^*)\boldsymbol{x} \geq \lambda_{\min}\boldsymbol{x}^\top \boldsymbol{x} \quad \forall \boldsymbol{x} \in \mathbb{R}^n.$$

The Taylor series (12.17) with $\nabla f(\boldsymbol{x}^*) = 0$ then implies

$$f(\boldsymbol{x}^* + \boldsymbol{x}) - f(\boldsymbol{x}^*) \geq \left(\frac{1}{2}\lambda_{\min} + r(\boldsymbol{x}) \right) \|\boldsymbol{x}\|^2.$$

Now if $x^\top H(x^*)x > 0$, $\lambda_{\min} > 0$, and hence $f(x^*+x) > f(x^*)$ for sufficiently small but nonzero x, which means x^* is a strict local minimum.

To show the necessary condition, we apply the one-dimensional result (12.16) to the function $g(t) = f(x^* + tx)$. If f has a local minimum at x^*, then g also has a local minimum at $t = 0$, and hence (12.16) implies on the one hand

$$g'(0) = 0 \implies (f(x^*))^\top x = 0 \ \forall x \in \mathbb{R}^n \implies \nabla f(x^*) = 0,$$

and on the other hand

$$g''(0) \geq 0 \implies x^\top H(x^*)x \geq 0 \ \forall x \in \mathbb{R}^n,$$

which concludes the proof. $\square$

12.2.2 Constrained minima and Lagrange multipliers

We start again by an example: suppose we want to minimize $f = x_1 x_2$ on the manifold $\mathcal{M} = \{(x_1, x_2) | c(x_1, x_2) = x_1^2 + x_2^2 - 1 = 0\}$. The manifold is a circle, and it is instructive to look at the contours of f together with the circle, as shown in Figure 12.10, which was obtained with the MAPLE

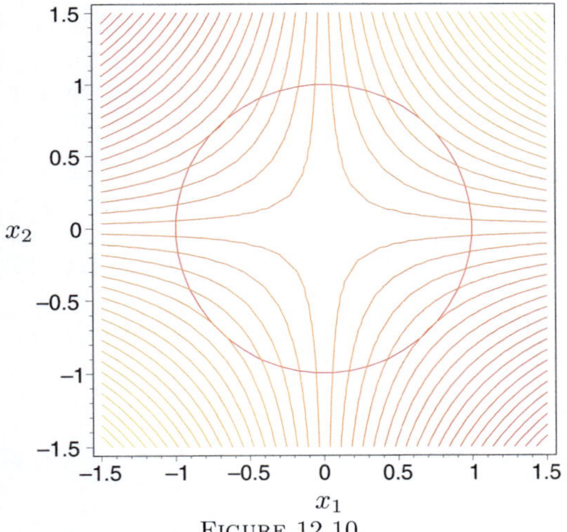

FIGURE 12.10.
Minimization of $f = x_1 x_2$ on the unit circle.

commands

```
> f:=x1*x2;
> c:=x1^2+x2^2-1;
> with(plots):
> P1:=contourplot(f,x1=-1.5..1.5,x2=-1.5..1.5,contours=40,axes=boxed):
> P2:=contourplot(c,x1=-1.5..1.5,x2=-1.5..1.5,contours=[0],axes=boxed):
> display([P1,P2]);
```

Since f is positive and increasing in the first and third quadrant, and negative and decreasing in the second and fourth, we can see from the plot that the minima on the circle are attained at $(x_1, x_2) = (-\frac{1}{\sqrt{2}}, \frac{1}{\sqrt{2}})$ and $(x_1, x_2) = (\frac{1}{\sqrt{2}}, -\frac{1}{\sqrt{2}})$.

We now present a geometric and an algebraic argument for finding these points. Geometrically, we see from Figure 12.10 that the extremal points of the constrained optimization problem are characterized by tangent level set curves of f to the constraint curve defined by c. To understand this, we note that the gradient of a function is always orthogonal to its level sets, as one can see for example from the Taylor expansion,

$$f(\boldsymbol{x}_0 + \boldsymbol{x}) - f(\boldsymbol{x}_0) = (\nabla f(\boldsymbol{x}_0))^\top \boldsymbol{x} + O(\|\boldsymbol{x}\|^2).$$

Now the minimum cannot be at a non-tangential intersection of the level sets of f and c, since one could then immediately decrease f by continuing a bit on the zero level set of c. Hence the level sets must be tangential at a minimum, which implies that the *gradient of f must be parallel to the gradient of c*. In other words, at the solutions of this problem, there must be a parameter $\lambda \in \mathbb{R}$ such that

$$\nabla f(\boldsymbol{x}^*) = \lambda \nabla c(\boldsymbol{x}^*). \tag{12.19}$$

In addition, we must be on the manifold, i.e. $g(\boldsymbol{x}^*) = 0$, which together with (12.19) forms a system of equations for the unknowns $\boldsymbol{x}^*$ and λ. In our example, we obtain with the MAPLE commands

```
> with(VectorCalculus):
> fp:=Gradient(f,[x1,x2]);
> gp:=Gradient(c,[x1,x2]);
> sols:=solve({seq(fp[i]-lambda*gp[i],i=1..2),c},{x1,x2,lambda});
> allvalues(sols[1]);
> allvalues(sols[2])
```

the solutions

$$\left\{\lambda = \frac{1}{2}, x1 = \frac{1}{2}\sqrt{2}, x2 = \frac{1}{2}\sqrt{2}\right\}, \left\{\lambda = \frac{1}{2}, x1 = -\frac{1}{2}\sqrt{2}, x2 = -\frac{1}{2}\sqrt{2}\right\}$$

$$\left\{\lambda = -\frac{1}{2}, x1 = -\frac{1}{2}\sqrt{2}, x2 = \frac{1}{2}\sqrt{2}\right\}, \left\{\lambda = -\frac{1}{2}, x1 = \frac{1}{2}\sqrt{2}, x2 = -\frac{1}{2}\sqrt{2}\right\}$$

as expected. Note that from these first order conditions, we cannot infer which of the solutions represent actual minima.

To find an algebraic argument, it is easiest to start with a parametric representation of the manifold. In our example, the circle can be represented by the function

$$\varphi(t) = \left(\begin{array}{c} \cos(t) \\ \sin(t) \end{array} \right).$$

Now the optimization problem becomes an unconstrained optimization problem by substituting the parametrization into the function and then minimizing, i.e. we minimize the function

$$F(t) := f(\boldsymbol{\varphi}(t)) = \sin(t)\cos(t) = \frac{1}{2}\sin(2t).$$

Computing the first and second derivatives, we find

$$F'(t) = \cos(2t), \quad F''(t) = -2\sin(2t),$$

and hence the minima are at $\frac{3\pi}{4}$ and $\frac{7\pi}{4}$, as expected.

For a general function $f : \mathbb{R}^2 \longrightarrow \mathbb{R}$ and a one-dimensional manifold $\mathcal{M}$ given by the parametrization $\boldsymbol{\varphi} : \mathbb{R} \longrightarrow \mathbb{R}^2$, we obtain at the extrema

$$F'(t) = (f(\boldsymbol{\varphi}(t)))' = (\nabla f(\boldsymbol{\varphi}(t)))^\top \boldsymbol{\varphi}'(t) = 0.$$

At a minimum $\boldsymbol{x}^*$, we therefore have

$$\nabla f(\boldsymbol{x}^*) \perp \boldsymbol{\varphi}'(t^*),$$

where $\boldsymbol{\varphi}'(t^*)$ is the tangent to the manifold $\mathcal{M}$ at $\boldsymbol{x}^*$.

THEOREM 12.2. (CONSTRAINED SUFFICIENT AND NECESSARY OPTIMALITY CONDITIONS 1) *Let* $f : \mathbb{R}^n \longrightarrow \mathbb{R}$ *be twice continuously differentiable, and let* $\boldsymbol{\varphi} : \mathbb{R}^k \longrightarrow \mathbb{R}^n$ *be a local parametrization of the manifold* $\mathcal{M}$ *in a neighborhood of* $\boldsymbol{x}^* \in \mathbb{R}^n$, $\boldsymbol{\varphi}(\boldsymbol{z}^*) = \boldsymbol{x}^*$. *Then*

$$\left\{ \begin{array}{c} (\nabla f(\boldsymbol{x}^*))^\top \boldsymbol{\varphi}'(\boldsymbol{z}^*) = 0 \\ \boldsymbol{z}^\top H(\boldsymbol{z}^*)\boldsymbol{z} > 0 \\ \textit{for all } \boldsymbol{z} \in \mathbb{R}^k, \ \boldsymbol{z} \neq 0 \end{array} \right\} \Longrightarrow \left\{ \begin{array}{c} f|_{\mathcal{M}} \textit{ has a} \\ \textit{local minimum} \\ \textit{at } \boldsymbol{x}^* \end{array} \right\}$$

$$\Longrightarrow \left\{ \begin{array}{c} (\nabla f(\boldsymbol{x}^*))^\top \boldsymbol{\varphi}'(\boldsymbol{z}^*) = 0 \\ \boldsymbol{z}^\top H(\boldsymbol{z}^*)\boldsymbol{z} \geq 0 \\ \textit{for all } \boldsymbol{z} \in \mathbb{R}^k \end{array} \right\}, \qquad (12.20)$$

where $H(\boldsymbol{z})$ *is the Hessian matrix of* $\boldsymbol{z} \mapsto f(\boldsymbol{\varphi}(\boldsymbol{z}))$.

PROOF. It suffices to apply Theorem 12.1 for the unconstrained case to the function $F(\boldsymbol{z}) := f(\boldsymbol{\varphi}(\boldsymbol{z}))$. $\qquad\square$

The condition $(\nabla f(\boldsymbol{x}^*))^\top \boldsymbol{\varphi}'(\boldsymbol{z}^*) = 0$ means that $\nabla f(\boldsymbol{x}^*)$ is orthogonal to the tangent space of the manifold $\mathcal{M}$ at $\boldsymbol{x}^*$, $T_{\boldsymbol{x}^*}\mathcal{M}$. If the manifold is given by a set of equations $\boldsymbol{c}(\boldsymbol{x}) = 0$, the tangent space $T_{\boldsymbol{x}^*}\mathcal{M}$ is $\ker(\boldsymbol{c}'(\boldsymbol{x}^*))$, where $\boldsymbol{c}'$ denotes the *Jacobian matrix* of $\boldsymbol{c}$ and $\ker(\boldsymbol{c}')$ is its *null space*. Now

$$\boldsymbol{x} \in \ker A \iff A\boldsymbol{x} = 0 \iff \boldsymbol{y}^\top A\boldsymbol{x} = 0, \ \forall \boldsymbol{y}$$

$$\iff \boldsymbol{x} \perp A^\top \boldsymbol{y} \iff \boldsymbol{x} \perp \text{range}(A^\top),$$

where $\text{range}(A^\top)$ is the *range* of $A^\top$, and hence

$$\nabla f(\boldsymbol{x}^*) \perp \ker(\boldsymbol{c}'(\boldsymbol{x}^*)) \iff \nabla f(\boldsymbol{x}^*) \in \text{range}((\boldsymbol{c}'(\boldsymbol{x}^*))^\top),$$

or in other words, there exists a vector $\boldsymbol{\lambda} = (\lambda_1, \ldots, \lambda_m)^\top$ such that

$$\nabla f(\boldsymbol{x}^*) = (\boldsymbol{c}'(\boldsymbol{x}^*))^\top \boldsymbol{\lambda}. \tag{12.21}$$

The λ_j, $j = 1, \ldots, m$ are called *Lagrange multipliers*. Defining the *Lagrangian* as

$$\mathcal{L}(\boldsymbol{x}, \boldsymbol{\lambda}) := f(\boldsymbol{x}) - (\boldsymbol{c}(\boldsymbol{x}))^\top \boldsymbol{\lambda}, \tag{12.22}$$

Equation (12.21) is equivalent to $\nabla_{\boldsymbol{x}}\mathcal{L}(\boldsymbol{x}, \boldsymbol{\lambda}) = 0$, while $\boldsymbol{c}(\boldsymbol{x}) = \boldsymbol{0}$ is equivalent to $\nabla_{\boldsymbol{\lambda}}\mathcal{L}(\boldsymbol{x}, \boldsymbol{\lambda}) = 0$. We therefore have the necessary condition $\nabla\mathcal{L}(\boldsymbol{x}, \boldsymbol{\lambda}) = 0$ for a minimum of f on the manifold defined by $\boldsymbol{c}(\boldsymbol{x}) = \boldsymbol{0}$.

THEOREM 12.3. (CONSTRAINED SUFFICIENT AND NECESSARY OPTI-MALITY CONDITIONS 2) *Let* $f : \mathbb{R}^n \longrightarrow \mathbb{R}$ *be twice continuously differentiable, and let* $\mathcal{M} = \{\boldsymbol{x} \in \mathbb{R}^n | \boldsymbol{c}(\boldsymbol{x}) = 0\}$ *be a manifold,* $\boldsymbol{x}^* \in \mathcal{M}$. *Then, with the Lagrangian (12.22), we have*

$$\left\{ \begin{array}{c} \text{There exists } \boldsymbol{\lambda} \in \mathbb{R}^m \text{ s.t.} \\ \nabla\mathcal{L}(\boldsymbol{x}^*, \boldsymbol{\lambda}) = 0 \text{ and} \\ \boldsymbol{w}^\top H(\boldsymbol{x}^*, \boldsymbol{\lambda})\boldsymbol{w} > 0 \\ \text{for all } \boldsymbol{w} \in T_{\boldsymbol{x}^*}\mathcal{M}, \ \boldsymbol{w} \neq \boldsymbol{0} \end{array} \right\} \implies \left\{ \begin{array}{c} f|_{\mathcal{M}} \text{ has a} \\ \text{local minimum} \\ \text{at } \boldsymbol{x}^* \end{array} \right\}$$

$$\implies \left\{ \begin{array}{c} \text{There exists } \boldsymbol{\lambda} \in \mathbb{R}^m \text{ s.t.} \\ \nabla\mathcal{L}(\boldsymbol{x}^*, \boldsymbol{\lambda}) = 0 \text{ and} \\ \boldsymbol{w}^\top H(\boldsymbol{x}^*, \boldsymbol{\lambda})\boldsymbol{w} \geq 0 \\ \text{for all } \boldsymbol{w} \in T_{\boldsymbol{x}^*}\mathcal{M} \end{array} \right\}, \tag{12.23}$$

where H *is the Hessian of* $\mathcal{L}$ *with respect to the variable* $\boldsymbol{x}$ *only.*
    PROOF. See Problem 12.10. $\square$

12.2.3   Equality and Inequality Constraints

For continuously differentiable functions $f : \mathbb{R}^n \longrightarrow \mathbb{R}$, $\boldsymbol{c}_E : \mathbb{R}^n \longrightarrow \mathbb{R}^m$ and $\boldsymbol{c}_I : \mathbb{R}^n \longrightarrow \mathbb{R}^k$, we look for local minima of the problem

$$\begin{aligned} f(\boldsymbol{x}) &\longrightarrow \text{min,} \\ \boldsymbol{c}_E(\boldsymbol{x}) &= \boldsymbol{0}, \\ \boldsymbol{c}_I(\boldsymbol{x}) &\geq \boldsymbol{0}. \end{aligned} \tag{12.24}$$

While such a problem is difficult to solve analytically in general, we show in the following example that in principle it could be treated analytically.

    EXAMPLE 12.1. *Let* $f(\boldsymbol{x}) = x_1^2 + x_2^2 + x_3^2$, $c_E(\boldsymbol{x}) = x_1 - x_2 - x_3$ *and* $\boldsymbol{c}_I(\boldsymbol{x}) = (x_1 - x_2 + \frac{1}{10}, x_2 - (x_3 + x_2 - 1)^2)^\top$. *We first use the equality constraint to eliminate the variable* $x_3 = x_1 - x_2$, *which leads to the simplified problem with* $f(\boldsymbol{x}) = 2(x_1^2 - x_1 x_2 + x_2^2)$ *and* $\boldsymbol{c}_I(\boldsymbol{x}) = (x_1 - x_2 + \frac{1}{10}, x_2 - (x_1 - 1)^2)^\top$. *Using the* MAPLE *commands*

```
> f:=x1^2+x2^2+x3^2;
```

```
> cE:=x1-x2-x3;
> cI1:=x1-x2+1/10;
> cI2:=x2-(x3+x2-1)^2;
> x3:=solve(cE,x3);
> with(plots):
> P1:=contourplot(f,x1=0..3,x2=0..3,contours=20,axes=boxed):
> P2:=contourplot(cI1,x1=0..3,x2=0..3,contours=[0],axes=boxed):
> P3:=contourplot(cI2,x1=0..3,x2=0..3,contours=[0],axes=boxed):
> display([P1,P2,P3]);
```

we obtain Figure 12.11.

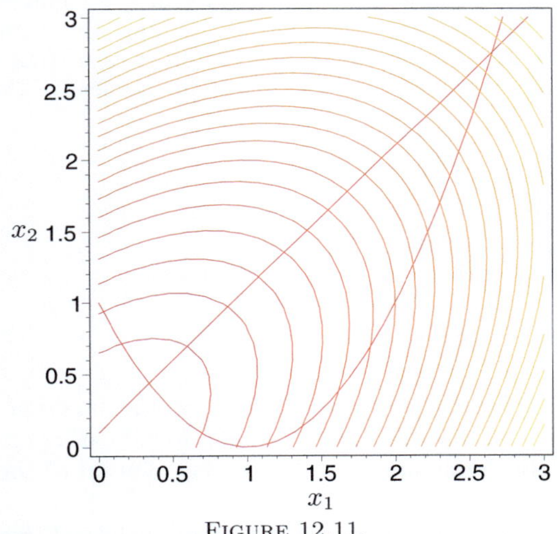

FIGURE 12.11.
Level sets of the function to minimize, and zero level sets
of the two inequality constraints in Example 12.1.

From this figure, we can see that in the interior of the domain bounded by the inequality constraints, f does not have any extrema, as one can readily test with the MAPLE commands

```
> with(VectorCalculus):
> fp:=Gradient(f,[x1,x2]);
> solve({fp[1],fp[2]},{x1,x2});
```

which gives only the zero solution. We therefore need to search for extrema on the boundary of the compact set. Using Lagrange multipliers, we obtain with MAPLE on the straight line constraint

```
> L1:=f-lambda*cI1;
> L1p:=Gradient(L1,[x1,x2,lambda]);
> solve({seq(L1p[i],i=1..3)},{x1,x2,lambda});
```

which gives the only solution $(x_1, x_2) = (-\frac{1}{20}, \frac{1}{20})$, outside of the compact set, which is also easily identified by looking at Figure 12.11, slightly outside on the left. On the second inequality constraint, we obtain

```
> L2:=f-lambda*cI2;
> L2p:=Gradient(L2,[x1,x2,lambda]);
> solve({seq(L2p[i],i=1..3)},{x1,x2,lambda});
> sols:=allvalues(%);
> evalf(sols[1]);
> assign(%);
> f;
```

and hence there is an extrema at $(x_1, x_2) \approx (0.3929927044, 0.368457857)$, which is indeed a minimum, on the boundary of the compact set, and the value of f at this point is .2908064168. Using the commands `evalf (sols[2]);` and `evalf(sols[3]);` one readily checks that the two solutions are complex. We finally need to check the corners of the boundary of the compact set,

```
> x1:='x1';x2:='x2';
> x2c1:=solve(cI1,x2);
> x2c2:=solve(cI2,x2);
> x1sols:=solve(x2c2=x2c1,x1);
> x1:=x1sols[1];x2:=x2c1;
> evalf(f);
> x1:=x1sols[2];x2:=x2c1;
> evalf(f);
```

which gives 14.72374902 and 0.3162509758 for the function values at the corners. We conclude that the local minimum found earlier on the boundary is indeed the global minimum, and hence the solution of the optimization problem.

As one can see from this simple example, an analytical treatment of such optimization problems is hopeless for realistic high-dimensional problems. We will therefore in the sequel consider numerical techniques for finding local minima.

As we have seen in subsection 12.2.2, equality constraints can be dealt with using Lagrange multipliers, and Theorem 12.2 gave necessary and sufficient conditions for optimality in that case. It is natural, for a problem with inequality constraints, that only the constraints active at a local optimum play a role in the optimality conditions, i.e the constraints for which the relation $(c_I)_i(x^*) = 0$ holds at the optimum x^* . We call the set $A = \{i \mid (c_I)_i(x^*) = 0\}$ the *active set for the local optimum* x^*.

THEOREM 12.4. (KARUSH-KUHN-TUCKER (KKT)) *Suppose that* $f : \mathbb{R}^n \longrightarrow \mathbb{R}$ *and* $c : \mathbb{R}^n \longrightarrow \mathbb{R}^m$ *are twice continuously differentiable and that* $x^* \in \mathbb{R}^n$ *is a local minimum of*

$$f(x) \longrightarrow \min$$
$$c(x) \geq 0.$$

Then there exists $\boldsymbol{\lambda} \in \mathbb{R}^m$ such that

$$
\begin{aligned}
\boldsymbol{c}(\boldsymbol{x}^*) &\geq 0, \\
\nabla f(\boldsymbol{x}^*) - (\nabla \boldsymbol{c}(\boldsymbol{x}^*))^\top \boldsymbol{\lambda} &= 0, \quad \text{with } \boldsymbol{\lambda} \geq 0 \\
c_i(\boldsymbol{x}^*)\lambda_i &= 0, \quad \forall i.
\end{aligned}
$$

PROOF. We consider a perturbation around $\boldsymbol{x}^*$ that satisfies the constraints. Since any constraint that is inactive at $\boldsymbol{x}^*$, i.e. $c_i(\boldsymbol{x}^*) > 0$, remains inactive in a neighborhood of $\boldsymbol{x}^*$, it suffices to consider the active constraints only, $c_i(\boldsymbol{x}^*) = 0$ for $i \in A$. If $c_i(\boldsymbol{x}^*) > 0$, $i \neq A$ is an inactive constraint, then we put $\lambda_i = 0$ to remove its contribution from the equation $\nabla f(\boldsymbol{x}^*) - (\nabla \boldsymbol{c}(\boldsymbol{x}^*))^\top \boldsymbol{\lambda} = 0$. Thus, it remains to show the positivity of λ_i for *active constraints*, i.e., for $i \in A$.

Let $\boldsymbol{x}(\alpha)$, $\alpha \geq 0$ be a twice continuously differentiable path with $\boldsymbol{x}(0) = \boldsymbol{x}^*$, i.e.

$$
\boldsymbol{x}(\alpha) = \boldsymbol{x}^* + \alpha \boldsymbol{s} + \frac{1}{2}\alpha^2 \boldsymbol{p} + O(\alpha^3),
$$

along which the active constraints remain satisfied, i.e. $c_i(\boldsymbol{x}(\alpha)) \geq 0$ for $i \in A$. This implies that

$$
\begin{aligned}
0 &\leq c_i(\boldsymbol{x}(\alpha)) \\
&= c_i(\boldsymbol{x}^* + \alpha \boldsymbol{s} + \frac{1}{2}\alpha^2 \boldsymbol{p} + O(\alpha^3)) \\
&= c_i(\boldsymbol{x}^*) + (\alpha \boldsymbol{s} + \frac{1}{2}\alpha^2 \boldsymbol{p})^\top \nabla c_i(\boldsymbol{x}^*) + \frac{1}{2}\alpha^2 \boldsymbol{s}^\top H_i(\boldsymbol{x}^*)\boldsymbol{s} + O(\alpha^3) \\
&= \alpha \boldsymbol{s}^\top \nabla c_i(\boldsymbol{x}^*) + \frac{1}{2}\alpha^2 \left(\boldsymbol{p}^\top \nabla c_i(\boldsymbol{x}^*) + \boldsymbol{s}^\top H_i(\boldsymbol{x}^*)\boldsymbol{s}\right) + O(\alpha^3),
\end{aligned}
$$

where $H_i(\boldsymbol{x})$ denotes the Hessian of $c_i(\boldsymbol{x})$. Hence for α small, we must either have

$$
\boldsymbol{s}^\top \nabla c_i(\boldsymbol{x}^*) > 0,
$$

or, if $\boldsymbol{s}^\top \nabla c_i(\boldsymbol{x}^*) = 0$, we must have

$$
\boldsymbol{p}^\top \nabla c_i(\boldsymbol{x}^*) + \boldsymbol{s}^\top H_i(\boldsymbol{x}^*)\boldsymbol{s} \geq 0
$$

for all $i \in A$. If we expand the function $f(\boldsymbol{x}(\alpha))$, we obtain

$$
\begin{aligned}
f(\boldsymbol{x}(\alpha)) &= f(\boldsymbol{x}^* + \alpha \boldsymbol{s} + \frac{1}{2}\alpha^2 \boldsymbol{p} + O(\alpha^3)) \\
&= f(\boldsymbol{x}^*) + (\alpha \boldsymbol{s} + \frac{1}{2}\alpha^2 \boldsymbol{p})^\top \nabla f(\boldsymbol{x}^*) + \frac{1}{2}\alpha^2 \boldsymbol{s}^\top H(\boldsymbol{x}^*)\boldsymbol{s} + O(\alpha^3) \\
&= f(\boldsymbol{x}^*) + \alpha \boldsymbol{s}^\top \nabla f(\boldsymbol{x}^*) + \frac{1}{2}\alpha^2 \left(\boldsymbol{p}^\top \nabla f(\boldsymbol{x}^*) + \boldsymbol{s}^\top H(\boldsymbol{x}^*)\boldsymbol{s}\right) + O(\alpha^3),
\end{aligned}
$$

where $H(\boldsymbol{x})$ denotes the Hessian of $f(\boldsymbol{x})$. Therefore, $\boldsymbol{x}^*$ can only be a local minimum if the set

$$S := \{\boldsymbol{s} \mid \boldsymbol{s}^\top \nabla f(\boldsymbol{x}^*) < 0 \text{ and } \boldsymbol{s}^\top \nabla c_i(\boldsymbol{x}^*) \geq 0 \text{ for } i \in A\}$$

is empty. Using now Farkas' Lemma below, the result follows. $\qquad\square$

LEMMA 12.1. (FARKAS) *Let $\boldsymbol{u} \in \mathbb{R}^n$, $\boldsymbol{v}_i \in \mathbb{R}^n$ for $i \in A$. Then the set*

$$S := \{\boldsymbol{s} \mid \boldsymbol{s}^\top \boldsymbol{u} < 0 \text{ and } \boldsymbol{s}^\top \boldsymbol{v}_i \geq 0 \text{ for } i \in A\}$$

is empty if and only if there exist $\lambda_i \geq 0$, $i \in A$ such that

$$\boldsymbol{u} = \sum_{i \in A} \lambda_i \boldsymbol{v}_i.$$

PROOF. See Problem 12.13. $\qquad\square$

12.3   Unconstrained Optimization

For a function $f : \mathbb{R}^n \longrightarrow \mathbb{R}$, we want to find a local minimum,

$$f(\boldsymbol{x}) \longrightarrow \min.$$

We have seen that mathematically, one simply searches for zeros of the gradient and verifies that the Hessian is positive definite in order to find local minima. However, finding a zero of a nonlinear system of equations is already a highly non-trivial task, as we have seen in Chapter 5, and the numerical methods can have severe convergence problems. In optimization, instead of simply trying to solve a nonlinear system of equations given by $\nabla f(\boldsymbol{x}) = 0$, one immediately tries to use the fact that one searches for a minimum of a function by requiring the value of the objective function to *decrease* at each iteration. In general, these methods only converge to a local minimum of f, except if f satisfies certain stringent requirements.

12.3.1   Line Search Methods

Line search methods construct a sequence of iterates $\boldsymbol{x}_k$ by the formula

$$\boldsymbol{x}_{k+1} = \boldsymbol{x}_k + \alpha_k \boldsymbol{p}_k,$$

where $\boldsymbol{p}_k$ is a search direction, and the scalar parameter $\alpha_k > 0$ is a step length. Typically, one requires $\boldsymbol{p}_k^\top \nabla f(\boldsymbol{x}_k) < 0$ to guarantee that it is a descent direction.

A first idea for determining the step length α_k is to simply minimize f in the search direction $\boldsymbol{p}_k$, which is a one-dimensional minimization problem, i.e. we compute a zero of the derivative with respect to α,

$$\frac{d}{d\alpha} f(\boldsymbol{x}_k + \alpha \boldsymbol{p}_k) = \boldsymbol{p}_k^\top \nabla f(\boldsymbol{x}_k + \alpha \boldsymbol{p}_k) = 0.$$

We have seen this approach already in Chapter 11, when we introduced the method of *steepest descent* to solve a symmetric positive definite system of linear equations $A\boldsymbol{x} = \boldsymbol{b}$. Unfortunately this method was not very good, so we introduced the method of *conjugate gradients*, a Krylov method, with much better performance. Conjugate gradients can be considered as an optimization algorithm for unconstrained problems of the form $f(\boldsymbol{x}) = \frac{1}{2}\boldsymbol{x}^\top A\boldsymbol{x} - \boldsymbol{b}^\top \boldsymbol{x}$, whose first order optimality condition is given by the linear system $A\boldsymbol{x} = \boldsymbol{b}$.

Minimizing with respect to α is called an exact line search, and is in general far too expensive to perform. Today, there are many inexact line search methods whose goal is to find a step size α_k that is neither too short nor too long. This is critical, as one can see with the following two examples.

EXAMPLE 12.2. *Suppose we need to minimize $f(x) = x^2$ using a line search algorithm starting at $x_0 = 2$. If we choose the descent direction to be $p_k = (-1)^{k+1}$, and the step length $\alpha_k = 2 + \frac{3}{2^{k+1}}$, the result is shown in Figure 12.12 on the left, which was obtained with the* MATLAB *commands*

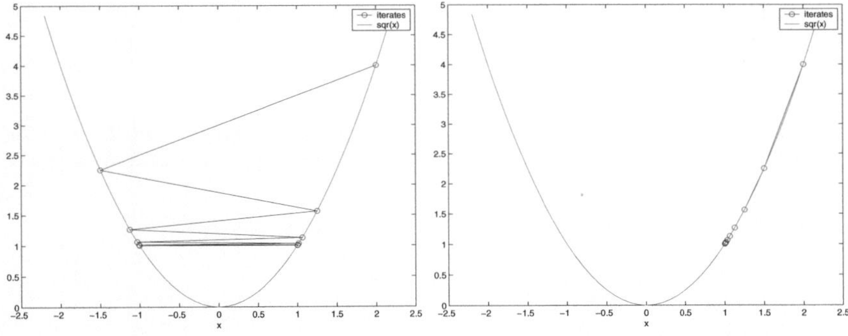

FIGURE 12.12.
On the left a case where the step length in the line search is too long, and on the right one where the step length is too short. In both cases, the line search methods do not converge to a minimum.

```
x(1)=2;
for k=1:10
  x(k+1)=x(k)+(2+3/(2^k))*(-1)^k;
end
xx=-2.2:0.1:2.2;
plot(x,x.^2,'-o',xx,xx.^2,'-');
xlabel('x');
legend('iterates','sqr(x)');
```

where we needed to pay attention to the fact that array indices in MATLAB *always start at 1. Clearly the method fails, since the step length is too large, and the method becomes trapped in a two-cycle. The cycle can easily be computed using the* MAPLE *recurrence relation solver,*

```
> rsolve({x(k+1)=x(k)+(2+3/2^(k+1))*(-1)^(k+1),x(0)=2},{x});
```

$$\{x(k) = (-1)^k + (\frac{-1}{2})^k\}$$

and thus shows convergence to an oscillation between 1 *and* -1.

EXAMPLE 12.3. *Suppose we need to minimize the same function* $f(x) = x^2$, *but now we choose the descent direction* $p_k = -1$, *and the step length* $\alpha_k = \frac{1}{2^{k+1}}$. *The result is shown in Figure 12.12 on the right, obtained with the* MATLAB *commands*

```
x(1)=2;
for k=1:10
  x(k+1)=x(k)-1/(2^k);
end
xx=-2.2:0.1:2.2;
plot(x,x.^2,'-o',xx,xx.^2,'-');
xlabel('x');
legend('iterates','sqr(x)');
```

Again the method fails: this time the step length is too short, and the method stagnates before reaching the minimum. We can also solve this recurrence relation with MAPLE,

```
> rsolve({x(k+1)=x(k)-1/2^(k+1),x(0)=2},{x});
```

$$\{x(k) = 1 + (\frac{1}{2})^k\}$$

which shows convergence to 1.

These two examples illustrate how important it is to choose a line search parameter that is neither too large nor too small.

Armijo Backtracking Line Search

Among the many line search techniques, one of the most successful ones is the *Armijo backtracking line search*, which starts with a reasonably large line search parameter α_{init}, and then reduces it until the function value at the new position is sufficiently reduced relative to the value at the old position:

ALGORITHM 12.5. *Generic line search method*

---

for $\boldsymbol{x}$ and $\boldsymbol{p}$ given;
$\alpha = \alpha_{\text{init}}$;
while $f(\boldsymbol{x} + \alpha\boldsymbol{p}) > f(\boldsymbol{x}) + \alpha\beta\boldsymbol{p}^{\top}\nabla f(\boldsymbol{x})$
   $\alpha = \tau\alpha$;
end;

Here α_{init} is an initial guess for the line search parameter, for example $\alpha_{\text{init}} = 1$, $\beta \in (0, 1)$ is a parameter to control the required amount of decrease of the function (note that $\boldsymbol{p}^\top \nabla f(\boldsymbol{x}) < 0$ for a descent direction), and $\tau \leq 1$ is the factor by which α is diminished each time, for example one half. The Armijo backtracking line search avoids steps that are too small, since the method accepts the first step that gives the required decrease in the function value. Overly large steps are also avoided, since one requires a sufficient decrease in the function value. Naturally, one might wonder if the Armijo backtracking strategy always finds a suitable value for the line search parameter α. The following theorem and its corollary give an affirmative answer.

THEOREM 12.5. *Let $f : \mathbb{R}^n \longrightarrow \mathbb{R}$ be continuously differentiable, and let ∇f be locally Lipschitz continuous at $\boldsymbol{x}_k$ with Lipschitz constant $L(\boldsymbol{x}_k)$,*

$$\|\nabla f(\boldsymbol{x}_k + \boldsymbol{x}) - \nabla f(\boldsymbol{x}_k)\|_2 \leq L(\boldsymbol{x}_k)\|\boldsymbol{x}\|_2. \tag{12.25}$$

If $\beta \in (0, 1)$ and $\boldsymbol{p}_k$ is a descent direction of f at $\boldsymbol{x}_k$, then the Armijo condition

$$f(\boldsymbol{x}_k + \alpha \boldsymbol{p}_k) \leq f(\boldsymbol{x}_k) + \alpha \beta \boldsymbol{p}_k^\top \nabla f(\boldsymbol{x}_k) \tag{12.26}$$

is satisfied for all $\alpha \in [0, \alpha_{\max}(\boldsymbol{x}_k, \boldsymbol{p}_k)]$, where

$$\alpha_{\max}(\boldsymbol{x}_k, \boldsymbol{p}_k) = \frac{2(\beta - 1)\boldsymbol{p}_k^\top \nabla f(\boldsymbol{x}_k)}{L(\boldsymbol{x}_k)\|\boldsymbol{p}_k\|_2^2}. \tag{12.27}$$

PROOF. Using integration, we find

$$
\begin{aligned}
f(\boldsymbol{x}_k + \alpha \boldsymbol{p}_k) &- f(\boldsymbol{x}_k) - \alpha \boldsymbol{p}_k^\top \nabla f(\boldsymbol{x}_k) \\
&= \int_0^1 \alpha \boldsymbol{p}_k^\top \nabla f(\boldsymbol{x}_k + \tau \alpha \boldsymbol{p}_k)d\tau - \int_0^1 \alpha \boldsymbol{p}_k^\top \nabla f(\boldsymbol{x}_k)d\tau \\
&= \alpha \int_0^1 \boldsymbol{p}_k^\top (\nabla f(\boldsymbol{x}_k + \tau \alpha \boldsymbol{p}_k) - \nabla f(\boldsymbol{x}_k))d\tau \\
&\leq \alpha \|\boldsymbol{p}_k\|_2 \int_0^1 \|\nabla f(\boldsymbol{x}_k + \tau \alpha \boldsymbol{p}_k) - \nabla f(\boldsymbol{x}_k)\|_2 d\tau \\
&\leq \alpha^2 \|\boldsymbol{p}_k\|_2^2 L(\boldsymbol{x}_k) \int_0^1 \tau d\tau \\
&= \frac{\alpha^2}{2} L(\boldsymbol{x}_k)\|\boldsymbol{p}_k\|_2^2,
\end{aligned}
$$

where we used the Cauchy–Schwarz inequality for vectors in the first inequality, and the Lipschitz condition (12.25) in the second one. Now using that $\alpha \leq \alpha_{\max}$, we can replace one factor α in the α^2 on the right hand side to obtain

$$
\begin{aligned}
f(\boldsymbol{x}_k + \alpha \boldsymbol{p}_k) &\leq f(\boldsymbol{x}_k) + \alpha \boldsymbol{p}_k^\top \nabla f(\boldsymbol{x}_k) + \alpha(\beta - 1)\boldsymbol{p}_k^\top \nabla f(\boldsymbol{x}_k) \\
&= f(\boldsymbol{x}_k) + \alpha \beta \boldsymbol{p}_k^\top \nabla f(\boldsymbol{x}_k),
\end{aligned}
$$

which concludes the proof. $\qquad\qquad\qquad\qquad\qquad\qquad\qquad\qquad\square$

COROLLARY 12.1. *Under the conditions of the previous theorem, the backtracking Armijo line search terminates with*

$$\alpha_k \geq \min(\alpha_{\text{init}}, \frac{2\tau(\beta - 1)\boldsymbol{p}_k^\top \nabla f(\boldsymbol{x}_k)}{L(\boldsymbol{x}_k)\|\boldsymbol{p}_k\|_2^2}). \qquad (12.28)$$

PROOF. Either α_{init} already satisfies the Armijo condition, or there is a second-to-last step in the Armijo backtracking algorithm which does not yet satisfy the Armijo condition. In this case, the next step will multiply this second-to-last one by τ, which then satisfies the Armijo condition, and the algorithm stops with α_k satisfying (12.28). $\qquad\qquad\square$

In general, it would be very difficult to get an estimate of the local Lipschitz constant $L(\boldsymbol{x}_k)$, so the backtracking Armijo search is precisely the tool that finds a suitable line search parameter without knowing this quantity.

Generic Line Search Method with Armijo Backtracking

A generic minimization procedure with the Armijo backtracking line search is

---

ALGORITHM 12.6.
Generic minimization with Armijo line search

```
x = x₀;
k = 0;
while not converged
    find a suitable descent direction p;
    α = αinit;
    while f(x + αp) > f(x) + αβpᵀ∇f(x)
        α = τα;
    end;
    x = x + αp;
    k = k + 1;
end;
```

THEOREM 12.6. *Let $f : \mathbb{R}^n \longrightarrow \mathbb{R}$ be continuously differentiable, and let ∇f be locally Lipschitz at $\boldsymbol{x}_k$ with Lipschitz constant $L(\boldsymbol{x}_k)$. Then the generic minimization algorithm with backtracking Armijo line search leads to one of the following results:*

1. $\nabla f(\boldsymbol{x}_k) = 0$ for some $k \geq 0$.

2. $\lim_{k\to\infty} f(\boldsymbol{x}_k) = -\infty$.

3. $\lim_{k\to\infty} \min(|\boldsymbol{p}_k^\top \nabla f(\boldsymbol{x}_k)|, \frac{|\boldsymbol{p}_k^\top \nabla f(\boldsymbol{x}_k)|}{\|\boldsymbol{p}_k\|_2}) = 0$.

PROOF. Case 1 is a lucky hit of a stationary point, and Case 2 can happen when the function is not bounded from below. Assuming that neither of these cases happen, i.e. $\nabla f(\boldsymbol{x}_k) \neq 0$ for all $k \geq 0$, and $\lim_{k\to\infty} f(\boldsymbol{x}_k) > -\infty$, the Armijo condition gives for all $k \geq 0$

$$f(\boldsymbol{x}_{k+1}) - f(\boldsymbol{x}_k) \leq \alpha_k \beta \boldsymbol{p}_k^\top \nabla f(\boldsymbol{x}_k),$$

and summing over k, we obtain

$$f(\boldsymbol{x}_{k+1}) - f(\boldsymbol{x}_0) \leq \sum_{j=0}^{k} \alpha_j \beta \boldsymbol{p}_j^\top \nabla f(\boldsymbol{x}_j),$$

which shows that the sum remains bounded from below, since $f(\boldsymbol{x}_{k+1}) > -\infty$. Now all the terms in the sum are negative, since the $\boldsymbol{p}_j$ are descent directions, and therefore the terms must converge to zero,

$$\lim_{k\to\infty} \alpha_k \boldsymbol{p}_k^\top \nabla f(\boldsymbol{x}_k) = 0.$$

We now define the two sets

$$K_1 := \{k | \alpha_{\mathrm{init}} > \frac{2\tau(\beta-1)\boldsymbol{p}_k^\top \nabla f(\boldsymbol{x}_k)}{L(\boldsymbol{x}_k)\|\boldsymbol{p}_k\|_2^2}\}, \quad K_2 := \{k | \alpha_{\mathrm{init}} \leq \frac{2\tau(\beta-1)\boldsymbol{p}_k^\top \nabla f(\boldsymbol{x}_k)}{L(\boldsymbol{x}_k)\|\boldsymbol{p}_k\|_2^2}\}.$$

If $k \in K_1$, we have from the corollary

$$\alpha_k \geq \frac{2\tau(\beta-1)\boldsymbol{p}_k^\top \nabla f(\boldsymbol{x}_k)}{L(\boldsymbol{x}_k)\|\boldsymbol{p}_k\|_2^2}$$

and therefore

$$\alpha_k \boldsymbol{p}_k^\top \nabla f(\boldsymbol{x}_k) \leq \frac{2\tau(\beta-1)}{L(\boldsymbol{x}_k)}\left(\frac{\boldsymbol{p}_k^\top \nabla f(\boldsymbol{x}_k)}{\|\boldsymbol{p}_k\|_2}\right)^2 < 0,$$

which implies

$$\lim_{k\in K_1 \to\infty} \frac{|\boldsymbol{p}_k^\top \nabla f(\boldsymbol{x}_k)|}{\|\boldsymbol{p}_k\|_2} = 0.$$

For $k \in K_2$, we have from the corollary

$$\alpha_k \geq \alpha_{\mathrm{init}},$$

which implies

$$\lim_{k\in K_2 \to\infty} |\boldsymbol{p}_k^\top \nabla f(\boldsymbol{x}_k)| = 0,$$

and hence concludes the proof. $\square$

Unfortunately, this first convergence result does not imply that the algorithm converges to a stationary point where the gradient vanishes, since in Case 3, the gradient can simply become more and more orthogonal to the search direction. For a successful algorithm, one therefore needs to demand more from the search direction than simply that it be a descent direction.

Steepest Descent

The classical choice is to go along the direction of *steepest descent*, i.e. $\boldsymbol{p}_k :=$ $-\nabla f(\boldsymbol{x}_k)$, which leads to a convergent algorithm:

COROLLARY 12.2. *Let $f : \mathbb{R}^n \longrightarrow \mathbb{R}$ be continuously differentiable, and let ∇f be locally Lipschitz at $\boldsymbol{x}_k$ with Lipschitz constant $L(\boldsymbol{x}_k)$. Then the generic minimization algorithm with backtracking Armijo line search and steepest descent search direction leads to one of the following results:*

1. *$\nabla f(\boldsymbol{x}_k) = 0$ for some $k \geq 0$.*

2. *$\lim_{k \to \infty} f(\boldsymbol{x}_k) = -\infty$.*

3. *$\lim_{k \to \infty} \nabla f(\boldsymbol{x}_k) = \mathbf{0}$.*

PROOF. The two special cases 1 and 2 are as in Theorem 12.6. For Case 3, we obtain from Theorem 12.6 that

$$\lim_{k \to \infty} \min(|\boldsymbol{p}_k^\top \nabla f(\boldsymbol{x}_k)|, \frac{|\boldsymbol{p}_k^\top \nabla f(\boldsymbol{x}_k)|}{\|\boldsymbol{p}_k\|_2}) = \lim_{k \to \infty} \min(\|\nabla f(\boldsymbol{x}_k)\|_2^2, \|\nabla f(\boldsymbol{x}_k)\|_2) = 0,$$

which concludes the proof. $\qquad\qquad\square$

Here is a MATLAB implementation of the optimization algorithm with Armijo line search and steepest descent direction:

ALGORITHM 12.7. *Minimization with Steepest Descent*

```
function [x,xk]=SteepestDescent(f,fp,x0,tol,maxiter,tau,be,alinit)
% STEEPESTDESCENT steepest descent min. search with Armijo line search
%    [x,xk]=SteepestDescent(f,fp,x0,tol,maxiter,tau,be,alinit) finds an
%    approximate minimum of the function f with gradient fp, starting
%    at the initial guess x0. The remaining parameters are optional and
%    default values are used if they are omited. xk contains all the
%    iterates of the method.

if nargin<8, alinit=1; end;
if nargin<7, be=0.1; end;
if nargin<6, tau=0.5; end;
if nargin<5, maxiter=100; end;
if nargin<4, tol=1e-6; end;

x=x0;
xk=x0;
p=-feval(fp,x);
k=0;
while norm(p)>tol & k<maxiter
  al=alinit;
  while feval(f,x+al*p)>feval(f,x)-al*be*p'*p
```

```
    al=tau*al;
  end;
  x=x+al*p
  p=-feval(fp,x);
  k=k+1;
  xk(:,k+1)=x;
end;
```

As we have seen in Chapter 11 on linear equations, steepest descent is not necessarily the best choice, and the same holds true for optimization, as the example in Figure 12.13 shows, which was generated using the MATLAB commands

```
f=@(x) 10*(x(2)-x(1)^2)^2+(x(1)-1)^2;
fp=@(x) [-40*(x(2)-x(1)^2)*x(1)+2*(x(1)-1); 20*(x(2)-x(1)^2)];
[x,xk]=SteepestDescent(f,fp,[-1.2;1])
F=@(x,y) 10*(y-x.^2).^2+(x-1).^2;         % for plotting purposes
Fp1=@(x,y) -40*(y-x.^2).*x+2*(x-1);
Fp2=@(x,y) 20*(y-x.^2);
[X,Y]=meshgrid(-1.5:0.1:1,-0.5:0.1:1.5);
contour(X,Y,F(X,Y),50)
hold on
quiver(X,Y,Fp1(X,Y),Fp2(X,Y),5)
plot(xk(1,:),xk(2,:),'-o')
hold off
```

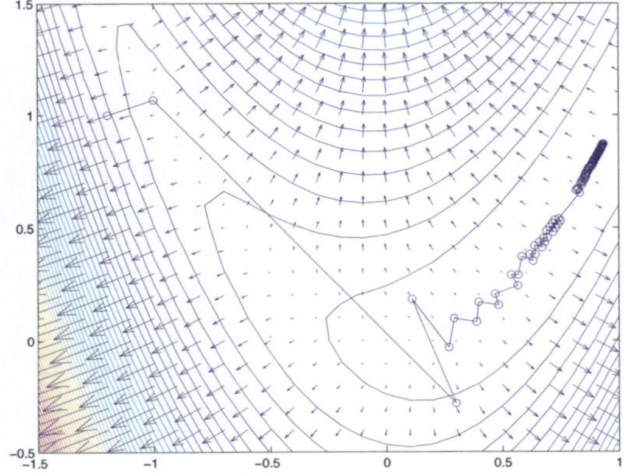

FIGURE 12.13.
Steepest Descent method for a model problem showing convergence problems of the method.

However, even though there are better choices than steepest descent for the search direction, as we will see in what follows, many modern codes resort to steepest descent as the search direction if convergence problems are detected in the code.

Newton Search Direction

To obtain a better search direction, we first notice that for any symmetric positive definite matrix B_k, the search direction

$$\boldsymbol{p}_k := -B_k^{-1}\nabla f(\boldsymbol{x}_k)$$

is a descent direction, since

$$\boldsymbol{p}_k^\top \nabla f(\boldsymbol{x}_k) = -(\nabla f(\boldsymbol{x}_k))^\top B_k^{-\top} \nabla f(\boldsymbol{x}_k) < 0,$$

because the fact that B_k is positive definite implies that $B_k^{-\top}$ is positive definite as well. In addition, this particular search direction leads directly to a stationary point of the quadratic problem

$$g(\boldsymbol{x}_k + \boldsymbol{p}) := f(\boldsymbol{x}_k) + \boldsymbol{p}^\top \nabla f(\boldsymbol{x}_k) + \frac{1}{2}\boldsymbol{p}^\top B_k \boldsymbol{p},$$

which can be considered as a quadratic approximation of the function f in the neighborhood of $\boldsymbol{x}_k$. In particular, if $B_k = H(\boldsymbol{x}_k)$, the Hessian of f at $\boldsymbol{x}_k$, then g is none other than a second-order Taylor approximation of f about $\boldsymbol{x}_k$. In this case, one calls the search direction $\boldsymbol{p}_k = -H(\boldsymbol{x}_k)^{-1}\nabla f(\boldsymbol{x}_k)$ the Newton direction, because this is the step the classical Newton method (c.f. Chapter 5, Section 5.4.3) would take when trying to find a root of the equation $\nabla f(\boldsymbol{x}) = \boldsymbol{0}$. However, in optimization, this direction is only useful if the Hessian $H(\boldsymbol{x}_k)$ is positive definite, since the Newton direction could otherwise be an ascent direction.

THEOREM 12.7. *Let* $f : \mathbb{R}^n \longrightarrow \mathbb{R}$ *be continuously differentiable and* ∇f *be Lipschitz in* $\mathbb{R}^n$. *Suppose the generic minimization algorithm with the back-tracking Armijo line search is used with descent direction* $\boldsymbol{p}_k = -B_k \nabla f(\boldsymbol{x}_k)$, *where the* B_k *are symmetric positive definite matrices with* $\lambda_{\max}(B_k) \leq \lambda_{\max} < \infty$ *and* $\lambda_{\min}(B_k) \geq \lambda_{\min} > 0$ *for all* k. *Then the iterates* $\boldsymbol{x}_k$ *satisfy one of the following statements:*

1. $\nabla f(\boldsymbol{x}_k) = 0$ *for some* $k \geq 0$.

2. $\lim_{k \to \infty} f(\boldsymbol{x}_k) = -\infty$.

3. $\lim_{k \to \infty} \nabla f(\boldsymbol{x}_k) = 0$.

PROOF. The points 1 and 2 are as in Theorem 12.6. For 3, we have for all $\boldsymbol{x} \neq 0$ that

$$\lambda_{\min} \leq \lambda_{\min}(B_k) \leq \frac{\boldsymbol{x}^\top B_k \boldsymbol{x}}{\|\boldsymbol{x}\|_2} \leq \lambda_{\max}(B_k) \leq \lambda_{\max},$$

and therefore the two estimates

$$|\boldsymbol{p}_k^\top \nabla f(\boldsymbol{x}_k)| = |\nabla f(\boldsymbol{x}_k)^\top B_k^{-T} \nabla f(\boldsymbol{x}_k)| \geq \lambda_{\min}(B_k^{-1})\|\nabla f(\boldsymbol{x}_k)\|_2^2 \geq \frac{1}{\lambda_{\max}}\|\nabla f(\boldsymbol{x}_k)\|_2^2$$

and

$$\begin{aligned}
\|\boldsymbol{p}_k\|_2^2 &= \nabla f(\boldsymbol{x}_k)^\top B_k^{-T} B_k^{-1} \nabla f(\boldsymbol{x}_k) \\
&= \nabla f(\boldsymbol{x}_k)^\top B_k^{-2} \nabla f(\boldsymbol{x}_k) \\
&\leq \lambda_{\max}(B_k^{-2}) \|\nabla f(\boldsymbol{x}_k)\|_2^2 \\
&\leq \frac{1}{\lambda_{\min}^2} \|\nabla f(\boldsymbol{x}_k)\|_2^2
\end{aligned}$$

hold. Using them together leads to

$$\frac{|\boldsymbol{p}_k^\top \nabla f(\boldsymbol{x}_k)|}{\|\boldsymbol{p}_k\|_2} \geq \frac{\lambda_{\min}}{\lambda_{\max}} \|\nabla f(\boldsymbol{x}_k)\|_2.$$

Therefore, the important quantity converging to zero in Theorem 12.6 is an upper bound,

$$\min(|\boldsymbol{p}_k^\top \nabla f(\boldsymbol{x}_k)|, \frac{|\boldsymbol{p}_k^\top \nabla f(\boldsymbol{x}_k)|}{\|\boldsymbol{p}_k\|_2}) \geq \frac{1}{\lambda_{\max}} \|\nabla f(\boldsymbol{x}_k)\|_2 \min(\|\nabla f(\boldsymbol{x}_k)\|_2, \lambda_{\min}),$$

and since $\lambda_{\min} > 0$ and $\lambda_{\max}$ is finite, $\|\nabla f(\boldsymbol{x}_k)\|_2$ must converge to zero as k goes to infinity, which concludes the proof. $\square$

Here is a MATLAB implementation of the Armijo backtracking line search using the Newton search direction:

ALGORITHM 12.8. *Minimization with Newton Direction*

```
function [x,xk]=Newton(f,fp,fpp,x0,tol,maxiter,tau,be,alinit)
% NEWTON Minimization with Newton descent and Armijo line search
%    [x,xk]=Newton(f,fp,fpp,x0,tol,maxiter,tau,be,alinit) finds an
%    approximate minimum of the function f with gradient fp and Hessian
%    fpp, starting at the initial guess x0. The remaining parameters are
%    optional and default values are used if they are omited. xk
%    contains all the iterates of the method.

if nargin<9, alinit=1; end;
if nargin<8, be=0.1; end;
if nargin<7, tau=0.5; end;
if nargin<6, maxiter=100; end;
if nargin<5, tol=1e-6; end;

x=x0;
xk=x0;
p=-feval(fpp,x)\feval(fp,x);
k=0;
while norm(feval(fp,x))>tol & k<maxiter
  al=alinit;
  while feval(f,x+al*p)>feval(f,x)-al*be*p'*p
```

```
   al=tau*al;
 end;
 x=x+al*p
 p=-feval(fpp,x)\feval(fp,x);
 k=k+1;
 xk(:,k+1)=x;
end;
```

For the same example as in Figure 12.13, we show in Figure 12.14 how much better the Newton step performs on this problem. The following the-

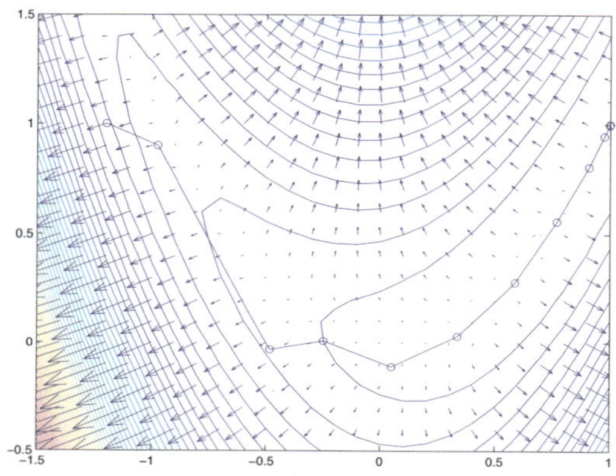

FIGURE 12.14.
Backtracking Armijo line search with Newton search
direction for a model problem showing much better
convergence behavior than the steepest descent direction
from Figure 12.13.

orem gives a convergence estimate for the backtracking Armijo line search with Newton direction.

THEOREM 12.8. *Let $f : \mathbb{R}^n \longrightarrow \mathbb{R}$ be twice continuously differentiable, and its Hessian matrix $H : \mathbb{R}^n \longrightarrow \mathbb{R}^{n \times n}$ be Lipschitz in $\mathbb{R}^n$ with Lipschitz constant L. If the sequence $\{x_k\}$, $k = 1, 2, \ldots$ generated by the generic minimization algorithm with backtracking Armijo line search and $\alpha_{init} = 1$, $\beta \in (0, \frac{1}{2})$ and*

$$p_k = \begin{cases} -H(x_k)^{-1}\nabla f(x_k) & \text{if } H(x_k) \text{ is positive definite} \\ -B_k^{-1}\nabla f(x_k) & B_k \text{ as in Theorem 12.7,} \end{cases}$$

has a limit point x^* with $H(x^*)$ positive definite, then

1. *$\alpha_k = 1$ for k large,*

2. *the entire sequence $\{x_k\}$ converges to x^*, and*

3. *the convergence rate is quadratic, i.e.*

$$\lim_{k\to\infty} \frac{\|\boldsymbol{x}_{k+1} - \boldsymbol{x}^*\|_2}{\|\boldsymbol{x}_k - \boldsymbol{x}^*\|_2^2} \leq C, \quad C > 0 \ a \ constant.$$

PROOF. We start by considering a subsequence $\{\boldsymbol{x}_{k'}\}$ that converges to $\boldsymbol{x}^*$. By continuity, $H(\boldsymbol{x}_{k'})$ is positive definite for k' large enough, and there exists a k'_0 such that for all $k' \geq k'_0$ we have

$$\boldsymbol{p}_{k'}^\top H(\boldsymbol{x}_{k'})\boldsymbol{p}_{k'} \geq \frac{1}{2}\lambda_{\min}(H(\boldsymbol{x}^*))\|\boldsymbol{p}_{k'}\|_2^2. \tag{12.29}$$

This implies with the Newton step $H(\boldsymbol{x}_{k'})\boldsymbol{p}_{k'} = -\nabla f(\boldsymbol{x}_{k'})$

$$|\boldsymbol{p}_{k'}^\top \nabla f(\boldsymbol{x}_{k'})| = -\boldsymbol{p}_{k'}^\top \nabla f(\boldsymbol{x}_{k'}) = \boldsymbol{p}_{k'}^\top H(\boldsymbol{x}_{k'})\boldsymbol{p}_{k'} \geq \frac{1}{2}\lambda_{\min}(H(\boldsymbol{x}^*))\|\boldsymbol{p}_{k'}\|_2^2, \tag{12.30}$$

and $\lim_{k'\to\infty} \boldsymbol{p}_{k'} = 0$ by Theorem 12.8.

Now from Taylor's Theorem with integral remainder term, we obtain

$$f(\boldsymbol{x}_k + \boldsymbol{p}_k) = f(\boldsymbol{x}_k) + \boldsymbol{p}_k^\top \nabla f(\boldsymbol{x}_k) + \int_0^1 (1-t)\boldsymbol{p}_k^\top H(\boldsymbol{x}_k + t\boldsymbol{p}_k)\boldsymbol{p}_k dt$$

$$= f(\boldsymbol{x}_k) + \boldsymbol{p}_k^\top \nabla f(\boldsymbol{x}_k) + \boldsymbol{p}_k^\top H(\boldsymbol{x}_k + \tau\boldsymbol{p}_k)\boldsymbol{p}_k \int_0^1 (1-t)dt, \ 0 \leq \tau \leq 1$$

$$= f(\boldsymbol{x}_k) + \boldsymbol{p}_k^\top \nabla f(\boldsymbol{x}_k) + \frac{1}{2}\boldsymbol{p}_k^\top H(\boldsymbol{x}_k + \tau\boldsymbol{p}_k)\boldsymbol{p}_k$$

and we can therefore estimate

$$\begin{aligned}
f(\boldsymbol{x}_k + \boldsymbol{p}_k) - f(\boldsymbol{x}_k) - \frac{1}{2}\boldsymbol{p}_k^\top \nabla f(\boldsymbol{x}_k) &= \frac{1}{2}\left(\boldsymbol{p}_k^\top \nabla f(\boldsymbol{x}_k) + \boldsymbol{p}_k^\top H(\boldsymbol{x}_k + \tau\boldsymbol{p}_k)\boldsymbol{p}_k\right) \\
&= \frac{1}{2}\left(\boldsymbol{p}_k^\top \nabla f(\boldsymbol{x}_k) + \boldsymbol{p}_k^\top H(\boldsymbol{x}_k)\boldsymbol{p}_k\right) \\
&+ \frac{1}{2}\left(\boldsymbol{p}_k^\top (H(\boldsymbol{x}_k + \tau\boldsymbol{p}_k) - H(\boldsymbol{x}_k))\boldsymbol{p}_k\right),
\end{aligned}$$

and since $H(\boldsymbol{x}_k)\boldsymbol{p}_k = -\nabla f(\boldsymbol{x}_k)$, the first term on the right hand side is identically zero. Using the Lipschitz condition $H(\boldsymbol{x}_k)$ satisfies, and that $0 \leq \tau \leq 1$, we obtain

$$f(\boldsymbol{x}_k + \boldsymbol{p}_k) - f(\boldsymbol{x}_k) - \frac{1}{2}\boldsymbol{p}_k^\top \nabla f(\boldsymbol{x}_k) \leq \frac{1}{2}L\|\boldsymbol{p}_k\|_2^3. \tag{12.31}$$

Hence, choosing k' large enough so that

$$L\|\boldsymbol{p}_{k'}\|_2 \leq \frac{1}{2}\lambda_{\min}(H(\boldsymbol{x}^*))(1 - 2\beta),$$

we obtain first using (12.31) and then (12.29) that

$$
\begin{aligned}
f(\boldsymbol{x}_{k'} + \boldsymbol{p}_{k'}) - f(\boldsymbol{x}_{k'}) \;&\leq\; \frac{1}{2}\boldsymbol{p}_{k'}^{\top}\nabla f(\boldsymbol{x}_{k'}) + \frac{1}{4}\lambda_{\min}(H(\boldsymbol{x}^*))(1 - 2\beta)\|\boldsymbol{p}_{k'}\|_2^2 \\
&\leq\; \frac{1}{2}(1 - (1 - 2\beta))\boldsymbol{p}_{k'}^{\top}\nabla f(\boldsymbol{x}_{k'}) \;=\; \beta\boldsymbol{p}_{k'}^{\top}\nabla f(\boldsymbol{x}_{k'}),
\end{aligned}
$$

which shows that for k' large enough, the Armijo condition is satisfied with $\alpha = 1$. In addition, we have for k' large enough $\|H(\boldsymbol{x}_k)^{-1}\|_2 \leq \frac{2}{\lambda_{\min}(H(\boldsymbol{x}^*))}$, and

$$
\begin{aligned}
\boldsymbol{x}_{k'+1} - \boldsymbol{x}^* &= \boldsymbol{x}_{k'} - H(\boldsymbol{x}_{k'})^{-1}\nabla f(\boldsymbol{x}_{k'}) - \boldsymbol{x}^* \\
&= \boldsymbol{x}_{k'} - H(\boldsymbol{x}_{k'})^{-1}(\nabla f(\boldsymbol{x}_{k'}) - \nabla f(\boldsymbol{x}^*)) - \boldsymbol{x}^* \quad (\text{note } \nabla f(\boldsymbol{x}^*) = 0) \\
&= H(\boldsymbol{x}_{k'})^{-1}(\nabla f(\boldsymbol{x}_{k'}) - \nabla f(\boldsymbol{x}^*) - H(\boldsymbol{x}_{k'})(\boldsymbol{x}^* - \boldsymbol{x}_{k'})),
\end{aligned}
$$

which allows us to estimate the difference in norm,

$$
\begin{aligned}
\|\boldsymbol{x}_{k'+1} - \boldsymbol{x}^*\|_2 &\leq \|H(\boldsymbol{x}_{k'})^{-1}\|_2\|(\nabla f(\boldsymbol{x}_{k'}) - \nabla f(\boldsymbol{x}^*) - H(\boldsymbol{x}_{k'})(\boldsymbol{x}^* - \boldsymbol{x}_{k'}))\|_2 \\
&= \|H(\boldsymbol{x}_{k'})^{-1}\|_2\Big\| \int_0^1 H(\boldsymbol{x}_{k'} + \tau(\boldsymbol{x}^* - \boldsymbol{x}_{k'}))(\boldsymbol{x}^* - \boldsymbol{x}_{k'})d\tau - H(\boldsymbol{x}_{k'})(\boldsymbol{x}^* - \boldsymbol{x}_{k'})\Big\|_2 \\
&\leq \|H(\boldsymbol{x}_{k'})^{-1}\|_2 \int_0^1 \|H(\boldsymbol{x}_{k'} + \tau(\boldsymbol{x}^* - \boldsymbol{x}_{k'})) - H(\boldsymbol{x}_{k'})\|_2\|\boldsymbol{x}^* - \boldsymbol{x}_{k'}\|_2 d\tau \\
&\leq \|H(\boldsymbol{x}_{k'})^{-1}\|_2 \frac{1}{2}L\|\boldsymbol{x}^* - \boldsymbol{x}_{k'}\|_2^2,
\end{aligned}
$$

and hence for k' large enough, we have

$$
\frac{\|\boldsymbol{x}_{k'+1} - \boldsymbol{x}^*\|_2}{\|\boldsymbol{x}_{k'} - \boldsymbol{x}^*\|_2^2} \leq \frac{L}{2}\|H(\boldsymbol{x}_{k'})^{-1}\|_2 \leq \frac{L}{\lambda_{\min}(H(\boldsymbol{x}^*))}
$$

which proves item 3 for $k = k'$ with $C = \frac{L}{\lambda_{\min}(H(\boldsymbol{x}^*))}$. In addition, this implies that for k' large enough, $k'+1$ is also an index in the convergent subsequence, which proves item 2, and thus item 1 and 3 for all k large enough. $\qquad\square$

Modified Newton Methods

In *modified Newton methods*, one replaces the Newton search direction determined by

$$
H(\boldsymbol{x}_k)\boldsymbol{p}_k = -\nabla f(\boldsymbol{x}_k)
$$

by a modified search direction determined by

$$
(H(\boldsymbol{x}_k) + M_k)\boldsymbol{p}_k = -\nabla f(\boldsymbol{x}_k),
$$

where the matrix M_k is chosen to guarantee that the matrix $H(\boldsymbol{x}_k) + M_k$ is sufficiently positive definite. There are several possibilities to obtain such a matrix M_k:

1. If one has an eigenvalue decomposition of the Hessian, $H(\boldsymbol{x}_k){=}Q_k^{\top}\Lambda_k Q_k$, one can choose

$$
H(\boldsymbol{x}_k) + M_k = Q_k^{\top}\max(\varepsilon I, \Lambda_k)Q_k,
$$

i.e., one shifts the eigenvalues that are too small or negative to $\varepsilon > 0$. The computation of the eigendecomposition can however be very expensive at each step.

2. If one has an estimate of $\lambda_{\min}(H(\boldsymbol{x}_k))$, the smallest eigenvalue of the Hessian, one can choose

$$M_k = \max(0, \varepsilon - \lambda_{\min}(H(\boldsymbol{x}_k))) \cdot I,$$

which shifts the whole spectrum whenever small or negative eigenvalues are present. This has the disadvantage that one large negative eigenvalue can distort the entire problem.

3. One can try to compute the Cholesky decomposition of the Hessian, which is needed anyway to solve the system defining the search direction $\boldsymbol{p}_k$, see Subsection 3.4. If the Cholesky decomposition fails due to small or negative eigenvalues, there are techniques to adapt it to obtain a decomposition of the form (see for example [30])

$$H(\boldsymbol{x}_k) + M_k = L_k L_k^\top.$$

Quasi-Newton Methods

Quasi-Newton methods were developed in order to reduce the cost of computing the Hessian, which is nowadays less of an issue. There are two main techniques:

1. Approximation of the Hessian by finite differences, i.e. for the i-th unit vector $\boldsymbol{e}_i$, one computes

$$H(\boldsymbol{x}_k)\boldsymbol{e}_i \approx \frac{\nabla f(\boldsymbol{x}_k + h\boldsymbol{e}_i) - \nabla f(\boldsymbol{x}_k)}{h}$$

and h is chosen approximately to be the square root of `macheps` to balance roundoff and approximation errors, see Section 8.2.

2. Using a secant approximation: one searches for a matrix B_k satisfying the so-called secant condition,

$$B_{k+1}(\boldsymbol{x}_{k+1} - \boldsymbol{x}_k) = \nabla f(\boldsymbol{x}_{k+1}) - \nabla f(\boldsymbol{x}_k),$$

which is motivated by the fact that the real Hessian matrix satisfies the secant condition approximately by Taylor's theorem. One has a lot of freedom in choosing a matrix B_k that satisfies the secant condition. Two of the most successful choices are

Symmetric rank-one method: Starting with a given matrix B_0, one computes recursively the rank-one updates

$$B_{k+1} = B_k + \frac{(\boldsymbol{y}_k - B_k \boldsymbol{s}_k)(\boldsymbol{y}_k - B_k \boldsymbol{s}_k)^\top}{\boldsymbol{s}_k^\top (\boldsymbol{y}_k - B_k \boldsymbol{s}_k)},$$

where $\boldsymbol{y}_k := \nabla f(\boldsymbol{x}_{k+1}) - \nabla f(\boldsymbol{x}_k)$ and $\boldsymbol{s}_k := \boldsymbol{x}_{k+1} - \boldsymbol{x}_k$. This approach can unfortunately lead to indefinite matrices B_k, or even fail with a division by zero.

Broyden–Fletcher–Goldfarb–Shanno (BFGS) method: One computes recursively

$$B_{k+1} = B_k + \frac{\boldsymbol{y}_k \boldsymbol{y}_k^\top}{\boldsymbol{s}_k^\top \boldsymbol{y}_k} - \frac{B_k \boldsymbol{s}_k \boldsymbol{s}_k^\top B_k}{\boldsymbol{s}_k^\top B_k \boldsymbol{s}_k}.$$

It is easy to show (see Problem 12.18) that if one starts with a matrix B_0 which is symmetric and positive definite, then all the B_k are symmetric and positive definite, provided the $\boldsymbol{s}_k^\top \boldsymbol{y}_k$ are positive. This latter condition can be guaranteed by a slight modification, see for example [19].

In general, the finite difference approximations of the Hessian in 1 are more expensive than the secant condition updates in 2.

Truncated Newton Methods

Truncated Newton methods are especially useful if the problem is of very high dimension. In that case, solving for the Newton direction using the linear system

$$H(\boldsymbol{x}_k)\boldsymbol{p}_k = -\nabla f(\boldsymbol{x}_k)$$

can be prohibitively expensive. The idea of truncated Newton methods is to solve the linear system approximately using the conjugate gradient method, see Section 11.7.1. If one starts the conjugate gradient method with the zero vector, the first iteration gives an approximate search direction $\boldsymbol{p}_k$ that is identical to steepest descent. If one iterates with the conjugate gradient method to convergence, one obtains the Newton search direction. Stopping anywhere in between gives an approximate search direction between steepest descent and the Newton direction.

12.3.2   Trust Region Methods

Trust region methods construct a local model $m(\boldsymbol{s})$ of the function $f(\boldsymbol{x}_k + \boldsymbol{s})$ and then accept a step $\boldsymbol{x}_{k+1} = \boldsymbol{x}_k + \boldsymbol{s}_k$, if $f(\boldsymbol{x}_k + \boldsymbol{s}_k)$ gives a similar decrease compared to the decrease predicted by the local model $m(\boldsymbol{s}_k)$. The two most commonly used models are

Linear model: $m_l(\boldsymbol{s}) := f(\boldsymbol{x}_k) + \boldsymbol{s}^\top \nabla f(\boldsymbol{x}_k)$

Quadratic model: $m_q(\boldsymbol{s}) := f(\boldsymbol{x}_k) + \boldsymbol{s}^\top \nabla f(\boldsymbol{x}_k) + \frac{1}{2}\boldsymbol{s}^\top B_k \boldsymbol{s}$, where the matrix B_k is an approximation to the Hessian at $\boldsymbol{x}_k$.

Since these models cannot be accurate for large s (the linear model is not even bounded from below), one imposes in addition the so-called trust region constraint,

$$\|s\| \le R_k, \tag{12.32}$$

and one computes at each step of the trust region algorithm

$$\min_{\|s\| \le R_k} m(s).$$

A generic *trust region algorithm* has the following form:

ALGORITHM 12.9. *Generic trust region algorithm*

For given x_0 and R_0;
$k = 0$;
while not converged
    $s_k = \operatorname*{argmin}\limits_{\|s\| \le R_k} m(s)$;
    $\rho = (f(x_k) - f(x_k + s_k))/(f(x_k) - m(s_k))$;
    if $\rho \ge \rho_s$
        $x_{k+1} = x_k + s_k$;
        if $\rho \ge \rho_v$
            $R_{k+1} = \gamma_i R_k$;
        else
            $R_{k+1} = R_k$;
        end;
    else
        $x_{k+1} = x_k$;
        $R_{k+1} = \gamma_d R_k$;
    end;
    $k = k + 1$;
end

After solving the trust region subproblem, the algorithm computes ρ, the ratio between the actual and predicted decrease. If this ratio is bigger than ρ_s, e.g. $\rho_s = 0.1$, a threshold for success, then the step is accepted. If the ratio is even bigger than ρ_v, $0 < \rho_s < \rho_v < 1$, e.g. $\rho_v = 0.9$, then the step was very successful, and we increase the trust region by the factor $\gamma_i \ge 1$, for example $\gamma_i = 2$. Otherwise, the size of the trust region is kept the same. If, however, the step was not a success, then we stay at the present position and decrease the trust region size by the factor γ_d, e.g. $\gamma_d = 0.5$.

One can show under certain hypotheses that the Trust Region Algorithm 12.9 leads to one of the following results, just like the line search algorithm:

1. $\nabla f(x_k) = 0$ for some $k \ge 0$.

2. $\lim_{k \to \infty} f(\boldsymbol{x}_k) = -\infty$.

3. $\lim_{k \to \infty} \nabla f(\boldsymbol{x}_k) = 0$.

To solve the trust region problem at each step, one can simply use an analytical solution in the case of the linear model, whereas for the quadratic model, there exist elegant numerical solutions, see for example [107].

Here is a MATLAB implementation of the trust region algorithm with a linear model:

ALGORITHM 12.10.
Trust Region Algorithm with a Linear Model

```
function [x,xk,Rk]=TrustRegionLinear(f,fp,x0,tol,maxiter,R0,rs,rv,gi,gd)
% TRUSTREGIONLINEAR trust region method with linear model
%    [x,xk,Rk]=TrustRegionLinear(f,fp,x0,tol,maxiter,r0,rs,rv,gi,gd)
%    finds an approximate minimum of the function f with gradient fp,
%    starting at the initial guess x0. The remaining parameters are
%    optional and default values are used if they are omited. xk
%    contains all the iterates of the method and rk the trust region
%    radii of all iterations.

if nargin<10, gd=0.5; end;
if nargin<9, gi=2; end;
if nargin<8, rv=0.9; end;
if nargin<7, rs=0.1; end;
if nargin<6, R0=1; end;
if nargin<5, maxiter=100; end;
if nargin<4, tol=1e-6; end;

x=x0;
xk=x0;
R=R0;
Rk=R0;
p=-feval(fp,x);
k=0;
while norm(p)>tol & k<maxiter
  s=R*p/norm(p);
  r=(feval(f,x)-feval(f,x+s))/(s'*p);
  if r>=rs
    x=x+s;
    p=-feval(fp,x);
    if r>-rv
      R=gi*R;
    end
  else
    R=gd*R
  end
  k=k+1;
```

```
  xk(:,k+1)=x;
  Rk(:,k+1)=R;
end;
```

The results for the same model problem as in Figure 12.13 are shown in Figure 12.15 on the left. We observe that the trust region method with a

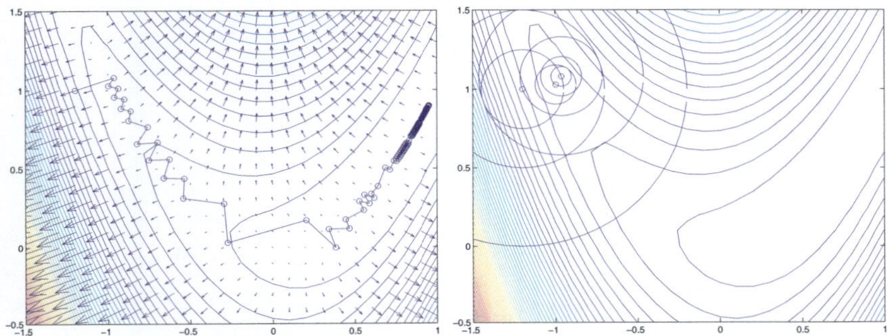

FIGURE 12.15.
*Trust region method with a linear model for an example,
showing that the method with the linear model has
similar problems as line search with steepest descent.*

linear model has a similar behavior to a line search method with steepest descent: the search can slow down dramatically in the bottom of a valley. On the right in Figure 12.15, we show for the first three steps how the trust region was adjusted to find an acceptable step.

12.3.3 Direct Methods

Direct optimization methods are methods that do not use derivative information, either explicitly or implicitly. We have seen an example of such a method in the introduction: Algorithm 12.1 uses the golden section to find the minimum of a scalar function, in the same way bisection is used for finding the zero of a scalar function.

Probably the most successful direct method is the *Nelder–Mead Algorithm*, invented in 1965. This algorithm uses a simplex, or a triangle in two dimensions, to explore the function $f : \mathbb{R}^n \to \mathbb{R}$ to be minimized. To start the algorithm, an initial simplex given by its corner points $\boldsymbol{x}_j, j = 1, 2, \ldots, n+1$ is needed. The first step in each iteration of the algorithm is to sort the corners such that $\boldsymbol{x}_1$ is the corner with the smallest, and $\boldsymbol{x}_{n+1}$ is the corner with the largest function value. Then the algorithm computes the center of gravity on the hypersurface described by all but $\boldsymbol{x}_{n+1}$,

$$\bar{\boldsymbol{x}} = \frac{1}{n} \sum_{j=1}^{n} \boldsymbol{x}_j,$$

and then the reflected point

$$x_r = \bar{x} + \rho(\bar{x} - x_{n+1}),$$

where ρ is a parameter, usually $\rho = 1$. An example for $n = 2$ is shown in Figure 12.16. Now the algorithm makes a first decision: if $f(x_r) < f(x_1)$,

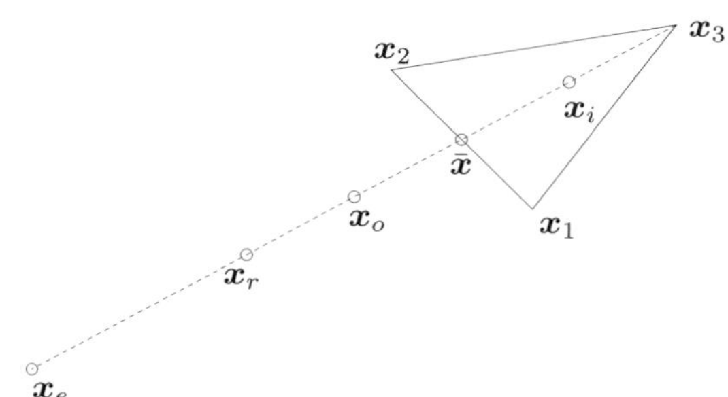

FIGURE 12.16.
The center of gravity $\bar{x}$, the reflection point x_r, the expansion point x_e and the two contraction points x_i and x_o, computed by the Nelder–Mead algorithm.

the direction x_r seems very promising, and the expansion point

$$x_e = \bar{x} + \chi(x_r - \bar{x})$$

is computed, with the parameter $\chi = 2$ usually, as shown in Figure 12.16. Now if $f(x_e) < f(x_r)$, then the expansion is accepted, and x_e replaces the worst point x_{n+1} to obtain a new simplex for the next iteration. Otherwise, if $f(x_r) < f(x_n)$, then the reflected point x_r is accepted and replaces the worst point x_{n+1} to form a new simplex for the next iteration. Finally, if $f(x_r) \geq f(x_n)$, the reflected point is not such a good direction, and a contraction step is tried. To do so, an inner and outer contraction point is computed,

$$x_i = \bar{x} - \gamma(x_r - \bar{x}), \quad x_o = \bar{x} + \gamma(x_r - \bar{x}),$$

with the usual parameter choice $\gamma = \frac{1}{2}$. This leads to the two points shown in Figure 12.16. Now if $f(x_r) < f(x_{n+1})$ and $f(x_o) < f(x_r)$, then x_o replaces x_{n+1} for a new simplex, and if $f(x_r) \geq f(x_{n+1})$ and $f(x_i) < f(x_{n+1})$, then x_i replaces x_{n+1} for a new simplex to restart the iteration. If neither of these conditions holds, a shrinking step is performed, in which all the vertices get closer to x_1, i.e.

$$x_j = x_1 + \sigma(x_j - x_1), \quad j = 2, \ldots, n+1,$$

with the parameter σ usually chosen to be $\frac{1}{2}$.

Here is a MATLAB implementation of the Nelder–Mead algorithm:

ALGORITHM 12.11. *Optimization with Nelder–Mead*

```
function x=NelderMead(f,x0,tol,maxiter,r,c,g,s)
% NELDERMEAD direct minimization algorithm by Nelder Mead
%   x=NelderMead(f,x0,tol,maxiter,r,c,g,s) tries to find a minimum of
%   f, starting at x0. The remaining parameters r, c, g, and s are
%   optional, and correspond to the parameters rho, chi, gamma, and
%   sigma; defaults are chosen, if they are not given.

if nargin<8, s=0.5; end;
if nargin<7, g=0.5; end;
if nargin<6, c=2; end;
if nargin<5, r=1; end;
if nargin<4, maxiter=100; end;
if nargin<3, tol=1e-6; end;

x=x0; k=0; fk=feval(f,x);
while k<maxiter
  [fk,id]=sort(fk); x=x(id,:);
  xb=mean(x(1:end-1,:)); xr=xb+r*(xb-x(end,:));
  fr=feval(f,xr);
  if fr>=fk(1) & fr<fk(end-1)                     % reflection step
    x(end,:)=xr; fk(end)=fr;
  elseif fr<fk(1)
    xe=xb+c*(xr-xb); fe=feval(f,xe);
    if fe<fr
      x(end,:)=xe; fk(end)=fe;
    else
      x(end,:)=xr; fk(end)=fr;
    end;
  else                                            % contraction step
    xi=xb-g*(xr-xb); fi=feval(f,xi);
    xo=xb+g*(xr-xb); fo=feval(f,xo);
    if fr<fk(end) & fo<fr
      x(end,:)=xo; fk(end)=fo;
    elseif fr>=fk(end) & fi<fk(end)
      x(end,:)=xi; fk(end)=fi;
    else                                          % shrinking step
      for j=2:size(x,1)
        x(j,:)=x(1,:)+s*(x(j,:)-x(1,:)); fk(j)=feval(f,x(j,:));
      end;
    end;
  end;
  k=k+1;
end;
```

Figure 12.17 shows how the Nelder–Mead algorithm approaches the minimum in the model problem of the previous sections. It is impressive how well

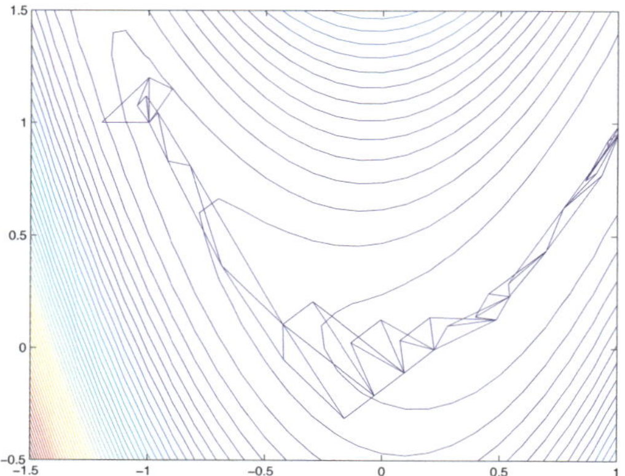

FIGURE 12.17. *Nelder–Mead algorithm for an example.*

the algorithm works on the model problem, and it has proved to work very well for innumerable applications. There is however no mathematical foundation for the Nelder–Mead algorithm: the only convergence result known is for strictly convex functions in one (!) dimension (see [81]). For two dimensional problems, there exist examples due to McKinnon (1998), that even in the strictly convex case, the Nelder–Mead algorithm can fail and converge to a degenerate simplex which does not indicate a minimum of the function. One does not even know if there exists a function in two dimensions for which Nelder–Mead works for arbitrary initial guesses. It currently remains a mystery why Nelder–Mead is so successful for so many practical problems.

12.4　Constrained Optimization

Constraint optimization problems contain a special subclass of problems, namely linear problems with linear constraints, for which Dantzig invented in 1951 a very elegant solution algorithm [23]. We start this section by explaining this well known *simplex algorithm.* We then treat the general case of constrained optimization problems with penalty and barrier functions, and also explain the revolutionary interior point methods, invented by Karmarkar in 1984 first for linear programming problems [79]. We close this section with sequential quadratic programming.

12.4.1　Linear Programming

Linear programming refers to the solution of constrained optimization problems in which both the objective function and the equality and inequality constraints in (12.13) are linear (or more precisely affine). After elimination of the equality constraints, we obtain the so-called linear program

$$
\begin{array}{rcrcccrcl}
\tilde{c}_1\tilde{x}_1 & + & \tilde{c}_2\tilde{x}_2 & + & \ldots & + & \tilde{c}_{\tilde{n}}\tilde{x}_{\tilde{n}} & \longrightarrow & \min, \\
\tilde{a}_{11}\tilde{x}_1 & + & \tilde{a}_{12}\tilde{x}_2 & + & \ldots & + & \tilde{a}_{1\tilde{n}}\tilde{x}_{\tilde{n}} & \leq & b_1, \\
\tilde{a}_{21}\tilde{x}_1 & + & \tilde{a}_{22}\tilde{x}_2 & + & \ldots & + & \tilde{a}_{2\tilde{n}}\tilde{x}_{\tilde{n}} & \leq & b_2, \\
\vdots & & \vdots & & & & \vdots & & \vdots \\
\tilde{a}_{m1}\tilde{x}_1 & + & \tilde{a}_{m2}\tilde{x}_2 & + & \ldots & + & \tilde{a}_{m\tilde{n}}\tilde{x}_{\tilde{n}} & \leq & b_m, \\
& & & & & & \tilde{x}_i & \geq & 0,
\end{array}
\tag{12.33}
$$

which one can write in more compact matrix notation,

$$
\begin{array}{rcl}
\tilde{c}^{\top}\tilde{x} & \longrightarrow & \min, \\
\tilde{A}\tilde{x} & \leq & b, \\
\tilde{x} & \geq & 0.
\end{array}
\tag{12.34}
$$

If some of the $\tilde{x}_i$ do not need to satisfy a sign constraint, one can artificially introduce two new variables $\tilde{x}_i^+ \geq 0$ and $\tilde{x}_i^- \geq 0$ and decompose $\tilde{x}_i = \tilde{x}_i^+ - \tilde{x}_i^-$ to fit the framework above. Dantzig invented in 1951 an elegant algorithm [23], the simplex algorithm, to solve problems of the form (12.34), which led to a real boom and the formation of entire operations research departments at universities. The simplex algorithm is listed as one of the *top ten algorithms of the last century* [27]. We now derive the simplex algorithm and illustrate it with the model problem from the introduction, subsection 12.1.3 on operations research.

The first step is the introduction of slack variables y_i, to transform the inequality constraints in (12.34) into equality constraints. This is achieved by setting $y = b - A\tilde{x}$, and the inequality constraint in (12.34) shows that the slack variables satisfy $y_i \geq 0$. This leads to the new equivalent problem

$$
\begin{array}{rcl}
\tilde{c}^{\top}\tilde{x} & \longrightarrow & \min, \\
\tilde{A}\tilde{x} + y & = & b, \\
\tilde{x} & \geq & 0, \\
y & \geq & 0.
\end{array}
\tag{12.35}
$$

It is worthwhile to interpret this problem before continuing: if one of the slack variables is zero, the corresponding underlying inequality constraints in (12.33) holds with equality, which implies that we are on the boundary of the region of admissible values of $\tilde{x}$; refer to Figure 12.8 for the example in the introduction. In this example, one would have four slack variables, and setting one of them equal to zero gives one of the four lines that form the boundaries of the region of admissible solution points. Two more lines are obtained by looking at $\tilde{x}_i = 0$ from the inequality constraints on $\tilde{x}_i$. Hence the slack variables and the real variables play a very similar role: setting one equal to zero gives a hyperplane which can be part of the boundary of the set of admissible solution points. If we combine in (12.35) the two vectors

$\tilde{x} \in \mathbb{R}^{\tilde{n}}$ and $y \in \mathbb{R}^m$ into the larger vector $x \in \mathbb{R}^n$, $n = \tilde{n} + m$, extend $\tilde{c}$ by zeros to obtain $c \in \mathbb{R}^n$, and form a new matrix $A \in \mathbb{R}^{m \times n}$ from $\tilde{A}$ and the identity matrix of size $m \times m$, we obtain in the new, more compact notation

$$
\begin{aligned}
c^\top x &\longrightarrow \text{min}, \\
Ax &= b, \\
x &\geq 0.
\end{aligned}
\tag{12.36}
$$

Again, setting any of the components x_i equal to zero means we are on a hyperplane which can be part of the boundary of the set of admissible solutions, and setting $\tilde{n}$ components equal to zero yields a corner point, provided the hyperplanes all intersect. This is the case when the remaining matrix after deleting the corresponding columns in A is invertible, so that the linear system in (12.36) has a unique solution after the corresponding x_i have been set to zero. Furthermore, if this solution is non-negative, it is a corner on the boundary of the set of admissible solutions, and thus a candidate for the solution of the optimization problem.

For our model problem (12.12) from Subsection 12.1.3, we need to introduce four slack variables: x_3, x_4, x_5 and x_6. The problem then becomes

$$
(5, 4, 0, 0, 0, 0)^\top x \rightarrow \text{min},
$$

$$
\begin{pmatrix}
1 & 1 & 1 & 0 & 0 & 0 \\
-1 & -1 & 0 & 1 & 0 & 0 \\
1 & 0 & 0 & 0 & 1 & 0 \\
0 & 1 & 0 & 0 & 0 & 1
\end{pmatrix}
x =
\begin{pmatrix}
8 \\
-5 \\
4 \\
7
\end{pmatrix},
\tag{12.37}
$$

$$
x_i \geq 0.
$$

Choosing for example $x_3 = x_5 = 0$ leads to an invertible linear system for the remaining variables,

$$
\begin{pmatrix}
1 & 1 & 0 & 0 \\
-1 & -1 & 1 & 0 \\
1 & 0 & 0 & 0 \\
0 & 1 & 0 & 1
\end{pmatrix}
\begin{pmatrix}
x_1 \\
x_2 \\
x_4 \\
x_6
\end{pmatrix}
=
\begin{pmatrix}
8 \\
-5 \\
4 \\
7
\end{pmatrix},
\tag{12.38}
$$

whose solution is $(x_1, x_2, x_4, x_6) = (4, 4, 3, 3) \geq 0$, and hence is the upper right-hand corner $(4, 4)$ on the Figure 12.8, the intersection of the hyperplanes $x_3 = x_5 = 0$.

Without loss of generality, we assume in the sequel that A has full rank, since otherwise there is either no solution, or one can eliminate equations from the linear system in (12.36) without changing the solutions. Motivated by the previous interpretation of (12.36) when setting some components equal to zero, we introduce the partitioning of the columns of the matrix A into matrices A_B and A_R using the index sets $B = \{b_1, \ldots, b_m\}$ and $R = \{r_1, \ldots, r_{n-m}\}$, $B \cup R = \{1, 2, \ldots, n\}$, $B \cap R = \emptyset$,

$$
A = (A_1, A_2, \cdots, A_n), \quad A_B = (A_{b_1}, A_{b_2}, \cdots, A_{b_m}), \quad A_R = (A_{r_1}, A_{r_2}, \cdots, A_{r_{n-m}}),
$$

and similarly we also partition the vectors,

$$
\boldsymbol{x}_B = \begin{pmatrix} x_{b_1} \\ x_{b_2} \\ \vdots \\ x_{b_m} \end{pmatrix}, \quad
\boldsymbol{x}_R = \begin{pmatrix} x_{r_1} \\ x_{r_2} \\ \vdots \\ x_{r_{n-m}} \end{pmatrix}, \quad
\boldsymbol{c}_B = \begin{pmatrix} c_{b_1} \\ c_{b_2} \\ \vdots \\ c_{b_m} \end{pmatrix}, \quad
\boldsymbol{c}_R = \begin{pmatrix} c_{r_1} \\ c_{r_2} \\ \vdots \\ c_{r_{n-m}} \end{pmatrix}.
$$

In this new notation, the linear program (12.36) becomes

$$
\begin{aligned}
\boldsymbol{c}_B^\top \boldsymbol{x}_B + \boldsymbol{c}_R^\top \boldsymbol{x}_R &\longrightarrow \quad \min, \\
A_B \boldsymbol{x}_B + A_R \boldsymbol{x}_R &= \quad \boldsymbol{b}, \\
\boldsymbol{x}_B &\geq \quad \boldsymbol{0}, \\
\boldsymbol{x}_R &\geq \quad \boldsymbol{0},
\end{aligned}
\tag{12.39}
$$

and setting $\boldsymbol{x}_R$ equal to zero corresponds to choosing a corner of the set of admissible solutions, provided that A_B is invertible, and $A_B^{-1}\boldsymbol{b} \geq 0$. This observation motivates the following definition:

DEFINITION 12.2. (BASIS) *The set B is a basis if A_B is invertible, and the associated vector*

$$
\boldsymbol{x} = \begin{pmatrix} \boldsymbol{x}_B \\ \boldsymbol{x}_R \end{pmatrix} = \begin{pmatrix} A_B^{-1}\boldsymbol{b} \\ \boldsymbol{0} \end{pmatrix}
$$

is called the basis solution. It is an admissible basis if $\boldsymbol{x} \geq 0$.

If the matrix A_B is invertible, we can eliminate the unknowns $\boldsymbol{x}_B$ from the problem (12.39) using the relation $\boldsymbol{x}_B = A_B^{-1}(\boldsymbol{b} - A_R\boldsymbol{x}_R)$, which leads to the new problem

$$
\begin{aligned}
\boldsymbol{u}^\top \boldsymbol{x}_R + z &\longrightarrow \quad \min, \\
S\boldsymbol{x}_R &\leq \quad \boldsymbol{t}, \\
\boldsymbol{x}_R &\geq \quad 0,
\end{aligned}
\tag{12.40}
$$

where $\boldsymbol{u}^\top = \boldsymbol{c}_R^\top - \boldsymbol{c}_B^\top A_B^{-1} A_R$, $z = \boldsymbol{c}_B^\top A_B^{-1}\boldsymbol{b}$, $S = A_B^{-1} A_R$ and $\boldsymbol{t} = A_B^{-1}\boldsymbol{b}$. Dantzig collected all the relevant quantities in the *simplex table*

$$
\begin{array}{|c|c|c|}
\hline
 & R & \\
\hline
B & S & \boldsymbol{t} \\
\hline
 & \boldsymbol{u}^\top & z \\
\hline
\end{array}
\tag{12.41}
$$

THEOREM 12.9. (SIMPLEX CRITERION) *Let B be a basis of the linear program (12.36) and let (12.41) be the associated simplex table. If $\boldsymbol{t} \geq 0$ and $\boldsymbol{u} \geq 0$, then $\boldsymbol{x}_B = \boldsymbol{t}$ and $\boldsymbol{x}_R = 0$ is the solution of the linear program (12.36).*

PROOF. The basis B is admissible, since $\boldsymbol{t} = A_B^{-1}\boldsymbol{b} \geq 0$. In addition, $\boldsymbol{u} \geq 0$ for a point $\boldsymbol{x}$ which satisfies $A\boldsymbol{x} = \boldsymbol{b}$ and $\boldsymbol{x} \geq 0$ implies that

$$
\boldsymbol{c}^\top \boldsymbol{x} = \boldsymbol{u}^\top \boldsymbol{x}_R + z \geq z = \boldsymbol{c}^\top \begin{pmatrix} \boldsymbol{t} \\ \boldsymbol{0} \end{pmatrix},
$$

which shows that there is no other solution with a smaller objective function value, and hence concludes the proof. $\qquad\square$

The simplex algorithm by Dantzig is a directed search along the edges of the set of admissible solutions to find the corner which minimizes the objective function. It starts with an admissible basis solution and its associated simplex table, and then constructs iteratively new admissible basis solutions and simplex tables with smaller and smaller values of the objective function, until the simplex criterion is satisfied. To transform one simplex table into another, the following theorem is useful.

THEOREM 12.10. (SIMPLEX TABLE TRANSFORM) *Let B be a basis, $R = \{1,\ldots,n\}\backslash B$, $k \in B$ and $l \in R$, with the associated simplex table*

$$
\begin{array}{|c|c|c|} \hline & R & \\ \hline B & S & t \\ \hline & \boldsymbol{u}^\top & z \\ \hline \end{array}
=
\begin{array}{|c|c|c|c|} \hline & l & j & \\ \hline k & s_{kl} & s_{kj} & t_k \\ \hline i & s_{il} & s_{ij} & t_i \\ \hline & u_l & u_j & z \\ \hline \end{array}
\, , \quad i \in B\backslash\{k\}, \; j \in R\backslash\{l\}.
$$

$$(12.42)$$

If the so-called pivot $s_{k,l}$ is non-zero, then

1. *$\tilde{B} := (B \cup \{l\})\backslash\{k\}$ is also a basis,*

2. *the corresponding new simplex table is*

$$
\begin{array}{|c|c|c|} \hline & \tilde{R} & \\ \hline \tilde{B} & \tilde{S} & \tilde{t} \\ \hline & \tilde{\boldsymbol{u}}^\top & \tilde{z} \\ \hline \end{array}
=
\begin{array}{|c|c|c|c|} \hline & k & j & \\ \hline l & \dfrac{1}{s_{kl}} & \dfrac{s_{kj}}{s_{kl}} & \dfrac{t_k}{s_{kl}} \\ \hline i & -\dfrac{s_{il}}{s_{kl}} & s_{ij} - \dfrac{s_{kj}s_{il}}{s_{kl}} & t_i - \dfrac{t_k s_{il}}{s_{kl}} \\ \hline & -\dfrac{u_l}{s_{kl}} & u_j - \dfrac{s_{kj}u_l}{s_{kl}} & z + \dfrac{t_k u_l}{s_{kl}} \\ \hline \end{array} \; .
$$

$$(12.43)$$

PROOF. The k-th component of the equation $\boldsymbol{x}_B + S\boldsymbol{x}_R = \boldsymbol{t}$ is

$$
x_k + \sum_{j \in R} s_{kj}x_j = t_k.
$$

Since the pivot $s_{kl} \neq 0$, we can solve this equation for x_l for $l \in R$, and obtain

$$
x_l + \frac{1}{s_{kl}}x_k + \sum_{j \in R\backslash\{l\}} \frac{s_{kj}}{s_{kl}}x_j = \frac{t_k}{s_{kl}},
$$

$$(12.44)$$

which corresponds to the l-th row in the new simplex table (12.43). Replacing now x_l from (12.44) into the i-th component of the equation $\boldsymbol{x}_B + S\boldsymbol{x}_R = \boldsymbol{t}$, $i \in B\backslash\{k\}$, we obtain

$$
x_i + s_{il}\left(-\frac{1}{s_{kl}}x_k - \sum_{j \in R\backslash\{l\}} \frac{s_{kj}}{s_{kl}}x_j + \frac{t_k}{s_{kl}} \right) + \sum_{j \in R\backslash\{l\}} s_{ij}x_j = t_i,
$$

which becomes after rearranging

$$x_i - \frac{s_{il}}{s_{kl}}x_k + \sum_{j \in R\backslash\{l\}} \left(s_{ij} - \frac{s_{il}s_{kj}}{s_{kl}}\right)x_j = t_i - \frac{t_k s_{il}}{s_{kl}},$$

which corresponds to the i-th row in the new simplex table (12.43). Similarly, one also obtains the last row of the new simplex table, which concludes the proof. $\square$

Theorem 12.10 allows us to easily compute a new simplex table from a previous one:

```
while there exists l in R with u(l)<0
  if s(i,l)<=0 for all i in B
    error('no solution, the objective function is unbounded');
  else
    choose k in B such that t(k)/s(k,l) is minimal
            for k in B with s(k,l)>0
    exchange k and l using the Theorem
  end;
end;
```

We explain each line in the simplex algorithm separately:

- The algorithm stops as soon as the simplex criterion from Theorem 12.9 is satisfied, $\boldsymbol{u} \geq 0$, which means the solution has been found.

- If for some $l \in R$, we have $u_l < 0$, and if $s_{il} \leq 0$ for all $i \in B$, then the objective function is unbounded from below on the set of admissible points, as one can see as follows: let $\boldsymbol{x}_R(\alpha)$ be defined by $x_l(\alpha) = \alpha$ for $\alpha \geq 0$, and $x_j(\alpha) = 0$ for $j \in R\backslash\{l\}$. Then $\boldsymbol{x}_R(\alpha) \geq 0$ and

$$S\boldsymbol{x}_R(\alpha) = \alpha \begin{pmatrix} s_{1l} \\ s_{2l} \\ \vdots \\ s_{ml} \end{pmatrix} \leq \boldsymbol{0} \leq \boldsymbol{t}.$$

Hence, the points $\boldsymbol{x}_R(\alpha)$ are admissible for all $\alpha \geq 0$. However, the objective function can then be made arbitrarily large negative, since

$$\boldsymbol{u}^\top \boldsymbol{x}_R(\alpha) + z = u_l\alpha + z \to -\infty, \quad \text{for } \alpha \to \infty,$$

so there is no minimum.

- The particular choice of l and k in the algorithm leads to a new admissible simplex table, $\tilde{\boldsymbol{t}} \geq 0$, with improved objective $-\tilde{z} \leq -z$, as one can see as follows: the new components of $\boldsymbol{t}$ for $i \in B\backslash\{k\}$ are $\tilde{t}_i = t_i - \frac{t_k s_{il}}{s_{kl}}$, which are bigger than or equal to $t_i > 0$ if $s_{il} \leq 0$, since

$t_k \geq 0$ and $s_{kl} > 0$. If $s_{il} > 0$, then $\tilde{t}_i = s_{il}\left(\frac{t_i}{s_{il}} - \frac{t_k}{s_{kl}}\right) \geq 0$, since k has been chosen such that the ratio was the smallest in the algorithm.

Now for the objective, we have $\tilde{z} = z + \frac{t_k u_l}{s_{kl}} \leq z$, since $u_l < 0$, $t_k \geq 0$ and $s_{kl} > 0$, and thus the new objective is at least as good or better than the old one.

Here is an implementation of the simplex algorithm in MATLAB:

ALGORITHM 12.12.
*Simplex Algorithm to Solve Linear Programming
Problem*

```
function R=Simplex(B,R,S,t,u,z);
% SIMPLEX Simplex algorithm starting with a simplex table
%    R=Simplex(B,R,S,t,u,z); implements the simplex algorithm starting
%    with a given simplex table consisting of B,R,S,t,u and z and finds
%    the optimal solution.

l=find(u<0);
while length(l)>0
  l=l(1);
  i=find(S(:,l)>0);
  if length(i)<1
    error('solution is unbounded');
  else
    [d,k]=min(t(i)./S(i,l));k=i(k);
    t([1:k-1 k+1:end])=t([1:k-1 k+1:end])-...
                    t(k)*S([1:k-1 k+1:end],l)/S(k,l);
    z=z+t(k)*u(l)/S(k,l);
    t(k)=t(k)/S(k,l);
    u([1:l-1 l+1:end])=u([1:l-1 l+1:end])-...
                    S(k,[1:l-1 l+1:end])*u(l)/S(k,l);
    u(l)=-u(l)/S(k,l);
    S([1:k-1 k+1:end],[1:l-1 l+1:end])=...
                    S([1:k-1 k+1:end],[1:l-1 l+1:end])...
      -S([1:k-1 k+1:end],l)*S(k,[1:l-1 l+1:end])/S(k,l);
    S(k,[1:l-1 l+1:end])=S(k,[1:l-1 l+1:end])/S(k,l);
    S([1:k-1 k+1:end],l)=-S([1:k-1 k+1:end],l)/S(k,l);
    S(k,l)=1/S(k,l);
    d=R(l);R(l)=B(k);B(k)=d;
    l=find(u<0);
    [0 R 0
     B' S t
     0 u z]
    pause
  end;
end;
```

The three lines before the command `pause` display the simplex tables generated and are for illustrative purposes only. To see how the simplex algorithm works, we apply it to the model problem (12.12) from Subsection 12.1.3. From the reformulation in (12.37), we define in MATLAB

```
A=[ 1  1 1 0 0 0
   -1 -1 0 1 0 0
    1  0 0 0 1 0
    0  1 0 0 0 1];
b=[8;-5;4;7];
c=[5;4;0;0;0;0];
```

and to start with the admissible basis solution found in (12.38), we define the sets B and R to be

```
B=[1 2 4 6]; R=[3 5];
```

We can now compute the corresponding simplex table, take a look at it, and then start the simplex algorithm:

```
S=A(:,B)\A(:,R);
t=A(:,B)\b;
u=c(R)'-c(B)'*(A(:,B)\A(:,R));
z=c(B)'*(A(:,B)\b);
[ 0  R 0
   B' S t
   0  u z]
Simplex(B,R,S,t,u,z);
```

The algorithm computes from the starting simplex table two new ones and then stops:

```
0   3   5   0        0   4   5   0        0   4   1   0
1   0   1   4        1   0   1   4        5   0   1   4
2   1  -1   4        2  -1  -1   1        2  -1   1   5
4   1   0   3        3   1   0   3        3   1   0   3
6  -1   1   3        6   1   1   6        6   1  -1   2
0  -4  -1  36        0   4  -1  24        0   4   1  20
```

Finding an admissible basis solution

The program `Simplex` will find the optimal solution provided the initial basis B yields an admissible solution. Unfortunately, the choice of this initial basis is not always obvious. For instance, if we had chosen R = [1,2], B = [3,4,5,6], the solution would not be admissible: indeed, looking at Figure 12.8, we see that $(x_1, x_2) = (0,0)$ is not within the shaded area, and hence is not an admissible solution. In other cases, an admissible basis may not even exist, e.g., if the set defined by the constraints turns out to be empty. The question becomes: how can one decide whether an admissible basis exists, and if it does, how can one compute it?

Consider once again the linear program whose constraints are

$$Sx_R \leq t, \qquad\qquad y + Sx_R = t,$$
$$x_R \geq 0 \qquad \Longleftrightarrow \qquad x_R \geq 0, \quad y \geq 0.$$

Here, we have omitted the objective function, since it does not affect whether a basis is admissible or not. If $t \geq 0$, i.e., if all components of t are non-negative, then setting $x_R = 0$ yields an admissible solution, so the basis $x_B = y$ is admissible. On the other hand, if $t_i < 0$ for some i, then the i-th constraint renders the basis inadmissible. To make the basis admissible again, we consider a *modified* problem in which we introduce one *gap variable* $z_i \geq 0$ per negative t_i, and replace the offending constraint by

$$y_i + \sum_{j \in \mathbb{R}} s_{ij} x_j = t_i + z_i, \qquad \text{whenever } t_i < 0,$$

i.e., we replace t_i by $t_i + z_i$. Thus, z_i is the gap between the original and modified constraints. Notice that for large enough z_i, we have $t_i + z_i \geq 0$, so it suffices to choose z_i instead of y_i as the basis variable to obtain an admissible solution. To see if the original problem has an admissible solution (and a corresponding basis), we want to minimize the gap. Thus, we obtain the modified problem

$$\sum_{t_i < 0} z_i \quad \longrightarrow \quad \min,$$
$$y_i + \sum_{j \in R} s_{ij} x_j \quad = \quad t_i, \qquad \text{if } t_i \geq 0,$$
$$z_i - y_i - \sum_{j \in R} s_{ij} x_j \quad = \quad -t_i, \qquad \text{if } t_i < 0,$$

for which $B = \{y_i \mid t_i \geq 0\} \cup \{z_i \mid t_i < 0\}$ is an admissible basis. After solving this linear program, if we obtain $\sum z_i = 0$, then the z_i must be zero individually. In other words, the z_i are redundant, and the corresponding basis is in fact an admissible basis for the original problem. If, on the other hand, we have $\sum z_i > 0$, then at least one of the gap variables is essential to the existence of a solution, so the original problem has no solution. The following algorithm computes an *admissible basis* for a given linear program.

ALGORITHM 12.13. *Computing an Admissible Basis*

```
function [B,R]=AdmissibleBasis(B,R,S,t)
% ADMISSIBLEBASIS Finds an admissible basis for a linear program
%    [B,R]=AdmissibleBasis(B,R,S,t) computes the sets of basis and
%    slack variables that yield an admissible basis solution. This is
%    done by solving an auxiliary linear program using SIMPLEX. The
%    result can then be used by SIMPLEX to find the minimum of the
```

```
%    original linear program.  A warning is produced if no admissible
%    solution exists.

u=find(t>=0);
l=find(t<0);
A=zeros(length(t),length(B)+length(R)+length(l));
[m,n]=size(A);

A(:,R)=S;
A(:,B)=eye(length(B));
A(l,:)=-A(l,:);
A(l,n-length(l)+1:end)=eye(length(l));
t(l)=-t(l);

Ba=[B(u),n-length(l)+1:n];
Ra=setdiff(1:n,Ba);
Sa=A(:,Ba)\A(:,Ra);
ta=A(:,Ba)\t;
ca=[zeros(n-length(l),1);ones(length(l),1)];
ua=ca(Ra)'-ca(Ba)'*(A(:,Ba)\A(:,Ra));
za=ca(Ba)'*(A(:,Ba)\t);
[0 Ra 0
 Ba' Sa ta
 0 ua za]
Ra=Simplex(Ba,Ra,Sa,ta,ua,za);
Ba=setdiff(1:n,Ra);
z=ca(Ba)'*(A(:,Ba)\t);
if (z>0),
  warning('no admissible solution exists');
else
  R=setdiff(Ra,(n-length(l)+1:n));
  B=Ba;
end;
```

Consider once again the linear program

```
A=[ 1   1 1 0 0 0
   -1 -1 0 1 0 0
    1  0 0 0 1 0
    0  1 0 0 0 1];
b=[8;-5;4;7];
c=[5;4;0;0;0;0];
```

This time, assume that we have not yet found an admissible basis, so we choose the slack variables instead as a basis:

```
B=[3 4 5 6]; R=[1 2];
```

Now the call

```
[Ba,Ra]=AdmissibleBasis(B,R,A(:,R),b)
```

uses `Simplex` as a subroutine to produce the basis

```
Ba =
     1     2     3     6
Ra =
     4     5
```

which also appears as a basis in the output of `Simplex` (specifically, in the middle table) on page 869.

12.4.2   Penalty and Barrier Functions

The simplex algorithm is a very elegant method that marches along the edges of the polytope described by the linear constraints toward the node with the smallest value of the objective function. The method can however be very slow, as illustrated in the example in Figure 12.18. Clearly it would be more

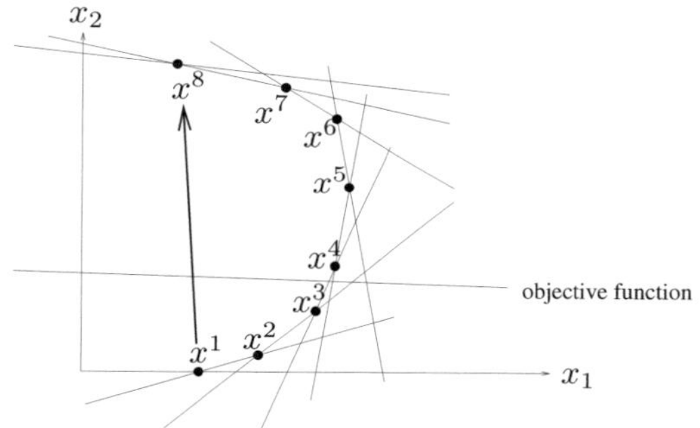

FIGURE 12.18.
An example where the simplex method could take many
steps.

efficient to follow a more direct path, as indicated by the bold arrow. Karmarkar from the AT&T Bell Labs revolutionized linear programming in 1984, when he presented a so-called *interior point method* which precisely followed a straighter path in the interior of the domain bounded by the constraints, instead of moving a long the boundary as the simplex algorithm does. Subsequently, AT&T developed the KORBX system, a hardware-software implementation of various interior point methods, and offered it for sale in 1989 for $8.9 million. Public domain versions of such interior point methods quickly followed, since their complexity is provably better than the worst-case complexity of the simplex algorithm.

Nonetheless, the fundamental idea of interior point methods is not new: in an optimization problem with equality constraints,

$$
\begin{aligned}
f(\boldsymbol{x}) &\longrightarrow \quad \min \\
\boldsymbol{c}_E(\boldsymbol{x}) &= \quad 0,
\end{aligned}
$$

we can replace the constraint by a *penalty function*, and then solve the unconstrained optimization problem

$$\phi(\boldsymbol{x}) := f(\boldsymbol{x}) + \frac{1}{2\mu}\|\boldsymbol{c}_E(\boldsymbol{x})\|_2^2 \longrightarrow \min,$$

for smaller and smaller values of the parameter $\mu > 0$. The penalty function we used here is quadratic, but there are many other possibilities. One can show that as μ converges to zero from above, some of the minima of ϕ converge to solutions of the constrained optimization problem. Unfortunately, it is also possible that ϕ has other stationary points as μ tends towards zero.

Similarly, in the optimization problem with inequality constraints

$$\begin{aligned} f(\boldsymbol{x}) &\longrightarrow \min \\ \boldsymbol{c}_I(\boldsymbol{x}) &\geq 0, \end{aligned}$$

we can replace the constraint by a *barrier function*, and then solve the unconstrained optimization problem

$$\phi(\boldsymbol{x}) := f(\boldsymbol{x}) - \mu \sum_{i=1}^{m} \log((\boldsymbol{c}_I)_i(\boldsymbol{x})) \longrightarrow \min, \tag{12.45}$$

for smaller and smaller values of the parameter $\mu > 0$. The log barrier function is the one most often used. One can again show that as μ approaches zero from above, some of the minima of ϕ converge to solutions of the constrained optimization problem, but once again it is possible that ϕ has other stationary points when μ converges to zero. Figure 12.19 shows two simple examples of the sequence of unconstrained problems obtained from a constrained one using the log barrier function: on the left, we want to minimize $f(x) = x$ under the constraint $x \geq 0$, and on the right $f(\boldsymbol{x}) = x_1^2 + x_2^2$ under the constraint $x_1 + x_2^2 \geq 1$. One can clearly see in these two examples how the minimum of the unconstrained problem with the log barrier function approaches the minimum of the constrained problem. The graphs on the right in Figure 12.19 were obtained for various values of μ with the MAPLE commands

```
f:=x1^2+x2^2;
phi:=f-mu*log(x1+x2^2-1);
mu:=1;
P1:=plots[contourplot](phi,x1=0..2,x2=
    max(0,sign(1-x1^2)*sqrt(abs(1-x1^2)))..2,
    axes=boxed,contours=[seq(i/4,i=4..36)],grid=[200,200]):
P2:=plot(sqrt(1-x1),x1=0..2):
plots[display](P1,P2);
```

12.4.3 Interior Point Methods

A generic *interior point method* uses a barrier function to convert a minimization problem with inequality constraints into a sequence of unconstrained

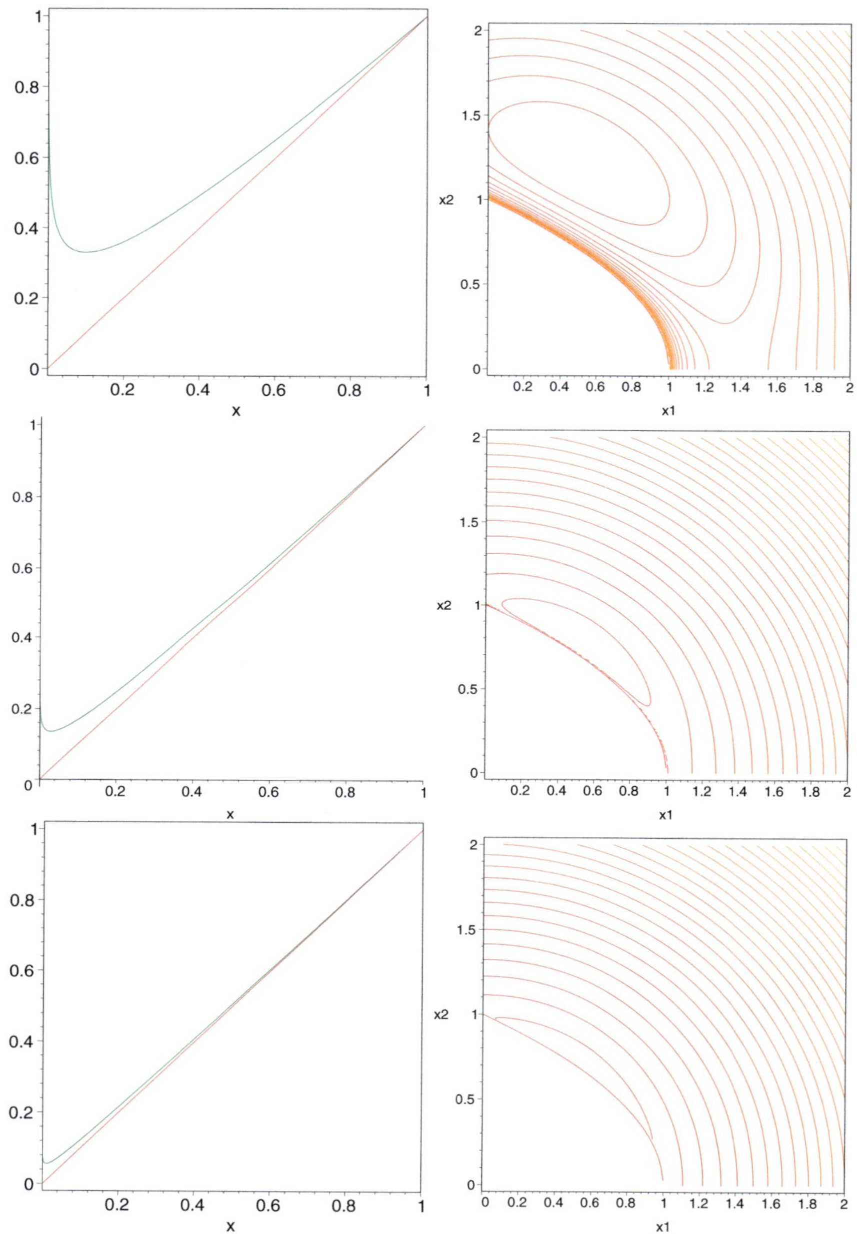

FIGURE 12.19.

Minimizing $f(x) = x$ under the constraint $x \geq 0$ using the log barrier function with $\mu = 0.1, 0.03, 0.01$ on the left, and $f(\boldsymbol{x}) = x_1^2 + x_2^2$ under the constraint $x_1 + x_2^2 \geq 1$ with the log barrier function and $\mu = 1, 0.1, 0.01$ on the right.

minimization problems, which are then solved to get better and better approximations of the original problem:

ALGORITHM 12.14. *Generic interior point method*

```
For given μ₀ > 0;
k = 0;
while not converged
    find xₖˢ with c(xₖˢ) > 0;      % e.g. xₖˢ = x*ₖ₋₁
    starting with xₖˢ, compute an approximate minimum
    xₖ of φ(x, μₖ);
    compute μₖ₊₁ > 0 smaller than μₖ,
    s.t. limₖ→∞ μₖ = 0;           e.g. μₖ₊₁ = 0.1μₖ or μₖ₊₁ = μₖ²
    k = k + 1;
end
```

The following theorem shows that under suitable conditions, algorithm 12.14 produces a better and better approximations to a point satisfying the first order optimality conditions of the original problem, given in Theorem 12.4.

THEOREM 12.11. *For $f : \mathbb{R}^n \to \mathbb{R}$ and $c : \mathbb{R}^n \to \mathbb{R}^m$ twice continuously differentiable, let*

$$(\lambda_k)_i := \frac{\mu_k}{c_i(\boldsymbol{x}_k)} \quad i = 1, 2, \ldots, m, \tag{12.46}$$

and assume that

$$\|\nabla_x \phi(\boldsymbol{x}_k, \mu_k)\|_2 \leq \epsilon_k, \tag{12.47}$$

where $\epsilon_k, \mu_k \to 0$ as $k \to \infty$. Assume further that $\boldsymbol{x}_k \longrightarrow \boldsymbol{x}^*$ and that the vectors $\nabla c_i(\boldsymbol{x}^*)$, corresponding to the set A of active constraints at $\boldsymbol{x}^*$, are linearly independent. Then $\boldsymbol{x}^*$ satisfies the first order optimality conditions of Theorem 12.4, and $\lambda_k \to \lambda$, the associated Lagrange multipliers.

PROOF. Let I be the complement of A at $\boldsymbol{x}^*$, and let $G(\boldsymbol{x}) := \nabla c(\boldsymbol{x})^\top$ be the Jacobian of c, with $G_A(\boldsymbol{x})$ being the rows of $\nabla c_i(\boldsymbol{x})^\top$ with $i \in A$ and $G_I(\boldsymbol{x})$ the rows $\nabla c_i(\boldsymbol{x})^\top$ with $i \in I$. By the assumption that $G_A(\boldsymbol{x}^*)$ has full rank, we have that in a neighborhood of $\boldsymbol{x}^*$, the pseudo-inverse (see Subsection 6.3.1)

$$G_A^+(\boldsymbol{x}) := (G_A(\boldsymbol{x})G_A(\boldsymbol{x})^\top)^{-1}G_A(\boldsymbol{x})$$

is well defined. Now let $\boldsymbol{\lambda}_A := G_A^+(\boldsymbol{x}^*)\nabla f(\boldsymbol{x}^*)$ and $\boldsymbol{\lambda}_I := 0$. If the set I is not empty, then using the definition (12.46), we can estimate for k large enough

$$
\begin{aligned}
\|(\boldsymbol{\lambda}_k)_I\|_2 &= \sqrt{\sum_{i\in I}\left(\frac{\mu_k}{c_i(\boldsymbol{x}_k)}\right)^2} \\
&\leq \mu_k\sqrt{\sum_{i\in I}\left(\frac{1}{\min_{j\in I}|c_j(\boldsymbol{x}_k)|}\right)^2} \\
&= \frac{\mu_k}{\min_{j\in I}|c_j(\boldsymbol{x}_k)|}\sqrt{|I|} \\
&\leq \frac{2\mu_k\sqrt{|I|}}{\min_{j\in I}|c_j(\boldsymbol{x}^*)|}.
\end{aligned}
\tag{12.48}
$$

Here, we have replaced $c_j(\boldsymbol{x}_k)$ by $c_j(\boldsymbol{x}^*)$ in the denominator on the last line and added a factor 2. The inequality then follows from the fact that $|c_j(\boldsymbol{x}^*)| \geq \frac{1}{2}|c_j(\boldsymbol{x}_k)|$ for k large enough, since the c_j are continuous and $\boldsymbol{x}_k \to \boldsymbol{x}^*$.

Next, we estimate

$$
\|\nabla f(\boldsymbol{x}_k) - G_A^\top(\boldsymbol{x}_k)(\boldsymbol{\lambda}_k)_A\|_2 \leq \|\nabla f(\boldsymbol{x}_k) - G^\top(\boldsymbol{x}_k)\boldsymbol{\lambda}_k\|_2 + \|G_I^\top(\boldsymbol{x}_k)(\boldsymbol{\lambda}_k)_I\|_2.
\tag{12.49}
$$

Now, by the definition of $\boldsymbol{\lambda}_k$, we have

$$
\begin{aligned}
(\nabla f(\boldsymbol{x}_k) - G^\top(\boldsymbol{x}_k)\boldsymbol{\lambda}_k)_i &= \frac{\partial f}{\partial x_i} - \sum_{j=1}^m (\lambda_k)_j \frac{\partial c_j}{\partial x_i}(\boldsymbol{x}_k) \\
&= \frac{\partial f}{\partial x_i} - \sum_{j=1}^m \frac{\mu_k}{c_j(\boldsymbol{x}_k)} \frac{\partial c_j}{\partial x_i}(\boldsymbol{x}_k),
\end{aligned}
$$

which we recognize as $\frac{\partial \phi}{\partial x_i}(\boldsymbol{x}_k)$ from (12.45). Thus, assumption (12.47) shows that

$$
\|\nabla f(\boldsymbol{x}_k) - G^\top(\boldsymbol{x}_k)\boldsymbol{\lambda}_k\|_2 \leq \epsilon_k.
\tag{12.50}
$$

Next, we combine this estimate, (12.48) and (12.49) to obtain

$$
\begin{aligned}
\|\nabla f(\boldsymbol{x}_k) - G_A^\top(\boldsymbol{x}_k)(\boldsymbol{\lambda}_k)_A\|_2 &\leq \|\nabla f(\boldsymbol{x}_k) - G^\top(\boldsymbol{x}_k)\boldsymbol{\lambda}_k\|_2 + \|G_I^\top(\boldsymbol{x}_k)(\boldsymbol{\lambda}_k)_I\|_2 \\
&\leq \epsilon_k + \frac{2\mu_k\|G_I^\top(\boldsymbol{x}_k)\|_2\sqrt{|I|}}{\min_{j\in I}|c_j(\boldsymbol{x}^*)|} =: \tilde{\epsilon}_k.
\end{aligned}
\tag{12.51}
$$

Now, since $G_A^+(\boldsymbol{x}_k)G_A^\top(\boldsymbol{x}_k) = I$, we have

$$
\begin{aligned}
\|G_A^+(\boldsymbol{x}_k)\nabla f(\boldsymbol{x}_k) - (\boldsymbol{\lambda}_k)_A\|_2 &= \|G_A^+(\boldsymbol{x}_k)(\nabla f(\boldsymbol{x}_k) - G_A^\top(\boldsymbol{x}_k)(\boldsymbol{\lambda}_k)_A)\|_2 \\
&\leq 2\|G_A^+(\boldsymbol{x}^*)\|_2\tilde{\epsilon}_k
\end{aligned}
\tag{12.52}
$$

for k big enough, again using the fact that $\nabla \boldsymbol{c}$ is continuous and $\boldsymbol{x}_k \to \boldsymbol{x}^*$. This allows us to estimate the convergence of the active λ_i using

$$
\|\boldsymbol{\lambda}_A - (\boldsymbol{\lambda}_k)_A\|_2 \leq \|G_A^+(\boldsymbol{x}^*)\nabla f(\boldsymbol{x}^*) - G_A^+(\boldsymbol{x}_k)\nabla f(\boldsymbol{x}_k)\| + \|G_A^+(\boldsymbol{x}_k)\nabla f(\boldsymbol{x}_k) - (\boldsymbol{\lambda}_k)_A\|_2.
$$

The first norm tends to zero as $\boldsymbol{x}_k \to \boldsymbol{x}^*$ because of the continuity of ∇f and $\nabla \boldsymbol{c}$. For the second norm, we observe from (12.51) that both ϵ_k and

μ_k approach zero as $k \to \infty$, meaning $\tilde{\epsilon}_k \to 0$ as well. Thus, by (12.52), the second norm tends to zero as well. In addition, $(\boldsymbol{\lambda}_k)_I \to 0$ because of (12.48) and $\boldsymbol{\lambda}_I$ is defined to be zero. Thus, both the active and inactive parts converge to the exact $\boldsymbol{\lambda}$, i.e., we have shown that $\boldsymbol{\lambda}_k \to \boldsymbol{\lambda}$ as $k \to \infty$. Furthermore, continuity of the gradients and (12.50) imply that

$$\nabla f(\boldsymbol{x}^*) - \nabla \boldsymbol{c}^\top(\boldsymbol{x}^*)\boldsymbol{\lambda} = 0,$$

and since $\boldsymbol{c}(\boldsymbol{x}_k) > 0$ for all k, we have $\boldsymbol{c}(\boldsymbol{x}^*) \geq 0$. Finally, the definition of $\boldsymbol{\lambda}_k$ in (12.46) gives

$$c_i(\boldsymbol{x}_k)(\boldsymbol{\lambda}_k)_i = \mu_k,$$

and with $\boldsymbol{\lambda}_k \to \boldsymbol{\lambda}$, we obtain

$$c_i(\boldsymbol{x}^*)\lambda_i = 0,$$

which concludes the proof. $\qquad\qquad\qquad\qquad\qquad\qquad\qquad\qquad\square$

It is remarkable that the generic interior point method also computes an estimate for the Lagrange multipliers, as shown in the previous theorem. While this theorem establishes a basic convergence result for the generic interior point method, there are several issues that need to be addressed to make this a useful algorithm:

1. A fundamental problem in the algorithm is that the Hessian of the unconstrained problem to be solved in each iteration becomes more and more ill conditioned as the algorithm progresses: One can show that its condition number is proportional to $\frac{1}{\mu_k}$. The ill-conditioning is evident for the example on the right in Figure 12.19, where the curvature is becoming extreme in the direction where the minimum is squeezed toward the boundary, whereas in the other direction, the curvature remains small.

2. Another important problem is the starting point $\boldsymbol{x}_k^s$. Using $\boldsymbol{x}_{k-1}^*$ can be a very unlucky choice, since one can show that from this point, a Newton step becomes asymptotically infeasible if $\mu_{k+1} < \frac{1}{2}\mu_k$.

These two problems can however be addressed using a perturbation argument in the optimality conditions, which leads to the powerful class of primal-dual interior point methods with both excellent theoretical and practical properties.

12.4.4   Sequential Quadratic Programming

Sequential quadratic programming (SQP) are methods for solving optimization problems with equality (or sometimes inequality) constraints based on Lagrange multipliers. The general minimization problem

$$\begin{array}{llll} f(\boldsymbol{x}) & \to & \min & f : \mathbb{R}^n \to \mathbb{R}, \\ \boldsymbol{c}(\boldsymbol{x}) & = & 0 & \boldsymbol{c} : \mathbb{R}^n \to \mathbb{R}^m, \end{array} \qquad (12.53)$$

has the Lagrangian

$$\mathcal{L}(\boldsymbol{x}, \boldsymbol{\lambda}) = f(\boldsymbol{x}) - \boldsymbol{c}^{\top}(\boldsymbol{x})\boldsymbol{\lambda}.$$

The system of first-order optimality conditions

$$\begin{aligned} \nabla_x \mathcal{L}(\boldsymbol{x}, \boldsymbol{\lambda}) = \nabla f(\boldsymbol{x}) - (\nabla \boldsymbol{c}(\boldsymbol{x}))^{\top} \boldsymbol{\lambda} &= 0 \\ \boldsymbol{c}(\boldsymbol{x}) &= 0 \end{aligned} \tag{12.54}$$

has $m+n$ equations for $m+n$ unknowns in $\boldsymbol{x}$ and $\boldsymbol{\lambda}$. Using Newton's method to compute an approximate solution of (12.54), we obtain the iteration

$$\begin{pmatrix} \boldsymbol{x}^{k+1} \\ \boldsymbol{\lambda}^{k+1} \end{pmatrix} = \begin{pmatrix} \boldsymbol{x}^k \\ \boldsymbol{\lambda}^k \end{pmatrix} - \begin{pmatrix} H(\boldsymbol{x}^k, \boldsymbol{\lambda}^k) & -\nabla \boldsymbol{c}(\boldsymbol{x}^k)^{\top} \\ \nabla \boldsymbol{c}(\boldsymbol{x}^k) & 0 \end{pmatrix}^{-1} \begin{pmatrix} \nabla_x \mathcal{L}(\boldsymbol{x}^k, \boldsymbol{\lambda}^k) \\ \boldsymbol{c}(\boldsymbol{x}^k) \end{pmatrix},$$

where $H(\boldsymbol{x}, \boldsymbol{\lambda})$ denotes the Hessian with respect to $\boldsymbol{x}$ of the Lagrangian. Therefore, at each iteration of Newton's method, one has to solve the linear system of equations

$$\begin{pmatrix} H(\boldsymbol{x}^k, \boldsymbol{\lambda}^k) & -\nabla \boldsymbol{c}(\boldsymbol{x}^k)^{\top} \\ \nabla \boldsymbol{c}(\boldsymbol{x}^k) & 0 \end{pmatrix} \begin{pmatrix} \Delta \boldsymbol{x} \\ \Delta \boldsymbol{\lambda} \end{pmatrix} = - \begin{pmatrix} \nabla_x \mathcal{L}(\boldsymbol{x}^k, \boldsymbol{\lambda}^k) \\ \boldsymbol{c}(\boldsymbol{x}^k) \end{pmatrix}.$$

This system can easily be symmetrized, by writing it in the form

$$\begin{pmatrix} H(\boldsymbol{x}^k, \boldsymbol{\lambda}^k) & \nabla \boldsymbol{c}(\boldsymbol{x}^k)^{\top} \\ \nabla \boldsymbol{c}(\boldsymbol{x}^k) & 0 \end{pmatrix} \begin{pmatrix} \Delta \boldsymbol{x} \\ -\Delta \boldsymbol{\lambda} \end{pmatrix} = - \begin{pmatrix} \nabla_x \mathcal{L}(\boldsymbol{x}^k, \boldsymbol{\lambda}^k) \\ \boldsymbol{c}(\boldsymbol{x}^k) \end{pmatrix},$$

and there are special techniques to solve such *saddle point problems*, see [89].

By linearity, we obtain from the first part of the system,

$$H(\boldsymbol{x}^k, \boldsymbol{\lambda}^k)\Delta \boldsymbol{x} - \nabla \boldsymbol{c}(\boldsymbol{x}^k)^{\top} \Delta \boldsymbol{\lambda} = -\nabla_x \mathcal{L}(\boldsymbol{x}^k, \boldsymbol{\lambda}^k) = -\nabla f(\boldsymbol{x}^k) + (\nabla \boldsymbol{c}(\boldsymbol{x}^k))^{\top} \boldsymbol{\lambda}^k,$$

or an equivalent form,

$$H(\boldsymbol{x}^k, \boldsymbol{\lambda}^k)\Delta \boldsymbol{x} - (\nabla \boldsymbol{c}(\boldsymbol{x}^k))^{\top} (\boldsymbol{\lambda}^k + \Delta \boldsymbol{\lambda}) = -\nabla f(\boldsymbol{x}^k),$$

and hence with the new variable $\bar{\boldsymbol{\lambda}} = \boldsymbol{\lambda}^k + \Delta \boldsymbol{\lambda}$, the system in Newton's method is also equivalent to

$$\begin{pmatrix} H(\boldsymbol{x}^k, \boldsymbol{\lambda}^k) & \nabla \boldsymbol{c}(\boldsymbol{x}^k)^{\top} \\ \nabla \boldsymbol{c}(\boldsymbol{x}^k) & 0 \end{pmatrix} \begin{pmatrix} \Delta \boldsymbol{x} \\ -\bar{\boldsymbol{\lambda}} \end{pmatrix} = - \begin{pmatrix} \nabla f(\boldsymbol{x}^k) \\ \boldsymbol{c}(\boldsymbol{x}^k) \end{pmatrix}, \tag{12.55}$$

which will be useful later. Finally, one would often use an approximation B^k of the Hessian $H(\boldsymbol{x}^k, \boldsymbol{\lambda}^k)$, and thus solve the system

$$\begin{pmatrix} B^k & \nabla \boldsymbol{c}(\boldsymbol{x}^k)^{\top} \\ \nabla \boldsymbol{c}(\boldsymbol{x}^k) & 0 \end{pmatrix} \begin{pmatrix} \Delta \boldsymbol{x} \\ -\bar{\boldsymbol{\lambda}} \end{pmatrix} = - \begin{pmatrix} \nabla f(\boldsymbol{x}^k) \\ \boldsymbol{c}(\boldsymbol{x}^k) \end{pmatrix}. \tag{12.56}$$

If B^k is invertible, one can solve (12.56) by using a Schur complement approach.

Where in all this does the name SQP come from? To understand this, we need to consider the *quadratic programming problem* with linear constraint

$$\begin{aligned} \tfrac{1}{2}s^\top B^k s + s^\top \nabla f(x) &\longrightarrow \text{min}, \\ \nabla c(x)s &= -c(x) \end{aligned} \tag{12.57}$$

where B^k is a symmetric matrix. If $B^k \approx H(x)$, this is a quadratic model of the original problem (12.53)

$$\begin{aligned} f(x + s) &\longrightarrow \text{min}, \\ c(x + s) &= 0, \end{aligned}$$

since we can approximate the objective function by truncating the Taylor expansion

$$f(x + s) = f(x) + s^\top \nabla f(x) + \frac{1}{2}s^\top B^k s + \ldots,$$

and the constraint $c(x + s) = 0$ can be approximated to first order by

$$c(x + s) = c(x) + (\nabla c(x))^\top s + \ldots.$$

which means an approximation of our original problem The Lagrangian for the model (12.57) is

$$\mathcal{L}^a(s, \bar{\lambda}) = \frac{1}{2}s^\top B^k s + s^\top \nabla f(x) - (\nabla c(x)s + c(x))^\top \bar{\lambda},$$

and the first-order optimality conditions are

$$\begin{aligned} B^k s + \nabla f(x) - \nabla(c(x))^\top \bar{\lambda} &= 0 \\ \nabla c(x)s &= -c(x), \end{aligned}$$

which is precisely the system (12.56) solved at each iteration by Newton's methods applied to the first order optimality conditions of the original problem (12.53), provided B^k is an approximation of the Hessian of the Lagrangian, and not of the objective function only. One can therefore interpret Newton's method as a method that solves a sequence of quadratic optimization problems with linear constraints, which explains the name sequential quadratic programming. A generic SQP method is thus given by

---

ALGORITHM 12.15. *Generic interior point method*

For given x^0 and λ^0;
$k = 0$;
while not converged
    compute $B^k \approx H(x^k, \lambda^k)$;
    $s^k = \text{argmin}_{s \ \text{s.t.} \ \nabla c(x^k)s = -c(x^k)} \frac{1}{2}s^\top B^k s + s^\top \nabla f(x^k)$;
    $x^{k+1} = x^k + s^k$;
    k=k+1;
end

This method is simple and fast: if $B^k = H(\boldsymbol{x}^k, \boldsymbol{\lambda}^k)$, then convergence is ultimately quadratic, as in Newton's method.

If the underlying problem had been an optimization problem with inequality constraints, the subproblem in the generic SQP algorithm would simply have to be replaced by a subproblem of the form

$$\begin{aligned} \tfrac{1}{2}\boldsymbol{s}^\top B^k \boldsymbol{s} + \boldsymbol{s}^\top \nabla f(\boldsymbol{x}) \quad &\to \quad \min, \qquad (B^k)^\top = B^k, \\ \nabla \boldsymbol{c}(\boldsymbol{x})\boldsymbol{s} \quad &\geq \quad -\boldsymbol{c}(\boldsymbol{x}). \end{aligned} \qquad (12.58)$$

To obtain a more robust method, one adds in general a line search method,

$$\boldsymbol{x}^{k+1} = \boldsymbol{x}^k + \alpha^k \boldsymbol{s}^k,$$

and chooses the line search parameter α^k such that a penalty function

$$\phi(\boldsymbol{x}, \mu) := f(\boldsymbol{x}) + \frac{1}{\mu}\|\boldsymbol{c}(\boldsymbol{x})\|$$

satisfies a criterion of sufficient decrease. As an alternative, one can also use a trust region approach.

12.5 Problems

PROBLEM 12.1.

1. *Implement the minimization algorithm with golden section search.*

2. *Use both the bisection algorithm from Section 5.2.1 in Chapter 5 and the function* `Minimize` *to solve the following problems:*

   - *Find the minimum of the cosine function between* 2.5 *and* 4. *Which of the two algorithms is more precise ? Explain why.*

   - *Solve the problem from the introduction on how much daily exercise is optimal.*

   - *An enterprise wants to maximize the income from the sale of a product. We assume that if p is the price of the product, $C\frac{\mathrm{e}^{-p}}{1+\mathrm{e}^{-p}}$ units of the product will be sold, where C is a constant, which means the income is*

   $$Cp\frac{\mathrm{e}^{-p}}{1+\mathrm{e}^{-p}}.$$

   - *A simple model of how to maximize halieutic resources (fishing). We suppose that the evolution of a fish population $p(t)$ is governed by*

   $$\frac{\mathrm{d}p}{\mathrm{d}t} = a(1-\delta)p - bp^2,$$

where a and b are given constants, and δ represents the fishing rate, $\delta \in [0,1]$. Solve the differential equation by separation of variables, which gives for the total amount of fish being fished over a time period $[0,T]$ the quantity

$$\int_0^T a\delta p(t)\mathrm{d}t = \int_0^T \frac{a\delta(1-\delta)\mathrm{d}t}{bp_0 + (a(1-\delta) - bp_0)\exp(-a(1-\delta)t)}.$$

Maximize this function using δ, for $p_0 = 1$, $a = 0.34$, $b = 0.01$ and $T = 20$.

*(**Hint:** Use a quadrature rule from the Chapter 9 in order to integrate the total amount numerically)*

PROBLEM 12.2. *Let $A \in \mathbb{C}^{n \times n}$ be a Hermitian matrix, i.e $A^* = A$. Show that all eigenvalues λ of A are real.*

PROBLEM 12.3. *Let Q be a unitary matrix in $\mathbb{C}^n$, i.e. $Q^*Q = I$. Prove that*

$$\|Q\boldsymbol{x}\|_2 = \|\boldsymbol{x}\|_2, \qquad\qquad \|Q^*BQ\|_2 = \|B\|_2.$$

for any vector x in $\mathbb{C}^n$ and any matrix B in $\mathbb{C}^{n \times n}$.

PROBLEM 12.4. *A wine farmer produces two different types of wine: a white wine of very good quality, and a red wine of regular quality. The white wine sells for \$15 a bottle, and the red for \$11. The wine farmer possesses 400 hectares of land, and can produce on each hectare either 50 liters of white wine or 75 liters of red wine. The maximum demand for red wine in his shop is 25000 liters. With all the seasonal helpers, the wine farmer has in total 4200 man-hours of labor available. To produce 100 liters of red wine, 12 man-hours are needed, whereas the same quantity of white wine requires 28 man-hours. The wine farmer would like to maximize his revenue. Solve this optimization problem graphically.*

PROBLEM 12.5. *A mobile phone network operator has a collection of fixed antennas with interference matrix $G = (g_{ij})$, $g_{ij} \geq 0$ for all i,j, $g_{ii} = 0$. He also owns two specific frequencies on which he can operate his network. As shown in Subsection 12.1.2, he tries to assign frequencies to his antennas in order to minimize interference, i.e. he is looking for a permutation matrix Π and an integer p,*

$$\Pi^T G \Pi = \begin{bmatrix} A & B \\ C & D \end{bmatrix} \qquad\qquad A \in \mathbb{R}^{p \times p},$$

such that $\max(\lambda_{\max}(A), \lambda_{\max}(D))$ is minimized.

1. *Implement the brute force search strategy to find the optimum in MATLAB, as shown in Subsection 12.1.2.*

2. *Implement the approximate minimization algorithm using the Frobenius norm, as shown in Subsection 12.1.2.*

3. *Test the two algorithms on the matrices* **G1** *and* **G2** *obtained by*

```
>> rand('seed',900);
>> [G1,x1,y1]=GenerateProblem(4,4,0.1);
>> [G2,x2,y2]=GenerateProblem(4,4,0.4);
```

PROBLEM 12.6. *For each of the following functions from $\mathbb{R}^2$ to $\mathbb{R}$, compute the location (x, y) of the global minimum:*

1. $f_1(x, y) = x^2 + y^2 - xy$.

2. $f_2(x, y) = x^4 + y^4 - \frac{3}{2}x^2y^2$.

3. $f_3(x, y) = 3x^2 - 3y^2 + 8xy$.

4. $f_4(x, y) = \exp(\sin(50x)) + \sin(60\exp(y)) + \sin(70\sin(x)) + \sin(\sin(80y)) - \sin(10(x+y)) + \frac{x^2+y^2}{4}$ *(this last one from Nick Trefethen is challenging)*

PROBLEM 12.7. *For n cities in a flat country, we want to build the shortest possible road network to connect them. This problem goes back to a problem Gauss had worked on, namely to connect four important cities in Germany by the shortest possible railroad network; but it is even older, see [38]. For a given network, let p be the number of cross points (a point different from a city where at least two roads come together), so that the network contains a total of $n + p$ nodes (a node can be either a city or a cross point).*

1. *Between two nodes, what is the form of the shortest road network connecting them ?*

2. *Prove that at a cross point, at least three roads must come together.* (**Hint:** *Show that if not, the cross point can be eliminated*)

3. *Using the previous result, show that the number of roads must be larger than $\frac{3p+n}{2}$.*

4. *Prove that if there are $p + n$ nodes, the shortest network has at most $p + n - 1$ roads.*

5. *Conclude from the previous result that the shortest network between n cities has at most $n - 2$ cross points, i.e $p \leq n - 2$.*

6. *Consider now 4 cities placed in the corners of a square:*

   (a) *Find the shortest road network with $p = 0$ (easy).*

   (b) *Find the shortest road network with $p = 1$ (less easy).*

(c) What is the shortest road network you can imagine with $p = 2$?

(d) What is the shortest road network?

PROBLEM 12.8. *We would like to build a window frame in the form of a rectangle, with a semi-circle on top. Suppose we have a total length l of wood available. Compute the height h of the rectangle and the radius r of the semi-circle such that the window lets the most light into the house, i.e. the surface of the window is maximal, using all the available wood for the rim of the window.*

PROBLEM 12.9. *A financial investor wants to invest \$1000 into three different types of investment portfolios, which have a return given by the random variables R_a, R_b and R_c. The expectations are $\mathbb{E}[R_a] = \mu_a$, $\mathbb{E}[R_b] = \mu_b$ and $\mathbb{E}[R_c] = \mu_c$, and suppose the covariances $\sigma_{ij} = \mathbb{E}[(R_i - \mathbb{E}[R_i])(R_j - \mathbb{E}[R_j])]$ are also known.*

The investor wants to invest his \$1000 in \$1000$\pi_a \geq 0$ of portfolio a, in \$1000$\pi_b \geq 0$ of portfolio b, and in \$1000$\pi_c \geq 0$ of portfolio c, $\pi_a + \pi_b + \pi_c = 1$, such that his expected return is at least 2%, minimizing the associated risk, i.e. $\mathrm{Var}(\sum_{i=a,b,c} \pi_i R_i) = \sqrt{\sum_{i,j=a,b,c} \sigma_{ij}\pi_i\pi_j}$.

With the given data $\mu_a = 0.01$ (1%), $\mu_b = 0.022$ (2.2%), $\mu_c = 0.027$ (2.7%) and

$$(\sigma_{ij})_{i,j=a,b,c} = \begin{bmatrix} 0.001 & 0.0002 & 0.0005 \\ 0.0002 & 0.005 & 0.0008 \\ 0.0005 & 0.0008 & 0.02 \end{bmatrix},$$

compute the optimal choice of π_a, π_b, π_c for which the minimum return is 2%.

PROBLEM 12.10. *Let $f : \mathbb{R}^n \to \mathbb{R}$ be twice continuously differentiable, and let $\mathcal{M}$ be the sub-manifold of $\mathbb{R}^n$ with dimension $n - m$ defined by $\mathcal{M} = \{x, g(x) = 0\}$, $g : \mathbb{R}^n \to \mathbb{R}^m$. Let $x^* \in \mathcal{M}$ and $\lambda \in \mathbb{R}^m$. The Lagrangian $\mathcal{L}$ is then defined by $\mathcal{L}(x, \lambda) = f(x) - g^T(x)\lambda$.*

Using a parametrization ψ of the sub-manifold $\mathcal{M}$, prove that

$$\exists \lambda \in \mathbb{R}^m \begin{cases} \nabla\mathcal{L}(x^*, \lambda) = 0 \\ w^T H(x^*, \lambda)w > 0 \quad \forall w \in T_{x^*}\mathcal{M}, \ w \neq 0 \end{cases}$$
$$\implies \quad f \text{ has a local minimum } x^* \text{ on } \mathcal{M}$$
$$\implies \quad \exists \lambda \in \mathbb{R}^m \begin{cases} \nabla\mathcal{L}(x^*, \lambda) = 0 \\ w^T H(x^*, \lambda)w \geq 0 \quad \forall w \in T_{x^*}\mathcal{M} \end{cases}$$

where H is the Hessian of $\mathcal{L}$ with respect to the variable x only.

PROBLEM 12.11.

1. Show that for a matrix to be positive definite, it is necessary, but not sufficient that all the diagonal elements are strictly positive.

2. *Show that the matrix*

$$\mathbf{A} = \begin{bmatrix} 3 & -2 & 0 \\ -2 & 2 & 1 \\ 0 & 1 & 2 \end{bmatrix},$$

is positive definite by

- *computing explicitly the eigenvalues,*
- *only computing determinants of sub-matrices.*

PROBLEM 12.12.

1. *Prove the Taylor formula with integral remainder term: for all f in $C^{n+1}(\mathbb{R})$ with values in $\mathbb{R}$,*

$$f(x+h) = \sum_{j=0}^{n} f^{(j)}(x)\frac{h^j}{j!} + \int_0^h \frac{(h-t)^n}{n!} f^{(n+1)}(x+t)\mathrm{d}t.$$

(Hint: *Use integration by parts with a well chosen constant in the integration)*

2. *Let $g : \mathbb{R}^n \to \mathbb{R}$ be in C^{n+1}. Show that*

$$g(\boldsymbol{x}+\boldsymbol{h}) = \sum_{|\alpha|\leq n} \partial^\alpha g(\boldsymbol{x})\frac{\boldsymbol{h}^\alpha}{\alpha!} + \sum_{|\alpha|=n+1} \int_0^1 \frac{n+1}{\alpha!}(1-t)^n \partial^\alpha g(\boldsymbol{x}+t\boldsymbol{h})\boldsymbol{h}^\alpha \mathrm{d}t,$$

where α represents a multi-index $(\alpha_1, \alpha_2, \ldots, \alpha_n)$ $\partial^\alpha = \partial_{x_1}^{\alpha_1}\partial_{x_2}^{\alpha_2}\ldots\partial_{x_n}^{\alpha_n}$, $\boldsymbol{h}^\alpha = h_1^{\alpha_1}h_2^{\alpha_2}\ldots, h_n^{\alpha_n}$, $|\alpha| = \sum_i |\alpha_i|$, and $\alpha! = \prod_{i=1}^n \alpha_i!$.

(Hint: *Consider $f : \mathbb{R} \to \mathbb{R}$, $t \mapsto g(\boldsymbol{x}+t\boldsymbol{h})$)*

3. *Let $\mathcal{S}$ be an open set of $\mathbb{R}^n$. Let $\boldsymbol{s} \in \mathbb{R}^n$ be such that $\boldsymbol{x}+\theta\boldsymbol{s}$ is in $\mathcal{S}$ for all θ in $[0,1]$:*

   (a) *Let $f : \mathbb{R}^n \to \mathbb{R}$ be continuously differentiable on $\mathcal{S}$. We suppose that ∇f is Lipschitz continuous in $\boldsymbol{x}$ with Lipshitz constant $L(\boldsymbol{x})$. Prove that*

   $$|f(\boldsymbol{x}+\boldsymbol{s}) - f(\boldsymbol{x}) - \nabla f(\boldsymbol{x})\cdot\boldsymbol{s}| \leq \frac{1}{2}L(\boldsymbol{x})\|s\|_2^2.$$

   (b) *Let $f : \mathbb{R}^n \to \mathbb{R}$ be twice continuously differentiable on $\mathcal{S}$. We suppose that the Hessian matrix of f, denoted by H, is Lipschitz continuous in $\boldsymbol{x}$ with Lipshitz constant $L_2(\boldsymbol{x})$. Prove that*

   $$|f(\boldsymbol{x}+\boldsymbol{s}) - f(\boldsymbol{x}) - \nabla f(\boldsymbol{x})\cdot\boldsymbol{s} - \frac{1}{2}(\boldsymbol{s}, H(\boldsymbol{x})\boldsymbol{s})| \leq \frac{1}{6}L_2(\boldsymbol{x})\|s\|_2^3.$$

PROBLEM 12.13. *Prove Farkas' Lemma, see the end of Subsection 12.2.3.*

PROBLEM 12.14. *Let β be in $(0,1)$. Let $f : \mathbb{R}^n \to \mathbb{R}$ be in $\mathcal{C}^1$ such that ∇f is Lipschitz with constant $L(\boldsymbol{x_k})$ for $\boldsymbol{x_k}$. Let p_k be a descent direction corresponding to the gradient. Prove that the step obtained by the Armijo line search backtracking algorithm verifies*

$$\alpha_k \geq \min(\alpha_{init}, \frac{2\tau(\beta-1)p_k^T \nabla f(\boldsymbol{x_k})}{L(\boldsymbol{x_k})p_k^T p_k}).$$

PROBLEM 12.15.

1. *Implement the steepest descent algorithm with Armijo line search backtracking shown in Subsection 12.3.1.*

2. *Test your algorithm on the function from Problem 12.6. What do you observe ?*

PROBLEM 12.16.

1. *Show that for a square invertible matrix A*

$$(A^{\mathrm{T}})^{-1} = (A^{-1})^{\mathrm{T}}.$$

2. *Prove that if the matrix A est symmetric, then A^{T}, A^{-1} and $(A^{\mathrm{T}})^{-1}$ are also symmetric.*

3. *Let A be a symmetric matrix, and $\boldsymbol{b}$ a vector. We define the function*

$$f : \mathbb{R}^n \to \mathbb{R},$$
$$\boldsymbol{x} \mapsto \frac{1}{2}\boldsymbol{x}^{\mathrm{T}} A\boldsymbol{x} + x^{\mathrm{T}}b + c.$$

Prove that $\nabla f(\boldsymbol{x}) = A\boldsymbol{x} + b$.

4. *Prove that if A is symmetric, then*

$$\lambda_{\min}(A) \leq \frac{\boldsymbol{x}^{\mathrm{T}} A\boldsymbol{x}}{\|\boldsymbol{x}\|_2^2} \leq \lambda_{\max}(A).$$

5. *Prove that if A is symmetric, then*

$$\lambda_{\min}(A^{-1}) = \frac{1}{\lambda_{\max}(A^{-1})}$$
$$\lambda_{\max}(A^{-2}) = \frac{1}{\lambda_{\min}(A^{-1})}$$

PROBLEM 12.17. *Implement Newton's method for optimization as described in Subsection 12.3.1.*

PROBLEM 12.18. *Consider the BFGS update*

$$B_{k+1} = B_k + \frac{\boldsymbol{y}_k \boldsymbol{y}_k^\top}{\boldsymbol{s}_k^\top \boldsymbol{y}_k} - \frac{B_k \boldsymbol{s}_k \boldsymbol{s}_k^\top B_k}{\boldsymbol{s}_k^\top B_k \boldsymbol{s}_k}$$

Show that if B_k is positive definite and $\boldsymbol{s}_k^\top \boldsymbol{y}_k > 0$, then B_{k+1} is also positive definite. **Hint:** *Expand $\boldsymbol{v}^T B_{k+1} \boldsymbol{v}$ using the update formula, then notice that a Cauchy–Schwarz inequality holds for the inner product $\langle \boldsymbol{v}, \boldsymbol{w} \rangle := \boldsymbol{v}^\top B_k \boldsymbol{w}$ when B_k is positive definite.*

PROBLEM 12.19. *Solve the trust region sub-problems for the linear and quadratic models:*

1. *Let $R > 0$, $\boldsymbol{x}_0$ be a point in $\mathbb{R}^m$, y be in $\mathbb{R}$ and $\boldsymbol{d}$ be a vector of $\mathbb{R}^m$. Let*

$$f_{lin} : \mathbb{R}^m \to \mathbb{R}$$
$$\boldsymbol{x} \mapsto y + \boldsymbol{d}^\top (\boldsymbol{x} - \boldsymbol{x}_0).$$

   Compute the point $\boldsymbol{x}$ where f_{lin} attains its minimum in the closed ball $B(\boldsymbol{x}_0, R)$.

2. *Let $R > 0$, $\boldsymbol{x}_0$ be a point in $\mathbb{R}^m$, y be in $\mathbb{R}$, $\boldsymbol{d}$ be a vector in $\mathbb{R}^m$ and H be a square symmetric matrix in $\mathbb{R}^{m \times m}$. Let*

$$f_{quad} : \mathbb{R}^m \to \mathbb{R}$$
$$\boldsymbol{x} \mapsto y + \boldsymbol{d}^\top (\boldsymbol{x} - \boldsymbol{x}_0) + (\boldsymbol{x} - \boldsymbol{x}_0)^\top H (\boldsymbol{x} - \boldsymbol{x}_0).$$

   Devise a numerical method to compute the point $\boldsymbol{x}$ where f_{quad} attains its minimum in the closed ball $B(\boldsymbol{x}_0, R)$.

PROBLEM 12.20.

1. *Implement the trust region methods described in Subsection 12.3.2 with a linear model using the exact solution of the sub problem from Problem 12.19.*

2. *Implement the trust region methods described in Subsection 12.3.2 with a quadratic model.*

3. *Compare the performance of the trust region methods with the line search algorithms steepest descent of Problem 12.15. What do you observe ?*

Bibliography

[1] Milton Abramowitz and Irene A. Stegun. *Handbook of Mathematical Functions with Formulas, Graphs, and Mathematical Tables*. Dover, 1970.

[2] Alexander Aitken. On Bernoulli's numerical solution of algebraic equations. *Proceedings of the Royal Society of Edinburgh*, 46:289–305, 1926.

[3] Robert S. Anderssen and Gene H. Golub. Richardson's non-stationary matrix iterative procedure. Technical Report CS-TR-72-304, Computer Science Department, Stanford University, 1972.

[4] Bengt Aspvall and John R. Gilbert. Graph coloring using eigenvalue decomposition. *SIAM J. Alg. Disc. Meth.*, 5(4), 1984.

[5] Forbes A. B. Geometric tolerance assessment. *TR DITC 210/92, National Physical Laboratory,Teddington*, 1992.

[6] Pierre Baldi and Edward Posner. Graph coloring bounds for cellular radio. *Computers Math. Applic.*, 19(10):91–97, 1990.

[7] Richard Barrett, Michael Berry, Tony F. Chan, James Demmel, June Donato, Jack Dongarra, Victor Eijkhout, Roldan Pozo, Charles Romine, and Henk van der Vorst. *Templates for the Solution of Linear Systmes: Buliding Blocks for Iterative Methods*. SIAM, 1994.

[8] M. Benzi. Splittings of symmetric matrices and a question of Ortega. *Linear Algebra and its Applications*, 429:2340–2343, 2008.

[9] Å. Björck. *Numerical Methods for Least Squares Problems*. SIAM, 1996.

[10] Susanne C Brenner and Larkin Ridgway Scott. *The mathematical theory of finite element methods*, volume 15. Springer, 2008.

[11] C. Brezinski. *Accélération de la Convergence en Analyse Numérique*. Number 584 in Lecture notes in Mathematics 584. Springer, 1977.

[12] M. Bronstein. *Symbolic Integration I : Transcendental Functions*, volume 1 of *Algorithms and Computation in Mathematics*. Springer, 1996.

[13] J. C. Butcher. On runge-kutta processes of high order, part 2. *J. Austral. Math. Soc.*, IV:179–194, 1964.

[14] John C Butcher. *Numerical methods for ordinary differential equations*. John Wiley & Sons, 2008.

W. Gander et al., *Scientific Computing - An Introduction using Maple and MATLAB*,
Texts in Computational Science and Engineering 11,
DOI 10.1007/978-3-319-04325-8,
© Springer International Publishing Switzerland 2014

[15] B. P. Butler, A. B. Forbes, and P. M. Harris. Algorithms for geometric tolerance assessment. *TR DITC 228/94, National Physical Laboratory, Teddington*, 1994.

[16] G. Cardano. *Ars Magna or The Rules of Algebra, 1545*. MIT, 1968.

[17] I. B. Cohen. *Howard Aiken : Portrait of a Computer Planner*. MIT Press, 1999.

[18] James W. Cooley and John W. Tukey. An algorithm for the machine calculation of complex fourier series. *Math. Comput.*, 19:297–301, 1965.

[19] I.D. Coope and C.J. Price. A modified BFGS formula maintaining positive definiteness with Armijo-Goldstein steplengths. *Journal of Compuational Mathematics*, 13:156–160, 2008.

[20] C.F. Curtiss and J.O. Hirschfelder. Integration of stiff equations. *Proc. Nat. Acad. Sci. U.S.A.*, 38:235–243, 1952.

[21] Germund Dahlquist. Convergence and stability in the numerical integration of ordinary differential equations. *Math. Scand.*, 4:33–53, 1956.

[22] Germund Dahlquist. A special stability problem for linear multistep methods. *BIT*, 3:27–43, 1963.

[23] G.B. Dantzig. Maximization of a linear function of variables subject to linear inequalities. In T.C. Koopmans, editor, *Activity Analysis of Production and Allocation*, pages 339–347. Wiley, New York, 1951.

[24] C. de Boor. *A Practical Guide to Splines*. Springer, 1978.

[25] Carl de Boor. *Splinefunktionen*. Birkhäuser, 1990.

[26] J. Dongarra, C. Moler, J. Bunch, and G. W. Stewart. *LINPACK Users' Guide*. SIAM, 1979.

[27] J. Dongarra and F. Sullivan. The top 10 algorithms. *Computing in Science and Engineering*, 2(1):22–23, 2000.

[28] Michael Drexler and Martin J Gander. Circular billiard. *SIAM review*, 40(2):315–323, 1998.

[29] L. Euler. *Institutionum Calculi Integralis. Volumen Primum*, volume XI of *Opera OMNIA*. Birkhäuser, 1768.

[30] Haw-ren Fang and Dianne P. O'Leary. Modified Cholesky algorithms: a catalog with new approaches. *Math. Program., Ser. A*, 115:319–349, 2008.

[31] K. Fernando and B. Parlett. Accurate singular values and differential qd algorithms. *Numerische Mathematik*, 67:191–229, 1994.

[32] R. Fletcher. Conjugate gradient methods for indefinite systems. In G. Alistair Watson, editor, *Numerical Analysis– Dundee 1975*, volume 506, pages 73–89. Lecture Notes in Mathematics, Springer-Verlag, Heidelberg, 1976.

[33] G. Forsythe and P. Henrici. The cyclic jacobi method for computing the principal values of acomplex matrix. *Trans. Amer. Math. Soc.*, 94:1–23, 1960.

[34] J. G. F. Francis. The qr transformation, a unitary analoque to the lr transformation – part 1. *The Computer Journal*, 4:265–271, 1961.

[35] Roland Freund and Noël Nachtigal. QMR: A quasi-minimal residual method for non-Hermitian linear systems. *Numerische Mathematik*, 60:315–339, 1991.

[36] A. Gamst and W. Rave. On the frequency assignment in mobile automatic telephone systems. In *Proc. of GLOBECOM 82*, pages 309–315, 1982.

[37] Martin J. Gander and Felix Kwok. Chladni figures and the Tacoma bridge: motivating PDE eigenvalues problems via vibrating plates. *SIAM Rev.*, 54:573–596, 2012.

[38] M.J. Gander, K. Santugini, and A. Steiner. La rete stradale piu breve che collega le citta (shortest road network connecting cities). *Bolettino dei docenti di matematica*, 56:9–19, 2008.

[39] W. Gander. On Halley's iteration method. *The American Mathematical Monthly*, 92(2), 1985.

[40] W. Gander. Algorithms for the polar decomposition. *SIAM J. on Sci. and Stat. Comp.*, 11(6), 1990.

[41] W. Gander. *Computermathematik*. Birkhäuser, 1992.

[42] W. Gander. Zeros of determinants of λ-matrices. In *Proceedings in Applied Mathematics and Mechanics*, volume 8, pages 10811–10814. Wiley, 2008.

[43] W. Gander and W. Gautschi. Adaptive quadrature - revisited. *BIT*, 40(1):84–101, 2000.

[44] W. Gander and D. Gruntz. The billiard problem. *International Journal of Mathematical Education in SCIENCE and Technology*, 23(6):825–830, 1992.

[45] W. Gander and J. Hřebíček. *Solving Problems in Scientific Computing using Maple AND Matlab*. Springer, 3rd edition, 1997.

[46] Walter Gander and Dominik Gruntz. Derivation of numerical methods using computer algebra. *SIAM Review*, 41(3), 1999.

[47] Garbow, Boyle, Dongarra, and Moler. *EISPACK Guide Extension*. Lecture Notes in Computer Science. Springer, 1977.

[48] Walter Gautschi, Ronald S FRIEDMAN, Jonathan BURNS, Rajan DAR-JEE, and Andrew MCINTOSH. *Orthogonal Polynomials: Computation and Approximation, Numerical Mathematics and Scientific Computation Series*. Oxford University Press, Oxford, 2004.

[49] Philip E Gill, Walter Murray, and Margaret H Wright. Practical optimization. 1981.

[50] L. Giraud, J. Langou, M. Rozlŏzník, and J. van den Eshof. Rounding error analysis of the classical gram-schmidt orthogonalization process. *Numerische Mathematik*, 101:87–100, 2005.

[51] G. H. Golub and C. F. Van Loan. *Matrix Computations*. Johns Hopkins Studies in the Mathematical Sciences. Johns Hopkins University Press, Baltimore, MD, 1996.

[52] G. H. Golub and J. H. Welsch. Calculation of gauss quadrature rules. *Math. Comp.*, 23:221–230, 1969.

[53] Gene Golub and Christian Reinsch. Singular value decomposition and least squares solutions. *Numerische Mathematik*, 14:403–420, 1970.

[54] Gene H. Golub. *The Use of Chebyshev Matrix Polynomials in the ITERA-TIVE Solution of Linear Equations Compared to THE Method of Successive*

Overrelaxation. PhD thesis, University of Illionis at Urbana-Champaign, 1959.

[55] Gene H. Golub and William Kahan. Calculating the singular values and pseudo-inverse of a matrix. *SIAM J. Numer. Anal.*, 2:205—224, 1965.

[56] Gene H. Golub and Victor Pereyra. The differentiation of pseudoinverses and nonlinear least squares problems whose variables separate. *SIAM J. Numer. Anal.*, 10:416–432, 1973.

[57] N. I. M. Gould and S. Leyffer. An introduction to algorithms for nonlinear optimization. In *Frontiers in Numerical Analysis*, pages 109–197. Springer Verlag, 2003.

[58] Andreas Griewank and Andrea Walther. *Evaluating Derivatives: Principles and Techniques of ALGORITHMIC Differentiation.* SIAM, 2008.

[59] Martin Gutknecht. Lanczos-type solvers for nonsymmetric linear systems of equations. *Acta Numerica*, pages 271–397, 1997.

[60] J. Hadamard. Sur les problèmes aux dérivées partielles et leur signification physique. *Princeton University Bulletin*, 13:49–52, 1902.

[61] E. Hairer, Ch. Lubich, and G. Wanner. *Geometric Numerical Integration: Structure-Preserving Algorithms for Ordinary Differential Equations.* Springer-Verlag, 2002.

[62] E. Hairer, S. P. Nørsett, and G. Wanner. *Solving Ordinary Differential Equations I, Nonstiff PROBLEMS.* Springer-Verlag, 2nd revised edition, 1993.

[63] E. Hairer and G. Wanner. *Solving Ordinary Differential Equations II, the Stiff CASE.* Springer-Verlag, 2nd revised edition, 1994.

[64] E. Hairer and G. Wanner. *Analysis by Its History.* Springer, New York, 1996.

[65] R. Hanson and M. Norris. Analysis of measurements based on the singular value decomposition. *SIAM J. on Sci. and Stat. Comp.*, 2(3), 1981.

[66] Jin-Kao Hao and Raphaël Dorne. Study of genetic search for the frequency assignment problem. In *Artificial Evolution European Conference*, pages 333–344. Springer-Verlag, 1996.

[67] P. Henrici. On the speed of convergence of cyclic and quasicyclic jacobi methods for computing eigenvalues of hermitian matrices. *J. Soc. Indust. Appl. Math.*, 6:144–162, 1958.

[68] P. Henrici. *Discrete variable methods in ordinary differential equations.* Wiley, 1962.

[69] P. Henrici. *Elements of Numerical Analysis.* Wiley, 1964.

[70] M. Hestenes and E. Stiefel. Methods of conjugate gradients for solving linear systems. *J. Research Nat. Bur. Standards*, 49:409–436, 1952.

[71] Nicholas J. Higham. *Accuracy and stability of numerical algorithms.* SIAM, 2002.

[72] A.S. Householder. On the convergence of matrix iterations. Technical Report 1883, Oak Ridge National Laboratory, 1955.

[73] Thomas JR Hughes. *The finite element method: linear static and dynamic finite element analysis.* DoverPublications. com, 2012.

[74] C. G. J. Jacobi. über ein leichtes verfahren, die in der theorie der säkularstörungen vorkommenden gleichungen numerisch aufzulösen. *Crelle's Journal*, 30:51–94, 1846.

[75] F. John. *Advanced Numerical Methods*. Lecture Notes, Department of Mathematics, 1956.

[76] Johan Joss. *Algorithmishes Differenzieren*. PhD thesis, Eidgenoessische Technische Hochschule, Zürich, Switzerland, 1976.

[77] William Kahan. *Gauss-Seidel Methods of Solving Large Systems of Linear EQUATIONS*. PhD thesis, University of Toronto, 1958.

[78] D. K. Kahaner. Comparison of numerical quadratur formulas. *Mathematical Software*, pages 229–269, 1971.

[79] Narendra Karmarkar. A new polynomial-time algorithm for linear programming. In *Proceedings of the sixteenth annual ACM symposium on Theory of computing*, pages 302–311. ACM, 1984.

[80] Daniel Kressner. *Numerical methods for general and structured eigenvalue problems*, volume 46. Springer, 2005.

[81] Jeffrey C. Lagarias, James A. Reeds, Margaret H. Wright, and Paul E. Wright. Convergence properties of the Nelder-Mead simplex method in low dimensions. *SIAM Journal of Optimization*, 9:112–147, 1998.

[82] Cornelius Lanczos. An iterative method for the solution of the eigenvalue problem of linear differential and integral operators. *J. Res. Nat. Bur. Standards, Sect. B.*, 45:225–280, 1950.

[83] Cornelius Lanczos. Solution of systems of linear equations by minimized iterations. *J. Res. Nat. Bur. Standards, Sect. B.*, 49:33–53, 1952.

[84] Ben Leimkuhler and Sebastian Reich. *Simulating Hamiltonian Dynamics*. Cambridge University Press, 2005.

[85] Steven J. Leon. *Linear Algebra with Applications*. Pearson, 2010.

[86] Jörg Liesen and Zdenek Strakos. *Krylov subspace methods: principles and analysis*. Oxford University Press, 2012.

[87] J. Liouville. Remarques nouvelles sur l'équation de riccati. *J. des Math. pures et appl.*, 6:1–13, 1841.

[88] James N Lyness and Cleve B Moler. Numerical differentiation of analytic functions. *SIAM Journal on Numerical Analysis*, 4(2):202–210, 1967.

[89] J. Liesen M.Benzi, G.H.Golub. Numerical solution of saddle point problems. *Acta Numerica*, 14:1–137, 2005.

[90] JB McLeod. A note on the ε-algorithm. *Computing*, 7(1-2):17–24, 1971.

[91] J. C. P. Miller. *Neville's and Romberg's Processes: A fresh Appraisal with Extensions*, volume 263 of *Series A, Mathematical and Physical Sciences*. Philosophical Transactions of the Royal Society of London, 1969.

[92] C. Moler. The qr algorithm – striving for infallibility. *MathWorks Newsletter*, 1995.

[93] Cleve Moler and Charles Van Loan. Nineteen dubious ways to compute the exponential of a matrix. *SIAM Review*, 20(4), 1978.

[94] A. M. Mood and F. A. Graybill. *Introduction to the Theory of Statistics.* McGraw-Hill Book Company, New York, 2nd edition, 1963.

[95] I. Newton. *Methodus Fluxionum et Serierum INFINITARUM,* volume 1 of *Opuscula mathematica.* edita Londini, 1736. Traduit en franç ais par M. de Buffon, Paris MDCCXL.

[96] Alan V. Oppenheim and Alan S. Willsky. *Signals & Systems.* Prentice Hall, 1996.

[97] J örg Waldvogel. Fast construction of the fejér and clenshaw-curtis quadrature rules. *BIT,* 46(1):195–202, 2006.

[98] AM Ostrowski. On the linear iteration procedures for symmetric matrices. *Rend. Mat. Appl,* 14:140–163, 1954.

[99] A.M. Ostrowski. *Solution of Equations and Systems of Equations.* Academic Press, 1973.

[100] M. L. Overton. *Numerical Computing with IEEE Floating Point Arithmetic.* SIAM, 2001.

[101] Chris C. Paige and Michael A. Saunders. Solution of sparse indefinite systems of linear equations. *SIAM J. Numer. Anal.,* 12:617–629, 1975.

[102] Chris C. Paige and Michael A. Saunders. LSQR: An algorithm for sparse linear equations and sparse least squares. *ACM Trans. Math. Soft.,* 8:43–71, 1982.

[103] Christopher Conway Paige. *The computation of eigenvalues and eigenvectors of very large sparse matrices.* PhD thesis, University of London, 1971.

[104] B. N. Parlett. *The Symmetric Eigenvalue Problem.* Classics in Applied Mathematics. SIAM, 2nd edition, 1998.

[105] Edgar Reich. On the convergence of the classical iterative method of solving linear simultaneous equations. *The Annals of Mathematical Statistics,* 20(3):448–451, 1949.

[106] Lewis Fry Richardson. On the approximate arithmetical solution by finite differences of physical problems involving differential equations, with an application to the stresses in a masonry dam. *Proceedings of the Royal Society of London. Series A,* 83(563):335–336, 1910.

[107] M. Rojas, S.A. Santos, and D.C. Sorensen. A new matrix-free algorithm for the large-scale trust-region subproblem. *SIAM J. Optim.,* 11(3):611–646, 2000.

[108] Walter Rudin. *Real and Complex Analysis.* McGraw-Hill, International Edition, 1987.

[109] H. Rutishauser. Der quotienten-differenzen-algorithmus. *Z. Angew. Math. Physik,* 5(1):233–251, 1954.

[110] H. Rutishauser. *Der Quotienten-Differenzen-Algorithmus.* Birkhäuser, 1957.

[111] H. Rutishauser. *Description of ALGOL 60.* Springer, 1967.

[112] H. Rutishauser. *Lectures on Numerical Mathematics.* Birkhäuser, 1990.

[113] H. Rutishauser and H. R. Schwarz. The lr transformation method for symmetric matrices. *Numerische Mathematik,* 5:273–289, 1963.

[114] Heinz Rutishauser. Über die Instabilität von Methoden zur Integration gewöhnlicher Differentialgleichungen. *ZAMP*, 3:65–74, 1952.

[115] Heinz Rutishauser. Beiträge zur kenntnis des biorthogonalisierungs- algorithmus von lanczos. *Zeitschrift für angewandte Mathematik und Physik (ZAMP)*, 4(1):35–56, 1953.

[116] W. Gander S. J. Leon, Å. Björck. Gram-schmidt orthogonalization: 100 years and more. *Numer. Linear Algebra Appl.*, 20:492–532, 2013.

[117] M. H. Saad, Y. & Schultz. Gmres: A generalized minimal residual algorithm for solving nonsymmetric linear systems. *SIAM J. Sci. Statist. Comput.*, 7:856–869, 1986.

[118] Yousef Saad. *Iterative Methods for Sparse Linear Systems*. SIAM, 2003.

[119] J. M. Sanz-Serna. Two topics on nonlinear stability. *Advances in Numerical Analysis (W. Light ed.)*, I:147–174, 1991.

[120] Ernst Schröder. über iterirte functionen. *Mathematische Annalen*, 3(2):296–322, 1870.

[121] Ernst Schröder. Über unendlich viele algorithmen zur auflösung der gleichungen. *Mathematische Annalen*, 2(2):317–365, 1870.

[122] H. R. Schwarz. *Numerik symmetrischer Matrizen*. Teubner, 1972.

[123] H. Schwetlik and T. Schütze. Least squares approximation by splines with free knots. *BIT*, 35(3):361–384, 1995.

[124] Daniel Shanks. Non-linear transformation of divergent and slowly convergent sequences. *Journal of Mathematics and Physics*, 34:1–42, 1955.

[125] Avram Sidi. Efficient implementation of minimal polynomial and reduced rank extrapolation methods. *J. of Comp. and Appl. Math.*, 36:305–337, 1991.

[126] Avram Sidi and Jacob Bridger. Convergence and stability analyses for some vector extrapolation methods in the presence of defective iteration matrices. *J. of Comp. and Appl. Math.*, 22:35–61, 1988.

[127] H. Späth. Orthogonal least squares fitting with linear manifolds. *Numer. Math.*, 48:441–445, 1986.

[128] William Squire and George Trapp. Using complex variables to estimate derivatives of real functions. *Siam Review*, 40(1):110–112, 1998.

[129] A. Steiner and M. Arrigoni. Die lösung gewisser räuber-beute-systeme. *Studia Biophysica*, 123(2), 1988.

[130] G. W. Stewart. The economical storage of plane rotations. *Numerische Mathematik*, 25:137–138, 1976.

[131] E. Stiefel. *Einführung in die numerische Mathematik*. Teubner, 1976.

[132] J. Stoer and R. Bulirsch. *Introduction to Numerical Analysis*. Springer, 1991.

[133] Gilbert Strang and George J Fix. *An analysis of the finite element method*, volume 212. Prentice-Hall Englewood Cliffs, 1973.

[134] A.M. Stuart and A.R. Humphries. *Dynamical Systems and Numerical Analysis*. Mathematics. Cambridge University Press, 1998.

[135] V. Szebehely. *Theory of Orbits, The restricted problem of three BODIES.* Acad. Press, New York, 1967.

[136] R.C.E. Tan. Implementation of the topological ε-algorithm. *SIAM J. Sci. Stat. Comput.*, 9(5), 1988.

[137] F. Tisseur and K. Meerbergen. The quadratic eigenvalue problem. *SIAM. Rev*, 43:234–286, 2001.

[138] Joseph F. Traub. *Iterative Methods for the Solution of Equations.* Algorithms and Computation in Mathematics. Prentice Hall, 1964.

[139] L. N. Trefethen. The definition of numerical analysis. SIAM News, November 1992. Available online at `http://people.maths.ox.ac.uk/trefethen/publication/PDF/1992_55.pdf`.

[140] L. N. Trefethen and D. Bau III. *Numerical Linear Algebra.* SIAM, 1997.

[141] Henk van der Vorst. Bi-CGSTAB: A fast and smoothly converging variant of Bi-CG for the solution of nonsymmetric linear systems. *SIAM J. Sci. Statist. Comput.*, 13:631–644, 1992.

[142] Richard S. Varga. *Matrix Iterative Analysis.* Prentice Hall, first edition, 1962.

[143] Vito Volterra. *Leçon sur la théorie mathématique de la lutte pour la vie.* Cahiers Scientifiques, Paris, 1931.

[144] Urs von Matt. The orthogonal qd-algorithm. *SIAM Journal on Scientific Computing*, 18:1163–1186, 1997.

[145] J. von Neumann and H.H. Goldstine. Numerical inverting of matrices of high order. *Bull. Amer. Math. Soc.*, 53:479–557, 1947.

[146] Gene H. Golub Walter Gander and Rolf Strebel. Least-squares fitting of circles and ellipses. *BIT*, 34:558–578, 1994.

[147] H. Weyl. Das asymptotische Verteilungsgesetz der Eigenwerte linearer partieller Differentialgleichungen (mit einer Anwendung auf die Theorie der Hohlraumstrahlung). *Mathematische Annalen*, 71:441–479, 1912.

[148] J. Wilkinson and C. Reinsch. *Linear Algebra.* Springer, 1971.

[149] J. H. Wilkinson. Error analysis of direct methods of matrix inversion. *Journal of the ACM (JACM)*, Volume 8 Issue 3:281–330, 1961.

[150] J. H. Wilkinson. *The Algebraic Eigenvalue Problem.* Monographs on Numerical Analysis. Oxford Science Publications, 1965.

[151] P. Wynn. On a device for computing the $e_m(s_n)$-transformation. *MTAC*, 10:91–96, 1956.

[152] P. Wynn. General purpose vector epsilon algorithm algol procedures. *Numerische Mathematik*, 6:22–36, 1964.

[153] David M. Young. *Iterative Methods for Solving Partial Difference EQUATIONS of Elliptic Type.* PhD thesis, Harvard University, May 1950.

[154] K. Zuse. *The computer, my life.* Springer, 1993.

Index

W. Gander et al., *Scientific Computing - An Introduction using Maple and MATLAB*,
Texts in Computational Science and Engineering 11,
DOI 10.1007/978-3-319-04325-8,
© Springer International Publishing Switzerland 2014

Series Editors

Timothy J. Barth
NASA Ames Research Center
NAS Division
Moffett Field, CA 94035, USA
barth@nas.nasa.gov

Michael Griebel
Institut für Numerische Simulation
der Universität Bonn
Wegelerstr. 6
53115 Bonn, Germany
griebel@ins.uni-bonn.de

David E. Keyes
Mathematical and Computer Sciences
and Engineering
King Abdullah University of Science
and Technology
P.O. Box 55455
Jeddah 21534, Saudi Arabia
david.keyes@kaust.edu.sa

and

Department of Applied Physics
and Applied Mathematics
Columbia University
500 W. 120th Street
New York, NY 10027, USA
kd2112@columbia.edu

Risto M. Nieminen
Department of Applied Physics
Aalto University School of Science
and Technology
00076 Aalto, Finland
risto.nieminen@aalto.fi

Dirk Roose
Department of Computer Science
Katholieke Universiteit Leuven
Celestijnenlaan 200A
3001 Leuven-Heverlee, Belgium
dirk.roose@cs.kuleuven.be

Tamar Schlick
Department of Chemistry
and Courant Institute
of Mathematical Sciences
New York University
251 Mercer Street
New York, NY 10012, USA
schlick@nyu.edu

Editor for Computational Science
and Engineering at Springer:
Martin Peters
Springer-Verlag
Mathematics Editorial IV
Tiergartenstrasse 17
69121 Heidelberg, Germany
martin.peters@springer.com

Texts in Computational Science and Engineering

1. H. P. Langtangen, *Computational Partial Differential Equations.* Numerical Methods and Diffpack Programming, 2nd Edition.
2. A. Quarteroni, F. Saleri, P. Gervasio, *Scientific Computing with MATLAB and Octave*, 3rd Edition.
3. H. P. Langtangen, *Python Scripting for Computational Science*, 3rd Edition.
4. H. Gardner, G. Manduchi, *Design Patterns for e-Science.*
5. M. Griebel, S. Knapek, G. Zumbusch, *Numerical Simulation in Molecular Dynamics.*
6. H. P. Langtangen, *A Primer on Scientific Programming with Python*, 3rd Edition.
7. A. Tveito, H. P. Langtangen, B. F. Nielsen, X. Cai, *Elements of Scientific Computing.*
8. B. Gustafsson, *Fundamentals of Scientific Computing.*
9. M. Bader, *Space-Filling Curves.*
10. M.G. Larson, F. Bengzon, *The Finite Element Method: Theory, Implementation, and Applications.*
11. W. Gander, M.J. Gander, F. Kwok, *Scientific Computing.* An Introduction using Maple and MATLAB.

For further information on these books please have a look at our mathematics catalogue at the following URL: www.springer.com/series/5151

Monographs in Computational Science and Engineering

1. J. Sundnes, G.T. Lines, X. Cai, B.F. Nielsen, K.-A. Mardal, A. Tveito, *Computing the Electrical Activity in the Heart.*

For further information on this book, please have a look at our mathematics catalogue at the following URL: www.springer.com/series/7417

Lecture Notes in Computational Science and Engineering

1. D. Funaro, *Spectral Elements for Transport-Dominated Equations.*
2. H.P. Langtangen, *Computational Partial Differential Equations.* Numerical Methods and Diffpack Programming.
3. W. Hackbusch, G. Wittum (eds.), *Multigrid Methods V.*
4. P. Deuflhard, J. Hermans, B. Leimkuhler, A.E. Mark, S. Reich, R.D. Skeel (eds.), *Computational Molecular Dynamics: Challenges, Methods, Ideas.*

5. D. Kröner, M. Ohlberger, C. Rohde (eds.), *An Introduction to Recent Developments in Theory and Numerics for Conservation Laws.*

6. S. Turek, *Efficient Solvers for Incompressible Flow Problems.* An Algorithmic and Computational Approach.

7. R. von Schwerin, *Multi Body System SIMulation.* Numerical Methods, Algorithms, and Software.

8. H.-J. Bungartz, F. Durst, C. Zenger (eds.), *High Performance Scientific and Engineering Computing.*

9. T.J. Barth, H. Deconinck (eds.), *High-Order Methods for Computational Physics.*

10. H.P. Langtangen, A.M. Bruaset, E. Quak (eds.), *Advances in Software Tools for Scientific Computing.*

11. B. Cockburn, G.E. Karniadakis, C.-W. Shu (eds.), *Discontinuous Galerkin Methods.* Theory, Computation and Applications.

12. U. van Rienen, *Numerical Methods in Computational Electrodynamics.* Linear Systems in Practical Applications.

13. B. Engquist, L. Johnsson, M. Hammill, F. Short (eds.), *Simulation and Visualization on the Grid.*

14. E. Dick, K. Riemslagh, J. Vierendeels (eds.), *Multigrid Methods VI.*

15. A. Frommer, T. Lippert, B. Medeke, K. Schilling (eds.), *Numerical Challenges in Lattice Quantum Chromodynamics.*

16. J. Lang, *Adaptive Multilevel Solution of Nonlinear Parabolic PDE Systems.* Theory, Algorithm, and Applications.

17. B.I. Wohlmuth, *Discretization Methods and Iterative Solvers Based on Domain Decomposition.*

18. U. van Rienen, M. Günther, D. Hecht (eds.), *Scientific Computing in Electrical Engineering.*

19. I. Babuška, P.G. Ciarlet, T. Miyoshi (eds.), *Mathematical Modeling and Numerical Simulation in Continuum Mechanics.*

20. T.J. Barth, T. Chan, R. Haimes (eds.), *Multiscale and Multiresolution Methods.* Theory and Applications.

21. M. Breuer, F. Durst, C. Zenger (eds.), *High Performance Scientific and Engineering Computing.*

22. K. Urban, *Wavelets in Numerical Simulation.* Problem Adapted Construction and Applications.

23. L.F. Pavarino, A. Toselli (eds.), *Recent Developments in Domain Decomposition Methods.*

24. T. Schlick, H.H. Gan (eds.), *Computational Methods for Macromolecules: Challenges and Applications.*

25. T.J. Barth, H. Deconinck (eds.), *Error Estimation and Adaptive Discretization Methods in Computational Fluid Dynamics.*

26. M. Griebel, M.A. Schweitzer (eds.), *Meshfree Methods for Partial Differential Equations.*

27. S. Müller, *Adaptive Multiscale Schemes for Conservation Laws.*

28. C. Carstensen, S. Funken, W. Hackbusch, R.H.W. Hoppe, P. Monk (eds.), *Computational Electromagnetics.*

29. M.A. Schweitzer, *A Parallel Multilevel Partition of Unity Method for Elliptic Partial Differential Equations.*

30. T. Biegler, O. Ghattas, M. Heinkenschloss, B. van Bloemen Waanders (eds.), *Large-Scale PDE-Constrained Optimization.*

31. M. Ainsworth, P. Davies, D. Duncan, P. Martin, B. Rynne (eds.), *Topics in Computational Wave Propagation.* Direct and Inverse Problems.

32. H. Emmerich, B. Nestler, M. Schreckenberg (eds.), *Interface and Transport Dynamics.* Computa- tional Modelling.

33. H.P. Langtangen, A. Tveito (eds.), *Advanced Topics in Computational Partial Differential Equations.* Numerical Methods and Diffpack Programming.

34. V. John, *Large Eddy Simulation of Turbulent Incompressible Flows.* Analytical and Numerical Results for a Class of LES Models.

35. E. Bänsch (ed.), *Challenges in Scientific Computing - CISC 2002.*

36. B.N. Khoromskij, G. Wittum, *Numerical Solution of Elliptic Differential Equations by Reduction to the Interface.*

37. A. Iske, *Multiresolution Methods in Scattered Data Modelling.*

38. S.-I. Niculescu, K. Gu (eds.), *Advances in Time-Delay Systems.*

39. S. Attinger, P. Koumoutsakos (eds.), *Multiscale Modelling and Simulation.*

40. R. Kornhuber, R. Hoppe, J. Périaux, O. Pironneau, O. Wildlund, J. Xu (eds.), *Domain Decomposition Methods in Science and Engineering.*

41. T. Plewa, T. Linde, V.G. Weirs (eds.), *Adaptive Mesh Refinement – Theory and Applications.*

42. A. Schmidt, K.G. Siebert, *Design of Adaptive Finite Element Software.* The Finite Element Toolbox ALBERTA.

43. M. Griebel, M.A. Schweitzer (eds.), *Meshfree Methods for Partial Differential Equations II.*

44. B. Engquist, P. Lötstedt, O. Runborg (eds.), *Multiscale Methods in Science and Engineering.*

45. P. Benner, V. Mehrmann, D.C. Sorensen (eds.), *Dimension Reduction of Large-Scale Systems.*

46. D. Kressner, *Numerical Methods for General and Structured Eigenvalue Problems.*

47. A. Boriçi, A. Frommer, B. Joó, A. Kennedy, B. Pendleton (eds.), *QCD and Numerical Analysis III.*

48. F. Graziani (ed.), *Computational Methods in Transport.*

49. B. Leimkuhler, C. Chipot, R. Elber, A. Laaksonen, A. Mark, T. Schlick, C. Schütte, R. Skeel (eds.), *New Algorithms for Macromolecular Simulation.*

50. M. Bücker, G. Corliss, P. Hovland, U. Naumann, B. Norris (eds.), *Automatic Differentiation: Applications, Theory, and Implementations.*

51. A.M. Bruaset, A. Tveito (eds.), *Numerical Solution of Partial Differential Equations on Parallel Computers.*

52. K.H. Hoffmann, A. Meyer (eds.), *Parallel Algorithms and Cluster Computing.*

53. H.-J. Bungartz, M. Schäfer (eds.), *Fluid-Structure Interaction.*

54. J. Behrens, *Adaptive Atmospheric Modeling.*

55. O. Widlund, D. Keyes (eds.), *Domain Decomposition Methods in Science and Engineering XVI.*

56. S. Kassinos, C. Langer, G. Iaccarino, P. Moin (eds.), *Complex Effects in Large Eddy Simulations.*

57. M. Griebel, M.A Schweitzer (eds.), *Meshfree Methods for Partial Differential Equations III.*

58. A.N. Gorban, B. Kégl, D.C. Wunsch, A. Zinovyev (eds.), *Principal Manifolds for Data Visualization and Dimension Reduction.*

59. H. Ammari (ed.), *Modeling and Computations in Electromagnetics: A Volume Dedicated to Jean-Claude Nédélec.*

60. U. Langer, M. Discacciati, D. Keyes, O. Widlund, W. Zulehner (eds.), *Domain Decomposition Methods in Science and Engineering XVII.*

61. T. Mathew, *Domain Decomposition Methods for the Numerical Solution of Partial Differential Equations.*

62. F. Graziani (ed.), *Computational Methods in Transport: Verification and Validation.*

63. M. Bebendorf, *Hierarchical Matrices.* A Means to Efficiently Solve Elliptic Boundary Value Problems.

64. C.H. Bischof, H.M. Bücker, P. Hovland, U. Naumann, J. Utke (eds.), *Advances in Automatic Differentiation.*

65. M. Griebel, M.A. Schweitzer (eds.), *Meshfree Methods for Partial Differential Equations IV.*

66. B. Engquist, P. Lötstedt, O. Runborg (eds.), *Multiscale Modeling and Simulation in Science.*

67. I.H. Tuncer, Ü. Gülcat, D.R. Emerson, K. Matsuno (eds.), *Parallel Computational Fluid Dynamics 2007.*

68. S. Yip, T. Diaz de la Rubia (eds.), *Scientific Modeling and Simulations.*

69. A. Hegarty, N. Kopteva, E. O'Riordan, M. Stynes (eds.), *BAIL 2008 – Boundary and Interior Layers.*

70. M. Bercovier, M.J. Gander, R. Kornhuber, O. Widlund (eds.), *Domain Decomposition Methods in Science and Engineering XVIII.*

71. B. Koren, C. Vuik (eds.), *Advanced Computational Methods in Science and Engineering.*

72. M. Peters (ed.), *Computational Fluid Dynamics for Sport Simulation.*

73. H.-J. Bungartz, M. Mehl, M. Schäfer (eds.), *Fluid Structure Interaction II – Modelling, Simulation, Optimization.*

74. D. Tromeur-Dervout, G. Brenner, D.R. Emerson, J. Erhel (eds.), *Parallel Computational Fluid Dynamics 2008.*

75. A.N. Gorban, D. Roose (eds.), *Coping with Complexity: Model Reduction and Data Analysis.*

76. J.S. Hesthaven, E.M. Rønquist (eds.), *Spectral and High Order Methods for Partial Differential Equations.*

77. M. Holtz, *Sparse Grid Quadrature in High Dimensions with Applications in Finance and Insurance.*

78. Y. Huang, R. Kornhuber, O. Widlund, J. Xu (eds.), *Domain Decomposition Methods in Science and Engineering XIX.*

79. M. Griebel, M.A. Schweitzer (eds.), *Meshfree Methods for Partial Differential Equations V.*

80. P.H. Lauritzen, C. Jablonowski, M.A. Taylor, R.D. Nair (eds.), *Numerical Techniques for Global Atmospheric Models.*

81. C. Clavero, J.L. Gracia, F. Lisbona (eds.), *BAIL 2010 – Boundary and Interior Layers, Computational and Asymptotic Methods.*

82. B. Engquist, O. Runborg, Y.R. Tsai (eds.), *Numerical Analysis and Multiscale Computations.*

83. I.G. Graham, T.Y. Hou, O. Lakkis, R. Scheichl (eds.), *Numerical Analysis of Multiscale Problems.*

84. A. Logg, K.-A. Mardal, G. Wells (eds.), *Automated Solution of Differential Equations by the Finite Element Method.*

85. J. Blowey, M. Jensen (eds.), *Frontiers in Numerical Analysis - Durham 2010.*

86. O. Kolditz, U.-J. Gorke, H. Shao, W. Wang (eds.), *Thermo-Hydro-Mechanical-Chemical Processes in Fractured Porous Media - Benchmarks and Examples.*

87. S. Forth, P. Hovland, E. Phipps, J. Utke, A. Walther (eds.), *Recent Advances in Algorithmic Differentiation.*

88. J. Garcke, M. Griebel (eds.), *Sparse Grids and Applications.*

89. M. Griebel, M. A. Schweitzer (eds.), *Meshfree Methods for Partial Differential Equations VI.*

90. C. Pechstein, *Finite and Boundary Element Tearing and Interconnecting Solvers for Multiscale Problems.*

91. R. Bank, M. Holst, O. Widlund, J. Xu (eds.), *Domain Decomposition Methods in Science and Engineering XX.*

92. H. Bijl, D. Lucor, S. Mishra, C. Schwab (eds.), *Uncertainty Quantification in Computational Fluid Dynamics.*

93. M. Bader, H.-J. Bungartz, T. Weinzierl (eds.), *Advanced Computing.*

94. M. Ehrhardt, T. Koprucki (eds.), *Advanced Mathematical Models and Numerical Techniques for Multi-Band Effective Mass Approximations.*

95. M. Azaïez, H. El Fekih, J.S. Hesthaven (eds.), *Spectral and High Order Methods for Partial Differential Equations ICOSAHOM 2012.*

96. M.P. Desjarlais, F. Graziani, R. Redmer, S.B. Trickey (eds.), *Frontiers and Challenges in Warm Dense Matter.*

97. J. Garcke, D. Pflüger (eds.), *Sparse Grids and Applications - Munich 2012.*

98. J. Erhel, M. Gander, L. Halpern, G. Pichot, T. Sassi, O. Widlund (eds.), *Domain Decomposition Methods in Science and Engineering XXI.*

99. R. Abgrall, H. Beaugendre, P.M. Congedo, C. Dobrzynski, M. Ricchiuto, V. Perrier (eds.), *High Order Nonlinear Numerical Methods for Evolutionary PDEs - HONOM 2013.*

For further information on these books please have a look at our mathematics catalogue at the following URL: www.springer.com/series/3527

FSC
www.fsc.org
MIX
Papier aus verantwortungsvollen Quellen
Paper from responsible sources
FSC® C105338